(Continued on page I10 following the Subject Index)

Finite Mathematics and Applied Calculus

Frank C. Wilson

Green River Community College

HOUGHTON MIFFLIN COMPANY

Boston   New York

Vice President and Publisher: Jack Shira
Sponsoring Editor: Leonid Tunik
Associate Editor: Jennifer King
Senior Project Editor: Tamela Ambush
Editorial Assistant: Sage Anderson
Senior Manufacturing Buyer: Renee Ostrowski
Executive Marketing Manager: Brenda Bravener-Greville
Senior Marketing Manager: Jennifer Jones

Cover photograph: © Shirichi Maruyama/Photonica, © Ed Young/CORBIS, © Royalty-Free/CORBIS, © Richard T. Norvitz/CORBIS.

Photo Credits: Chapter 1: p. 1, © Lou Dematteis/Reuther/CORBIS; p. 43, © Richard Cummins/CORBIS; **Chapter 2:** p. 53, © Wolfgang Kaehler/CORBIS; p.57, Author; **Chapter 3:** p. 111, © Larry Williams and Associates/CORBIS; p. 132, © MacDuff/Everton/CORBIS; **Chapter 4:** p.171, © Brownie Harris Productions/CORBIS; p. 215, © Skjold Productions/PhotoEdit, Inc.; p. 265, © Connie Gardner AFP Getty Images; **Chapter 5:** p. 281, © Frederic J. Brown/Getty Images; p.330, © Reuters/CORBIS; p. 382, Bureau of Labor Statistics; **Chapter 6:** p. 383, © Tom and Dee Ann McCarthy/CORBIS; p. 418, © Getty Images News; **Chapter 7:** p. 440, © Strauss/Curtis/CORBIS; p. 458, © Chris Trotman/Duomo/CORBIS; **Chapter 8:** p. 494, © Darren Modricker/CORBIS; p. 518, © Digital Images/Getty Images; p. 536, © Animals/Animals; **Chapter 9:** p. 599, © Alex Fezer/CORBIS; p. 610, © Tony Duffy/Getty Images; **Chapter 10:** p. 660, © Royalty Free/CORBIS; p. 677, © Taxi/Getty Images; **Chapter 11:** p. 705, © Ariel Skelley/CORBIS; p. 770, © Stone/Getty Images; **Chapter 12:** p. 782, © Karen Kasmauski/CORBIS; p. 799, © Reuters/CORBIS; **Chapter 13:** p. 850, © Royalty Free/CORBIS; p. 886, © Arthur Morris/CORBIS; p. 893, © Lawson Wood/CORBIS; p. 893, © Jeffrey L. Rotman/CORBIS; p.893, © Stuart Westmoreland/CORBIS; p. 900, © David Fleetham/CORBIS; **Chapter 14:** p. 903, © Royalty Free/CORBIS; p. 921, Author; p. 921, Author; p. 921, Author; p. 921, Author.

Printed in the U.S.A.

Library of Congress Control Number: 2005934027

Instructor's exam copy:
ISBN 13: 978-0-618-73265-4
ISBN 10: 0-618-73265-9

For orders, use student text ISBNs:
ISBN 13: 978-0-618-33291-5
ISBN 10: 0-618-33291-X

1 2 3 4 5 6 7 8 9 –CRK– 10 09 08 07 06

Contents

Preface

To the Student

Have you ever asked, "When am I ever going to use this?" or "Why do I care?" after learning a new mathematical procedure? Many students have. This book seeks to answer those questions by teaching mathematical skills which, when applied, can improve your quality of life. Whether calculating how long it will take to pay off a car loan or predicting how much tuition will be next year, examples and exercises in this book are based on interesting and engaging real-life data. Consequently, when you find a mathematical solution, you will learn something new about the world in which we live. To make real-life data analysis even more meaningful, you will become skilled in collecting and analyzing data from your own life through the *Make It Real* projects. The analytical skills learned through these projects will remain with you long after you have left this course.

This book is written in a casual, reader-friendly style. Although key mathematical terms and concepts are appropriately addressed, the focus is conceptual understanding not mathematical jargon. If you are pursuing a degree in business, social science, or a similar field, this book is written specifically for you.

You will find that your understanding of key concepts will be enhanced through the use of a graphing calculator. The TI-83 Plus (or the TI-83 or TI-84) is the ideal calculator for this course. Because learning how to use the calculator is a challenge for many students, Technology Tips are scattered throughout the text allowing you to learn a new calculator technique when it is needed. These tips detail how to graph a function, solve an equation, find the maximum value of a function, and so on. Rather than giving a broad overview of a procedure, the tips take you through the actual keystrokes on the TI-83 Plus and show you calculator screenshots so that you can verify that you're doing each step correctly.

I have learned much about teaching from my students and deeply value their input. Likewise, I'm interested in hearing from you. Let me know how this book works for you and feel free to share any feedback you may have on how to improve this text. I may be reached by e-mail by clicking the Contact the Author icon on the text's website: **college.hmco.com/pic/wilsonACFM1e.**

To the Instructor

Thank you for selecting this book. I believe you will find its approach refreshing and its content interesting to you and your students. It is written specifically for students pursuing business, social science, or a related field. Many of these students do not enjoy mathematics and are taking this course only because it is required for their major. (Many business schools require a finite math—applied calculus—statistics sequence.) Several features are included in this text to make the course content more accessible to these students including:

- an informal writing style that addresses key concepts appropriately without becoming bogged down in mathematical jargon

- examples and exercises throughout the text based on interesting and engaging real-life data
- *Make It Real* projects which allow students to collect and analyze data relevant to their personal lives
- detailed *Technology Tips* (including screenshots) which teach students how to use the TI-83 Plus calculator as a tool to analyze real-life data
- a bare-bones approach to course content. Topics covered in detail include those that are most relevant to everyday people

I believe that candid feedback from colleagues is helpful in enhancing teaching effectiveness. Please feel free to contact me with any recommendations, comments, or other feedback that you feel will enhance the effectiveness of future editions of this text. I may be reached by e-mail by clicking the Contact the Author icon on the text's website: **college.hmco.com/pic/wilsonACFM1e.**

Disclaimer

In this book, I have attempted to incorporate real-world data from the financial markets to the medical field. In each case, I have done my best to present the data accurately and interpret the data realistically. However, I do not claim to be an expert in financial, medical, and other similar fields. My interpretations of real-world data and my associated conclusions may not adequately consider all relevant factors. Therefore, readers are encouraged to seek professional advice from experts in the appropriate fields before making decisions related to the topics addressed herein.

Despite the usefulness of mathematical models as representations of real-world data sets, most mathematical models have a certain level of error. It is common for model results to differ from raw data set values. Consequently, conclusions drawn from a mathematical model may differ (sometimes dramatically) from conclusions drawn by looking at raw data sets. Readers are encouraged to interpret model results with this understanding.

Acknowledgments

This textbook would not have been possible without the contributions of many colleagues. I greatly appreciate all of the people who contributed time and talent to bring this book to fruition.

The feedback from the following reviewers was invaluable and helped to shape the final form of the text: Bill Ardis, *Collin County Community College, TX;* James J. Ball, *Indiana State University, IN;* Michael L. Berry, *West Virginia Wesleyan College, WV;* Marcelle Bessman, *Jacksonville University, FL;* Mike Bosch, *Iowa Lakes Community College, IA;* Emily Bronstein, *Prince George's Community College, MD;* Dean S. Burbank, *Gulf Coast Community College, FL;* Andra Buxkemper, *Bunn College, TX;* Roxanne Byrne, *University of Colorado—Denver, CO;* Scott A. Clary, *Florida Institute of Technology, FL;* David Collingwood, *University of Washington, WA;* Mark A. Crawford Jr., *Western Michigan University, MI;* Khaled Dib, *University of Minnesota Duluth, MN;* Lance D. Drager, *Texas Tech University, TX;* Klara Grodzinsky, *Georgia Institute of Technology, GA;* Lucy L. Hanks, *Virginia Polytechnic Institute and State University, VA;* Jean B. Harper, *State Univiversity of N.Y. – College at Fredonia, NY;* Kevin M. Jenerette, *Coastal Carolina University, SC;* Cynthia Kaus, *Metropolitan State University, MN;* Michael LaValle, *Rochester Community and Technical College, MN;* Roger D. Lee, *Salt Lake Community College, UT;* Lia Liu, *Univer-*

sity of Illinois Chicago, IL; Alan Mabry, *Unitersity of Texas at El Paso;* Quincy Magby, *Arizona Western College, AZ;* Mary M. Marco, *Bucks County Community College, PA;* Nicholas Martin, *Shepherd College, WV;* William C. McClure, *Orange Coast College, CA;* James McGlothin, *Lower Columbia College, WA;* Victoria Neagoe, *Goldey Beacom College;* David W. Nelson, *Green River Community College;* Ralph W. Oberste-Vorth, *Marshall University, WV;* Armando I. Perez, *Laredo Community College;* Cyril Petras, *Lord Fairfax Community College, VA;* Mihaela Poplicher, *University of Cincinnati;* John E. Porter, *Murray State University, KY;* David W. Roach, *Murray State University, KY;* R. A. Rock, *Daniel Webster College, NH;* Arthur Rosenthal, *Salem State College, MA;* Kimmo I. Rosenthal, *Union College, NY;* Sharon Mayhew Saxton, *Cascadia Community College, WA;* Edwin Shapiro, *University of San Francisco, CA;* Denise Szecsei, *Stetson University;* Abolhassan S. Taghavy, *Richard J. Daley College, IL;* Muhammad Usman, *University of Cincinnati, OH;* Jorge R. Viramontes Olivas, *University of Texas at El Paso, TX;* Beverly Vredevelt, *Spokane Falls Community College, WA;* Michael L. Wright, *Cossatot Community College, AR.*

I extend special thanks to Cindy Harvey, who provided detailed recommendations on how to improve an early draft of this text, to Helen Medley, who reviewed the final draft text for accuracy, and to Paul Lorczak, who conducted the accuracy review for the typeset manuscript. Their suggestions greatly enhanced the quality of the text.

My appreciation goes to Rob Jonas, who wrote most of the solutions to the calculus portion of the text, and to Donnie Hallstone, who worked many of the solutions to the probability and statistics portion of the text.

I appreciate the work of my exceptional student, Jon Austin, who developed some of the mathematical models used in the text. I value the recommendations from the math instructors who class-tested this book, including:

Beverly A. Vredevelt, *Spokane Falls Community College*
Frank "Bud" Wright, *Green River Community College*
David Nelson, *Green River Community College*
Keith Alford, *Green River Community College*

Additional comments from students using a draft copy of the text at Green River Community College and Spokane Falls Community College were also helpful.

I'm thankful for Erica Carlson, who assisted with the art manuscript and for Uli Gersiek, who created most of the artwork in the text.

The Houghton Mifflin Company team was phenomenal to work with. I extend gratitude to Lauren Schultz, who first believed in the project; Jennifer King and Marika Hoe, who kept things moving; Kasey McGarrigle, who provided behind the scenes support; and Tamela Ambush and Sage Anderson, who ensured the quality of the published text. Additionally, I'm grateful for the management and support staff at Houghton Mifflin who made the production of this text a reality.

I also thank Texas Instruments for providing a way to enhance student learning by creating the TI-83 Plus calculator. Without it, the Technology Tips would not have been possible.

On a personal note, I could not have written this text without the tireless support of my wife, Shelley Wilson. She worked overtime caring for our home and children for the three years I worked on the text. Despite this challenge, she never faltered in offering me encouragement, support, and love. For this, I am eternally grateful and indebted to her.

Frank C. Wilson

Supplements for the Instructor

Eduspace®

Powerful, customizable, and interactive, Eduspace, powered by Blackboard®, is Houghton Mifflin's online learning tool. By pairing the widely recognized tools of Blackboard with quality, text-specific content from Houghton Mifflin, Eduspace makes it easy for instructors to create all or part of a course online. Homework exercises, quizzes, tests, tutorials, and supplemental study materials all come ready-to-use or can be modified by the instructor. Visit **www.eduspace.com** for more information.

HM ClassPrep™ with HM Testing (powered by Diploma™)

HM ClassPrep offers text-specific resources for the instructor. *HM Testing* offers instructors a flexible and powerful tool for test generation and test management. Now supported by the Brownstone Research Group's market-leading *Diploma™* software, this new version of *HM Testing* significantly improves ease of use.

Blackboard®, WebCT®, and eCollege®

Houghton Mifflin can provide you with valuable content to include in your existing Blackboard, WebCT, and eCollege systems. This text-specific content enables instructors to teach all or part of their course online.

Supplements for the Student

Eduspace®

Powerful, customizable, and interactive, Eduspace®, powered by Blackboard®, is Houghton Mifflin's online learning tool for instructors and students. Eduspace is a text-specific, web-based learning environment that your instructor can use to offer students a combination of practice exercises, multimedia tutorials, video explanations, online algorithmic homework, and more. Specific content is available 24 hours a day to help you succeed in your course.

SMARTHINKING®

Houghton Mifflin's unique partnership with SMARTHINKING brings students real-time, online tutorial support when they need it most. Using state-of-the-art whiteboard technology and feedback tools, students interact and communicate with "e-structors." Tutors guide students through the learning and problem solving process without providing answers or rewriting a student's work. Visit **www.smarthinking.com** for more information.

HM mathSpace® Student Tutorial CD-ROM

For students who prefer the portability of a CD-ROM, this tutorial provides opportunities for self-paced review and practice with algorithmically generated exercises and step-by-step solutions.

Houghton Mifflin Instructional DVDs

Hosted by Dana Mosely and professionally produced, these DVDs cover topics in the text. Ideal for promoting individual study and review, these comprehensive DVDs also support students in online courses or those who have missed a lecture.

Online Learning Center and Online Teaching Center

The free Houghton Mifflin websites at **college.hmco.com/pic/wilsonACFM1e** contain an abundance of resources.

Student Solutions Guide (ISBN 0-618-33300-2) contains solutions to the odd-numbered exercises from all Exercise Sets in the book.

Excel Guide for Finite Mathematics and Applied Calculus by Revathi Narasimhan (ISBN 0-618-67691-0) provides lists of exercises from the text that can be completed after each stepped out Excel example. No prior knowledge of Excel is necessary.

Functions and Linear Models

Mathematical functions are a powerful tool used to model real-world phenomena. Whether simple or complex, functions give us a way to forecast expected results. Remarkably, anything that has a constant rate of change may be accurately modeled with a linear function. For example, the cost of filling your car's gas tank is a linear function of the number of gallons purchased.

1.1 Functions
- Distinguish between functions and nonfunctions in tables, graphs, and words
- Use function notation
- Graph functions using technology
- Determine the domain of a function

1.2 Linear Functions
- Calculate and interpret the meaning of the slope of a linear function
- Interpret the physical and graphical meaning of x- and y-intercepts
- Formulate the equation of a line given two points
- Recognize the slope-intercept, point-slope, and standard forms of a line

1.3 Linear Models
- Use technology to model linear and near-linear data
- Use a linear equation to describe the relationship between directly proportional quantities
- Recognize and model naturally occurring linear, near-linear, and piecewise linear relationships

1.1 Functions

- Distinguish between functions and nonfunctions in tables, graphs, and words
- Use function notation
- Graph functions using technology
- Determine the domain of a function

GETTING STARTED   Our society is a complex system of relationships between people, places, and things. Many of these relationships are interconnected. In mathematics, we often model the relationship between two or more interdependent quantities by using a **function**. In this section, we will show how to distinguish between functions and nonfunctions and will practice using function notation. We will also demonstrate how to use technology to draw a function graph, and discuss how to find the domain of a function.

DEFINITION: FUNCTION

A function is a rule that associates each input with *exactly one* output.

Often the rule is represented in a table of data with the inputs on the left-hand side and the outputs on the right-hand side. For example, the amount of money we pay to fill up our gas tank is a function of the number of gallons pumped (see Table 1.1).

TABLE 1.1

Gallons Pumped	Total Fuel Cost
10	$15.99
15	$23.99
20	$31.98

In this case, the input to the function is *gallons pumped* and the output of the function is *total fuel cost*. Fuel cost is a function of the number of gallons pumped because each input has exactly one output.

Similarly, the weekly wage of a service station employee is a function of the number of hours worked. The *number of hours worked* is the input to the function, and the *weekly wage* is the output of the function. Since weekly wage is a function of the number of hours worked, an employee who works 40 hours expects to be paid the same wage each time she works that amount of time.

EXAMPLE 1   Determining If a Table of Data Represents a Function

A car dealer tracks the number of blue cars in each of three shipments and records the data in a table (see Table 1.2). Is the number of blue cars a function of the number of cars in the shipment?

TABLE 1.2

Number of Cars in the Shipment	Number of Blue Cars
22	6
24	7
24	5

same input { 24, 24 }   different outputs { 7, 5 }

SOLUTION According to the definition of a function, each input must have exactly one output. The input value 24 has two different outputs: 5 and 7. Since the input 24 has more than one output, the number of blue cars is *not* a function of the number of cars in the shipment.

Function Notation

When we encounter functions in real life, they are often expressed in words. To make functions easier to work with, we typically use symbolic notation to represent the relationship between the input and the output. Let's return to the fuel cost table introduced previously (Table 1.3).

TABLE 1.3

Gallons Pumped	Total Fuel Cost
10	$15.99
15	$23.99
20	$31.98

Observe that the fuel cost is equal to $1.599 times the number of gallons pumped. We represent this symbolically as $C(g) = 1.599g$, where $C(g)$ represents the total fuel cost when g gallons are pumped. [$C(g)$ is read "C of g"]. The letter C is used to represent the name of the rule, and the letter g in the parentheses indicates that the rule works with different values of g (see Figure 1.1).

$$g \longrightarrow \boxed{C} \longrightarrow C(g)$$

input value   function name   output value

FIGURE 1.1

We call the output variable of a function the **dependent variable** because the value of the output variable *depends* upon the value of the input variable. The input variable is called the **independent variable.** (One way to remember the meaning of the terms is to observe that both *input* and *independent* begin with *in.*) From the table, we see that

$$C(10) = 15.99$$
$$C(15) = 23.99$$
$$C(20) = 31.98$$

For this function, the independent variable took on the values 10, 15, and 20, and the dependent variable assumed the values 15.99, 23.99, and 31.98.

EXAMPLE 2   **Determining a Linear Model from a Verbal Description**

An electronics store employee earns $8.50 per hour. Write an equation for the employee's earnings as a function of the hours worked. Then calculate the amount of money the employee earns (in dollars) by working 30 hours.

SOLUTION   Since the employee earns $8.50 for each hour worked, the employee's total earnings are equal to $8.50 times the total number of hours worked. That is,

$$E(h) = 8.50h$$

where E is the employee's earnings (in dollars) and h is the number of hours worked. To calculate the amount of money earned by working 30 hours, we evaluate this function at $h = 30$.

$$E(30) = 8.5(30)$$
$$= 255$$

The employee earns $255 for 30 hours of work.

Function notation is extremely versatile. Suppose we are given the function $f(x) = x^2 - 2x + 1$. We may evaluate the function using either numerical values or nonnumerical values. For example,

$$f(2) = (2)^2 - 2(2) + 1 \qquad\qquad f(\triangle) = (\triangle)^2 - 2(\triangle) + 1$$
$$= 4 - 4 + 1$$
$$= 1$$

$$f(a + 2) = (a + 2)^2 - 2(a + 2) + 1$$
$$= (a^2 + 4a + 4) - 2a - 4 + 1$$
$$= a^2 + 2a + 1$$

In each case, we replaced the value of x in the function $f(x) = x^2 - 2x + 1$ with the quantity in the parentheses. Whether the independent variable value was 2, $\triangle$, or $a + 2$, the process was the same.

EXAMPLE 3   **Evaluating a Function Using Function Notation**

Evaluate the function $s(t) = t^3 + 4t$ at $t = 3$, $t = \triangle$, and $t = a^2$.

SOLUTION

$$s(3) = (3)^3 + 4(3) \qquad\qquad s(\triangle) = (\triangle)^3 + 4(\triangle)$$
$$= 27 + 12$$
$$= 39$$

$$s(a^2) = (a^2)^3 + 4(a^2)$$
$$= a^6 + 4a^2$$

Graphs of Functions

Functions are represented visually by plotting points on a Cartesian coordinate system (see Figure 1.2). The horizontal axis shows the value of the independent variable (in this case, x), and the vertical axis shows the value of the dependent variable (in this case, y).

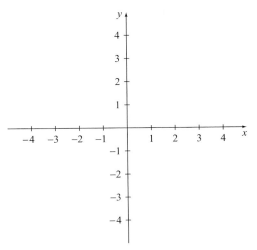

FIGURE 1.2

When using the coordinate system, y is typically used in place of the function notation $f(x)$. That is, $y = f(x)$. This is true even if the function has a name other than f.

    The point of intersection of the horizontal and vertical axes is referred to as the **origin** and is represented by the ordered pair $(0, 0)$. To graph an ordered pair (a, b), we move from the origin $|a|$ units horizontally and $|b|$ units vertically and draw a point. If $a > 0$, we move to the right. If $a < 0$, we move to the left. Similarly, if $b > 0$, we move up, and if $b < 0$, we move down. For example, consider the table of values with its associated interpretation in Table 1.4 and the graph in Figure 1.3.

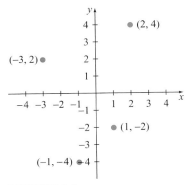

FIGURE 1.3

TABLE 1.4

x	y	Horizontal	Vertical
-3	2	left 3	up 2
-1	-4	left 1	down 4
1	-2	right 1	down 2
2	4	right 2	up 4

When we are given the equation of a function, we can generate a table of values and then plot the corresponding points. Once we have drawn a sufficient number of points to be able to determine the basic shape of the graph, we typically connect the points with a smooth curve. For example, the function $y = x^3 - 9x$ has the table of values and graph shown in Figure 1.4.

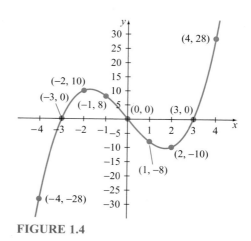

x	y
-4	-28
-3	0
-2	10
-1	8
0	0
1	-8
2	-10
3	0
4	28

FIGURE 1.4

EXAMPLE **4**

Estimating Function Values from a Graph

Estimate $f(-3)$ and $f(2)$ using the graph of $f(x) = x^3 - 16x$ shown in Figure 1.5.

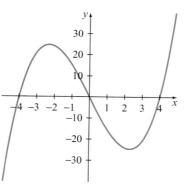

FIGURE 1.5

SOLUTION It appears from Figure 1.6 that $f(-3) \approx 20$ and $f(2) \approx -25$. Calculating these values with the algebraic equation, we see that

$$f(-3) = (-3)^3 - 16(-3)$$
$$= -27 + 48$$
$$= 21$$

and

$$f(2) = (2)^3 - 16(2)$$
$$= 8 - 32$$
$$= -24$$

FIGURE 1.6

One drawback of using a graph to determine the values of a function is that it is difficult to be precise. For this reason, algebraic methods are typically preferred when precision is important.

Not all data sets represent functions. If any value of the independent variable is associated with more than one value of the dependent variable, the table of data and its associated graph will not represent a function. For example, consider $y = \pm\sqrt{2x}$ (see Figure 1.7).

x	y
0	0
0.5	−1
0.5	1
2	−2
2	2
4.5	−3
4.5	3
8	−4
8	4

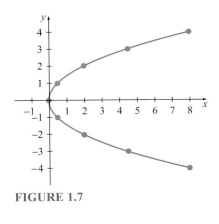

FIGURE 1.7

Each positive value of x is associated with two different values of y. We can easily determine this from the graph by observing that a vertical line drawn through any positive value of x will cross the graph twice. This observation leads us to the Vertical Line Test.

VERTICAL LINE TEST

If every vertical line drawn on a graph intersects the graph in at most one place, then the graph is the graph of a function. Otherwise, the graph is not the graph of a function.

EXAMPLE 5

Determining If a Graph Represents a Function

The graph of $y = x^2$ is shown in Figure 1.8. Does the graph represent a function?

FIGURE 1.8

SOLUTION

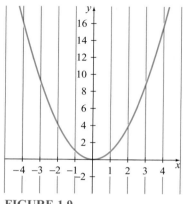

FIGURE 1.9

Each of the vertical lines drawn crosses the graph in exactly one place, as shown in Figure 1.9. Therefore, the graph passes the Vertical Line Test and y is a function of x.

EXAMPLE 6

Determining If a Graph Represents a Function

The graph of $y = \pm\sqrt{x}$ is shown in Figure 1.10. Does the graph represent a function?

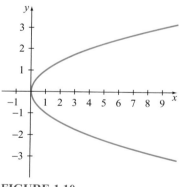

FIGURE 1.10

SOLUTION   Notice that many of the vertical lines in Figure 1.11 cross the graph in two places. The graph does not pass the Vertical Line Test, so y is not a function of x. Each positive value of x corresponds with two values of y.

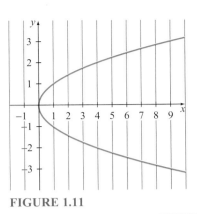

FIGURE 1.11

The graphs shown thus far have been continuous. That is, each graph could be drawn without lifting the pencil. However, some function graphs are *discontinuous*. A discontinuous graph has a break in the graph. The break in the graph is referred to as a **discontinuity.**

EXAMPLE 7

Determining If a Discontinuous Graph Represents a Function

The graph of $h(x) = \begin{cases} x \text{ if } x \leq 1 \\ 2 \text{ if } x > 1 \end{cases}$ is shown in Figure 1.12. Determine if the graph represents a function.

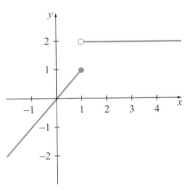

FIGURE 1.12

SOLUTION The graph is discontinuous at $x = 1$. The value of the function at $x = 1$ is 1. [The open dot at $(1, 2)$ indicates that the point $(1, 2)$ is not part of the graph.] Although the graph is discontinuous, each value of the independent variable is associated with exactly one value of the dependent variable. Therefore, the graph represents a function.

Using Technology to Graph a Function

Graphing calculators, such as the TI-83 Plus, can quickly draw the graph of a function as shown in the following Technology Tip.

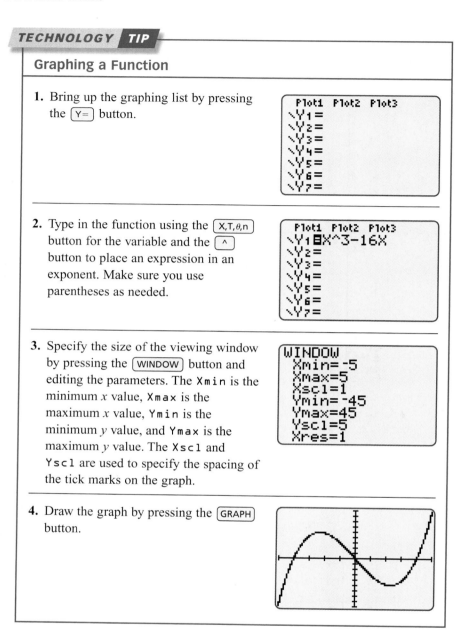

Graphing a Function

1. Bring up the graphing list by pressing the [Y=] button.

2. Type in the function using the [X,T,θ,n] button for the variable and the [^] button to place an expression in an exponent. Make sure you use parentheses as needed.

3. Specify the size of the viewing window by pressing the [WINDOW] button and editing the parameters. The Xmin is the minimum *x* value, Xmax is the maximum *x* value, Ymin is the minimum *y* value, and Ymax is the maximum *y* value. The Xscl and Yscl are used to specify the spacing of the tick marks on the graph.

4. Draw the graph by pressing the [GRAPH] button.

Domain and Range

We are often interested in the set of all possible values of the independent variable and the set of all possible values of the dependent variable.

DOMAIN AND RANGE

The set of all possible values of the independent variable of a function is called the **domain**. The set of all possible values of the dependent variable of a function is called the **range**.

It is easy to remember the meaning of the terms if we observe that ***input, independent variable,*** and ***domain*** all contain the word ***in***.

Consider a common kitchen blender. If we put an orange into a blender and turn the blender on, the orange is transformed into orange juice. We say that *orange* is in the domain of the blender function and *orange juice* is in the range of the blender function. On the other hand, if we put a rock into the blender and turn it on, the blender will self-destruct. We say that *rock* is not in the domain of the blender function, since blenders are unable to process rocks.

Finding the domain of a function frequently involves solving an equation or an inequality. Recall that *solving an equation* means finding the value of the variable that makes the equation a true statement.

The domain of most frequently used mathematical functions is the set of all real numbers. However, there are three common situations in which the domain of a function is restricted to a subset of the real numbers. They are:

1. A zero in the denominator
2. A negative value under a square root symbol (radical)
3. The context of a word problem

Let's look at an example for each of the situations.

EXAMPLE 8

Determining the Domain of a Function

What is the domain of $g(x) = \dfrac{3x - 1}{2x + 6}$?

SOLUTION We know that the value in the denominator must be nonzero, since division by zero is undefined. That is, $2x + 6 \neq 0$. To find the value of x that must be excluded from the domain, we must solve the following equation:

$$2x + 6 = 0$$
$$2x + 6 - 6 = 0 - 6 \qquad \text{Subtract 6 from both sides}$$
$$2x = -6$$
$$\frac{2x}{2} = \frac{-6}{2} \qquad \text{Divide both sides by 2}$$
$$x = -3$$

The domain of the function is all real numbers except -3. We may rewrite the function equation with the domain restriction as follows:

$$g(x) = \frac{3x - 1}{2x + 6}, \qquad x \neq -3$$

EXAMPLE 9

Determining the Domain of a Function

What is the domain of $y = \sqrt{x - 3}$?

SOLUTION We know that the value underneath the radical must be nonnegative, since the square root of a negative number is undefined in the real number system. Therefore, $x - 3 \geq 0$. Solving for x, we get

$$x - 3 \geq 0$$

$$x - 3 + 3 \geq 0 + 3 \qquad \text{Add 3 to both sides}$$

$$x \geq 3$$

The domain of the function is all real numbers greater than or equal to 3.

EXAMPLE 10   ## Determining the Domain of a Function

The revenue R from the sale of x gallons of gasoline is given by the equation

$$R(x) = 1.379x$$

What is the domain of the function?

SOLUTION   In the context of this problem, it doesn't make sense to talk about selling a negative number of gallons of gas. So $x \geq 0$. The domain of the function is all nonnegative real numbers.

1.1 Summary

In this section, you learned to distinguish between functions and nonfunctions, and you practiced using function notation. You learned how to use technology to graph a function, and you discovered how to find the domain of a function. Mastering each of these techniques will help you understand the subsequent concepts covered in this chapter.

1.1 Exercises

In Exercises 1–6, determine whether the output is a function of the input.

1. $W(a)$ = your weight in pounds when you were a years old.

2. $S(n)$ = your score on test number n in a finite math course. (Assume you could only take the test once.)

3.

Time of Day	Temperature (°F)
11:00 a.m.	68
1:00 p.m.	73
3:00 p.m.	75

4.

Time of Day	Temperature (°F)
10:00 a.m.	66
12:00 p.m.	74
2:00 p.m.	74

5.

Fish Caught	Salmon in Catch
169	24
182	32
182	47

6.

Fish Caught	Salmon in Catch
252	74
276	74
301	92

In Exercises 7–13, calculate the value of the function at the designated input and interpret the result.

7. $C(x) = 39.95x$ at $x = 4$, where $C(x)$ is the cost of buying x pairs of shoes.

8. $C(x) = 29.95x + 200$ at $x = 300$, where $C(x)$ is the cost of making x pairs of shoes.

9. $H(t) = -16t^2 + 120$ at $t = 2$, where $H(t)$ is the height of a cliff diver above the water t seconds after he jumped from a 120-foot cliff.

10. 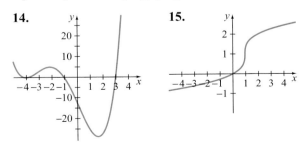 **Revenue**   Find $R(t)$ at $t = 4$, where $R(t)$ is the quarterly revenue for an international tortilla producer t quarters after December 1999 as shown below.

Quarter	Revenue (in millions of dollars)
1	460.0
2	452.0
3	474.0
4	483.8
5	466.0

Source: Gruma, S.A. de C.V.

11. 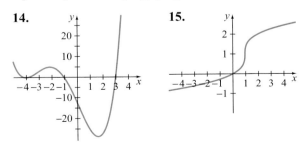 **Earnings per Share**   Find $E(t)$ at $t = 4$, where $E(t)$ is the quarterly earnings per share for an international tortilla producer t quarters after December 1999.

Quarter	Earnings per Share (in dollars)
1	0.13
2	−0.11
3	0.15
4	0.06
5	0.03
6	0.06

Source: Gruma, S.A. de C.V.

12. **Stock Price**   Find $P(t)$ at $t = 3$, where $P(t)$ is the stock price of an e-learning company at the end of the day and t is the number of days after November 4, 2001.

Source: Digital Think Corporation.

13. 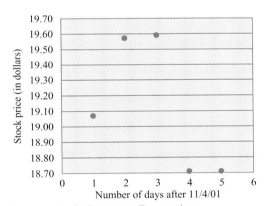 **Stock Price**   Find $P(t)$ at $t = 4$, where $P(t)$ is the stock price of a computer company at the end of the day and t is the number of days after November 4, 2001.

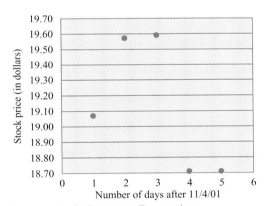

Source: Apple Computer Corporation.

In Exercises 14–19, determine whether the graphs represent functions by applying the Vertical Line Test.

14.

15.

16.

17.

18.

19.

In Exercises 20–23, graph the function on your graphing calculator, using the specified viewing window. Note that a ≤ x ≤ b means x min = a and x max = b. Similarly, c ≤ y ≤ d means y min = c and y max = d.

 20. $y = x^2 - 5x; -3 \le x \le 7, -8 \le y \le 8$

 21. $y = -x^2 - 4; -3 \le x \le 3, -10 \le y \le 1$

 22. $y = -x + 2; -3 \le x \le 3, -2 \le y \le 6$

23. $y = -2x^2 + 1; -2 \le x \le 2, -3 \le y \le 2$

In Exercises 24–27, use the graph to estimate the value of the function at the indicated x value. Then calculate the exact value algebraically.

24. $y = 0.25(x + 1)^2(x - 3); x = 4$

25. $y = 5|x| - x^2; x = 1$

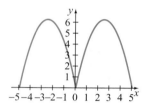

26. $y = \dfrac{|x + 2|}{x^2 + 1}; x = 3$

27. $y = -\dfrac{4(x^2 - 4)}{x^3 + 2x^2 + 3x + 6}; x = -2$

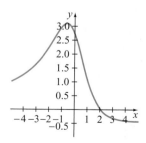

In Exercises 28–41, determine the domain of the function.

28. $f(x) = 5x^3 - 5x$ **29.** $f(p) = p^2 - 2p + 1$

30. $C(r) = \dfrac{r^2 - 1}{r + 1}$ **31.** $h(r) = \dfrac{r^2 - 1}{r^2 + 1}$

32. $S(t) = \dfrac{t - 3}{4t + 1}$

33. $g(t) = \dfrac{3t - 3}{4t - 4}$

34. $H(x) = \sqrt{x - 5}$

35. $f(a) = a^3 - \sqrt{a + 1}$

36. $f(x) = \dfrac{\sqrt{x + 5}}{x + 2}$

37. $f(x) = \dfrac{\sqrt{2x + 6}}{x^2 + 3}$

38. $C(n) =$ the cost of buying n apples

39. $P(n) =$ the profit earned from the sale of n bags of candy

40. $W(n) =$ the average birth weight of someone born in year n

41. $H(n) =$ the average height of someone born in year n

Exercises 42–46 are intended to challenge your understanding of functions and graphs.

 42. Graph the function $f(x) = \dfrac{x - 1}{x^2 - 1}$ on your calculator. What do you think is the domain of f?

43. Determine the domain of $f(x) = \dfrac{x - 1}{x^2 - 1}$ algebraically and compare your answer to the solution of Exercise 42.

44. Are all lines functions? Explain.

45. Does the table of data represent a function? Explain.

x	y
0	5
1	6
1	6
3	0
4	2

46. For a given date and location, is air temperature a function of the time of day? Explain.

1.2 **Linear Functions**

- Calculate and interpret the meaning of the slope of a linear function
- Interpret the physical and graphical meaning of *x*- and *y*-intercepts
- Formulate the equation of a line given two points
- Recognize the slope-intercept, point-slope, and standard forms of a line

GETTING **STARTED**   The cost of filling our car's gas tank, the number of calories we consume by eating a few bags of fruit snacks, and the amount of sales tax we pay when we buy new clothes are examples of linear functions. In this section, we will show how to calculate the slope of a linear function. We will discuss how to interpret the physical and graphical meaning of slope, *x*-intercept, and *y*-intercept. We will also show three different ways to write a linear equation and demonstrate how to find the equation of a linear function from a data set.

DEFINITION: GRAPH OF A LINEAR FUNCTION

The line passing through any two points (x_1, y_1) and (x_2, y_2) with $x_1 \neq x_2$ is referred to as the **graph of a linear function.**

Linear functions are characterized by a constant rate of change. That is, increasing a domain value by one unit will always change the corresponding range value by a constant amount. The converse is also true. Any table of data with a constant rate of change represents a linear function. The constant rate of change is referred to as the **slope** of the linear function.

DEFINITION: SLOPE

The slope of a linear function is the change in the output that occurs when the input is increased by one unit. The slope m may be calculated by dividing the difference of two outputs by the difference in the corresponding inputs. That is,

$$m = \frac{y_2 - y_1}{x_2 - x_1}$$

where (x_1, y_1) and (x_2, y_2) are data points of the linear function.

The number of calories C in n bags of fruit snacks is shown in Table 1.5.

TABLE 1.5

Bags (n)	Calories (C)
1	100
2	200
3	300

Source: Fruit Smiles 9-oz. box label

Notice that the calorie count increases by 100 calories for each additional bag. Since the dependent variable C is changing at a constant rate (100 calories per bag), the data may be modeled by a linear function. In this case, the linear function is $C = 100n$.

To calculate the slope of the function, we may use any two data points. Using the data points $(x_1, y_1) = (1, 100)$ and $(x_2, y_2) = (3, 300)$, the slope is

$$m = \frac{300 - 100}{3 - 1} \frac{\text{calories}}{\text{bags}}$$

$$= \frac{200}{2} \frac{\text{calories}}{\text{bags}}$$

$$= 100 \text{ calories per bag}$$

A one-bag increase in the input results in a 100-calorie increase in the output.

Will we get the same result if we use different points? Let's check using $(1, 100)$ and $(2, 200)$ (see Figure 1.13).

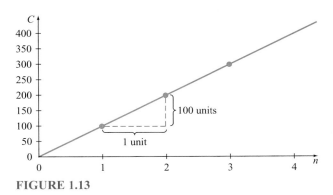

FIGURE 1.13

UNITS

$$m = \frac{100 - 200 \text{ calories}}{1 - 2 \text{ bags}}$$

$$= \frac{-100}{-1} \frac{\text{calories}}{\text{bag}}$$

$$= 100 \text{ calories per bag}$$

The result is the same! With linear functions, we may use any two points to calculate the slope.

Not all lines are functions. Vertical lines fail to pass the Vertical Line Test, so they are not functions. As shown in Example 1, vertical lines have an undefined slope.

EXAMPLE 1

Determining the Slope of a Line from Two Points on the Line

What is the slope of the line going through $(2, 4)$ and $(2, 8)$ (Figure 1.14)?

FIGURE 1.14

SOLUTION

$$m = \frac{8 - 4}{2 - 2}$$

$$= \frac{4}{0}$$

$$= \text{undefined}$$

Since division by zero is not defined, the line has an undefined slope. Vertical lines are the only lines that have an undefined slope.

EXAMPLE 2   **Determining the Slope of a Line from Two Points on the Line**

What is the slope of the line going through (2, 4) and (5, 4) in Figure 1.15?

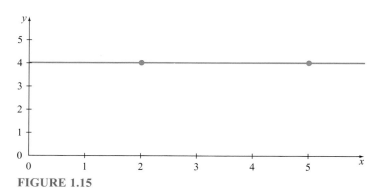

FIGURE 1.15

SOLUTION

$$m = \frac{4 - 4}{5 - 2}$$

$$= \frac{0}{3}$$

$$= 0$$

Any line with a zero slope is a horizontal line.

The absolute value of the slope is referred to as the **magnitude** of the slope. In a general sense, slope is a measure of steepness: The greater the magnitude of the slope, the greater the steepness of the line. The graph of a line with a negative slope falls as the independent variable increases. The graph of a line with a positive slope rises as the independent variable increases.

EXAMPLE 3   **Determining the Sign and Magnitude of a Line's Slope from a Graph**

Determine from the graph in Figure 1.16 which lines have a negative slope and which lines have a positive slope. Then identify the line whose slope has the greatest magnitude.

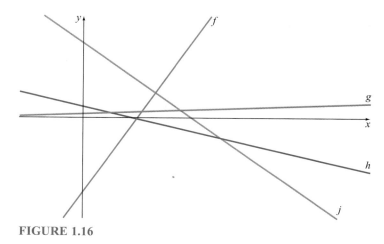

FIGURE 1.16

SOLUTION   Since lines h and j fall as the value of x increases, they have negative slopes. Since lines f and g rise as the value of x increases, they have positive slopes.

The steepest line is the line whose slope has the greatest magnitude. Although both f and j are steep, line f is steeper. Therefore, f has the slope with the greatest magnitude.

Intercepts

In discussing linear functions, it is often useful to talk about where the line crosses the x and y axes. Knowing these **intercepts** helps us to determine the equation of the line.

DEFINITION: y-INTERCEPT

The y-intercept is the point on the graph where the function intersects the y axis. It occurs when the value of the independent variable is 0. It is formally written as an ordered pair $(0, b)$, but b itself is often called the y-intercept.

EXAMPLE 4

Finding the y-intercept of the Graph of a Linear Function

What is the y-intercept of the linear function $y = 3x + 5$?

SOLUTION   At the y-intercept, the x coordinate is 0.

$$y = 3(0) + 5 \qquad \text{Substitute 0 for } x$$
$$= 5$$

So the y-intercept of the function is $(0, 5)$.

DEFINITION: x-INTERCEPT

The x-intercept is the point on the graph where the function intersects the x axis. It occurs when the value of the dependent variable is 0. It is formally written as an ordered pair $(a, 0)$, but a itself is often called the x-intercept.

EXAMPLE 5

Finding the x-intercept of the Graph of a Linear Function

What is the x-intercept of the linear function $y = 3x + 5$?

SOLUTION   At the x-intercept, the y coordinate is 0.

$$y = 3x + 5$$
$$0 = 3x + 5 \qquad \text{Substitute 0 for } y$$
$$-3x = 5$$
$$x = -\frac{5}{3}$$

So the x-intercept of the function is $\left(-\frac{5}{3}, 0\right)$.

EXAMPLE 6

Determining the Slope and Intercepts of the Graph of a Linear Function

Determine the slope, the y-intercept, and the x-intercept of the linear function from its graph (Figure 1.17).

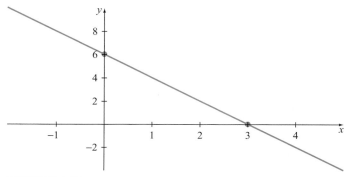

FIGURE 1.17

SOLUTION   Since the graph of the function crosses the y axis at $(0, 6)$, the y-intercept is $(0, 6)$. Since the graph crosses the x axis at $(3, 0)$, the x-intercept is $(3, 0)$.

Since the line is falling as x increases, the slope will be negative. The slope of the line is

$$m = \frac{y_2 - y_1}{x_2 - x_1}$$

$$= \frac{6 - 0}{0 - 3}$$

$$= \frac{6}{-3}$$

$$= -2$$

Linear Equations

The graph of any line may be represented by a linear equation. Vertical and horizontal lines have the simplest linear equations.

EQUATIONS OF VERTICAL AND HORIZONTAL LINES

- The equation of a vertical line passing through a point (a, b) is $x = a$.
- The equation of a horizontal line passing through a point (a, b) is $y = b$.

If we know the slope and y-intercept of a linear function, we are able to easily determine the slope-intercept form of the linear equation.

SLOPE-INTERCEPT FORM OF A LINE

A linear function with slope m and y-intercept $(0, b)$ has the equation

$$y = mx + b$$

EXAMPLE 7   Determining the Slope-Intercept Form of a Line

What is the slope-intercept form of the linear function with slope 5 and y-intercept $(0, 4)$?

SOLUTION   Since $m = 5$ and $b = 4$,

$$y = 5x + 4$$

is the slope-intercept form of the line.

EXAMPLE 8   Finding a Linear Model from a Verbal Description

For breakfast, we decide to eat an apple containing 5.7 grams of dietary fiber and a number of servings of Cheerios™, each containing 3 grams of dietary fiber. (**Source:** Cheerios box label.) Write the equation for our dietary fiber intake as a function of number of servings of cereal.

SOLUTION   We know that if we don't eat any cereal, we will consume 5.7 grams of dietary fiber (from the apple). So the y-intercept of the fiber function is $(0, 5.7)$. Since each serving of Cheerios contains the same amount of fiber, we know that the function is linear. The slope of the function is 3 grams per serving. So the equation of the fiber function is

$$F(n) = 3n + 5.7$$

where $F(n)$ is the amount of dietary fiber (in grams) and n is the number of servings of cereal.

To check our work, we directly calculate the number of grams of dietary fiber in a breakfast containing two servings of cereal and one apple.

$$\text{Fiber} = 3 + 3 + 5.7 = 11.7 \text{ grams}$$

Using the fiber function formula, we get

$$F(2) = 3(2) + 5.7$$
$$= 11.7 \text{ grams}$$

The results are the same, so we are confident that our formula is correct.

EXAMPLE 9   Determining a Linear Model from a Table of Data

The amount of sales tax paid on a clothing purchase in Seattle is a function of the sales price of the clothes, as shown in Table 1.6.

TABLE 1.6

Sales Price (p)	Tax (T)
$20.00	$1.72
$30.00	$2.58
$40.00	$3.44

Source: www.cityofseattle.net.

If the function is linear, write the equation for the sales tax as a function of the sales price.

 UNITS

SOLUTION We must first determine if the function is linear. Since each $10 increase in sales price increases the sales tax by a constant $0.86, we conclude that the function is linear. The slope of the function is

$$m = \frac{2.58 - 1.72 \text{ dollars}}{30 - 20 \text{ dollars}}$$

$$= \frac{0.86 \quad \text{dollar}}{10 \quad \text{dollars}}$$

$$= 0.086 \text{ tax dollar per sales price dollar}$$

(In other words, for each dollar increase in the sales price, the sales tax increases by 8.6¢.) Since a sale of $0 results in $0 sales tax, we know that the y-intercept is $(0, 0)$. The equation of the function is

$$T = 0.086p$$

To check our work, we substitute the point $(40, 3.44)$ into the equation.

$$3.44 = 0.086(40) \qquad p = 40 \text{ and } T = 3.44$$

$$3.44 = 3.44$$

The statement is true, so we are confident that our equation is correct.

EXAMPLE 10   Finding a Linear Model from the Graph of a Linear Function

The graph in Figure 1.18 shows the balance of a checking account as a function of the number of ATM withdrawals from the account.

FIGURE 1.18

(a) Identify the *x*-intercept and *y*-intercept and interpret their physical meaning.

(b) Write the linear equation for the function.

SOLUTION

(a) The *x*-intercept is $(10, 0)$. When there have been 10 withdrawals, the account balance is $0. The *y*-intercept is $(0, 200)$. When there have been no withdrawals, the account balance is $200.

(b) The slope of the function is

$$m = \frac{200 - 0 \text{ dollars}}{0 - 10 \text{ withdrawals}}$$

$$= -\$20 \text{ per withdrawal}$$

So the slope-intercept form of the linear equation is

$$B = -20w + 200$$

where B is the account balance and w is the number of withdrawals.

Finding the Equation of a Line

As demonstrated in Example 10, to find the slope-intercept form of a line from two points, we need to calculate the slope and the *y*-intercept of the function. To do this, we proceed as follows:

1. Calculate the slope
2. Write the function in slope-intercept form, substituting the slope for *m*.
3. Select one of the points and substitute the output value for *y* and the input value for *x*.
4. Solve for *b*.
5. Write the function in slope-intercept form, substituting the *y*-intercept for *b*.

EXAMPLE 11 **Finding the Slope-Intercept Form of a Line from Two Points**

Find the equation of the line passing through the points $(3, 5)$ and $(7, 1)$.

SOLUTION

$$m = \frac{5 - 1}{3 - 7}$$

$$= \frac{4}{-4}$$

$$= -1$$

So the slope is -1. Substituting in the slope and the point $(3, 5)$, we get

$$y = -1 \cdot x + b$$
$$5 = -1(3) + b$$
$$b = 8$$

The *y*-intercept is $(0, 8)$.

The slope-intercept form of the line is $y = -1 \cdot x + 8$ or $y = -x + 8$.

Other Forms of Linear Equations

There are two additional forms of linear equations that are commonly used: standard form and point-slope form.

STANDARD FORM OF A LINE

A linear equation may be written as

$$ax + by = c$$

where a, b, and c are real numbers. If $a = 0$, the graph of the equation is a horizontal line. If $b = 0$, the graph of the equation is a vertical line. In the equation, a and b cannot both be zero.

The standard form of a linear equation is extremely useful when working with systems of equations or solving linear programming problems. We will discuss the standard form of a line further when we introduce these topics.

The point-slope form of a line is especially useful when we know a line's slope and the coordinates of a point on the line.

POINT-SLOPE FORM OF A LINE

A linear function written as

$$y - y_1 = m(x - x_1)$$

has slope m and passes through the point (x_1, y_1).

EXAMPLE 12

Finding a Linear Model from a Verbal Description

Based on data from 1980 to 1999, per capita consumption of milk as a beverage has been decreasing by approximately 0.219 gallon per year. In 1997, the per capita consumption of milk as a beverage was 24.0 gallons. (**Source:** *Statistical Abstract of the United States, 2001*, Table 202, p 129.) Find an equation for the per capita milk consumption as a function of years since 1980.

SOLUTION The slope of the line is $m = -0.219$. The point $(17, 24.0)$ lies on the line, since 1997 is 17 years after 1980. The point-slope form of the line is

$$y - 24.0 = -0.219(t - 17)$$

If preferred, the equation may be rewritten in slope-intercept form,

$$y = -0.219(t - 17) + 24.0$$
$$y = -0.219t + 3.723 + 24.0$$
$$y = -0.219t + 27.723$$

or standard form,

$$0.219t + y = 27.723$$

Graphing Linear Functions

To graph a linear function, we first generate a table of values by substituting different values of x into the equation and calculating the corresponding value of y. For example, if we are given the linear equation $y = 4x - 8$, we may choose to evaluate the function at $x = -1$, $x = 0$, $x = 1$, $x = 2$, and $x = 3$, as shown in Table 1.7.

TABLE 1.7

x	y
-1	-12
0	-8
1	-4
2	0
3	4

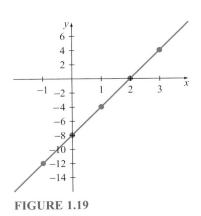

FIGURE 1.19

We then plot the points and connect them with a straight line, as shown in Figure 1.19. (Although we plotted multiple points, only two points are necessary to determine the line.)

EXAMPLE 13   **Graphing a Linear Function**

Graph the function $y = 2x - 3$.

SOLUTION   From the equation, we see that the y-intercept is $(0, -3)$. We need to find only one more point. Evaluating the function at $x = 2$ yields $y = 1$, so $(2, 1)$ is a point on the line. We plot each point and connect the points with a straight line, as shown in Figure 1.20.

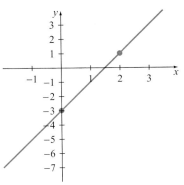

FIGURE 1.20

Recall that the standard form of a line is $ax + by = c$. Graphing a line from standard form is remarkably easy. The y-intercept of the graph occurs when the x value is equal to zero. Similarly, the x-intercept of the graph occurs when the

y value is equal to zero. Using these facts, we can quickly determine the x- and y-intercepts of the function. To find the x-intercept, we set $y = 0$.

$$ax + by = c$$
$$ax + b(0) = c$$
$$ax = c$$
$$x = \frac{c}{a}$$

Notice that the x coordinate of the x-intercept is the constant term divided by the coefficient on the x term.

To find the y-intercept, we set $x = 0$.

$$ax + by = c$$
$$a(0) + by = c$$
$$by = c$$
$$y = \frac{c}{b}$$

Notice that the y coordinate of the y-intercept is the constant term divided by the coefficient on the y term. Using this procedure to find intercepts will allow us to graph linear equations in standard form quickly by hand.

EXAMPLE 14

Graphing a Linear Function

Graph the linear function $2x + y = 4$.

SOLUTION   The x-intercept is found by dividing the constant term by the coefficient on the x term.

$$x = \frac{4}{2}$$
$$= 2$$

The point $(2, 0)$ is the x-intercept.

The y-intercept is found by dividing the constant term by the coefficient on the y term.

$$y = \frac{4}{1}$$
$$= 4$$

The point $(0, 4)$ is the y-intercept. We graph the x- and y-intercepts and then draw the line through the intercepts, as shown in Figure 1.21.

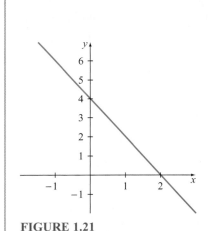

FIGURE 1.21

1.2 Summary

In this section, you learned how to calculate the slope of a linear function and how to interpret the physical and graphical meaning of slope, x-intercept, and y-intercept. You also discovered how to calculate the equation of a linear function from a set of data and learned three different ways to write a linear equation.

1.2 **Exercises**

In Exercises 1–6, calculate the slope of the linear function passing through the points.

1. (2, 5) and (4, 3)   **2.** (−3, 4) and (0, −2)

3. (1.2, 3.4) and (2.7, 3.1)

4. (7, 11) and (9, 2)   **5.** (2, 2) and (5, 2)

6. (4, 3) and (4, 7)

In Exercises 7–12, find the x-intercept and y-intercept of the linear function.

7. $y = 5x + 10$ **8.** $y = -3x + 9$

9. $y = 2x + 11$ **10.** $y = \dfrac{1}{2}x - 2$

11. $3x - y = 4$ **12.** $4x - 2y = 5$

In Exercises 13–18, write the equation of the linear function passing through the points in slope-intercept form, point-slope form, and standard form.

13. (2, 5) and (4, 3)

14. (−3, 4) and (0, −2)

15. (1.2, 3.4) and (2.7, 3.1)

16. (7, 11) and (9, 2)

17. (−2, 2) and (5, 2)

18. (−3.1, 4.5) and (2.1, −3.4)

In Exercises 19–25, graph the line.

19. $y = 4x - 2$ **20.** $x - y = 3$

21. $y - 4 = 0.5(x - 2)$ **22.** $y = -5x + 10$

23. $2x - 3y = 5$ **24.** $y - 9 = -3(x + 2)$

25. $y = -\dfrac{2}{3}x + \dfrac{4}{3}$

In Exercises 26–30, use the graph to determine the equation of the line.

26.

27.

28.

29.

30.

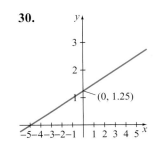

In Exercises 31–35, determine if the table of data represents a linear function. If so, calculate the slope and interpret its real-life significance.

31.   **Average Personal Income**

Year	U.S. Average Personal Income (in terms of year 2000 dollars)
1989	18,593
1999	28,525

Source: www.census.gov.

32.   **Internet Access**

Year	Number of People with Internet Access at Home or at Work (in thousands)
1997	46,305
1998	62,273
1999	83,677
2000	112,949

Source: www.census.gov.

33.   **Take-Home Pay**

Months (since Sept. 2001)	Take-Home Pay (dollars)
0	3,167.30
1	4,350.31

Source: Employee pay stubs.

34.   **Museum Admission Cost**

People in Group	Total Cost of Admission for a Group Visiting the Experience Music Project
3	$59.85
4	$79.80
5	$99.75
6	$119.70

Source: www.emplive.com.

35.   **Solid Waste Disposal**

Clean Wood (pounds)	Cost to Dispose of Clean Wood at Enumclaw Transfer Station
500	$18.75
700	$26.25
900	$33.75
1,000	$37.50

Source: www.dnr.metrokc.gov.

In Exercises 36–38, find the linear equation that models the data and answer the specified questions.

36.   **Nutrition**   The Recommended Daily Allowance for dietary fiber is 25 grams for a person on a 2,000-calorie-per-day diet. A $\frac{3}{4}$-cup serving of Post® Fruity Pebbles® contains 0.2 gram of dietary fiber. (**Source:** Package labeling.) A 1-cup serving of 2 percent milk fortified with vitamin A doesn't contain any dietary fiber. A large banana $\left(8 \text{ to } 8\frac{7}{8} \text{ inches long}\right)$

contains 3.3 grams of dietary fiber. (**Source:** www.nutri-facts.com.) How many servings of Fruity Pebbles with milk would you have to eat along with a large banana to consume 8 grams of dietary fiber? (Round up to the nearest number of servings.)

37.   **Nutrition**   The Recommended Daily Allowance for dietary fiber is 25 grams for a person on a 2,000-calorie-per-day diet. A $\frac{3}{4}$-cup serving of General Mills Wheaties® contains 2.1 grams of dietary fiber. (**Source:** Package labeling.) A 1-cup serving of 2 percent milk fortified with vitamin A doesn't contain any dietary fiber. A large $\left(8 \text{ to } 8\frac{7}{8} \text{ inches long}\right)$ banana contains 3.3 grams of dietary fiber. (**Source:** www.nutri-facts.com.) How many servings of Wheaties with milk would you have to eat along with a large banana to consume 8 grams of dietary fiber? (Round up to the nearest number of servings.)

38.   **Nutrition**   The Recommended Daily Allowance for dietary fiber is 30 grams for a person on a 2,500-calorie-per-day diet. A large apple $\left(3\frac{1}{4} \text{ inches in diameter}\right)$ contains 5.7 grams of dietary fiber. A large banana (8 to $\left(8\frac{7}{8} \text{ inches long}\right)$ contains 3.3 grams of dietary fiber. (**Source:** www.nutri-facts.com.) How many apples would you have to eat along with a banana to consume 30 grams of dietary fiber?

Exercises 39–51 are intended to challenge your understanding of linear equations and their graphs.

39. What is the equation of the line that passes through $(4, 7)$ and $(4, -1)$?

40. Do the two equations represent the same line? Explain.

$$3x + 4y = 12; \quad y + 3 = -0.75(x - 8)$$

41. Explain why the equation of a vertical line cannot be written in slope-intercept or point-slope form.

42.   **Alaska Population**   The U.S. Census Bureau projects that the population of Alaska will grow at an approximate rate of 9,321 people per year between 1995 and 2025. It estimates that 791,000 people will be living in Alaska in 2025. (**Source:** www.census.gov.) What is the projected population of Alaska in 2035?

43. Are all lines functions? Explain.

44. Two lines are said to be parallel if they do not intersect. If $f(x) = mx + b$ and $g(x) = nx + c$ are parallel lines, what conclusions can you draw about b, c, m, and n?

45. Two lines are perpendicular if the slope of one line is the negative reciprocal of the slope of the other line. Find the equation of the line perpendicular to $y = 4$ that passes through the point $(3, -2)$.

46. **M** **World Population** Based on data from 1980 to 2001, the population of the world may be modeled by $P = (81.3t + 4460)$ million people, where t is the number of years since 1980. (**Source:** Modeled from *Statistical Abstract of the United States, 2001,* Table 1324, p. 829.) According to the model, how quickly will the world population be increasing in 2010?

47. The equation of a nonvertical line is given by $ax + by = c$. What are the slope, x-intercept, and y-intercept of the line?

48. Is the function shown in the table linear? Explain.

x	y
2	11
5	17
8	23
10	29
14	35

49. In order for the function shown in the table to be linear, what must be the value of b?

x	y
2	7
5	13
8	19
10	b
14	31

50. In order for the function shown in the table to be linear, what must be the value of a?

x	y
2	6
5	12
8	18
a	26
14	30

51. What is the point of intersection of the lines $x = a$ and $y = b$?

1.3 **Linear Models**

- Use technology to model linear and near-linear data
- Use a linear equation to describe the relationship between directly proportional quantities
- Recognize and model naturally occurring linear, near-linear, and piecewise linear relationships

GETTING **STARTED** The amount of money deducted from a paycheck for Social Security, the annual income of a sales representative, and the profit from a club car wash may be modeled by linear functions. In this section, we will look at a wide variety of information encountered in everyday life and learn how to determine if the data are linear, near-linear, or piecewise linear. We will also demonstrate how to use technology to determine the linear equation that best fits a data set. Additionally, we will show how to find linear equations to model real-life phenomena.

Direct Proportions

Two quantities are said to be **directly proportional** if the ratio of the output to the input is a constant. Quantities that are directly proportional to each other may be modeled by a linear equation of the form

$$y = kx$$

The constant $k = \dfrac{y}{x}$ and is called the **constant of proportionality.**

EXAMPLE 1 **Determining If a Data Set Is Directly Proportional**

For the data in Table 1.8, determine whether the employee's weekly income is directly proportional to the number of hours worked. If so, find a linear function that models the data.

TABLE 1.8

Hours Worked (h)	Income (I)
35.0	$210.00
38.5	$231.00
42.3	$253.80

SOLUTION Calculate the ratio $\dfrac{I}{h}$ for each ordered pair.

UNITS

$$\frac{\$210}{35 \text{ hours}} = \$6/\text{hr}$$

$$\frac{\$231}{38.5 \text{ hours}} = \$6/\text{hr}$$

$$\frac{\$253.80}{42.3 \text{ hours}} = \$6/\text{hr}$$

Since $\dfrac{I}{h} = 6$ for each ordered pair, the weekly income is directly proportional to the hours worked with a constant of proportionality $k = 6$. The linear function is

$$I = 6h$$

EXAMPLE 2 **Determining If a Data Set Is Directly Proportional**

Determine whether the cost of filling a gas tank as given in Table 1.9 is directly proportional to the number of gallons put into the tank. If so, find a linear function that models the data.

Number of Gallons (g)	Cost (C)
3.50	$5.25
6.28	$9.41
17.34	$25.99

SOLUTION   Calculate the ratio $\dfrac{C}{g}$ for each ordered pair.

UNITS

$$\dfrac{\$5.25}{3.50 \text{ gallons}} = 1.500 \text{ dollars per gallon}$$

$$\dfrac{\$9.41}{6.28 \text{ gallons}} \approx 1.498 \text{ dollars per gallon}$$

$$\dfrac{\$25.99}{17.34 \text{ gallons}} \approx 1.499 \text{ dollars per gallon}$$

The ratios are close to constant, but they are different. But wait! Although gas prices are given to the tenth of a cent (e.g., $1.599), the cost of buying gasoline is rounded to the nearest cent. Could the difference in ratios be due to round-off error? Let's see. It looks as if $k \approx 1.499$. The associated linear model is

$$C = 1.499g$$

How good is the model?

TABLE 1.10

Number of Gallons	Cost	Model Estimate
3.50	$5.25	$5.2465
6.28	$9.41	$9.4137
17.34	$25.99	$25.993

As we see in Table 1.10, if we round to the nearest cent, the model fits the data perfectly. From our personal experience purchasing gasoline, we know that the cost of filling the gas tank is directly proportional to the number of gallons we put in. Our personal experience validates the accuracy of the model.

EXAMPLE   3

Determining If a Data Set Is Directly Proportional

A teacher's salary may fluctuate over the course of the year. (Teaching contracts typically cover a nine-month period.) A teacher recorded his paycheck amount and the OASI (Social Security) deduction over a three-month period in Table 1.11.

TABLE 1.11

Paycheck Amount (p)	OASI Deduction (S)
$1,474.77	$91.44
$2,386.17	$145.58
$2,795.74	$170.98

Is the OASI deduction directly proportional to the paycheck amount? If so, find a linear function that models the data.

SOLUTION Calculate the ratio $\dfrac{S}{p}$ for each ordered pair.

$$\frac{91.44}{1474.77} \approx 0.06200 \text{ OASI dollar per paycheck dollar}$$

$$\frac{145.58}{2386.17} \approx 0.06101 \text{ OASI dollar per paycheck dollar}$$

$$\frac{170.98}{2795.74} \approx 0.06116 \text{ OASI dollar per paycheck dollar}$$

The ratios are not constant, so, technically, we may conclude that the data are not directly proportional. (The tax law requires that 6.2 percent of the gross pay amount reduced by the cost of medical premiums be paid into OASI. In the first check, no amount of money was paid toward the teacher's medical plan. For the second two checks, the gross pay amount was reduced by the cost of the medical premiums before calculating the 6.2 percent OASI amount.)

Although the data in Example 3 are not linear, the fact that the ratios are nearly constant indicates that the data are near-linear. It looks as if $k \approx 0.0615$. A linear model for the data is

$$S = 0.0615p$$

How good is the model? Let's check.

$$S = 0.0615(1474.77) = 90.70$$
$$S = 0.0615(2386.17) = 146.75$$
$$S = 0.0615(2795.74) = 171.94$$

Table 1.12 summarizes the results.

TABLE 1.12

Paycheck Amount (p)	OASI Deduction (S)	Model Estimate (M)	Error ($S - M$)
$1,474.77	$91.44	$90.70	0.74
$2,386.17	$145.58	$146.75	−1.17
$2,795.74	$170.98	$171.94	−0.96

The error represents the difference between the actual data and the model estimate. The closer the magnitude of the error is to 0, the better the model fits the data. Is there a linear equation that better fits the data? Using technology, we can find the linear equation that best fits the data.

Linear Regression and the Line of Best Fit

Any number of lines may be used to model a near-linear set of data. We want to find the **line of best fit,** the line that best fits the data. To do this, we minimize the sum of the squares of the errors. Figure 1.22 shows three data points and a line used to model the data.

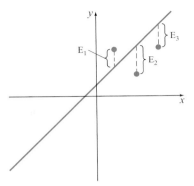

FIGURE 1.22

The vertical lines show the difference between the data output and the model estimate. The difference between the actual value and the estimated value of the individual data points (the error) is designated by $E1$, $E2$, and $E3$. To find the line of best fit, we square each of the errors, add the results together, and then minimize the sum using advanced calculus techniques. That is, we minimize $E = (E_1)^2 + (E_2)^2 + (E_3)^2$. This process is called **linear regression.** Fortunately, the TI-83 Plus calculator has a linear regression feature that does the necessary calculations automatically.

TECHNOLOGY **TIP**

Linear Regression

1. Press (2nd) then (0), scroll to `DiagnosticOn`, and press (ENTER) twice. This will ensure that the correlation coefficient r and the coefficient of determination r^2 will appear.

```
CATALOG
 DependAuto
 det(
 DiagnosticOff
▶DiagnosticOn
 dim(
 Disp
 DispGraph
```

2. Bring up the Statistics Menu by pressing the (STAT) button.

```
EDIT CALC TESTS
1:Edit…
2:SortA(
3:SortD(
4:ClrList
5:SetUpEditor
```

3. Bring up the List Editor by selecting `1:Edit…` and pressing (ENTER).

```
L1    L2    L3    2
1     9     ------
2     7
3     5
----- -----
L2(4) =
```

(Continued)

4. Clear the lists. If there are data in the list, use the arrows to move the cursor to the list heading L1. Press the (CLEAR) button and press (ENTER). This clears all of the list data. Repeat for each list with data. (*Warning:* Be sure to use (CLEAR) instead of (DELETE). (DELETE) removes the entire column.)

5. Enter the numeric values of the *inputs* in list L1, pressing (ENTER) after each entry.

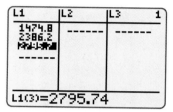

6. Enter the numeric values of the *outputs* in list L2, pressing (ENTER) after each entry.

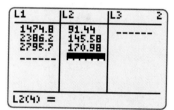

7. Return to the Statistics Menu by pressing the (STAT) button.

8. Bring up the Calculate Menu by using the arrows to select CALC.

9. Calculate the linear equation of the model by selecting 4:LinReg(ax+b) and pressing (ENTER) twice. The line of best fit is $y = 0.06008x + 2.6928$ and has correlation coefficient $r = 0.9999$.

```
LinReg
 y=ax+b
 a=.060078251
 b=2.692769342
 r²=.9998949304
 r=.9999474638
```

The correlation coefficient is a number r between -1 and 1 that measures the degree to which two variables are linearly related. If the line of best fit has a positive slope, $r > 0$. If the line of best fit has a negative slope, $r < 0$. If there is a perfect linear relationship between the two variables, $|r| = 1$. The more closely related the variables, the closer the correlation coefficient will be to ± 1. If the correlation coefficient is 0, the variables are not linearly related.

The coefficient of determination is equal to r^2. For linear functions, the coefficient of determination is the square of the correlation coefficient. (The formulas for calculating the correlation coefficient and coefficient of determination will not be shown here but may be found in many college-level statistics textbooks.)

In Example 3, the correlation coefficient was $r = 0.9999474638$. This number is extremely close to 1, so the line of best fit fits the data well. The linear function that best fits the data from Example 3 is $S = 0.06008p + 2.6928$. We can readily see from Table 1.13 that the model we found using linear regression fits the data better than the first model.

TABLE 1.13

Paycheck Amount (p)	OASI Deduction (S)	First Model $S = 0.0615p$	Linear Regression Model $S = 0.06008p + 2.6928$
$1,474.77	$91.44	$90.70	$91.30
$2,386.17	$145.58	$146.75	$146.05
$2,795.74	$170.98	$171.94	$170.66

Sometimes it is useful to graph the model. One method of doing this is detailed in the following Technology Tip.

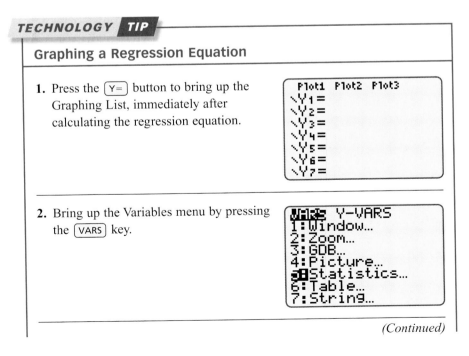

TECHNOLOGY TIP

Graphing a Regression Equation

1. Press the [Y=] button to bring up the Graphing List, immediately after calculating the regression equation.

2. Bring up the Variables menu by pressing the [VARS] key.

(Continued)

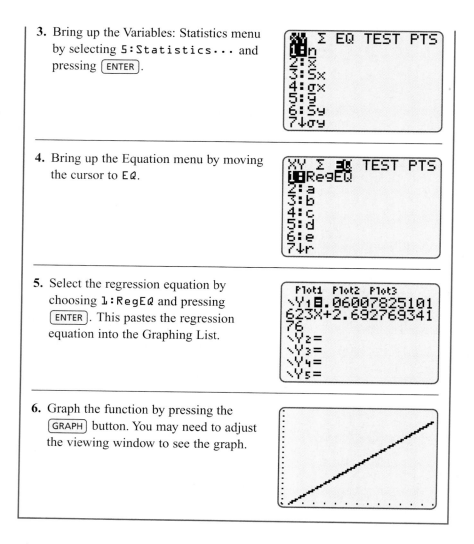

3. Bring up the Variables: Statistics menu by selecting 5:Statistics... and pressing ENTER.

4. Bring up the Equation menu by moving the cursor to EQ.

5. Select the regression equation by choosing 1:RegEQ and pressing ENTER. This pastes the regression equation into the Graphing List.

6. Graph the function by pressing the GRAPH button. You may need to adjust the viewing window to see the graph.

The graph of a set of discrete data is called a **scatter plot.** We can draw a scatter plot by hand, using graph paper, or we can use a calculator to draw the scatter plot.

TECHNOLOGY **TIP**

Drawing a Scatter Plot

1. Bring up the Statistics Plot menu by pressing the [2nd] button, then the [Y=] button.

2. Open Plot1 by pressing [ENTER].

3. Turn on Plot1 by moving the cursor to On and pressing [ENTER]. Confirm that the other menu entries are as shown.

4. Graph the scatter plot by pressing [ZOOM] and scrolling to 9:ZoomStat. Press [ENTER]. This will graph the entire scatter plot along with any functions in the Graphing List. The ZoomStat feature automatically adjusts the viewing window so that all of the data points are visible.

The graph shows the data points along with the linear model. We see that the linear model fits the data well.

EXAMPLE 4   **Using the Line of Best Fit to Forecast Sales**

The annual sales of McDonalds Corporation are shown in Table 1.14.

TABLE 1.14

Years Since 1990 (t)	Franchised Sales (S) (millions of dollars)
1	12,959
2	14,474
3	15,756
4	17,146
5	19,123
6	19,969
7	20,863
8	22,330
9	23,830
10	24,463
11	24,838

Source: www.mcdonalds.com.

(a) Draw a scatter plot of the data. Will a linear function fit the data well?
(b) Find the equation of the linear model that best fits the data.
(c) Graph the linear model simultaneously with the scatter plot. How well does the model predict the data?
(d) What does the slope of the function tell us about the company's sales?

SOLUTION

(a) Figure 1.23 shows a scatter plot of the data.

FIGURE 1.23

The sales revenue increases every year, but the amount by which it increases varies from year to year. Since the rate of increase is not constant, a linear function won't fit the data perfectly. However, the data do appear to be near-linear, so a linear model may fit the data well.

(b) Using linear regression, we determine that the linear model for sales is $S = 1233.5t + 12{,}213$ million dollars where t is the number of years since the end of 1990.

(c)

FIGURE 1.24

As shown in Figure 1.24 and Table 1.15, the linear model fits the data fairly well.

TABLE 1.15

Years Since 1990 (t)	Franchised Sales (S) (millions of dollars)	Estimated Sales (E) (millions of dollars)	Error ($S - E$) (millions of dollars)	% Error $\left(\dfrac{S - E}{S}\right)$
1	12,959	13,447	−488	−3.76%
2	14,474	14,680	−206	−1.42%
3	15,756	15,914	−158	−1.00%
4	17,146	17,147	−1	−0.01%
5	19,123	18,381	743	3.88%
6	19,969	19,614	355	1.78%
7	20,863	20,848	16	0.07%
8	22,330	22,081	249	1.12%
9	23,830	23,315	516	2.16%
10	24,463	24,548	−85	−0.35%
11	24,838	25,782	−944	−3.80%

Although the actual sales and the estimated sales from the model differed by as much as 944 million dollars, the model projections were within 3.88 percent of the actual values.

(d) The slope of the sales function is 1233.5 million dollars per year. According to the model, we expect that, on average, the sales will increase by about 1233.5 million dollars per year. We may use this information to help us determine whether to buy stock in the company. We should be aware that past performance is no guarantee of future results. Before making a stock purchase, we would want to look at other indicators, such as earnings per share, recent news, analyst ratings, and so on.

Finding Linear Functions to Model Real-Life Phenomena

Not all of the data we encounter in the real world are laid out nicely in an input-output table. Often we have to ferret out the information needed to formulate a model. The next two examples will illustrate real-life situations where using mathematics can give us financial leverage.

EXAMPLE 5   **Using Linear Models to Make Business Decisions**

Many companies offer their salespeople a base salary plus a commission based on the employee's sales. In November 2001, a Nebraska firm posted an ad on an electronic bulletin board advertising a sales representative position paying a $36,000 base salary plus commission. (**Source:** www.dice.com.) Based on her past sales experience, a saleswoman estimates that she can generate $600,000 in sales annually. To maintain her standard of living, she needs to earn a total of $63,000 annually.

(a) What commission rate (as a percentage of sales) must she earn to maintain her standard of living?

(b) Write the equation for her annual income as a function of her annual sales dollar volume.

(c) If she wants to increase her annual income to $75,000, by how much will she have to increase her annual sales?

(d) If she wanted to increase her annual income to $75,000 without increasing her sales volume, what commission rate must she receive?

(e) How could the answers to the previous questions help her during salary negotiations?

SOLUTION

(a) We have

$$I = ms + b$$

where I is the annual income (in thousands of dollars) and s is the annual sales (in thousands of dollars). If she doesn't sell anything (assuming she doesn't get fired), she will still earn $36,000, so $(0, 36)$ is the y-intercept. Thus

$$I = ms + 36$$

When $s = 600$, $I = 63$, so

$$63 = m \cdot 600 + 36$$
$$27 = 600m$$
$$m = 0.045$$

She must earn a 4.5 percent commission to maintain her standard of living.

(b) The equation for her annual income is $I = 0.045s + 36$ thousand dollars, where s is the sales volume (in thousands of dollars).

(c) Setting $I = 75$ and solving for s, we get

$$75 = 0.045s + 36$$
$$39 = 0.045s$$
$$s = 866.667$$

She must increase her annual sales from $600,000 to $866,667 to increase her annual income to $75,000.

(d) Letting m be the variable and setting I equal to 75, we get

$$75 = m \cdot 600 + 36$$
$$39 = 600m$$
$$m = 0.065$$

So to increase her income to $75,000 without increasing sales, she must earn a 6.5 percent commission.

(e) We know that her minimum acceptable commission rate is 4.5 percent. To increase her salary without having to increase her workload, she will have to convince the employer to increase her commission rate to 6.5 percent. If the employer is unwilling to raise the commission rate, she will have to bring in an additional $266,667 in sales to earn $75,000. This information can give her leverage in salary negotiations.

EXAMPLE 6

Using Linear Models to Make Business Decisions

In 2001, BeautiControl Cosmetics sold a new consultant a demonstration package for $119.82. (**Source:** BeautiControl Cosmetics consultant.) The consultant's director told her that the average revenue from each in-home party is $250. The consultant's cost for the cosmetics is 50 percent of the sales price.

(a) Find the linear model for the consultant's cost as a function of the dollar volume of the cosmetics sold.
(b) Find the linear model for the consultant's revenue as a function of the dollar volume of cosmetics sold.
(c) Find the linear model for the consultant's profit as a function of the dollar volume of cosmetics sold.
(d) How many dollars of cosmetics must the consultant sell before she begins to make a profit?
(e) How many in-home parties must the consultant hold in order to begin making a profit?

SOLUTION

(a) Let x be the sales volume in dollars. Then the cost function C is

$$C(x) = 0.5x + 119.82$$

The cost of the demonstration package, $119.82, is the fixed cost. The consultant's variable cost, $0.5x$ dollars, is dependent upon her sales dollar volume.

(b) The revenue function R is

$$R(x) = x$$

The consultant's revenue is equal to her sales dollar volume.

(c) The profit function is the difference between the revenue and cost functions.

$$P(x) = R(x) - C(x)$$
$$= x - (0.5x + 119.82)$$
$$= 0.5x - 119.82$$

(d) We need to find the break-even point (the sales dollar volume that makes the profit equal zero). Setting the profit equal to zero, we get

$$0 = 0.5x - 119.82$$

$$119.82 = 0.5x$$

$$x = 239.64$$

So to break even, she must sell \$239.64 worth of cosmetics.

(e) Dividing the sales dollar volume needed to break even by the amount of revenue per party, we get

UNITS

$$\frac{239.64 \text{ dollars}}{250.00 \frac{\text{dollars}}{\text{party}}} = 0.95856 \text{ parties}$$

It doesn't make sense to talk about fractions of a party, so we round up. The consultant should begin to make a profit at her first party.

Piecewise Linear Models

A function of the form

$$f(x) = \begin{cases} ax + b & x \le c \\ dx + g & x > c \end{cases}$$

is called a **piecewise linear function.** The rule used to calculate $f(x)$ depends upon the value of x. In this case, if x is less than or equal to c, the rule is $f(x) = ax + b$. If x is greater than c, the rule is $f(x) = dx + g$.

A common piecewise linear function is the absolute value function $f(x) = |x|$. This function makes negative numbers positive and leaves positive numbers unchanged. In practical terms, it "erases" the negative sign of any negative number. The absolute value function is formally defined as a piecewise linear function.

$$f(x) = |x|$$
$$= \begin{cases} x & x \ge 0 \\ -x & x < 0 \end{cases}$$

What is $|-5|$? Since $-5 < 0$, we apply the second rule of the piecewise function.

$$|-5| = -(-5)$$
$$= 5$$

Piecewise linear functions are very common in a variety of real-life situations, such as the ones given in Examples 7 and 8.

EXAMPLE 7

Using a Piecewise Linear Model to Forecast Monthly Payments

An orthodontist and the parents of a girl scheduled to receive braces agreed upon the following payment plan: a $550 down payment when treatment begins and $100 per month until the treatment is paid in full. The total cost for the treatment is $1250, and the treatment is expected to take 18 months to complete. (**Source:** Author's personal payment plan.)

(a) How many months after treatment begins will the treatment be paid in full?

(b) Write the equation for the balance due as a function of the number of months since treatment began.

SOLUTION

(a) The patient's parents pay $550 when treatment begins, so $700 remains for the balance to be paid in full. Since $100 is paid each month, the treatment will be paid in full 7 months later.

(b)

$$B(t) = \begin{cases} -100t + 700 & 0 \le t \le 7 \\ 0 & 7 < t \le 18 \end{cases}$$

where $B(t)$ is the balance of the account t months after treatment begins. Since payments are made monthly, the function is defined for whole number values of t.

 We initially use a linear function for $B(t)$, since the account balance starts at $700 and is reduced by a constant $100 per month. However, after 7 months, the account balance has reached $0. If we continued to use the linear function, the balance would end up being a negative number, which doesn't make sense in the context of the problem. Therefore, no further payments are made after the seventh month, and the account balance remains at $0 from the end of the seventh month through the eighteenth month.

EXAMPLE 8

Using a Piecewise Linear Model to Make Spending Decisions

Experience Music Project, an interactive music museum in Seattle, charges adult visitors a $19.95 admission fee. Groups of 15 or more may enter for $14.50 per person. (**Source:** www.emplive.com.)

(a) Draw a scatter plot for group cost as a function of group size.

(b) Write the total cost of admission as a function of the number of people in a group. Confirm that the function and the scatter plot are in agreement.

(c) Calculate the cost of admitting a group of 14 people and the cost of admitting a group of 15 people.

(d) Determine the largest group we could bring in and still remain below the cost of a 14-person group.

SOLUTION

(a)

FIGURE 1.25

(b)

$$C(x) = \begin{cases} 19.95x & x < 15 \\ 14.5x & x \ge 15 \end{cases}$$

where x is the number of people in the group and C is the total cost in dollars.

(c) $C(14) = 19.95(14) = \$279.30$ is the cost for a group of 14 people.

$C(15) = 14.5(15) = \$217.50$ is the cost for a group of 15 people.

(d) From the scatter plot, it looks as if the maximum number of people we could bring in and still remain below the 14-person group price is 19 people. We can confirm the result algebraically.

$$279.30 = 14.5x$$

$$x \approx 19.26$$

Since it doesn't make sense to talk about 0.26 of a person, we round down to 19.

$$C(19) = 14.5(19)$$

$$= 275.50$$

The cost for a 19-person group is $275.50. This is slightly less than the $279.30 cost for a 14-person group.

Determining Which Modeling Method to Use

Although linear regression is a powerful tool for creating linear models, many linear models can be constructed algebraically, as shown in Examples 5 through 8. It is good practice to first attempt to find the linear model algebraically. A common pitfall of beginning students is to use the calculator to find the linear model when it is actually easier to find the model algebraically.

Whenever the data set contains exactly two points, we should find the linear model algebraically. In addition, there are other key phrases that alert us to the fact that a linear model may be constructed algebraically. Table 1.16 details how to interpret the mathematical meaning of some commonly occurring phrases.

TABLE 1.16

Phrase	Mathematical Meaning
constant rate of change	slope
increasing at a rate of 20 people per year	The slope of the linear function is $20\dfrac{\text{people}}{\text{year}}$.
tickets cost $37 per person	The slope is $37\dfrac{\text{dollars}}{\text{person}}$.
sales increase by 100 units for every $1 decrease in price	The slope is $$m = \dfrac{100\ \text{units}}{-1\ \text{dollar}}$$ $$= -100\ \text{units per dollar}.$$
initial price of $1.25	The y-intercept is 1.25.
There are 350 chairs today. The number of chairs is increasing by 10 chairs per month.	The y-intercept is 350 and the slope is 10.
The price is $2.25 and is decreasing at a constant rate of $0.02 per day.	The y-intercept is 2.25 and the slope is -0.02.

By applying the techniques demonstrated in the table, you will increase your comfort level in working with real-world problems.

1.3 Summary

In this section, you looked at a wide variety of information encountered in everyday life, and you learned how to model linear, near linear, and piecewise linear data sets. You also learned how to use technology to determine the line of best fit to model real-life phenomena.

1.3 Exercises

In Exercises 1–5, do the following:

(a) Draw the scatter plot.
(b) Find the equation of the line of best fit.
(c) Interpret the meaning of the slope and the y-intercept of the model.
(d) Explain why you do or do not believe that the model would be a useful tool for businesses and/or consumers.

1.   **Share Price**

Month in 2000	Harbor Capital Appreciation Fund Share Price on Last Day of Month
10	$48.16
11	$41.50
12	$35.58

Source: www.quicken.com.

2. Share Price

Month in 2001	CREF Stock Retirement Annuity Share Price on Last Day of Month
3	$164.5760
6	$173.1588
9	$146.4102

Source: TIAA-CREF statement.

  3. University Enrollment

Washington State Public University Enrollment	
Years Since 1990 (t)	Students (S)
0	81,401
1	81,882
2	83,052
3	84,713
4	85,523
5	86,080
6	87,309
7	89,365
8	90,189
9	91,543
10	92,821

Source: Washington State Higher Education Coordinating Board.

 4. Per Capita Personal Income

Per Capita Personal Income—Utah	
Years Since 1993 (t)	Personal Income (P) (dollars)
0	16,830
1	17,638
2	18,508
3	19,514
4	20,613
5	21,594
6	22,305
7	23,436

Source: Bureau of Economic Analysis (www.bea.gov).

  5. Per Capita Income Ranking

Years since 1995	North Carolina Per Capita Income Ranking (out of 50 states)
0	42
1	34
2	32
3	28
4	28

Source: http://www.census.gov/statab.

In Exercises 6–16, formulate the linear or piecewise linear model for the real-life scenario. If a linear model represents a directly proportional relationship between the independent and dependent variables, so state.

6.   **Personal Income** In November 2001, Hall Kinion advertised an Account Executive position in Savannah, Georgia, for a salesperson with experience selling Internet services. The job advertised a $35K base salary and $24K to $48K in commissions in the first year. (**Source:** www.dice.com.)

(a) Assuming a 20 percent commission rate, how many dollars in sales would you have to generate to earn $64K annually?

(b) If you were in the sales field, why would this type of information be valuable to you?

7. 🌐 **Solid Waste Disposal** King County charges $82.50 per ton to dispose of garbage at area transfer stations. A minimum disposal fee of $13.72 per entry is charged for vehicles disposing of trash. In addition to the disposal fee, a moderate risk waste (MRW) fee of $2.61 per ton is assessed. A minimum MRW fee of $1 is charged on all transactions. Tax is charged on the combined price of the disposal and moderate risk waste fees. The tax rate is 3.6 percent. The weight of the trash is rounded to the nearest 20 pounds before the fees are calculated. (**Source:** www.dnr.metrokc.gov.)

(a) Write the equation for disposal cost as a function of the number of pounds of refuse.

(b) How much will it cost to drop off 230 pounds of trash?

(c) How much will it cost to drop off 513 pounds of trash?

(d) If you ran a construction company, how could you use the results of this exercise?

8. 🌐 **Cellular Phone Plan** In 2002, Qwest offered a 150-minute cellular plan including free long distance and free roaming to its Seattle-area customers for $29.99 per month. Additional time cost $0.35 per minute or portion of a minute. (**Source:** Qwest customer monthly statement.)

(a) Write the cellular phone cost equation as a function of the number of minutes used.

(b) How could you use the cost equation if you were enrolled in the plan?

9. 🌐 **Cellular Phone Plan** Sprint PCS offered a service plan to its Seattle-area customers for $29.99 per month that included 200 anytime minutes. The plan also included free long distance. Additional minutes or portions of a minute were $0.40 each. (**Source:** www1.sprintpcs.com.)

(a) Determine the cost equation for the Sprint PCS plan as a function of the number of anytime minutes used.

(b) Is the Sprint plan or the Qwest plan (from Exercise 8) a better deal for a consumer who uses 300 anytime minutes monthly? Explain.

10. 🌐 **Solid Waste Disposal** The Enumclaw Transfer/Recycling Station accepts clean wood (stumps, branches, etc.) in addition to household garbage. The station charges $75 per ton with a minimum fee of $12.75 per entry for vehicles disposing of clean wood. (**Source:** www.dnr.metrokc.gov.)

(a) How many pounds of clean wood could be dropped off at the station without exceeding the $12.75 minimum fee?

(b) Write the equation for the disposal cost as a function of the number of pounds of clean wood.

(c) If you ran a landscaping business, how could you use the results of parts (a) and (b)?

11. 🌐 **Used Car Value** In 2001, the average retail price of a 2000 Toyota Land Cruiser 4-Wheel Drive was $43,650. The average retail price of a 1995 Toyota Land Cruiser 4-Wheel Drive was $21,125. (**Source:** www.nadaguides.com.)

(a) Find a linear model for the value of a Toyota Land Cruiser in 2001 as a function of its production year.

(b) Use your linear model to predict the value of a 1992 Toyota Land Cruiser in 2001.

(c) A 1992 Toyota Land Cruiser had an average retail price of $14,325 in 2001. How good was your linear model at predicting the value of the vehicle?

12. 🌐 **Nutrition** The Recommended Daily Allowance (RDA) for fat for a person on a 2,000-calorie-per-day diet is less than 65 grams. A McDonalds Big Mac® sandwich contains 34 grams of fat. A large order of French fries contains 26 grams of fat. (**Source:** www.mcdonalds.com.)

(a) Write the equation for fat grams consumed as a function of large orders of French fries.

(b) Write the equation for fat grams consumed as a function of Big Macs.

(c) How many large orders of French fries can you eat without exceeding the RDA for fat?

(d) How many Big Macs can you eat without exceeding the RDA for fat?

(e) How many combination meals (Big Mac and French fries) can you eat without exceeding the RDA for fat?

13.   **Nutrition**   The Recommended Daily Allowance (RDA) for fat for a person on a 2,500-calorie-per-day diet is less than 80 grams. A McSalad Shaker Chef Salad® contains 8 grams of fat. A package (44.4 ml) of ranch salad dressing contains 18 grams of fat. (**Source:** www.mcdonalds.com.)

 (a) Write the equation for fat grams as a function of Chef Salads.

 (b) How many salads can you eat and remain below the RDA for fat?

 (c) If you use one package of ranch salad dressing, how many salads can you eat and remain below the RDA for fat?

 (d) The RDA for fat for a person on a 2000-calorie diet is 65 grams. If a person on a 2,000-calorie diet uses one package of ranch salad dressing, how many salads can that person eat and remain below the RDA for fat?

14.   **Sales Tax**   The Picture People photography studio in Tacoma, Washington, charges $40 for one of its promotional photo packages. The total purchase price for the promotional package, including tax, is $43.20. (**Source:** Picture People sales receipt.)

 (a) Write the equation for Tacoma sales tax as a function of the pretax purchase price.

 (b) A family has $60 cash to pay for pictures. What is the maximum pretax purchase price that the family can afford?

15.   **Airline Ticket Cost**   The total cost of an airline ticket includes the published fare, a federal flight segment tax, a federal security fee, and an airport Passenger Facility Charge (PFC). The flight segment tax is $3 on each segment of the itinerary. (A segment is a takeoff and landing.) The PFC varies from airport to airport; it ranges from $3 to $4.50, and it is imposed by an airport on enplaning passengers. When a passenger is departing from or connecting at any airport, that airport's PFC will apply in addition to the fare. Likewise, a $2.50 security fee is imposed on all enplaning passengers.

 A Southwest Airlines refundable airline ticket from Seattle to Phoenix is advertised at $222. Write the maximum total cost of a Southwest (**Source:** www.southwest.com.) airline ticket from Seattle to Phoenix as a function of the number of

segments. (Assume that a passenger travels exactly one segment per airplane. That is, if the plane stops, the passenger changes planes.)

16.   **Airline Ticket Cost**   The total cost of an airline ticket includes the published fare, a federal flight segment tax, a federal security fee, and an airport Passenger Facility Charge (PFC). The flight segment tax is $3 on each segment of the itinerary. (A segment is a takeoff and landing.) The PFC varies from airport to airport; it ranges from $3 to $4.50, and it is imposed by an airport on enplaning passengers. When a passenger is departing from or connecting at any airport, that airport's PFC will apply in addition to the fare. Likewise, a $2.50 security fee is imposed on all enplaning passengers.

 A Southwest Airlines refundable round-trip airline ticket from Seattle to Phoenix is advertised at $444. (**Source:** www.southwest.com.) Write the maximum total cost of a round-trip airline ticket from Seattle to Phoenix as a function of the number of segments.

Exercises 17–25 are intended to challenge your understanding of linear models.

17. The standard form of a linear function in three dimensions is given by $ax + by + cz = d$, where x, y, and z are variables and a, b, c, and d are constants. A business determines that the ***revenue*** generated by selling x cups of coffee, y bagels, and z muffins is given by

$$R = ax + by + cz$$

What is the practical meaning of the constants a, b, and c?

18. A business determines that the ***cost*** of producing x cups of coffee, y bagels, and z muffins is given by

$$C = ax + by + cz + d$$

What is the practical meaning of the constants a, b, c, and d?

19. The equation of the line of best fit for the data in the table is $y = 0$.

x	y
0	−1
1	1
2	0
3	0
4	1
5	−1
6	0

This model passes through three of the seven points shown in the table. Is this model a good fit for the data? Justify your answer.

20. Are production costs directly proportional to the number of items produced for most businesses? Explain.

21. Does the fact that a linear model passes through the point $(0, 0)$ imply that the dependent variable of the original data is directly proportional to the independent variable of the original data? Explain.

22. A linear model has a correlation coefficient of $r = -1$. Does the fact that the linear model passes through the point $(0, 0)$ imply that the output of the original data is directly proportional to the input of the original data? Explain.

23. Use your calculator to find the line of best fit for the data in the table.

x	y
0	1
1	1
2	1
3	1

Why do you think the correlation coefficient is undefined? Does this mean that the model doesn't fit the data? Justify your answer.

24. Find the piecewise linear function that best fits the data in the table.

x	y
0	2
1	4
2	6
3	7
4	8
5	9
6	10

25. Find the piecewise linear function that best fits the data in the table.

x	y
−2.0	14
1.0	8
2.5	5
4.0	2
8.0	4
10.0	5

Chapter 1 Review Exercises

Section 1.1   *In Exercises 1–3, calculate the value of the function at the designated input and interpret the result.*

1. $C(x) = 49.95x$ at $x = 2$, where C is the cost (in dollars) of buying x pairs of shoes.

2. $C = 9.95x + 1200$ at $x = 400$, where C is the cost of making x pairs of shoes.

3. $H = -16t^2 + 100$ at $t = 2$, where H is the height of a cliff diver above the water t seconds after he jumped from a 100-foot cliff.

In Exercises 4–7, determine the domain of the function.

4. $f(x) = 4x^3 - 10x$

5. $f(p) = p^2 - 9p + 15$

6. $C = \dfrac{r^2 - 1}{2r + 1}$ **7.** $C = \dfrac{r^2 + 1}{r^2 - 1}$

In Exercises 8–9, interpret the meaning of the indicated point on the graph.

8.   **Stock Price**   Find P at $t = 4$, where P is the stock price of an e-learning company at the end of the day and t is the number of days after December 16, 2001.

y-axis: Stock price (in dollars), values 2.00 to 3.40
x-axis: Number of days after 12/16/01, values 0 to 6

Source: Click2Learn Corporation.

9. 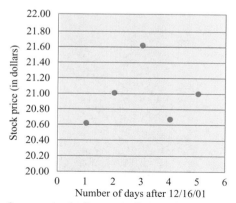   **Stock Price**   Find P at $t = 2$, where P is a computer company's stock price at the end of the day, t days after December 16, 2001.

y-axis: Stock price (in dollars), values 20.00 to 22.00
x-axis: Number of days after 12/16/01, values 0 to 6

Source: Apple Computer Corporation.

In Exercises 10–11, determine whether the graphs represent functions by applying the Vertical Line Test.

10.

11.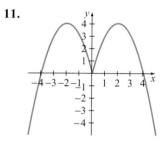

Section 1.2   *In Exercises 12–15, calculate the slope of the linear function passing through the points.*

12. $(2, 9)$ and $(4, 3)$

13. $(-3, -4)$ and $(0, -2)$

14. $(1.3, 3.5)$ and $(3.5, 1.3)$

15. $(7, 11)$ and $(8, 0)$

In Exercises 16–18, find the x-intercept and the y-intercept of the linear function.

16. $y = -5x + 10$ **17.** $y = -3x + 18$

18. $y = 2x - 12$

In Exercises 19–20, write the equation of the linear function passing through the points in slope-intercept form and in standard form.

19. $(2, 5)$ and $(4, 3)$ **20.** $(-3, 4)$ and $(0, -2)$

Section 1.3   *In Exercises 21–23, find a model for the data and answer the given questions.*

21.   **Nutrition**   The Recommended Daily Allowance (RDA) for fat for a person on a 2,000-calorie-per-day diet is less than 65 grams.

A McDonalds Big N' Tasty® sandwich contains 32 grams of fat. A super size order of French fries contains 29 grams of fat. (Source: www.mcdonalds.com.)

(a) Write the equation for fat grams consumed as a function of large orders of French fries eaten.

(b) Write the equation for fat grams consumed as a function of Big N' Tasty sandwiches eaten.

(c) How many combination meals (Big N' Tasty sandwich and super size order of French fries) can you eat without exceeding the RDA for fat?

22. **Solid Waste Disposal** The Enumclaw Transfer/Recycling Station accepts clean wood (stumps, branches, etc.) in addition to household garbage. The station charges $75 per ton with a minimum fee of $12.75 per entry for vehicles disposing of clean wood. (Source: www.dnr.metrokc.gov.)

(a) Write a piecewise linear function to model the cost of disposing of x pounds of clean wood. (Assume that the entire amount of wood is delivered in a single load.)

(b) How much does it cost to dispose of 300 pounds of clean wood? 700 pounds?

23. **Used Car Value** In 2001, the average retail price of a 1998 Mercedes-Benz Roadster two-door SL500 was $51,400. The average retail price of a 2000 Mercedes-Benz Roadster two-door SL500 was $66,025. (Source: www.nadaguides.com.)

(a) Find a linear model for the value of a Mercedes-Benz Roadster two-door SL500 in 2001 as a function of its production year.

(b) Use your linear model to predict the 2001 value of a 1999 Mercedes-Benz Roadster two-door SL500.

(c) A 1999 Mercedes-Benz Roadster two-door SL500 had an average retail price of $58,500 in 2001. How good was your linear model at predicting the value of the vehicle?

Make It Real

What to do

1. Find a set of at least six data points from an area of personal interest.
2. Draw a scatter plot of the data and explain why you do or do not believe that a linear model would fit the data well.
3. Find the equation of the line of best fit for the data.
4. Interpret the physical meaning of the slope and y-intercept of the model.
5. Use the model to predict the value of the function at an unknown point and explain why you do or do not think the prediction is accurate.
6. Explain how a consumer and/or a businessperson could benefit from the model.
7. Present your findings to the class and defend your conclusions.

Where to look for data

Box Office Guru
www.boxofficeguru.com
Look at historical data on movie revenues.

Nutri-Facts
www.nutri-facts.com
Compare the nutritional content of common foods based on serving size.

Quantitative Environmental Learning Project
www.seattlecentral.org/qelp
Look at environmental information in easy-to-access charts and tables.

U.S. Census Bureau
www.census.gov
Look at data on U.S. residents ranging from Internet usage to family size.

Local Gas Station or Supermarket
Track an item's price daily for a week.

School Registrar
Ask for historical tuition data.

Utility Bills
Look at electricity, water, or gas usage.

Employee Pay Statements
Look at take-home pay or taxes.

Body Weight Scale
Track your body weight for a week.

Chapter

2

Systems of Linear Equations

New companies often lose money for a few years before becoming profitable. Business owners may forecast profitability by modeling business costs and revenue with mathematical functions. Systems of linear equations are commonly used to determine when a linear revenue function is first expected to exceed a linear cost function. That is, systems of equations may be used to forecast the profitability of a company.

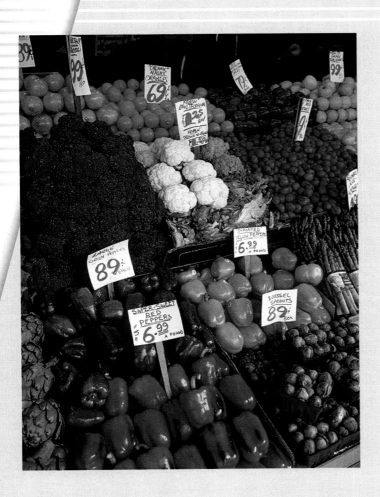

2.1 Systems of Linear Equations
- Graphically find the point of intersection of two lines
- Solve systems of linear equations using algebraic methods
- Interpret the meaning of solutions to linear systems

2.2 Using Matrices to Solve Linear Systems of Equations
- Represent systems of linear equations with augmented matrices
- Use row operations to put augmented matrices in reduced row echelon form

2.3 Linear System Applications
- Formulate linear systems from real-life data
- Interpret the meaning of solutions to linear systems

2.1 Systems of Linear Equations

- Graphically find the point of intersection of two lines
- Solve systems of linear equations using algebraic methods
- Interpret the meaning of solutions to linear systems

GETTING STARTED   In order to be profitable, a company's revenue must exceed its costs. Analysts often try to predict when a new company will begin to turn a profit. When both revenue and expenditures may be modeled by linear functions, a system of linear equations may be used to determine the **break-even point,** the point at which revenue and costs are equal. In this section, we will show how to find the break-even point of a company. Additionally, we will demonstrate how to find the point of intersection of any two lines graphically and algebraically.

SYSTEM OF LINEAR EQUATIONS IN TWO VARIABLES

A **system of linear equations in two variables** is a collection of two or more linear equations. The solution to the system of equations, if it exists, is the point (or points) of intersection of the graphs of the equations. In other terms, a solution is an ordered pair that satisfies each of the equations in the system.

In reality, systems of linear equations may actually contain hundreds of variables and hundreds of equations. However, increasing the number of variables beyond three makes a graphical interpretation of the solution impossible. Even a three-variable system of equations may be difficult to represent graphically. For this reason, our graphical discussion will focus on systems of linear equations in two variables.

The Graphical Method

An estimate of the solution to a system of linear equations may easily be obtained using the graphical method.

EXAMPLE 1

Finding a Solution to a System of Linear Equations Using Technology

Find the solution to the system of linear equations. (That is, find the point of intersection of the graphs of the two lines.)

$$y = -1.5x + 8$$
$$y = x + 1$$

SOLUTION Plotting the linear equations yields the graph in Figure 2.1.

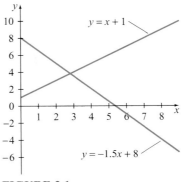

FIGURE 2.1

It appears from the graph that the lines intersect at or near (3, 4). One of the limitations of graphical analysis is that it is difficult to be precise; however, the TI-83 calculator has a built-in tool that calculates the point of intersection of two graphs. Using the technique demonstrated in the following Technology Tip, we can determine that the point of intersection of the two lines is (2.8, 3.8).

When using the calculator to determine the point of intersection, it is important to note that the calculator may not give the *exact* answer. It will, however, return the best possible decimal approximation of the exact answer. For example, if the point of intersection is $(\sqrt{3}, \sqrt{2})$, the calculator will return (1.732050808, 1.414213562).

TECHNOLOGY **TIP**

Finding a Point of Intersection

1. Simultaneously graph both functions. Adjust the window as necessary to make the point of intersection visible. (If you've forgotten how to do this, reread the "Graphing a Function" technology tip in Section 1.1.)

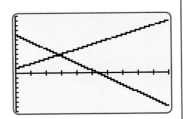

2. Press [2nd] [TRACE] to bring up the Calculate Menu.

(Continued)

3. Select `5:intersect`. The calculator asks "`First curve?`" Use the blue arrow buttons to select either curve, then press `ENTER`. The calculator asks "`Second curve?`" Use the arrow buttons to select the second line and press `ENTER`.

4. The calculator asks "`Guess?`" If there is more than one point of intersection, move the cursor near the desired point of intersection and press `ENTER`. Otherwise, just press `ENTER`. The point of intersection is highlighted on the graph and its coordinates are displayed.

EXAMPLE 2 **Finding a Solution to a System of Linear Equations Graphically**

Solve the system of equations graphically.

$$2x - 3y = 6$$
$$3x + 2y = 48$$

SOLUTION We begin by writing each of the lines in slope-intercept form.

$$2x - 3y = 6 \qquad\qquad 3x + 2y = 48$$
$$-3y = -2x + 6 \qquad\qquad 2y = -3x + 48$$
$$y = \frac{2}{3}x - 2 \qquad\qquad y = -\frac{3}{2}x + 24$$

Then we graph the functions, as shown in Figure 2.2.

It appears that the graphs intersect at $(12, 6)$. We can verify this by using the graphing calculator or by substituting the values of x and y into each of the original equations. We will do the latter.

$$2x - 3y = 6$$
$$2(12) - 3(6) = 6 \qquad \text{Substituting in } (12, 6)$$
$$24 - 18 = 6$$
$$6 = 6$$

$$3x + 2y = 48$$
$$3(12) + 2(6) = 48 \qquad \text{Substituting in } (12, 6)$$
$$36 + 12 = 48$$
$$48 = 48$$

Since the point $(12, 6)$ satisfies both equations, $x = 12$ and $y = 6$ is the solution to the system of equations.

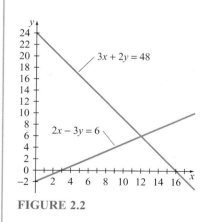

FIGURE 2.2

Business analysts are often interested in the profit, revenue, and costs of a company. In evaluating a business plan, it is important to know at what sales level revenue and costs are expected to be equal. This sales level is referred to as the **break-even point** and is the point at which the company begins to turn a profit.

Any business has two types of costs: **fixed costs** and **variable costs.** Fixed costs are those that remain constant independent of production levels. For example, building rent, product research and development, and advertising costs are often viewed as fixed costs. Variable costs are those that vary depending upon the level of production. For example, raw materials and production-line workers' wages may be viewed as variable costs. If no items are produced, no raw materials are purchased and no production-line workers are needed. As the number of items produced increases, the raw material and production-line labor costs also increase.

Break-even analysis may be applied to large companies or small home-based businesses. An example of one such business is given in Example 3.

Using a System of Linear Equations to Find the Break-Even Point

An artisan is considering selling her handmade craft angels online. She estimates her material cost for each angel to be $3.50. An online merchant, craftmall.com, offers to advertise her product for a $24.95 monthly fee. Comparing her craft to similar crafts on the market, she estimates that she can sell the craft angel for $9.95. How many angels will she have to sell each month to break even? At that production level, what will be her production cost, revenue, and profit?

SOLUTION The cost equation for the craft angels is the sum of the variable cost, $3.50 per angel, and the fixed cost, $24.95. Let a be the number of angels sold. The cost equation is

$$y = 3.50a + 24.95 \text{ dollars}$$

The revenue equation is

$$y = 9.95a \text{ dollars}$$

She must determine when her revenue will equal her cost. In other words, she must solve the following system of equations.

$$y = 3.50a + 24.95$$
$$y = 9.95a$$

Graphing the two functions simultaneously results in the graph shown in Figure 2.3.

It appears that the graphs intersect near $(4, 38)$. At the intersection point, the revenue and cost functions are equal. We can find the exact point of intersection by setting the revenue and cost expressions equal to each other and solving algebraically.

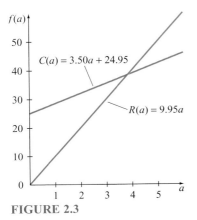

FIGURE 2.3

$$9.95a = 3.50a + 24.95 \quad \text{Revenue expression} = \text{cost expression}$$
$$6.45a = 24.95$$
$$a = \frac{24.95}{6.45}$$
$$a \approx 3.868$$
$$a \approx 4 \text{ angels}$$

The break-even point is (3.868, 38.49); however, in the context of the problem, it doesn't make sense to talk about 3.868 angels. We conclude that she must sell four craft angels each month in order to cover her costs.

The cost to produce four angels is

$$y = 3.50(4) + 24.95$$
$$= \$38.95$$

The revenue from the sale of four angels is

$$y = 9.95(4)$$
$$= \$39.80$$

She will have a profit of $0.85 if she sells four angels.

When solving an application problem (such as the one shown in Example 3), it is essential to make sure that the solution makes sense in the context of the problem. A common error among students is to accept a mathematically correct answer [i.e., (3.868, 38.49)] without verifying that it makes sense in the context of the problem.

In Example 3, we found the intersection of two lines algebraically. Two common methods used to solve systems of linear equations algebraically are the substitution method and the elimination method.

HOW TO **The Substitution Method**

To solve a system of linear equations

$$ax + by = c$$
$$dx + ey = f$$

do the following:

1. Solve the first equation for y in terms of x.

2. Replace y in the second equation with the result from Step 1.

3. Solve the second equation for x.

4. Substitute the value of x back into the first equation and solve for y.

If the system of equations models real-life data, you should also complete Steps 5 and 6.

5. Ask yourself if the mathematical solution makes sense in its real world context.

6. Reevaluate the functions, as necessary, and verbally express the real-world meaning of the result.

Although we stated that you should solve the first equation for y in terms of x, you may equivalently solve for x in terms of y and obtain the same result.

EXAMPLE **4**   ### Solving a System of Linear Equations with the Substitution Method

Solve the system of equations.

$$5x + 2y = 12$$
$$4x - y = 7$$

SOLUTION We may solve either equation for y. Since the second equation looks easier, we will solve it for y.

$$4x - y = 7$$
$$4x = y + 7$$
$$y = 4x - 7$$

Substituting this result in for y in the equation $5x + 2y = 12$ yields

$$5x + 2(4x - 7) = 12$$
$$5x + 8x - 14 = 12$$
$$13x = 26$$
$$x = 2$$

Since $y = 4x - 7$,

$$y = 4(2) - 7 \quad \text{Substitute } x = 2$$
$$y = 1$$

Therefore, the solution to the system of equations is $x = 2$ and $y = 1$. We can verify the accuracy of this answer by substituting these values back into each of the original equations.

$$5x + 2y = 12 \qquad 4x - y = 7$$
$$5(2) + 2(1) = 12 \qquad 4(2) - (1) = 7$$
$$12 = 12 \qquad\qquad 7 = 7$$

EXAMPLE **5**   ### Solving a System of Linear Equations with the Substitution Method

Solve the system of equations.

$$2x - 7y = -4$$
$$x + 12y = 29$$

SOLUTION Since the coefficient on the x term in the second equation is a 1, it will be easier to solve the equation for x instead of y.

$$x + 12y = 29$$
$$x = -12y + 29$$

Substituting this value of x into the first equation yields

$$2x - 7y = -4$$
$$2(-12y + 29) - 7y = -4$$
$$-24y + 58 - 7y = -4$$
$$-31y = -62$$
$$y = 2$$

To find the corresponding value of x, we back-substitute $y = 2$.

$$x = -12y + 29$$
$$x = -12(2) + 29$$
$$= 5$$

The solution to the system of equations is $x = 5$ and $y = 2$.

Solving a system of equations by substitution is especially easy when the linear equations are written in slope-intercept form, $y = mx + b$. Since y is already written as a function of x, we can proceed directly to Step 2 of the substitution process.

EXAMPLE 6   **Solving a System of Linear Equations with the Substitution Method**

Based on data from 1992 to 2000, the annual value of printing products shipped in the United States may be modeled by

$$V(t) = 3203.0t + 80{,}732 \text{ million dollars}$$

and the annual value of furniture and related products shipped in the United States may be modeled by

$$V(t) = 3511.7t + 48{,}294 \text{ million dollars}$$

where t is the number of years since the end of 1992. According to the models, when will the annual value of products shipped by the printing industry and the furniture industry be the same?

SOLUTION   We will replace the V in the second equation with the value of V from the first equation and solve for t.

$$3203.0t + 80{,}732 = 3511.7t + 48{,}294$$
$$32{,}438 = 308.7t$$
$$t \approx 105.08$$

We solve for the value of $V(t)$ by substituting $t = 105.08$ into either equation.

$$V(105.08) = 3203.0(105.08) + 80{,}732$$
$$= 417{,}300 \text{ (accurate to five significant digits)}$$

According to the model, at the end of 2097 ($t = 105$), the annual value of products shipped by the printing and furniture industries will be approximately equal ($417,300 million). Although we found a mathematical solution, we are a bit skeptical of the real-life accuracy of the result, since it is so far outside of the original data set.

EXAMPLE 7

Solving a System of Linear Equations with No Solution

Find the solution to the system of linear equations.

$$4x - 5y = 20$$
$$8x - 10y = 30$$

SOLUTION We begin by solving the first equation for y.

$$4x - 5y = 20$$
$$5y = 4x - 20$$
$$y = 0.8x - 4$$

We substitute this value of y into the second equation and solve for x.

$$8x - 10y = 30$$
$$8x - 10(0.8x - 4) = 30$$
$$8x - 8x + 40 = 30$$
$$40 = 30$$

But $40 \neq 30$. Since the system of equations led to a contradiction, the system of equations does not have a solution. This can readily be seen by writing both equations in slope-intercept form.

$$y = 0.8x - 4$$
$$y = 0.8x - 3$$

The lines are parallel, since they have the same slope but different y-intercepts (see Figure 2.4). Therefore, the two lines do not intersect and the system of equations does not have a solution.

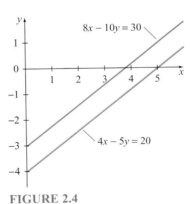

FIGURE 2.4

EXAMPLE 8

Solving a System of Linear Equations with Infinitely Many Solutions

Find the solution to the system of linear equations.

$$12x - 15y = 60$$
$$8x - 10y = 40$$

SOLUTION We begin by solving the first equation for y.

$$12x - 15y = 60$$
$$15y = 12x - 60$$
$$y = 0.8x - 4$$

We substitute this value of y into the second equation and solve for x.

$$8x - 10y = 40$$
$$8x - 10(0.8x - 4) = 40$$
$$8x - 8x + 40 = 40$$
$$40 = 40$$

As in Example 7, the x was eliminated; however, in this case, the resulting statement was true. Therefore, the system of equations has infinitely many solutions. The slope-intercept form of each equation is $y = 0.8x - 4$. Any point on this line is a solution to the system of equations.

Systems of linear equations will have 0, 1, or infinitely many solutions. A system of equations without a solution is said to be **inconsistent.** A system of equations with infinitely many solutions is said to be **dependent** (see Figure 2.5).

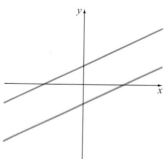

inconsistent system
two parallel lines
no solution

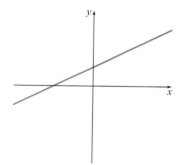

dependent system
both linear equations give the same line
infinitely many solutions

FIGURE 2.5

In order for a system of three or more linear equations to have a solution, all of the lines must intersect at the same point. The fact that two lines intersect at a point (a, b) does not ensure that a third line will intersect the first two lines at the same point.

EXAMPLE 9

Solving a System of Equations with More Equations than Unknowns

Solve the system of equations.

$$2x - y = 5$$
$$3x + 2y = 11$$
$$4x - 4y = 8$$

SOLUTION We will find the point of intersection of the first two lines and then check to see if the point is a solution to the third equation. The first equation may be written as $y = 2x - 5$. Substituting this value in for y in the second equation and solving for x yields

$$3x + 2y = 11$$
$$3x + 2(2x - 5) = 11 \qquad \text{Substitute } y = 2x - 5$$
$$3x + 4x - 10 = 11$$
$$7x = 21$$
$$x = 3$$

Since $y = 2x - 5$,

$$y = 2(3) - 5$$
$$= 1$$

The point of intersection of the first two lines is $(3, 1)$. We will check to see if this point satisfies the third equation.

$$4x - 4y = 8$$
$$4(3) - 4(1) = 8 \qquad \text{Substitute } x = 3 \text{ and } y = 1$$
$$12 - 4 = 8$$
$$8 = 8$$

Since the resulting statement was true, the solution to the system of equations is $x = 3$ and $y = 1$. (A false statement would have implied that the system was inconsistent.) We graphically confirm the result in Figure 2.6.

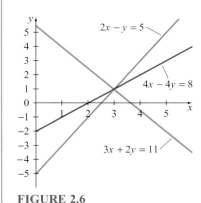

FIGURE 2.6

EXAMPLE 10

Solving an Inconsistent System of Linear Equations

Solve the system of equations.

$$5x - 3y = -2$$
$$-x + y = 2$$
$$-3x - 2y = -19$$

SOLUTION We will find the point of intersection of the first two lines and then check to see if the point is a solution to the third equation. The second equation may be written as $y = x + 2$. Substituting this value in for y in the first equation and solving for x yields

$$5x - 3y = -2$$
$$5x - 3(x + 2) = -2 \qquad \text{Substitute } y = x + 2$$
$$5x - 3x - 6 = -2$$
$$2x = 4$$
$$x = 2$$

Back-substituting $x = 2$ into $y = x + 2$ yields

$$y = (2) + 2$$
$$= 4$$

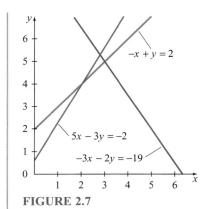

FIGURE 2.7

The point of intersection of the first two lines is $(2, 4)$. We will now determine if this point is on the third line.

$$-3x - 2y = -19$$
$$-3(2) - 2(4) = -19 \qquad \text{Substitute } x = 2 \text{ and } y = 4$$
$$-6 - 8 = -19$$
$$-14 = -19$$

Since $-14 \neq -19$, the statement is false and $(2, 4)$ is not on the third line. Therefore the system is inconsistent. Graphing the lines confirms our result, as shown in Figure 2.7.

The Elimination Method

Although the elimination method is most commonly used when a system of linear equations contains three or more variables, it may also be used to solve two-variable systems. This method uses the properties of equivalent systems of equations to find the solution.

> **EQUIVALENT SYSTEMS OF EQUATIONS**
>
> The following operations on a system of equations result in a system of equations with the same solution as the original system:
>
> 1. Interchange (change the position of) two equations.
> 2. Multiply an equation by a nonzero number.
> 3. Add a nonzero multiple of one equation to a nonzero multiple of another equation and replace either equation with the result.

EXAMPLE 11 **Finding an Equivalent System of Equations**

Given the system of equations

$$2x + 3y - z = 6$$
$$3x - 6y + 2z = 12$$

find an equivalent system of equations by doing the following:

1. Multiply the first equation by 2 and write the result as the first equation.
2. Add the first equation to the second and write the result as the second equation.

SOLUTION We first multiply the first equation $2x + 3y - z = 6$ by two.

$$4x + 6y - 2z = 12$$
$$3x - 6y + 2z = 12$$

We add the two equations together and write the result as the second equation.

$$4x + 6y - 2z = 12$$
$$7x + 0y + 0z = 24$$

HOW TO The Elimination Method

To solve a system of equations using the elimination method, do the following:

1. Write each equation in standard form ($Ax + By + Cz + \cdots = K$).

2. Vertically align the variables in each equation with the corresponding variables in the other equations.

3. Use the equivalency operations to eliminate the x variable in all but the first equation.

4. Use the equivalency operations to eliminate the y variable in all but the second equation.

5. Continue to simplify the system in a like manner for any remaining variables.

6. Simplify the system to the form

$$x = a$$
$$y = b$$
$$z = c$$
$$\vdots$$

EXAMPLE 12 Solving a System of Linear Equations with the Elimination Method

Find the solution to the system of linear equations.

$$2x - 9y = 29$$
$$3x + 8y = 22$$

SOLUTION If we multiply the first equation by 3 and the second equation by -2, we get an equivalent system.

$$6x - 27y = 87 \qquad \text{Multiply by 3}$$
$$-6x - 16y = -44 \qquad \text{Multiply by } -2$$

Now that the x terms have the opposite coefficients, we can eliminate the x terms by adding the second equation to the first equation.

$$6x - 27y = 87$$
$$+(-6x - 16y = -44)$$
$$\overline{0x - 43y = 43}$$

We place this result in the position of the second equation.

$$6x - 27y = 87$$
$$-43y = 43$$

Multiplying the first equation by $\frac{1}{3}$ and the second equation by $-\frac{1}{43}$ yields

$$2x - 9y = 29 \qquad \text{Multiply by } \frac{1}{3}$$

$$y = -1 \qquad \text{Multiply } -\frac{1}{43}$$

We want to eliminate the y term in the first equation. If we add nine times the second equation to the first equation, we will eliminate the y term in the first equation.

$$2x - 9y = 29$$
$$+ \qquad 9y = -9$$
$$\overline{2x + 0y = 20}$$

We place this result in the position of the first equation.

$$2x = 20$$
$$y = -1$$

We finish by multiplying the first equation by $\frac{1}{2}$.

$$x = 10 \qquad \text{Multiply by } \frac{1}{2}$$

$$y = -1$$

The solution is $x = 10$ and $y = -1$.

EXAMPLE 13 **Solving a System of Linear Equations with the Elimination Method**

Find the solution to the system of linear equations.

$$3x - 2y + z = -2$$
$$x + 2y - z = 6$$
$$-x + \ y + z = 0$$

SOLUTION A system of linear equations in two variables may be interpreted graphically as a group of lines drawn in two dimensions. A system of linear equations in three variables may be interpreted graphically as a group of planes drawn in three dimensions. The three-dimensional graph of the given system is shown in Figure 2.8.

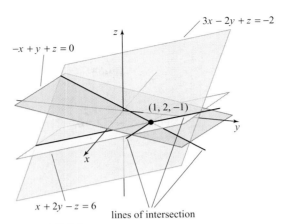

$-x + y + z = 0$

$3x - 2y + z = -2$

$(1, 2, -1)$

$x + 2y - z = 6$

lines of intersection

FIGURE 2.8

Notice that any two planes intersect in a line. The point where all three planes intersect will be the solution of the system of equations. In solving linear systems in three variables, we rarely draw the graphs because of the difficulty of three-dimensional graphing. However, a graphical understanding of what is going on helps to add meaning to the numerical solution.

We will proceed to solve the system algebraically.

$$(1) \qquad 3x - 2y + z = -2$$
$$(2) \qquad x + 2y - z = 6$$
$$(3) \qquad -x + y + z = 0$$

We have labeled the equations (1), (2), and (3) to make it easier to keep track of our work. Our strategy is to eliminate variables by adding a multiple of one equation to a multiple of another equation in such a way that one or more variables are eliminated. We will place the resultant equation in the position of either of the original equations. It is computationally convenient to have the coefficient of the x variable in the first equation equal to 1. We could divide equation (1) by 3 to achieve this result; however, that would create fractional coefficients on the y and z variables. We will instead interchange equations (1) and (2). After making the switch, we relabel the result.

$$(1) \qquad x + 2y - z = 6$$
$$(2) \qquad 3x - 2y + z = -2$$
$$(3) \qquad -x + y + z = 0$$

We observe that if we multiply equation (1) by -3 and add it to equation (2), the x variable in equation (2) will be eliminated. We will place the result back in equation (2). Symbolically, we can represent the operation as $-3E_1 + E_2 \rightarrow E_2$, where E_1 represents equation (1) and E_2 represents equation (2). We multiply each term in equation (1) by -3 and add it to the corresponding term in equation (2).

$$
\begin{array}{r}
-3x - 6y + 3z = -18 \\
+ \quad (3x - 2y + z = -2) \\
\hline
0x - 8y + 4z = -20
\end{array}
$$

We write the new equation back in (2).

$$
\begin{aligned}
&(1) \quad & x + 2y - z &= 6 \\
&(2) \quad & -8y + 4z &= -20 \qquad -3E_1 + E_2 \rightarrow E_2 \\
&(3) \quad & -x + y + z &= 0
\end{aligned}
$$

We next want to eliminate the x in the third equation. We observe that adding equations (1) and (3) will achieve the desired result. Symbolically, we write $E_1 + E_3 \rightarrow E_3$.

$$
\begin{aligned}
x + 2y - z &= 6 \\
+ (-x + y + z &= 0) \\
\hline
0x + 3y + 0z &= 6
\end{aligned}
$$

We write the new equation back in (3).

$$
\begin{aligned}
&(1) \quad & x + 2y - z &= 6 \\
&(2) \quad & -8y + 4z &= -20 \\
&(3) \quad & 3y &= 6 \qquad E_1 + E_3 \rightarrow E_3
\end{aligned}
$$

Observe that we can solve for y by multiplying (3) by $\frac{1}{3}$. Symbolically, we write $\frac{1}{3}E_3 \rightarrow E_3$.

$$
\begin{aligned}
&(1) \quad & x + 2y - z &= 6 \\
&(2) \quad & -8y + 4z &= -20 \\
&(3) \quad & y &= 2 \qquad \frac{1}{3}E_3 \rightarrow E_3
\end{aligned}
$$

We will interchange equations (2) and (3) so that the middle equation is of the form $y = b$. Symbolically, we write $E_2 \leftrightarrow E_3$.

$$
\begin{aligned}
&(1) \quad & x + 2y - z &= 6 \\
&(2) \quad & y &= 2 \qquad E_2 \leftrightarrow E_3 \\
&(3) \quad & -8y + 4z &= -20 \qquad E_2 \leftrightarrow E_3
\end{aligned}
$$

We will now eliminate the y term in equation (3). Observe that $8E_2 + E_3 \rightarrow E_3$ will do this.

$$
\begin{aligned}
0x + 8y + 0z &= 16 \\
+ (0x - 8y + 4z &= -20) \\
\hline
0x + 0y + 4z &= -4
\end{aligned}
$$

We write the new equation back in equation (3).

$$
\begin{aligned}
&(1) \quad & x + 2y - z &= 6 \\
&(2) \quad & y &= 2 \\
&(3) \quad & 4z &= -4 \qquad 8E_2 + E_3 \rightarrow E_3
\end{aligned}
$$

Using $\frac{1}{4}E_3 \rightarrow E_3$, we solve for z.

$$
\begin{aligned}
&(1) \quad & x + 2y - z &= 6 \\
&(2) \quad & y &= 2 \\
&(3) \quad & z &= -1 \qquad \frac{1}{4}E_3 \rightarrow E_3
\end{aligned}
$$

Using $E_1 + E_3 \rightarrow E_1$, we eliminate the z term in equation (1).

$$\begin{array}{r} x + 2y - z = 6 \\ + (0x + 0y + z = -1) \\ \hline x + 2y + 0z = 5 \end{array}$$

We write the new equation back in equation (1).

$$\begin{array}{lll} (1) & x + 2y \quad\;\; = 5 & E_1 + E_3 \rightarrow E_1 \\ (2) & \qquad y \quad\;\; = 2 & \\ (3) & \qquad\qquad z = -1 & \end{array}$$

Using $E_1 - 2E_2 \rightarrow E_1$, we eliminate the y term in equation (1).

$$\begin{array}{r} x + 2y + 0z = 5 \\ - (0x + 2y + 0z = 4) \\ \hline x + 0y + 0z = 1 \end{array}$$

We write the new equation back in equation (1).

$$\begin{array}{lll} (1) & x \qquad\qquad = 1 & E_1 - 2E_2 \rightarrow E_1 \\ (2) & \quad y \qquad\; = 2 & \\ (3) & \qquad\; z = -1 & \end{array}$$

The solution to the system of equations is $x = 1$, $y = 2$, and $z = -1$.

To be certain we're correct, we will check our work by plugging our solution into each of the original equations.

$$\begin{array}{ll} (1) & 3(1) - 2(2) + (-1) = -2 \\ (2) & (1) + 2(2) - (-1) = 6 \\ (3) & -1(1) + (2) + (-1) = 0 \end{array}$$

Simplifying, we get

$$\begin{array}{ll} (1) & -2 = -2 \\ (2) & 6 = 6 \\ (3) & 0 = 0 \end{array}$$

Each statement is true so we are certain that our solution is correct.

EXAMPLE 14 **Solving a System of Linear Equations with the Elimination Method**

Solve the system of equations.

$$\begin{array}{r} 4x + 5y - 2z = 30 \\ 3x - 2y + 7z = -43 \\ -2x + 4y + 3z = 1 \end{array}$$

SOLUTION

$$\begin{array}{ll} (1) & 4x + 5y - 2z = 30 \\ (2) & 3x - 2y + 7z = -43 \\ (3) & -2x + 4y + 3z = 1 \end{array}$$

We want to make the coefficient on the x term in equation (1) equal 1. The operation $E_1 - E_2 \rightarrow E_1$ will achieve this objective. The new system is

(1) $\quad x + 7y - 9z = 73$ $\qquad E_1 - E_2 \rightarrow E_1$

(2) $\quad 3x - 2y + 7z = -43$

(3) $-2x + 4y + 3z = 1$

Next we will eliminate the x term in equation (2) with the operation $3E_1 - E_2 \rightarrow E_2$.

(1) $\quad x + 7y - 9z = 73$

(2) $\qquad\quad 23y - 34z = 262$ $\qquad 3E_1 - E_2 \rightarrow E_2$

(3) $-2x + 4y + 3z = 1$

Next, we will eliminate the x term in equation (3) with the operation $2E_1 + E_3 \rightarrow E_3$.

(1) $\quad x + 7y - 9z = 73$

(2) $\qquad\quad 23y - 34z = 262$

(3) $\qquad\quad 18y - 15z = 147$ $\qquad 2E_1 + E_3 \rightarrow E_3$

We now want to eliminate the y term in equation (3). We will do this with the operation $18E_2 - 23E_3 \rightarrow E_3$. Since this computation is somewhat complex, we will show each step.

$$18(23y - 34z = 262)$$
$$\underline{-23(18y - 15z = 147)}$$

$$414y - 612z = 4716$$
$$\underline{-(414y - 345z = 3381)}$$
$$0y - 267z = 1335$$

We write the new equation back in equation (3).

(1) $\quad x + 7y - 9z = 73$

(2) $\qquad\quad 23y - 34z = 262$

(3) $\qquad\qquad -267z = 1335$ $\qquad 18E_2 - 23E_3 \rightarrow E_3$

Solving equation (3) for z with the operation $-\dfrac{1}{267}E_3 \rightarrow E_3$ yields

(1) $\quad x + 7y - 9z = 73$

(2) $\qquad\quad 23y - 34z = 262$

(3) $\qquad\qquad\qquad z = -5$ $\qquad -\dfrac{1}{267}E_3 \rightarrow E_3$

We will now eliminate the z term in equation (2) with $E_2 + 34E_3 \rightarrow E_2$.

(1) $\quad x + 7y - 9z = 73$

(2) $\qquad\quad 23y \qquad\; = 92$ $\qquad E_2 + 34E_3 \rightarrow E_2$

(3) $\qquad\qquad\qquad z = -5$

Then we eliminate the z term in equation (1) with $E_1 + 9E_3 \rightarrow E_1$.

(1) $\quad x + 7y \qquad = 28 \qquad E_1 + 9E_3 \rightarrow E_1$

(2) $\qquad\quad 23y \qquad = 92$

(3) $\qquad\qquad\quad z = -5$

Solving equation (2) for y with the operation $\frac{1}{23}E_2 \rightarrow E_2$ yields

(1) $\quad x + 7y \qquad = 28$

(2) $\qquad\quad y \qquad = 4 \qquad \frac{1}{23}E_2 \rightarrow E_2$

(3) $\qquad\qquad z = -5$

We now eliminate the y term from equation (1) with $E_1 - 7E_2 \rightarrow E_1$.

(1) $\quad x \qquad\qquad = 0 \qquad E_1 - 7E_2 \rightarrow E_1$

(2) $\qquad\quad y \qquad = 4$

(3) $\qquad\qquad z = -5$

The solution to the system of equations is $x = 0$, $y = 4$, and $z = -5$.

EXAMPLE 15

Solving a Dependent System of Equations with the Elimination Method

Solve the system of equations.

$$2x + 6y - 4z = 42$$
$$4x - 3y - 2z = -42$$
$$6x + 3y - 6z = 0$$

SOLUTION We first obtain a coefficient of 1 on the x term in equation (1) by using the operation $\frac{1}{2}E_1 \rightarrow E_1$.

$$x + 3y - 2z = 21$$
$$4x - 3y - 2z = -42$$
$$6x + 3y - 6z = 0$$

We then eliminate the x terms in equations (1) and (2) with the operations $4E_1 - E_2 \rightarrow E_2$ and $6E_1 - E_3 \rightarrow E_3$.

$$x + 3y - 2z = 21$$
$$15y - 6z = 126$$
$$15y - 6z = 126$$

Observe that equations (2) and (3) are identical. In fact, the operation $E_2 - E_3 \rightarrow E_3$ eliminates all variables from the third equation.

$$x + 3y - 2z = 21$$
$$15y - 6z = 126$$
$$0 = 0$$

When the number of equations equals the number of variables, a unique solution may exist. However, if one of the equations "drops out" (as shown in this example), the system is either dependent or inconsistent. We will proceed to eliminate the y from equation (1) with the operation $5E_1 - E_2 \to E_1$ before discussing the final solution.

$$5x \qquad - 4z = -21$$
$$15y - 6z = 126$$
$$0 = 0$$

We will now solve each equation for x or y in terms of z.

$$5x - 4z = -21 \qquad\qquad 15y - 6z = 126$$
$$5x = 4z - 21 \qquad\qquad 15y = 6z + 126$$
$$x = 0.8z - 4.2 \qquad\qquad y = 0.4z + 8.4$$

If we let $z = t$ (where t is any real number), then $x = 0.8t - 4.2$ and $y = 0.4t + 8.4$. The system is a dependent system of equations and thus has infinitely many solutions. Each solution is of the form $x = 0.8t - 4.2$, $y = 0.4t + 8.4$, and $z = t$. For example, if we let $t = 0$, then the solution is $(-4.2, 8.4, 0)$. If we let $t = 1$, the solution is $(-3.4, 8.8, 1)$.

In graphical terms, a dependent system of three linear equations with three variables is a set of three planes that intersect in a common line instead of a single point. The graph for the system of equations given in this example is shown in Figure 2.9.

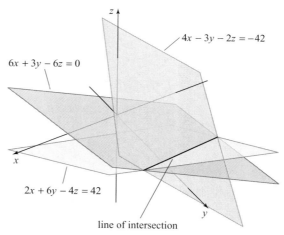

FIGURE 2.9

In the next section, we will introduce a process for solving systems of equations with three or more variables that, when mastered, is easier than the approach shown in Example 15.

2.1 Summary

In this section, you learned how to find the solution to a system of linear equations both graphically and algebraically. You learned two algebraic methods for solving linear systems of equations: the substitution method and the elimination method.

2.1 Exercises

In Exercises 1–10, use your graphing calculator to find the point(s) of intersection of the lines. If the lines do not intersect at a common point, state that a solution does not exist.

1. $y = 5x - 9$
$y = 2x - 3$

2. $-5x + 2y = 8$
$y = -3x + 26$

3. $y = 5$
$y = -3.21x + 4.32$

4. $-26x + 10y = 112$
$21x + 100y = 1000$

5. $-98x + 10y = -76$
$-98x + 10y = -28$

6. $y = 0.5x - 9$
$y = 0.5x + 10$

7. $y = 2x - 9$
$y = -3x - 9$
$y = \dfrac{1}{2}x - 9$

8. $y = 0.25x + 1$
$4x + y = 18$
$y = 2$

9. $y = 4.8x + 6$
$y = 1.9x - 2.1$
$y = 2x + 1.6$

10. $y = 11x - 12$
$y = 15x - 21$
$y = 15x - 12$

In Exercises 11–20, solve the system of equations algebraically. Compare your solutions to those obtained in Exercises 1–10.

11. $y = 5x - 9$
$y = 2x - 3$

12. $-5x + 2y = 8$
$y = -3x + 26$

13. $y = 5$
$y = -3.21x + 4.32$

14. $-26x + 10y = 112$
$21x + 100y = 1000$

15. $-98x + 10y = -76$
$-98x + 10y = -28$

16. $y = 0.5x - 9$
$y = 0.5x + 10$

17. $y = 2x - 9$
$y = -3x - 9$
$y = \dfrac{1}{2}x - 9$

18. $y = 0.25x + 1$
$4x + y = 18$
$y = 2$

19. $y = 4.8x + 6$
$y = 1.9x - 2.1$
$y = 2x + 1.6$

20. $y = 11x - 12$
$y = 15x - 21$
$y = 15x - 12$

In Exercises 21–30, determine the solution by setting up and solving a linear system of equations.

21. **M** **Diabetes** The incidence of diabetes in men and women aged 45–64 years increased between 1994 and 1999. The incidence for women may be modeled by

$$F(t) = 3.75t + 131 \text{ people per thousand}$$

and the incidence for men may be modeled by

$$M(t) = 20t + 109 \text{ people per thousand}$$

where t is the number of years after 1994. (**Source:** Modeled from *Statistical Abstract of the United States, 2001,* Table 159, p. 109.) According to the models, when will the incidence for men and women be the same?

22. **M** **Gasoline Prices** Based on data from 1990 to 2000, the retail cost of unleaded regular gasoline may be modeled by

$$R(t) = 0.035t + 1.16 \text{ dollars per gallon}$$

and the retail cost of unleaded premium gasoline may be modeled by

$$P(t) = 0.034t + 1.35 \text{ dollars per gallon}$$

where t is the number of years after 1990. (**Source:** Modeled from *Statistical Abstract of the United States, 2001,* Table 704, p. 467.) According to the models, when will the cost of unleaded regular and unleaded premium gasoline be the same? Based on your experience of purchasing gasoline, does your solution seem reasonable? Explain.

23. **M** **Weekly Food Cost** Based on data from 1990 and 2000, the weekly food cost for a 15- to 19-year-old male in a four-person family t years after 1990 may be modeled by one of the following functions based on the family's food-spending behavior.

Thrifty Plan:
$$T(t) = 0.56t + 21.40 \text{ dollars}$$

Moderate–Cost Plan:
$$M(t) = 1.06t + 36.80 \text{ dollars}$$

Liberal Plan:
$$L(t) = 1.21t + 42.60 \text{ dollars}$$

(**Source:** Modeled from *Statistical Abstract of the United States, 2001,* Table 705, p. 467.)

According to the models, will the food cost of all three plans ever be the same? Explain.

24. **M** **Weekly Food Cost** Based on data from 1990 and 2000, the weekly food cost for a 12- to 19-year-old female in a four-person family t years after 1990 may be modeled by one of the following functions based on the family's food-spending behavior. (**Source:** Modeled from *Statistical Abstract of the United States, 2001,* Table 705, p. 467.)

Thrifty Plan:
$$T(t) = 0.55t + 20.80 \text{ dollars}$$

Moderate–Cost Plan:
$$M(t) = 0.85t + 30.10 \text{ dollars}$$

Liberal Plan:
$$L(t) = 1.04t + 36.30 \text{ dollars}$$

According to the models, will the food cost of all three plans ever be the same? Explain.

25. **M** **Bread and Gasoline Prices** Based on data from 1990 to 2000, the retail cost of unleaded regular gasoline may be modeled by

$$R(t) = 0.035t + 1.16 \text{ dollars per gallon}$$

where t is the number of years since 1990. Based on data from 1993 to 2000, the price of a loaf of whole wheat bread may be modeled by

$$C(t) = 0.0404t + 0.991$$

where t is the number of years since 1990.

(**Source:** Modeled from *Statistical Abstract of the United States, 2001,* Tables 704 and 706, pp. 467–468.) Rounded up to the nearest year, when will a loaf of whole wheat bread cost more than a gallon of unleaded regular gasoline?

26. **M** **Entertainment Revenues** Based on data from 1998 and 1999, the amount of revenue brought in by amusement and theme parks may be modeled by

$$A(t) = 177t + 7335 \text{ million dollars}$$

where t is the number of years since the end of 1998. The amount of revenue brought in by racetracks may be modeled by

$$R(t) = 507t + 4599 \text{ million dollars}$$

According to the models, in what year will racetrack revenues surpass amusement park revenues? (**Source:** Modeled from *Statistical Abstract of the United States, 2001,* Table 1231, pp. 753–754.)

27. **M** **Artist and Athlete Earnings** In 1999 the number of paid employees in the independent artist, writer, and performer industry surpassed the number of paid employees in the sports team industry; however, the annual payroll for the artist industry still trailed that for the sports team industry by $4.1 billion. Based on data from 1998 and 1999, the payroll for the sports team industry may be modeled by

$$S(t) = 990t + 5718 \text{ million dollars}$$

and the payroll for the independent artist industry may be modeled by

$$A(t) = -35t + 3494 \text{ million dollars}$$

where t is the number of years since 1998. (**Source:** Modeled from *Statistical Abstract of the United States, 2001,* Table 1232, p. 754.) According to the models, when were the payrolls for the two industries equal? Does this solution seem reasonable? Explain.

28. **M** **High School Athletics** From 1971 to 1999, participation in high school athletics among women increased by 809 percent. In contrast, participation among men increased by a mere 5 percent. Based on data from 1991 to 2000, the number of women in high school sports may be modeled by

$$F(t) = 0.104t + 1.83 \text{ million}$$

and the number of men in high school sports may be modeled by

$$M(t) = 0.0606t + 3.33 \text{ million}$$

where t is the number of years since the 1990–1991 school year. (**Source:** Modeled from *Statistical Abstract of the United States, 2001,* Table 1247, p. 764.) According to the models, in what school year will the number of female athletes surpass the number of male athletes?

29. **M** **Chicken and Fish Consumption** Using data from 1970 to 1999, the per capita consumption of chicken may be modeled by

$$C(t) = 0.982t + 23.7 \text{ pounds}$$

and the per capita consumption of fish may be modeled by

$$F(t) = 0.137t + 11.7 \text{ pounds.}$$

where t is the number of years since 1970. (**Source:** Modeled from *Statistical Abstract of the United States, 2001,* Table 202, p. 129.) According to the models, will the per capita consumption of fish ever exceed the per capita consumption of chicken? Explain.

30. **M** **Pork and Beef Consumption** Using data from 1981 to 2001, the demand for pork may be modeled by

$$q = -21.2p + 95.2 \text{ pounds per person}$$

where p is the constant-demand price of pork in dollars per pound. The demand for beef over the same period may be modeled by

$$q = -18.1p + 128 \text{ pounds per person}$$

where p is the constant-demand price of beef in dollars per pound. (**Source:** Modeled from the Research Institute on Livestock Pricing.) Is there a price at which the demand for beef and pork will be the same?

In Exercises 31–40, solve the system of equations using the elimination method. If the system of equations is inconsistent or dependent, so state.

31. $x - y + z = 6$
$x + y + z = 8$
$x - y - z = -4$

32. $3x - y \quad\quad = 4$
$2x \quad\quad + z = 5$
$x + y + z = 5$

33. $6x - 3y + 3z = 6$
$x + 5y + 10z = 1$
$4x - 5y - z = 4$

34. $-2x - y + 4z = 8$
$2x + y - 4z = -8$
$x + y + z = 6$

35. $2x + 2y + 2z = 0$
$3x - 5y + 4z = -90$
$5x - 3y + 6z = -90$

36. $3x - y \quad\quad = 2$
$x \quad\quad + z = 2$
$2x - y - z = 2$

37. $3x - 3y + z = 5$
$x + y + z = 11$
$2x \quad\quad + 2z = 10$

38. $-2x - 7y + 3z = 0$
$5x + 4y - 2z = 0$
$4x + 3y + 9z = 0$

39. $9x - 8y - 7z = 16$
$7x - 5y + 3z = 4$
$5x - 3y - z = 6$

40. $10x - 5y \quad\quad = 0$
$20x + 5y + z = 0$
$30x \quad\quad + z = 0$

Exercises 41–45 are intended to challenge your understanding of systems of linear equations.

41. Solve the system of linear equations.

$$x + y + z + w = 2$$
$$x - y + z - w = 2$$
$$x - y - z + w = 0$$
$$x - y - z - w = 0$$

42. Solve the system of linear equations.

$$x + y + z + w = 1$$
$$x - y + z - w = 2$$
$$x - y - z + w = 3$$
$$x - y - z - w = 4$$

43. A failing business sells three colors of hair dye: fluorescent green, magenta, and sky blue. Its supplier charges $3 per bottle of fluorescent green, $2 per bottle of magenta, and $1 per bottle of sky blue. Regardless of the color, the business sells a bottle of hair dye to consumers for $5. The business has budgeted $300 to spend on hair dye and hopes to generate sufficient income from hair dye sales to pay its hair dye supplier and pay the $900 lease on the building that houses the store. It has sufficient space to store 250 bottles. Is it mathematically possible for the business to achieve its goal? Explain.

44. A system of equations has more equations than it has variables. Is it possible for such a system to have exactly one solution? Explain.

45. A system of equations has more variables than it has equations. Is it possible for such a system to have exactly one solution? Explain.

2.2 Using Matrices to Solve Linear Systems of Equations

- Represent systems of linear equations with augmented matrices
- Use row operations to put augmented matrices in reduced row echelon form

GETTING STARTED   In the previous section, we solved linear systems of equations in two and three variables using the graphical, substitution, and elimination methods. For equations containing more than three variables, the elimination method is the procedure of choice. In this section, we will introduce matrix notation and demonstrate how matrices are used to represent and solve systems of linear equations. We will also show how to use technology to simplify an augmented matrix to reduced row echelon form.

Matrix Notation

AN $m \times n$ MATRIX

An $m \times n$ matrix A is an array of numbers with m rows and n columns. The plural of matrix is *matrices*.

Capital letters are typically used to represent matrices. The following examples show matrices of various dimensions.

$$A = \begin{bmatrix} 2 & 4 & 6 \\ 5 & 3 & 1 \end{bmatrix} \qquad B = \begin{bmatrix} 3 \\ -1 \\ 6 \end{bmatrix} \qquad C = \begin{bmatrix} 2.1 & 1.9 \\ 0.1 & 0.8 \end{bmatrix} \qquad D = \begin{bmatrix} 2 & 7 \end{bmatrix}$$

Matrix A is a 2×3 matrix because it has two rows and three columns. Similarly, B is a 3×1 matrix, C is a 2×2 matrix, and D is a 1×2 matrix.

A matrix that consists of a single row of numbers is called a **row matrix.** Matrix D is a row matrix. A matrix that consists of a single column of numbers is called a **column matrix.** Matrix B is a column matrix. A matrix that has the same number of rows as columns is called a **square matrix.** Matrix C is a square matrix.

The numbers (or variables) inside the matrix are called the **entries** of the matrix. The entries of an $m \times n$ matrix A may be represented as

$$A = \begin{bmatrix} a_{11} & a_{12} & \cdots & a_{1n} \\ a_{21} & a_{22} & \cdots & a_{2n} \\ \vdots & \vdots & \vdots & \vdots \\ a_{m1} & a_{m2} & \cdots & a_{mn} \end{bmatrix}$$

The subscripts of each entry indicate its row and column position. An entry a_{ij} is the term in row i and column j. For example, a_{21} refers to the entry in the second row and first column.

EXAMPLE 1

Determining the Dimensions and Entry Values of a Matrix

Determine the dimensions of the matrix A and the value of the entries a_{12}, a_{21}, and a_{24}.

$$A = \begin{bmatrix} 1 & 2 & 3 & 4 \\ 5 & 6 & 7 & 8 \\ 9 & 10 & 11 & 12 \end{bmatrix}$$

SOLUTION The matrix has three rows and four columns, so it is a 3×4 matrix. For this matrix, $a_{12} = 2$, $a_{21} = 5$, and $a_{24} = 8$.

Representing Systems of Linear Equations with Augmented Matrices

A system of linear equations may be represented with an augmented matrix. For example, the system

$$2x + 3y = 8$$
$$4x - y = 2$$

is written as

$$\begin{bmatrix} 2 & 3 & 8 \\ 4 & -1 & 2 \end{bmatrix}$$

The matrix is called an **augmented matrix** because the coefficient matrix, $\begin{bmatrix} 2 & 3 \\ 4 & -1 \end{bmatrix}$, is augmented (added onto) with the column matrix, $\begin{bmatrix} 8 \\ 2 \end{bmatrix}$. (Recall that the term *coefficient* refers to the numeric value written in front of a given variable. For example, the coefficient of the term $2x$ is 2.) The vertical bar between the last two columns of numbers indicates that the matrix is an augmented matrix. The first column of the matrix contains the coefficients of the x terms. In this case, the x terms $2x$ and $4x$ have coefficients 2 and 4, respectively. The second column contains the coefficients of the y terms. In this case, the y terms $3y$ and $-y$ have coefficients 3 and -1, respectively.

EXAMPLE 2

Writing a Systems of Equations as an Augmented Matrix

Rewrite the system of equations as an augmented matrix.

$$x - 4y - z = -5$$
$$3y + z = 7$$
$$2x + y = 10$$

SOLUTION Before converting a system of equations to an augmented matrix, it is useful to write the equations so that each of the variables lines up vertically. That is, we rewrite the system as

$$x - 4y - z = -5$$
$$3y + z = 7$$
$$2x + y \quad = 10$$

The resultant augmented matrix is

$$\begin{bmatrix} 1 & -4 & -1 & -5 \\ 0 & 3 & 1 & 7 \\ 2 & 1 & 0 & 10 \end{bmatrix}$$

The first column represents the x variable, the second column represents the y variable, and the third column represents the z variable. Since the second equation didn't contain an x variable, we concluded that the coefficient on the second equation's x variable was zero. Similarly, the coefficient on the z variable in the third equation was zero.

Row Operations

When we introduced the elimination method, we identified three operations that yielded an equivalent system of equations. The operations were

1. Interchange (change the position of) two equations.
2. Multiply an equation by a nonzero number.
3. Add a nonzero multiple of one equation to a nonzero multiple of another equation.

Similar operations, called **row operations**, may be applied to augmented matrices.

ROW OPERATIONS

For any augmented matrix of a system of equations, the following row operations yield an augmented matrix of an equivalent system of equations:

1. Interchange (change the position of) two rows.
2. Multiply a row by a nonzero number.
3. Add a nonzero multiple of one row to a nonzero multiple of another row and replace either row with the result.

We will use the following notation to specify which operation we are applying.

Operation	Stand Alone Notation	Matrix Notation
Interchange Row 1 and Row 2	$R_1 \leftrightarrow R_2$	$\begin{bmatrix} a_{11} & a_{12} & a_{13} \\ a_{21} & a_{22} & a_{23} \\ a_{31} & a_{32} & a_{33} \end{bmatrix} \begin{matrix} R_2 \\ R_1 \\ \\ \end{matrix}$
Multiply Row 1 by a and place back in Row 1	$aR_1 \rightarrow R_1$	$\begin{bmatrix} a_{11} & a_{12} & a_{13} \\ a_{21} & a_{22} & a_{23} \\ a_{31} & a_{32} & a_{33} \end{bmatrix} \begin{matrix} aR_1 \\ \\ \\ \end{matrix}$
a times Row 1 is added to b times Row 2 and placed in Row 2	$aR_1 + bR_2 \rightarrow R_2$	$\begin{bmatrix} a_{11} & a_{12} & a_{13} \\ a_{21} & a_{22} & a_{23} \\ a_{31} & a_{32} & a_{33} \end{bmatrix} \begin{matrix} \\ aR_1 + bR_2 \\ \\ \end{matrix}$

In matrix notation, the vertical position of the operation indicates how the entries of the adjacent row were calculated.

The process of simplifying the matrix is almost identical to the elimination method illustrated in the previous section. We will first seek to obtain a 1 in the a_{11} entry of the matrix. We will then seek to obtain zeros below the 1 in the first column. We will simplify the remaining columns in a similar manner until the matrix is in **reduced row echelon form.**

REDUCED ROW ECHELON FORM

An augmented matrix A is said to be in reduced row echelon form if it satisfies each of the following conditions:

1. The leading entry (first nonzero entry) in each row is a 1.
2. The leading entry in each row is the only nonzero entry in its corresponding column.
3. The leading entry in each row is to the right of the leading entry in the row above it.
4. All rows of zeros are at the bottom of the matrix.

A matrix in reduced row echelon form is also referred to as a **reduced matrix.**

EXAMPLE **3**

Determining If an Augmented Matrix Is in Reduced Row Echelon Form

Which of the augmented matrices are in reduced row echelon form? For each matrix not in reduced row echelon form, specify the operations needed to produce a reduced matrix.

$$A = \begin{bmatrix} 1 & 0 & 1 & \vdots & 5 \\ 0 & 1 & 2 & \vdots & 2 \\ 0 & 0 & 0 & \vdots & 0 \end{bmatrix} \quad B = \begin{bmatrix} 0 & 1 & 0 & \vdots & 5 \\ 1 & 0 & 0 & \vdots & 7 \\ 0 & 0 & 1 & \vdots & 2 \end{bmatrix}$$

$$C = \begin{bmatrix} 1 & 0 & 0 & \vdots & 12 \\ 0 & 1 & 0 & \vdots & -2 \\ 0 & 0 & 6 & \vdots & 12 \end{bmatrix} \quad D = \begin{bmatrix} 1 & 2 & 0 & \vdots & 4 \\ 0 & 1 & 0 & \vdots & 3 \\ 0 & 0 & 1 & \vdots & 2 \end{bmatrix}$$

SOLUTION Matrix A is the only matrix in reduced row echelon form.

In matrix B, the leading entry in Row 2 is to the left instead of to the right of the leading entry in Row 1. Switching Row 1 with Row 2 would result in a reduced matrix.

In matrix C, the leading entry in the third row is a 6. Multiplying Row 3 by $\frac{1}{6}$ would result in a reduced matrix.

In matrix D, the leading entry in Row 2 contains a 2 above it instead of a 0. Multiplying Row 2 by -2, adding Row 2 to Row 1, and writing the final result in Row 1 would result in a reduced matrix.

EXAMPLE **4**

Rewriting an Augmented Matrix in Reduced Row Echelon Form

Rewrite the system of equations as an augmented matrix, then simplify the matrix to reduced row echelon form.

$$\begin{aligned} x - 4y - z &= -5 \\ 3y + z &= 7 \\ 2x + y \quad\;\; &= 10 \end{aligned}$$

SOLUTION

$$\begin{bmatrix} 1 & -4 & -1 & \vdots & -5 \\ 0 & 3 & 1 & \vdots & 7 \\ 2 & 1 & 0 & \vdots & 10 \end{bmatrix}$$

We will simplify the matrix entry by entry, beginning with the first column. Our goal is to make the matrix look like

$$\begin{bmatrix} 1 & 0 & 0 & \vdots & a \\ 0 & 1 & 0 & \vdots & b \\ 0 & 0 & 1 & \vdots & c \end{bmatrix}$$

We have a leading 1 in Row 1. We will use that entry to eliminate the leading entry in Row 3. If we multiply Row 1 by 2 and subtract Row 3, we can eliminate the leading 2 in Row 3. We will place the result of the operation in Row 3. Symbolically, we represent this as $2R_1 - R_3 \rightarrow R_3$.

We calculate the new values for Row 3 one entry at a time as follows:

$$a_{31} = 2(1) - 2 = 0$$
$$a_{32} = 2(-4) - 1 = -9$$
$$a_{33} = 2(-1) - 0 = -2$$
$$a_{34} = 2(-5) - 10 = -20$$

Observe that the numbers in the parentheses are the entries of Row 1 and the numbers following the minus sign are the entries of Row 3. The resultant matrix is

$$\begin{bmatrix} 1 & -4 & -1 & \vdots & -5 \\ 0 & 3 & 1 & \vdots & 7 \\ 0 & -9 & -2 & \vdots & -20 \end{bmatrix} \quad 2R_1 - R_3$$

Next we will eliminate the -9 in the third row using the following operation:

$$3R_2 + R_3 \rightarrow R_3$$

We calculate the new values for Row 3 one entry at a time as follows:

$$a_{31} = 3(0) + 0 = 0$$
$$a_{32} = 3(3) + (-9) = 0$$
$$a_{33} = 3(1) + (-2) = 1$$
$$a_{34} = 3(7) + (-20) = 1$$

The resultant matrix is

$$\begin{bmatrix} 1 & -4 & -1 & \vdots & -5 \\ 0 & 3 & 1 & \vdots & 7 \\ 0 & 0 & 1 & \vdots & 1 \end{bmatrix} \quad 3R_2 + R_3$$

Notice that Row 3 contains only zeros in Columns 1 and 2. This means that adding a multiple of Row 3 to any of the other rows will leave the first two columns unchanged. We will use this fact to zero out the entries a_{13} and a_{23}. Using the operations

$$R_2 - R_3 \rightarrow R_2 \qquad \text{and} \qquad R_1 + R_3 \rightarrow R_1$$

we obtain the matrix

$$\begin{bmatrix} 1 & -4 & 0 & \vdots & -4 \\ 0 & 3 & 0 & \vdots & 6 \\ 0 & 0 & 1 & \vdots & 1 \end{bmatrix} \quad \begin{matrix} R_1 + R_3 \\ R_2 - R_3 \end{matrix}$$

Next we need to get a leading 1 in Row 2. Using the operation

$$\frac{1}{3}R_2 \rightarrow R_2$$

yields the matrix

$$\begin{bmatrix} 1 & -4 & 0 & \vdots & -4 \\ 0 & 1 & 0 & \vdots & 2 \\ 0 & 0 & 1 & \vdots & 1 \end{bmatrix} \quad \frac{1}{3}R_2$$

To finish, we zero out the a_{12} entry with the operation

$$R_1 + 4R_2 \rightarrow R_1$$

resulting in the matrix

$$\begin{bmatrix} 1 & 0 & 0 & 4 \\ 0 & 1 & 0 & 2 \\ 0 & 0 & 1 & 1 \end{bmatrix} \qquad R_1 + 4R_2$$

The matrix is in reduced row echelon form and yields the solution to the system of equations.

$$x = 4$$
$$y = 2$$
$$z = 1$$

In Example 4, we simplified the matrix in the following order:

1. Make Column 1 look like $\begin{bmatrix} 1 \\ 0 \\ 0 \end{bmatrix}$ and the resultant matrix look like

$$\begin{bmatrix} 1 & a & d & g \\ 0 & b & e & h \\ 0 & c & f & i \end{bmatrix}.$$

2. Make Row 3 look like $\begin{bmatrix} 0 & 0 & 1 \end{bmatrix}$ and the resultant matrix look like

$$\begin{bmatrix} 1 & a & c & e \\ 0 & b & d & f \\ 0 & 0 & 1 & g \end{bmatrix}.$$

3. Make Column 3 look like $\begin{bmatrix} 0 \\ 0 \\ 1 \end{bmatrix}$ and the resultant matrix look like

$$\begin{bmatrix} 1 & a & 0 & c \\ 0 & b & 0 & d \\ 0 & 0 & 1 & e \end{bmatrix}.$$

4. Make Column 2 look like $\begin{bmatrix} 0 \\ 1 \\ 0 \end{bmatrix}$ and the resultant matrix look like

$$\begin{bmatrix} 1 & 0 & 0 & a \\ 0 & 1 & 0 & b \\ 0 & 0 & 1 & c \end{bmatrix}.$$

Note: The variables *a* through *i* do not represent the same values in each step.

The idea is to first create a "triangle" of zeroes below the main diagonal of the matrix and then create a "triangle" of zeroes above the main diagonal of the matrix. We believe this approach is the simplest; however, regardless of the order in which you simplify the columns and rows of the matrix, the reduced row echelon form of the matrix will be the same.

When the number of columns on the coefficient side of the augmented matrix exceeds the number of rows of the matrix, the matrix corresponds to a system of equations with more variables than equations. For this type of matrix, there will be multiple solutions to the system of equations (if a solution exists).

If the number of rows exceeds the number of columns on the coefficient side of the augmented matrix, the matrix corresponds to a system of equations with more equations than variables. If a solution to the system exists, the simplified matrix will contain at least one row of zeros. Examples 5 and 6 demonstrate these concepts.

EXAMPLE 5

Rewriting an Augmented Matrix in Reduced Row Echelon Form

Rewrite the system of equations as an augmented matrix, then simplify the matrix to reduced row echelon form. Identify the solution(s) to the system of equations.

$$x + y + 6z = 16$$
$$2x - y \quad\quad = 2$$

SOLUTION

$$\begin{bmatrix} 1 & 1 & 6 & \vdots & 16 \\ 2 & -1 & 0 & \vdots & 2 \end{bmatrix}$$

The operation $2R_1 - R_2 \rightarrow R_2$ will zero out the a_{21} entry.

$$\begin{bmatrix} 1 & 1 & 6 & \vdots & 16 \\ 0 & 3 & 12 & \vdots & 30 \end{bmatrix} \quad 2R_1 - R_2$$

The operation $\frac{1}{3}R_2 \rightarrow R_2$ will give us a leading 1 in the a_{22} entry.

$$\begin{bmatrix} 1 & 1 & 6 & \vdots & 16 \\ 0 & 1 & 4 & \vdots & 10 \end{bmatrix} \quad \frac{1}{3}R_2$$

The operation $R_1 - R_2 \rightarrow R_1$ will zero out the a_{12} entry.

$$\begin{bmatrix} 1 & 0 & 2 & \vdots & 6 \\ 0 & 1 & 4 & \vdots & 10 \end{bmatrix} \quad R_1 - R_2$$

The matrix is in reduced row echelon form. We have the equations

$$x + 2z = 6$$
$$y + 4z = 10$$

Solving the equations for x and y, respectively, yields

$$x = -2z + 6$$
$$y = -4z + 10$$

Notice that we have x as a function of z and y as a function of z. We can pick z to be any number we want and then calculate the corresponding values of x and y. To represent this idea, we set z equal to an arbitrary variable, say t. The final solution is

$$x = -2t + 6$$
$$y = -4t + 10$$
$$z = t$$

There are an infinite number of points (x, y, z) that satisfy the system of equations, so the system of equations is dependent. We can calculate a particular solution by picking a value of t. For example, let $t = 0$. Then $x = 6$, $y = 10$, and $z = 0$.

EXAMPLE 6

Rewriting an Augmented Matrix in Reduced Row Echelon Form

Rewrite the system of equations as an augmented matrix, then simplify the matrix to reduced row echelon form. Identify the solution(s) to the system of equations.

$$x + 2y = 5$$
$$3x - y = 11.5$$
$$5x + 3y = 21.5$$

SOLUTION

$$\begin{bmatrix} 1 & 2 & | & 5 \\ 3 & -1 & | & 11.5 \\ 5 & 3 & | & 21.5 \end{bmatrix}$$

The operation $3R_1 - R_2 \rightarrow R_2$ will zero out the a_{21} entry, and the operation $5R_1 - R_3 \rightarrow R_3$ will zero out the a_{31} entry. The resultant matrix is

$$\begin{bmatrix} 1 & 2 & | & 5 \\ 0 & 7 & | & 3.5 \\ 0 & 7 & | & 3.5 \end{bmatrix} \quad \begin{matrix} 3R_1 - R_2 \\ 5R_1 - R_3 \end{matrix}$$

Observe that Row 2 and Row 3 are identical. The operation $R_2 - R_3 \rightarrow R_3$ will zero out Row 3.

$$\begin{bmatrix} 1 & 2 & | & 5 \\ 0 & 7 & | & 3.5 \\ 0 & 0 & | & 0 \end{bmatrix} \quad R_2 - R_3$$

The operation $\frac{1}{7}R_2 \rightarrow R_2$ will give us a leading 1 in Row 2.

$$\begin{bmatrix} 1 & 2 & | & 5 \\ 0 & 1 & | & 0.5 \\ 0 & 0 & | & 0 \end{bmatrix} \quad \frac{1}{7}R_2$$

The operation $R_1 - 2R_2 \rightarrow R_1$ will zero out the a_{12} entry.

$$\begin{bmatrix} 1 & 0 & | & 4 \\ 0 & 1 & | & 0.5 \\ 0 & 0 & | & 0 \end{bmatrix} \quad R_1 - 2R_2$$

The matrix is in reduced row echelon form. The solution to the system of equations is $x = 4$ and $y = 0.5$.

Not every system of equations has a solution. When a system of equation has no solution, a contradiction will occur in the augmented matrix. This case is demonstrated in Examples 7 and 8.

EXAMPLE 7

Using Matrices to Solve an Inconsistent System of Equations

Rewrite the system of equations as an augmented matrix, then simplify the matrix to reduced row echelon form. Identify the solution(s) to the system of equations.

$$x + 2y = 5$$
$$3x - y = 2$$
$$4x + y = 8$$

SOLUTION

$$\begin{bmatrix} 1 & 2 & \vdots & 5 \\ 3 & -1 & \vdots & 2 \\ 4 & 1 & \vdots & 8 \end{bmatrix}$$

The operation $3R_1 - R_2 \rightarrow R_2$ will zero out the a_{21} entry, and the operation $4R_1 - R_3 \rightarrow R_3$ will zero out the a_{31} entry. The resultant matrix is

$$\begin{bmatrix} 1 & 2 & \vdots & 5 \\ 0 & 7 & \vdots & 13 \\ 0 & 7 & \vdots & 12 \end{bmatrix} \quad \begin{matrix} \\ 3R_1 - R_2 \\ 4R_1 - R_3 \end{matrix}$$

The operation $R_2 - R_3 \rightarrow R_3$ will zero out the a_{32} entry.

$$\begin{bmatrix} 1 & 2 & \vdots & 5 \\ 0 & 7 & \vdots & 13 \\ 0 & 0 & \vdots & 1 \end{bmatrix} \quad \begin{matrix} \\ \\ R_2 - R_3 \end{matrix}$$

Row 3 is equivalent to

$$0x + 0y = 1$$
$$0 = 1$$

But $0 \neq 1$ so the system of equations is inconsistent (has no solution).

EXAMPLE 8

Using Matrices to Solve an Inconsistent System of Equations

Rewrite the system of equations as an augmented matrix, then simplify the matrix to reduced row echelon form. Identify the solution(s) to the system of equations.

$$3x + 3y \quad = 9$$
$$2x + 3y + z = 6$$
$$x \quad - z = 1$$

SOLUTION

$$\begin{bmatrix} 3 & 3 & 0 & \vdots & 9 \\ 2 & 3 & 1 & \vdots & 6 \\ 1 & 0 & -1 & \vdots & 1 \end{bmatrix}$$

The operation $R_1 \leftrightarrow R_3$ places a 1 in the upper left-hand corner of the matrix.

$$\begin{bmatrix} 1 & 0 & -1 & | & 1 \\ 2 & 3 & 1 & | & 6 \\ 3 & 3 & 0 & | & 9 \end{bmatrix} \quad \begin{matrix} R_3 \\ \\ R_1 \end{matrix}$$

The operation $2R_1 - R_2 \rightarrow R_2$ will zero out the a_{21} entry. The operation $3R_1 - R_3 \rightarrow R_3$ will zero out the a_{31} entry.

$$\begin{bmatrix} 1 & 0 & -1 & | & 1 \\ 0 & -3 & -3 & | & -4 \\ 0 & -3 & -3 & | & -6 \end{bmatrix} \quad \begin{matrix} \\ 2R_1 - R_2 \\ 3R_1 - R_3 \end{matrix}$$

The operation $R_2 - R_3 \rightarrow R_3$ will zero out the a_{32} entry.

$$\begin{bmatrix} 1 & 0 & -1 & | & 1 \\ 0 & -3 & -3 & | & -4 \\ 0 & 0 & 0 & | & 2 \end{bmatrix} \quad \begin{matrix} \\ \\ R_2 - R_3 \end{matrix}$$

Row 3 is equivalent to

$$0x + 0y + 0z = 2$$
$$0 = 2$$

But $0 \neq 2$. Therefore, the system of equations is inconsistent.

In each of the examples, we have looked at systems of equations with integer coefficients and rational solutions. Although the elimination process may be used on any augmented matrix, computations may become particularly messy when the system of equations has noninteger coefficients. Fortunately, the TI-83 Plus has a powerful feature that will simplify augmented matrices to reduced row echelon form.

EXAMPLE 9

Using Technology to Find the Reduced Row Echelon Form of a Matrix

Rewrite the system of equations as an augmented matrix, then simplify the matrix to reduced row echelon form. Identify the solution(s) to the system of equations.

$$0.31x + 6.23y + 2.15z = 5.61$$
$$0.07x - 7.21y + 1.15z = 2.23$$
$$4.11x - 1.23y - 3.91z = 8.99$$

SOLUTION

$$\begin{bmatrix} 0.31 & 6.23 & 2.15 & | & 5.61 \\ 0.07 & -7.21 & 1.15 & | & 2.23 \\ 4.11 & -1.23 & -3.91 & | & 8.99 \end{bmatrix}$$

Using the techniques shown in the following Technology Tips, the reduced matrix is approximately equal to

$$\begin{bmatrix} 1 & 0 & 0 & \vdots & 4.03 \\ 0 & 1 & 0 & \vdots & 0.0365 \\ 0 & 0 & 1 & \vdots & 1.92 \end{bmatrix}$$

The solution to the system of equations is $x \approx 4.03$, $y \approx 0.0365$, and $z \approx 1.92$.

TECHNOLOGY **TIP**

Entering a Matrix

1. Activate the Matrix Menu by pressing
   [2nd] [x^{-1}]. You may or may not have
   some matrices displayed in the
   name list.

   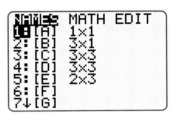

2. Create the augmented matrix A by
   moving the cursor to EDIT and pressing
   [ENTER].

   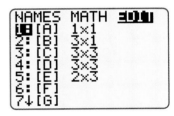

3. Enter the dimensions of the matrix. Type
   the number of rows and press [ENTER],
   then type the number of columns and
   press [ENTER]. The example matrix is a
   3×4 augmented matrix.

4. Enter the individual entries of the
   matrix. Use the blue arrow buttons to
   move from one entry to another. When
   you've entered all of the entries, press
   [2nd] [MODE] to exit the Matrix Editor.

   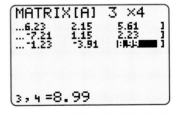

TECHNOLOGY **TIP**

Finding the Reduced Row Echelon Form of a Matrix *A*

1. Activate the Matrix Math Menu by pressing [2nd] [x⁻¹] and moving the cursor to the MATH menu item.

```
NAMES MATH EDIT
1:det(
2:ᵀ
3:dim(
4:Fill(
5:identity(
6:randM(
7↓augment(
```

2. Select the rref operation by scrolling down the list to item B:rref(. Press [ENTER] to place the operation on the home screen. (The rref(function will convert a matrix to reduced row echelon form.)

```
rref(
```

3. Select the matrix *A* from the Matrix Names Menu by pressing [2nd] [x⁻¹], moving the cursor to matrix *A*, and pressing [ENTER]. This places the matrix *A* on the home screen.

```
rref([A]
```

4. Calculate the reduced row echelon form of the matrix by pressing [ENTER].

```
rref([A]
[[1 0 0 4.02752…
 [0 1 0 .036500…
 [0 0 1 1.92282…
```

2.2 Summary

In this section, you learned basic matrix notation and saw how to use matrices to represent and solve systems of linear equations. You also learned how to use technology to simplify an augmented matrix to reduced row echelon form.

2.2 Exercises

In Exercises 1–5, determine the dimensions of the matrix A.

1. $A = \begin{bmatrix} 19 & 5 & 1 \\ 2 & 8 & 18 \end{bmatrix}$

2. $A = \begin{bmatrix} -10 \\ 42 \\ 17 \end{bmatrix}$

3. $A = \begin{bmatrix} 14 & 0 & -5 \end{bmatrix}$

4. $A = \begin{bmatrix} -4 & 5 & 8 & 11 \\ 11 & 5 & 4 & -6 \end{bmatrix}$

5. $A = \begin{bmatrix} 17 \end{bmatrix}$

In Exercises 6–10, write the augmented matrix as a system of linear equations, using the variables x, y, and z as appropriate.

6. $\left[\begin{array}{ccc:c} 2 & 1 & 0 & 11 \\ 1 & 0 & 4 & 0 \end{array}\right]$

7. $\left[\begin{array}{cc:c} 1 & 3 & 12 \\ 4 & 1 & 7 \\ -1 & 6 & 9 \end{array}\right]$

8. $\left[\begin{array}{ccc:c} 1 & 0 & 4 & 0 \\ 0 & 1 & 2 & 19 \\ 6 & 0 & 3 & 22 \end{array}\right]$

9. $\left[\begin{array}{ccc:c} 3 & -1 & 9 & 13 \\ -3 & 1 & 0 & -31 \end{array}\right]$

10. $\left[\begin{array}{ccc:c} 5 & 2 & 4 & 11 \\ -1 & 2 & 2 & 3 \\ 5 & 10 & 3 & 18 \end{array}\right]$

In Exercises 11–20, determine if the augmented matrix is in reduced row echelon form. If it is not reduced, specify which of the four criteria for reduced row echelon form the matrix fails to meet.

11. $\left[\begin{array}{ccc:c} 1 & 0 & 0 & 3 \\ 0 & 0 & 1 & 8 \\ 0 & 1 & 0 & -2 \end{array}\right]$

12. $\left[\begin{array}{ccc:c} 1 & 0 & 0 & 0 \\ 0 & 0 & 1 & 11 \\ 0 & 0 & 0 & 0 \end{array}\right]$

13. $\left[\begin{array}{ccc:c} 1 & 0 & 0 & 0 \\ 0 & 3 & 0 & 12 \\ 0 & 0 & 1 & 5 \end{array}\right]$

14. $\left[\begin{array}{ccc:c} 1 & 0 & 0 & 0 \\ 0 & 1 & 2 & 14 \\ 0 & 0 & 1 & 52 \end{array}\right]$

15. $\left[\begin{array}{ccc:c} 1 & 0 & 0 & 2.7 \\ 0 & 0 & 1 & 4.2 \end{array}\right]$

16. $\left[\begin{array}{ccc:c} 1 & 1 & 0 & 12.1 \\ 0 & 0 & 1 & -34.2 \end{array}\right]$

17. $\left[\begin{array}{ccc:c} 0 & 1 & 0 & 11 \\ 1 & 0 & 1 & 0 \end{array}\right]$

18. $\left[\begin{array}{ccc:c} 1 & 2 & 9 & 13 \\ 0 & 0 & 0 & 0 \end{array}\right]$

19. $\left[\begin{array}{cc:c} 1 & 0 & 12 \\ 0 & 1 & 7 \\ 0 & 1 & 9 \end{array}\right]$

20. $\left[\begin{array}{cc:c} 2 & 0 & 12 \\ 0 & 1 & 7 \\ 0 & 0 & 0 \end{array}\right]$

In Exercises 21–40, rewrite the system of equations as an augmented matrix. Then simplify the matrix to reduced row echelon form. Identify the solution(s) to the system of equations. If the system is inconsistent or dependent, so state.

21. $2x + 5y = 2$
$3x - 5y = 3$

22. $6x + 2y = 10$
$-x - 2y = -5$

23. $-2x + 6y + 4z = 10$
$4x - 12y + 2z = -20$
$3x + 4y - z = 11$

24. $10x - y + z = 7$
$9x - 2y + 4z = 7$
$x + 2y - 4z = -17$

25. $3x - 2y + z = 6$
$11x - 20y - z = 0$
$y + z = 3$

26. $9x - 6y \qquad = 0$
$\quad 4x + 5y \qquad = 23$
$\quad x \qquad - z = 6$

27. $\quad x - y = -5$
$\quad 9x + y = 25$
$\quad 29x + y = 65$

28. $\quad x - 2y = -7$
$\quad 6x + 5y = 94$
$\quad 10x + 3y = 114$

29. $\quad 2x - 4y = 16$
$\quad 9x + y = -4$
$\quad -3x + 6y = -24$

30. $x - y = 3$
$\quad 6x + 7y = 44$
$\quad 6x - 7y = 16$

31. $2x - 2y = 0$
$\quad 3x + y = 4$
$\quad 5x - y = 5$

32. $8x - y = 3$
$\quad x + 2y = 11$
$\quad 9x + y = 14$

33. $x + y + z = 6$
$\quad 2x - y + z = 3$

34. $2x - 4y + 2z = 6$
$\quad 5x - 5y + z = 1$

35. $3x + 5y + z = 6$
$\quad 6x + 10y + 2z = 10$

36. $x + y + z = 6$
$\quad 2x - y + z = 12$

37. $x + y + z = 6$
$\quad 2x - y + z = 3$
$\quad 4x - 2y + 3z = 9$

38. $5x + y + z = 1$
$\quad 4x - 2y + 5z = -2$
$\quad x - 7y + 6z = -7$

39. $3x + 2y + 3z = 6$
$\quad 2x - 5y + z = -11$
$\quad 4x + 2y + 3z = 3$

40. $x + y + z = 3$
$\quad 2x - y + z = 2$
$\quad 4x - 2y + 3z = 5$

In Exercises 41–50, use technology to write the matrix in reduced row echelon form. Then interpret the result.

41. $\begin{bmatrix} 0.1 & 0.3 & 1.2 & | & 2.1 \\ 0.7 & -2.8 & 3.5 & | & 1.2 \\ 8.2 & -7.3 & -1.1 & | & 3.9 \end{bmatrix}$

42. $\begin{bmatrix} 2.6 & 0.0 & 0.2 & | & 9.6 \\ 3.1 & 4.0 & 7.5 & | & 0 \\ 9.1 & -2.1 & -0.8 & | & 9.9 \end{bmatrix}$

43. $\begin{bmatrix} 0.11 & 0.99 & 1.25 & | & 2.61 \\ 0.97 & -2.28 & 2.15 & | & 11.23 \\ 5.12 & -4.43 & -1.81 & | & 3.29 \end{bmatrix}$

44. $\begin{bmatrix} 1.2 & 2.3 & 3.4 & | & 1.0 \\ 2.3 & 3.4 & 4.5 & | & 2.0 \\ 3.4 & 4.5 & 5.6 & | & 3.0 \end{bmatrix}$

45. $\begin{bmatrix} 0 & 2.3 & 4.4 & | & 8.0 \\ 2.3 & 0 & 6.5 & | & 6.0 \\ 4.4 & 6.5 & 0 & | & 4.0 \end{bmatrix}$

46. $\begin{bmatrix} 5.7 & 7.8 & 9.4 & | & 8.2 \\ 1.3 & 3.0 & 0.5 & | & 5.9 \\ 4.7 & 0.5 & 2.0 & | & 3.1 \end{bmatrix}$

47. $\begin{bmatrix} 1.2 & 3.4 & 5.6 & | & 7.8 \\ 0.1 & 2.3 & 4.5 & | & 6.7 \\ 9.8 & 7.6 & 5.4 & | & 3.2 \end{bmatrix}$

48. $\begin{bmatrix} 1.7 & 2.8 & 6.4 & | & 8.2 \\ 1.6 & 3.6 & 2.5 & | & 5.9 \\ 4.9 & 0.5 & 2.0 & | & 3.1 \end{bmatrix}$

49. $\begin{bmatrix} 0.50 & 0.25 & 0.25 & | & 1 \\ 0.25 & 0.50 & 0.25 & | & 1 \\ 0.25 & 0.25 & 0.50 & | & 1 \end{bmatrix}$

50. $\begin{bmatrix} 1.9 & 2.8 & 3.7 & | & 4.6 \\ 2.5 & 3.6 & 4.7 & | & 5.8 \\ 3.7 & 0.3 & 2.0 & | & 3.3 \end{bmatrix}$

2.3 Linear System Applications

- Formulate linear systems from real-life data
- Interpret the meaning of solutions to linear systems

GETTING STARTED The per capita consumption of sugar rose from 62.7 pounds in 1985 to 67.9 pounds in 1999. Over the same period, the per capita consumption of high-fructose corn syrup (used to sweeten soft drinks, juices, and other such products) rose from 45.2 pounds to 66.2 pounds. (**Source:** *Statistical Abstract of the United States, 2001,* Table 202, p. 129.) Based on the trends in sugar and corn syrup consumption, producers may want to adjust their strategic (long-range) production plan. In this section, we will show how to formulate linear systems from real-life data and use the solutions to these systems to make business decisions. We will also show some of the strengths and limitations of mathematical models.

EXAMPLE 1 **Using a Linear Model to Forecast Demand for a Product**

The per capita consumption of sugar and high-fructose corn syrup are shown in Table 2.1.

TABLE 2.1

Years since 1985 (t)	Per Capita Sugar Consumption (pounds) (S)	Per Capita High-Fructose Corn Syrup Consumption (pounds) (C)
0	62.7	45.2
5	64.4	49.6
10	65.5	58.4
12	66.5	62.4
13	66.9	63.8
14	67.9	66.2

(a) Use linear functions to model the consumption of each of the sweeteners.
(b) Predict the year in which the consumption of each type of sweetener will be equal.
(c) Explain how the results of your analysis could benefit a sweetener producer.

SOLUTION

(a) Using linear regression on each data set, we obtain the models

Sugar: $S = 0.339t + 62.6$ pounds
Corn syrup: $C = 1.53t + 43.8$ pounds

For both models, t is the number of years after 1985.

(b) We want to find when $S = C$. Setting $S = C$ and moving all of the variables to the left-hand side of the equal sign we obtain the system of equations

$$C - 0.339t = 62.6$$
$$C - 1.53t = 43.8$$

and the corresponding augmented matrix

$$\begin{bmatrix} 1 & -0.339 & \vdots & 62.6 \\ 1 & -1.53 & \vdots & 43.8 \end{bmatrix}$$

Using technology to reduce the matrix yields

$$\begin{bmatrix} 1 & 0 & \vdots & 68.0 \\ 0 & 1 & \vdots & 15.8 \end{bmatrix}$$

and the solution $C = 68.0$ and $t = 15.8$. That is, about 15.8 years after 1985, the per capita consumption of both sugar and high-fructose corn syrup were projected to be about 68 pounds. So in the latter part of 2001, sugar and high-fructose corn syrup consumption were projected to be equal.

(c) Based on the slopes of the linear models, we see that sugar consumption between 1985 and 1999 was increasing by 0.339 pound per year while corn syrup consumption was increasing by 1.53 pounds per year. A producer could use this information to adjust production levels of corn syrup and sugar. Recognizing that the demand for high-fructose corn syrup would surpass the demand for sugar in 2001, the producer could adjust her marketing strategy to take advantage of the increased consumer demand.

It is essential to recognize that mathematical models rarely yield exact results. Nevertheless, they are of value because they give us an idea of what may happen if current trends continue. They also offer a mechanism for making an educated guess concerning yet unknown future results.

In Example 1, we used linear regression to model the data. Linear regression is just one of the many different types of regression that may be used to find a model for a data set. In Chapter 5, we will demonstrate other types of regression, including quadratic, cubic, quartic, exponential, and logarithmic. In deciding which type of regression to use, we select the type of function that we believe will best fit the data.

EXAMPLE 2

Using a Linear Model to Forecast School Enrollment

School district facilities managers are often asked to provide information to district boards regarding projected growth in enrollment within the district. The boards use this information to determine whether the existing education infrastructure (buildings, buses, and so on) can accommodate the projected need. These projections may go out more than 25 years, but they are frequently adjusted to improve accuracy as more data become available.

In January 2002, the White River School District facilities manager published a student enrollment report including the data shown in Table 2.2.

TABLE 2.2

Years Since 1998 (t)	Projected Student Enrollment, Grade 3 (students) (T)	Projected Student Enrollment, Grade 4 (students) (F)
0	273	298
1	311	282
2	321	327
3	312	339
4	311	319
5	337	317
6	309	340
7	365	312
8	372	368
9	379	375
10	387	380

(a) Find the linear models that best fit the data for enrollment in grades 3 and 4.
(b) Use the models (not the table) to predict the years in which grade 3 enrollment is projected to exceed grade 4 enrollment.
(c) Explain why your result in part (b) is different from that shown in the table.
(d) Explain how the results of your analysis could benefit the school district.

SOLUTION

(a) Using linear regression on each data set, we obtain the models

Grade 3: $T = 9.99t + 284$ students
Grade 4: $F = 7.93t + 293$ students

For both models, t is the number of years after 1998.
  We note that the correlation coefficients for the models are 0.908 and 0.830, respectively. The models don't fit the data sets as well as we would have liked. (We would prefer to see a correlation coefficient closer to 1.)

(b) We want to find when $T = F$. Setting $T = F$ and moving all of the variables to the left-hand side of the equal sign, we obtain the system of equations

$$F - 9.99t = 284$$
$$F - 7.93t = 293$$

and the corresponding augmented matrix

$$\begin{bmatrix} 1 & -9.99 & \vdots & 284 \\ 1 & -7.93 & \vdots & 293 \end{bmatrix}$$

Writing the matrix in reduced row echelon form and rounding to three significant digits yields

$$\begin{bmatrix} 1 & 0 & \vdots & 328 \\ 0 & 1 & \vdots & 4.37 \end{bmatrix}$$

and the solution $F = 328$ and $t = 4.37$. That is, about 4.37 years after 1998, enrollment in both grade 3 and grade 4 is projected to be 328 students. Looking at the slopes of our linear models, we see that grade 3 is increasing by about 10 students per year, while grade 4 is increasing by about 8 students per year. Consequently, grade 3 is expected to have a greater enrollment for the fifth and subsequent years after 1998.

(c) In the table, we see that the facilities manager projects that grade 3 will have greater enrollment 1, 5, 7, 8, 9, and 10 years after 1998. The facilities manager's projections take into consideration the fact that this year's grade 3 will be the basis for next year's grade 4, a fact not considered in our models.

(d) Our results could be used to determine if existing classrooms are sufficient to accommodate the growing third- and fourth-grade population. The results could also be used to justify hiring additional elementary school teachers.

EXAMPLE 3 **Using Linear Models to Make Investment Decisions**

An investor with $30,000 in an existing mutual fund IRA wants to roll over her investment into a new retirement plan with TIAA-CREF. The company offers a variety of accounts with varying returns, as shown in Table 2.3.

TABLE 2.3

CREF Variable Annuity Accounts as of 5/31/04	Unit Value	1-year Return	5-year Return	10-year Return	Return Since Inception
Bond Market	$71.01	−0.65%	6.70%	7.05%	7.63%
Equity Index	$71.94	19.16%	−0.75%	10.83%	10.89%
Global Equities	$70.45	24.50%	−0.84%	6.96%	8.52%
Growth	$55.93	16.01%	−6.72%	8.14%	8.23%
Inflation-Linked Bond	$42.14	2.11%	9.15%	—	7.53%
Money Market	$21.86	0.67%	3.16%	4.25%	4.95%
Social Choice	$101.39	11.07%	2.49%	10.01%	10.27%
Stock	$174.37	21.18%	−0.62%	9.53%	10.44%

Source: www.tiaa-cref.com.

The investor looks at the long-term performance of each account as an indicator of the likely long-term return. She would like to invest her money in the Bond Market, Equity Index, and Social Choice accounts. After analyzing her risk tolerance, she decides to invest twice as much money in the Bond Market account as in the Equity Index account. She wants to earn a 9 percent return on her investments. Assuming that she will be able to earn the 10-year average annual return, how much money should she invest in each account? [For ease of computation, round the 10-year rate to the nearest whole-number percent (e.g., 7.05 percent is rounded to 7 percent)].

SOLUTION Let x be the amount of money invested in the Bond Market account, y be the amount invested in the Equity Index account, and z be the amount invested in the Social Choice account. Since the investor has $30,000 to invest, we know that

$$x + y + z = 30,000 \qquad \text{Total amount of money to be invested}$$

Since she plans to invest twice as much money in the Bond Market account as in the Equity Index account, we know that

$$x = 2y \qquad \text{The Bond Market investment is twice the Equity Index investment}$$

$$x - 2y = 0$$

Rounding to the nearest whole-number percentage, the Bond Market account has a 7 percent return rate, the Equity Index account has an 11 percent return rate, and the Social Choice Stock account has a 10 percent return rate. Since she wants to earn a 9 percent return on her total investment, we know that

$$0.7x + 0.11y + 0.10z = 0.09(30,000) \qquad \text{Total annual return on investment}$$

$$0.07x + 0.11y + 0.10z = 2700$$

We need to solve the following system of equations:

$$
\begin{aligned}
x + \quad y + \quad z &= 30,000 \\
x - \quad 2y \qquad\quad &= 0 \\
0.07x + 0.11y + 0.10z &= 2700
\end{aligned}
$$

The corresponding augmented matrix is

$$
\begin{bmatrix}
1 & 1 & 1 & 30,000 \\
1 & -2 & 0 & 0 \\
0.07 & 0.11 & 0.10 & 2,700
\end{bmatrix}
$$

Entering the matrix into the TI-83 Plus and using the **rref** feature gives the results in Figure 2.10.
    The reduced matrix is

$$
\begin{bmatrix}
1 & 0 & 0 & 12,000 \\
0 & 1 & 0 & 6,000 \\
0 & 0 & 1 & 12,000
\end{bmatrix}
$$

FIGURE 2.10

and corresponds with the system of equations

$$x = 12,000$$
$$y = 6000$$
$$z = 12,000$$

She should invest $12,000 in the Bond Market account, $6000 in the Equity Index account, and $12,000 in the Social Choice account. By diversifying her investments into different types of accounts, she is able to reduce her overall risk.

EXAMPLE 4

Using a System of Linear Equations to Find a Quadratic Model

According to its 2001 annual report, the gross revenue of Johnson & Johnson and its subsidiaries increased in 1999, 2000, and 2001, as shown in Table 2.4.

TABLE 2.4   Johnson & Johnson and Subsidiaries Gross Revenue

Years Since the End of 1999 (t)	Revenue from Sales (millions of dollars) (R)
0	27,357
1	29,172
2	32,317

Source: Johnson & Johnson 2001 annual report, p. 6.

Use matrices to find a quadratic equation that models the company's gross revenue.

SOLUTION A quadratic equation will have the form $R(t) = at^2 + bt + c$, where $R(t)$ is the gross revenue (in millions of dollars) and t is the number of years since the end of 1999. Each of the data points must satisfy the quadratic equation. That is,

$$27{,}357 = a(0)^2 + b(0) + c \qquad \text{Using } t = 0 \text{ and } R = 27{,}357$$
$$29{,}172 = a(1)^2 + b(1) + c \qquad \text{Using } t = 1 \text{ and } R = 29{,}172$$
$$32{,}317 = a(2)^2 + b(2) + c \qquad \text{Using } t = 2 \text{ and } R = 32{,}317$$

Reducing each equation yields a system of linear equations

$$c = 27{,}357$$
$$a + b + c = 29{,}172$$
$$4a + 2b + c = 32{,}317$$

and its corresponding augmented matrix

$$\begin{bmatrix} 0 & 0 & 1 & \vdots & 27{,}357 \\ 1 & 1 & 1 & \vdots & 29{,}172 \\ 4 & 2 & 1 & \vdots & 32{,}317 \end{bmatrix}$$

We can readily see that $c = 27{,}357$. We will use the TI-83 Plus to find the values of a and b. The reduced matrix

$$\begin{bmatrix} 1 & 0 & 0 & \vdots & 665 \\ 0 & 1 & 0 & \vdots & 1{,}150 \\ 0 & 0 & 1 & \vdots & 27{,}357 \end{bmatrix}$$

yields the corresponding solution to the system

$$a = 665$$
$$b = 1150$$
$$c = 27{,}357$$

The quadratic model for the gross revenue of Johnson & Johnson and its subsidiaries is

$$R(t) = 665t^2 + 1150t + 27{,}357 \text{ million dollars}$$

EXAMPLE 5 **Using a System of Equations to Determine Business Needs**

A chef has been commissioned to create a party mix containing pretzels, bagel chips, and Chex® cereal for a corporate gathering of 180 people. He estimates that on average each person will consume 1/2 cup of the party mix. The Original Chex Party Mix recipe calls for 9 cups of Chex cereal, 1 cup of pretzels, and 1 cup of bagel chips. (**Source:** www.chex.com.) However, the chef plans to modify the recipe so that there are half as many bagel chips as pretzels. How many cups of each of the ingredients will be needed to make the party mix?

SOLUTION Let x be the number of cups of pretzels, y be the number of cups of bagel chips, and z be the number of cups of Chex cereal. Since each of the 180 people is expected to consume 1/2 cup of party mix, a total of 90 cups of the party mix is needed. Therefore,

$$x + y + z = 90 \qquad \text{The quantity of the party mix is the sum of the ingredients}$$

Since there are to be half as many bagel chips as pretzels, we know that

$$y = 0.5x \qquad \text{There are half as many bagel chips as pretzels}$$
$$2y = x$$
$$-x + 2y = 0$$

No other constraints are given, so the system of equations is

$$x + y + z = 90$$
$$-x + 2y \phantom{{}+ z} = 0$$

with the corresponding matrix

$$\begin{bmatrix} 1 & 1 & 1 & \vdots & 90 \\ -1 & 2 & 0 & \vdots & 0 \end{bmatrix}$$

Using the elimination method, we obtain

$$\begin{bmatrix} 1 & 1 & 1 & \vdots & 90 \\ 0 & 3 & 1 & \vdots & 90 \end{bmatrix} \qquad R_1 + R_2$$

$$\begin{bmatrix} 3 & 0 & 2 & \vdots & 180 \\ 0 & 3 & 1 & \vdots & 90 \end{bmatrix} \qquad 3R_1 - R_2$$

$$\begin{bmatrix} 1 & 0 & \frac{2}{3} & \vdots & 60 \\ 0 & 1 & \frac{1}{3} & \vdots & 30 \end{bmatrix} \qquad \begin{matrix} \frac{1}{3}R_1 \\ \frac{1}{3}R_2 \end{matrix}$$

The system of linear equations is reduced to

$$x + \frac{2}{3}z = 60 \qquad\qquad y + \frac{1}{3}z = 30$$
$$\text{and}$$
$$x = 60 - \frac{2}{3}z \qquad\qquad y = 30 - \frac{1}{3}z$$

Since all quantities must be nonnegative, $x \geq 0$, $y \geq 0$, and $z \geq 0$. Furthermore,

$$x \geq 0 \qquad\qquad y \geq 0$$

$$60 - \frac{2}{3}z \geq 0 \qquad\qquad 30 - \frac{1}{3}z \geq 0$$

and

$$60 \geq \frac{2}{3}z \qquad\qquad 30 \geq \frac{1}{3}z$$

$$90 \geq z \qquad\qquad 90 \geq z$$

So $0 \leq z \leq 90$. The same constraints hold for each of the other ingredients as well, since the total quantity of the mix cannot exceed 90 cups.

Let $z = t$. Then the solution to the system is

$$x = 60 - \frac{2}{3}t \text{ cups of pretzels}$$

$$y = 30 - \frac{1}{3}t \text{ cups of bagel chips}$$

$$z = t \text{ cups of Chex cereal}$$

The chef has a variety of solutions. If he chooses to use 48 cups of Chex cereal, then he will need

$$x = 60 - \frac{2}{3}t \qquad\qquad y = 30 - \frac{1}{3}t$$

$$x = 60 - \frac{2}{3}(48) \qquad\qquad y = 30 - \frac{1}{3}(48)$$

$$x = 60 - 32 \qquad\qquad y = 30 - 16$$

$$x = 28 \text{ cups of pretzels} \qquad y = 14 \text{ cups of bagel chips}$$

If he chooses to use 60 cups of Chex cereal, he will need 20 cups of pretzels and 10 cups of bagel chips. In the original recipe, the quantity of cereal was nine times as much as each of the other ingredients. If the chef wants the final product to be similar to the Original Chex Party Mix, roughly 82 percent of the mix should be cereal. If 75 cups of Chex cereal are used, then 10 cups of pretzels and 5 cups of bagel chips will be required. Using this combination of ingredients, 83 percent $\left(\frac{75}{90} = 0.8333 \right)$ of the mixture is cereal.

2.3 Summary

In this section, you learned how to formulate linear systems from real-life data and use the solutions to these systems to make business decisions. You also discovered some of the strengths and limitations of mathematical models.

2.3 Exercises

For Exercises 1–18, do the following:

(a) Find the linear function that best fits each data set as a function of time.

(b) Determine if and when the output of the two models will be equal.

(c) Interpret the results in part (b). Do they make sense in their real-world context?

(d) Explain how the results of your analysis could be used to improve business or society.

  1. School Enrollment

Projected Student Enrollment

Years Since 1998 (t)	Grade 2 (students) (S)	Grade 3 (students) (T)
0	299	273
1	312	311
2	301	321
3	282	312
4	307	311
5	283	337
6	338	309
7	343	365
8	350	372
9	359	379
10	369	387

Source: White River School District.

  2. Margarine and Cheese Consumption

Per Capita Consumption

Years Since 1980 (t)	Margarine (pounds) (M)	Mozzarella Cheese (pounds) (C)
0	11.3	3.0
5	10.8	4.6
10	10.9	6.9
15	9.2	8.1
17	8.6	8.4
18	8.3	8.8
19	8.1	9.2

Source: *Statistical Abstract of the United States,* *2001,* Table 202, p. 129.

  3. Fish and Chicken Consumption

Per Capita Consumption

Years Since 1970 (t)	Chicken (pounds) (C)	Fish and Shellfish (Boneless, Trimmed Weight) (pounds) (F)
0	27.4	11.7
5	26.4	12.1
10	32.7	12.4
15	36.4	15.0
20	42.4	15.0
25	48.8	14.9
29	54.2	15.2

Source: *Statistical Abstract of the United States,* *2001,* Table 202, p. 129.

 4. Diabetes Incidence

Incidence of Diabetes in Adults Ages 45–64

Years Since 1995 (t)	Male Diabetes Rate per 1,000 (M)	Female Diabetes Rate per 1,000 (F)
0	109	131
4	189	146

Source: *Statistical Abstract of the United States, 2001,* Table 159, p. 109.

5.   **Baseball Game Attendance**

Major League Baseball Attendance

Years Since 1994 (t)	National League Games (millions of people) (N)	American League Games (millions of people) (A)
0	25.8	24.2
1	25.1	25.4
2	30.4	29.7
3	31.9	31.3
4	38.4	31.9
5	38.3	31.8

Source: *Statistical Abstract of the United States, 2001,* Table 1241, p. 759.

6.   **Population of Argentina and Canada**

Population of Two Countries

Years Since 1980 (t)	Argentina (thousands) (A)	Canada (thousands) (C)
0	28,237	24,593
10	32,634	27,791
15	34,818	29,619
20	36,955	31,278
21	37,385	31,593

Source: *Statistical Abstract of the United States, 2001,* Table 1327, p. 831.

7.   **Births to Unmarried Women**

Births to Unmarried Women in the USA and UK

Years Since 1980 (t)	United States (percent of all births) (U)	United Kingdom (percent of all births) (K)
0	18	12
18	33	38

Source: *Statistical Abstract of the United States, 2001,* Table 1331, p. 836.

8.   **Number of Medical Doctors**

Medical Doctors per 1,000 Population

Years Since 1990 (t)	United States (U)	United Kingdom (K)
0	2.4	1.4
7	2.7	1.7

Source: *Statistical Abstract of the United States, 2001,* Table 1332, p. 836.

9.   **Postal Service Rates**

U.S. Postal Service Rates

Years Since 1985 (t)	1-Ounce Letter (cents) (L)	Postcard (cents) (P)
0	22	14
3	25	15
6	29	19
10	32	20
14	33	20
16	34	20

Source: *Statistical Abstract of the United States, 2001,* Table 1117, p. 697.

10.   **Professional Athlete Salaries**

Average Major League Athlete Salaries

Years Since 1980 (t)	Baseball ($1,000s) ($B$)	Football ($1,000s) ($F$)
0	144	79
5	371	194
10	598	352
15	1,111	714
16	1,120	791
17	1,337	725
18	1,399	1,138

Source: *Statistical Abstract of the United States, 2001,* Table 1241, p. 759.

11. **Annual Earnings by Industry**

Paper Versus Publishing Industry

Years Since 1995 (t)	Paper Manufacturing Employee Earnings (dollars) (E)	Printing and Publishing Employee Earnings (dollars) (P)
0	39,458	34,539
1	40,718	35,897
2	42,129	37,427
3	43,185	39,256
4	44,900	41,083

Source: *Statistical Abstract of the United States, 2001,* Table 979, p. 622.

 12. **Annual Earnings by Industry**

Plastic Versus Apparel Industry

Years Since 1995 (t)	Rubber and Plastics Manufacturing Industry Employee Earnings (dollars) (E)	Apparel and Other Textiles Manufacturing Industry Employee Earnings (dollars) (A)
0	29,867	18,800
1	30,898	19,832
2	32,237	20,838
3	33,574	22,103
4	34,508	23,255

Source: *Statistical Abstract of the United States, 2001,* Table 979, p. 622.

13. **Clothing and Food Stores**

Retail Establishments

Year Since End of 1998 (t)	Clothing and Accessories Stores (C)	Food and Beverage Stores (F)
0	152,603	147,652
1	151,674	151,506

Source: *Statistical Abstract of the United States, 2001,* Table 1017, p. 641.

14. **Gasoline Stations and Parts Dealers**

Retail Establishments

Year Since End of 1998 (t)	Gasoline Stations (G)	Motor Vehicle and Parts Dealers (M)
0	123,894	123,359
1	121,095	123,855

Source: *Statistical Abstract of the United States, 2001,* Table 1017, p. 641.

15. **Specialized Clothing Stores**

Retail Establishments

Year Since End of 1998 (t)	Men's Clothing Stores (M)	Children's Clothing Stores (C)
0	11,861	5,165
1	11,445	5,333

Source: *Statistical Abstract of the United States, 2001,* Table 1017, p. 641.

16. **New and Used Vehicle Dealers**

Retail Establishments

Year Since End of 1998 (t)	New Vehicle Dealers (N)	Used Vehicle Dealers (U)
0	26,216	23,651
1	26,117	23,998

Source: *Statistical Abstract of the United States, 2001,* Table 1017, p. 641.

17.   **Retail Sales**

Sales (in millions of dollars)

Year Since End of 1992 (t)	Building Materials and Supplies Stores (B)	Department Stores (D)
0	160,171	181,255
1	171,733	192,292
2	190,817	205,302
3	199,068	212,759
4	212,759	218,740
5	229,489	225,062
6	243,490	226,024
7	263,958	231,236
8	277,185	241,958

Source: *Statistical Abstract of the United States, 2001,* Table 1020, p. 644.

18.   **Dining Retail Sales**

Sales (in millions of dollars)

Year Since End of 1992 (t)	Full-Service Restaurants (R)	Limited-Service Eating Places (L)
0	86,493	87,433
1	91,476	94,736
2	97,117	98,446
3	99,430	103,143
4	104,514	106,192
5	114,591	109,298
6	119,663	116,836
7	124,463	123,081
8	134,363	127,504

Source: *Statistical Abstract of the United States, 2001,* Table 1020, p. 644.

In Exercises 19–35, set up the system of linear equations and solve it using matrices.

19.   Investment Choices   The following table shows the average annual rate of return on a variety of TIAA-CREF investment accounts over a 10-year period.

CREF Variable Annuity Accounts as of 5/31/04	10-Year Average
Bond Market	7.05%
Equity Index	10.83%
Global Equities	6.96%
Growth	8.14%
Money Market	4.25%
Social Choice	10.01%
Stock	9.53%

Source: www.tiaa-cref.com.

An investor chooses to invest $3000 in the Bond Market, Growth, and Stock accounts. He assumes that he will be able to get a return equal to the 10-year average, and he wants the total return on his investment to be 9 percent. He wants to invest three times as much money in the Bond Market account as in the Growth account. How much money should he invest in each account? [For ease of computation, round each percentage to the nearest whole-number percent (e.g., 9.53 percent = 10 percent.]

20.   Investment Choices   An investor chooses to invest $5000 in the Global Equities, Money Market, and Social Choice accounts shown in Exercise 19. She wants to put five times as much money in the Money Market account as in the Global Equities account. She assumes that she will be able to get a return equal to the 10-year average, and she wants the total return on her investment to be 7 percent.

How much money should she invest in each account? [For ease of computation, round each percentage to the nearest whole-number percent (e.g., 6.96 percent = 7 percent).]

21. 🌐 **Vehicle Sales** A company has budgeted $970,000 to purchase new vehicles. The vehicles are offered as incentives to top salespeople and include the following models:

- 2004 Toyota Corolla 4-door sedan LE Manual, Manufacturer's Suggested Retail Price $14,780

- 2004 Toyota Camry 4-door sedan XLE V6 Auto, Manufacturer's Suggested Retail Price $25,405

- 2004 Prius 5-door hatchback, Manufacturer's Suggested Retail Price $20,295

(**Source:** www.nada.com.)

The company expects to be able to negotiate lower sale prices on each of the vehicles. For budgeting purposes, the company estimates the total cost of each vehicle including tax to be as follows:

- Corolla $14,000

- Camry $25,000

- Prius $20,000

Company staff anticipate that they will give away three times as many Corollas as Camrys and half as many Prius cars as Corollas. How many of each type of car should the company buy if it plans to spend the entire new vehicle budget?

22. 🌐 **Vehicle Sales** Because of low sales, the company in Exercise 21 revises its ordering criteria midway through the year. The company decides to order 30 cars, including five times as many Corollas as Camrys.

Ignoring the original budget limitation, how many of each type of car should the company buy? Find three or more solutions.

23. 🌐 **Johnson & Johnson Costs** According to the company's 2001 annual report, the cost of goods sold by Johnson & Johnson and subsidiaries increased in 1999, 2000, and 2001 as shown in the table.

Johnson & Johnson and Subsidiaries Cost of Goods Sold

Years Since the End of 1999 (t)	Cost of Goods Sold (millions of dollars) (C)
0	8,539
1	8,957
2	9,581

Source: Johnson & Johnson 2001 annual report, p. 6.

Use matrices to find a quadratic equation that models the company's cost of goods sold.

24. 🌐 **Frito Lay Profit** According to the PepsiCo 2001 annual report, the operating profit of Frito-Lay North America increased in 1999, 2000, and 2001 as shown in the table.

Frito-Lay North America Operating Profit

Years Since the End of 1999 (t)	Operating Profit (millions of dollars) (P)
0	1,679
1	1,915
2	2,056

Source: PepsiCo 2001 annual report, pp. 23, 44.

Use matrices to find a quadratic equation that models the company's operating profit.

25. 🌐 **Gatorade/Tropicana Sales** According to the PepsiCo 2001 annual report, the net sales of Gatorade/Tropicana North America increased in 1999, 2000, and 2001 as shown in the table.

Gatorade/Tropicana North America Net Sales

Years Since the End of 1999 (t)	Net Sales (millions of dollars) (R)
0	3,452
1	3,841
2	4,016

Source: PepsiCo 2001 annual report, p. 44.

Use matrices to find a quadratic equation that models the company's net sales.

26. 🌐 **Gatorade/Tropicana Profit** According to the PepsiCo 2001 annual report, operating profit of Gatorade/Tropicana North America increased in 1999, 2000, and 2001 as shown in the table.

Gatorade/Tropicana North America Operating Profit

Years Since the End of 1999 (t)	Operating Profit (millions of dollars) (P)
0	433
1	500
2	530

Source: PepsiCo 2001 annual report, p. 44.

Use matrices to find a quadratic equation that models the company's operating profit.

27. **Resource Allocation: Sandwiches** A plain hamburger requires one ground beef patty and a bun. A cheeseburger requires one ground beef patty, one slice of cheese, and a bun. A double cheeseburger requires two ground beef patties, two slices of cheese, and a bun.

Frozen hamburger patties are typically sold in packs of 12; hamburger buns, in packs of 8; and cheese slices, in packs of 24.

A family is in charge of providing burgers for a neighborhood block party. It has purchased 13 packs of buns, 11 packs of hamburger patties, and 3 packs of cheese slices. How many of each type of sandwich should the family make if the goal is to use up all of the buns, patties, and cheese slices?

28. **Pet Nutrition: Food Cost** PETsMART.com sold the following varieties of dog food in June 2003. The price shown is for an eight-pound bag.

Pro Plan Adult Chicken & Rice Formula, 25 percent protein, 3 percent fiber, $7.99
Pro Plan Adult Lamb & Rice Formula, 28 percent protein, 3 percent fiber, $7.99
Pro Plan Adult Turkey & Barley Formula, 26 percent protein, 3 percent fiber, $8.49
(**Source:** www.petsmart.com.)

A dog breeder wants to make 120 pounds of a mix containing 27 percent protein and 3 percent fiber. How many 8-pound bags of each dog food variety should the breeder buy? What is the breeder's food cost?

29. **Pet Nutrition: Food Cost** PETsMART.com sold the following varieties of dog food in June 2003.

Nature's Recipe Venison Meal & Rice Canine, 20 percent protein, $21.99 per 20-pound bag
Nutro Max Natural Dog Food, 27 percent protein, $12.99 per 17.5-pound bag
PETsMART Premier Oven Baked Lamb Recipe, 25 percent protein, $22.99 per 30-pound bag
(**Source:** www.petsmart.com.)

A dog breeder wants to make 300 pounds of a mix containing 22 percent protein. How many bags of each dog food variety should the breeder buy? (*Hint:* Note that each bag is a different weight. Fractions of bags may not be purchased.)

30. **Utilization of Ingredients** A custard recipe calls for 3 eggs and 2.5 cups of milk. A vanilla pudding recipe calls for 2 eggs and 2 cups of milk. A bread pudding recipe calls for 2 eggs, 2 cups of milk, and 8 slices of bread. (**Source:** *Betty Crocker's Cookbook*)

A stocked kitchen contains 18 eggs, 1 gallon of milk, and 24 slices of bread. How many batches of each recipe should a chef make in order to use up all of the ingredients?

31. **First-Aid Kit Supplies** Safetymax.com sells first-aid and emergency preparedness supplies to businesses. A company that assembles first-aid kits for consumers purchases 3500 1″ × 3″ plastic adhesive bandages, 1800 alcohol wipes, and 220 tubes of antibiotic ointment from Safetymax.com.

The company assembles compact, standard, and deluxe first-aid kits for sale to consumers. A compact first-aid kit contains 20 plastic adhesive bandages, 8 alcohol wipes, and 1 tube of antibiotic ointment. A standard first-aid kit contains 40 plastic adhesive bandages, 20 alcohol wipes, and 2 tubes of antibiotic ointment. A deluxe first-aid kit contains 50 plastic adhesive bandages, 28 alcohol wipes, and 4 tubes of antibiotic ointment.

How many of each type of kit should the company assemble in order to use up all of the bandages, wipes, and antibiotics ordered?

32. **Concert Ticket Sales** On the weekend of July 23–25, 2004, the House of Blues Sunset Strip in Hollywood, California, hosted three concerts: Saves the Day, Jet, and The Bodeans. Saves the Day tickets cost $15, Jet tickets cost $20, and The Bodeans tickets cost $22. (**Source:** www.ticketmaster.com.)

If concert planners expected that a total of 1200 tickets would be sold over the weekend, how many tickets for each concert needed to be sold in order to bring in $23,500? Find three different solutions.

33.   **Concert Ticket Sales**   On July 7, 2004, Shania Twain was scheduled to perform at the TD Waterhouse Centre in Orlando, Florida. The center offered 18,039 seats for the concert in three seating classifications: floor, lower, and upper. Based upon their location, tickets were offered at three different prices: $80, $65, and $45. (**Source:** www.ticketmaster.com.) Suppose that the average price of a floor ticket was $80, the average price of a lower ticket was $65, and the average price of an upper ticket was $45. If concert planners expected that a total of 12,000 tickets would be sold, including five times as many tickets on the lower level as on the floor, how many of each type of ticket would have to be sold in order to obtain $675,000 in ticket revenue?

34.   **Concert Ticket Sales**   On June 24, 2004, Madonna was scheduled to perform at Madison Square Garden in New York City. Tickets were offered at four different prices: $49.50, $94.50, $154.50, and $304.50. (**Source:** www.ticketmaster.com.) If concert planners expected that a total of 25,000 tickets would be sold, including ten times as many of the least expensive ticket as of the most expensive ticket and one-third as many $154.50 tickets as $94.50 tickets, how many of each type of ticket would need to be sold in order to earn $2,130,000 in revenue?

35.   **Furniture Production**   Based on data from 1991–2001, approximately 32 percent of the production value of a piece of furniture is the result of the labor cost associated with producing the piece. (**Source:** www.bls.gov.)

In June 2004, an online furniture retailer offered the following items at the indicated prices.

Teak Double Rocker: $745
Avalon Teak Armchair: $378
Teak Tennis Bench: $124

(**Source:** www.outdoordecor.com)

Suppose that the number of hours required to produce each item is as shown in the following table.

	Cut	Finish	Package
Rocker	4	7	1
Armchair	2	4	1
Bench	1	3	1

If the company has 360 labor hours available in the Cutting Department, 720 labor hours available in the Finishing Department, and 160 labor hours available in the Packaging Department, how many of each type of item should the company produce in order to use all of the available labor hours?

Exercises 36–40 are intended to challenge your understanding of linear system applications.

36. **Amusement Park Rides**   Rides at an amusement park require three, four, five, or six tickets. A family purchases 75 tickets and receives two tickets from a park guest who had leftover tickets. The family wants to use all its tickets and wants the sum of five- and six-ticket rides to be twice as much as the sum of three- and four-ticket rides. What is the largest possible number of six-ticket rides that the family can go on subject to these constraints?

37. **Assortment of Coins**   A coin purse contains 64 coins including pennies, nickels, dimes, and quarters. The total value of the coins is $4.94. There is the same number of nickels as dimes. How many of each type of coin is in the bag?

38.   **Investments**   The following table shows the average annual rate of return on a variety of TIAA-CREF investment accounts over a 10-year period.

CREF Variable Annuity Accounts (as of 5/31/04)	10 year
Bond Market	7.05%
Equity Index	10.83%
Global Equities	6.96%
Growth	8.14%
Money Market	4.25%
Social Choice	10.01%
Stock	9.53%

Source: www.tiaa-cref.com.

An investor wants to earn an 8 percent annual return on a $37,625 investment. The investor expects that the annual return on each account will be equal to the 10-year rate rounded to the nearest whole-number percentage (e.g., 10.83 percent = 11 percent).

The investor wants to invest the same amount of money in the Bond Market account as in the Global Equities account and twice as much in the Social Choice account as in the Growth account. How much money should the investor place in each account?

39. Traffic Flow The figure shows the flow of traffic at four city intersections.

For each intersection, the number of cars entering the intersection must equal the number of cars leaving the intersection. For example, the number of cars entering the intersection of North and West Sts. is $x_3 + 300$. The number of cars leaving the intersection is $x_4 + 150$. Therefore, $x_3 + 300 = x_4 + 150$.

Find two separate set of values x_1, x_2, x_3, and x_4 that work in the traffic flow system.

40. Traffic Flow Repeat Exercise 39 for the following traffic flow diagram.

Then explain who could benefit from this type of traffic flow analysis.

Chapter 2 Review Exercises

Section 2.1 *In Exercises 1–8, use your graphing calculator to find the point(s) of intersection of the lines. If the lines do not intersect at a common point, state that a solution does not exist.*

1. $y = 3x + 4$
$y = 1.7x + 5.3$

2. $y = 4.6x + 10$
$y = -10x + 44$

3. $y = 7.1x - 41.5$
$y = -1.9x + 3.5$

4. $y = 2.5x - 1$
$y = -0.25x + 10$

5. $y = 7x - 1$
$y = 7x + 1$

6. $y = 5.5x - 1.2$
$y = -2.5x + 5.5$

7. $y = 3.3x - 3.1$
$y = -3.1x + 3.3$

8. $y = -12.5x - 11.5$
$y = -6.2x + 10.4$

In Exercises 9–16, solve the system of equations algebraically. Compare your solutions to those obtained in Exercises 1–8.

9. $y = 3x + 4$
$y = 1.7x + 5.3$

10. $y = 4.6x + 10$
$y = -10x + 44$

11. $y = 7.1x - 41.5$
$y = -1.9x + 3.5$

12. $y = 2.5x - 1$
$y = -0.25x + 10$

13. $y = 7x - 1$
$y = 7x + 1$

14. $y = 5.5x - 1.2$
$y = -2.5x + 5.5$

15. $y = 3.3x - 3.1$
$y = -3.1x + 3.3$

16. $y = -12.5x - 11.5$
$y = -6.2x + 10.4$

Section 2.2 *In Exercises 17–24, determine if the augmented matrix is in reduced row echelon form. If it is not reduced, specify which of the four criteria for reduced row echelon form the matrix fails to meet.*

17. $\begin{bmatrix} 1 & 0 & 0 & | & 3 \\ 0 & 1 & 1 & | & 7 \\ 0 & 1 & 0 & | & -2 \end{bmatrix}$ **18.** $\begin{bmatrix} 1 & 0 & 0 & | & 9 \\ 0 & 1 & 0 & | & 11 \\ 0 & 0 & 1 & | & 2 \end{bmatrix}$

19. $\begin{bmatrix} 0 & 0 & 0 & | & 0 \\ 0 & 1 & 0 & | & 13 \\ 1 & 0 & 0 & | & 27 \end{bmatrix}$ **20.** $\begin{bmatrix} 1 & 0 & 1 & | & -9 \\ 0 & 1 & 0 & | & 14 \\ 0 & 0 & 1 & | & 2.1 \end{bmatrix}$

21. $\begin{bmatrix} 1 & 1 & 2 & | & 0 \\ 0 & 0 & 0 & | & 1 \\ 0 & 0 & 0 & | & 0 \end{bmatrix}$ **22.** $\begin{bmatrix} 1 & 0 & | & 5 \\ 0 & 1 & | & 9 \\ 0 & 0 & | & 0 \end{bmatrix}$

23. $\begin{bmatrix} 1 & 0 & 1 & 0 & | & -2 \\ 0 & 1 & 0 & 0 & | & 4 \\ 0 & 0 & 0 & 1 & | & 0 \end{bmatrix}$ **24.** $\begin{bmatrix} 1 & 0 & 0 & 1 & | & 3 \\ 0 & 1 & 1 & 0 & | & -2 \\ 0 & 0 & 0 & 1 & | & 5 \end{bmatrix}$

In Exercises 25–32, rewrite the system of equations as an augmented matrix, then simplify the matrix to reduced row echelon form. Identify the solution(s) to the system of equations.

25. $2.1x - y = -8$
$3.4x + y = 8$

26. $5.2x - 1.3y = 12$
$-10.4x + 2.6y = 24$

27. $4x - 8y = -16$
$x + y = 5$

28. $2.9x - 8.1y = 4$
$3.7x + 16.2y = 1.5$

29. $x + y + z = 6$
$2x - y + z = 3$
$3x + 2z = 9$

30. $6x - 2y = 6$
$4x + 2y = 14$
$x + y = 5$

31. $x - y + 3z = 5$
$2x + 4y + z = -1$

32. $3x - 2y + z = 11$
$x + 2y - 3z = -1$

In Exercises 33–36, use technology to write the matrix in reduced row echelon form.

33. $\begin{bmatrix} 8.1 & 4.3 & 6.2 & | & 3.1 \\ 2.7 & -2.8 & 7.5 & | & 7.2 \\ 8.2 & -5.3 & -1.1 & | & 5.9 \end{bmatrix}$

34. $\begin{bmatrix} 2.6 & 3.0 & 0.2 & | & 9.6 \\ 6.1 & 4.0 & 6.5 & | & 0.8 \\ 9.1 & -5.0 & -0.8 & | & 9.9 \end{bmatrix}$

35. $\begin{bmatrix} 6.2 & 5.3 & -0.2 & | & 3.3 \\ 1.7 & -2.6 & 2.5 & | & 2.2 \\ 4.5 & 7.9 & -2.7 & | & 1.1 \end{bmatrix}$

36. $\begin{bmatrix} 4.5 & 3.9 & 1.2 & | & -1.7 \\ 1.5 & 1.3 & 2.5 & | & 0.3 \\ 9.0 & 7.8 & -9.8 & | & 7.9 \end{bmatrix}$

Section 2.3 *For Exercises 37–40, do the following:*

(a) Find the linear function that best models each data set as a function of time.

(b) Determine if and when the output of the two models will be equal.

(c) Interpret the results in part (b). Do they make sense in their real-world context?

(d) Explain how the results of your analysis could be used to improve business or society.

37. **Births to Unmarried Women**

Births to Unmarried Women in the UK and Iceland

Years Since 1980 (t)	United Kingdom (percent of all births) (K)	Iceland (percent of all births) (I)
0	12	40
18	38	64

Source: *Statistical Abstract of the United States, 2001,* Table 1331, p. 836.

38. **Fruit and Vegetable Production**

Worldwide Produce Production

Years Since 1990 (t)	Fruit (millions of metric tons) (F)	Vegetables (millions of metric tons) (V)
0	352.2	461.4
2	379.0	478.6
4	391.0	532.1
6	427.3	591.2
8	432.8	626.0
10	459.2	670.1

Source: *Statistical Abstract of the United States, 2001,* Table 1357, p. 849.

39. **Basketball Game Attendance**

NCAA Basketball Game Attendance

Years Since 1985 (t)	Men's Games (thousands) (M)	Women's Games (thousands) (W)
0	26,584	2,072
5	28,741	2,777
10	28,548	4,962
14	29,025	8,698

Source: *Statistical Abstract of the United States, 2001*, Table 1241, p. 759.

 40.   **Education and Crime**

Years Since 1980 (t)	Private College Students (thousands) (S)	Adults in Prison, in Jail, on Probation, or on Parole (thousands) (A)
0	2,640	1,840
2	2,730	2,193
4	2,765	2,689
6	2,790	3,239
8	2,894	3,714
10	2,974	4,348
12	3,103	4,763
14	3,145	5,141
16	3,247	5,483
18	3,373	6,126

Source: *Statistical Abstract of the United States, 2001*, Tables 205, 335, pp. 133, 202.

In Exercises 41–43, set up the system of linear equations and solve it using matrices.

41. **Company Sales** The following table shows the sales revenue for the Starbucks Corporation from 1993 to 2002.

Starbucks Corporation Sales

Years Since 9/93	Sales Revenue (millions of dollars)
0	163.5
1	284.9
2	465.2
3	696.5
4	966.9
5	1,308.7
6	1,680.1
7	2,169.2
8	2,649.0
9	3,288.9

Source: moneycentral.msn.com.

Use matrices and the sales data from 1994, 1998, and 2002 to find a quadratic model for the income from sales. Then use the model to forecast the 2004 sales.

42. **Music Market Size** According to the Recording Industry Association of America, the size of the music market fluctuated between 1997 and 2002 as shown in the following table.

Music Market Size

Years Since 1997 (t)	Dollar Volume (millions) (M)
0	$12,236.80
1	$13,723.50
2	$14,584.50
3	$14,323.00
4	$13,740.89
5	$12,614.21

Source: www.riaa.com.

Use matrices and the dollar volume data from 1997, 1998, and 2002 to find a quadratic model for the dollar volume of the music market. Then use the model to forecast the market size in 2003.

43.   **First-Aid Kit Supplies**   Safetymax.com sells first-aid and emergency preparedness supplies to businesses. A company that assembles first-aid kits for consumers purchases 3275 $1'' \times 3''$ plastic adhesive bandages, 1300 alcohol wipes, and 185 tubes of antibiotic ointment from Safetymax.com.

The company assembles compact, standard, and deluxe first-aid kits for sale to consumers. A compact first-aid kit contains 25 plastic adhesive bandages, 10 alcohol wipes, and 1 tube of antibiotic ointment. A standard first-aid kit contains 40 plastic adhesive bandages, 15 alcohol wipes, and 2 tubes of antibiotic ointment. A deluxe first-aid kit contains 60 plastic adhesive bandages, 25 alcohol wipes, and 4 tubes of antibiotic ointment.

How many of each type of kit should the company assemble in order to use up all of the bandages, wipes, and antibiotics ordered?

Make It Real

What to do

1. Find two sets of data from an area of personal interest that appear to be increasing or decreasing at a nearly constant rate. Both data sets should have the same domain (input values).

2. Find the linear function that best models each data set.

3. Determine if and when the output of the two models will be equal. (Find the point of intersection of their graphs.)

4. Interpret the results from Step 3. Do they make sense in their real-world context?

5. Explain how the results of your analysis could be used to improve business or society.

Where to look for data

U.S. Census Bureau
www.census.gov
Look at data on U.S. residents ranging from Internet usage to family size.

NADA
www.nada.com
Check out current trade-in and retail values of used cars, trucks, and vans. Compare the retail value of a model to its trade-in value over time.

State Office of Financial Management
Go to your state's web site and search for the Office of Financial Management. This office typically publishes city and county population data. Compare the growth of two nearby cities over the past decade.

Chapter 3

Matrix Algebra and Applications

As states faced budget crises in the early 2000s, citizens rallied to let state legislatures know which programs they saw as a priority. In Washington State, voters overwhelmingly approved an initiative in 2000 granting annual cost-of-living increases to K through 12 teachers and community college employees. Determining a new pay scale resulting from an across-the-board percentage increase in salaries, such as that approved by Washington voters, is easy to do using matrix algebra.

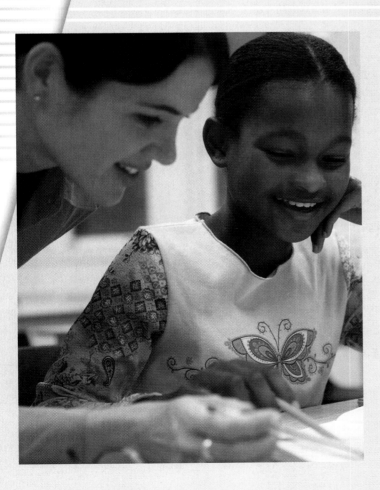

3.1 Matrix Addition and Scalar Multiplication
- Add and subtract matrices
- Do scalar multiplication of matrices
- Use matrices to analyze and interpret real-life data

3.2 Matrix Multiplication and Inverses
- Do matrix multiplication
- Find the inverse of an invertible matrix with and without technology
- Use matrices to analyze and interpret real-life data

3.3 Solving Matrix Equations
- Use matrix algebra to solve systems of linear equations
- Use matrices to analyze and interpret real-life data
- Find the inverse of an invertible matrix with and without technology

3.4 Leontief Input-Output Models
- Do basic input-output analysis using Leontief input-output models

3.1 Matrix Addition and Scalar Multiplication

- Add and subtract matrices
- Do scalar multiplication of matrices
- Use matrices to analyze and interpret real-life data

GETTING STARTED   Financial conflict is one of the leading causes of marital discord. To help alleviate future financial stress, many engaged couples take marriage preparation classes to help them confront marital issues, including finances, before they tie the knot.

In Chapter 2, we introduced augmented matrices as a tool for solving systems of linear equations. In this section, we will further develop the concept of matrices by demonstrating matrix addition and scalar multiplication. We will also demonstrate how to use these techniques to analyze real-life financial situations, such as a forthcoming marriage.

Matrix Addition

Two matrices, A and B, of the same dimension may be added together to form a new matrix $C = A + B$ by adding their corresponding entries. For example, if the 2×2 matrices $A = \begin{bmatrix} 1 & 2 \\ 3 & 4 \end{bmatrix}$ and $B = \begin{bmatrix} 4 & -2 \\ -3 & 1 \end{bmatrix}$ are added together, we get

$$
\begin{aligned}
C &= A + B \\
&= \begin{bmatrix} 1 & 2 \\ 3 & 4 \end{bmatrix} + \begin{bmatrix} 4 & -2 \\ -3 & 1 \end{bmatrix} \\
&= \begin{bmatrix} 1 + 4 & 2 + (-2) \\ 3 + (-3) & 4 + 1 \end{bmatrix} \quad \text{Add corresponding entries} \\
&= \begin{bmatrix} 5 & 0 \\ 0 & 5 \end{bmatrix}
\end{aligned}
$$

MATRIX ADDITION

An $m \times n$ matrix A and an $m \times n$ matrix B may be added together to form a new $m \times n$ matrix, C. The value of the entry in the ith row and jth column of C is $c_{ij} = a_{ij} + b_{ij}$.

If A and B are not of the same dimension, matrix addition is undefined.

EXAMPLE 1   ## Calculating the Sum of Two Matrices

Calculate $C = A + B$ given $A = \begin{bmatrix} 2 & 0 \\ -1 & 6 \\ 4 & 5 \end{bmatrix}$ and $B = \begin{bmatrix} 9 & 2 \\ -7 & -6 \\ 8 & -1 \end{bmatrix}$.

SOLUTION

$$C = A + B$$

$$= \begin{bmatrix} 2 & 0 \\ -1 & 6 \\ 4 & 5 \end{bmatrix} + \begin{bmatrix} 9 & 2 \\ -7 & -6 \\ 8 & -1 \end{bmatrix}$$

$$= \begin{bmatrix} 2 + 9 & 0 + 2 \\ -1 + (-7) & 6 + (-6) \\ 4 + 8 & 5 + (-1) \end{bmatrix} \quad \text{Add corresponding entries}$$

$$= \begin{bmatrix} 11 & 2 \\ -8 & 0 \\ 12 & 4 \end{bmatrix}$$

TECHNOLOGY **TIP**

Adding Two Matrices

1. Enter matrix A and matrix B into the calculator using the Matrix Editor.

```
MATRIX[B] 3 x2
[ 9     2      ]
[ -7    -6     ]
[ 8     ▉      ]


3,2=-1
```

2. Use the Matrix Names menu to place matrix A on the home screen. Press the ⌈ + ⌋ key. Then place matrix B on the home screen.

```
[A]+[B]
```

3. Press ⌈ENTER⌋ to display the sum of the two matrices.

```
[A]+[B]
              [[11 2]
               [-8 0]
               [12 4]]
```

Error Alert:
If you try to add two matrices of different dimensions, you will get a dimension mismatch error.

```
ERR:DIM MISMATCH
1█Quit
2:Goto
```

EXAMPLE 2

Using Matrix Addition to Solve Real-World Problems

Jamaal and Shina, an engaged couple, are discussing each of their financial obligations. Jamaal has the following debts:

$5000 student loan with Back Bay Credit Union
$2000 student loan with Aspen Bank
$7000 car loan with IntelliBank
$1500 credit card balance with IntelliBank
$800 credit card balance with Aspen Bank

Shina has the following debts:

$8000 student loan with Back Bay Credit Union
$4500 car loan with Back Bay Credit Union
$700 credit card balance with Aspen Bank
$1200 credit card balance with IntelliBank

Use matrix addition to determine their combined debts of each type at the various financial institutions.

SOLUTION Jamaal's debts (in thousands of dollars) may be represented as shown in Table 3.1.

TABLE 3.1

	Car Loan	Credit Card	Student Loan
Aspen Bank	0	0.8	2
Back Bay Credit Union	0	0	5
IntelliBank	7	1.5	0

The information may be further represented as a matrix whose rows and columns have the same meaning as the rows and columns of the table.

$$J = \begin{bmatrix} 0 & 0.8 & 2 \\ 0 & 0 & 5 \\ 7 & 1.5 & 0 \end{bmatrix}$$

Similarly, Shina's debts (in thousands of dollars) may be represented as shown in Table 3.2

TABLE 3.2

	Car Loan	Credit Card	Student Loan
Aspen Bank	0	0.7	0
Back Bay Credit Union	4.5	0	8
IntelliBank	0	1.2	0

and the matrix

$$S = \begin{bmatrix} 0 & 0.7 & 0 \\ 4.5 & 0 & 8 \\ 0 & 1.2 & 0 \end{bmatrix}$$

Their combined debt D is the sum of the two matrices.

$$D = J + S$$

$$= \begin{bmatrix} 0 & 0.8 & 2 \\ 0 & 0 & 5 \\ 7 & 1.5 & 0 \end{bmatrix} + \begin{bmatrix} 0 & 0.7 & 0 \\ 4.5 & 0 & 8 \\ 0 & 1.2 & 0 \end{bmatrix}$$

$$= \begin{bmatrix} 0 & 1.5 & 2 \\ 4.5 & 0 & 13 \\ 7 & 2.7 & 0 \end{bmatrix}$$

The result shows the types of debt owed to each institution (see Table 3.3).

TABLE 3.3

	Car Loan	Credit Card	Student Loan
Aspen Bank	0	1.5	2
Back Bay Credit Union	4.5	0	13
IntelliBank	7	2.7	0

To determine the total amount owed to each institution, the row entries should be summed. To determine the total amount of each type of debt, the column entries should be summed.

Using matrices to calculate the combined debts in Example 2 may have seemed trivial. Wouldn't it have been easier just to add the various quantities without introducing matrix notation? Perhaps. But what if we were looking at the merger of two companies with 40 creditors and 20 different types of debt? Would we want to calculate all 800 entries manually? Probably not. The notion of matrix addition, a feature embedded in most types of spreadsheet software, would greatly facilitate the calculation.

Scalar Multiplication

There are two types of multiplication that may be used with matrices: **scalar multiplication** and **matrix multiplication.** We will address scalar multiplication here and matrix multiplication in Section 3.2. The term **scalar** means *constant* or *number.* Scalar multiplication "scales" the entries of a matrix by making them larger or smaller by a given factor.

For example, if $A = \begin{bmatrix} 1 & 3 & 0 \\ -2 & 4 & 6 \\ 5 & 7 & 9 \end{bmatrix}$, then

$$5A = 5 \begin{bmatrix} 1 & 3 & 0 \\ -2 & 4 & 6 \\ 5 & 7 & 9 \end{bmatrix}$$

$$= \begin{bmatrix} 5 \cdot 1 & 5 \cdot 3 & 5 \cdot 0 \\ 5 \cdot (-2) & 5 \cdot 4 & 5 \cdot 6 \\ 5 \cdot 5 & 5 \cdot 7 & 5 \cdot 9 \end{bmatrix}$$ Multiply each entry by the scalar 5

$$= \begin{bmatrix} 5 & 15 & 0 \\ -10 & 20 & 30 \\ 25 & 35 & 45 \end{bmatrix}$$

In this example, the scalar 5 increased the magnitude of each entry by a factor of five. Any real number may be used as a scalar.

SCALAR MULTIPLICATION

An $m \times n$ matrix A and a real number k may be multiplied together to form a new $m \times n$ matrix, C. The value of the entry in the ith row and jth column of C is $c_{ij} = k \cdot a_{ij}$.

EXAMPLE 3   **Multiplying by a Scalar**

Calculate $C = -\dfrac{1}{2}A$ given $A = \begin{bmatrix} 2 & 0 \\ -4 & 6 \\ 1 & 8 \end{bmatrix}$.

SOLUTION

$$C = -\frac{1}{2}A$$

$$= -\frac{1}{2} \begin{bmatrix} 2 & 0 \\ -4 & 6 \\ 1 & 8 \end{bmatrix}$$

$$= \begin{bmatrix} -\dfrac{1}{2} \cdot 2 & -\dfrac{1}{2} \cdot 0 \\ -\dfrac{1}{2} \cdot (-4) & -\dfrac{1}{2} \cdot 6 \\ -\dfrac{1}{2} \cdot 1 & -\dfrac{1}{2} \cdot 8 \end{bmatrix}$$ Multiply each entry by the scalar $-\dfrac{1}{2}$

$$= \begin{bmatrix} -1 & 0 \\ 2 & -3 \\ -\dfrac{1}{2} & -4 \end{bmatrix}$$

TECHNOLOGY TIP

Scalar Multiplication

1. Enter matrix *A* into the calculator using the Matrix Editor.

```
MATRIX[A] 3 ×2
[ 2      0       ]
[ -4     6       ]
[ 1     I:■■■■■   ]


3,2=8
```

2. Type in the scalar on the home screen, press the ⌧ key, and then use the Matrix Names menu to place matrix *A* on the home screen.

```
-1/2*[A]
```

3. Press ENTER to display the product of the scalar and the matrix.

```
-1/2*[A]
      [[-1  0 ]
       [2   -3]
       [-.5 -4]]
```

4. If you would like to convert decimal entries to fractions, press MATH, select 1:▶Frac, then press ENTER.

```
        [[-1    0 ]
         [2     -3]
         [-.5  -4]]
Ans▶Frac
        [[-1    0 ]
         [2     -3]
         [-1/2 -4]]
```

EXAMPLE 4

Using Scalar Multiplication to Adjust a Pay Scale

In 2000, Washington State voters overwhelmingly approved an initiative to give state K through 12 and community college educators annual cost-of-living increases. (**Source:** I-732 Text.) Faculty at Green River Community College received a 3.7 percent salary increase for the 2001–2002 school year.

Each faculty member's base salary is calculated based on teaching experience and educational/professional development credits as shown in Table 3.4.

TABLE 3.4   2000–2001 Pay Scale

	Credits 240	Credits 300	Credits 360
Teaching Level 1	$33,878	$37,360	$40,843
Teaching Level 2	$35,695	$39,177	$42,660
Teaching Level 3	$37,512	$40,994	$44,476
Teaching Level 4	$39,329	$42,811	$46,293

Source: Green River Community College.

Determine the 2001–2002 pay scale for Green River Community College faculty.

SOLUTION   We can solve the problem by writing the table as a matrix and multiplying by a scalar. Since salaries are to increase by 3.7 percent, the 2001–2002 salaries will be 103.7 percent of the 2000–2001 salaries.

$$103.7\% = \frac{103.7}{100}$$
$$= 1.037$$

We will multiply the matrix by 1.037. This will increase each entry by 3.7 percent.

$$S = \begin{bmatrix} 33,878 & 37,360 & 40,843 \\ 35,695 & 39,177 & 42,660 \\ 37,512 & 40,994 & 44,476 \\ 39,329 & 42,811 & 46,293 \end{bmatrix}$$

$$1.037S = 1.037 \begin{bmatrix} 33,878 & 37,360 & 40,843 \\ 35,695 & 39,177 & 42,660 \\ 37,512 & 40,994 & 44,476 \\ 39,329 & 42,811 & 46,293 \end{bmatrix}$$

$$\approx \begin{bmatrix} 35,131 & 38,742 & 42,354 \\ 37,016 & 40,627 & 44,238 \\ 38,900 & 42,511 & 46,122 \\ 40,784 & 44,395 & 48,006 \end{bmatrix}$$

The new matrix represents the 2001–2002 pay scale for Green River Community College faculty, as shown in Table 3.5.

TABLE 3.5   2001–2002 Pay Scale

	Credits 240	Credits 300	Credits 360
Teaching Level 1	$35,131	38,742	$42,354
Teaching Level 2	$37,016	$40,627	$44,238
Teaching Level 3	$38,900	$42,511	$46,122
Teaching Level 4	$40,784	$44,395	$48,006

Matrix Addition and Scalar Multiplication Properties

Let A, B, and C be $m \times n$ matrices and let c and k be real numbers. Let O be the $m \times n$ **zero matrix** (a matrix with entries of all zeros.) The following properties hold:

Matrix Addition Properties

1. Additive Associative $A + (B + C) = (A + B) + C$
2. Additive Commutative $A + B = B + A$
3. Additive Identity $A + O = O + A = A$
4. Additive Inverse $(-A) + A = A + (-A) = O$

Scalar Multiplication Properties

5. Distributive $c(A + B) = cA + cB$
6. Distributive $(c + k)A = cA + kA$
7. Multiplicative Associative $c(kA) = (ck)A$
8. Scalar Unit 1 $1 \cdot A = A$
9. Scalar Unit 0 $0 \cdot A = O$

You may have noticed that no matrix subtraction properties were listed. This is because we view matrix subtraction as a combination of scalar multiplication and matrix addition. The matrix expression $A - B$ is equivalent to $A + (-1)B$. As shown in Example 5, in practice we typically simplify the matrix expression $A - B$ by subtracting the entries of B from the corresponding entries of A.

EXAMPLE 5

Calculating the Difference of Two Matrices

Let $A = \begin{bmatrix} 4 & -1 & 2 \\ 3 & -5 & 0 \\ 9 & 11 & 20 \end{bmatrix}$ and $B = \begin{bmatrix} -4 & 7 & 3 \\ 10 & -9 & 10 \\ -2 & 4 & 8 \end{bmatrix}$. Show that $A - B$ is equivalent to $A + (-1)B$.

SOLUTION

$$A - B = \begin{bmatrix} 4 & -1 & 2 \\ 3 & -5 & 0 \\ 9 & 11 & 20 \end{bmatrix} - \begin{bmatrix} -4 & 7 & 3 \\ 10 & -9 & 10 \\ -2 & 4 & 8 \end{bmatrix}$$

$$= \begin{bmatrix} 4 - (-4) & -1 - 7 & 2 - 3 \\ 3 - 10 & -5 - (-9) & 0 - 10 \\ 9 - (-2) & 11 - 4 & 20 - 8 \end{bmatrix} \quad \begin{array}{l} \text{Subtract the entries of } B \\ \text{from the corresponding} \\ \text{entries of } A \end{array}$$

$$= \begin{bmatrix} 8 & -8 & -1 \\ -7 & 4 & -10 \\ 11 & 7 & 12 \end{bmatrix}$$

$$A + (-1)B = \begin{bmatrix} 4 & -1 & 2 \\ 3 & -5 & 0 \\ 9 & 11 & 20 \end{bmatrix} + (-1)\begin{bmatrix} -4 & 7 & 3 \\ 10 & -9 & 10 \\ -2 & 4 & 8 \end{bmatrix}$$

$$= \begin{bmatrix} 4 & -1 & 2 \\ 3 & -5 & 0 \\ 9 & 11 & 20 \end{bmatrix} + \begin{bmatrix} (-1)(-4) & (-1)7 & (-1)3 \\ (-1)10 & (-1)(-9) & (-1)10 \\ (-1)(-2) & (-1)4 & (-1)8 \end{bmatrix}$$

Multiply the entries of B by the scalar − 1

$$= \begin{bmatrix} 4 & -1 & 2 \\ 3 & -5 & 0 \\ 9 & 11 & 20 \end{bmatrix} + \begin{bmatrix} 4 & -7 & -3 \\ -10 & 9 & -10 \\ 2 & -4 & -8 \end{bmatrix}$$

$$= \begin{bmatrix} 8 & -8 & -1 \\ -7 & 4 & -10 \\ 11 & 7 & 12 \end{bmatrix}$$

Matrix expressions may combine one or more matrix operations. Example 6 includes matrix addition and scalar multiplication with three different matrices, including the zero matrix.

EXAMPLE 6

Solving a Matrix Algebra Problem Involving the Zero Matrix

Let $A = \begin{bmatrix} 1 & 0 & 4 \\ -1 & 2 & -5 \end{bmatrix}$, $B = \begin{bmatrix} 5 & 7 & 0 \\ 3 & -2 & -4 \end{bmatrix}$, and $O = \begin{bmatrix} 0 & 0 & 0 \\ 0 & 0 & 0 \end{bmatrix}$. Calculate $A - (2B + O)$.

SOLUTION

$$A - (2B + O) = \begin{bmatrix} 1 & 0 & 4 \\ -1 & 2 & -5 \end{bmatrix} - \left(2 \cdot \begin{bmatrix} 5 & 7 & 0 \\ 3 & -2 & -4 \end{bmatrix} + \begin{bmatrix} 0 & 0 & 0 \\ 0 & 0 & 0 \end{bmatrix}\right)$$

$$= \begin{bmatrix} 1 & 0 & 4 \\ -1 & 2 & -5 \end{bmatrix} + (-1)\left(\begin{bmatrix} 10 & 14 & 0 \\ 6 & -4 & -8 \end{bmatrix} + \begin{bmatrix} 0 & 0 & 0 \\ 0 & 0 & 0 \end{bmatrix}\right)$$

Multiply B by 2. Rewrite "−" as "+".

$$= \begin{bmatrix} 1 & 0 & 4 \\ -1 & 2 & -5 \end{bmatrix} + (-1)\begin{bmatrix} 10 & 14 & 0 \\ 6 & -4 & -8 \end{bmatrix}$$

$$= \begin{bmatrix} 1 & 0 & 4 \\ -1 & 2 & -5 \end{bmatrix} + \begin{bmatrix} -10 & -14 & 0 \\ -6 & 4 & 8 \end{bmatrix} \qquad \text{Multiply } 2B \text{ by } -1$$

$$= \begin{bmatrix} -9 & -14 & 4 \\ -7 & 6 & 3 \end{bmatrix}$$

3.1 Summary

In this section, you learned how to do matrix addition and scalar multiplication. You also saw how these techniques could be used to analyze real-life financial situations in life and in business.

3.1 Exercises

In Exercises 1–10, perform the indicated matrix operation, if possible, given the following matrices. Solve these problems without technology.

$$A = \begin{bmatrix} 1 & 6 \\ 3 & 4 \\ 5 & 2 \end{bmatrix}, \quad B = \begin{bmatrix} -5 & 0 \\ 7 & 8 \\ -9 & 1 \end{bmatrix},$$

$$C = \begin{bmatrix} 2 & 3 & 0 \\ -1 & 4 & 1 \\ 5 & -2 & 0 \end{bmatrix}, \quad D = \begin{bmatrix} 1 & -3 & 4 \\ 0 & 5 & 7 \\ 9 & 8 & 2 \end{bmatrix}$$

1. $A + B$ **2.** $A - B$ **3.** $B + C$

4. $C - D$ **5.** $D + C$ **6.** $2A$

7. $3B$ **8.** $2A + 3B$ **9.** $-2C + 2D$

10. $2A + 2C$

In Exercises 11–20, use technology to simplify the matrix expressions, given the following matrices.

$$A = \begin{bmatrix} 1.2 & 6.3 & 0.4 \\ -9.1 & 4.2 & 1.7 \\ 0.9 & -2.0 & 0.3 \end{bmatrix}$$

$$B = \begin{bmatrix} 1.4 & -0.3 & 0.4 \\ -2.8 & 5.5 & 7.1 \\ 9.2 & 8.6 & 2.0 \end{bmatrix}$$

11. $1.2A$ **12.** $-2.3B$

13. $1.2A - 2.3B$ **14.** $4.1A + 0.1B$

15. $-1.1A + 2.9B$ **16.** $2.9A + 0.1B$

17. $-8.7A + 8.7B$ **18.** $-A - 9.2B$

19. $7.8A + 9.9B$ **20.** $-1.7A - 2.1B$

In Exercises 21–36, use matrix addition and/or scalar multiplication to find the solution.

21. Personal Debt A couple is planning to get married. He has the following debts: a $2700 consumer loan at Loan Shark Larry's, a $26,500 car loan and an $8200 credit card balance at Risky Bank, and a $2700 consumer loan at Mastercraft Jewelers (for the ring). She has the following debts: a $1200 car loan, a $82,500 home mortgage, a $200 credit card balance at Risky Bank, and a $250 consumer loan at Mastercraft Jewelers. Determine their combined debts of each type at each of the various financial institutions.

22. **Faculty Salaries** For the 2002–2003 academic year, Green River Community College faculty received a 3.432 percent cost-of-living increase. (**Source:** Green River Community College.) The 2001–2002 pay scale is shown in the following table. Determine the 2002–2003 pay scale.

2001–2002 Pay Scale

	240 Credits ($)	300 Credits ($)	360 Credits ($)
Level 1	35,131	38,742	42,354
Level 2	37,016	40,627	44,238
Level 3	38,900	42,511	46,122
Level 4	40,784	44,395	48,006

23. **Faculty Salaries** If the Green River Community College faculty in Exercise 22 get a 3.432 percent raise annually, determine the 2005–2006 pay scale.

24. **Condiment Prices** On July 20, 2002, Albertsons.com advertised the following items at the indicated prices.

	Albertsons	Hunts	Kraft
Ketchup, 24 oz	$0.69	$1.89	
Barbecue sauce, 18 oz		$0.99	$0.99
Mayonnaise, 32 oz	$2.19		$3.19

If the prices are affected only by inflation and the annual rate of inflation is 3 percent, determine the price of each of the items on July 20, 2006.

25. **Soda Prices** On July 20, 2002, Albertsons.com advertised the following items at the indicated prices.

	Albertsons	A&W	Henry Weinhards
Root beer, 6-pack	$1.59	$1.67	$4.50
Cream soda, 12-pack	$3.18	$3.34	$9.00
Club soda, 2 liters	$0.99		

If the prices are affected only by inflation and the annual rate of inflation is 3 percent, determine the price of each of the items on July 20, 2006.

26. **Auto Prices** The average trade-in value of a Volkswagen New Beetle and Volkswagen Golf in July 2002 is shown in the first table. The average retail value of the two vehicles is shown in the second table.

Average Trade-in Value

	Golf	New Beetle
2000 model	$11,000	$11,850
2001 model	$11,875	$13,175

Average Retail Value

	Golf	New Beetle
2000 model	$13,050	$14,000
2001 model	$14,025	$15,475

Source: www.nada.com.

Use matrices to create a table that shows the average dealer markup for each of the vehicles.

27. **Auto Prices** The average trade-in value of a Honda Civic and a Honda Accord in July 2002 is shown in the first table. The average retail value of the two vehicles is shown in the second table.

Average Trade-in Value

	Accord	Civic
2000 model	$14,800	$8,925
2001 model	$16,575	$9,850

Average Retail Value

	Accord	Civic
2000 model	$17,100	$10,800
2001 model	$18,975	$11,825

Source: www.nada.com.

Use matrices to create a table that shows the average dealer markup for each of the vehicles.

28. **Food Supply** Bed-and-breakfasts attract visitors with their intimate ambiance and their delicious cuisine. Many establishments serve wedding brunches in addition to providing lodging services.

A bed-and-breakfast hostess is planning an upcoming wedding brunch. She plans to make four dozen muffins (two dozen apple and two dozen blueberry) and six fruit crisps (three apple and three blueberry). The amount of flour, sugar, and fruit required to make a single fruit crisp is shown in the first table, and the amount required to make a dozen muffins is shown in the second table.

Fruit Crisp

	Flour	Sugar	Fruit
Apple	1/2 cup	3/4 cup	3 cups
Blueberry	1/2 cup	1 cup	4 cups

Muffins

	Flour	Sugar	Fruit
Apple	2 cups	1/4 cup	3/4 cup
Blueberry	2 cups	1/3 cup	1 cup

Use matrices to create a table that shows how much flour, sugar, and fruit will be required for the apple desserts and the blueberry desserts.

29. **Energy Usage** The amount of natural gas and coal energy produced and consumed in the United States is shown in the following tables. Use matrix operations to create a table that shows the difference between energy production and energy consumption.

Energy Production

Years Since 1960	Natural Gas (quadrillion BTUs)	Coal (quadrillion BTUs)
0	12.66	10.82
10	21.67	14.61
20	19.91	18.60
30	18.36	22.46
40	19.74	22.66

Energy Consumption

Years Since 1960	Natural Gas (quadrillion BTUs)	Coal (quadrillion BTUs)
0	12.39	9.84
10	21.80	12.27
20	20.39	15.42
30	19.30	19.25
40	23.33	22.41

Source: *Statistical Abstract of the United States, 2001*, Table 891, p. 569.

30.   **Energy Policy** If you were a lobbyist for the natural gas industry, how would you use the results of Exercise 29 to persuade legislators to support further natural gas exploration?

31.   **Energy Usage** The amount of nuclear electric power and coal energy produced and consumed in the United States is shown in the following tables. Use matrix operations to create a table that shows the difference between energy production and energy consumption.

Energy Production

Years Since 1960	Nuclear Electric Power (quadrillion BTUs)	Coal (quadrillion BTUs)
0	0.01	10.82
10	0.24	14.61
20	2.74	18.60
30	6.16	22.46
40	8.01	22.66

Energy Consumption

Years Since 1960	Nuclear Electric Power (quadrillion BTUs)	Coal (quadrillion BTUs)
0	0.01	9.84
10	0.24	12.27
20	2.74	15.42
30	6.16	19.25
40	8.01	22.41

Source: *Statistical Abstract of the United States, 2001*, Table 891, p. 569.

32.   **Energy Policy** An environmentalist believes that the United States should produce only as much energy as it consumes. Based on the results of Exercise 31, which energy technology (coal or nuclear electric power) seems to best support the environmentalist's position? Defend your conclusion.

33.   **Renewable Energy Consumption** The amount of renewable energy consumed in the United States in various years is shown in the following tables. Use matrices to construct a table showing the total amount of renewable energy consumed between the beginning of 1997 and the end of 1999.

1997 Renewable Energy Consumption

Renewable Energy Type	Energy Consumed (quadrillion BTUs)
Conventional hydroelectric power	3.94
Geothermal energy	0.33
Biomass	2.98
Solar energy	0.07
Wind energy	0.03

1998 Renewable Energy Consumption

Renewable Energy Type	Energy Consumed (Quadrillion BTUs)
Conventional hydroelectric power	3.55
Geothermal energy	0.34
Biomass	2.99
Solar energy	0.07
Wind energy	0.03

1999 Renewable Energy Consumption

Renewable Energy Type	Energy Consumed (quadrillion BTUs)
Conventional hydroelectric power	3.42
Geothermal energy	0.33
Biomass	3.51
Solar energy	0.08
Wind energy	0.04

Source: *Statistical Abstract of the United States, 2001*, Table 896, p. 572.

34. **Average Energy Consumption** Using the results from Exercise 33, use matrix operations to construct a table showing the average amount of energy consumed annually between the start of 1997 and the end of 1999.

35. **Energy Usage Trends** Using the results from Exercise 34 and the data tables from Exercise 33, construct a table showing the difference between the 1999 energy consumption and the average annual energy consumption. What conclusions can you draw from this table?

36. **Organized Physical Activity** The following tables show the percentage of high school students involved in physical education classes. Based on the information given, can you determine the percentage of ninth graders who exercise 20 or more minutes per class? Justify your answer.

Males Enrolled in a Physical Education Class

Grade	Enrolled in a P.E. Class (percent)	Exercised 20 Minutes or More per Class (percent)
9	60.7	82.1
10	82.3	84.4
11	65.3	79.4
12	44.6	82.0

Females Enrolled in a Physical Education Class

Grade	Enrolled in a P.E. Class (percent)	Exercised 20 Minutes or More per Class (percent)
9	51.5	69.6
10	75.6	72.5
11	56.6	70.2
12	36.8	68.0

Source: *Statistical Abstract of the United States, 2001*, Table 1246, p. 764.

Exercises 37–40 are intended to challenge your ability to apply matrix addition and scalar multiplication properties. For each exercise, $A = \begin{bmatrix} 2 & 15 & 0 \\ 6 & -4 & -9 \\ -8 & \frac{1}{2} & \frac{2}{3} \end{bmatrix}$

and $C = \begin{bmatrix} 12 & 1 & 4 \\ 3 & 0 & 5 \\ -2 & \frac{3}{2} & -\frac{1}{3} \end{bmatrix}.$

37. Solve the matrix equation for B.
$$5A - B = C$$

38. Solve the matrix equation for B.
$$-2A + 3B = C$$

39. Solve the matrix equation for B.
$$\frac{1}{2}A + \frac{2}{3}B = \frac{1}{6}C$$

40. Solve the matrix equation for B.
$$-100A - 200B = 330C$$

3.2 Matrix Multiplication and Inverses

- Do matrix multiplication
- Find the inverse of an invertible matrix with and without technology
- Use matrices to analyze and interpret real-life data

GETTING STARTED A theater charges $8.50 per adult, $5.50 per child, and $6.50 per senior. What will be the total admission cost for a group of 12 adults, 16 children, and 4 seniors? This question may be answered using matrix multiplication.

In this section, we will introduce some special types of matrices, demonstrate how to do matrix multiplication, and show how to find the inverse of a 2×2 matrix. We will also look at real-life applications of matrix multiplication.

Special Types of Matrices

A $1 \times n$ **row matrix** is a matrix that consists of a single row and n columns. For example, $A = \begin{bmatrix} 2 & 5 & 3 \end{bmatrix}$ is a 1×3 row matrix and $B = \begin{bmatrix} 3 & 1 & 0 & 9 \end{bmatrix}$ is a 1×4 row matrix.

An $n \times 1$ **column matrix** is a matrix that consists of n rows and a single column. For example, $C = \begin{bmatrix} 2 \\ 1 \end{bmatrix}$ is a 2×1 column matrix and $D = \begin{bmatrix} -2 \\ 7 \\ 5 \end{bmatrix}$ is a 3×1 column matrix.

An $n \times n$ **square matrix** is a matrix with n rows and n columns. For example, $E = \begin{bmatrix} 0 & -1 & 4 \\ -5 & 5 & 7 \\ 9 & 4 & 1 \end{bmatrix}$ is a 3×3 square matrix and $F = \begin{bmatrix} 1 & 9 \\ 9 & 8 \end{bmatrix}$ is a 2×2 square matrix. We will use each of these types of matrices as we discuss matrix multiplication.

Matrix Multiplication

In the last section, we introduced the concept of scalar multiplication—the multiplication of a number and a matrix. There is another type of multiplication involving matrices—matrix multiplication. Since matrix multiplication is a somewhat strange process, we will introduce the concept with a brief example before giving a formal definition.

EXAMPLE 1 Using Matrix Multiplication to Calculate Theater Admission Cost

A theater charges $8.50 per adult, $5.50 per child, and $6.50 per senior. What will be the total admission cost for a group of 12 adults, 16 children, and 4 seniors?

SOLUTION We know that the total cost is given by

$$\text{Total cost} = 8.50(12) + 5.50(16) + 6.50(4)$$
$$\phantom{\text{Total cost} =}\ \underset{\text{Adults}}{} \quad \underset{\text{Children}}{} \quad \underset{\text{Seniors}}{}$$

$$= \quad 102 \quad + \quad 88 \quad + \quad 26$$

$$= 216$$

The total admission cost for the group is $216.

The same result may be obtained by representing the individual admission cost as a row matrix and the number of guests as a column matrix.

$$C = \begin{bmatrix} 8.50 & 5.50 & 6.50 \end{bmatrix} \qquad \text{Individual admission cost matrix}$$

$$N = \begin{bmatrix} 12 \\ 16 \\ 4 \end{bmatrix} \qquad \text{Number of guests matrix}$$

The product of the 1×3 matrix C and the 3×1 matrix N is given by

$$CN = \begin{bmatrix} 8.50 & 5.50 & 6.50 \end{bmatrix} \begin{bmatrix} 12 \\ 16 \\ 4 \end{bmatrix}$$

From our initial solution, we know that the total cost is given by the expression

$$8.50(12) + 5.50(16) + 6.50(4)$$

How can we combine the elements of each matrix so that we end up with the desired expression? Observe that if we multiply the first entry in the row matrix by the first entry in the column matrix, we obtain $8.50(12)$, the first term of the expression. Similarly, multiplying the second entry in the row matrix by the second entry in the column matrix yields $5.50(16)$, the second term of the expression. The third term of the expression, $6.50(4)$, is obtained by multiplying the third term in the row matrix by the third term in the column matrix. The final result is obtained by summing the individual terms. Therefore,

$$CN = \begin{bmatrix} 8.50 & 5.50 & 6.50 \end{bmatrix} \begin{bmatrix} 12 \\ 16 \\ 4 \end{bmatrix}$$

$$= \begin{bmatrix} 8.50(12) + 5.50(16) + 6.50(4) \end{bmatrix}$$
$$= \begin{bmatrix} 102 + 88 + 26 \end{bmatrix}$$
$$= \begin{bmatrix} 216 \end{bmatrix}$$

The total ticket cost is $216.

How did we know the units of the 1×1 matrix? The units of C were dollars per person, since the ticket prices given were individual ticket prices. The units of N were people or persons. The units of the product of the matrices are the product of the units of the matrices, in the same order. Therefore,

$$\text{Units of } CN = \left(\frac{\text{dollars}}{\text{person}} \right) (\text{persons})$$

$$= \left(\frac{\text{dollars}}{\text{person}} \right) (\text{persons})$$

$$= \text{dollars}$$

In Example 1, we demonstrated how to multiply a row matrix and a column matrix. The process is summarized as follows.

THE PRODUCT OF A ROW MATRIX AND A COLUMN MATRIX

The product of a $1 \times n$ row matrix A and an $n \times 1$ column matrix B is the 1×1 square matrix given by

$$AB = \begin{bmatrix} a_1 & a_2 & \cdots & a_n \end{bmatrix} \begin{bmatrix} b_1 \\ b_2 \\ \vdots \\ b_n \end{bmatrix}$$

$$= \begin{bmatrix} a_1 b_1 + a_2 b_2 + \cdots + a_n b_n \end{bmatrix}$$

EXAMPLE 2

Determining the Product of a Row Matrix and a Column Matrix

Calculate AB given $A = \begin{bmatrix} 1 & 2 \end{bmatrix}$ and $B = \begin{bmatrix} 3 \\ 4 \end{bmatrix}$.

SOLUTION

$$AB = \begin{bmatrix} 1 & 2 \end{bmatrix} \begin{bmatrix} 3 \\ 4 \end{bmatrix}$$

$$= \begin{bmatrix} 1(3) + 2(4) \end{bmatrix}$$

$$= \begin{bmatrix} 3 + 8 \end{bmatrix}$$

$$= \begin{bmatrix} 11 \end{bmatrix}$$

The solution is the 1×1 matrix $\begin{bmatrix} 11 \end{bmatrix}$. Note that this is a matrix, not the number 11.

EXAMPLE 3

Determining the Product of a Row Matrix and a Column Matrix

Calculate AB given $A = \begin{bmatrix} 2 & 0 & -1 & 8 \end{bmatrix}$ and $B = \begin{bmatrix} 5 \\ 7 \\ 4 \\ -2 \end{bmatrix}$.

SOLUTION

$$AB = \begin{bmatrix} 2 & 0 & -1 & 8 \end{bmatrix} \begin{bmatrix} 5 \\ 7 \\ 4 \\ -2 \end{bmatrix}$$

$$= \begin{bmatrix} 2(5) + 0(7) + (-1)(4) + 8(-2) \end{bmatrix}$$

$$= \begin{bmatrix} 10 + 0 - 4 - 16 \end{bmatrix}$$

$$= \begin{bmatrix} -10 \end{bmatrix}$$

The method used to calculate the product of a $1 \times n$ row matrix and an $n \times 1$ column matrix may be used to calculate the entries of the product of any two matrices (provided the product exists).

MATRIX MULTIPLICATION

An $m \times n$ matrix A and an $n \times p$ matrix B may be multiplied together to form a new $m \times p$ matrix, C. The value of the entry in the ith row and the jth column of C is the product of the ith row of A and the jth column of B.

It is significant to note that matrix multiplication can be performed only if the number of columns of the first matrix is equal to the number of rows of the second matrix. Consider the $m \times n$ matrix A and the $n \times p$ matrix B shown here. We have boxed the ith row of A and the jth column of B.

$$C = AB$$

$$= \begin{bmatrix} a_{11} & a_{12} & \cdots & a_{1n} \\ a_{21} & a_{22} & \cdots & a_{2n} \\ \vdots & \vdots & \ddots & \vdots \\ \boxed{a_{i1} \quad a_{i2} \quad \cdots \quad a_{in}} \\ \vdots & \vdots & \ddots & \vdots \\ a_{m1} & a_{m2} & \cdots & a_{mn} \end{bmatrix} \cdot \begin{bmatrix} b_{11} & b_{12} & \cdots & \boxed{b_{1j}} & \cdots & b_{1p} \\ b_{21} & b_{22} & \cdots & \boxed{b_{2j}} & \cdots & b_{2p} \\ \vdots & \vdots & \ddots & \vdots & \ddots & \vdots \\ b_{n1} & b_{n2} & \cdots & \boxed{b_{nj}} & \cdots & b_{np} \end{bmatrix}$$

The entry c_{ij} of matrix C is determined by calculating

$$[c_{ij}] = [a_{i1} \quad a_{i2} \quad \cdots \quad a_{in}] \begin{bmatrix} b_{1j} \\ b_{2j} \\ \vdots \\ b_{nj} \end{bmatrix}$$

$$= [a_{i1}b_{1j} + a_{i2}b_{2j} + \cdots + a_{in}b_{nj}]$$

EXAMPLE 4 **Determining the Product of Two Matrices**

Find $C = AB$ given $A = \begin{bmatrix} 1 & 0 \\ 3 & 7 \\ 4 & -2 \end{bmatrix}$ and $B = \begin{bmatrix} 5 \\ 6 \end{bmatrix}$.

SOLUTION A is a 3×2 matrix and B is a 2×1 matrix. Since the "inside" dimensions match (both equal 2), the matrix $C = AB$ will take on the "outside" dimensions. That is, C will be a 3×1 matrix.

$$C = AB$$

$$= \begin{bmatrix} 1 & 0 \\ 3 & 7 \\ 4 & -2 \end{bmatrix} \begin{bmatrix} 5 \\ 6 \end{bmatrix}$$

$$= \begin{bmatrix} 1(5) + 0(6) \\ 3(5) + 7(6) \\ 4(5) + (-2)(6) \end{bmatrix} \qquad \begin{array}{l} \text{Row 1 of } A \text{ times Column 1 of } B \\ \text{Row 2 of } A \text{ times Column 1 of } B \\ \text{Row 3 of } A \text{ times Column 1 of } B \end{array}$$

$$= \begin{bmatrix} 5 + 0 \\ 15 + 42 \\ 20 - 12 \end{bmatrix}$$

$$= \begin{bmatrix} 5 \\ 57 \\ 8 \end{bmatrix}$$

Notice that the entries of C are the product of the rows of A and the column of B. For example, c_{21} is the product of the second row of A, $\begin{bmatrix} 3 & 7 \end{bmatrix}$, and the first column of B, $\begin{bmatrix} 5 \\ 6 \end{bmatrix}$.

EXAMPLE 5

Determining the Product of Two Matrices

Find $C = DA$ given $D = \begin{bmatrix} 2 & 5 \\ 7 & 6 \end{bmatrix}$ and $A = \begin{bmatrix} 1 & 0 \\ 3 & 7 \\ 4 & -2 \end{bmatrix}$.

SOLUTION D is a 2×2 matrix and A is a 3×2 matrix. The matrix $C = DA$ cannot be computed, since the number of columns of D (two) does not equal the number of rows of A (three). Let's try to do the multiplication anyway, just to see what happens. To calculate the entry c_{21}, we need to multiply the second row of D by the first column of A.

$$[c_{21}] = \begin{bmatrix} 7 & 6 \end{bmatrix} \begin{bmatrix} 1 \\ 3 \\ 4 \end{bmatrix}$$

$$= [7(1) + 6(3) + ?(4)]$$

We can't perform the calculation. The conflict shown here will appear anytime you try to multiply two matrices whose rows and columns don't match up.

We will use the matrices from Example 5 in Example 6. However, this time we will reverse the order in which the matrices are to be multiplied.

EXAMPLE 6 **Determining the Product of Two Matrices**

Find $C = AD$ given $A = \begin{bmatrix} 1 & 0 \\ 3 & 7 \\ 4 & -2 \end{bmatrix}$ and $D = \begin{bmatrix} 2 & 5 \\ 7 & 6 \end{bmatrix}$.

SOLUTION A is a 3×2 matrix and D is a 2×2 matrix. The matrix $C = AD$ is a 3×2 matrix.

$$C = AD$$

$$= \begin{bmatrix} 1 & 0 \\ 3 & 7 \\ 4 & -2 \end{bmatrix} \begin{bmatrix} 2 & 5 \\ 7 & 6 \end{bmatrix}$$

$$= \begin{bmatrix} 1(2) + 0(7) & 1(5) + 0(6) \\ 3(2) + 7(7) & 3(5) + 7(6) \\ 4(2) + (-2)(7) & 4(5) + (-2)(6) \end{bmatrix}$$

$$= \begin{bmatrix} 2 & 5 \\ 55 & 57 \\ -6 & 8 \end{bmatrix}$$

From Examples 5 and 6, we see that matrix multiplication is not commutative. That is, in general, $AB \neq BA$.

EXAMPLE 7 **Using Technology to Determine the Product of Two Matrices**

Find $C = AB$ given $A = \begin{bmatrix} 1 & 2 & 3 & 4 \\ 0 & 1 & 2 & 3 \\ 4 & 5 & 6 & 7 \\ 5 & 6 & 7 & 0 \end{bmatrix}$ and $B = \begin{bmatrix} -1.1 & 1.8 \\ 0.4 & 4.2 \\ -0.2 & 1.0 \\ -3.9 & 1.1 \end{bmatrix}$.

SOLUTION A is a 4×4 matrix and B is a 4×2 matrix, so AB will be a 4×2 matrix. Because of the size of the matrices and the complexity of the entries in matrix B, we will use technology to calculate the product, as demonstrated in the following Technology Tip. The result is

$$C = AB$$

$$= \begin{bmatrix} -16.5 & 17.6 \\ -11.7 & 9.5 \\ -30.9 & 41.9 \\ -4.5 & 41.2 \end{bmatrix}$$

TECHNOLOGY **TIP**

Matrix Multiplication

1. Enter matrix A and matrix B into the calculator, using the Matrix Editor. Use the Matrix Names menu to place matrix A on the home screen. Then place matrix B on the home screen.

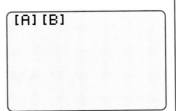

2. Press [ENTER] to display the product of the two matrices.

Error Alert:
If a dimension mismatch error occurs, double-check to make sure you have entered your matrices correctly. If you have, this error tells you that matrix multiplication is not possible because the number of columns of the first matrix is not the same as the number of rows of the second matrix.

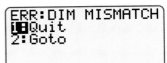

Matrix Multiplication Properties

Let A, B, and C be matrices. Let I be an identity matrix (see page 135), and let O be a zero matrix. Given that the dimensions of the matrices allow each of the operations to be performed, the following properties hold:

1. Multiplicative Associative $\quad A(BC) = (AB)C$
2. Multiplicative Identity $\quad AI = IA = A$
3. Distributive $\quad A(B + C) = AB + AC$
4. Distributive $\quad (A + B)C = AC + BC$
5. Multiplication by a Zero Matrix $\quad OA = AO = O$

Real-Life Applications of Matrix Multiplication

Up to this point, we have focused on the mathematical skill of matrix multiplication without demonstrating its usefulness in addressing real-world problems. In the next two examples, we will illustrate real-world uses of matrix multiplication.

EXAMPLE 8

Using Matrix Multiplication to Analyze Livestock Inventory Values

The U.S. Department of Agriculture monitors livestock inventories and their values. In a 1999 report on New York state hog and pig production, the USDA provided the data in Table 3.6.

TABLE 3.6   All Hogs and Pigs

Year	Number (thousands)	Value per Head (dollars)
1992	105	72.00
1993	90	77.00
1994	72	54.00
1995	66	71.00
1996	82	92.00
1997	79	81.00
1998	60	45.00

Source: www. nass.usda.gov.

From 1992 through 1998, what was the cumulative value of hogs and pigs in New York state?

SOLUTION   To determine the annual value of hog and pig inventories in a particular year, we simply need to multiply the number of hogs and pigs by the value per head. To calculate the cumulative value of the hog and pig inventories, we need to add up the annual inventory values. These operations may be performed using matrix multiplication.

We'll first define a column matrix N. The rows of N represent the years 1992 through 1998, and the column of N represents the number of hogs and pigs (in thousands):

$$N = \begin{bmatrix} 105 \\ 90 \\ 72 \\ 66 \\ 82 \\ 79 \\ 60 \end{bmatrix}$$

We will next define a row matrix V. The row of V represents the value per head, and the columns of V represent the years 1992 through 1998.

$$V = \begin{bmatrix} 72 & 77 & 54 & 71 & 92 & 81 & 45 \end{bmatrix}$$

We now must determine how to multiply the matrices so that the results make sense. Let's calculate VN. The dimensions of V are 1 value per head (in dollars) $\times$ 7 years. The dimensions of N are 7 years $\times$ 1 number of hogs and pigs (in thousands). Since the columns of V and the rows of N represent the same thing, matrix multiplication is possible and meaningful. The dimensions of VN will be 1×1.

$$VN = \begin{bmatrix} 72 & 77 & 54 & 71 & 92 & 81 & 45 \end{bmatrix} \begin{bmatrix} 105 \\ 90 \\ 72 \\ 66 \\ 82 \\ 79 \\ 60 \end{bmatrix}$$

$$= \begin{bmatrix} 72(105) + 77(90) + 54(72) + 71(66) + 92(82) + 81(79) + 45(60) \end{bmatrix}$$

$$= \begin{bmatrix} 39{,}707 \end{bmatrix}$$

What are the units of VN? We are multiplying value per head (in dollars) by number of head of hogs and pigs (in thousands).

UNITS

$$\frac{\text{dollars}}{\text{pig or hog}} \cdot \text{thousand of pigs or hogs} = \frac{\text{dollars}}{\text{pig or hog}} \cdot \text{thousand } \cancel{\text{of pigs or hogs}}$$

$$= \text{thousand dollars}$$

The cumulative value of the hog and pig inventory in New York from 1992 through 1998 is \$39,707,000.

Let's now look at NV. The dimensions of N are 7 years $\times$ 1 number of hogs and pigs (in thousands). The dimensions of V are 1 value per head (in dollars) $\times$ 7 years. Since the number of columns of N is the same as the number of rows of V, matrix multiplication is possible. However, since the meaning of the column of N (number of hogs and pigs) is not the same as the meaning of the rows of V (value per head), NV is meaningless in a real-world context.

EXAMPLE 9

Using Matrix Multiplication to Find the Nutritional Content of a Mix

Breakfast cereal connoisseurs often enjoy mixing cereals to create a new breakfast taste. A connoisseur working at a bed-and-breakfast wants to report the nutritional content of various mixtures of Honey Nut Cheerios®, Rice Crunch-Ems!, and Corn Crunch-Ems! to his health-conscious guests. From the package labeling, he determines the nutritional content of each cereal and records it in Table 3.7.

TABLE 3.7

	Honey Nut Cheerios	Rice Crunch-Ems!	Corn Crunch-Ems!
Protein	3 grams/cup	1.6 grams/cup	2 grams/cup
Carbohydrates	24 grams/cup	20.8 grams/cup	27 grams/cup
Fat	1.5 grams/cup	0 grams/cup	0 grams/cup

Source: Health Valley Rice Crunch-Ems! and Corn Crunch-Ems! labels and General Mills Honey Nut Cheerios label.

His first mixture will contain 1 cup of Honey Nut Cheerios, 2 cups of Rice Crunch-Ems!, and 1 cup of Corn Crunch-Ems! His second mixture will contain 2 cups of Honey Nut Cheerios, 4 cups of Rice Crunch-Ems!, and 3 cups of Corn Crunch-Ems!. Determine the amount of protein, carbohydrates, and fat in a 1-cup serving of each mixture.

SOLUTION   The nutrition content table may be represented by the matrix

$$N = \begin{matrix} & \text{Honey} & \text{Rice} & \text{Corn} \\ & \begin{bmatrix} 3.0 & 1.6 & 2.0 \\ 24.0 & 20.8 & 27.0 \\ 1.5 & 0.0 & 0.0 \end{bmatrix} & & \begin{matrix} \text{Protein} \\ \text{Carbs} \\ \text{Fat} \end{matrix} \end{matrix}$$

The mixture ingredients may be represented by Table 3.8

TABLE 3.8

	Mixture 1	Mixture 2
Honey Nut Cheerios	1.0 cup	2.0 cups
Rice Crunch-Ems!	2.0 cups	4.0 cups
Corn Crunch-Ems!	1.0 cup	3.0 cups

and the corresponding matrix

$$M = \begin{matrix} & \text{Mix 1} & \text{Mix 2} \\ & \begin{bmatrix} 1 & 2 \\ 2 & 4 \\ 1 & 3 \end{bmatrix} & \begin{matrix} \text{Honey} \\ \text{Rice} \\ \text{Corn} \end{matrix} \end{matrix}$$

Notice that the columns of N and the rows of M represent the same cereals (Honey Nut Cheerios, Rice Crunch-Ems!, and Corn Crunch-Ems!). The matrix NM will be a 3×2 matrix with rows representing protein, carbohydrates, and fat and columns representing Mixture 1 and Mixture 2.

$$NM = \begin{bmatrix} 3.0 & 1.6 & 2.0 \\ 24.0 & 20.8 & 27.0 \\ 1.5 & 0.0 & 0.0 \end{bmatrix} \begin{bmatrix} 1 & 2 \\ 2 & 4 \\ 1 & 3 \end{bmatrix}$$

$$= \begin{matrix} & \text{Mix 1} & \text{Mix 2} \\ & \begin{bmatrix} 8.2 & 18.4 \\ 92.6 & 212.2 \\ 1.5 & 3.0 \end{bmatrix} & \begin{matrix} \text{Protein} \\ \text{Carbs} \\ \text{Fat} \end{matrix} \end{matrix}$$

What are the units of the entries in matrix NM? The units of N are grams/cup, and the units of M are cups.

UNITS

$$\frac{\text{grams}}{\text{cup}} \cdot \text{cup} = \frac{\text{grams}}{\cancel{\text{cup}}} \cdot \cancel{\text{cup}}$$

$$= \text{grams}$$

So the units of the product is grams.

Converting the matrix back to a table, we have Table 3.9.

TABLE 3.9

	Mixture 1	Mixture 2
Protein	8.2 grams	18.4 grams
Carbohydrates	92.6 grams	212.2 grams
Fat	1.5 grams	3.0 grams

The table shows the total amount of protein, carbohydrates, and fat in each mixture. However, the first mixture contains 4 cups of cereal, and the second contains 9 cups of cereal.

Dividing the terms in the first column by 4 cups and the entries in the second column by 9 cups, we get the results in Table 3.10.

TABLE 3.10

	Mixture 1	Mixture 2
Protein	2.05 grams/cup	2.04 grams/cup
Carbohydrates	23.15 grams/cup	23.58 grams/cup
Fat	0.375 gram/cup	0.33 gram/cup

Since the original data were accurate to one decimal place, we will round the table entries as shown in Table 3.11.

TABLE 3.11

	Mixture 1	Mixture 2
Protein	2.1 grams/cup	2.0 grams/cup
Carbohydrates	23.2 grams/cup	23.6 grams/cup
Fat	0.4 gram/cup	0.3 gram/cup

Table 3.11 shows the nutritional content of each of the mixtures.

Inverse Matrices

The **identity matrix** I_n is the $n \times n$ square matrix with 1s along the main diagonal of the matrix and 0s elsewhere. For example, $I_3 = \begin{bmatrix} 1 & 0 & 0 \\ 0 & 1 & 0 \\ 0 & 0 & 1 \end{bmatrix}$, $I_2 = \begin{bmatrix} 1 & 0 \\ 0 & 1 \end{bmatrix}$, and $I_1 = [1]$ are identity matrices. (The subscript on the I is often omitted, but the dimensions of I are typically implied by the context of the problem.) We are often interested in matrices that have the property that $AB = BA = I$. These types of matrices are often used to solve systems of linear equations. In general, $AB \neq BA$, since matrix multiplication is not commutative. However, if two $n \times n$ matrices are inverses of each other, then $AB = BA = I$.

INVERSE MATRICES

An $n \times n$ matrix A and an $n \times n$ matrix B are inverses of each other if and only if $AB = BA = I_n$. We say that $B = A^{-1}$ (read "A inverse").

A matrix with an inverse is said to be **invertible.** A matrix without an inverse is said to be **singular.**

EXAMPLE 10   **Determining If a Matrix Is the Inverse of Another Matrix**

Let $A = \begin{bmatrix} 3 & 5 \\ 1 & 2 \end{bmatrix}$ and $B = \begin{bmatrix} 2 & -5 \\ -1 & 3 \end{bmatrix}$. Determine if matrix B is the inverse of matrix A.

SOLUTION

$$AB = \begin{bmatrix} 3 & 5 \\ 1 & 2 \end{bmatrix}\begin{bmatrix} 2 & -5 \\ -1 & 3 \end{bmatrix}$$
$$= \begin{bmatrix} 6-5 & -15+15 \\ 2-2 & -5+6 \end{bmatrix}$$
$$= \begin{bmatrix} 1 & 0 \\ 0 & 1 \end{bmatrix}$$
$$= I_2$$

$$BA = \begin{bmatrix} 2 & -5 \\ -1 & 3 \end{bmatrix}\begin{bmatrix} 3 & 5 \\ 1 & 2 \end{bmatrix}$$
$$= \begin{bmatrix} 6-5 & 10-10 \\ -3+3 & -5+6 \end{bmatrix}$$
$$= \begin{bmatrix} 1 & 0 \\ 0 & 1 \end{bmatrix}$$
$$= I_2$$

Since $AB = BA = I_2$, $B = A^{-1}$.

In Example 10, the matrices looked remarkably similar to each other. In fact, for a 2×2 invertible matrix, it is fairly simple to calculate the inverse.

INVERSE OF A 2 × 2 MATRIX

The inverse of an invertible 2×2 matrix $A = \begin{bmatrix} a & b \\ c & d \end{bmatrix}$ is

$$A^{-1} = \frac{1}{ad-bc}\begin{bmatrix} d & -b \\ -c & a \end{bmatrix}$$

The quantity $ad - bc$ is called the **determinant** of the matrix A and is often written $\det(A)$. If $\det(A) = 0$, then $\frac{1}{ad-bc}$ is undefined and A is singular.

This method works only for 2×2 matrices. For larger matrices, finding the inverse requires a bit more effort. We will demonstrate how to find the inverse of a 3×3 matrix algebraically in Section 3.3.

EXAMPLE 11   **Finding the Inverse of a 2 × 2 Matrix**

Find the inverse of matrix $A = \begin{bmatrix} 4 & 2 \\ 5 & 3 \end{bmatrix}$, if it exists.

SOLUTION

$$\det(A) = 4(3) - 2(5)$$
$$\det(A) = 2$$

Since the determinant of A is not zero, the inverse of A exists.

$$A^{-1} = \frac{1}{2} \begin{bmatrix} 3 & -2 \\ -5 & 4 \end{bmatrix}$$
$$= \begin{bmatrix} 1.5 & -1 \\ -2.5 & 2 \end{bmatrix}$$

EXAMPLE 12   **Using the Determinant to Determine If a Matrix Is Singular**

Find the inverse of matrix $A = \begin{bmatrix} 2 & 6 \\ 1 & 3 \end{bmatrix}$, if it exists.

SOLUTION   Since $\det(A) = 2(3) - 6(1) = 0$, A is singular. That is, A does not have an inverse.

Notice that Row 1 is equal to twice Row 2. Whenever one row of a matrix is a multiple of another row, the determinant of the matrix will be zero.

3.2 Summary

In this section, you learned about row, column, square, identity, and inverse matrices. You discovered matrix multiplication and learned techniques for finding the inverse of an invertible 2×2 matrix. You were also introduced to real-world applications of matrix multiplication.

3.2 Exercises

In Exercises 1–20, perform the indicated operation. As appropriate, use the matrices

$$A = \begin{bmatrix} 3 & 4 \\ -1 & -2 \end{bmatrix}, B = \begin{bmatrix} 2 & -4 \\ 1 & 2 \end{bmatrix},$$

$$C = \begin{bmatrix} 8 & 4 \\ 6 & 1 \\ 7 & 3 \end{bmatrix}, \text{ and } D = \begin{bmatrix} -2 & 0 & 3 \\ 0 & 4 & 3 \\ 1 & 7 & 0 \end{bmatrix}.$$

If the specified operation is undefined, so state.

1. $\begin{bmatrix} 3 & 9 & -2 \end{bmatrix} \begin{bmatrix} 2 \\ 2 \\ 12 \end{bmatrix}$ **2.** $\begin{bmatrix} -2 & 9 & 0.5 \end{bmatrix} \begin{bmatrix} 1.5 \\ 4 \\ -6 \end{bmatrix}$

3. $A + B$ **4.** $A - B$

5. $4A$ **6.** $-2B$

7. $4A - 2B$ **8.** $C + D$

9. AB **10.** BA

11. CD **12.** DC

13. CA **14.** AC

15. CB **16.** A^{-1}

17. B^{-1} **18.** $A^{-1}A$

19. AA^{-1} **20.** $DC + C$

In Exercises 21–30, use the determinant formula to determine if the matrix is invertible or singular. If the matrix is invertible, find its inverse.

21. $A = \begin{bmatrix} 2 & 3 \\ 1 & 2 \end{bmatrix}$ **22.** $B = \begin{bmatrix} -2 & 4 \\ 1 & 2 \end{bmatrix}$

23. $C = \begin{bmatrix} 9 & 6 \\ 3 & 2 \end{bmatrix}$ **24.** $D = \begin{bmatrix} 0 & -1 \\ 4 & 4 \end{bmatrix}$

25. $E = \begin{bmatrix} -0.5 & -0.7 \\ 4.0 & 3.4 \end{bmatrix}$ **26.** $A = \begin{bmatrix} -1 & 1 \\ 1 & 1 \end{bmatrix}$

27. $B = \begin{bmatrix} 5 & 10 \\ 2 & 4 \end{bmatrix}$ **28.** $C = \begin{bmatrix} 0.1 & 0.6 \\ 2.0 & 1.2 \end{bmatrix}$

29. $D = \begin{bmatrix} 1.0 & -1.1 \\ 0.9 & 1.0 \end{bmatrix}$ **30.** $E = \begin{bmatrix} -0.9 & -0.9 \\ -0.9 & 0.9 \end{bmatrix}$

In Exercises 31–40, use matrix multiplication to find the answer to each question.

31. **Lumber Manufacturing Payroll**   The number of employees working in the lumber and wood products manufacturing industry is shown in the following table.

Manufacturing Full-Time Employees: Lumber and Wood Products

Years Since 1995 (t)	Employees (E) (thousands)
0	772
1	782
2	794
3	816
4	843

Source: *Statistical Abstract of the United States, 2001,* Table 979, p. 622.

The average annual wage/salary of a full-time employee working in the lumber and wood products manufacturing industry is shown in the following table.

Manufacturing Salaries: Lumber and Wood Products

Years Since 1995 (t)	Average Annual Wage/Salary (S) (dollars)
0	25,110
1	26,148
2	27,382
3	28,278
4	29,040

Source: *Statistical Abstract of the United States, 2001,* Table 979, p. 622.

From 1995 through 1999, what was the total amount of money spent on employee wages and salaries?

32. 🌐 **Rubber and Plastics Payroll** The number of employees working in the rubber and plastics manufacturing industry is shown in the following table.

Manufacturing Full-Time Employees: Rubber and Plastics Industry

Years Since 1995 (t)	Employees (N) (thousands)
0	963
1	965
2	984
3	998
4	994

Source: *Statistical Abstract of the United States, 2001*, Table 979, p. 622.

The average annual wage/salary of a full-time employee working in the rubber and plastics manufacturing industry is shown in the following table.

Manufacturing Rubber and Plastics Industry Employee Average Earnings

Years Since 1995 (t)	Earnings (E) (dollars)
0	29,867
1	30,898
2	32,237
3	33,574
4	34,508

Source: *Statistical Abstract of the United States, 2001*, Table 979, p. 622.

From 1995 through 1999, what was the total amount of money spent on employee earnings?

33. 🌐 **Fruit Smoothie Nutritional Content**
The Vita-Mix® Super 5000 is a blender-like kitchen appliance with a 2+-horsepower motor. The Vita-Mix blade tips move at up to 240 miles per hour, easily converting whole fruits into luscious smoothies or grinding wheat kernels into fine flour. (*Source:* www.vita-mix.com.)

The author uses his Vita-Mix regularly to make breakfast beverages and was curious about the nutritional content of two different types of whole fruit smoothies. The ingredients of the first smoothie are 1 large apple, 1 large orange, 1 large banana, and 1 cup of water. The ingredients of the second smoothie are 2 large oranges, 1 large banana, and 1 cup of water.

Assuming that each fruit yields a 1-cup serving, we have the following nutrition information. A large orange contains 53.2 milligrams (mg) of vitamin C (ascorbic acid), 40 mg of calcium, and 2.4 grams (g) of fiber. A large banana contains 9.1 mg of vitamin C, 6 mg of calcium, and 2.4 g of fiber. A large apple contains 5.7 mg of vitamin C, 7 mg of calcium, and 2.7 g of fiber. (*Source:* USDA.) How much vitamin C, calcium, and fiber are in each smoothie?

34. 🌐 **Fruit Smoothie Nutritional Content**
One cup of flaxseed contains 1.3 mg of vitamin C, 199 mg of calcium, and 27.9 g of fiber. If one tablespoon of flaxseed is added to each of the smoothies in Exercise 33, how much vitamin C, calcium, and fiber are in each smoothie? (*Hint:* There are 16 tablespoons in a cup.)

35. 🌐 **Fruit Smoothie Nutritional Content** A tropical fruit smoothie is made from 2 cups of chopped mango, 1 cup of pineapple, 1 cup of coconut, and 2 cups of ice. A pina colada smoothie is made from 2 cups of pineapple, 1 cup of coconut, and 2 cups of ice.

A cup of fresh mango contains 10 mg of calcium, 27.7 mg of vitamin C, and 1.8 g of fiber. A cup of fresh pineapple contains 7 mg of calcium, 15.4 mg of vitamin C, and 1.2 g of fiber. A cup of raw coconut contains 14 mg of calcium, 3.3 mg of vitamin C, and 9 g of fiber. (*Source:* USDA.) How much vitamin C, calcium, and fiber are in each smoothie?

36. **Alcohol-Related Homicides** The following table shows the annual number of homicides resulting from an alcohol-related brawl.

Homicides Resulting from an Alcohol-Related Brawl

Years Since 1990 (t)	Homicides (H)
1	500
2	429
3	383
4	316
5	254
6	256
7	239
8	213
9	203
10	181

Source: Federal Bureau of Investigation.

Write the number of homicides as a 10×1 column matrix H. Let

$$A = [1 \quad 1 \quad 1 \quad 1 \quad 1 \quad 1 \quad 1 \quad 1 \quad 1 \quad 1]$$

Calculate and interpret the meaning of AH.

37. **Military Personnel** The table shows the number of military personnel in 1996 and 1998.

Year (t)	All Military Personnel (m) (thousands)	Army Personnel (a) (thousands)
1996	1,056	491
1998	1,004	484

Source: Statistical Abstract of the United States, 2001, Tables 499 and 500, pp. 328–329.

We can create a 2×2 matrix P with rows representing 1996 and 1998 and columns representing the number of military personnel.

$$P = \begin{bmatrix} 1056 & 491 \\ 1004 & 484 \end{bmatrix} \begin{matrix} \text{Year 1996} \\ \text{Year 1998} \end{matrix}$$

Let $R = [-1 \quad 1]$ and $C = \begin{bmatrix} 1 \\ -1 \end{bmatrix}$. Calculate and interpret the meaning of RP and PC.

38. **Movie Theater Revenue** The following table shows the average movie theater ticket price and total theater attendance from 1995 to 1999.

Years Since 1995	Price (dollars per person)	Attendance (millions of people)
0	4.35	1,263
1	4.42	1,339
2	4.59	1,388
3	4.69	1,481
4	5.08	1,465

Source: Statistical Abstract of the United States, 2001, Table 1244, p. 761.

Use matrices to calculate the accumulated movie theater revenue from ticket sales from the start of 1995 through 1999.

39. **Company Profit** The following table shows the net sales and cost of goods sold for the Kellogg Company from 1999 through 2001.

Year	Net Sales (millions of dollars)	Cost of Goods Sold (millions of dollars)
1999	6,984.2	3,325.1
2000	6,954.7	3,327.0
2001	8,853.3	4,128.5

Source: Kellogg Company 2001 Annual Report, pp. 7, 27.

Use matrices to calculate the accumulated net sales, accumulated cost of goods sold, and accumulated profit for the time period 1999 through 2001.

40. **Company Profit** The following table shows the net revenue and cost of goods sold for the Coca-Cola Company from 1999 through 2001.

Year	Net Revenue (millions of dollars)	Cost of Goods Sold (millions of dollars)
1999	19,284	6,009
2000	19,889	6,204
2001	20,092	6,044

Source: Coca-Cola Company 2001 Annual Report, p. 57.

Use matrices to calculate the accumulated net revenue, accumulated cost of goods sold, and accumulated profit for the time period 1999 through 2001.

Exercises 41–45 are intended to challenge your understanding of matrix multiplication and matrix inverses.

41. Find three different 2×2 singular matrices.

42. For real numbers a and b,
$(a + b)^2 = a^2 + 2ab + b^2$. For $n \times n$ matrices A and B, does $(A + B)^2 = A^2 + 2AB + B^2$? Explain.

43. A **diagonal matrix** A is a matrix with the property that $a_{ij} = 0$ when $i \neq j$. Show that if A is invertible, the inverse of the diagonal matrix

$$A = \begin{bmatrix} a & 0 & 0 \\ 0 & b & 0 \\ 0 & 0 & c \end{bmatrix} \text{ is } A^{-1} = \begin{bmatrix} \frac{1}{a} & 0 & 0 \\ 0 & \frac{1}{b} & 0 \\ 0 & 0 & \frac{1}{c} \end{bmatrix}. \text{ Under}$$

what conditions is A singular?

44. Find three 3×3 diagonal matrices that are their own inverses.

45. A **symmetric matrix** A is a matrix with the property that $a_{ij} = a_{ji}$. Show that if A is a 3×3 symmetric matrix, then A^2 is a 3×3 symmetric matrix. (*Hint:* $A^2 = A \cdot A$.)

3.3 Solving Matrix Equations

- Use matrix algebra to solve systems of linear equations
- Use matrices to analyze and interpret real-life data
- Find the inverse of an invertible matrix with and without technology

GETTING STARTED Many students seek to maintain or increase their GPA to be eligible for academic recognition, including scholarships, dean's lists, honor societies, and so on. A common question that students ask is, "If I get straight A's from now on, how high can I raise my GPA?" By using the techniques demonstrated in this section, you will be able to determine your highest possible cumulative GPA.

In this section, we will demonstrate how to find the inverse of a square matrix algebraically. We will also show how to use the inverse and matrix algebra to solve systems of equations. We will conclude by further investigating how matrix algebra may be used in business and consumer applications.

Finding the Inverse of a Matrix Algebraically

Recall that a matrix A is said to be invertible if there exists a matrix A^{-1} such that $AA^{-1} = A^{-1}A = I$. In Section 3.2, we introduced a quick method for finding the inverse of a 2×2 matrix; however, we did not explain how the method was derived. The inverse of an invertible matrix of any size may be determined algebraically by augmenting the matrix with the identity matrix and then row reducing the matrix. For example, if $A = \begin{bmatrix} 1 & 2 \\ 3 & 4 \end{bmatrix}$, we add on the identity matrix $I = \begin{bmatrix} 1 & 0 \\ 0 & 1 \end{bmatrix}$.

$$\begin{bmatrix} 1 & 2 & \vdots & 1 & 0 \\ 3 & 4 & \vdots & 0 & 1 \end{bmatrix} \qquad \text{Augment } A \text{ with } I$$

$$\begin{bmatrix} 1 & 2 & \vdots & 1 & 0 \\ 0 & 2 & \vdots & 3 & -1 \end{bmatrix} \qquad 3R_1 - R_2$$

$$\begin{bmatrix} 1 & 0 & \vdots & -2 & 1 \\ 0 & 2 & \vdots & 3 & -1 \end{bmatrix} \qquad R_1 - R_2$$

$$\begin{bmatrix} 1 & 0 & \vdots & -2 & 1 \\ 0 & 1 & \vdots & \frac{3}{2} & -\frac{1}{2} \end{bmatrix} \qquad \frac{1}{2}R_2$$

When the left-hand side of the augmented matrix is reduced to the identity matrix, the matrix on the right-hand side of the augmented matrix is A^{-1}. We may verify the accuracy of our result by multiplying A by A^{-1}:

$$AA^{-1} = \begin{bmatrix} 1 & 2 \\ 3 & 4 \end{bmatrix} \begin{bmatrix} -2 & 1 \\ \frac{3}{2} & -\frac{1}{2} \end{bmatrix}$$

$$= \begin{bmatrix} 1(-2) + 2\left(\frac{3}{2}\right) & 1(1) + 2\left(-\frac{1}{2}\right) \\ 3(-2) + 4\left(\frac{3}{2}\right) & 3(1) + 4\left(-\frac{1}{2}\right) \end{bmatrix}$$

$$= \begin{bmatrix} -2 + 3 & 1 + (-1) \\ -6 + 6 & 3 + (-2) \end{bmatrix}$$

$$= \begin{bmatrix} 1 & 0 \\ 0 & 1 \end{bmatrix}$$

$$= I$$

Multiplying A^{-1} by A will yield the same result.

We may determine the inverse matrix for all invertible 2×2 matrices by augmenting a generic matrix $A = \begin{bmatrix} a & b \\ c & d \end{bmatrix}$ with the identity matrix $I = \begin{bmatrix} 1 & 0 \\ 0 & 1 \end{bmatrix}$ and row reducing.

$$\begin{bmatrix} a & b & \vdots & 1 & 0 \\ c & d & \vdots & 0 & 1 \end{bmatrix} = \begin{bmatrix} a & b & \vdots & 1 & 0 \\ 0 & ad-bc & \vdots & -c & a \end{bmatrix} \qquad -cR_1 + aR_2$$

$$= \begin{bmatrix} a & b & \vdots & 1 & 0 \\ 0 & 1 & \vdots & \dfrac{-c}{ad-bc} & \dfrac{a}{ad-bc} \end{bmatrix} \qquad \tfrac{1}{ad-bc}R_2$$

$$= \begin{bmatrix} a & 0 & \vdots & 1 - b\left(\dfrac{-c}{ad-bc}\right) & -b\left(\dfrac{a}{ad-bc}\right) \\ 0 & 1 & \vdots & \dfrac{-c}{ad-bc} & \dfrac{a}{ad-bc} \end{bmatrix} \qquad R_1 - bR_2$$

$$= \begin{bmatrix} a & 0 & \vdots & 1 + \dfrac{bc}{ad-bc} & \dfrac{-ab}{ad-bc} \\ 0 & 1 & \vdots & \dfrac{-c}{ad-bc} & \dfrac{a}{ad-bc} \end{bmatrix} \qquad \text{Simplify}$$

$$= \begin{bmatrix} 1 & 0 & \vdots & \dfrac{1}{a}\left(1 + \dfrac{bc}{ad-bc}\right) & \dfrac{1}{a}\left(\dfrac{-ab}{ad-bc}\right) \\ 0 & 1 & \vdots & \dfrac{-c}{ad-bc} & \dfrac{a}{ad-bc} \end{bmatrix} \qquad \tfrac{1}{a}R_1$$

Although the left-hand side of the augmented matrix is I, the right-hand side doesn't yet look like $\dfrac{1}{ad-bc}\begin{bmatrix} d & -b \\ -c & a \end{bmatrix}$. However, we can show that our matrix is equivalent to the general 2×2 inverse matrix with a few additional steps.

$$A^{-1} = \begin{bmatrix} \dfrac{1}{a}\left(1 + \dfrac{bc}{ad-bc}\right) & \dfrac{1}{a}\left(\dfrac{-ab}{ad-bc}\right) \\ \dfrac{-c}{ad-bc} & \dfrac{a}{ad-bc} \end{bmatrix}$$

$$= \begin{bmatrix} \dfrac{1}{a}\left(\dfrac{ad-bc}{ad-bc} + \dfrac{bc}{ad-bc}\right) & \dfrac{-ab}{a(ad-bc)} \\ \dfrac{-c}{ad-bc} & \dfrac{a}{ad-bc} \end{bmatrix}$$

$$= \dfrac{1}{ad-bc}\begin{bmatrix} \dfrac{ad-bc+bc}{a} & \dfrac{-ab}{a} \\ -c & a \end{bmatrix}$$

$$= \dfrac{1}{ad-bc}\begin{bmatrix} d & -b \\ -c & a \end{bmatrix}$$

The process of augmenting a matrix with the identity and row reducing may be used for any size square matrix.

Consider the 3×3 matrix $A = \begin{bmatrix} 2 & 3 & 4 \\ 3 & 4 & 5 \\ 4 & 0 & 6 \end{bmatrix}$. We will calculate the inverse of matrix A by creating the augmented matrix $[A \vdots I]$. Then, using the row operations demonstrated in Section 2.2, we will reduce the left-hand side of the augmented matrix to reduced row echelon form. If the matrix is invertible, the reduced matrix will be $[I \mid A^{-1}]$.

$$[A \mid I] = \begin{bmatrix} 2 & 3 & 4 & 1 & 0 & 0 \\ 3 & 4 & 5 & 0 & 1 & 0 \\ 4 & 0 & 6 & 0 & 0 & 1 \end{bmatrix}$$

We would like to have a leading 1 in column 1 to use to zero out the remaining entries in that column. Observe that $R_2 - R_1 \to R_1$ will give us a leading 1 in row 1.

$$\begin{bmatrix} 1 & 1 & 1 & -1 & 1 & 0 \\ 3 & 4 & 5 & 0 & 1 & 0 \\ 4 & 0 & 6 & 0 & 0 & 1 \end{bmatrix} \qquad R_2 - R_1$$

Note that we performed the row operation on all six entries of row 1. We will now zero out the entries a_{21} and a_{31} with the operations $3R_1 - R_2 \to R_2$ and $4R_1 - R_3 \to R_3$.

$$\begin{bmatrix} 1 & 1 & 1 & -1 & 1 & 0 \\ 0 & -1 & -2 & -3 & 2 & 0 \\ 0 & 4 & -2 & -4 & 4 & -1 \end{bmatrix} \qquad \begin{matrix} 3R_1 - R_2 \\ 4R_1 - R_3 \end{matrix}$$

Recall that we are trying to make the left-hand side look like the identity matrix. Next we will use $-R_2 \to R_2$ to get a 1 in a_{22}.

$$\begin{bmatrix} 1 & 1 & 1 & -1 & 1 & 0 \\ 0 & 1 & 2 & 3 & -2 & 0 \\ 0 & 4 & -2 & -4 & 4 & -1 \end{bmatrix} \qquad -R_2$$

We will now zero out a_{32} with the operation $4R_2 - R_3 \to R_3$.

$$\begin{bmatrix} 1 & 1 & 1 & -1 & 1 & 0 \\ 0 & 1 & 2 & 3 & -2 & 0 \\ 0 & 0 & 10 & 16 & -12 & 1 \end{bmatrix} \qquad 4R_2 - R_3$$

We can create a 1 in a_{33} by multiplying row 3 by 0.1; however, that will introduce decimals on the right-hand side of the matrix, which will make mental computations more difficult. Consequently, we will reduce the matrix as far as possible before creating decimal entries. We will next zero out a_{12} with the operation $R_1 - R_2 \to R_1$.

$$\begin{bmatrix} 1 & 0 & -1 & -4 & 3 & 0 \\ 0 & 1 & 2 & 3 & -2 & 0 \\ 0 & 0 & 10 & 16 & -12 & 1 \end{bmatrix} \qquad R_1 - R_2$$

Observe that the first two columns of the augmented matrix are identical to the first two columns of the identity matrix I_3. We cannot simplify the matrix any further without creating decimal entries. We will proceed by creating a 1 in a_{33} with the operation $0.1R_3 \to R_3$.

$$\begin{bmatrix} 1 & 0 & -1 & -4 & 3 & 0 \\ 0 & 1 & 2 & 3 & -2 & 0 \\ 0 & 0 & 1 & 1.6 & -1.2 & 0.1 \end{bmatrix} \qquad 0.1R_3$$

Finally, we will zero out a_{13} and a_{23} with the operations $R_1 + R_3 \to R_1$ and $R_2 - 2R_3 \to R_2$.

$$\left[\begin{array}{ccc:ccc} 1 & 0 & 0 & -2.4 & 1.8 & 0.1 \\ 0 & 1 & 0 & -0.2 & 0.4 & -0.2 \\ 0 & 0 & 1 & 1.6 & -1.2 & 0.1 \end{array}\right] \begin{array}{l} R_1 + R_3 \\ R_2 - 2R_3 \\ {} \end{array}$$

The matrix on the left is I_3, and the matrix on the right is A^{-1}. That is,

$$A^{-1} = \begin{bmatrix} -2.4 & 1.8 & 0.1 \\ -0.2 & 0.4 & -0.2 \\ 1.6 & -1.2 & 0.1 \end{bmatrix}$$

which may alternatively be written as

$$A^{-1} = 0.1\begin{bmatrix} -24 & 18 & 1 \\ -2 & 4 & -2 \\ 16 & -12 & 1 \end{bmatrix}$$

We can check our work by calculating $A^{-1}A$.

$$A^{-1}A = 0.1\begin{bmatrix} -24 & 18 & 1 \\ -2 & 4 & -2 \\ 16 & -12 & 1 \end{bmatrix}\begin{bmatrix} 2 & 3 & 4 \\ 3 & 4 & 5 \\ 4 & 0 & 6 \end{bmatrix}$$

$$= 0.1\begin{bmatrix} 10 & 0 & 0 \\ 0 & 10 & 0 \\ 0 & 0 & 10 \end{bmatrix} = \begin{bmatrix} 1 & 0 & 0 \\ 0 & 1 & 0 \\ 0 & 0 & 1 \end{bmatrix}$$

Since $A^{-1}A = I_3$, the matrices are inverses of each other.

The demonstrated method will work for any size square matrix. Let's recap the method.

HOW TO Finding the Inverse of a Matrix *A*

1. Augment the matrix A with the identity matrix I.
2. Use row operations to reduce the left-hand side of the augmented matrix to reduced row echelon form.
3. If the left-hand side of the reduced matrix is the identity matrix, the right-hand side is A^{-1}. If the left-hand side of the reduced matrix contains a row of zeros, the matrix A is singular.

EXAMPLE 1

Finding the Inverse of a 3 × 3 Matrix

Find the inverse of the matrix $A = \begin{bmatrix} 3 & 4 & 2 \\ -1 & 0 & -1 \\ 5 & 2 & 5 \end{bmatrix}$, if it exists.

SOLUTION

$$\begin{bmatrix} 3 & 4 & 2 & \vdots & 1 & 0 & 0 \\ -1 & 0 & -1 & \vdots & 0 & 1 & 0 \\ 5 & 2 & 5 & \vdots & 0 & 0 & 1 \end{bmatrix} \qquad \text{Augment } A \text{ with } I$$

$$= \begin{bmatrix} -1 & 0 & -1 & \vdots & 0 & 1 & 0 \\ 3 & 4 & 2 & \vdots & 1 & 0 & 0 \\ 5 & 2 & 5 & \vdots & 0 & 0 & 1 \end{bmatrix} \qquad \begin{matrix} R_2 \\ R_1 \\ {} \end{matrix}$$

$$= \begin{bmatrix} 1 & 0 & 1 & \vdots & 0 & -1 & 0 \\ 0 & 4 & -1 & \vdots & 1 & 3 & 0 \\ 0 & 2 & 0 & \vdots & 0 & 5 & 1 \end{bmatrix} \qquad \begin{matrix} -R_1 \\ 3R_1 + R_2 \\ 5R_1 + R_3 \end{matrix}$$

$$= \begin{bmatrix} 1 & 0 & 1 & \vdots & 0 & -1 & 0 \\ 0 & 0 & -1 & \vdots & 1 & -7 & -2 \\ 0 & 2 & 0 & \vdots & 0 & 5 & 1 \end{bmatrix} \qquad \begin{matrix} {} \\ R_2 - 2R_3 \\ {} \end{matrix}$$

$$= \begin{bmatrix} 1 & 0 & 1 & \vdots & 0 & -1 & 0 \\ 0 & 2 & 0 & \vdots & 0 & 5 & 1 \\ 0 & 0 & -1 & \vdots & 1 & -7 & -2 \end{bmatrix} \qquad \begin{matrix} {} \\ R_3 \\ R_2 \end{matrix}$$

$$= \begin{bmatrix} 1 & 0 & 0 & \vdots & 1 & -8 & -2 \\ 0 & 1 & 0 & \vdots & 0 & 2.5 & 0.5 \\ 0 & 0 & 1 & \vdots & -1 & 7 & 2 \end{bmatrix} \qquad \begin{matrix} R_1 + R_3 \\ 0.5R_2 \\ -R_3 \end{matrix}$$

$$= [I \vdots A^{-1}]$$

$$\text{So } A^{-1} = \begin{bmatrix} 1 & -8 & -2 \\ 0 & 2.5 & 0.5 \\ -1 & 7 & 2 \end{bmatrix}.$$

EXAMPLE 2   **Determining If a 3 × 3 Matrix Is Singular**

Find the inverse of the matrix $A = \begin{bmatrix} 5 & 2 & 6 \\ -2 & 2 & -4 \\ 1 & 6 & -2 \end{bmatrix}$, if it exists.

SOLUTION

$$\begin{bmatrix} 5 & 2 & 6 & \vdots & 1 & 0 & 0 \\ -2 & 2 & -4 & \vdots & 0 & 1 & 0 \\ 1 & 6 & -2 & \vdots & 0 & 0 & 1 \end{bmatrix} \qquad \text{Augment } A \text{ with } I$$

$$= \begin{bmatrix} 1 & 6 & -2 & \vdots & 0 & 0 & 1 \\ -2 & 2 & -4 & \vdots & 0 & 1 & 0 \\ 5 & 2 & 6 & \vdots & 1 & 0 & 0 \end{bmatrix} \qquad \begin{matrix} R_3 \\ {} \\ R_1 \end{matrix}$$

$$= \begin{bmatrix} 1 & 6 & -2 & \vdots & 0 & 0 & 1 \\ 0 & 14 & -8 & \vdots & 0 & 1 & 2 \\ 0 & 28 & -16 & \vdots & -1 & 0 & 5 \end{bmatrix} \qquad \begin{matrix} {} \\ 2R_1 + R_2 \\ 5R_1 - R_3 \end{matrix}$$

$$= \begin{bmatrix} 1 & 6 & -2 & \vdots & 0 & 0 & 1 \\ 0 & 14 & -8 & \vdots & 0 & 1 & 2 \\ 0 & 0 & 0 & \vdots & 1 & 2 & -1 \end{bmatrix} \qquad \begin{matrix} {} \\ {} \\ 2R_2 - R_3 \end{matrix}$$

The left-hand side of the augmented matrix has a row of zeros. As a result, in reduced row echelon form, it will not be the identity matrix. Consequently, A is singular.

The TI-83 Plus calculator has a built-in function that allows you to easily calculate the inverse of an invertible matrix. This feature should be used primarily to check your work for accuracy in larger matrices (3×3 and above). For the Technology Tip, we will find the inverse of $A = \begin{bmatrix} 6 & 14 & 16 \\ 4 & 1 & 14 \\ 1 & 4 & 6 \end{bmatrix}$.

TECHNOLOGY **TIP**

Finding the Inverse of a Matrix

1. Enter matrix A into the calculator, using the Matrix Editor. Use the Matrix Names menu to place matrix A on the home screen. Then press the $\boxed{x^{-1}}$ button.

```
[A]⁻¹
```

2. Press $\boxed{\text{ENTER}}$ to display the inverse of the matrix.

```
[A]⁻¹
[[.25    .1   -.9…
 [.05   -.1  .1…
 [-.075 .05  .25…
```

3. To convert the entries to fractions, press $\boxed{\text{MATH}}$, then 1:▶Frac. Press $\boxed{\text{ENTER}}$.

```
[[.25    .1   -.9…
 [.05   -.1  .1…
 [-.075 .05  .25…
Ans▶Frac
[[1/4    1/10  -…
 [1/20  -1/10 1…
 [-3/40 1/20  1…
```

Error Alert:
If, when attempting to find the inverse of a matrix, a singular matrix error occurs, double-check to make sure that you have entered the matrix correctly. If you have, this error tells you that the matrix is singular (not invertible).

```
ERR:SINGULAR MAT
1:Quit
2:Goto
```

EXAMPLE 3

Using Technology to Find the Inverse of a Square Matrix

Use technology to find the inverse of the matrix $A = \begin{bmatrix} 6 & 14 & 16 \\ 4 & 1 & 14 \\ 1 & 4 & 6 \end{bmatrix}$.

SOLUTION

Using technology, we determine that $A^{-1} = \begin{bmatrix} 0.25 & 0.1 & -0.9 \\ 0.05 & -0.1 & 0.1 \\ -0.075 & 0.05 & 0.25 \end{bmatrix}$. The in-

verse matrix may alternatively be written as

$$A^{-1} = \begin{bmatrix} \frac{1}{4} & \frac{1}{10} & -\frac{9}{10} \\ \frac{1}{20} & -\frac{1}{10} & \frac{1}{10} \\ -\frac{3}{40} & \frac{1}{20} & \frac{1}{4} \end{bmatrix} \quad \text{or} \quad A^{-1} = \frac{1}{40}\begin{bmatrix} 10 & 4 & -36 \\ 2 & -4 & 4 \\ -3 & 2 & 10 \end{bmatrix}$$

Solving Systems of Linear Equations Using Matrix Equations

We will now return to the process of solving systems of linear equations, introduced in Chapter 2, with the added capability of matrix algebra. Consider the system of linear equations

$$\begin{aligned} x - 4y - z &= -5 \\ 3y + z &= 7 \\ 2x + y \quad &= 10 \end{aligned}$$

Let's define A to be the coefficient matrix of the system. That is, the entries of A are the coefficients of the variables of the system.

$$A = \begin{bmatrix} 1 & -4 & -1 \\ 0 & 3 & 1 \\ 2 & 1 & 0 \end{bmatrix}$$

Let's define a column matrix X to be the variable matrix of the system. That is, the entries of X are the variables of the system.

$$X = \begin{bmatrix} x \\ y \\ z \end{bmatrix}$$

Finally, let's define the column matrix B to be the constant matrix of the system. That is, the entries of B are the constants from the right-hand side of the equal sign of the system:

$$B = \begin{bmatrix} -5 \\ 7 \\ 10 \end{bmatrix}$$

Consider the matrix product AX.

$$AX = \begin{bmatrix} 1 & -4 & -1 \\ 0 & 3 & 1 \\ 2 & 1 & 0 \end{bmatrix} \begin{bmatrix} x \\ y \\ z \end{bmatrix}$$

$$= \begin{bmatrix} 1x - 4y - 1z \\ 0x + 3y + 1z \\ 2x + 1y + 0z \end{bmatrix}$$

But from the system of equations we know that

$$1x - 4y - 1z = -5$$
$$0x + 3y + 1z = 7$$
$$2x + 1y + 0z = 10$$

So

$$AX = \begin{bmatrix} -5 \\ 7 \\ 10 \end{bmatrix}$$

$$= B$$

Therefore, a system of linear equations may be represented by the matrix equation $AX = B$. The solution to the system of equations is given by the matrix X. Is there a way to solve the matrix equation for X? Let's left multiply both sides by the matrix A^{-1}. (Since matrix multiplication is not commutative, we use the terms *left multiply* and *right multiply* to designate on which side to place the matrix being inserted into the equation.)

$$AX = B$$
$$A^{-1}(AX) = A^{-1}B \qquad \text{Left multiply by } A^{-1}$$
$$(A^{-1}A)X = A^{-1}B \qquad \text{Associative Property}$$
$$IX = A^{-1}B \qquad A^{-1}A = I$$
$$X = A^{-1}B \qquad IX = X$$

So the product of the inverse of the coefficient matrix and the constant matrix is the solution matrix. Using the algebraic or technological methods previously introduced, it may be shown that the inverse of $A = \begin{bmatrix} 1 & -4 & -1 \\ 0 & 3 & 1 \\ 2 & 1 & 0 \end{bmatrix}$ is

$A^{-1} = \begin{bmatrix} \frac{1}{3} & \frac{1}{3} & \frac{1}{3} \\ -\frac{2}{3} & -\frac{2}{3} & \frac{1}{3} \\ 2 & 3 & -1 \end{bmatrix}$. Thus, the solution to the matrix equation

$$\begin{bmatrix} 1 & -4 & -1 \\ 0 & 3 & 1 \\ 2 & 1 & 0 \end{bmatrix} \begin{bmatrix} x \\ y \\ z \end{bmatrix} = \begin{bmatrix} -5 \\ 7 \\ 10 \end{bmatrix} \text{ is the product of } A^{-1} \text{ and } B.$$

$$X = A^{-1}B$$

$$= \begin{bmatrix} \frac{1}{3} & \frac{1}{3} & \frac{1}{3} \\ -\frac{2}{3} & -\frac{2}{3} & \frac{1}{3} \\ 2 & 3 & -1 \end{bmatrix} \begin{bmatrix} -5 \\ 7 \\ 10 \end{bmatrix}$$

$$= \begin{bmatrix} 4 \\ 2 \\ 1 \end{bmatrix}$$

So $x = 4$, $y = 2$, and $z = 1$.

This method of solving systems of linear equations is extremely efficient and works well for large systems of linear equations with unique solutions. Unfortunately, if a system has multiple solutions, this method cannot be used because the coefficient matrix will not be invertible.

EXAMPLE **4**   **Solving a System of Equations Using Matrix Algebra**

Write the system of equations as a matrix equation and solve.

$$\begin{aligned} x + \ y + \ z &= 7 \\ 3x - \ y + \ z &= 21 \\ -x + 2y + 2z &= 2 \end{aligned}$$

SOLUTION The system of equations is equivalent to the matrix equation

$$\begin{bmatrix} 1 & 1 & 1 \\ 3 & -1 & 1 \\ -1 & 2 & 2 \end{bmatrix} \begin{bmatrix} x \\ y \\ z \end{bmatrix} = \begin{bmatrix} 7 \\ 21 \\ 2 \end{bmatrix}$$

The solution to the system is given by

$$\begin{bmatrix} x \\ y \\ z \end{bmatrix} = \begin{bmatrix} 1 & 1 & 1 \\ 3 & -1 & 1 \\ -1 & 2 & 2 \end{bmatrix}^{-1} \begin{bmatrix} 7 \\ 21 \\ 2 \end{bmatrix}$$

Using technology, we determine that

$$\begin{bmatrix} 1 & 1 & 1 \\ 3 & -1 & 1 \\ -1 & 2 & 2 \end{bmatrix}^{-1} = \begin{bmatrix} \frac{2}{3} & 0 & -\frac{1}{3} \\ \frac{7}{6} & -\frac{1}{2} & -\frac{1}{3} \\ -\frac{5}{6} & \frac{1}{2} & \frac{2}{3} \end{bmatrix}$$

Therefore,

$$
\begin{bmatrix} x \\ y \\ z \end{bmatrix} = \begin{bmatrix} 1 & 1 & 1 \\ 3 & -1 & 1 \\ -1 & 2 & 2 \end{bmatrix}^{-1} \begin{bmatrix} 7 \\ 21 \\ 2 \end{bmatrix}
$$

$$
= \begin{bmatrix} \frac{2}{3} & 0 & -\frac{1}{3} \\ \frac{7}{6} & -\frac{1}{2} & -\frac{1}{3} \\ -\frac{5}{6} & \frac{1}{2} & \frac{2}{3} \end{bmatrix} \begin{bmatrix} 7 \\ 21 \\ 2 \end{bmatrix}
$$

$$
= \begin{bmatrix} \frac{14}{3} + 0 + \left(-\frac{2}{3}\right) \\ \frac{49}{6} + \left(-\frac{21}{2}\right) + \left(-\frac{2}{3}\right) \\ -\frac{35}{6} + \frac{21}{2} + \frac{4}{3} \end{bmatrix}
$$

$$
= \begin{bmatrix} \frac{12}{3} \\ \frac{49}{6} + \left(-\frac{63}{6}\right) + \left(-\frac{4}{6}\right) \\ -\frac{35}{6} + \frac{63}{6} + \frac{8}{6} \end{bmatrix}
$$

$$
= \begin{bmatrix} 4 \\ -\frac{18}{6} \\ \frac{36}{6} \end{bmatrix}
$$

$$
= \begin{bmatrix} 4 \\ -3 \\ 6 \end{bmatrix}
$$

So $x = 4$, $y = -3$, and $z = 6$ is the solution to the system of equations.

Analyzing Real-Life Data with Matrices

Throughout this chapter, we have focused on developing your matrix algebra skills. With a well-developed set of matrix algebra skills, you will be able to set up matrix equations with relative ease. Since the entries in the matrices in many real-life situations are decimal numbers, we will often use technology to determine the desired solution.

EXAMPLE 5

Using Matrix Algebra to Make Informed Investment Decisions

Financial advisors often counsel their clients to diversify their investments into a variety of accounts with different levels of performance and risk. The average annual return (over the 10-year period prior to June 30, 2002) of two mutual funds offered by Harbor Fund is shown in Table 3.12.

TABLE 3.12

	Average Annual Return
Capital Appreciation Fund	12.69%
Bond Fund	7.97%

Source: Harbor Fund account statement.

High-performance accounts typically have greater volatility than lower-performance accounts. For example, although the Capital Appreciation Fund has a higher average annual return over the 10-year period, it earned -23.42 percent in a recent year. The Bond Fund has a lower average annual return over the 10-year period, but it earned 10.86 percent in the same year that the Capital Appreciation Fund suffered the 23.42 percent loss.

An investor has $1,000 to invest in the two accounts. Assuming that the accounts will earn the returns specified in the table over the next year, how much should she invest in each account if she wants to earn 8 percent, 10 percent, or 12 percent?

SOLUTION Let x be the amount invested in the Capital Appreciation Fund and y be the amount invested in the Bond Fund. Since the sum of the individual investments is $1000, we have

$$x + y = 1000$$

The annual return on each account is the product of the rate of return and the amount of money invested in the account. For the Capital Appreciation Fund, the annual return is given by $0.1269x$. For the Bond Fund, the annual return is given by $0.0797y$. The combined return on the two accounts is given by the expression $0.1269x + 0.0797y$. If we let r be the desired rate of return on the $1000 investment, then $1000r$ is the dollar amount of the return. Since these two expressions must be equal, we have

$$0.1269x + 0.0797y = 1000r$$

For the 8 percent return, $r = 0.08$. Therefore,

$$0.1269x + 0.0797y = 1000(0.08)$$
$$0.1269x + 0.0797y = 80$$

Consequently, we have the system of equations

$$x + y = 1000 \qquad \text{Total investment}$$
$$0.1269x + 0.0797y = 80 \qquad \text{Total return on investment}$$

and the corresponding matrix equation

$$
\begin{matrix} A & X & B \end{matrix}
$$
$$
\begin{bmatrix} 1 & 1 \\ 0.1269 & 0.0797 \end{bmatrix} \begin{bmatrix} x \\ y \end{bmatrix} = \begin{bmatrix} 1000 \\ 80 \end{bmatrix}
$$

Similarly, for the 10 percent return, we have the system of equations

$$x + y = 1000 \qquad \text{Total investment}$$
$$0.1269x + 0.0797y = 100 \qquad \text{Total return on investment}$$

and the corresponding matrix equation

$$
\begin{matrix} A & X & B \end{matrix}
$$
$$
\begin{bmatrix} 1 & 1 \\ 0.1269 & 0.0797 \end{bmatrix} \begin{bmatrix} x \\ y \end{bmatrix} = \begin{bmatrix} 1000 \\ 100 \end{bmatrix}
$$

Likewise, for the 12 percent return, we have the system of equations

$$x + y = 1000 \qquad \text{Total investment}$$
$$0.1269x + 0.0797y = 120 \qquad \text{Total return on investment}$$

and the corresponding matrix equation

$$\begin{matrix} A & X & B \end{matrix}$$

$$\begin{bmatrix} 1 & 1 \\ 0.1269 & 0.0797 \end{bmatrix} \begin{bmatrix} x \\ y \end{bmatrix} = \begin{bmatrix} 1000 \\ 120 \end{bmatrix}$$

Notice that although the constant matrices of the three matrix equations differ, each of the equations has the same coefficient matrix A. We know that the solution to the matrix equation $AX = B$ is $X = A^{-1}B$. The inverse of A is given by

$$A^{-1} \approx \begin{bmatrix} -1.689 & 21.19 \\ 2.689 & -21.19 \end{bmatrix}$$

We will use the inverse matrix shown on our calculator, instead of the rounded matrix entries shown here, to increase the accuracy of our results.

Therefore, for the 8 percent return, we have

$$X = A^{-1}B$$

$$= \begin{bmatrix} -1.689 & 21.19 \\ 2.689 & -21.19 \end{bmatrix} \begin{bmatrix} 1000 \\ 80 \end{bmatrix}$$

$$= \begin{bmatrix} 6.36 \\ 993.64 \end{bmatrix}$$

She should invest $6.36 in the Capital Appreciation Fund and $993.64 in the Bond Fund in order to earn an 8 percent return.

For the 10 percent return, we have

$$X = A^{-1}B$$

$$= \begin{bmatrix} -1.689 & 21.19 \\ 2.689 & -21.19 \end{bmatrix} \begin{bmatrix} 1000 \\ 100 \end{bmatrix}$$

$$= \begin{bmatrix} 430.08 \\ 569.92 \end{bmatrix}$$

She should invest $430.08 in the Capital Appreciation Fund and $569.92 in the Bond Fund in order to earn a 10 percent return.

For the 12 percent return, we have

$$X = \begin{bmatrix} -1.689 & 21.19 \\ 2.689 & -21.19 \end{bmatrix} \begin{bmatrix} 1000 \\ 120 \end{bmatrix}$$

$$= \begin{bmatrix} 853.81 \\ 146.19 \end{bmatrix}$$

She should invest $853.81 in the Capital Appreciation Fund and $146.19 in the Bond Fund in order to earn a 12 percent return.

Recall that we *assumed* that she would earn the 10-year average annual return on all of her investments. Since none of the returns are guaranteed, which blend of investments she decides to choose will depend upon her risk tolerance.

In Example 5, we could have solved each matrix equation individually without using the inverse; however, since each equation had the same coefficient matrix, we greatly reduced our computations by using the inverse of the coefficient matrix to help find the solution.

Using Matrix Algebra to Forecast a Grade Point Average

Students at Green River Community College must earn 90 credits to obtain an Associate of Arts degree. Three friends discover that they all have the same grade point average (GPA), even though they have earned different numbers of credits. They each have a 3.42 GPA, and they hope to raise their GPAs to 3.70 by the time they have earned 90 credits. The first student has earned 45 credits; the second student, 65 credits; and the third student, 75 credits. Is it possible for all of the students to increase their cumulative GPAs to 3.7 by the time they obtain 90 credits?

SOLUTION   Let x be the number of credits a student has earned and y be the number of credits remaining. We know that for each student,

$$x + y = 90$$

In order to have a 3.7 cumulative GPA, each student must earn $3.7(90) = 333$ grade points. Each student has earned $3.42x$ grade points from the courses already taken. Assuming that 4.00 is the highest grade a student can earn in any course, the maximum number of grade points that can be earned on the remaining credits is $4.00y$. Combining these two expressions, we have

$$3.42x + 4.00y = 333$$

Setting up and solving the matrix equation for the system of equations, we get

$$x + y = 90 \qquad \text{Total number of credits}$$
$$3.42x + 4.00y = 333 \qquad \text{Total number of grade points}$$

$$\begin{bmatrix} 1 & 1 \\ 3.42 & 4.00 \end{bmatrix} \begin{bmatrix} x \\ y \end{bmatrix} = \begin{bmatrix} 90 \\ 333 \end{bmatrix}$$

$$\begin{bmatrix} x \\ y \end{bmatrix} = \begin{bmatrix} 1 & 1 \\ 3.42 & 4.00 \end{bmatrix}^{-1} \begin{bmatrix} 90 \\ 333 \end{bmatrix}$$

$$\begin{bmatrix} x \\ y \end{bmatrix} \approx \begin{bmatrix} 6.90 & -1.72 \\ -5.90 & 1.72 \end{bmatrix} \begin{bmatrix} 90 \\ 333 \end{bmatrix}$$

$$\begin{bmatrix} x \\ y \end{bmatrix} \approx \begin{bmatrix} 46.55 \\ 43.45 \end{bmatrix}$$

(Although we rounded A^{-1} when we wrote it down, we used the more accurate calculator result in our computations to come up with $X = \begin{bmatrix} 46.55 \\ 43.45 \end{bmatrix}$.)

If a student has earned fewer than 46.55 credits, it is possible for her to raise her GPA to 3.7 by the time she has earned 90 credits. If she has earned more than 46.55 credits, it is mathematically impossible for her to raise her GPA to 3.7 by the time she has earned 90 credits. In the case of the three friends, only the student with 45 credits could conceivably raise her GPA to 3.7. To do so, she must earn a 4.0 grade on the majority of her remaining coursework.

EXAMPLE 7   **Using Matrix Algebra to Calculate Management Bonuses**

A small company rewards its upper management by offering annual bonuses. Each executive receives a percentage of the profits that remain after the bonuses of all the executives have been deducted from the company's profits. The CEO receives 4 percent; the CFO, 4 percent; and the vice president, 2 percent. What will be the bonus amount of each executive if the company's annual profit is $200,000? If it is $300,000? If it is $500,000?

SOLUTION Let x be the CEO's bonus, y be the CFO's bonus, z be the vice president's bonus, and P be the annual profit of the company. We know that the amount of profit remaining after the executives' bonuses are deducted is $P - (x + y + z)$. Since each executive receives a different percentage of this quantity, we end up with three different equations.

$$x = 0.04[P - (x + y + z)] \qquad \text{CEO receives 4.0\%}$$
$$y = 0.04[P - (x + y + z)] \qquad \text{CFO receives 4.0\%}$$
$$z = 0.02[P - (x + y + z)] \qquad \text{Vice president receives 2.0\%}$$

Each equation must be simplified before we can write the matrix equation.

$$x = 0.04[P - (x + y + z)]$$
$$x = 0.04P - 0.04x - 0.04y - 0.04z$$
$$1.04x + 0.04y + 0.04z = 0.04P$$
$$\frac{1.04}{0.04}x + y + z = P$$
$$26x + y + z = P$$

$$y = 0.04[P - (x + y + z)]$$
$$y = 0.04P - 0.04x - 0.04y - 0.04z$$
$$0.04x + 1.04y + 0.04z = 0.04P$$
$$x + \frac{1.04}{0.04}y + z = P$$
$$x + 26y + z = P$$

$$z = 0.02[P - (x + y + z)]$$
$$z = 0.02P - 0.02x - 0.02y - 0.02z$$
$$0.02x + 0.02y + 1.02z = 0.02P$$
$$x + y + \frac{1.02}{0.02}z = P$$
$$x + y + 51z = P$$

Combining the simplified equations yields the system of equations

$$26x + y + z = P$$
$$x + 26y + z = P$$
$$x + y + 51z = P$$

and the corresponding matrix equation

$$
\begin{array}{ccc}
A & X & B
\end{array}
$$

$$
\begin{bmatrix} 26 & 1 & 1 \\ 1 & 26 & 1 \\ 1 & 1 & 51 \end{bmatrix} \begin{bmatrix} x \\ y \\ z \end{bmatrix} = \begin{bmatrix} P \\ P \\ P \end{bmatrix}
$$

Using technology, we determine that

$$
A^{-1} = \begin{bmatrix} 0.03855 & -0.001455 & -0.0007273 \\ -0.001455 & 0.03855 & -0.0007273 \\ -0.0007273 & -0.0007273 & 0.01964 \end{bmatrix}
$$

To minimize round-off error, we store this matrix on our calculator before proceeding further.

The solution to the matrix equation is

$$
X = A^{-1}B
$$

$$
= \begin{bmatrix} 0.03855 & -0.001455 & -0.0007273 \\ -0.001455 & 0.03855 & -0.0007273 \\ -0.0007273 & -0.0007273 & 0.01964 \end{bmatrix} \begin{bmatrix} P \\ P \\ P \end{bmatrix}
$$

$$
= P \begin{bmatrix} 0.03855 & -0.001455 & -0.0007273 \\ -0.001455 & 0.03855 & -0.0007273 \\ -0.0007273 & -0.0007273 & 0.01964 \end{bmatrix} \begin{bmatrix} 1 \\ 1 \\ 1 \end{bmatrix} \qquad \text{Since } \begin{bmatrix} P \\ P \\ P \end{bmatrix} = P \begin{bmatrix} 1 \\ 1 \\ 1 \end{bmatrix}
$$

$$
= P \begin{bmatrix} 0.03636 \\ 0.03636 \\ 0.01818 \end{bmatrix} = P \begin{bmatrix} \frac{2}{55} \\ \frac{2}{55} \\ \frac{1}{55} \end{bmatrix} = \begin{bmatrix} \frac{2}{55}P \\ \frac{2}{55}P \\ \frac{1}{55}P \end{bmatrix}
$$

The CEO and CFO each receive $\frac{2}{55}$ of the annual profits. The vice president receives $\frac{1}{55}$ of the annual profits. Table 3.13 shows each bonus amount for the identified profit levels.

TABLE 3.13

Profit (P)	$\frac{2}{55}P$	$\frac{1}{55}P$
$200,000	$7,272.73	$3,636.36
$300,000	$10,909.09	$5,454.55
$500,000	$18,181.82	$9,090.91

3.3 Summary

In this section, you learned how to find the inverse of a matrix algebraically. You also used the inverse matrix to solve systems of equations, and you saw how matrix algebra may be used in both business and consumer applications.

3.3 Exercises

In Exercises 1–10, find the inverse of the matrix A algebraically. If A is singular, so state.

1. $A = \begin{bmatrix} 2 & 5 & 3 \\ 1 & 4 & 2 \\ 0 & 2 & 4 \end{bmatrix}$
2. $A = \begin{bmatrix} 4 & 2 & 1 \\ 5 & 3 & 2 \\ 0 & 2 & 3 \end{bmatrix}$

3. $A = \begin{bmatrix} 4 & 0 & 2 \\ 2 & 1 & 2 \\ 3 & 2 & 4 \end{bmatrix}$
4. $A = \begin{bmatrix} 3 & 5 & 1 \\ 0 & 3 & 1 \\ 2 & 9 & 3 \end{bmatrix}$

5. $A = \begin{bmatrix} 6 & 1 & 0 \\ 5 & 4 & 8 \\ 0 & 1 & 2 \end{bmatrix}$
6. $A = \begin{bmatrix} -1 & 0 & 0 \\ -1 & 1 & 1 \\ 0 & 1 & 0 \end{bmatrix}$

7. $A = \begin{bmatrix} 1 & 1 & 0 \\ 0 & 1 & 1 \\ 1 & 0 & -1 \end{bmatrix}$
8. $A = \begin{bmatrix} 2 & 3 & 3 \\ 8 & 2 & 1 \\ 7 & 1 & 0 \end{bmatrix}$

9. $A = \begin{bmatrix} 2 & 4 & 6 \\ 3 & 6 & 9 \\ 1 & 7 & 5 \end{bmatrix}$
10. $A = \begin{bmatrix} 2 & 0 & 0 \\ 8 & 1 & 0 \\ 4 & 1 & 1 \end{bmatrix}$

In Exercises 11–20, write the system of equations as a matrix equation, AX = B, and solve the equation, either algebraically or by using technology.

11.
$\begin{aligned} x + y + z &= 6 \\ 2x - y + z &= 3 \\ 4x - 2y + 3z &= 9 \end{aligned}$
12.
$\begin{aligned} 5x + y + z &= 1 \\ 4x - 2y + 5z &= -2 \\ x - 7y + 6z &= -7 \end{aligned}$

13.
$\begin{aligned} 3x + 2y + 3z &= 6 \\ 2x - 5y + z &= -11 \\ 4x + 2y + 3z &= 3 \end{aligned}$
14.
$\begin{aligned} x + y + z &= 1 \\ 2x - y + z &= 4 \\ 4x - 2y + 3z &= 9 \end{aligned}$

15.
$\begin{aligned} x + y + z &= 3 \\ 2x - y + z &= 13 \\ 4x - 2y + 3z &= 28 \end{aligned}$
16.
$\begin{aligned} 3x - y + 2z &= 17 \\ 4x + 3y + 5z &= 12 \\ 6x + 8y + 4z &= 4 \end{aligned}$

17.
$\begin{aligned} 3x - y + 2z &= 12 \\ 4x + 3y + 5z &= 45 \\ 6x + 8y + 4z &= 64 \end{aligned}$
18.
$\begin{aligned} 2x + 5y + 3z &= 10 \\ x + 4y + 2z &= 7 \\ 0x + 2y + 4z &= 6 \end{aligned}$

19.
$\begin{aligned} 2x + 5y + 3z &= 0 \\ x + 4y + 2z &= -1 \\ 0x + 2y + 4z &= 2 \end{aligned}$
20.
$\begin{aligned} x + y + z &= 1 \\ 2x - y + z &= 1 \\ 4x - 2y + 3z &= 3 \end{aligned}$

In Exercises 21–30, use technology to find the inverse of the matrix, if it exists. If the matrix is singular, so state.

21. $A = \begin{bmatrix} 1 & 1 & 2 \\ 3 & 5 & 8 \\ 13 & 21 & 34 \end{bmatrix}$

22. $B = \begin{bmatrix} 1.0 & 5.2 & -2.8 \\ 0 & 4.0 & 9.4 \\ 0 & 0 & 0.5 \end{bmatrix}$

23. $C = \begin{bmatrix} 3.4 & 1.2 & 0.5 \\ 1.8 & 4.2 & 0 \\ 0.3 & 0 & 0 \end{bmatrix}$

24. $D = \begin{bmatrix} 1 & 1 & 0 \\ 0 & 1 & 1 \\ 1 & 0 & 1 \end{bmatrix}$

25. $E = \begin{bmatrix} 1 & 2 & 3 & 4 \\ 2 & 3 & 4 & 5 \\ 3 & 0 & 5 & 6 \\ 4 & 0 & 0 & 7 \end{bmatrix}$

26. $A = \begin{bmatrix} 3 & 2 & 1 \\ 2 & 1 & 3 \\ 1 & 3 & 2 \end{bmatrix}$

27. $B = \begin{bmatrix} 2.0 & 9.2 & -2.9 \\ 6.1 & 4.7 & 1.4 \\ 0.0 & 2.0 & 3.5 \end{bmatrix}$

28. $D = \begin{bmatrix} 1.1 & 2.2 & 3.3 \\ -1.1 & 2.2 & 1.1 \\ 0.0 & 4.4 & 1.1 \end{bmatrix}$

29. $D = \begin{bmatrix} 1 & 1 & 0 & 2 \\ 0 & 1 & 1 & 2 \\ 1 & 0 & 1 & 2 \\ 2 & 1 & 0 & 3 \end{bmatrix}$

30. $E = \begin{bmatrix} 1 & 7 & 8 & 0 \\ 7 & 2 & 6 & 9 \\ 8 & 6 & 3 & 5 \\ 0 & 9 & 5 & 4 \end{bmatrix}$

In Exercises 31–36, determine the solution by setting up and solving the matrix equation.

31.   **Nutritional Content**   A nut distributor wants to determine the nutritional content of various mixtures of pecans (oil-roasted, salted), cashews (dry-roasted, salted), and almonds (honey-roasted, unblanched). Her supplier has provided the following nutrition information:

	Almonds	Cashews	Pecans
Protein	26.2 grams/ cup	21.0 grams/ cup	10.1 grams/ cup
Carbohydrates	40.2 grams/ cup	44.8 grams/ cup	14.3 grams/ cup
Fat	71.9 grams/ cup	63.5 grams/ cup	82.8 grams/ cup

Source: www.Nutri-facts.com.

Her first mixture, Protein Blend, contains 6 cups of almonds, 3 cups of cashews, and 1 cup of pecans. Her second mixture, Low-Fat Mix, contains 3 cups of almonds, 6 cups of cashews, and 1 cup of pecans. Her final mixture, Low-Carb Mix, contains 3 cups of almonds, 1 cup of cashews, and 6 cups of pecans. Determine the amount of protein, carbohydrates, and fat in a 1-cup serving of each of the mixtures.

32.   **Floral Costs**   A florist purchases her flowers from an online flower wholesaler. White daisies are $3.38 per bunch, football mums are $7.40 per bunch, super blue purple statice is $4.25 per bunch, and misty blue limonium is $5.25 per bunch. (**Source:** www.FlowerSales.com.) From these flowers, she will make three types of bouquets.

	Type 1	Type 2	Type 3
Daisies	1 bunch	1 bunch	2 bunches
Mums	1 bunch	1 bunch	None
Statice	1/2 bunch	None	1/2 bunch
Limonium	None	1/2 bunch	None

What is her flower cost for each type of bouquet? How much should she charge for each bouquet if her markup is 50 percent of her flower cost?

33.   **Return on Investments**   The average annual return (over the 10-year period prior to June 30, 2002) of two mutual funds offered by Harbor Fund is shown in the table.

	Average Annual Return
Money Market	4.50%
Large-Cap Value	11.01%

Source: Harbor Fund.

Suppose you have $2,000 to invest in these two accounts. Assuming that the accounts will earn the returns specified in the table over the next year, how much should you invest in each account if you want to earn 6 percent, 8 percent, or 10 percent?

34.   **Nutritional Content**   A nut distributor wants to determine the nutritional content of various mixtures of pecans (oil-roasted, salted), cashews (dry-roasted, salted), and almonds (honey-roasted, unblanched). Her supplier has provided the following nutrition information.

	Almonds	Cashews	Pecans
Protein	26.2 grams/ cup	21.0 grams/ cup	10.1 grams/ cup
Sugars	20.5 grams/ cup	40.7 grams/ cup	3.9 grams/ cup
Fiber	19.7 grams/ cup	4.1 grams/ cup	10.4 grams/ cup

Source: www.Nutri-facts.com.

Her first mixture, Protein Blend, contains 6 cups of almonds, 3 cups of cashews, and 1 cup of pecans. Her second mixture, Low-Sugar Mix, contains 2 cups of almonds, 1 cup of cashews, and 7 cups of pecans. Her final mixture, High-Fiber Mix, contains 5 cups of almonds, 1 cup of cashews, and 4 cups of pecans. Determine the amount of protein, sugar, and fiber in each 1-cup serving of the mixtures.

35.   **Grade Point Average**   A student at Green River Community College must earn 90 credits to obtain an Associate of Arts degree. Three students with a 2.9 existing GPA hope to increase their cumulative GPA to 3.5. The first student has earned 30 credits; the second, 45 credits; and the third, 55 credits. Is it possible for all of the students to increase their cumulative GPA to a 3.5 GPA by the time they obtain 90 credits?

36. **Executive Bonus Plan**   A small company rewards its upper management by offering annual bonuses. Each executive receives a percentage of the profits that remain after the bonuses of all of the executives have been deducted from the company's profits. The CEO receives 5 percent; the CFO, 4 percent; and the vice president, 3 percent. What will be the bonus amount of each executive if the company's annual profit is $500,000? $800,000? $1,000,000?

Exercises 37– 46 are intended to challenge your understanding of the process of finding matrix inverses algebraically.

37. Find the inverse of the matrix
$$A = \begin{bmatrix} 1 & 0 & 0 & a \\ 0 & 1 & 1 & 0 \\ 0 & 1 & 0 & 1 \\ 0 & 0 & 1 & 1 \end{bmatrix}.$$

38. Under what conditions is the matrix
$$A = \begin{bmatrix} 1 & 0 & a \\ 0 & 1 & 0 \\ a & 0 & a \end{bmatrix} \text{ singular?}$$

39. Determine if the following statement is true or false: The matrix $A = \begin{bmatrix} 1 & a & 1 \\ 0 & a & 0 \\ 1 & -a & 1 \end{bmatrix}$ is singular for any value of a.

40. Determine if the following statement is true or false: The matrix $A = \begin{bmatrix} 1 & a & b \\ 0 & 1 & c \\ 0 & 0 & 1 \end{bmatrix}$ is invertible for all values of a, b, and c.

41. Determine if the following statement is true or false: The matrix $A = \begin{bmatrix} a & 1 & b \\ 0 & 1 & c \\ 0 & 0 & 1 \end{bmatrix}$ is invertible for all values of a, b, and c.

42. Show that if A is invertible, then A^2 is invertible. [*Hint:* $(AB)^{-1} = B^{-1}A^{-1}.$]

43. Show that if A and B are invertible and AB is defined, then AB is invertible. [*Hint:* $(AB)^{-1} = B^{-1}A^{-1}.$]

44. Find three different 2×2 matrices A with the property that $A^2 = I$.

45. Find two different 3×3 matrices that are their own inverses. That is, $A \cdot A = I$.

46. Given $A^2 = I$, write A^{-1} in terms of A.

3.4 Leontief Input-Output Models

- Do basic input-output analysis using Leontief input-output models

GETTING STARTED   Economists typically divide the economies they analyze into industrial sectors. For example, a primitive economy may consist of energy, manufacturing, and agriculture sectors. All of the sectors are interrelated. The energy industry must produce enough energy to meet its own needs as well as the energy needs of the manufacturing and agriculture sectors. Similarly, the manufacturing industry must produce enough manufactured goods to meet its own needs as well as the manufactured goods needs of the energy and agriculture industries.

The economist Wassily Leontief earned the Nobel Prize in Economics in 1973, in part, for his innovative approach to analyzing economies. Leontief studied the inter-relationship among 500 different sectors in the U.S. economy and laid the groundwork for the field of mathematics referred to as **input-output analysis.** In this section, we will demonstrate how to do basic input-output analysis.

A primitive economy consists of two sectors: energy and manufacturing. A unit of a product is interpreted to be $1 worth of that product. To produce 1 unit of energy requires 0.1 unit of energy and 0.3 unit of manufactured products. To produce 1 unit of manufactured products requires 0.3 unit of energy and 0.6 unit of manufactured products. These are the economy's *internal demands*. The economy also demands a surplus of energy and manufactured products. This demand is referred to as an *external demand*. A $27,000 surplus of energy and an $81,000 surplus of manufactured goods are required. How many units of energy and manufacturing does the economy need to produce in order to meet its internal and external demands?

Let x be the number of units of energy produced and y be the number of units of manufactured products produced. To produce $1 of energy costs $0.10 in energy and $0.30 in manufactured products. That is, $0.1x + 0.3y$ is the input required to meet the internal demand for energy. Similarly, to produce $1 of manufactured products costs $0.30 of energy and $0.60 of manufactured goods. That is, $0.3x + 0.6y$ is the input required to meet the internal demand for manufactured goods. In addition to meeting the internal demands, the energy and manufacturing sectors must meet the external demands for their products. We have

Total output = internal demand + external demand

$$x = (0.1x + 0.3y) + 27,000 \qquad \text{Energy sector}$$
$$y = (0.3x + 0.6y) + 81,000 \qquad \text{Manufacturing sector}$$

This system may be alternatively written as a matrix equation.

$$\begin{bmatrix} x \\ y \end{bmatrix} = \begin{bmatrix} 0.1 & 0.3 \\ 0.3 & 0.6 \end{bmatrix} \begin{bmatrix} x \\ y \end{bmatrix} + \begin{bmatrix} 27,000 \\ 81,000 \end{bmatrix}$$

Total output = internal demand + external demand

The matrix $X = \begin{bmatrix} x \\ y \end{bmatrix}$ is called the **output matrix** (or **production matrix**), and the

matrix $D = \begin{bmatrix} 27,000 \\ 81,000 \end{bmatrix}$ is called the **external demand matrix** (or **final demand**

matrix). The matrix $T = \begin{bmatrix} 0.1 & 0.3 \\ 0.3 & 0.6 \end{bmatrix}$ is called the **technology matrix.** Each entry

of the technology matrix has a specific meaning.

$$T = \begin{bmatrix} \boxed{\begin{array}{l}\text{Input from energy} \\ \text{sector required to} \\ \text{produce \$1 of energy}\end{array}} & \boxed{\begin{array}{l}\text{Input from energy sector} \\ \text{required to produce \$1 of} \\ \text{manufactured goods}\end{array}} \\ \boxed{\begin{array}{l}\text{Input from manufacturing} \\ \text{sector required to produce} \\ \text{\$1 of energy}\end{array}} & \boxed{\begin{array}{l}\text{Input from manufacturing} \\ \text{sector required to produce} \\ \text{\$1 of manufactured goods}\end{array}} \end{bmatrix}$$

Using the matrix names $X, D,$ and $T,$ we can write the matrix equation as $X = TX + D$ and solve for X as shown.

$$\begin{bmatrix} x \\ y \end{bmatrix} = \begin{bmatrix} 0.1 & 0.3 \\ 0.3 & 0.6 \end{bmatrix} \begin{bmatrix} x \\ y \end{bmatrix} + \begin{bmatrix} 27{,}000 \\ 81{,}000 \end{bmatrix}$$

$$X = TX + D$$

$IX = TX + D$ \qquad Recall that I is the identity matrix,

$$I = \begin{bmatrix} 1 & 0 \\ 0 & 1 \end{bmatrix}$$

$IX - TX = D$ \qquad Subtract TX from both sides

$(I - T)X = D$ \qquad Factor out X

$(I - T)^{-1}(I - T)X = (I - T)^{-1}D$ \qquad Left multiply both sides by the inverse of $I - T$

$IX = (I - T)^{-1}D$ \qquad The product of a matrix and its inverse is I

$X = (I - T)^{-1}D$ \qquad The product of a matrix X and matrix I is X

Note that the general solution $X = (I - T)^{-1}D$ is not dependent upon the specific values of $X, T,$ and D used in this problem. Consequently, the solution to all problems of the form $X = TX + D$ will be $X = (I - T)^{-1}D$, provided that $I - T$ is invertible. Returning to our specific problem, we have

$$I - T = \begin{bmatrix} 1 & 0 \\ 0 & 1 \end{bmatrix} - \begin{bmatrix} 0.1 & 0.3 \\ 0.3 & 0.6 \end{bmatrix}$$

$$= \begin{bmatrix} 0.9 & -0.3 \\ -0.3 & 0.4 \end{bmatrix}$$

Recall that the inverse of a 2×2 matrix $A = \begin{bmatrix} a & b \\ c & d \end{bmatrix}$ is

$$A^{-1} = \frac{1}{ad - bc} \begin{bmatrix} d & -b \\ -c & a \end{bmatrix}$$

Consequently,

$$(I - T)^{-1} = \frac{1}{(0.9)(0.4) - (-0.3)(-0.3)} \begin{bmatrix} 0.4 & 0.3 \\ 0.3 & 0.9 \end{bmatrix}$$

$$= \frac{1}{0.36 - 0.09} \begin{bmatrix} 0.4 & 0.3 \\ 0.3 & 0.9 \end{bmatrix}$$

$$= \frac{1}{0.27} \begin{bmatrix} 0.4 & 0.3 \\ 0.3 & 0.9 \end{bmatrix}$$

$$= \begin{bmatrix} \dfrac{40}{27} & \dfrac{10}{9} \\ \dfrac{10}{9} & \dfrac{10}{3} \end{bmatrix}$$

Therefore,

$$X = (I - T)^{-1}D$$

$$X = \begin{bmatrix} \frac{40}{27} & \frac{10}{9} \\ \frac{10}{9} & \frac{10}{3} \end{bmatrix} \begin{bmatrix} 27{,}000 \\ 81{,}000 \end{bmatrix}$$

$$\begin{bmatrix} x \\ y \end{bmatrix} = \begin{bmatrix} 130{,}000 \\ 300{,}000 \end{bmatrix}$$

The economy needs to produce \$130,000 worth of energy and \$300,000 worth of manufactured goods in order to meet its internal and external demands. We can check our work by verifying that $X = TX + D$.

$$X = TX + D$$

$$\begin{bmatrix} 130{,}000 \\ 300{,}000 \end{bmatrix} \overset{?}{=} \begin{bmatrix} 0.1 & 0.3 \\ 0.3 & 0.6 \end{bmatrix} \begin{bmatrix} 130{,}000 \\ 300{,}000 \end{bmatrix} + \begin{bmatrix} 27{,}000 \\ 81{,}000 \end{bmatrix}$$

$$\begin{bmatrix} 130{,}000 \\ 300{,}000 \end{bmatrix} \overset{?}{=} \begin{bmatrix} 103{,}000 \\ 219{,}000 \end{bmatrix} + \begin{bmatrix} 27{,}000 \\ 81{,}000 \end{bmatrix}$$

$$\begin{bmatrix} 130{,}000 \\ 300{,}000 \end{bmatrix} = \begin{bmatrix} 130{,}000 \\ 300{,}000 \end{bmatrix}$$

The result checks out, so we are confident that we have solved the problem correctly.

Although we developed the theory of solving input-output problems using an economy containing exactly two sectors, the same problem-solving approach works for economies with any number of sectors.

INPUT-OUTPUT PROBLEM SOLUTION

An economy with a technology matrix T and an external demand matrix D has an output matrix X. The matrix X satisfies the input-output equation

$$X = TX + D$$

The input-output equation has the solution

$$X = (I - T)^{-1}D$$

provided that $I - T$ is invertible.

The entry t_{ij} of the technology matrix T represents the input from sector i required to produce 1 unit of output in sector j.

Within the U.S. government, the Bureau of Economic Analysis (www.bea.gov) is responsible for conducting analyses of the economy. The bureau publishes annual input-output tables for various sectors of the economy. Many of the examples and exercises throughout this section are based on data from the 2002 input-output tables published by the bureau. We will typically focus on four or fewer sectors of the economy in a given problem. This approach will not give a complete picture of the economy because it assumes that the output of the industries in a given problem is dependent only upon the sectors given in the problem. However, it will give us an idea of how various sectors of the economy are interrelated.

EXAMPLE 1

Using Input-Output Analysis to Determine the Relationship Between Two Industries

The *motion picture and sound recording* (movie) industry and the *broadcasting and telecommunications* (telecom) industry are interrelated. Based on data from 2002, it requires roughly $0.27 from the movie industry and $0.01 from the telecom industry to produce $1 of movie industry products. Similarly, it requires $0.03 from the movie industry and $0.24 from the telecom industry to produce $1 of telecom products. If the external demand for movie products is $42 billion and the external demand for telecom products is $381 billion, what output from each sector is required to meet the demand for movie and telecom products?

SOLUTION Let x be the output of the movie industry and y be the output of the telecom industry, both in billions of dollars. The total output of the sectors is $X = \begin{bmatrix} x \\ y \end{bmatrix}$. The technology matrix T is given by

$$
\begin{array}{cc}
\text{Movie} & \text{Telecom} \\
\text{output} & \text{output}
\end{array}
$$

$$
T = \begin{bmatrix} 0.27 & 0.03 \\ 0.01 & 0.24 \end{bmatrix} \quad \begin{array}{l} \text{Movie input} \\ \text{Telecom input} \end{array}
$$

The external demand matrix is $D = \begin{bmatrix} 42 \\ 381 \end{bmatrix}$. From our earlier discussion, we know that $X = (I - T)^{-1}D$.

$$
\begin{aligned}
X &= (I - T)^{-1}D \\
&= \left(\begin{bmatrix} 1 & 0 \\ 0 & 1 \end{bmatrix} - \begin{bmatrix} 0.27 & 0.03 \\ 0.01 & 0.24 \end{bmatrix} \right)^{-1} \begin{bmatrix} 42 \\ 381 \end{bmatrix} \\
&= \begin{bmatrix} 0.73 & -0.03 \\ -0.01 & 0.76 \end{bmatrix}^{-1} \begin{bmatrix} 42 \\ 381 \end{bmatrix} \\
&\approx \begin{bmatrix} 1.37 & 0.054 \\ 0.018 & 1.32 \end{bmatrix} \begin{bmatrix} 42 \\ 381 \end{bmatrix} \quad \text{We calculated the inverse using technology} \\
&\approx \begin{bmatrix} 78 \\ 502 \end{bmatrix}
\end{aligned}
$$

We estimate that $78 billion of movie products and $502 billion of telecom products need to be produced to meet the demand.

EXAMPLE 2

Using Input-Output Analysis to Determine the Relationship Between Three Industries

The *wood products* (wood), *paper products* (paper), and *printing and related support activities* (print) industries are interrelated. The following technology matrix quantifies the relationship between the industries:

$$
\begin{array}{ccc}
\text{Wood} & \text{Paper} & \text{Print} \\
\text{output} & \text{output} & \text{output}
\end{array}
$$

$$
T = \begin{bmatrix} 0.216 & 0.023 & 0.000 \\ 0.004 & 0.219 & 0.241 \\ 0.000 & 0.001 & 0.060 \end{bmatrix} \quad \begin{array}{l} \text{Wood input} \\ \text{Paper input} \\ \text{Print input} \end{array}
$$

Explain the meaning of the entries t_{11}, t_{23}, and t_{31}. Then determine how much each industry needs to produce in order to meet internal and external demands, given that the external demand for wood products, paper products, and print products are $64 billion, $96 billion, and $64 billion, respectively.

SOLUTION t_{11} is the entry in the first row and first column of the technology matrix. It represents the amount of wood products input that is required to produce $1 of wood products. It requires nearly $0.22 ($0.216) of wood products to produce $1 of wood products.

t_{23} is the entry in the second row and third column of the technology matrix. It represents the amount of paper products input that is required to produce $1 of print products. It requires roughly $0.24 ($0.241) of paper products to produce $1 of print products.

t_{31} is the entry in the third row and first column of the technology matrix. It represents the amount of print products input that is required to produce $1 of wood products. Since the entry is equal to 0, no print products are required to produce $1 of wood products.

To determine the total demand for wood, paper, and print products, we calculate $X = (I - T)^{-1}D$.

$$X = (I - T)^{-1}D$$

$$= \left(\begin{bmatrix} 1 & 0 & 0 \\ 0 & 1 & 0 \\ 0 & 0 & 1 \end{bmatrix} - \begin{bmatrix} 0.216 & 0.023 & 0.000 \\ 0.004 & 0.219 & 0.241 \\ 0.000 & 0.001 & 0.060 \end{bmatrix} \right)^{-1} \begin{bmatrix} 64 \\ 96 \\ 64 \end{bmatrix}$$

$$= \begin{bmatrix} 0.784 & -0.023 & 0.000 \\ -0.004 & 0.781 & -0.241 \\ 0.000 & -0.001 & 0.940 \end{bmatrix}^{-1} \begin{bmatrix} 64 \\ 96 \\ 64 \end{bmatrix}$$

$$\approx \begin{bmatrix} 1.276 & 0.038 & 0.010 \\ 0.007 & 1.281 & 0.328 \\ 0.000 & 0.001 & 1.064 \end{bmatrix} \begin{bmatrix} 64 \\ 96 \\ 64 \end{bmatrix}$$

$$\approx \begin{bmatrix} 86 \\ 144 \\ 68 \end{bmatrix}$$

In order to satisfy internal and external demands, $86 billion of wood products, $144 billion of paper products, and $68 billion of print products need to be produced.

3.4 Summary

In this section, you learned how to do basic input-output analysis. These skills allow you to analyze the interrelationships among sectors of the economy.

3.4 Exercises

In Exercises 1–4, interpret the meaning of the indicated entries in the technology matrix.

1. Entries t_{11}, t_{12}, t_{21}, and t_{22}.

$$T = \begin{bmatrix} 0.2 & 0.5 \\ 0.4 & 0.8 \end{bmatrix} \begin{matrix} \text{Sector A} \\ \text{Sector B} \end{matrix}$$

2. Entries t_{11}, t_{12}, t_{21}, and t_{22}.

$$T = \begin{bmatrix} 0.27 & 0.03 \\ 0.01 & 0.24 \end{bmatrix} \begin{matrix} \text{Movie} \\ \text{Telecom} \end{matrix}$$

3. Entries t_{12}, t_{21}, t_{23}, and t_{31}.

$$T = \begin{bmatrix} 0.14 & 0.01 & 0.00 \\ 0.00 & 0.10 & 0.00 \\ 0.02 & 0.02 & 0.01 \end{bmatrix} \begin{matrix} \text{Oil} \\ \text{Mining} \\ \text{Support} \end{matrix}$$

4. Entries t_{13}, t_{22}, t_{32}, and t_{33}.

$$T = \begin{bmatrix} 0.01 & 0.01 & 0.00 \\ 0.00 & 0.00 & 0.02 \\ 0.03 & 0.00 & 0.06 \end{bmatrix} \begin{matrix} \text{Computer} \\ \text{Management} \\ \text{Support} \end{matrix}$$

In Exercises 5–10, use the given technology and external demand matrices to find the answers to the questions. The demand matrices are in millions of dollars.

$$T = \begin{bmatrix} 0.2 & 0.1 \\ 0.6 & 0.3 \end{bmatrix} \begin{matrix} \text{Agriculture} \\ \text{Manufacturing} \end{matrix}$$

$$D_1 = \begin{bmatrix} 210 \\ 300 \end{bmatrix} \quad D_2 = \begin{bmatrix} 240 \\ 360 \end{bmatrix}$$

5. How much agriculture and manufacturing input is required to produce $1 of manufacturing?

6. How much agriculture and manufacturing input is required to produce $1 of agriculture?

7. Calculate $I - T$ and $(I - T)^{-1}$.

8. How much agriculture and manufacturing input is required to meet the internal demands of each industry and the external demands given in D_1 ?

9. How much agriculture and manufacturing input is required to meet the internal demands of each industry and the external demands given in D_2 ?

10. Given that the external demand for agriculture and manufacturing products is given by $D = \begin{bmatrix} d_1 \\ d_2 \end{bmatrix}$, how much agriculture and manufacturing input is required to meet the internal demands of each industry and the external demands given in D ?

Exercises 11–20 are derived from economic data for 2002 published by the U.S. Bureau of Economic Analysis.

11. **Industry Analysis: Two Industries** The *textile mills* and *textile product mills* (textile) industry and the *apparel and leather and allied products* (apparel) industry are interrelated. Based on data from 2002, it requires roughly $0.23 from the textile industry and $0.01 from the apparel industry to produce $1 of textile products. Similarly, it requires $0.20 from the textile industry and $0.10 from the apparel industry to produce $1 of apparel products. If the external demand for textile products is $43 billion and the external demand for apparel products is $48 billion, what output from each sector is required to meet the demand for textile and apparel products?

12. **Industry Analysis: Two Industries** The *fabricated metal products* (metal) industry and the *machinery* (machine) industry are interrelated. Based on data from 2002, it requires roughly $0.11 from the metal industry and $0.01 from the machine industry to produce $1 of metal products. Similarly, it requires $0.10 from the metal industry and $0.09 from the machine industry to produce $1 of machine products. If the external demand for metal products is $185 billion and the external demand for machine products is $210 billion, what output from each sector is required to meet the demand for metal and machine products?

13. **Industry Analysis: Two Industries** The *motor vehicles, bodies and trailers, and parts* (vehicle) industry and the *other transportation equipment* (transportation) industry are interrelated. Based on data from 2002, it requires roughly $0.31 from the vehicle industry and $0.00 from the transportation industry to produce $1 of vehicle products. Similarly, it requires $0.01 from the vehicle industry and $0.18 from the transportation industry to produce $1 of transportation products. If the external demand for vehicle products is $296 billion and the external demand for transportation products is $134 billion, what output from each sector is required to meet the demand for vehicle and transportation products?

14.   **Industry Analysis: Two Industries**   The *computer and electronic products* (computer) industry and the *electrical equipment, appliances, and components* (equipment) industry are interrelated. Based on data from 2002, it requires roughly $0.22 from the computer industry and $0.02 from the equipment industry to produce $1 of computer products. Similarly, it requires $0.03 from the computer industry and $0.06 from the equipment industry to produce $1 of equipment products. If the external demand for computer products is $284 billion and the external demand for equipment products is $82 billion, what output from each sector is required to meet the demand for computer and equipment products?

15.   **Industry Analysis: Three Industries**   The *primary metals* (metals), *fabricated metal products* (products), and *machinery* (machinery) industries are interrelated. Based on data from 2002, it requires $0.26 of metals, $0.02 of products, and $0.01 of machinery to produce $1 of metals. Similarly, it requires $0.17 of metals, $0.11 of products, and $0.01 of machinery to produce $1 of products. Likewise, it requires $0.09 of metals, $0.10 of products, and $0.09 of machinery to produce $1 of machinery. If the external demand for metals is $37 billion, the external demand for products is $181 billion, and the external demand for machinery is $208 billion, what output from each sector is required to meet the demand for metals, products, and machinery?

16.   **Industry Analysis: Three Industries**   The *oil and gas extraction* (oil), *mining, except oil and gas* (mining), and *support activities for mining* (support) industries are interrelated. Based on data from 2002, it requires $0.14 of oil, $0.00 of mining, and $0.02 of support to produce $1 of oil. Similarly, it requires $0.01 of oil, $0.10 of mining, and $0.02 of support to produce $1 of mining. Likewise, it requires $0.00 of oil, $0.00 of mining, and $0.01 of support to produce $1 of support. If the external demand for oil is $87 billion, the external demand for mining is $43 billion, and the external demand for support is $26 billion, what output from each sector is required to meet the demand for oil, mining, and support?

17.   **Industry Analysis: Three Industries**   The *computer systems design and related services* (computer), *management of companies*

and enterprises (management), and *administrative and support services* (support) industries are interrelated. Based on data from 2002, $0.01 of computer, $0.00 of management, and $0.03 of support is required to produce $1 of computer. To produce $1 of management requires $0.01 of computer, $0.00 of management, and $0.00 of support. To produce $1 of support requires $0.00 of computer, $0.02 of management, and $0.06 of support. If the external demand for computer products is $139 billion, the external demand for management is $276 billion, and the external demand for support is $401 billion, what output from each sector is required to meet the demand?

18.   **Industry Analysis: Four Industries**   The *petroleum and coal products* (petrol), *chemical products* (chem), *plastics and rubber products* (plastic), and *nonmetallic mineral products* (mineral) industries are interrelated as shown in the given technology matrix.

$$T = \begin{bmatrix} 0.09 & 0.02 & 0.00 & 0.00 \\ 0.01 & 0.21 & 0.23 & 0.03 \\ 0.00 & 0.02 & 0.07 & 001 \\ 0.00 & 0.00 & 0.01 & 0.10 \end{bmatrix} \begin{matrix} \text{Petrol} \\ \text{Chem} \\ \text{Plastic} \\ \text{Mineral} \end{matrix}$$

The estimated external demand for each of these products in 2002 is given in the table.

Industry	External Demand
Petrol	$168 billion
Chem	$268 billion
Plastic	$149 billion
Mineral	$74 billion

What production level in each industry is required in order to satisfy internal and external demands?

19.   **Industry Analysis: Four Industries**   The *Federal Reserve banks, credit intermediation, and related activities* (banks), *securities, commodity contracts, and investments* (securities), *insurance carriers and related activities* (insurance), and *funds, trusts, and other financial vehicles* (funds) industries are interrelated as shown in the given technology matrix.

$$T = \begin{bmatrix} 0.08 & 0.03 & 0.01 & 0.03 \\ 0.05 & 0.12 & 0.01 & 0.54 \\ 0.00 & 0.00 & 0.38 & 0.02 \\ 0.00 & 0.00 & 0.01 & 0.00 \end{bmatrix} \begin{matrix} \text{Banks} \\ \text{Securities} \\ \text{Insurance} \\ \text{Funds} \end{matrix}$$

The estimated external demand for each of these products in 2002 is given in the table.

Industry	External Demand
Banks	$453 billion
Securities	$129 billion
Insurance	$281 billion
Funds	$67 billion

What production levels are required to meet product demand?

20. **Industry Analysis: Four Industries**   The *publishing (includes software)* (publish), *motion picture and sound recording* (movie), *broadcasting and telecommunications* (telecom), and *information and data processing services* (data) industries are interrelated as shown in the given technology matrix.

$$T = \begin{bmatrix} 0.09 & 0.00 & 0.00 & 0.00 \\ 0.00 & 0.27 & 0.03 & 0.00 \\ 0.03 & 0.01 & 0.24 & 0.06 \\ 0.02 & 0.00 & 0.00 & 0.02 \end{bmatrix} \begin{matrix} \text{Publish} \\ \text{Movie} \\ \text{Telecom} \\ \text{Data} \end{matrix}$$

The estimated external demand for each of these products in 2002 is given in the table.

Industry	External Demand
Publish	$144 billion
Movie	$42 billion
Telecom	$370 billion
Data	$88 billion

What production levels are required to meet internal and external demands?

Exercises 21–26 are intended to challenge your understanding of input-output tables.

21. A classmate presents you with the following "technology" matrix.

$$T = \begin{bmatrix} 0.6 & 0.3 \\ 0.5 & 0.9 \end{bmatrix}$$

What causes you to immediately question his results?

22. The following input-output problem has no solution. Explain why.

$$x = 0.5x + 0.4y + 2000$$
$$y = 0.5x + 0.6y + 3000$$

23. The relationship between three sectors of an economy are summarized in the given technology matrix.

$$T = \begin{bmatrix} 0.2 & 0.0 & 0.0 \\ 0.0 & 0.3 & 0.4 \\ 0.0 & 0.2 & 0.1 \end{bmatrix} \begin{matrix} \text{Sector A} \\ \text{Sector B} \\ \text{Sector C} \end{matrix}$$

Assume that the output of each sector is dependent only upon input from the indicated sectors. If production in Sector A is halted because of an employee strike, how will that affect Sectors B and C?

24. Referring to the technology matrix in Exercise 23, explain how Sectors A and B will be affected if production in Sector C is halted.

25. Let T be any technology matrix. For a functioning economy, between what values must the sum of the entries in a given column of T range?

26. Let T be any technology matrix. What numerical restrictions, if any, are placed upon the entries in T?

Chapter 3 Review Exercises

Section 3.1   *In Exercises 1–5, perform the indicated matrix operation, if possible, given the following matrices. Solve these problems without technology.*

$$A = \begin{bmatrix} 2 & 3 \\ -3 & -4 \\ 0 & 7 \end{bmatrix}, \ B = \begin{bmatrix} 5 & 6 \\ 3 & -4 \\ 9 & 2 \end{bmatrix}$$

1. $A + B$

2. $A - B$

3. $2A$

4. $3B$

5. $2A + 3B$

In Exercises 6–10, use technology to simplify the matrix expressions, given

$$A = \begin{bmatrix} 1.2 & 6.1 & -0.4 \\ -9.1 & 4.2 & 1.7 \\ 0.9 & 3.0 & 3.3 \end{bmatrix} \text{ and}$$

$$B = \begin{bmatrix} 3.4 & -0.3 & -0.4 \\ -3.8 & 5.6 & 7.2 \\ 2.2 & 2.6 & 2.0 \end{bmatrix}$$

 6. $3.2A$ **7.** $5.3B$ **8.** $3.2A - 5.3B$

9. $4.1A + 0.1B$ **10.** $-1.1A + 2.9B$

11. **Auto Prices** The average *trade-in value* of a Toyota Celica and a Toyota MR2 Spyder in July 2002 is shown in the first table. The average *retail value* of the two vehicles is shown in the second table.

Average Trade-In

	Celica	MR2 Spyder
2000 Model	$13,025	$18,750
2001 Model	$14,850	$20,100

Average Retail

	Celica	MR2 Spyder
2000 Model	$15,300	$21,300
2001 Model	$17,250	$22,725

Source: www.nada.com.

Use matrices to create a table that shows the average dealer markup for each of the vehicles.

Section 3.2 *In Exercises 12–20, perform the indicated operation. As appropriate, use the matrices*

$$A = \begin{bmatrix} 5 & 8 \\ 1 & 2 \end{bmatrix}, B = \begin{bmatrix} 4 & 5 \\ 2 & 2 \end{bmatrix},$$

$$C = \begin{bmatrix} 2 & 1 \\ 5 & 4 \\ 8 & 7 \end{bmatrix}, \text{ and } D = \begin{bmatrix} -1 & 0 & 4 \\ -2 & 2 & -2 \\ -6 & 4 & 1 \end{bmatrix}.$$

If the specified operation is undefined, so state.

12. AB **13.** BA **14.** CD

15. DC **16.** CA **17.** AC

18. A^{-1} **19.** B^{-1} **20.** D^{-1}

In Exercises 21–22, use the determinant formula to determine if the matrix is invertible or singular.

21. $A = \begin{bmatrix} 9 & 6 \\ -3 & 2 \end{bmatrix}$ **22.** $B = \begin{bmatrix} -8 & -4 \\ 5 & 3 \end{bmatrix}$

In Exercises 23–24, use technology to find the inverse of the matrix, if it exists. If the matrix is singular, so state.

23. $B = \begin{bmatrix} 2.0 & 6.2 & -0.8 \\ 0 & 3.2 & 5.4 \\ 0 & 0 & -0.5 \end{bmatrix}$

24. $B = \begin{bmatrix} 2.0 & 6.2 & -0.8 \\ 4.2 & 3.2 & 5.4 \\ 6.2 & 9.4 & 4.6 \end{bmatrix}$

Section 3.3 *In Exercises 25–27, find the inverse of the matrix A algebraically. If A is singular, so state.*

25. $A = \begin{bmatrix} 6 & -2 & 1 \\ 3 & -1 & 1 \\ 0 & 2 & 4 \end{bmatrix}$ **26.** $A = \begin{bmatrix} 9 & 7 & 8 \\ 3 & 3 & 3 \\ 6 & 10 & 11 \end{bmatrix}$

27. $A = \begin{bmatrix} 9 & 8 & 7 \\ 6 & 5 & 4 \\ -2 & 0 & 1 \end{bmatrix}$

In Exercises 28–30, write the system of equations as a matrix equation, AX = B, and solve.

28. $\begin{aligned} x + y + z &= 2 \\ 2x - y + z &= -6 \\ 4x - 2y + 3z &= -17 \end{aligned}$

29. $\begin{aligned} x + y + z &= 14 \\ x - y + z &= 26 \\ x - y - z &= 2 \end{aligned}$

30. $\begin{aligned} 3x + 2y + 3z &= 0 \\ 2x - 5y + z &= 0 \\ 4x + 2y + 3z &= 0 \end{aligned}$

In Exercises 31–33, determine the solution by setting up and solving the matrix equation.

31. **Return on Investment** The average annual return (over the 10-year period prior to June 30, 2002) of two mutual funds offered by Harbor Fund is shown in the following table.

	Average Annual Return
Growth Fund	5.84%
Capital Appreciation Fund	12.69%

Source: Harbor Fund account statement.

Suppose you have $2,000 to invest in these two accounts. Assuming that the accounts will earn the returns specified in the table over the next year, how much should you invest in each account if you want to earn 7 percent, 9 percent, or 11 percent?

32. Grade Point Average A student at Green River Community College must earn 90 credits to obtain an Associate of Arts degree. A student with a 3.1 existing GPA hopes to increase her cumulative GPA to 3.4. If she has 36 credits now and anticipates that she will be able to earn a 3.7 GPA on her remaining coursework, is it possible for her to increase her cumulative GPA to 3.4 by the time she obtains 90 credits?

33. Floral Costs A florist purchases her flowers from an online flower wholesaler. White daisies are $3.38 per bunch, football mums are $7.40 per bunch, and super blue purple statice is $4.25 per bunch. (**Source:** www.FlowerSales.com.) From these flowers she will make three types of jumbo bouquets:

	Type 1	Type 2	Type 3
Daisies	2 bunches	3 bunches	2 bunches
Mums	1 bunch	2 bunches	2 bunches
Statice	1 bunch	None	1 bunch

What is her flower cost for each type of bouquet? How much should she charge for each bouquet if her markup is 50 percent of her flower cost?

Exercises 3.4 *Exercises 34–35 are derived from economic data for 2002 published by the U.S. Bureau of Economic Analysis.*

34. Industry Analysis: Three Industries The *farms* (farm), *food and beverage and tobacco products* (food), and *food services and drinking places* (dining) industries are interrelated. Based on data from 2002, it requires $0.14 of farm products, $0.09 of food products, and $0.00 of dining products to produce $1 of farm products. Similarly, it requires $0.19 of farm products, $0.13 of food products, and $0.00 of dining products to produce $1 of food products. Likewise, it requires $0.01 of farm products, $0.18 of food products, and $0.01 of dining products to produce $1 of dining products. If the external demand for farm products is $53 billion, the external demand for food products is $422 billion, and the external demand for dining products is $379 billion, what output from each sector is required to meet the demand for farm, food, and dining products?

35. Industry Analysis: Two Industries The *machinery* (machine) and *motor vehicles, bodies and trailers, and parts* (auto) industries are interrelated. Based on data from 2002, it requires roughly $0.09 from the machine industry and $0.02 from the auto industry to produce $1 of machine products. Similarly, it requires $0.02 from the machine industry and $0.31 from the auto industry to produce $1 of auto products. If the external demand for machine products is $202 billion and the external demand for auto products is $293 billion, what output from each sector is required to meet the demand for machine and auto products?

Make It Real

What to do

1. Find out your current cumulative grade point average and your total number of graded credits.

2. Determine how many credits are required for your degree program.

3. Find two or three scholarships that require a minimum GPA.

4. If your cumulative GPA is below the required minimum for these scholarships, complete Step 5. Otherwise, complete Step 6.

5. Set up and solve a system of equations to determine how many credits with an "A" grade (4.0) a student with your cumulative GPA must earn in order to meet the GPA requirement. To do this, let x be the number of credits already earned and y be the number of credits left to be earned. Proceed to Step 7.

6. Set up and solve a system of equations to determine how many credits with a "C" grade (2.0) a student with your cumulative GPA can earn and still meet the GPA requirement. To do this, let x be the number of credits already earned and y be the number of credits left to be earned.

7. Suppose that one of your friends has the same GPA as you but has earned 12 credits fewer than you. Based upon your previous calculations, determine if it is mathematically possible for both you and your friend to meet the grade point average requirement for each of the scholarships by the time you have earned the required number of credits.

Linear Programming

Businesses seek to maximize their profits while operating under budget, supply, labor, and space constraints. Determining which combination of variables will result in the maximum profit may be done through the use of linear programming. Although many factors affect a business's profitability, linear programming can help a business owner determine the "ideal" conditions for business success.

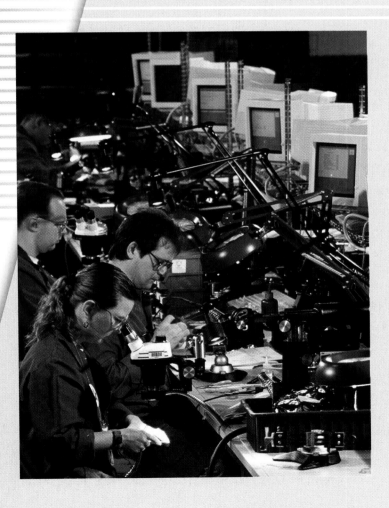

4.1 Graphing Linear Inequalities

- Graph linear inequalities
- Set up and solve systems of linear inequalities

GETTING **STARTED** Many students work multiple part-time jobs to finance their education. Often the jobs pay different wages and offer varying hours. Suppose that a student earns $10.50 per hour delivering pizza and $8.00 per hour working in a campus computer lab. If the student has only 30 hours per week to work and must earn $252 during that period, how many hours must he spend at each job in order to meet his earnings goal? In this section, we will explain how linear inequalities may be used to answer this question. We will demonstrate how to graph linear inequalities and show that the solution region of a system of linear inequalities is the intersection of the graphs of the individual inequalities.

Linear Inequalities

In many real-life applications, we are interested in a range of possible solutions instead of a single solution. For example, when you prepare to buy a house, a lender will calculate the maximum amount of money it is willing to lend you; however, the lender doesn't require you to borrow the maximum amount. You may borrow any amount of money up to the maximum. Recall that in mathematics, we use inequalities to represent the range of possible solutions that meet the given criteria.

INEQUALITY NOTATION

$x \leq y$ is the set of all values of x less than or equal to y.

$x \geq y$ is the set of all values of x greater than or equal to y.

$x < y$ is the set of all values of x less than but not equal to y.

$x > y$ is the set of all values of x greater than but not equal to y.

The inequalities $x < y$ and $x > y$ are called **strict inequalities** because the two variables cannot ever be equal. Although strict inequalities have many useful applications, we will focus on the nonstrict inequalities in this chapter.

An easy way to keep track of the meaning of an inequality is to remember that the inequality sign always points toward the smaller number. Consider these everyday examples of inequalities:

You must be at least 16 years old to get a driver's license. ($16 \leq a$) or ($a \geq 16$).

You must be at least 21 years old to legally buy alcohol. ($21 \leq a$) or ($a \geq 21$).

The maximum fine for littering is $200. ($200 \geq f$) or ($f \leq 200$).

Your carry-on bag must be no more than 22 inches long. ($22 \geq l$) or ($l \leq 22$).

A linear inequality looks like a linear equation with an inequality sign in the place of the equal sign. Recall that linear inequalities may be manipulated algebraically in the same way as linear equations, with one major exception: When we multiply or divide both sides of an inequality by a negative number, we must reverse the direction of the inequality sign. For example, if we multiply both sides of $-3x + 2y \leq 10$ by -1, we get $3x - 2y \geq -10$, not $3x - 2y \leq -10$.

Graphing Linear Inequalities

The graph of a linear inequality is a region bordered by a line called a **boundary line.** The **solution region** of a linear inequality is the set of all points (including the boundary line) that satisfy the inequality.

Consider the inequality $x + 2y \geq 4$ (see Figure 4.1). The boundary line of the solution region is $x + 2y = 4$, since the points satisfying this linear equation are on the border of the solution region.

We need to find all points (x, y) that satisfy the inequality. We know that all points on the line satisfy the inequality. Which points off the line satisfy the inequality? Let's pick a few points off the line (see Table 4.1) and test them to see if they satisfy the inequality. In order to satisfy the inequality, $x + 2y$ must be at least 4.

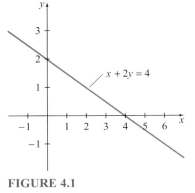

FIGURE 4.1

TABLE 4.1

x	y	$x + 2y$	In Solution Region?
-1	1	1	No
0	1	2	No
1	3	7	Yes
2	2	6	Yes
3	0	3	No
5	2	9	Yes
6	1	8	Yes

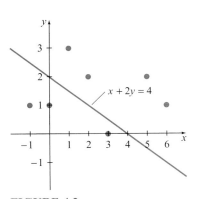

FIGURE 4.2

Graphically speaking, what do the four points in the solution region have in common (see Figure 4.2)?

They are all on the same side of the boundary line! In fact, all points on or above this boundary line satisfy the inequality. We represent this notion by shading the region above the boundary line, as shown in Figure 4.3.

Although we checked multiple points in this problem, we need to check only one point off the boundary line in order to determine which region to shade.

The linear inequality graphing process is summarized as follows.

FIGURE 4.3

HOW TO   Linear Inequality Graphing Technique

To graph the solution region of the linear inequality $ax + by \leq c$ (or $ax + by \geq c$), do the following:

1. Graph the boundary line $ax + by = c$.

2. Select a point on one side of the line. [If the line doesn't pass through the origin, $(0, 0)$ is an excellent choice for easy computations.]

3. Substitute the point into the linear inequality and simplify. If the simplified statement is true, the selected point and all other points on the same side of the line are in the solution region. If the simplified statement is false, all points on the opposite side of the line are in the solution region.

4. Shade the solution region.

EXAMPLE 1   Graphing the Solution Region of a Linear Inequality

Graph the solution region of the linear inequality $2x + y \leq 4$.

SOLUTION   As shown in Chapter 1, the x-intercept is easily found by dividing the constant term by the coefficient on the x term.

$$x = \frac{4}{2}$$
$$= 2$$

The point $(2, 0)$ is the x-intercept.

The y-intercept is found by dividing the constant term by the coefficient on the y term.

$$y = \frac{4}{1}$$
$$= 4$$

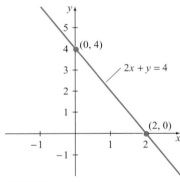

FIGURE 4.4

The point $(0, 4)$ is the y-intercept. We graph the x- and y-intercepts and then draw the line through the intercepts, as shown in Figure 4.4.

Next, we will pick the point $(0, 0)$ to plug into the inequality.

$$2(0) + 0 \leq 4$$
$$0 \leq 4$$

The statement is true, so all points on the same side of the line as the origin are in the solution region. We shade the solution region (see Figure 4.5).

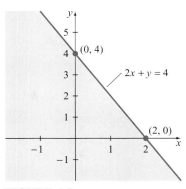

FIGURE 4.5

If you choose to convert lines from standard to slope-intercept form before graphing them, the following properties will help you to quickly identify the solution region without having to check a point.

SOLUTION REGION OF A LINEAR INEQUALITY

The solution region of a linear inequality $y \geq mx + b$ contains the line $y = mx + b$ and the shaded region **above** the line.

The solution region of a linear inequality $y \leq mx + b$ contains the line $y = mx + b$ and the shaded region **below** the line.

Graphing Systems of Linear Inequalities

Just as we can graph systems of linear equations, we can graph systems of linear inequalities. The solution region of a system of linear inequalities is the intersection of the solution regions of the individual inequalities. When we graph a solution region by hand, we will typically place arrows on the boundary lines to indicate which side of the lines satisfies the given inequality. Once all of the linear inequality graphs have been drawn, we will shade the region that has arrows pointing into the interior of the region from all sides.

EXAMPLE 2

Graphing the Solution Region of a System of Linear Inequalities

Graph the solution region of the system of linear inequalities.

$$3x + 2y \leq 5$$
$$x \geq 0$$
$$y \geq 0$$

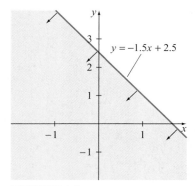

FIGURE 4.6

SOLUTION We first rewrite the linear inequality as a linear inequality in slope-intercept form.

$$3x + 2y \leq 5$$

$$2y \leq -3x + 5 \qquad \text{Subtract } 3x \text{ from both sides}$$

$$y \leq -\frac{3}{2}x + \frac{5}{2} \qquad \text{Divide both sides by 2}$$

$$y \leq -1.5x + 2.5 \qquad \text{Write as a decimal (optional)}$$

The boundary line is a line with slope -1.5 and y-intercept $(0, 2.5)$. Since y is less than or equal to the expression $-1.5x + 2.5$, we will shade the region below the line, as shown in Figure 4.6.

The next two inequalities $(x \geq 0, y \geq 0)$ limit the solution region to positive values of x and y. The line $x = 0$ is the y axis. The line $y = 0$ is the x axis. Therefore, the solution region of the system of inequalities is the triangular region to the right of the line $x = 0$, above the line $y = 0$, and below the line $y = -1.5x + 2.5$ (see Figure 4.7).

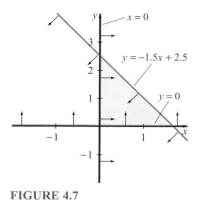

FIGURE 4.7

If it is possible to draw a circle around the solution region, the solution region is **bounded.** If no circle can be drawn that will enclose the entire solution region, the solution region is **unbounded.** The solution region in Example 2 was bounded. The solution region in Example 3 will be unbounded.

EXAMPLE **3**

Graphing the Solution Region of a System of Linear Inequalities

Graph the solution region of the system of linear inequalities.

$$4x + y \geq 4$$

$$-x + y \geq 1$$

SOLUTION The x-intercept of the boundary line $4x + y = 4$ is $(1, 0)$, and the y-intercept is $(0, 4)$. Plugging in the point $(0, 0)$, we get

$$4(0) + (0) \geq 4$$

$$0 \geq 4$$

Since the statement is false, we graph $4x + y = 4$ and place arrows on the side of the line not containing the origin, as shown in Figure 4.8.

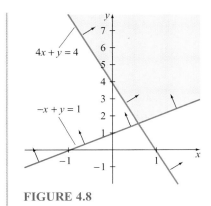

4x + y = 4

-x + y = 1

FIGURE 4.8

The x-intercept of the boundary line $-x + y = 1$ is $(-1, 0)$, and the y-intercept is $(0, 1)$. Plugging in the point $(0, 0)$, we get

$$-(0) + (0) \geq 1$$
$$0 \geq 1$$

Since the statement is false, we graph $-x + y = 1$ and place arrows on the side of the line not containing the origin, as shown in Figure 4.8.

The solution region is not bounded above the line $y = x + 1$ or above the line $y = -4x + 4$. Consequently, the shaded solution region is unbounded.

EXAMPLE 4

Graphing a System of Linear Inequalities with an Empty Solution Region

Graph the solution region of the system of linear inequalities.

$$-2x + 2y \geq 6$$
$$-x + y \leq 1$$

SOLUTION We will graph the boundary lines by first rewriting them in slope-intercept form.

Solving the first inequality for y, we get $y \geq x + 3$ and draw arrows pointing to the region above the line $y = x + 3$. Solving the second inequality for y, we get $y \leq x + 1$ and draw arrows pointing to the region below the line $y = x + 1$ (see Figure 4.9).

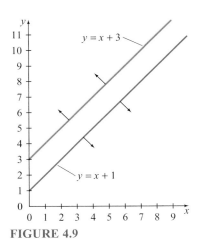

$y = x + 3$

$y = x + 1$

FIGURE 4.9

Because the lines have the same slope ($m = 1$), they are parallel. Consequently, the lines will never intersect. As seen in the graph, the two regions also will never intersect; they will always be separated by the region between the two lines. Therefore, this system of linear inequalities does not have a solution. That is, there is no ordered pair (x, y) that can satisfy both inequalities simultaneously.

The corners of a solution region are called **corner points.** To find the coordinates of each corner point, we solve the system of equations formed by the two intersecting boundary lines that create the corner.

EXAMPLE 5

Finding the Corner Points of a Solution Region

Graph the solution region of the system of linear inequalities and determine the coordinates of each corner point.

$$2x + y \leq 6$$
$$-x + y \geq 0$$
$$x \geq 0$$

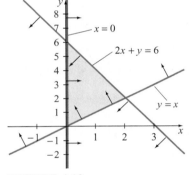

FIGURE 4.10

SOLUTION We will graph the boundary line $2x + y = 6$ using its x- and y-intercepts. The x-intercept is $(3, 0)$, since $\frac{6}{2} = 3$. The function has y-intercept $(0, 6)$, since $\frac{6}{1} = 6$. Plugging in the point $(0, 0)$, we get

$$2(0) + (0) \leq 6$$
$$0 \leq 6$$

The statement is true, so we shade the side of the line containing the point $(0, 0)$. Solving the second inequality for y, we get $y \geq x$, and we shade the region above the line $y = x$. The third inequality restricts x to all positive values, so we shade the region to the right of the y axis.

The triangular region shown in Figure 4.10 is the intersection of the three regions and is the solution region of the system of linear inequalities. From the graph, it appears that the solution region has corner points at or near $(0, 0)$, $(0, 6)$, and $(2, 2)$. We will verify these results algebraically.

The coordinates of the first corner point may be found by solving the system of equations formed by the boundary lines that make up the corner.

$$-x + y = 0$$
$$x \quad\quad = 0$$

Adding the first equation to the second equation yields $y = 0$. Since the second equation tells us that $x = 0$, the coordinates of the first corner point are $(0, 0)$.

The coordinates of the second corner point may be found by solving the system of equations formed by the boundary lines that make up the corner.

$$2x + y = 6$$
$$x \quad\quad = 0$$

The second equation tells us that $x = 0$. Substituting this value of x into the first equation yields

$$2x + y = 6$$
$$2(0) + y = 6 \quad\quad \text{Substitute } x = 0$$
$$y = 6$$

The coordinates of the second corner point are $(0, 6)$.

The coordinates of the third corner point may be found by solving the system of equations formed by the boundary lines that make up the corner.

$$2x + y = 6$$
$$-x + y = 0$$

The second equation may be rewritten as $y = x$. Substituting this result into the first equation yields

$$2x + y = 6$$
$$2x + (x) = 6 \qquad \text{Since } y = x$$
$$3x = 6$$
$$x = 2$$

Since $x = y$, the coordinates of the corner point are $(2, 2)$.

It may have seemed superfluous to calculate the coordinates of the corner points algebraically in Example 5 when the coordinates were readily apparent from the graph of the solution region. Despite the apparent redundancy of the procedure, it is a necessary step. Example 6 illustrates the hazards of relying solely upon a graph for the coordinates of the corner points.

EXAMPLE **6** **Finding the Corner Points of a Solution Region**

Graph the solution region for the system of inequalities and determine the coordinates of the corner points.

$$x + y \leq 5$$
$$-5x + 5y \leq 6$$
$$y \geq 2$$

SOLUTION The graph of the solution region is shown in Figure 4.11.
From the graph, it appears that the corner points of the region are at or near $(2, 3)$, $(3, 2)$, and $(0.75, 2)$.
The coordinates of the first corner point may be found by solving the system of equations formed by the boundary lines that make up the corner.

$$x + y = 5$$
$$-5x + 5y = 6$$

Adding five times the first equation to the second equation yields

$$0x + 10y = 31$$
$$y = 3.1$$

Substituting this result back into the first equation yields

$$x + y = 5$$
$$x + (3.1) = 5$$
$$x = 1.9$$

The coordinates of the first corner point are $(1.9, 3.1)$. [From the graph, it looked as if the corner point was $(2, 3)$.]

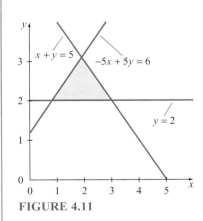

FIGURE 4.11

The coordinates of the second corner point may be found by solving the system of equations formed by the boundary lines that make up the corner.

$$x + y = 5$$
$$y = 2$$

Since the second equation tells us that $y = 2$, we may substitute this value into the first equation.

$$x + y = 5$$
$$x + (2) = 5 \qquad \text{Since } y = 2$$
$$x = 3$$

The coordinates of the second corner point are $(3, 2)$. Unlike the first corner point, this result agrees with our graphical conclusion.

The coordinates of the third corner point may be found by solving the system of equations formed by the boundary lines that make up the corner.

$$-5x + 5y = 6$$
$$y = 2$$

Since the second equation tells us that $y = 2$, we may substitute this value into the first equation.

$$-5x + 5y = 6$$
$$-5x + 5(2) = 6 \qquad \text{Since } y = 2$$
$$-5x + 10 = 6$$
$$-5x = -4$$
$$x = 0.8$$

The coordinates of the third corner point are $(0.8, 2)$. [From the graph, it looked as if the corner point was $(0.75, 2)$.]

Using Technology to Graph Linear Inequalities

Many graphing calculators can draw the graphs of linear inequalities, as detailed in the following Technology Tip. Often, however, it is quicker to draw the graphs by hand.

TECHNOLOGY TIP

Graphing a System of Linear Inequalities

1. Enter the linear equations associated with each inequality by using the $\boxed{Y=}$ editor. (We will use the system $y \leq -3x + 6$ and $y \geq 2x + 4$ for this example.)

(Continued)

2. Move the cursor to the \ to the left of Y1 and press ENTER repeatedly. This will cycle through several graphing options. We want to shade the region below the line, so we will pick the lower triangular option.

3. Move the cursor to the \ to the left of Y2 and press ENTER repeatedly. We want to shade the region above the line, so we will pick the upper triangular option.

4. Press GRAPH to draw each of the shaded regions. The region with the crisscross pattern is the solution region.

Linear Inequality Applications

Many real-life problems are subject to multiple constraints, such as budget, staffing, resources, and so on. Yet even when subjected to these constraints, the problems often have multiple solutions. As will be shown in Example 7, the linear inequality graphing techniques introduced earlier in this section may often be used to find solutions to real-life problems.

EXAMPLE 7

Using a Linear System of Inequalities to Find the Ideal Work Schedule

A student earns $8.00 per hour working in a campus computer lab and $10.50 per hour delivering pizza. If he has only 30 hours per week to work and must earn at least $252 during that period, how many hours can he spend at each job in order to earn at least $252?

SOLUTION Let c be the number of hours the student works in the computer lab and p be the number of hours he works delivering pizza. He can work at most 30 hours. This is represented by the inequality

$$c + p \leq 30 \qquad \text{The maximum number of work hours is 30}$$

The amount he earns working in the lab is $8.00c$, and the amount of money he earns delivering pizza is $10.50p$. His total income must be at least $252. That is,

$$8c + 10.5p \geq 252 \qquad \text{The minimum amount of income is \$252}$$

Solving the inequalities for p in terms of c, we get the following system of inequalities and its associated graph (Figure 4.12). (We add the restrictions $p \geq 0$ and $c \geq 0$, since it doesn't make sense to work a negative number of hours at either job.)

$$p \leq -c + 30$$

$$p \geq -\frac{16}{21}c + 24 \qquad \text{(In decimal form, } p \geq -0.7619c + 24 \text{ approximately)}$$

$$p \geq 0, c \geq 0$$

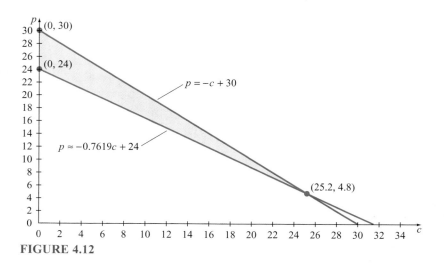

FIGURE 4.12

Every point of the solution region represents a combination of hours at the two jobs that will result in earnings of at least $252.

The corner points of the solution region are $(0, 24)$, $(0, 30)$, and $(25.2, 4.8)$. (The last point is found by calculating the intersection of the two boundary lines.) Let's calculate the student's weekly earnings at the corner points and some of the other points in the region (see Table 4.2).

TABLE 4.2

	Lab Hours	Pizza Hours	Weekly Earnings
Corner point	0	24	$252.00
Corner point	0	30	$315.00
Corner point	25.2	4.8	$252.00
Interior point	5	25	$302.50
Interior point	10	20	$290.00
Interior point	15	14	$267.00

The weekly earnings vary; however, in every case the number of work hours is less than or equal to 30 hours and the earnings are greater than or equal to $252.

4.1 Summary

In this section, you learned how to graph linear inequalities. You discovered that the solution to a system of linear inequalities is the intersection of the graphs of the solution regions of the individual inequalities.

4.1 Exercises

In Exercises 1–10, graph the solution region of the linear inequality. Then use the graph to determine if point P is in the solution region.

1. $2x + y \leq 6$;
$P = (2, 4)$

2. $4x + y \leq 0$;
$P = (1, 1)$

3. $x + 5y \leq 10$;
$P = (0, 0)$

4. $5x + 6y \leq 30$;
$P = (0, 5)$

5. $-2x + 4y \geq -2$;
$P = (1, 2)$

6. $x - y \leq 10$;
$P = (5, -5)$

7. $5x - 4y \leq 0$;
$P = (1, 0)$

8. $-3x - 3y \leq 9$;
$P = (2, -1)$

9. $2x - y \geq 8$;
$P = (-3, 2)$

10. $7x - 6y \geq 12$;
$P = (0, -1)$

In Exercises 11–25, graph the solution region of the system of linear equations. If there is no solution, explain why.

11. $-4x + y \geq 2$
$-2x + y \geq 1$
$x \leq 0$

12. $-5x + y \geq 0$
$2x + y \leq 4$
$y \geq 0$

13. $-2x + 6y \leq 8$
$4x - 12y \leq -6$

14. $10x - y \geq 12$
$9x - 2y \geq 2$

15. $3x - 2y \leq 4$
$11x - 20y \geq 2$

16. $9x - 6y \leq 0$
$4x + 5y \leq 23$

17. $x - y \leq -5$
$9x + y \leq 25$

18. $2x + 5y \leq 2$
$3x - 5y \leq 3$

19. $6x + 2y \leq 10$
$-x - 2y \geq -5$
$x \geq 0$
$y \geq 0$

20. $x - y \geq 3$
$6x + 7y \leq 44$
$6x - 7y \leq 16$

21. $2x - 4y \geq 16$
$9x + y \leq -4$
$-3x + 6y \leq -24$

22. $2x - 2y \geq 0$
$3x + y \leq 4$
$5x - y \geq 5$

23. $8x - y \geq 3$
$x + 2y \leq 11$
$9x + y \leq 14$

24. $-4x + y \geq 2$
$-2x + y \geq 1$
$y \geq 1$
$x \leq 1$

25. $-5x + y \geq 0$
$2x + y \leq 4$
$y \leq 1$
$x \leq 1$

In Exercises 26–30, set up the system of linear inequalities that can be used to solve the problem. Then graph the system of equations and solve the problem.

26. **Nutritional Content**   A 32-gram serving of Skippy® Creamy Peanut Butter contains 150 milligrams of sodium and 17 grams of fat. A 56-gram serving of Bumble Bee® Chunk Light Tuna in Water contains 250 milligrams of sodium and 0.5 gram of fat. (**Source:** Product labeling.) Some health professionals advise that a person on a 2500-calorie diet should consume no more than 2400 mg of sodium and 80 grams of fat. Graph the region showing all possible serving combinations of peanut butter and tuna that a person could eat and still meet the dietary guidelines.

27. **Nutritional Content**   A Nature Valley® Strawberry Yogurt Chewy Granola Bar contains 130 milligrams of sodium and 3.5 grams of fat. A Nature's Choice® Multigrain Strawberry Cereal Bar contains 65 milligrams of sodium and 1.5 grams of fat. (**Source:** Product labeling.) Some health professionals advise that a person on a 2500-calorie diet should consume no more than 2400 mg of sodium and 80 grams of fat. Graph the region showing all possible serving combinations of granola bars and cereal bars that a person could eat and still meet the dietary guidelines.

28. **Student Wages** A student earns $15.00 per hour designing web pages and $9.00 per hour supervising a campus tutoring center. She has at most 30 hours per week to work, and she needs to earn at least $300. Graph the region showing all possible work-hour allocations that meet her time and income requirements.

29. **Wages** A salaried employee earns $900 per week managing a copy center. He is required to work a minimum of 35 hours but no more than 45 hours weekly. As a side business, he earns $25 per hour designing brochures for local business clients. In order to maintain his standard of living, he must earn $1100 per week. In order to maintain his quality of life, he limits his workload to 50 hours per week. Given that he has no control over the number of hours he has to work managing the copy center, will he be able to consistently meet his workload and income goals? Explain.

30. **Commodity Prices** Today's Market Prices (www.todaymarket.com) is a daily fruit and vegetable wholesale market price service. Produce retailers who subscribe to the service can use the wholesale prices to aid them in setting retail prices for the fruits and vegetables they sell.

A 25-pound carton of peaches holds 60 medium peaches or 70 small peaches. In August 2002, the wholesale price for local peaches in Los Angeles was $9.00 per carton for medium peaches and $10.00 per carton for small peaches. (**Source:** Today's Market Prices.) A fruit vendor has budgeted up to $100 to spend on peaches. He estimates that weekly demand for peaches is at least 420 peaches but no more than 630 peaches. He wants to buy enough peaches to meet the minimum estimated demand but no more than the maximum estimated demand. Graph the region showing which small- and medium-size peach carton combinations meet his demand and budget restrictions.

Exercises 31–40 are intended to challenge your understanding of the graphs of linear inequalities.

31. Graph the solution region of the system of linear inequalities and identify the coordinates of the corner points.

$$2x + 3y \le 6$$
$$-2x + 4y \ge 4$$
$$-5x + y \le 15$$
$$x \le 5$$
$$y \ge 2$$

32. Graph the solution region of the system of linear inequalities and identify the coordinates of the corner points.

$$-2x + y \le 4$$
$$7x + 2y \ge 8$$
$$x \le 0$$

33. Graph the solution region of the system of linear inequalities and identify the coordinates of the corner points.

$$-x + y \le 0$$
$$-x - y \ge -4$$
$$y \ge 2$$

34. Write a system of inequalities whose solution region has the corner points $(0, 0)$, $(1, 3)$, $(3, 5)$, and $(2, 1)$.

35. Write a system of inequalities whose solution region has the corner points $(1, 1)$, $(1, 3)$, $(5, 3)$, and $(2, 1)$.

36. Write a system of inequalities whose *unbounded* solution region has the corner points $(0, 5)$, $(2, 1)$, and $(5, 0)$.

37. Write a system of inequalities whose *unbounded* solution region has the corner points $(0, 5)$, $(4, 4)$, and $(5, 0)$.

38. A student concludes that the corner points of a solution region defined by a system of linear inequalities are $(0, 0)$, $(1, 1)$, $(0, 2)$, and $(2, 2)$. After looking at the graph of the region, the instructor immediately concludes that the student is incorrect. How did the instructor know?

39. Is it possible to have a bounded solution region with exactly one corner point? If so, give a system of inequalities whose solution region is bounded and has exactly one corner point.

40. Is it possible to have an *unbounded* solution region with exactly one corner point? If so, give a system of inequalities whose solution region is unbounded and has exactly one corner point.

4.2 Solving Linear Programming Problems Graphically

- Determine the feasible region of a linear programming problem
- Solve linear and integer programming problems in two variables

GETTING **STARTED** Many products, such as printer ink, are sold to business customers at a discount if large quantities are ordered. Profitable businesses want to minimize their supply costs yet have sufficient ink on hand to fulfill their printing requirements. How much ink should they order? The process of acquiring, producing, and distributing supplies can often be made more efficient by setting up and solving systems of linear inequalities.

In this section, we show how a mathematical method called **linear programming** can help businesses determine the most cost-effective way to manage their resources. We will demonstrate how linear programming is used to optimize an objective function subject to a set of linear constraints. We will also reveal how to find the whole-number solution of an integer programming problem. We will begin our discussion with the following set of definitions.

LINEAR PROGRAMMING PROBLEM

A linear equation $z = ax + by$, called an **objective function,** may be maximized or minimized subject to a set of linear **constraints** of the form

$$cx + dy \le f \quad \text{or} \quad cx + dy \ge f$$

where x and y are variables (called **decision variables**) and a, b, c, d and f are real numbers. A problem consisting of an objective function and a set of linear constraints is called a **linear programming problem.** The values of x and y that optimize (maximize or minimize) the value of the objective function are called the **optimal solution.** A linear programming problem with the additional constraint that x and y are integers is called an **integer programming problem.**

In Section 4.1, we gave the example of a student earning \$10.50 an hour delivering pizza and \$8.00 an hour working in a campus computer lab. He had only 30 hours per week to work, and he had to earn at least \$252 during that period. We let p be the number of hours he spent delivering pizza and c be the number of hours he spent working in the computer lab. We had the constraints

$$c + p \le 30 \qquad \text{He can work at most 30 hours}$$
$$8c + 10.5p \ge 252 \qquad \text{He must earn at least \$252}$$
$$p \ge 0, c \ge 0 \qquad \text{He must work a nonnegative number of hours at each job}$$

which we rewrote as

$$p \le -c + 30$$
$$p \ge -\frac{16}{21}c + 24$$
$$p \ge 0, c \ge 0$$

From this scenario, we can set up linear programming problems to address each of the following questions:

1. What is the largest amount of money he can earn?

Objective function: Maximize $z = 8c + 10.5p$ Total amount earned

Subject to
$$
\begin{cases}
p \leq -c + 30 \\
p \geq -\dfrac{16}{21}c + 24 \\
p \geq 0, c \geq 0
\end{cases}
$$

2. What is the least number of hours he can work?

Objective function: Minimize $z = c + p$ Total hours worked

Subject to
$$
\begin{cases}
p \leq -c + 30 \\
p \geq -\dfrac{16}{21}c + 24 \\
p \geq 0, c \geq 0
\end{cases}
$$

3. What is the maximum number of hours he can work in the computer lab?

Objective function: Maximize $z = c$ Total computer lab hours

Subject to
$$
\begin{cases}
p \leq -c + 30 \\
p \geq -\dfrac{16}{21}c + 24 \\
p \geq 0, c \geq 0
\end{cases}
$$

Although each of the three objective functions has the same constraints, the optimal solution to each linear programming problem will differ based upon the objective function. However, all solutions will lie within the solution region of the system of constraints. In the context of linear programming, we call the solution region of the system of constraints the **feasible region** and the points within the region **feasible points.** Which of all the feasible points will optimize each of the objective functions? Testing all of the points in the feasible region would be an impossible task! Fortunately, we don't have to. The Fundamental Theorem of Linear Programming limits the number of points we have to test.

FUNDAMENTAL THEOREM OF LINEAR PROGRAMMING

1. If the solution to a linear programming problem exists, it will occur at a corner point.

2. If two adjacent corner points are optimal solutions, then all points on the line segment between them are also optimal solutions.

3. Linear programming problems with bounded feasible regions will always have optimal solutions.

4. Linear programming problems with unbounded feasible regions may or may not have optimal solutions.

Recall that the graph of the feasible region for the computer lab and pizza delivery problem had the corner points $(0, 24)$, $(0, 30)$, and $(25.2, 4.8)$, as shown in Figure 4.13.

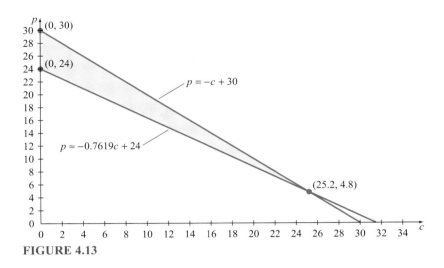

FIGURE 4.13

The solution to each one of the linear programming problems will occur at one of these points.

   1. What is the largest amount of money he can earn?

$$\text{Maximize} \quad z = 8c + 10.5p.$$

We evaluate the objective function at each corner point (see Table 4.3).

TABLE 4.3

Corner Point		Objective Function
Computer Lab Hours (c)	Pizza Delivery Hours (p)	Total Earnings (in dollars) ($z = 8c + 10.5p$)
0	24	252
0	30	315
25.2	4.8	252

The maximum value of z occurs at corner point $(0, 30)$. The optimal solution is $c = 0$ and $p = 30$. To maximize his earnings, he should work 0 hours in the lab and 30 hours delivering pizza.

   2. What is the least number of hours he can work?

$$\text{Minimize} \quad z = c + p.$$

We evaluate the objective function at each corner point (see Table 4.4).

TABLE 4.4

Corner Point		Objective Function
Computer Lab Hours (c)	Pizza Delivery Hours (p)	Total Work Hours ($z = c + p$)
0	24	24
0	30	30
25.2	4.8	30

The minimum value of z occurs at corner point (0, 24). The optimal solution is $c = 0$ and $p = 24$. To minimize his work hours, he should work 0 hours in the lab and 24 hours delivering pizza.

3. What is the maximum number of hours he can work in the computer lab?

$$\text{Maximize} \quad z = c.$$

We evaluate the objective function at each corner point (see Table 4.5).

TABLE 4.5

Corner Point		Objective Function
Computer Lab Hours (c)	Pizza Delivery Hours (p)	Lab Work Hours ($z = c$)
0	24	0
0	30	0
25.2	4.8	25.2

The maximum value of z occurs at corner point (25.2, 4.8). The optimal solution is $c = 25.2$ and $p = 4.8$. The maximum amount of time he can work in the computer lab is 25.2 hours. He will still have to work 4.8 hours delivering pizza to reach his earnings goal.

Why do we have to check only the corner points of the feasible region? We'll address this question by considering a "family" of objective functions. Suppose we are asked to maximize the objective function $z = 3x + y$ subject to the following constraints:

$$4x + y \leq 12$$
$$2x + y \leq 8$$
$$x \geq 0$$
$$y \geq 0$$

The graph of the feasible region and the coordinates of the corner points are shown in Figure 4.14.

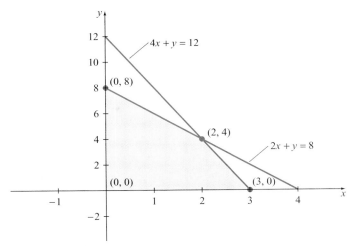

FIGURE 4.14

If we set the variable z in the objective function $z = 3x + y$ to a fixed value c, then the graph of the line $3x + y = c$ will pass through all points (x, y) that satisfy the equation $3x + y = c$. If we repeat this for different values of c, we will end up with a "family" of objective function lines. The lines will be parallel with different y-intercepts. (In Figure 4.15, we set z equal to the following constant values: 2, 4, 6, 8, 10, 12.)

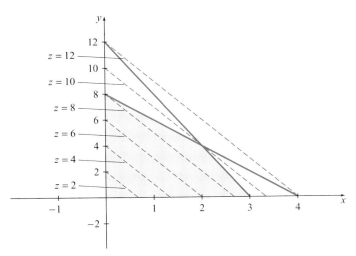

FIGURE 4.15

Recall that we want to maximize the objective function subject to the constraints. Observe that although $z = 12$ is the largest value of z shown in the figure, the line $3x + y = 12$ does not intersect the feasible region. Consequently, no feasible solution lies on the line $3x + y = 12$.

The first line above the feasible region that intersects the feasible region is the line $3x + y = 10$. This line intersects the feasible region at a single point: the corner point $(2, 4)$. At this corner point, $z = 10$. All other objective function lines that cross the feasible region take on values of z less than 10. Therefore, the objective function $z = 3x + y$ has an optimal solution at $(2, 4)$. At this point, the objective function takes on its maximum value: $z = 10$. This value of the objective function is referred to as the **optimal value** for the linear programming problem. Regardless of the objective function, the maximum (or minimum) value of the objective function will occur at a corner point of the feasible region. An argument similar to that given here can be made for any objective function and any feasible region.

The process of solving linear programming problems graphically is summarized in the following box.

HOW TO **Graphical Method for Solving Linear Programming (LP) Problems**

1. Graph the feasible region determined by the constraints.
2. Find the corner points of the feasible region.
3. Find the value of the objective function at each of the corner points.
4. If the feasible region is bounded, the maximum or minimum value of the objective function will occur at one of the corner points.
5. If the feasible region is an unbounded region in the first quadrant and the coefficients of the objective function are positive, then the objective function has a minimum value at a corner point. The objective function will not have a maximum value.

EXAMPLE 1   **Solving a Linear Programming Problem Graphically**

Solve the linear programming problem:

$$\text{Maximize} \quad z = 6x + 2y$$

$$\text{Subject to} \quad \begin{cases} -3x + y \geq 2 \\ x + y \leq 10 \\ x \geq 0 \\ y \geq 0 \end{cases}$$

SOLUTION   We begin by solving each inequality for y.

$$-3x + y \geq 2 \qquad\qquad x + y \leq 10$$
$$y \geq 3x + 2 \qquad\qquad y \leq -x + 10$$

Graphing the feasible region yields the graph shown in Figure 4.16.

The corner points occur where the boundary lines intersect. The corner points of the feasible region are (0, 2), (0, 10), and (2, 8). Substituting each of these points into the objective function, $z = 6x + 2y$, we get the results in Table 4.6.

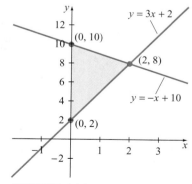

FIGURE 4.16

TABLE 4.6

Corner Point		Objective Function
x	y	$z = 6x + 2y$
0	2	4
0	10	20
2	8	28

The objective function $z = 6x + 2y$ is maximized when $x = 2$ and $y = 8$. At that point, $z = 28$.

EXAMPLE 2

Solving an LP Problem with an Unbounded Feasible Region

Solve the linear programming problem:

$$\text{Minimize} \quad z = 2x + 5y$$

$$\text{Subject to} \quad \begin{cases} 4x + y \geq 4 \\ -x + y \geq 1 \\ x \geq 0 \\ y \geq 0 \end{cases}$$

SOLUTION The objective function is

$$z = 2x + 5y$$

The constraints are

$$4x + y \geq 4 \qquad x \geq 0$$
$$-x + y \geq 1 \qquad y \geq 0$$

We draw each boundary line and shade the feasible region, as shown in Figure 4.17.

The region is unbounded and has two corner points. Since the coefficients of the objective function are both positive, the objective function will have a minimum. The first corner point of the feasible region is $(0, 4)$, the y-intercept of one of the constraints. We determine the coordinates of the second corner point by finding the intersection point of the boundary lines, $y = x + 1$ and $y = -4x + 4$.

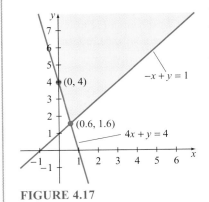

FIGURE 4.17

$$-4x + 4 = x + 1 \qquad \text{Solve the system of equations by the substitution method}$$
$$-5x = -3$$
$$x = 0.6$$
$$y = (0.6) + 1 \qquad \text{Substitute } x = 0.6 \text{ into } y = x + 1$$
$$y = 1.6$$

The second corner point is $(0.6, 1.6)$. Substituting each of the points into the objective function, $z = 2x + 5y$, we obtain the values in Table 4.7.

TABLE 4.7

Corner Point		Objective Function
x	y	$z = 2x + 5y$
0	4	20
0.6	1.6	9.2

The objective function has a minimum at $(0.6, 1.6)$. At that point, $z = 9.2$.

In Example 2, we had an unbounded feasible region. We were able to find a solution because we were asked to minimize the objective function $z = 2x + 5y$. If we had been asked to maximize the objective function, the problem would have had no solution. For a linear programming problem with an unbounded feasible region and an objective function with positive coefficients, no matter what "optimal solution" we pick, we will always be able to find a point (x, y) in the feasible region that yields a greater "optimal value."

EXAMPLE 3 Solving a Linear Programming Problem Graphically

Solve the linear programming problem:

$$\text{Minimize} \quad z = -5x + 3y$$

$$\text{Subject to} \quad \begin{cases} 6x + y \geq 6 \\ -2x + y \geq 1 \\ x \leq 2 \\ y \geq 1 \\ x \geq 0 \end{cases}$$

SOLUTION After graphing each of the five boundary lines, we will use arrows to show which side of the boundary line will be shaded, as shown in Figure 4.18. This technique is especially helpful when working with a large number of constraints.

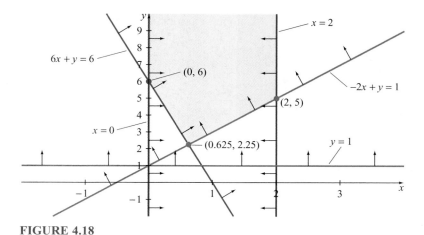

FIGURE 4.18

We shade the region that has boundary lines with all arrows pointing to the interior of the region. Notice that the line $y = 1$ does not form a boundary line of the feasible region. This is okay so long as the feasible region is on the appropriate side of $y = 1$. Since the arrows on the line $y = 1$ point toward the side that contains the feasible region, the constraint $y \geq 1$ is satisfied.

The first corner point $(0, 6)$ is easily determined, since it is the y-intercept of the boundary line $6x + y = 6$. The second corner point occurs at the intersection of $6x + y = 6$ and $-2x + y = 1$. We must solve the system of equations

$$6x + y = 6$$
$$-2x + y = 1$$

Subtracting the second equation from the first equation yields

$$8x = 5$$
$$x = 0.625$$

To determine the value of y, we substitute the x value back into the equation $-2x + y = 1$.

$$-2x + y = 1$$
$$-2(0.625) + y = 1 \qquad \text{Since } x = 0.625$$
$$-1.25 + y = 1$$
$$y = 2.25$$

The second corner point is $(0.625, 2.25)$.

The third point occurs at the intersection of $x = 2$ and $-2x + y = 1$. The x coordinate of the corner point is $x = 2$. To determine the y coordinate, we substitute this result into $-2x + y = 1$.

$$-2x + y = 1$$
$$-2(2) + y = 1 \qquad \text{Since } x = 2$$
$$-4 + y = 1$$
$$y = 5$$

The third corner point is $(2, 5)$.

With the corner points identified, we are ready to evaluate the objective function $z = -5x + 3y$ at each corner point, as shown in Table 4.8.

TABLE 4.8

Corner Point		Objective Function
x	y	$z = -5x + 3y$
0	6	18
0.625	2.25	3.625
2	5	5

Since we are looking for the minimum value of the objective function, the optimal solution is $(0.625, 2.25)$ and the optimal value is 3.625.

Real-Life Applications

As shown in Examples 4 and 5, many real-life problems can be analyzed using the techniques of linear programming.

EXAMPLE 4

Using Linear Programming to Do Investment Analysis

Table 4.9 shows the average annual rate of return on two TIAA-CREF investment accounts over a 10-year period.

TABLE 4.9

As of 6/30/04	
CREF Variable Annuity Accounts	10-Year Average
Bond Market	7.15%
Social Choice	10.31%

Source: www.tiaa-cref.com.

An investor wants to invest at least $3000 in the Bond Market and Social Choice accounts. He assumes that he will be able to get a return equal to the 10-year average, and he wants the total return on his investment to be at least 9 percent. He assigns each share in the Bond Market account a risk rating of 2 and each share in the Social Choice account a risk rating of 4. The approximate share price at the end of June 2004 was $74 per share for the Bond Market account and $102 per share for the Social Choice account. He will use these prices in his analysis. How many shares of each account should he buy in order to minimize his overall risk? [Note that fractions of shares may be purchased. Also, to make computations easier, round each percentage to the nearest whole-number percent (i.e., 10.31 percent = 10 percent).]

SOLUTION Let x be the number of shares in the Bond Market account and y be the number of shares in the Social Choice account. The rounded rate of return for the Bond Market account is 7 percent, and that for the Social Choice account is 10 percent. The share price for the Bond Market account is $74, and that for the Social Choice account is $102. Each Bond Market share has a risk rating of 2, and each Social Choice share has a risk rating of 4. We want to minimize the overall risk. That is, we want to minimize $z = 2x + 4y$.

The amount of money invested in each account is equal to the share price times the number of shares. The amount invested is $74x$ for the Bond Market account and $102y$ for the Social Choice account. The total amount invested is given by the equation

$$74x + 102y \geq 3000 \qquad \text{The total amount invested is at least \$3000}$$

The dollar amount of the return on the investment is equal to the product of the rate and the amount invested. The return on the Bond Market account is $(0.07)(74x) = 5.18x$, and the return on the Social Choice account is $(0.10)(102y) = 10.20y$. Since we want to earn at least 9 percent on the total amount of money invested $(74x + 102y)$, the dollar amount of the minimum combined return is $(0.09)(74x + 102y) = 6.66x + 9.18y$. The combined return is given by the equation

$$5.18x + 10.20y \geq 6.66x + 9.18y \qquad \text{The combined return is at least 9 percent of the amount invested}$$

$$-1.48x + 1.02y \geq 0$$

Combining the objective function and each of the constraints yields the following linear programming problem:

Minimize $z = 2x + 4y$

Subject to
$$\begin{cases} 74x + 102y \geq 3000 & \text{The total amount invested is at least} \\ & \$3000 \\ -1.48x + 1.02y \geq 0 & \text{The combined return is at least 9 per-} \\ & \text{cent of the amount invested.} \\ x \geq 0 \\ y \geq 0 \end{cases}$$

The graph of the feasible region is shown in Figure 4.19.

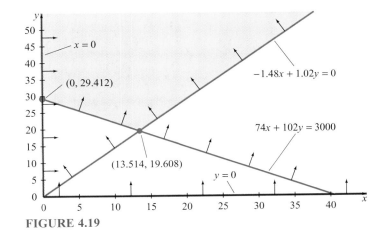

FIGURE 4.19

The first corner point is the y-intercept of the line $74x + 102y = 3000$. Since $\frac{3000}{102} \approx 29.412$, the corner point is $(0, 29.412)$.

The second corner point is the point of intersection of $74x + 102y = 3000$ and $-1.48x + 1.02y = 0$. We must solve the system of equations

$$74x + 102y = 3000$$
$$-1.48x + 1.02y = 0$$

We will find the solution using matrices.

$$\begin{bmatrix} 74 & 102 & \vdots & 3000 \\ -1.48 & 1.02 & \vdots & 0 \end{bmatrix}$$

$$\begin{bmatrix} 74 & 102 & \vdots & 3000 \\ 222 & 0 & \vdots & 3000 \end{bmatrix} \quad R_1 - 100R_2$$

$$\begin{bmatrix} 0 & 306 & \vdots & 6000 \\ 222 & 0 & \vdots & 3000 \end{bmatrix} \quad 3R_1 - R_2$$

$$\begin{bmatrix} 0 & 1 & \vdots & 19.608 \\ 1 & 0 & \vdots & 13.514 \end{bmatrix} \quad \begin{matrix} 1/306\, R_1 \\ 1/222\, R_2 \end{matrix}$$

$$\begin{bmatrix} 1 & 0 & \vdots & 13.514 \\ 0 & 1 & \vdots & 19.608 \end{bmatrix} \quad \begin{matrix} R_2 \\ R_1 \end{matrix}$$

The second corner point is $(13.514, 19.608)$.

We evaluate the objective function $z = 2x + 4y$ at each corner point, as shown in Table 4.10.

TABLE 4.10

Corner Point		Objective Function
x	y	$z = 2x + 4y$
0	29.412	117.648
13.514	19.608	105.460

The optimal solution is $(13.514, 19.608)$, and the optimal value is 105.460.

The risk is minimized when 13.514 shares of the Bond Market account and 19.608 shares of the Social Choice account are purchased.

Did the rounding of the percentages and share prices affect the solution? Yes. However, since the 10-year average rate of return on each account was not guaranteed, the investor decided that rounding the percentages to whole-number percents and rounding the share prices to whole-dollar amounts was good enough for modeling purposes.

Integer Programming Problems

In many real-applications, we have the additional constraint that the objective function input values must be whole numbers. Problems of this type are called **integer programming problems.**

INTEGER PROGRAMMING PROBLEM

A linear programming problem with the additional constraint that the decision variables are integers is called an **integer programming problem.**

Integer programming problems containing many input variables and constraints can be very difficult to solve. Even in the two-decision-variable case, solving these types of problems requires several extra steps. Although we won't require you to solve any of these types of problems in the exercises, we will work an example here to illustrate the errors that may arise when the optimal solution of a linear programming problem is rounded to whole-number values.

EXAMPLE 5

Using Linear Programming to Make Business Decisions

Marlborough Printer Supplies sells generic replacement ink cartridges for a variety of printers through its eBay store. Shipping is free when 10 or more cartridges are ordered. In 2002, a single black ink cartridge for the Epson Color Stylus 660 printer cost $2.50, and a three-pack of black ink cartridges cost $6. (**Source:** eBay online store.) The owner of a small business needs to purchase at least 20 ink cartridges and wants to minimize her overall cost. How many single cartridges and how many three-packs should she buy?

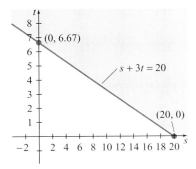

FIGURE 4.20

SOLUTION Let s be the number of single cartridges and t be the number of three-packs. Since 10 or more cartridges will be ordered, shipping will be free. Consequently, the equation of the objective function (the cost function) is

$$C = 2.5s + 6t$$

We have the constraints

$$s + 3t \geq 20 \qquad \text{A minimum of 20 cartridges are ordered}$$
$$s \geq 0, t \geq 0 \qquad \text{The number of each type ordered is nonnegative}$$

The feasible region is unbounded and has corner points $\left(0, 6\frac{2}{3}\right)$ and $(20, 0)$, as shown in Figure 4.20.

Evaluating the objective function $C = 2.5s + 6t$ at the two corner points yields the data in Table 4.11.

TABLE 4.11

Corner Point		Objective Function
Single (s)	Three-Pack (t)	$C = 2.5s + 6t$
20	0	50
0	$6\frac{2}{3}$	40

Since the smallest value of the objective function is 40, the optimal solution is $\left(0, 6\frac{2}{3}\right)$ and the optimal value is 40. However, since we can't order a fraction of a three-pack, we must find the whole-number solution. Our natural tendency might be to round the optimal solution to $(0, 7)$. However, doing so also alters the optimal value, as shown in Table 4.12.

TABLE 4.12

Rounded Optimal Solution		Objective Function
Single (s)	Three-Pack (t)	$C = 2.5s + 6t$
0	7	42

Using whole-number values for s and t, is it possible to further reduce the cost? We will investigate this question by adding additional constraints. We create Subproblem 1 by adding the constraint $t \leq 6$ and Subproblem 2 by adding the constraint $t \geq 7$, as shown in Figure 4.21. (These are the whole-number values on either side of $6\frac{2}{3}$.)

These new constraints split the feasible region into two separate regions. We will solve Subproblem 1 first. The unbounded feasible region of Subproblem 1 has corner points $(20, 0)$ and $(2, 6)$.

FIGURE 4.21

TABLE 4.13

Corner Point of Subproblem 1		Objective Function
Single (s)	Three-Pack (t)	$C = 2.5s + 6t$
20	0	50
2	6	41

The optimal solution for Subproblem 1 is (2, 6), as shown in Table 4.13. Since this is a whole-number solution, it makes sense in the context of the problem. When two single cartridges and six three-packs are purchased, the total ink cost is $41.

The feasible region of Subproblem 2 has the corner point (0, 7) (see Table 4.14).

TABLE 4.14

Corner Point of Subproblem 2		Objective Function
Single (s)	Three-Pack (t)	$C = 2.5s + 6t$
0	7	42

The optimal solution for Subproblem 2 is (0, 7). Since this is also a whole-number solution, it makes sense in the context of the problem. When no single cartridges and seven three-packs are purchased, the total ink cost is $42.

The optimal whole-number solution for the entire linear programming problem will be the subproblem solution that yields the smallest value of the objective function. The whole-number solution for Subproblem 1, (2, 6), had optimal value $41. The whole-number solution for Subproblem 2, (0, 7), had optimal value $42. Comparing the results of Subproblems 1 and 2, we conclude that the optimal whole-number solution of the entire problem is (2, 6). When two single cartridges and six three-packs are ordered, the overall cost is minimized. (It is important to note that if the business owner spent the extra dollar and ordered no single cartridges and seven three-packs, she would get 21 cartridges instead of 20 cartridges. She may decide that the extra cartridge is worth the extra dollar.)

As shown in Example 5, we must be aware that rounding an optimal solution to whole-number values does not guarantee that we have found the optimal whole-number solution.

4.2 Summary

In this section, you learned how a mathematical method called linear programming can help businesses determine the most cost-effective way to manage their resources. You used linear programming to optimize an objective function subject to a set of linear constraints.

4.2 Exercises

In Exercises 1–20, find the optimal solution and optimal value of the linear programming problem. If a solution does not exist, explain why.

1. Minimize $z = 3x + 7y$

Subject to $\begin{cases} 4x + y \geq 4 \\ -x + y \geq 1 \\ x \geq 0 \\ y \geq 0 \end{cases}$

2. Minimize $z = 6x + 2y$

Subject to $\begin{cases} 6x + y \geq 16 \\ -2x + y \geq 0 \\ x \geq 0 \\ y \geq 0 \end{cases}$

3. Minimize $z = 9x + y$

Subject to $\begin{cases} 6x + y \geq 16 \\ -2x + y \geq 0 \\ x \geq 0 \\ y \geq 0 \end{cases}$

4. Maximize $z = 9x + y$

Subject to $\begin{cases} 6x + y \leq 16 \\ -2x + y \leq 0 \\ x \geq 0 \\ y \geq 0 \end{cases}$

5. Maximize $z = x + 10y$

Subject to $\begin{cases} 6x + y \leq 16 \\ -2x + y \leq 0 \\ x \geq 0 \\ y \geq 0 \end{cases}$

6. Minimize $z = 2x - 5y$

Subject to $\begin{cases} 4x + y \leq 12 \\ -6x + 2y \leq 24 \\ x \geq 0 \\ y \geq 0 \end{cases}$

7. Maximize $z = 2x - 5y$

Subject to $\begin{cases} 4x + y \leq 12 \\ -6x + 2y \leq 24 \\ x \geq 0 \\ y \geq 0 \end{cases}$

8. Minimize $z = x - y$

Subject to $\begin{cases} -4x + y \geq 8 \\ -3x + y \leq 6 \\ x \geq 0 \\ y \geq 0 \end{cases}$

9. Maximize $z = -2x - y$

Subject to $\begin{cases} -4x + y \geq 8 \\ -3x + y \leq 6 \\ x \leq 4 \\ x \geq 0 \\ y \geq 0 \end{cases}$

10. Minimize $z = 5x - y$

Subject to $\begin{cases} -3x + y \geq 9 \\ -2x + y \leq 6 \\ x \leq 3 \\ x \geq 0 \\ y \geq 0 \end{cases}$

11. Minimize $z = -2x + 7y$

Subject to $\begin{cases} -x + 4y \geq 4 \\ -x + y \geq 1 \\ x \geq 0 \\ y \geq 0 \end{cases}$

12. Minimize $z = 3x + 5y$

Subject to $\begin{cases} 6x + y \geq 21 \\ -2x + y \geq 1 \\ x \geq 3 \\ y \geq 0 \end{cases}$

13. Minimize $z = 11x + 9y$

Subject to $\begin{cases} 6x + y \geq 16 \\ -2x + y \geq 0 \\ x \geq 0 \\ y \geq 2 \end{cases}$

14. Maximize $z = 11x + 9y$

Subject to $\begin{cases} 6x + y \geq 16 \\ -2x + y \geq 0 \\ x \geq 0 \\ y \geq 2 \end{cases}$

15. Maximize $z = x + 10y$

Subject to
$$\begin{cases} 6x + y \le 29 \\ -2x + y \le -3 \\ y \le 5 \\ x \ge 0 \\ y \ge 0 \end{cases}$$

16. Minimize $z = 20x - y$

Subject to
$$\begin{cases} 4x + y \le 20 \\ -6x + 2y \le 40 \\ x \ge 0 \\ y \ge 0 \end{cases}$$

17. Maximize $z = 20x - y$

Subject to
$$\begin{cases} 4x + y \le 20 \\ -6x + 2y \le 40 \\ x \ge 0 \\ y \ge 0 \end{cases}$$

18. Minimize $z = x - y$

Subject to
$$\begin{cases} 2x + y \ge 0 \\ 3x - y \le 0 \\ x \ge 1 \\ y \ge 1 \end{cases}$$

19. Maximize $z = -2x + y$

Subject to
$$\begin{cases} -4x + 3y \ge 0 \\ -3x + 4y \ge -1 \\ x \le 4 \\ x \ge 0 \\ y \ge 2 \end{cases}$$

20. Minimize $z = x + y$

Subject to
$$\begin{cases} 10x + 2y \ge 8 \\ -20x - 4y \le -16 \\ x \le 4 \\ x \ge 0 \\ y \ge 0 \end{cases}$$

For Exercises 21–32, identify the objective function and constraints of the linear programming problem. Then solve the problem and interpret the real-world meaning of the results.

21. **Minimum Commodity Cost** Today's Market Prices (www.todaymarket.com) is a daily fruit and vegetable wholesale market price service. Produce retailers who subscribe to the service can use wholesale prices to aid them in setting retail prices for the fruits and vegetables they sell.

A 25-pound carton of peaches holds 60 medium peaches or 70 small peaches. In August 2002, the wholesale price for local peaches in Los Angeles was $9.00 per carton for medium peaches and $10.00 per carton for small peaches. (**Source:** Today's Market Prices.) A fruit vendor sells the medium peaches for $0.50 each and the small peaches for $0.45 each. He estimates that weekly demand for peaches is at least 420 peaches but no more than 630 peaches. He wants to buy enough peaches to meet the minimum estimated demand, but no more than the maximum estimated demand. How many boxes of each size of peaches should he buy if he wants to minimize his wholesale cost?

22. **Painkiller Costs** An online drugstore sells Tylenol Extra Strength in a variety of bottle sizes. The 250-caplet bottle costs $\approx$ $15, and the 150-caplet bottle costs $\approx$ $12. (**Source:** www.drugstore.com.) A family wants to order a supply of at least 750 caplets. How many 150-caplet bottles and how many 250-caplet bottles should the family order if it wants to minimize costs?

23. **Investment Choices** The following table shows the average annual rate of return on a variety of TIAA-CREF investment accounts over a 10-year period.

As of 6/30/04	
CREF Variable Annuity Accounts	10-Year Average
Bond Market	7.15%
Equity Index	11.34%
Global Equities	7.24%
Growth	8.56%
Money Market	4.22%
Social Choice	10.31%
Stock	9.97%

Source: www.tiaa-cref.com.

An investor wants to invest at least $2000 in the Stock and Growth accounts. He assumes that he will be able to get a return equal to the 10-year average, and he wants the total return on his investment to be at least 9 percent. He assigns each share in the Stock account a risk rating of 6 and each share in the Growth account a risk rating of 7. The approximate share price at the end of June 2004 was $174 per share for the Stock account and $55 per share for the Growth account. He will use these prices in his analysis. How many shares of each account should he buy in order to minimize his overall risk? [Note that fractions of shares may be purchased. Also, to make computations easier, round each percentage to the nearest whole-number percent (i.e., for 9.97 percent, use 10 percent).]

24. **Investment Choices** An investor wants to invest at least $5000 in the Global Equities and Equity Index accounts shown in Exercise 23. She assumes that she will be able to get a return equal to the 10-year average, and she wants the total return on her investment to be at least 10 percent. She assigns each share in the Global Equities account a risk rating of 6 and each share in the Equity Index account a risk rating of 5. The approximate share price at the end of June 2004 was $70 per share for the Global Equities account and $72 per share for the Equity Index account. She will use these prices in her analysis. How many shares of each account should she buy in order to minimize her overall risk? [Note that fractions of shares may be purchased. Also, to make computations easier, round each percentage to the nearest whole-number percent (i.e., for 11.34 percent, use 11 percent).]

25. **Pet Nutrition: Food Cost** PETsMART.com sold the following varieties of dog food in June 2003:

    Nature's Recipe Venison Meal & Rice Canine, 20 percent protein, $21.99 per 20-pound bag
    PETsMART Premier Oven Baked Lamb Recipe, 25 percent protein, $22.99 per 30-pound bag.
    (**Source:** www.petsmart.com.)

    A dog breeder wants to make at least 300 pounds of a mix containing at most 22 percent protein. How many bags of each dog food variety should the breeder buy in order to minimize cost? (*Hint:* Note that each bag is a different weight.)

26. **First Aid Kit Supplies** Safetymax.com sells first aid supplies to businesses. A company that assembles first aid kits for consumers purchases 3500 $1'' \times 3''$ plastic adhesive bandages and 1800 alcohol wipes from Safetymax.com.

The company assembles standard and deluxe first aid kits for sale to consumers. A **standard** first aid kit contains 40 plastic adhesive bandages and 20 alcohol wipes. A **deluxe** first aid kit contains 50 plastic adhesive bandages and 28 alcohol wipes.

The company makes a profit of $3 from each standard first aid kit sold and $4 from each deluxe first aid kit sold. Assuming that every kit produced will be sold, how many of each type of kit should the company assemble in order to maximize profit?

27. **Furniture Production** In June 2004, an online furniture retailer offered the following items at the indicated prices:

    Teak Double Rocker, $745
    Teak Tennis Bench, $124
    (**Source:** www.outdoordecor.com.)

    Suppose that the number of hours required to produce each item is as shown in the following table.

	Cut	Finish	Package
Rocker	4	7	1
Bench	1	3	1

    The company has a maximum of 360 labor hours available in the Cutting Department, a maximum of 730 labor hours available in the Finishing Department, and a maximum of 150 labor hours available in the Packaging Department. Suppose that the company makes a profit of $314 from the sale of each rocker and $57 from the sale of each bench. Assuming that all items produced are sold, how many rockers and how many benches should the company produce in order to maximize profit?

28. **Furniture Production** In June 2004, an online furniture retailer offered the following items at the indicated prices:

    Avalon Teak Armchair, $378
    Teak Tennis Bench, $124
    (**Source:** www.outdoordecor.com.)

    Suppose that the number of hours required to produce each item is as shown in the following table.

	Cut	Finish	Package
Armchair	2	4	1
Bench	1	3	1

The company has a maximum of 200 labor hours available in the Cutting Department, a maximum of 480 labor hours available in the Finishing Department, and a maximum of 150 labor hours available in the Packaging Department. Suppose the company makes a profit of $181 from the sale of each armchair and $57 from the sale of each bench. Assuming that all items produced are sold, how many armchairs and how many benches should the company produce in order to maximize profit?

29. **Transportation Costs** A high school PTA in southern Florida is planning an overnight trip to Orlando, Florida, for its graduating class. A Plus Transportation, a local charter transportation company, offers the following rates (as of December 2003):

Vehicle Capacity	Overnight Charter Rate (for 2 Days of Service)
29	$1000
49	$1800

Source: www.buscharter.net.

The school anticipates that 135 students will go on the trip. Each 29-passenger vehicle requires two chaperones, and each 49-passenger vehicle requires four chaperones. (The chaperones will be traveling with the students.) At most 16 chaperones are available to go on the trip. How many of each type of vehicle should the PTA charter in order to minimize transportation costs?

30. **Transportation Costs** A high school PTA in southern Florida is planning an overnight trip to Orlando, Florida, for its graduating class. A Plus Transportation, a local charter transportation company, offers the following rates (as of December 2003).

Vehicle Capacity	Overnight Charter Rate (for 2 Days of Service)
10	$625
57	$1995

Source: www.buscharter.net.

The school anticipates that 153 students will go on the trip. Each 10-passenger vehicle requires one chaperone, and each 57-passenger vehicle requires six chaperones. (The chaperones will be traveling with the students.) At most 24 chaperones are available to go on the trip. How many of each type of vehicle should the PTA charter in order to minimize transportation costs?

31. **Television Advertising** For the week of July 5-July 11, 2004, a national media research company estimated that 14,834,000 viewers watched *CSI* and 10,557,000 viewers watched *Law and Order*. (**Source:** www.nielsenmedia .com.)

The amount of money a network can charge for advertising is based in part on the size of the viewing audience. Suppose that a 30-second commercial running on *CSI* costs $3100 per spot and a 30-second commercial running on *Law and Order* costs $2500 per spot. A beverage company is willing to spend up to $87,000 for commercials run during episodes of the two programs. The company requires at least 10 spots to be run on each program. How many spots on each program should be purchased in order to maximize the number of viewers?

32. **Television Advertising** For the week of July 5-July 11, 2004, a national media research company estimated that 14,834,000 viewers watched *CSI* and 10,557,000 viewers watched *Law and Order*. (**Source:** www .nielsenmedia.com.)The amount of money a network can charge for advertising is based in part on the size of the viewing audience. Suppose that a 30-second commercial running on *CSI* costs $3100 per spot and a 30-second commercial running on *Law and Order* costs $2500 per spot. An athletic gear company is willing to spend up to $68,500 for commercials run during episodes of the two programs. The company requires at least 10 spots to be run on *CSI* and at least 15 spots on *Law and Order*. How many spots on each program should be purchased in order to maximize the number of viewers?

For Exercises 33–35, use the following data for publicly traded recreational vehicle companies. The information was accurate as of July 16, 2004.

Company	Share Price (dollars)	Earnings/ Share (dollars)	Dividend/ Share (dollars)
Harley-Davidson, Inc. (HDI)	62.70	2.56	0.40
Polaris Industries Inc. (PII)	50.50	2.50	0.92
Winnebago Industries Inc. (WGO)	33.42	1.80	0.20

Source: moneycentral.msn.com.

33. **Investment Choices**   An investor has up to $4000 to invest in Harley-Davidson, Inc., and Polaris Industries, Inc. The investor wants to earn at least $60 in dividends while maximizing total earnings. How many shares of each company's stock should the investor buy? (Assume that portions of shares may be purchased.)

34. **Investment Choices**   An investor has up to $10,000 to invest in Harley-Davidson, Inc., and Winnebago, Inc. The investor wants to earn at least $50 in dividends while maximizing total earnings. How many shares of each company's stock should the investor buy? (Assume that portions of shares may be purchased.)

35. **Investment Choices**   An investor has up to $17,000 to invest in Polaris Industries, Inc., and Winnebago, Inc. The investor wants to have total earnings of at least $900 while

maximizing total dividends. How many shares of each company's stock should the investor buy? (Assume that portions of shares may be purchased.)

Exercises 36–40 are intended to challenge your understanding of linear programming.

36. Consider the following linear programming problem.

$$\text{Maximize} \quad P = 3x + 4y$$

$$\text{Subject to} \quad \begin{cases} x + y \le 2 \\ 3x + 5y \le 9 \\ x \ge 0, y \ge 0 \end{cases}$$

Can a whole-number solution to a linear programming problem be obtained by simply rounding the noninteger solution to whole-number values? Explain.

37. Is it possible for a corner point (a, b) to simultaneously minimize and maximize an objective function? If yes, give an example.

38. Give an example of a linear programming problem that does not have a solution.

39. Is it possible for the feasible region of a linear programming problem not to have any corner points? If yes, give an example.

40. Given the following linear programming problem, what is the maximum possible number of corner points the feasible region could have?

$$\text{Maximize} \quad P = x + y$$

$$\text{Subject to} \quad \begin{cases} -ax + by \le c \\ -dx + fy \le g \\ x \ge 0, y \ge 0 \end{cases}$$

Assume $a, b, c, d, f,$ and g are positive.

4.3 Solving Standard Maximization Problems with the Simplex Method

- Apply the simplex method to solve multivariable standard maximization problems

GETTING STARTED Suppose you are a wholesale fruit buyer for a grocery store. You have been given the assignment to purchase five varieties of apples, two varieties of peaches, and three varieties of pears. The grocery store has known budget and space constraints. You are asked to maximize the number of pieces of fruit purchased. This linear programming problem has 10 decision variables and cannot be solved graphically. Although the graphical method works well for linear programming problems with two decision variables, it does not work for problems containing more than two decision variables. In this section, we will introduce an alternative method that may be used to solve linear programming problems with any number of decision variables.

In 1947, George B. Dantzig developed the **simplex method** to solve linear programming problems. His method has been used to solve linear programming problems with hundreds of decision variables and hundreds of constraints. We will use the method to solve **standard maximization problems.**

STANDARD MAXIMIZATION PROBLEM

A **standard maximization problem** is a linear programming problem with an objective function that is to be maximized. The objective function is of the form

$$P = ax + by + cz + \cdots$$

where $a, b, c, \ldots$ are real numbers and $x, y, z, \ldots$ are decision variables.

The decision variables are constrained to nonnegative values. Additional constraints are of the form

$$Ax + By + Cz + \cdots \leq M$$

where $A, B, C, \ldots$ are real numbers and M is nonnegative.

Observe that substituting $x = 0$, $y = 0$, $z = 0, \ldots$ into each constraint inequality $Ax + By + Cz + \cdots \leq M$ yields $A(0) + B(0) + C(0) + \cdots \leq M$, which simplifies to $0 \leq M$. Since the inequality $0 \leq M$ is valid for all nonnegative values of M, the origin $(0, 0, \ldots, 0)$ is contained in the solution region of each constraint and, consequently, in the feasible region of the standard maximization problem. Furthermore, if at least one of the constraints $Ax + By + Cz + \cdots \leq M$ has all nonnegative coefficients, we are guaranteed that the feasible region is bounded. For the vast majority of our real-life applications, the constraints will have nonnegative coefficients.

Graphically speaking, the simplex method starts at the origin and moves from corner point to corner point of the feasible region, each time increasing the value of the objective function until it attains its maximum value. Remarkably, the method doesn't require that all corner points be tested, a fact that is appreciated by those solving linear programming problems with hundreds of corner points. Although the simplex method may be used for linear programming problems with any number of decision variables, we will restrict our use to problems with four variables or less. Larger problems are typically solved using computers.

We will introduce the simplex method with a two-variable example before formally listing the steps of the method.

Suppose we are asked to use the simplex method to maximize $P = 6x + 2y$ subject to the constraints

$$2x + y \leq 10$$
$$x + y \leq 8$$
$$x \geq 0$$
$$y \geq 0$$

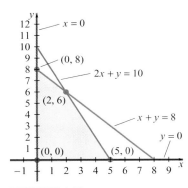

FIGURE 4.22

Since the decision variables are nonnegative and additional constraints are of the form $Ax + By + Cz + \cdots \leq M$ with M nonnegative, this linear programming problem is a standard maximization problem. Consequently, we will be able to maximize the objective function using the simplex method. Although this is not part of the simplex method, we will draw the graph of the feasible region to give you a frame of reference as we work through the problem. The feasible region has corner points $(0, 0)$, $(0, 8)$, $(2, 6)$, and $(5, 0)$ and is shown in Figure 4.22.

The value of the objective function, $P = 6x + 2y$, varies depending upon the corner point. Substituting each corner point (x, y) into $P = 6x + 2y$ and solving for P yields Table 4.15.

TABLE 4.15

Corner Point		Objective Function Value
x	y	$P = 6x + 2y$
0	0	0
0	8	16
2	6	24
5	0	30

We see that the objective function $P = 6x + 2y$ attains a maximum value of 30 at the corner point $(5, 0)$. Using the simplex method to solve the standard maximization problem will yield the same result.

Recall that we are to solve the following standard maximization problem:

$$\text{Maximize} \quad P = 6x + 2y$$

$$\text{Subject to} \quad \begin{cases} 2x + y \leq 10 \\ x + y \leq 8 \\ x \geq 0 \\ y \geq 0 \end{cases}$$

The first step of the simplex method requires us to convert each of the constraint inequalities into an equation. Since $2x + y \leq 10$, there is some value $s \geq 0$ such that $2x + y + s = 10$. The variable s is called a **slack variable** because it "takes up the slack." That is, s adds in whatever value is necessary to make the left-hand side of the equation equal 10. The value of s will vary depending upon the value of x and y (see Table 4.16).

TABLE 4.16

Corner Point		Slack Variable Value
x	y	$s = 10 - 2x - y$
0	0	10
0	8	2
2	6	0
5	0	0

For the constraint inequality $x + y \leq 8$, we observe that there is some value t such that $x + y + t = 8$. Again, the value of t will vary depending upon the value of x and y (see Table 4.17).

TABLE 4.17

Corner Point		Slack Variable Value
x	y	$t = 8 - x - y$
0	0	8
0	8	0
2	6	0
5	0	3

We rewrite the objective function equation $P = 6x + 2y$ by moving all terms to the left-hand side of the equation, which yields the equation $-6x - 2y + P = 0$. The system of equations representing the linear programming problem created by combining the constraint equations and the objective function equation is

$$
\begin{array}{ll}
2x + y + s = 10 & \text{Constraint 1} \\
x + y + t = 8 & \text{Constraint 2} \\
-6x - 2y + P = 0 & \text{Objective function}
\end{array}
$$

This is a system of three equations and five unknowns (x, y, s, t, P). Since the number of unknowns exceeds the number of equations, there will be more than one solution to the system, if a solution exists.

We can represent the system of equations using matrix notation. The augmented matrix is called the **initial simplex tableau.**

$$
\begin{array}{ccccc}
x & y & s & t & P
\end{array}
$$
$$
\left[\begin{array}{ccccc|c}
2 & 1 & 1 & 0 & 0 & 10 \\
1 & 1 & 0 & 1 & 0 & 8 \\
\hline
-6 & -2 & 0 & 0 & 1 & 0
\end{array} \right]
\begin{array}{l}
\text{Constraint 1} \\
\text{Constraint 2} \\
\text{Objective function}
\end{array}
$$

The vertical and horizontal lines in the tableau are used to separate the bottom row and rightmost column from the rest of the matrix. This separation will become important when we calculate test ratios (as will be explained later).

The variables of the columns of the simplex tableau that contain exactly one nonzero entry are called **active (basic)** variables. The variables of the columns that contain more than one nonzero entry are called **inactive (nonbasic)** variables. For the initial tableau, the variables s, t, and P are the active variables, since their corresponding columns in the tableau contain exactly one nonzero value. The variables x and y are the inactive variables in the initial simplex tableau, since their corresponding columns contain more than one nonzero value.

$$
\begin{array}{ccccc}
x & y & s & t & P \\
\end{array}
$$

$$
\left[
\begin{array}{ccccc|c}
2 & 1 & 1 & 0 & 0 & 10 \\
1 & 1 & 0 & 1 & 0 & 8 \\
-6 & -2 & 0 & 0 & 1 & 0 \\
\end{array}
\right]
\quad
\begin{array}{l}
\text{Constraint 1} \\
\text{Constraint 2} \\
\text{Objective function}
\end{array}
$$

The basic feasible solution of the tableau always corresponds with the origin. The solution is obtained by substituting $x = 0$ and $y = 0$ into the equations that generated the tableau.

$$
\begin{array}{ll}
2(0) + (0) + s = 10 & \text{Constraint 1} \\
(0) + (0) + t = 8 & \text{Constraint 2} \\
-6(0) - 2(0) + P = 0 & \text{Objective function}
\end{array}
$$

Simplifying yields

$$
\begin{array}{ll}
s = 10 & \text{Constraint 1} \\
t = 8 & \text{Constraint 2} \\
P = 0 & \text{Objective function}
\end{array}
$$

These results are summarized in Table 4.18.

TABLE 4.18

Corner Point		Objective Function Value $P = 6x + 2y$	Slack Variable Value	
x	y		s	t
0	0	0	10	8

The objective function $P = 6x + 2y$ has a value of 0 at the corner point $(0, 0)$.

An alternative and quicker approach to finding the solution may be obtained from the initial tableau itself by crossing out each inactive variable column and reading the result from the tableau. Crossing out an inactive variable column is equivalent to setting the column variable to zero. For the initial simplex tableau, we have

$$
\begin{array}{ccccc}
x & y & s & t & P \\
\end{array}
$$

$$
\left[
\begin{array}{ccccc|c}
2 & 1 & 1 & 0 & 0 & 10 \\
1 & 1 & 0 & 1 & 0 & 8 \\
-6 & -2 & 0 & 0 & 1 & 0 \\
\end{array}
\right]
\quad
\begin{array}{l}
\text{Constraint 1} \\
\text{Constraint 2} \\
\text{Objective function}
\end{array}
$$

The first row of the resultant matrix is equivalent to

$$
1s + 0t + 0P = 10
$$

$$
s = 10
$$

The second row is equivalent to

$$0s + 1t + 0P = 8$$
$$t = 8$$

The bottom row is equivalent to

$$0s + 0t + 1P = 0$$
$$P = 0$$

These results are the same as those shown in Table 4.18. Clearly, the corner point $(0, 0)$ does not maximize the value of the objective function.

How can we make P as large as possible? The coefficients on the x and y terms of the objective function $P = 6x + 2y$ give us a clue. Since the coefficient on the x term is 6, each 1-unit increase in x will result in a 6-unit increase in P. Similarly, since the coefficient on the y term is 2, each 1-unit increase in y will result in a 2-unit increase in P. Consequently, our initial strategy is to make the x term as big as possible, since it contributes the most to the value of P. The pivoting process detailed next will achieve this objective.

We will now use row operations to convert the initial simplex tableau into a new tableau. We will first select a **pivot column.** Then, using row operations, we will make the pivot column look like one of the columns of the 3 × 3 identity matrix. To choose the pivot column, we find the negative number in the bottom row that is furthest away from zero. The corresponding column is the pivot column. In this tableau, the pivot column will be the x column, since -6 is further away from 0 than -2.

$$
\begin{array}{ccccc}
x & y & s & t & P \\
\end{array}
$$
$$
\left[
\begin{array}{ccccc|c}
2 & 1 & 1 & 0 & 0 & 10 \\
1 & 1 & 0 & 1 & 0 & 8 \\
\hline
-6 & -2 & 0 & 0 & 1 & 0 \\
\end{array}
\right]
$$

Pivot column

Why do we pick the column with the negative number that has the largest magnitude? Recall that we are trying to maximize the objective function $P = 6x + 2y$. As was pointed out earlier, the value of x contributes more to P than the value of y because of the differences in their coefficients. Picking the column with the negative number of the largest magnitude guarantees that we are making the variable that contributes the most to the value of P as large as possible.

The **pivot** is the nonzero entry in the pivot column that will remain after we zero out the other column entries. It must be a positive number. If the pivot column contains more than one positive number, we will select the pivot by using a **test ratio.** To calculate the test ratio, we divide the rightmost entry of each row (excluding the bottom row) by the positive entry in the pivot column of that row.

$$
\begin{array}{ccccc}
x & y & s & t & P \\
\end{array}
$$
$$
\left[
\begin{array}{ccccc|c}
2 & 1 & 1 & 0 & 0 & 10 \\
1 & 1 & 0 & 1 & 0 & 8 \\
\hline
-6 & -2 & 0 & 0 & 1 & 0 \\
\end{array}
\right]
$$

Pivot column

$10/2 = 5 \leftarrow$ Test ratio for first row
$8/1 = 8 \leftarrow$ Test ratio for second row

Since the x column is the pivot column, these test ratios correspond geometrically with the x-intercepts of the constraints (see Figure 4.23).

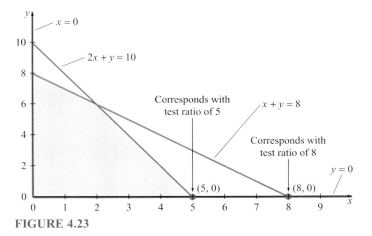

FIGURE 4.23

We select as the pivot the pivot-column entry that has the smallest test ratio. That is, we pick the x-intercept on the graph that is closest to the origin to be the pivot. This ensures that we remain in the feasible region. In this case, the first row of the tableau had the smallest test ratio. Consequently, the pivot will be the 2 in the first row and first column.

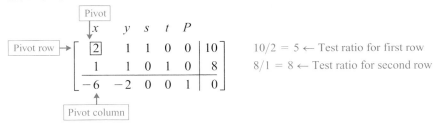

Why do we calculate test ratios only for positive column entries? When we calculate test ratios for the x column, we are actually determining the values of the x-intercepts of the constraints. If the column entry is a negative number, the x-intercept of the constraint will be to the left of the origin. However, we are constrained to nonnegative values for both x and y. Therefore, determining the value of the x-intercepts is necessary only for positive values. (If the column entry is zero, the corresponding line is horizontal and doesn't have an x-intercept.)

Recall that the pivot for this tableau is the 2 in the x column and the first row.

$$
\begin{array}{ccccc}
x & y & s & t & P \\
\end{array}
$$
$$
\left[\begin{array}{ccccc|c}
\boxed{2} & 1 & 1 & 0 & 0 & 10 \\
1 & 1 & 0 & 1 & 0 & 8 \\
\hline
-6 & -2 & 0 & 0 & 1 & 0
\end{array}\right]
$$

We will use the pivot to zero out the remaining entries in the column. However, to ensure that all active variables remain nonnegative, row operations must be of the form

$$aR_c \pm bR_p \rightarrow R_c \text{ with } a > 0 \text{ and } b > 0$$

where R_c is the row to be changed and R_p is the row containing the pivot. Using the row operations

$$2R_2 - R_1 \rightarrow R_2$$

and

$$R_3 + 3R_1 \rightarrow R_3$$

we get the new tableau

$$
\begin{array}{ccccc}
x & y & s & t & P \\
\end{array}
$$
$$
\left[
\begin{array}{ccccc|c}
\boxed{2} & 1 & 1 & 0 & 0 & 10 \\
0 & 1 & -1 & 2 & 0 & 6 \\
0 & 1 & 3 & 0 & 1 & 30
\end{array}
\right]
\begin{array}{c}
\\ 2R_2 - R_1 \\ R_3 + {}_3R_1
\end{array}
$$

Columns x and t almost look like columns of the identity matrix; however, their nonzero entry is a 2 instead of a 1. Using the row operations

$$\frac{1}{2}R_1 \rightarrow R_1$$

and

$$\frac{1}{2}R_2 \rightarrow R_2$$

we get

$$
\begin{array}{ccccc}
x & y & s & t & P \\
\end{array}
$$
$$
\left[
\begin{array}{ccccc|c}
\boxed{1} & \frac{1}{2} & \frac{1}{2} & 0 & 0 & 5 \\
0 & \frac{1}{2} & -\frac{1}{2} & 1 & 0 & 3 \\
0 & 1 & 3 & 0 & 1 & 30
\end{array}
\right]
\begin{array}{c}
1/2\,R_1 \\ 1/2\,R_2 \\ \\
\end{array}
$$

The three columns that look like columns from the 3×3 identity matrix are x, t, and P. Each of these columns has exactly one nonzero entry, so their corresponding variables are the active variables. The remaining columns, y and s, have more than one nonzero entry, so their corresponding variables are the inactive variables. The system of equations represented by the tableau is

$$x + \frac{1}{2}y + \frac{1}{2}s = 5$$

$$\frac{1}{2}y - \frac{1}{2}s + t = 3$$

$$1y + 3s + P = 30$$

Setting the inactive variables y and s to zero yields the system of equations

$$x + \frac{1}{2}(0) + \frac{1}{2}(0) = 5$$

$$\frac{1}{2}(0) - \frac{1}{2}(0) + t = 3$$

$$1(0) + 3(0) + P = 30$$

which simplifies to $x = 5$, $t = 3$, and $P = 30$. The inactive variables y and s are both zero. As noted earlier, a quick way to see the values of the active variables is to cross out the columns of the inactive variables and read the resultant values of the active variables directly from the tableau.

$$
\begin{bmatrix}
x & y & s & t & P & \\
1 & \frac{1}{2} & \frac{1}{2} & 0 & 0 & 5 \\
0 & \frac{1}{2} & -\frac{1}{2} & 1 & 0 & 3 \\
0 & 1 & 3 & 0 & 1 & 30
\end{bmatrix}
$$

Reading from the tableau, we have

$$
x = 5
$$
$$
t = 3
$$
$$
P = 30
$$

The inactive variables y and s are both equal to zero. Since $x = 5$ and $y = 0$, this solution corresponds with the corner point $(5, 0)$ of the feasible region. At $(5, 0)$, $P = 30$.

If there were negative entries in the bottom row of the tableau, we would select a new pivot column and repeat the process. However, since the bottom row contains only nonnegative entries, we are done. The objective function is maximized at $(5, 0)$. This corroborates our earlier conclusion, which was based on Table 4.15.

The simplex method is not intuitive and takes some effort to learn. We will summarize the steps of the process and do several more examples to help you master the method.

HOW TO The Simplex Method

1. Verify that the linear programming problem is a standard maximization problem.

2. Add slack variables to convert the constraints into linear equations.

3. Rewrite the objective function in the form $ax + by + \cdots + P = 0$, where $a, b, \ldots$ are real numbers and $x, y, \ldots$ are decision variables. (Make sure the coefficient on P is positive one.)

4. Set up the initial tableau from the system of equations generated in Steps 2 and 3. Be sure to put the objective function equation from Step 3 in the bottom row of the tableau.

5. Select the pivot column by identifying the negative entry in the bottom row with the largest magnitude.

6. Select the pivot by calculating test ratios for the positive entries in the pivot column and choosing the entry with the smallest test ratio. The test ratio of a row is calculated by dividing the last entry of the row by the entry in the pivot column of the row.

7. Use the pivot to zero out the remaining entries in the pivot column. All row operations must be of the form $aR_c \pm bR_p \rightarrow R_c$ with $a > 0$ and $b > 0$, where R_c is the row to be changed and R_p is the pivot row. As needed, multiply a row by a positive number to convert the pivot to a 1.

8. If the bottom row contains all nonnegative entries, cross out the columns of the inactive variables and read the solution from the tableau. If the bottom row contains negative entries, return to Step 5 and repeat the process for the new tableau.

EXAMPLE 1 **Using the Simplex Method to Solve a Standard Maximization Problem**

Maximize $P = 4x + 3y$ subject to

$$3x + 2y \leq 12$$
$$x + y \leq 5$$
$$x \geq 0, y \geq 0$$

SOLUTION We confirm that the problem is a standard maximization problem. Converting the linear programming problem into a system of linear equations, we get

$$3x + 2y + s = 12$$
$$x + y + t = 5$$
$$-4x - 3y + P = 0$$

and the initial simplex tableau

$$
\begin{array}{ccccc}
x & y & s & t & P \\
\end{array}
$$
$$
\left[
\begin{array}{ccccc|c}
3 & 2 & 1 & 0 & 0 & 12 \\
1 & 1 & 0 & 1 & 0 & 5 \\
\hline
-4 & -3 & 0 & 0 & 1 & 0 \\
\end{array}
\right]
$$

Since -4 is the negative number in the bottom row that has the largest magnitude, we select the x column as the pivot column. Because the entries in the first and second rows of the pivot column are positive, we must calculate the test ratios.

$$
\begin{array}{ccccc}
x & y & s & t & P \\
\end{array}
$$
$$
\left[
\begin{array}{ccccc|c}
3 & 2 & 1 & 0 & 0 & 12 \\
1 & 1 & 0 & 1 & 0 & 5 \\
\hline
-4 & -3 & 0 & 0 & 1 & 0 \\
\end{array}
\right]
\qquad
\begin{array}{l}
12/3 = 4 \\
5/1 = 5 \\
\end{array}
$$

Pivot column

Since the x column is the pivot column, the test ratios give us the x-intercepts of the constraint equations, as shown in Figure 4.24.

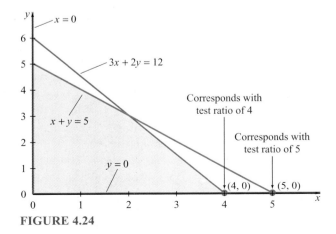

FIGURE 4.24

By picking the column entry with the smallest test ratio, we guarantee that our computations will yield a result that is in the feasible region. The pivot entry is the 3 in the first row and the x column. We zero out the remaining entries in the pivot column with the indicated operations. The resultant tableau is

$$
\begin{array}{ccccc}
x & y & s & t & P \\
\end{array}
$$
$$
\left[\begin{array}{ccccc|c}
\boxed{3} & 2 & 1 & 0 & 0 & 12 \\
0 & 1 & -1 & 3 & 0 & 3 \\
0 & -1 & 4 & 0 & 3 & 48
\end{array}\right]
\begin{array}{l}
\\
3R_2 - R_1 \\
3R_3 + 4R_1
\end{array}
$$

Further simplifying the tableau yields

$$
\begin{array}{ccccc}
x & y & s & t & P \\
\end{array}
$$
$$
\left[\begin{array}{ccccc|c}
1 & \frac{2}{3} & \frac{1}{3} & 0 & 0 & 4 \\
0 & \frac{1}{3} & -\frac{1}{3} & 1 & 0 & 1 \\
0 & -\frac{1}{3} & \frac{4}{3} & 0 & 1 & 16
\end{array}\right]
\begin{array}{l}
\frac{1}{3}R_1 \\
\frac{1}{3}R_2 \\
\frac{1}{3}R_3
\end{array}
$$

We cross out the columns of the inactive variables and read the result:

$$
\begin{array}{ccccc}
x & y & s & t & P \\
\end{array}
$$
$$
\left[\begin{array}{ccccc|c}
1 & \frac{2}{3} & \frac{1}{3} & 0 & 0 & 4 \\
0 & \frac{1}{3} & -\frac{1}{3} & 1 & 0 & 1 \\
0 & -\frac{1}{3} & \frac{4}{3} & 0 & 1 & 16
\end{array}\right]
$$

$x = 4$, $t = 1$, and $P = 16$. The inactive variables y and s are all zero. The corner point $(4, 0)$ yields an objective function value of $P = 16$.

Observe that the fractional entries in the inactive variable columns of the tableau were eliminated when we set the inactive variables to zero. To avoid spending the time calculating fractional entries that will be eliminated anyway, an alternative approach is to cross out the inactive variable columns before converting any active variable column entries to a 1. We then read the resultant equations from the tableau and solve.

$$
\begin{array}{ccccc}
x & y & s & t & P \\
\end{array}
$$
$$
\left[\begin{array}{ccccc|c}
3 & 2 & 1 & 0 & 0 & 12 \\
0 & 1 & -1 & 3 & 0 & 3 \\
0 & -1 & 4 & 0 & 3 & 48
\end{array}\right]
$$

$$3x = 12 \qquad 3t = 3 \qquad 3P = 48$$
$$x = 4 \qquad t = 1 \qquad P = 16$$

Observe that this approach yields the same result and requires fewer computations.

The bottom row of the tableau is equivalent to the equation

$$-y + 4s + 3P = 48$$
$$3P = y - 4s + 48$$
$$P = \frac{1}{3}y - \frac{4}{3}s + 16$$

Observe that since the coefficient on the y term is positive, increasing the value of y while leaving the value of s unchanged will yield a value of P greater than 16. Consequently, the tableau

$$
\begin{array}{ccccc}
x & y & s & t & P \\
\end{array}
$$
$$
\left[\begin{array}{ccccc|c}
3 & 2 & 1 & 0 & 0 & 12 \\
0 & 1 & -1 & 3 & 0 & 3 \\
\hline
0 & -1 & 4 & 0 & 3 & 48
\end{array}\right]
$$

is **not** the final simplex tableau. The negative value in the bottom row indicates that the value of P may be increased.

We will repeat the simplex method for this new tableau.

$$
\begin{array}{ccccc}
x & y & s & t & P \\
\end{array}
$$
$$
\left[\begin{array}{ccccc|c}
3 & 2 & 1 & 0 & 0 & 12 \\
0 & 1 & -1 & 3 & 0 & 3 \\
\hline
0 & -1 & 4 & 0 & 3 & 48
\end{array}\right]
$$

$\uparrow$

New pivot column

The y column is the pivot column, since it is the column with a negative entry in the bottom row. Because the entries in the first and second rows of the pivot column are positive, we must calculate the test ratios.

$$
\begin{array}{ccccc}
x & y & s & t & P \\
\end{array}
$$
$$
\left[\begin{array}{ccccc|c}
3 & \boxed{2} & 1 & 0 & 0 & 12 \\
0 & \boxed{1} & -1 & 3 & 0 & 3 \\
\hline
0 & -1 & 4 & 0 & 3 & 48
\end{array}\right]
\qquad
\begin{array}{l}
12/2 = 6 \\
3/1 = 3
\end{array}
$$

The 1 in the second row is the pivot. We simplify the tableau using the indicated row operations.

$$
\begin{array}{ccccc}
x & y & s & t & P \\
\end{array}
$$
$$
\left[\begin{array}{ccccc|c}
3 & 0 & 3 & -6 & 0 & 6 \\
0 & 1 & -1 & 3 & 0 & 3 \\
\hline
0 & 0 & 3 & 3 & 3 & 51
\end{array}\right]
\qquad
\begin{array}{l}
R_1 - 2R_2 \\
\\
R_3 + R_2
\end{array}
$$

$$
\begin{array}{lll}
3x = 6 & y = 3 & 3P = 51 \\
x = 2 & & P = 17
\end{array}
$$

Because the bottom row of the tableau is entirely nonnegative, we do not need to repeat the pivoting process. The corner point $(2, 3)$ yields a maximum objective function value of $P = 17$.

Observe that each new simplex tableau corresponds with a different corner point. The initial simplex tableau corresponded with $(0, 0)$; the second tableau, with $(4, 0)$; and the final tableau, with $(2, 3)$, as shown in Figure 4.25.

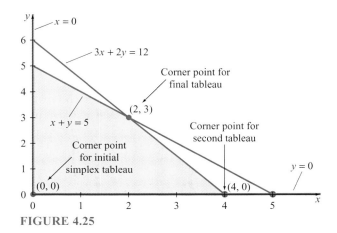

FIGURE 4.25

Graphically speaking, the simplex method moves from a corner point to an adjacent corner point, each time increasing the value of the objective function.

We can verify the accuracy of our result by calculating the value of the objective function at each corner point (see Table 4.19).

TABLE 4.19

Corner Points		Objective Function
x	y	$P = 4x + 3y$
0	0	0
4	0	16
2	3	17
0	5	15

EXAMPLE 2

Using the Simplex Method to Make Business Decisions

A store can buy at most 60 packages of three different brands of cheese. The first brand costs $2 per package, the second brand costs $3 per package, and the third brand costs $4 per package. The store can spend up to $120 on cheese. The first brand resells for $5 per package, the second brand resells for $6 per package, and the third brand resells for $8 per package. How many packages of each brand of cheese should the store buy in order to maximize profit? (*Note:* Profit per item = revenue per item − cost per item.)

SOLUTION Let x be the number of packages of the first brand, y be the number of packages of the second brand, and z be the number of packages of the third brand. The total revenue from cheese sales is $R = 5x + 6y + 8z$. The total cost of the cheese is $C = 2x + 3y + 4z$. Since profit is the difference between revenue and cost,

$$P = R - C$$
$$= (5x + 6y + 8z) - (2x + 3y + 4z)$$
$$= 3x + 3y + 4z$$

Since we want to maximize profit, we must maximize $P = 3x + 3y + 4z$ subject to

$$x + y + z \le 60 \qquad \text{There are at most 60 packages}$$
$$2x + 3y + 4z \le 120 \qquad \text{At most, \$120 is spent on cheese}$$
$$x \ge 0, y \ge 0, z \ge 0 \qquad \text{Nonnegative number of each type of cheese is purchased}$$

The problem is a standard maximization problem. Converting the linear programming problem into a system of linear equations, we get

$$x + \quad y + \quad z + s = 60 \qquad \text{Package quantity constraint}$$
$$2x + 3y + 4z + t = 120 \qquad \text{Budget constraint}$$
$$-3x - 3y - 4z + P = 0 \qquad \text{Objective function}$$

and the initial simplex tableau

$$
\begin{array}{cccccc}
x & y & z & s & t & P \\
\end{array}
$$
$$
\left[
\begin{array}{cccccc|c}
1 & 1 & 1 & 1 & 0 & 0 & 60 \\
2 & 3 & 4 & 0 & 1 & 0 & 120 \\
\hline
-3 & -3 & -4 & 0 & 0 & 1 & 0 \\
\end{array}
\right]
$$

$\uparrow$ Pivot column

Since -4 is the negative number in the bottom row that has the largest magnitude, the z column is the pivot column. Because the entries in the first and second rows of the pivot column are positive, we must calculate the test ratios.

$$
\begin{array}{cccccc}
x & y & z & s & t & P \\
\end{array}
$$
$$
\left[
\begin{array}{cccccc|c}
1 & 1 & 1 & 1 & 0 & 0 & 60 \\
2 & 3 & \boxed{4} & 0 & 1 & 0 & 120 \\
\hline
-3 & -3 & -4 & 0 & 0 & 1 & 0 \\
\end{array}
\right]
\qquad
\begin{array}{l}
60/1 = 60 \\
120/4 = 30 \\
\end{array}
$$

The pivot entry is 4, since the second row had the smallest test ratio. We zero out the remaining entries in the pivot column with the indicated row operations. The resultant tableau is

$$
\begin{array}{cccccc}
x & y & z & s & t & P \\
\end{array}
$$
$$
\left[
\begin{array}{cccccc|c}
2 & 1 & 0 & 4 & -1 & 0 & 120 \\
2 & 3 & 4 & 0 & 1 & 0 & 120 \\
\hline
-1 & 0 & 0 & 0 & 1 & 1 & 120 \\
\end{array}
\right]
\qquad
\begin{array}{l}
4R_1 - R_2 \\
\\
R_3 + R_2 \\
\end{array}
$$

We could convert the nonzero entries in the active columns to a 1 by multiplying Rows 1 and 2 by $\frac{1}{4}$; however, that would result in messy fractional entries in the tableau. We will hold off on the reduction until we have obtained the final simplex tableau. Since there is still a negative value in the bottom row, we must again select a pivot column and entry. The pivot column x has two positive entries, so we must compute the test ratios.

$$
\begin{array}{cccccc}
x & y & z & s & t & P \\
\end{array}
$$
$$
\left[
\begin{array}{cccccc|c}
\boxed{2} & 1 & 0 & 4 & -1 & 0 & 120 \\
2 & 3 & 4 & 0 & 1 & 0 & 120 \\
\hline
-1 & 0 & 0 & 0 & 1 & 1 & 120 \\
\end{array}
\right]
\qquad
\begin{array}{l}
120/2 = 60 \\
120/2 = 60 \\
\end{array}
$$

$\uparrow$ Pivot column

Since both entries have the same test ratio, we may pick either entry to be the pivot. We select the entry in Row 1 as the pivot and zero out the remaining entries in the pivot column with the indicated row operations. The resultant tableau is

$$
\begin{array}{c}
\begin{array}{cccccc} x & y & z & s & t & P \end{array} \\
\left[\begin{array}{cccccc|c}
2 & 1 & 0 & 4 & -1 & 0 & 120 \\
0 & 2 & 4 & -4 & 2 & 0 & 0 \\
0 & 1 & 0 & 4 & 1 & 2 & 360
\end{array}\right]
\begin{array}{c}
\\
R_2 - R_1 \\
2R_3 + R_1
\end{array}
\end{array}
$$

Since the bottom row of the tableau is entirely nonnegative, it is unnecessary to repeat the pivot selection process. Crossing out the inactive variable columns and reading the result yields

$$
\begin{array}{c}
\begin{array}{cccccc} x & y & z & s & t & P \end{array} \\
\left[\begin{array}{cccccc|c}
2 & 1 & 0 & 4 & -1 & 0 & 120 \\
0 & 2 & 4 & -4 & 2 & 0 & 0 \\
0 & 1 & 0 & 4 & 1 & 2 & 360
\end{array}\right]
\end{array}
$$

$$2x = 120 \qquad 4z = 0 \qquad 2P = 360$$
$$x = 60 \qquad z = 0 \qquad P = 180$$

The corner point (x, y, z) that maximizes the objective function is $(60, 0, 0)$. It is interesting to note that although z is an active variable, $z = 0$. Active variables may equal zero, but inactive variables must equal zero.

In the context of the scenario, the store should order 60 packages of the first brand to earn the maximum profit of $180. Even though the profit per package for the third brand ($4 per package) was higher than the profit per package for the first brand ($3 per package), the higher wholesale cost of the third brand ($4 versus $2) limited the number of third-brand items available for sale. Ironically, the lowest-priced brand of cheese resulted in the greatest profit.

In Examples 1 and 2, there was a unique solution to the standard maximization problem; however, on occasion you'll encounter a standard maximization problem with infinitely many solutions. Example 3 is such a case.

EXAMPLE 3   **Using the Simplex Method to Make Business Decisions**

A store can purchase three brands of binders. The first brand, A, costs $2 per binder. The second brand, B, costs $1 per binder. The third brand, C, costs $2 per binder. The store can spend up to $100 and purchase up to 50 binders. It resells binder A for $5 per binder, binder B for $3 per binder, and binder C for $5 per binder. Recall that profit per binder is revenue per binder minus cost per binder. How many of each type of binder should the store buy if it wants to maximize its profit?

SOLUTION   Let x be the number of binders of the first brand, y be the number of binders of the second brand, and z be the number of binders of the third brand. The total revenue from binder sales is $R = 5x + 3y + 5z$. The total cost of the binders is $C = 2x + 1y + 2z$. Since profit is the difference between revenue and cost,

$$
\begin{aligned}
P &= R - C \\
&= (5x + 3y + 5z) - (2x + 1y + 2z) \\
&= 3x + 2y + 3z
\end{aligned}
$$

Since we want to maximize profit, we must maximize $P = 3x + 2y + 3z$ subject to

$$x + y + z \le 50 \qquad \text{At most 50 binders are purchased}$$
$$2x + y + 2z \le 100 \qquad \text{At most \$100 is spent on binders}$$
$$x \ge 0, y \ge 0, z \ge 0 \qquad \text{A nonnegative number of each brand of binder is purchased}$$

The problem is a standard maximization problem. Converting the linear programming problem into a system of linear equations, we get

$$x + y + z + s = 50 \qquad \text{Binder quantity constraint}$$
$$2x + y + 2z + t = 100 \qquad \text{Budget constraint}$$
$$-3x - 2y - 3z + P = 0 \qquad \text{Objective function}$$

and the initial simplex tableau

$$
\begin{array}{cccccc}
x & y & z & s & t & P \\
\end{array}
$$
$$
\left[
\begin{array}{ccccccc}
1 & 1 & 1 & 1 & 0 & 0 & 50 \\
2 & 1 & 2 & 0 & 1 & 0 & 100 \\
\hline
-3 & -2 & -3 & 0 & 0 & 1 & 0
\end{array}
\right]
$$

$$\uparrow$$
$$\boxed{\text{Pivot column}}$$

Since -3 is the negative number in the bottom row that has the largest magnitude, we may select either column x or column z as the pivot column. We select column z. Because the entries in the first and second rows of the pivot column are positive, we must calculate the test ratios.

$$
\begin{array}{cccccc}
x & y & z & s & t & P \\
\end{array}
$$
$$
\left[
\begin{array}{ccccccc}
1 & 1 & \boxed{1} & 1 & 0 & 0 & 50 \\
2 & 1 & 2 & 0 & 1 & 0 & 100 \\
\hline
-3 & -2 & -3 & 0 & 0 & 1 & 0
\end{array}
\right]
\qquad
\begin{array}{l}
50/1 = 50 \\
100/2 = 50
\end{array}
$$

Since the entries have the same test ratio, we may select either one as the pivot. We select the entry in the first row. We zero out the remaining entries in the pivot column with the indicated row operations and obtain the tableau

$$
\begin{array}{cccccc}
x & y & z & s & t & P \\
\end{array}
$$
$$
\left[
\begin{array}{ccccccc}
1 & 1 & 1 & 1 & 0 & 0 & 50 \\
0 & -1 & 0 & -2 & 1 & 0 & 0 \\
\hline
0 & 1 & 0 & 3 & 0 & 1 & 150
\end{array}
\right]
\qquad
\begin{array}{l}
R_2 - 2R_1 \\
R_3 + 3R_1
\end{array}
$$

Crossing out the inactive variable columns and reading the resultant equations, we get

$$
\begin{array}{cccccc}
x & y & z & s & t & P \\
\end{array}
$$
$$
\left[
\begin{array}{ccccccc}
1 & 1 & 1 & 1 & 0 & 0 & 50 \\
0 & -1 & 0 & -2 & 1 & 0 & 0 \\
\hline
0 & 1 & 0 & 3 & 0 & 1 & 150
\end{array}
\right]
$$

$$x + z = 50 \qquad\qquad t = 0 \qquad P = 150$$
$$z = -x + 50$$

There are infinitely many solutions (x, y, z) that maximize the objective function. Each solution is of the form $(n, 0, -n + 50)$ with $0 \leq n \leq 50$.

In the context of the real-world scenario, the store may purchase any combination of 50 binders from brands A and C and still maximize profits. This is because both brands have the same wholesale cost and the same retail price. As a result, the cost, revenue, and profit for each item are the same.

When a standard maximization problem has more than two constraints, additional slack variables will be needed. This will result in additional columns in the initial simplex tableau. We typically use the letters u, v, and so on for slack variables in addition to the letters s and t. Even with additional rows or columns, the simplex method is effective in finding the optimal solution to any standard maximization problem.

EXAMPLE **4**

Using the Simplex Method to Make Business Decisions

Clothing retailers typically mark up the price of apparel by 100 percent. That is, the retail price is typically 200 percent of the wholesale cost. WholesaleFashion.com claims to be the number one Internet business-to-business source for fashion retailers of women's apparel and shoes. It sells directly to businesses instead of to consumers. In January 2003, WholesaleFashion.com offered a 3/4 Sleeves Stretch Top for $9.50, a Long Sleeve Turtle Neck Top for $12.50, a Heart Neck Tank Top for $9.25, and an American Flag Tank Top for $9.75. (**Source:** www.wholesalefashion.com.)

A women's apparel store has up to $1000 to spend on the four items. It has enough rack space for up to 60 tops. It anticipates the demand for tank tops to be greater than the demand for the other tops, so it wants to order at least twice as many tank tops as sleeved tops. The store intends to mark up all of the items by 100 percent. Assuming that all of the items that the store orders will sell, how many of each item should it order if it wants to maximize revenue?

SOLUTION Let x be the number of 3/4 Sleeves Stretch Tops, y be the number of Long Sleeve Turtle Neck Tops, z be the number of Heart Neck Tank Tops, and w be the number of American Flag Tank Tops. The number of sleeved tops is $x + y$, and the number of tank tops is $z + w$. Since the number of tank tops is to be at least twice the number of sleeved tops, we have

$$z + w \geq 2(x + y)$$
$$z + w \geq 2x + 2y$$
$$-2x - 2y + z + w \geq 0$$
$$2x + 2y - z - w \leq 0 \qquad \text{Dividing by } -1 \text{ reverses the inequality sign}$$

We are asked to maximize the revenue from sales

$$R = 19x + 25y + 18.50z + 19.50w$$

subject to the constraints

$$x + y + z + w \leq 60 \qquad \text{At most 60 tops are purchased}$$
$$2x + 2y - z - w \leq 0 \qquad \text{Twice as many tank tops are purchased}$$
$$9.5x + 12.5y + 9.25z + 9.75w \leq 1000 \qquad \text{At most \$1000 is spent on tops}$$
$$x \geq 0, y \geq 0, z \geq 0, w \geq 0 \qquad \text{A nonnegative number of each type is purchased}$$

Rewriting the objective function and adding the slack variables to the constraint inequalities yields the system of equations

$$
\begin{aligned}
x + \quad y + \quad z + \quad w + s &= 60 \\
2x + \quad 2y - \quad z - \quad w \quad + t &= 0 \\
9.5x + 12.5y + \quad 9.25z + \quad 9.75w \quad\quad + u &= 1000 \\
-19x - \quad 25y - 18.50z - 19.50w \quad\quad\quad + R &= 0
\end{aligned}
$$

and the associated initial simplex tableau

$$
\begin{array}{ccccccccc|c}
x & y & z & w & s & t & u & R & \\
1 & 1 & 1 & 1 & 1 & 0 & 0 & 0 & 60 \\
2 & 2 & -1 & -1 & 0 & 1 & 0 & 0 & 0 \\
9.5 & 12.5 & 9.25 & 9.75 & 0 & 0 & 1 & 0 & 1000 \\
-19 & -25 & -18.5 & -19.5 & 0 & 0 & 0 & 1 & 0
\end{array}
$$

Pivot column (under y)

The y column is the pivot column, since -25 is the negative entry in the bottom row with the greatest magnitude. Since the first three entries in the y column are positive, we must calculate three test ratios.

$$
\begin{array}{ccccccccc|c}
x & y & z & w & s & t & u & R & \\
1 & 1 & 1 & 1 & 1 & 0 & 0 & 0 & 60 \\
2 & \boxed{2} & -1 & -1 & 0 & 1 & 0 & 0 & 0 \\
9.5 & 12.5 & 9.25 & 9.75 & 0 & 0 & 1 & 0 & 1000 \\
-19 & -25 & -18.5 & -19.5 & 0 & 0 & 0 & 1 & 0
\end{array}
\quad
\begin{array}{l}
60/1 = 60 \\
0/2 = 0 \\
1000/12.5 = 80
\end{array}
$$

The pivot entry is the 2 in the second column and second row, since the second row has the smallest test ratio. We zero out the remaining entries in the second column with the indicated operations.

Because of the numerical complexity of the entries in the tableau, we will use the Technology Tips following this example to do the row operations. The resultant tableau is

$$
\begin{array}{ccccccccc|c}
x & y & z & w & s & t & u & R & \\
0 & 0 & 3 & 3 & 2 & -1 & 0 & 0 & 120 \\
2 & 2 & -1 & -1 & 0 & 1 & 0 & 0 & 0 \\
-6 & 0 & 31 & 32 & 0 & -12.5 & 2 & 0 & 2000 \\
12 & 0 & -62 & -64 & 0 & 25 & 0 & 2 & 0
\end{array}
\quad
\begin{array}{l}
2R_1 - R_2 \\
\\
2R_3 - 12.5R_2 \\
2R_4 + 25R_2
\end{array}
$$

New pivot column (under w)

The w column is the new pivot column, since -64 is the negative entry in the bottom row that has the largest magnitude. Since the only positive entries in the w column are in the first and third rows, we need to calculate only two test ratios.

$$
\begin{array}{ccccccccc|c}
x & y & z & w & s & t & u & R & \\
0 & 0 & 3 & \boxed{3} & 2 & -1 & 0 & 0 & 120 \\
2 & 2 & -1 & -1 & 0 & 1 & 0 & 0 & 0 \\
-6 & 0 & 31 & 32 & 0 & -12.5 & 2 & 0 & 2000 \\
12 & 0 & -62 & -64 & 0 & 25 & 0 & 2 & 0
\end{array}
\quad
\begin{array}{l}
120/3 = 40 \\
\\
2000/32 = 62.5 \\
\end{array}
$$

Since the entry in the second row of the pivot column was negative, we did not calculate a test ratio for the second row. The pivot entry is the 3 in the w column and first row. We will zero out the remaining entries in the w column with the indicated operations. The resultant tableau is

$$
\begin{array}{cccccccc}
x & y & z & w & s & t & u & R \\
\end{array}
$$

$$
\left[
\begin{array}{ccccccc|c}
0 & 0 & 3 & 3 & 2 & -1 & 0 & 0 & 120 \\
6 & 6 & 0 & 0 & 2 & 2 & 0 & 0 & 120 \\
-18 & 0 & -3 & 0 & -64 & -5.5 & 6 & 0 & 2160 \\
\hline
36 & 0 & 6 & 0 & 128 & 11 & 0 & 6 & 7680
\end{array}
\right]
\begin{array}{l}
\\ 3R_2 + R_1 \\ 3R_3 - 32R_1 \\ 3R_4 + 64R_1
\end{array}
$$

Since all entries in the bottom row are nonnegative, this is the final simplex tableau. Crossing out the inactive variable columns and reading the resultant equations, we get

$$
\begin{array}{cccccccc}
x & y & z & w & s & t & u & R \\
\end{array}
$$

$$
\left[
\begin{array}{ccccccc|c}
0 & 0 & 3 & 3 & 2 & -1 & 0 & 0 & 120 \\
6 & 6 & 0 & 0 & 2 & 2 & 0 & 0 & 120 \\
-18 & 0 & -3 & 0 & -64 & -5.5 & 6 & 0 & 2160 \\
\hline
36 & 0 & 6 & 0 & 128 & 11 & 0 & 6 & 7680
\end{array}
\right]
$$

$$x = 0 \qquad 6y = 120 \qquad z = 0 \qquad 3w = 120$$
$$y = 20 \qquad\qquad w = 40$$
$$s = 0 \qquad t = 0 \qquad 6u = 2160 \qquad 6R = 7680$$
$$u = 360 \qquad R = 1280$$

The maximum revenue that can be attained is $1280. This will be achieved when 20 Long Sleeve Turtle Neck Tops and 40 American Flag Tank Tops are purchased. Recall that u was the slack variable associated with the equation for the total cost of the tops. Since $u = 360$, the amount spent by the store is $360 less than the maximum amount allowed. That is, $640 is spent on the tops. The additional money remains unspent. It was the space constraint (a maximum of 60 tops), not the budget constraint (a maximum of $1000), that limited how many tops could be ordered.

For large tableaus or tableaus with noninteger entries, it is often helpful to do row operations on the calculator. The next four Technology Tips detail how to do this on the TI-83 Plus.

TECHNOLOGY **TIP**

Interchanging Two Rows

1. Enter a matrix or simplex tableau using the Matrix Editor. (Press (2nd) (x⁻¹) to access the Matrix Editor.)

```
[A]
[[1   2   1   0   0   6...
 [2   1   0   1   0   4...
 [-5  -3  0   0   1   0...
```

(Continued)

2. Close the Matrix Editor to store the matrix, then reopen the Matrix Editor and move the cursor to the MATH menu.

```
NAMES MATH EDIT
1▪det(
2:ᵀ
3:dim(
4:Fill(
5:identity(
6:randM(
7↓augment(
```

3. Scroll to C:rowSwap(and press ENTER. This operation is used to interchange one row with another row.

```
NAMES MATH EDIT
7↑augment(
8:Matr▸list(
9:List▸matr(
0:cumSum(
A:ref(
B:rref(
█:rowSwap(
```

4. Type in the matrix name from the Matrix Menu, the first row number, and the second row number. For example, rowSwap([A],1,2) is equivalent to $R_1 \leftrightarrow R_2$.

```
rowSwap([A],1,2)
```

5. Press ENTER to display the new matrix or tableau.

```
rowSwap([A],1,2)
[[2   1   0  1  0  4…
 [1   2   1  0  0  6…
 [-5 -3   0  0  1  0…
```

TECHNOLOGY **TIP**

Adding One Row to Another Row

1. Enter a matrix or simplex tableau using the Matrix Editor. (Press 2nd x⁻¹ to access the Matrix Editor.)

```
[A]
[[1   2   1  0  0  6…
 [2   1   0  1  0  4…
 [-5 -3   0  0  1  0…
```

(Continued)

2. Close the Matrix Editor to store the matrix, then reopen the Matrix Editor and move the cursor to the MATH menu.

3. Scroll to D:row+(and press ⟨ENTER⟩. This operation is used to add one row to another row.

4. Type in the matrix name from the Matrix Menu, the first row number, and the second row number. The result will be placed in the last row listed. For example, row+([A],2,3) is equivalent to $R_2 + R_3 \rightarrow R_3$.

5. Press ⟨ENTER⟩ to display the new matrix or tableau.

Multiplying a Row by a Nonzero Constant

1. Enter a matrix or simplex tableau using the Matrix Editor. (Press ⟨2nd⟩ ⟨x⁻¹⟩ to access the Matrix Editor.)

(Continued)

2. Close the Matrix Editor to store the matrix, then reopen the Matrix Editor and move the cursor to the MATH menu.

```
NAMES MATH EDIT
1█det(
2:T
3:dim(
4:Fill(
5:identity(
6:randM(
7↓augment(
```

3. Scroll to E:*row(and press ENTER. This operation is used to multiply a row by a nonzero constant.

```
NAMES MATH EDIT
0↑cumSum(
A:ref(
B:rref(
C:rowSwap(
D:row+(
█:*row(
F:*row+(
```

4. Type in the multiplier value, the matrix name from the Matrix Menu, and the row to be multiplied. The result will be placed in the last row listed. For example, *row(-2,[A],1) means $-2R_1 \to R_1$.

```
*row(-2,[A],1)
```

5. Press ENTER to display the new matrix or tableau.

```
*row(-2,[A],1)
[[-2 -4 -2 0 0 …
 [2  1  0  1 0 …
 [-5 -3  0  0 1 …
```

TECHNOLOGY **TIP**

Adding a Multiple of One Row to Another Row

1. Enter a matrix or simplex tableau using the Matrix Editor. (Press 2nd x⁻¹ to access the Matrix Editor.)

```
[A]
[[1   2  1 0 0 6…
 [2   1  0 1 0 4…
 [-5 -3  0 0 1 0…
```

(Continued)

2. Close the Matrix Editor to store the matrix, then reopen the Matrix Editor and move the cursor to the MATH menu.

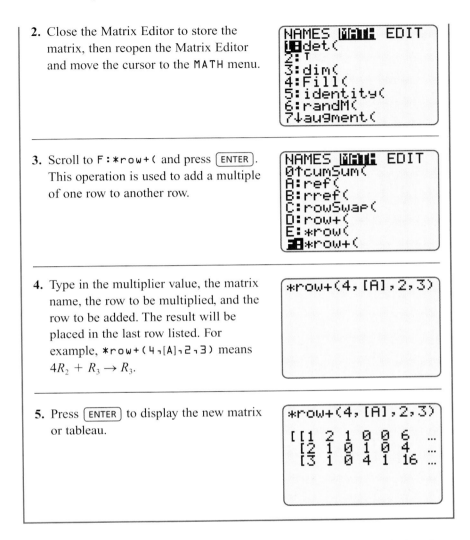

3. Scroll to F:*row+(and press [ENTER]. This operation is used to add a multiple of one row to another row.

4. Type in the multiplier value, the matrix name, the row to be multiplied, and the row to be added. The result will be placed in the last row listed. For example, *row+(4,[A],2,3) means $4R_2 + R_3 \rightarrow R_3$.

5. Press [ENTER] to display the new matrix or tableau.

Recall that simplex method row operations must be of the form $aR_c \pm bR_p \rightarrow R_c$, with both a and b positive. Alternatively, we may write this expression as $\pm bR_p + aR_c \rightarrow R_c$. The latter form is more consistent with the calculator functionality and may make using the calculator easier.

EXAMPLE 5

Using the Simplex Method to Make Business Decisions

In January 2003, WholesaleFashion.com offered a Long Halter Dress for $25.75, a Long Open Back Dress for $21.75, a pair of Thai Silk Pump shoes for $17.75, and a pair of Glitter Evening High Heels for $17.50.
(**Source:** www.wholesalefashion.com.)

A women's apparel store has up to $1000 to spend on the four items. It will mark up the dresses 100 percent but mark up the shoes only 50 percent. The store anticipates that it will sell 80 percent of each type of dress at full price and sell the remaining 20 percent of each type of dress at its cost. It expects to sell 90 percent of each type of shoes at full price and the remaining 10 percent of each shoe type at its cost. It expects to sell a pair of shoes with each dress it sells. (The store will market a dress and shoes as a package.) Additionally, it may sell some shoes separately. How many of each item should the store order to maximize its profit?

SOLUTION Let x be the number of Long Halter Dresses, y be the number of Long Open Back Dresses, z be the number of pairs of Thai Silk Pump shoes, and w be the number of pairs of Glitter Evening High Heels.

The revenue generated by selling 80 percent of each type of dress at full retail price is given by

$$R = 51.50(0.8x) + 43.50(0.8y)$$
$$= 41.20x + 34.80y$$

The revenue generated by selling the remaining 20 percent of each type of dress at the wholesale cost is given by

$$R = 25.75(0.2x) + 21.75(0.2y)$$
$$= 5.15x + 4.35y$$

The combined revenue from dress sales is

$$\text{Dress revenue} = (41.20x + 34.80y) + (5.15x + 4.35y)$$
$$= 46.35x + 39.15y$$

The retail price of each shoe variety is 150 percent of the wholesale price.

$$\text{Thai Silk Pumps} = 1.5(17.75)$$
$$= 26.625$$
$$\approx 26.63 \qquad \text{Rounded to the nearest cent}$$

$$\text{Glitter Evening High Heels} = 1.5(17.50)$$
$$= 26.25$$

The revenue generated by selling 90 percent of each shoe variety at full retail price is given by

$$R = 26.63(0.9z) + 26.25(0.9w)$$
$$\approx 23.97z + 23.63w \qquad \text{Rounded to the nearest cent}$$

The revenue generated by selling 10 percent of each shoe variety at wholesale cost is

$$R = 17.75(0.1z) + 17.50(0.1w)$$
$$\approx 1.78z + 1.75w \qquad \text{Rounded to the nearest cent}$$

The combined revenue from shoe sales is

$$\text{Shoe revenue} = (23.97z + 23.63w) + (1.78z + 1.75w)$$
$$\approx 25.75z + 25.38w$$

The combined revenue from dresses and shoes is the sum of the dress revenue and the shoe revenue.

$$\text{Combined revenue} = \text{dress revenue} + \text{shoe revenue}$$
$$\approx (46.35x + 39.15y) + (25.75z + 25.38w)$$
$$\approx 46.35x + 39.15y + 25.75z + 25.38w$$

The total cost of the dresses and shoes is

$$C = 25.75x + 21.75y + 17.75z + 17.5w$$

Since profit is the difference in revenue and cost,

$$P = R - C$$
$$\approx (46.35x + 39.15y + 25.75z + 25.38w)$$
$$- (25.75x + 21.75y + 17.75z + 17.50w)$$
$$\approx 20.60x + 17.40y + 8.00z + 7.88w$$

We need to order at least as many pairs of shoes as dresses, so

Number of dresses $\leq$ number of pairs of shoes

$$(x + y) \leq (z + w)$$
$$x + y - z - w \leq 0$$

We are asked to maximize the profit $P = 20.6x + 17.4y + 8.00z + 7.88w$ subject to the constraints

$$x + y - z - w \leq 0 \qquad \text{Quantity constraint}$$
$$25.75x + 21.75y + 17.75z + 17.5w \leq 1000 \qquad \text{Budget constraint}$$
$$x \geq 0, y \geq 0, z \geq 0, w \geq 0$$

Rewriting the objective function and adding the slack variables to the constraint inequalities yields the following system of equations:

$$x + \quad y - \quad z - \quad w + s \qquad\qquad = 0$$
$$25.75x + 21.75y + 17.75z + 17.5w \qquad + t \qquad = 1000$$
$$-20.60x - 17.40y - 8.00z - 7.88w \qquad\qquad + P = 0$$

and the associated initial simplex tableau

$$
\begin{array}{ccccccc}
x & y & z & w & s & t & P \\
\end{array}
$$
$$
\left[
\begin{array}{ccccccc|c}
1 & 1 & -1 & -1 & 1 & 0 & 0 & 0 \\
25.75 & 21.75 & 17.75 & 17.50 & 0 & 1 & 0 & 1000 \\
-20.60 & -17.40 & -8.00 & -7.88 & 0 & 0 & 1 & 0 \\
\end{array}
\right]
$$

Pivot column

The x column is the pivot column, since -20.60 is the negative entry in the bottom row that has the largest magnitude. We must calculate test ratios for the first two rows.

$$
\begin{array}{ccccccc}
x & y & z & w & s & t & P \\
\end{array}
$$
$$
\left[
\begin{array}{ccccccc|c}
\boxed{1} & 1 & -1 & -1 & 1 & 0 & 0 & 0 \\
25.75 & 21.75 & 17.75 & 17.50 & 0 & 1 & 0 & 1000 \\
-20.60 & -17.40 & -8.00 & -7.88 & 0 & 0 & 1 & 0 \\
\end{array}
\right]
\begin{array}{l}
0/1 = 0 \\
1000/25.75 \approx 38.83 \\
\\
\end{array}
$$

The pivot entry is the 1 in the x column and the first row. We will zero out the remaining entries in the first column with the indicated operations. Because of the numerical complexity of the entries in the tableau, we will use the Technology Tips described in this section to do the row operations.

The resulting tableau is

$$
\begin{array}{ccccccc}
x & y & z & w & s & t & P \\
\end{array}
$$

$$
\left[
\begin{array}{ccccccc|c}
1 & 1 & -1 & -1 & 1 & 0 & 0 & 0 \\
0 & -4 & \boxed{43.5} & 43.25 & -25.75 & 1 & 0 & 1000 \\
0 & 3.2 & -28.60 & -28.48 & 20.60 & 0 & 1 & 0 \\
\end{array}
\right]
\begin{array}{l}
\\
-25.75R_1 + R_2 \\
20.60\,R_1 + R_3
\end{array}
$$

↑ New pivot column

(Note: We wrote the row operations in the alternative form $\pm bR_p + aR_c \to R_c$.)

The z column is the new pivot column, since -28.60 is the negative entry in the bottom row that has the largest magnitude. Since there is only one positive entry in the z column above the bottom row, that entry is the pivot. The pivot entry is the 43.5 in the z column and the second row. We will zero out the remaining entries in the z column with the following row operations written in the alternative form:

$$R_2 + 43.5R_1 \to R_1$$

$$28.6R_2 + 43.5R_3 \to R_3$$

To do this on the calculator requires a sequence of operations. To perform the operation $R_2 + 43.5R_1 \to R_1$, we do the following to the tableau that is on the screen of our calculator:

Step 1: *row(43.5, *ans*, 1)	Multiplies R_1 by 43.5
Step 2: ENTER	Returns the modified matrix
Step 3: row + (*ans*, 2, 1)	Adds the first two rows of the modified matrix
Step 4: ENTER	Completes the operation

To perform the operation $28.6R_2 + 43.5R_3 \to R_3$, we do the following to the tableau that is on the screen of our calculator:

Step 1: *row(43.5, *ans*, 3)	Multiplies R_3 by 43.5
Step 2: ENTER	Returns the modified matrix
Step 3: *row + (28.6, *ans*, 2, 3)	Multiplies R_2 by 28.6 and adds it to the modified R_3
Step 4: ENTER	Completes the operation

The resultant tableau is

$$
\begin{array}{ccccccc}
x & y & z & w & s & t & P \\
\end{array}
$$

$$
\left[
\begin{array}{ccccccc|c}
43.5 & 39.5 & 0 & -0.25 & 17.75 & 1 & 0 & 1000 \\
0 & -4 & 43.5 & \boxed{43.25} & -25.75 & 1 & 0 & 1000 \\
0 & 24.8 & 0 & -1.93 & 159.65 & 28.60 & 43.5 & 28600 \\
\end{array}
\right]
$$

↑ New pivot column

The bottom row contains a negative entry in the w column, so we must repeat the pivoting process. Since the only positive entry in the pivot column is 43.25, it is the new pivot. The indicated row operations (in alternative form) will zero out the additional terms in the pivot column.

$$0.25R_2 + 43.25R_1 \to R_1$$

$$1.93R_2 + 43.25R_3 \to R_3$$

To perform the operation $0.25R_2 + 43.25R_1 \rightarrow R_1$, we do the following to the tableau that is on the screen of our calculator:

Step 1: *row(43.25, *ans*, 1)   Multiplies R_1 by 43.25
Step 2: ENTER   Returns the modified matrix
Step 3: *row + (0.25, *ans*, 2, 1)   Multiplies R_2 by 0.25 and adds it to the modifed R_1
Step 4: ENTER   Completes the operation

To perform the operation $1.93R_2 + 43.25R_3 \rightarrow R_3$, we do the following to the tableau that is on the screen of our calculator:

Step 1: *row(43.25, *ans*, 3)   Multiplies R_3 by 43.25
Step 2: ENTER   Returns the modified matrix
Step 3: *row + (1.93, *ans*, 2, 3)   Multiplies R_2 by 1.93 and adds it to the modified R_3
Step 4: ENTER   Completes the operation

The resultant simplex tableau is

x	y	z	w	s	t	P	
1881.38	1707.38	10.88	0	761.25	43.5	0	43500
0	-4	43.5	43.25	-25.75	1	0	1000
0	1064.88	83.96	0	6855.17	1238.88	1881.38	1238880

Crossing out the inactive variable columns and reading the resultant equations, we get

x	y	z	w	s	t	P	
1881.38	1707.38	10.88	0	761.25	43.5	0	43500
0	-4	43.5	43.25	-25.75	1	0	1000
0	1064.88	83.96	0	6855.17	1238.88	1881.38	1238880

Reading from the tableau, we see that

$$1881.38x = 43500 \qquad y = 0 \qquad z = 0 \qquad 43.25w = 1000$$
$$x = 23.12 \qquad\qquad\qquad\qquad\qquad w = 23.12$$
$$s = 0 \qquad t = 0 \qquad 1881.38P = 1238880$$
$$P = 658.50$$

Since it doesn't make sense to talk about a fraction of a dress or pair of shoes, we will round the values to the nearest whole number. Rounding to whole numbers doesn't guarantee the maximum whole number solution. However, if the solution isn't the maximum whole number solution, it will be close to the optimal solution. We estimate that the maximum profit will be achieved when 23 Long Halter Dresses and 23 pairs of Glitter Evening High Heels are purchased. Since $t = 0$, the amount spent by the store appears to be exactly $1000. (Actually, the cost is $994.75, since we rounded the number of dresses and shoes.) The maximum profit is expected to be $658.50. (Since we rounded down the number of shoes and dresses, the exact profit is $655.04.)

Even though Example 5 required some rounding to make sense of the results, the numerical analysis provided some valuable input to help us make a sound business decision.

4.3 Summary

In this section, you learned the simplex method for solving standard maximization problems. You learned that this is the method of choice in solving problems with more than two decision variables.

4.3 Exercises

In Exercises 1–10, determine whether the problem is a standard maximization problem. If it isn't, explain why.

1. Maximize $P = 9x + 8y$

Subject to $\begin{cases} 2x + y \le 10 \\ -x + y \le 1 \\ x \ge 0 \\ y \ge 0 \end{cases}$

2. Maximize $P = -2x - 5y$

Subject to $\begin{cases} 4x + 9y \le 10 \\ -11x + y \le -21 \\ x \ge 0 \\ y \ge 0 \end{cases}$

3. Maximize $P = 6xy$

Subject to $\begin{cases} 6x + 7y \le 13 \\ -8x - 4y \le 12 \\ x \ge 0 \\ y \ge 0 \end{cases}$

4. Maximize $P = 4x - 2y + z$

Subject to $\begin{cases} 2x + y + 4z \le 24 \\ -2x + 3y + 3z \le 150 \\ x \ge 0 \\ y \ge 0 \\ z \ge 0 \end{cases}$

5. Maximize $P = -1.2x - 2.8y + 4.3z$

Subject to $\begin{cases} 3.2x + 1.5y + 7.4z \le 249.8 \\ -2.7x + 3.4y + 3.9z \le 190.1 \end{cases}$

6. Maximize $P = 4x + 4y + 9z$

Subject to $\begin{cases} -6x - y + 4z \le -24 \\ -9x - 1.5y + 6z \le 36 \\ x \ge 0 \\ y \ge 0 \\ z \ge 0 \end{cases}$

7. Maximize $P = x + 9y + 8z$

Subject to $\begin{cases} -6x - y + 4z \le -24 \\ -9x - 1.5y + 6z \le 36 \\ x \ge 0 \\ y \ge 0 \\ z \ge 0 \end{cases}$

8. Maximize $P = 4x + 4y + 9z$

Subject to $\begin{cases} -3x - 2y + 4z \ge -22 \\ 5x - 2.5y + 6z \le 36 \\ x \ge 0 \\ y \ge 0 \\ z \ge 0 \end{cases}$

9. Maximize $P = -8x + 2y - z$

Subject to $\begin{cases} z \le 4 \\ x - y \le 3 \\ 2x - z \le 5 \\ x + y + z \le 0 \end{cases}$

10. Maximize $P = 5x + 7z$

Subject to $\begin{cases} 4x + 2y \le 16 \\ 4x - 2y \le 3 \\ x \ge 0 \\ y \ge 0 \end{cases}$

In Exercises 11–20, solve the standard maximization problems by using the simplex method. Check your answer by graphing the feasible region and calculating the value of the objective function at each of the corner points.

11. Maximize $P = 9x + 8y$

Subject to $\begin{cases} 2x + y \le 10 \\ -x + y \le 1 \\ x \ge 0 \\ y \ge 0 \end{cases}$

12. Maximize $P = -2x + 4y$

Subject to $\begin{cases} 2x + y \le 6 \\ x + y \le 4 \\ x \ge 0 \\ y \ge 0 \end{cases}$

13. Maximize $P = 5x - 2y$

Subject to $\begin{cases} -3x + y \le 0 \\ 3x + y \le 12 \\ x \ge 0 \\ y \ge 0 \end{cases}$

14. Maximize $P = 7x + 10y$

Subject to $\begin{cases} x + y \le 4 \\ 5x + y \le 12 \\ x \ge 0 \\ y \ge 0 \end{cases}$

15. Maximize $P = 4x + 4y$

Subject to $\begin{cases} 4x + 3y \le 20 \\ -x + 2y \le 6 \\ x \ge 0 \\ y \ge 0 \end{cases}$

16. Maximize $P = 5x - 2y$

Subject to $\begin{cases} 4x + 5y \le 28 \\ -3x + 2y \le 2 \\ x \ge 0 \\ y \ge 0 \end{cases}$

17. Maximize $P = 6x + 5y$

Subject to $\begin{cases} 3x + 2y \le 12 \\ x + y \le 5 \\ x \ge 0 \\ y \ge 0 \end{cases}$

18. Maximize $P = 6x + 2y$

Subject to $\begin{cases} x + y \le 10 \\ 2x + y \le 16 \\ -2x + y \le 0 \\ x \ge 0 \\ y \ge 0 \end{cases}$

19. Maximize $P = 7x - 5y$

Subject to $\begin{cases} 6x - 5y \le 30 \\ 2x + 2y \le 10 \\ -5x + 3y \le 7 \\ x \ge 0 \\ y \ge 0 \end{cases}$

20. Maximize $P = x + y$

Subject to $\begin{cases} 5x + y \le 20 \\ -x + 2y \le 29 \\ 10x + y \le 30 \\ x \ge 0 \\ y \ge 0 \end{cases}$

In Exercises 21–30, solve the standard maximization problems by using the simplex method.

21. Maximize $P = 5x + 6y + 6z$

Subject to $\begin{cases} 2x + 3y + 2z \le 120 \\ -x + y + z \le 60 \\ x \ge 0 \\ y \ge 0 \\ z \ge 0 \end{cases}$

22. Maximize $P = 4x + 6y + 5z$

Subject to $\begin{cases} 2x + 3y + 3z \le 210 \\ x + y + z \le 100 \\ x \ge 0 \\ y \ge 0 \\ z \ge 0 \end{cases}$

23. Maximize $P = 10x + 6y + 12z$

Subject to $\begin{cases} 8x + 5y + 9z \le 360 \\ x + y + z \le 50 \\ x \ge 0 \\ y \ge 0 \\ z \ge 0 \end{cases}$

24. Maximize $P = x - 2y + 4z$

Subject to $\begin{cases} 6x - 5y + 10z \le 300 \\ 2x + 5y + 2z \le 500 \\ x \ge 0 \\ y \ge 0 \\ z \ge 0 \end{cases}$

25. Maximize $P = 5x - 2y + 4z$

Subject to $\begin{cases} 8x + 5y + 9z \le 360 \\ x + y + z \le 50 \\ x \ge 0 \\ y \ge 0 \\ z \ge 0 \end{cases}$

26. Maximize $P = 4x + 4y - 10z$

Subject to $\begin{cases} 3x + 3y + 6z \le 42 \\ 2x + y + z \le 8 \\ 3x + 2y + 5z \le 23 \\ x \ge 0 \\ y \ge 0 \\ z \ge 0 \end{cases}$

27. Maximize $P = -x - y + 10z$

Subject to $\begin{cases} 2x + 2y + 2z \le 14 \\ 4x + 2y + 2z \le 16 \\ 3x + 2y + 5z \le 23 \\ x \ge 0 \\ y \ge 0 \\ z \ge 0 \end{cases}$

28. Maximize $P = x + 2y + 3z$

Subject to $\begin{cases} x + z \le 20 \\ y + 2z \le 30 \\ x + 2y + 3z \le 60 \\ x \ge 0 \\ y \ge 0 \\ z \ge 0 \end{cases}$

29. Maximize $P = x + y + z$

Subject to $\begin{cases} x + z \le 0 \\ y + 2z \le 5 \\ x + 2y + 3z \le 8 \\ x \ge 0 \\ y \ge 0 \\ z \ge 0 \end{cases}$

30. Maximize $P = x + 2y + 3z$

Subject to $\begin{cases} x - z \le 20 \\ y + z \le 30 \\ x - 2y + z \le 40 \\ x \ge 0 \\ y \ge 0 \\ z \ge 0 \end{cases}$

In Exercises 31–36, set up and solve the standard maximization problem using the simplex method.

31. **Resource Allocation: Beverages**   In the mood for a new beverage recipe? iDrink.com provides you with recipes for alcoholic and nonalcoholic drinks based on the ingredients you have on hand.

The Cranberry Cooler calls for 2 ounces of lemon-lime soda and 4 ounces of cranberry juice in addition to other ingredients. Nancy's Party Punch calls for 32 ounces of cranberry juice and no lemon-lime soda. Jimmy Wallbanger calls for 6 ounces of lemon-lime soda and no cranberry juice. (**Source:** www.idrink.com.) The Cranberry Cooler recipe yields 1 serving, the Nancy's Party Punch recipe yields 9 servings, and the Jimmy Wallbanger recipe yields 1 serving.

If you have 2 quarts (64 ounces) of lemon-lime soda and 1 gallon (128 ounces) of cranberry juice, at most how many drink servings can you make?

32. **Resource Allocation: Sandwiches**   A plain hamburger requires one ground beef patty and a bun. A cheeseburger requires one ground beef patty, one slice of cheese, and a bun. A double cheeseburger requires two ground beef patties, two slices of cheese, and a bun.

Frozen hamburger patties are typically sold in packs of 12; hamburger buns, in packs of 8; and cheese slices, in packs of 24.

A family is in charge of providing burgers for a neighborhood block party. The family members have purchased 13 packs of buns, 11 packs of hamburger patties, and 3 packs of cheese slices. How many of each type of sandwich should the family prepare if its members want to maximize the number of burgers with cheese?

33. **Resource Allocation: Sandwiches**   Repeat Exercise 32, except maximize the number of hamburger patties used.

34. **Resource Allocation: Sandwiches**   Repeat Exercise 32. except maximize the number of double cheeseburgers.

35. **Battery Sales**   AAA Alkaline Discount Batteries sells low-cost batteries to consumers. In August 2002, the company offered AA batteries at the following prices: 50-pack for $10.00, 100-pack for $18.00, and 600-pack for $96.00. (**Source:** www.aaaalkalinediscountbatteries .com.)

An electronics store owner wants to buy at most 1000 batteries and spend at most $175. She expects that she'll be able to resell all of the batteries she orders for $4.00 per 4-pack. How many packs (50-packs, 100-packs, or 600-packs) should she order if she wants to maximize her revenue? What is her maximum revenue?

36. **Battery Sales**   Repeat Exercise 35, except maximize profit. Assume that her only cost is the cost of the batteries. What is her maximum profit?

Exercises 37–38 are intended to expand your understanding of the simplex method.

37. Given the standard maximization problem

Maximize $z = 4x + 4y$

Subject to $\begin{cases} -x + y \le 4 \\ x + 2y \le 14 \\ 5x + 2y \le 30 \\ x \ge 0, y \ge 0 \end{cases}$

do the following:

(a) Graph the feasible region and find the coordinates of the corner points.

(b) Solve the standard maximization problem using the simplex method.

(c) Find the feasible solution associated with each simplex tableau and label the point on the graph that is associated with each feasible solution.

(d) Explain what the simplex method does in terms of the graph of the feasible region.

38. Repeat Exercise 37; however, this time pick a different column to be the first pivot column. Compare and contrast your results with those of Exercise 37.

4.4 Solving Standard Minimization Problems with the Dual

■ Find the solution to a standard minimization problem by solving the dual problem with the simplex method

GETTING STARTED A food distributor has two production facilities: one in Portland, Oregon, and the other in Spokane, Washington. Subject to production limitations and market demands, what shipment plan will minimize the distributor's shipment costs? Questions such as this may often be answered by solving a standard minimization problem.

In this section, we will introduce standard minimization problems and show how finding the maximum solution for a dual problem leads us to the minimum solution of the minimization problem.

STANDARD MINIMIZATION PROBLEM

A **standard minimization problem** is a linear programming problem with an objective function that is to be minimized. The objective function is of the form

$$P = ax + by + cz + \cdots$$

where $a, b, c, \ldots$ are real numbers and $x, y, z, \ldots$ are decision variables.

The decision variables are constrained to nonnegative values. Additional constraints are of the form

$$Ax + By + Cz + \cdots \geq M$$

where $A, B, C, \ldots$ are real numbers and M is nonnegative.

We are told that M is nonnegative. That is, $M \geq 0$. Is the origin in the feasible region? Substituting the origin $x = 0, \quad y = 0, \quad z = 0, \ldots$ into each constraint inequality $Ax + By + Cz + \cdots \geq M$ yields the inequality $A(0) + B(0) + C(0) + \cdots \geq M$ which simplifies to $0 \geq M$. Since M is nonnegative, the inequality $0 \geq M$ is satisfied if and only if $M = 0$. That is, the origin is in the feasible region if and only if $M = 0$ for every constraint. If $M \neq 0$ for one or more constraints, the origin is not in the feasible region.

EXAMPLE 1

Determining If a Linear Programming Problem Is a Standard Minimization Problem

Determine if the linear programming problem is a standard minimization problem.

$$\text{Minimize} \quad P = 2x + 3y$$

$$\text{Subject to} \begin{cases} 3x + y \geq 6 \\ 2x - y \geq 4 \\ \quad\quad x \geq 0, y \geq 0 \end{cases}$$

SOLUTION The objective function is to be minimized, the constraints are of the form $Ax + By + Cz + \cdots \geq M$ with M nonnegative, and the decision variables are nonnegative. Therefore, the problem is a standard minimization problem.

EXAMPLE 2

Determining If a Linear Programming Problem Is a Standard Minimization Problem

Determine if the linear programming problem is a standard minimization problem.

$$\text{Minimize} \quad P = 4x + 2y$$

$$\text{Subject to} \begin{cases} -3x + 7y \leq -9 \\ 7x + 5y \geq 4 \\ \quad\quad x \geq 0, y \geq 0 \end{cases}$$

SOLUTION At first glance, the problem doesn't look like a standard minimization problem. The first constraint has a less than or equal sign, and the constant term is negative. However, if we multiply the first constraint by -1, it becomes $3x - 7y \geq 9$. The problem may then be rewritten as

$$\text{Minimize} \quad P = 4x + 2y$$

$$\text{Subject to} \begin{cases} 3x - 7y \geq 9 \\ 7x + 5y \geq 4 \\ \quad\quad x \geq 0, y \geq 0 \end{cases}$$

Therefore, the linear programming problem is a standard minimization problem.

When setting up a linear programming problem, if we make sure that the constant term of each constraint is nonnegative, then we will be able to readily see if the problem is a standard maximization or minimization problem.

The Dual

For a standard minimization problem whose objective function has nonnegative coefficients, we may construct a standard maximization problem called the **dual problem.** By solving the dual problem, we can find the solution to the standard minimization problem. Consider the standard minimization problem,

$$\text{Minimize} \quad P = 6x + 5y$$

$$\text{Subject to} \begin{cases} x + y \geq 2 \\ 2x + y \geq 3 \\ \quad\quad x \geq 0, y \geq 0 \end{cases}$$

We first construct a matrix for the problem as shown.

$$A = \begin{bmatrix} 1 & 1 & | & 2 \\ 2 & 1 & | & 3 \\ \hline 6 & 5 & | & 1 \end{bmatrix} \qquad \begin{matrix} x + y \geq 2 \\ 2x + y \geq 3 \\ 6x + 5y = P \end{matrix}$$

The constraints are placed in the first two rows of the matrix, and the coefficients of the objective function are placed in the last row of the matrix. The matrix for the dual problem is found by transposing the matrix. The **transpose** of a matrix A, which is written A^T, is created by switching the rows and columns of A. In this case, the first row of A is $[1 \quad 1 \quad 2]$, so the first column of A^T is $\begin{bmatrix} 1 \\ 1 \\ 2 \end{bmatrix}$. Repeating the process for the additional columns of A^T yields $A^T = \begin{bmatrix} 1 & 2 & | & 6 \\ 1 & 1 & | & 5 \\ 2 & 3 & | & 1 \end{bmatrix}$. We construct the dual problem from A^T. We first extract the constraints and the objective function from the matrix.

$$A = \begin{bmatrix} 1 & 2 & | & 6 \\ 1 & 1 & | & 5 \\ \hline 2 & 3 & | & 1 \end{bmatrix} \qquad \begin{matrix} x + 2y \leq 6 \\ x + y \leq 5 \\ 2x + 3y = P \end{matrix}$$

The dual problem is

$$\text{Maximize} \quad P = 2x + 3y$$

$$\text{Subject to} \begin{cases} x + 2y \leq 6 \\ x + y \leq 5 \\ x \geq 0, y \geq 0 \end{cases}$$

How are the optimal values of the standard minimization problem and its dual problem related? Let's look at the graphs of the feasible regions (Figures 4.26 and 4.27).

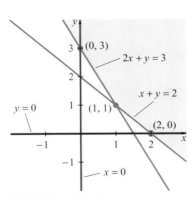

FIGURE 4.26 Standard Minimization Problem
Corner points:
$(0, 3), (1, 1), (2, 0)$

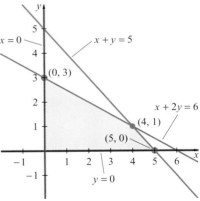

FIGURE 4.27 Standard Maximization Problem
Corner Points:
$(0, 0), (0, 3), (4, 1), (5, 0)$

For each problem, we calculate the value of the objective function at each corner point (see Tables 4.20 and 4.21).

TABLE 4.20 Standard Minimization Problem

x	y	P = 6x + 5y
0	3	15
1	1	**11**
2	0	12

TABLE 4.21 Standard Maximization Problem

x	y	P = 2x + 3y
0	0	0
0	3	9
4	1	**11**
5	0	10

Observe that the minimum value of the *minimization* problem objective function is 11. Similarly, the maximum value of the *maximization* problem objective function is 11. This leads us to the Fundamental Principle of Duality.

FUNDAMENTAL PRINCIPLE OF DUALITY

A standard minimization problem has a solution if and only if its dual problem has a solution. If a solution exists, the optimal value of the standard minimization problem and the optimal value of the dual problem are equal.

Although we already know the optimal value of the dual problem, we will demonstrate how using the simplex method to find the maximum value of the dual problem helps us determine the minimum value of the standard minimization problem. Recall that the dual problem is

$$\text{Maximize} \quad P = 2x + 3y$$
$$\text{Subject to} \begin{cases} x + 2y \le 6 \\ x + y \le 5 \\ x \ge 0, y \ge 0 \end{cases}$$

Adding in the slack variables and rewriting the objective function yield the system of equations

$$x + 2y + s \qquad = 6$$
$$x + y \qquad + t \quad = 5$$
$$-2x - 3y \qquad + P = 0$$

and the corresponding initial simplex tableau

$$\begin{array}{ccccc} x & y & s & t & P \\ \left[\begin{array}{ccccc|c} 1 & 2 & 1 & 0 & 0 & 6 \\ 1 & 1 & 0 & 1 & 0 & 5 \\ \hline -2 & -3 & 0 & 0 & 1 & 0 \end{array}\right] \end{array}$$

The y column is the pivot column, since it contains the negative entry in the bottom row with the largest magnitude. We calculate the test ratios and identify the pivot.

$$
\begin{array}{ccccc}
x & y & s & t & P \\
\end{array}
$$

$$
\left[\begin{array}{ccccc|c}
1 & \boxed{2} & 1 & 0 & 0 & 6 \\
1 & 1 & 0 & 1 & 0 & 5 \\
\hline
-2 & -3 & 0 & 0 & 1 & 0
\end{array}\right]
\quad
\begin{array}{l}
6/2 = 3 \\
5/1 = 5 \\
\end{array}
$$

The 2 in the first row of the y column is the pivot, since the first row has the smallest test ratio. We zero out the remaining terms in the y column using the indicated operations.

$$
\begin{array}{ccccc}
x & y & s & t & P \\
\end{array}
$$

$$
\left[\begin{array}{ccccc|c}
1 & 2 & 1 & 0 & 0 & 6 \\
1 & 0 & -1 & 2 & 0 & 4 \\
\hline
-1 & 0 & 3 & 0 & 2 & 18
\end{array}\right]
\quad
\begin{array}{l}
\\
2R_2 - R_1 \\
2R_3 + 3R_1 \\
\end{array}
$$

The x column is the new pivot column, since it is the only column with a negative entry in the bottom row. We calculate the test ratios and identify the pivot.

$$
\begin{array}{ccccc}
x & y & s & t & P \\
\end{array}
$$

$$
\left[\begin{array}{ccccc|c}
1 & 2 & 1 & 0 & 0 & 6 \\
\boxed{1} & 0 & -1 & 2 & 0 & 4 \\
\hline
-1 & 0 & 3 & 0 & 2 & 18
\end{array}\right]
\quad
\begin{array}{l}
6/1 = 6 \\
4/1 = 4 \\
\end{array}
$$

The 1 in the second row of the x column is the pivot, since the second row has the smallest test ratio. We zero out the remaining entries in the x column with the indicated operations.

$$
\begin{array}{ccccc}
x & y & s & t & P \\
\end{array}
$$

$$
\left[\begin{array}{ccccc|c}
0 & 2 & 2 & -2 & 0 & 2 \\
1 & 0 & -1 & 2 & 0 & 4 \\
\hline
0 & 0 & 2 & 2 & 2 & 22
\end{array}\right]
\quad
\begin{array}{l}
R_1 - R_2 \\
\\
R_3 + R_2 \\
\end{array}
$$

We must perform one additional step to make the nonzero term in each active column equal 1.

$$
\begin{array}{ccccc}
x & y & s & t & P \\
\end{array}
$$

$$
\left[\begin{array}{ccccc|c}
0 & 1 & 1 & -1 & 0 & 1 \\
1 & 0 & -1 & 2 & 0 & 4 \\
\hline
0 & 0 & 1 & 1 & 1 & 11
\end{array}\right]
\quad
\begin{array}{l}
1/2\,R_1 \\
\\
1/2\,R_3 \\
\end{array}
$$

Reading from the tableau, we have

$$
x = 4 \qquad y = 1 \qquad P = 11
$$

The corner point $(4, 1)$ yields the maximum value of the objective function of the dual problem $P = 2x + 3y$.

  Ironically, the optimal solution of the standard minimization problem is also nested in the final simplex tableau. With all active variable columns containing only 1s and 0s, the values in the bottom row of the s and t columns correspond with the x and y values, respectively, of the optimal solution of the standard minimization problem. That is, $x = 1$ and $y = 1$. The optimal value is the same as that of the dual problem: $P = 11$. We can verify that the corner point $(1, 1)$ of the standard minimization problem yields an objective function value of 11 by substituting the point into the objective function of the standard minimization problem $P = 6x + 5y$.

$$P = 6x + 5y$$
$$11 \overset{?}{=} 6(1) + 5(1)$$
$$11 = 11$$

The relationship between the standard minimization problem and its dual problem is somewhat remarkable, and the problem-solving method is straightforward.

> **HOW TO** Using Duals to Solve Standard Minimization Problems
>
> 1. Find the dual standard maximization problem.
> 2. Use the simplex method to solve the maximization problem.
> 3. The maximum value of the objective function of the dual problem is the minimum value of the objective function of the minimization problem.
> 4. The optimum solution of the minimization problem is given in the entries of the bottom row of the final tableau corresponding with the columns of the slack variables (as long as the entry in the P column equals 1).

EXAMPLE 3 Solving a Standard Minimization Problem

Solve the standard minimization problem.

$$\text{Minimize} \quad P = 4x + 5y + 6z$$
$$\text{Subject to} \begin{cases} x + y + z \geq 3 \\ 2x - z \geq 0 \\ x \geq 0, y \geq 0, z \geq 0 \end{cases}$$

SOLUTION We begin by constructing a matrix from the constraints and objective function.

$$A = \begin{matrix} & x & y & z & \\ & \begin{bmatrix} 1 & 1 & 1 & 3 \\ 2 & 0 & -1 & 0 \\ \hline 4 & 5 & 6 & 1 \end{bmatrix} \end{matrix} \quad \begin{matrix} x + y + z \geq 3 \\ 2x + 0y - z \geq 0 \\ 4x + 5y + 6z = P \end{matrix}$$

Next we find A^T and extract the objective function and constraints. Note that the constraints are of the form $ax + by \leq M$ with M nonnegative.

$$A^T = \begin{bmatrix} 1 & 2 & 4 \\ 1 & 0 & 5 \\ 1 & -1 & 6 \\ \hline 3 & 0 & 1 \end{bmatrix} \quad \begin{matrix} x + 2y \leq 4 \\ x + 0y \leq 5 \\ x - y \leq 6 \\ 3x + 0y = P \end{matrix}$$

The dual problem is

$$\text{Maximize} \quad P = 3x$$

$$\text{Subject to} \quad \begin{cases} x + 2y \le 4 \\ x \le 5 \\ x - y \le 6 \\ x \ge 0, y \ge 0, z \ge 0 \end{cases}$$

The dual is a standard maximization problem. We solve the problem using the simplex method. We begin by adding slack variables to the constraints.

$$x + 2y + s = 4$$
$$x + t = 5$$
$$x - y + u = 6$$
$$x \ge 0, y \ge 0, z \ge 0$$

Rewriting the objective function yields $-3x + P = 0$. The initial simplex tableau is

$$
\begin{array}{cccccc}
x & y & s & t & u & P \\
\end{array}
$$

$$
\begin{bmatrix}
\boxed{1} & 2 & 1 & 0 & 0 & 0 & 4 \\
1 & 0 & 0 & 1 & 0 & 0 & 5 \\
1 & -1 & 0 & 0 & 1 & 0 & 6 \\
-3 & 0 & 0 & 0 & 0 & 1 & 0
\end{bmatrix}
$$

The x column is the pivot column, since it is the only column with a negative entry in the bottom row. Mentally calculating the test ratios, we see that the entry in the first row and the x column is the pivot.

The final tableau is obtained by using the indicated row operations.

$$
\begin{array}{cccccc}
x & y & s & t & u & P \\
\end{array}
$$

$$
\begin{bmatrix}
1 & 2 & 1 & 0 & 0 & 0 & 4 \\
0 & -2 & -1 & 1 & 0 & 0 & 1 \\
0 & -3 & -1 & 0 & 1 & 0 & 2 \\
0 & 6 & 3 & 0 & 0 & 1 & 12
\end{bmatrix}
\quad
\begin{array}{l}
\\
R_2 - R_1 \\
R_3 - R_1 \\
R_4 + 3R_1
\end{array}
$$

Reading from the tableau, we see that the objective function of the *maximization* problem has a maximum value of 12. This occurs when $x = 4$ and $y = 0$. According to the fundamental principle of duality, the minimum value of the objective function of the *minimization* problem is also 12. The solution to the *minimization* problem may be found by looking at the slack variable columns in the bottom row of the final tableau. The s, t, and u columns of the final simplex tableau are the x, y, and z values of the solution to the minimization problem. That is, $(3, 0, 0)$ yields an objective function value of 12. This can be verified by substituting $x = 3, y = 0$, and $z = 0$ into $P = 4x + 5y + 6z$.

$$P = 4(3) + 5(0) + 6(0)$$
$$= 12$$

It is important to note that a standard minimization problem and its dual problem may have different numbers of decision variables. In Example 3, the standard minimization problem had three decision variables, x, y, and z, whereas the dual problem had only two decision variables, x and y.

EXAMPLE 4

Solving a Standard Minimization Problem

Solve the standard minimization problem.

$$\text{Minimize} \quad P = 2x + y + 2z$$

$$\text{Subject to} \quad \begin{cases} x + 2y + z \geq 6 \\ \quad\quad 3y + 2z \geq 12 \\ x \geq 0, y \geq 0, z \geq 0 \end{cases}$$

SOLUTION   We have

$$A = \begin{bmatrix} 1 & 2 & 1 & 6 \\ 0 & 3 & 2 & 12 \\ 2 & 1 & 2 & 1 \end{bmatrix} \qquad \begin{aligned} x + 2y + z &\geq 6 \\ 3y + 2z &\geq 12 \\ 2x + y + 2z &= P \end{aligned}$$

Therefore,

$$A^T = \begin{bmatrix} 1 & 0 & 2 \\ 2 & 3 & 1 \\ 1 & 2 & 2 \\ 6 & 12 & 1 \end{bmatrix} \quad \text{and} \quad \begin{aligned} x &\leq 2 \\ 2x + 3y &\leq 1 \\ x + 2y &\leq 2 \\ 6x + 12y &= P \end{aligned}$$

The dual problem is

$$\text{Maximize} \quad P = 6x + 12y$$

$$\text{Subject to} \quad \begin{cases} x \leq 2 \\ 2x + 3y \leq 1 \\ x + 2y \leq 2 \\ x \geq 0, y \geq 0 \end{cases}$$

Adding the slack variables and rewriting the objective function yields

$$x + s = 2$$
$$2x + 3y + t = 1$$
$$x + 2y + u = 2$$
$$-6x - 12y + P = 0$$

The initial simplex tableau is

$$\begin{array}{cccccc} x & y & s & t & u & P \\ \end{array}$$
$$\begin{bmatrix} 1 & 0 & 1 & 0 & 0 & 0 & 2 \\ 2 & \boxed{3} & 0 & 1 & 0 & 0 & 1 \\ 1 & 2 & 0 & 0 & 1 & 0 & 2 \\ \hline -6 & -12 & 0 & 0 & 0 & 1 & 0 \end{bmatrix}$$

The y column is the pivot, since it contains the negative entry in the bottom row with the largest magnitude. The pivot is the 3 in the second row and the y column because the second row's test ratio of $\frac{1}{3}$ is smaller than the third row's test ratio of $\frac{2}{2}$. Using the indicated row operations yields

$$
\begin{array}{ccccccc}
 & x & y & s & t & u & P \\
\end{array}
$$

$$
\left[\begin{array}{cccccc|c}
1 & 0 & 1 & 0 & 0 & 0 & 2 \\
2 & 3 & 0 & 1 & 0 & 0 & 1 \\
-1 & 0 & 0 & -2 & 3 & 0 & 4 \\
2 & 0 & 0 & 4 & 0 & 1 & 4
\end{array}\right]
\quad
\begin{array}{l}
3R_3 - 2R_2 \\
R_4 + 4R_2
\end{array}
$$

To obtain the final simplex tableau, we must ensure that the nonzero entry of each active variable column is equal to 1.

$$
\begin{array}{ccccccc}
 & x & y & s & t & u & P \\
\end{array}
$$

$$
\left[\begin{array}{cccccc|c}
1 & 0 & 1 & 0 & 0 & 0 & 2 \\
\frac{2}{3} & 1 & 0 & \frac{1}{3} & 0 & 0 & \frac{1}{3} \\
-\frac{1}{3} & 0 & 0 & -\frac{2}{3} & 1 & 0 & \frac{4}{3} \\
2 & 0 & 0 & 4 & 0 & 1 & 4
\end{array}\right]
\quad
\begin{array}{l}
\frac{1}{3}R_2 \\
\\
\frac{1}{3}R_3
\end{array}
$$

The solution to the dual problem is $x = 0$ and $y = \frac{1}{3}$. The optimal value is 4.

From the bottom row of the final tableau, we read the solution of the minimization problem from the slack variable columns: $x = 0$, $y = 4$, and $z = 0$. The minimal value of the objective function is 4. This can be verified by substituting the solution into the objective function equation of the standard minimization problem.

$$
\begin{aligned}
P &= 2x + y + 2z \\
&= 2(0) + (4) + 2(0) \\
&= 4
\end{aligned}
$$

Substituting the point $(0, 4, 0)$ back into the constraints of the standard minimization problem yields

$$
\begin{array}{ll}
x + 2y + z \geq 6 & 3y + 2z \geq 12 \\
(0) + 2(4) + (0) \geq 6 & 3(4) + 2(0) \geq 12 \\
8 \geq 6 & 12 \geq 12
\end{array}
$$

We are further convinced that we have found the optimal solution, since the solution satisfies both constraints (as it should). Also, $3y + 2z$ was made as small as the constraint would allow.

EXAMPLE 5

Using a Standard Minimization Problem to Minimize Shipping Costs

A northwest region canned food distributor has two production facilities: one in Spokane, Washington, and one in Portland, Oregon. For two supermarkets, the Portland facility produces a minimum of 700 cases of canned food per week and the Spokane facility produces a minimum of 300 cases per week. A Seattle-area supermarket requires at least 600 cases per week, and a supermarket in Ellensburg, Washington, requires at least 400 cases per week. Shipping costs (based on a rate of $0.36 per mile) vary based on the point of origin and destination as shown in Table 4.22.

TABLE 4.22

		Destination	
		Seattle	Ellensburg
Origin	Portland	$0.62 per case	$0.80 per case
	Spokane	$1.01 per case	$0.63 per case

What shipping schedule will minimize the food distributor's shipping costs?

SOLUTION

Let x be the number of cases shipped from Portland to Seattle.
Let y be the number of cases shipped from Spokane to Seattle.
Let z be the number of cases shipped from Portland to Ellensburg.
Let w be the number of cases shipped from Spokane to Ellensburg.

We need to minimize the shipment cost function,
$C = 0.62x + 1.01y + 0.80z + 0.63w$. We have the following constraints:

$$x + z \geq 700 \qquad \text{The Portland facility produces at least 700 cases}$$

$$y + w \geq 300 \qquad \text{The Spokane facility produces at least 300 cases}$$

$$x + y \geq 600 \qquad \text{The Seattle-area supermarket needs at least 600 cases}$$

$$z + w \geq 400 \qquad \text{The Ellensburg supermarket needs at least 400 cases}$$

$$x \geq 0, y \geq 0, z \geq 0, w \geq 0 \qquad \text{The number of cases shipped is nonnegative}$$

This is a standard minimization problem. We will solve it by using the dual and the simplex method. We have

$$A = \begin{bmatrix} 1 & 0 & 1 & 0 & 700 \\ 0 & 1 & 0 & 1 & 300 \\ 1 & 1 & 0 & 0 & 600 \\ 0 & 0 & 1 & 1 & 400 \\ \hline 0.62 & 1.01 & 0.80 & 0.63 & 1 \end{bmatrix}$$

Thus

$$A^T = \begin{bmatrix} 1 & 0 & 1 & 0 & 0.62 \\ 0 & 1 & 1 & 0 & 1.01 \\ 1 & 0 & 0 & 1 & 0.80 \\ 0 & 1 & 0 & 1 & 0.63 \\ \hline 700 & 300 & 600 & 400 & 1 \end{bmatrix}$$

The dual problem is

Maximize $\quad P = 700x + 300y + 600z + 400w$

$$\text{Subject to} \begin{cases} x + z \leq 0.62 \\ y + z \leq 1.01 \\ x + w \leq 0.80 \\ y + w \leq 0.63 \\ \quad x \geq 0, y \geq 0, z \geq 0, w \geq 0 \end{cases}$$

The initial tableau is given by

$$
\begin{array}{ccccccccc}
x & y & z & w & s & t & u & v & P \\
\end{array}
$$

$$
\left[
\begin{array}{ccccccccc|c}
\boxed{1} & 0 & 1 & 0 & 1 & 0 & 0 & 0 & 0 & 0.62 \\
0 & 1 & 1 & 0 & 0 & 1 & 0 & 0 & 0 & 1.01 \\
1 & 0 & 0 & 1 & 0 & 0 & 1 & 0 & 0 & 0.80 \\
0 & 1 & 0 & 1 & 0 & 0 & 0 & 1 & 0 & 0.63 \\
\hline
-700 & -300 & -600 & -400 & 0 & 0 & 0 & 0 & 1 & 0
\end{array}
\right]
$$

The pivot is the 1 in the first row of the x column of the tableau. The second tableau is generated with the indicated row operations. The 1 in the third row of the w column is the new pivot.

$$
\begin{array}{ccccccccc}
x & y & z & w & s & t & u & v & P \\
\end{array}
$$

$$
\left[
\begin{array}{ccccccccc|c}
1 & 0 & 1 & 0 & 1 & 0 & 0 & 0 & 0 & 0.62 \\
0 & 1 & 1 & 0 & 0 & 1 & 0 & 0 & 0 & 1.01 \\
0 & 0 & -1 & \boxed{1} & -1 & 0 & 1 & 0 & 0 & 0.18 \\
0 & 1 & 0 & 1 & 0 & 0 & 0 & 1 & 0 & 0.63 \\
\hline
0 & -300 & 100 & -400 & 700 & 0 & 0 & 0 & 1 & 434
\end{array}
\right]
\begin{array}{l}
\\
\\
R_3 - R_1 \\
\\
R_5 + 700R_1
\end{array}
$$

The third tableau is generated with the indicated row operations. We may choose either the y column or the z column to be the new pivot column. We pick the 1 in the fourth row of the y column to be the new pivot.

$$
\begin{array}{ccccccccc}
x & y & z & w & s & t & u & v & P \\
\end{array}
$$

$$
\left[
\begin{array}{ccccccccc|c}
1 & 0 & 1 & 0 & 1 & 0 & 0 & 0 & 0 & 0.62 \\
0 & 1 & 1 & 0 & 0 & 1 & 0 & 0 & 0 & 1.01 \\
0 & 0 & -1 & 1 & -1 & 0 & 1 & 0 & 0 & 0.18 \\
0 & \boxed{1} & 1 & 0 & 1 & 0 & -1 & 1 & 0 & 0.45 \\
\hline
0 & -300 & -300 & 0 & 300 & 0 & 400 & 0 & 1 & 506
\end{array}
\right]
\begin{array}{l}
\\
\\
\\
R_4 - R_3 \\
R_5 + 400R_3
\end{array}
$$

The final simplex tableau is generated with the indicated row operations.

$$
\begin{array}{ccccccccc}
x & y & z & w & s & t & u & v & P \\
\end{array}
$$

$$
\left[
\begin{array}{ccccccccc|c}
1 & 0 & 1 & 0 & 1 & 0 & 0 & 0 & 0 & 0.62 \\
0 & 0 & 0 & 0 & -1 & 1 & 1 & -1 & 0 & 0.56 \\
0 & 0 & -1 & 1 & -1 & 0 & 1 & 0 & 0 & 0.18 \\
0 & 1 & 1 & 0 & 1 & 0 & -1 & 1 & 0 & 0.45 \\
\hline
0 & 0 & 0 & 0 & 600 & 0 & 100 & 300 & 1 & 641
\end{array}
\right]
\begin{array}{l}
\\
R_2 - R_4 \\
\\
\\
R_5 + 300R_4
\end{array}
$$

The optimal solution of the dual problem is $x = 0.62$, $y = 0.45$, $z = 0$, and $w = 0.18$ with optimal value $P = 641$. We read the optimal solution for the minimization problem from the bottom row of the final tableau. When $x = 600$, $y = 0$, $z = 100$, and $w = 300$, the optimal value of 641 is obtained. In the context of the problem, the total shipping cost is minimized when 600 cases are shipped from Portland to Seattle, no cases are shipped from Spokane to Seattle, 100 cases are shipped from Portland to Ellensburg, and 300 cases are shipped from Spokane to Ellensburg. The total shipping cost is $641. (*Note:* This result assumes that the minimum number of cases were produced at each facility.)

EXAMPLE 6

Using a Standard Minimization Problem to Minimize Training Costs

Many companies with large government contracts use a competitive bidding process to hire subcontractors to do much of the work, A company has a contract to create at least 1000 hours of training and plans to hire three subcontractors (Trainum, Teachum, and Schoolum) to help with the work. Each subcontractor requires a contract for at least 200 hours of training development. Trainum charges $250 per hour of training development, Teachum charges $300 per hour of training development, and Schoolum charges $250 per hour of training development. How many hours should the company allocate to each subcontractor in order to minimize costs?

SOLUTION

Let x be the number of hours allocated to Trainum.
Let y be the number hours allocated to Teachum.
Let z be the number of hours allocated to Schoolum.

We need to minimize the training development cost function,
$C = 250x + 300y + 250z$, subject to the following constraints:

$$x + y + z \geq 1000 \qquad \text{At least 1000 training hours are produced}$$

$$x \geq 200 \qquad \text{Trainum is allocated at least 200 hours}$$

$$y \geq 200 \qquad \text{Teachum is allocated at least 200 hours}$$

$$z \geq 200 \qquad \text{Schoolum is allocated at least 200 hours}$$

This is a standard minimization problem. We will solve it by using the dual and the simplex method.

$$A = \begin{bmatrix} 1 & 1 & 1 & | & 1000 \\ 1 & 0 & 0 & | & 200 \\ 0 & 1 & 0 & | & 200 \\ 0 & 0 & 1 & | & 200 \\ \hline 250 & 300 & 250 & | & 1 \end{bmatrix} \quad \text{and} \quad A^T = \begin{bmatrix} 1 & 1 & 0 & 0 & | & 250 \\ 1 & 0 & 1 & 0 & | & 300 \\ 1 & 0 & 0 & 1 & | & 250 \\ \hline 1000 & 200 & 200 & 200 & | & 1 \end{bmatrix}$$

The dual problem is

$$\text{Maximize} \quad P = 1000x + 200y + 200z + 200w$$

$$\text{Subject to} \quad \begin{cases} x + y \leq 250 \\ x + z \leq 300 \\ x + w \leq 250 \\ x \geq 0, y \geq 0, z \geq 0, w \geq 0 \end{cases}$$

The initial simplex tableau is

$$\begin{array}{cccccccc} x & y & z & w & s & t & u & P \\ \begin{bmatrix} \boxed{1} & 1 & 0 & 0 & 1 & 0 & 0 & 0 & | & 250 \\ 1 & 0 & 1 & 0 & 0 & 1 & 0 & 0 & | & 300 \\ 1 & 0 & 0 & 1 & 0 & 0 & 1 & 0 & | & 250 \\ \hline -1000 & -200 & -200 & -200 & 0 & 0 & 0 & 1 & | & 0 \end{bmatrix} \end{array}$$

The pivot column is the x column, since -1000 is the negative entry in the bottom row with the largest magnitude. The test ratios of the first and third rows are both equal to 250. Since this is smaller than the test ratio for the second row

(300), either the first or the third entry in the x column may be selected as the pivot. We pick the first entry. Using the indicated row operations, we obtain the second tableau.

$$
\begin{array}{cccccccc|c}
x & y & z & w & s & t & u & P & \\
\hline
1 & 1 & 0 & 0 & 1 & 0 & 0 & 0 & 250 \\
0 & -1 & 1 & 0 & -1 & 1 & 0 & 0 & 50 \\
0 & -1 & 0 & \boxed{1} & -1 & 0 & 1 & 0 & 0 \\
\hline
0 & 800 & -200 & -200 & 1000 & 0 & 0 & 1 & 250000
\end{array}
\quad
\begin{array}{l}
\\
R_2 - R_1 \\
R_3 - R_1 \\
R_4 + 1000R_1
\end{array}
$$

Since the magnitude of the negative entries in the bottom row is the same, we may pick either the z column or the w column to be the pivot column. We pick the w column. The pivot is the 1 in the third row. Using the indicated row operation, we get the third tableau. The new pivot is the entry in the second row of the z column.

$$
\begin{array}{cccccccc|c}
x & y & z & w & s & t & u & P & \\
\hline
1 & 1 & 0 & 0 & 1 & 0 & 0 & 0 & 250 \\
0 & -1 & \boxed{1} & 0 & -1 & 1 & 0 & 0 & 50 \\
0 & -1 & 0 & 1 & -1 & 0 & 1 & 0 & 0 \\
\hline
0 & 600 & -200 & 0 & 800 & 0 & 200 & 1 & 250000
\end{array}
\quad
\begin{array}{l}
\\
\\
\\
R_4 + 200R_3
\end{array}
$$

Using the indicated row operation, we obtain the final tableau.

$$
\begin{array}{cccccccc|c}
x & y & z & w & s & t & u & P & \\
\hline
1 & 1 & 0 & 0 & 1 & 0 & 0 & 0 & 250 \\
0 & -1 & 1 & 0 & -1 & 1 & 0 & 0 & 50 \\
0 & -1 & 0 & 1 & -1 & 0 & 1 & 0 & 0 \\
\hline
0 & 400 & 0 & 0 & 600 & 200 & 200 & 1 & 260000
\end{array}
\quad
\begin{array}{l}
\\
\\
\\
R_4 + 200R_2
\end{array}
$$

The optimal solution for the dual problem is $x = 250$, $y = 0$, $z = 50$, and $w = 0$. The optimal value is 260,000.

An optimal solution for the minimization problem is $x = 600$, $y = 200$, and $z = 200$ with optimal value 260,000. In the context of the problem, the training development costs are minimized if Trainum develops 600 hours of training, Teachum develops 200 hours of training, and Schoolum develops 200 hours of training. The minimum training costs are $260,000. Since the hourly training development costs for Trainum and Schoolum are the same, another optimal solution is $(200, 200, 600)$. In fact, any point of the form $(t, 200, 800 - t)$ for $0 \le t \le 800$ will yield an optimal solution.

As shown in Example 6, a standard minimization problem may have more than one optimal solution. Each solution, however, will generate the same optimal value. The method of using the dual guarantees that we will find *an* optimal solution, if there is one.

EXAMPLE 7

Using a Standard Minimization Problem to Minimize Food Costs

PETsMART.com, an online retailer, allows pet owners to quickly compare nutrition and pricing for different brands of pet food sold on its web site. A June 2003 query revealed the data given in Table 4.23.

TABLE 4.23

Brand	Protein (percent)	Fiber (percent)	Price (dollars per 20-lb bag)
Eukanuba Adult Maintenance Formula	25	5	18.99
Nutro Natural Choice Plus	27	3	17.99
Science Diet Active Formula Canine Maintenance	26.5	3.5	17.99

Source: www.petsmart.com.

A dog breeder wants to create at least 300 pounds of a dog food mix that is at least 26 percent protein and 4 percent fiber while minimizing dog food cost. How many 20-pound bags of each type of dog food should the breeder buy?

SOLUTION

Let x be the number of bags of the Eukanuba brand.
Let y be the number of bags of the Nutro brand.
Let z be the number of bags of the Science Diet brand.

To make computations easier, we will convert the percentages per bag to pounds per bag by multiplying the percentage per bag by 20 pounds and dividing by 100, as shown in Table 4.24.

TABLE 4.24

Brand	Protein (pounds)	Fiber (pounds)
Eukanuba Adult Maintenance Formula	5.0	1.0
Nutro Natural Choice Plus	5.4	0.6
Science Diet Active Formula Canine Maintenance	5.3	0.7

The total weight of the mix is $20x + 20y + 20z$ pounds.

The total amount of protein in the mix is $5.0x + 5.4y + 5.3z$ pounds. We want the mix to be at least 26 percent protein. Thus we require that

$$0.26(20x + 20y + 20z) \leq 5.0x + 5.4y + 5.3z$$

$$5.2x + 5.2y + 5.2z \leq 5.0x + 5.4y + 5.3z$$

$$0.2x - 0.2y - 0.1z \leq 0$$

$$-0.2x + 0.2y + 0.1z \geq 0$$

The total amount of fiber in the mix is $1.0x + 0.6y + 0.7z$ pounds. We want the mix to be at least 4 percent fiber. Thus we require that

$$0.04(20x + 20y + 20z) \leq 1.0x + 0.6y + 0.7z$$
$$0.8x + 0.8y + 0.8z \leq 1.0x + 0.6y + 0.7z$$
$$-0.2x + 0.2y + 0.1z \leq 0$$
$$0.2x - 0.2y - 0.1z \geq 0$$

For computational ease, we round the prices to the nearest dollar. We must minimize $C = 19x + 18y + 18z$ subject to the following constraints:

$$20x + 20y + 20z \geq 300 \qquad \text{At least 300 pounds must be purchased}$$
$$-0.2x + 0.2y + 0.1z \geq 0 \qquad \text{The mix is at least 26 percent protein}$$
$$0.2x - 0.2y - 0.1z \geq 0 \qquad \text{The mix is at least 4 percent fiber}$$
$$x \geq 0,\, y \geq 0,\, z \geq 0 \qquad \text{A nonnegative number of bags are purchased}$$

We have

$$A = \begin{bmatrix} 20 & 20 & 20 & 300 \\ -0.2 & 0.2 & 0.1 & 0 \\ 0.2 & -0.2 & -0.1 & 0 \\ \hline 19 & 18 & 18 & 1 \end{bmatrix} \quad \text{and} \quad A^T = \begin{bmatrix} 20 & -0.2 & 0.2 & 19 \\ 20 & 0.2 & -0.2 & 18 \\ 20 & 0.1 & -0.1 & 18 \\ \hline 300 & 0 & 0 & 1 \end{bmatrix}$$

The dual problem is

Maximize $P = 300x + 0y + 0z$

Subject to $\begin{cases} 20x - 0.2y + 0.2z \leq 19 \\ 20x + 0.2y - 0.2z \leq 18 \\ 20x + 0.1y - 0.1z \leq 18 \\ x \geq 0,\, y \geq 0,\, z \geq 0 \end{cases}$

with the corresponding initial tableau

$$\begin{bmatrix} x & y & z & s & t & u & P & \\ 20 & -0.2 & 0.2 & 1 & 0 & 0 & 0 & 19 \\ \boxed{20} & 0.2 & -0.2 & 0 & 1 & 0 & 0 & 18 \\ 20 & 0.1 & -0.1 & 0 & 0 & 1 & 0 & 18 \\ \hline -300 & 0 & 0 & 0 & 0 & 0 & 1 & 0 \end{bmatrix}$$

The x column is the pivot column. Either the second or the third entry in the x column may be used as a pivot. Using the entry in the second row of the x column as a pivot and the indicated row operations, we obtain the second tableau. The entry in the third row of the z column is the new pivot, since it has the smallest test ratio.

$$\begin{bmatrix} x & y & z & s & t & u & P & \\ 0 & -0.4 & 0.4 & 1 & -1 & 0 & 0 & 1 \\ 20 & 0.2 & -0.2 & 0 & 1 & 0 & 0 & 18 \\ 0 & -0.1 & \boxed{0.1} & 0 & -1 & 1 & 0 & 0 \\ \hline 0 & 3 & -3 & 0 & 15 & 0 & 1 & 270 \end{bmatrix} \begin{matrix} R_1 - R_2 \\ \\ R_3 - R_2 \\ R_4 + 15R_2 \end{matrix}$$

The indicated row operations give the next tableau. The new pivot is the 3 in the first row of the t column.

$$
\begin{array}{ccccccc|c}
x & y & z & s & t & u & P & \\
\hline
0 & 0 & 0 & 1 & \boxed{3} & -4 & 0 & 1 \\
20 & 0 & 0 & 0 & -1 & 2 & 0 & 18 \\
0 & -0.1 & 0.1 & 0 & -1 & 1 & 0 & 0 \\
\hline
0 & 0 & 0 & 0 & -15 & 30 & 1 & 270
\end{array}
\qquad
\begin{array}{l}
R_1 - 4R_3 \\
R_2 + 2R_3 \\
\\
R_4 + 30R_3
\end{array}
$$

The indicated row operations give the next tableau.

$$
\begin{array}{ccccccc|c}
x & y & z & s & t & u & P & \\
\hline
0 & 0 & 0 & 1 & 3 & -4 & 0 & 1 \\
60 & 0 & 0 & 1 & 0 & 2 & 0 & 55 \\
0 & -0.3 & 0.3 & 1 & 0 & -1 & 0 & 1 \\
\hline
0 & 0 & 0 & 5 & 0 & 10 & 1 & 275
\end{array}
\qquad
\begin{array}{l}
\\
3R_2 + R_1 \\
3R_3 + R_1 \\
R_4 + 5R_1
\end{array}
$$

The final simplex tableau is obtained by converting the nonzero entry of each column with a single nonzero entry to a 1 by using the indicated row operations.

$$
\begin{array}{ccccccc|c}
x & y & z & s & t & u & P & \\
\hline
0 & 0 & 0 & \frac{1}{3} & 1 & -\frac{4}{3} & 0 & \frac{1}{3} \\
1 & 0 & 0 & \frac{1}{60} & 0 & \frac{1}{30} & 0 & \frac{11}{12} \\
0 & -1 & 1 & \frac{10}{3} & 0 & -\frac{10}{3} & 0 & \frac{10}{3} \\
\hline
0 & 0 & 0 & 5 & 0 & 10 & 1 & 275
\end{array}
\qquad
\begin{array}{l}
\frac{1}{3}R_1 \\
\frac{1}{60}R_2 \\
\frac{1}{0.3}R_3 \\
\\
\end{array}
$$

Setting the inactive variable columns s and u to 0 yields

$$
x = \frac{11}{12} \qquad -y + z = \frac{10}{3} \qquad t = \frac{1}{3} \qquad P = 275
$$

$$
z = y + \frac{10}{3}
$$

Any solution of the form $\left(\frac{11}{12}, r, r + \frac{10}{3}\right)$ for nonnegative r maximizes the objective function of the dual problem. The optimal value is 275.

The minimization problem has the solution $x = 5$, $y = 0$, and $z = 10$. When 5 bags of the Eukanuba brand and 10 bags of the Science Diet brand are purchased, the total cost is $275. This is the minimum cost for a mix that meets the breeder's nutrition requirements.

4.4 Summary

In this section, you learned how to solve standard minimization problems using the notion of the dual problem. You discovered that finding the maximum solution of a dual problem leads us to the minimum solution of the minimization problem.

4.4 Exercises

In Exercises 1–10, determine if the problem is a standard minimization problem. If it isn't, explain why.

1. Minimize $P = 9x + 8y$

Subject to $\begin{cases} 2x + y \geq 10 \\ -x + y \geq 1 \\ x \geq 0 \\ y \geq 0 \end{cases}$

2. Minimize $P = -2x - 5y$

Subject to $\begin{cases} 4x + 9y \geq 10 \\ -11x + y \geq -21 \\ x \geq 0 \\ y \geq 0 \end{cases}$

3. Minimize $P = 6xy$

Subject to $\begin{cases} 6x + 7y \geq 13 \\ -8x - 4y \geq 12 \\ x \geq 0 \\ y \geq 0 \end{cases}$

4. Minimize $P = 4x - 2y + z$

Subject to $\begin{cases} 2x + y + 4z \geq 24 \\ -2x + 3y + 3z \geq 150 \\ x \geq 0 \\ y \geq 0 \\ z \geq 0 \end{cases}$

5. Minimize $P = -1.2x - 2.8y + 4.3z$

Subject to $\begin{cases} 3.2x + 1.5y + 7.4z \geq 249.8 \\ -2.7x + 3.4y + 3.9z \geq 190.1 \end{cases}$

6. Minimize $P = 4x + 4y + 9z$

Subject to $\begin{cases} -6x - y + 4z \leq -24 \\ -9x - 1.5y + 6z \geq 36 \\ x \geq 0 \\ y \geq 0 \\ z \geq 0 \end{cases}$

7. Minimize $P = x + 9y + 8z$

Subject to $\begin{cases} -6x - y + 4z \geq -24 \\ -9x - 1.5y + 6z \geq 36 \\ x \geq 0 \\ y \geq 0 \\ z \geq 0 \end{cases}$

8. Minimize $P = 4x + 4y + 9z$

Subject to $\begin{cases} -3x - 2y + 4z \leq -22 \\ 5x - 2.5y + 6z \geq 36 \\ x \geq 0 \\ y \geq 0 \\ z \geq 0 \end{cases}$

9. Minimize $P = -8x + 2y - z$

Subject to $\begin{cases} z \geq 4 \\ x - y \geq 3 \\ 2x - z \geq 5 \\ x + y + z \leq 0 \end{cases}$

10. Minimize $P = 5x + 7z$

Subject to $\begin{cases} 4x + 2y \geq 16 \\ 4x - 2y \geq 3 \\ x \geq 0 \\ y \geq 0 \end{cases}$

In Exercises 11–20, find the transpose of the given matrix.

11. $A = \begin{bmatrix} 2 & 2 & 1 & 4 \\ 5 & 9 & 7 & 3 \\ 8 & 1 & 0 & 1 \end{bmatrix}$ **12.** $A = \begin{bmatrix} 1 & 1 & 1 & 4 \\ 1 & 1 & 1 & 5 \\ 4 & 5 & 1 & 1 \end{bmatrix}$

13. $A = \begin{bmatrix} 1 & 2 & 4 \\ 6 & 8 & 1 \end{bmatrix}$ **14.** $A = \begin{bmatrix} 8 & 2 & 1 \\ 2 & 5 & 7 \\ 1 & 7 & 1 \end{bmatrix}$

15. $A = \begin{bmatrix} 9 & 3 & -1 & 4 \\ 7 & 2 & -2 & 1 \end{bmatrix}$

16. $A = \begin{bmatrix} 1 & -2 & 1 & 4 & 11 \\ -5 & 4 & -7 & 3 & 9 \\ 3 & 7 & 9 & 3 & 1 \end{bmatrix}$

17. $A = \begin{bmatrix} 9 & 2 & 18 \\ -3 & -2 & 6 \\ 4 & 0 & 0 \\ 5 & 6 & 1 \end{bmatrix}$

18. $A = \begin{bmatrix} 6 & 5 & 4 & 3 & 2 \\ 5 & 6 & 7 & 8 & 9 \\ 4 & 5 & 6 & 5 & 1 \end{bmatrix}$

19. $A = \begin{bmatrix} 0 & 2 & 1 \\ -5 & 4 & 7 \\ 8 & 1 & 0 \\ -1 & 2 & 1 \\ 4 & 2 & 1 \end{bmatrix}$

20. $A = \begin{bmatrix} 2 & 4 & 6 & 5 & 1 \end{bmatrix}$

In Exercises 21–30, do the following:

(i)   Write the dual problem for the given standard minimization problem.

(ii) Solve the dual problem using the simplex method.

(iii) Use the final simplex tableau of the dual problem to solve the standard minimization problem.

(iv) Check your answer by graphing the feasible region of the standard minimization problem and calculating the value of the objective function at each of the corner points.

21. Minimize $P = x + 2y$

Subject to $\begin{cases} 3x + y \geq 6 \\ 2x + 3y \geq 11 \\ x \geq 0, y \geq 0 \end{cases}$

22. Minimize $P = 4x + 2y$

Subject to $\begin{cases} 5x + 7y \geq 19 \\ -2x + 3y \geq 5 \\ x \geq 0, y \geq 0 \end{cases}$

23. Minimize $P = 9x + 6y$

Subject to $\begin{cases} 5x + 2y \geq 30 \\ 10x - 5y \geq 15 \\ x \geq 0, y \geq 0 \end{cases}$

24. Minimize $P = 4x + 4y$

Subject to $\begin{cases} 3x + 4y \geq 18 \\ 5x + 3y \geq 19 \\ x \geq 0, y \geq 0 \end{cases}$

25. Minimize $P = 5x + 20y$

Subject to $\begin{cases} -5x + 2y \geq 10 \\ 5x + y \geq 25 \\ x \geq 0, y \geq 0 \end{cases}$

26. Minimize $P = 2x + 5y$

Subject to $\begin{cases} x + y \geq 9 \\ 2x - 2y \geq 6 \\ x \geq 0, y \geq 0 \end{cases}$

27. Minimize $P = 3x + 2y$

Subject to $\begin{cases} x + y \geq 5 \\ 3x + 2y \geq 11 \\ 4x - y \geq 0 \\ x \geq 0, y \geq 0 \end{cases}$

28. Minimize $P = 9x + 7y$

Subject to $\begin{cases} 5x + 3y \geq 8 \\ 3x + 5y \geq 8 \\ 8x - 8y \geq 0 \\ x \geq 0, y \geq 0 \end{cases}$

29. Minimize $P = 6x + 5y$

Subject to $\begin{cases} x + y \geq 4 \\ x - y \geq 2 \\ 2x - y \geq 6 \\ x \geq 0, y \geq 0 \end{cases}$

30. Minimize $P = 5x + 7y$

Subject to $\begin{cases} x + y \geq 4 \\ x - y \geq 2 \\ 2x - y \geq 6 \\ x \geq 0, y \geq 0 \end{cases}$

In Exercises 31–40, set up and solve the standard minimization problem.

31.   **Pet Nutrition: Food Cost**
PETsMART.com sold the following varieties of dog food in June 2003. The price shown is for an 8-pound bag.

Pro Plan Adult Chicken & Rice Formula,
   25 percent protein, 3 percent fiber, $7.99
Pro Plan Adult Lamb & Rice Formula,
   28 percent protein, 3 percent fiber, $7.99
Pro Plan Adult Turkey & Barley Formula,
   26 percent protein, 3 percent fiber, $8.49

(**Source:** www.petsmart.com.)

   A dog breeder wants to make at least 120 pounds of a mix containing at least 27 percent protein and at least 3 percent fiber. How many 8-pound bags of each dog food variety should the breeder buy in order to minimize cost? (Round prices up to the nearest dollar.)

32. 🌐 **Pet Nutrition: Food Cost**
PETsMART.com sold the following varieties of dog food in June 2003.

Authority Chicken Adult Formula, 32 percent protein, 3 percent fiber, $19.99 per 33-pound bag

Bil-Jac Select Dog Food, 27 percent protein, 4 percent fiber, $18.99 per 18-pound bag

Iams Minichuncks, 26 percent protein, 4 percent fiber, $8.99 per 8-pound bag

(**Source:** www.petsmart.com.)

A dog kennel wants to make at least 2178 pounds of a mix containing at least 29 percent protein and at least 3.5 percent fiber. How many bags of each dog food variety should the kennel buy in order to minimize cost? (Round prices up to the nearest dollar.)

33. 🌐 **Pet Nutrition: Food Cost**
PETsMART.com sold the following varieties of dog food in June 2003.

Nature's Recipe Venison Meal & Rice Canine, 20 percent protein, 10 percent fat, $21.99 per 20-pound bag

Nutro Max Natural Dog Food, 27 percent protein, 16 percent fat, $12.99 per 17.5-pound bag

PETsMART Premier Oven Baked Lamb Recipe, 25 percent protein, 14 percent fat, $22.99 per 30-pound bag

(**Source:** www.petsmart.com.)

A dog breeder wants to make at least 175 pounds of a mix containing at least 25 percent protein and at least 14 percent fat. How many bags of each dog food variety should the breeder buy in order to minimize cost? (Round prices *up* to the nearest dollar.)

34. 🌐 **Pet Nutrition: Fat Content**
PETsMART.com sold the following varieties of dog food in June 2003.

Nature's Recipe Venison Meal & Rice Canine, 20 percent protein, 10 percent fat, $21.99 per 20-pound bag

Nutro Max Natural Dog Food, 27 percent protein, 16 percent fat, $12.99 per 17.5-pound bag

PETsMART Premier Oven Baked Lamb Recipe, 25 percent protein, 14 percent fat, $22.99 per 30-pound bag

(**Source:** www.petsmart.com.)

A dog breeder wants to make at least 210 pounds of a mix containing at least 25 percent protein. How many bags of each dog food variety should the breeder buy in order to minimize fat content?

35. 🌐 **Food Distribution Cost** Wal-Mart Stores, Inc., has food distribution centers in Monroe, Georgia, and Shelbyville, Tennessee, and Wal-Mart Supercenters in Birmingham, Alabama, and Scottsboro, Alabama. (**Source:** www.walmart.com.) Suppose that the Monroe distribution center must ship at least 600 cases of peanut butter weekly and the Shelbyville distribution center must ship at least 400 cases of peanut butter weekly.* If the Birmingham store requires at least 700 cases of peanut butter weekly and the Scottsboro store requires at least 300 cases of peanut butter weekly, what shipment plan will minimize the distribution cost?

Estimated Distribution Cost per Case

	Monroe	Shelbyville
Birmingham	$0.68	$0.57
Scottsboro	$0.65	$0.32

*Distribution costs and amounts are hypothetical.

36. 🌐 **Food Distribution Cost** Wal-Mart Stores, Inc., has food distribution centers in Monroe, Georgia, and Shelbyville, Tennessee, and Wal-Mart Supercenters in Birmingham, Alabama, and Calhoun, Georgia. (**Source:** www.walmart.com.) Suppose that the Monroe distribution center must ship at least 600 cases of pickles weekly and the Shelbyville distribution center must ship at least 700 cases of pickles weekly.* If the Birmingham store requires at least 1000 cases of pickles weekly and the Calhoun store requires at least 300 cases of pickles weekly, what shipment plan will minimize the distribution cost?

Estimated Distribution Cost per Case

	Monroe	Shelbyville
Birmingham	$0.68	$0.57
Calhoun	$0.39	$0.50

*Distribution costs and amounts are hypothetical.

37. 🌐 **Food Distribution Cost** Wal-Mart Stores, Inc., has food distribution centers in Monroe, Georgia, and Shelbyville, Tennessee, and Wal-Mart Supercenters in Scottsboro, Alabama, and Calhoun, Georgia. (**Source:** www.walmart.com.) Suppose that the Monroe distribution center must ship at least 400 cases of potato chips weekly and the Shelbyville distribution center must ship at least 200 cases of potato chips weekly.* If the Scottsboro store requires at least 300 cases of potato chips weekly and the Calhoun store requires at least 300 cases of potato chips weekly, what shipment plan will minimize the distribution cost?

Estimated Distribution Cost per Case

	Monroe	Shelbyville
Calhoun	$0.39	$0.50
Scottsboro	$0.65	$0.32

*Distribution costs and amounts are hypothetical.

38. Food and Entertainment It costs $35 for a family to dine out and $10 for the family to eat at home. Eating out has a fun rating of 10 points, while eating at home has a fun rating of 2 points.

It costs $32 for a family to play a game of miniature golf and $30 to watch a movie. Miniature golf has a fun rating of 8 points, and watching a movie has a fun rating of 6 points.

The family must eat at least 21 meals weekly and must go out for food or miniature golf at least five times weekly. The family wants to earn at least 82 fun points per week. How many times a week should the family participate in each activity in order to minimize food and entertainment costs?

39. Marital Harmony It costs a couple 3 hours and $200 to go clothes shopping together. She gives shopping a fun rating of 10 points, while he gives it a fun rating of 1 point (a total of 11 fun points for the couple).

It costs a couple 4 hours and $250 to go to a major league baseball playoff game. She gives the game a fun rating of 4 points, while he gives it a fun rating of 10 points (a total of 14 fun points for the couple).

The couple wants to spend at least 10 hours together (shopping and watching baseball) while earning at least 36 fun points. She insists that they go shopping at least twice. How shall they spend their time if they want to minimize their financial costs?

40. Marital Discord It costs a couple 3 hours to go clothes shopping together. He gives shopping with her a fun rating of 6 points while she gives it a fun rating of 1 point (a total of 7 fun points for the couple).

It costs a couple 4 hours to go to a major league baseball game. She gives going to a game with him a fun rating of 7 points while he gives it a fun rating of 2 points (a total of 9 fun points for the couple).

The couple wants to earn at least 30 fun points. She refuses to go shopping with him unless he goes with her to at least one game. How shall they spend their time if they want to minimize their time together? (Things are not going well in the relationship.)

4.5 Solving General Linear Programming Problems with the Simplex Method

- Solve general linear programming problems with the simplex method

GETTING STARTED An investor plans to invest at most $3000 in the three publicly traded recreational vehicle companies shown in Table 4.25.

TABLE 4.25

Company	Share Price (dollars)	Dividend/Share (dollars)
Harley- Davidson, Inc. (HDI)	62.70	0.40
Polaris Industries Inc. (PII)	50.50	0.92
Winnebago Industries Inc. (WGO)	33.42	0.20

Source: moneycentral.msn.com. (Accurate as of July 16, 2004.)

He wants to earn at least $50 in dividends while maximizing the number of shares purchased. How many shares of each company should he purchase?

At first glance, this looks like a standard maximization problem. However, a closer analysis reveals that it is not. While the investment constraint is of the form $Ax + By + Cz + \cdots \leq M$, the dividend constraint is of the form $Ax + By + Cz + \cdots \geq M$ instead of $Ax + By + Cz + \cdots \leq M$. This is a **general linear programming problem** or a linear programming problem with *mixed constraints*.

In this section, we will show how to solve general linear programming problems. We will demonstrate how minimization problems may be solved by maximizing the negative of the objective function. We will return to the recreational vehicle investment problem in Example 3.

Recall that the feasible region of any standard maximization problem always includes the origin. Graphically speaking, the simplex method starts at the origin and moves from corner point to adjacent corner point, each time increasing the value of the objective function until the maximum value is reached. The simplex method works with standard maximization problems because we are guaranteed that the origin is in the feasible region. But what if the origin is outside of the feasible region? The feasible region of a linear programming problem with mixed constraints often does not contain the origin.

A linear programming problem with mixed constraints has constraints in two or more of the following forms: $Ax + By + Cz + \cdots \leq M$, $Ax + By + Cz + \cdots \geq M$, or $Ax + By + Cz + \cdots = M$. In all cases, $M \geq 0$. Consider the following linear programming problem with mixed constraints.

$$\text{Maximize} \quad P = 3x + 2y$$

$$\text{Subject to} \quad \begin{cases} -x + 2y \leq 4 \\ 2x - y \leq 4 \\ x + y \geq 5 \\ x \geq 0, y \geq 0 \end{cases}$$

Since the constraint $x + y \geq 5$ contains a $\geq$ sign instead of a $\leq$ sign, this is not a standard maximization problem. It is a problem with mixed constraints. The graph of the feasible region is shown in Figure 4.28.

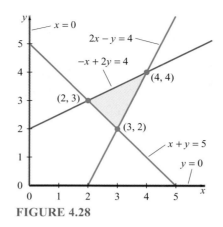

FIGURE 4.28

Notice that the origin is not contained within the feasible region. Our strategy for solving linear programming problems with mixed constraints is a two-stage process.

Stage 1: Get inside the feasible region.
Stage 2: Solve the problem with the simplex method.

Since our problem contains only two decision variables, we have the luxury of observing how the process works graphically. This is not the case in problems involving three or more decision variables.

We set up the problem for the simplex method as usual by adding in slack variables to constraints of the form $Ax + By + Cz + \cdots \leq M$.

$$-x + 2y + s = 4$$
$$2x - y + t = 4$$

Can we add in slack variables to constraints of the form $Ax + By + Cz + \cdots \geq M$? Let's see. We have the constraint $x + y \geq 5$. Recall that slack variables add in what is necessary to make the inequality an equality. By definition, slack variables must be nonnegative. What must we do to the inequality $x + y \geq 5$ to make it an equality? Adding a slack variable to the inequality will increase instead of decrease the value of the left-hand side. Since the left-hand side of the inequality is greater than or equal to 5, we must subtract some nonnegative value u in order to make $x + y$ equal to 5. That is,

$$x + y - u = 5$$

Since this variable takes away the surplus, it is referred to as a **surplus variable.** We have

$$-x + 2y + s = 4$$
$$2x - y + t = 4$$
$$x + y - u = 5$$
$$-3x - 2y + P = 0$$

and the corresponding initial tableau

$$
\begin{array}{cccccc}
 & x & y & s & t & u & P \\
\left[\begin{array}{cccccc|c}
-1 & 2 & 1 & 0 & 0 & 0 & 4 \\
2 & -1 & 0 & 1 & 0 & 0 & 4 \\
1 & 1 & 0 & 0 & -1 & 0 & 5 \\
\hline
-3 & -2 & 0 & 0 & 0 & 1 & 0
\end{array}\right]
\end{array}
$$

Setting the x and y variables equal to zero yields the solution to this tableau.

$$x = 0 \qquad y = 0 \qquad s = 4 \qquad t = 4 \qquad -u = 5 \qquad P = 0$$
$$u = -5$$

Recall that all decision, slack, and surplus variables are required to be nonnegative. Since $u < 0$, this solution is not in the feasible region. This may be easily seen graphically (Figure 4.29). The origin is not in the feasible region.

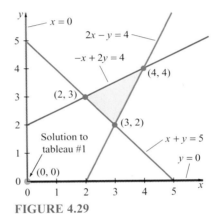

FIGURE 4.29

In the initial tableau, columns s, t, u, and P were the active variable columns. Since the active variable u was negative, we want to modify the tableau in such a way as to make u an inactive variable. We place a * on the left of the tableau to highlight the rows above the bottom row that resulted in negative variable values. In this case, the third row was the only one with a negative variable value.

$$
\begin{array}{cccccc}
 & x & y & s & t & u & P \\
\left[\begin{array}{cccccc|c}
-1 & 2 & 1 & 0 & 0 & 0 & 4 \\
2 & -1 & 0 & 1 & 0 & 0 & 4 \\
1 & 1 & 0 & 0 & -1 & 0 & 5 \\
\hline
-3 & -2 & 0 & 0 & 0 & 1 & 0
\end{array}\right]
\end{array}
$$

Since all surplus variables in the initial tableau are negative, each row of the initial tableau containing surplus variables will be starred. We select the column with the largest positive entry in the starred row as the pivot column. Since the positive values in the starred row are equal, we may select either the x column or the y column to be the pivot column. We select the x column. To determine the pivot, we calculate the test ratios and select the row with the smallest test ratio. This will ensure that all nonnegative variables will remain nonnegative.

$$
\begin{array}{ccccccc}
 & x & y & s & t & u & P & \\
 & \begin{bmatrix} -1 & 2 & 1 & 0 & 0 & 0 & 4 \\ \boxed{2} & -1 & 0 & 1 & 0 & 0 & 4 \\ 1 & 1 & 0 & 0 & -1 & 0 & 5 \\ \hline -3 & -2 & 0 & 0 & 0 & 1 & 0 \end{bmatrix} & & & & & & \begin{array}{l} 4/2 = 2 \\ 5/1 = 5 \end{array}
\end{array}
$$

* (stars next to row 3)

↑ Pivot column

The 2 in the second row of the x column is the pivot. We modify the initial tableau with the indicated operations and star any rows corresponding with negative active variables.

$$
\begin{array}{ccccccc}
 & x & y & s & t & u & P & \\
 & \begin{bmatrix} 0 & 3 & 2 & 1 & 0 & 0 & 12 \\ \boxed{2} & -1 & 0 & 1 & 0 & 0 & 4 \\ 0 & 3 & 0 & -1 & -2 & 0 & 6 \\ 0 & -7 & 0 & 3 & 0 & 2 & 12 \end{bmatrix} & & & & & & \begin{array}{l} 2R_1 + R_2 \\ \\ 2R_3 - R_2 \\ 2R_4 + 3R_2 \end{array}
\end{array}
$$

* (star next to row 3)

$$2x = 4 \qquad y = 0 \qquad 2s = 12 \qquad t = 0 \qquad -2u = 6 \qquad 2P = 12$$
$$x = 2 \qquad\qquad\qquad s = 6 \qquad\qquad\qquad u = -3 \qquad\quad P = 6$$

We starred the third row, since the value of u is negative. Although u is still negative, it is less negative than it was before. We will continue the process until u is nonnegative. However, before continuing, let's examine what is happening graphically (Figure 4.30).

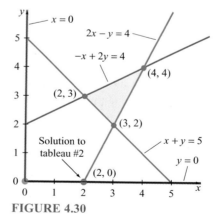

FIGURE 4.30

We have moved from the origin to the x-intercept of one of the constraints. However, we still are outside of the feasible region. This is due to the fact that the surplus variable u is still negative. We will pivot again and further increase the value of u.

The largest value in the starred row is the 3 in the y column. The y column is the new pivot column.

$$
\begin{array}{cccccc}
x & y & s & t & u & P \\
\end{array}
$$

$$
\begin{bmatrix}
0 & 3 & 2 & 1 & 0 & 0 & | & 12 \\
2 & -1 & 0 & 1 & 0 & 0 & | & 4 \\
0 & \boxed{3} & 0 & -1 & -2 & 0 & | & 6 \\
0 & -7 & 0 & 3 & 0 & 2 & | & 12
\end{bmatrix}
\begin{array}{l}
12/3 = 4 \\
\\
6/3 = 2 \\
\\
\end{array}
$$

* (marks the third row)

Pivot column (points to y column)

The pivot is the 3 in the third row of the y column, since that row has the smallest test ratio. We zero out the remaining terms in the column with the indicated operations and read the solution from the tableau.

$$
\begin{array}{cccccc}
x & y & s & t & u & P \\
\end{array}
$$

$$
\begin{bmatrix}
0 & 0 & 2 & 2 & 2 & 0 & | & 6 \\
6 & 0 & 0 & 2 & -2 & 0 & | & 18 \\
0 & 3 & 0 & -1 & -2 & 0 & | & 6 \\
0 & 0 & 0 & 2 & -14 & 6 & | & 78
\end{bmatrix}
\begin{array}{l}
R_1 - R_3 \\
3R_2 + R_3 \\
\\
3R_4 + 7R_3
\end{array}
$$

$$
6x = 18 \quad 3y = 6 \quad 2s = 6 \quad t = 0 \quad u = 0 \quad 6P = 78
$$

$$
x = 3 \quad y = 2 \quad s = 3 \qquad\qquad\qquad P = 13
$$

Since none of the variables are negative, this solution is in the feasible region (Figure 4.31). Since we are now in the feasible region, this completes Stage 1 of the process.

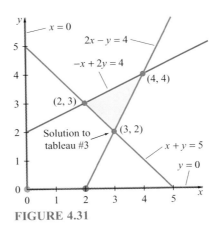

FIGURE 4.31

Stage 2 allows us to apply the simplex method to find the optimal solution. The u column is the pivot column. The 2 in the first row of that column is the pivot, since it is the only nonnegative value in the column.

$$
\begin{array}{cccccc}
x & y & s & t & u & P \\
\end{array}
$$

$$
\begin{bmatrix}
0 & 0 & 2 & 2 & \boxed{2} & 0 & | & 6 \\
6 & 0 & 0 & 2 & -2 & 0 & | & 18 \\
0 & 3 & 0 & -1 & -2 & 0 & | & 6 \\
0 & 0 & 0 & 2 & -14 & 6 & | & 78
\end{bmatrix}
$$

Pivot column (points to u column)

We zero out the remaining terms in the u column using the indicated operations.

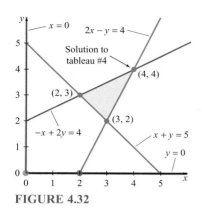

FIGURE 4.32

$$
\begin{array}{cccccc|c}
x & y & s & t & u & P & \\
0 & 0 & 2 & 2 & 2 & 0 & 6 \\
6 & 0 & 2 & 4 & 0 & 0 & 24 \\
0 & 3 & 2 & 1 & 0 & 0 & 12 \\
0 & 0 & 14 & 16 & 0 & 6 & 120
\end{array}
\quad
\begin{array}{l}
\\
R_2 + R_1 \\
R_3 + R_1 \\
R_4 + 7R_1
\end{array}
$$

$$
\begin{array}{cccccc}
6x = 24 & 3y = 12 & s = 0 & t = 0 & 2u = 6 & 6P = 120 \\
x = 4 & y = 4 & & & u = 3 & P = 20
\end{array}
$$

Since the bottom row of the tableau does not contain any negative values, this is the final simplex tableau. The solution that maximizes the objective function $P = 3x + 2y$ is $x = 4$ and $y = 4$ (Figure 4.32). The maximum value of the objective function is $P = 20$.

The following box details the steps used to solve the general linear programming problem.

HOW TO **Solving General Linear Programming Problems**

Stage 1: Get inside the feasible region

1. Star all rows that correspond with a negative value of a decision, slack, or surplus variable.

2. Identify the largest positive entry in the starred row. The corresponding column is the pivot column.

3. Calculate the test ratios for the positive entries above the bottom row of the pivot column.

4. Pick the pivot column entry with the smallest test ratio as the pivot.

5. Row reduce the tableau using operations of the form $aR_c \pm bR_p \rightarrow R_c$ with a and b positive. (Recall that R_c is the row we want to change and R_p is the pivot row.)

6. From the tableau, calculate the value of all decision, slack, and surplus variables. If any of the variables are negative, repeat Steps 1 through 5 for the new tableau. Otherwise, go to Stage 2.

Stage 2: Solve the maximization problem with the simplex method.

EXAMPLE 1 **Solving a Linear Programming Problem with No Optimal Solution**

Graph the feasible region associated with the given linear programming problem. Then solve the problem using the two-stage method described previously, indicating on the graph the solution that corresponds with each tableau.

$$
\begin{aligned}
\text{Maximize} \quad & P = 3x + 10y \\
\text{Subject to} \quad &
\begin{cases}
4x - y \geq 11 \\
x + 2y \geq 5 \\
-x + y \leq 1 \\
\qquad x \geq 0, y \geq 0
\end{cases}
\end{aligned}
$$

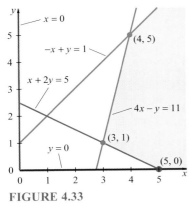

FIGURE 4.33

SOLUTION The graph of the feasible region is shown in Figure 4.33.

We can see from the graph that the feasible region is unbounded. Since the objective function has positive coefficients and the region is unbounded, the objective function will not have a maximum value. Nevertheless, we will proceed with the two-stage method to discover what indicators in the tableau let us know that there is no objective function maximum.

We add the slack and surplus variables and rewrite the objective function. We have

$$4x - y - s = 11$$
$$x + 2y - t = 5$$
$$-x + y + u = 1$$
$$-3x - 10y + P = 0$$

and the corresponding initial tableau

$$\begin{array}{cccccc}
x & y & s & t & u & P \\
\left[\begin{array}{cccccc|c}
4 & -1 & -1 & 0 & 0 & 0 & 11 \\
1 & 2 & 0 & -1 & 0 & 0 & 5 \\
-1 & 1 & 0 & 0 & 1 & 0 & 1 \\
\hline
-3 & -10 & 0 & 0 & 0 & 1 & 0
\end{array}\right]
\end{array}$$

Reading from the tableau, we have

$$x = 0 \qquad y = 0 \qquad -s = 11 \qquad -t = 5 \qquad u = 1 \qquad P = 0$$
$$s = -11 \qquad t = -5$$

Since both s and t are negative, we will star the corresponding rows.

$$\begin{array}{cccccc}
 & x & y & s & t & u & P \\
* & \left[\begin{array}{cccccc|c}
4 & -1 & -1 & 0 & 0 & 0 & 11 \\
1 & 2 & 0 & -1 & 0 & 0 & 5 \\
-1 & 1 & 0 & 0 & 1 & 0 & 1 \\
\hline
-3 & -10 & 0 & 0 & 0 & 1 & 0
\end{array}\right]
\end{array}$$

We may choose either starred row to work with. We pick the row corresponding with the variable t (second row). The largest value in the labeled columns of the second row is the 2 in the y column. Consequently, the y column is the pivot column. We calculate the test ratios and locate the pivot.

$$\begin{array}{cccccc}
 & x & y & s & t & u & P \\
* & \left[\begin{array}{cccccc|c}
4 & -1 & -1 & 0 & 0 & 0 & 11 \\
1 & 2 & 0 & -1 & 0 & 0 & 5 \\
-1 & \boxed{1} & 0 & 0 & 1 & 0 & 1 \\
\hline
-3 & -10 & 0 & 0 & 0 & 1 & 0
\end{array}\right] & \begin{array}{l} \\ 5/2 = 2.5 \\ 1/1 = 1 \\ \end{array}
\end{array}$$

Pivot column

We zero out the remaining terms in the pivot column using the indicated operations and star the rows that correspond with negative variable values.

$$
\begin{array}{c}
\begin{array}{cccccc}
x & y & s & t & u & P
\end{array}\\
\begin{array}{c}
* \\
* \\
\ \\
\
\end{array}
\left[
\begin{array}{cccccc|c}
3 & 0 & -1 & 0 & 1 & 0 & 12 \\
3 & 0 & 0 & -1 & -2 & 0 & 3 \\
-1 & 1 & 0 & 0 & 1 & 0 & 1 \\
-13 & 0 & 0 & 0 & 10 & 1 & 10
\end{array}
\right]
\begin{array}{c}
R_1 + R_3 \\
R_2 - 2R_3 \\
\ \\
R_4 + 10R_3
\end{array}
\end{array}
$$

$$x = 0 \qquad y = 1 \qquad -s = 12 \qquad -t = 3 \qquad u = 0 \qquad P = 10$$
$$s = -12 \qquad t = -3$$

We observe that although t remains negative, it has become less negative. We will again select the largest value in a labeled column of the second row. The 3 in the x column is the largest value. Consequently, the x column is the pivot column.

$$
\begin{array}{c}
\begin{array}{cccccc}
x & y & s & t & u & P
\end{array}\\
\begin{array}{c}
* \\
* \\
\ \\
\
\end{array}
\left[
\begin{array}{cccccc|c}
3 & 0 & -1 & 0 & 1 & 0 & 12 \\
\boxed{3} & 0 & 0 & -1 & -2 & 0 & 3 \\
-1 & 1 & 0 & 0 & 1 & 0 & 1 \\
-13 & 0 & 0 & 0 & 10 & 1 & 10
\end{array}
\right]
\begin{array}{c}
12/3 = 4 \\
3/3 = 1 \\
\ \\
\
\end{array}
\end{array}
$$

Pivot column

The 3 in the second row of the x column is the pivot, since the second row has the smallest test ratio. We zero out the remaining entries in the x column using the indicated operations. We then star the row corresponding to the negative variable.

$$
\begin{array}{c}
\begin{array}{cccccc}
x & y & s & t & u & P
\end{array}\\
\begin{array}{c}
* \\
\ \\
\ \\
\
\end{array}
\left[
\begin{array}{cccccc|c}
0 & 0 & -1 & 1 & 3 & 0 & 9 \\
3 & 0 & 0 & -1 & -2 & 0 & 3 \\
0 & 3 & 0 & -1 & 1 & 0 & 6 \\
0 & 0 & 0 & -13 & 4 & 3 & 69
\end{array}
\right]
\begin{array}{c}
R_1 - R_2 \\
\ \\
3R_3 + R_2 \\
3R_4 + 13R_2
\end{array}
\end{array}
$$

$$3x = 3 \qquad 3y = 6 \qquad -s = 9 \qquad t = 0 \qquad u = 0 \qquad 3P = 69$$
$$x = 1 \qquad y = 2 \qquad s = -9 \qquad \qquad \qquad P = 23$$

Observe that the surplus variable t is now nonnegative. We need to continue the process to make s nonnegative. The largest value in the labeled columns of the starred row is the 3 in the u column. Consequently, the u column is the pivot column.

$$
\begin{array}{c}
\begin{array}{cccccc}
x & y & s & t & u & P
\end{array}\\
\begin{array}{c}
* \\
\ \\
\ \\
\
\end{array}
\left[
\begin{array}{cccccc|c}
0 & 0 & -1 & 1 & \boxed{3} & 0 & 9 \\
3 & 0 & 0 & -1 & -2 & 0 & 3 \\
0 & 3 & 0 & -1 & 1 & 0 & 6 \\
0 & 0 & 0 & -13 & 4 & 3 & 69
\end{array}
\right]
\begin{array}{c}
9/3 = 3 \\
\ \\
6/1 = 6 \\
\
\end{array}
\end{array}
$$

Pivot column

Using the 3 in the first row of the u column as a pivot, we zero out the remaining column entries using the indicated operations.

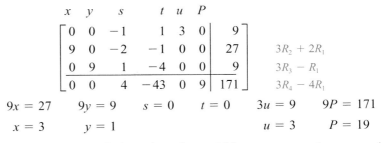

$$9x = 27 \qquad 9y = 9 \qquad s = 0 \qquad t = 0 \qquad 3u = 9 \qquad 9P = 171$$
$$x = 3 \qquad\quad y = 1 \qquad\qquad\qquad\qquad\qquad u = 3 \qquad P = 19$$

Since all decision, slack, and surplus variables are nonnegative, we are in the feasible region. We can now proceed to Stage 2 of the process: applying the simplex method. Before going on, let's observe what has happened graphically (see Figure 4.34).

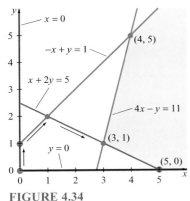

FIGURE 4.34

The blue dots on the graph indicate the solutions to the various tableaus. We started at $(0, 0)$, proceeded to $(0, 1)$, then to $(1, 2)$, and finally to $(3, 1)$.

Returning to the tableau, we apply the simplex method. The only negative entry in the bottom row is in the t column, so the t column is our pivot column. The only positive entry in the t column is the 1 in the first row, so that is our pivot.

$$
\begin{array}{cccccc|c}
x & y & s & t & u & P & \\
\hline
0 & 0 & -1 & \boxed{1} & 3 & 0 & 9 \\
9 & 0 & -2 & -1 & 0 & 0 & 27 \\
0 & 9 & 1 & -4 & 0 & 0 & 9 \\
\hline
0 & 0 & 4 & -43 & 0 & 9 & 171
\end{array}
$$

$$\uparrow$$
$$\boxed{\text{Pivot column}}$$

We zero out the remaining entries in the t column with the indicated operations.

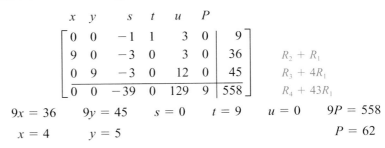

$$9x = 36 \qquad 9y = 45 \qquad s = 0 \qquad t = 9 \qquad u = 0 \qquad 9P = 558$$
$$x = 4 \qquad\quad y = 5 \qquad\qquad\qquad\qquad\qquad\qquad\qquad P = 62$$

The s column is the new pivot column. However, we are unable to select a pivot because every entry in the s column is negative. This signifies that the objective function has no optimal solution. Figure 4.35 shows what has happened graphically.

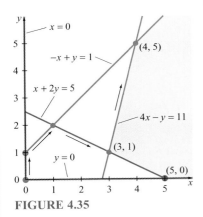

FIGURE 4.35

We increased the value of the objective function by moving from the corner point $(3, 1)$ to the corner point $(4, 5)$. Out of all of the corner points, this is the corner point that yields the largest value of the objective function. However, since the region is unbounded, we can continue to increase the value of the objective function indefinitely.

> **LINEAR PROGRAMMING PROBLEMS WITH NO OPTIMAL SOLUTIONS**
>
> If none of the entries in the pivot column of a simplex tableau are positive, the corresponding feasible region is unbounded and the objective function has no optimal solution.

EXAMPLE 2   **Solving a General Linear Programming Problem**

Solve the general linear programming problem:

$$\text{Maximize} \quad P = 4x + 3y + 2z$$

$$\text{Subject to} \begin{cases} -x - y + z \geq 1 \\ -2x + y + z \geq 3 \\ -x + y + z \leq 3 \\ x \geq 0, y \geq 0, z \geq 0 \end{cases}$$

SOLUTION   We add the slack and surplus variables and rewrite the objective function.

$$-x - y + z - s = 1$$
$$-2x + y + z - t = 3$$
$$-x + y + z + u = 3$$
$$-4x - 3y - 2z + P = 0$$

The corresponding initial tableau is shown. We calculate the basic solution and star the rows with negative variable values.

$$\begin{array}{c} \\ * \\ * \\ \\ \\ \end{array} \begin{array}{ccccccc} x & y & z & s & t & u & P \\ \left[\begin{array}{ccccccc|c} -1 & -1 & 1 & -1 & 0 & 0 & 0 & 1 \\ -2 & 1 & 1 & 0 & -1 & 0 & 0 & 3 \\ -1 & 1 & 1 & 0 & 0 & 1 & 0 & 3 \\ \hline -4 & -3 & -2 & 0 & 0 & 0 & 1 & 0 \end{array}\right] \end{array}$$

$$x = 0 \qquad y = 0 \qquad z = 0 \qquad \begin{array}{l} -s = 1 \\ s = -1 \end{array} \qquad \begin{array}{l} -t = 3 \\ t = -3 \end{array} \qquad u = 3 \qquad P = 0$$

We may work with either starred row. We choose the second row. (This row corresponds to $t = -3$.) The largest positive entries in the labeled columns of the second row are the 1 in the y column and the 1 in the z column. We may pick either column as the pivot column. We pick the y column. We calculate the test ratios and locate the pivot. The test ratios are equal, so we may pick either the second or the third row as the pivot row. We choose the second row.

$$\begin{array}{ccccccc} x & y & z & s & t & u & P \\ \left[\begin{array}{ccccccc|c} -1 & -1 & 1 & -1 & 0 & 0 & 0 & 1 \\ -2 & \boxed{1} & 1 & 0 & -1 & 0 & 0 & 3 \\ -1 & 1 & 1 & 0 & 0 & 1 & 0 & 3 \\ \hline -4 & -3 & -2 & 0 & 0 & 0 & 1 & 0 \end{array}\right] \end{array} \quad \begin{array}{l} \\ \\ 3/1 = 3 \\ 3/1 = 3 \\ \\ \end{array}$$

$$\uparrow$$
$$\boxed{\text{Pivot column}}$$

We zero out the y column using the indicated operations and star the row corresponding with a negative variable value.

$$
\begin{array}{ccccccc}
 & x & y & z & s & t & u & P \\
* & \begin{bmatrix} -3 & 0 & 2 & -1 & -1 & 0 & 0 \\ -2 & 1 & 1 & 0 & -1 & 0 & 0 \\ 1 & 0 & 0 & 0 & 1 & 1 & 0 \\ -10 & 0 & 1 & 0 & -3 & 0 & 1 \end{bmatrix} & \begin{matrix} 4 \\ 3 \\ 0 \\ 9 \end{matrix} & \begin{matrix} R_1 + R_2 \\ \\ R_3 - R_2 \\ R_4 + 3R_2 \end{matrix}
\end{array}
$$

$$x = 0 \qquad y = 3 \qquad z = 0 \qquad -s = 4 \qquad t = 0 \qquad u = 0 \qquad P = 9$$

$$s = -4$$

The 2 in the z column is the largest positive value in the starred row. Consequently, the z column is the pivot column. We calculate the test ratios and locate the pivot.

$$
\begin{array}{ccccccc}
x & y & z & s & t & u & P \\
\begin{bmatrix} -3 & 0 & \boxed{2} & -1 & -1 & 0 & 0 \\ -2 & 1 & 1 & 0 & -1 & 0 & 0 \\ 1 & 0 & 0 & 0 & 1 & 1 & 0 \\ -10 & 0 & 1 & 0 & -3 & 0 & 1 \end{bmatrix} & \begin{matrix} 4 \\ 3 \\ 0 \\ 9 \end{matrix} & \begin{matrix} 4/2 = 2 \\ 3/1 = 3 \end{matrix}
\end{array}
$$

$$\uparrow$$
$$\boxed{\text{Pivot column}}$$

We zero out the z column with the indicated operations.

$$
\begin{array}{ccccccc}
x & y & z & s & t & u & P \\
\begin{bmatrix} -3 & 0 & 2 & -1 & -1 & 0 & 0 \\ -1 & 2 & 0 & 1 & -1 & 0 & 0 \\ 1 & 0 & 0 & 0 & 1 & 1 & 0 \\ -17 & 0 & 0 & 1 & -5 & 0 & 2 \end{bmatrix} & \begin{matrix} 4 \\ 2 \\ 0 \\ 14 \end{matrix} & \begin{matrix} \\ 2R_2 - R_1 \\ \\ 2R_4 - R_1 \end{matrix}
\end{array}
$$

$$x = 0 \qquad 2y = 2 \qquad 2z = 4 \qquad s = 0 \qquad t = 0 \qquad u = 0 \qquad 2P = 14$$

$$y = 1 \qquad z = 2 \qquad\qquad\qquad\qquad\qquad\qquad\qquad P = 7$$

Since all of the decision, slack, and surplus variables are positive, we are in the feasible region. We may now move to Stage 2 and apply the simplex method. Since the x column contains the negative value in the bottom row with the largest magnitude, it is the pivot column. Since there is only one positive entry in the x column, it is the pivot.

$$
\begin{array}{ccccccc}
x & y & z & s & t & u & P \\
\begin{bmatrix} -3 & 0 & 2 & -1 & -1 & 0 & 0 \\ -1 & 2 & 0 & 1 & -1 & 0 & 0 \\ \boxed{1} & 0 & 0 & 0 & 1 & 1 & 0 \\ -17 & 0 & 0 & 1 & -5 & 0 & 2 \end{bmatrix} & \begin{matrix} 4 \\ 2 \\ 0 \\ 14 \end{matrix}
\end{array}
$$

$$\uparrow$$
$$\boxed{\text{Pivot column}}$$

We zero out the remaining terms in the x column using the indicated operations.

$$
\begin{array}{ccccccc|c}
x & y & z & s & t & u & P & \\
\hline
0 & 0 & 2 & -1 & 2 & 3 & 0 & 4 \\
0 & 2 & 0 & 1 & 0 & 1 & 0 & 2 \\
1 & 0 & 0 & 0 & 1 & 1 & 0 & 0 \\
\hline
0 & 0 & 0 & 1 & 12 & 17 & 2 & 14
\end{array}
\qquad
\begin{array}{l}
R_1 + 3R_3 \\
R_2 + R_3 \\
\\
R_4 + 17R_3
\end{array}
$$

$$x = 0 \qquad 2y = 2 \qquad 2z = 4 \qquad s = 0 \qquad t = 0 \qquad u = 0 \qquad 2P = 14$$
$$\qquad\qquad y = 1 \qquad z = 2 \qquad\qquad\qquad\qquad\qquad\qquad\qquad P = 7$$

Since all of the entries in the bottom row of the tableau are nonnegative, this solution is the optimal solution. When $x = 0$, $y = 1$, and $z = 2$, the objective function $P = 4x + 3y + 2z$ attains a maximum value of 7.

Figure 4.36 shows the graph of the constraints and the optimal solution. Each constraint is a plane. The planes intersect at $(0, 1, 2)$.

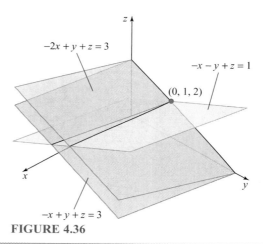

FIGURE 4.36

EXAMPLE 3 **Using Linear Programming to Make Investment Decisions**

An investor plans to invest at most $3000 in the three publicly traded recreational vehicle companies shown in Table 4.26. (Share prices are rounded to the nearest dollar, and dividends per share are rounded to the nearest dime.)

TABLE 4.26

Company	Share Price (dollars)	Dividends/Share (dollars)
Harley-Davidson, Inc. (HDI)	63	0.40
Polaris Industries Inc. (PII)	51	0.90
Winnebago Industries Inc. (WGO)	33	0.20

Source: moneycentral.msn.com. (Accurate as of July 16, 2004.)

He wants to earn at least \$50 in dividends while maximizing the number of shares purchased. How many shares of each company should he purchase?

SOLUTION   Let x be the number of shares of Harley-Davidson, y be the number of shares of Polaris Industries, and z be the number of shares of Winnebago Industries. We must maximize $P = x + y + z$ subject to

$$63x + 51y + 33z \leq 3000 \qquad \text{The total amount invested is at most \$3000}$$
$$0.4x + 0.9y + 0.2z \geq 50 \qquad \text{Total dividends are at least \$50}$$
$$x \geq 0, y \geq 0, z \geq 0$$

We add in the slack and surplus variables and rewrite the objective function.

$$63x + 51y + 33z + s = 3000$$
$$0.4x + 0.9y + 0.2z - t = 50$$
$$-x - y - z + P = 0$$

We then write the initial tableau and star the row corresponding with a negative variable value.

$$
\begin{array}{cccccc}
x & y & z & s & t & P \\
\end{array}
$$

$$
\begin{bmatrix}
63 & 51 & 33 & 1 & 0 & 0 & 3000 \\
0.4 & 0.9 & 0.2 & 0 & -1 & 0 & 50 \\
-1 & -1 & -1 & 0 & 0 & 1 & 0
\end{bmatrix}
$$
(with $*$ marking the middle row)

$$x = 0 \qquad y = 0 \qquad z = 0 \qquad s = 3000 \qquad -t = 50 \qquad P = 0$$
$$t = -50$$

The largest entry in the starred row is the 0.9 in the y column; hence, the y column is the pivot column. We calculate the test ratios and locate the pivot.

$$
\begin{array}{cccccc}
x & y & z & s & t & P \\
\end{array}
$$

$$
\begin{bmatrix}
63 & \boxed{51} & 33 & 1 & 0 & 0 & 3000 \\
0.4 & \boxed{0.9} & 0.2 & 0 & -1 & 0 & 50 \\
-1 & -1 & -1 & 0 & 0 & 1 & 0
\end{bmatrix}
\qquad
\begin{array}{l}
3000/51 \approx 58.82 \\
50/0.9 \approx 55.56
\end{array}
$$

We zero out the y column with the indicated operations and star the row corresponding with a negative variable value.

$$
\begin{array}{cccccc}
 & x & y & z & s & t & P \\
\left[\begin{array}{cccccc|c}
36.3 & 0 & 19.5 & 0.9 & 51 & 0 & 150 \\
0.4 & 0.9 & 0.2 & 0 & -1 & 0 & 50 \\
\hline
-0.5 & 0 & -0.7 & 0 & -1 & 0.9 & 50
\end{array}\right]
\end{array}
\quad
\begin{array}{l}
0.9R_1 - 51R_2 \\
\\
0.9R_3 + R_2
\end{array}
$$

$$x = 0 \qquad 0.9y = 50 \qquad z = 0 \qquad 0.9s = 150 \qquad t = 0 \qquad 0.9P = 50$$
$$y \approx 55.56 \qquad\qquad s \approx 166.67 \qquad\qquad P \approx 55.56$$

Since all of the decision, slack, and surplus variables are nonnegative, we are in the feasible region and may move to Stage 2.

The negative entry in the bottom row of a labeled column with the largest magnitude is the -1 in the t column. Consequently, the t column is the pivot column. Since 51 is the only positive entry in the t column, it is the pivot. We zero out the t column with the indicated operations.

$$
\begin{array}{cccccc}
 & x & y & z & s & t & P \\
\left[\begin{array}{cccccc|c}
36.3 & 0 & 19.5 & 0.9 & \boxed{51} & 0 & 150 \\
56.7 & 45.9 & 29.7 & 0.9 & 0 & 0 & 2700 \\
\hline
10.8 & 0 & -16.2 & 0.9 & 0 & 45.9 & 2700
\end{array}\right]
\end{array}
\quad
\begin{array}{l}
\\
51R_2 + R_1 \\
51R_3 + R_1
\end{array}
$$

$$x = 0 \qquad 45.9y = 2700 \qquad z = 0 \qquad s = 0 \qquad 51t = 150 \qquad 45.9P = 2700$$
$$y \approx 58.82 \qquad\qquad\qquad t \approx 2.94 \qquad P \approx 58.82$$

The z column is the new pivot column, since it contains a negative entry in the bottom row. We calculate the test ratios and identify the pivot.

$$
\begin{array}{cccccc}
 & x & y & z & s & t & P \\
\left[\begin{array}{cccccc|c}
36.3 & 0 & \boxed{19.5} & 0.9 & 51 & 0 & 150 \\
56.7 & 45.9 & 29.7 & 0.9 & 0 & 0 & 2700 \\
\hline
10.8 & 0 & -16.2 & 0.9 & 0 & 45.9 & 2700
\end{array}\right]
\end{array}
\quad
\begin{array}{l}
150/19.5 \approx 7.69 \\
2700/29.7 \approx 90.91
\end{array}
$$

We zero out the z column with the indicated operations.

$$
\begin{array}{cccccc}
 & x & y & z & s & t & P \\
\left[\begin{array}{cccccc|c}
36.3 & 0 & 19.5 & 0.9 & 51 & 0 & 150 \\
27.54 & 895.05 & 0 & -9.18 & -1514.7 & 0 & 48195 \\
\hline
798.66 & 0 & 0 & 32.13 & 826.2 & 895.05 & 55080
\end{array}\right]
\end{array}
\quad
\begin{array}{l}
\\
19.5R_2 - 29.7R_1 \\
19.5R_3 + 16.2R_1
\end{array}
$$

$$x = 0 \qquad 895.05y = 48195 \qquad 19.5z = 150 \qquad s = 0 \qquad t = 0 \qquad 895.05P = 55080$$
$$y \approx 53.85 \qquad\qquad z \approx 7.69 \qquad\qquad\qquad P \approx 61.54$$

The investor should purchase 53.85 shares of Polaris Industries and 7.69 shares of Winnebago Industries in order to maximize the number of shares while simultaneously earning dividends of at least $50.

Minimization Problems

Examples 1, 2, and 3 showed how to *maximize* an objective function of a linear programming problem with mixed constraints. What if we want to *minimize* the objective function? Fortunately, with one minor modification to the objective function, the same procedure works. We begin by observing the relationship between a function f and the function $-f = -1 \cdot f$ shown in Figure 4.37.

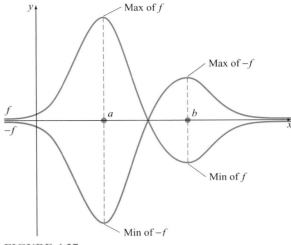

FIGURE 4.37

Observe that the maximum of f and the minimum of $-f$ both occur at $x = a$. Similarly, the minimum of f and the maximum of $-f$ both occur at $x = b$. This relationship holds true for all functions. Consequently, if we want to *minimize* an objective function P, we need only *maximize* the function $-P$.

EXAMPLE 4

Solving a General Linear Programming Problem

Solve the general linear programming problem:

$$\text{Minimize} \quad P = 3x + 5y + 2z$$

$$\text{Subject to} \quad \begin{cases} x + 2y - z \le 0 \\ -2x + 4y + z \ge 3 \\ -x + y + z \le 3 \\ x \ge 0, y \ge 0, z \ge 0 \end{cases}$$

SOLUTION We first introduce a new function $C = -P$.

$$\begin{aligned} C &= -P \\ &= -(3x + 5y + 2z) \\ &= -3x - 5y - 2z \end{aligned}$$

The solution that maximizes C will minimize P. Therefore, the linear programming problem may be rewritten as

$$\text{Maximize} \quad C = -3x - 5y - 2z$$

$$\text{Subject to} \quad \begin{cases} x + 2y - z \le 0 \\ -2x + 4y + z \ge 3 \\ -x + y + z \le 3 \\ x \ge 0, y \ge 0, z \ge 0 \end{cases}$$

We add in the slack and surplus variables and rewrite the objective function. The system of equations is

$$x + 2y - z + s = 0$$
$$-2x + 4y + z - t = 3$$
$$-x + y + z + u = 3$$
$$3x + 5y + 2z + C = 0$$

and has the corresponding tableau

$$
\begin{array}{ccccccc}
x & y & z & s & t & u & C \\
\end{array}
$$

$$
\left[
\begin{array}{ccccccc|c}
1 & 2 & -1 & 1 & 0 & 0 & 0 & 0 \\
-2 & 4 & 1 & 0 & -1 & 0 & 0 & 3 \\
-1 & 1 & 1 & 0 & 0 & 1 & 0 & 3 \\
\hline
3 & 5 & 2 & 0 & 0 & 0 & 1 & 0 \\
\end{array}
\right]
$$

$$x = 0 \quad y = 0 \quad z = 0 \quad s = 0 \quad -t = 3 \quad u = 3 \quad C = 0$$
$$t = -3$$

Since the surplus variable t is negative, we star the corresponding row. The y column is the pivot column because it contains the largest positive entry in the starred row. Calculating the test ratios, we select the pivot.

$$
\begin{array}{ccccccc}
x & y & z & s & t & u & C \\
\end{array}
$$

$$
*\left[
\begin{array}{ccccccc|c}
1 & \boxed{2} & -1 & 1 & 0 & 0 & 0 & 0 \\
-2 & 4 & 1 & 0 & -1 & 0 & 0 & 3 \\
-1 & 1 & 1 & 0 & 0 & 1 & 0 & 3 \\
\hline
3 & 5 & 2 & 0 & 0 & 0 & 1 & 0 \\
\end{array}
\right]
\quad
\begin{array}{l}
0/2 = 0 \\
3/4 = 0.75 \\
3/1 = 3 \\
\end{array}
$$

We zero out the remaining terms in the y column using the indicated operations and star the row corresponding with the negative variable value.

$$
\begin{array}{ccccccc}
x & y & z & s & t & u & C \\
\end{array}
$$

$$
*\left[
\begin{array}{ccccccc|c}
1 & 2 & -1 & 1 & 0 & 0 & 0 & 0 \\
-4 & 0 & 3 & -2 & -1 & 0 & 0 & 3 \\
-3 & 0 & 3 & -1 & 0 & 2 & 0 & 6 \\
\hline
1 & 0 & 9 & -5 & 0 & 0 & 2 & 0 \\
\end{array}
\right]
\quad
\begin{array}{l}
\\
R_2 - 2R_1 \\
2R_3 - R_1 \\
2R_4 - 5R_1 \\
\end{array}
$$

$$x = 0 \quad 2y = 0 \quad z = 0 \quad s = 0 \quad -t = 3 \quad 2u = 6 \quad 2C = 0$$
$$y = 0 \qquad\qquad\qquad\qquad\qquad t = -3 \quad u = 3 \quad C = 0$$

The largest positive value in the starred row is the 3 in the z column. Consequently, the z column is the pivot column. We calculate the test ratios and select the pivot.

$$
\begin{array}{ccccccc}
x & y & z & s & t & u & C \\
\end{array}
$$

$$
\left[
\begin{array}{ccccccc|c}
1 & 2 & -1 & 1 & 0 & 0 & 0 & 0 \\
-4 & 0 & \boxed{3} & -2 & -1 & 0 & 0 & 3 \\
-3 & 0 & 3 & -1 & 0 & 2 & 0 & 6 \\
\hline
1 & 0 & 9 & -5 & 0 & 0 & 2 & 0 \\
\end{array}
\right]
\quad
\begin{array}{l}
\\
3/3 = 1 \\
6/3 = 2 \\
\end{array}
$$

We zero out the remaining terms in the z column with the indicated operations.

$$
\begin{array}{ccccccc}
x & y & z & s & t & u & C \\
\end{array}
$$

$$
\left[
\begin{array}{ccccccc|c}
-1 & 6 & 0 & 1 & -1 & 0 & 0 & 3 \\
-4 & 0 & 3 & -2 & -1 & 0 & 0 & 3 \\
1 & 0 & 0 & 1 & 1 & 2 & 0 & 3 \\
\hline
13 & 0 & 0 & 1 & 3 & 0 & 2 & -9
\end{array}
\right]
\begin{array}{l}
3R_1 + R_2 \\
\\
R_3 - R_2 \\
R_4 - 3R_2
\end{array}
$$

$$x = 0 \qquad 6y = 3 \qquad 3z = 3 \qquad s = 0 \qquad t = 0 \qquad 2u = 3 \qquad 2C = -9$$

$$y = \frac{1}{2} \qquad z = 1 \qquad\qquad\qquad\qquad\qquad u = \frac{3}{2} \qquad C = -\frac{9}{2}$$

Since all decision, slack, and surplus variables are nonnegative, we are now in the feasible region. (There is no restriction on the sign of the objective function. In this case, C is negative.) We may now move to Stage 2 of the process. However, since no entry in the bottom row of a labeled column is negative, this is the final tableau. The objective function $C = -3x - 5y - 2z$ is maximized at $\left(0, \frac{1}{2}, 1\right)$.

Consequently, the objective function $P = 3x + 5y + 2z$ is minimized at $\left(0, \frac{1}{2}, 1\right)$.

The maximum value of C is $-\frac{9}{2}$. Since $P = -C$, the minimum value of P is $\frac{9}{2}$.

So far, the general linear programming problems we have demonstrated have included only linear inequalities as constraints. However, some linear programming problems have linear equations as constraints. Any linear equation may be written as a system of linear inequalities. For example, consider the linear equation $2x + 3y = 5$. This equation is equivalent to the following system of linear inequalities:

$$2x + 3y \le 5$$
$$2x + 3y \ge 5$$

When we are given a linear equation as a constraint, we will rewrite it as a system of linear inequalities, as demonstrated in Example 5.

EXAMPLE 5

Using Linear Programming to Make Investment Decisions

An investor has \$5000 to invest in three mutual funds: Bond, Index, and Growth. The Bond fund is expected to earn 7 percent; the Index fund, 11 percent; and the Growth fund, 9 percent. The investor wants to earn an annual return of at least 10 percent while minimizing the amount invested in the Index fund. How much should she invest in each account?

SOLUTION   Let x be the amount invested in the Bond fund, y be the amount invested in the Index fund, and z be the amount invested in the Growth fund. We must solve the following linear programming problem.

Minimize $P = y$

Subject to
$$\begin{cases} x + y + z = 5000 & \text{\$5000 is invested} \\ 0.07x + 0.11y + 0.09z \ge 0.1(5000) & \text{The return is at least 10\%} \\ & \text{of \$5000} \\ x \ge 0, y \ge 0, z \ge 0 & \end{cases}$$

We rewrite the first constraint as two inequalities and simplify the second constraint. We then create a new objective function $C = -P$. Since $P = y$, $C = -y$. The new linear programming problem is

$$\text{Maximize} \quad C = -y$$

$$\text{Subject to} \begin{cases} x + y + z \leq 5000 \\ x + y + z \geq 5000 \\ 0.07x + 0.11y + 0.09z \geq 5 \\ x \geq 0, y \geq 0, z \geq 0 \end{cases}$$

We add the surplus and slack variables and rewrite the objective function.

$$x + y + z + s = 5000$$

$$x + y + z - t = 5000$$

$$0.07x + 0.11y + 0.09z - u = 500$$

$$y + C = 0$$

We write the initial tableau and star the appropriate rows.

$$\begin{array}{c} \\ \\ * \\ * \\ \end{array} \begin{array}{cccccccc} x & y & z & s & t & u & C & \\ \end{array}$$

$$\begin{bmatrix} 1 & 1 & 1 & 1 & 0 & 0 & 0 & 5000 \\ 1 & 1 & 1 & 0 & -1 & 0 & 0 & 5000 \\ 0.07 & 0.11 & 0.09 & 0 & 0 & -1 & 0 & 500 \\ \hline 0 & 1 & 0 & 0 & 0 & 0 & 1 & 0 \end{bmatrix}$$

$$x = 0 \quad y = 0 \quad z = 0 \quad s = 5000 \quad -t = 5000 \quad -u = 500 \quad C = 0$$
$$t = -5000 \quad u = -500$$

Although we may use either starred row, we pick the second row of the tableau because the entries are simpler. We may pick the x, y, or z column as the pivot column. We pick the x column. We calculate the test ratios and identify the pivot.

$$\begin{array}{cccccccc} x & y & z & s & t & u & C & \\ \end{array}$$

$$\begin{bmatrix} \boxed{1} & 1 & 1 & 1 & 0 & 0 & 0 & 5000 \\ 1 & 1 & 1 & 0 & -1 & 0 & 0 & 5000 \\ 0.07 & 0.11 & 0.09 & 0 & 0 & -1 & 0 & 500 \\ \hline 0 & 1 & 0 & 0 & 0 & 0 & 1 & 0 \end{bmatrix} \quad \begin{array}{l} 5000/1 = 5000 \\ 5000/1 = 5000 \\ 500/0.07 \approx 7143 \end{array}$$

We may pick either the first or the second term in the x column to be the pivot. We pick the first term and zero out the remaining entries with the indicated operations. We then star the appropriate row.

$$\begin{array}{c} \\ \\ \\ * \\ \\ \end{array} \begin{array}{cccccccc} x & y & z & s & t & u & C & \\ \end{array}$$

$$\begin{bmatrix} 1 & 1 & 1 & 1 & 0 & 0 & 0 & 5000 \\ 0 & 0 & 0 & -1 & -1 & 0 & 0 & 0 \\ 0 & 0.04 & 0.02 & -0.07 & 0 & -1 & 0 & 150 \\ \hline 0 & 1 & 0 & 0 & 0 & 0 & 1 & 0 \end{bmatrix} \quad \begin{array}{l} \\ R_2 - R_1 \\ R_3 - 0.07R_1 \\ \\ \end{array}$$

$$x = 5000 \quad y = 0 \quad z = 0 \quad s = 0 \quad -t = 0 \quad -u = 150 \quad C = 0$$
$$t = 0 \quad u = -150$$

The y column is the pivot column, since the largest positive entry in the starred row is the 0.04 in the y column. We calculate the test ratios and locate the pivot.

$$
\begin{array}{c}
\begin{array}{ccccccc}
x & y & z & s & t & u & C
\end{array} \\
\left[\begin{array}{ccccccc|c}
1 & 1 & 1 & 1 & 0 & 0 & 0 & 5000 \\
0 & 0 & 0 & -1 & -1 & 0 & 0 & 0 \\
0 & \boxed{0.04} & 0.02 & -0.07 & 0 & -1 & 0 & 150 \\
0 & 1 & 0 & 0 & 0 & 0 & 1 & 0
\end{array}\right]
\end{array}
\qquad
\begin{array}{l}
5000/1 = 5000 \\
\\
150/0.04 = 3750 \\
\\
\end{array}
$$

(The third row is marked with $*$.)

We zero out the y column with the indicated operations.

$$
\begin{array}{c}
\begin{array}{ccccccc}
x & y & z & s & t & u & C
\end{array} \\
\left[\begin{array}{ccccccc|c}
0.04 & 0 & 0.02 & 0.11 & 0 & 1 & 0 & 50 \\
0 & 0 & 0 & -1 & -1 & 0 & 0 & 0 \\
0 & 0.04 & 0.02 & -0.07 & 0 & -1 & 0 & 150 \\
0 & 0 & -0.02 & 0.07 & 0 & 1 & 0.04 & -150
\end{array}\right]
\end{array}
\qquad
\begin{array}{l}
0.04R_1 - R_3 \\
\\
\\
0.04R_4 - R_3
\end{array}
$$

$$
\begin{array}{llllllll}
0.04x = 50 & 0.04y = 150 & z = 0 & s = 0 & -t = 0 & u = 0 & 0.04C = -150 \\
x = 1250 & y = 3750 & & & t = 0 & & C = -3750
\end{array}
$$

Since all decision, slack, and surplus variables are nonnegative, we are in the feasible region and may proceed to Stage 2.

Since the only negative entry in the bottom row of a labeled column is the -0.02 in the z column, the z column is the pivot column. We calculate the test ratios and identify the pivot.

$$
\begin{array}{c}
\begin{array}{ccccccc}
x & y & z & s & t & u & C
\end{array} \\
\left[\begin{array}{ccccccc|c}
0.04 & 0 & \boxed{0.02} & 0.11 & 0 & 1 & 0 & 50 \\
0 & 0 & 0 & -1 & -1 & 0 & 0 & 0 \\
0 & 0.04 & 0.02 & -0.07 & 0 & -1 & 0 & 150 \\
0 & 0 & -0.02 & 0.07 & 0 & 1 & 0.04 & -150
\end{array}\right]
\end{array}
\qquad
\begin{array}{l}
50/0.02 = 2500 \\
\\
150/0.04 = 3750 \\
\\
\end{array}
$$

We zero out the remaining terms in the pivot column with the indicated operations.

$$
\begin{array}{c}
\begin{array}{ccccccc}
x & y & z & s & t & u & C
\end{array} \\
\left[\begin{array}{ccccccc|c}
0.04 & 0 & 0.02 & 0.11 & 0 & 1 & 0 & 50 \\
0 & 0 & 0 & -1 & -1 & 0 & 0 & 0 \\
-0.04 & 0.04 & 0 & -0.18 & 0 & -2 & 0 & 100 \\
0.04 & 0 & 0 & 0.18 & 0 & 2 & 0.04 & -100
\end{array}\right]
\end{array}
\qquad
\begin{array}{l}
\\
\\
R_3 - R_1 \\
R_4 + R_1
\end{array}
$$

$$
\begin{array}{llllllll}
x = 0 & 0.04y = 100 & 0.02z = 50 & s = 0 & -t = 0 & u = 0 & 0.04C = -100 \\
& y = 2500 & z = 2500 & & t = 0 & & C = -2500
\end{array}
$$

Since all entries in the bottom row are positive, this is the final tableau. The objective function C reaches its maximum value when $x = 0$, $y = 2500$, and $z = 2500$. The maximum value is $C = -2500$. The objective function $P = -C$ attains its minimum value of 2500 at the same point. In the context of the problem, a 10 percent return is earned when \$0 is invested in the Bond fund, \$2500 is invested in the Index fund, and \$2500 is invested in the Growth fund. This solution minimizes the amount of money invested in the Index fund.

4.5 Summary

In this section, you learned how to solve general linear programming problems. You discovered that minimization problems may be solved by maximizing the negative of the objective function.

4.5 Exercises

In Exercises 1–5, rewrite the general linear programming problem as a system of equations with slack and surplus variables and a rewritten objective function.

1. Maximize $P = 4x + 2y + 3z$

Subject to $\begin{cases} 2x + y + z \le 10 \\ x + z \ge 3 \\ x - 2z \ge 1 \\ x \ge 0, y \ge 0, z \ge 0 \end{cases}$

2. Maximize $P = 2x + 10y + z$

Subject to $\begin{cases} -x + 3y \le 2 \\ 3x + z \ge 4 \\ 5x - 4z \ge 1 \\ x \ge 0, y \ge 0, z \ge 0 \end{cases}$

3. Maximize $P = 3x - y + 9z$

Subject to $\begin{cases} -12x + 8y + 2z \le 15 \\ 8x + 5y - 2z \le 10 \\ 3x - 2y + 6z \ge 5 \\ x \ge 0, y \ge 0, z \ge 0 \end{cases}$

4. Maximize $P = x - z$

Subject to $\begin{cases} -x + y + z \le 12 \\ x + 3y - 5z \ge 5 \\ 2x + y - 2z \le 1 \\ x \ge 0, y \ge 0, z \ge 0 \end{cases}$

5. Maximize $P = 10x + 9y + z$

Subject to $\begin{cases} 4x - 3y - 2z \le 12 \\ -2x + 4y + 6z \le 12 \\ 3x - 2y + 4z \ge 12 \\ x \ge 0, y \ge 0, z \ge 0 \end{cases}$

In Exercises 6–10, star the rows of the tableau that correspond with negative variable values.

6.

x	y	s	t	u	P	
-1	2	1	0	0	0	4
2	-3	0	1	0	0	6
4	1	0	0	-1	0	5
-6	-2	0	0	0	1	0

7.

x	y	s	t	u	P	
9	4	-1	0	0	0	10
-2	5	0	1	0	0	8
1	1	0	0	-1	0	12
5	4	0	0	0	1	0

8.

x	y	s	t	u	P	
-1	2	-1	0	0	0	4
2	0	-2	1	0	0	6
4	0	1	0	-5	0	5
-6	0	3	0	0	4	24

9.

x	y	s	t	u	P	
0	6	-2	0	0	0	16
5	1	0	1	0	0	1
0	3	0	4	-4	0	6
0	10	0	5	0	3	9

10.

x	y	s	t	u	P	
0	3	0	0	3	0	15
0	0	1	1	2	0	1
2	0	0	-2	-4	0	4
0	0	0	6	3	1	22

In Exercises 11–20, graph the feasible region of the general linear programming problem. Then solve the problem by using the methods demonstrated in the section. For each tableau, indicate on the graph of the feasible region the point that corresponds with the solution of the tableau.

11. Maximize $P = 2x - 3y$

Subject to $\begin{cases} 2x + y \le 10 \\ x + y \ge 3 \\ 3x - y \ge 1 \\ x \ge 0, y \ge 0 \end{cases}$

12. Maximize $P = 4x + y$

Subject to $\begin{cases} 2x + y \le 10 \\ 2x + 3y \ge 14 \\ -2x + y \le 2 \\ x \ge 0, y \ge 0 \end{cases}$

13. Maximize $P = 5x - 8y$

Subject to $\begin{cases} 2x + y \le 13 \\ 9x - 3y \le 6 \\ -x + 2y \ge 6 \\ x \ge 0, y \ge 0 \end{cases}$

14. Maximize $P = 6x + 3y$

Subject to $\begin{cases} -x + y \ge 0 \\ 2x + y \ge 9 \\ x + 2y \ge 6 \\ x \ge 0, y \ge 0 \end{cases}$

15. Maximize $P = -3x + 4y$

Subject to $\begin{cases} 3x + 12y \ge 12 \\ 2x + y \le 8 \\ x + y \le 6 \\ x \ge 0, y \ge 0 \end{cases}$

16. Minimize $P = 3x + 5y$

Subject to $\begin{cases} -5x + 2y \le 2 \\ -x + y \le 4 \\ -7x + 4y \ge 4 \\ x \ge 0, y \ge 0 \end{cases}$

17. Minimize $P = 7x + 4y$

Subject to $\begin{cases} 3x + y \ge 10 \\ x + y \ge 6 \\ 3x + 2y \le 17 \\ x \ge 0, y \ge 0 \end{cases}$

18. Minimize $P = -3x + 8y$

Subject to $\begin{cases} y \le 6 \\ -2x + y \ge 0 \\ 2x + y \ge 8 \\ x \ge 0, y \ge 0 \end{cases}$

19. Minimize $P = -x + y$

Subject to $\begin{cases} x + y \ge 1 \\ 2x - 3y \ge 2 \\ x + 4y \ge 4 \\ x \ge 0, y \ge 0 \end{cases}$

20. Maximize $P = 4x + 2y$

Subject to $\begin{cases} 5x + 2y \ge 17 \\ x - 2y \le 1 \\ x + y \ge 7 \\ x \ge 0, y \ge 0 \end{cases}$

In Exercises 21–35, solve the general linear programming problem. If there is no solution, so state.

21. Maximize $P = 3x + 2y + z$

Subject to $\begin{cases} x + y \le 10 \\ x + z \le 5 \\ y + z \ge 4 \\ x \ge 0, y \ge 0, z \ge 0 \end{cases}$

22. Maximize $P = 5x + 6y + 2z$

Subject to $\begin{cases} x + 2y \le 10 \\ 2x + z \le 8 \\ y + 2z \ge 6 \\ x \ge 0, y \ge 0, z \ge 0 \end{cases}$

23. Maximize $P = x - y + z$

Subject to $\begin{cases} 4x + 3y + 2z \ge 12 \\ 2x + 2y + z \le 8 \\ x + y + 4z \ge 4 \\ x \ge 0, y \ge 0, z \ge 0 \end{cases}$

24. Maximize $P = 6x + y$

Subject to $\begin{cases} 5x + 2y + 4z \le 23 \\ -x + y + z \ge 4 \\ x - y + z \ge 4 \\ x \ge 0, y \ge 0, z \ge 0 \end{cases}$

25. Maximize $P = 2z$

Subject to $\begin{cases} x + 2y + 3z \ge 12 \\ 3x + y + 2z \le 12 \\ 2x + 3y + z \le 12 \\ x \ge 0, y \ge 0, z \ge 0 \end{cases}$

26. Maximize $P = 2x + y + z$

Subject to $\begin{cases} -x + y + z \le 5 \\ x - y + z \le 7 \\ -x - y + z \ge 3 \\ x \ge 0, y \ge 0, z \ge 0 \end{cases}$

27. Maximize $P = 6x - 2y + 6z$

Subject to $\begin{cases} -x - y + z \le 1 \\ x + y + z \le 7 \\ 2x - 3y + z \ge 12 \\ x \ge 0, y \ge 0, z \ge 0 \end{cases}$

28. Minimize $P = 4y + 9z$

Subject to $\begin{cases} -2x - 3y + z \ge 0 \\ 4x + y + z \ge 10 \\ -4x + 2y + z \ge 3 \\ x \ge 0, y \ge 0, z \ge 0 \end{cases}$

29. Minimize $P = -x - y + 4z$

Subject to $\begin{cases} -2x + 3y + z \ge 4 \\ 2x + 4y + z \le 4 \\ -4x + y + z \le 12 \\ x \ge 0, y \ge 0, z \ge 0 \end{cases}$

30. Minimize $P = 5x + 4y + 3z$

Subject to $\begin{cases} 2x - 5y + z \ge 4 \\ 2x + 4y + z \le 13 \\ z \ge 3 \\ x \ge 0, y \ge 0, z \ge 0 \end{cases}$

31. Minimize $P = 9x - 3y - 3z$

Subject to $\begin{cases} -3x - 4y + z \le 4 \\ 2x - 5y + z \le 14 \\ 3x - 2y + z \ge 8 \\ x \ge 0, y \ge 0, z \ge 0 \end{cases}$

32. Minimize $P = x - y + 3z$

Subject to $\begin{cases} 2x + 2y - z \le 4 \\ 5x - 4y + z \ge 16 \\ x - 4y + z \ge 5 \\ x \ge 0, y \ge 0, z \ge 0 \end{cases}$

33. Minimize $P = 6x - 2y + 4z$

Subject to $\begin{cases} 8x + y - z \le 8 \\ x - y + z \ge 1 \\ x - y + z \le 10 \\ x \ge 0, y \ge 0, z \ge 0 \end{cases}$

34. Minimize $P = x + 2y + 3z$

Subject to $\begin{cases} x + y + z \ge 6 \\ x - y + z \ge 2 \\ x + z \le 4 \\ x \ge 0, y \ge 0, z \ge 0 \end{cases}$

35. Maximize $P = 4x + 3y + 2z$

Subject to $\begin{cases} x + y + z \ge 8 \\ -2x + z \ge 1 \\ 3y + 5z \le 27 \\ x \ge 0, y \ge 0, z \ge 0 \end{cases}$

In Exercises 36–45, use the techniques demonstrated in the section to set up and solve each problem.

36. **Commodity Prices** Today's Market Prices (www.todaymarket.com) is a daily fruit and vegetable wholesale market price service. Produce retailers who subscribe to the service can use wholesale prices to aid them in setting retail prices for the fruits and vegetables they sell.

A 25-pound carton of peaches holds 60 medium peaches or 70 small peaches. In August 2002, the wholesale price for local peaches in Los Angeles was $9.00 per carton for medium peaches and $10.00 per carton for small peaches. (**Source:** Today's Market Prices.) A fruit vendor has budgeted up to $100 to spend on peaches. He estimates that weekly demand for peaches is no more than 660 peaches. He wants to buy at least four boxes of each size of peach. Subject to these constraints, how many boxes of each size of peach should he buy in order to maximize the number of peaches available for sale?

37. **Battery Sales** AAA Alkaline Discount Batteries sells low-cost batteries to consumers. In August 2002, the firm offered AA batteries at the following prices: 50-pack for $10.00, 100-pack for $18.00, and 600-pack for $96.00. (**Source:** www.aaaalkalinediscountbatteries.com.)

An electronics store owner wants to buy at least 900 batteries and spend at most $150. She expects that she'll be able to resell all of the batteries she orders for $1.50 each. How many packs (50-packs, 100-packs, or 600-packs) should she order if she wants to maximize her revenue? What is her maximum revenue? (*Hint:* At most one 600-pack may be purchased without exceeding the $150 limit.)

38. Resource Allocation: Food An ice cream parlor wants to make three different types of ice cream: vanilla, strawberry, and peach-cherry. The parlor has 120 cups of cream, 48 eggs, and 32 cups of sugar on hand. The vanilla ice cream recipe calls for 4 cups of cream, 1 egg, and 0.75 cup of sugar. The strawberry ice cream recipe calls for 2 cups of cream, 2 eggs, and 0.75 cup of sugar. The peach-cherry ice cream recipe calls for 4 cups of cream, 1 egg, and 1.25 cups of sugar. Each recipe yields 1.5 quarts of ice cream. The parlor needs at least 18 quarts of vanilla and at least 6 quarts of each of the other varieties. The parlor wants to maximize the amount of ice cream produced. How many batches of each variety of ice cream does the parlor need to produce?

39. Resource Allocation: Food A delicatessen makes three types of pudding: rice, tapioca, and vanilla. The deli has 108 cups of milk, 150 cups of sugar, and 84 eggs on hand. The rice pudding recipe requires 12 cups of milk, 1.5 cups of sugar, and 9 eggs and yields 24 servings. The tapioca pudding recipe requires 12 cups of milk, 1.5 cups of sugar, and 9 eggs and yields 18 servings. The vanilla pudding recipe requires 6 cups of milk, 1.5 cups of sugar, and 6 eggs and yields 12 servings. The deli requires that the sum of the number of batches of tapioca pudding and the number of batches of rice pudding be at least two. How many batches of each recipe must the deli produce in order to maximize the number of servings?

40. Advertising A large company advertises through magazines, radio, and television. For every $10,000 spent on magazine advertising, the company estimates that it reaches 150,000 people. For every $10,000 spent on radio advertising, the company estimates that it reaches 250,000 people. For every $10,000 spent on television advertising, the company estimates that it reaches 300,000 people. The company has at most $2.5 million to spend on advertising. It requires that at least twice as much be spent on radio as on television and that the amount spent on magazines be at least $150,000 more than the amount spent on radio. How much should the company spend on each type of advertising in order to maximize the number of people reached?

41. Nutrition The nutritional content of canned beans varies based on the type of

bean and the manufacturer. A 1/2-cup serving of Fred Meyer Pinto Beans contains 7 grams of fiber, 1 gram of sugar, and 6 grams of protein. A 1/2-cup serving of Fred Meyer Kidney Beans contains 11 grams of fiber, 1 gram of sugar, and 8 grams of protein. A 1/2-cup serving of Trader Joe's Cuban Style Black Beans contains 2 grams of fiber, 1 gram of sugar, and 6 grams of protein. (**Source:** Package labeling.) A bean dish is to be made using the three bean varieties. The dish can contain at most 3 cups of beans but must include at least 39 grams of fiber and 42 grams of protein. How much of each type of bean should be included in the dish in order to minimize the amount of kidney beans used? (Fractions of cups may be used.)

42. Nutrition The nutritional content of canned vegetables varies based on the type of vegetable and the manufacturer. A 1/2-cup serving of S&W Cut Blue Lake Green Beans contains 2 grams of fiber, 2 grams of sugar, and 20 calories. A 1/2-cup serving of Safeway Golden Sweet Whole Kernel Corn contains 2 grams of fiber, 6 grams of sugar, and 80 calories. A 1/2-cup serving of Pot O'Gold Sliced Carrots contains 2 grams of fiber, 4 grams of sugar, and 30 calories. (**Source:** Package labeling.) A dish is to be made using the three types of vegetables. The dish can contain at most 5 cups of vegetables but must include at least 18 grams of fiber and at least 32 grams of sugar (for flavor). How much of each type of vegetable should be included in the dish in order to minimize the amount of calories in the dish?

43. Resource Allocation: Schools A city has two elementary schools. The first school has a maximum enrollment of 500 students, and the second school has a maximum enrollment of 420 students. The city is divided into two regions: North and South. There are at least 400 students in the North region and at least 430 students in the South region. The annual transportation cost varies by region as shown in the table.

Transportation Costs

	School 1 (cost per student)	School 2 (cost per student)
North	120	180
South	100	150

Based on these constraints, what is the minimum possible transportation cost, and under what conditions does it occur?

44. Resource Allocation: Schools A city has two high schools. The first school has a maximum enrollment of 900 students, and the second school has a maximum enrollment of 550 students. The city is divided into two regions: Inner City and Suburbs. There are at least 500 students in the Inner City region and at least 800 students in the Suburbs region. The annual transportation cost varies by region as shown in the table.

Transportation Costs

	School 1 (cost per student)	School 2 (cost per student)
Inner City	220	260
Suburbs	200	280

Based on these constraints, what is the minimum possible transportation cost, and under what conditions does it occur?

45. 🪐 **Transportation Costs** A furniture company has warehouses in Phoenix, Arizona, and Las Vegas, Nevada. The company has customers in Kingman, Arizona, and Flagstaff, Arizona. Its Phoenix warehouse has 500 desks in stock, and its Las Vegas warehouse has 200 desks in stock. Its Kingman customer needs at least 250 desks, and its Flagstaff customer needs at least 400 desks. Based on a rate of $0.01 per mile per item, the cost of delivery per item is shown in the table.

	Phoenix	Las Vegas
Kingman	$1.85	$1.05
Flagstaff	$1.45	$2.51

Subject to these constraints, under what conditions are the company's delivery costs minimized?

Exercises 46–50 are intended to challenge your understanding of general linear programming problems.

46. Maximize $P = x + y + z + w$

Subject to $\begin{cases} x + y + z \leq 9 \\ y + z + w \leq 8 \\ x + z + w \geq 7 \\ x + y + w \geq 6 \\ x \geq 0, y \geq 0, z \geq 0, w \geq 0 \end{cases}$

47. Minimize $P = x + y + z + w$

Subject to $\begin{cases} x + y + z \leq 6 \\ y + z + w \leq 8 \\ x + z + w \geq 9 \\ x + y + w \geq 7 \\ x \geq 0, y \geq 0, z \geq 0, w \geq 0 \end{cases}$

48. Create a two-variable linear programming problem with three or more mixed constraints with the property that the feasible region consists of exactly one point. Provide evidence that the feasible region contains exactly one point.

49. Create a three-variable linear programming problem with three or more mixed constraints with the property that the feasible region is empty. Provide evidence that the feasible region is empty.

50. Explain what happens graphically in Stage 1 of the general linear programming problem-solving process.

Chapter 4 Review Exercises

Section 4.1 *In Exercises 1–4, graph the solution region of the linear inequality. Then use the graph to determine if point P is a solution.*

1. $4x - 2y \leq 6$; $P = (2, 9)$

2. $3x + 5y \leq 0$; $P = (1, 7)$

3. $9x - 8y \leq 12$; $P = (8, 9)$

4. $7x + 6y \leq 42$; $P = (3, 4)$

In Exercises 5–8, graph the solution region of the system of linear equations. If there is no solution, explain why.

5. $-4x + 3y \geq 2$
$-3x + 2y \geq 1$

6. $-10x + y \geq 0$
$2x + y \leq 4$

7. $2x + 4y \leq 8$
$6x - 2y \leq -6$

8. $x + y \leq 8$
$-x + y \leq 0$
$4x - 2y \leq 8$

In Exercise 9, set up and graphically solve the system of linear inequalities.

9. Wages A salaried employee earns $800 per week managing a copy center. She is required to work a minimum of 40 hours but no more than 50 hours weekly. As a side business, she earns $30 per hour designing brochures for local business clients. In order to maintain her standard of living, she must earn $1000 per week. In order to maintain her quality of life, she limits her workload to 50 hours per week. Given that she has no control over the number of hours she will have to work managing the copy center, will she be able to consistently meet her workload and income goals? Explain.

Section 4.2 *In Exercises 10–13, find the optimal solution to the linear programming problem, if it exists. If a solution does not exist, explain why.*

10. Minimize $z = 5x - 7y$

Subject to $\begin{cases} 4x + y \geq 4 \\ -x + y \geq 1 \\ x \geq 0 \\ y \geq 0 \end{cases}$

11. Minimize $z = 9x + 4y$

Subject to $\begin{cases} 6x + y \geq 16 \\ -2x + y \geq 0 \\ x \geq 0 \\ y \geq 0 \end{cases}$

12. Maximize $z = 6x + 10y$

Subject to $\begin{cases} 6x + y \leq 16 \\ -2x + y \leq 0 \\ x \geq 0 \\ y \geq 0 \end{cases}$

13. Maximize $z = 5x - y$

Subject to $\begin{cases} 6x + 2y \leq 16 \\ -3x - y \leq 10 \\ x \geq 0 \\ y \geq 0 \end{cases}$

For Exercise 14, identify the objective function and constraints of the linear programming problem. Then solve the problem and interpret the real-world meaning of your results.

14. **Family Food Storage** A family wants to purchase at least 75 pounds of beans. A #10 can of pinto beans weighs 5.0 pounds and costs $2.75. A #10 can of white beans weighs 5.3 pounds and costs $2.88. (**Source:** Kent Washington Cannery.) The family wants to buy at least 25 pounds of pinto beans and at least 53 pounds of white beans. It has budgeted $50 and, because of limited storage space, wants to minimize the number of cans purchased. How many cans of each type of beans should the family purchase?

Section 4.3 *In Exercises 15–16, determine if the problem is a standard maximization problem. If it isn't, explain why.*

15. Maximize $P = -2x + 8y$

Subject to $\begin{cases} 2x + y \leq 1 \\ -x + -y \leq -20 \end{cases}$

16. Maximize $P = -9x + 8y$

Subject to $\begin{cases} 42x + 19y \leq 10 \\ -11x + 19y \leq 21 \\ x \geq 0 \\ y \geq 0 \end{cases}$

In Exercises 17–18, solve the standard maximization problems by using the simplex method. Check your answer by graphing the feasible region and calculating the value of the objective function at each of the corner points.

17. Maximize $P = -2x + 10y$

Subject to $\begin{cases} 2x + y \leq 10 \\ -x + y \leq 1 \\ x \geq 0 \\ y \geq 0 \end{cases}$

18. Maximize $P = 5x - 2y$

Subject to $\begin{cases} 2x + y \le 6 \\ 2x + 2y \le 8 \\ x \ge 0 \\ y \ge 0 \end{cases}$

In Exercises 19–20, solve the standard maximization problems by using the simplex method.

19. Maximize $P = 6x + 4y + 5z$

Subject to $\begin{cases} 2x + 3y + 2z \le 120 \\ x + y + z \le 60 \\ x \ge 0 \\ y \ge 0 \\ z \ge 0 \end{cases}$

20. Maximize $P = 4x - y + z$

Subject to $\begin{cases} 2x + 3y + 3z \le 210 \\ x + y + z \le 100 \\ x \ge 0 \\ y \ge 0 \\ z \ge 0 \end{cases}$

In Exercises 21–22, set up and solve the standard maximization problem using the simplex method.

21. **Battery Sales** AAA Alkaline Discount Batteries sells low-cost batteries to consumers. In August 2002, the company offered AAA batteries at the following prices: 100-pack for $18.00, 600-pack for $96.00, and 1200-pack for $180.00.
(**Source:** www.aaaalkalinediscountbatteries.com.)

An electronics store owner wants to buy at most 1800 batteries and spend at most $300. She expects that she'll be able to resell all of the batteries she orders for $1.00 each. How many packs (100-packs, 600-packs, or 1200-packs) should she order if she wants to maximize her revenue? What is her maximum revenue?

22. **Battery Sales** Repeat Exercise 21 except maximize profit. Assume that her only cost is the cost of the batteries. What is her maximum profit?

Section 4.4 *In Exercises 23–25, determine if the problem is a standard minimization problem. If it isn't, explain why.*

23. Minimize $P = 11x + 8y + 2z$

Subject to $\begin{cases} 2x + y + z \ge 10 \\ -x + y - z \ge 1 \\ x \ge 0 \\ y \ge 0 \\ z \ge 0 \end{cases}$

24. Maximize $P = -2x - 5y + 7z$

Subject to $\begin{cases} x + 2y + z \ge 11 \\ -11x + y \le -21 \\ x \ge 0 \\ y \ge 0 \\ z \ge 0 \end{cases}$

25. Maximize $P = 4x - 2y$

Subject to $\begin{cases} 6x + 7y \ge 13 \\ -8x - 4y \ge -12 \\ x \ge 0 \\ y \ge 0 \end{cases}$

In Exercises 26–28, find the transpose of the given matrix.

26. $A = \begin{bmatrix} 7 & -2 & 0 & 4 \\ 5 & 3 & 7 & 6 \\ -1 & 1 & 2 & 1 \end{bmatrix}$

27. $A = \begin{bmatrix} 12 & 10 & 8 \\ 14 & 12 & 10 \\ 16 & 14 & 12 \end{bmatrix}$

28. $A = \begin{bmatrix} 1 & 0 & 1 & 0 & 2 \\ 0 & 1 & 0 & 1 & 9 \\ 1 & 1 & 1 & 1 & 11 \end{bmatrix}$

In Exercises 29–32, do the following:

(i) Write the dual problem for the given standard minimization problem.

(ii) Solve the dual problem using the simplex method.

(iii) Use the final simplex tableau of the dual problem to solve the standard minimization problem.

(iv) Check your answer by graphing the feasible region of the standard minimization problem and calculating the value of the objective function at each of the corner points.

29. Minimize $P = 4x + 2y$

Subject to $\begin{cases} 4x + 5y \geq 20 \\ 2x + 3y \geq 11 \\ \quad x \geq 0, y \geq 0 \end{cases}$

30. Minimize $P = x + 6y$

Subject to $\begin{cases} 2x + 9y \geq 19 \\ -x + 10y \geq 5 \\ \quad x \geq 0, y \geq 0 \end{cases}$

31. Minimize $P = 2x + 3y$

Subject to $\begin{cases} 5x + 2y \geq 26 \\ 10x - 5y \geq 25 \\ \quad x \geq 0, y \geq 0 \end{cases}$

32. Minimize $P = 3x + 5y$

Subject to $\begin{cases} 9x + 4y \geq 36 \\ 3x + 2y \geq 18 \\ \quad x \geq 0, y \geq 0 \end{cases}$

In Exercises 33–34, set up and solve the standard minimization problem.

33. **Pet Nutrition: Food Cost**
PETsMART.com sold the following varieties of dog food in June 2003.

Authority Chicken Adult Formula, 32 percent protein, 3 percent fiber, $19.99 per 33-pound bag

Bil-Jac Select Dog Food, 27 percent protein, 4 percent fiber, $18.99 per 18-pound bag

Iams Minichuncks, 26 percent protein, 4 percent fiber, $8.99 per 8-pound bag

(**Source:** www.petsmart.com.)

A dog kennel wants to make at least 330 pounds of a mix containing at least 30 percent protein and at least 3 percent fiber. How many bags of each dog food variety should the kennel buy in order to minimize cost? (To make computations easier, round the price of each bag to the nearest dollar.)

34. **Pet Nutrition: Fat Content**
PETsMART.com sold the following varieties of dog food in June 2003.

Nature's Recipe Venison Meal & Rice Canine, 20 percent protein, 10 percent fat, $21.99 per 20-pound bag

Nutro Max Natural Dog Food, 27 percent protein, 16 percent fat, $12.99 per 17.5-pound bag

PETsMART Premier Oven Baked Lamb Recipe, 25 percent protein, 14 percent fat, $22.99 per 30-pound bag

(**Source:** www.petsmart.com.)

A dog breeder wants to make at least 300 pounds of a mix containing at least 24 percent protein. How many bags of each dog food variety should the breeder buy in order to minimize fat content?

Section 4.5 *In Exercises 35–36, rewrite the general linear programming problem as a system of equations with slack and surplus variables and a rewritten objective function.*

35. Maximize $P = 2x + 3y + z$

Subject to $\begin{cases} 2x + 4y + z \leq 10 \\ x + 2z \geq 3 \\ 3x - 2z \geq 1 \\ x \geq 0, y \geq 0, z \geq 0 \end{cases}$

36. Maximize $P = x - 10y + 6z$

Subject to $\begin{cases} -2x + 2y + 3z \leq 2 \\ 5x + 2y + 3z \geq 4 \\ 7x - 3y + 2z \geq 1 \\ x \geq 0, y \geq 0, z \geq 0 \end{cases}$

In Exercises 37–38, solve the general linear programming problem. If there is no solution, so state.

37. Maximize $P = 4x + y + 10z$

Subject to $\begin{cases} x + 3y + 2z \leq 5 \\ 2x + y + 2x \geq 4 \\ 2y + 3z \leq 6 \\ x \geq 0, y \geq 0, z \geq 0 \end{cases}$

38. Minimize $P = 4x + y + 10z$

Subject to $\begin{cases} x + 3y + 2z \leq 5 \\ 2x + y + 2x \geq 4 \\ 2y + 3z \leq 6 \\ x \geq 0, y \geq 0, z \geq 0 \end{cases}$

Make It Real

What to do

1. Find the wholesale and retail prices of three items of personal interest.

2. Estimate the number of cubic feet each item (or box of items) occupies.

3. Find a place in your home or workplace where you could store the items. Calculate the number of cubic feet in the storage area.

4. Assume that you may spend up to 100 times the retail price of the most expensive item in purchasing a combination of the three items.

5. Assume that you will be able to sell at retail price all of the items you order at wholesale. Determine how many of the items you should order if you want to maximize your profit subject to the spending and storage area constraints.

Where to look for prices

Wholesale Prices

Fruits: www.todaymarket.com
Vegetables: www.todaymarket.com
Flowers: www.flowersales.com
Computer parts: www.tcwo.com
Batteries: www.aaaalkalinediscountbatteries.com
Various items: Costco, Sam's Club, etc.
Search the Internet at www.yahoo.com and type in the key word
"wholesale".

Retail Prices

Local grocery stores, clothing stores, etc.
Advertisements from newspapers
Search the Internet and type in a key word for your product of interest.

Chapter

5

Nonlinear Models

Mathematical functions are commonly used to model real-world data. Although every model has its limitations, models are often used to forecast expected results. Selecting which mathematical model to use is relatively easy once you become familiar with the basic types of mathematical functions. Remarkably, these basic functions may be used to effectively model many real-world data sets. For example, based on data from 1993 to 2002, the sales revenue from Starbucks stores may be effectively modeled by a quadratic function.

5.1 Quadratic Function Models
- Use nonlinear functions to model real-life phenomena
- Interpret the meaning of mathematical models in their real-world context
- Model real-life data with quadratic functions

5.2 Higher-Order Polynomial Function Models
- Use nonlinear functions to model real-life phenomena
- Interpret the meaning of mathematical models in their real-world context
- Model real-life data with higher-order polynomial functions

5.3 Exponential Function Models
- Use nonlinear functions to model real-life phenomena
- Interpret the meaning of mathematical models in their real-world context
- Graph exponential functions
- Find the equation of an exponential function from a table
- Model real-life data with exponential functions

5.4 Logarithmic Function Models
- Use nonlinear functions to model real-life phenomena
- Interpret the meaning of mathematical models in their real-world context
- Graph logarithmic functions
- Model real-life data with logarithmic functions
- Apply the rules of logarithms to simplify logarithmic expressions
- Solve logarithmic equations

5.5 Choosing a Mathematical Model
- Use critical thinking skills in selecting a mathematical model

5.1 Quadratic Function Models

- Use nonlinear functions to model real-life phenomena
- Interpret the meaning of mathematical models in their real-world context
- Model real-life data with quadratic functions

GETTING **STARTED** Many college-bound high school students enroll in Advanced Placement Program (AP®) courses to earn college credit while still in high school. The College Board® offers 35 courses in 19 disciplines ranging from Studio Art: 3D Design to Calculus AB. (**Source:** The College Board®.) The number of students taking the Calculus AB test has increased dramatically since 1969, as shown in Figure 5.1.

FIGURE 5.1

How many students will take the test in 2005? Nobody knows; however, using a mathematical model, we can predict what may happen.

In this section, we will discuss the identifying features of quadratic functions. We will then demonstrate how to use algebraic methods and quadratic regression to model data sets whose graphs open upward or downward over their domain.

A polynomial is a function that is the sum of terms of the form ax^n, where a is a real number and n is a nonnegative integer. For example, each of the following functions is a polynomial.

$$f(x) = 6x^4 - 3x^3 + 5x^2 - 2x + 9$$
$$g(x) = 4x^2 + 2x$$
$$h(x) = -1.3x + 9.7$$
$$j(x) = x^2 - 2x + 6$$

May a constant term like 9 be written in the form ax^n? Yes! Since $x^0 = 1$ for nonzero x, $9x^0 = 9$. Consequently, a constant function $s(x) = 9$ is also a polynomial.

The **degree of a polynomial** is the value of its largest exponent. For example, the degree of the polynomial $f(x) = 6x^4 - 3x^3 + 5x^2 - 2x + 9$ is 4, since 4 is the largest exponent. Since the equation of any line may be written as $y = ax^1 + b$, lines are polynomials of degree 1. We've worked extensively with first-degree polynomials (lines) in the preceding chapters. In this section, we are interested in polynomials of degree 2. Polynomials of degree 2 are called **quadratic functions.**

QUADRATIC FUNCTION

A polynomial function of the form $f(x) = ax^2 + bx + c$ with $a \neq 0$ is called a **quadratic function.** The graph of a quadratic function is a **parabola.**

When a parabola opens upward, we say that it is "concave up." When the parabola opens downward, we say that it is "concave down." The steepness of the sides of the parabola and its concavity are controlled by the value of a, the coefficient on the x^2 term in its equation. If $a > 0$, the graph is concave up. If $a < 0$, the graph is concave down. As the magnitude of a increases, the steepness of the graph increases. (For $a > 0$, the magnitude of a is a. For $a < 0$, the magnitude of a is $-a$.) Consider the graphs in Figures 5.2 and 5.3. These graphs have the same values for b and c ($b = -2$ and $c = 3$) but differing values of a.

In Figure 5.2, since a is the coefficient on the x^2 term, $a = 1$. The magnitude of a is 1. Since $a > 0$, the graph is concave up.

In Figure 5.3, since a is the coefficient on the x^2 term, $a = 2$. The magnitude of a is 2. Increasing the magnitude of a increased the graph's steepness.

Consider the graphs in Figures 5.4 and 5.5. These graphs have the same values for b and c ($b = 4$ and $c = 0$) but differing values of a.

FIGURE 5.2

FIGURE 5.3

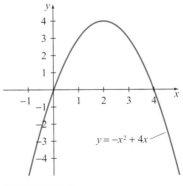

FIGURE 5.4

In Figure 5.4, since a is the coefficient on the x^2 term, $a = -1$. The magnitude of a is 1. Since $a < 0$, the graph is concave down.

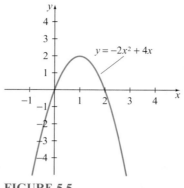

FIGURE 5.5

In Figure 5.5, since a is the coefficient on the x^2 term, $a = -2$. The magnitude of a is 2. Increasing the magnitude of a from 1 to 2 increased the graph's steepness.

The coefficient b in the equation affects the horizontal and vertical placement of the parabola. The constant term c in the equation indicates that the point $(0, c)$ is the y-intercept of the graph.

Recall that the graph of a function is said to be increasing if the value of y gets bigger as the value of x increases. Similarly, the graph of a function is said to be decreasing if the value of y gets smaller as the value of x increases. The vertex of a parabola is the point on the graph of a quadratic function where the curve changes from decreasing to increasing (or vice versa). The minimum y value of a concave up parabola occurs at the vertex (see Figure 5.6a). Similarly, the maximum y value of a concave down parabola occurs at the vertex (see Figure 5.6b).

For parabolas with x-intercepts, the x coordinate of the vertex always lies halfway between the x-intercepts. Recall that as a result of the quadratic formula, we know that the x coordinate of the vertex is $x = \dfrac{-b}{2a}$. The y coordinate of the vertex is obtained by evaluating the quadratic function at this x value. For example, for the quadratic function $y = x^2 - 2x + 3$, we know that $a = 1$ and $b = -2$. The x coordinate of the vertex is

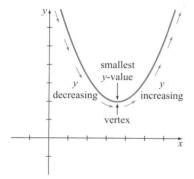

FIGURE 5.6a Concave Up Parabola

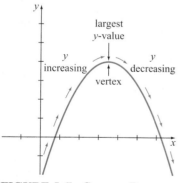

FIGURE 5.6b Concave Down Parabola

$$x = \frac{-b}{2a}$$
$$= \frac{-(-2)}{2(1)}$$
$$= \frac{2}{2}$$
$$= 1$$

Evaluating the function at $x = 1$ yields

$$y = x^2 - 2x + 3$$
$$= (1)^2 - 2(1) + 3$$
$$= 1 - 2 + 3$$
$$= 2$$

Therefore, the vertex of the function is $(1, 2)$.

Parabolas are symmetrical. That is, if we draw a vertical line through the vertex of the parabola, the portion of the graph on the left of the line is the mirror image of the portion of the graph on the right of the line. The line is referred to as the **axis of symmetry** (see Figure 5.7).

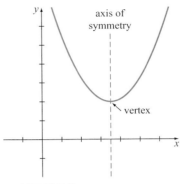

FIGURE 5.7

Since the axis of symmetry is a vertical line passing through the vertex, the equation of the axis of symmetry is $x = \dfrac{-b}{2a}$.

EXAMPLE 1 | ### Describing the Graph of a Quadratic Function from Its Equation

Determine the concavity, the y-intercept, and the vertex of the quadratic function $y = 3x^2 + 6x - 1$.

SOLUTION We have $a = 3$, $b = 6$, and $c = -1$. Since $a > 0$, the parabola is concave up. Since $c = -1$, the y-intercept is $(0, -1)$. The x coordinate of the vertex is given by

$$x = \frac{-b}{2a}$$
$$= \frac{-6}{2(3)}$$
$$= \frac{-6}{6}$$
$$= -1$$

The y coordinate of the vertex is obtained by evaluating the function at $x = -1$.

$$y = 3x^2 + 6x - 1$$
$$= 3(-1)^2 + 6(-1) - 1$$
$$= 3 - 6 - 1$$
$$= -4$$

The vertex of the parabola is $(-1, -4)$.

EXAMPLE 2   ## Determining the Equation of a Parabola from Its Graph

Determine the equation of the parabola shown in Figure 5.8.

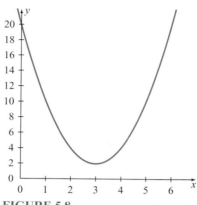

FIGURE 5.8

SOLUTION   The y-intercept is $(0, 20)$, so $c = 20$. The vertex is $(3, 2)$. Since the x coordinate of the vertex is $\dfrac{-b}{2a}$, we know that

$$\frac{-b}{2a} = 3$$
$$-b = 6a$$
$$b = -6a$$

Since $f(x) = ax^2 + bx + c$, we have

$$f(x) = ax^2 + (-6a)x + (20) \qquad \text{Since } b = -6a$$
$$= ax^2 - 6ax + 20$$

The vertex is $(3, 2)$, so $f(3) = 2$. Therefore,

$$f(x) = ax^2 - 6ax + 20$$
$$2 = a(3)^2 - 6a(3) + 20 \qquad \text{Substitute in } (3, 2)$$
$$2 = 9a - 18a + 20$$
$$-18 = -9a$$
$$a = 2$$

Since $b = -6a$, $b = -12$. The equation of the parabola is $f(x) = 2x^2 - 12x + 20$.

We can check our work by substituting in a different point from the graph. The parabola passes through the point $(4, 4)$. Consequently, this point should satisfy the equation $f(x) = 2x^2 - 12x + 20$.

$$f(x) = 2x^2 - 12x + 20$$
$$4 \stackrel{?}{=} 2(4)^2 - 12(4) + 20 \qquad \text{Substitute in } (4, 4)$$
$$4 \stackrel{?}{=} 2(16) - 48 + 20$$
$$4 \stackrel{?}{=} 32 - 48 + 20$$
$$4 = 4$$

Recognizing the relationship between the graph of a parabola and its corresponding quadratic function provides a quick way to evaluate the accuracy of a quadratic model. Quadratic models may be determined algebraically or by using quadratic regression. We will demonstrate both methods in the next several examples.

Let's return to the AP Calculus exam data introduced at the beginning of the section. At first glance, the Calculus AB exam data don't look at all like a parabola; however, we do observe that the data appear to be increasing at an ever-increasing rate. (It took six years [1980 to 1986] for the number of exams to increase from 20,000 to 40,000, but it took only three years [1986 to 1989] for the number of exams to increase from 40,000 to 60,000.) Plotting the quadratic equation $f(t) = 157.8486t^2 - 622,378t + 613,500,479$ together with the data set, we observe that, in fact, a quadratic equation fits the data very well (Figure 5.9). (**Source:** The College Board®.) This equation is found by performing quadratic regression on the data, a process that we will detail in the forthcoming Technology Tip.

FIGURE 5.9

Observe that the coefficients of the variables in the quadratic equation are very large. We can come up with an equation with smaller coefficients by **aligning the data.** We'll let $t = 0$ in 1969, $t = 1$ in 1970, and so on. Doing quadratic regression on the aligned data yields the equation $f(t) = 157.85t^2 - 770.64t + 10,268$ with the coefficient of determination $r^2 = 0.9979$. Recall that the coefficient of determination is a measure of how well the model fits the data. The closer r^2 is to 1, the better the model fits the data. Although both of the models fit the data, the second model will make computations easier because of the smaller coefficients.

Using the model, we predict how many tests will be administered in 2005. In 2005, $t = 36$.

$$f(36) = 157.85(36)^2 - 770.64(36) + 10,268$$
$$= 187,099$$

We estimate that 187,099 AP Calculus AB exams will be administered in 2005.

A quadratic function model for a data set may be generated by using the quadratic regression feature on a graphing calculator, as demonstrated in the following Technology Tip. However, the fact that the calculator can create a quadratic model does not guarantee that the model will be a good fit for the data.

TECHNOLOGY **TIP**

Quadratic Regression

1. Enter the data using the Statistics Menu List Editor. (Refer to Section 1.3 if you've forgotten how to do this.)

2. Bring up the Statistics Menu Calculate feature by pressing [STAT] and using the blue arrows to move to the CALC menu. Then select item 5:QuadReg and press [ENTER].

3. If you want to automatically paste the regression equation into the [Y=] editor, press the key sequence [VARS] Y-Vars; 1:Function; 1:Y1 and press [ENTER]. Otherwise press [ENTER].

Although the quadratic model fits the AP Calculus AB exam data very well from 1969 to 2002, we must be cautious in using it to predict future behavior. (Predicting the output value for an input value outside of the interval of the input data is called **extrapolation.**) For this data set, we may feel reasonably comfortable with an estimate two or three years beyond the last data point; however, we would doubt the accuracy of the model 100 years beyond 2002. For example, in 2102 ($t = 133$), the estimated number of exams is 2,702,587. This figure is more than 17 times greater than the maximum number of exams that have ever been administered!

We also need to look at the population of students who could possibly take the exam. Between 1970 and 1999, the number of twelfth graders (those who typically take the AP Calculus exam) fluctuated from a high of 3,026,000 in 1977 to a low of 2,381,000 in 1990. (**Source:** *Statistical Abstract of the United States, 2001,* Table 232, p. 149.) If the number of twelfth graders continues to fluctuate between these two values, at some point our model estimate for the exams would exceed the number of people who could conceivably take the exam. This illustrates the necessity to verify that a model makes sense in its real-world context.

It is often possible to model a data set with more than one mathematical model. When selecting a model, we should consider the following:

1. The graphical fit of the model to the data
2. The correlation coefficient (r) or the coefficient of determination (r^2)
3. The known behavior of the thing being modeled

Recall that the closer the correlation coefficient is to 1 or -1, the better the model fits the data. Similarly, the closer the coefficient of determination is to 1, the better the model fits the data.

EXAMPLE 3

Using Quadratic Regression to Forecast Prescription Drug Sales

Retail prescription drug sales in the United States increased from 1995 to 2000 as shown in Table 5.1.

TABLE 5.1

Years Since 1995 (t)	Retail Sales (billions of dollars) [$S(t)$]
0	68.6
2	89.1
3	103.0
4	121.7
5	140.7

Source: *Statistical Abstract of the United States, 2001,* Table 127, p. 94.

Model the data using a quadratic function. Then use the model to predict retail prescription drug sales in 1996 and 2001.

SOLUTION We observe from the scatter plot that the data appear concave up everywhere (Figure 5.10).

$S(t) = 1.411t^2 + 7.441t + 68.55$

FIGURE 5.10

A quadratic function may fit the data well. We use quadratic regression to find the quadratic model that best fits the data.

Based on data from 1995 to 2000, retail prescription drug sales in the United States may be modeled by

$$S(t) = 1.411t^2 + 7.441t + 68.55 \text{ billion dollars}$$

where t is the number of years since 1995.

The coefficient of determination ($r^2 = 0.9997$) is extremely close to 1. The graph also appears to "touch" each data point. The model appears to fit the data well.

In 1996, $t = 1$, and in 2001, $t = 6$. Evaluating the function at each t value, we get

$$S(1) = 1.411(1)^2 + 7.441(1) + 68.55$$
$$= 77.402$$
$$S(6) = 1.411(6)^2 + 7.441(6) + 68.55$$
$$= 163.992$$

Since the original data were accurate to only one decimal place, we will round our solutions to one decimal place as well. We estimate that prescription drug sales were $77.4 billion in 1996 and $164.0 billion in 2001.

You may ask, "Is there a way to find a quadratic model without using quadratic regression?" There is. The model may not be the model of best fit, but it may still model the data effectively. In Example 4, we repeat Example 3 using an algebraic method to find a quadratic model.

EXAMPLE 4

Using Algebraic Methods to Model Prescription Drug Sales

Retail prescription drug sales in the United States increased from 1995 to 2000 as shown in Table 5.2.

TABLE 5.2

Years Since 1995 (t)	Retail Sales (billions of dollars) [$S(t)$]
0	68.6
2	89.1
3	103.0
4	121.7
5	140.7

Source: *Statistical Abstract of the United States, 2001*, Table 127, p. 94.

Model the data using a quadratic function. Then use the model to predict retail prescription drug sales in 1996 and 2001.

SOLUTION Given any three data points, we can find a quadratic function that passes through the points, provided that the points define a nonlinear function. We will pick the points $(0, 68.6), (3, 103.0),$ and $(5, 140.7)$ from the table. A quadratic function is of the form $S(t) = at^2 + bt + c$. Each of the points must satisfy this equation.

$$68.6 = a(0)^2 + b(0) + c \qquad \text{Substitute } t = 0,\ S(t) = 68.6$$
$$c = 68.6$$

$$103.0 = a(3)^2 + b(3) + c \qquad \text{Substitute } t = 3,\ S(t) = 103.0$$
$$103.0 = 9a + 3b + c$$

$$140.7 = a(5)^2 + b(5) + c \qquad \text{Substitute } t = 5,\ S(t) = 140.7$$
$$140.7 = 25a + 5b + c$$

Since we know $c = 68.6$, we can simplify the last two equations.

$$103.0 = 9a + 3b + 68.6$$
$$34.4 = 9a + 3b$$

$$140.7 = 25a + 5b + 68.6$$
$$72.1 = 25a + 5b$$

We can now find the values of a and b by solving the system of equations.

$$9a + 3b = 34.4$$
$$25a + 5b = 72.1$$

$$\begin{bmatrix} 9 & 3 & \vdots & 34.4 \\ 25 & 5 & \vdots & 72.1 \end{bmatrix}$$

$$\begin{bmatrix} 9 & 3 & \vdots & 34.4 \\ -30 & 0 & \vdots & -44.3 \end{bmatrix} \qquad 5R_1 - 3R_2$$

$$\begin{bmatrix} 0 & 30 & \vdots & 211.1 \\ -30 & 0 & \vdots & -44.3 \end{bmatrix} \qquad 10R_1 + 3R_2$$

$$-30a = -44.3 \qquad\qquad 30b = 211.1$$
$$a = 1.477 \qquad\qquad b = 7.037$$

A quadratic model for the data is

$$S(t) = 1.477t^2 + 7.037t + 68.6$$

Graphing this model with the data shows that it fits the data relatively well (Figure 5.11).

FIGURE 5.11

In 1996, $t = 1$, and in 2001, $t = 6$. Evaluating the function at each t value, we get

$$S(1) = 1.477(1)^2 + 7.037(1) + 68.6 \qquad S(6) = 1.477(6)^2 + 7.037(6) + 68.6$$
$$\approx 77.1 \text{ billion} \qquad\qquad\qquad \approx 164.0 \text{ billion}$$

These estimates are close to the estimates from Example 3. In Example 3, we estimated that prescription drug sales were $77.1 billion in 1996 and $164.0 billion in 2001.

EXAMPLE 5

Using Quadratic Regression to Forecast Nursing Home Care Costs

Because of their need for constant medical care, many elderly Americans are placed in nursing homes by their families. The amount of money spent on nursing home care has risen substantially since 1960, as shown in Table 5.3.

TABLE 5.3

Years Since 1960 (t)	Nursing Home Care (billions of dollars) [$N(t)$]
0	1
10	4
20	18
30	53
40	96

Source: *Statistical Abstract of the United States, 2001,* Table 119, p. 91.

The U.S. Centers for Medicare and Medicaid Services predict that by 2010 the cost of nursing home care will reach $183 billion.

Model the data with a quadratic function and calculate the cost of nursing home care in 2010. Compare your result to the U.S. Centers for Medicare & Medicaid Services estimate.

SOLUTION Using quadratic regression, we determine that the quadratic model of best fit is

$$N(t) = 0.07214t^2 - 0.4957t + 1.029 \text{ billion dollars}$$

where t is the number of years since 1960.

Since the coefficient of determination ($r^2 = 0.9987$) is extremely close to 1, we anticipate that the model fits the data well. Graphing the data and the model together yields Figure 5.12.

FIGURE 5.12

The graph passes near each data point. The model appears to fit the data fairly well. Zooming in, however, we see that the graph decreases from 1 billion to 0 billion between 1960 and 1963 (Figure 5.13). (We rounded to whole numbers, since the original data were in whole numbers.)

FIGURE 5.13

In reality, nursing home expenditures have increased every year since 1960. Nevertheless, despite this limitation in our model, we will use it to predict nursing home expenditures in 2010. In 2010, $t = 50$.

$$N(50) = 0.721(50)^2 - 0.4957(50) + 1.0286$$
$$= 156.60$$

We estimate that in 2010, $157 billion will be spent on nursing home care. Our model estimate is substantially less than the $183 billion that the U.S. Centers for Medicare & Medicaid Services estimate. Their estimate probably anticipated the health care needs of the aging population of baby boomers, while ours did not.

Models projecting health care costs are useful for legislators, insurance companies, and consumers. By considering future costs, people can prepare for the future and avert financial crises.

EXAMPLE **6** **Using Algebraic Methods to Find a Quadratic Model for Net Sales**

Based on the data in Table 5.4, find a quadratic model for the net sales of the Kellogg Company algebraically.

TABLE 5.4

Years Since 1999 (t)	Kellogg Company Net Sales (millions of dollars) $R(t)$
0	6,984.2
1	6,954.7
2	8,853.3

Source: Kellogg Company 2001 Annual Report, pp. 7, 27.

SOLUTION Since we are given the y-intercept, $(0, 6984.2)$, we know that $c = 6984.2$. We have

$$R(t) = at^2 + bt + c$$
$$= at^2 + bt + 6984.2$$

$$R(1) = a(1)^2 + b(1) + 6984.2$$
$$6954.7 = a + b + 6984.2$$
$$-29.5 = a + b$$

$$R(2) = a(2)^2 + b(2) + 6984.2$$
$$8853.3 = 4a + 2b + 6984.2$$
$$1869.1 = 4a + 2b$$

We must solve the system of equations

$$a + b = -29.5$$
$$4a + 2b = 1869.1$$

We will solve the system using the substitution method. Solving the first equation for a yields $a = -b - 29.5$. Substituting this result into the second equation $4a + 2b = 1869.1$ yields

$$4(-b - 29.5) + 2b = 1869.1 \qquad \text{Since } a = -b - 29.5$$
$$-4b - 118 + 2b = 1869.1$$
$$-2b = 1987.1$$
$$b = -993.55$$

Since $a = -b - 29.5$,

$$a = -(-993.55) - 29.5$$
$$= 964.05$$

The quadratic function that models the revenue of the Kellogg Company is $R(t) = 964.05t^2 - 993.55t + 6984.2$, where t is the number of years since the end of 1999 and $R(t)$ is the revenue from sales in millions of dollars.

In each of the preceding examples, the quadratic model fit the data well. This is not always the case, as demonstrated in Example 7.

EXAMPLE 7 **Determining When a Quadratic Model Should Not Be Used**

The per capita consumption of ready-to-eat and ready-to-cook breakfast cereal is shown in Table 5.5.

TABLE 5.5

Years Since 1980 (t)	Cereal Consumption (pounds) $[C(t)]$	Years Since 1980 (t)	Cereal Consumption (pounds) $[C(t)]$
0	12	10	15.4
1	12	11	16.1
2	11.9	12	16.6
3	12.2	13	17.3
4	12.5	14	17.4
5	12.8	15	17.1
6	13.1	16	16.6
7	13.3	17	16.3
8	14.2	18	15.6
9	14.9	19	15.5

Source: *Statistical Abstract of the United States, 2001*, Table 202, p. 129.

Explain why you do or do not believe that a quadratic function will model the data set well.

SOLUTION We first draw the scatter plot of the data set (Figure 5.14).

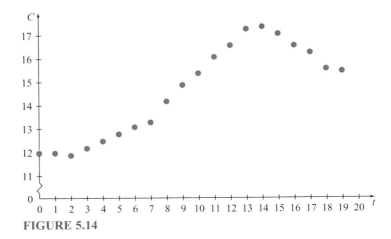

FIGURE 5.14

(*Note:* We have adjusted the viewing window so that we may better analyze the data.

Recall that a parabola is either concave up everywhere or concave down everywhere. This scatter plot appears to be concave up between $t = 1$ and $t = 10$, concave down between $t = 10$ and $t = 17$, and concave up between $t = 17$ and $t = 19$. The fact that the scatter plot changes concavity causes us to doubt that a quadratic model will fit the data well. In short, the scatter plot doesn't look like a parabola or a portion of a parabola.

The quadratic model that best fits the data is $C(t) = -0.0179t^2 + 0.637t + 10.8$ and is shown in Figure 5.15.

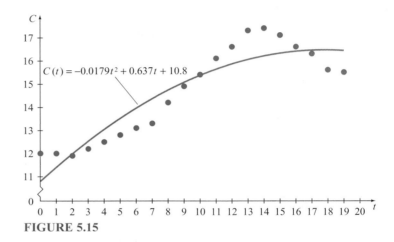

FIGURE 5.15

The coefficient of determination ($r^2 = 0.853$) is not as close to 1 as in our previous examples. Additionally, we can see visually that the model does not fit the data well.

Once a model has been created, we can often use it to make educated guesses about what may happen in the future. Forecasting the future is a key function for many businesses.

EXAMPLE **8**

Using Quadratic Regression to Forecast Fuel Consumption

As vans, pickups, and SUVs have increased in popularity, the total fuel consumption of these types of vehicles has also increased (Table 5.6).

TABLE 5.6　Motor Fuel Consumption of Vans, Pickups, and SUVs

Years Since 1980 (t)	Fuel Consumption (billions of gallons) (F)
0	23.8
5	27.4
10	35.6
15	45.6
19	52.8

Source: *Statistical Abstract of the United States, 2001*, Table 1105, p. 691.

Model the fuel consumption with a quadratic function and forecast the year in which fuel consumption will reach 81.0 billion gallons.

SOLUTION　Using quadratic regression, we determine the model to be

$$F(t) = 0.0407t^2 + 0.809t + 23.3 \text{ billion gallons}$$

where t is the number of years since the end of 1980. We want to know at what value of t does $F(t) = 81.0$. This problem may be solved algebraically or graphically using technology. We will solve the problem twice (once with each method) and allow you to use the method of your choice in the exercises.

ALGEBRAIC SOLUTION

$$81.0 = 0.0407t^2 + 0.809t + 23.3$$
$$0 = 0.0407t^2 + 0.809t - 57.7$$

This is a quadratic function with $a = 0.0407$, $b = 0.809$, and $c = -57.7$.

Recall that the solution to a quadratic equation of the form $at^2 + bt + c = 0$ is given by the Quadratic Formula,

$$t = \frac{-b \pm \sqrt{b^2 - 4ac}}{2a}$$

Substituting our values of a, b, and c into the Quadratic Formula yields

$$t = \frac{-0.809 \pm \sqrt{(-0.809)^2 - 4(0.0407)(-57.7)}}{2(0.0407)}$$

$$= \frac{-0.809 \pm \sqrt{10.0}}{0.0814}$$

$$= \frac{-0.809 \pm 3.17}{0.0814}$$

In the context of the problem, we know that t must be nonnegative. Consequently, we will calculate only the nonnegative solution.

$$t = \frac{-0.809 + 3.17}{0.0814}$$

$$= \frac{2.36}{0.0814}$$

$$= 29.0$$

We anticipate that at the end of 2009 (29 years after the end of 1980), the fuel consumption will have reached 81.0 billion gallons.

GRAPHICAL SOLUTION

We graph the function $F(t) = 0.0407t^2 + 0.809t + 23.3$ and the horizontal line $y = 81.0$ simultaneously (Figure 5.16). $F(t) = 81.0$ at the point at which these two functions intersect.

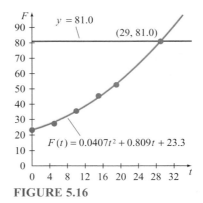

FIGURE 5.16

Using the intersect feature on the graphing calculator, we determine that the point of intersection is $(29, 81.0)$. (The technique for finding a point of intersection on a graphing calculator is detailed in Section 2.1.) The interpretation of the solution is the same as that given in the discussion of the algebraic method.

5.1 Summary

In this section, you learned how to use quadratic regression to model a data set. You also discovered the importance of analyzing a mathematical model before using it to calculate unknown values.

5.1 Exercises

In Exercises 1–5, determine the concavity, y-intercept, and vertex of the quadratic equation.

1. $y = x^2 - 2x + 1$ **2.** $f(x) = -2x^2 + 4$

3. $g(x) = 3x^2 + 3x$

4. $h(t) = -1.2t^2 + 2.4t + 4.5$

5. $f(t) = 2.8t^2 - 1.4t + 2.1$

In Exercises 6–10, determine the equation of the parabola from the graph.

6.

7.

8.

9.

10.

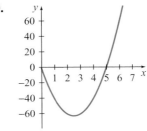

In Exercises 11–15, do the following:

(a) Find a quadratic model for the data algebraically.

(b) Graph the model together with the scatter plot.

(c) Explain why you do or do not believe that the function is a good model.

11. **Basketball Game Attendance**

NCAA Basketball Game Attendance

Years Since 1985 (t)	Women's Games (thousands) (W)
0	2,072
5	2,777
10	4,962
14	8,698

Source: *Statistical Abstract of the United States, 2001*, Table 1241, p. 759.

12.   **Ice Cream Cost**

Prepackaged Ice Cream Cost

Years Since 1990 (t)	Ice Cream Cost (dollars per 1/2 gallon) (C)
0	2.54
3	2.59
4	2.62
5	2.68
6	2.94
7	3.02
8	3.30
9	3.40
10	3.66

Source: *Statistical Abstract of the United States, 2001*, Table 706, p. 468.

13.   **DVD Player Shipments**

DVD Player Shipments in North America

Years Since 1997 (t)	Shipments (thousands) (D)
1	1,080
2	4,100
3	9,700
4	16,700

Source: DVD Entertainment Group.

14.   **Health Care Spending**

Health Care Spending by Insurance Companies

Years Since 1960 (t)	Spending (billions of dollars) (I)
0	6
10	16
20	68
30	233
40	438

Source: *Statistical Abstract of the United States, 2001*, Table 119, p. 91.

15.   **VHS Tape and CD-R Sales**

VHS Tape Versus CD-R Disk Sales, 2000-2002

Millions of Blank VHS Tapes Sold Worldwide (v)	Millions of CD-R Disks Sold Worldwide (C)
1,409	2,730
1,322	3,730
1,195	4,225

Source: International Recording Media Association.

In Exercises 16–50, do the following:

(a) Draw a scatter plot for the data.
(b) Explain why a quadratic function might or might not fit the data.

If it appears that a quadratic function may fit the data:

(c) Use quadratic regression to find the quadratic function that best fits the data.
(d) Graph the model together with the scatter plot.
(e) Explain why you do or do not believe that the function is a good model.
(f) Answer any additional questions that may be given.

16. College Freshmen

Male College Freshmen

Years Since 1970 (t)	Percentage of Freshmen That Are Male (F)
0	52.1
10	48.8
15	48.9
20	46.9
25	45.6
27	45.5
28	45.5
29	45.3
30	45.2

According to the model, in what year were there an equal number of male and female freshmen?

  17. **Personal Income** Because of inflation, the buying power of a dollar decreases over time. As a result, your personal income may increase while your buying power decreases. In the following table, personal income per capita is listed using constant (1996) dollars. That is, the actual income figures are adjusted for inflation.

Personal Income per Capita, Connecticut, Ranked #1 in 2000

Years Since 1980 (t)	Income (dollars) (I)
0	22,530
10	31,223
20	37,854

Source: *Statistical Abstract of the United States, 2001*, Table 652, p. 426.

  18. **Personal Income** In the following table, personal income per capita is listed using constant (1996) dollars. That is, the actual income figures are adjusted for inflation.

Personal Income per Capita, Mississippi, ranked #50 in 2000

Years Since 1980 (t)	Income (dollars) (I)
0	12,817
10	15,373
20	19,554

Source: *Statistical Abstract of the United States, 2001*, Table 652, p. 426.

What is the vertex of the quadratic model, and what does it mean in the context of the problem?

  19. Game Sales

U.S. Video Game Industry Sales

Years Since 1997 (t)	Sales (billions of dollars) (S)
0	5.1
1	6.2
2	6.9
3	6.6
4	9.4

Source: www.npdfunworld.com.

20.   **Milk Consumption**

**Per Capita Milk Beverage
Consumption**

Years Since 1980 (t)	Consumption (gallons) (M)
0	27.6
5	26.7
10	25.7
15	24.3
17	24.0
18	23.7
19	23.6

Source: *Statistical Abstract of the
United States, 2001*, Table 202,
p. 129.

 21.   **Conventional Mortgages**

**Percentage of New Privately Owned
One-Family Houses Financed with
a Conventional Mortgage**

Years Since 1970 (t)	Conventional Mortgages (percent) (M)
0	47
10	55
20	62
25	74
29	79

Source: *Statistical Abstract of the
United States, 2001*, Table 938,
p. 597.

According to the model, in what year were
50 percent of the new privately owned homes
financed with a conventional mortgage?

 22.   **Home Square Footage**

**Percentage of New One-Family
Houses Under 1,200 Square Feet**

Years Since 1970 (t)	Percentage of New Houses (percent) (H)
0	36
10	21
20	11
25	10
29	7

Source: *Statistical Abstract of the
United States, 2001*, Table 938,
p. 597.

What is the concavity of the graph of the
model, and what does it mean in the context of
the problem?

 23.   **Home Square Footage**

**New One-Family Homes: Average
Number of Square Feet**

Years Since 1970 (t)	Houses (square feet) (H)
0	1,500
10	1,740
20	2,080
25	2,095
29	2,225

Source: *Statistical Abstract of
the United States, 2001*, Table
938, p. 597.

24. **Air Conditioning in Homes**

**New One-Family Homes
with Central Air Conditioning**

Years Since 1970 (t)	Percentage of New Houses (percent) (H)
0	34
10	63
20	76
25	80
29	84

Source: *Statistical Abstract of the United States, 2001*, Table 938, p. 597.

 25. **Homes with Garages**

**New One-Family Homes with
a Garage**

Years Since 1970 (t)	Percentage of New Houses (percent) (H)
0	58
10	69
20	82
25	84
29	87

Source: *Statistical Abstract of the United States, 2001*, Table 938, p. 597.

In what year were 90 percent of new one-family homes expected to have a garage?

 26. **Home Sales Price**

**Median Sales Price of a New
One-Family House in the
Southern United States**

Years Since 1980 (t)	Price (thousands of dollars) (P)
0	59.6
5	75.0
10	99.0
15	124.5
20	148.0

Source: *Statistical Abstract of the United States, 2001,* Table 940, p. 598.

 27. **Home Sales Price**

**Median Sales Price of a New
One-Family House in the
Northeastern United States**

Years Since 1980 (t)	Price (thousands of dollars) (P)
0	69.5
5	103.3
10	159.0
15	180.0
20	227.4

Source: *Statistical Abstract of the United States, 2001,* Table 940, p. 598.

28. **Home Sales Price**

Median Sales Price of a New One-Family House in the Western United States

Years Since 1980 (t)	Price (thousands of dollars) (P)
0	72.3
5	92.6
10	147.5
15	141.4
20	196.4

Source: *Statistical Abstract of the United States, 2001*, Table 940, p. 598.

  29. **Vehicle Leasing**

Percentage of Households Leasing Vehicles

Years Since 1989 (t)	Percentage of Households (percent) (P)
0	2.5
3	2.9
6	4.5
9	6.4

Source: *Statistical Abstract of the United States, 2001*, Table 1086, p. 680.

In what year did the percentage of households leasing vehicles reach 5 percent?

 30. **Fuel Consumption**

Motor Fuel Consumption of Vans, Trucks, and SUVs

Years Since 1980 (t)	Fuel Consumption (billions of gallons) (F)
0	23.8
5	27.4
10	35.6
15	45.6
19	52.8

Source: *Statistical Abstract of the United States, 2001*, Table 1105, p. 691.

 31. **Coffee Sales**

Starbucks Corporation Sales

Years Since 09/93 (t)	Income from Sales (millions of dollars) (S)
0	163.5
1	284.9
2	465.2
3	696.5
4	966.9
5	1,308.7
6	1,680.1
7	2,169.2
8	2,649.0
9	3,288.9

Source: moneycentral.msn.com.

According to the model, when did Starbucks Corporation sales reach $5000 million?

32. **Organic Cropland**

Certified Organic Cropland

Years Since 1992 (t)	Acres (thousands) (A)
0	403
1	465
2	557
3	639
4	850

Source: *Statistical Abstract of the United States, 2001*, Table 805, p. 526.

33. **Poultry Pricing**

Average Retail Price of Fresh Whole Chicken

Years Since 1985 (t)	Price (dollars per pound) (P)
0	0.78
5	0.86
6	0.86
7	0.88
8	0.91
9	0.90
10	0.94
11	1.00
12	1.00
13	1.06
14	1.05
15	1.08

Source: *Statistical Abstract of the United States, 2001*, Table 706, p. 468.

34. **Prison Rate**

Federal and State Prison Rate

Years Since 1980 (t)	Rate (prisoners per 100,000 people) (R)
0	139
2	171
4	188
6	217
8	247
10	297
12	332
14	389
16	427
18	461

Source: *Statistical Abstract of the United States, 2001*, Table 332, p. 200.

35. **College Enrollment**

Private College Enrollment

Years Since 1980 (t)	Students (in thousands) (S)
0	2,640
2	2,730
4	2,765
6	2,790
8	2,894
10	2,974
12	3,103
14	3,145
16	3,247
18	3,373

Source: *Statistical Abstract of the United States, 2001*, Table 205, p. 133.

36. **Military Personnel**

Active-Duty Military Personnel

Years Since 1990 (t)	Personnel (thousands) (P)
1	1,263
2	1,214
3	1,171
4	1,131
5	1,085
6	1,056
7	1,045
8	1,004
9	1,003

Source: *Statistical Abstract of the United States, 2001*, Table 499, p. 328.

   37. **Personal Income**

Per Capita Personal Income: California

Years Since 1993 (t)	Personal Income (dollars) (P)
0	22,833
1	23,348
2	24,339
3	25,373
4	26,521
5	28,240
6	29,772
7	32,149

Source: Bureau of Economic Analysis (www.bea.gov).

  38. **Personal Income**

Per Capita Personal Income: Colorado

Years Since 1993 (t)	Personal Income (dollars) (P)
0	22,196
1	23,055
2	24,289
3	25,514
4	27,067
5	28,764
6	30,206
7	32,434

Source: Bureau of Economic Analysis (www.bea.gov).

  39. **Personal Income**

Per Capita Personal Income: Louisiana

Years Since 1993 (t)	Personal Income (dollars) (P)
0	17,587
1	18,602
2	19,314
3	19,978
4	20,874
5	21,948
6	22,274
7	23,090

Source: Bureau of Economic Analysis (www.bea.gov).

40.   **Personal Income**

Per Capita Personal Income: Minnesota

Years Since 1993 (t)	Personal Income (dollars) (P)
0	21,903
1	23,241
2	24,295
3	25,904
4	27,086
5	29,092
6	30,105
7	31,935

Source: Bureau of Economic Analysis (www.bea.gov).

41.   **Pharmaceutical Company Profit**

Johnson & Johnson Pharmaceutical Operating Profit

Years Since 1997 (t)	Operating Profit (millions of dollars) (P)
0	2,332
1	3,114
2	3,735
3	4,394
4	4,928

Source: Johnson & Johnson Annual Report, 2001, p. 8.

In what year were Johnson & Johnson profits expected to reach $6000 million?

42.   **Pharmaceutical Income**

Johnson & Johnson Net Income

Years Since 1990 (t)	Net Income (millions of dollars) (I)
0	1,195
1	1,441
2	1,572
3	1,786
4	1,998
5	2,418
6	2,958
7	3,385
8	3,798
9	4,348
10	4,998
11	5,899

Source: Johnson & Johnson Annual Report, 2001, p. 18.

In what year was Johnson & Johnson net income projected to reach $7000 million?

43. **Company Payroll**

Payroll of Private-Employer Firms*

Years Since 1990 (t)	Payroll (billions of dollars) (F)
0	2,104
1	2,145
2	2,272
3	2,363
4	2,488
5	2,666
6	2,849
7	3,048
8	3,309

*Firms are an aggregation of all establishments owned by a parent company.

Source: *Statistical Abstract of the United States, 2001*, Table 726, p. 486.

 44. **Public-School Teachers**

Projected Number of Public-School Teachers

Years Since 2000 (t)	Public-School Teachers (thousands) (G)
0	2,850
1	2,865
2	2,877
3	2,891
4	2,905
5	2,914
6	2,919
7	2,927
8	2,932
9	2,937
10	2,940

Source: *Statistical Abstract of the United States, 2001*, Table 207, p. 134.

According to the model, in what year is the number of public-school teachers expected to reach 3 million?

 45. **Private-School Teachers**

Projected Number of Private-School Teachers

Years Since 2000 (t)	Private-School Teachers (thousands) (G)
0	402
1	403
2	404
3	405
4	407
5	408
6	409
7	410
8	411
9	411
10	412

Source: *Statistical Abstract of the United States, 2001*, Table 207, p. 134.

46. **Apparel Production Wages**

Apparel Production Workers' Wages

Years Since 1980 (t)	Wage of Men's and Boys' Furnishings Production Workers (dollars per hour) (M)
0	4.23
10	6.06
15	7.19
16	7.40
17	7.72
18	7.97
19	8.27
20	8.54

Source: *Statistical Abstract of the United States, 2001*, Table 609, p. 394.

 47. **Apparel Production Wages**

Apparel Production Workers' Wages

Years Since 1980 (t)	Wage of Women's and Misses' Outerwear Production Workers (dollars per hour) (W)
0	4.61
10	6.26
15	7.27
16	7.49
17	7.84
18	8.15
19	8.41
20	8.40

Source: *Statistical Abstract of the United States, 2001*, Table 609, p. 394.

 48. **McDonalds Restaurant Sales**

McDonalds Franchised Sales

Years Since 1990 (t)	Franchised Sales (millions of dollars) (S)
1	12,959
2	14,474
3	15,756
4	17,146
5	19,123
6	19,969
7	20,863
8	22,330
9	23,830
10	24,463
11	24,838

Source: www.mcdonalds.com.

 49. **Advertising Expenditures**

Advertising Expenditures: Billboards

Years Since 1990 (t)	Advertising Expenditures (millions of dollars) (A)
0	1,084
1	1,077
2	1,030
3	1,090
4	1,167
5	1,263
6	1,339
7	1,455
8	1,576
9	1,725
10	1,870

Source: *Statistical Abstract of the United States, 2001*, Table 1272, p. 777.

In what year were billboard advertising expenditures projected to reach $2 billion?

 50. **Advertising Expenditures**

Advertising Expenditures: Direct Mail

Years Since 1990 (t)	Advertising Expenditures (millions of dollars) (A)
0	23,370
1	24,460
2	25,392
3	27,266
4	29,638
5	32,866
6	34,509
7	36,890
8	39,620
9	41,403
10	44,715

Source: *Statistical Abstract of the United States, 2001*, Table 1272, p. 777.

In what year were direct mail advertising expenditures expected to reach $50 billion?

Exercises 51–55 are intended to challenge your understanding of quadratic functions.

51. Find two different quadratic functions that pass through the points $(0, 0)$ and $(4, 0)$.

52. A classmate claims that any group of three points defines a unique quadratic function. Is your classmate correct? Explain.

53. Is there a single quadratic function that passes through each of the following points: $(1, 3), (2, 5), (4, 9)$? Explain.

54. What is the graphical significance of the point with x coordinate $x = \dfrac{-b + \sqrt{b^2 - 4ac}}{2a}$ on the graph of $y = ax^2 + bx + c$?

55. If $ax^2 + bx + c > 3$ for all values of x, what do we know about a and the graph of $y = ax^2 + bx + c$?

5.2 Higher-Order Polynomial Function Models

- Use nonlinear functions to model real-life phenomena
- Interpret the meaning of mathematical models in their real-world context
- Model real-life data with higher-order polynomial functions

GETTING STARTED Since 1975, the average wage of sailors working on the East Coast of the United States has trailed that of sailors working on the West Coast (Table 5.7). At one point, West Coast sailors made nearly $1\frac{1}{2}$ times the wage of East Coast sailors.

TABLE 5.7   **Typical Basic Monthly Wage for Able-Bodied Seamen in Addition to Room and Board**

Years Since 1980 (t)	East Coast Wage (dollars) (E)	West Coast Wage (dollars) (W)
0	967	1,414
5	1,419	2,029
10	1,505	2,218
13	1,721	2,438
14	1,790	2,536
15	1,918	2,637
16	2,014	2,769
17	2,094	2,879
18	2,178	2,994
19	2,265	3,114
20	2,453	3,114

Source: *Statistical Abstract of the United States, 2001*, Table 1072, p. 674.

Salaries of both groups of sailors have increased substantially since 1980, although pay increases have been somewhat irregular. Will East Coast sailors ever make as much as West Coast sailors?

In this section, we will use *cubic* and *quartic* polynomials to model data. To keep computations simple, we use these models only if we're dissatisfied with the fit of simpler models.

Cubic Functions

A constant function is a polynomial of degree 0. A nonconstant linear function is a polynomial of degree 1. A quadratic function is a polynomial of degree 2. A **cubic function** is a polynomial of degree 3 and has the form $f(x) = ax^3 + bx^2 + cx + d$. The graph of a cubic function has exactly one **inflection point** (a point where the graph of the function changes concavity). The term **concavity** refers to the curvature of the graph. Recall that when the graph curves upward, we say that it is "concave up," and when the graph curves downward, we say that it is "concave down." The following rhyme is helpful in remembering the meaning of the terms.

Concave up is like a cup.
Concave down is like a frown.

The graphs in Figure 5.17 are concave up.

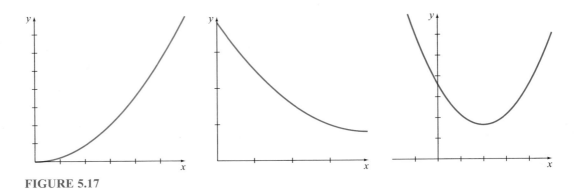

FIGURE 5.17

The graphs in Figure 5.18 are concave down.

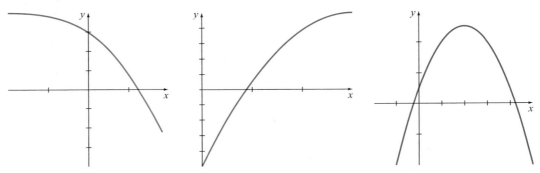

FIGURE 5.18

Cubic function graphs consist of two pieces: a concave up piece and a concave down piece. If $a > 0$, the graph of the cubic function is first concave down and then concave up. The graph of the cubic function will take on one of the basic shapes shown in Figure 5.19.

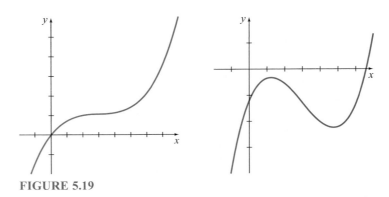

FIGURE 5.19

The size of the "bends" in the graph will vary depending upon the values of b and c. When a is positive, the graph of a cubic function will be increasing (rising from left to right) for sufficiently large values of x.

If $a < 0$, the graph of the cubic function is first concave up and then concave down. The graph of the cubic function will take on one of the shapes shown in Figure 5.20.

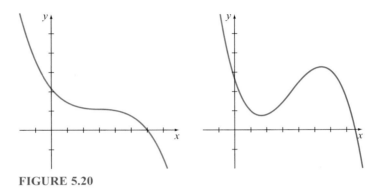

FIGURE 5.20

Again, the size of the "bends" in the graph will vary depending upon the values of b and c. When a is negative, the graph of a cubic function will be decreasing (falling from left to right) for sufficiently large values of x.

Returning to the sailors' wages, we can draw some conclusions by looking at a scatter plot of the data. Since a portion of the scatter plot appears linear, we first attempt to model the data with a linear function (Figure 5.21).

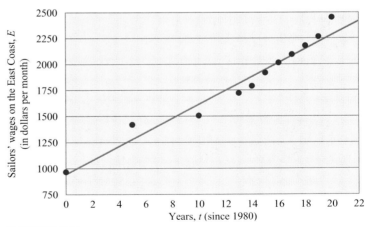

FIGURE 5.21

The linear model does not fit the data as well as we would like. We also note that from 1980 to 1990, the graph appears to be concave down, and from 1990 to 2000, the graph appears to be slightly concave up. We expect that a cubic model with $a > 0$ may fit the data well. Using the cubic regression feature on our calculator, we determine that the cubic function that best fits the data is

$$E(t) = 0.2792t^3 - 7.037t^2 + 102.6t + 983.6 \text{ dollars per month}$$

where t is the number of years since 1980. The model graph is shown in Figure 5.22.

FIGURE 5.22

The model fits the data fairly well, especially after 1993. The coefficient of determination of $r^2 = 0.9905$ further increases our confidence in the model. Let's now look at the scatter plot of the data from the West Coast (Figure 5.23).

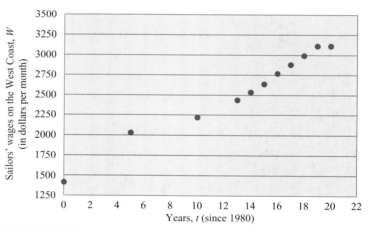

FIGURE 5.23

Again, the graph appears to be concave down from 1980 to 1990. From 1990 to 1999, the graph appears to be concave up. From 1999 to 2000, the wage did not change. Using the cubic regression feature on our calculator, we determine that the cubic equation that best fits the data is

$$W(t) = 0.2546t^3 - 7.277t^2 + 132.2t + 1437 \text{ dollars per month}$$

where t is the number of years since 1980. The model graph is shown in Figure 5.24.

FIGURE 5.24

Will the two models ever be equal? In 2017, the graphs intersect with a monthly wage of nearly $9100 (Figure 5.25)! Since 2017 is 17 years beyond the last point in our data set, we are somewhat skeptical of the result. Nevertheless, we can conclude that the wage gap isn't likely to be eliminated in the next couple of years.

FIGURE 5.25

The following Technology Tip details how to use your TI-83 Plus calculator to do cubic and quartic regression.

TECHNOLOGY **TIP**

Cubic and Quartic Regression

1. Enter the data using the Statistics Menu List Editor. (Refer to Section 1.3 if you've forgotten how to do this.)

L1	L2	L3	1
0	612	------	
5	967		
10	1419		
15	1505		
18	1721		
19	1790		
20	1918		

L1(1) = 0

2. Bring up the Statistics Menu Calculate feature by pressing [STAT] and using the blue arrows to move to the CALC menu. Then select item 6:CubicReg or item 7:QuadReg and press [ENTER].

```
EDIT CALC TESTS
1:1-Var Stats
2:2-Var Stats
3:Med-Med
4:LinReg(ax+b)
5:QuadReg
6:CubicReg
7↓QuartReg
```

3. If you want to automatically paste the regression equation into Y1 in the [Y=] editor, press the key sequence [VARS] Y-VARS; 1:Function; 1:Y1 and press [ENTER]. Otherwise, press [ENTER].

```
CubicReg
y=ax³+bx²+cx+d
a=.1643275034
b=-5.657940597
c=112.3153143
d=594.9549798
R²=.9927303408
```

EXAMPLE **1**

Choosing a Polynomial Function to Model a Company's Revenue

Electronic Arts, Inc., is one of the premier producers of interactive electronic games playable on game platforms such as the Sony PlayStation 2®, Microsoft Xbox®, Nintendo GameCube™, and Sony PlayStation. Despite the sluggish economy in the early 2000s, Electronic Arts showed an overall strong growth in revenue (Table 5.8).

TABLE 5.8

Years Since End of Fiscal Year 1998 (t)	Net Revenues (millions of dollars) $[R(t)]$
0	908.852
1	1221.863
2	1420.011
3	1322.273
4	1724.675
5	2482.244

Source: Electronic Arts, Inc., Annual Reports, March 31, 2002 and 2004.

Find a polynomial model for the annual revenue of Electronic Arts, Inc., and forecast in what year the revenue will reach $3 billion.

SOLUTION From the scatter plot, it appears that a cubic function will best fit the data (Figure 5.26). Using cubic regression, we determine that the cubic model that best fits the data is

$$R(t) = 43.87t^3 - 274.8t^2 + 594.7t + 900.0 \text{ million dollars}$$

where t is the number of years since the end of fiscal year 1998.

FIGURE 5.26

Since $3 billion is the same as $3000 million, we must determine when $R(t) = 3000$. To do this, we simultaneously graph $R(t) = 43.87t^3 - 274.8t^2 + 594.7t + 900.0$ and $y = 3000$. Then, using the intersection feature on our graphing calculator, we determine when the two functions will be equal (Figure 5.27).

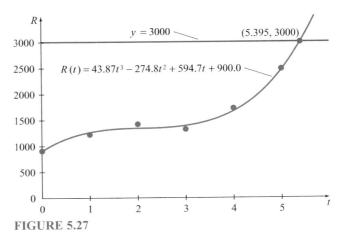

FIGURE 5.27

According to the model, 5.395 years after the end of fiscal year 1998, the annual revenue is expected to reach $3 billion. To add meaning to the result, we convert 0.395 years into months.

$$0.395 \text{ years} \cdot \frac{12 \text{ months}}{1 \text{ year}} \approx 5 \text{ months}$$

During the one-year period prior to the end of the fifth month of fiscal year 2004, we anticipate that $3 billion in revenues will be earned. (*Note:* Since $t = 5$ is the end of fiscal year 2003, $t = 5.395$ is in fiscal year 2004.)

Quartic Functions

A fourth-degree polynomial is a function of the form $f(x) = ax^4 + bx^3 + cx^2 + dx + e$ and is called a **quartic** function. A quartic function may have zero or two inflection points. A quartic graph with $a > 0$ will have one of the basic shapes shown in Figure 5.28.

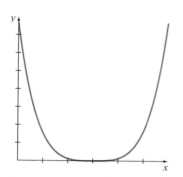

Concave up everywhere
No inflection points
FIGURE 5.28

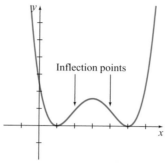

Concave up, then down, then up
Two inflection points

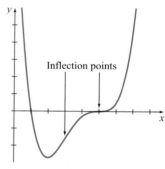

Concave up, then down, then up
Two inflection points

Notice that each graph opens upward. The size of the "bends" in the graph will vary depending upon the values of b, c, and d. When a is positive, the graph of a quartic function will be increasing (rising from left to right) for sufficiently large values of x.

A quartic graph with $a < 0$ will have one of the basic shapes shown in Figure 5.29.

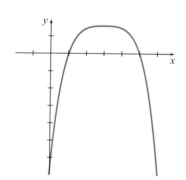

Concave down everywhere
No inflection points
FIGURE 5.29

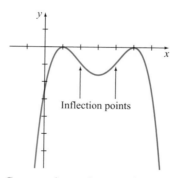

Concave down, then up, then down
Two inflection points

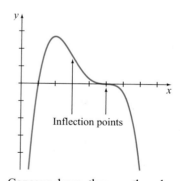

Concave down, then up, then down
Two inflection points

Notice that each graph opens downward. Again, the size of the "bends" in the graph will vary depending upon the values of b, c, and d. When a is negative, the graph of a quartic function will be decreasing (falling from left to right) for sufficiently large values of x.

EXAMPLE **2**

Choosing a Nonlinear Function to Model Interest Rates

The average fixed interest rate for a conventional mortgage on a new home fluctuated between 1985 and 1995 as shown in Table 5.9.

TABLE 5.9   Average Fixed Interest Rates on New Home Conventional Mortgages

Years Since 1985 (t)	Percentage Rate (percent) (R)
0	11.90
2	9.50
4	10.20
6	9.32
8	7.27
10	7.95

Source: *Statistical Abstract of the United States, 2001*, Table 1185, p. 733.

Model the data with a nonlinear function. Do you think the model will be a good indicator of future interest rates? Explain.

SOLUTION

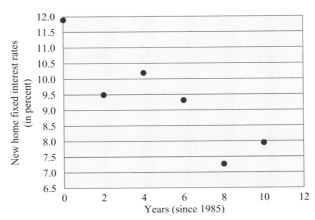

FIGURE 5.30

As shown in Figure 5.30, the graph appears to be concave up between 1985 and 1988, concave down from 1988 to 1992, and concave up from 1992 to 1995. A quartic model may work well (Figure 5.31).

FIGURE 5.31

Using the quartic regression feature on our calculator, we determine that fixed interest rates for mortgages on new homes between 1985 and 1995 may be modeled by the equation

$$R(t) = 0.0112t^4 - 0.2242t^3 + 1.403t^2 - 3.1796t + 11.89 \text{ percent}$$

where t is the number of years since the end of 1985.

The model shows that interest rates will rise rapidly every year after 1995. Let's calculate the predicted rate at the end of 1997 ($t = 12$).

$$R(12) = 0.0112(12)^4 - 0.2242(12)^3 + 1.403(12)^2 - 3.1796(12) + 11.89$$
$$= 20.59$$

According to the model, interest rates reached 20.59 percent in 1997. This figure seems a bit unrealistic. (Between 1970 and 2000, interest rates remained below 15 percent.) The model may be a better indicator for intermediate years, such as 1992 or 1994.

EXAMPLE   3

Choosing a Function to Model the Number of Mutual Funds

The number of mutual funds available to investors has increased dramatically since 1970, as shown in Table 5.10.

TABLE 5.10   Number of Mutual Funds

Years Since 1970 (t)	Mutual Funds (M)
0	361
2	410
4	431
6	452
8	505
10	564
12	857
14	1,241
16	1,835
18	2,708
20	3,081
22	3,826
24	5,330
26	6,254
28	7,314
30	8,171

Source: *Statistical Abstract of the United States, 2001*, Table 1214, p. 744.

Find a function that models the data and estimate the number of mutual funds in 2005.

SOLUTION We could model the data with a quadratic function (Figure 5.32); however, even though the correlation coefficient is close to 1, the quadratic model doesn't fit the data very well between 1970 and 1980.

$y = 12.65x^2 - 117.4x + 545.3$

FIGURE 5.32

A quartic model will give a better fit (Figure 5.33).

FIGURE 5.33

The number of mutual funds is modeled by the quartic equation

$$M(t) = -0.02524t^4 + 1.519t^3 - 16.18t^2 + 62.46t + 351.2 \text{ mutual funds}$$

where t is the number of years since 1970. For the year 2005, $t = 35$. To estimate the number of mutual funds in 2005, we evaluate $M(t)$ at $t = 35$.

$$M(35) = -0.02524(35)^4 + 1.519(35)^3 - 16.18(35)^2 + 62.46(35) + 351.2$$
$$= 9968$$

According to the model, in 2005 ($t = 35$), there will be 9968 mutual funds. Given the historically rapid growth in the number of mutual funds, this estimate seems reasonable.

5.2 Summary

In this section, you learned how to use cubic and quartic polynomials to model data. You also discovered the importance of being cautious in using model predictions.

5.2 Exercises

In Exercises 1–20, find the quadratic, cubic, or quartic model that best fits the data. If multiple models appear to fit the data well, use the simplest model to answer the questions.

  1. **Public Ninth- to Twelfth-Grade Enrollment**

Years Since 1984 (t)	Students (in thousands) (S)
0	12,304
1	12,388
2	12,333
3	12,076
4	11,687
5	11,390
6	11,338
7	11,541
8	11,735
9	11,961
10	12,213
11	12,500
12	12,847
13	13,054
14	13,193
15	13,369

Source: *Statistical Abstract of the United States, 2001,* Table 232, p. 149.

According to your model, how many students were enrolled in high school (ninth to twelfth grades) in 2000?

  2. **Student-to-Teacher Ratio at Private Elementary and Secondary Schools**

Years Since 1960 (t)	Students per Teacher (R)
0	30.7
5	28.3
10	23.0
11	22.6
12	21.6
13	21.2
14	20.4
15	19.6
20	17.7
25	16.2
26	15.7
27	15.5
28	15.2
29	14.2
30	14.7
31	14.7
32	14.6
33	14.8
34	14.9
35	14.9
36	14.9
37	15.1
38	15.2

Source: *Statistical Abstract of the United States, 2001,* Table 235, p. 150.

According to your model, what will be the student-to-teacher ratio in 2006? Does this seem reasonable? Explain.

3. **Public-School Teachers**

**Public Elementary- and
Secondary-School Teachers**

Years Since 1980 (t)	Teachers (thousands) (R)
0	2,211
1	2,192
2	2,158
3	2,134
4	2,144
5	2,175
6	2,215
7	2,249
8	2,282
9	2,324
10	2,362
11	2,409
12	2,429
13	2,466
14	2,512
15	2,568
16	2,607
17	2,673
18	2,745
19	2,813
20	2,886

Source: *Statistical Abstract of
the United States, 2001,*
Table 237, p. 151.

According to the model, in what year is the number of teachers expected to reach 3 million?

 4. **Principals' Salaries**

**High School Principals'
Salaries**

Years Since 1980 (t)	Salary (dollars) (S)
0	29,207
1	32,231
2	34,776
3	37,602
4	39,334
5	42,094
6	44,986
7	47,896
8	50,512
9	52,987
10	55,722
11	59,106
12	61,768
13	63,054
14	64,993
15	66,596
16	69,277
17	72,410
18	74,380
19	76,768
20	79,839

Source: *Statistical Abstract of
the United States, 2001,*
Table 238, p. 152.

According to the model, when will the average high school principal's salary reach $100,000?

5. **School Internet Access**

Percentage of Public-School Classrooms with Internet Access

Years Since 1994 (t)	Percentage of Classrooms (percent) (C)
0	3
1	8
2	14
3	27
4	51
5	64
6	77

Source: *Statistical Abstract of the United States, 2001*, Table 243, p. 155.

According to the model, what percentage of classrooms had Internet access in 2001? In 2002? Do the model estimates seem reasonable? Explain.

 6. **Banks**

Number of Different Banks (Not Bank Branches)

Years Since 1984 (t)	Banks (B)
0	17,900
1	18,033
2	17,876
3	17,325
4	16,562
5	15,829
6	15,192
7	14,517
8	13,891
9	13,261
10	12,641
11	12,002
12	11,478
13	10,923
14	10,463
15	10,221
16	9,908

Source: *Statistical Abstract of the United States, 2001*, Table 1173, p. 728.

According to the model, when will the number of banks drop below 9000? What do you think is causing the number of different banking companies to decrease?

7. **Debit Cards**

Number of Debit Cards

Years Since 1990 (t)	Cards (millions) (C)
0	164
5	201
6	205
7	211
8	217
9	228

Source: *Statistical Abstract of the United States, 2001,* Table 1189, p. 734.

A consulting firm has projected that the number of debit cards will reach 270 million by 2005. Does this projection agree with your model?

 8. **NASDAQ Volume**

NASDAQ Average Daily Volume

Years Since 1980 (t)	Shares (millions) (S)
0	27
2	33
4	60
6	114
8	123
10	132
12	191
14	295
16	544
18	802
20	1,757

Source: *Statistical Abstract of the United States, 2001,* Table 1205, p. 741.

According to your model, when will the NASDAQ market have an average daily volume of 2 billion shares? Visit www.marketdata.nasdaq.com to see if your model estimate is accurate.

 9. **Stock Trading**

New York Stock Exchange Shares Traded

Years Since 1980 (t)	Shares (millions) (S)
0	11,562
2	16,669
4	23,309
6	36,009
8	41,118
10	39,946
12	51,826
14	74,003
16	105,477
18	171,188
20	265,499

Source: *Statistical Abstract of the United States, 2001,* Table 1207, p. 742.

According to the model, how many shares will be traded in 2005? Does this seem reasonable? Explain.

10. **Listed Stocks**

**Companies with Stocks
Listed on the NYSE**

Years Since 1980 (t)	Companies (C)
0	1,570
2	1,526
4	1,543
6	1,575
8	1,681
10	1,774
12	2,088
14	2,570
16	2,907
18	3,114
20	2,862

Source: *Statistical Abstract of the United States, 2001,* Table 1207, p. 742.

According to the model, in which years is the number of companies listed on the New York Stock Exchange expected to exceed 2700 companies?

 11. **Computer Retailers' Wages**

**Average Annual Wage per
Worker for Computer and
Equipment Retailers**

Years Since 1992 (t)	Wages (dollars) (W)
1	32,200
2	30,500
3	32,100
4	33,800
5	35,000
6	37,300

Source: *Statistical Abstract of the United States, 2001,* Table 1123, p. 703.

In what year do you expect the average wage to exceed $40,000?

 12. **Computer Wholesalers' Wages**

**Average Annual Wage per
Worker for Computer and
Equipment Wholesalers**

Years Since 1992 (t)	Wages (dollars) (W)
1	52,500
2	52,900
3	52,900
4	54,300
5	56,700
6	62,200

Source: *Statistical Abstract of the United States, 2001,* Table 1123, p. 703.

Calculate the average wage in 2000. Does this seem reasonable? Explain.

 13. **Software Developers' Wages**

**Average Wage per Worker
at Prepackaged Software
Development Firms**

Years Since 1992 (t)	Wages (dollars) (W)
0	57,000
1	54,500
2	57,000
3	63,700
4	70,100
5	79,200
6	94,100

Source: *Statistical Abstract of the United States, 2001,* Table 1123, p. 703.

In what year will the wage reach $120,000?

14. Software Retailers' Wages

**Average Wage per Worker at
Prepackaged Software Retailers**

Years Since 1992 (*t*)	Wages (dollars) (*W*)
0	32,200
1	30,500
2	32,100
3	33,800
4	35,000
5	37,300
6	40,400

Source: *Statistical Abstract of
the United States, 2001,*
Table 1123, p. 703.

What do you expect the average worker's
wage was in 1999?

  15. Software Wholesalers' Wages

**Average Wage per Worker at
Prepackaged Software Wholesalers**

Years Since 1992 (*t*)	Wages (dollars) (*W*)
0	52,500
1	52,900
2	52,900
3	54,300
4	56,700
5	62,200
6	69,700

Source: *Statistical Abstract of
the United States, 2001,*
Table 1123, p. 703.

Compare the wholesalers' average wage
model with the retailers' average wage model
from Exercise 14. According to the models, will
wages at retailers ever catch up with wages at
wholesalers?

  16. Radio Broadcasters' Wages

**Average Annual Wage per Worker
in Radio Broadcasting**

Years Since 1992 (*t*)	Annual Wage (dollars) (*W*)
0	23,500
1	24,300
2	26,000
3	27,200
4	29,300
5	31,300
6	34,200

Source: *Statistical Abstract of the
United States, 2001,* Table 1123,
p. 703.

According to the model, what was the average
wage of a radio broadcasting employee in 2006?
Does this estimate seem reasonable? Explain.

  17. Television Broadcasters' Wages

**Average Annual Wage per Worker
in Television Broadcasting**

Years Since 1992 (*t*)	Annual Wage (dollars) (*W*)
0	41,400
1	42,200
2	43,700
3	47,200
4	51,100
5	51,000
6	54,600

Source: *Statistical Abstract of the
United States, 2001,* Table 1123,
p. 703.

According to the model, when will the
average annual wage for television broadcasting
employees exceed $60,000? Does this estimate
seem reasonable? Explain.

18. **Internet Usage**

Projected Internet Usage, Hours per Person per Year (Based on 1995–1999 Data)

Years Since 1995 (t)	Usage per Person (hours per year) (H)
0	5
1	10
2	34
3	61
4	99
5	135
6	162
7	187
8	208
9	228

Source: *Statistical Abstract of the United States, 2001*, Table 1125, p. 704.

What do you project the Internet usage will be in 2006? Do you believe your projection is a good estimate? Explain.

  19. **Homes with Cable**

Percentage of TV Homes with Cable

Years Since 1970 (t)	TV Homes with Cable (percent) (C)
0	6.7
5	12.6
10	19.9
15	42.8
16	45.6
17	47.7
18	49.4
19	52.8
20	56.4
21	58.9

(Continued)

(Continued)

22	60.2
23	61.4
24	62.4
25	63.4
26	65.3
27	66.5
28	67.2
29	67.5

Source: *Statistical Abstract of the United States, 2001*, Table 1126, p. 705.

According to the model, what percentage of TV homes had cable in 2002? Does this seem reasonable? Explain.

20. **Homes with VCRs**

Percentage of TV Homes with a VCR

Years Since 1980 (t)	Homes with a VCR (percent) (C)
0	1.1
5	20.8
6	36.0
7	48.7
8	58.0
9	64.6
10	68.6
11	71.9
12	75.0
13	77.1
14	79.0
15	81.0
16	82.2
17	84.2
18	84.6
19	84.6

Source: *Statistical Abstract of the United States, 2001*, Table 1126, p. 705.

Using the model from Exercise 19, determine in what year the percentage of homes with VCRs and the percentage of homes with cable was the same. (*Hint:* The meaning of t in the two exercises is different.)

Exercises 21–25 are intended to challenge your understanding of higher-order polynomial functions.

21. A "bend" in a graph occurs when the graph changes from increasing to decreasing or from decreasing to increasing. A quadratic function has exactly one bend. A cubic function has zero or two bends. A quartic function has one or three bends. In general, how many bends may the graph of a polynomial of degree n have?

22. Can the number of x-intercepts of a graph of a polynomial function ever exceed the degree of the polynomial? Explain.

23. A function of the form $f(x) = x^n$ is called a **power function.** If n is a positive even integer, what is the relationship between $f(x)$ and $f(-x)$?

24. Find the equation of the cubic function that passes through the points $(-1, 0)$, $(0, 0)$, $(1, 0)$, and $(3, 24)$.

25. Find the equation of the quartic function that passes through the points $(-1, 0)$, $(0, 0)$, $(1, 0)$, $(2, 0)$, and $(3, 24)$.

5.3 Exponential Function Models

- Use nonlinear functions to model real-life phenomena
- Interpret the meaning of mathematical models in their real-world context
- Graph exponential functions
- Find the equation of an exponential function from a table
- Model real-life data with exponential functions

GETTING STARTED Michael Jordan is arguably the most renowned athlete of all time. With millions of dollars in endorsements in addition to his multimillion-dollar salary, he is one of the wealthiest athletes in history. Since his entry into the NBA in 1984, the average salary for NBA players has increased *exponentially*.

In this section, we will demonstrate how exponential functions can be used to model rapidly increasing data sets such as the average salary for NBA players. We will develop exponential models from tables of data and verbal descriptions. In addition, we will show what an exponential function graph looks like.

The average annual salary of an NBA player increased from $170 thousand in 1980 to $2.6 million in 1998, as shown in Table 5.11.

If we plot the NBA salary data together with the graph of the function $S(t) = 161.4(1.169)^t$, we see that the graph of S fits the data fairly well (Figure 5.34).

TABLE 5.11   Average NBA Athlete's Salary

Years Since 1980 (t)	Annual Salary (thousands of dollars) (S)
0	170
5	325
10	750
15	1,900
16	2,000
17	2,200
18	2,600

Source: *Statistical Abstract of the United States, 2001,* Table 1324, p. 829.

FIGURE 5.34

Notice that the independent variable, t, appears as an exponent. The function S is called an **exponential function.**

EXPONENTIAL FUNCTION

If a and b are real numbers with $a \neq 0$, $b > 0$, and $b \neq 1$, then the function

$$y = ab^x$$

is called an **exponential function.** The value b is called the **base** of the exponential function.

Why must b be positive? Consider the function $y = (-1)^x$. For integer values, y oscillates between -1 and 1 (Table 5.12). However, the function is undefined for numerous noninteger values of x (Table 5.13).

TABLE 5.12

x	y
-2	1
-1	-1
0	1
1	-1
2	1

TABLE 5.13

x	y
-0.5	Undefined
-0.3	Undefined
0.1	Undefined
1.4	-1
1.5	Undefined

On the other hand, if $b > 0$, then the value of b^x is defined for *all* integer and noninteger values of x.

Why don't we allow b to be 1? If $b = 1$, then

$$y = ab^x$$

$$y = a(1)^x$$

$$y = a \qquad \text{Since } (1)^x = 1 \text{ for all } x$$

The graph of the function $y = a$ is a horizontal line and does not exhibit the same graphical behavior as all other functions of the form $y = ab^x$ with $b > 0$. By eliminating the case of $b = 1$, we are able to talk about a family of like functions.

Exponential functions are used frequently to model growth and decay situations, such as growth in population, depreciation of a vehicle, or growth in a retirement account. When the growth or decay of a quantity is modeled by an exponential function $y = ab^x$, the independent variable is frequently time. The beginning value of the quantity at time $x = 0$ is referred to as the **initial value** of the function and, as demonstrated below, is equal to a.

$$y = ab^0$$

$$= a(1) \qquad \text{Since } b^0 = 1 \text{ for all nonzero real numbers } b$$

$$= a$$

Graphically speaking, the constant a in $y = ab^x$ corresponds with the y-intercept of the exponential graph. In most real-life applications, a will be positive. The initial value of the NBA player salary function $S(t) = 161.4(1.169)^t$ is 161.4. That is, the model estimates that the average NBA player's salary was about \$161 thousand in the year $t = 0$ (1980).

The base b of $y = ab^x$ is often referred to as the **growth factor** of the function. Increasing the value of x by one unit increases y by a factor of b. The growth factor of the NBA player salary function $S(t) = 161.4(1.169)^t$ is 1.169, since $b = 1.169$. That is, the average NBA player's salary is increasing by a factor of 1.169 annually. Next year's salary is forecast to be 1.169 times this year's salary.

Recall that the x-intercept of a function occurs when $y = 0$. For what values of x does $ab^x = 0$? Let's consider the exponential function $y = 2^x$. We'll generate a table of values and plot a few points to get an idea about what is happening graphically (Figure 5.35).

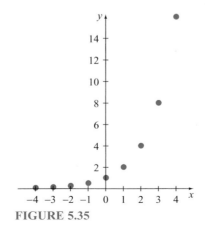

FIGURE 5.35

x	$y = 2^x$
-4	$\dfrac{1}{16}$
-3	$\dfrac{1}{8}$
-2	$\dfrac{1}{4}$
-1	$\dfrac{1}{2}$
0	1
1	2
2	4
3	8
4	16

Observe that as x increases, y also increases. Also, observe that y is positive for all values of x. That is, *there are no x-intercepts*. This remains true even if we pick a negative number with a larger magnitude, say $x = -33$.

$$2^{-33} = \frac{1}{2^{33}}$$

$$2^{-33} = \frac{1}{8,589,934,592}$$

We call the line $y = 0$ a **horizontal asymptote** of the function $y = 2^x$. For sufficiently small values of x, the graph of $y = 2^x$ approaches the graph of the line $y = 0$. All exponential functions of the form $y = ab^x$ have a horizontal asymptote at $y = 0$.

Exponential Function Graphs

Exponential function graphs will take on one of the four basic shapes specified in Table 5.14. Each graph has a horizontal asymptote at $y = 0$ and a y-intercept at $(0, a)$.

TABLE 5.14 Exponential Function Graphs: $y = ab^x$

Value of a	Value of b	Concavity of Graph	Increasing/ Decreasing	Sample Graph
$a > 0$	$b > 1$	Concave up	Increasing	
$a > 0$	$0 < b < 1$	Concave up	Decreasing	
$a < 0$	$b > 1$	Concave down	Decreasing	
$a < 0$	$0 < b < 1$	Concave down	Increasing	

EXAMPLE **1** **Comparing Exponential Graphs**

Compare and contrast the graphs of $f(x) = 3(2)^x$ and $g(x) = 4(0.5)^x$. (You may sketch the graphs by hand by plotting several points, or you may use technology to graph the functions.)

SOLUTION Both f and g are exponential functions. In the function equation of f, $a = 3$ and $b = 2$. The graph of f is concave up, since $a > 0$, and is increasing, since $b > 1$. The graph of f has a y-intercept at $(0, 3)$.

In the function equation of g, $a = 4$ and $b = 0.5$. The graph of g is concave up, since $a > 0$, and is decreasing, since $b < 1$. The graph of g has a y-intercept at $(0, 4)$. The graphs of both functions have a horizontal asymptote at $y = 0$ (Figure 5.36).

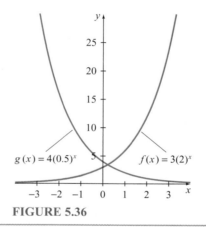

FIGURE 5.36

Properties of Exponents

As seen earlier in this section, it is often necessary to apply the properties of exponents when working with exponential functions. Since these properties are typically covered in depth in an algebra course, we will only summarize them here.

PROPERTIES OF EXPONENTS

If b, m, and n are real numbers with $b > 0$, the following properties hold.

Property	Example
1. $b^{-n} = \dfrac{1}{b^n}$	1. $2^{-3} = \dfrac{1}{2^3}$
2. $b^m \cdot b^n = b^{m+n}$	2. $3^2 \cdot 3^4 = 3^6$
3. $\dfrac{b^m}{b^n} = b^{m-n}$	3. $\dfrac{5^6}{5^4} = 5^2$
4. $b^{mn} = (b^m)^n = (b^n)^m$	4. $6^{2 \cdot 3} = (6^2)^3 = (6^3)^2$

Finding an Exponential Function from a Table

We can easily determine if a table of values models an exponential function by calculating the ratio of consecutive outputs for evenly spaced inputs. The ratio will be constant if the table of values models an exponential function. Consider Table 5.15.

The domain values are equally spaced (two units apart). Calculating the ratios of the consecutive range values, we get

TABLE 5.15

x	y
0	6
2	24
4	96
6	384

$$\frac{24}{6} = 4 \qquad \frac{96}{24} = 4 \qquad \frac{384}{96} = 4$$

In each case, the ratio was 4. Therefore, the table of values models an exponential function. But what is the equation of the function?

We can find the equation of the function algebraically. We know that an exponential function must be of the form $y = ab^x$. Substituting the point $(0, 6)$ into the equation and solving, we get

$$y = ab^x$$
$$6 = ab^0 \qquad \text{Substitute } x = 0 \text{ and } y = 6$$
$$6 = a$$

Since $a = 6$, $y = 6b^x$. Substituting the point $(2, 24)$ into the equation $y = 6b^x$, we get

$$y = 6b^x$$
$$24 = 6b^2 \qquad \text{Substitute } x = 2 \text{ and } y = 24$$
$$4 = b^2$$
$$b = 2 \qquad b \neq -2 \text{ since the base of an exponential function} \\ \text{must be positive}$$

Therefore, the exponential function that models the table data is $y = 6(2)^x$.

EXAMPLE 2

Finding an Exponential Equation from a Table

Find the equation of the exponential function modeled by Table 5.16.

TABLE 5.16

x	y
1	21
2	63
3	189
4	567

SOLUTION   We know that an exponential function must be of the form $y = ab^x$. Substituting the point $(2, 63)$ into the equation, we get

$$y = ab^x$$
$$63 = ab^2 \qquad \text{Substitute } x = 2 \text{ and } y = 63$$

Substituting the point $(1, 21)$ into the equation $y = ab^x$, we get

$$y = ab^x$$
$$21 = ab^1 \qquad \text{Substitute } x = 1 \text{ and } y = 21$$

We can eliminate the a variable by dividing the first equation by the second equation. (*Note:* What we are really doing is dividing both sides of the first equation by the same nonzero quantity, expressed in two different forms.)

$$\frac{63}{21} = \frac{ab^2}{ab^1}$$

$$3 = \frac{a}{a} \cdot b^{2-1} \qquad \text{Since } \frac{b^m}{b^n} = b^{m-n}$$

$$3 = 1 \cdot b$$

$$b = 3$$

We may then substitute the value of b into either equation to find a.

$$21 = ab^1$$

$$21 = a \cdot 3^1 \qquad \text{Substitute } b = 3$$

$$\frac{21}{3} = a$$

$$a = 7$$

The exponential function that models the table data is $y = 7(3)^x$.

When using the method of dividing one equation by the other, computations will tend to be easier if we divide the equation with the largest exponent by the equation with the smallest exponent.

The technique shown in Example 2 may be generalized for all exponential functions. Using the result from the generalized solution will allow us to determine the exponential equation more quickly.

Consider an exponential function $y = ab^x$ whose graph goes through the points (x_1, y_1) and (x_2, y_2). We have

$$y = ab^x$$

$$y_1 = ab^{x_1} \qquad \text{Substitute } x = x_1 \text{ and } y = y_1$$

and

$$y = ab^x$$

$$y_2 = ab^{x_2} \qquad \text{Substitute } x = x_2 \text{ and } y = y_2$$

Dividing the second equation by the first equation yields a quick way to calculate b.

$$\frac{y_2}{y_1} = \frac{ab^{x_2}}{ab^{x_1}}$$

$$\frac{y_2}{y_1} = \frac{a}{a} b^{x_2 - x_1}$$

$$\frac{y_2}{y_1} = b^{x_2 - x_1}$$

$$\left(\frac{y_2}{y_1}\right)^{\frac{1}{x_2 - x_1}} = \left(b^{x_2 - x_1}\right)^{\frac{1}{x_2 - x_1}}$$

$$\left(\frac{y_2}{y_1}\right)^{\frac{1}{x_2 - x_1}} = b^{\frac{x_2 - x_1}{x_2 - x_1}}$$

$$b = \left(\frac{y_2}{y_1}\right)^{\frac{1}{x_2 - x_1}}$$

Using the points $(1, 21)$ and $(3, 189)$ from Example 2, we get

$$b = \left(\frac{y_2}{y_1}\right)^{\frac{1}{x_2 - x_1}}$$

$$b = \left(\frac{189}{21}\right)^{\frac{1}{3-1}} \qquad \text{Substitute } x_1 = 0, x_2 = 2, y_1 = 21, \text{ and } y_2 = 189$$

$$= 9^{1/2}$$

$$= \sqrt{9} \qquad \text{Recall that for } b \geq 0, b^{1/2} = \sqrt{b}$$

$$= 3$$

This method is especially useful when checking to see if a data table with unequally spaced inputs is an exponential function. If it is an exponential function, the value of b will be constant regardless of which two points we substitute into the formula.

Using Exponential Regression to Model Data

Data sets with near-constant ratios of change may be modeled using the exponential regression feature on our graphing calculator. It is often helpful to do a scatter plot of the data first to see if the graph looks like an exponential function.

EXAMPLE 3

Using Exponential Regression to Model the Population of Akron, Ohio

The population of Akron, Ohio, is shown in Table 5.17. Model the population with an exponential function and forecast the population of Akron in 2010.

TABLE 5.17

Years Since 1970 (x)	Population (thousands) (y)
0	275
10	237
20	223
30	217

Source: *Statistical Abstract of the United States, 2001,* Table 34, p. 34.

SOLUTION We first draw a scatter plot so that we can visually predict whether an exponential function will fit the data well (Figure 5.37). (Note that because we have zoomed in on the data, the "x axis" is given by $y = 210$ instead of $y = 0$.)

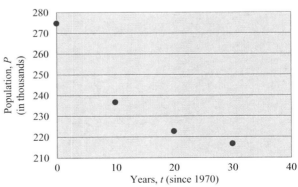

FIGURE 5.37

Since the scatter plot is decreasing and concave up, an exponential function may fit the data well. Using techniques given in the Technology Tip following this example, we can calculate an exponential model for the data set (Figure 5.38).

FIGURE 5.38

TABLE 5.18

Year (x)	Population −210 (thousands) (y)
0	65
10	27
20	13
30	7

The model $y = 266.1(0.9923)^x$ is a terrible fit for the data. What happened? Recall that an exponential function of the form $y = ab^x$ has a horizontal asymptote at $y = 0$. From the scatter plot of the data, it appears that the population is approaching a constant value of $y = 210$ instead of $y = 0$. (This value is not precise. We are *guessing* what the horizontal asymptote would be based on the scatter plot of the data set.) If we subtract 210 from each of the y values, we get a new data set with a horizontal asymptote of $y = 0$. This is referred to as an **aligned data set** (Table 5.18).

Using exponential regression to find a model for the aligned data set yields $y = 60.81(0.9285)^x$, a function that fits the aligned data well (Figure 5.39).

FIGURE 5.39

Subtracting 210 from each of the y values of the original data set had the graphical implication of moving the data down 210 units. To return the aligned data model to the position of the original data, we need to move the model up 210 units. To do this, we add back the 210.

$$P(x) = 60.81(0.9285)^x + 210 \qquad \text{Add 210 to shift the data graph upward}$$

The model created from the aligned data fits those data much better than the unaligned data model (Figure 5.40).

FIGURE 5.40

To estimate the population of Akron, Ohio, in 2010, we evaluate $P(x) = 60.81(0.9285)^x + 210$ at $x = 40$.

$$P(40) = 60.81(0.9285)^{40} + 210$$

$$\approx 213$$

We forecast that the population of Akron, Ohio, in 2010 will be 213,000.

The methods of aligning data and of performing exponential regression are detailed in the following Technology Tips.

TECHNOLOGY TIP

Aligning a Data Set

1. Enter the data using the Statistics Menu List Editor. (Refer to Section 1.3 if you've forgotten how to do this.)

2. Move the cursor to the top of L3. We want the entries in L3 to equal the entries in L2 minus the amount of the vertical shift (in this case 210). To do this, we must enter the equation L3=L2-210.

(Continued)

3. Press 2nd then 2 to place L2 on the equation line at the bottom of the viewing window. Then press − and 2 1 0 to subtract 210.

L1	L2	■3	3
0	275	------	
10	237		
20	223		
30	217		
------	------		

L3 =L2−210

4. Press ENTER to display the list of aligned values in L3.

L1	L2	L3	3
0	275	65	
10	237	27	
20	223	13	
30	217	7	
------	------	▄▄▄▄▄▄	

L3(5) =

TECHNOLOGY TIP

Exponential Regression

1. Enter the data using the Statistics Menu List Editor. (Refer to Section 1.3 if you've forgotten how to do this.)

L1	L2	L3	3
0	275	65	
10	237	27	
20	223	13	
30	217	7	
------	------	▄▄▄▄▄▄	

L3(5) =

2. Bring up the Statistics Menu Calculate feature by pressing STAT and using the blue arrows to move to the CALC menu. Then select item Ø:ExpReg and press ENTER.

```
EDIT CALC TESTS
7↑QuartReg
8:LinReg(a+bx)
9:LnReg
ØEExpReg
A:PwrReg
B:Logistic
C:SinReg
```

3. If the data to be evaluated are in L1 and L2, press ENTER. Otherwise, go to step 4. (If you want to automatically paste the regression equation into the Y= editor, press the key sequence VARS Y-VARS; 1:Function; 1:Y1 before pressing ENTER.)

```
ExpReg
y=a×b^x
a=266.0545948
b=.9923145889
r²=.8858866457
r=-.941215515
```

(Continued)

4. If the data to be evaluated are in L1 and L3, press the key sequence (2nd) (1) (,) (2nd) (3) to place the entries L1 and L3 on the home screen.

---

5. Press (ENTER) to display the exponential model. (If you want to automatically paste the regression equation into the (Y=) editor, press the key sequence (VARS) Y-VARS; 1:Function; 1:Y1 before pressing (ENTER).)

```
ExpReg
y=a*b^x
a=60.8078408
b=.9285201575
r²=.9938920452
r=-.9969413449
```

Finding an Exponential Function from a Verbal Description

Much of the information we encounter in the media is given to us in terms of percentages. Consider these typical news headlines. "Gasoline Prices Flare Up 9 Percent" and "Home Values Increase by 6 Percent." Anything that increases or decreases at a constant percentage rate may be modeled with an exponential function. The value of the growth factor b is $1 + r$, where r is the decimal form of the percentage (i.e., 5 percent = 0.05). Note that if $r < 0$, then $b < 1$. A negative rate of growth is referred to as **depreciation,** while a positive rate of growth is referred to as **appreciation.**

ANNUAL GROWTH RATE

Let $y = ab^t$ model the amount of a quantity at time t years. The annual growth rate r of the quantity y is given by $r = b - 1$. Similarly, the annual growth factor is $b = r + 1$.

For example, the author ordered an appraisal of his home in August 2004. In her report, the appraiser asserted that home values in the area were increasing at a rate of 5 percent per year. In other words, $r = 0.05$. The corresponding growth factor is $b = 1.05$. To forecast the value of his home in August 2005, the author multiplied the appraised value by the growth factor.

EXAMPLE 4

Using an Exponential Model to Forecast a Car's Value

New cars typically lose 50 percent of their value in the first three years. Many cars lose more than 25 percent of their value in the first year alone. (**Source:** Runzheimer International.) Depreciating approximately 13.4 percent annually, the Saturn

Sedan 2 holds its value better than most vehicles. Estimate the value of a $12,810 Saturn Sedan 2 three years after it is purchased.

SOLUTION Since the car is depreciating at a constant percentage rate, an exponential function may be used to model the data. The initial value of the car is $12,810, so $a = 12,810$. The car is losing value at 13.4 percent per year, so $r = -0.134$. Since $b = 1 + r$, the exponential function is

$$V(t) = 12,810(b)^t$$
$$= 12,810[1 + (-0.134)]^t \qquad \text{Since } b = 1 + (-0.134)$$
$$= 12,810(0.866)^t \text{ dollars}$$

where t is the age of the car in years.

We want to know the value of the car after three years, so $t = 3$.

$$V(3) = 12,810(0.866)^3$$
$$= 12,810(0.649)$$
$$= 8320$$

We estimate that the car will be valued at $8320 when it is three years old.

EXAMPLE 5

Using an Exponential Function to Model Inflation

Inflation (rising prices) causes money to lose its buying power. In the United States, inflation hovers around 3 percent annually.

(a) If a candy bar costs $0.59 today, what will it cost 10 years from now?
(b) When will a candy bar cost $1.00?

SOLUTION Since we're assuming a constant percentage increase in the price of the candy bar, an exponential model should be used. The initial value is $0.59, so $a = 0.59$. The price of the candy bar is increasing by 3 percent annually, so $r = 0.03$. The cost of the candy bar is given by

$$C(t) = 0.59(1 + 0.03)^t \text{ dollars}$$
$$= 0.59(1.03)^t \text{ dollars}$$

where t is the number of years from today.

(a) Evaluating the function at $t = 10$, we get

$$C(10) = 0.59(1.03)^{10}$$
$$= 0.59(1.344)$$
$$= 0.79$$

We estimate that the candy bar will cost $0.79 ten years from now.

(b) We want to know when $C(t) = 1$.

$$1 = 0.59(1.03)^t$$
$$\frac{1}{0.59} = 1.03^t$$
$$1.695 = 1.03^t$$

At this point, we are stuck. We haven't yet developed the mathematical machinery (logarithms) to get the variable out of the exponent. We can, however, attack the problem graphically by breaking the equation $1.695 = 1.03^t$ into two separate functions: $y_1 = 1.695$ and $y_2 = 1.03^t$. By graphing these functions simultaneously, we can determine the point of intersection of the two functions using our graphing calculator.

Using the graphing calculator, we generate the graph in Figure 5.41 and determine the point of intersection.

FIGURE 5.41

The functions intersect when $t = 17.85$. In about 18 years, we estimate that the cost of a candy bar will be $1.00.

We also could have used the guess-and-check method to approximate the result by evaluating $C(t) = 0.59(1.03)^t$ at various values of t. Since

$$C(17) = 0.59(1.03)^{17} \qquad \text{and} \qquad C(18) = 0.59(1.03)^{18}$$
$$= 0.9752 \qquad\qquad\qquad\qquad = 1.004$$

we determine that the price reaches $1.00 between the seventeenth and eighteenth years.

5.3 Summary

In this section, you learned how exponential functions can be used to model rapidly increasing (or decreasing) data sets. You graphed exponential functions and developed exponential models from tables of data and verbal descriptions.

5.3 Exercises

In Exercises 1–10, do the following:

(a) Determine if the graph of the function is increasing or decreasing.

(b) Determine if the graph of the function is concave up or concave down.

(c) Identify the coordinate of the y-intercept.

(d) Graph the function to verify your conclusions.

1. $y = 4(0.25)^x$ **2.** $y = -2(0.5)^x$

3. $y = 0.5(2)^x$ **4.** $y = 6(0.1)^x$

5. $y = 0.4(5)^x$ **6.** $y = -0.1(0.2)^x$

7. $y = -1.2(2.3)^x$ **8.** $y = 5(0.4)^x$

9. $y = 3(0.9)^x$ **10.** $y = -5(3)^x$

In Exercises 11–20, use algebraic methods to find the equation of the exponential function that fits the data in the table.

11.

x	y
0	2
1	6
2	18
3	54

12.

x	y
0	3
1	12
2	48
3	192

13.

x	y
1	10
2	20
3	40
4	80

14.

x	y
1	10
2	50
3	250
4	1,250

15.

x	y
2	16
4	4
6	1
8	0.25

16.

x	y
2	100
4	1
6	0.01
8	0.0001

17.

x	y
1	1
2	4
4	64
5	256

18.

x	y
2	4
5	32
9	512
11	2,048

19.

x	y
0	256
5	8
7	2
10	0.25

20.

x	y
0	10,000
2	900
5	24.3
7	2.187

In Exercises 21–25, use exponential regression to model the data in the table. Use the model to predict the value of the function when t = 25, and interpret the real-world meaning of the result.

   The Consumer Price Index is used to measure the increase in prices over time. In each of the following tables, the index is assumed to have the value 100 in the year 1984.

  21. **Dental Prices**

Price of Dental Services

Years Since 1980 (t)	Price Index (I)
0	78.9
5	114.2
10	155.8
15	206.8
20	258.5

Source: *Statistical Abstract of the United States, 2001,* Table 694, p. 455.

22. **Wine Prices**

Price of Wine Consumed at Home

Years Since 1980 (t)	Price Index (I)
0	89.5
5	100.2
10	114.4
15	133.6
20	151.6

Source: *Statistical Abstract of the United States, 2001,* Table 694, p. 455.

  23. **Television Prices**

Price of a Television Set

Years Since 1980 (t)	Price Index (I)
0	104.6
5	88.7
10	74.6
15	68.1
20	49.9

Source: *Statistical Abstract of the United States, 2001,* Table 694, p. 455.

  24. **Alcohol Prices**

Price of Distilled Spirits Consumed at Home

Years Since 1980 (t)	Price Index (I)
0	89.8
5	105.3
10	125.7
15	145.7
20	162.3

Source: *Statistical Abstract of the United States, 2001,* Table 694, p. 455.

  25. **Entertainment Prices**

Price of Admission to Entertainment Venues

Years Since 1980 (t)	Price Index (I)
0	83.8
5	112.8
10	151.2
15	181.5
20	230.5

Source: *Statistical Abstract of the United States, 2001,* Table 694, p. 455.

In Exercises 26–30, find the exponential function that fits the verbal description.

26. Salaries Instructors' salaries are $45,000 per year and are expected to increase by 3.5 percent annually. What will instructors' salaries be five years from now?

27. Savings Account A savings account balance is currently $235 and is earning 2.32 percent per year. When will the balance reach $250?

28. **Television Price** The cost of a 27-inch flat-screen television in 2002 was $599.99. (**Source:** www.bestbuy.com.) Television prices are expected to decrease by 16 percent per year. How much is the flat-screen television expected to cost in 2007?

29. **Concert Admission** Reserved seating at a Dave Matthews Band concert at the Gorge Amphitheater in George, Washington, cost $59.90 in September 2004. (**Source:** www.ticketmaster.com.) If admission fees for concerts are expected to increase by 28 percent per year, what is the expected price of a Dave Matthews Band ticket in September 2006?

30. Tuition Tuition is currently $2024 per year and is increasing by 12 percent annually. In how many years from now will tuition reach $3567?

In Exercises 31–35, find the solution by solving the equation graphically.

 31. $(1.2)^x = 5$ **32.** $(0.9)^x = 0.5$

33. $(1.05)^x = 2$ **34.** $(1.12)^x = 2$

35. $(0.3)^x = 0.09$

Exercises 36–40 are intended to challenge your understanding of exponential functions.

36. **Retirement Investments** An investor has $3000 to invest in two different investment accounts: CREF Social Choice and TIAA Retirement Annuity. Based on data from January 1, 1993, through December 31, 2002, the average annual return on the CREF Social Choice account was 8.67 percent, and the average annual return on the TIAA Retirement Annuity was 6.93 percent. (**Source:** TIAA-CREF.) The investor plans to invest twice as much in the Retirement Annuity as in the Social Choice account. Assuming that the future rates of return will be the same as the past rates of return, do the following:

(a) Find the equation of the function for the predicted value of the $3000 investment *t* years from now.

(b) What is the predicted value of the investment 20 years from now?

37. **Retirement Investments** Repeat Exercise 36, except this time invest twice as much in the Social Choice account as in the Retirement Annuity. What factors should the investor consider before changing the amount invested in each of the accounts?

38. **Retirement Investments** Based on data from January 1, 2002, through December 31, 2002, the annual rate of return on the CREF Social Choice account was −2.98 percent, substantially lower than the 10-year average return of 8.67 percent. Over the same period, the CREF Stock account earned −20.73 percent, substantially lower than its 10-year average return of 7.69 percent. (**Source:** TIAA-CREF.)

An investor invested $1000 in each of the two accounts at the start of 2002.

(a) What was the value of the investment in each account at the end of 2002?

(b) Assuming that in each year following 2002, the investments earned an annual return equal to the 10-year average, predict how long it will take for the combined value of the accounts to reach $2000.

39. **Retirement Investments** Based on data from January 1, 2002, through December 31, 2002, the annual rate of return of the CREF Growth account was −30.06 percent, substantially lower than the five-year average return of −5.48 percent. Over the same period, the CREF Inflation-linked Bond account earned 16.32 percent, substantially higher than its five-year average return of 8.33 percent. (**Source:** TIAA-CREF.)

An investor invested $1000 in each of the two accounts at the start of 2002.

(a) What was the value of the investment in each account at the end of 2002?

(b) Assuming that in each year following 2002 the investments earned an annual return equal to the five-year average, predict how long it will take for the combined $2000 investment in the accounts to double.

40. **Investments** How long will it take for a $1000 investment earning 10 percent annually to equal the value of a $2000 investment earning 6 percent annually?

5.4 Logarithmic Function Models

- Use nonlinear functions to model real-life phenomena
- Interpret the meaning of mathematical models in their real-world context
- Graph logarithmic functions
- Model real-life data with logarithmic functions
- Apply the rules of logarithms to simplify logarithmic expressions
- Solve logarithmic equations

GETTING **STARTED**   In Section 5.3, we showed that the NBA players' average salary could be modeled by

$$S(t) = 161.4(1.169)^t \text{ thousand dollars}$$

where t is the number of years since 1980. We could use the model to estimate the average salary in any year t. But what if we wanted to algebraically determine the year in which the salary would be $1,000,000?

In this section, we will demonstrate how logarithmic functions can be used to model data sets that are increasing at a decreasing rate. We will show how to graph logarithmic functions and will illustrate graphically the inverse relationship between logarithmic and exponential functions. We will explain how the rules of logarithms can be used to simplify logarithmic expressions and solve logarithmic equations. We will also show how to develop logarithmic models from tables of data, such as NBA players' salaries, by using logarithmic regression.

We want to determine the year algebraically in which the NBA players' average salary reached $1 million. We can estimate the year by looking at the data in Table 5.19.

TABLE 5.19   NBA Players' Average Salary

Salary (thousands of dollars) (s)	Years Since 1980 (t)
170	0
325	5
750	10
1,900	15
2,000	16
2,200	17
2,600	18

Source: *Statistical Abstract of the United States, 2001*, Table 1324, p. 829.

Notice that *salary* is the independent variable and *years* is the dependent variable. The average salary was $750,000 in 1990 and $1,900,000 in 1995, so we know that $1,000,000 was reached sometime between 1990 and 1995.

Returning to the exponential model for the NBA players' average salary, we have

$$s(t) = 161.4(1.169)^t \text{ thousand dollars}$$

where t is the number of years since 1980. Recall that $1,000,000 is equal to $1000 thousand. We must determine when $s(t) = 1000$. In other words, we must solve the equation

$$1000 = 161.4(1.169)^t \qquad \text{Since the units of the salary are thousands of dollars}$$

$$\frac{1000}{161.4} = \frac{161.4(1.169)^t}{161.4}$$

$$6.196 = 1.169^t$$

If we want to solve this equation algebraically, at this point we are stuck. We must come up with some way to get the t out of the exponent. Observe that t is the exponent that we place on 1.169 in order to get 6.196. We need to create a symbol to represent the phrase "is the exponent we place on." The chosen symbol, as bizarre as it may seem, is **log**.

$$\underbrace{t \text{ is}}_{t\,=} \quad \underbrace{\text{the exponent we place on}}_{\log} \quad \underbrace{1.169}_{1.169} \quad \underbrace{\text{in order to get } 6.196}_{6.196}$$

Compressing these terms into a single equation yields

$$t = \log_{1.169} 6.196$$

Using the calculator ⌊ LOG ⌋ button and some log rules that will be introduced later, we determine that $t \approx 11.7$ years. That is, between 1991 ($t = 11$) and 1992 ($t = 12$), the NBA players' average salary reached $1 million.

Logarithmic Functions

The notation **log** is short for **logarithm.** It is helpful to think of **log** as representing the phrase "the exponent we place on." The equation $y = \log_b(x)$, which is read "y equals log base b of x," means "y is the exponent we place on b in order to get x." For example, $y = \log_2(8)$ means "y is the exponent we place on 2 in order to get 8." That is, $2^y = 8$. Since $2^3 = 8$, we conclude that $y = 3$.

EXAMPLE 1 **Calculating a Logarithm**

Find the value of y given $y = \log_3(9)$.

SOLUTION $y = \log_3(9)$ means "y is the exponent we place on 3 in order to get 9." That is, $3^y = 9$. Since $3^2 = 9$, we conclude that $y = 2$.

> **LOGARITHMIC FUNCTION**
>
> Let b and x be real numbers with $b > 0$ and $b \neq 1$ and $x > 0$. The function
>
> $$y = \log_b(x)$$
>
> is called a **logarithmic function.** The value b is called the **base** of the logarithmic function. We read the expression $\log_b(x)$ as "log base b of x."

The use of parentheses around the x in $y = \log_b(x)$ is optional. That is, $y = \log_b(x)$ is equivalent to $y = \log_b x$. However, we will consistently use parentheses throughout this text to remind ourselves that the expression "$\log_b$" is meaningless by itself.

Although any positive number not equal to 1 may be used as the base of a logarithmic function, the most commonly used bases are 10 and e. The irrational number e is approximately equal to 2.71828. The number e, like π, has many wonderful uses. Although we will see some of the power of e in this chapter, we will witness its full strength when we move into calculus. The number e is generated by calculating $\left(1 + \frac{1}{n}\right)^n$ for infinitely large n. You might think that $\left(1 + \frac{1}{n}\right)^n$ will get infinitely big as n goes to infinity. Surprisingly, $\left(1 + \frac{1}{n}\right)^n \approx 2.718281828$ for large values of n (see Table 5.20).

TABLE 5.20

n	$\left(1 + \dfrac{1}{n}\right)^n$
1	2.000
10	2.594
100	2.705
1,000	2.717
10,000	2.718
100,000	2.718

The value of e, 2.718281828. . . , may be accessed on the TI-83 Plus calculator by pressing the key sequence [2nd] and [÷].

The expression $\log_{10}(x)$ is often written $\log(x)$ and is called the **common log.** We will assume that the base is 10 when no base is written.

The expression $\log_e(x)$ is typically written $\ln(x)$ and is called the **natural log.**

Graphs of Logarithmic Functions

Like polynomial and exponential function graphs, logarithmic function graphs share many common characteristics. Consider the logarithmic functions $f(x) = \log_2(x)$, $g(x) = \log_3(x)$, and $h(x) = \log_4(x)$ (Figure 5.42).

Each graph is concave down and increasing. Each graph passes through the point $(1, 0)$ and appears to have a vertical asymptote at $x = 0$ (the y axis). In fact, for all positive values of b not equal to 1, the graph of $y = \log_b(x)$ has a vertical asymptote at $x = 0$ and an x-intercept at $(1, 0)$. Logarithmic functions are undefined for $x < 0$. The shape of the graph is determined by b. The base b controls how rapidly the graph increases (or decreases). For each of the graphs in Figure 5.42, $b > 1$. What if $b < 1$? Consider the logarithmic functions $f(x) = \log_{0.3}(x)$, $g(x) = \log_{0.5}(x)$, and $h(x) = \log_{0.7}(x)$ (Figure 5.43).

FIGURE 5.42

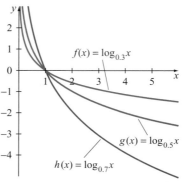

FIGURE 5.43

Each graph is concave up and decreasing. However, these graphs also have a vertical asymptote at $x = 0$ and pass through the point $(1,0)$. Our graphical observations are summarized in Table 5.21.

TABLE 5.21   Logarithmic Function Graphs: $y = \log_b(x)$

Value of b	Concavity	Increasing/Decreasing	Graph
$b > 1$	Concave down	Increasing	
$0 < b < 1$	Concave up	Decreasing	

EXAMPLE 2 **Determining the Shape of a Logarithmic Function**

Determine the concavity and increasing/decreasing behavior of $y = \log_2(x)$ and $y = \log_{0.5}(x)$. Then graph both functions to verify your results.

SOLUTION   Since $2 > 1$, $y = \log_2(x)$ will be concave down and increasing. In order to graph the function by hand, we must generate a table of values for the function. You may find it helpful to write $y = \log_2(x)$ as $x = 2^y$, then select the values of y and calculate the values of x (Table 5.22). Plotting the points and connecting the dots yields the graph shown in Figure 5.44.

TABLE 5.22

$x = 2^y$	y
1	0
2	1
4	2
8	3
16	4

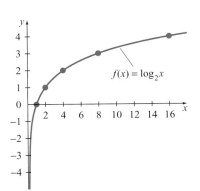

FIGURE 5.44

Since $0.5 < 1$, $y = \log_{0.5}(x)$ will be concave up and decreasing. You may find it helpful to write $y = \log_{0.5}(x)$ as $x = (0.5)^y$, then select the values of y and calculate the values of x (Table 5.23). Plotting the points and connecting the dots yields the graph shown in Figure 5.45.

TABLE 5.23

$x = (0.5)^y$	y
16	-4
8	-3
4	-2
2	-1
1	0

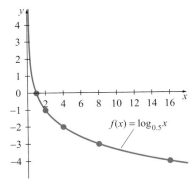

FIGURE 5.45

EXAMPLE 3

Comparing the Graphs of Logarithmic Functions

Graph $f(x) = \ln(x)$ and $g(x) = \log(x)$ simultaneously. Then determine where $f(x) < g(x)$ and where $g(x) > f(x)$.

SOLUTION

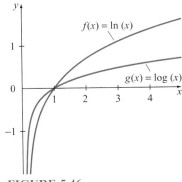

FIGURE 5.46

(To graph the function on the TI-83 Plus, use the [LOG] and [LN] keys as appropriate.) Based on the graphs in Figure 5.46, we conclude that for $x < 1$, $\ln(x) < \log(x)$. For $x > 1$, $\ln(x) > \log(x)$.

Relationship Between Logarithmic and Exponential Functions

Exponential and logarithmic functions are intimately related. Any logarithmic function may be rewritten as an exponential function (and vice versa).

LOGARITHMIC AND EXPONENTIAL FUNCTION RELATIONSHIP

For $b > 0$ with $b \neq 1$ and $x > 0$, the following statements are equivalent:

1. $y = \log_b(x)$

2. $b^y = x$

Consider tables of values for the functions $f(x) = 10^x$ and $g(x) = \log(x)$ (Table 5.24).

TABLE 5.24

x	$f(x) = 10^x$	x	$g(x) = \log(x)$
−2	0.01	0.01	−2
−1	0.1	0.1	−1
0	1	1	0
1	10	10	1
2	100	100	2

Notice that the *input* of the exponential function is the *output* of the logarithmic function. Similarly, the *output* of the exponential function is the *input* of the logarithmic function. A point (x_1, y_1) is on the graph of f if and only if the point (y_1, x_1) is on the graph of g. Functions with this property are called **inverse** functions. Exponential functions and logarithmic functions are inverse functions of each other. The inverse relationship of logarithmic and exponential functions may also be seen by graphing $y = \ln(x)$ and $y = e^x$ together with the graph of $y = x$ (Figure 5.47). The graph of $y = \ln(x)$ is the reflection of $y = e^x$ about the line $y = x$. That is, if we folded our paper along the line $y = x$, the graphs of the two functions would lie directly on top of each other.

When we solve a logarithmic function $y = \log_b(x)$ for x, we get the exponential function, $x = b^y$. This relationship between logarithmic and exponential functions will be critical to solving the logarithmic and exponential equations.

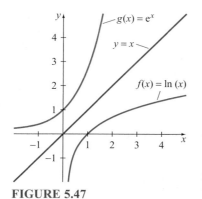

FIGURE 5.47

Rules of Logarithms

There are several rules that are used to manipulate logarithms. Each rule is important; however, we will use Rule 7 most extensively. It is so essential that we comically call it "the mother of all log rules."

RULES OF LOGARITHMS

If x, y, and n are real numbers with $x > 0$ and $y > 0$, then the following rules apply.

Common Logarithm	Natural Logarithm
1. $\log(1) = 0$	1. $\ln(1) = 0$
2. $\log(10) = 1$	2. $\ln(e) = 1$
3. $\log(10^x) = x$	3. $\ln(e^x) = x$
4. $10^{\log(x)} = x$	4. $e^{\ln(x)} = x$
5. $\log(xy) = \log(x) + \log(y)$	5. $\ln(xy) = \ln(x) + \ln(y)$
6. $\log\left(\dfrac{x}{y}\right) = \log(x) - \log(y)$	6. $\ln\left(\dfrac{x}{y}\right) = \ln(x) - \ln(y)$
7. $\log(x^p) = p \log(x)$	7. $\ln(x^p) = p \ln(x)$

In fact, Rules 1, 5, 6, and 7 apply for any base, not just 10 and e. Rules similar to Rules 2 through 4 may easily be formulated for any other base. In our applications, we will be using base 10 and base e predominantly. This is because the following change-of-base formula allows us to convert any logarithm into a logarithm with one of these two bases.

CHANGE-OF-BASE FORMULA

$$\log_b(a) = \frac{\log(a)}{\log(b)} = \frac{\ln(a)}{\ln(b)}$$

The change-of-base formula is easily derived.

Let $y = \log_b(a)$	
$b^y = a$	Rewrite in exponential form
$\log(b^y) = \log(a)$	Take the log of both sides
$y \log(b) = \log(a)$	Apply log Rule 7
$y = \dfrac{\log(a)}{\log(b)}$	Divide both sides by $\log(b)$
But $y = \log_b(a)$	

so

$$\log_b(a) = \frac{\log(a)}{\log(b)}$$

The ability to apply the log rules to manipulate expressions and equations is one of the critical skills required to be successful in solving exponential and logarithmic equations. To assist you in mastering those skills, we will do several "skill-and-drill" examples.

EXAMPLE 4

Using Log Rules to Simplify a Logarithmic Expression

Use the log rules to rewrite the logarithmic expression as a single term.

$$\log(3) - \log(27) + \log(12)$$

SOLUTION

$$\log(3) - \log(27) + \log(12) = \log\left(\frac{3}{27}\right) + \log(12) \qquad \text{Rule 6}$$

$$= \log\left(\frac{3}{27} \cdot 12\right) \qquad \text{Rule 5}$$

$$= \log\left(\frac{36}{27}\right)$$

$$= \log\left(\frac{4}{3}\right)$$

EXAMPLE 5

Writing a Logarithmic Expression as a Single Logarithm

Rewrite the logarithmic expression as a single logarithm.

$$\log(x) + \log(x^3) - 4\log(x^2)$$

SOLUTION

$$\log(x) + \log(x^3) - 4\log(x^2)$$

$$= \log(x) + 3\log(x) - 2 \cdot 4\log(x) \qquad \text{Rule 7}$$

$$= 4\log(x) - 8\log(x) \qquad \text{Combine like terms}$$

$$= -4\log(x) \qquad \text{Combine like terms}$$

The solution may also be written as

$$-\log(x^4), \log(x^{-4}), \text{ or } \log\left(\frac{1}{x^4}\right)$$

When checking a solution with the answers in the back of the book, it is important to recognize that the correct solution may be written in a variety of forms.

EXAMPLE 6

Writing a Logarithmic Expression as a Single Logarithm

Rewrite the expression $2\log\left(\frac{2}{3x}\right) - \log(x^3)$ as a single logarithm.

SOLUTION

$$2\log\left(\frac{2}{3x}\right) - \log(x^3)$$

$$= \log\left(\frac{2}{3x}\right)^2 - \log(x^3) \qquad \text{Rule 7}$$

$$= \log\left[\frac{(2)^2}{(3x)^2}\right] + \log(x^3)^{-1} \qquad \text{Rule 7, rules of exponents}$$

$$= \log\left(\frac{4}{9x^2}\right) + \log(x^{-3}) \qquad \text{Rules of exponents}$$

$$= \log\left(\frac{4}{9x^2} \cdot x^{-3}\right) \qquad \text{Rule 5}$$

$$= \log\left(\frac{4}{9x^2} \cdot \frac{1}{x^3}\right) \qquad \text{Rules of exponents}$$

$$= \log\left(\frac{4}{9x^5}\right) \qquad \text{Rules of exponents and fractions}$$

EXAMPLE 7

Solving a Logarithmic Equation

Solve the equation $\log_3(x) = 4$ for x.

SOLUTION

$$\log_3(x) = 4$$
$$3^4 = x \qquad \text{Rewrite in exponential form}$$
$$x = 81$$

EXAMPLE 8

Solving an Exponential Equation

Solve the equation $3^x = \frac{1}{12}$ for x.

SOLUTION

$$3^x = \frac{1}{12}$$

$$\ln(3^x) = \ln\left(\frac{1}{12}\right) \qquad \text{Take the natural log of both sides}$$

$$x \ln(3) = \ln(1) - \ln(12) \qquad \text{Rule 7 and Rule 6}$$
$$x \ln(3) = 0 - \ln(12) \qquad \text{Rule 1}$$
$$x \ln(3) = -\ln(12)$$
$$x = -\frac{\ln(12)}{\ln(3)} \qquad \text{Divided by } \ln(3)$$
$$x \approx -2.262$$

It is important to note that $-\frac{\ln(12)}{\ln(3)} \neq -\ln\left(\frac{12}{3}\right)$.

Logarithmic Models

Functions that are increasing at a decreasing rate may often be modeled by logarithmic functions. Using the logarithmic regression feature, **LnReg,** on our calculator returns a function of the form $y = a + b \ln(x)$. The a shifts the graph of $y = b \ln(x)$ vertically by $|a|$ units. The resultant graph passes through the point $(1, a)$ instead of $(1, 0)$.

EXAMPLE 9

Using Logarithmic Regression to Forecast the Population of Hawaii

TABLE 5.25 Projected Population of Hawaii

People (thousands) (p)	Years Since 1995 (T)
1,187	0
1,257	5
1,342	10
1,553	20
1,812	30

Source: www.census.gov.

Find the logarithmic model that best fits the data in Table 5.25. Then evaluate the function at $p = 1700$ and interpret your result.

SOLUTION Using the logarithmic regression feature on our calculator (as shown in the following Technology Tip), we get

$$y = -496.0 + 70.17 \ln(x)$$

or, in terms of our variables,

$$T(p) = -496.0 + 70.17 \ln(p)$$

Evaluating the function when $p = 1700$, we get

$$T(1700) = -496.0 + 70.17 \ln(1700)$$
$$= -496.0 + 70.17(7.438)$$
$$= -496.0 + 522.0$$
$$= 26.0$$

The population of Hawaii is projected to reach 1,700,000 in 2021 (26 years after 1995).

TECHNOLOGY **TIP**

Logarithmic Regression

1. Enter the data using the Statistics Menu List Editor. (Refer to Section 1.3 if you've forgotten how to do this.)

```
L1      L2      L3      3
1187    0       ------
1257    5
1342    10
1553    20
1812    30
------  ------
L3(1) =
```

2. Bring up the Statistics Menu Calculate feature by pressing [STAT] and using the blue arrows to move to the CALC menu. Then select item 9:LnReg and press [ENTER].

```
EDIT CALC TESTS
7↑QuartReg
8:LinReg(a+bx)
9:LnReg
0:ExpReg
A:PwrReg
B:Logistic
C:SinReg
```

3. If you want to automatically paste the regression equation into the [Y=] editor, press the key sequence [VARS] Y-VARS; 1:Function; 1:Y1 and press [ENTER]. Otherwise, press [ENTER].

```
LnReg
y=a+blnx
a=-496.0111246
b=70.17413909
r²=.9976542757
r=-.9988264492
```

5.4 Summary

In this section, you learned how to graph logarithmic functions. You discovered the inverse relationship between logarithmic and exponential functions and practiced using rules of logarithms to simplify logarithmic expressions and solve logarithmic equations. You also developed logarithmic models from data tables by using logarithmic regression.

5.4 Exercises

In Exercises 1–5, solve the logarithmic equation for y without using a calculator.

1. $y = \log_5(25)$

2. $y = \log_3(81)$

3. $y = \log_2(64)$

4. $y = \log_2(4^{-1})$

5. $y = \log_3\left(\dfrac{1}{9}\right)$

In Exercises 6–10, determine the concavity and increasing/decreasing behavior of the graph of the function. Then graph the function. (Hint: You will have to use the change-of-base formula if you graph the function on your calculator. That is,

$y = \log_b(x) \Rightarrow y = \dfrac{\ln(x)}{\ln(b)}.$)

6. $y = \log_2(x)$

7. $y = \log_4(x)$

8. $y = \log_{0.2}(x)$

9. $y = \log_{0.7}x$

10. $y = \log_{0.8}x$

In Exercises 11–20, use the inverse relationship between logarithmic and exponential functions to solve the equations for x. Simplify your answers.

11. $2^x = 64$

12. $5^x = 125$

13. $2^x = \dfrac{1}{2}$

14. $3^x = \dfrac{1}{9}$

15. $4^x = 64$

16. $\log_5(x) = 3$

17. $\log_4(x) = -2$

18. $\log_2(x) = 6$

19. $\log_3(x) = 5$

20. $\log_4(x) = \dfrac{1}{2}$

In Exercises 21–40, use the rules of logarithms to rewrite each expression as a single logarithmic expression.

21. $\log(2) + \log(8)$

22. $\log(2) - \log(8)$

23. $\log(2x) + \log(x^3)$

24. $\log(5x) + \log(20)$

25. $\log(j) + \log(a) + \log(m)$

26. $3\log(x^2) - 4\log(x^2) + \log(2x^3)$

27. $7\log(x^{-1}) - 4\log(3x^2) - \log(4x)$

28. $2\log(3x)^2 + \log(x)^{-12}$

29. $4\log\left(\dfrac{1}{x}\right) + 3\log(x)$

30. $3\log\left(\dfrac{2}{5x}\right) - \log(x^3)$

31. $\ln(x) - \ln\left(\dfrac{1}{x}\right)$

32. $2\ln(2x)^2 + \ln\left(\dfrac{1}{x^2}\right)$

33. $\ln(1) - \log(1) + \ln(x^2)$

34. $-2\ln\left(\dfrac{x}{2}\right) + \ln(2)$

35. $3\ln(3x^2) - \ln(x^6)$

36. $2\ln(3x) - \ln(3x^2) + 2\ln(3)$

37. $-\ln(4x^2)^3 + 2\ln(x^6) - 6\ln(x)$

38. $-\ln(3x)^2 - 3\ln(x^{-2}) - 5\ln(x)$

39. $4\ln(3x) + \ln(81x^2) - \ln(9x)$

40. $\ln(3x) - \ln(9x^2) + \ln(3) - \ln(x)$

In Exercises 41–45, use logarithmic regression to model the data in the table. Use the model to predict the value of the function when i = 125, and interpret the real-world meaning of the result.

The Consumer Price Index is used to measure the increase in prices over time. In each of the following tables, the index is assumed to have the value 100 in the year 1984.

  41. **Dental Services**

Price of Dental Services

Price Index (i)	Years Since 1980 (t)
78.9	0
114.2	5
155.8	10
206.8	15
258.5	20

Source: *Statistical Abstract of the United States, 2001,* Table 694, p. 455.

   42. **Price of Wine**

Price of Wine Consumed at Home

Price Index (i)	Years Since 1980 (t)
89.5	0
100.2	5
114.4	10
133.6	15
151.6	20

Source: *Statistical Abstract of the United States, 2001,* Table 694, p. 455.

  43. **Price of a TV**

Price of a Television Set

Price Index (i)	Years Since 1980 (t)
104.6	0
88.7	5
74.6	10
68.1	15
49.9	20

Source: *Statistical Abstract of the United States, 2001,* Table 694, p. 455.

  44. **Price of Alcohol**

Price of Distilled Spirits Consumed at Home

Price Index (i)	Years Since 1980 (t)
89.8	0
105.3	5
125.7	10
145.7	15
162.3	20

Source: *Statistical Abstract of the United States, 2001,* Table 694, p. 455.

  45. **Entertainment Admission Price**

Price of Admission to Entertainment Venues

Price Index (i)	Years Since 1980 (t)
83.8	0
112.8	5
151.2	10
181.5	15
230.5	20

Source: *Statistical Abstract of the United States, 2001,* Table 694, p. 455.

Exercises 46–50 are intended to challenge your understanding of logarithmic and exponential functions.

46. **Homes with VCRs**   Given the following table, perform each of the tasks identified as (**a**) through (**c**).

Percentage of TV Homes with a VCR

Years Since 1984 (t)	Homes with a VCR (percent) (V)
1	20.8
2	36.0
3	48.7
4	58.0
5	64.6
6	68.6
7	71.9
8	75.0
9	77.1
10	79.0
11	81.0
12	82.2
13	84.2
14	84.6

Source: *Statistical Abstract of the United States, 2001*, Table 1126, p. 705.

(**a**) Use logarithmic regression to find the logarithmic function that best fits the data.

(**b**) Solve the equation for the input variable, t.

(**c**) Determine the year in which 90 percent of TV homes were expected to have VCRs. Do you think this prediction will be accurate? Explain.

47. The domain of the function $y = \log_b(x)$ is the set of all positive real numbers. Explain why the negative real numbers and zero are not in the domain of the function.

48. If $b > c > 1$, for what values of x is $\log_b(x) > \log_c(x)$?

49. Does the equation $e^x = \ln(x)$ have a solution? Explain.

50. In order for the equation $\log_b(x) = 2\log_c(x)$ to be true for all positive values of x, what must be the relationship between b and c?

5.5 Choosing a Mathematical Model

- Use critical thinking skills in selecting a mathematical model

GETTING STARTED   Many data sets may be effectively fitted with more than one mathematical model. Determining which model to use requires critical thinking. In this section, we will discuss strategies and techniques for choosing a mathematical model. However, before delving into model selection strategies, we will *briefly* introduce one other type of mathematical model: the logistic function. More detailed coverage of logistic functions is given in Chapter 14 of the *Finite Math and Applied Calculus* version of this text.

Logistic Functions

A logistic function graph is an *s*-shaped curve that is bounded above and below by horizontal asymptotes (Figures 5.48 and 5.49). It may be either an increasing function or a decreasing function.

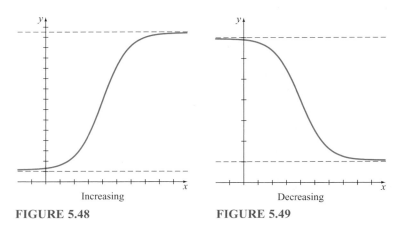

Increasing	Decreasing
FIGURE 5.48	**FIGURE 5.49**

The equation of a logistic function is of the form

$$f(x) = \frac{c}{1 + ae^{-bx}} + k$$

where *a, b, c,* and *k* are all constants. Logistic functions exhibit slow growth, then rapid growth, followed by slow growth or slow decay, then rapid decay, followed by slow decay.

Selecting a Mathematical Model

Graphing calculators and computer spreadsheets are extremely useful tools for mathematical modeling. When you are given a table of data to model, use the following model selection strategies.

HOW TO **Model Selection Strategies to Use When Given a Table of Data**

1. Draw a scatter plot.
2. Determine whether the scatter plot exhibits the behavior of the graph of one or more of the standard mathematical functions: linear, quadratic, cubic, quartic, exponential, logarithmic, or logistic.
3. Find a mathematical model for each function type selected in Step 2.
4. Use all available information to anticipate the expected behavior of the thing being modeled outside of the data set. Eliminate models that don't exhibit the expected behavior. (Sometimes it is convenient to switch the order of Steps 3 and 4.)
5. Choose the simplest model from among the models that meet your criteria.

In accomplishing Step 2, it is helpful to recognize key graphical features exhibited by the data set. Often we can eliminate one or more model types by noting the concavity of the scatter plot. Table 5.26 identifies key graphical features to look for.

TABLE 5.26

Function Type	Model Equation	Key Graphical Features
Linear	$y = mx + b$	Straight line
Quadratic	$y = ax^2 + bx + c$	Concave up everywhere or concave down everywhere
Cubic	$y = ax^3 + bx^2 + cx + d$	Changes concavity exactly once; no horizontal asymptotes
Quartic	$y = ax^4 + bx^3 + cx^2 + dx + f$	Changes concavity zero or two times; no horizontal asymptotes
Exponential	$y = ab^x$ with $a > 0$	Concave up; horizontal asymptote at $y = 0$
Logarithmic	$y = a + b \ln(x)$ with $b > 0$	Concave down; vertical asymptote at $x = 0$
Logistic	$y = \dfrac{c}{1 + ae^{-bx}} + k$	Changes concavity exactly once; horizontal asymptotes at $y = k$ and $y = c + k$

EXAMPLE   1

Choosing a Mathematical Model to Forecast a Room Rate

As shown in Table 5.27, the average room rate for a hotel/motel increased between 1990 and 1999. Find a mathematical model for the data and forecast the average hotel/motel room rate for 2004.

TABLE 5.27   Average Hotel/Motel Room Rate

Years Since 1990 (t)	Room Rate (dollars) (R)
0	57.96
1	58.08
2	58.91
3	60.53
4	62.86
5	66.65
6	70.93
7	75.31
8	78.62
9	81.33

Source: *Statistical Abstract of the United States, 2001,* Table 1266, p. 774.

SOLUTION We first draw a scatter plot of the data on our graphing calculator (see Figure 5.50).

FIGURE 5.50

The graph is initially concave up, but it changes concavity near $t = 6$. Since the graph changes concavity exactly once, a cubic or a logistic model (Figure 5.51) may fit the data.

R as a function of *t*

$$R(t) = -0.04815t^3 + 0.8868t^2 - 1.495t + 58.32$$

R as a function of *t*

$$R(t) = \frac{27.80}{(1 + 47.23e^{-0.6425t})} + 57$$

FIGURE 5.51

Both the cubic model and the logistic model appear to fit the data extremely well. Which model will be best for forecasting *R* in 2004 ($t = 14$)? To answer this question, we extend the viewing rectangle for each graph to include the interval $[-1, 15]$ (see Figure 5.52).

R as a function of *t*

$$R(t) = -0.04815t^3 + 0.8868t^2 - 1.495t + 58.32$$

R as a function of *t*

$$R(t) = \frac{27.80}{(1 + 47.23e^{-0.6425t})} + 57$$

FIGURE 5.52

The cubic model forecasts a decrease in the room rate by 2004 ($t = 14$). According to the cubic model, the rate will drop to $79.07. Since room rates have increased every year since 1990, it is unlikely that the room rate will drop at any point in the future.

The logistic model forecasts a leveling off of the room rate. According to the logistic model, the 2004 room rate will be $84.64. This forecast is likely to be more accurate than the cubic model projection.

It is important to note that there is a certain degree of uncertainty when picking a model to use from a group of models that seem to fit the data. It is possible that two different people may select different models as the "best" model. For this reason, it is important to always explain the reasoning behind the selection of a particular model.

EXAMPLE 2

Choosing a Mathematical Model to Forecast a Movie Ticket Price

As shown in Table 5.28, the average price of a movie ticket increased between 1975 and 1999. Find a mathematical model for the data and forecast the average price of a movie ticket in 2004.

TABLE 5.28

Years Since 1975 (t)	Movie Ticket Price (dollars) (P)
0	2.05
5	2.69
10	3.55
15	4.23
20	4.35
24	5.08

Source: *Statistical Abstract of the United States, 2001*, Table 1244, p. 761.

SOLUTION We first draw the scatter plot (Figure 5.53).

FIGURE 5.53

The first four data points appear to be somewhat linear, so our initial impression is that a linear model may fit the data well. However, the fifth data point is not aligned with the first four, so we know that the data set isn't perfectly linear. However, a linear model may still fit the data fairly well.

Alternatively, we may look at the scatter plot and conclude that the function is concave down on $[0, 15]$ and concave up on $[15, 24]$. Since the scatter plot appears to change concavity once and does not have any horizontal asymptotes, a cubic model may work well. Figure 5.54 shows the two models.

P as a function of t

$P(t) = 0.1224t + 2.148$

P as a function of t

$P(t) = 0.00007426t^3 - 0.004278t^2 + 0.1846t + 2.001$

FIGURE 5.54

In 2004, $t = 29$. Evaluating each function at $t = 29$ yields the following:

$$P(29) = 0.1224(29) + 2.148$$
$$\approx \$5.70$$

$$P(29) = 0.00007426(29)^3 - 0.004278(29)^2 + 0.1846(29) + 2.001$$
$$\approx \$5.57$$

It is difficult to know which of these estimates of the average price of a movie ticket is more accurate. The models seem to fit the data equally well. Furthermore, we have no other information that leads us to believe that one model would be better than the other. Therefore, we pick the simpler model: $P(t) = 0.1224t + 2.148$.

Certain verbal descriptions often hint at the particular mathematical model to use. By watching for key phrases, we can narrow the model selection process. Table 5.29 presents some examples of typical phrases, their interpretation, and possible models.

TABLE 5.29   Finding a Mathematical Model from a Verbal Description

Phrase	Interpretation	Possible Model
Salaries are projected to increase by $800 per year for the next several years.	The graph of the salary function will have a constant slope, $m = 800$.	Linear $S(t) = 800t + b$
The company's revenue has decreased by 5 percent annually for the past six years.	Since the annual decrease is a *percentage* of the previous year's revenue, the dollar amount by which revenue decreases annually is decreasing. The revenue graph is decreasing and concave up.	Exponential $R(t) = a(1 - 0.05)^t$ $\quad\;\; = a(0.95)^t$
Product sales were initially slow when the product was introduced, but sales increased rapidly as the popularity of the product increased. Sales are continuing to increase, but not as quickly as before.	The graph of the sales function may have a horizontal asymptote at or near $y = 0$ and a horizontal asymptote slightly above the maximum projected sales amount.	Logistic
Enrollments have been dropping for years. Each year we lose more students than we did the year before.	The graph of the enrollment function is decreasing, since enrollments are dropping. It will also be concave down, since the rate at which students are dropping continues to increase in magnitude.	Quadratic
Company profits increased rapidly in the early 1990s but leveled off in the late 1990s. In the early 2000s, profits again increased rapidly.	The graph of the profit function increases rapidly, then levels off, then increases rapidly again.	Cubic

EXAMPLE 3   Choosing a Mathematical Model to Forecast Coca-Cola Production

In its 2001 Annual Report, the Coca-Cola Company reported the following:

> Our worldwide unit case volume increased 4 percent in 2001, on top of a 4 percent increase in 2000. The increase in unit case volume reflects consistent performance across certain key operations despite difficult global economic conditions. Our business system sold 17.8 billion unit cases in 2001.
>
> (**Source:** Coca-Cola Company 2001 Annual Report, p. 46.)

Find a mathematical model for the unit case volume of the Coca-Cola Company.

SOLUTION Since the unit case volume is increasing at a constant percentage rate (4 percent), an exponential model will fit the data well. Furthermore, the initial number of unit cases sold was 17.8 billion.

$$V(t) = ab^t \text{ where } t \text{ is the number of years since the end of 2001}$$
$$= 17.8(1 + 0.04)^t \text{ billion unit cases}$$
$$= 17.8(1.04)^t \text{ billion unit cases}$$

The mathematical model for the unit case volume is $V(t) = 17.8(1.04)^t$.

Sometimes a data set may not be effectively modeled by any of the aforementioned functions. In these cases, we look to see if we can model the data with a piecewise function. That is, we use one function to model a portion of the data and a different function to model another portion of the data. Even with this approach, there are some data sets (e.g., daily stock prices) that can rarely be modeled by one of the standard functions.

EXAMPLE 4

Finding a Piecewise Model for AIDS Deaths in the United States

From 1981 to 1995, the number of adult and adolescent AIDS deaths in the United States increased dramatically. However, from 1995 to 2001, the annual death rate plummeted, as shown in Table 5.30.

TABLE 5.30 Adult and Adolescent AIDS Deaths in the United States

Years Since 1981 (t)	Number of Deaths During Year (D)
0	122
1	453
2	1,481
3	3,474
4	6,877
5	12,016
6	16,194
7	20,922
8	27,680
9	31,436
10	36,708

(Continued)

(Continued)

Years Since 1981 (t)	Number of Deaths During Year (D)
11	41,424
12	45,187
13	50,071
14	50,876
15	37,646
16	21,630
17	18,028
18	16,648
19	14,433
20	8,963

Source: Centers for Disease Control and Prevention, "HIV/AIDS Surveillance Report," December 2001, p. 30.

Find the mathematical model that best models the data and forecast the number of adult and adolescent AIDS deaths in 2004.

SOLUTION We first draw a scatter plot of the data (Figure 5.55).

FIGURE 5.55

The data set appears to exhibit logistic behavior up until 1995. After 1995, the graph appears to exhibit cubic behavior. We will use a piecewise function to model the data set. We determine each of the model pieces by using logistic and cubic regression. (Although the data point associated with $t = 14$ was used in finding both pieces of the model, we must assign the domain value $t = 14$ to one piece or the other. We choose to assign $t = 14$ to the logistic piece of the model.)

$$P(t) = \begin{cases} \dfrac{53,955}{1 + 38.834e^{-0.45127t}} & 0 \le t \le 14 \\ -381.06t^3 + 20,770t^2 - 379,469t + 2,339,211 & t > 14 \end{cases}$$

FIGURE 5.56

Our piecewise model (Figure 5.56) appears to fit the data set very well. We are asked to forecast the number of AIDS deaths in 2004 ($t = 23$). Since $23 > 14$, we will use the second function in the piecewise model.

$$P(23) = -381.06(23)^3 + 20{,}770(23)^2 - 379{,}469(23) + 2{,}339{,}211$$
$$= -37{,}603 \text{ adult and adolescent AIDS deaths}$$

It is impossible to have a negative number of deaths! Despite the fact that the model fit the data well, using the model to forecast the 2004 mortality rate yielded an unreasonable result. Returning to the data set, we estimate that the number of AIDS deaths in the years beyond 2001 will be somewhere between 0 and 8963 (the 2001 figure).

If a data set cannot be effectively modeled by one of the standard mathematical functions or a piecewise function, we may conclude that we don't know how to effectively model the data. In such cases, we may estimate a future result by identifying a range of seemingly reasonable values.

EXAMPLE 5

Analyzing Data Not Easily Modeled with a Common Function

The number of firearms detected during airport passenger screening is shown in Table 5.31.

TABLE 5.31

Years Since 1980 (t)	Firearms Detected (F)	Years Since 1980 (t)	Firearms Detected (F)
0	1,914	14	2,994
5	2,913	15	2,390
10	2,549	16	2,155
11	1,644	17	2,067
12	2,608	18	1,515
13	2,798	19	1,552

Source: *Statistical Abstract of the United States, 2001,* Table 1062, p. 669.

Estimate the number of firearms detected by airport screeners in 2002.

SOLUTION

FIGURE 5.57

The scatter plot (Figure 5.57) does not resemble any of the standard mathematical functions. It also does not appear that a piecewise model will fit the data well. The number of firearms detected in a given year appears to be somewhat random, ranging from about 1500 firearms to roughly 3000 firearms.

After September 11, 2001, airline screening became much more thorough. The increased security may be having a deterrent effect. According to the Bureau of Transportation Statistics (www.bts.gov), there were 1071 firearms detected in 2001. As of March 2004, airline screening data for 2002 had not yet been published on the bureau's web site. We estimate that the number of firearms detected in 2002 will be in the 1000–2000 range. Our estimate is based in part on the additional data we discovered through research.

5.5 Summary

In this section, you learned strategies and techniques for selecting mathematical models. You also discovered that oftentimes more than one function may be used to model the same data set.

5.5 Exercises

In Exercises 1–10, find the equation of the mathematical model (if possible) that you believe will most accurately forecast the indicated result. Justify your conclusions.

1.   **Gaming Hardware Sales**

Electronic Gaming Hardware Factory Sales

Years Since 1990 (t)	Sales (millions of dollars) (S)
0	975
1	1,275
2	1,575
3	1,650
4	1,575
5	1,500
6	1,600
7	1,650
8	1,980
9	2,250

Source: *Statistical Abstract of the United States, 2001*, Table 1005, p. 634.

    Forecast gaming hardware sales for 2002.

2.   **Number of Farms**

Years Since 1978 (t)	Farms (thousands) (f)
0	1,015
4	987
9	965
14	946
19	932

Source: *Statistical Abstract of the United States, 2001*, Table 803, p. 523.

    Forecast the number of farms in 2000.

3.   **Community College Education Costs**

Maricopa Community College District Tuition and Fees

Years Since 1997–98 (t)	Cost per Credit (dollars) (C)
0	37
1	38
2	40
3	41
4	43
5	46
6	51

Source: www.dist.maricopa.edu.

    Forecast the cost per credit for students attending college in the district in 2005–06.

4.   **Breakfast Cereal Consumption**

Per Capita Consumption of Breakfast Cereals

Years Since 1980 (t)	Consumption (pounds) (C)
0	12
1	12
2	11.9
3	12.2
4	12.5
5	12.8
6	13.1
7	13.3
8	14.2
9	14.9
10	15.4
11	16.1

(Continued)

(Continued)

12	16.6
13	17.3
14	17.4
15	17.1
16	16.6
17	16.3
18	15.6
19	15.5

Source: *Statistical Abstract of the United States, 2001*, Table 202, p. 129.

Forecast the per capita consumption of breakfast cereal in 2001.

5.   **State University Enrollment**

Washington State Public University Enrollment

Years Since 1990 (t)	Students (S)
0	81,401
1	81,882
2	83,052
3	84,713
4	85,523
5	86,080
6	87,309
7	89,365
8	90,189
9	91,543
10	92,821

Source: Washington State Higher Education Coordinating Board, Higher Education Statistics, September 2001.

Forecast the Washington state public university enrollment in 2010.

6.   **Community College Education Costs**

Average Annual Undergraduate Tuition and Fees at Washington State Community Colleges

Years Since 1984-85 (t)	Tuition and Fees (dollars) (F)
0	581
1	699
2	699
3	759
4	780
5	822
6	867
7	945
8	999
9	1,125
10	1,296
11	1,350
12	1,401
13	1,458
14	1,515
15	1,584
16	1,641
17	1,743

Source: Washington State Higher Education Coordinating Board, Higher Education Statistics, September 2001.

Forecast the annual tuition and fees at a Washington state community college in 2005.

7. **AIDS Incidence in Children**

**Estimated Pediatric AIDS Incidence
(United States only)**

Years Since 1992 (t)	Number of Cases (C)
0	954
1	927
2	821
3	687
4	515
5	329
6	235
7	179
8	120
9	101

Source: Centers for Disease Control and
Prevention, "HIV/AIDS Surveillance
Report," December 2001, p. 36.

Forecast the number of pediatric AIDS cases
in the United States in 2005.

8. **Per Capita Personal Income**

Per Capita Personal Income-Florida

Years Since 1993 (t)	Personal Income (dollars) (P)
0	21,320
1	21,905
2	22,942
3	23,909
4	24,869
5	26,161
6	26,593
7	27,764

Source: Bureau of Economic Analysis
(www.bea.gov).

Forecast the per capita personal income in
Florida in 2004.

9. **Aviation**

Air Carrier Accidents

Year (t)	Accidents (A)
1992	18
1993	23
1994	23
1995	36
1996	37
1997	49
1998	50
1999	52
2000	54

Source: Statistical Abstract
of the United States, 2001,
Table 1063, p. 669.

Forecast the number of air carrier accidents in
2002.

10. **Federal Funds for Elections**

**Federal Funds for Presidential Election
Campaigns**

Years Since 1980 (t)	Federal Funds (millions of dollars) (F)
0	62.7
4	80.3
8	92.2
12	110.4
16	152.6

Source: Statistical Abstract of the United States,
2001, Table 409, p. 255.

Forecast the amount of federal funds spent in
the 2004 presidential election.

*In Exercises 11–20, find mathematical models for
each of the verbal descriptions.*

11. Candy Bar Prices Candy bars currently cost
$0.60 each. The price of a candy bar is expected
to increase by 3 percent per year in the future.

12. **Housing Prices** On March 12, 2004, a builder priced a new home in Queen Creek, Arizona, at $198.9K. The builder's sales representative told the author that the price for that home style would increase on March 16, March 30, and April 13 to $205.9K, $209.9K, and $212.9K, respectively. (**Source:** Fulton Homes.)

Find a mathematical model for the price of the new home style.

13. **Calculator Prices** A calculator currently costs $87. The price of the calculator is expected to increase by $3 per year.

Find a mathematical model for the price of the calculator.

14. **Mortality Rates** There are presently 95 members of a high school graduating class who are still living. The number of surviving class members is decreasing at a rate of 4 percent per year.

Find a mathematical model for the number of surviving class members.

15. **Product Sales Growth** We are introducing a new product next year. We anticipate that sales will initially be slow but will increase rapidly once people become aware of our product. We anticipate that our monthly sales will start to level off in 18 months at about $200,000. We predict that sales for the first two months will be $12,000 and $19,000, respectively.

Develop a mathematical model to forecast monthly product sales.

16. **Club Membership** We are concerned about the decreasing number of members of our business club. Two years ago, we had 200 members. Last year we had 165 members, and this year we have 110 members. If something doesn't change, we expect to lose even more members next year than we lost this year.

Develop a mathematical model for the club membership.

17. **Population Growth** The town of Queen Creek, Arizona, was founded in 1989. In 1990, there were 2667 people living in the town. The town grew rapidly in the 1990s, in large part because of new home construction in the area. There were 4316 people living in the town in 2000 and 4940 people in 2001. The

Arizona Department of Commerce estimated the 2002 population of Queen Creek at 5555 people.

If a logistic model is used to model the population of Queen Creek, what is the maximum projected population of the city? (*Hint:* First align the data.)

The Arizona Department of Commerce estimated the 2003 population of Queen Creek at 7480. In light of this additional information, is there a different type of model that would have better predicted the 2003 population of Queen Creek? Explain.

18. **Housing Prices** On March 12, 2004, a builder priced a new home in Queen Creek, Arizona, at $143.9K. The builder's sales representative told the author that the price for that home style would increase on March 16, March 30, and April 13 to $148.9K, $151.9K, and $153.9K, respectively. (**Source:** Fulton Homes.)

Find a mathematical model for the price of the new home style.

19. **Federal Tax Rates** Federal income tax rates are dependent upon the amount of taxable income received. In 2003, federal income taxes were calculated as follows. For single filers, the first $7000 earned was taxed at 10 percent. The next $21,400 earned was taxed at 15 percent. The next $40,400 earned was taxed at 25 percent.

For example, the tax of a single woman who earned $25,000 would be calculated as follows:

10% tax on the first $7,000

$7,000 \times 0.10 = $700

Amount to be taxed at a higher rate

$25,000 - $7000 = $18,000

15% tax on the next $18,000

$18,000 \times 0.15 = $2,700

The person's total tax is

$700 + $2,700 = $3,400

Find a mathematical model for income tax as a function of taxable income for single filers.

20. **Federal Tax Rates** In 2003, federal income taxes for married individuals filing jointly were calculated as follows. The first $14,000 was taxed at 10 percent. The next $42,800 was taxed at 15 percent. The next $57,850 was taxed at 25 percent.

The tax of a married couple who earned $65,000 would be calculated as follows:

10% tax on the first $14,000

$14,000 × 0.10 = $1,400

Amount to be taxed at a higher rate

$65,000 − $14,000 = $51,000

15% tax on the next $42,800

$42,800 × 0.15 = $6420

Amount to be taxed at a higher rate

$51,000 − $42,800 = $8,200

25% tax on next $8,200

$8,200 × 0.25 = $2,050

The couple's total tax is

$1,400 + $6,420 + $2,050 = $9,870

Find a mathematical model for income tax as a function of taxable income for couples filing jointly.

Exercises 21–25 are intended to challenge your understanding of mathematical modeling.

21. The graph of a mathematical model passes through all of the points of a data set. A student claims that the model is a perfect forecaster of future results. How would you respond?

22. A scatter plot is concave up and increasing on [0, 5], concave down and increasing on [5, 8], and decreasing at a constant rate on [8, 15]. Describe two different mathematical models that may fit the data set.

23. Describe how a business owner can benefit from mathematical modeling, despite the imprecision of a model's results.

24. Daily fluctuations in the stock market make the share price of a stock very difficult to model. What approach would you take if you wanted to model the long-term performance of a particular stock?

25. You are asked by your boss to model the data shown in the following scatter plot. How would you respond?

Chapter 5 Review Exercises

Section 5.1 *In Exercises 1–13, find the model that best fits the data. Use the model to answer the given questions.*

1.   **Advertising Expenditures: Magazines**

Years Since 1990 (t)	Advertising Expenditures (millions of dollars) (A)
0	6,803
1	6,524
2	7,000
3	7,357
4	7,916
5	8,580
6	9,010
7	9,821
8	10,518
9	11,433
10	12,348

Source: *Statistical Abstract of the United States, 2001,* Table 1272, p. 777.

According to the model, how much money was spent on magazine advertising in 2002?

2.

Advertising Expenditures: Cable TV

Years Since 1990 (t)	Advertising Expenditures (millions of dollars) (A)
0	2,457
1	2,728
2	3,201
3	3,678
4	4,302
5	5,108
6	6,438
7	7,237
8	8,301
9	10,429
10	12,364

Source: *Statistical Abstract of the United States, 2001,* Table 1272, p. 777.

According to the model, how much money was spent on cable television advertising in 2002?

3. Using the models from Exercises 1 and 2, determine in what year cable television advertising expenditures are expected to exceed magazine advertising expenditures.

4. **Advertising Expenditures: Radio**

Years Since 1990 (t)	Advertising Expenditures (millions of dollars) (A)
0	8,726
1	8,476
2	8,654
3	9,457
4	10,529
5	11,338
6	12,269
7	13,491
8	15,073
9	17,215
10	19,585

Source: *Statistical Abstract of the United States, 2001,* Table 1272, p. 777.

According to the model, when will radio advertising exceed $25 billion?

5. **Advertising Expenditures: Yellow Pages**

Years Since 1990 (t)	Advertising Expenditures (millions of dollars) (A)
0	8,926
1	9,182
2	9,320
3	9,517
4	9,825
5	10,236
6	10,849
7	11,423
8	11,990
9	12,652
10	13,367

Source: *Statistical Abstract of the United States, 2001,* Table 1272, p. 777.

According to the model, when will Yellow Pages advertising exceed $15 billion?

Section 5.2

6. **Federal Credit Unions**

Years Since 1975 (t)	Credit Unions (C)
0	12,737
5	12,440
10	10,125
15	8,511
20	7,329
25	6,336

Source: *Statistical Abstract of the United States, 2001,* Table 1184, p. 732.

According to the model, how many federal credit unions were there in 1999? Does this seem reasonable? Explain.

7. **Advertising Expenditures: Newspapers**

Years Since 1990 (t)	Advertising Expenditures (millions of dollars) (A)
0	32,281
1	30,409
2	30,737
3	32,025
4	34,356
5	36,317
6	38,402
7	41,670
8	44,292
9	46,648
10	49,246

Source: *Statistical Abstract of the United States, 2001,* Table 1272, p. 777.

According to the model, how much money was spent on newspaper advertising in 2001?

8. **Full-Service Restaurant Sales**

Years Since 1980 (t)	Sales (millions of dollars) (S)
0	39,307
2	46,443
4	54,815
6	61,474
8	69,356
10	77,811
12	83,561
14	91,457
16	100,830
18	117,774
20	134,461

Source: *Statistical Abstract of the United States, 2001,* Table 1268, p. 775.

According to the model, when will full-service restaurant sales exceed $150 billion?

9. **Advertising Expenditures: Broadcast TV**

Years Since 1990 (t)	Advertising Expenditures (millions of dollars) (A)
0	26,616
1	25,461
2	27,249
3	28,020
4	31,133
5	32,720
6	36,046
7	36,893
8	39,173
9	40,011
10	44,438

Source: *Statistical Abstract of the United States, 2001,* Table 1272, p. 777.

According to the model, when will broadcast television advertising expenditures exceed $50 billion?

10. **Manufacturing Full-Time Employees: Leather and Leather Products**

Years Since 1995 (t)	Employees (thousands) (N)
0	106
1	95
2	89
3	84
4	76

Source: *Statistical Abstract of the United States, 2001,* Table 979, p. 622.

According to the model, in what year will the number of full-time employees in the leather and leather products manufacturing industry drop below 65,000?

11.   **Wages**

Average Hourly Earnings in Manufacturing Industries: Michigan

Years Since 1980 (t)	Average Earnings (dollars per hour) (E)
0	9.52
1	10.53
2	11.18
3	11.62
4	12.18
5	12.64
6	12.80
7	12.97
8	13.31
9	13.51
10	13.86
11	14.52
12	14.81
13	15.36
14	16.13
15	16.31
16	16.67
17	17.18
18	17.61
19	18.38
20	19.20

Source: *Statistical Abstract of the United States, 2001*, Table 978, p. 622.

According to the model, what will the average hourly wage in Michigan manufacturing industries be in 2003?

12. **Wages**

Average Hourly Earnings in Manufacturing Industries: Florida

Years Since 1980 (t)	Average Earnings (dollars per hour) (E)
0	5.98
1	6.53
2	7.02
3	7.33
4	7.62
5	7.86
6	8.02
7	8.16
8	8.39
9	8.67
10	8.98
11	9.30
12	9.59
13	9.76
14	9.97
15	10.18
16	10.55
17	10.95
18	11.43
19	11.83
20	12.28

Source: *Statistical Abstract of the United States, 2001*, Table 978, p. 622.

According to the model, what will the average hourly wage in Florida manufacturing industries be in 2003?

13. Based on the wage data in Exercises 11–12, do you think it would be better to start up a manufacturing business in Florida or in Michigan? Justify your answer and explain what other issues might affect your decision.

Section 5.3 *In Exercises 14–15, determine if the graph of the function is increasing or decreasing and if it is concave up or concave down. Identify the coordinates of the y-intercept. Then graph the function to verify your conclusions.*

14. $y = 2(0.75)^x$ **15.** $y = -0.3(2.8)^x$

In Exercises 16–17, find the equation of the exponential function that fits the data in the table algebraically.

16.

x	y
0	3
1	15
2	75
3	375

17.

x	y
2	36
4	144
6	576
8	2,304

In Exercises 18–19, use exponential regression to model the data in the table. Use the model to predict the value of the function when t = 10, and interpret the real-world meaning of the result.

18.

Number of Subway Restaurants

Years Since 1996 (t)	Restaurants (N)
0	12,516
1	13,066
2	13,600
3	14,162
4	14,662

Source: www.subway.com.

19.

Number of McDonalds Restaurants

Years Since 1997 (t)	Restaurants (N)
0	22,928
1	24,513
2	26,309
3	28,707
4	30,093

Source: www.mcdonalds.com.

In Exercises 20–21, find the exponential function that fits the verbal description. Calculate the value of the function five years from now and interpret its real-world meaning.

20. My car is depreciating at a rate of 17 percent per year. It is currently valued at $6000.

21. My monthly household expenses are increasing by 2.3 percent annually. It currently costs $4500 per month to maintain my household.

In Exercises 22–23, find the solution by graphically solving the equation.

22. $1.9(2.6)^x = 10$ **23.** $9.7(0.4)^x = 2$

Section 5.4 *In Exercises 24–25, solve the logarithmic equation for y without using a calculator.*

24. $y = \log_6(36)$ **25.** $y = \log_5(0.2)$

In Exercises 26–27, determine the concavity and increasing/decreasing behavior of the graph of the function. Then graph the function to verify your results. (Hint: You will have to use the change-of-base formula if you graph the function on your calculator. That is, $y = \log_b(x) \Rightarrow y = \frac{\ln(x)}{\ln(b)}$)

26. $y = \log_5(x)$ **27.** $y = \log_{0.4}(x)$

In Exercises 28–31, use the inverse relationship between logarithmic and exponential functions to solve the equations for x. Simplify your answers.

28. $\left(\frac{1}{2}\right)^x = 64$ **29.** $5^x = 625$

30. $\log_3(x) = 4$ **31.** $\log_5(x) = -2$

In Exercises 32–35, use the rules of logarithms to rewrite each expression as a single logarithm.

32. $\log(5) + \log(3)$ **33.** $2\log(4x) - \log(8)$

34. $3\log(2x)^2 - \log(2x^3)$

35. $-\log(3x)^2 + \log(3x^2)$

In Exercises 36–37, use logarithmic regression to model the data in the table. Use the model to predict the value of the function when t = 10, and interpret the real-world meaning of the result.

  36. **McDonalds' Systemwide Sales**

Years Since 1997 (*t*)	Sales (millions of dollars) (*S*)
1	33,638
2	35,979
3	38,491
4	40,181
5	40,630

Source: www.mcdonalds.com.

37. **Non-Alcohol-Related Auto Accident Fatalities**

Years Since 1989 (*t*)	Fatalities (percentage) (*F*)
1	50.5
3	54.5
4	56.5
5	59.3
6	58.8
7	59.1
8	61.5
9	61.4
10	61.7

Source: *Statistical Abstract of the United States, 2001,* Table 1099, p. 688.

Does the model estimate at *t* = 10 agree with the raw data value?

Section 5.5 *In Exercises 38–40, find the equation of the mathematical model (if possible) that you believe will most accurately forecast the future behavior of the thing being modeled. Justify your conclusions.*

38. **Projected Teacher Salaries**

Years Since 1990 (*t*)	Public School Teacher Average Annual Salary (thousands of dollars) (*S*)
0	31.4
1	33.1
2	34.1
3	35.0
4	35.7
5	36.7
6	37.7
7	38.5
8	39.5
9	40.6
10	41.7

Source: *Statistical Abstract of the United States, 2001,* Table 237, p. 151.

39.   **Cassette Tape Market Share**

Cassette Tape Sales

Years Since 1993 (t)	Percent of Music Market (percentage points) (P)
0	38.0
1	32.1
2	25.1
3	19.3
4	18.2
5	14.8
6	8.0
7	4.9
8	3.4
9	2.4

Source: Recording Industry Association of America.

40.   **Music Market Size**

Music Market Size

Years Since 1997 (t)	Dollar Volume (millions) (P)
0	$12,236.80
1	$13,723.50
2	$14,584.50
3	$14,323.00
4	$13,740.89
5	$12,614.21

Source: Recording Industry Association of America.

What to do

1. Visit the Bureau of Labor Statistics web site (www.bls.gov/cpi) and access the most recent Consumer Price Index news report.

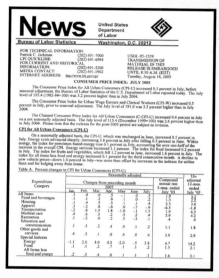

2. Select an expenditure category.

3. Record the annual percentage rate of change in that category. (This is called the compound annual rate in the report.)

4. Find the price of a product that belongs to the category you selected at a local retailer.

5. Using the Consumer Price Index information, estimate the price of the item five years from now.

Consumer Price Index explained

"The Consumer Price Index (CPI) is a measure of the average change in prices over time of goods and services purchased by households. The CPI for All Urban Consumers (CPI-U) and the Chained CPI for All Urban Consumers (C-CPI-U), which cover approximately 87 percent of the total population and include in addition to wage earners and clerical worker households, groups such as professional, managerial, and technical workers, the self-employed, short-term workers, the unemployed, and retirees and others not in the labor force." (**Source:** Bureau of Labor Statistics News, July 2002, USDL-020-480.)

Mathematics of Finance

Effective retirement planning can help ensure peace and prosperity in one's golden years. One of the greatest benefits young workers have is time. By investing wisely early in their careers, young workers have more time for their investments to grow before retirement than their older counterparts have. By understanding compound interest and annuities, you can better plan for your retirement.

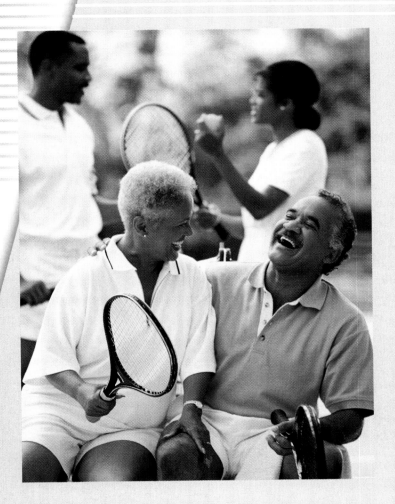

6.1 Solving Exponential Equations
- Solve exponential equations algebraically

6.2 Simple and Compound Interest
- Solve logarithmic equations algebraically
- Use simple and compound interest formulas to find the future value of an investment

6.3 Future Value of an Increasing Annuity
- Use the future value of an increasing annuity formula to answer questions related to accounts with systematic investments, such as retirement savings plans

6.4 Present Value of a Decreasing Annuity
- Use the present value of a decreasing annuity formula to answer questions related to accounts with systematic payments, such as auto loans

6.1 Solving Exponential Equations

■ Solve exponential
equations algebraically

GETTING **STARTED** The price of whole chicken between 1990 and 2000
may be modeled by

$$P(t) = 0.8443(1.025)^t \text{ dollars per pound}$$

where t is the number of years since the end of 1990. (**Source:** Modeled from
Statistical Abstract of the United States, 2001, Table 707, p. 468.) A catering business
is forecasting future food costs. It plans to increase the price of its Chicken Divan
entrée when the cost of chicken reaches $1.25 per pound. When should the cater-
ing business increase the price of the entrée? This question can be answered by
using logarithms to solve an exponential equation.

In this section, we will refine our technique of solving exponential equations algebraically. When algebraic methods fail to isolate the variable, we will use graphical
methods to estimate the solution.

Returning to the whole chicken price problem, we want to know when
$P = 1.25$. That is, we must solve

$$1.25 = 0.8443(1.025)^t$$
$$1.481 = 1.025^t \qquad \text{Divide both sides by 0.8443}$$

To get the t out of the exponent, we will take the natural log of both sides and
apply Log Rule 7.

$$\ln(1.481) = \ln(1.025)^t \qquad \text{Take the natural log of both sides}$$
$$\ln(1.481) = t\ln(1.025) \qquad \text{Rule 7}$$
$$t = \frac{\ln(1.481)}{\ln(1.025)} \qquad \text{Divide both sides by } \ln(1.025)$$
$$t \approx 15.9 \qquad \text{Evaluate on calculator}$$

The price of whole chicken is expected to reach $1.25 per pound by 2006
(16 years after 1990). The catering business should raise its prices by 2006.

The following process for solving exponential equations minimizes the
number of log rules we have to use.

HOW TO **Method for Solving Exponential Equations of the
Form $y = ab^x$**

To solve an exponential equation $y = ab^x$, do the following:

1. Divide both sides by a.

$$\frac{y}{a} = b^x$$

(Continued)

2. Take the log or natural log of both sides.

$$\log\left(\frac{y}{a}\right) = \log(b)^x \text{ or } \ln\left(\frac{y}{a}\right) = \ln(b)^x$$

3. Bring the x down in front of the log or natural log.

$$\log\left(\frac{y}{a}\right) = x \log(b) \text{ or } \ln\left(\frac{y}{a}\right) = x \ln(b)$$

4. Divide both sides by $\log(b)$ or $\ln(b)$, respectively.

$$x = \frac{\log\left(\frac{y}{a}\right)}{\log(b)} \text{ or } x = \frac{\ln\left(\frac{y}{a}\right)}{\ln(b)}$$

We will repeat this basic process in the next couple of examples to give you a feel for the method.

EXAMPLE 1   Determining the Doubling Time for Prescription Drug Spending

The amount of money spent on prescription drugs in the United States may be modeled by

$$D(t) = 2.333(1.099)^t \text{ billion dollars}$$

where t is the number of years since the end of 1960. (**Source:** Modeled from *Statistical Abstract of the United States, 2001*, Table 119, p. 91.) According to the model, how many years does it take for the amount of money spent on prescription drugs to double?

SOLUTION According to the model, $2.333 billion was spent on prescription drugs in 1960. Double the initial amount is $4.666 billion.

$$4.666 = 2.333(1.099)^t$$

$$\frac{4.666}{2.333} = (1.099)^t \qquad \text{Divide both sides by 2.333}$$

$$2 = (1.099)^t \qquad \text{Simplify}$$

$$\ln(2) = \ln(1.099)^t \qquad \text{Take the natural log of both sides}$$

$$\ln(2) = t \ln(1.099) \qquad \text{Rule 7}$$

$$t = \frac{\ln(2)}{\ln(1.099)} \qquad \text{Divide both sides by } \ln(1.099)$$

$$t \approx 7.34 \qquad \text{Evaluate on calculator}$$

According to the model, prescription drug spending doubles every seven years or so. So why do we care? Increases in drug costs are typically passed on to the consumer through higher premiums or higher co-payments.

| EXAMPLE 2 | ## Forecasting Tuition Costs with an Exponential Function |

The cost of full-time resident *quarterly* tuition at Green River Community College may be modeled by

$$E(t) = 430.6(1.042)^t \text{ dollars}$$

where t is the number of years since the end of 1994. (**Source:** Modeled from Green River Community College data.)

When their daughter was born in 1994, a couple invested $2000 in a college savings account earning 3 percent annually. They hope to use the money to pay for the first year of their daughter's college tuition, but, with rising tuition costs, they are uncertain whether their savings will cover the annual cost of the tuition. How long do they have before the annual cost of tuition (three quarters) will exceed the amount of their savings?

SOLUTION The value of the couple's investment can be modeled by

$$A(t) = 2000(1.03)^t \text{ dollars}$$

where t is the number of years since the end of 1994.

The annual cost of tuition is $3E$, since three quarters is considered a full academic year. We want to find out when $3E = A$.

$$3E = A$$

$$3[430.6(1.042)^t] = 2000(1.03)^t$$

$1291.8(1.042)^t = 2000(1.03)^t$ Simplify

$\dfrac{(1.042)^t}{(1.03)^t} = \dfrac{2000}{1291.8}$ Divide both sides by $(1.03)^t$ and by 1291.8

$\left(\dfrac{1.042}{1.03}\right)^t = 1.548$ Simplify

$(1.012)^t = 1.548$ Simplify

$\ln(1.012)^t = \ln(1.548)$ Take the natural log of both sides

$t \ln(1.012) = \ln(1.548)$ Rule 7

$t = \dfrac{\ln(1.548)}{\ln(1.012)}$ Divide both sides by $\ln(1.012)$

$t \approx 37.74$ Evaluate on calculator

(*Note:* As in other examples and exercises in this text, we are keeping the actual values in our calculator even though we round the values when we write them down. By waiting to round until we have our final answer, we obtain a more accurate result.)

According to the model, the annual cost of tuition will exceed the balance of the savings account in 2032 (38 years after 1994). What will the annual cost of tuition be that year? Evaluating either function at $t = 37.74$, we find that the *annual* tuition is $6102. That is, in 2032, we expect the *quarterly* tuition to be about $2034! Fortunately, the couple's savings will be sufficient to cover a year of tuition provided that their daughter goes to college before she's 38 years old.

Example 2 was a bit more challenging because it required us to solve an exponential equation of the form $ab^t = cd^t$. An equation of this form can always be reduced as detailed in the following box.

HOW TO Method for Solving Exponential Equations of the Form $ab^x = cd^x$

To solve an exponential equation $ab^x = cd^x$, do the following:

1. Divide both sides by a.

$$b^x = \frac{cd^x}{a}$$

2. Divide both sides by d^x.

$$\frac{b^x}{d^x} = \frac{c}{a}$$

3. Rewrite $\dfrac{b^x}{d^x}$ as $\left(\dfrac{b}{d}\right)^x$.

$$\left(\frac{b}{d}\right)^x = \frac{c}{a}$$

4. Solve using the method for solving exponential equations of the form $y = b^x$.

$$\ln\left(\frac{b}{d}\right)^x = \ln\left(\frac{c}{a}\right)$$

$$x \ln\left(\frac{b}{d}\right) = \ln\left(\frac{c}{a}\right)$$

$$x = \frac{\ln\left(\dfrac{c}{a}\right)}{\ln\left(\dfrac{b}{d}\right)}$$

EXAMPLE 3 Solving a System of Equations Graphically

Health care expenditures by insurance companies and patients have risen sharply over the past four decades. Based on data from 1960 to 2000, patient out-of-pocket expenditures can be modeled by

$$P(t) = 13.16(1.075)^t \text{ billion dollars}$$

and insurance company expenditures can be modeled by

$$I(t) = 0.3593t^2 - 3.561t + 7.857 \text{ billion dollars}$$

where t is the number of years since the end of 1960. (**Source:** Modeled from *Statistical Abstract of the United States, 2001*, Table 119, p. 91.)

Patient spending initially exceeded insurance company spending. In what year, were insurance company and patient out-of-pocket spending the same?

SOLUTION   We must find when $P(t) = I(t)$.

$$P(t) = I(t)$$

$$13.16(1.075)^t = 0.3593t^2 - 3.561t + 7.857$$

Observe that the left-hand side of the equation has a variable in the exponent. The only way we will be able to get the variable out of the exponent is by taking the log of both sides. However, if we take the log of both sides, the equation becomes

$$\log[13.16(1.075)^t] = \log(0.3593t^2 - 3.561t + 7.857)$$

leaving us with the variable trapped inside the logarithm on the right-hand side. Because of this dilemma, we are unable to solve the problem algebraically. However, we can use a graphing calculator to graph $P(t) = 13.16(1.075)^t$ and $I(t) = 0.3593t^2 - 3.561t + 7.857$ and find their points of intersection. Doing so yields Figure 6.1.

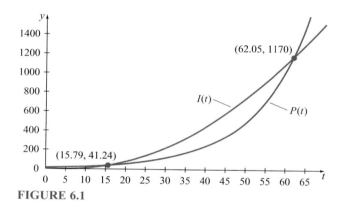

FIGURE 6.1

The graphs intersect at (15.79, 41.24) and (62.05, 1170). [The graphs also intersect at $(-1.089, 12.16)$; however, we are interested only in positive values of t.] According to the models, insurance spending first surpassed out-of-pocket expenditures in 1976 ($t = 16$). However, out-of pocket expenditures are expected to exceed insurance spending in 2022 ($t = 62$). Since our model was based on data from $t = 0$ to $t = 40$, we are a bit skeptical about the accuracy of the second projection because $t = 62$ is relatively far away from the data values used to create the model. On the other hand, we feel very comfortable with our first projection, since it is within the range of data values used to create the model. Referring back to the source data used to create the model (not included here), we discover that in 1977, out-of-pocket and insurance expenditures were both equal to $45 billion. Our projection of $41 billion in 1976 was off by about a year and about $4 billion.

Predicting the output value for an input value outside of the interval of the input data is called **extrapolation.** Predicting the output value for an input value inside the interval of the input data is called **interpolation.** We must use extreme caution when considering the extrapolated values of a model. Typically, the further away the input value is from the domain of the data set, the less likely it is that the extrapolated value is accurate in modeling the real-life scenario.

6.1 Summary

In this section, you refined your ability to solve exponential equations algebraically. When algebraic methods failed to isolate the variable, you used graphical methods to estimate the solution.

6.1 Exercises

In Exercises 1–10, solve the exponential equation algebraically.

1. $2(3^x) = 162$

2. $3(2^x) = 768$

3. $1.1^x = 1.21$

4. $5(0.5)^x = 0.125$

5. $3(2^x) = 4^x$

6. $4(3^x) = 10(2^x)$

7. $2(1.09)^t = 3(2.11)^t$

8. $4(3.9)^t = 1.7$

9. $2(3.1)^t = 3.6$

10. $3(7.1)^t = 99$

In Exercises 11–20, solve the problems algebraically.

11. **M** **Church Growth** The number of religious congregations of the Church of Jesus Christ of Latter-Day Saints in the United States may be modeled by
$$C(t) = 8679(1.023)^t \text{ congregations}$$
where *t* is the number of years since the end of 1988. (**Source:** Modeled from data compiled by Mark Davies, Brigham Young University.)
In what year is the number of congregations expected to reach 15,000?

12. **M** **Video Game Sales** In the first quarter of 2001, U.S. video game industry sales totaled $1.6 billion. Sales increased by 20 percent in the first quarter of 2002. (**Source:** www.npdfunworld.com.) If this pattern continues, video game industry sales in the first quarter, *S*, may be modeled by
$$S(t) = 1.6(1.2)^t \text{ billion dollars}$$
where *t* is the number of years since the end of 2001.
In what year are first-quarter sales expected to exceed $3.2 billion?

13. **M** **Land in Farms** The amount of land in U.S. farms may be modeled by
$$F(t) = 115.1(0.9352)^t + 900 \text{ million acres}$$
where *t* is the number of years since the end of 1978. (**Source:** Modeled from *Statistical Abstract of the United States, 2001*, Table 796, p. 523.)

According to the model, in what year will the amount of land in U.S. farms drop below 910 million acres?

14. **M** **Milk Beverage Consumption** The annual per capita milk beverage consumption may be modeled by
$$M(t) = 27.76(0.9914)^t \text{ gallons}$$
where *t* is the number of years since the end of 1980. (**Source:** Modeled from *Statistical Abstract of the United States, 2001*, Table 202, p. 129.)
According to the model, in what year will the average person drink 20 gallons of milk?

15. **M** **Bottled Water Consumption** The annual per capita bottled water consumption may be modeled by
$$W(t) = 2.593(1.106)^t \text{ gallons}$$
where *t* is the number of years since the end of 1980. (**Source:** Modeled from *Statistical Abstract of the United States, 2001*, Table 204, p. 130.)
According to the model, in what year will the average person drink 20 gallons of bottled water?

16. **M** **Beverage Consumption** The annual per capita milk beverage consumption may be modeled by
$$M(t) = 27.76(0.9914)^t \text{ gallons}$$
and the annual per capita bottled water consumption may be modeled by
$$W(t) = 2.593(1.106)^t \text{ gallons}$$
where *t* is the number of years since the end of 1980. (**Source:** Modeled from *Statistical Abstract of the United States, 2001*, Tables 202, 204, pp. 129–130.)
A dairy farmer discovered a freshwater spring on his property and, recognizing the decreasing demand for milk and the rising demand for bottled water, is considering producing bottled water instead of milk. As a part of his business

plan, he wants to know when the consumption of bottled water is projected to exceed the consumption of milk. According to the models, when will bottled water consumption surpass milk beverage consumption?

17. **Church Growth**   Based on data from 1990 to 2001, the number of American adults who consider themselves Lutherans may be modeled by

$$L(t) = 9110(1.005)^t \text{ thousand adults}$$

and the number who consider themselves nondenominational Christians may be modeled by

$$C(t) = 8073(1.053)^t \text{ thousand adults}$$

where t is the number of years since the end of 1990. (**Source:** Modeled from American Religious Identification Survey.)

According to the models, in what year will the number of adults in each religious group be the same?

18. **Bread Prices**   The price of white bread has been increasing exponentially since 1980. The consumer price index for white bread may be modeled by

$$I(t) = 86.56(1.043)^t \text{ index points}$$

where t is the number of years since the end of 1980. (**Source:** Modeled from *Statistical Abstract of the United States, 2001*, Table 694, p. 455.)

In 1984, the index was 100. When the index reaches 300, the price of bread will have tripled over its 1984 price. In what year is that expected to occur?

19. **Population Growth**   The projected population of Ohio may be modeled by

$$O(t) = 11{,}160(1.003)^t \text{ thousand people}$$

and the projected population of Illinois may be modeled by

$$I(t) = 12{,}050(1.002)^t \text{ thousand people}$$

where t is the number of years since the end of 1995.

The number of representatives a state has in Congress is based in part on the state's population. According to the models, in what year will the population of the two states (and therefore their representation in Congress) be equal?

20. **Prescription Drug Expenditures**   Based on data from 1960 to 2000, prescription drug expenditures in the United States may be modeled by

$$D(t) = 2.333(1.099)^t \text{ billion dollars}$$

and out-of-pocket health care expenditures by consumers may be modeled by

$$P(t) = 13.16(1.075)^t \text{ billion dollars}$$

where t is the number of years since the end of 1960. (**Source:** Modeled from *Statistical Abstract of the United States, 2001*, Table 119, p. 91.)

According to the models, in what year will out-of-pocket health care spending and prescription drug spending be equal?

In Exercises 21–25, solve the problem graphically.

21. $2(3^x) = x^2 + 4$ **22.** $2(3^{-x}) = -x^2 + 2$

23. **Teacher Salaries**   Based on data from 1990 to 2000, the average salary of public elementary and secondary school teachers may be modeled by

$$S(t) = 32.05(1.027)^t \text{ thousand dollars}$$

where t is the number of years since the end of 1990. (**Source:** Modeled from *Statistical Abstract of the United States, 2001*, Table 237, p. 151.)

Based on data from 1998 to 2000, the average salary of an assistant professor at a public university may be modeled by

$$P(x) = 1.7x + 42.4 \text{ thousand dollars}$$

where x is the number of years since the end of 1998. (**Source:** Modeled from *Statistical Abstract of the United States, 2001*, Table 280, p. 173.)

In what year are the salaries of public elementary and secondary school teachers and public university assistant professors expected to be equal? (*Hint:* Before graphing the functions, make sure they are in terms of the same variable, t or x.)

24. **Beverage Consumption**   The annual per capita milk beverage consumption may be modeled by

$$M(t) = -0.219t + 27.7 \text{ gallons}$$

and the annual per capita bottled water consumption may be modeled by

$$W(t) = 2.593(1.106)^t \text{ gallons}$$

where t is the number of years since the end of 1980. (**Source:** Modeled from *Statistical Abstract of the United States, 2001*, Tables 202, 204, pp. 129–130.)

In what year is the consumption of milk and bottled water expected to be the same?

Compare your results for this exercise to the solution of Exercise 16. Did the exponential and linear models for milk consumption yield the same result?

25. **M** **Coca-Cola Company Sales** Based on data from 1999-2001, the number of unit cases of Coca-Cola brand beverages sold worldwide may be modeled by

$$S(t) = 16.5(1.04)^t \text{ billion unit cases}$$

and the net operating revenue of the Coca-Cola Company may be modeled by

$$R(t) = -0.2t^2 + 0.8t + 19.3 \text{ billion dollars}$$

where t is the number of years since the end of 1999. (**Source:** Modeled from Coca-Cola Company 2001 Annual Report, pp. 46, 57.)

When is the number of unit cases sold (in billions) expected to equal the number of dollars of net operating revenue (in billions)?

In Exercises 26–30, use algebraic or graphical methods to find the solution.

26. **Investments** Between August 1, 1952, and December 31, 2002, the CREF Stock Account earned an average annual return of 10.11 percent. (**Source:** TIAA-CREF.) If an investor invests $500 in an account earning 10.11 percent annually, how long will it take for the investment to grow to $1000?

27. **Investments** Between October 2, 1995, and December 31, 2002, the TIAA Real Estate Account earned an average annual return of 7.74 percent. (**Source:** TIAA-CREF.) If an investor invests $500 in an account earning 7.74 percent annually, how long will it take for the investment to grow to $1000?

28. **Investments** If $500 is invested in an account earning 9.2 percent annually and $1000 is invested in an account earning 5.3 percent annually, how long will it take for the values of the two accounts to be equal?

29. **Investments** An investor puts $500 into an account earning 8 percent annually and $1000 into an account losing 8 percent annually.

(a) How long will it take for the value of the accounts to be equal?

(b) When the value of the two accounts are equal, what is the combined value of the investments?

(c) Thirty years after the initial investment is made, what will be the combined value of the investments?

(d) If the investor had invested the initial $1500 in an account paying an annual rate of 4.15 percent annually, what would the account value have been at the end of 30 years?

(e) Compare the results of parts (c) and (d). As an investor, how could you use this information?

30. **Investments** If $500 is invested in an account earning 10 percent annually and $500 is invested in an account losing 10 percent annually, will the combined account value remain at a constant $1000? Explain.

Exercises 31–40 are intended to challenge your understanding of exponential equations.

31. **Investments** If $100 is invested in an account earning 5 percent annually and $100 is invested in an account losing 6 percent annually, will the combined value of the accounts increase over time? Explain.

32. Does the equation

$$3^x = -x^2 + 4x - 2$$

have a solution? Explain.

33. Solve the exponential equation algebraically.

$$3(2^x) = 3^{2x}$$

34. Solve the exponential equation algebraically.

$$5(2^{x+1}) = 2^{2x-1}$$

35. For what values of a will the equation

$$-x^2 = a(2^x)$$

have a solution?

36. For what values of b will the equation

$$x^2 = 2(b^x)$$

have a solution?

37. At least how many solutions will the given equation have?

$$a^{3x} = b^{2x}; \quad a > 0, b > 0$$

38. For what values of a and b will the given equation have an infinite number of solutions?

$$a^{3x} = b^{2x}; \ a > 0, b > 0$$

39. For what values of a and b do the graphs of $g(x) = a^x$ and $f(x) = b^x$ intersect in more than one place? Explain.

40. Given that $ac \neq bd$, do the graphs of $g(x) = ca^x$ and $f(x) = db^x$ intersect in more than one place? Explain.

6.2 Simple and Compound Interest

- Solve logarithmic equations algebraically
- Use the simple and compound interest formulas to find the future value of an investment

GETTING STARTED Inflation diminishes the buying power of money over time. A common challenge faced by all investors is earning a high enough return on an investment to stay ahead of inflation. A commonly used, conservative investment is the certificate of deposit (CD). CDs are FDIC-insured up to $100,000 per person. They have a fixed maturity date, usually from three months to five years, and they typically pay higher interest than a savings account. A penalty is often charged for withdrawing funds before the maturity date.

Bankrate.com compiles advertised rates from financial institutions across the nation and allows consumers to search for the best rates locally or nationwide. In this section, we will demonstrate how to use the simple and compound interest formulas to compare investment options such as those found on Bankrate.com.

Simple Interest

An investment paying an annual rate of simple interest pays a fixed amount annually. This amount, called **interest,** is calculated as a percentage of the initial investment. The simple interest earned after t years on an investment of P dollars earning an interest rate r annually is $I = Prt$. The initial investment, P, is commonly called the **principal.** The value of the investment after t years is referred to as the **future value** of the investment.

SIMPLE INTEREST

The future value A of an initial investment P earning a simple interest rate r is given by

$$A = P + Prt$$
$$= P(1 + rt)$$

where t is the number of years after the initial investment is made.

The rate, r, is the decimal form of the percentage rate. That is, 5 percent is written as 0.05. The amount of interest earned after t years is Prt.

EXAMPLE 1

Determining the Future Value of a Certificate of Deposit

On August 30, 2002, the one-year CD with the highest advertised rate nationally was a CD paying an annual rate of 3 percent simple interest annually, offered by ING DIRECT. (**Source:** www.bankrate.com.) What will be the value of a $200 investment when it matures?

SOLUTION We have $P = 200$ dollars, $r = 0.03$, and $t = 1$ year (since it is a one-year CD). The future value of the investment is

$$A = P + Prt$$
$$= 200 + 200(0.03)(1)$$
$$= 200 + 6$$
$$= 206$$

When the CD matures, it will be worth $206.

EXAMPLE 2

Determining a Simple Interest Rate

How high a return would we have to earn if we wanted a $200 simple-interest investment to be valued at $300 after five years?

SOLUTION We have $A = 300$ dollars, $P = 200$ dollars, and $t = 5$ years.

$$A = P + Prt$$
$$300 = 200 + 200(r)(5)$$
$$100 = 1000r$$
$$r = 0.1$$

We would have to earn an annual rate of 10 percent simple interest in order for a $200 investment to reach $300 in five years.

Compound Interest

Simple interest investments earn interest only on the initial amount invested. On the other hand, compound interest investments earn interest on the initial amount invested and any previously earned interest. Consequently, simple interest investments grow linearly, while compound interest investments grow exponentially.

For example, consider a $1000 investment into an account paying a 5 percent interest rate compounded quarterly. What will be the future value of the account after one year? As demonstrated in Chapter 5, an annual rate of 5 percent corresponds with an annual growth factor of 1.05 (recall that $b = 1 + r$). However, since interest is paid quarterly, not annually, we must determine the quarterly rate. Since there are four quarters in one year, the quarterly interest rate is calculated by dividing the annual interest rate by 4. The quarterly rate is $\frac{5 \text{ percent}}{4} = 1.25$ percent. The quarterly growth factor is 1.0125. So to determine the value of our investment after interest is paid at the end of the first quarter, we multiply our $1000 investment by 1.0125.

$$1000(1.0125) = \$1012.50$$

Over the course of the first quarter, $12.50 in interest was earned.

What will the value of the account be after interest is paid at the end of the *second* quarter? Since the rate and **compounding frequency** (the number of times interest is paid each year) remain unchanged, the quarterly growth factor is still 1.0125. Therefore, if we multiply the $1012.50 account value by the growth factor, we will obtain the investment value after interest is paid at the end of the second quarter.

$$1012.50(1.0125) = \$1025.16$$

Over the course of the second quarter, $12.66 in interest was earned. Why was this amount greater than the interest for the first quarter? Because the first-quarter interest earned interest itself during the second quarter.

What will the value of the account be after interest is paid at the end of the *third* quarter? Since the rate and compounding frequency remain unchanged, the quarterly growth factor is still 1.0125. Therefore, if we multiply the $1025.16 account value by the growth factor, we will obtain the investment value after interest is paid at the end of the third quarter.

$$1025.16(1.0125) = \$1037.97$$

Over the course of the third quarter, $12.81 in interest was earned.

What will the value of the account be after interest is paid at the end of the *fourth* quarter? Again, we multiply $1037.97 by the growth factor of 1.0125. The investment value after interest is paid at the end of the fourth quarter is given by

$$1037.97(1.0125) = \$1050.95$$

Over the course of the fourth quarter, $12.98 in interest was earned.

Observe that for each quarter, we calculated the future value of the account by multiplying by the growth factor one or more times.

$$1000(1.0125)^1 = 1012.50$$
$$1000(1.0125)^2 = 1025.16$$
$$1000(1.0125)^3 = 1037.97$$
$$1000(1.0125)^4 = 1050.95$$

Observe that for each subsequent quarter, the exponent on the growth factor is increased by 1. If we want to know the account value at the end of t years, we must first convert the years to quarters. Since there are 4 quarters in a year, there are $4t$ quarters in t years.

Recall that the growth factor itself was determined by adding 1 to the annual percentage rate divided by the number of times interest was paid each year.

$$\left(1 + \frac{0.05}{4}\right) = 1.0125$$

Combining all of these observations, we can find an equation for calculating the future value A of the account after t years.

$$A = 1000\left(1 + \frac{0.05}{4}\right)^{4t}$$

This equation is the compound interest formula for the account earning 5 percent interest compounded quarterly with an initial investment of \$1000.

COMPOUND INTEREST

The future value A of an initial investment P earning a compound interest rate $100r$ percent is given by

$$A = P\left(1 + \frac{r}{n}\right)^{nt}$$

where n is the number of times interest is paid annually and t is the number of years after the initial investment is made. As with simple interest, the rate r is the decimal form of the percentage rate.

As noted earlier, the number of times interest is paid annually, n, is called the *compounding frequency*. In solving interest rate problems, we must convert the verbal description of the compounding frequency into a numeric value, as illustrated in Table 6.1.

TABLE 6.1

Interest Is Compounded	Compounding Frequency n
Annually	1
Semiannually	2
Quarterly	4
Monthly	12
Daily	365

Using the Compound Interest Formula to Forecast Investment Value

On August 30, 2002, the one-year CD with the second highest advertised rate nationally was a CD offered by DeepGreen Bank paying an annual rate of 2.95 percent interest compounded daily. A $1000 minimum deposit was required. (**Source:** www.bankrate.com.) What will be the value of a $1000 investment when it matures?

SOLUTION We have $P = 1000$ dollars, $r = 0.0295$, $t = 1$ year (since it is a one-year CD), and $n = 365$ (since interest is calculated daily). The future value of the investment is

$$A = P\left(1 + \frac{r}{n}\right)^{nt}$$

$$= 1000\left(1 + \frac{0.0295}{365}\right)^{365(1)}$$

$$= 1000(1.0000808)^{365}$$

$$= 1000(1.0299)$$

$$= 1029.94$$

When the CD matures, it will be worth $1029.94.

Observe that $1029.94 = 1000\left(1 + \frac{0.02994}{1}\right)^{1(1)}$. That is, earning 2.95 percent interest compounded daily for one year is equivalent to earning 2.994 percent interest compounded annually for one year.

The **annual percentage yield** of an account is the percentage rate at which interest would need to be paid if interest were calculated and paid only once a year. The annual percentage yield is the percentage that reflects the total interest to be received for a 365-day year based on an institution's compounding method. To protect consumers, Truth-in-Savings regulations require that the annual percentage yield on interest-bearing accounts be publicly disclosed.

ANNUAL PERCENTAGE YIELD

The annual percentage yield, APY, for an account earning $100r$ percent compounded n times per year is

$$APY = \left(1 + \frac{r}{n}\right)^{n} - 1$$

For a simple interest account,

$$APY = r$$

It is informative to look deeper into the origin of the annual percentage yield formulas. Observe that the compound interest formula

$$A = P\left(1 + \frac{r}{n}\right)^{nt}$$

may be written as

$$A = P\left[\left(1 + \frac{r}{n}\right)^n\right]^t$$

If r and n are known values, then $b = \left(1 + \frac{r}{n}\right)^n$ is a constant. The compound interest formula may then be rewritten as an exponential function:

$$A = P(b)^t$$

As discussed earlier, the annual rate of return of a quantity with growth factor b is equal to $b - 1$. In terms of investments, the annual percentage yield, APY, is given by

$$APY = b - 1$$

$$= \left(1 + \frac{r}{n}\right)^n - 1 \qquad \text{Since } b = \left(1 + \frac{r}{n}\right)^n$$

For *simple interest* investments, the annual percentage yield is equivalent to the simple interest rate.

EXAMPLE 4

Comparing Investments with Different Interest Payment Methods

On August 30, 2002, NetBank offered a five-year CD paying an annual rate of 4.74 percent compounded daily and ING DIRECT offered a five-year CD paying an annual rate of 4.80 percent simple interest. An investor has $1000 to invest. Which CD will have a higher value when it matures?

SOLUTION

NetBank:

$$A = P\left(1 + \frac{r}{n}\right)^{nt}$$

$$= 1000\left(1 + \frac{0.0474}{365}\right)^{365(5)}$$

$$= 1000(1.0001299)^{1825}$$

$$= 1000(1.26742)$$

$$= 1267.42$$

The NetBank CD will be worth $1267.42.

ING DIRECT:

$$A = P + Prt$$

$$= 1000 + 1000(0.0480)(5)$$

$$= 1000 + 240$$

$$= 1240$$

The ING DIRECT CD will be worth $1240.00.

The NetBank CD will be worth $27.42 more than the ING DIRECT CD. Surprisingly, the CD with the lower rate earned a greater return. The interest calculation method used made a substantial difference.

EXAMPLE   5

Determining the Annual Percentage Yield of an Investment

Returning to the investments introduced in Example 4, determine the annual percentage yield (*APY*) for each CD. Recall that NetBank offered a five-year CD paying an annual rate of 4.74 percent compounded daily and ING DIRECT offered a five-year CD paying an annual rate of 4.80 percent simple interest.

SOLUTION Since the annual percentage yield is the simple interest rate on a one-year investment, the *APY* for ING DIRECT is 4.80 percent. For NetBank,

$$APY = \left(1 + \frac{r}{n}\right)^n - 1$$

$$= \left(1 + \frac{0.0474}{365}\right)^{365} - 1$$

$$= 1.0485 - 1$$

$$= 0.0485$$

The *APY* for NetBank is 4.85 percent.

When comparing savings options, it is essential to look at the *APY* in addition to the advertised rate. As we saw in Example 5, sometimes the investment with the higher annual rate has a lower *APY*.

Continuous Compound Interest

What happens to the *APY* as the compounding frequency increases? Will it increase dramatically if interest is calculated every hour ($n = 8760$), every minute ($n = 525,600$), or every second ($n = 31,536,000$)? Consider an account earning an annual rate of 6 percent interest (see Table 6.2).

TABLE 6.2

Compounding Frequency	Number of Times Compounded Each Year (n)	APY $\left(1 + \dfrac{0.06}{n}\right)^n - 1$
Annually	1	0.06000
Quarterly	4	0.06136
Monthly	12	0.06168
Daily	365	0.06183
Hourly	8,760	0.06184
Every minute	525,600	0.06184
Every second	31,536,000	0.06184

It appears that as the compounding frequency increases, the annual percentage yield approaches a constant value of 0.06184. Recall from Section 5.4 that the expression $\left(1 + \dfrac{1}{n}\right)^n$ was observed to approach the irrational number

$e = 2.718281828 \ldots$ as n approached infinity. It is interesting to note that

$$\left(1 + \frac{0.06}{n}\right)^n \approx e^{0.06}$$
$$\approx 1.06184$$

In fact, as the compounding frequency increases, $\left(1 + \frac{r}{n}\right)^n \to e^r$ for any r. (The symbol $\to$ means "approaches" and is commonly used when one quantity is getting infinitely close to another quantity. When we use the phrase *infinitely close,* we mean that we can make the first value get as close to the second value as we would like.) This result is not immediately obvious; however, we can easily convince ourselves of its validity.

Recall that $\left(1 + \frac{1}{x}\right)^x \to e$ as $x \to \infty$. Let $x = \frac{n}{r}$. Observe that if r is any positive constant, then $x \to \infty$ as $n \to \infty$. Since $x = \frac{n}{r}$, $n = rx$. Let's return to the compound interest formula, $A = P\left(1 + \frac{r}{n}\right)^{nt}$. This may be rewritten as

$$A = P\left[\left(1 + \frac{r}{n}\right)^n\right]^t$$
$$= P\left[\left(1 + \frac{1}{x}\right)^{rx}\right]^t \qquad \text{Since } \frac{r}{n} = \frac{1}{x} \text{ and } n = rx$$
$$= P\left[\left(1 + \frac{1}{x}\right)^x\right]^{rt}$$

But $\left(1 + \frac{1}{x}\right)^x \to e$ as $x \to \infty$. So, as the compounding frequency n gets infinitely large, the compound interest formula becomes

$$A = Pe^{rt}$$

We represent the notion of infinitely large n by using the term **continuous compounding.**

CONTINUOUS COMPOUND INTEREST

The future value A of an initial investment P earning a continuous compound interest rate $100r$ percent is given by

$$A = Pe^{rt}$$

where t is the number of years after the initial investment is made. The annual percentage yield is $APY = e^r - 1$.

To understand how we determined the APY, observe that the continuous compound interest formula may be converted to the compound interest formula as shown here.

$$A = Pe^{rt}$$
$$= P(e^r)^t$$
$$= Pb^t \qquad \text{Let } b = e^r$$
$$= P(1 + APY)^t \qquad \text{Since } b = 1 + APY$$

Notice that $e^r = 1 + APY$ or, in other terms, $APY = e^r - 1$.

EXAMPLE 6

Determining the Effect of Increasing the Compounding Frequency

Verify that the annual percentage yield of an account earning a compound interest rate of 5.00 percent approaches $e^{0.05} - 1$ as the compounding frequency increases.

SOLUTION

TABLE 6.3

Compounding Frequency	n	$\left(1 + \dfrac{0.05}{n}\right)^n - 1$
Annually	1	0.05000
Monthly	12	0.05116
Daily	365	0.05127
Hourly	8,760	0.05127
Continuous	$n \to \infty$	$e^{0.05} - 1 = 0.05127$

EXAMPLE 7

Converting a Quarterly Rate to a Continuous Rate

An account earning an annual interest rate of 12.00 percent compounded quarterly earns the same as an account earning what continuous rate?

SOLUTION Let r be the rate (as a decimal) for the continuously compounding account. The compound interest formula for each account may be written in the form $A = Pb^t$. For the 12.00 percent account, we have

$$b = \left(1 + \frac{0.12}{4}\right)^4$$
$$= (1 + 0.03)^4$$
$$= (1.03)^4$$
$$= 1.1255$$

For the continuously compounding account, we have $b = e^r$. Since $b = e^r$, $\ln(b) = r$.

$$r = \ln(1.1255)$$
$$= 0.1182$$

A 12.00 percent interest rate compounded quarterly yields the same as an 11.82 percent rate compounded continuously.

EXAMPLE 8   Comparing the Values of Different Types of Investments

How many years will it take for a $300 investment earning 5 percent compounded continuously to be worth more than a $500 investment earning an annual interest rate of 4 percent compounded quarterly?

SOLUTION

Continuous: $A = 300e^{0.05t}$

Quarterly: $A = 500\left(1 + \dfrac{0.04}{4}\right)^{4t}$

We want to know when the accounts will have the same balance, so we set them equal to each other and solve.

$$300e^{0.05t} = 500\left(1 + \frac{0.04}{4}\right)^{4t}$$

$$300e^{0.05t} = 500(1.01)^{4t}$$

$0.6e^{0.05t} = (1.01)^{4t}$	Divide both sides by 500
$0.6(1.0513)^t = (1.0406)^t$	Evaluate the exponents
$0.6 = \left(\dfrac{1.0406}{1.0513}\right)^t$	Divide both sides by $(1.0513)^t$
$0.6 = (0.9899)^t$	Simplify
$\ln(0.6) = t\,\ln(0.9899)$	Take the natural log of both sides
$t = \dfrac{\ln(0.6)}{\ln(0.9899)}$	Divide both sides by $\ln(0.9899)$
$t \approx 50.09$	Evaluate

It will take a little more than 50 years for the accounts to reach the same balance. At the end of the first quarter of the 51st year ($t = 50.25$), the account paying an annual rate of interest compounded quarterly will have a balance of $3694.59 and the account paying an annual rate of continuously compounded interest will have a balance of $3700.72. It is important to note that although the mathematical solution is 50.09, the balance of the account paying an annual rate of interest compounded quarterly increases only at the end of a quarter (i.e., 50.00, 50.25, 50.50, etc.). Therefore, the quarterly-interest formula gives an accurate value for the account only if the input is one of these quarterly values.

EXAMPLE 9   Using Compound Interest to Analyze Payday Loans

On February 17, 2003, mycashnow.com published the following loan terms:

> Loan amount: $500
> Loan term: 7–14 days
> Finance charge: $93.10

If a consumer borrows the money for seven days, what is the annual percentage yield on the investment for the lender? (Assume that the finance charge is entirely interest.)

SOLUTION In this problem, the lender may be viewed as the investor and the consumer is the one paying interest. The lender's initial investment is $500. At the end of the seven-day period, the lender gets back its $500 investment plus $93.10 in interest. We have

$$A = P\left(1 + \frac{r}{n}\right)^{nt}$$

$$593.10 = 500\left(1 + \frac{r}{n}\right)^{n\left(\frac{7}{365}\right)} \qquad \text{Since 7 days is } \frac{7}{365} \text{ of a year}$$

$$593.10 = 500(b)^{\left(\frac{7}{365}\right)} \qquad \text{Let } b \text{ equal } \left(1 + \frac{r}{n}\right)^{n}$$

$$1.1862 = b^{\frac{7}{365}} \qquad \text{Divide both sides by 500}$$

$$(1.1862)^{\frac{365}{7}} = \left(b^{\frac{7}{365}}\right)^{\frac{365}{7}} \qquad \text{Raise both sides to the } \frac{365}{7}$$

$$7358.78 = b \qquad \text{Evaluate}$$

Since $APY = b - 1$, $APY = 7357.78$. Converting this to a percentage gives an annual percentage yield of 735,778 percent! To put this in perspective, a $500 investment earning 735,778 percent annually would be worth $3,679,390 after one year.

6.2 Summary

In this section, you learned how to use the simple and compound interest formulas to compare investment options. You also learned the importance of looking at the annual percentage yield in addition to the advertised rate.

6.2 Exercises

In Exercises 1–10, answer each of the questions using the simple and/or compound interest formulas. For each exercise, quoted rates are from Bankrate.com and are accurate as of August 2002.

1. **Certificates of Deposit** First NB of Baldwin County offers a five-year CD paying an annual rate of 4.5 percent simple interest with a minimum investment of $500. How much will the CD be worth at maturity if the minimum amount is invested?

2. **Certificates of Deposit** State Bank of Texas offers a five-year CD paying an annual rate of 4.25 percent simple interest with a minimum investment of $1000. How much will the CD be worth at maturity if the minimum amount is invested?

3. **Certificates of Deposit** New South Federal Savings offers a one-year CD paying an annual rate of 2.75 percent simple interest with a minimum investment of $5000. How much will the CD be worth at maturity if the minimum amount is invested?

4. **Doubling Time** What simple interest rate would an investor need to earn in order to double his or her investment in seven years?

5. **Certificates of Deposit** Ascencia Bank offers a five-year CD paying an annual rate of 4.70 percent interest compounded monthly with a minimum investment of $500. How much will the CD be worth at maturity if the minimum amount is invested?

6. **Certificates of Deposit** Countrywide Bank offers a five-year CD paying an

annual rate of 4.69 percent interest compounded monthly with a minimum investment of $1000. How much will the CD be worth at maturity if the minimum amount is invested?

7. **Certificates of Deposit** ING DIRECT offers a five-year CD paying an annual rate of 4.80 percent simple interest. Is this CD a better investment than the Ascencia Bank CD in Exercise 5? Explain.

8. **Certificates of Deposit** Resource Bank offers a three-month CD paying an annual rate of 2.59 percent compounded quarterly with a minimum investment of $10,000. What will the value of the CD be at maturity?

9. **Certificates of Deposit** Medford Savings Bank offers a three-month CD paying an annual rate of 2.00 percent simple interest. Everbank.com offers a three-month CD paying an annual rate of 1.98 percent compounded daily. Which account will be worth more at maturity?

10. **Certificates of Deposit** ING DIRECT, Ascencia Bank, and interState NetBank offer five-year CDs with a 4.80 percent *APY*. ING DIRECT pays simple interest, Ascencia Bank pays 4.70 percent compounded monthly, and interState NetBank pays 4.69 percent compounded daily. Which account will yield the highest return at maturity? Explain.

In Exercises 11–15, calculate the annual percentage yield for each investment.

11. A savings account paying an annual rate of 3.50 percent compounded quarterly.

12. A checking account paying an annual rate of 1.25 percent compounded daily.

13. A three-year certificate of deposit paying an annual rate of 3.43 percent compounded continuously.

14. A three-year certificate of deposit paying an annual rate of 2.45 percent simple interest.

15. A savings account paying an annual rate of 2.35 percent compounded continuously.

In Exercises 16–20, determine how many years it will take for the two accounts to reach the same balance.

16. A $1000 investment in an account paying an annual rate of 3.00 percent compounded monthly

and an $1100 investment in an account paying an annual rate of 3.02 percent compounded annually.

17. A $200 investment in an account paying an annual rate of 5.00 percent compounded continuously and a $300 investment in an account paying an annual rate of 6.00 percent compounded semiannually (twice a year).

18. A $2000 investment in an account paying an annual rate of 4.00 percent compounded continuously and a $3000 investment in an account paying an annual rate of 5.00 percent compounded monthly.

19. A $1000 investment in an account paying an annual rate of 5.25 percent compounded continuously and an $800 investment in an account paying an annual rate of 6.25 percent compounded continuously.

20. A $2000 investment at a simple interest rate of 12.00 percent and a $1500 investment at an interest rate of 12.00 percent compounded monthly.

In Exercises 21–28, do the following. For continuous rates, determine the annual rate that has the same APY as the continuous rate. For annual rates, determine the continuous rate that has the same APY as the annual rate.

21. Continuous rate: 7.23 percent

22. Continuous rate: 2.98 percent

23. Annual rate: 3.29 percent

24. Annual rate: 4.75 percent

25. Continuous rate: 11.02 percent

26. Annual rate: 6.54 percent

27. Annual rate: 17.20 percent

28. Continuous rate: 1.00 percent

Exercises 29–32 deal with payday loans. Some consumers turn to payday loans for needed cash. The lender gives the consumers an advance on their forthcoming paycheck and, when the consumers get paid, they pay off the lender. Although many payday lenders don't consider the finance charges they assess to be "interest", we will treat the charges as interest.

29. **Payday Loans** On February 17, 2003, mycashnow.com published the following loan terms:

Loan amount: $100
Loan term: 7–14 days
Finance charge: $18.62

If the consumer borrows the money for seven days, what is the annual percentage rate on the loan?

30. **Payday Loans** On February 17, 2003, mycashnow.com published the following loan terms:

Loan amount: $500
Loan term: 7–14 days
Finance charge: $93.10

If the consumer borrows the money for 14 days, what is the annual percentage rate on the loan?

31. **Payday Loans** On February 17, 2003, mycashnow.com published the following loan terms:

Loan amount: $200
Loan term: 7–14 days
Finance charge: $37.24

If the consumer borrows the money for 14 days, what is the annual percentage rate on the loan?

32. **Payday Loans** On February 17, 2003, mycashnow.com published the following loan terms:

Loan amount: $300
Loan term: 7–14 days
Finance charge: $55.86

If the consumer borrows the money for 10 days, what is the annual percentage rate on the loan?

Exercises 33–36 deal with inflation. Inflation is defined to be the percentage rate at which prices are increasing. The Consumer Price Index is often used to measure inflation.

33. **White Bread Prices** The average cost of a loaf of bread in 1983 was assigned a Consumer Price Index value of 100. By 2000, the index number had increased to 199.1, indicating that the price had nearly doubled. (**Source:** *Statistical Abstract of the United States, 2001*, Table 694, p. 454.)

(a) On average, at what annual percentage rate was the cost of white bread increasing between 1983 and 2000?

(b) If the average price of a loaf of bread was $1.12 in 1993 and prices continue to increase at the same rate, what will be the cost of a loaf of bread in 2008?

34. **Tomato Prices** The average cost of a pound of tomatoes in 1980 was assigned a Consumer Price Index value of 81.9. By 2000, the index number had increased to 234.7, indicating that the price had nearly tripled. (**Source:** *Statistical Abstract of the United States, 2001*, Table 694, p. 454.)

(a) On average, at what annual percentage rate was the cost of tomatoes increasing between 1980 and 2000?

(b) In 1990, the average price of field-grown tomatoes was $0.86 per pound. (**Source:** *Statistical Abstract of the United States, 2001*, Table 706, p. 468.) If prices continue to increase at the same percentage rate, what will be the cost of a pound of tomatoes in 2010?

35. **Rental Costs** The average cost of renting a place to live was assigned a Consumer Price Index value of 103.0 in 1983. By 1997, the index number had increased to 186.4. (**Source:** *Statistical Abstract of the United States, 2001*, Table 694, p. 454.)

(a) On average, at what annual percentage rate was the cost of rent increasing between 1983 and 1997?

(b) If an apartment cost $220 a month to rent in 1983, in what year is it expected to cost $500 per month?

36. **Interstate Long Distance Costs** The average cost per minute to make an interstate long distance phone call was assigned a Consumer Price Index value of 101.5 in 1983. By 2000, the index number had fallen to 68.0. (**Source:** *Statistical Abstract of the United States, 2001*, Table 694, p. 454.)

(a) On average, at what annual percentage rate was the cost of interstate long distance calls decreasing between 1983 and 2000?

(b) What do you think may have been the cause of the price reduction?

Exercises 37–43 deal with savings bonds. The U.S. government issues savings bonds to consumers at half of their face value. For example, a $50 savings bond is issued to a consumer for $25. The bond is guaranteed to be worth its face value no later than its original maturity date.

37. **Savings Bonds** Series E/EE bonds issued on 12/01/81 have an original maturity date of 12/01/89. (**Source:** www.publicdebt.treas.gov.)

What is the minimum annual percentage yield earned by the bonds from the date of issue to the original maturity date?

38. **Savings Bonds** Series E/EE bonds issued on 1/1/93 have an original maturity date of 1/1/05.
(**Source:** www.publicdebt.treas.gov.)
What is the minimum annual percentage yield that will be earned by the bonds from the date of issue to the original maturity date?

39. **Savings Bonds** Series E/EE bonds issued on 1/1/96 have an original maturity date of 1/1/13.
(**Source:** www.publicdebt.treas.gov.)
What is the minimum annual percentage yield earned by the bonds from the date of issue to the original maturity date?

40. **Savings Bonds** A $500 EE savings bond issued for $250 in July 2000 was worth $283.60 in July 2003.
(**Source:** www.publicdebt.treas.gov.)
What is the annual percentage yield on the bond?

41. **Savings Bonds** A $50 EE savings bond issued for $25 in January 2000 was worth $28.68 in February 2003.
(**Source:** www.publicdebt.treas.gov.)
What is the annual percentage yield on the bond?

42. **Savings Bonds** A $50 EE savings bond issued for $25 in January 1980 was worth $125.62 in February 2003.
(**Source:** www.publicdebt.treas.gov.)
What is the annual percentage yield on the bond?

43. **Savings Bonds** A $50 EE savings bond issued for $25 in January 1990 was worth $52.88 in February 2003.
(**Source:** www.publicdebt.treas.gov.)
What is the annual percentage yield on the bond?

Exercises 44–48 are intended to challenge your understanding of simple and compound interest.

44. Explain the practical meaning of $\frac{r}{12}$ from the compound interest formula $A = P\left(1 + \frac{r}{12}\right)^{12t}$.

45. A $1000 simple interest investment is earning twice the interest rate of a $1000 investment earning interest compounded annually. If, after *two* years, the investments have the same value, what is the compound interest rate?

46. A $1000 simple interest investment is earning twice the interest rate of a $1000 investment earning interest compounded annually. If, after *three* years, the investments have the same value, what is the compound interest rate?

47. You decide to open up a payday loan company. You charge 20 percent of the loan amount for a one-month loan. Treating the finance charge as interest, what is your annual percentage yield on a $500 loan?

48. Explain how an understanding of the compound interest and simple interest formulas can benefit you personally.

6.3 Future Value of an Increasing Annuity

- Use the future value of an increasing annuity formula to answer questions related to accounts with systematic investments, such as retirement savings plans

GETTING STARTED You can be a millionaire in 40 years. Some of you will become millionaires much sooner. Becoming a millionaire is not a rich man's game. With a little financial discipline and a basic knowledge of annuities, you can make your money work for you, rather than always working for money.

In this section, we will discuss how to use the future value of an increasing annuity formula to help prepare for a comfortable retirement. Using this formula, we can analyze a variety of financial scenarios with relative ease.

Financial instruments that require a series of payments of set size and frequency are called **annuities.** Suppose we want to have $1,000,000 in our retirement account after 40 years of investing. We plan on making monthly contributions to an account earning 12 percent interest compounded monthly. How much should we invest each month? The monthly periodic interest rate is $i = \frac{0.12}{12} = 0.01$.

Since we plan on making monthly contributions at the end of each month for 40 years, we will make a total of 480 contributions. Since interest is also paid at the end of each month, our first payment will earn interest for 479 months. (It didn't earn any interest the first month, since the contribution was made at the end of the month.) Our second payment will earn interest for 478 months. Our third payment will earn interest for 477 months, and so on. Our second to last payment will earn interest for one month, and our final payment, made at the end of the 480th month, will not earn any interest at all. If R is the amount of the monthly contribution, then the compound interest formula tells us that the future value of a monthly contribution after m months is

$$FV = R(1 + 0.01)^m$$
$$= R(1.01)^m$$

The future value of our entire investment is the sum of the future values of each monthly contribution (see Table 6.4).

TABLE 6.4

Contribution Number	Contribution Amount	Future Value
1	R	$FV = R(1.01)^{479}$
2	R	$FV = R(1.01)^{478}$
3	R	$FV = R(1.01)^{477}$
...	...	...
479	R	$FV = R(1.01)^1$
480	R	$FV = R(1.01)^0$

We want the future value of our investment to be $1,000,000.

$$1,000,000 = R(1.01)^{479} + R(1.01)^{478} + \cdots + R(1.01)^1 + R(1.01)^0$$
$$1,000,000 = R[(1.01)^{479} + (1.01)^{478} + \cdots + (1.01)^1 + 1]$$

Calculating the value of the right-hand side of the equation looks like an arduous task; however, we will be able to greatly simplify it with some clever algebraic manipulation. We will begin by multiplying the equation by 1.01. The resultant equation is

$$1.01(1,000,000) = 1.01R[(1.01)^{479} + (1.01)^{478} + \cdots + (1.01)^1 + 1]$$
$$1,010,000 = R[(1.01)^{480} + (1.01)^{479} + \cdots + (1.01)^2 + (1.01)^1]$$

We will now subtract the original equation $1{,}000{,}000 = R\left[(1.01)^{479} + (1.01)^{478} + \cdots + (1.01)^{1} + 1\right]$ from the new equation $1{,}010{,}000 = R\left[(1.01)^{480} + (1.01)^{479} + \cdots + (1.01)^{2} + (1.01)^{1}\right]$.

$$
\begin{array}{r}
1{,}010{,}000 = R\left[(1.01)^{480} + (1.01)^{479} + \cdots + (1.01)^{2} + (1.01)^{1}\right] \\
-\ 1{,}000{,}000 = R\left[(1.01)^{479} + (1.01)^{478} + \cdots + (1.01)^{1} + 1\right] \\
\hline
10{,}000 = R\left[(1.01)^{480} \qquad\qquad - 1\right]
\end{array}
$$

Solving for R, we get

$$
R = \frac{10{,}000}{(1.01)^{480} - 1}
$$
$$
= \$85.00
$$

You will be a millionaire in 40 years if you invest \$85.00 a month in an account earning an annual rate of 12 percent interest compounded monthly. Isn't that amazing? Most of us spend three to four times that much on car payments alone.

Generalizing the formula used in the million-dollar example, we have

$$
R = \frac{FV(i)}{(1 + i)^{m} - 1}
$$

where $m = nt$ is the number of payments to be made, $i = \frac{r}{n}$ is the periodic interest rate, FV is the future value of the annuity, and R is the payment amount. It is common to represent the payment amount R by PMT. Solving the formula for FV, we get the future value of an increasing annuity formula.

FUTURE VALUE OF AN INCREASING ANNUITY (WITH A ZERO PRESENT VALUE)

The future value FV of an increasing annuity with an initial balance of 0 dollars is given by

$$
FV = PMT\,\frac{(1 + i)^{m} - 1}{i}
$$

where $i = \frac{r}{n}$ is the periodic interest rate, $m = nt$ is the number of payments, and PMT is the payment amount.

If the annuity has a nonzero present value (a balance), the present value will grow in accordance with the compound interest formula. The future value of the entire investment will be the sum of the two quantities.

> **FUTURE VALUE OF AN INCREASING ANNUITY (WITH A NONZERO PRESENT VALUE)**
>
> The future value FV of an annuity with a present value PV is given by
>
> $$FV = PV(1 + i)^m + PMT \frac{(1 + i)^m - 1}{i}$$
>
> where $i = \frac{r}{n}$ is the periodic interest rate, $m = nt$ is the number of payments, and PMT is the payment amount.

The future value of an increasing annuity formulas are typically used with sinking funds. A **sinking fund** is an account established for the purpose of accumulating money to pay off future debts or obligations.

EXAMPLE 1

Forecasting Retirement Savings with an Increasing Annuity

Since its inception on August 1, 1952, the CREF Stock Account has performed very well. As of April 30, 2005, it had earned an average of 10.39 percent annually. (**Source:** TIAA-CREF.) A 35-year-old college professor contributes $200 to her CREF Stock Account twice a month. She hopes to retire in 25 years with a retirement account worth $800,000 or more. Her current account balance is $15,000. Assuming that she will be able to earn an annual rate of 10.39 percent compounded semimonthly, will she be able to reach her retirement goal?

SOLUTION   The professor's retirement account is a sinking fund, since it will be used to pay for future living expenses after she retires. The future value of her current account balance may be calculated using the compound interest formula. Since interest is compounded twice a month, $n = 24$.

$$A = P\left(1 + \frac{r}{n}\right)^{nt}$$

$$= 15,000\left(1 + \frac{0.1039}{24}\right)^{24(25)}$$

$$= 15,000(13.3550)$$

$$= 200,325.57$$

The current account balance will grow to $200,325.57 over the next 25 years.

The future value of her future contributions may be calculated using the future value of an increasing annuity formula. For her scenario,

$m = 24 \cdot 25 = 600$ and $i = \frac{0.1039}{24} = 0.004329$.

$$FV = PMT \frac{(1 + i)^m - 1}{i}$$

$$= 200 \frac{(1 + 0.004329)^{600} - 1}{0.004329}$$

$$= 200(2853.9067)$$

$$= 570,781.34$$

In 25 years, her future contributions will grow to $570,781.34

As shown in the future value of an increasing annuity (with a nonzero balance) formula, her final account balance can be determined by summing the two amounts.

$$200,325.57 + 570,781.34 = 771,106.91$$

Her final account balance will be $771,106.91. According to the scenario, she will be roughly $29,000 short of her $800,000 goal.

EXAMPLE 2

Using an Increasing Annuity to Plan for Retirement

In Example 1, the college professor ended up short of her investment goal. To meet her goal, she will need to either increase her payment amount or increase the number of payments.

(a) Determine the payment amount that will allow her to reach her goal in 25 years.

(b) Determine the number of $200 payments that will allow her to reach her goal.

SOLUTION

(a) Since she is not increasing the length of her investment, her current balance will still grow to $200,325.57. She needs to earn an additional $599,674.43 from her future contributions in order to reach her $800,000 goal. We can determine the payment amount by solving the future value of an increasing annuity formula for *PMT*.

$$FV = PMT \, \frac{(1 + i)^m - 1}{i}$$

$$PMT = FV \frac{i}{(1 + i)^m - 1}$$

$$PMT = 599{,}674.43 \left[\frac{0.004329}{(1 + 0.004329)^{600} - 1} \right]$$

$$= 599{,}674.43(0.0003504)$$

$$= 210.13$$

Increasing the payment amount by $10.13 will increase the final account balance by $28,910.07, slightly more than the $28,893.09 we needed. A $10.12 increase would increase the final account balance by $28,881.53, slightly less than the $28,893.09 we needed.

(b) Since she is increasing the amount of time her investment will earn interest, growth in both the current balance and in future contributions will be affected. We need the sum of her current and future investments to total $800,000.

$$FV = P\left(1 + \frac{r}{n}\right)^{nt} + PMT\left[\frac{\left(1 + \frac{r}{n}\right)^{nt} - 1}{\frac{r}{n}}\right]$$

$$800{,}000 = 15{,}000(1.004329)^{24t} + 200\left[\frac{(1.004329)^{24t} - 1}{0.004329}\right]$$

$$800{,}000 = 15{,}000(1.1092)^{t} + 200\left[\frac{(1.1092)^{t} - 1}{0.004329}\right] \qquad \text{Evaluate exponents}$$

$$800{,}000 = 15{,}000(1.1092)^{t} + 46{,}198.27[(1.1092)^{t} - 1] \qquad \text{Divide 200 by 0.004329}$$

$$800{,}000 = 15{,}000(1.1092)^{t} + 46{,}198.27(1.1092)^{t} - 46{,}198.27 \qquad \text{Distribute 46,198.27}$$

$$846{,}198.27 = 15{,}000(1.1092)^{t} + 46{,}198.27(1.1092)^{t} \qquad \text{Add 46,198.27}$$

$$846{,}198.27 = 61{,}198.27(1.1092)^{t} \qquad \text{Combine like terms}$$

$$13.8272 = (1.1092)^{t} \qquad \text{Divide by 61,198.27}$$

$$\ln(13.8272) = t\ln(1.1092) \qquad \text{Log both sides; Rule 7}$$

$$t = \frac{\ln(13.8272)}{\ln(1.1092)} \qquad \text{Divide by ln (1.1092)}$$

$$t \approx 25.34 \qquad \text{Evaluate on calculator}$$

She needs to make contributions for an additional 0.34 year (4.1 months). Since she makes contributions twice a month, she needs to make only nine more payments to reach her goal.

In Example 2, we saw how a small change in the payment amount or in the length of the investment may substantially increase the value of the investment. Likewise, seemingly small changes in the interest rate can have a dramatic impact on the growth of an investment. The magnifying factor is time. This phenomenon is commonly referred to as the **time value of money.** A dollar today is worth more than a dollar in the future, even after adjusting for inflation, because a dollar invested today can earn interest until the time the dollar in the future is received. By increasing her semimonthly payment by $10.13 she contributed only an additional $6078 to her retirement account. However, over the 25-year period, that additional investment grew to nearly $29,000.

The future value of an increasing annuity formula is so applicable to everyday life that the TI-83 Plus has a built-in Time Value of Money Solver that allows us to solve annuity problems quickly and easily.

TECHNOLOGY **TIP**

Using the TVM Solver

1. Press the blue APPS button to bring up a list of available applications. Select **1:Finance···** from the list and press ENTER

2. Select **1:TVM Solver** from the list of available financial applications and press ENTER.

3. The TVM Solver displays several variables. **N** is the number of payments, **I%** is the annual interest rate (not the periodic rate) written as a percent, **PV** is the negative of the present value of the investment, **PMT** is the negative of the payment amount, **FV** is the future value of the investment, **P/Y** is the number of payments per year, **C/Y** is the number of compounding periods per year, and **PMT: END BEGIN** tells the calculator whether to calculate interest at the beginning or end of the compounding period.

4. To solve for any quantity, place the cursor on the unknown quantity and press ALPHA then ENTER.

Since the TVM Solver is designed for decreasing annuities such as auto and home loans, we had to make the present value and payment amounts negative in order to use it with our increasing annuity problem. Admittedly, this is a bit counterintuitive. We will run through a couple of examples to allow you to practice using the TVM Solver.

EXAMPLE 3

Determining the Future Value of an Investment

You're contributing $100 per month to a retirement plan earning 6.31 percent compounded monthly. Your retirement account has a current balance of $2309.33. How long will it take for your investment to reach $10,000?

SOLUTION Entering each of the known values, we get the result shown in Figure 6.2.

FIGURE 6.2

Notice that the value of N is still 604.6393625 from a previous problem. We haven't solved for N yet. As soon as we do, the calculator will replace the value with the correct value of N. Solving for N, we get the result shown in Figure 6.3.

FIGURE 6.3

Notice that the calculator placed a black square to the left of N. This is the calculator's way of indicating that the value of N is a calculated value.

It will take 59 monthly payments for the account balance to reach $10,000. That is, it will take four years and eleven months for the account balance to reach $10,000.

EXAMPLE 4

Determining the Future Value of an Investment

Returning to the scenario in Example 3, what interest rate would you have to earn if you wanted your account balance to reach $10,000 in four years?

SOLUTION Four years of monthly payments equals 48 payments. Updating the TVM Solver, we have the result shown in Figure 6.4.

FIGURE 6.4

Solving for I%, we get the result shown in Figure 6.5.

FIGURE 6.5

Notice that the calculator placed a black square to the left of the I%. This is the calculator's way of indicating that the value of I% is a calculated value.

You must earn an annual rate of 12.56 percent compounded monthly if you want the balance to grow to $10,000 in four years.

EXAMPLE 5

Comparing the Future Values of Two Investment Accounts

You plan on investing $50 per month for five years into one of two accounts. The first account pays 2.95 percent interest compounded daily. The second account pays 3.00 percent interest compounded quarterly. Which account will have the greatest value at the end of five years?

SOLUTION Notice that the compounding frequency and the number of payments are not equal, as in previous examples.

First Account:

Since interest is being compounded daily, C/Y=365.

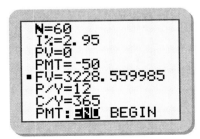

FIGURE 6.6

The future value will be $3,228.56, as shown in Figure 6.6.

Second Account:

Since interest is being compounded quarterly, C/Y=4.

FIGURE 6.7

The future value of the account will be $3,231.73, as shown in Figure 6.7. The second account will yield a higher return than the first account.

6.3 Summary

In this section, you learned how to use the future value of an increasing annuity formula to help you prepare for a comfortable retirement. You also learned how to use the TVM Solver to solve increasing annuity problems quickly.

6.3 Exercises

In Exercises 1–5, solve for the indicated value algebraically by using one of the future value of an increasing annuity formulas.

1. An investor opened a new account earning 6 percent interest compounded monthly. She plans to make a $250 contribution each month. How long will it take for her account to be valued at $10,000?

2. If an investor contributes $100 a month to an account earning 4 percent interest compounded monthly, what will be the value of the account in five years?

3. The present value of an account earning 8 percent interest compounded monthly is $2400. An investor needs the account to be worth $4800 two years from now. How much must he contribute monthly?

4. How long will it take a $2000 account balance earning 2 percent compounded monthly with monthly contributions of $50 to grow to $3000?

5. How much must you invest each month into an account earning 4 percent interest compounded monthly if you want the account balance to be at least $2000 after two years?

In Exercises 6–15, retirement plan scenarios are set up using mutual funds, which, by their very nature, do not guarantee a particular rate of return. Mutual funds are used in these scenarios instead of savings accounts because they typically outperform savings accounts in the long run and are considered by many to be a better long-term investment. It is meaningful to look at the long-range performance of a mutual fund as an indicator of what its future performance might *be. Performance data are accurate as of August 31, 2002.*

Solve each of the increasing annuity problems algebraically. Then check your solution using the TVM Solver on your calculator.

6. **Mutual Funds** The CREF Global Equities Fund has earned an average of 8.37 percent since May 1, 1992. (Source: TIAA-CREF.) Assuming that the account will earn an annual rate of 8.37 percent compounded monthly, what will be the future value of the account in 10 years if $150 contributions are made monthly?

7. **Mutual Funds** The CREF Growth Fund has earned an average of 8.16 percent since April 29, 1994. (Source: TIAA-CREF.) Assuming that the account will earn an annual rate of 8.16 percent compounded monthly, what will be the future value of the account in 30 years if $200 contributions are made monthly?

8. **Mutual Funds** The CREF Equity Index Fund has earned an average of 11.33 percent since April 29, 1994. (Source: TIAA-CREF.) Assuming that the account will earn an annual rate of 11.33 percent compounded monthly, what will be the future value of the account in 25 years if $200 contributions are made monthly?

9. **Mutual Funds** The CREF Social Choice Fund has earned an average of 10.51 percent since March 1, 1990. (Source: TIAA-CREF.) Assuming that the account will earn an annual rate of 10.51 percent compounded monthly, what will be the future value of the account in 20 years if $300 contributions are made monthly?

10. **Mutual Funds** The CREF Bond Market Fund has earned an average of 8.00 percent since March 1, 1990. (Source: TIAA-CREF.) Assuming that the account will earn an annual rate of 8.00 percent compounded monthly, how many years will it take for the account to grow to $250,000 if $300 contributions are made monthly?

11. **Mutual Funds** The CREF Inflation-linked Bond Fund has earned an average of 6.88 percent since May 1, 1997. (Source: TIAA-CREF.) Assuming that the account will earn an annual rate of 6.88 percent compounded monthly, what will the future value of an account with a

present value of $2500 be in 20 years if $300 contributions are made monthly?

12. **Mutual Funds** The TIAA Real Estate Account has earned an average of 8.11 percent since October 2, 1995. (Source: TIAA-CREF.) Assuming that the account will earn an annual rate of 8.11 percent compounded monthly, what will the future value of an account with a present value of $3200 be in 25 years if $125 contributions are made monthly?

13. **Mutual Funds** The CREF Money Market Account has earned an average of 5.51 percent since April 1, 1988. (Source: TIAA-CREF.) Assuming that the account will earn an annual rate of 5.51 percent compounded monthly, what will the future value of an account with a present value of $2000 be in 15 years if $100 contributions are made monthly?

14. **Mutual Funds** The Harbor Capital Appreciation Fund has earned an average of 13.16 percent since December 29, 1987. (Source: Harbor Funds.) Assuming that the account will earn an annual rate of 13.16 percent compounded monthly, what will the future value of an account with a present value of $6200 be in 30 years if $250 contributions are made monthly?

15. **Mutual Funds** The Large Cap Value Fund has earned an average of 11.38 percent since December 29, 1987. (Source: Harbor Funds.) Assuming that the account will earn an annual rate of 11.38 percent compounded monthly and has a present value of $6200, how long will it take to reach an account value of $124,000 if $250 contributions are made monthly?

In Exercises 16–29, solve each problem using the TVM Solver.

16. **Retirement Account** An investor hopes to increase his retirement account to $950,000 by the time he retires 10 years from now. The present value of the account is $250,259.12. He expects his investment to earn an annual rate of 13.16 percent compounded monthly. What must his monthly investment be in order for him to reach his goal?

17. **Retirement Account** An investor hopes to increase her retirement account to $1,250,000 by the time she retires 15 years from now. The present value of the account is $219,976.31. She expects her investment to earn an annual rate of 11.38 percent compounded monthly. What must her monthly investment be in order for her to reach her goal?

18. Retirement Account An investor hopes to increase her retirement account to $1,000,000 by the time she retires 35 years from now. The present value of her account is $2498.39. She expects her investment to earn an annual rate of 6.88 percent compounded monthly. What must her monthly investment be in order for her to reach her goal?

19. Retirement Account An investor hopes to increase his retirement account to $500,000 by the time he retires two years from now. The present value of his account is $425,524.21. He expects his investment to earn an annual rate of 5.51 percent compounded monthly. What must his monthly investment be in order for him to reach his goal?

20. Retirement Account An investor hopes to increase his retirement account to $500,000 by the time he retires two years from now. The present value of his account is $425,524.21. He can afford to make $350 contributions monthly. Assuming that interest is compounded monthly, what interest rate must he earn in order for him to reach his goal?

21. Retirement Account An investor hopes to increase his retirement account to $700,000 by the time he retires 11 years from now. The present value of his account is $332,185.22. He can afford to make $450 contributions monthly. Assuming that interest is compounded monthly, what interest rate must he earn in order for him to reach his goal?

22. Retirement Account An investor hopes to increase her retirement account to $2,900,000 by the time she retires 21 years from now. The present value of her account is $10,185.22. She can afford to make $650 contributions monthly. Assuming that interest is compounded monthly, what interest rate must she earn in order for her to reach her goal?

23. Retirement Account An investor hopes to increase her retirement account to $1,000,000 by the time she retires 44 years from now. The present value of her account is $250. She can afford to make $75 contributions monthly.

Assuming that interest is compounded monthly, what interest rate must she earn in order for her to reach her goal?

24. Buying a Home A recent college graduate is saving up for the down payment on a home. She currently has $2025 in a savings account earning an annual rate of 2.35 percent compounded *quarterly*. She contributes $200 a month to the account. She anticipates that she will be able to afford a $125,000 home. She intends to finance the home through the government FHA loan program, which requires a 5 percent down payment instead of the typical 20 percent down payment. How long will it take her to save up enough money to make a down payment that is 5 percent of the price of the home?

25. Buying a Home A grocery store manager is saving up for the down payment on a home. He and his wife currently have $1200 in a savings account earning an annual rate of 3.25 percent compounded *quarterly*. They contribute $500 a month to the account. They anticipate that they will be able to afford a $195,000 home. They intend to finance the home through the government FHA loan program, which requires a 5 percent down payment instead of the typical 20 percent down payment. How long will it take them to save up enough money to make a down payment that is 5 percent of the price of the home?

26. Buying a Home A hotel catering manager is saving up for the down payment on a home. She and her husband currently have $502.02 in a savings account earning an annual rate of 1.25 percent compounded monthly. She contributes her $500.00 quarterly bonus to the account each quarter. They anticipate that they will be able to afford a $105,000 home. How long will it take them to save up enough money to make a down payment that is 5 percent of the price of the home?

27. Buying a Home A steelworker is saving up for the down payment on his next home. He expects to make a profit of $25,000 from the sale of his current home. He wants to move into a $183,000 home and finance it with a conventional loan, which requires a 20 percent down payment. He currently has $4200 in a savings account earning an annual rate of 2.33 percent compounded monthly. He contributes $200 to the account each time he gets his biweekly paycheck. Assuming that he will use the profit from the sale of his current home as part of the down payment, how long will it take him to save up enough money to make a down payment that is 20 percent of the price of the home?

28. **Buying a Home** An expert video-game programmer is saving up for the down payment on his next home. He expects to make a profit of $20,000 from the sale of his current home. He wants to move into a $525,000 home and finance it with a conventional loan, which requires a 20 percent down payment. He currently has $24,200 in a conservative investment account earning an annual rate of 4.69 percent compounded daily. He contributes $1,000 to the account monthly. How long will it take him to save up enough money to make a down payment that is 20 percent of the price of the home? Assume that he will use the profit from the sale of his first home as part of the down payment.

29. **Buying a Home** A freelance illustrator is saving up for the down payment on a condominium overlooking the ocean. She contributes $15,000 semiannually to a savings account earning an annual rate of 3.89 percent compounded quarterly. (Artists and authors earning royalties on their work are typically paid twice a year.) She currently has $15,290.23 in the account. Since she is uncertain of her future income, she doesn't want to finance any more than 25 percent of the price of the condominium. How long will it take her to save up enough money to make a 75 percent down payment on a $110,000 condominium?

Exercises 30–41 are intended to challenge your ability to work with increasing annuities.

30. **Investment Value** Investment account A has a present value of $1200 and is earning an annual rate of 6 percent interest compounded monthly. Investment account B has a present value of $500 and is earning an annual rate of 6 percent interest compounded monthly. Monthly contributions of $100 are made to investment account B.

How long will it take for the two accounts to have the same value?

31. **Investment Value** Investment account A has a present value of $8000 and is earning an annual rate of 8 percent interest compounded quarterly.

Investment account B has a present value of $2000 and is earning an annual rate of 6 percent interest compounded monthly. Monthly contributions of $100 are made to investment account B.

How long will it take for the two accounts to have the same value?

32. **Investment Value** A wise investor began investing $50 a month in an account earning an annual rate of 4 percent interest compounded monthly on her 20th birthday. Her twin brother began investing $100 a month into an account earning an annual rate of 4 percent compounded monthly 10 years later. How old will the twins be when the account values are equal?

33. **Investment Value** A wise investor began investing $200 a month in an account earning an annual rate of 6 percent interest compounded monthly on his 18th birthday. His twin brother began investing $200 a month into an account earning an annual rate of 12 percent compounded monthly 10 years later. How old will the twins be when the account values are equal?

34. **Solve the equation**

$$FV = PV(1 + i)^m + PMT \frac{(1 + i)^m - 1}{i}$$

for *PV.* Simplify the result as much as possible.

35. Give an example of a case in which the result of Exercise 34 would be useful.

36. **Investment Value** An investment account earns 1 percent interest each *month.* (This is *not* the same as 1 percent interest compounded monthly.) If $100 is contributed to the account monthly, what will be the value of the account after 10 payments are made? (Assume that deposits are made and interest is paid at the end of the month. An investment must be in the account for a full month before it earns interest.)

37. **Investment Value** An account earns 2 percent interest each *quarter.* (This is *not* the same as 2 percent interest compounded quarterly.) If $200 is contributed to the account quarterly, what will be the value of the account after six payments are made? (Assume that deposits are made and interest is paid at the end of the quarter. An investment must be in the account for a full quarter before it earns interest.)

38. Investment Value   An account earns 3 percent interest every six months. (This is *not* the same as 3 percent interest compounded semiannually.) If $500 is contributed to the account every six months, what will be the value of the account after two years? (Assume that deposits are made and interest is paid at the end of the six-month period. An investment must be in the account for the full six months before it earns interest.)

 39. Graph the functions

$$f(t) = \frac{100[(1.005)^{12t} - 1]}{0.005} \quad \text{and}$$

$$g(t) = \frac{30[(1.003)^{24t} - 1]}{0.003}$$

Find all points of intersection of the graphs. If f and g represent the value of two investment accounts t years from now, how should we interpret the meaning of the intersection point(s)?

 40. Graph the functions

$$f(t) = \frac{100[(1.005)^{12t} - 1]}{0.005} \quad \text{and}$$

$$g(t) = \frac{60[(1.006)^{6t} - 1]}{0.006}$$

Find all points of intersection of the graphs. If f and g represent the value of two investment accounts t years from now, how should we interpret the meaning of the intersection point(s)?

41. Investment Value   Two different investment accounts earn an annual rate of 6 percent interest compounded monthly over a period of 40 years. At the end of 40 years, which account will have a higher value?

Account A: $100 monthly investments during the first 20 years and no investments during the last 20 years

Account B: No investments during the first 20 years and $300 monthly investments during the last 20 years.

6.4 Present Value of a Decreasing Annuity

- Use the present value of a decreasing annuity formula to answer questions related to accounts with systematic payments, such as auto loans

GETTING STARTED   A college student is planning on purchasing a two-year-old Honda Civic coupe. The average retail price of the used car is $10,800. (**Source:** www.nadaguides.com.) The student has $2000 for a down payment and wants to finance the balance of the loan over three years. She has been approved for a 5.24 percent loan by VirtualBank. (In September 2002, VirtualBank was identified by Bankrate.com as having one of the best auto loan rates nationally.) What will her monthly payments be? Including interest charges, how much will she actually pay for the car? Questions such as these can be answered by using the present value of a decreasing annuity formula.

In this section, we will discuss how to use the present value of a decreasing annuity formula as a tool in making financial decisions such as borrowing money for an auto or a home. We will also look at the power and peril of credit cards.

Amortization

A familiar decreasing annuity is the auto loan. Payments reduce the **principal** (amount owed) and cover interest charges. This process of paying off a loan is called **amortization.** An amortization table shows the lender and the loan recipient what portion of each payment goes to principal and what portion goes to interest. It also details how many payments are necessary to pay off the loan. Before signing any loan documents, wise consumers should make sure they understand the full cost of taking on a loan. The amortization table for the auto loan introduced at the start of this section is shown in Table 6.5. We generated the table using a spreadsheet.

TABLE 6.5

Payment Number	Principal (Present Value)	Interest (0.0044 × Principal)	Payment	New Balance (Principal + Interest − Payment)
1	$8,800.00	$38.43	$264.69	$8,573.74
2	$8,573.74	$37.44	$264.69	$8,346.49
3	$8,346.49	$36.45	$264.69	$8,118.24
4	$8,118.24	$35.45	$264.69	$7,889.00
5	$7,889.00	$34.45	$264.69	$7,658.76
6	$7,658.76	$33.44	$264.69	$7,427.51
7	$7,427.51	$32.43	$264.69	$7,195.26
8	$7,195.26	$31.42	$264.69	$6,961.99
9	$6,961.99	$30.40	$264.69	$6,727.70
10	$6,727.70	$29.38	$264.69	$6,492.38
11	$6,492.38	$28.35	$264.69	$6,256.04
12	$6,256.04	$27.32	$264.69	$6,018.67
13	$6,018.67	$26.28	$264.69	$5,780.26
14	$5,780.26	$25.24	$264.69	$5540.81
15	$5,540.81	$24.19	$264.69	$5,300.32
16	$5,300.32	$23.14	$264.69	$5,058.77
17	$5,058.77	$22.09	$264.69	$4,816.17
18	$4,816.17	$21.03	$264.69	$4,572.51
19	$4,572.51	$19.97	$264.69	$4,327.79

(Continued)

TABLE 6.5 *(Continued)*

Payment Number	Principal (Present Value)	Interest (0.0044 × Principal)	Payment	New Balance (Principal + Interest − Payment)
20	$4,327.79	$18.90	$264.69	$4,082.00
21	$4,082.00	$17.82	$264.69	$3,835.13
22	$3,835.13	$16.75	$264.69	$3,587.19
23	$3,587.19	$15.66	$264.69	$3,338.16
24	$3,338.16	$14.58	$264.69	$3,088.05
25	$3,088.05	$13.48	$264.69	$2,836.85
26	$2,836.85	$12.39	$264.69	$2,584.54
27	$2,584.54	$11.29	$264.69	$2,331.14
28	$2,331.14	$10.18	$264.69	$2,076.63
29	$2,076.63	$9.07	$264.69	$1,821.01
30	$1,821.01	$7.95	$264.69	$1,564.27
31	$1,564.27	$6.83	$264.69	$1,306.41
32	$1,306.41	$5.70	$264.69	$1,047.42
33	$1,047.42	$4.57	$264.69	$787.31
34	$787.31	$3.44	$264.69	$526.06
35	$526.06	$2.30	$264.69	$263.66
36	$263.66	$1.15	$264.81	$0.00

A graph of the loan balance as a function of the payment number is shown in Figure 6.8.

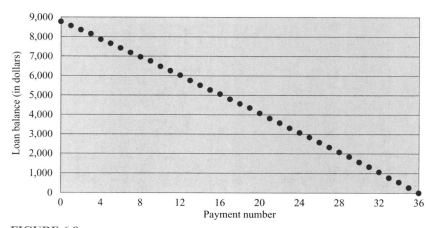

FIGURE 6.8

Since the interest rate is 5.24 percent compounded monthly, we needed to determine the monthly percentage rate in order to calculate the monthly interest.

$$\text{Monthly rate} = \frac{\text{annual rate}}{12}$$

$$= \frac{0.0524}{12}$$

$$\approx 0.0044$$

Monthly interest is about 0.44 percent of the principal. Since the first loan payment is not due until the month after the student acquired the loan, she will pay interest on the full loan amount in her first payment. The new principal balance ($8573.74) is calculated by adding the interest cost ($38.43) to the initial principal amount ($8800) and then subtracting the payment amount ($264.69). This process of calculating interest is repeated for the new principal balance. Since the principal balance has been reduced, the amount of interest due with the second payment is reduced to $37.44. By the 36th payment, interest is a mere $1.15. You may be wondering how we knew that the monthly payment should be $264.69. As you may have guessed, there is also a formula for decreasing annuities such as auto loans. The formula may be used to determine the amount of the monthly car payment.

PRESENT VALUE OF A DECREASING ANNUITY (WITH A FUTURE VALUE OF ZERO)

The present value PV of a decreasing annuity with a future value of zero is given by

$$PV = PMT \left[\frac{1 - (1 + i)^{-m}}{i} \right]$$

where $i = \frac{r}{n}$ is the periodic interest rate, $m = nt$ is the number of payments, and PMT is the payment amount.

The **present value** of an annuity is the amount of money that would need to be invested today in order to provide the desired number of equal periodic payments. We will demonstrate how the formula was developed by using an example. Suppose we want to withdraw $3000 per month from our retirement account for the next 25 years, beginning at the end of the month. The account is earning an annual rate of 6 percent interest compounded monthly. We want to know if we have enough money in the account to last us 25 years.

Since the payments will be withdrawn over a period of 25 years and will continue to earn interest until they are withdrawn, the present value of each payment is less than $3000. We can calculate the present value of each payment by using a slightly modified version of the compound interest formula.

$$A = P\left(1 + \frac{r}{n}\right)^{nt}$$

$$A = PMT\left(1 + \frac{r}{n}\right)^{m} \qquad \text{Let } P = PMT \text{ and } m = nt$$

$$3000 = PMT\left(1 + \frac{0.06}{12}\right)^{m}$$

$$3000 = PMT(1 + 0.005)^{m}$$

$$3000 = PMT(1.005)^{m}$$

$$PMT = \frac{3000}{(1.005)^{m}}$$

We can calculate the present value of the first payment by substituting in $m = 1$.

$$PMT = \frac{3000}{(1.005)^{1}}$$

$$= 2985.07$$

The present value of the first payment is $2985.07, since it will earn interest for only a month before it is withdrawn. Calculating the present value of the second payment, we get

$$PMT = \frac{3000}{(1.005)^{2}}$$

$$= 2970.22$$

The present value of the second payment is $2970.22, since it will earn interest for two months before it is withdrawn. We can calculate the present value of any of the 300 payments.

$$PMT = \frac{3000}{(1.005)^{300}}$$

$$= 671.90$$

The present value of the 300th payment is $671.90. We can calculate the present value of the entire investment by adding up the present values of the individual payments.

$$PV = \frac{3000}{(1.005)^{300}} + \frac{3000}{(1.005)^{299}} + \cdots + \frac{3000}{(1.005)^{2}} + \frac{3000}{(1.005)^{1}}$$

Multiplying both sides of the equation by $(1.005)^{300}$, we get

$$(1.005)^{300} \cdot PV = (1.005)^{300}\left[\frac{3000}{(1.005)^{300}} + \frac{3000}{(1.005)^{299}} + \cdots + \frac{3000}{(1.005)^{2}} + \frac{3000}{(1.005)^{1}}\right]$$

$$(1.005)^{300} \cdot PV = 3000\left[\frac{(1.005)^{300}}{(1.005)^{300}} + \frac{(1.005)^{300}}{(1.005)^{299}} + \cdots + \frac{(1.005)^{300}}{(1.005)^{2}} + \frac{(1.005)^{300}}{(1.005)^{1}}\right] \qquad \text{Factor out 3000} \\ \text{Distribute } (1.005)^{300}$$

$$(1.005)^{300} \cdot PV = 3000[1 + (1.005)^{1} + \cdots + (1.005)^{298} + (1.005)^{299}] \qquad \text{Reduce quotients}$$

One amazing property of algebra is that $1 + x + x^2 + \cdots + x^{n-1} = \dfrac{x^n - 1}{x - 1}$. Consequently, $[1 + (1.005)^1 + \cdots + (1.005)^{298} + (1.005)^{299}]$ is equivalent to $\dfrac{1.005^{300} - 1}{1.005 - 1}$. Therefore, the equation may be rewritten as

$$(1.005)^{300} \cdot PV = 3000 \left[\frac{(1.005)^{300} - 1}{1.005 - 1} \right]$$

$$PV = \frac{3000}{(1.005)^{300}} \left[\frac{(1.005)^{300} - 1}{0.005} \right] \qquad \text{Divide both sides by } (1.005)^{300}$$

$$= 3000 \left[\frac{(1.005)^{300} - 1}{(1.005)^{300} 0.005} \right] \qquad \text{Distribute } \frac{1}{(1.005)^{300}}$$

$$= 3000 \left[\frac{1 - (1.005)^{-300}}{0.005} \right] \qquad \text{Divide numerator by } (1.005)^{300}$$

Observe that this formula is of the form $PV = PMT \left[\dfrac{1 - (1 + i)^{-m}}{i} \right]$, that is, the present value of a decreasing annuity formula introduced earlier. Continuing with the computation, we get

$$PV = 3000(155.21)$$
$$= 465,620.59$$

If we have \$465,620.59 in our retirement account today, we will be able to withdraw \$3000 per month for the next 25 years. At the end of that time, our account balance will be zero.

EXAMPLE 1

Determining the Monthly Payment on an Auto Loan

A student intends to buy a two-year-old Honda Civic for \$10,800. She plans to put \$2000 down and finance the remaining balance on a 36-month loan charging 5.24 percent interest compounded monthly. What will be her monthly payment?

SOLUTION Since $10,800 - 2000 = 8800$, her initial loan amount is \$8800. Since payments are made monthly, her periodic interest rate is

$$i = \frac{0.0524}{12}$$
$$= 0.004367$$

She will make 36 payments, so $m = 36$. Substituting these values into the present value of a decreasing annuity formula, we get

$$PV = PMT \left[\frac{1 - (1 + i)^{-m}}{i} \right]$$

$$8800 = PMT \left[\frac{1 - (1 + 0.004367)^{-36}}{0.004367} \right]$$

$$8800 = PMT(33.24)$$

$$PMT = 264.69$$

Her monthly payment is \$264.69.

EXAMPLE 2

Determining the Loan Period on an Auto Loan

The boyfriend of the student in Example 1 wants her to buy a brand new $13,270 Civic coupe instead of the used model, even though she's told him that she can't afford more than a $264.69 monthly payment. He assures her that she can still make the same payment; she will just have to extend the length of her loan by a few months. How much longer will it take her to pay off the loan if she buys the new car instead of the used model?

SOLUTION She will still make the $2000 down payment, so she needs to finance only $11,270. The monthly periodic rate and payment amount remain the same. Substituting these values into the present value of a decreasing annuity formula, we get

$$PV = PMT\left[\frac{1 - (1 + i)^{-m}}{i}\right]$$

$$11,270 = 264.69\left[\frac{1 - (1 + 0.004367)^{-m}}{0.004367}\right]$$

$11,270 = 60,616.03[1 - (1.004367)^{-m}]$	Divide 264.69 by 0.004367
$0.1859 = 1 - (1.004367)^{-m}$	Divide both sides by 60,616.03
$-0.8141 = -(1.004367)^{-m}$	Subtract 1 from both sides
$0.8141 = (1.004367)^{-m}$	Multiply both sides by -1
$\ln(0.8141) = -m \ln(1.004367)$	Log both sides; Log Rule 7
$m = -\dfrac{\ln(0.8141)}{\ln(1.004367)}$	Divide both sides by $\ln(1.004367)$
$= 47.21$	Evaluate on calculator

She will have to make 48 payments to pay off the loan. (The last payment will be smaller than the first 47.) It will take her one year longer (48 months versus 36 months) to pay for the new car than to pay for the used car.

We can determine the actual amount paid for the car (including interest charges) by multiplying the payment amount by the number of payments. In Example 1, the total cost of the car was approximately

$$P = 2000 + 36(264.69)$$
$$= \$11,528.84$$

In Example 2, the total cost of the car was approximately

$$P = 2000 + 48(264.69)$$
$$= \$14,705.12$$

These are both approximations, since the final payment on a loan often differs slightly from the rest of the payments. This is due to the fact that we round payment amounts to the nearest cent instead of the nearest fraction of a penny. The final payment is used to compensate for the error introduced by rounding. In the case of Example 2, the final payment will be substantially different from the rest of the payments, since only 0.21 of a normal payment is necessary to pay off the loan.

Although we can use the present value of a decreasing annuity formula to determine the amount of any loan payment, one way to reduce the number of computations in future problems is to solve the present value of a decreasing annuity formula $PV = PMT \left[\dfrac{1 - (1 + i)^{-m}}{i} \right]$ for PMT.

$$PV = PMT \left[\frac{1 - (1 + i)^{-m}}{i} \right]$$

$$i(PV) = PMT \left[1 - (1 + i)^{-m} \right] \qquad \text{Multiply both sides by } i$$

$$\frac{i(PV)}{1 - (1 + i)^{-m}} = PMT \qquad \qquad \text{Divide both sides by } 1 - (1 + i)^{-m}$$

This is referred to as the **amortization formula.**

AMORTIZATION

The payment PMT required to amortize (pay off) a debt of PV dollars is given by

$$PMT = \frac{i(PV)}{1 - (1 + i)^{-m}}$$

where $i = \dfrac{r}{n}$ is the periodic interest rate and $m = nt$ is the number of equal periodic payments.

EXAMPLE 3

Determining the Mortgage Payment for a Home Loan

Homeowners typically finance their homes using a 15-year or 30-year mortgage. Interest rates are typically lower for 15-year mortgages than for 30-year mortgages; however, the monthly payments are substantially larger. On September 4, 2002, mortgage rates hit their lowest point since the mid-1960s. The average 30-year fixed mortgage rate was 6.17 percent, and the average 15-year fixed mortgage rate was 5.62 percent. (**Source:** www.bankrate.com.) What would the mortgage payment be for a couple purchasing a $233,750 home given that they made a 20 percent down payment? Calculate the payment amount for both mortgage options.

SOLUTION Since the couple is making a 20 percent down payment, only 80 percent of the purchase price will be financed. Since $0.8(233,750) = 187,000$, the amount financed is $187,000.

15-year mortgage:

The monthly periodic rate is $i = \dfrac{0.0562}{12} = 0.004683$. The number of payments is $m = 15(12) = 180$. Applying the amortization formula, we have

$$PMT = \frac{i(PV)}{1 - (1 + i)^{-m}}$$

$$= \frac{0.004683(187,000)}{1 - (1 + 0.004683)^{-180}}$$

$$= \frac{875.78}{0.5687}$$

$$= 1539.88$$

The monthly principal and interest payment for the 15-year mortgage is $1539.88. (Although we round values before we write them down, we keep the actual values in our calculator. This ensures that our final answer is not adversely affected by round-off error.)

30-year mortgage:

The monthly periodic rate is $i = \frac{0.0617}{12} = 0.005142$. The number of payments is $m = 30(12) = 360$. Applying the amortization formula, we have

$$PMT = \frac{i(PV)}{1 - (1 + i)^{-m}}$$

$$= \frac{0.005142(187{,}000)}{1 - (1 + 0.005142)^{-360}}$$

$$= \frac{961.49}{0.8422}$$

$$= 1141.68$$

The monthly principal and interest payment for the 30-year mortgage is $1141.68.

The TVM Solver on our calculator may also be used to solve decreasing annuity problems; however, this time the value of *PMT* will be positive instead of negative. The present value amount will still be negative. We will illustrate how to use the TVM Solver for a decreasing annuity in the next example.

EXAMPLE   4   **Determining How to Reduce the Length of a Loan**

Suppose the couple in Example 3 chose the 30-year mortgage. If they increase their monthly payment to $1200, how long will it take them to pay off the loan? If they want to pay off the loan in 20 years, how much should they pay each month?

SOLUTION   Using the TVM Solver, we get the results shown in Figure 6.9.

FIGURE 6.9

If they increase their monthly payment to $1200, it will take about 315 monthly payments (26.25 years) for them to pay off the loan. The $58.32 increase in their payment knocked more than three years off the loan.

Using the TVM Solver to solve for paying off the loan in 20 years, we get the results in Figure 6.10.

N=240
I%=6.17
PV=-187000
▪PMT=1358.130416
FV=0
P/Y=12
C/Y=12
PMT: **END** BEGIN

FIGURE 6.10

If they increase their payment to $1358.13, they'll be able to pay off the loan in 20 years.

Many first-time home buyers are surprised when they learn how much their home (including interest) actually costs. Truth-in-Lending laws require mortgage companies to disclose the total cost to home buyers before loan documents are signed.

The couple in Example 3 paid $1141.68 a month for 30 years. The total cost of their $233,750 home (including the $46,750 down payment) was $457,754.80, nearly double the value of their home! If they increased their payment to $1358.13, as shown in Example 4, their total home cost would be $372,701.20, an $85,053.60 savings. Fortunately, because of inflation, most homes appreciate over time. If housing prices increase by 2 percent annually, the $233,750 home will be worth over $423,000 thirty years later.

Sometimes we are interested in knowing when a decreasing annuity will reach a certain nonzero value. Since the interest rate on a 15-year mortgage is typically less than the rate on a 30-year mortgage, many homeowners refinance their homes with a 15-year mortgage once their principal balance reaches a certain level. The present value of a decreasing annuity (with a nonzero future value) formula may be used to address this situation.

PRESENT VALUE OF A DECREASING ANNUITY (WITH A NONZERO FUTURE VALUE)

The present value PV of a decreasing annuity with a nonzero future value FV is given by

$$PV = FV(1 + i)^{-m} + PMT \left[\frac{1 - (1 + i)^{-m}}{i} \right]$$

where $i = \frac{r}{n}$ is the periodic interest rate, $m = nt$ is the number of payments, and PMT is the payment amount.

EXAMPLE 5

Determining When a Loan Will Reach a Designated Value

A couple has purchased a home with a 30-year, $150,000 mortgage charging 6 percent interest compounded monthly. Their monthly principal and interest payment is $899.33. They want to refinance their home when the loan balance is $75,000. How long will it take for the account value to drop below $75,000?

SOLUTION   The present value is $150,000, and the future value is $75,000. The loan rate is 6 percent, so $i = \frac{0.06}{12} = 0.005$. Since payments are made monthly, the total number of payments is given by $m = 12t$. We have

$$PV = FV(1 + i)^{-m} + PMT\left[\frac{1 - (1 + i)^{-m}}{i}\right]$$

$$150{,}000 = 75{,}000(1 + 0.005)^{-12t} + 899.33\left[\frac{1 - (1 + 0.005)^{-12t}}{0.005}\right]$$

$$150{,}000 = 75{,}000(0.9419)^t + 899.33\left[\frac{1 - (0.9419)^t}{0.005}\right] \qquad \text{Evaluate exponents}$$

$$(0.005)150{,}000 = (0.005)\left\{75{,}000(0.9419)^t + 899.33\left[\frac{1 - (0.9419)^t}{0.005}\right]\right\} \qquad \text{Multiply by 0.005}$$

$$750 = 375(0.9419)^t + 899.33[1 - (0.9419)^t] \qquad \text{Distribute 0.005}$$

$$750 = 375(0.9419)^t + 899.33 - 899.33(0.9419)^t \qquad \text{Distribute 899.33}$$

$$-149.33 = 375(0.9419)^t - 899.33(0.9419)^t \qquad \text{Subtract 899.33}$$

$$-149.33 = -524.33(0.9419)^t \qquad \text{Group like terms}$$

$$0.2848 = (0.9419)^t \qquad \text{Divide by } -524.33$$

$$\ln(0.2848) = \ln(0.9419)^t \qquad \text{Log both sides}$$

$$\ln(0.2848) = t \ln(0.9419) \qquad \text{Log Rule 7}$$

$$t = \frac{\ln(0.2848)}{\ln(0.9419)} \qquad \text{Divide by } \ln(0.9419)$$

$$\approx 20.983 \text{years} \qquad \text{Evaluate on calculator}$$

In 21 years (252 monthly payments), the account balance will drop below $75,000.

Since we will solve multiple problems of this type in the exercises, it will be helpful to solve the present value of a decreasing annuity formula for m. This will make our computations in the exercises easier.

$$PV = FV(1 + i)^{-m} + PMT\left[\frac{1 - (1 + i)^{-m}}{i}\right]$$

$$PV = FV(1 + i)^{-m} + \frac{PMT}{i}[1 - (1 + i)^{-m}]$$

$$PV = FV(1 + i)^{-m} + \frac{PMT}{i} - \frac{PMT}{i}(1 + i)^{-m} \qquad \text{Distribute } \frac{PMT}{i}$$

$$PV - \frac{PMT}{i} = FV(1 + i)^{-m} - \frac{PMT}{i}(1 + i)^{-m} \qquad \text{Subtract } \frac{PMT}{i}$$

$$PV - \frac{PMT}{i} = \left(FV - \frac{PMT}{i}\right)(1 + i)^{-m} \qquad \text{Group like terms}$$

$$\frac{PV - \dfrac{PMT}{i}}{FV - \dfrac{PMT}{i}} = (1 + i)^{-m} \qquad \text{Divide by } \left(FV - \frac{PMT}{i}\right)$$

$$\frac{\dfrac{iPV - PMT}{i}}{\dfrac{iFV - PMT}{i}} = (1 + i)^{-m} \qquad \text{Rewrite rational expressions}$$

$$\frac{iPV - PMT}{iFV - PMT} = (1 + i)^{-m} \qquad \text{Rewrite rational expressions}$$

$$\ln\left(\frac{iPV - PMT}{iFV - PMT}\right) = \ln(1 + i)^{-m} \qquad \text{Log both sides}$$

$$\ln\left(\frac{iPV - PMT}{iFV - PMT}\right) = -m \ln(1 + i) \qquad \text{Log Rule 7}$$

$$m = \frac{\ln\left(\dfrac{iPV - PMT}{iFV - PMT}\right)}{-\ln(1 + i)} \qquad \text{Divide both sides by } -\ln(1 + i)$$

When using this form of the formula, it is important to remember that m represents the number of payments and, consequently, must be rounded to a whole number.

EXAMPLE 6

Determining When a Retirement Account Will Reach a Certain Value

After working hard for 45 years, an investor has accumulated $450,000 in her retirement account. She has placed the money into an investment that guarantees a 5 percent interest rate compounded monthly. She plans to withdraw $5000 per month. How many monthly withdrawals will she be able to make before the account value drops below $100,000?

SOLUTION We have

$$i = \frac{0.05}{12}$$

$$= 0.004167$$

Using the modified form of the present value of a decreasing annuity formula, we get

$$m = \frac{\ln\left(\dfrac{iPV - PMT}{iFV - PMT}\right)}{-\ln(1 + i)}$$

$$= \frac{\ln\left[\dfrac{0.004167(450{,}000) - 5000}{0.004167(100{,}000) - 5000}\right]}{-\ln(1 + 0.004167)}$$

$$= \frac{\ln\left(\dfrac{-3125}{-4583.33}\right)}{-\ln(1 + 0.004167)}$$

$$= \frac{\ln(0.6818)}{-\ln(1.004167)}$$

$$= 92.11$$

She will be able to make 92 monthly withdrawals before the account balance drops below $100,000.

Credit Cards

Many credit card companies target college students as a source of new customers. They offer free T-shirts, low introductory rates, and other gimmicks to attract potential patrons. Many college students need to develop a good credit history, and using a credit card responsibly can aid in that endeavor. Credit cards, however, are not annuities. The minimum payment required is typically a percentage of the account balance.

The author purchased a bunk bed (for his children) for $532.22. He charged the purchase to his credit card. When he received his credit card statement, the minimum payment due was $12.00, a mere 2.25 percent of the balance (Figure 6.11).

Furniture WHOLESALE   Customer Service (Servicio al Cliente)

Days in Billing Cycle: 31   **Statement Date: 08/26/2002**

Account Summary

PAYMENT DUE DATE:	TOTAL MINIMUM PAYMENT DUE	NEW BALANCE	PAST DUE AMOUNT	AVAILABLE CREDIT
09/20/2002	$12.00	$532.22	$0.00	$2,468.00

To avoid additional finance charges, pay by the Payment Due date above: 1) the New Balances of your Regular and Reduced Rate Credit Plans, plus 2) the New Balance of any Same As Cash or Waived Finance Charge Credit Plan that has expired or that will expire in the current billing cycle, plus 3) any Minimum Payments due on an unexpired Same As Cash or Waived Finance Charge Credit Plan.

Transactions

Transaction Date	Transaction Detail	Promo. Type/Credit Plan	Amount
06/27/2002	Previous Balance		$0.00
08/21/2002	Warehouse	Regular Purchase	$532.22
08/26/2002	New Balance		$532.22

Finance Charge Summary

Promotion Type/Credit Plan	Purchase Date	Promotion Expiration Date	Previous Balance	Average Daily Balance	Daily Periodic Rate	Corresponding APR	ANNUAL PERCENTAGE RATE (APR)	FINANCE CHARGES at Periodic Rate	Deferred FINANCE CHARGES	New Balance	Minimum Payment Due
Regular Purchase 00007-01	N/A	N/A	$0.00	$68.71	0.04287%	15.65%	15.65%	$0.00	$1.00	$532.22	$12.00

FIGURE 6.11

Suppose that he pays 2.25 percent of the balance (rounded up to the nearest dollar) every month. The card has a 15.65 percent annual percentage rate (APR). (APR is similar to annual percentage yield.)

For illustration purposes, we will assume that he makes his payment at the same time every month. Since the account balance will change once a month, we can convert the annual percentage rate into a monthly periodic rate. It appears that Costco calculates the APR by multiplying the daily periodic rate by 365. Therefore, we will calculate the monthly periodic rate by dividing the APR by 12.

$$\text{Monthly periodic rate} = \frac{0.1565}{12}$$

$$\approx 0.0130$$

Using a spreadsheet, we set up an amortization table for his credit card. How long do you think it will take him to pay off the card? Five years? Ten years? Because of the size of the amortization table (Table 6.6), we will only show a few of the payments made over the next 21.25 years! If he makes only the minimum payment, he will pay $569.41 in interest!

TABLE 6.6

Payment Number	Principal	Interest (1.30% of Principal)	Payment (2.25% of Principal rounded up)	New Balance (Principal + Interest − Payment)
1	$532.22	$6.94	$12.00	$527.16
2	$527.16	$6.88	$12.00	$522.04
3	$522.04	$6.81	$12.00	$516.84
4	$516.84	$6.74	$12.00	$511.58
5	$511.58	$6.67	$12.00	$506.26
7	$500.86	$6.53	$12.00	$495.39
9	$489.85	$6.39	$12.00	$484.24
10	$484.24	$6.32	$11.00	$479.56
50	$315.22	$4.11	$8.00	$311.33
100	$175.97	$2.29	$4.00	$174.26
200	$39.38	$0.51	$1.00	$38.90
254	$1.60	$0.02	$1.00	$0.62
255	$0.62	$0.01	$0.63	$0.00

Can we find a formula for the balance on a credit card with a minimum payment that is a percentage of the principal? Let i be the monthly periodic rate and $100r$ percent be the percentage of the account balance B that is required for the minimum payment. Then we have

Balance after first payment:

$$B_1 = B + iB - rB$$
$$= B(1 + i - r)$$

Balance after second payment:

$$B_2 = B(1 + i - r) + i[B(1 + i - r)] - r[B(1 + i - r)]$$
$$= B(1 + i - r)(1 + i - r)$$
$$= B(1 + i - r)^2$$

The pattern continues, and we get

Balance after nth payment:

$$B_n = B(1 + i - r)^n$$

ESTIMATE OF THE FUTURE BALANCE OF A CREDIT CARD

The balance on a credit card, B_n, after n minimum payments have been made may be estimated by

$$B_n = B(1 + i - r)^n$$

where B is the initial balance, i is the monthly periodic interest rate, and r is the percentage of the balance that is the minimum amount required to be paid.

This formula will give us an estimate of the balance after making n payments. It will not be precise, however, because many credit card companies impose additional rules on finance charges and payments, such as requiring a minimum $0.50 finance charge or rounding up the minimum payment to the nearest dollar. One of the reasons they do so is to ensure that the debt is eventually paid off. Note that in the formula, $B_n \neq 0$ for any n. That is, the balance will never be paid off. The credit card companies also need to ensure that the interest they collect is sufficient to make billing the customer profitable.

EXAMPLE 7

Estimating the Future Balance of a Credit Card

A credit card has a $532.22 balance and a 15.65 percent APR. The minimum payment required is 2.25 percent of the balance. Estimate the credit card balance after 50 payments have been made.

SOLUTION We have $B = 532.22$, $i = \frac{0.1565}{12} = 0.0130$, $r = 0.0225$, and $n = 50$.

$$B_{50} = 532.22(1 + 0.0130 - 0.0225)^{50}$$
$$= 532.22(0.9905)^{50}$$
$$= 532.22(0.6218)$$
$$= 330.92$$

We estimate that after 50 minimum payments, we will still owe $330.92. Referring back to the amortization table presented earlier in this section, we see that the actual amount owed after the 50th payment was $311.33. Even though the amortization table had the additional requirement of rounding up the minimum payment to the nearest dollar, the estimate was still close to the actual value.

6.4 Summary

In this section, you learned how to use the present value of a decreasing annuity formula as a tool in making financial decisions such as borrowing money to purchase an auto or a home. You also learned how to estimate a future credit card balance.

6.4 Exercises

In Exercises 1–20, use the present value of a decreasing annuity formula to solve the problem algebraically. Then check your work using the TVM Solver on your calculator. All quoted prices are accurate as of September 2002. Assume that interest is compounded monthly.

1. **Auto Loan** A new 2002 Chevrolet Tahoe 1500 LS has a manufacturer's suggested retail price (MSRP) of $32,954. (Source: www.car-prices-costs.com.) Determine the amount of the car payment if a buyer finances the entire retail price with an 8.99 percent, 48-month loan.

2. **Auto Loan** A new 2002 Volkswagen New Beetle GL has a manufacturer's suggested retail price (MSRP) of $15,900. (Source: www.car-prices-costs.com.) Determine the amount of the car payment if a buyer finances the entire retail price with a 3.95 percent, 48-month loan.

3. **Auto Loan** A new 2002 Chevrolet Corvette convertible has a manufacturer's suggested retail price (MSRP) of $48,205.00. (Source: www.car-prices-costs.com.) Determine the amount of the car payment if a buyer finances the entire retail price with a 6.99 percent, 60-month loan.

4. **Auto Loan** A new 2002 Toyota Camry Solara SE has a manufacturer's suggested retail price (MSRP) of $19,365. (Source: www.car-prices-costs.com.) Determine the amount of the car payment if a buyer finances the entire retail price with a 1.5 percent, 36-month loan.

5. **Auto Loan** A new 2002 Toyota Sequoia Limited 4 × 4 has a manufacturer's suggested retail price (MSRP) of $42,725. (Source: www.car-prices-costs.com.) A buyer gets a $7500 credit for her trade-in and finances the balance of the price at her local credit union with a 60-month, 5.99 percent loan. Determine the amount of the car payment.

6. **Auto Loan** A new 2002 Jaguar S-Type 4.0-Liter has a manufacturer's suggested retail price (MSRP) of $49,330. (Source: www.car-prices-costs.com.) A buyer gets a $18,500 credit for her trade-in and finances the balance of the price at her local credit union with a 48-month, 6.99 percent loan. Determine the amount of the car payment.

7. **Auto Loan** A new 2002 Jeep Grand Cherokee Laredo has a manufacturer's suggested retail price (MSRP) of $25,665. (Source: www.car-prices-costs.com.) A buyer gets an $8500 credit for her trade-in and finances the balance of the price at her local bank with a 72-month, 8.99 percent loan. Determine the amount of the car payment.

8. **Auto Loan** A new 2002 Ford Windstar LTD has a manufacturer's suggested retail price (MSRP) of $34,040. (Source: www.car-prices-costs.com.) A buyer gets a $17,200 credit for his trade-in and finances the balance of the price with a 60-month, 2.90 percent loan. Find the amount of the car payment.

9. **Auto Loan** A family wants to buy a 2002 Toyota Sequoia SR5 but wants to keep the monthly payments to $400 or less. If the family keeps the length of the loan to 48 months or less, the credit union will give them a loan with a 2.90 percent APR. The SUV retails for $31,256. (Source: www.car-prices-costs.com.) How much of a down payment does the family need to make?

10. **Auto Loan** A couple wants to buy a 2002 Ford Windstar LTD, but wants to keep the monthly payments to $450 or less. If

they keep the length of the loan to 36 months or less, the dealer will give them a loan with a 1.90 percent APR. The minivan retails for $34,040. (**Source:** www.car-prices-costs.com.) How much of a down payment does the couple need to make?

11. **Timeshare Condominiums**
Worldmark, the Club by Trendwest, a popular timeshare condominium company, sells vacation credit packages to consumers. Buyers can cash in credits to stay in four-star accommodations around the globe. In 2002, a perpetual 6000-credit-per-year package cost $9300. In a sales presentation, a company representative indicated that with a 10 percent down payment, the company would finance the balance over seven years at a 13.9 percent interest rate. (**Source:** Worldmark, the Club.)

If the buyer finances the package with the company, what will be the total cost of the package (including interest)?

12. **Timeshare Condominiums** After hearing the sales presentation referred to in Exercise 11, the author went to ebay.com and found a number of people selling their Worldmark, the Club memberships at a discount. One person offered the same 6000-credit package for $5100. However, the eBay seller indicated that the buyer would have to pay a $150 document fee to change the ownership of the package officially. (The author confirmed with Worldmark, the Club that memberships are transferable.) If the buyer finances the entire cost on his 9.90 percent APR credit card and makes $133.13 payments monthly, what is the actual cost of the package (including interest)?

13. **Home Mortgage** A 2608-square-foot, four-bedroom, two-bath home in Baker County, Florida, was advertised for $265,000 in September 2002. (**Source:** www.realtor.com.)
What will the monthly payment on a 15-year, 7.25 percent mortgage be if the buyer makes a 20 percent down payment?

14. **Home Mortgage** A three-bedroom home on 41.6 acres of land in Kittitas County, Washington, was advertised for $220,000 in September 2002. (**Source:** www.realtor.com.)
What will the monthly payment on a 30-year, 6.85 percent mortgage be if the buyer makes a 20 percent down payment?

15. **Home Mortgage** A three-bedroom home on five acres of land in Oregon was advertised for $89,900 in September 2002. (**Source:** www.realtor.com.)
What will the monthly payment on a 15-year, 6.50 percent mortgage be if the buyer makes a 5 percent down payment?

16. **Home Mortgage** A two-bedroom, two-bath home in New York, New York, was advertised for $1,677,545 in September 2002. (**Source:** www.realtor.com.)
What will the monthly payment on a 30-year, 7.75 percent mortgage be if the buyer makes a 20 percent down payment?

17. **Home Mortgage** A three-bedroom, two-bath home in Waukesha County, Wisconsin, was advertised for $249,900 in September 2002. (**Source:** www.realtor.com.)
What will the monthly payment on a 20-year, 6.00 percent mortgage be if the buyer makes a 10 percent down payment?

18. **Home Mortgage** A 2992-square-foot, four-bedroom, 2.5-bath home in Lancaster County, Nebraska, was advertised for $149,900 in September 2002. (**Source:** www.realtor.com.)
A couple has been prequalified for a 6.75 percent, 15-year VA loan. (VA loans, which may be obtained by military veterans, don't require a down payment.) The maximum monthly payment the couple can afford is $1000. Can the couple afford this home?

19. **Home Mortgage** A 3824-square-foot, six-bedroom, three-bath home in Queen Creek, Arizona, was advertised for $302,691 in February 2005. (**Source:** www.fultonhomes.com.)
A family finances 80 percent of the purchase price with a 5.75 percent loan. If the family makes $1500 principal and interest payments monthly, how long will it take to pay off the loan?

20. **Home Mortgage** A three-bedroom, three-bath home on six acres in Rensselaer County, New York, was advertised for $895,000 in September 2002. (**Source:** www.realtor.com.)
A retired executive has been prequalified for a 5.25 percent, 30-year loan. She intends to make a 20 percent down payment. The maximum monthly payment she can afford is $4000. Can she afford this home?

In Exercises 21–25, determine the amount of time it will take for the loan balance to reach the indicated value. Assume that interest is compounded monthly.

21. **Loan Balance**   How long will it take for a 5 percent interest, $250,000 loan with a monthly payment of $1000 to reach a loan balance below $100,000?

22. **Loan Balance**   How long will it take for an 8 percent interest, $100,000 loan with a monthly payment of $500 to reach a loan balance below $50,000?

23. **Loan Balance**   How long will it take for a 9 percent interest, $13,000 loan with a monthly payment of $375 to have a loan balance below $5000?

24. **Loan Balance**   How long will it take for an 18 percent interest, $9000 loan with a monthly payment of $400 to have a loan balance below $4000?

25. **Loan Balance**   How long will it take for an 11 percent interest, $10,000 loan with a monthly payment of $200 to have a loan balance below $2000?

In Exercises 26–30, determine the amount of time it will take for the retirement account balance to reach the indicated value.

26. **Retirement Account Balance**   How long will it take for a $300,000 retirement account earning an annual rate of 5 percent interest compounded annually to drop below $200,000 if $50,000 withdrawals are taken annually?

27. **Retirement Account Balance**   How long will it take for a $190,000 retirement account earning 3 percent interest compounded annually to drop below $100,000 if $30,000 withdrawals are taken annually?

28. **Retirement Account Balance**   How long will it take for a $800,000 retirement account earning 6 percent interest compounded annually to drop below $300,000 if $80,000 withdrawals are taken annually?

29. **Retirement Account Balance**   How long will it take for a $400,000 retirement account earning an annual rate of 4 percent interest compounded monthly to drop below $50,000 if $5000 withdrawals are taken monthly?

30. **Retirement Account Balance**   How long will it take for a $300,000 retirement account earning an annual rate of 6 percent interest compounded monthly to drop below $60,000 if $4000 withdrawals are taken monthly?

In Exercises 31–35, estimate the credit card balance after the indicated number of payments has been made. In each case, assume that the minimum payment is 2.25 percent of the current balance.

31. **Credit Cards**   The initial balance on a credit card with an 18.9 percent APR is $2505.98. What will the estimated balance be after 20 payments have been made?

32. **Credit Cards**   The initial balance on a credit card with a 21.9 percent APR is $1890.25. What will the estimated balance be after two years of payments have been made?

33. **Credit Cards**   The initial balance on a credit card with a 15.67 percent APR is $632.10. What will the estimated balance be after six payments have been made? Create an amortization table to check the accuracy of your estimate.

34. **Credit Cards**   The initial balance on a credit card with a 9.90 percent APR is $981.52. What will the estimated balance be after 36 payments have been made?

35. **Credit Cards**   The initial balance on a credit card with a 13.35 percent APR is $1225.00. What will the estimated balance be after 15 payments have been made?

Exercises 36–40 are intended to challenge your understanding of decreasing annuities.

36. **Retirement Planning**   An employee plans on working and contributing to her retirement account monthly for 44 years and then plans to live off her retirement savings for the next 25 years. She expects her retirement account to earn an annual rate of 8 percent compounded monthly. If she wants to make $3500 monthly withdrawals from her account during retirement, how much should she contribute to her retirement account monthly during her working years?

37. **Retirement Planning**   An employee plans on working and contributing to his retirement account monthly for 30 years and then plans to live off his retirement savings for the next 20

years. He expects his retirement account to earn an annual rate of 6 percent compounded monthly. If he wants to make $3000 monthly withdrawals from his account during retirement, how much should he contribute to his retirement account monthly during his working years?

38. Retirement Planning An employee plans on working and contributing to his retirement account monthly for 40 years and then plans to live off his retirement savings for the next 20 years. He expects his retirement account to earn an annual rate of 6 percent compounded monthly. He can afford to make $300 monthly contributions to his retirement account. How much money can he withdraw from the account monthly during the 20 years of his retirement?

39. Retirement Planning An employee plans on working and contributing to her retirement account monthly for 30 years and then plans to live off her retirement savings for the next 35 years. She expects her retirement account to earn an annual rate of 7 percent compounded monthly. She can afford to make $600 monthly contributions to her retirement account. How much money can she withdraw from the account monthly during the 35 years of her retirement?

40. Mortgage Payoff A couple has a $200,000 mortgage with a 7 percent annual interest rate. If they pay $1500 a month for 120 months and then $2000 a month until the loan is paid off, how long will it take them to pay off the loan?

Chapter 6 Review Exercises

Section 6.1 *In Exercises 1–4, solve the exponential equation algebraically.*

1. $1.3^x = 1.69$

2. $3^x = 243$

3. $1.2^x = 1.44$

4. $(0.5)^x = 4$

In Exercises 5–6, solve the problems algebraically.

5. **Auto Fatalities** The percentage of auto accident fatalities that are not alcohol-related may be modeled by

$$F(t) = 49.96 + 5.112 \ln{(t)} \text{ percent}$$

where t is the number of years since the end of 1989. (**Source:** Modeled from *Statistical Abstract of the United States, 2001*, Table 1099, p. 688.) According to the model, when will less than 25 percent of auto accident fatalities be alcohol-related?

6. **Video-Game Industry** While much of the U.S. economy was struggling in fiscal year 2002, Electronic Arts Inc., which is known for its array of video-game software, earned more than $917 million in profits. Stockholders' equity in the company (total assets minus liabilities, preferred stock, and intangible assets) may be modeled by

$$S(t) = 566.9(1.225)^t \text{ million dollars}$$

where t is the number of years since fiscal year 1998. (**Source:** Modeled from Electronic Arts Inc Annual Report, March 31, 2002.) According to the model, how long will it take for stockholders' equity to reach $1 billion?

Section 6.2 *In Exercises 7–10, answer each of the questions using the simple and/or compound interest formulas. For each exercise, quoted rates are from Bankrate.com and are accurate as of August 2002.*

7. **Certificates of Deposit** NBC Bank offers a three-month CD paying an annual rate of 2.27 percent simple interest with a minimum investment of $5000. How much will the CD be worth at maturity if the minimum amount is invested?

8. **Certificates of Deposit** Beal Bank offers a three-month CD paying an annual rate of 2.23 percent compounded quarterly with a minimum investment of $1000. How much will the CD be worth at maturity if the minimum amount is invested?

9. **Certificates of Deposit** State Farm Bank offers a five-year CD paying an annual rate of 4.63 percent interest compounded daily. Stonebridge Bank offers a five-year CD paying an annual rate of 4.65 percent compounded monthly. Assuming that the same amount of money is invested in each account, which account will be worth more at maturity?

10. **Certificates of Deposit** Providian National Bank offers a five-year CD paying an annual rate of 4.45 percent interest compounded daily. Eastern Savings Bank offers a five-year CD paying an annual rate of 4.46 percent compounded monthly. Assuming that the same amount of money is invested in each account, which account will be worth more at maturity?

In Exercises 11–12, calculate the annual percentage yield for each investment.

11. A savings account paying an annual rate of 2.57 percent compounded monthly.

12. A checking account paying an annual rate of 1.09 percent compounded daily.

In Exercises 13–14, determine how many years it will take for the two accounts to reach the same balance.

13. A $2000 investment in an account paying an annual rate of 2.75 percent compounded monthly and a $1800 investment in an account paying an annual rate of 3.02 percent compounded daily.

14. A $750 investment in an account paying an annual rate of 2.84 percent compounded quarterly and a $700 investment in an account paying an annual rate of 2.98 percent compounded daily.

In Exercises 15–16, convert the annual rate into a continuous rate or the continuous rate into an annual rate as appropriate.

15. Annual rate: 10.25 percent

16. Continuous rate: 3.23 percent

Section 6.3 *In Exercises 17–20, solve each of the increasing annuity problems algebraically. Check your solution using the TVM Solver on your calculator.*

17. **Mutual Funds** The Harbor Capital Appreciation fund earned an average of 11.51 percent in the 10-year period prior to August 31, 2002. (**Source:** Harbor Funds.) Assuming that the account will earn an annual rate of 11.51 percent compounded monthly, what will the future value of the account be in 10 years if $250 contributions are made monthly?

18. **Mutual Funds** The Harbor Growth fund earned an average of 4.14 percent annually in the 10-year period prior to August 31, 2002. (**Source:** Harbor Funds.) Assuming that the account will earn an annual rate of 4.14 percent compounded monthly, what will the future value of the account be in 10 years if $250 contributions are made monthly?

19. **Mutual Funds** The Harbor International fund earned an average of 10.05 percent in the 10-year period prior to August 31, 2002. (**Source:** Harbor Funds.) Assuming that the account has a current balance of $2320.00 and will earn an annual rate of 10.05 percent compounded monthly, what will the future value of the account be in 20 years if $300 contributions are made monthly?

20. **Mutual Funds** The Large Cap Value Fund earned an average of 10.12 percent in the 10-year period prior to August 31, 2002. (**Source:** Harbor Funds.) Assuming that the account has a current balance of $12,650.00 and will earn an annual rate of 10.12 percent compounded monthly, what will the future value of the account be in 30 years if $400 contributions are made monthly?

In Exercises 21–22, solve each problem using the TVM Solver.

21. **Retirement Savings** An investor hopes to increase his retirement account to $800,000 by the time he retires 15 years from now. The present value of the account is $35,289.19. He expects his investment to earn an annual rate of 9.89 percent compounded monthly. What must his monthly investment be in order for him to reach his goal?

22. **Home Buying** A rock star's ex-wife is saving up for a down payment on her next home. She wants to move into a $425,000 condominium in Paris with the former drummer of the band, and to finance it with a conventional loan requiring a 20 percent down payment. She currently has $34,820.21 in an offshore investment account earning an annual rate of 8.69 percent compounded daily. She contributes $3000 to the account monthly from her online sales of items containing her ex-husband's autograph. She wants to buy the condo within a year. Will she have enough money saved to make the down payment in 12 months?

Section 6.4

In Exercises 23–26, use the present value of a decreasing annuity formula to solve the problem algebraically. Then check your work using the TVM Solver on your calculator. All quoted prices are accurate as of September 2002.

23. **Retirement Savings** A retired couple has $450,000 in an investment account earning an annual rate of 5 percent interest compounded monthly. Starting at the end of the month, they'll begin withdrawing $4000 a month from the account. How long will it take for them to deplete all of the money in the account?

24. **Auto Loan** All used cars from Hunk-a-Heap Motors cost under $1000. A young driver wants to buy a $950 car he has nicknamed "the Bomb." He will pay 29 percent interest compounded daily on a two-year loan as a part of the dealer's "first-time buyer" package. What will his monthly payment be?

25. **Investment Property** A 4400-square-foot, ten-bedroom, six-bath duplex in Utah County, Utah, was advertised for $309,000 in September 2002. (**Source:** www.realtor.com.)

    A real estate investor has the required 20 percent down payment and is prequalified for a 6.25 percent, 15-year loan. She can afford a $300 monthly payment and expects to bring in an additional $2400 a month by renting out the rooms to local students. With her personal funds and the rent monies, will she have enough to make the monthly payment?

26. **Investment Property** A 1272-square-foot, three-bedroom, two-bath duplex in Utah County, Utah, was advertised for $124,900 in September 2002. (**Source:** www.realtor.com.)

    A real estate investor has the required 20 percent down payment and is prequalified for a 6.25 percent, 15-year loan. She can afford a $300 monthly payment and expects to bring in an additional $1000 a month by renting out the rooms to students at Brigham Young University and Utah Valley State College. With her personal funds and the rent monies, will she have enough to make the monthly payment?

In Exercises 27–28, estimate the credit card balance after the indicated number of payments has been made. In each case, assume that the minimum payment is 2.25 percent of the current balance.

27. The initial balance on a credit card with a 13.9 percent APR is $3795.99. What will the estimated balance be after 15 payments have been made?

28. The initial balance on a credit card with a 10.9 percent APR is $862.88. What will the estimated balance be after two years of payments have been made?

Make It Real

What to do

1. Find a home in your area that you would like to buy.

2. Open a savings account.

3. Calculate how much money you'll need for a 5 percent down payment on the house.

4. Calculate how long it will take you to save enough for the down payment if you contribute 5 percent of your monthly income (or $100 if you earn less than $2000 per month) to the savings account.

5. Find the best 30-year FHA mortgage rate offered in your area.

6. Find the monthly mortgage payment for the house from Step 1 using the rate from Step 5, assuming that you made a 5 percent down payment on the house.

7. Set a personal goal of getting a home mortgage within the next 10 years.

Where to look for data

Homes for Sale
www.realtor.com
www.homes.com
www.homebuilder.com
Local real estate guides (typically found outside supermarkets)

Mortgage Rates
www.bankrate.com
www.lendingtree.com
Wall Street Journal
Local banks and credit unions

Savings Account Rates
www.bankrate.com
Local banks and credit unions

Chapter 7

Sets and Probability

Although we typically try to control as many factors as possible, many things in life are influenced by an element of chance. If we understand basic probability concepts, we can estimate the likelihood of a certain event's occurrence. For example, did you know that in any class with 23 students or more, it is likely that two students in the class have the same birthday?

7.1 Introduction to Sets
- Use set notation to represent relationships between groups

7.2 Cardinality and the Addition and Multiplication Principles
- Count the number of items in a finite set

7.3 Permutations and Combinations
- Use permutations and combinations to determine the number of possible outcomes of a series of choices

7.4 Introduction to Probability
- Use probability terminology to describe real-life scenarios
- Use probability concepts to calculate probabilities

7.5 Basic Probability Concepts
- Use probability concepts to calculate probabilities

7.1 Introduction to Sets

- Use set notation to represent relationships between groups

GETTING STARTED   Human resources departments have the arduous task of sorting through job applications and determining which applicants should be interviewed. One sorting strategy is to eliminate all applications that do not meet the advertised criteria for the job. The mathematical notion of sets is useful in developing the sorting process.

In this section, we will introduce set notation and Venn diagrams. We will show how sets can be used to clarify complex relationships.

In September 2002, Levi Strauss & Co. requested applications for the position of Director of Strategic Development in its San Francisco office. The director's duties included performing complex business analysis and providing a clear understanding of supply chain dynamics, the competition, and the company's marketplace performance to senior executives. In addition to other skills, the position required a bachelor's degree and a minimum of 10 years of business experience. (**Source:** www.levistrauss.com.)

Suppose the company's human resource department received applications from the nine candidates listed in Table 7.1.

TABLE 7.1

	Name	Education	Experience
1	Chen	M.B.A.	1 year internship at Bugle Boy
2	Smythe	High school diploma	12 years managing a clothing retail store
3	Rodriguez	B.S.	10 years in marketing; various positions
4	Lewis	B.A.	5 years market research and analysis
5	Black	B.S.	15 years teaching high school science
6	Nguyen	B.S.	11 years project management
7	Romero	M.B.A.	21 years marketing and management
8	Alexander	High school diploma	25 years in retail distribution
9	Farris	High school diploma	12 years as an actor/artist

The department wants to interview only those candidates who meet the company's minimum criteria. Candidates 1, 4, 5, and 9 lack the required business experience. Candidates 2, 8, and 9 lack the required educational background. The only candidates meeting both criteria are Candidates 3, 6, and 7.

An alternative way to represent the information is by using rectangles and circles in a **Venn diagram.** We will represent the set of all candidates with rectangle U, the **universal set.** We will represent the set of candidates with at least a bachelor's degree with circle A and the set of candidates with 10 or more years of business experience with circle B. Figure 7.1 represents the situation graphically.

FIGURE 7.1

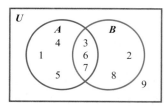

FIGURE 7.2

The set $A = \{1,3,4,5,6,7\}$. The set $B = \{2,3,6,7,8\}$. The entries inside the braces are called the **elements** of the set. We will now place the elements of each set in their appropriate location on the Venn diagram (Figure 7.2).

The candidates who meet the education requirement are in set A. The candidates who meet the business requirement are in set B. Candidates who meet both requirements are written in the **intersection of A and B,** the region that is common to both A and B. Candidate 9 doesn't meet either of the requirements and is written outside of both sets A and B.

EXAMPLE 1

Using a Venn Diagram to Sort Data

In September 2002, Electronic Arts advertised a Games Technical Support Representative position in Redwood Shores, California. Minimum qualifications included "avid computer and video gamer," "knowledge of Windows 95/98/Me Operating Systems in relation to the gaming environment (knowledge of MacOS and Windows XP is a plus)," and "self motivated and strong work ethics." (**Source:** jobs.ea.com.)

Use a Venn diagram to classify the hypothetical applicants in Table 7.2 according to their qualifications.

TABLE 7.2

	Name	Gaming Experience	Operating Systems	Work Ethic
1	Draxx	Video-game expert; plays 12 hours a day, knows 150 different games	Win 95/98/Me, Win XP, and MacOS	Fired from last two jobs for not coming to work
2	Argo	Six hours weekly on computer/video games for past 5 years	None	Successful athlete, average student, works in family business
3	Vogel	Five hours weekly on computer/video games for past 5 years	Win 95/98/Me	Eagle Scout, excellent student, successful athlete
4	Stewart	Risk, Monopoly, and other board games; hosts monthly bridge parties	None	Manages local retail store; active in a variety of professional organizations
5	Wilson	15 hours monthly on computer games	Win 95/98/Me, Win XP, and MacOS	Received outstanding employee of the year award

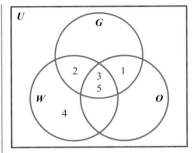

FIGURE 7.3

SOLUTION Let G be the set of applicants who are avid computer/video gamers, O be the set of applicants who know Win 95/98/Me, and W be the set of applicants with a strong work ethic. We'll consider anyone who plays computer games 15 or more hours a month to be an avid gamer. The Venn diagram is shown in Figure 7.3.

All of the applicants met at least one of the criteria; however, only applicants 3 and 5 met all three criteria.

To further our discussion of sets, it is necessary to develop some notation and terminology that will allow us to represent such concepts as the intersection of two sets symbolically.

SET TERMINOLOGY AND NOTATION

1. A set A is represented by separating its elements with commas and writing the list of elements within braces. For example, $A = \{1, 3, 5, 8\}$.

2. We write $x \in A$ and read "x is an element of A" if x is in A. For example, $3 \in A$.

3. We write $x \notin A$ and read "x is not an element of A" if x is not in A. For example, $4 \notin A$.

4. The set that contains no elements is called the **empty set** and is written $\varnothing$ or $\{\}$.

5. The set of all elements under consideration is called the **universal set** and is written U.

6. The set of all elements contained in U but not in A is called A **complement** and is written A'.

7. The set of all elements contained in A *and* in B is called the **intersection of A and B** and is written $A \cap B$. $A \cap B$ is read "A intersect B."

8. The set of all elements contained in A *or* in B *or* in both is called the **union of A and B** and is written $A \cup B$. $A \cup B$ is read "A union B."

9. A set A and a set B contain the same elements if and only if $A = B$.

10. If a set A is contained entirely in a set B, we say that "A is a **subset** of B" and write $A \subseteq B$.

11. If a set A is contained entirely in a set B, and $A \neq B$, then we say that "A is a **proper subset** of B" and write $A \subset B$.

EXAMPLE **2**

Using Set Notation

Write each of the sentences using set notation, using Figure 7.4 as a reference.

(a) Two is an element of the intersection of G and W.

(b) The union of G and W contains the elements 1, 2, 3, 4, and 5.

(c) Four is not in O.

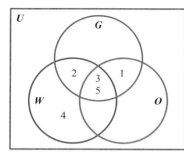

FIGURE 7.4

(d) Four is outside of G.

(e) Two is outside of the intersection of G and O.

(f) O is a proper subset of G.

SOLUTION

(a) $2 \in G \cap W$

(b) $G \cup W = \{1,2,3,4,5\}$

(c) $4 \notin O$ or $4 \in O'$

(d) $4 \in G'$ or $4 \notin G$

(e) $2 \notin (G \cap O)$ or $2 \in (G \cap O)'$

(f) $O \subset G$

EXAMPLE 3

Using a Venn Diagram to Interpret Set Notation

Use the Venn diagram in Figure 7.5 to find the elements in each of the sets. Then interpret the meaning of the results.

(a) $A \cap B$

(b) $A \cup B$

(c) U

(d) B'

(e) $A \cap B'$

FIGURE 7.5

SOLUTION

(a) $A \cap B = \{1, 5\}$ Everything that is in both A and B

(b) $A \cup B = \{1, 2, 4, 5, 6\}$ Everything that is in A or B or both

(c) $U = \{1, 2, 3, 4, 5, 6, 7\}$ Everything in the rectangle

(d) $B' = \{2, 3, 4, 7\}$ Everything not in B

(e) $A \cap B' = \{2, 4\}$ Everything in A that's not in B

When two sets do not intersect, we call them **disjoint sets.** For example, the set of all licensed drivers and the set of unlicensed drivers are disjoint sets. The set of all blondes and the set of all people with Ph.D.'s are not disjoint sets because some blondes have Ph.D.'s.

When a set has a large number of elements, it is cumbersome to write down each element. Suppose C is a set containing all positive even numbers less than or equal to 100,000. Listing all 50,000 elements would require a visible effort and a lot of paper. Often it is much easier to define a set by a rule. In this case, **set-builder notation** is extremely useful.

SET-BUILDER NOTATION

The set A of all x such that x meets specific criteria may be written as

$$A = \{x \mid x \text{ meets these criteria}\}$$

and is read A is the set of all x such that x meets these criteria.

EXAMPLE 4

Using Set-Builder Notation

The set C is a set containing all positive even numbers less than or equal to 100,000. Write C using set-builder notation.

SOLUTION $C = \{x \mid x \text{ is a positive even number and } x \leq 100,000\}$; this is read "$C$ is the set of all x such that x is a positive even number and is less than or equal to 100,000."

EXAMPLE 5

Using Set Notation and a Venn Diagram

Many people buy life insurance so that their families will be financially secure in the event of the unexpected death of an income-earning family member. Life insurance premiums are determined in part by actuaries who analyze mortality (death) data. Term life insurance provides protection only for a specified period of time, whereas whole-life insurance provides coverage for an individual's whole life. Premiums for life insurance vary dramatically based upon the insured person's use of tobacco. In September 2002, a 30-year, \$100,000 term life policy for a 20-year-old male nonsmoker cost \$162 a year. The same policy for a 20-year-old male smoker cost \$425 a year. (**Source:** Beneficial Life Insurance Company.)

A life insurance salesperson keeps a database of her customers on her home office computer. Her company has just reduced the rates for 20-40-year-old males who don't use tobacco. She plans on calling each of her 20-40-year-old male, nonsmoker clients to see if they want to increase their insurance coverage now that rates have dropped. She will use her database filter to create a list of the clients she needs to call. Use set notation to specify what search criteria she should use. Then draw a Venn diagram and shade the solution region.

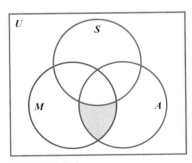

FIGURE 7.6

SOLUTION Let $S = \{x \mid x \text{ is a smoker}\}$, $M = \{x \mid x \text{ is a male}\}$, and $A = \{x \mid 20 \leq x \leq 40 \text{ years old}\}$. The list of clients she wants to call is given by $(S' \cap M) \cap A$ and is shown as the shaded region in Figure 7.6.

EXAMPLE **6** **Interpreting Set Notation**

The results of a U.S. Department of Education study of instructional faculty and staff at two-year institutions is shown in Table 7.3.

TABLE 7.3

Employment Status and Years of Experience on Current Job	Instructional Faculty and Staff (in thousands)	Highest Credential Attained (percent)		
		Bachelor's or Less	Master's Degree	Ph.D. or First Professional
Total full time	94.9	17.5	63.7	18.8
Years of experience on current job				
Less than 10 years	46.0	21.6	60.8	17.6
10–19 years	25.8	19.2	63.6	17.3
20 or more years	23.2	7.5	69.6	22.9
Total part time	153.1	33.3	53.3	13.4
Years of experience on current job				
Less than 10 years	122.2	34.6	52.6	12.9
10–19 years	24.5	29.7	55.1	15.2
20 or more years	6.4	22.4	60.0	17.6

Source: U.S. Department of Education, National Center for Education Statistics, 1993 National Study of Postsecondary Faculty (NSOPF: 1993).

Let F be the group of full-time instructors and E be the group of people who have 20 or more years of teaching experience. Interpret the meaning of $F \cap E'$ and determine the number of people in that group.

SOLUTION E' is the group of people who have less than 20 years of teaching experience. $F \cap E'$ is the group of full-time faculty with less than 20 years of teaching experience. From the full-time section of the table, we see that 46.0 thousand instructors have less than 10 years' experience and 25.8 thousand have 10–19 years' experience. Since $F \cap E'$ is the union of these two groups, there are 71.8 thousand full-time instructors with less than 20 years of teaching experience.

7.1 Summary

In this section, you learned how to use set notation and Venn diagrams. You discovered how sets can be used in real life to clarify complex relationships.

7.1 Exercises

In Exercises 1–10, copy the Venn diagram given here and shade the region defined in each problem, if possible. Then list the elements of the shaded region.

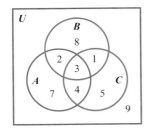

1. A'

2. $A \cap B$

3. $B \cup C$

4. $B' \cup C$

5. $(A \cap B)'$

6. $(A \cap B) \cup C$

7. $A \cap A'$

8. $(A \cap B) \cap C'$

9. $(A \cup B) \cap C'$

10. $((A \cap C) \cup (B \cap C))'$

In Exercises 11–15, let x represent a person.
$S = \{x \mid x \text{ is currently single}\}$,
$D = \{x \mid x \text{ never divorced}\}$, *and*
$M = \{x \mid x \text{ is or has been married}\}$.
Because of the complexity of multiple marriages, assume that each person in the universal set U has married at most once. Use set notation to represent the following groups of people.

11. Divorcees

12. Widows and widowers

13. People who never married

14. People who are currently married

15. Divorcees, widows, and widowers

In Exercises 16–18, use set notation to describe the relationship between A and B represented by the Venn diagram.

16.

17.

18.

In Exercises 19–24, write the set using set-builder notation.

19. The set of all positive numbers less than 10

20. The set of all integers that are a multiple of 3

21. The set of numbers that are perfect squares (*Hint:* A perfect square is a number that is the product of two identical integer factors.)

22. The set of birth years for people who were 21 years old or younger at the end of 2006.

23. The set of numbers that are factors of 24

24. The set of negative integers

For Exercises 25–36, six people are categorized by their gender, household income, and employment status.

> 1: Male, $67K, employed
> 2: Male, $43K, employed
> 3: Female, $52K, employed
> 4: Female, $25K, unemployed
> 5: Male, $31K, unemployed
> 6: Female, $73K, unemployed

The group is further classified, with M being the set of all men, R being the set of people with household income more than $50,000, and E being the set of employed people. Explain the real-life meaning of the set specified in each exercise. Then specify which people are elements of the set. (Hint: It may be helpful to draw a Venn diagram.)

25. $M \cap R$

26. $E \cup R$

27. $E' \cap R$

28. $(R \cap M)'$

29. $(E \cap M)' \cap R$

30. $(E' \cap R') \cup (M' \cap R)$

31. $(E' \cup R') \cap (M' \cup R)$

32. $(E \cup R') \cup (M' \cap R)$

33. $(E \cap R') \cup M$ **34.** $(E \cap R') \cap M$

35. $(M \cap R')'$ **36.** $(M \cup R') \cup E$

For Exercises 37–41, the results of a U.S. Department of Education study of instructional faculty and staff at two-year institutions are shown in the table. Let F be the group of full-time instructors and E be the group of people who have 20 or more years of teaching experience.

Employment Status and Years of Experience on Current Job	Instructional Faculty and Staff (in thousands)	Highest Credential Attained (percent)		
		Bachelor's or Less	Master's Degree	Ph.D. or First Professional
Total full time	94.9	17.5	63.7	18.8
Years of experience on current job				
Less than 10 years	46.0	21.6	60.8	17.6
10–19 years	25.8	19.2	63.6	17.3
20 or more years	23.2	7.5	69.6	22.9
Total part time	153.1	33.3	53.3	13.4
Years of experience on current job				
Less than 10 years	122.2	34.6	52.6	12.9
10–19 years	24.5	29.7	55.1	15.2
20 or more years	6.4	22.4	60.0	17.6

Source: U.S. Department of Education, National Center for Education Statistics, 1993 National Study of Postsecondary Faculty (NSOPF: 1993).

37. Interpret the meaning of $F \cup E$ and determine the number of people in that group.

38. Interpret the meaning of $F' \cap E$ and determine the number of people in that group.

39. Interpret the meaning of $F' \cap E'$ and determine the number of people in that group.

40. Interpret the meaning of $F' \cup E'$ and determine the number of people in that group.

41. Interpret the meaning of $(F \cap E')'$ and determine the number of people in that group.

Exercises 42–46 are intended to challenge your understanding of the concepts of sets and set notation.

42. Explain how the concepts of set intersection and union could be used by a business that tracks the type and number of products purchased by each customer.

43. DeMorgan's Law states that

$(A \cap B)' = A' \cup B'$ and $(A \cup B)' = A' \cap B'$.

For each statement, construct a Venn diagram for each side of the equal sign. Do you think DeMorgan's Law works for all sets A and B? Explain.

44. What is an equivalent way in which the statement $(A \cap B') \cup A'$ could be written?

45. What is an equivalent way in which the statement $(A \cap B') \cap (A' \cap B)$ could be written?

46. Are the two statements $(A \cap B') \cup B$ and $A \cap (B' \cup B)$ equivalent? Show the steps that lead to your conclusion.

7.2 Cardinality and the Addition and Multiplication Principles

- Count the number of items in a finite set

GETTING STARTED In the early 1940s, Marie Callender made a name for herself with her mouthwatering pies known for their light, flaky crusts and tasty fresh fruit fillings. Demand for her pies was so great that by 1948, the Callender family opened a wholesale bakery. The first Marie Callender's restaurant was opened in 1964, and by 2002, there were restaurants in more than 160 locations.

Marie Callender's offers seven pies called "Marie's Classics." Three of the pies list cream cheese as a main ingredient. Three of the pies list chocolate as an ingredient. Two pies don't list cream cheese or chocolate as an ingredient. (**Source:** www.mariecallender.com.) How many pies contain both cream cheese and chocolate? Questions such as this can be answered using the notion of cardinality.

In this section, we will show how to find the **cardinality of a set** (number of items in the set). We will also introduce the addition and multiplication principles for counting items in a set. We will use these concepts to analyze real-world data such as classifying Marie Callender's scrumptious pies.

CARDINALITY

If A is a set with a finite number of elements, then

$$n(A) = \text{the number of elements in } A$$

$n(A)$ is read "the cardinality of A."

Let U be the set of all Marie Callender's classic pies, C be the set of classic pies with cream cheese, and D be the set of classic pies with chocolate.

We have $n(U) = 7$, $n(C) = 3$, $n(D) = 3$, and $n((C \cup D)') = 2$. We want to find $n(C \cap D)$, the number of pies with both cream cheese and chocolate. To solve the problem, we need to know how to find the cardinality of a union and the cardinality of a complement.

CARDINALITY OF A UNION

If A and B are sets with a finite number of elements, then

$$n(A \cup B) = n(A) + n(B) - n(A \cap B)$$

Since the items in the intersection of A and B were "double counted" [once in $n(A)$ and once in $n(B)$], it was necessary to subtract $n(A \cap B)$.

CARDINALITY OF A COMPLEMENT

If U is a universal set with a finite number of elements and A is a subset of U, then

$$n(A') = n(U) - n(A) \text{ and } n(A) = n(U) - n(A')$$

From the cardinality of a complement formula, we know that

$$n(C \cup D) = n(U) - n((C \cup D)')$$
$$= 7 - 2$$
$$= 5$$

There are five pies with cream cheese or chocolate.

From the cardinality of a union formula, we know that

$$n(C \cup D) = n(C) + n(D) - n(C \cap D)$$
$$5 = 3 + 3 - n(C \cap D)$$
$$-1 = -n(C \cap D)$$
$$n(C \cap D) = 1$$

One pie has both cream cheese and chocolate listed as ingredients. Each of the seven pies is identified in the Venn diagram in Figure 7.7.

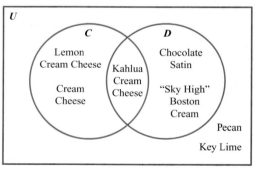

FIGURE 7.7

EXAMPLE 1

Finding the Cardinality of a Set

Wheaties has been known as "The Breakfast of Champions" since 1933. Packages featuring Babe Ruth, Mary Lou Retton, Michael Jordan, and Tiger Woods are among the top ten most popular Wheaties packages. Of the top ten packages, four packages feature people playing baseball and eight feature people in sports other than football. How many packages feature athletes not playing baseball or football?

SOLUTION Let U be the set of all top ten packages, B be the set of packages featuring baseball players, and F be the set of packages with football players. We must find $n((B \cup F)')$. We have $n(U) = 10$, $n(B) = 4$, and $n(F') = 8$.

We know that

$$n(F) = n(U) - n(F')$$
$$= 10 - 8$$
$$= 2$$

Since each athlete is featured playing only one sport, B and F are disjoint sets and $n(B \cap F) = 0$.

From the cardinality of a union formula, we know that

$$n(B \cup F) = n(B) + n(F) - n(B \cap F)$$
$$= 4 + 2 - 0$$
$$= 6$$

Six athletes played baseball or football.

From the cardinality of a complement formula, we know that

$$n((B \cup F)') = n(U) - n(B \cup F)$$
$$= 10 - 6$$
$$= 4$$

Four athletes playing sports other than baseball or football are featured. They include the Men's U.S. Gold Medal Hockey Team, Michael Jordan, Mary Lou Retton, and Tiger Woods.

Mary Lou Retton (featured in 1984) was the first female athlete to be a Wheaties spokesperson. Since then, the Women's U.S. Gold Medal Gymnastics Team, Amy Van Dyken (swimming), the Women's U.S. Gold Medal Hockey Team, and Sarah Hughes (figure skating) have joined the ranks of featured women.

Cartesian Products

Many sets are defined by linking two different quantities. The Cartesian product is used to maintain the linked relationship.

CARTESIAN PRODUCT

If A and B are sets with a finite number of elements, then the Cartesian product is

$$A \times B = \{(a, b) | a \in A \text{ and } b \in B\}$$

The elements of the set $A \times B$ are ordered pairs with the first value from set A and the second value from set B.

EXAMPLE 2

Determining the Outcomes of a Cartesian Product

In a game, you flip a coin and roll a die. What are the possible results?

SOLUTION

The coin has two possible outcomes: heads or tails.

The die has six possible outcomes: 1, 2, 3, 4, 5, 6.

Let $C = \{H, T\}$ be the set of possible outcomes for the coins and $D = \{1, 2, 3, 4, 5, 6\}$ be the set of possible outcomes for the die.

Then $C \times D = \begin{Bmatrix} (H,1), & (H,2), & (H,3), & (H,4), & (H,5), & (H,6), \\ (T,1), & (T,2), & (T,3), & (T,4), & (T,5), & (T,6), \end{Bmatrix}$ is the set of possible results.

CARDINALITY OF A CARTESIAN PRODUCT

If A and B are sets with a finite number of elements, then

$$n(A \times B) = n(A) \cdot n(B)$$

From Example 2, we have $n(C) = 2$ and $n(D) = 6$. It follows that

$$n(C \times D) = n(C) \cdot n(D)$$
$$= 2 \cdot 6$$
$$= 12$$

This solution matches the result in Example 2.

EXAMPLE 3

Finding the Cardinality of a Cartesian Product

Two siblings have been accepted by Harvard, Yale, UCLA, and the University of Washington. The first sibling has also been accepted by the University of Michigan. How many different enrollment combinations are possible for the two siblings?

SOLUTION   Let F be the number of enrollment choices for the first sibling and S be the number of enrollment choices for the second sibling. We have $n(F) = 5$ and $n(S) = 4$. So

$$n(F \times S) = 5 \cdot 4$$
$$= 20$$

There are 20 possible enrollment combinations for the siblings.

Addition Principle

Marie Callender's classifies pies into five different categories: Famous Cream Pies, Callender's Double Cream Pies, Seasonal Fresh Fruit Pies, Marie's Classics, and Traditional Favorites. No pie is listed in more than one category (**Source:** www.mariecallender.com.) How many varieties of pie are there to pick from? Questions such as these may be answered using the Addition Principle.

ADDITION PRINCIPLE

When choosing an element from among n disjoint sets, the total number of elements to pick from is the *sum* of the cardinalities of each set.

There are four Famous Cream Pies, three Callender's Double Cream Pies, six Seasonal Fresh Fruit Pies (including the "no sugar added" varieties), seven Marie's Classics, and sixteen Traditional Favorites. The total number of pies to choose from is $4 + 3 + 6 + 7 + 16 = 36$.

EXAMPLE 4

Using the Addition Principle

As of September 10, 2002, Toyota of Puyallup had several used Toyota Tacoma SR5 trucks for sale, including four 2001 models, one 2000 model, and six 1999 models. (**Source:** ToyotaofPuyallup.com.) How many Toyota Tacoma trucks were for sale?

SOLUTION Since $4 + 1 + 6 = 11$, there were 11 Toyota Tacoma trucks for sale.

Multiplication Principle

The Royal Argosy dinner cruise is a popular attraction for visitors in the Seattle area. The cruise ship spends three hours on beautiful Elliot Bay while guests enjoy an exquisite dinner prepared by some of the area's premier chefs. The views of the Seattle skyline from the ship are phenomenal. The author and his wife dined on the Royal Argosy while celebrating their anniversary. They were given the following menu options:

Welcome Aboard Pre-Appetizer
Assorted Artisan Bread Basket

Starters
Hearts of Baby Romaine
Elliot's Dungeness Crab and Corn Bisque
Gathered Greens

Entrees
Macadamia Nut Halibut
Wild Chinook Salmon
New York Strip Loin
Filet Mignon Royal Argosy
Grilled Portobello

Desserts
Burnt Cream
Dark Chocolate Tower of Decadence
Mixed Berry Shortcake

Source: www.royalargosy.com.

How many different four-course meals can be ordered from the menu? Questions such as this can be answered using the Multiplication Principle.

MULTIPLICATION PRINCIPLE

The total number of possible selections that include one element from each of r sets is the *product* of the cardinalities of each set. That is, given sets $A_1, A_2, \ldots, A_r$ the total number of possible selections including exactly one element from each set is

$$n(A_1) \cdot n(A_2) \cdot \cdots \cdot n(A_r)$$

The dinner cruise offered one pre-appetizer, three starters, five entrees, and three desserts. The total number of possible dinner selections is $1 \cdot 3 \cdot 5 \cdot 3 = 45$.

EXAMPLE 5

Using the Multiplication Principle

Cars.com allows you to pick and price the features of your new car. In completing the online form for a 2002 Toyota Sequoia SR5 V8, potential buyers were given options in the following categories:

> Styles (2 choices)
> Drive type (2 choices)
> Body color (9 choices)
> Interior color (2 choices)
> Option groups (3 choices)
> Wheels (3 choices)
> Tires (2 choices)
> Audio (3 choices)
> Other options (14 choices)

If one choice is made from each category, how many different Toyota Sequoias can be made?

SOLUTION Assuming that the sets are disjoint (as they appear to be), the multiplication principle may be used. There are $2 \cdot 2 \cdot 9 \cdot 2 \cdot 3 \cdot 3 \cdot 2 \cdot 3 \cdot 14 = 54{,}432$ different cars possible.

Decision Trees

Decision trees are a graphical representation of each of the possible outcomes that result from a series of choices. In a sense, they are a visual interpretation of the Multiplication Principle. They are most useful for a series of four or fewer choices from sets containing five or fewer elements. They can be used for larger sets, but they tend to become unwieldy. We will demonstrate how to use a decision tree with an example.

A student needs to enroll in three online classes next quarter: English, math, and art. There are two online English courses: Basic Composition and Basic Literature; there are three online math courses: Finite Math, Consumer Math, and Precalculus; and there are two online art courses: Art History and Design. We will construct the decision tree by making each of the choices in sequence. We must first choose an English class (see Figure 7.8).

FIGURE 7.8

We will next choose a math class (see Figure 7.9).

FIGURE 7.9

So far there are six different course schedules available to us. We will now pick the art class (see Figure 7.10).

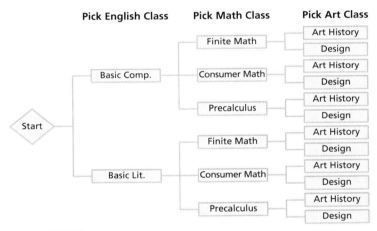

FIGURE 7.10

There are 12 schedules available to us. Reading from the decision tree, the first choice is Basic Composition, Finite Math, and Art History. The second choice is Basic Composition, Finite Math, and Design. The third choice is Basic Composition, Consumer Math, and Art History. The process of reading the diagram continues until you reach the final choice of Basic Literature, Precalculus, and Design.

7.2 **Summary**

In this section, you learned how to find the cardinality of a set and how to find a Cartesian product. You saw how the addition and multiplication principles are used in counting the items in a set. You finished the section by learning how to construct a decision tree.

7.2 Exercises

In Exercises 1–10, A and B are finite subsets of the universal set U with $n(U) = 15$, $n(A) = 10$, $n(A \cap B) = 4$, and $n(A \cup B) = 13$. Find the cardinality of each of the specified sets.

1. B

2. A'

3. B'

4. $A' \cup B'$

5. $A' \cap B'$

6. $(A \cup B)'$

7. $(A \cap B)'$

8. $A \cap B'$

9. $A' \cap B$

10. $A \cup B'$

In Exercises 11–15, calculate the cardinality of the Cartesian product $A \times B$ and write each of the elements of $A \times B$.

11. Let $A = \{$red, green, yellow$\}$ and $B = \{$apple, tomato$\}$.

12. Let $A = \{25, 36, 49\}$ and $B = \{x|x$ is a positive integer less than 6$\}$.

13. Let $A = \{$chocolate, vanilla$\}$ and $B = \{$ice cream, pie, shake, cookie$\}$.

14. Let $A = \{$a, e, i, o, u$\}$ and $B = \{$n, s, t$\}$.

15. Let $A = \{$Heads, Tails$\}$ and $B = \{x|x$ is a positive factor of 6$\}$.

In Exercises 16–25, use the Addition or Multiplication Principle as appropriate to determine the cardinality of the set.

16. **Online Class Schedules** Washington Online is a virtual campus of the Washington State two-year colleges. Students may enroll in a variety of courses at area colleges. For the fall quarter of 2002, South Seattle Community College offered three English courses, ten computer-related courses, and six other types of courses. (**Source:** www.waol.org.) How many schedule choices does a student who intends to enroll in one course from each category have?

17. **Online Class Schedules** For the fall quarter of 2002, Green River Community College offered two English courses, four social science courses, and four other types of online courses. (**Source:** www.waol.org.) How many schedule choices does a student who intends to enroll in one course from each category have?

18. **Ice Cream Varieties** The Tillamook Creamery Association, a farmer-owned cooperative in Oregon, operates the Tillamook Cheese creamery. Nearly a million tourists visit the creamery each year to watch the cheese makers at work and to partake of the rich Tillamook ice cream. On a hot summer day, the ice cream counter serves roughly 4000 ice cream cones.

The creamery offers 4 types of cones (wafer cone, sugar cone, waffle cone, and waffle dish) and 45 varieties of ice cream, sherbet, and frozen yogurt. Cones may be served with one, two, or three scoops. (**Source:** www.tillamookcheese.com.) How many ice cream cone choices are there?

19. **Frozen Dessert Varieties** The Tillamook Cheese creamery sells 39 flavors of 13.5 percent butterfat ice cream, 3 varieties of frozen yogurt, 1 flavor of sherbet, and 2 varieties of reduced-fat ice cream. (**Source:** www.tillamookcheese.com.) How many different varieties of frozen dairy desserts does the company sell?

20. **Lodging and Dining** An executive is flying a potential business client to her Boston office. She has asked her administrative assistant to book a room for her client in a five-star hotel and to make dinner reservations for them at a top-rated restaurant. Her computer-savvy assistant surfs to www.find-luxury-hotels.com and pulls up a list of three five-star hotels in the Boston area: *Lenox Hotel, Fairmont Copley Plaza, and Millennium Bostonian*. He next goes to www.diningguide.net and finds two fine dining restaurants in downtown Boston: *Julien* and *Rowes Wharf Restaurant*. How many different dining/lodging options does the assistant have? Draw a decision tree showing the possible options.

21. **Lodging and Dining** If the administrative assistant from Exercise 20 includes 4-star and 4½-star hotels, there are 17 hotels to choose from. If he includes restaurants outside of the Boston downtown area, there are 8 fine dining restaurants to choose from. How many different dining/lodging options does he have?

22. 🌐 **Menu Options** The Royal Argosy lunch cruise offers the following menu options:

Welcome Aboard Pre-Appetizer
Assorted Artisan Bread Basket

Starters
Elliot's Clam Chowder
Gathered Greens

Entrees
Hearts of Romaine & Alaskan Shrimp
Dungeness Crab Cakes
Wild Chinook Salmon
Herb Crusted Local Farm Chicken
Butternut Squash Ravioli

Desserts
Burnt Cream
Dark Chocolate Tower of Decadence
Warm Pineapple Upside-down Cake

Source: www.royalargosy.com.

How many different meals can be created from the menu, assuming that one item is selected from each category?

23. 🌐 A calorie-conscious guest wants to forgo dessert on the lunch cruise in Exercise 22. How many meal options are there if dessert is omitted?

24. 🌐 A calorie-conscious guest who has discovered that Burnt Cream (Crème Brulée) is the most exquisite dessert ever invented decides to enjoy a serving of Burnt Cream and start her diet tomorrow. Using the menu from Exercise 22, how many meal combinations are possible with Burnt Cream as a dessert?

25. 🌐 Using the menu in Exercise 22, how many different dish choices (not meals) are there?

For Exercises 26–30, you are given the sets $R = \{\text{marble, block}\}$, $C = \{\text{black, red, blue}\}$, $S = \{\text{small, medium, large}\}$, *and* $M = \{\text{glass, wood, plastic}\}$. *Draw a decision tree to show all possible outcomes for the specified criteria.*

26. The outcomes with a single element from each of the sets R, C, and S.

27. The outcomes with a single element from each of the sets R, C, and M.

28. The outcomes with a single element from each of the sets R, S, and M.

29. The outcomes with a single element from each of the sets R, C, M, and S.

30. The outcomes with a single element from each of the sets R, C, M, and S that contain the elements *red* and *marble*.

For Exercises 31–35, refer to the accompanying table, which depicts the highest educational level attained by U.S. Air Force officers. Data were accurate as of May 2003.

Educational Level	Males (M)	Females (F)	Total
B.A./B.S. (B)	25,900	6,235	32,135
M.A./M.S. (S)	24,694	4,514	29,208
Ph.D. (D)	852	155	1,007
Professional Degree (P)	4,518	1,199	5,717
Unknown (N)	3,313	902	4,215
Total	59,277	13,005	72,282

Source: www.afpc.randolph.af.mil.

31. 🌐 Calculate $n(D \cap F)$.

32. 🌐 Calculate $n((M \cup F) \cap P)$.

33. 🌐 Calculate $n(B' \cap F)$.

34. 🌐 Calculate $n((P' \cap F) \cup (M \cap N'))$.

35. 🌐 Calculate $n((N' \cap M) \cup F)$.

Exercises 36–40 are intended to challenge your understanding of the concepts covered in this section.

36. If $n(A \cap B) = n(A \cup B)$, what can you conclude about sets A and B?

37. If $n(A \cap B) \neq n(A \cup B)$, what can you conclude about sets A and B?

38. **Class Schedule** A student needs to select three courses for next semester: a math class, an English class, and a class from some other category. There are five math courses, four English courses, three P.E. courses, and four art courses available. How many different course schedules can be created?

39. **Sandwiches** Of the sandwiches listed on the McDonalds web site, nine contain beef, four contain chicken, and one includes fish. Of the fourteen different sandwiches, seven contain lettuce and seven contain cheese. All sandwiches include a bun. (**Source:** www.mcdonalds.com.)

How many different types of meat sandwiches could be made from the ingredients beef, chicken, fish, bun, lettuce, and cheese? (Assume that each sandwich contains only one type of meat and one bun. Additional ingredients are optional.)

40. Explain why the answer to Exercise 39 is less than 14.

7.3 **Permutations and Combinations**

- Use permutations and combinations to determine the number of possible outcomes of a series of choices

GETTING **STARTED** In an upset victory, Sarah Hughes won the women's figure skating gold medal in the 2002 Winter Olympics. Sarah, in fourth place after the short program, skated an impeccable long program that landed her the gold

medal. Since the games, she has reportedly received millions of dollars in endorsements and has been featured on the Wheaties cereal package. Fellow competitor Irina Slutskaya (Russia) won the silver medal, and teammate Michelle Kwan (United States) won the bronze. How many different ways could the top three skaters have placed? Questions such as these may be answered using permutations and combinations.

In this section, we will demonstrate how permutations and combinations can be used to count the number of possible results of an event. These concepts will lay the groundwork for our investigation into probability.

How many different ways could the three medalists have placed? Let's look at Table 7.4.

TABLE 7.4

	1	2	3	4	5	6
Gold	Hughes	Hughes	Slutskaya	Slutskaya	Kwan	Kwan
Silver	Slutskaya	Kwan	Hughes	Kwan	Hughes	Slutskaya
Bronze	Kwan	Slutskaya	Kwan	Hughes	Slutskaya	Hughes

There were six different ways the medalists could have placed. An alternative way to look at the ordering of the medalists is to look at the number of choices available for each medal and use the multiplication principle.

$$\overline{\text{Gold}} \quad \overline{\text{Silver}} \quad \overline{\text{Bronze}}$$

From the set $W = \{\text{Hughes, Slutskaya, Kwan}\}$, we have three choices for the gold medal. After choosing the gold medal winner, we have only two choices left for the silver medal. After choosing the silver medalist, only one choice remains for the bronze medal.

$$\frac{3}{\text{Gold}} \cdot \frac{2}{\text{Silver}} \cdot \frac{1}{\text{Bronze}}$$

There are $3 \cdot 2 \cdot 1 = 6$ possible ways to order the elements of the set W.

How many different ways could the top five contestants have placed? Using the same argument, there are $5 \cdot 4 \cdot 3 \cdot 2 \cdot 1 = 120$ possible ways the top five contestants could have placed.

How many ways could the top 10 contestants have placed? There are $10 \cdot 9 \cdot 8 \cdot 7 \cdot 6 \cdot 5 \cdot 4 \cdot 3 \cdot 2 \cdot 1 = 3,628,800$ possible ways they could have placed. As you can see, it becomes increasingly difficult to write out the product as the number of different items in the set increases. To facilitate this process, we introduce **factorial** notation.

FACTORIAL

For nonnegative integers,

$$n! = n \cdot (n-1) \cdot (n-2) \cdot \cdots \cdot 2 \cdot 1$$

$n!$ is read "n factorial." By definition, $0! = 1$.

For example,

$$10! = 10 \cdot 9 \cdot 8 \cdot 7 \cdot 6 \cdot 5 \cdot 4 \cdot 3 \cdot 2 \cdot 1$$
$$= 3,628,800$$

Factorial notation is used extensively when determining the number of different ways to sequence a set of distinct items.

PERMUTATION OF *n* ITEMS

A **permutation of *n* items** is an ordered sequence of those items. The number of possible permutations of *n* items is

$$n! = n \cdot (n-1) \cdot (n-2) \cdot \cdots \cdot 2 \cdot 1$$

EXAMPLE 1 **Counting the Number of Ways to Rank the Members of a Group**

How many ways can a class of 15 students be ranked?

SOLUTION We must order the 15 students. There are 15! ways to rank them.

$$15! = 15 \cdot 14 \cdot 13 \cdot 12 \cdot 11 \cdot 10 \cdot 9 \cdot 8 \cdot 7 \cdot 6 \cdot 5 \cdot 4 \cdot 3 \cdot 2 \cdot 1$$
$$= 1{,}307{,}674{,}368{,}000$$

There are more than a trillion ways to rank order a class of 15 students!

Calculating *n*! by hand can be cumbersome for even small values of *n*. The following Technology Tip details how to use a graphing calculator to calculate *n*!.

TECHNOLOGY TIP

Calculating *n*!

1. Type the value of *n* in the home screen.

> 15

2. Press (MATH), use the blue arrows to move to the PRB menu, and select item 4 : !. Press (ENTER).

> MATH NUM CPX **PRB**
> 1:rand
> 2:nPr
> 3:nCr
> **4**:!
> 5:randInt(
> 6:randNorm(
> 7:randBin(

3. Press (ENTER) to display the result.

> 15!
> 1.307674368ε12

Returning to the Olympic figure skating event, we didn't know who would win medals in 2002, even after the short program. Michelle Kwan was favored to win, and Sarah Hughes was in fourth place. Out of the top 15 skaters, how many different ways could the contestants have won medals?

Let S be the set of the top 15 skaters at the end of the short program. We know that $n(S) = 15$. We still have only three medals to award. How many choices do we have for the gold medal? There are 15 contestants, so we have 15 choices. After selecting the gold medalist, how many choices do we have for the silver medal? We can choose from among the remaining 14 contestants. After selecting the gold and silver medalists, we have 13 contestants from which to pick the bronze medalist.

$$\underset{\text{Gold}}{15} \cdot \underset{\text{Silver}}{14} \cdot \underset{\text{Bronze}}{13}$$

There are $15 \cdot 14 \cdot 13 = 2730$. possible ways to pick three elements from the set W. That is, there were 2730 different ways the medals could have been awarded to one of the top 15 short program contestants.

Observe that

$$15 \cdot 14 \cdot 13 = 15 \cdot 14 \cdot 13 \cdot \left(\frac{12 \cdot 11 \cdot 10 \cdot 9 \cdot 8 \cdot 7 \cdot 6 \cdot 5 \cdot 4 \cdot 3 \cdot 2 \cdot 1}{12 \cdot 11 \cdot 10 \cdot 9 \cdot 8 \cdot 7 \cdot 6 \cdot 5 \cdot 4 \cdot 3 \cdot 2 \cdot 1} \right)$$

$$= \frac{15 \cdot 14 \cdot 13 \cdot 12 \cdot 11 \cdot 10 \cdot 9 \cdot 8 \cdot 7 \cdot 6 \cdot 5 \cdot 4 \cdot 3 \cdot 2 \cdot 1}{12 \cdot 11 \cdot 10 \cdot 9 \cdot 8 \cdot 7 \cdot 6 \cdot 5 \cdot 4 \cdot 3 \cdot 2 \cdot 1}$$

$$= \frac{15!}{12!}$$

$$= \frac{15!}{(15 - 3)!}$$

We recognize the 15 as the number of unique elements in the set. We recognize the 3 as the number of medals to be awarded. Selecting 3 medalists from a group of 15 contestants is an example of a permutation of 15 items taken 3 at a time.

PERMUTATIONS OF n ITEMS TAKEN r AT A TIME

A **permutation of n items taken r at a time** is an ordered sequence of r items chosen from a set of n distinct items. The number of possible permutations of n items taken r at a time is

$$P(n, r) = \frac{n!}{(n - r)!}$$

EXAMPLE 2

Using Permutations in the Real World

In the early 1980s, Tommy Tutone made it big with the hit "867-5309 Jenny." The song tells of a guy trying to get up the nerve to call "Jenny," whose phone number he presumably found scrawled on a bathroom wall. The single hit number 4 on the charts in early 1982. (**Source:** musicfinder.yahoo.com.)

The number 8675309 is fascinating in its own right. First, no digit is repeated in the number. Second, the number is divisible only by 1 and itself.

How many different seven-digit phone numbers can be created without repeating any of the digits? Remember that phone numbers can't start with a 1 or a 0.

SOLUTION We must select each digit from the set $N = \{0, 1, 2, 3, 4, 5, 6, 7, 8, 9\}$. We know that $n(N) = 10$. Once a digit has been selected, it cannot be reused. We need seven digits to make the number.

$$P(n, r) = \frac{n!}{(n - r)!}$$

$$P(10, 7) = \frac{10!}{(10 - 7)!}$$

$$= \frac{10!}{3!}$$

$$= \frac{10 \cdot 9 \cdot 8 \cdot 7 \cdot 6 \cdot 5 \cdot 4 \cdot 3!}{3!}$$

$$= 604,800$$

There are 604,800 seven-digit numbers that don't repeat any digits. However, phone numbers can't start with a 0 or a 1, so we need to subtract all numbers that start with a 0 or a 1. Let's first consider the seven-digit numbers that start with a 0. We need to select six more digits from the remaining nine numbers.

$$P(n, r) = \frac{n!}{(n - r)!}$$

$$P(9, 6) = \frac{9!}{(9 - 6)!}$$

$$= \frac{9!}{3!}$$

$$= \frac{9 \cdot 8 \cdot 7 \cdot 6 \cdot 5 \cdot 4 \cdot 3!}{3!}$$

$$= 60,480$$

Similarly, there are 60,480 numbers that start with a 1. The number of seven-digit phone numbers is given by

$$604,800 - 2(60,480) = 483,840$$

Depending on what region of the country you live in, phone numbers beginning with certain other prefixes (e.g., 911, 611, or 411) also may not be used. The actual number of seven-digit phone numbers for a particular region will be less than the calculated 483,840.

Combinations

When the order in which items are selected from a set is unimportant, **combinations** are used instead of permutations. For example, suppose you plan on driving to a Norah Jones concert. You have five friends you would like to invite to go with you; however, your car can carry only three passengers in addition to you. How many different groupings of three friends do you have to pick from? For visual purposes, let's suppose your friends are named Alex, Ben, Carmen, Dina, and Ernesto. In Table 7.5, we represent the friends by the first initial of their names. If the order in which the group members are picked matters, there are $P(5, 3) = 60$ possible groupings (Table 7.5).

TABLE 7.5

ABC	ABD	ABE	ACD	ACE	ADE	BCD	BCE	BDE	CDE
ACB	ADB	AEB	ADC	AEC	AED	BDC	BEC	BED	CED
BAC	BAD	BAE	CAD	CAE	DAE	CBD	CBE	DBE	DCE
BCA	BDA	BEA	CDA	CEA	DEA	CDB	CEB	DEB	DEC
CAB	DAB	EAB	DAC	EAC	EAD	DBC	EBC	EBD	ECD
CBA	DBA	EBA	DCA	ECA	EDA	DCB	ECB	EDB	EDC

Notice that each group listed in the first column of the table contains Alex, Ben, and Carmen. If the order in which each friend is picked is unimportant, the six groups in the first column are equivalent. Likewise, the six groups in the second column are equivalent. The same is true for the remaining columns in the table. Therefore, there are $\frac{60}{6} = 10$ possible groupings if the order in which group members are picked is unimportant. There are 10 different combinations of three-person groups that can be picked from among the five friends.

COMBINATIONS OF _n_ ITEMS TAKEN _r_ AT A TIME

A **combination of _n_ items taken _r_ at a time** is an unordered sequence of r items chosen from a set of n distinct items. The number of possible combinations of n items taken r at a time is

$$C(n, r) = \frac{n!}{(n - r)! \cdot r!}$$

For all n and r, $C(n, r) \leq P(n, r)$.

The origin of the equation for calculating combinations is not immediately obvious. As we will show, $C(n, r) = \frac{P(n, r)}{r!}$.

$$\frac{P(n, r)}{r!} = \frac{\text{number of ways to pick } r \text{ things from a set of } n}{\text{number of ways to order a set of } r \text{ things}}$$

$$= \frac{\frac{n!}{(n - r)!}}{r!} \qquad \text{Definition of permutation}$$

$$= \frac{n!}{(n - r)!} \cdot \frac{1}{r!} \qquad \text{Division of fractions}$$

$$= \frac{n!}{(n - r)!r!} \qquad \text{Product of fractions}$$

$$= C(n, r)$$

For example, a soccer league official is asked to identify how many different matches among the eight teams in the league are possible. The official returns with the following chart:

$$\begin{bmatrix} \cancel{11} & 12 & 13 & 14 & 15 & 16 & 17 & 18 \\ 21 & \cancel{22} & 23 & 24 & 25 & 26 & 27 & 28 \\ 31 & 32 & \cancel{33} & 34 & 35 & 36 & 37 & 38 \\ 41 & 42 & 43 & \cancel{44} & 45 & 46 & 47 & 48 \\ 51 & 52 & 53 & 54 & \cancel{55} & 56 & 57 & 58 \\ 61 & 62 & 63 & 64 & 65 & \cancel{66} & 67 & 68 \\ 71 & 72 & 73 & 74 & 75 & 76 & \cancel{77} & 78 \\ 81 & 82 & 83 & 84 & 85 & 86 & 87 & \cancel{88} \end{bmatrix}$$

He explains that the number *ij* means that Team *i* plays Team *j*. He tells the league committee that there are 56 different matches possible, since a team can't play itself. One committee member points out that 12 and 21 mean the same thing, since Team 1 playing Team 2 is the same as Team 2 playing Team 1. She tells him that to be accurate, duplicate matches need to be eliminated.

$$\begin{bmatrix} \cancel{11} & \cancel{12} & \cancel{13} & \cancel{14} & \cancel{15} & \cancel{16} & \cancel{17} & \cancel{18} \\ 21 & \cancel{22} & \cancel{23} & \cancel{24} & \cancel{25} & \cancel{26} & \cancel{27} & \cancel{28} \\ 31 & 32 & \cancel{33} & \cancel{34} & \cancel{35} & \cancel{36} & \cancel{37} & \cancel{38} \\ 41 & 42 & 43 & \cancel{44} & \cancel{45} & \cancel{46} & \cancel{47} & \cancel{48} \\ 51 & 52 & 53 & 54 & \cancel{55} & \cancel{56} & \cancel{57} & \cancel{58} \\ 61 & 62 & 63 & 64 & 65 & \cancel{66} & \cancel{67} & \cancel{68} \\ 71 & 72 & 73 & 74 & 75 & 76 & \cancel{77} & \cancel{78} \\ 81 & 82 & 83 & 84 & 85 & 86 & 87 & \cancel{88} \end{bmatrix}$$

He eliminates the duplicate matches and reports that there are 28 unique matches possible. His result can easily be confirmed by calculating $C(8, 2)$.

$$C(n, r) = \frac{n!}{(n - r)! \cdot r!}$$

$$C(8, 2) = \frac{8!}{(8 - 2)! \cdot 2!}$$

$$= \frac{8!}{6! \cdot 2!}$$

$$= \frac{8 \cdot 7 \cdot \cancel{6!}}{\cancel{6!} \cdot 2!}$$

$$= \frac{56}{2}$$

$$= 28$$

Graphing calculators can quickly and easily compute both combinations and permutations, as detailed in the following Technology Tip.

TECHNOLOGY **TIP**

Calculating $C(n, r)$ and $P(n, r)$

1. Type the value of n in the home screen.

2. Press MATH, use the blue arrows to move to the PRB menu, and select item 2:nPr for permutations or 3:nCr for combinations.

3. Type in the value of r and press ENTER to display the result.

EXAMPLE 3

Using Permutations in the Real World

A couple is expecting their third boy and is searching for a unique boy's name. The names of their first two boys contain five unique letters and start with J. Each name has a different vowel in the second and fourth positions and different consonants in the third and fifth positions. They want to maintain the pattern with their third son. How many names do they have to pick from?

SOLUTION There are 26 letters in the alphabet, 5 vowels and 21 consonants. (We will assume that y is a consonant.) The couple has used up J, so 20 consonants remain. Since the first letter is J, we need to pick only two vowels and two consonants. Does order matter? Since *Jared* is not the same as *Jerad*, order matters. The number of ways to pick two different vowels from the set of five vowels is given by

$$P(5, 2) = \frac{5!}{3!}$$

$$= 20$$

The number of ways to pick two different consonants (excluding j) is

$$P(20, 2) = \frac{20!}{18!}$$

$$= 380$$

The number of five-letter *J* names with alternating vowels and consonants is given by

$$P(5, 2) \cdot P(20, 2) = 20 \cdot 380$$
$$= 7600$$

There are 7600 such names.

An alternative way to look at the problem is to list the five slots and determine how many different letters may be used to fill each slot.

$$J\underline{5} \quad \underline{20} \quad \underline{4} \quad \underline{19}$$
$$5 \cdot 20 \cdot 4 \cdot 19 = 7600$$

They have already used two names for their first two sons, so there are 7598 names remaining. When this chapter was written, the couple's top name candidates were *Jarin, Jaden, Jaxon,* and *Javin.*

The Multiplication Principle Revisited

To use permutations and combinations, we assume that no element can be picked twice. That is, once an item has been picked, it is eliminated from the selection pool. If an item is placed back into the selection pool after it is picked, the number of possible outcomes increases dramatically.

SELECTING *r* ITEMS FROM A SET OF *n* ITEMS (WITH REPLACEMENT)

If *r* items are chosen from a set of *n* items and each item is returned to the selection pool after it is picked, then the total number of possible groupings is n^r.

EXAMPLE 4

Counting the Number of Possible Phone Numbers

The North American Number Plan Administration (NANPA) is a third-party administrator that works with the telecommunications industry and oversees the assignment of area codes. On April 1, 2002, NANPA issued a press release indicating that three counties in Central Florida would be receiving a new overlay area code. According to the report, the 407 and 321 area codes were expected to be exhausted by first quarter 2004. (**Source:** NANPA.)

Certain phone prefixes are restricted and may not be used, most notably 000, 911, and 411. Suppose that there are a total of 300 restricted prefixes in Central Florida, including those prefixes starting with a 1 or a 0. (The phone company would not disclose to the author the exact number of restricted prefixes.) Assuming that all other phone numbers in the 407 and 321 area codes were used, how many phone numbers were there in Central Florida when the area codes were exhausted?

SOLUTION We will first consider the total number of possible seven-digit numbers and then subtract the phone numbers with restricted prefixes. We have seven slots to fill.

— — — — — — —

We may select each of the seven digits from the following 10 numbers: 0, 1, 2, 3, 4, 5, 6, 7, 8, 9. For the first slot, we have 10 choices, and likewise for the second through seventh slots.

$$\underline{10} \quad \underline{10} \quad \underline{10} \quad \underline{10} \quad \underline{10} \quad \underline{10} \quad \underline{10}$$

Multiplying the quantities together, we have $10^7 = 10,000,000$ possible seven-digit numbers. We will now eliminate the restricted numbers. For each restricted prefix, we have

$$\underline{1} \quad \underline{1} \quad \underline{1} \quad \underline{10} \quad \underline{10} \quad \underline{10} \quad \underline{10}$$

Multiplying the quantities together, there are 10,000 seven-digit numbers that start with any given prefix. We have 300 restricted prefixes, so we have

$$300 \cdot 10,000 = 3,000,000 \text{ restricted numbers}$$

Subtracting the quantity of restricted numbers from the total quantity of seven-digit numbers, we determine that there are 7,000,000 phone numbers in each area code, or a total of 14,000,000 numbers in the two area codes.

7.3 Summary

In this section, you learned how permutations and combinations can be used to count the number of possible results of an event. You also learned that permutations and combinations may be used only if an item is eliminated from the selection pool after it is picked.

7.3 Exercises

In Exercises 1–10, calculate the value of the expression without using a calculator.

1. $P(5, 2)$ **2.** $P(7, 5)$

3. $P(5, 5)$ **4.** $P(4, 1)$

5. $P(9, 0)$ **6.** $C(5, 2)$

7. $C(7, 5)$ **8.** $C(5, 5)$

9. $C(4, 1)$ **10.** $C(9, 0)$

In Exercises 11–20, calculate the value of the expression using a calculator.

 11. $P(15, 7)$ **12.** $P(20, 3)$

 13. $P(19, 7)$ **14.** $P(39, 6)$

 15. $P(369, 3)$ **16.** $C(15, 7)$

 17. $C(20, 3)$ **18.** $C(19, 7)$

 19. $C(39, 6)$ **20.** $C(369, 3)$

In Exercises 21–35, use permutations or combinations as appropriate to calculate the total number of possible outcomes.

21. **Lottery** In POWERBALL Lotto, five numbered white balls are drawn out of a drum of 49 white balls and one numbered red ball is drawn out of a drum of 42 red balls. To win the jackpot, you must match all five white balls (in any order) and the red ball. (**Source:** www.musl.com.) On September 7, 2002, the POWERBALL Lotto jackpot was $80 million. (**Source:** www.lotteryusa.com.) The probability of winning the jackpot is one divided by the total number of possible outcomes. How many outcomes are possible?

22. **License Plates** Some states use three letters followed by three numbers on their license plates. How many such license plates can be made if numbers and letters may be repeated?

23. **License Plates** How many license plates with three letters followed by three numbers can be made if no letters or numbers are repeated?

24. **Phone Numbers** Assuming that no phone numbers are prohibited, what is the maximum number of seven-digit phone numbers that could be issued in area code 360 if digits may be repeated?

25. **Phone Numbers** Assuming that no phone numbers are prohibited, what is the maximum number of seven-digit phone numbers with no repeating digits that could be issued in area code 360?

26. **Zip Codes** The U.S. Postal Service delivers mail based on a five-digit zip code. At most how many zip codes can begin with "98" if numbers may be repeated?

27. **Zip Codes** The U.S. Postal Service delivers mail based on a five-digit zip code. At most how many zip codes can begin with "98" if numbers may *not* be repeated?

28. **Soccer Matches** A local soccer league has 20 teams. How many different matches are possible?

29. **Class Schedules** A student needs to take a math class, two business classes, and a P.E. class. There are four math classes, seven business classes, and five P.E. classes that fit into her schedule. How many different class schedules does she have to pick from?

30. **Pizza Varieties** In 1995, Papa Aldo's Pizza and Murphy's Pizza merged into Papa Murphy's Pizza International. At the time of the merger, there were 140 stores. As of September 2002, there were 628 Papa Murphy's® Take 'N' Bake Pizza stores in 21 states. Ninety-eight percent of the stores are franchised. (**Source:** www.papamurphys.com.)

According to the company's web site, Papa Murphy's offers six different meat toppings, ten

vegetable toppings, and four other toppings. Assuming that a topping is not used more than once on the same pizza, how many different types of three-topping pizzas can be made?

31. **Pizza Varieties** According to the company's web site, Papa Murphy's offers six different meat toppings, ten vegetable toppings, and four other toppings. How many different three-topping pizzas can be made if one topping is selected from each category?

32. **Pizza Varieties** According to the company's web site, Papa Murphy's offers six different meat toppings, ten vegetable toppings, and four other toppings. Assuming that a topping is not used more than once on the same pizza, how many different types of six-topping pizzas can be made?

33. **Pizza Varieties** According to the company's web site, Papa Murphy's offers six different meat toppings, ten vegetable toppings, and four other toppings. How many different types of six-topping pizzas can be made if two toppings are selected from each category?

34. **Radio Station Call Signs** Although there are 69 radio or TV stations with three-letter call signs, most stations are identified by a four-letter call sign. Stations to the west of the Mississippi River have call signs that begin with K, and stations to the east of the river have call signs that begin with W. (**Source:** www.earlyradiohistory.us.)

How many different four-letter radio station call signs can be made?

35. **Radio Station Call Signs** Radio station call signs were originally three letters long; however, the dramatic growth in the industry necessitated a move to a four-letter system in 1922. Four-letter call signs are required to begin with K or W. Furthermore, original regulations required that no letter could be repeated three times in a row. (**Source:** www.earlyradiohistory.us.)

How many four-letter call signs can be made without repeating a letter three times in a row?

Exercises 36–40 are intended to challenge your understanding of permutations and combinations.

36. How many three-digit numbers with no repeating digits are there that end with a 5, 7, or 9?

37. How many three-digit numbers are there with the property that each digit is greater than the digit to its right?

38. How many three-digit numbers are there with the property that each digit is less than or equal to the digit to its left?

39. A bag contains three types of coins: pennies, nickels, and dimes. The number of each type of coin is unknown; however, there is the same number of each type of coin (e.g., if there are five nickels, there are also five pennies and five dimes). There is a total of $3r$ coins in the bag. How many different ways are there to withdraw three coins from the bag whose combined value is

16 cents? (*Hint:* The answer will be in terms of r.) Assume that you can tell the difference between different coins of the same value.

40. A bag contains three types of coins: pennies, nickels, and dimes. The number of each type of coin is unknown; however, there is the same number of each type of coin (e.g., if there are five nickels, there are also five pennies and five dimes). There is a total of $3r$ coins in the bag. How many different ways are there to withdraw three coins from the bag whose combined value is *at most* 16 cents? (*Hint:* The answer will be in terms of r.) Assume that you can tell the difference between different coins of the same value.

7.4 Introduction to Probability

- Use probability terminology to describe real-life scenarios
- Use probability concepts to calculate probabilities

GETTING **STARTED** Many games include some element of chance. Players who understand the probability concepts associated with games of chance are better able to make wise decisions in their game play.

In this section, we will introduce the language and basic principles of probability. These will help us determine the likelihood of obtaining a desired result.

In discussing probability concepts, we use terms such as *experiment, trial, outcome,* and *sample space.* We begin by defining each term.

PROBABILITY TERMINOLOGY: EXPERIMENTS, TRIALS, OUTCOMES, AND SAMPLE SPACES

An **experiment** is an occurrence with an uncertain result. Each repetition of an experiment is called a **trial.** The possible results of each trial are called **outcomes.** The set of all possible outcomes of an experiment is called the **sample space** for the experiment.

To help you better understand the terminology, we will give several examples.

EXAMPLE 1 **Using Probability Terminology**

A coin is tossed. We are interested in the symbol on the face of the coin after it lands (heads or tails). Use probability terminology to describe the situation.

SOLUTION

Experiment: Toss a coin and observe the face of the coin after it lands

Trials: 1

Outcomes: Heads or tails

Sample space: $S = \{\text{heads, tails}\}$

EXAMPLE 2

Using Probability Terminology

Three people are picked from the audience at a U2 concert. We are interested in the gender of those selected, and we don't care about the order in which they are selected. That is, we consider picking a male first and then two females to be the same as picking two females first and then a male. Use probability terminology to describe the situation.

SOLUTION

Experiment: Randomly select three people from the U2 concert audience and observe their gender

Trials: 1

Outcomes: (M, M, M), (M, M, F), (M, F, F), (F, F, F)

Sample space: $S = \left\{ \begin{array}{l} (M, M, M), (M, M, F), (M, F, F), \\ (F, F, F) \end{array} \right\}$

If the order in which the people were selected was important to us, we would end up with a different sample space.

Sample space: $T = \left\{ \begin{array}{l} (M, M, M), (M, M, F), (M, F, M), (F, M, M), \\ (M, F, F), \ (F, M, F), \ (F, F, M), \ (F, F, F) \end{array} \right\}$

EXAMPLE 3

Using Probaility Terminology

In a game, a red die and a blue die are rolled. We're interested in the numbers shown on the dice. Use probability terminology to describe the situation.

SOLUTION

Experiment: Roll a red die and a blue die and observe the numbers shown

Trials: 1

Outcomes: (1, 1), (1, 2), (1, 3), . . . , (6, 3), (6, 4), (6, 5), (6, 6) (The first number is the red die value, and the second number is the blue die value.)

Sample space:
$$S = \left\{ \begin{array}{cccccc} (1,1) & (1,2) & (1,3) & (1,4) & (1,5) & (1,6) \\ (2,1) & (2,2) & (2,3) & (2,4) & (2,5) & (2,6) \\ (3,1) & (3,2) & (3,3) & (3,4) & (3,5) & (3,6) \\ (4,1) & (4,2) & (4,3) & (4,4) & (4,5) & (4,6) \\ (5,1) & (5,2) & (5,3) & (5,4) & (5,5) & (5,6) \\ (6,1) & (6,2) & (6,3) & (6,4) & (6,5) & (6,6) \end{array} \right\}$$

EXAMPLE 4

Using Probability Terminology

In a game, two six-sided dice are rolled. We're interested in the sum of the numbers shown. Use probability terminology to describe the situation.

SOLUTION

Experiment: Roll two dice and observe the sum of the numbers shown

Trials: 1

Outcomes: 2, 3, 4, 5, 6, 7, 8, 9, 10, 11, 12

Sample space: $S = \{2, 3, 4, 5, 6, 7, 8, 9, 10, 11, 12\}$

EXAMPLE 5

Using Probability Terminology

Pick two coins from a purse containing nickels, dimes, and pennies. Use probability terminology to describe the situation.

SOLUTION

Experiment: Pick two coins from a purse with nickels, dimes, and pennies and observe the type of coin

Trials: 1

Outcomes: (nickel, nickel), (dime, nickel), (penny, nickel), (nickel, dime), (dime, dime), (penny, dime), (nickel, penny), (dime, penny), (penny, penny)

Sample space: $S = \left\{\begin{array}{lll} (\text{nickel, nickel}) & (\text{dime, nickel}) & (\text{penny, nickel}) \\ (\text{nickel, dime}) & (\text{dime, dime}) & (\text{penny, dime}) \\ (\text{nickel, penny}) & (\text{dime, penny}) & (\text{penny, penny}) \end{array}\right\}$

Often we are interested in a particular outcome because obtaining that outcome leads to some favorable result, such as scoring points, winning money, and so on. We use the terms *event* and *favorable outcomes* to describe such results.

> **MORE PROBABILITY TERMINOLOGY: EVENTS AND FAVORABLE OUTCOMES**
>
> A subset E of a sample space S that meets a specified criterion is called an **event**. The elements of E are called the **favorable outcomes.**

EXAMPLE 6

Using Probability Terminology

Three people are randomly picked from the audience of a U2 concert. The order in which the people are picked is important to us. [That is, we treat (M, M, F) and (M, F, M) as different elements, even though they both contain two males and one female.] We are interested in the outcomes that contain at least one woman. Use the terms *sample space, event,* and *favorable outcomes* to describe the situation.

SOLUTION

Sample space:

$$S = \begin{cases} (M, M, M), (M, M, F), (M, F, M), (F, M, M), (M, F, F), \\ (F, M, F), \quad (F, F, M), \quad (F, F, F) \end{cases}$$

Event: $E = \begin{cases} (M, M, F), (M, F, M), (F, M, M), (M, F, F), \\ (F, M, F), \quad (F, F, M), \quad (F, F, F) \end{cases}$

Favorable outcomes:

$(M, M, F), (M, F, M), (F, M, M), (M, F, F), (F, M, F), (F, F, M), (F, F, F)$

EXAMPLE 7

Using Probability Terminology

Three people are randomly picked from a finite math class, and the gender of each person is recorded. We are interested in the outcomes that don't contain any men. Use the terms *sample space, event,* and *favorable outcomes* to describe the situation.

SOLUTION

Sample space:

$$S = \begin{cases} (M, M, M), (M, M, F), (M, F, M), (F, M, M), (M, F, F), \\ (F, M, F), \quad (F, F, M), \quad (F, F, F) \end{cases}$$

Event: $E = \{(F, F, F)\}$

Favorable outcomes: (F, F, F)

EXAMPLE 8

Using Probabilty Terminology

In a game, a red die and a blue die are rolled. We observe the numbers shown on the pair of dice each time they are rolled. We want to know the dice pairs that have a sum of 7 or 11. Use the terms *sample space, event,* and *favorable outcomes* to describe the situation.

SOLUTION

Sample space:

$$S = \begin{cases} (1,1) & (1,2) & (1,3) & (1,4) & (1,5) & (1,6) \\ (2,1) & (2,2) & (2,3) & (2,4) & (2,5) & (2,6) \\ (3,1) & (3,2) & (3,3) & (3,4) & (3,5) & (3,6) \\ (4,1) & (4,2) & (4,3) & (4,4) & (4,5) & (4,6) \\ (5,1) & (5,2) & (5,3) & (5,4) & (5,5) & (5,6) \\ (6,1) & (6,2) & (6,3) & (6,4) & (6,5) & (6,6) \end{cases}$$

Event:

$E = \{(1,6), (2,5), (3,4), (4,3), (5,2), (6,1), (5,6), (6,5)\}$

Favorable outcomes:

$(1,6), (2,5), (3,4), (4,3), (5,2), (6,1), (5,6), (6,5)$

When playing a game where scoring points is based on random events such as rolling dice or picking cards, we are often interested in the likelihood of obtaining a favorable outcome. The notion of *probability* is used to measure the likelihood of success or failure.

BASIC PRINCIPLE OF PROBABILITY

Let a sample space S have equally likely outcomes. Let E be a subset of S. The **probability that event E occurs** is given by

$$P(E) = \frac{n(E)}{n(S)}$$

For any event, $0 \le P(E) \le 1$.

EXAMPLE 9 **Calculating the Probability of an Event**

A coin purse contains a penny, a nickel, a dime, a quarter, a 50-cent piece, and a silver dollar. What is the probability that a randomly selected coin's value is a multiple of 10?

SOLUTION Let S be the sample space and E be the set of coins whose value is a multiple of 10.

$$S = \{\text{penny, nickel, dime, quarter, 50-cent piece, silver dollar}\}$$
$$n(S) = 6$$
$$E = \{\text{dime, 50-cent piece, silver dollar}\}$$
$$n(E) = 3$$
$$P(E) = \frac{n(E)}{n(S)}$$
$$= \frac{3}{6}$$
$$= 0.50$$

Assuming that there is an equally likely chance of picking any coin, there is a 50 percent chance that the selected coin's value will be a multiple of 10.

EXAMPLE 10 **Calculating the Probability of an Event**

In a game, a red die and a blue die are rolled. We observe the numbers shown on the pair of dice. What is the probability that the dice pair has a sum of 7 or 11?

SOLUTION

$$S = \begin{cases} (1,1) & (1,2) & (1,3) & (1,4) & (1,5) & (1,6) \\ (2,1) & (2,2) & (2,3) & (2,4) & (2,5) & (2,6) \\ (3,1) & (3,2) & (3,3) & (3,4) & (3,5) & (3,6) \\ (4,1) & (4,2) & (4,3) & (4,4) & (4,5) & (4,6) \\ (5,1) & (5,2) & (5,3) & (5,4) & (5,5) & (5,6) \\ (6,1) & (6,2) & (6,3) & (6,4) & (6,5) & (6,6) \end{cases}$$

$$n(S) = 36$$

Let the event E be the set of dice values that sum to 7 or 11.

$$E = \{(1, 6), \ (2, 5), \ (3, 4), \ (4, 3), \ (5, 2), \ (6, 1), \ (5, 6), \ (6, 5)\}$$
$$n(E) = 8$$

There are eight different dice rolls with a sum of 7 or 11.

$$P(E) = \frac{n(E)}{n(S)}$$
$$= \frac{8}{36}$$
$$= 0.2222$$

There is a 22.22 percent chance that the sum of the dice will be 7 or 11.

Up until this point, we have used sample spaces with relatively few elements in each of our examples. When sample spaces become increasingly large, it is helpful to use the counting techniques covered in the previous sections. For Example 11, we are selecting five numbers (with replacement) from the set $\{1, 2, 3, 4, 5, 6\}$

EXAMPLE 11

Calculating the Probability of an Event

What is the probability that five dice rolled will all show the same number? (In the game of Yahtzee, this is the highest-scoring dice roll.)

SOLUTION

$$S = \{x | x \text{ is a possible dice roll with 5 dice}\}$$
$$n(S) = 6^5$$
$$= 7776$$

$$E = \{x | x \text{ is a dice roll with dice all showing the same number}\}$$
$$n(E) = P(6, 1) \qquad \text{Pick one number from the possible six numbers: 1, 2, 3, 4, 5, 6}$$
$$= 6$$

$$P(E) = \frac{n(E)}{n(S)}$$
$$= \frac{P(6, 1)}{6^5}$$
$$= \frac{6}{6^5}$$
$$= \frac{1}{6^4}$$
$$= \frac{1}{1296}$$
$$= 0.0007716$$
$$= 0.07716\%$$

There is nearly a 0.08 percent chance that the dice will all show the same number.

EXAMPLE 12 Calculating the Probability of an Event

Four identical coins are tossed simultaneously. What is the probability that the faces of the coins will all be heads or all be tails when the coins land?

SOLUTION

$$S = \{x \mid x \text{ is a possible coin toss with 4 coins}\}$$

$$n(S) = 2^4$$

$$= 16$$

$$E = \{x \mid x \text{ is a coin toss showing all heads or all tails}\}$$

$$n(E) = P(2, 1) \quad \text{From the two possible coin faces, we must pick one: heads or tails}$$

$$= 2 \quad \text{One toss is all heads, the other toss is all tails}$$

$$P(E) = \frac{n(E)}{n(S)}$$

$$= \frac{2}{2^4}$$

$$= \frac{2}{16}$$

$$= 0.125$$

$$= 12.5\%$$

There is a 12.5 percent probability that the coin toss will show either all heads or all tails.

Odds

In gaming situations, it is common to speak of the **odds** for or against the event instead of the probability of the event. Probability and odds are closely related, as detailed here.

PROBABILITY AND ODDS

If the probability of an event E is $P(E)$ with $0 < P(E) < 1$, then the

1. **Odds for E** are given by Odds for $E = \dfrac{P(E)}{P(E')}$

2. **Odds against E** are given by

$$\text{Odds against } E = \frac{1}{\text{odds for } E}$$

$$= \frac{1}{\dfrac{P(E)}{P(E')}}$$

$$= \frac{P(E')}{P(E)}$$

It is customary to represent odds as ratios of whole numbers.

In Example 12, we showed that the probability $P(E)$ that the faces of four tossed coins would match was 12.5 percent. Consequently, the probability that the faces do not all match is given by

$$P(E') = 1 - P(E)$$
$$= 1 - 0.125$$
$$= 0.875$$
$$= 87.5\%$$

The odds for E are

$$\frac{P(E)}{P(E')} = \frac{0.125}{0.875}$$
$$= \frac{1}{7}$$

If having all of the coin faces match is considered a win, then the odds of winning are 1 to 7. This is often written as 1:7 instead of $\frac{1}{7}$. The odds against winning are 7 to 1 (or 7:1). In other words, the odds of losing are 7:1.

EXAMPLE 13

Determining Odds

The probability of winning a dice game is 25 percent. What are the odds of winning and the odds of losing?

SOLUTION   We have $P(E) = 0.25$. Therefore, $P(E') = 1 - 0.25 = 0.75$. The odds of winning are

$$\frac{P(E)}{P(E')} = \frac{0.25}{0.75}$$
$$= \frac{1}{3}$$

The odds of winning are 1 to 3 (or 1:3). Consequently, the odds of not winning are 3 to 1. Therefore, the odds of losing are 3:1.

7.4 Summary

In this section, you learned basic probability terms and the basic principle of probability. You saw how these could be used to determine the likelihood of a favorable event. You also learned the relationship between probability and odds.

7.4 Exercises

In Exercises 1–10, use the terms experiment, trials, outcomes, *and* sample space *to describe the situation.*

1. A person is randomly picked from the population of Ellensburg, Washington, and the person's gender is noted.

2. A pair of dice is rolled, and the product of the numbers shown is noted.

3. A pair of dice is rolled, and the difference between the larger number and the smaller number is noted.

4. A coin is flipped, and the image on the side facing up is noted. The coin is flipped three times.

5. A bag of marbles contains five red, three white, and two blue marbles. Three marbles are pulled from the bag, and their color is noted.

6. A student enrolled in a five-credit math class is selected at random, and the number of credits on the student's class schedule is noted. (Assume that the number of credits cannot exceed 21.)

7. A student enrolled in a two-credit P.E. class is selected at random, and the number of credits on the student's class schedule is noted. (Assume that the number of credits cannot exceed 21.)

8. A player is selected at random from the high school football team, and the player's grade in school (i.e., ninth grade) is noted.

9. A grocery store classifies its breakfast cereals as Kellogg's, General Mills, or Other Brand. A box is selected at random, and its brand is noted. This process is repeated 10 times.

10. A clothing store catalogs its men's denim pants as Levi's, Bugle Boy, Calvin Klein, and Other Brands. A pair of denim pants is selected, and its brand is noted. This process is repeated 150 times.

In Exercises 11–20, identify the sample space for the experiment. Then write the specified events in set notation.

11. **Scholarship Applicants**   Andrea (F), Miki (F), Miguel (M), Sara (F), and Sam (M) are scholarship contestants. (Their genders are noted in the parentheses.) Two of them are selected as finalists.
    (a) *Event:* At least one of the finalists is male.
    (b) *Event:* Both finalists are female.
    (c) *Event:* No more than one finalist is male.

12. **Scholarship Applicants**   Julie (3.95), Dawn (4.00), Andres (3.53), Layla (3.83), Ari (3.71), and Saul (4.00) are scholarship candidates. (Their G.P.A.s are noted in the parentheses.) Two of them are selected as finalists.
    (a) *Event:* At least one of the finalists has a 4.0 G.P.A.
    (b) *Event:* Both finalists have above a 3.90 G.P.A.
    (c) *Event:* At least one finalist has a G.P.A. between 3.50 and 3.75.

13. **Game of Dice**   In a game of dice, a red die and a blue die are rolled, and the numbers shown are noted.
    (a) *Event:* The sum of the numbers is even.
    (b) Event: The sum of the numbers is a multiple of 5.
    (c) *Event:* The sum of the numbers is not a multiple of 3.

14. **Game of Dice**   In a game of dice, a red die and a blue die are rolled, and the numbers shown are noted.
    (a) *Event:* The numbers are the same.
    (b) *Event:* The numbers are both even.
    (c) *Event:* The difference between the larger number and the smaller number is exactly 2.

15. **True/False Test**   A student takes a four-question true/false test, and her responses are noted. The correct answer to each question was *true*.
    (a) *Event:* She missed at most one question.
    (b) *Event:* She got at least two answers right.
    (c) *Event:* She got all of the questions wrong.

16. **True/False Test**   A student takes a four-question true/false test, and his responses are noted. The correct answers to the questions (in order) were *TFFT.*
    (a) *Event:* He missed at most one question.
    (b) *Event:* He got at least three answers right.
    (c) *Event:* He got all of the questions wrong.

17. **House of Representatives**   The two members of the U.S. House of Representatives from Idaho are selected in order of seniority, and their party affiliations (D, R, or I) are noted.
    (a) *Event:* More than half of them are Republicans (R).
    (b) *Event:* Democrats (D) outnumber Republicans (R).
    (c) *Event:* There are more non-Democrats (R and I) than Democrats (D).

18. House of Representatives The three members of the U.S. House of Representatives from Nebraska are selected in order of seniority, and their party affiliations (*D, R,* or *I*) are noted.

   (a) *Event:* At least half of them are Republicans (*R*).
   (b) *Event:* Democrats (*D*) outnumber Republicans (*R*).
   (c) *Event:* There are fewer non-Democrats (*I* and *R*) than Democrats (*D*).

19. House of Representatives The four members of the U.S. House of Representatives from Mississippi are selected in order of seniority, and their genders are noted.

   (a) *Event:* Females outnumber males.
   (b) *Event:* The least senior representative is female.
   (c) *Event:* The senior representative is male and the least senior representative is female.

20. House of Representatives The five members of the U.S. House of Representatives from Oregon are selected in order of seniority, and their genders are noted.

   (a) *Event:* Females outnumber males.
   (b) *Event:* The least senior representative is female.
   (c) *Event:* The senior representative is male and the least senior representative is female.

In Exercises 21–30, assuming that all outcomes are equally likely, calculate the probability of each event. Then calculate the odds for and against each event. You may find it helpful to refer to your work in Exercises 11–20.

21. **Scholarship Applicants** Andrea (F), Miki (F), Miguel (M), Sara (F), and Sam (M) are scholarship contestants. (Their genders are noted in the parentheses.) Two of them are selected as finalists.

   (a) *Event:* At least one of the finalists is male.
   (b) *Event:* Both finalists are female.
   (c) *Event:* No more than one finalist is male.

22. **Scholarship Applicants** Julie (3.95), Dawn (4.00), Andres (3.53), Layla (3.83), Ari (3.71), and Saul (4.00) are scholarship candidates. (Their G.P.A.s are noted in the parentheses.) Two of them are selected as finalists.

   (a) *Event:* At least one of the finalists has a 4.0 G.P.A.

   (b) *Event:* Both finalists have above a 3.90 G.P.A.
   (c) *Event:* At least one finalist has a G.P.A. between 3.50 and 3.75.

23. **Game of Dice** In a game of dice, a red die and a blue die are rolled, and the numbers shown are noted.

   (a) *Event:* The sum of the numbers is even.
   (b) *Event:* The sum of the numbers is a multiple of 5.
   (c) *Event:* The sum of the numbers is not a multiple of 3.

24. **Game of Dice** In a game of dice, a red die and a blue die are rolled, and the numbers shown are noted.

   (a) *Event:* The numbers are the same.
   (b) *Event:* The numbers are both even.
   (c) *Event:* The difference between the larger number and the smaller number is exactly 2.

25. **True/False Test** A student takes a four-question true/false test, and her responses are noted. The correct answer to each question was true.

   (a) *Event:* She missed at most one question.
   (b) *Event:* She got at least two answers right.
   (c) *Event:* She got all of the questions wrong.

26. **True/False Test** A student takes a four-question true/false test, and his responses are noted. The correct answers to the questions (in order) were *TFFT.*

   (a) *Event:* He missed at most one question.
   (b) *Event:* He got at least three answers right.
   (c) *Event:* He got all of the questions wrong.

27. House of Representatives The two members of the U.S. House of Representatives from Idaho are selected in order of seniority, and their party affiliations (*D, R,* or *I*) are noted.

   (a) *Event:* More than half of them are Republicans (*R*).
   (b) *Event:* Democrats (*D*) outnumber Republicans (*R*).
   (c) *Event:* There are more non-Democrats (*R* and *I*) than Democrats (*D*).

28. House of Representatives The three members of the U.S. House of Representatives from Nebraska are selected in order of seniority, and their party affiliations (*D, R,* or *I*) are noted.

(a) *Event:* At least half of them are Republicans (*R*).
(b) *Event:* Democrats (*D*) outnumber Republicans (*R*).
(c) *Event:* There are fewer non-Democrats (*I* and *R*) than Democrats (*D*).

29. **House of Representatives** The four members of the U.S. House of Representatives from Mississippi are selected in order of seniority, and their genders are noted.

(a) *Event:* Females outnumber males.
(b) *Event:* The least senior representative is female.
(c) *Event:* The senior representative is male and the least senior representative is female.

30. **House of Representatives** The five members of the U.S. House of Representatives from Oregon are selected in order of seniority, and their genders are noted.

(a) *Event:* Females outnumber males.
(b) *Event:* The least senior representative is female.
(c) *Event:* The senior representative is male and the least senior representative is female.

Exercises 31–35 are intended to challenge your understanding of basic probability concepts.

31. **Employment Status** In a five-person group, there are three women and two men. Two of the women and one of the men are employed. Assuming that all outcomes are equally likely, determine the probability of each event.

(a) *Event:* Two people selected from the group are both employed.
(b) *Event:* Two people selected from the group are both unemployed.
(c) *Event:* Two people selected from the group are employed women.

32. **Professions** In a thirty-person group, there are three doctors, five lawyers, and four teachers. Assuming that all outcomes are equally likely, determine the probability of each event.

(a) *Event:* Three people selected from the group are teachers.
(b) *Event:* Three people selected from the group are not lawyers.
(c) *Event:* At least one of three people selected from the group is a doctor.

33. **Dice Game** In a game of dice, three dice are rolled. The player wins if the sum *and* the product of the dice are even numbers. What is the probability that the player will win?

34. **Dice Game** In a game of dice, two dice are rolled. The player wins if the sum of the dice is greater than 5 *and* the product of the dice is less than 15. What is the probability that the player will win?

35. **Dice Game** In a game of dice, two dice are rolled. The player wins if the sum of the dice is less than 9 *and* the product of the dice is a multiple of 3. What is the probability that the player will win?

7.5 Basic Probability Concepts

▪ Use probability concepts to calculate probabilities

GETTING STARTED Political lobbyists representing various special interest groups seek to influence the decisions of members of Congress. Certain issues may receive support among Democrats, while other issues may receive support among women, regardless of their political affiliation. A lobbyist may use the concepts of probability to help him or her determine the likelihood of success with a particular issue.

In this section, we will discuss set operations for events and then introduce rules for the probability of a union and a complement. We will also introduce the idea of a probability distribution.

Often we are interested in the intersection, union, or complement of two events. As detailed here, the same set operations introduced earlier in the chapter may be used with events for a sample space.

SET OPERATIONS FOR EVENTS

Let E and F be events for a sample space S. Then

- $E \cap F$ is the set of outcomes in both E and F.
- $E \cup F$ is the set of all outcomes that are either in E or in F or in both E and F.
- E' is the set of all outcomes in S that are not in E.

EXAMPLE 1 **Using and Interpreting Set Notation**

A coin purse contains pennies, nickels, dimes, quarters, 50-cent pieces, and silver dollars. Pick a coin from a purse and note its value. Let E be the set of coins whose value is a multiple of 10. Let F be the set of coins whose value is a multiple of 25. Determine $E \cap F$, $E \cup F$, and E' and express in words the meaning of each set.

SOLUTION

$S = \{$penny, nickel, dime, quarter, 50-cent piece, silver dollar$\}$	
$E = \{$dime, 50-cent piece, silver dollar$\}$	Coins whose value is a multiple of 10
$F = \{$quarter, 50-cent piece, silver dollar$\}$	Coins whose value is a multiple of 25
$E \cap F = \{$50-cent piece, silver dollar$\}$	Coins whose value is a multiple of 10 and 25
$E \cup F = \{$dime, quarter, 50-cent piece, silver dollar$\}$	Coins whose value is a multiple of 10 or 25
$E' = \{$penny, nickel, quarter$\}$	Coins whose value is not a multiple of 10

Not all events share elements in common. Events without any shared elements are **mutually exclusive events.**

MUTUALLY EXCLUSIVE EVENTS

Two events E and F are mutually exclusive if $E \cap F = \varnothing$.

EXAMPLE 2

Determining If Sets Are Mutually Exclusive

Pick a person from the crowd and note whether the person is a citizen or a noncitizen of the United States. Let E be the set of citizens and F be the set of noncitizens. Determine if the sets E and F are mutually exclusive.

SOLUTION

$$S = \{\text{citizen, noncitizen}\}$$

$$E = \{\text{citizen}\} \qquad \text{People who are citizens}$$

$$F = \{\text{noncitizen}\} \qquad \text{People who are not citizens}$$

$$E \cap F = \varnothing \qquad \text{People who are both citizens and not citizens}$$

Since $E \cap F = \varnothing$, the events E and F are mutually exclusive.

When we talk about the probability of a union of two events E and F, we are looking for the likelihood that either event E **or** event F occurs. For example, a teenager may estimate that there is a 20 percent probability that Anastasia will ask him to the dance and a 40 percent probability that Marisol will ask him to the dance. The probability of the union of these two events is the probability that Anastasia **or** Marisol **or** both Anastasia and Marisol will ask him to the dance. To calculate the probability of a union, we use the probability of a union property.

PROBABILITY OF A UNION

Let E and F be events in a sample space S. Then

$$P(E \cup F) = P(E) + P(F) - P(E \cap F)$$

If the events E and F are mutually exclusive, then $P(E \cap F) = 0$, and the equation simplifies to

$$P(E \cup F) = P(E) + P(F)$$

EXAMPLE 3

Calculating the Probability of an Event

In 2002, the state of Missouri had nine representatives in the U.S. House of Representatives. Four of the representatives were Democrats, and two of the representatives were women. One of the representatives was a woman Democrat. (**Source:** www.clerk.house.gov.) If a political issue has strong support among both Democrats and women, and strong opposition from Republican men, what is the probability that a randomly selected representative from Missouri will support the issue?

SOLUTION The set $S = \{x | x$ is a representative from Missouri$\}$.

Let $E = \{x | x$ is a Democrat$\}$ and $F = \{x | x$ is a woman$\}$. We have $n(S) = 9$, $n(E) = 4$, $n(F) = 2$, and $n(E \cap F) = 1$.

We need to determine the probability that a representative is a woman **or** a Democrat. Recall that for an event E from a sample space S, $P(E) = \dfrac{n(E)}{n(S)}$.

$$P(E \cup F) = P(E) + P(F) - P(E \cap F)$$

$$= \frac{4}{9} + \frac{2}{9} - \frac{1}{9}$$

$$= \frac{5}{9}$$

$$= 0.5555$$

The probability that a representative is a Democrat or a woman is 55.55 percent. It is likely that the a randomly selected representative from the state of Missouri would support the issue.

EXAMPLE 4 Calculating the Probability of an Event

A coin is flipped three times, and the sequence of heads and tails is recorded. Let event E be the set of outcomes containing at least two heads. Let event F be the set of outcomes containing all heads or all tails. What is the probability that an outcome is in $E \cup F$?

SOLUTION

$$S = \{\text{HHH, HHT, HTH, THH, HTT, THT, TTH, TTT}\}$$
$$n(S) = 8$$

$$E = \{\text{HHH, HHT, HTH, THH}\} \quad \text{The set of outcomes with at least two heads}$$
$$n(E) = 4$$

$$F = \{\text{HHH, TTT}\} \quad \text{The set of outcomes containing all heads or all tails}$$
$$n(F) = 2$$

$$E \cap F = \{\text{HHH}\} \quad \text{The set of outcomes with at least two heads and all heads or all tails}$$

$$n(E \cap F) = 1$$

We must find $P(E \cup F)$.

$$P(E \cup F) = P(E) + P(F) - P(E \cap F)$$

$$= \frac{4}{8} + \frac{2}{8} - \frac{1}{8}$$

$$= \frac{5}{8}$$

$$= 0.6250$$

There is a 62.5 percent probability that an outcome is in $E \cup F$.

EXAMPLE 5

Calculating the Probability of an Event

Four men and two women sell high-quality kitchen knives. Three of them will be selected at random to work this weekend at a local fair. What is the probability that at most two men will have to work this weekend?

SOLUTION We must first determine the cardinality of our sample space S. The number of different groups of three people that can be selected from a group of six people is

$$n(S) = C(6, 3)$$
$$= \frac{6!}{3! \cdot 3!}$$
$$= \frac{6 \cdot 5 \cdot 4}{3 \cdot 2 \cdot 1}$$
$$= 20$$

Since there are only two women, at least one man will have to work this weekend. We will calculate the probability that a three-person group contains exactly one man and then calculate the probability that a three-person group contains exactly two men. The sum of these probabilities will give us the probability that at most two men will have to work this weekend.

Define event E to be the set of three-person groups containing exactly *one* man. The number of different groups of three people containing exactly one man is

$$n(E) = C(4, 1) \cdot C(2, 2)$$

Pick one out of four men.   Pick two out of two women.

$$n(E) = C(4, 1) \cdot C(2, 2)$$
$$= \frac{4!}{3!1!} \cdot \frac{2!}{2!0!}$$
$$= 4 \cdot 1$$
$$= 4$$

There are four three-person groups containing exactly one man.

$$P(E) = \frac{n(E)}{n(S)}$$
$$= \frac{4}{20}$$
$$= 0.20$$

There is a 20 percent probability that a three-person group will contain exactly one man.

Define event F to be the set of three-person groups containing exactly *two* men. The number of groups of three people containing exactly two men is

$$n(F) = C(4, 2) \cdot C(2, 1)$$

Pick two out of four men.   Pick one out of two women.

$$n(F) = C(4, 2) \cdot C(2, 1)$$

$$= \frac{4!}{2! \cdot 2!} \cdot \frac{2!}{1! \cdot 1!}$$

$$= \frac{4 \cdot 3}{2 \cdot 1} \cdot \frac{2}{1}$$

$$= 12$$

There are 12 three-person groups containing exactly two men.

$$P(F) = \frac{n(F)}{n(S)}$$

$$= \frac{12}{20}$$

$$= 0.60$$

There is a 60 percent probability that a three-person group will contain exactly two men.

Since E and F are mutually exclusive, $P(E \cap F) = 0$. Therefore, the probability that at most one or two men will have to work this weekend is given by

$$P(E \cup F) = P(E) + P(F) \qquad \text{Since } E \text{ and } F \text{ are mutually exclusive}$$

$$= 0.60 + 0.20$$

$$= 0.80$$

There is an 80 percent chance that at most two men will have to work this weekend.

The process of calculating probabilities can often be simplified by calculating the probability of a complement. The probability of a complement formula is derived by observing that there is a 100 percent probability of being in the sample space. That is,

$$P(S) = \frac{n(S)}{n(S)}$$

$$= 1$$

Note that $E \cup E' = S$. Since E and E' are mutually exclusive,

$$P(S) = P(E \cup E')$$

$$1 = P(E) + P(E') \qquad \text{Since } P(S) = 1$$

$$P(E') = 1 - P(E)$$

PROBABILITY OF A COMPLEMENT

Let E be an event in a sample space S. Then

$$P(E') = 1 - P(E)$$

In Example 6, we will rework Example 5 using the probability of a complement formula.

EXAMPLE 6

Calculating the Probability of a Complement

Four men and two women sell high-quality kitchen knives. Three of them will be selected at random to work this weekend at a local fair. What is the probability that at most two men will have to work this weekend?

SOLUTION We must first determine the cardinality of our sample space S. The number of different groups of three people that can be selected from a group of six people is

$$n(S) = C(6, 3)$$
$$= \frac{6!}{3! \cdot 3!}$$
$$= \frac{6 \cdot 5 \cdot 4}{3 \cdot 2 \cdot 1}$$
$$= 20$$

Define event E to be the set of three-person groups containing exactly *three* men. The number of groups of three people containing exactly three men is

$$n(E) = C(4, 3) \cdot C(2, 0)$$

| Pick three out of four men. | Pick zero out of two women. |

$$n(E) = C(4, 3) \cdot C(2, 0)$$
$$= \frac{4!}{3! \cdot 1!} \cdot \frac{2!}{2! \cdot 0!}$$
$$= 4 \cdot 1$$
$$= 4$$

There are four three-person groups containing exactly three men.

$$P(E) = \frac{n(E)}{n(S)}$$
$$= \frac{4}{20}$$
$$= 0.20$$

There is a 20 percent probability that a three-person group will contain exactly three men.

E' is the set of three-person groups not containing three men. In other words, it is the set of three-person groups containing at most two men.

$$P(E') = 1 - P(E)$$
$$= 1 - 0.20$$
$$= 0.80$$

There is an 80 percent chance that a three-person group will contain at most two men. This result is the same as that from Example 5 but required fewer computations.

EXAMPLE 7

Calculating the Probability of a Complement

Ten students are competing for a scholarship. Six of the students are women. What is the probability that a randomly selected application belongs to a man?

SOLUTION Since there are 10 applicants, $n(S) = 10$.

$$E = \{x \mid x \text{ is a female applicant}\}$$
$$n(E) = 6$$
$$E' = \{x \mid x \text{ is a male applicant}\}$$
$$n(E') = n(S) - n(E)$$
$$= 10 - 6$$
$$= 4$$

$$P(E') = \frac{n(E')}{n(S)}$$
$$= \frac{4}{10}$$
$$= 0.4$$

There is a 40 percent probability that the application belongs to a man.

When looking at the fairness of award processes, it is often helpful to look at the probability of obtaining an outcome that meets certain specified criteria. For example, suppose that the probability of a woman obtaining one of 10 identical scholarships is 55 percent. However, in actuality, men obtain all 10 of the scholarships. Since the actual result varies widely from our predicted result, we may want to analyze the awards process for gender bias.

EXAMPLE 8

Calculating the Probability of an Event

Ten students are competing for a scholarship. Exactly two of the men have a G.P.A. below 3.7. What is the probability that an applicant has a G.P.A above 3.7 or is a woman?

SOLUTION There are 10 students competing, so $n(S) = 10$. Let $E = \{x \mid x \text{ is a student who is a man}\}$ and $F = \{x \mid x \text{ is a student with a G.P.A below 3.7}\}$. The set of men with G.P.A.s below 3.7 is given by $E \cap F$. We know that $n(E \cap F) = 2$ and

$$P(E \cap F) = \frac{n(E \cap F)}{n(S)}$$
$$= \frac{2}{10}$$
$$= 0.2$$

There is a 20 percent probability that a randomly selected applicant is a male with a G.P.A. below 3.7.

We know that

$(E \cap F)' = \{x | x$ is a student who is not a man with a G.P.A below 3.7$\}$

$\qquad = \{x | x$ is a student who is a woman or has a G.P.A of 3.7 or better$\}$

$P(E \cap F)' = 1 - P(E \cap F)$

$\qquad = 1 - 0.2$

$\qquad = 0.8$

There is an 80 percent probability that a randomly selected applicant is a woman or has a G.P.A of 3.7 or better.

To help us better understand our society, researchers often classify populations by demographic factors such as ethnic origin, religion, or gender. Doing so allows us to investigate the fairness of a variety of societal processes, such as education, employment, and home buying. Whenever a population is divided into a group of disjoint sets, a probability distribution may be used to determine the likelihood that a randomly selected element belongs to any specified set.

PROBABILITY DISTRIBUTION

Let $E, F, G, H, \ldots$ be mutually exclusive events whose union contains all outcomes in a sample space S. Then

$$P(E) + P(F) + P(G) + P(H) + \cdots = 1$$

The probabilities $P(E), P(F), P(G), P(H), \ldots$ make up a **probability distribution** for S.

EXAMPLE 9

Finding a Probabilty Distribution

A class of students contains twenty-five students classified by hair color: six blondes, four brunettes, two with red hair, seven with black hair, four with brown hair, and two with hair of other colors. Find a probability distribution for the class based on hair color.

SOLUTION Let $S = \{x | x$ is a student in the class$\}$, $E = \{x | x$ is a blonde$\}$, $F = \{x | x$ is a brunette$\}$, $G = \{x | x$ has red hair$\}$, $H = \{x | x$ has black hair$\}$, $I = \{x | x$ has brown hair$\}$, and $J = \{x | x$ has hair of another color$\}$. These events are mutually exclusive.

$$P(E) = \frac{6}{25} = 0.24 \qquad P(F) = \frac{4}{25} = 0.16 \qquad P(G) = \frac{2}{25} = 0.08$$

$$P(H) = \frac{7}{25} = 0.28 \qquad P(I) = \frac{4}{25} = 0.16 \qquad P(J) = \frac{2}{25} = 0.08$$

Observe that $0.24 + 0.16 + 0.08 + 0.28 + 0.16 + 0.08 = 1$. There is a 24 percent chance that a randomly selected student is blonde, a 16 percent chance that the student is brunette, an 8 percent chance that the student has red hair, a 28 percent chance that the student has black hair, a 16 percent chance that the student has brown hair, and an 8 percent chance that the student has hair of another color.

EXAMPLE 10 **Finding a Probability Distribution**

In 2002, the state of Colorado had six representatives in the U.S. House of Representatives, four Republicans and two Democrats. Only one of the representatives was female. (**Source:** www.clerk.house.gov.) Make a probability distribution based on political party. Then make another probability distribution based on gender.

SOLUTION Note that for both probability distributions, the sample space S is the set of six Colorado representatives.

$$E = \{x|x \text{ is a Republican}\}$$
$$F = \{x|x \text{ is a Democrat}\}$$

The events are mutually exclusive.

$$P(E) = \frac{n(E)}{n(S)}$$
$$= \frac{4}{6}$$
$$= 0.6667$$

$$P(F) = \frac{n(F)}{n(S)}$$
$$= \frac{2}{6}$$
$$= 0.3333$$

Observe that $0.6667 + 0.3333 = 1$.

There is a 66.67 percent probability that a randomly selected representative is a Republican and a 33.33 percent probability that the representative is a Democrat.

$$E = \{x|x \text{ is a male}\}$$
$$F = \{x|x \text{ is a female}\}$$

The events are mutually exclusive.

$$P(E) = \frac{n(E)}{n(S)}$$
$$= \frac{5}{6}$$
$$= 0.8333$$

$$P(F) = \frac{n(F)}{n(S)}$$
$$= \frac{1}{6}$$
$$= 0.1667$$

Observe that $0.8333 + 0.1667 = 1$.

There is an 83.33 percent probability that a randomly selected representative is male and a 16.67 percent probability that the representative is female.

7.5 Summary

In this section, you learned the rules for the probability of a union and a complement. You also learned how to find a probability distribution.

7.5 Exercises

In Exercises 1–5, determine $E \cap F$, $E \cup F$, and E' and express in words the meaning of each set.

1. **Coin Purse** A coin purse contains pennies, nickels, dimes, quarters, 50-cent pieces, and silver dollars. Pick a coin from the purse and note its value. Let E be the set of coins whose value is a multiple of 5. Let F be the set of coins whose value is a multiple of 50.

2. **Recipe File** A recipe file includes recipes for blueberry cobbler, apple crisp, oatmeal cookies, chocolate-covered strawberries, strawberry ice cream, chocolate ice cream, and German chocolate cake. Let E be the set of recipes containing fruit and F be the set of recipes containing chocolate.

3. **Class Composition** Let E be the students in a class who are on scholarship and F be the students in the class who have a cumulative G.P.A of 3.5 or higher.

4. **Class Composition** Some students are unsuccessful the first time they take a particular course but are able to successfully pass the course after retaking it. Let E be the group of students who are enrolled in this course for the first time and F be the group of students who have previously earned a grade below a C in this course.

5. **United Nations Security Council** In 2004, the United Nations Security Council consisted of Chile, China, France, Germany, Pakistan, Philippines, Romania, Russian Federation, Spain, United Kingdom, United States, Algeria, Angola, Benin, and Brazil. (**Source:** www.un.org.) Let E be the nations from Europe and F be the nations who are permanent members of the Security Council.

In Exercises 6–10, the sample space S is created by rolling two dice (one red and one white) and observing the numbers shown. Calculate $P(E \cup F)$ and $P(E \cap F)$.

6. $E = \left\{ x \mid x \text{ is a dice roll with an even sum} \right\}$

   $F = \left\{ x \mid x \text{ is a dice roll with a sum less than 9} \right\}$

7. $E = \left\{ x \mid x \text{ is a dice roll with an odd sum} \right\}$

   $F = \left\{ x \mid x \text{ is a dice roll with a sum less than 4} \right\}$

8. $E = \left\{ x \mid x \text{ is a dice roll with both numbers less than 5} \right\}$

   $F = \left\{ x \mid x \text{ is a dice roll with both numbers even} \right\}$

9. $E = \left\{ x \mid x \text{ is a dice roll with both numbers more than 3} \right\}$

   $F = \left\{ x \mid x \text{ is a dice roll with an even sum} \right\}$

10. $E = \left\{ x \mid x \text{ is a dice roll with both numbers equal} \right\}$

    $F = \left\{ x \mid x \text{ is a dice roll with both numbers even} \right\}$

In Exercises 11–15, four women (Alicia, Britney, Carmela, and Dejanae) and two men (Ernesto and Frank) sell vegetable choppers at local fairs. Three of them will be selected at random to work this weekend. Find the probability of the specified event E.

11. $E = \left\{ x \mid x \text{ is a work group with more women than men} \right\}$

12. $E = \left\{ x \mid x \text{ is a work group with Alicia and Ernesto} \right\}$

13. $E = \left\{ x \mid x \text{ is a work group with both men} \right\}$

14. $E = \left\{ x \mid x \text{ is a work group with at most one man} \right\}$

15. $E = \left\{ x \mid x \text{ is a work group with exactly two women} \right\}$

In Exercises 16–20, three women (Alexis, Brianna, and Cali) and four men (Dakota, Eli, Fahim, and Gregorio) sell power blenders at local fairs. Four of them will be selected at random to work this weekend. Find the probability of the specified event E.

16. $E = \left\{ x \mid x \text{ is a work group with Brianna and Dakota} \right\}$

17. $E = \left\{ x \mid x \text{ is a work group without Alexis and Fahim} \right\}$

18. $E = \left\{ x \mid x \text{ is a work group with all men} \right\}$

19. $E = \left\{ x \mid x \text{ is a work group with women and men} \right\}$

20. $E = \left\{ x \mid x \text{ is a work group with Cali and Eli} \right\}$

21. Who would benefit from the results of Exercises 11–20? Explain.

In Exercises 22–30, find the probability distribution.

22. **House of Representatives**   In 2002, the U.S. House of Representatives had 435 members. There were 223 Republicans, 209 Democrats, 1 Independent, and 2 vacancies. Find a probability distribution for the House of Representatives.

23. **Senate**   In 2002, the U.S. Senate had 100 members, including 49 Republicans, 50 Democrats, and 1 Independent. Find a probability distribution for the Senate.

24. **Dice Game**   A pair of dice is rolled. Find a probability distribution for the sum of the dice.

25. **Citizenship**   A math class has 33 students, including 20 American citizens, 11 people with student visas, and 2 illegal aliens. Find a probability distribution for the legal status of the students in the class.

26. **Coin Collection**   A box of twenty coins contains four quarters, five dimes, three nickels, and eight pennies. Find a probability distribution for the box of coins.

27. **Educational Status**   A company with 40 employees tracks the highest earned degree of its employees. It has 13 employees with master's degrees, 17 employees with bachelor's degrees, and 10 employees with high school diplomas. Find a probability distribution for the educational status of the employees of the company.

28. **Dice Game**   A six-sided die (numbered from one to six) and a four-sided die (numbered from one to four) are rolled. Find a probability distribution for the absolute value of the difference of the numbers shown on the faces of the dice.

29. **Dice Game**   A six-sided die (numbered from one to six) and a four-sided die (numbered from one to four) are rolled. Find a probability distribution for the product of the numbers shown on the faces of the dice.

30. **Professions**   A dinner party of sixteen guests includes three doctors, one dentist, one physical therapist, two teachers, three training specialists, three mechanics, a bartender, a preacher who gives weekly sermons, and a cook. The guests may be categorized into groups of those who treat the human body, those who teach people, those who fix mechanical things, and those who make or serve foodstuffs. Find a probability distribution for the employment categories.

Exercises 31–35 are intended to challenge your understanding of basic probability concepts.

31. **Mixed Groups**   There are ten people in a courtroom, including six lawyers, three criminals, seven men, and one male lawyer who is a criminal. All of the women in the room, including the one who is a lawyer, aren't criminals. What is the probability that a person randomly selected from the group is a male lawyer?

32. **Doctor's Office**   There are twenty people in a room, including six pregnant women, eleven sick people, two doctors, and three nurses. None of the medical personnel are sick, but one of them is pregnant. What is the probability that a person randomly selected from the group is a pregnant women who is sick?

33. Dice Game   Two six-sided dice are rolled. Find a probability distribution for the product of the numbers shown minus the largest number shown on the faces of the dice.

34. Dice Game   Three four-sided dice are rolled. Find a probability distribution for the difference between the largest and the smallest value shown on the faces of the dice.

35. Dice Game   Based on the results of Exercise 34, is obtaining an even value or an odd value more likely?

Chapter 7 Review Exercises

Section 7.1   *In Exercises 1–4, copy the Venn diagram given here and shade the region defined in each problem, if possible. Then list the elements of the solution set.*

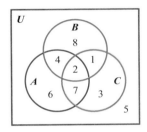

1. $(A \cap B)'$

2. $(A \cap B) \cup C$

3. $(A \cap B) \cap C'$

4. $(A \cup B) \cap C'$

For Exercises 5–8, six people are categorized by their gender, employment status, and highest academic degree as follows:

1: Male, employed, master's degree
2: Male, employed, high school diploma
3: Female, employed, bachelor's degree
4: Female, unemployed, bachelor's degree
5: Male, unemployed, Ph.D.
6: Female, unemployed, high school diploma

Further categorize the group by letting M be the set of all men, E be the set of employed people, and D be the set of people with a graduate degree (master's or Ph.D.). Explain the real-life meaning of the set specified in each problem. Then specify which people are elements of the set. (Hint: It may be helpful to draw a Venn diagram.)

5. $M \cap E$

6. $E \cup D$

7. $E' \cap D$

8. $(D \cap M)'$

Section 7.2   *In Exercises 9–12, A and B are finite subsets of the universal set U with $n(U) = 20$, $n(A) = 12$, $n(A \cap B) = 5$, and $n(A \cup B) = 17$. Find the cardinality of each of the specified sets.*

9. B'

10. $A' \cup B'$

11. $A' \cap B'$

12. $(A \cup B)'$

In Exercises 13–16, use the Addition or Multiplication Principle as appropriate to determine the cardinality of the set.

13. A restaurant offers 6 appetizers, 20 entrees, and 8 desserts. How many dish options are there?

14. A restaurant offers 6 appetizers, 20 entrees, and 8 desserts. How many three-course meal options are there?

15. A restaurant offers 7 appetizers, 25 entrees, and 4 desserts, including strawberry shortcake. How many three-course meal options are there with strawberry shortcake as dessert?

16. An office supply store offers three qualities of staplers: light-duty, medium-duty, and heavy-duty. The staplers come in grey, brown, or black. How many different types of staplers are there?

For Exercises 17–18, you are given the sets $R = \{truck, sport\ utility\ vehicle\}$, $C = \{black, red, blue\}$, $S = \{two-wheel\ drive, four-wheel\ drive\}$, and $M = \{new, used\}$. Draw a decision tree to show all possible outcomes for the specified criteria.

17. The outcomes with a single element from each of the sets R, C, and S.

18. The outcomes with a single element from each of the sets R, C, and M.

Section 7.3   *In Exercises 19–22, calculate the value of the expression without using a calculator.*

19. $P(9, 2)$ **20.** $P(15, 1)$

21. $C(6, 2)$ **22.** $C(10, 4)$

In Exercises 23–24, calculate the value of the expression using a calculator.

 23. $P(25, 7)$ **24.** $C(25, 7)$

In Exercises 25–28, use permutations or combinations as appropriate to calculate the total number of possible outcomes.

Pizza Hut offers 4 different types of crust (Pan Pizza, Stuffed Crust Pizza, The Big New Yorker Pizza, and Thin 'n Crispy Pizza) and 12 toppings (Italian Sausage, Pepperoni, Beef Topping, Pork Topping, Ham, Chicken, Mushrooms, Green Peppers, Red Onions, Tomatoes, Black Olives, and Bacon). (**Source:** www.pizzahut.com.)

25. How many different types of one-topping pizzas can be made?

26. How many different types of three-topping Stuffed Crust Pizzas can be made?

27. How many different types of three-topping pizzas can be made?

28. How many different types of three-topping pizzas can be made with the meatless toppings?

Section 7.4   *In Exercises 29–30, use the terms* experiment, trials, outcomes, *and* sample space *to explain the situation.*

29. A sweepstakes contestant is randomly picked from a box of entries and the last digit of the entry number is noted. (Assume that entries may end in any digit.)

30. A pair of dice is rolled twice. Each time the sum of the dice is recorded.

In Exercises 31–32, identify the sample space for the experiment. Then write the specified event in set notation.

31. Adi (M), Berto (M), Marsha (F), Mariah (F), and Liz (F) are scholarship contestants. (Their genders are noted in the parentheses.) Two of them are selected as finalists.

   (a) *Event:* At least one of the finalists is male.

   (b) *Event:* Both finalists are female.

   (c) *Event:* No more than one finalist is male.

32. Jorge (3.92), Shauna (3.97), Andrea (3.47), Mohammed (3.83), Teri (3.74), and Paul (4.00) are scholarship candidates. (Their G.P.A.s are noted in the parentheses.) Two of them are selected as finalists.

   (a) *Event:* At least one of the finalists has a 4.0 G.P.A.

   (b) *Event:* Both finalists have above a 3.90 G.P.A.

   (c) *Event:* At least one finalist has a G.P.A. between 3.40 and 3.75.

Section 7.5   *In Exercises 33–36, three women (Alena, Betty, and Carmen) and two men (Don and Ernesto) deliver pizza. Three of them will be selected at random to work this weekend. Find the probability of the specified event.*

33. $E = \left\{ x \mid x \text{ is a group with more women than men} \right\}$

34. $E = \left\{ x \mid x \text{ is a group with Alena and Ernesto} \right\}$

35. $E = \left\{ x \mid x \text{ is a group with Carmen and at least one man} \right\}$

36. $E = \left\{ x \mid x \text{ is a group with Alena but not Carmen} \right\}$

Make It Real

Dice games, including Casino Craps, Pig, Bunco, Zilch, and Yahtzee©, among others, are popular diversions for people seeking a bit of entertainment. In this project, you'll make up a simple dice game.

What to do

1. Find a pair of dice (or a paradise!).
2. Determine how you will interpret each dice roll (sum of dice, product of dice, compare dice values, and so on).
3. List all outcomes for each dice roll.
4. Define three or more events that will be scored (e.g., sum of dice is 7).
5. Find the probability distribution for the dice game.
6. Design a scoring system based on the probabilities (low probability offers high points).
7. Name your game.

Example dice game

Bonkaroo

1. Roll two dice and sum their values.
2. *Sample space:* {2, 3, 4, . . . ,11, 12}
3. *Events:* {2}, {7}, {11}, {not 2, 7, 11}
4. There are 36 possible rolls with two dice. The ways to get each event are

   2: (1,1)
   7: (1, 6), (2, 5), (3, 4), (4, 3), (5, 2), (6, 1)
   11: (5, 6), (6, 5)
   not 2, 7, 11: the remaining 27 rolls
   Probability Distribution

   2: $\dfrac{1}{36} = 0.0278$

   7: $\dfrac{6}{36} = 0.1667$

   11: $\dfrac{2}{36} = 0.0556$

   not 2, 7, 11: $\dfrac{27}{36} = 0.75$

5. A 2 is worth 30 points, a 7 is worth 5 points, and an 11 is worth 15 points.
6. Players take turns rolling the dice.
7. The first player to score a total of 100 points and yell *Bonkaroo!* wins.

Advanced Probability and Statistics

As diverse as we are as individuals, we still are remarkably similar to other people. Consider the physical feature of height. Ninety percent of Mexican American women are between 58 and 66 inches tall. (**Source:** www.halls.md.) If a woman is randomly selected from a large group of Mexican American women, how likely is it that she will be 70 inches tall? The statistical concept of a normal distribution may be used to calculate the probability of randomly selecting a woman of a particular height.

8.1 Conditional Probability
- Use conditional probability to determine if events are independent

8.2 Bayes' Theorem and Applications
- Apply Bayes' theorem to real-world applications
- Calculate the predictive-value positive and predictive-value negative of a test

8.3 Markov Chains
- Forecast future events with Markov chains

8.4 Random Variables and Expected Value
- Calculate the probability distribution of a random variable
- Draw a histogram
- Determine the expected value of a random variable

8.5 Measures of Central Tendency and Dispersion
- Calculate the mean, median, mode, and standard deviation of a data set

8.6 Normal Distributions
- Convert a normal distribution to a standard normal distribution
- Determine the area of a shaded region beneath a normal curve
- Calculate the probability of attaining a value within a designated number of standard deviations of the mean of a normal distribution

8.1 Conditional Probability

- Use conditional probability to determine if events are independent

GETTING STARTED In 2003, Washington state had nine representatives and two senators in Congress, all of whom were either Republicans or Democrats. Of the three female members of Congress, two were senators. The female representative in the House of Representatives was a Republican, and the female senators were Democrats. There were a total of eight congressional Democrats from Washington state. What is the probability that a randomly selected female member of Congress from Washington state is also a Democrat? What is the probability that a randomly selected member of the House of Representatives representing Washington state is female? Questions such as these can be answered using conditional probability.

In this section, we will show how conditional probability may be used when certain characteristics of a member of a set are known. We will also demonstrate how to create a probability tree and introduce the concept of independent and dependent events.

Let the sample space C represent the Washington state congressional delegation, the event D be the set of Washington state congressional Democrats, the event H be the set of Washington state Representatives, and the event F be the set of Washington state congressional women. Writing the cardinality of each set in the corresponding region on the Venn diagram yields Figure 8.1.

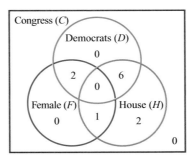

FIGURE 8.1

What is the probability that a randomly selected female member of Congress from Washington state is also a Democrat? In other words, what is the probability that a randomly selected member of Congress is a Democrat given that she is female? From the Venn diagram, we see that of the three women, two are Democrats. Since we know that the member of Congress is female, our sample space is restricted to the event F. We want to know $P(D)$, treating the event F as the entire sample space. In probability notation, we have

$$P(D) = \frac{n(F \cap D)}{n(F)}$$

$$= \frac{2}{3}$$

Therefore, the probability that a randomly selected female member of Congress from Washington state is a Democrat is $\frac{2}{3}$, or 66.67 percent. The idea of restricting the sample space to a particular event is an important one. In the example, we restricted the sample space to the set of female members of Congress. We represent the notion of restricting the sample space to a particular set with conditional probability notation.

CONDITIONAL PROBABILITY

The probability that event A occurs given that event B occurs is given by

$$P(A|B) = \frac{P(A \cap B)}{P(B)}$$

provided that $P(B) \neq 0$.

This equation looks somewhat different from our earlier calculations. Will it yield the same result? Let's see. We want to find $P(D|F)$. Our sample space is the entire Washington state congressional delegation C.

$$P(D|F) = \frac{P(D \cap F)}{P(F)}$$

$$= \frac{\dfrac{n(D \cap F)}{n(C)}}{\dfrac{n(F)}{n(C)}}$$

$$= \frac{\dfrac{2}{11}}{\dfrac{3}{11}}$$

$$= \frac{2}{3}$$

The result is the same as before.

EXAMPLE **1**   ## Using Conditional Probability

The makeup of the Washington state congressional delegation is shown in Figure 8.2. What is the probability that a congressman is a member of the House of Representatives given that the congressman is a Democrat?

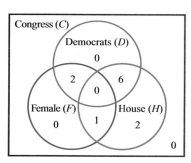

FIGURE 8.2

SOLUTION We are asked to find $P(H|D)$. From the conditional probability formula, we know that

$$P(H|D) = \frac{P(H \cap D)}{P(D)}$$

$$= \frac{\dfrac{n(H \cap D)}{n(C)}}{\dfrac{n(D)}{n(C)}}$$

$$= \frac{n(H \cap D)}{n(D)}$$

$$= \frac{6}{8}$$

$$= 75\%$$

Given that a member of Congress is a Democrat, there is a 75 percent probability that the member of Congress is a member of the House of Representatives.

EXAMPLE 2 **Using Conditional Probability**

An applied calculus class has 33 students enrolled, including 12 international students. There are 10 students in the class, 6 of whom are international students, who have Asian ancestors. What is the probability that a randomly selected student has Asian ancestors, given that the student is not an international student?

SOLUTION Let A be the set of students with Asian ancestors, I be the set of international students, and C be the set of students enrolled in the class. We are asked to find $P(A|I')$.

$$P(A|I') = \frac{P(A \cap I')}{P(I')}$$

$$= \frac{\dfrac{n(A \cap I')}{n(C)}}{\dfrac{n(I')}{n(C)}}$$

$$= \frac{n(A \cap I')}{n(I')}$$

$$= \frac{4}{21}$$

$$\approx 19\%$$

There is a 19 percent probability that a noninternational student has Asian ancestors.

EXAMPLE 3 Using Conditional Probability

Table 8.1 shows regional HIV/AIDS statistics for North America, the Caribbean, and the entire world as of the end of 2002. (**Source:** UNAIDS/WHO.)

TABLE 8.1

Region	People Living with HIV/AIDS (Infected Prior to 2002)	People Newly Infected with HIV in 2002	Total Number of People Infected with HIV/AIDS
North America	935,000	45,000	980,000
Caribbean	380,000	60,000	440,000
Worldwide	37,000,000	5,000,000	42,000,000

Is it more likely that a randomly selected person infected with HIV/AIDS lives in North America or lives in the Caribbean, given that the person was newly infected with HIV/AIDS in 2002?

SOLUTION Let C be the group of people with HIV/AIDS living in the Caribbean, let A be the group of people with HIV/AIDS living in North America, and let N be the group of people newly infected with HIV/AIDS. We are asked to compare $P(C|N)$ and $P(A|N)$

$$P(C|N) = \frac{P(C \cap N)}{P(N)}$$

$$= \frac{\dfrac{60,000}{42,000,000}}{\dfrac{5,000,000}{42,000,000}}$$

$$= 1.2\%$$

$$P(A|N) = \frac{P(A \cap N)}{P(N)}$$

$$= \frac{\dfrac{45,000}{42,000,000}}{\dfrac{5,000,000}{42,000,000}}$$

$$= 0.9\%$$

There is a 1.2 percent probability that a person infected with HIV/AIDS in 2002 lives in the Caribbean and a 0.9 percent probability that the person lives in North America. It is more likely that a person newly infected with HIV/AIDS in 2002 lives in the Caribbean than in North America.

We previously developed a formula to calculate the probability of the union of two events. In like manner, we can derive a formula for the probability of an intersection of two events. Recall that the conditional probability formula is given by $P(A|B) = \frac{P(A \cap B)}{P(B)}$. Since $P(B)$ is a real number, we can multiply both sides of the formula by $P(B)$. Doing so yields $P(A \cap B) = P(B) \cdot P(A|B)$. By a similar argument, $P(A \cap B) = P(A) \cdot P(B|A)$.

PRODUCT RULE OF PROBABILITY

Given events A and B,

$$P(A \cap B) = P(B) \cdot P(A|B) \text{ and } P(A \cap B) = P(A) \cdot P(B|A)$$

provided that $P(A) \neq 0$ and $P(B) \neq 0$. [If $P(A) = 0$ or $P(B) = 0$, then $P(A \cap B) = 0$.]

EXAMPLE 4 Using Conditional Probability in a Real-World Context

According to UNAIDS/WHO data, there is an 11.9 percent probability that a person with HIV/AIDS acquired the disease in 2002, given that the person lived in sub-Saharan Africa. There is a 70.0 percent probability that a person with HIV/AIDS in 2002 lived in sub-Saharan Africa. What is the probability that a randomly selected person who acquired HIV/AIDS in 2002 lived in sub-Saharan Africa?

SOLUTION Let S be the group of people with HIV/AIDS living in sub-Saharan Africa in 2002, and let N be the group of people newly infected with HIV in 2002. We are asked to find $P(S \cap N)$. We know $P(N|S) = 0.119$ and $P(S) = 0.700$.

$$P(S \cap N) = P(S) \cdot P(N|S)$$
$$= 0.700 \cdot 0.119$$
$$= 0.0833$$

There is an 8.33 percent probability that a randomly selected person with HIV/AIDS in 2002 acquired the disease in 2002 and lived in sub-Saharan Africa.

EXAMPLE 5 Finding the Probability of an Intersection of Sets

A couple has 11 grandchildren, including 5 girls and 6 boys. Two-thirds of the boys have curly hair. What is the probability that a randomly selected grandchild is a curly-haired boy?

SOLUTION Let B be the group of boys and C be the group of curly-haired grandchildren. We are asked to find $P(B \cap C)$. We know that $P(B) = \frac{6}{11}$ and $P(C|B) = \frac{2}{3}$.

$$P(B \cap C) = P(B) \cdot P(C|B)$$
$$= \frac{6}{11} \cdot \frac{2}{3}$$
$$= \frac{4}{11}$$
$$= 36.4\%$$

There is a 36.4 percent probability that a randomly selected grandchild is a curly-haired boy.

Probability Trees

The product rule of probability may also be represented through the use of a *probability tree*. A **probability tree** is a decision tree with the probability of the event annotated on each branch of the tree. In Section 7.2, we introduced the scenario of a student who needed to enroll in three online classes: English, math, and art. The student could choose from two online English courses: Basic Composition and Basic Literature; three online math courses: Finite Math, Consumer Math, and Precalculus; and two online art courses: Art History and Design. Based on these options, we constructed the decision tree shown in Figure 8.3).

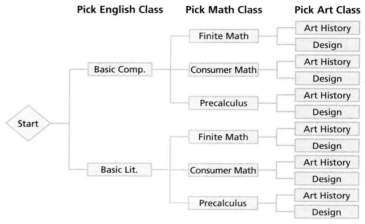

FIGURE 8.3

To convert the decision tree into a probability tree, we simply need to assign the probabilities to each event, as shown in Figure 8.4.

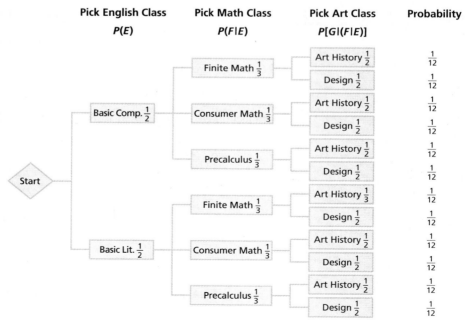

FIGURE 8.4

Since there are two English courses, the probability of randomly selecting a particular English course is $\frac{1}{2}$. Since there are three math courses, the probability of randomly selecting a particular math course is $\frac{1}{3}$. Likewise, since there are two art courses, the probability of randomly selecting a particular art course is $\frac{1}{2}$. Recall from our work in Chapter 7 that the total number of different three-course schedules is equal to the product of the number of choices for each course. That is, $2 \cdot 3 \cdot 2 = 12$. Since each of the schedules is different, the probability of randomly selecting one particular schedule is $\frac{1}{12}$. This same result may be obtained by multiplying the individual probabilities of each course selection as shown here. Assuming that the selection of any course in a given subject is equally likely, the probability of selecting any three-course schedule is $\frac{1}{12} = 8.33$ percent since $\frac{1}{2} \cdot \frac{1}{3} \cdot \frac{1}{2} = \frac{1}{12}$.

As just shown, the probability of choosing any of the 12 three-course schedules was the same for each schedule. However, if a choice alters the options for subsequent choices, the probabilities for different sequences of events often differ, as demonstrated in the following example.

EXAMPLE 6

Using a Probabilty Tree

In 2002, Missouri had nine members of the U.S. House of Representatives, five Republicans and four Democrats. Of the two women, one was a Democrat. (**Source:** clerk.house.gov.) Use a probability tree to determine the likelihood that a randomly selected representative is a female Democrat.

SOLUTION We will first choose the party of the representative and then select the gender (Figure 8.5).

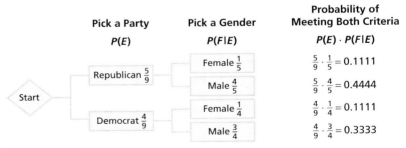

FIGURE 8.5

There is an 11.11 percent chance that a randomly selected Missouri representative is a female Democrat. It is most likely that a randomly selected Missouri representative is a male Republican (44.44 percent).

Independent Events

Observe that in the discussion of class schedules, $P(F|E) = P(F)$. That is, which English class was chosen did not alter the probability of choosing a given math class. However, in Example 6, $P(F|E) \neq P(F)$. That is, the probability of picking a male or female representative differed depending upon which political party was picked. If the probabilities of subsequent events are not altered by previous choices, the events are said to be **independent events.**

INDEPENDENT EVENTS

Events E and F are **independent events** if

$$P(E|F) = P(E) \text{ and } P(F|E) = P(F)$$

Events that are not independent are called **dependent events.** The events of selecting courses in the course schedule were independent events, whereas the events of selecting a representative's party affiliation and gender in Example 6 were dependent events.

EXAMPLE 7

Determining If Events Are Independent

The Federal Bureau of Investigation (FBI) collects crime statistics from cities across the nation and publishes the results regularly in the Uniform Crime Reports periodical. The crimes reported are murder, forcible rape, robbery, aggravated assault, burglary, larceny, and motor vehicle theft. In its December 16, 2002, report, the FBI published the crime statistics in Table 8.2.

TABLE 8.2 **2002 Crime Statistics**

City	Population (approx.)	Crimes
Anchorage, AK	260,300	6,325
Atlanta, GA	416,500	23,877
Boston, MA	589,100	17,463
Cleveland, OH	478,400	15,797
Green Bay, WI	102,300	1,585
Total	**1,846,600**	**65,047**

Source: FBI Uniform Crime Reports, January–June 2002, www.census.gov.

A family intends to move to one of the five cities. Assuming that it is equally likely that anyone in the selected city will be a crime victim and that crime rates are constant from year to year, determine if the likelihood of becoming a crime victim is independent of the city selected.

SOLUTION Let C be the event "the given city is selected" and V be the event "become a victim." We have the probability tree shown in Figure 8.6.

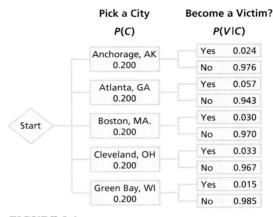

FIGURE 8.6

Based on the data for these five cities,

$$P(V) = \frac{65{,}047}{1{,}846{,}600}$$

$$= 0.0352$$

$$= 3.52\,\%$$

If the crime rate is independent of the city, then each of the cities should have a 3.52 percent crime rate. However,

$$P(V|\text{Anchorage}) = 2.4\%$$
$$P(V|\text{Atlanta}) = 5.7\%$$
$$P(V|\text{Boston}) = 3.0\%$$
$$P(V|\text{Cleveland}) = 3.3\%$$
$$P(V|\text{Green Bay}) = 1.5\%$$

Therefore, the incidence of crime is dependent upon the city selected.

8.1 Summary

In this section, you learned how conditional probability may be used when certain characteristics of a member of a set are known. You also saw how to create a probability tree and became familiar with the concept of independent and dependent events.

8.1 Exercises

In Exercises 1–15, determine the indicated probability.

1. $P(A \cap B) = 0.45$ and $P(B) = 0.5$. Find $P(A|B)$.

2. $P(A \cap B) = 0.10$ and $P(B) = 0.2$. Find $P(A|B)$.

3. $P(A \cap B) = 0.01$ and $P(A) = 0.10$. Find $P(B|A)$.

4. $P(A \cap B) = 0.5$ and $P(A) = 0.5$. Find $P(B|A)$.

5. $P(A|B) = 0.42$ and $P(B) = 0.13$. Find $P(A \cap B)$.

6. $P(A|B) = 0.96$ and $P(B) = 0.32$. Find $P(A \cap B)$.

7. $P(B|A) = 0.21$ and $P(A) = 0.60$. Find $P(A \cap B)$.

8. $P(B|A) = 0.44$ and $P(A) = 0.25$. Find $P(A \cap B)$.

9. Find $P(F|D)$.

10. Find $P(D|F)$.

11. Find $P(D'|H)$.

12. Find $P(D'|F')$.

13. Find $P(H'|F)$.

14. Find $P(F|H')$.

15. Find $P(H|F')$.

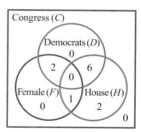

In Exercises 16–25, use the concept of conditional probability to answer the questions.

16. Gumballs   A gumball machine contains 10 red, 12 yellow, and 8 blue gumballs. Assuming that all of the gumballs are equally likely to be dispensed when the machine handle is turned, what is the probability that the second gumball dispensed from the machine is blue, given that the first gumball dispensed was blue?

17. **Gumballs**   A gumball machine contains 10 red, 12 yellow, and 8 blue gumballs. Assuming that all of the gumballs are equally likely to be dispensed when the machine handle is turned, what is the probability that the second gumball dispensed from the machine is yellow, given that the first gumball dispensed was red?

18. **Ethnic Background**   A finite math class has 32 students enrolled, including 13 international students. There are 9 students, 6 of whom are international students, who have Hispanic ancestors. What is the probability that a randomly selected student has Hispanic ancestors, given that the student is not an international student?

19. **Employment Status**   A finite math class has 28 students enrolled, including 15 women. There are 9 students, 6 of whom are women, who work full time. What is the probability that a randomly selected student doesn't work full time, given that the student is a woman?

20. **Demographics**   There were 35 job candidates who submitted applications for a management position. Of the 35 applicants, 20 were men and 6 were from racial minorities. Exactly 2 men from a racial minority applied for the position. What is the probability that a randomly selected candidate is from a racial minority, given that the candidate is female?

21. **Demographics**   There were 40 job candidates who submitted applications for a management position. Of the 40 applicants, 35 were women and 5 were from racial minorities. Exactly 2 men from a racial minority applied for the position. What is the probability that a randomly selected candidate is from a racial minority, given that the candidate is female?

22. **Demographics**   Twenty people, including twelve African American men, applied for a management position at a recording studio. Five applicants, all African American men, were well qualified for the job. What is the probability that a person randomly selected for the position is not an African American man, given that the candidate is well qualified for the job?

23. **Demographics**   Ninety people, including six women, applied for an advertising director position at an international shipping company. Twelve applicants, including five women, were well qualified for the job. What is the probability

that a randomly selected person is a woman, given that she was well qualified?

24. **Discrimination**   After being denied the position described in Exercise 23, a man sued the company, charging reverse discrimination. He argued that there was only a $\frac{6}{90} = 6.67$ percent probability that a woman would get the job and a 93.33 percent probability that a man would get the position. Considering the results of Exercise 23, how should the defense lawyer respond?

25. **Menu Choices**   A menu offers 30 different items. There are 15 entrées, 3 of which are vegetarian dishes. When salads and other side dishes are included, there are a total of 10 vegetarian items on the menu. What is the likelihood that a randomly selected item chosen from the menu is not a vegetarian dish, given that the item is not an entrée?

In Exercises 26–30, use a probability tree to determine the solution.

26. **House of Representatives**   In 2003, California had 53 members of the U.S. House of Representatives, 20 Republicans and 33 Democrats. Of the 17 female representatives, 1 was a Republican. (**Source:** clerk.house.gov.) What is the likelihood that a randomly selected representative is a female Democrat?

27. **House of Representatives**   In 2003, California had 53 members of the U.S. House of Representatives, 20 Republicans and 33 Democrats. Of the 17 female representatives, 1 was a Republican. (**Source:** clerk.house.gov.) What is the likelihood that a randomly selected representative is a male Republican?

28. **Storage Media**   Imation is a company that produces a variety of products, including CD-R and CD-RW storage media. According to its March 2003 product list, CD-R retail spindle packs are sold in packages containing 5, 10, 20, 25, 50, or 100 disks. CD-RW packs are sold in packages containing 1, 5, or 25 disks. (**Source:** www.imation.com.) What is the likelihood that a randomly selected item on the product list contains exactly 5 disks? Given that the selected disk is a CD-R disk, does the likelihood that it contains exactly 5 disks increase or decrease?

29. **Playing Cards** Suppose 3 cards are pulled from a 52-card deck containing 13 hearts, 13 diamonds, 13 clubs, and 13 spades. What is the probability that the third card drawn is a heart? Does the probability increase or decrease given that the first two cards drawn were hearts?

30. **Playing Cards** Suppose 3 cards are pulled from a 52-card deck containing 4 jacks. What is the probability that the third card drawn is a jack? Does the probability increase or decrease given that the first two cards drawn were jacks?

In Exercises 31–35, determine if the events E and F are independent or dependent.

31. Two coins are tossed.

$E = $ the first coin shows tails
$F = $ the second coin shows tails

32. Two dice are rolled.

$E = $ the first die shows a 5
$F = $ the second die shows a 1

33. Two coins are drawn from a bag containing five nickels, two dimes, and three pennies.

$E = $ the first coin is a dime
$F = $ the second coin is a nickel

34. Two coins are drawn from a bag containing five nickels, two dimes, and three pennies.

$E = $ the first coin is a dime
$F = $ the second coin is not a dime

35. A family has three children.

$E = $ the oldest child is a girl
$F = $ the second child is a girl

Exercises 36–38 are intended to challenge your understanding of probability concepts.

36. If $P(E|F) = P(F|E)$, what do you know about $P(E)$ and $P(F)$?

37. If $P(E|F) = P(F)$, are E and F independent? Explain.

38. Given $P(E \cap F) = P(F \cap G)$, show that $P(E \cap F \cap G') = P(E' \cap F \cap G)$.

8.2 Bayes' Theorem and Applications

- Apply Bayes' theorem to real-world applications
- Calculate the predictive-value positive and predictive-value negative of a test

GETTING STARTED One critical element of a disease control and prevention program is the ability to detect who has been infected by a disease. Blood screening and other tests are often used to identify infected people; however, sometimes these tests yield false-positive results. That is, the test indicates that the person is infected when in reality the person is not. Similarly, tests sometimes yield false-negative results. That is, the test indicates that the person is not infected when in reality the person is infected. One way to analyze the effectiveness of a test is by applying Bayes' theorem.

In this section, we will introduce Bayes' theorem and some of its applications. Bayes' theorem is used by the Centers for Disease Control and Prevention and others to investigate the accuracy of screening tests.

Our practical approach to Bayes' theorem differs from the approach taken by many other texts. We determined that it would be most beneficial for readers to learn how Bayes' theorem is actually used in medicine and industry. Many of the concepts covered herein may be applied by consumers in their personal lives.

When a disease detection test is administered, there are four possible outcomes: true positive, false positive, false negative, and true negative (see Table 8.3).

TABLE 8.3

Test Result	Actual Status		Total
	Disease Present	Disease Absent	
Positive	True positive (a)	False positive (b)	All positive tests ($a + b$)
Negative	False negative (c)	True negative (d)	All negative tests ($c + d$)
Total	All with disease ($a + c$)	All without disease ($b + d$)	Total tested ($a + b + c + d$)

Graphically, we may represent these results with a probability tree (Figure 8.7).

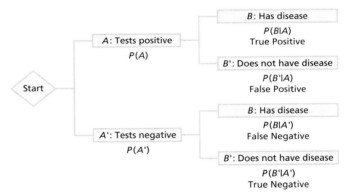

FIGURE 8.7

The probability that a test result will be positive when the test is administered to a person with the disease is referred to as the **sensitivity** of the test.

$$\text{Sensitivity} = \frac{a}{a + c} = \frac{\text{true positive}}{\text{all with disease}}$$

The probability that a test result will be negative when the test is administered to a person without the disease is referred to as the **specificity** of the test.

$$\text{Specificity} = \frac{d}{b + d} = \frac{\text{true negative}}{\text{all without disease}}$$

The probability that a person with a positive test result will actually have the disease is referred to as the **predictive-value positive (PVP)** of the test.

$$PVP = \frac{a}{a + b} = \frac{\text{true positive}}{\text{all with positive test}}$$

The probability that a person with a negative test result will not have the disease is referred to as the **predictive-value negative (PVN)** of the test.

$$PVN = \frac{d}{c + d} = \frac{\text{true negative}}{\text{all with negative test}}$$

The known percentage of people in the population that are infected is referred to as the **prevalence.**

EXAMPLE **1**

Determining the Accuracy of a Screening Test

In June 1984, five companies were licensed to produce a test kit for detecting the HIV antibody. In March 1985, the first enzyme-linked immunosorbent assay (EIA) test kit was approved by the Food and Drug Administration.

The Centers for Disease Control and Prevention offer this case study in their 2003 *Screening for Antibody to the Human Immunodeficiency Virus* instructor's guide. A blood bank has an EIA sensitivity of 95.0 percent and an EIA specificity of 98.0 percent. Of the 1 million blood donors, 0.04 percent actually have the HIV antibody. (**Source:** www.cdc.gov.) Using this information, complete Table 8.4. Then calculate the *PVP* and *PVN* for the EIA test.

TABLE 8.4

	Actual Status		
Test Result	Disease Present	Disease Absent	Total
Positive			
Negative			
Total			

SOLUTION The number of people infected is given by

$$(\text{Prevalence})(\text{population}) = 0.0004(1,000,000)$$
$$= 400$$

Therefore, the number of people not infected is given by

$$(1 - \text{prevalence})(\text{population}) = (1 - 0.0004)(1,000,000)$$
$$= (0.9996)(1,000,000)$$
$$= 999,600$$

(Alternatively, we could simply subtract the number infected from the total population: $1,000,000 - 400 = 999,600$.)

The number of true positives is given by

$$(\text{Sensitivity})(\text{infected}) = 0.95(400)$$
$$= 380$$

The number of true negatives is given by

$$(\text{Specificity})(\text{not infected}) = 0.98(999,600)$$
$$= 979,608$$

The number of false negatives is given by

$$\text{Infected} - \text{true positives} = 400 - 380$$
$$= 20$$

The number of false positives is given by

$$\text{not infected } - \text{ true negatives } = 999,600 - 979,608$$
$$= 19,992$$

The completed table is shown in Table 8.5.

TABLE 8.5

	Actual Status		
Test Result	Disease Present	Disease Absent	Total
Positive	380	19,992	20,372
Negative	20	979,608	979,628
Total	400	999,600	1,000,000

$$PVP = \frac{\text{true positive}}{\text{total positive}} \qquad PVN = \frac{\text{true negative}}{\text{total negative}}$$

$$= \frac{380}{20,372} \qquad\qquad = \frac{979,608}{979,628}$$

$$= 0.018653 \qquad\qquad = 0.99998$$

$$= 1.8653\% \qquad\qquad = 99.998\%$$

There is about a 1.9 percent probability that a person with a positive test result actually has the disease. Consequently, we are highly skeptical of a positive test result. On the other hand, there is nearly a 100 percent probability that a person with a negative result doesn't have the disease. We are highly confident that if the test result is negative, the person does not have the disease.

EXAMPLE 2

Determining the Accuracy of E-mail Filtering Software

Between December 30, 2003, and January 2, 2004, the author received 49 e-mails. His Internet security software tagged 13 e-mails as possible spam. All of the tagged e-mails were in fact spam. There were 25 e-mails that were not spam. Determine the sensitivity, specificity, predictive-value positive, and predictive-value negative of the spam filter. Interpret the real-world significance of the results.

SOLUTION Data for the solution are given in Table 8.6.

TABLE 8.6

	Actual Status		
Test Result	Spam	Not Spam	Total
Tagged	13	0	13
Not Tagged	11	25	36
Total	24	25	49

$$\text{Sensitivity} = \frac{\text{tagged spam}}{\text{total spam}}$$

$$= \frac{13}{24}$$

$$= 0.5417$$

$$= 54.17\%$$

$$\text{Specificity} = \frac{\text{not spam and not tagged}}{\text{not spam}}$$

$$= \frac{25}{25}$$

$$= 1$$

$$= 100\%$$

$$PVP = \frac{\text{tagged spam}}{\text{tagged}} \qquad PVN = \frac{\text{not spam and not tagged}}{\text{not tagged}}$$

$$= \frac{13}{13} \qquad\qquad = \frac{25}{36}$$

$$= 1 \qquad\qquad = 0.6944$$

$$= 100\% \qquad\qquad = 69.44\%$$

Based on this sample, the spam filter appears to tag about 54 percent of all spam. If the filter tagged an e-mail as possible spam, it was spam. If the filter did not tag an e-mail as possible spam, there was a 69 percent chance that the e-mail was not spam.

According to IT industry experts, spam cost American businesses $10 billion in 2003. As businesses have developed filtering software to block spam, however, they have run into a new challenge: false positives. Blocked legitimate e-mail cost U.S. businesses approximately $3.5 billion in 2003. (**Source:** itmanagement.earthweb.com.)

An alternative approach to these types of problems is through the use of set notation. Recall that the product rule of probability states that for events A and B, $P(A \cap B) = P(B) \cdot P(A|B)$ and $P(A \cap B) = P(A) \cdot P(B|A)$ [for $P(A) \neq 0$ and $P(B) \neq 0$]. Furthermore, observe that $B = (B \cap A) \cup (B \cap A')$. ($B \cap A$ is the set of items in B that are also in A, whereas $B \cap A'$ is the set of items in B that are outside of A.) Therefore,

$$P(B) = P[(B \cap A) \cup (B \cap A')]$$

$$= P(B \cap A) + P(B \cap A') \text{ since } (B \cap A) \cap (B \cap A') = \varnothing \qquad \text{Probability of a union}$$

$$= P(B|A)P(A) + P(B|A')P(A') \qquad\qquad\qquad\qquad\qquad \text{Product rule of probaility}$$

We know that

$$P(A|B) = \frac{P(A \cap B)}{P(B)} \qquad\qquad \text{Definition of conditional probability}$$

$$= \frac{P(B|A)P(A)}{P(B)} \qquad\qquad \text{Product rule of probability}$$

$$= \frac{P(B|A)P(A)}{P(B|A)P(A) + P(B|A')P(A')}$$

This result is referred to as Bayes' theorem.

BAYES' THEOREM: BRIEF FORM

Let A and B be events. Then

$$P(A|B) = \frac{P(B|A)P(A)}{P(B|A)P(A) + P(B|A')P(A')}.$$

We can further understand Bayes' theorem by looking again at the corresponding probability tree (Figure 8.8).

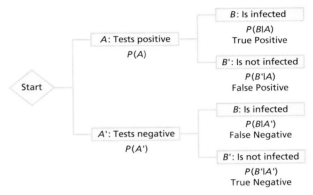

FIGURE 8.8

Bayes' theorem may be rewritten in terms of the probability tree.

BAYES' THEOREM: FROM A PROBABILITY TREE

Let A and B be events. Then

$$P(A|B) = \frac{\text{product of branch probabilities leading to } B \text{ through } A}{\text{sum of all branch probabilities leading to } B}$$

The numerator and denominator of the expression on the right-hand side of the equation in *Bayes' Theorem: From a Probability Tree* are equivalent to the numerator and denominator, respectively, of the expression on the right-hand side of the equation in Bayes' Theorem: Brief Form. That is, product of branch probabilities leading to B through $A = P(B|A)P(A)$ and sum of all branch probabilities leading to $B = P(B|A)P(A) + P(B|A')P(A')$. In Example 3, we investigate the scenario from Example 2 using the probability tree approach.

EXAMPLE 3

Applying Bayes' Theorem from a Probability Tree

Between December 30, 2003, and January 2, 2004, the author received 49 e-mails. His Internet security software tagged 13 e-mails as possible spam. All of the tagged e-mails were in fact spam. There were 25 e-mails that were not spam. What is the probability that a spam e-mail was tagged?

SOLUTION We use the given data to construct a probability tree (Figure 8.9).

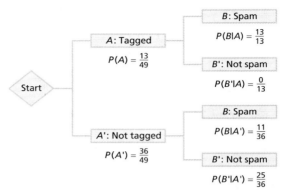

FIGURE 8.9

The probability that a spam e-mail was tagged is $P(A|B)$.

$$P(A|B) = \frac{\text{product of branch probabilities leading to } B \text{ through } A}{\text{sum of all branch probabilities leading to } B}$$

$$= \frac{\left(\dfrac{13}{49}\right)\left(\dfrac{13}{13}\right)}{\left(\dfrac{13}{49}\right)\left(\dfrac{13}{13}\right) + \left(\dfrac{36}{49}\right)\left(\dfrac{11}{36}\right)}$$

$$= \frac{\left(\dfrac{13}{49}\right)}{\left(\dfrac{13}{49}\right) + \left(\dfrac{11}{49}\right)} = \frac{\left(\dfrac{13}{49}\right)}{\left(\dfrac{24}{49}\right)}$$

$$= \frac{13}{24}$$

$$= 0.5417$$

There is a 54.17 percent probability that a spam e-mail was tagged.

EXAMPLE 4

Applying Bayes' Theorem in E-mail Filtering

Over a five-day period, a business's e-mail server received 2500 e-mails. Of these, 1100 were classified as spam and deleted before being delivered. Employees reported that 45 e-mails that were delivered were spam. Employees also reported that 10 legitimate e-mails that had been sent to the company were not delivered. (An IT investigation revealed that all 10 e-mails had been classified as spam when they reached the e-mail server.) What is the probability that a delivered e-mail was spam, and what is the probability that a legitimate e-mail was blocked?

SOLUTION   We define the following sets:

$$B = \{x\,|\,x \text{ is a blocked e-mail}\}$$
$$B' = \{x\,|\,x \text{ is a delivered e-mail}\}$$
$$S = \{x\,|\,x \text{ is spam}\}$$
$$S' = \{x\,|\,x \text{ is not spam}\}$$

We must determine $P(S|B')$ and $P(B|S')$. We have the results given in Table 8.7.

TABLE 8.7

Filter Result	Actual Status		Total
	Spam	Not Spam	
Blocked	1,100	10	1,110 $n(B)$
Delivered	45	1,345	1,390 $n(B')$
Total	1,145 $n(S)$	1,355 $n(S')$	2,500

Using this information, we can generate a Venn diagram showing the cardinality of each subset (Figure 8.10).

Observe that

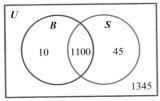

FIGURE 8.10

$$P(S) = \frac{1145}{2500}$$
$$= 0.458$$
$$= 45.8\%$$

$$P(S') = \frac{1355}{2500}$$
$$= 0.542$$
$$= 54.2\%$$

$$P(B|S) = \frac{1100}{1145} \quad \text{Sensitivity}$$
$$= 0.961$$
$$= 96.1\%$$

$$P(B|S') = \frac{10}{1355}$$
$$= 0.007$$
$$= 0.7\%$$

$$P(B'|S) = \frac{45}{1145}$$
$$= 0.039$$
$$= 3.9\%$$

$$P(B'|S') = \frac{1345}{1355} \quad \text{Specificity}$$
$$= 0.993$$
$$= 99.3\%$$

We have $P(B|S') = 0.7\%$, so there is a 0.7 percent probability that the filter will block a legitimate e-mail. To find $P(S|B')$, we will use Bayes' theorem.

From Bayes' theorem, we know that

$$P(S|B') = \frac{P(B'|S)P(S)}{P(B'|S)P(S) + P(B'|S')P(S')}$$

$$= \frac{\left(\dfrac{45}{1145}\right)\left(\dfrac{1145}{2500}\right)}{\left(\dfrac{45}{1145}\right)\left(\dfrac{1145}{2500}\right) + \left(\dfrac{1345}{1355}\right)\left(\dfrac{1355}{2500}\right)}$$

$$= \frac{\dfrac{45}{2500}}{\dfrac{45 + 1345}{2500}} = \frac{45}{1390}$$

$$= 0.032$$

$$= 3.2\%$$

There is a 3.2 percent probability that a delivered e-mail was spam.

Although it would have been quicker to calculate $P(S|B')$ from the Venn diagram in Example 4, we used Bayes' theorem to illustrate the relationship between sensitivity, specificity, and prevalence as related to the conditional probability.

$$P(S|B') = \frac{P(B'|S)P(S)}{P(B'|S)P(S) + P(B'|S')P(S')}$$

$$= \frac{(1 - \text{sensitivity})(\text{prevalence})}{(1 - \text{sensitivity})(\text{prevalence}) + (\text{specificity})(1 - \text{prevalence})}$$

In fact, we can calculate any of the conditional probabilities simply by knowing the sensitivity, specificity, and prevalence of a test. Since we are often most interested in knowing the predictive-value positive and the predictive-value negative of a test, we will use Bayes' theorem to rewrite these conditional probabilities in terms of sensitivity, specificity, and prevalence.

Let $A = \{x|x$ yields a positive test$\}$ and $B = \{x|x$ is infected$\}$. Then sensitivity $= P(A|B)$, specificity $= P(A'|B')$, prevalence $= P(B)$, $PVP = P(B|A)$, and $PVN = P(B'|A')$. Applying Bayes' theorem,

$$PVP = P(B|A)$$

$$= \frac{P(A|B)P(B)}{P(A|B)P(B) + P(A|B')P(B')} \qquad \text{Bayes' theorem}$$

$$= \frac{P(A|B)P(B)}{P(A|B)P(B) + [1 - P(A'|B')]P(B')} \qquad \text{Since } P(A|B') + P(A'|B') = 1$$

$$= \frac{P(A|B)P(B)}{P(A|B)P(B) + [1 - P(A'|B')][1 - P(B)]} \qquad \text{Since } P(B) + P(B') = 1$$

$$= \frac{(\text{sensitivity})(\text{prevalence})}{(\text{sensitivity})(\text{prevalence}) + (1 - \text{specificity})(1 - \text{prevalence})}$$

Similarly,

$$PVN = P(B'|A')$$

$$= \frac{P(A'|B')P(B')}{P(A'|B')P(B') + P(A'|B)P(B)} \qquad \text{Bayes' theorem}$$

$$= \frac{P(A'|B')[1 - P(B)]}{P(A'|B')[1 - P(B)] + P(A'|B)P(B)} \qquad \text{Since } P(B) + P(B') = 1$$

$$= \frac{P(A'|B')[1 - P(B)]}{P(A'|B')[1 - P(B)] + [1 - P(A|B)]P(B)} \qquad \text{Since } P(A|B) + P(A'|B') = 1$$

$$= \frac{(\text{specificity})(1 - \text{prevalence})}{(\text{specificity})(1 - \text{prevalence}) + (1 - \text{sensitivity})(\text{prevalence})}$$

Since many medical tests report their effectiveness in terms of sensitivity and specificity, we will calculate the predictive-value positive and predictive-value negative using these terms in the next few examples.

PREDICTIVE-VALUE POSITIVE AND PREDICTIVE-VALUE NEGATIVE

Let $A = \{x|x \text{ yields a positive test}\}$ and $B = \{x|x \text{ is infected}\}$. Then sensitivity $= P(A|B)$, specificity $= P(A'|B')$, and prevalence $= P(B)$. The predictive-value positive and predictive-value negative of the test are given by

$$PVP = \frac{(\text{sensitivity})(\text{prevalence})}{(\text{sensitivity})(\text{prevalence}) + (1 - \text{specificity})(1 - \text{prevalence})}$$

and

$$PVN = \frac{(\text{specificity})(1 - \text{prevalence})}{(\text{specificity})(1 - \text{prevalence}) + (1 - \text{sensitivity})(\text{prevalence})}$$

EXAMPLE 5

Applying Bayes' Theorem in Cancer Detection

Prostate cancer is the second leading cause of cancer death among American men. The incidence of prostate cancer increases substantially for men over 50 years old. According to the National Cancer Institute's December 2003 *Screening for Prostate Cancer* report, the age-adjusted incidence for African American men is 275.3 per 100,000, and that for white males is 172.9 per 100,000. (**Source:** www.cancer.gov.)

One common test to detect prostate cancer is the prostate-specific antigen (PSA) test. Many professionals consider a PSA value of 4.0 to be positive for prostate cancer. Although the sensitivity and specificity of the test vary depending upon the age of the individual and the type of cancer, medical experts estimate that the sensitivity for detection of cancer occurring within the first four years after the test is 73 percent. The overall specificity is estimated to be 91 percent. (**Source:** www.ncbi.nlm.nih.gov.) Determine the *PVP* and *PVN* for African American and Caucasian men and interpret the meaning of the result.

SOLUTION The prevalence for African American men is 0.2753 percent and that for Caucasian men is 0.1729 percent. The sensitivity of the test is 73 percent, and the specificity is 91 percent.

African American Men

$$
\begin{aligned}
PVP &= \frac{(\text{sensitivity})(\text{prevalence})}{(\text{sensitivity})(\text{prevalence}) + (1 - \text{specificity})(1 - \text{prevalence})} \\
&= \frac{(0.73)(0.002753)}{(0.73)(0.002753) + (1 - 0.91)(1 - 0.002753)} \\
&= 0.02190 \\
&= 2.19\%
\end{aligned}
$$

Of the African American men with a PSA value of 4.0 or more, 2.19 percent will develop prostate cancer within 4 years of the test.

$$
\begin{aligned}
PVN &= \frac{(\text{specificity})(1 - \text{prevalence})}{(\text{specificity})(1 - \text{prevalence}) + (1 - \text{sensitivity})(\text{prevalence})} \\
&= \frac{(0.91)(1 - 0.002753)}{(0.91)(1 - 0.002753) + (1 - 0.73)(0.002753)} \\
&= 0.9992 \\
&= 99.92\%
\end{aligned}
$$

Of the African American men with a PSA value less than 4.0, 99.92 percent won't develop prostate cancer within 4 years of the test.

Caucasian Men

$$PVP = \frac{(\text{sensitivity})(\text{prevalence})}{(\text{sensitivity})(\text{prevalence}) + (1 - \text{specificity})(1 - \text{prevalence})}$$

$$= \frac{(0.73)(0.001729)}{(0.73)(0.001729) + (1 - 0.91)(1 - 0.001729)}$$

$$= 0.01385$$

$$= 1.39\%$$

Of the Caucasian men with a PSA value of 4.0 or more, 1.39 percent will develop prostate cancer within 4 years of the test.

$$PVN = \frac{(\text{specificity})(1 - \text{prevalence})}{(\text{specificity})(1 - \text{prevalence}) + (1 - \text{sensitivity})(\text{prevalence})}$$

$$= \frac{(0.91)(1 - 0.001729)}{(0.91)(1 - 0.001729) + (1 - 0.73)(0.001729)}$$

$$= 0.9995$$

$$= 99.95\%$$

Of the Caucasian men with a PSA value less than 4.0, 99.95 percent won't develop prostate cancer within 4 years of the test.

EXAMPLE **6**

Applying Bayes' Theorem in Drug Testing

Screening for drug use by athletes has become a major issue in sports over the past several years. Laboratory urinalysis is the primary screening test used to detect drug use. Although the test is considered highly effective, the invasive nature of the urine collection procedure has concerned some privacy advocates. An alternative procedure, oral fluid analysis, shows promise as a less invasive yet highly effective drug screening test.

In a 2002 study of adult arrestees, it was estimated that the Intercept Oral Specimen Collection Device (IOSCD) was 95 percent sensitive and 98 percent specific for cocaine. (**Source:** www.criminology.fsu.edu.) In analyzing the effectiveness of the oral fluid collection method, it was assumed that the laboratory urinalysis approach accurately detected the use or nonuse of drugs.

Suppose that 8 percent of the athletes on a football team use cocaine. An athlete who tested positive for cocaine on the IOSCD test claims that he is not a drug user. If you were a lawyer assigned to represent the athlete, what analytical argument could you use to persuade officials that the athlete is telling the truth?

SOLUTION Since the athlete tested positive but claims not to be a drug user, we must determine the predictive-value positive.

$$PVP = \frac{(\text{sensitivity})(\text{prevalence})}{(\text{sensitivity})(\text{prevalence}) + (1 - \text{specificity})(1 - \text{prevalence})}$$

$$= \frac{(0.95)(0.08)}{(0.95)(0.08) + (1 - 0.98)(1 - 0.08)}$$

$$= 0.8051$$

$$= 80.51\%$$

We could argue that although the test theoretically correctly identifies 95 percent of the cocaine users and 98 percent of non-cocaine users, the probability that our client is a user is about 81 percent.

We could further argue that the sensitivity and specificity estimates are imprecise. We could cite additional studies that showed different sensitivities and specificities. For example, a separate study showed that the IOSCD test had an 82 percent sensitivity and a 96 percent specificity for cocaine. (**Source:** www.criminology.fsu.edu.) Using these values, the *PVP* is substantially lower.

$$PVP = \frac{(\text{sensitivity})(\text{prevalence})}{(\text{sensitivity})(\text{prevalence}) + (1 - \text{specificity})(1 - \text{prevalence})}$$

$$= \frac{(0.82)(0.08)}{(0.82)(0.08) + (1 - 0.96)(1 - 0.08)}$$

$$= 0.6406$$

$$= 64.06\%$$

Using the results from the second study, the probability that our client is a cocaine user is about 64 percent. The varying probabilities call into question the reliability of the test results. Although the analytical argument may add strength to the athlete's case, other factors, such as the testimony of witnesses, the athlete's typical behavior, and so on, should be considered as well.

In the preceding examples, we used the brief form of Bayes' theorem. An expanded form of the theorem is more complex yet highly useful. Recall that the denominator of the rational expression in Bayes' theorem was calculated as follows:

$$P(B) = P((B \cap A) \cup (B \cap A'))$$

$$= P(B \cap A) + P(B \cap A') \text{ since } (B \cap A) \cap (B \cap A') = \varnothing \qquad \text{Probability of a union}$$

$$= P(A)P(B|A) + P(A')P(B|A') \qquad\qquad\qquad \text{Product rule of probability}$$

In the second step, we made the observation that the sets $(B \cap A)$ and $(B \cap A')$ were disjoint, yet their union made up the entirety of B (Figure 8.11).

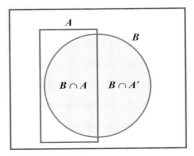

FIGURE 8.11

Although A and A' split B into only two regions, we may divide B into as many regions as we like. The division of a sample space into a group of disjoint sets whose union is the entire sample space is referred to as a **partition.** Figure 8.12 shows a partition of B into three disjoint sets.

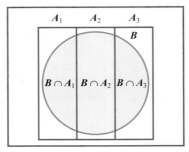

FIGURE 8.12

For the three-set partition of B, observe that $B = (B \cap A_1) \cup (B \cap A_2) \cup (B \cap A_3)$ and

$$
\begin{aligned}
P(B) &= P((B \cap A_1) \cup (B \cap A_2) \cup (B \cap A_3)) \\
&= P(B \cap A_1) + P(B \cap A_2) + P(B \cap A_3) && \text{Since } A_1, A_2, \text{ and } A_3 \text{ are disjoint} \\
&= P(B|A_1)P(A_1) + P(B|A_2)P(A_2) + P(B|A_3)P(A_3) && \text{Product rule of probability}
\end{aligned}
$$

The result leads us to the extended form of Bayes' theorem.

BAYES' THEOREM: EXTENDED FORM

Let events A_1, A_2, and A_3 form a partition of sample space B. Then

$$
P(A_1|B) = \frac{P(B|A_1)P(A_1)}{P(B|A_1)P(A_1) + P(B|A_2)P(A_2) + P(B|A_3)P(A_3)}
$$

A similar form would apply for subspaces with four or more partitions.

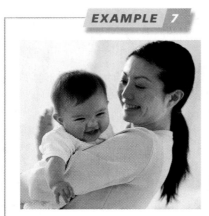

EXAMPLE 7

Applying Bayes' Theorem in Social Analysis

In 1999, 33.0 percent of live births in the United States were to unmarried mothers. Of mothers racially classified as Asian, 15.4 percent were unmarried. Of mothers racially classified as black, 68.9 percent were unmarried. Of mothers classified as other races, 27.2 percent were unmarried. Asian women made up 4.6 percent of the total number of women who gave birth in 1999, black women made up 15.3 percent, and women of other races made up 80.1 percent. (**Source:** *Statistical Abstract of the United States,* 2001, Table 827, p. 537.) What is the probability that an unmarried woman who gave birth in 1999 was Asian?

SOLUTION Define events $A_1 = \{x|x$ is a black woman$\}$, $A_2 = \{x|x$ is a black woman$\}$, $A_3 = \{x|x$ is a woman of another race$\}$, and $B = \{x|x$ is an unmarried mother$\}$. The events A_1, A_2, and A_3 form a partition of the racial classification. We know that $P(B) = 0.330$, $P(A_1) = 0.046$, $P(A_2) = 0.153$, $P(A_3) = 0.801$, $P(B|A_1) = 0.154$, $P(B|A_2) = 0.689$, and $P(B|A_3) = 0.272$. We are asked to find $P(A_1|B)$.

$$P(A_1|B) = \frac{P(B|A_1)P(A_1)}{P(B|A_1)P(A_1) + P(B|A_2)P(A_2) + P(B|A_3)P(A_3)}$$

$$= \frac{(0.154)(0.046)}{(0.154)(0.046) + (0.689)(0.153) + (0.272)(0.801)}$$

$$= 0.021$$

$$= 2.1\%$$

There is a 2.1 percent probability that a randomly selected unmarried woman who gave birth in 1999 was Asian.

Although we correctly applied Bayes' theorem in Example 7, we could have generated the result more quickly. Recall that the denominator of the expression is equal to $P(B)$. Since we knew that $P(B) = 0.33$, we could have calculated $P(A_1|B)$ as follows:

$$P(A_1|B) = \frac{P(B|A_1)P(A_1)}{P(B|A_1)P(A_1) + P(B|A_2)P(A_2) + P(B|A_3)P(A_3)}$$

$$= \frac{P(B|A_1)P(A_1)}{P(B)} = \frac{(0.154)(0.046)}{0.330}$$

$$= 0.021$$

$$= 2.1\%$$

Before diving into a problem with Bayes' theorem, it is wise to think through which approach will be most efficient. In the case of Example 7, simply applying the definition of conditional probability led us to our result more quickly. Bayes' theorem is most useful when we have incomplete information regarding the various probabilities.

EXAMPLE 8

Applying Bayes' Theorem in Educational Analysis

In an effort to assess the effectiveness of an intermediate algebra course in preparing students for college-level mathematics (precalculus, finite math, and contemporary math), the author collected the following statistics for Green River Community College. The statistics are based upon the number of students who passed the intermediate algebra course and enrolled in a college-level math course. Students are required to earn a grade of 2.0 or better in the intermediate algebra course, score well on a placement test, or obtain a waiver from a math instructor before they may enroll in the college-level math course. We restrict our analysis to students who took both intermediate algebra and a college level math course.

Of the students who earned a 2.0 or better in the intermediate algebra course on the first attempt, 80.8 percent earned a 2.0 or better in the college-level math course. Of the students who earned a 2.0 or better in intermediate algebra on the second attempt, 69.6 percent earned a 2.0 or better in the college-level math course. Of the students enrolled in the college-level math course, 96.1 percent earned a 2.0 or better in algebra on the first attempt and 3.9 percent earned a 2.0 or better in algebra on the second attempt. Further analyze the data and prepare a report for the Math Department.

SOLUTION We first acknowledge that the data won't give us a complete picture, since the sample does not include the students who took the intermediate algebra course but did not complete a college-level math course. However, our analysis will help us determine if the required 2.0 grade in intermediate algebra is a good predictor of performance in the college-level math course. We define events $A_1 = \{x \,|\, x$ is a student who earned 2.0 or better in algebra on the first attempt$\}$, $A_2 = \{x \,|\, x$ is a student who earned 2.0 or better in algebra on the second attempt$\}$, and $B = \{x \,|\, x$ is a student who earned 2.0 or better in a college-level math course$\}$. The events A_1 and A_2 form a partition of the population based on the number of times the student had to take the algebra course. We know that $P(B|A_1) = 0.808$, $P(B|A_2) = 0.696$, $P(A_1) = 0.961$, and $P(A_2) = 0.039$. We want to find answers to the following questions:

1. What is the probability that a student who earned a grade of 2.0 or better in a college-level math course earned a 2.0 or better in algebra on the first attempt? [Find $P(A_1|B)$.]

$$P(A_1|B) = \frac{P(B|A_1)P(A_1)}{P(B|A_1)P(A_1) + P(B|A_2)P(A_2)}$$

$$= \frac{(0.808)(0.961)}{(0.808)(0.961) + (0.696)(0.039)}$$

$$= 0.966$$

$$= 96.6\%$$

There is a 96.6 percent probability that a student who earned a grade of 2.0 or better in a college-level math course earned a 2.0 or better in intermediate algebra on the first attempt.

2. What is the probability that a student who earned a grade of 2.0 or better in a college-level math course earned a 2.0 or better in algebra on the second attempt? [Find $P(A_2|B)$.]

$$P(A_2|B) = \frac{P(B|A_2)P(A_2)}{P(B|A_2)P(A_2) + P(B|A_1)P(A_1)}$$

$$= \frac{(0.696)(0.039)}{(0.696)(0.039) + (0.808)(0.961)}$$

$$= 0.034$$

$$= 3.4\%$$

There is a 3.4 percent probability that a student who earned a grade of 2.0 or better in the college-level math course had to take intermediate algebra twice before obtaining a 2.0 in the algebra course.

The majority of students (96.6 percent) who did well in a college-level math course took the prerequisite intermediate algebra course only once. A small minority (3.4 percent) of those who did well in a college-level math course were required to take the intermediate algebra course twice. Students who had difficulty passing intermediate algebra on the first attempt did not do as well in the college-level math course. Our results indicate that 80.8 percent of the one-time algebra students passed the college-level math course, while only 69.6 percent of the repeating algebra students passed the college-level math course. These results could be shared with students during advising sessions to help them recognize the importance of putting forth their best effort in the intermediate algebra class the first time they enroll.

8.2 Summary

In this section, you learned how to apply Bayes' theorem in a variety of real-world applications ranging from drug test analysis to spam detection. You also discovered that Bayes' theorem is needed only when we have incomplete statistical information.

8.2 Exercises

In Exercises 1–10, calculate the prevalence, sensitivity, specificity, predictive-value positive, and predictive-value negative from the table. Then interpret the meaning of the results.

1.

Test Result	Actual Status	
	Disease Present	Disease Absent
Positive	125	18
Negative	20	822

2.

Test Result	Actual Status	
	Disease Present	Disease Absent
Positive	1,023	12
Negative	52	9,848

3.

Test Result	Actual Status	
	Spam	Not Spam
Positive	164	50
Negative	26	750

4.

Test Result	Actual Status	
	Spam	Not Spam
Positive	250	10
Negative	25	1,000

5.

	Will Pass Class	Will Fail Class
Studies	20	3
Doesn't Study	4	6

6.

	Will Pass Class	Will Fail Class
Studies	36	4
Doesn't Study	6	12

7.

	No Repairs Needed	Repairs Needed
Newer Home	56	14
Older Home	62	246

8.

	Has Children	Doesn't Have Children
Married	259	198
Single	123	326

9.

	Gets a Speeding Ticket	Doesn't Get a Speeding Ticket
Drives a Red Car	26	64
Drives a Non-red Car	39	189

10.

	Has Tongue Pierced	Doesn't Have Tongue Pierced
Has Tattoo	3	4
Doesn't Have Tattoo	5	22

In Exercises 11–20, complete the table by applying Bayes' theorem.

11.

	Percent
Prevalence	0.2
Sensitivity	98
Specificity	76
Predictive-value positive	
Predictive-value negative	

12.

	Percent
Prevalence	1
Sensitivity	87
Specificity	99
Predictive-value positive	
Predictive-value negative	

13.

	Percent
Prevalence	20
Sensitivity	50
Specificity	75
Predictive-value positive	
Predictive-value negative	

14.

	Percent
Prevalence	1.2
Sensitivity	88
Specificity	100
Predictive-value positive	
Predictive-value negative	

15.

	Percent
Prevalence	5
Sensitivity	85
Specificity	
Predictive-value positive	13
Predictive-value negative	

16.

	Percent
Prevalence	2
Sensitivity	60
Specificity	
Predictive-value positive	38
Predictive-value negative	

17.

	Percent
Prevalence	2
Sensitivity	92
Specificity	
Predictive-value positive	19
Predictive-value negative	

18.

	Percent
Prevalence	20
Sensitivity	
Specificity	62
Predictive-value positive	
Predictive-value negative	95

19.

	Percent
Prevalence	10
Sensitivity	
Specificity	98
Predictive-value positive	
Predictive-value negative	99

20.

	Percent
Prevalence	1
Sensitivity	
Specificity	50
Predictive-value positive	
Predictive-value negative	99

In Exercises 21–35, apply Bayes' theorem as appropriate in analyzing the real-world data.

21. **Skin Cancer Screening**   A common screening test for skin cancer is the total cutaneous examination (TCE). One study cited in the American College of Preventative Medicine policy statement on skin cancer reports that the TCE has a 93.3 percent sensitivity, 97.8 percent specificity, predictive-value positive of 54 percent, and predictive-value negative of 99.8 percent when the examination is conducted by dermatologists. (**Source:** www.acpm.org.)

Interpret the meaning of sensitivity, specificity, predictive-value positive, and predictive-value negative in the context of this study.

22. **Breast Cancer Screening**   The American Cancer Society and others recommend that women age 50 and older receive an annual mammogram, a breast cancer screening test. According to the American College of Preventative Medicine policy statement on breast cancer screening, the sensitivity of the mammogram is estimated to be 75 to 90 percent, and the specificity is estimated to be 90 to 95 percent. The predictive-value positive for women age 50 to 69 is estimated to be 60 to 80 percent. (**Source:** www.acpm.org.)

Explain how you could use these results when visiting a friend or relative who tested positive for breast cancer during a recent mammogram.

23. **Disease Detection**   Invasive aspergillosis commonly occurs in people whose immune systems are compromised by illness or chemotherapy. In 2003, the Food and Drug Administration approved the Platelia Aspergillus EIA test as a disease detection tool. Based on clinical studies, the sensitivity of the test was estimated to be 81 percent and the specificity was estimated to be 89 percent. (**Source:** www.accessdata.fda.gov.) Since only a few thousand cases of the disease occur in the United States each year, the prevalence of the disease is estimated to be 0.002 percent.

Calculate and interpret the meaning of the predictive-value positive and the predictive-value negative for the Platelia Aspergillus EIA test.

24. **Birth Defect Screening**   The incidence of Down's syndrome increases as a woman ages. When the mother is 35, roughly 1 out of 375 children born will have Down's syndrome. By age 45, the incidence increases to 1 out of 30 births. One screening test for Down's syndrome is a measurement of maternal serum alpha-fetoprotein (MSAFP). The MSAFP test has an estimated sensitivity of 25 percent and a specificity of about 90 percent. (**Source:** www.singhospi.com.)

Calculate the predictive-value positive and the predictive-value negative for MSAFP test for a 35-year-old and a 45-year-old woman. What conclusions can you draw from the results?

25. **Weather Forecasting**   A meteorologist tracks the accuracy of her daily forecasts. She determines that if she forecasts rain, her prediction is accurate 65 percent of the time. If she forecasts no rain, her prediction is accurate 80 percent of the time. She typically forecasts rain

73 days out of a year. Today it is raining. What is the probability that her forecast for today's weather was accurate?

26. Weather Forecasting A meteorologist tracks the accuracy of his daily forecasts. He determines that if he forecasts rain, his prediction is accurate 69 percent of the time. If he forecasts no rain, his prediction is accurate 78 percent of the time. He typically forecasts rain 30 days out of a year. Today it is not raining. What is the probability that his forecast for today's weather was not accurate?

27. Pregnancy Test Signify® Card Pregnancy Test is a test used to detect human chorionic gonadotropin, an early indicator of pregnancy. The test boasts a 99.1 percent sensitivity and a 100 percent specificity. (**Source:** www.abbottdiagnostics.com.)

Out of a group of 100 women undergoing treatment for infertility, 4 people test positive for pregnancy. What is the probability that a woman with a positive test is actually pregnant? What is the probability that a woman with a negative test is actually pregnant?

28. Out-of-Wedlock Births In 1999, the percentage of live births to unmarried mothers in the United States reached a new high. Of mothers racially classified as Asian, 15.4 percent were unmarried. Of mothers racially classified as black, 68.9 percent were unmarried. Of mothers classified as other races, 27.2 percent were unmarried. Asian women made up 4.6 percent of the total number of women who gave birth in 1999, black women made up 15.3 percent, and women of other races made of 80.1 percent. (**Source:** *Statistical Abstract of the United States, 2001*, Table 827, p 537.) What is the probability that an unmarried woman who gave birth in 1999 was black?

29. Law School Admission A user-compiled online database of law school admission statistics seeks to help law school applicants determine their likelihood of acceptance. As of January 12, 2004, fourteen of the applicants to the law school at University of California-Berkeley had received word of their admission status. Of those with a Law School Data Assembly Service (LSDAS) GPA of 3.8 or higher, 90.9 percent were admitted to the school. Of those with an LSDAS GPA of less than 3.8, 33.3 percent were rejected. Eleven of the fourteen students had an LSDAS GPA of 3.8 or higher. (**Source:** www.lawschoolnumbers.com.)

What is the probability that an applicant who was rejected had an LSDAS GPA of 3.8 or above? What is the probability that an applicant who was accepted had an LSDAS GPA below 3.8?

30. Law School Admission A user-compiled online database of law school admission statistics seeks to help law school applicants determine their likelihood of acceptance. As of January 12, 2004, twelve of the applicants to the law school at the University of Notre Dame had received word of their admission status. Of those with an LSAT score of 163 or higher, 87.5 percent were admitted to the school. Of those with an LSAT score of less than 163, 75 percent were rejected. Eight of the twelve students scored 163 or higher on the LSAT. (**Source:** www.lawschoolnumbers.com.)

What is the probability that an applicant who was rejected had a LSAT score of 163 or higher? What is the probability that an applicant who was accepted scored below 163?

31. Law School Admission A user-compiled online database of law school admission statistics seeks to help law school applicants determine their likelihood of acceptance. As of January 12, 2004, 35 of the applicants to the law school at the University of Virginia had received word of their admission status: 77.1 percent were accepted, 11.4 percent were wait-listed, and 11.4 percent were rejected. (The percentages add up to 99.9 percent because of rounding.) Of those accepted, 100 percent had an LSAT score of 163 or higher. Of those wait-listed, 50 percent had an LSAT score of 163 or higher. Of those rejected, 0 percent had an LSAT score of 163 or higher. (**Source:** www.lawschoolnumbers.com.)

What is the probability that a student who had an LSAT score of 163 or higher was wait-listed?

32. Mad Cow Disease In December 2003, the first U.S. case of mad cow disease was discovered in Mabton, Washington. The international reaction was swift and decisive. More than 30 countries placed an immediate ban on beef from the United States.

The Prionics®-Check LIA is a test used to detect bovine spongiform encephalopathy (mad cow disease). Based on a study of 1400 cattle, the sensitivity and specificity of the test were both determined to be 100 percent. (**Source:** www.prionics.ch.)

How confident should consumers be that meat determined safe by the Prionics-Check LIA test is actually safe to eat? Explain.

33. **Ethnicity and Graduation Rates**   The author determined the percentage of students taking courses at Green River Community College in 2000–2001 who graduated with an associate's degree by winter 2004. Treating the percentages as probabilities, he determined the following:

There was a 27.1 percent probability that a non-Latino/non-Hispanic student would earn a degree. There was a 17.3 percent probability that a Latino/Hispanic student would earn a degree. Latino/ Hispanic students made up 5.0 percent of the student population. According to this analysis, what is the probability that a student who earned a degree was Latino/Hispanic?

34. **Strep Throat Test**   Signify Strep A is a test used for the detection of group A streptococcal antigens from a throat swab. The test has sensitivity of 96 percent and a specificity of 97.8 percent. (**Source:** www.abbottdiagnostics.com.) The incidence of strep throat is estimated to be 5 out of 1000 people. (**Source:** www.healthcentral.com.)

What is the probability that someone who tests negative for strep throat actually has the illness? What is the probability that someone who tests positive for strep throat doesn't have the illness?

35. **Infectious Mononucleosis Test**
Signify Mono is a test used to detect infectious mononucleosis. The test boasts a 100 percent sensitivity and a 90.3 percent specificity. (**Source:** www.abbottdiagnostics.com.) The incidence of mononucleosis is estimated to be 8 out of 100,000 people. (**Source:** www.healthcentral.com.)

What is the probability that someone who tests negative for mono actually has the illness? What is the probability that someone who tests positive for mono doesn't have the illness?

Exercises 36–40 are intended to challenge your understanding of Bayes' theorem and its applications.

36. Assuming that the sensitivity and specificity of a disease screening test remain constant, what effect does an increase in the prevalence of the disease have on the predictive-value positive and predictive-value negative?

37. Let $A = \{x | x \text{ yields a positive test}\}$ and $B = \{x | x \text{ is infected}\}$. Then sensitivity $= P(A|B)$, specificity $= P(A'|B')$, and prevalence $= P(B)$. Write $P(A'|B)$ in terms of sensitivity, specificity, and/or prevalence.

38. **Drug Screening Test**   A drug-use screening test is given to a group of 1000 drug users. The test indicates that all members of the group are drug users. As a result, the manufacturer claims that the screening test is 100 percent accurate. Do you agree with the manufacturer? Explain.

39. The sensitivity and specificity of a test are often estimated. What factors influence the accuracy of these estimates?

40. Let $B = \{x | x \text{ is a person living in Maine}\}$, $A_1 = \{x | x \text{ is a person with blue eyes}\}$, $A_2 = \{x | x \text{ is a person with brown eyes}\}$, and $A_3 = \{x | x \text{ is a person with green eyes}\}$. A student claims that $P(B) = P(B|A_1)P(A_1) + P(B|A_2)P(A_2) + P(B|A_3)P(A_3)$. Do you agree? Justify your answer.

8.3 Markov Chains

- Forecast future events with Markov chains

GETTING STARTED   Business-savvy housing managers often analyze rental data to forecast occupancy rates in future years. A review of an apartment complex's lease records revealed that 85 percent of the 100 apartments vacated in 2005 had new occupants in 2006. Furthermore, 65 percent of the 400 apartments occupied in 2005 had the same occupants in 2006. (All occupants sign a one-year lease.)

Based on these results, the manager hypothesizes that the probability that a vacant apartment will be filled in the following year is 85 percent and the probability that a filled apartment will remain filled the following year is 65 percent. The housing manager can forecast the number of filled and vacant apartments in future years by using Markov chains.

In this section, we will demonstrate how to use Markov chains to address business concerns. Markov chains are frequently used to forecast consumer behavior. Markov chains are named for the Russian mathematician A. A. Markov, who developed many of the techniques covered in this section.

We summarize the apartment manager's hypothesis given at the beginning of this section in Table 8.8.

TABLE 8.8

		2006	
		Filled	Vacant
2005	Filled	0.65	0.35
	Vacant	0.85	0.15

Every apartment at the complex may be classified as either *filled* or *vacant*. These classifications are referred to as **states.** We are interested in the probability that an apartment *makes the transition* (moves) from one state to another. That is, what is the probability that a vacant apartment in 2005 will be filled in 2006? What is the probability that a filled apartment in 2005 will be vacant in 2006?

A **stochastic process** is a collection of possible states together with the transition probabilities between them. A square matrix with nonnegative entries is called a **stochastic matrix** or **transition matrix** if the sum of the entries in each row is 1. The probability table in Table 8.8 may be rewritten as a transition matrix.

$$\overbrace{\begin{matrix} S_1 & S_2 \end{matrix}}^{\text{Columns represent next state}}$$

Rows represent current state $\quad \begin{matrix} S_1 \\ S_2 \end{matrix}\begin{bmatrix} 0.65 & 0.35 \\ 0.85 & 0.15 \end{bmatrix} = P$

In this matrix, state 1 = filled and state 2 = vacant. The matrix entry p_{ij} = the probability of moving from state i to state j. p_{ij} is called the **transition probability.** For example, $p_{12} = 0.35$ means that the probability that a filled apartment is vacant a year later is 35 percent. Similarly, $p_{22} = 0.15$ means that the probability that a vacant apartment remains vacant a year later is 15 percent.

MARKOV CHAIN

A finite Markov chain is a set of states $S_1, S_2, \ldots, S_n$ together with an $n \times n$ transition matrix P such that the entry p_{ij} of the matrix P is the probability of making the transition from S_i to S_j.

EXAMPLE 1

Interpreting the Meaning of the Entries of a Transition Matrix

Given the transition matrix $P = \begin{bmatrix} 0.25 & 0.45 & 0.30 \\ 0.00 & 1.00 & 0.00 \\ 0.75 & 0.00 & 0.25 \end{bmatrix}$, interpret the meaning of p_{12}, p_{22}, and p_{31}.

SOLUTION Since $p_{12} = 0.45$, the probability of making a transition from state S_1 to state S_2 is 45 percent. Similarly, the probability of moving from state S_2 to state S_2 is 100 percent, since $p_{22} = 1.00$. (That is, we never leave state 2 after we reach it.) Likewise, since $p_{31} = 0.75$, the probability of moving from state S_3 to state S_1 is 75 percent.

An alternative way to represent a transition matrix is through a **state transition diagram.** The diagram shows each of the states together with the probability of moving from one state to another. We have the transition matrix

$$P = \begin{bmatrix} 0.25 & 0.45 & 0.30 \\ 0.00 & 1.00 & 0.00 \\ 0.75 & 0.00 & 0.25 \end{bmatrix}$$

Since P is a 3×3 matrix, the Markov chain has three states: S_1, S_2, and S_3. We draw a circle for each state. For each transition with a nonzero probability p_{ij}, we draw an arrow from state S_i to state S_j (see Figure 8.13).

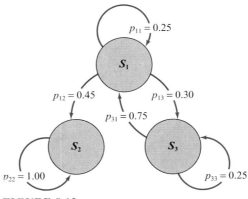

FIGURE 8.13

Observe that the only way to reach state S_2 from state S_3 is by passing through state S_1. Additionally, once state S_2 is reached, there is no leaving. We refer to S_2 as an **absorbing state.**

EXAMPLE 2

Drawing a State Transition Diagram

Draw the state transition diagram that corresponds with the given transition matrix. Identify which states, if any, are absorbing states.

$$P = \begin{bmatrix} 0.1 & 0.2 & 0.7 \\ 0.0 & 0.9 & 0.1 \\ 0.0 & 0.0 & 1.0 \end{bmatrix}$$

SOLUTION   Using the transition matrix, we construct the state transition diagram (Figure 8.14).

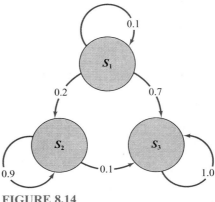

FIGURE 8.14

Observe that there are arrows leaving states S_1 and S_2; however, there are no arrows leaving state S_3. Thus, S_3 is the only absorbing state.

Powers of a Transition Matrix

In forecasting future events, we consider powers of the transition matrix P. Consider the apartment manager's transition matrix introduced earlier, $P = \begin{bmatrix} 0.65 & 0.35 \\ 0.85 & 0.15 \end{bmatrix}$. Since $p_{12} = 0.35$, there is a 35 percent probability of moving from state 1 to state 2 in exactly one step. In the context of the scenario, there is a 35 percent probability that a filled apartment will be vacant a year later. Now consider the matrix P^2.

$$P^2 = P \cdot P$$
$$= \begin{bmatrix} 0.65 & 0.35 \\ 0.85 & 0.15 \end{bmatrix}\begin{bmatrix} 0.65 & 0.35 \\ 0.85 & 0.15 \end{bmatrix}$$
$$= \begin{bmatrix} 0.72 & 0.28 \\ 0.68 & 0.32 \end{bmatrix}$$

What is the meaning of the entry p_{12} for the matrix P^2? This entry represents the probability of making a transition from state 1 to state 2 in exactly *two* steps. In this case, there is a 28 percent probability of making the transition from state 1 to state 2 in exactly *two* steps. In the context of the scenario, there is a 28 percent probability that a filled apartment will be vacant two years later.

What is the meaning of the entry p_{12} for the matrix P^3? This entry represents the probability of moving from state 1 to state 2 in exactly *three* steps.

$$P^3 = P^2 \cdot P$$

$$= \begin{bmatrix} 0.72 & 0.28 \\ 0.68 & 0.32 \end{bmatrix} \begin{bmatrix} 0.65 & 0.35 \\ 0.85 & 0.15 \end{bmatrix}$$

$$= \begin{bmatrix} 0.706 & 0.294 \\ 0.714 & 0.286 \end{bmatrix}$$

In this case, there is a 29.4 percent probability of moving from state 1 to state 2 in exactly *three* steps. In the context of the scenario, there is a 29.4 percent probability that a filled apartment will be vacant three years later. This pattern continues for any whole-number power of P.

POWERS OF A TRANSITION MATRIX

Let P be a transition matrix. Then the ijth entry of the matrix P^k is the probability of a transition from state i to state j in exactly k steps.

EXAMPLE **3**

Forecasting Future Results with a Markov Chain

A business-savvy travel agent conducts an annual review of her financial records to look for trends in client spending patterns. She determines that 65 percent of her clients who booked a trip in 2003 also booked a trip in 2004. Additionally, 40 percent of her clients who did not book a trip in 2003 booked a trip in 2004. Use a Markov chain to forecast the spending behavior of her clients in 2005, in 2007, and in 2009.

SOLUTION We identify the states and write the transition matrix for the Markov chain.

State 1: Books a trip

State 2: Does not book a trip

$$P = \begin{bmatrix} 0.65 & 0.35 \\ 0.40 & 0.60 \end{bmatrix}$$

Since 2005 is two years away from 2003, we square the transition matrix.

$$P^2 = \begin{bmatrix} 0.65 & 0.35 \\ 0.40 & 0.60 \end{bmatrix} \begin{bmatrix} 0.65 & 0.35 \\ 0.40 & 0.60 \end{bmatrix}$$

$$= \begin{bmatrix} 0.56 & 0.44 \\ 0.50 & 0.50 \end{bmatrix}$$

There is a 56 percent probability that a customer who books a trip in 2003 will book a trip in 2005. There is a 50 percent probability that a client who didn't book a trip in 2003 will book a trip in 2005.

Since 2007 is four years away from 2003, we raise the transition matrix to the fourth power.

$$P^4 = \begin{bmatrix} 0.65 & 0.35 \\ 0.40 & 0.60 \end{bmatrix}\begin{bmatrix} 0.65 & 0.35 \\ 0.40 & 0.60 \end{bmatrix}\begin{bmatrix} 0.65 & 0.35 \\ 0.40 & 0.60 \end{bmatrix}\begin{bmatrix} 0.65 & 0.35 \\ 0.40 & 0.60 \end{bmatrix}$$

$$= \begin{bmatrix} 0.54 & 0.46 \\ 0.53 & 0.47 \end{bmatrix}$$

There is a 54 percent probability that a customer who books a trip in 2003 will book a trip in 2007. There is a 53 percent probability that a client who didn't book a trip in 2003 will book a trip in 2007.

Since 2009 is six years away from 2003, we raise the transition matrix to the sixth power.

$$P^6 = \begin{bmatrix} 0.65 & 0.35 \\ 0.40 & 0.60 \end{bmatrix}\begin{bmatrix} 0.65 & 0.35 \\ 0.40 & 0.60 \end{bmatrix}\begin{bmatrix} 0.65 & 0.35 \\ 0.40 & 0.60 \end{bmatrix}\begin{bmatrix} 0.65 & 0.35 \\ 0.40 & 0.60 \end{bmatrix}\begin{bmatrix} 0.65 & 0.35 \\ 0.40 & 0.60 \end{bmatrix}\begin{bmatrix} 0.65 & 0.35 \\ 0.40 & 0.60 \end{bmatrix}$$

$$= \begin{bmatrix} 0.53 & 0.47 \\ 0.53 & 0.47 \end{bmatrix}$$

There is a 53 percent probability that a customer who books a trip in 2003 will book a trip in 2009. There is also a 53 percent probability that a client who didn't book a trip in 2003 will book a trip in 2009. In other words, there is a 53 percent probability that a randomly selected 2003 client will book a trip in 2009.

TECHNOLOGY TIP

Calculating Powers of a Square Matrix

1. Enter square matrix A into the calculator, using the Matrix Editor (see Section 2.2 for detailed steps). Then close the Matrix Editor.

2. Use the Matrix NAMES menu to place matrix A on the home screen.

3. To calculate A^n, press the key sequence ⌐^⌐ ⌐n⌐, where n is the numerical value of the exponent.

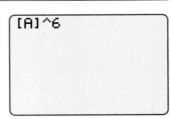

(Continued)

4. Press ENTER to display the result.

```
[A]^6
[[.5334472656  .…
 [.533203125   .…
```

5. To convert the matrix entries to rational numbers, press the key sequence MATH ENTER , ENTER . The calculator will convert the entries to rational numbers, if possible.

```
[A]^6
[[.5334472656  .…
 [.533203125   .…
Ans▶Frac
[[2185/4096  191…
 [273/512    239…
```

EXAMPLE 4

Forecasting Future Results with a Markov Chain

A survey of families belonging to the same church revealed the following. There was a 75 percent probability that children with mothers who were active in the church (attended church services weekly and volunteered service hours to the church) would be active in the church as adults. There was also a 75 percent probability that children with mothers who were less active in the church (did not attend church services weekly or did not volunteer service hours to the church) would be less active in the church as adults. (**Source:** Author's data.) What is the probability that the grandchild of an active mother is active in the church?

SOLUTION We identify the states and write the transition matrix for the Markov chain.

State 1: Active in the church

State 2: Less active in the church

$$P = \begin{bmatrix} 0.75 & 0.25 \\ 0.25 & 0.75 \end{bmatrix}$$

The time interval for this problem is one generation. Since we are interested in the active mother's grandchildren, we need to find P^2. (A grandchild is two generations away from its grandmother.)

$$P^2 = \begin{bmatrix} 0.75 & 0.25 \\ 0.25 & 0.75 \end{bmatrix} \begin{bmatrix} 0.75 & 0.25 \\ 0.25 & 0.75 \end{bmatrix}$$

$$= \begin{bmatrix} 0.625 & 0.375 \\ 0.375 & 0.625 \end{bmatrix}$$

There is a 62.5 percent probability that the grandchild of a woman who is active in the church will be active in the church.

State Distribution Vector

Returning to the apartment rental scenario given at the beginning of this section, we have a Markov chain with the transition matrix $P = \begin{bmatrix} 0.65 & 0.35 \\ 0.85 & 0.15 \end{bmatrix}$. We were told that in 2005, 400 of the 500 apartments were filled and the remaining 100 were vacant. In other words, 80 percent of the apartments were filled in 2005 and 20 percent were vacant. What percentage of the apartments will be filled in 2007?

$$\begin{aligned} \text{Rented in 2006} &= (\text{rented in 2005})(0.65) + (\text{vacant in 2005})(0.85) \\ &= (400)(0.65) + (100)(0.85) \\ &= 260 + 85 \\ &= 345 \end{aligned}$$

The manager forecasts that in 2006, 345 apartments will be filled and the remaining 155 will be vacant. In other words, 69 percent of the apartments will be filled and 31 percent will be vacant.

$$\begin{aligned} \text{Rented in 2007} &= (\text{rented in 2006})(0.65) + (\text{vacant in 2006})(0.85) \\ &= (345)(0.65) + (155)(0.85) \\ &= 224 + 132 \\ &= 356 \end{aligned}$$

The manager forecasts that in 2007, 356 apartments will be filled and the remaining 144 will be vacant. In other words, 71 percent of the apartments will be filled and 29 percent will be vacant.

Since 80 percent of the apartments were filled and 20 percent were vacant in 2005, we can alternatively represent the 2005 rental rates with the initial *state distribution vector* $D = \begin{bmatrix} 0.80 & 0.20 \end{bmatrix}$. The expected percentage of apartments rented in 2006 is given by

$$DP = \begin{bmatrix} \text{percentage filled} & \text{percentage vacant} \end{bmatrix} \begin{bmatrix} 0.65 & 0.35 \\ 0.85 & 0.15 \end{bmatrix}$$

$$= \begin{bmatrix} 0.80 & 0.20 \end{bmatrix} \begin{bmatrix} 0.65 & 0.35 \\ 0.85 & 0.15 \end{bmatrix}$$

$$= \begin{bmatrix} 0.69 & 0.31 \end{bmatrix}$$

The expected percentage of apartments rented in 2007 is given by

$$(DP)P = \begin{bmatrix} 0.69 & 0.31 \end{bmatrix} \begin{bmatrix} 0.65 & 0.35 \\ 0.25 & 0.75 \end{bmatrix}$$

$$= \begin{bmatrix} 0.71 & 0.29 \end{bmatrix}$$

These results agree with our earlier calculations.

STATE DISTRIBUTION VECTOR

A **state distribution vector** is a $1 \times n$ row matrix $D = \begin{bmatrix} d_1 & d_2 & \dots & d_n \end{bmatrix}$ whose entries are nonnegative and add to 1. If D is the current probability distribution for a Markov chain with transition matrix P, then the state distribution vector given by DP is the next probability distribution.

EXAMPLE 5

Forecasting Future Results with a Markov Chain

According to market analysts, Palm Incorporated had 30.6 percent of the personal data assistant (PDA) market share in the third quarter of 2002. (Source: www.palmpower.com.) Assume that by the third quarter of the following year, 20 percent of current Palm users will move to a competing brand and 10 percent of users of competing brands will move to a Palm PDA.

According to these assumptions, what is Palm's expected market share in the third quarter of 2003, 2004, and 2005?

SOLUTION We define $S_1 =$ Palm user and $S_2 =$ other brand PDA user. The initial state distribution vector is $D_0 = \begin{bmatrix} 0.306 & 0.694 \end{bmatrix}$. The transition matrix is $P = \begin{bmatrix} 0.80 & 0.20 \\ 0.10 & 0.90 \end{bmatrix}$.

$$D_1 = D_0 P$$

$$= \begin{bmatrix} 0.306 & 0.694 \end{bmatrix} \begin{bmatrix} 0.80 & 0.20 \\ 0.10 & 0.90 \end{bmatrix}$$

$$= \begin{bmatrix} 0.314 & 0.686 \end{bmatrix}$$

By the third quarter of 2003, Palm is projected to have a 31.4 percent market share.

$$D_2 = D_1 P$$

$$= \begin{bmatrix} 0.314 & 0.686 \end{bmatrix} \begin{bmatrix} 0.80 & 0.20 \\ 0.10 & 0.90 \end{bmatrix}$$

$$= \begin{bmatrix} 0.320 & 0.680 \end{bmatrix}$$

By the third quarter of 2004, Palm is projected to have a 32.0 percent market share.

$$D_3 = D_2 P$$

$$= \begin{bmatrix} 0.320 & 0.680 \end{bmatrix} \begin{bmatrix} 0.80 & 0.20 \\ 0.10 & 0.90 \end{bmatrix}$$

$$= \begin{bmatrix} 0.324 & 0.676 \end{bmatrix}$$

By the third quarter of 2005, Palm is projected to have a 32.4 percent market share.

---

kth STATE DISTRIBUTION VECTOR

Let D_0 be the initial state distribution vector of a Markov chain with transition matrix P. The entry in row i and column j of P^k is the probability of moving from state i to state j in exactly k transitions. The *kth* **state distribution vector** is given by $D_k = D_0 P^k$.

EXAMPLE 6

Finding State Distribution Vectors

Return to Example 5 and calculate the tenth, twentieth, and hundredth state distribution vectors. What do the results seem to infer?

SOLUTION

$$D_{10} = D_0 P^{10}$$

$$= \begin{bmatrix} 0.306 & 0.694 \end{bmatrix} \left(\begin{bmatrix} 0.80 & 0.20 \\ 0.10 & 0.90 \end{bmatrix} \right)^{10}$$

$$= \begin{bmatrix} 0.306 & 0.694 \end{bmatrix} \begin{bmatrix} 0.352 & 0.648 \\ 0.324 & 0.676 \end{bmatrix}$$

$$= \begin{bmatrix} 0.333 & 0.667 \end{bmatrix}$$

By the third quarter of 2012, Palm is projected to have a 33.3 percent market share.

$$D_{20} = D_0 P^{20}$$

$$= \begin{bmatrix} 0.306 & 0.694 \end{bmatrix} \left(\begin{bmatrix} 0.80 & 0.20 \\ 0.10 & 0.90 \end{bmatrix} \right)^{20}$$

$$= \begin{bmatrix} 0.306 & 0.694 \end{bmatrix} \begin{bmatrix} 0.334 & 0.666 \\ 0.333 & 0.667 \end{bmatrix}$$

$$= \begin{bmatrix} 0.333 & 0.667 \end{bmatrix}$$

By the third quarter of 2022, Palm is projected to have a 33.3 percent market share.

$$D_{100} = D_0 P^{100}$$

$$= \begin{bmatrix} 0.306 & 0.694 \end{bmatrix} \left(\begin{bmatrix} 0.80 & 0.20 \\ 0.10 & 0.90 \end{bmatrix} \right)^{100}$$

$$= \begin{bmatrix} 0.306 & 0.694 \end{bmatrix} \begin{bmatrix} 0.333 & 0.667 \\ 0.333 & 0.667 \end{bmatrix}$$

$$= \begin{bmatrix} 0.333 & 0.667 \end{bmatrix}$$

By the third quarter of 2102, Palm is projected to have a 33.3 percent market share.

Observe that the state distribution vector remained constant for values of k beyond 10. That is, the projected market share did not change after 2012. We refer to the state distribution vector $D = \begin{bmatrix} 0.333 & 0.667 \end{bmatrix}$ as a *steady state distribution vector* for the Markov chain, since multiplying this vector by P or any power of P leaves the vector unchanged.

STEADY STATE DISTRIBUTION VECTOR

A state distribution vector $D = \begin{bmatrix} d_1 & d_2 & \ldots & d_n \end{bmatrix}$ is a **steady state distribution vector** for a Markov chain with $n \times n$ transition matrix P if $DP = D$.

EXAMPLE 7

Finding a Steady State Distribution Vector

Find a steady state distribution vector (if it exists) for the Markov chain with transition matrix $P = \begin{bmatrix} 0.1 & 0.9 \\ 0.8 & 0.2 \end{bmatrix}$.

SOLUTION Define a steady state distribution vector $D = [d_1 \ d_2]$. We know that

$$DP = D$$

$$[d_1 \ d_2]\begin{bmatrix} 0.1 & 0.9 \\ 0.8 & 0.2 \end{bmatrix} = [d_1 \ d_2]$$

$$[0.1d_1 + 0.8d_2 \quad 0.9d_1 + 0.2d_2] = [d_1 \ d_2]$$

This result leads to the following system of equations:

$$0.1d_1 + 0.8d_2 = d_1$$
$$0.9d_1 + 0.2d_2 = d_2$$

Simplifying the system yields

$$-0.9d_1 + 0.8d_2 = 0$$
$$0.9d_1 - 0.8d_2 = 0$$

$$\begin{bmatrix} -0.9 & 0.8 & \vdots & 0 \\ 0.9 & -0.8 & \vdots & 0 \end{bmatrix}$$

$$\begin{bmatrix} -0.9 & 0.8 & \vdots & 0 \\ 0 & 0 & \vdots & 0 \end{bmatrix} \qquad R_1 + R_2$$

The system appears to be dependent. However, since D is a state distribution vector, we know that $d_1 + d_2 = 1$. Adding this information to the system yields

$$\begin{bmatrix} -0.9 & 0.8 & \vdots & 0 \\ 1 & 1 & \vdots & 1 \end{bmatrix}$$

$$\begin{bmatrix} 1 & 1 & \vdots & 1 \\ -0.9 & 0.8 & \vdots & 0 \end{bmatrix} \qquad \begin{matrix} R_2 \\ R_1 \end{matrix}$$

$$\begin{bmatrix} 1 & 1 & \vdots & 1 \\ 0 & 17 & \vdots & 9 \end{bmatrix} \qquad 9R_1 + 10R_2$$

$$\begin{bmatrix} 1 & 1 & \vdots & 1 \\ 0 & 1 & \vdots & \frac{9}{17} \end{bmatrix} \qquad \frac{1}{17}R_2$$

$$\begin{bmatrix} 1 & 0 & \vdots & \frac{8}{17} \\ 0 & 1 & \vdots & \frac{9}{17} \end{bmatrix} \qquad R_1 - R_2$$

The steady state distribution vector for the Markov chain with transition matrix $P = \begin{bmatrix} 0.1 & 0.9 \\ 0.8 & 0.2 \end{bmatrix}$ is $D = [\frac{8}{17} \ \frac{9}{17}]$.

EXAMPLE 8 **Finding a Steady State Distribution Vector**

Find a steady state distribution vector (if it exists) for the Markov chain with transition matrix $P = \begin{bmatrix} 0.25 & 0.75 \\ 0.10 & 0.90 \end{bmatrix}$.

SOLUTION
$$DP = D$$

$$[d_1 \ d_2]\begin{bmatrix} 0.25 & 0.75 \\ 0.10 & 0.90 \end{bmatrix} = [d_1 \ d_2]$$

$$[0.25d_1 + 0.10d_2 \quad 0.75d_1 + 0.90d_2] = [d_1 \ d_2]$$

This result leads to the following system of equations:

$$0.25d_1 + 0.10d_2 = d_1$$
$$0.75d_1 + 0.90d_2 = d_2$$

Simplifying the system yields

$$-0.75d_1 + 0.10d_2 = 0$$
$$0.75d_1 - 0.10d_2 = 0$$

$$\begin{bmatrix} -0.75 & 0.10 & \vdots & 0 \\ 0.75 & -0.10 & \vdots & 0 \end{bmatrix}$$

$$\begin{bmatrix} -0.75 & 0.10 & \vdots & 0 \\ 0 & 0 & \vdots & 0 \end{bmatrix} \quad R_1 + R_2$$

The system appears to be dependent. However, since D is a state distribution vector, we know that $d_1 + d_2 = 1$. Adding this information to the system yields

$$\begin{bmatrix} -0.75 & 0.10 & \vdots & 0 \\ 1 & 1 & \vdots & 1 \end{bmatrix}$$

$$\begin{bmatrix} 1 & 1 & \vdots & 1 \\ -0.75 & 0.10 & \vdots & 0 \end{bmatrix} \quad \begin{matrix} R_2 \\ R_1 \end{matrix}$$

$$\begin{bmatrix} 1 & 1 & \vdots & 1 \\ 0 & 85 & \vdots & 75 \end{bmatrix} \quad 75R_1 + 100R_2$$

$$\begin{bmatrix} 1 & 1 & \vdots & 1 \\ 0 & 1 & \vdots & \frac{75}{85} \end{bmatrix} \quad \frac{1}{85}R_2$$

$$\begin{bmatrix} 1 & 0 & \vdots & \frac{10}{85} \\ 0 & 1 & \vdots & \frac{75}{85} \end{bmatrix} \quad R_1 - R_2$$

$$\begin{bmatrix} 1 & 0 & \vdots & \frac{2}{17} \\ 0 & 1 & \vdots & \frac{15}{17} \end{bmatrix} \quad \begin{matrix} \text{Reduce fraction} \\ \text{Reduce fraction} \end{matrix}$$

The steady state distribution vector for the Markov chain with transition matrix $P = \begin{bmatrix} 0.25 & 0.75 \\ 0.10 & 0.90 \end{bmatrix}$ is $D = \begin{bmatrix} \frac{2}{17} & \frac{15}{17} \end{bmatrix}$.

EXAMPLE 9

Forecasting Genetic Makeup with a Markov Chain

The physical characteristics of a plant or animal's offspring are determined in large part by the genetic makeup of the parents. For some physical features, multiple genes from each parent determine the particular trait. For other physical features, a single gene from each parent determines the physical characteristic.

The North American Curly Horse is characterized by its curly coat. Whether the coat of a horse is curly or straight-haired is believed to be determined by a single gene from each parent. Mendelian genetics may be used to predict the coat of a foal based on its parents' genetic makeup.

Let C represent the curly-haired gene and c represent the straight-haired gene. Since curly hair is dominant, a horse with genetic makeup CC or Cc has curly hair. A horse with genetic makeup cc has straight hair. Horse breeders are often interested in predicting the genotype (genetic makeup) of a foal. If a mare of each genotype (CC, Cc, and cc) is bred with a stud of genotype Cc, the expected genotype of a foal may be determined by constructing a series of tables with column and row headings indicating the genotype of the parents and the entries in the table representing the expected genotype of the foal.

	C	c
C	CC	Cc
C	CC	Cc

	C	c
C	CC	Cc
c	Cc	cc

	C	c
c	Cc	cc
c	Cc	cc

Do the following:

1. Identify the genotype states.
2. Draw the transition diagram and its associated matrix.
3. Assuming that each mare of any future generation is bred with a stud of genotype Cc, determine the probability that a mare with genotype CC will have a great-grandchild with straight hair.
4. Suppose that the initial state distribution vector for the genotypes of a herd of horses is $D = \begin{bmatrix} 0.70 & 0.10 & 0.20 \end{bmatrix}$. Determine the state distribution for the fourth generation.

SOLUTION Let $S_1 = CC$, $S_2 = Cc$, and $S_3 = cc$. From the tables, we can see that if a CC is bred with a Cc, there is a 50 percent probability that the foal will be a CC and a 50 percent probability that the foal will be a Cc. If a Cc is bred with a Cc, there is a 25 percent probability that the foal will be a CC, a 50 percent probability that the foal will be a Cc, and a 25 percent probability that the foal will be a cc. If a cc is bred with a Cc, there is a 50 percent probability that the foal will be a Cc and a 50 percent probability that the foal will be a cc. The preceding results yield the transition diagram in Figure 8.15 and its associated transition matrix.

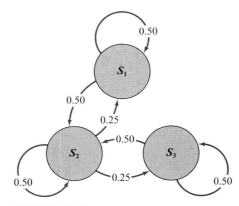

FIGURE 8.15

$$P = \begin{bmatrix} 0.50 & 0.50 & 0.00 \\ 0.25 & 0.50 & 0.25 \\ 0.00 & 0.50 & 0.50 \end{bmatrix}$$

To determine the probability that the great-grandchild of a *CC* mare is of geno-type *cc*, we look at the p_{13} entry of the matrix P^3. (A great-grandchild is three gen-erations away from its great-grandmother.)

$$P^3 = \begin{bmatrix} 0.50 & 0.50 & 0.00 \\ 0.25 & 0.50 & 0.25 \\ 0.00 & 0.50 & 0.50 \end{bmatrix} \begin{bmatrix} 0.50 & 0.50 & 0.00 \\ 0.25 & 0.50 & 0.25 \\ 0.00 & 0.50 & 0.50 \end{bmatrix} \begin{bmatrix} 0.50 & 0.50 & 0.00 \\ 0.25 & 0.50 & 0.25 \\ 0.00 & 0.50 & 0.50 \end{bmatrix}$$

$$= \begin{bmatrix} 0.31 & 0.50 & 0.19 \\ 0.25 & 0.50 & 0.25 \\ 0.19 & 0.50 & 0.31 \end{bmatrix}$$

There is a 19 percent probability that the great-grandchild of a mare of genotype *CC* will have straight hair, assuming that all mares of future generations are bred with a stud of genotype *Cc*.

The genotype state distribution for the fourth generation (three generations away from the first generation) is given by

$$DP^3 = \begin{bmatrix} 0.70 & 0.10 & 0.20 \end{bmatrix} \begin{bmatrix} 0.31 & 0.50 & 0.19 \\ 0.25 & 0.50 & 0.25 \\ 0.19 & 0.50 & 0.31 \end{bmatrix}$$

$$= \begin{bmatrix} 0.28 & 0.50 & 0.22 \end{bmatrix}$$

Of the fourth generation of horses, 28 percent are expected to have genotype *CC*, 50 percent are expected to have genotype *Cc*, and 22 percent are expected to have genotype *cc*.

8.3 Summary

In this section, you learned how to use Markov chains to forecast future results. You saw that the long-range behavior of a Markov chain often stabilizes over time.

8.3 Exercises

In Exercises 1–5, interpret the meaning of p_{13}, p_{21}, and p_{33} in the given transition matrix.

1. $P = \begin{bmatrix} 1 & 0 & 0 \\ 0 & 0.2 & 0.8 \\ 0 & 0.3 & 0.7 \end{bmatrix}$

2. $P = \begin{bmatrix} 0.7 & 0.0 & 0.3 \\ 0.1 & 0.4 & 0.5 \\ 0.2 & 0.8 & 0.0 \end{bmatrix}$

3. $P = \begin{bmatrix} 0.0 & 0.9 & 0.1 \\ 0.2 & 0.0 & 0.8 \\ 0.4 & 0.6 & 0.0 \end{bmatrix}$

4. $P = \begin{bmatrix} 0.1 & 0.2 & 0.7 \\ 0.2 & 0.3 & 0.5 \\ 0.3 & 0.4 & 0.3 \end{bmatrix}$

5. $P = \begin{bmatrix} 1 & 0 & 0 \\ 0 & 1 & 0 \\ 0.5 & 0 & 0.5 \end{bmatrix}$

In Exercises 6–15, draw the state transition diagram that corresponds to the given transition matrix. Identify which states, if any, are absorbing states.

6. $P = \begin{bmatrix} 0 & 1 \\ 1 & 0 \end{bmatrix}$

7. $P = \begin{bmatrix} 1 & 0 \\ 0 & 1 \end{bmatrix}$

8. $P = \begin{bmatrix} 0.1 & 0.9 \\ 0.4 & 0.6 \end{bmatrix}$

9. $P = \begin{bmatrix} 1.00 & 0.00 \\ 0.05 & 0.95 \end{bmatrix}$

10. $P = \begin{bmatrix} 1 & 0 & 0 \\ 0 & 0.2 & 0.8 \\ 0 & 0.3 & 0.7 \end{bmatrix}$

11. $P = \begin{bmatrix} 0.7 & 0.0 & 0.3 \\ 0.1 & 0.4 & 0.5 \\ 0.2 & 0.8 & 0.0 \end{bmatrix}$

12. $P = \begin{bmatrix} 0.0 & 0.9 & 0.1 \\ 0.2 & 0.0 & 0.8 \\ 0.4 & 0.6 & 0.0 \end{bmatrix}$

13. $P = \begin{bmatrix} 0.1 & 0.2 & 0.7 \\ 0.2 & 0.3 & 0.5 \\ 0.3 & 0.4 & 0.3 \end{bmatrix}$

14. $P = \begin{bmatrix} 0 & 1 & 0 & 0 \\ 0 & 0 & 1 & 0 \\ 0 & 0 & 0 & 1 \\ 1 & 0 & 0 & 0 \end{bmatrix}$

15. $P = \begin{bmatrix} 0 & 0.5 & 0.5 & 0 \\ 0 & 0 & 0.5 & 0.5 \\ 0 & 0 & 1 & 0 \\ 0 & 0 & 0 & 1 \end{bmatrix}$

In Exercises 16–20, write the transition matrix P that corresponds to the state transition diagram.

16.

17.

18.

19.

20.

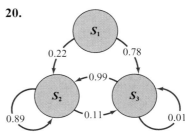

In Exercises 21–30, find a steady state distribution vector (if it exists) for each transition matrix.

21. $P = \begin{bmatrix} 0.2 & 0.8 \\ 0.3 & 0.7 \end{bmatrix}$

22. $P = \begin{bmatrix} 0.5 & 0.5 \\ 0.1 & 0.9 \end{bmatrix}$

23. $P = \begin{bmatrix} 0.1 & 0.9 \\ 0.4 & 0.6 \end{bmatrix}$

24. $P = \begin{bmatrix} 1.00 & 0.00 \\ 0.05 & 0.95 \end{bmatrix}$

25. $P = \begin{bmatrix} 0.64 & 0.36 \\ 0.25 & 0.75 \end{bmatrix}$

26. $P = \begin{bmatrix} 0.7 & 0.0 & 0.3 \\ 0.1 & 0.4 & 0.5 \\ 0.2 & 0.8 & 0.0 \end{bmatrix}$

27. $P = \begin{bmatrix} 0.0 & 0.9 & 0.1 \\ 0.2 & 0.0 & 0.8 \\ 0.4 & 0.6 & 0.0 \end{bmatrix}$

28. $P = \begin{bmatrix} 1.0 & 0.0 & 0.0 \\ 0.0 & 0.5 & 0.5 \\ 0.0 & 0.7 & 0.3 \end{bmatrix}$

29. $P = \begin{bmatrix} 0 & 1 & 0 \\ 0 & 0 & 1 \\ 1 & 0 & 0 \end{bmatrix}$

30. $P = \begin{bmatrix} 0 & 0.5 & 0.5 & 0 \\ 0 & 0 & 0.5 & 0.5 \\ 0 & 0 & 1 & 0 \\ 0 & 0 & 0 & 1 \end{bmatrix}$

In Exercises 31–35, calculate and interpret the meaning of P^2, P^5, and P^7 given a Markov chain with transition matrix P.

31. $P = \begin{bmatrix} 0.12 & 0.00 & 0.88 \\ 0.88 & 0.12 & 0.00 \\ 0.00 & 0.88 & 0.12 \end{bmatrix}$

32. $P = \begin{bmatrix} 0.5 & 0.5 & 0.0 \\ 0.0 & 0.5 & 0.5 \\ 0.0 & 0.0 & 1.0 \end{bmatrix}$

33. $P = \begin{bmatrix} 0.50 & 0.25 & 0.25 \\ 0.25 & 0.50 & 0.25 \\ 0.25 & 0.25 & 0.50 \end{bmatrix}$

34. $P = \begin{bmatrix} 0 & 1 & 0 \\ 0 & 0 & 1 \\ 1 & 0 & 0 \end{bmatrix}$

35. $P = \begin{bmatrix} 0.33 & 0.67 & 0.00 \\ 0.10 & 0.90 & 0.00 \\ 0.00 & 0.00 & 1.00 \end{bmatrix}$

In Exercises 36–40, determine the answer to each question by calculating the appropriate power of the transition matrix.

36. Family Business   In forecasting future workforce needs, the head of a small company estimates that there is a 20 percent probability that an employee who is working for the company today will not be working for the company one year from now. There is a 15 percent probability that an employee who used to work for the company will be employed by the company one year from now. Assuming that these estimates are accurate for future years, what is the probability that a current employee will not be employed by the company three years from now?

37. Coin Toss Game   A unique coin toss game is played as follows. To start the game, a coin is tossed and allowed to fall upon a tabletop. The face of the coin (heads or tails) is noted. The process is then repeated, with scoring as follows: If the coin shows heads or matches the previous coin toss, a win is counted. Otherwise, a loss is counted. Is it possible for a player to lose 25 rounds in a row? Explain. Then calculate the probability that a person who loses the first round will win the 25th round.

38. Deterioration of Theater Films   Each time a movie is shown in a theater, the actual film is exposed to the possibility of being scratched or of accumulating specks of dust. Consequently, the quality of the projected film image deteriorates over time.

Suppose that there is a 0.2 percent probability that an undamaged film frame will be damaged during the next showing of the film and that there is a 100 percent probability that a damaged film frame will remain damaged during the next showing of the film. What is the probability that an undamaged film frame will have become damaged after 200 additional showings of the film?

39. College Enrollment   The educational attainment of a student's parents has a dramatic effect on the educational attainment of the student. According to a 2001 Department of Education study, the percentage of high school graduates enrolled in college varied dramatically based upon the educational attainment of the parents. If the student's parents' highest level of education was a high school diploma or less, 59 percent of the students enrolled in college. If the parents' highest level of education was some college, 75 percent of the students enrolled in college. If the parents' highest level of education was a bachelor's degree or higher, 93 percent of the students enrolled in college. (**Source:** Special Analysis 2001, *Students Whose Parents Did Not Go to College: Postsecondary Access, Persistence, and Attainment,* U.S. Department of Education.) Based on these results and others presented within the study, we combine the last two percentages and estimate that 86 percent of students whose parents attended college also enrolled in college.

We define state 1 as "attended college" and state 2 as "did not attend college." Treating each of the percentages as a probability, what is the probability that the grandchild of a person who attended college also went to college?

40. College Enrollment   Referring to the scenario given in Exercise 39, explain whether or not a person's educational attainment substantially affects the likelihood that a great-great-grandchild of that person will enroll in college. (*Hint:* A great-great-grandchild is four generations away from its great-great-grandparent.)

In Exercises 41–46, do the following: (i) identify the states of the Markov chain, (ii) draw the transition diagram, (iii) write the transition matrix P, (iv) identify the initial state distribution vector D, and (v) calculate DP and interpret the meaning of the result.

41. **Software Market Share** According to market researchers, Mercury Interactive had a 55.2 percent market share of the automated software quality tools market in 2002. (**Source:** www.mercuryinteractive.com.) Assume that the probability that a Mercury Interactive customer will change to a competitor's program is 15 percent and that the probability that a competitor's customer will change to a Mercury Interactive program is 20 percent.

42. **Banking Deposit Market Share** According to market researchers, J. P. Morgan Chase had a 22.52 percent market share of the banking deposit market as of January 30, 2003. (**Source:** www.snl.com.) Assume that the probability that a J. P. Morgan Chase customer will transfer all of his or her deposits to a competitor's bank is 10 percent and that the probability that a competitor's customer will transfer all of his or her deposits to a J. P. Morgan Chase bank is 20 percent.

43. **Stock Prices** The closing share price of Apple Computer Corporation fluctuated during the 20 trading days in January 2004. On 11 of the trading days, the share price was higher than the preceding day's closing price. On the remaining trading days, the share price was lower than the preceding day's closing price. Further analysis of the data revealed that the probability of moving from a high price to a higher price on the next day was 45 percent. Similarly, the probability of moving from a low price to a lower price on the next day was 25 percent. On the first trading day of 2004, the share price was lower than on the preceding trading day.

44. **Employment and Mortality Rates** At the 10-year reunion of a high school graduating class, it is discovered that 94 percent of the graduates are employed, 5 percent of the graduates are unemployed, and 1 percent of the graduates are dead. Assume that the probability that an unemployed graduate will become employed by the 20-year reunion is 84 percent, and the probability that the graduate will be dead is 5 percent. The probability that an employed graduate will become unemployed by the 20-year reunion is 8 percent, and the probability that the graduate will be dead is 2 percent.

45. **Workout Schedules** In an effort to ensure that a fitness club has an adequate balance of cardiovascular and weight-training equipment, a manager monitors the workout behavior of the club's clients. The manager determines that a client who does weight training today is expected to do one of the following tomorrow:

Weight training (10 percent probability)
Cardiovascular (70 percent probability)
Not work out (20 percent probability)

A client who does cardiovascular training today is expected to do one of the following tomorrow:

Weight training (60 percent probability)
Cardiovascular (20 percent probability)
Not work out (20 percent probability)

A client who does not work out today is expected to do one of the following tomorrow:

Weight training (40 percent probability)
Cardiovascular (50 percent probability)
Not work out (10 percent probability)

Today 30 percent of the club's members did weight training, 25 percent of the members did cardiovascular training, and 45 percent of the members did not work out at the club.

46. **Gambling** A gambler predicts that if she wins the current game of poker, the probability that she will win the next game is 55 percent. She predicts that if she loses the current game of poker, there is a 40 percent probability that she will win the next game. So far this evening she has won 30 percent of her poker games.

In Exercises 47–55, use the Markov chain to answer the questions.

47. **Unemployment Rate** Unemployment rates fluctuate over time. In 1992, 7.5 percent of civilians 16 years old and older were unemployed. In 1996, 5.4 percent were unemployed. In 2000, 4.0 percent were unemployed. (**Source:** *Statistical Abstract of the United States, 2001*, Table 567, p. 367.) Based on these data, the unemployment rate may be modeled by a Markov chain with transition matrix

$$P = \begin{bmatrix} 0.671 & 0.329 \\ 0.004 & 0.996 \end{bmatrix}$$

and states S_1 = unemployed and S_2 = employed. The time interval for the matrix is four years. (That is, there are four years between transitions.) The initial state distribution vector is $D_0 = [0.075 \quad 0.925]$.

(a) Calculate and interpret the meaning of D_1 and D_2.

(b) Compare the results of part (a) with the reported unemployment data. Does the model seem to fit the data well? Explain.

(c) Use the model to predict the unemployment rate in 2004.

(d) According to this model, what percentage of the population is expected to be unemployed in the long run?

48. **Union Membership—Public Sector** Union membership rates tend to fluctuate over time. In 1990, 36.5 percent of the public-sector workers were members of a union. In 1995, 37.7 percent were union members. In 2000, 37.5 percent were union members. (**Source:** *Statistical Abstract of the United States, 2001*, Table 637, p. 411.) Based on these data, union membership rates for public-sector workers may be modeled by a Markov chain with transition matrix

$$P = \begin{bmatrix} 0.271 & 0.729 \\ 0.438 & 0.562 \end{bmatrix}$$

and states S_1 = union member and S_2 = not a union member. The time interval for the matrix is five years. (That is, there are five years between transitions.) The initial state distribution vector is $D_0 = [0.365 \quad 0.635]$.

(a) Calculate and interpret the meaning of D_1 and D_2.

(b) Compare the results of part (a) with the reported union membership data. Does the model seem to fit the data well? Explain.

(c) Use the model to predict the union membership rate in 2005.

(d) According to this model, what percentage of public-sector workers will be members of a union in the long run?

(e) How could union leaders use the results of part (d)?

49. **Union Membership—Private Sector** In 1990, 11.9 percent of the private-sector workers were members of a union. In 1995, 10.3 percent were union members. In 2000, 9.0 percent were union members. (**Source:** *Statistical*

Abstract of the United States, 2001, Table 637, p. 411.) Based on these data, union membership rates for private-sector workers may be modeled by a Markov chain with transition matrix

$$P = \begin{bmatrix} 0.819 & 0.181 \\ 0.006 & 0.994 \end{bmatrix}$$

and states S_1 = union member and S_2 = not a union member. The time interval for the matrix is five years. (That is, there are five years between transitions.) The initial state distribution vector is $D_0 = [0.119 \quad 0.881]$.

(a) Calculate and interpret the meaning of D_1 and D_2.

(b) Compare the results of part (a) with the reported union membership data. Does the model seem to fit the data well? Explain.

(c) Use the model to predict the union membership rate in 2005.

(d) According to this model, what percentage of private-sector workers will be members of a union in the long run?

(e) How could union leaders use the results of part (d)?

50. **African Population** In 1990, humans living in Africa made up 11.9 percent of the world's population. By 2000, the percentage had increased to 13.2 percent. According to Census Bureau estimates, Africans are expected to make up 14.4 percent of the world's population by 2010. (**Source:** *Statistical Abstract of the United States, 2001*, Table 1325, p. 829.) Based on these data, the percentage of humans living in Africa may be modeled by a Markov chain with transition matrix

$$P = \begin{bmatrix} 0.945 & 0.055 \\ 0.022 & 0.978 \end{bmatrix}$$

and states S_1 = living in Africa and S_2 = not living in Africa. The time interval for the matrix is 10 years. (That is, there are 10 years between transitions.) The initial state distribution vector is $D_0 = [0.119 \quad 0.881]$.

(a) Use the model to calculate the percentage of the human population expected to be living in Africa in 2050.

(b) The Census Bureau projects that by 2050, Africans will make up 20.3 percent of the world's human population. Does this estimate agree with the Markov chain model? Explain.

(c) According to the model, what percentage of the world's population is projected to live in Africa in the long run?

51. **Child Abuse and Neglect** In 1996, nearly 956,000 children in the United States were victims of child abuse. The majority of these children were neglected by their parents or guardians. In 1996, 51.6 percent of abused children were neglected. In 1997, 54.6 percent were neglected, and in 1998, 53.5 percent were neglected. (**Source:** *Statistical Abstract of the United States, 2001*, Table 328, p. 199.) Based on these data, the percentage of abused children who were neglected may be modeled by a Markov chain with transition matrix

$$P = \begin{bmatrix} 0.369 & 0.631 \\ 0.735 & 0.265 \end{bmatrix}$$

and states $S_1 =$ neglected and $S_2 =$ not neglected.

(a) According to the model, what percentage of abused children were neglected in 1999?

(b) According to the model, what percentage of abused children will be neglected in the long run?

(c) What can you do to help reduce child abuse? (*Hint:* Visit www.preventchildabuse.org.)

52. **Supermarkets with ATMs** In 1992, 28 percent of supermarkets had automatic teller machines. The percentage had increased to 60 percent by 1996 and to 64 percent by 2000. (**Source:** *Statistical Abstract of the United States, 2001*, Table 1033, p. 651.) Based on these data, the percentage of supermarkets with ATMs may be modeled by a Markov chain with transition matrix

$$P = \begin{bmatrix} 0.690 & 0.310 \\ 0.565 & 0.435 \end{bmatrix}$$

and states $S_1 =$ has ATM and $S_2 =$ doesn't have ATM. The time interval between transitions is four years.

(a) Does the model fit the data well? Justify your answer.

(b) Do the transition probabilities seem to make sense in the context of the problem?

(c) According to the model, will the percentage of supermarkets with ATM machines ever reach 80 percent? Justify your answer.

(d) Do you think that this model is an accurate predictor of the percentage of supermarkets with ATMs? Justify your answer.

53. **Supermarkets with a Pharmacy** In 1992, 18 percent of supermarkets had a pharmacy. The percentage had increased to 26 percent by 1996 and to 32 percent by 2000. (**Source:** *Statistical Abstract of the United States, 2001*, Table 1033, p. 651.) Based on these data, the percentage of supermarkets with a pharmacy may be modeled by a Markov chain with transition matrix

$$P = \begin{bmatrix} 0.875 & 0.125 \\ 0.125 & 0.875 \end{bmatrix}$$

and states $S_1 =$ has pharmacy and $S_2 =$ doesn't have pharmacy. The time interval between transitions is four years.

(a) Does the model fit the data well? Justify your answer.

(b) Do the transition probabilities seem to make sense in the context of the problem?

(c) According to the model, what is the maximum predicted percentage of supermarkets that will have a pharmacy?

(d) Do you think that this model is an accurate predictor of the percentage of supermarkets with a pharmacy? Justify your answer.

54. **Supermarkets with a Deli** In 1992, 78 percent of supermarkets had a deli. The percentage had increased to 80 percent by 1996 and to 81 percent by 2000. (**Source:** *Statistical Abstract of the United States, 2001*, Table 1033, p. 651.) Based on these data, the percentage of supermarkets with a deli may be modeled by a Markov chain with transition matrix

$$P = \begin{bmatrix} 0.91 & 0.09 \\ 0.41 & 0.59 \end{bmatrix}$$

and states $S_1 =$ has deli and $S_2 =$ doesn't have deli. The time interval between transitions is four years.

(a) Does the model fit the data well? Justify your answer.

(b) Do the transition probabilities seem to make sense in the context of the problem?

(c) According to the model, what is the maximum predicted percentage of supermarkets that will have a deli?

55. **Supermarkets with Service Fish Departments** In 1992, 41 percent of supermarkets had a service fish department. The percentage had increased to 46 percent by 1996 and then decreased to 45 percent by 2000. (**Source:** *Statistical Abstract of the United States, 2001,* Table 1033, p. 651.) Based on these data, the percentage of supermarkets with a service fish department may be modeled by a Markov chain with transition matrix

$$P = \begin{bmatrix} 0.342 & 0.658 \\ 0.542 & 0.458 \end{bmatrix}$$

and states $S_1 =$ has service fish department and $S_2 =$ doesn't have service fish department. The time interval between transitions is four years.

(a) Does the model fit the data well? Justify your answer.

(b) Do the transition probabilities seem to make sense in the context of the problem?

(c) According to the model, what percentage of supermarkets will have a service fish department in the long run?

Exercises 56–61 are intended to challenge your understanding of Markov chains.

56. Find three distinct steady state vectors for the transition matrix

$$P = \begin{bmatrix} 0.2 & 0.8 & 0.0 \\ 0.1 & 0.9 & 0.0 \\ 0.0 & 0.0 & 1.0 \end{bmatrix}$$

57. The vector $D = \begin{bmatrix} d_1 & d_2 & d_3 \end{bmatrix}$ is a steady state distribution vector for the transition matrix

$$P = \begin{bmatrix} 1 & 0 & 0 \\ 0.2 & 0.6 & 0.2 \\ 0 & 0 & 1 \end{bmatrix}$$

Determine the value of and/or relationship between d_1, d_2, and d_3. (*Hint:* We're looking for something more than just $d_1 + d_2 + d_3 = 1$.)

58. A Markov chain with a 5×5 transition matrix P has the property that its first four states are absorbing states. The probability of making a transition from the fifth state S_5 to state S_i is $ai\%$, where a is a real number. Write the numeric values of the transition matrix P.

59. Given the transition matrix

$$P = \begin{bmatrix} 0 & 0 & 1 \\ 1 & 0 & 0 \\ 0 & 1 & 0 \end{bmatrix}$$

determine for which values of k $P^k = P$. Show the work that leads to your conclusion.

60. Consider a Markov chain with transition matrix

$$P = \begin{bmatrix} 0 & 0 & 1 \\ 1 & 0 & 0 \\ 0 & 1 & 0 \end{bmatrix}$$

and initial state distribution vector $\begin{bmatrix} 0.1 & 0.3 & 0.6 \end{bmatrix}$. List all possible state distribution vectors for the Markov chain. What will the state probability distribution be after 291 transitions? (*Hint:* Refer to the results of Exercise 59.)

61. Suppose that a Markov chain with transition matrix $P = \begin{bmatrix} a & 1 - a \\ b & 1 - b \end{bmatrix}$ has state distribution vectors $d_0 = \begin{bmatrix} x & 1 - x \end{bmatrix}$, $d_1 = \begin{bmatrix} y & 1 - y \end{bmatrix}$, and $d_2 = \begin{bmatrix} z & 1 - z \end{bmatrix}$. Show that if $x \neq y$, $0 < x \leq 1$, and $0 < y \leq 1$, then $a = \dfrac{xz + y - y^2 - z}{x - y}$ and $b = \dfrac{xz - y^2}{x - y}$.

8.4 Random Variables and Expected Value

- Calculate the probability distribution of a random variable
- Draw a histogram
- Determine the expected value of a random variable

GETTING **STARTED** Many decision-making processes involve a certain element of chance. People often take risks after weighing the likelihood of a negative consequence. By using mathematical methods, we can quantify risk and make better-informed decisions.

In this section, we will demonstrate how to calculate the probability distribution of a *random variable*, draw a *histogram*, and determine the *expected value of a random variable.* These techniques provide important inputs into any decision-making process.

Random Variables and Histograms

RANDOM VARIABLE

A **random variable** is a function that assigns a number to each outcome in a sample space.

The term *random variable* is an odd name for this function, since it is neither random nor a variable. However, since the term is considered standard, we too will use it. Capital letters X and Y are often used to represent random variables.

Consider a simple coin toss. When a tossed coin lands, it will result in one of two outcomes: heads or tails. We may define a random variable X by assigning a value of 0 to *tails* and a value of 1 to *heads.* When $X = 0$, we know that a *tail* is displayed. When $X = 1$, we know that a *head* is displayed.

Random variables are not limited to experiments with just two possible outcomes. In many games of dice, we are interested in the sum of the face values of two dice. There are 36 outcomes in the sample space. We can define a random variable $X =$ sum of the face values of two six-sided dice. The value of the random variable is obtained by counting the dots on the face of each die (Figure 8.16).

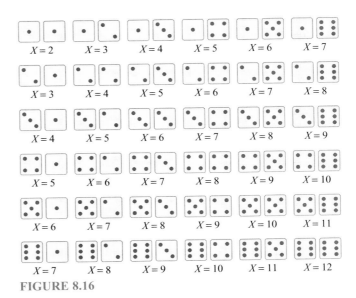

FIGURE 8.16

Some of the random variable values occur frequently (e.g., $X = 7$), while others occur only once (e.g., $X = 2$). We identify the number of times each value occurs in a **frequency distribution** (Table 8.9).

TABLE 8.9

x	2	3	4	5	6	7	8	9	10	11	12
Frequency	1	2	3	4	5	6	5	4	3	2	1

In games of chance, we are interested in the probability of a particular value occurring. To determine the probability of an event, we divide the frequency of the event by the total number of possible outcomes. We summarize all of the probabilities with a probability distribution.

PROBABILITY DISTRIBUTION OF A RANDOM VARIABLE

The **probability distribution of a random variable** X is denoted by

$$P(X = x) = p(x)$$

and represents the collection of probabilities for each possible value of x.

In the case of the dice sum, the random variable X may take on any of the whole-number values between 2 and 12. To calculate $P(X = x)$, we count the number of times X takes on the value of x and divide it by the total number of outcomes in the sample space (Table 8.10).

TABLE 8.10

x	2	3	4	5	6	7	8	9	10	11	12
$P(X = x)$	$\frac{1}{36}$	$\frac{2}{36}$	$\frac{3}{36}$	$\frac{4}{36}$	$\frac{5}{36}$	$\frac{6}{36}$	$\frac{5}{36}$	$\frac{4}{36}$	$\frac{3}{36}$	$\frac{2}{36}$	$\frac{1}{36}$

Since $P(X = x)$ is a probability distribution, the values in the table will always sum to 1. The graph of a probability distribution is referred to as a **histogram.** A histogram is a bar graph. The width of each bar is equal to 1 unit, and the height of each bar is the probability. The histogram for the random variable X is shown in Figure 8.17.

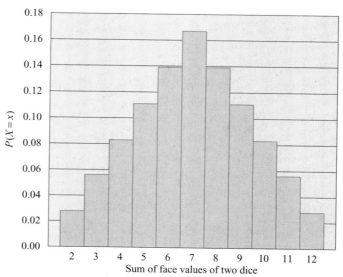

FIGURE 8.17

This particular histogram is symmetric about $X = 7$. We see visually that of all possible outcomes, that $X = 7$ is most likely and $X = 2$ or $X = 12$ is least likely.

EXAMPLE 1

Finding a Probabilty Distribution and Its Associated Histogram

The random variable Y is defined to be the absolute value of the difference in the face values of two dice. Determine the probability distribution for Y and draw the associated histogram.

SOLUTION The random variable Y may take on any of the whole-number values between 0 and 5. The frequency of each outcome varies as shown in Figure 8.18.

FIGURE 8.18

The probability distribution of Y is shown in Table 8.11 and the histogram in Figure 8.19.

TABLE 8.11

y	0	1	2	3	4	5
$P(Y = y)$	$\frac{6}{36}$	$\frac{10}{36}$	$\frac{8}{36}$	$\frac{6}{36}$	$\frac{4}{36}$	$\frac{2}{36}$

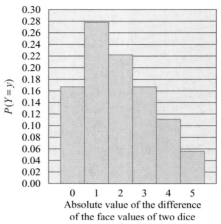

FIGURE 8.19

Unlike the previous histogram, the histogram in Figure 8.19 is not symmetric. The outcome $X = 1$ is the most likely, and the outcome $X = 5$ is the least likely.

The TI-83 Plus may be used to draw the histogram of a probability distribution, as illustrated in the following Technology Tip.

TECHNOLOGY TIP

Drawing a Histogram

1. Press [STAT] [ENTER] to open the Statistics Editor. Then enter the values of the random variable in L1 and the frequency of each value in L2.

2. Use the blue arrow to move the cursor to the L3 heading. Press the key sequence [2nd], [2], [÷], and the number of outcomes in the sample space. (This operation will divide the values in List 2 by the number of outcomes in the sample space.)

3. Press [ENTER] to display the probability distribution of the random variable in List 3.

(Continued)

4. Press the key sequence $\boxed{\text{2nd}}$, $\boxed{\text{Y=}}$, $\boxed{\text{ENTER}}$ to open Plot 1. Select the entries depicted in the figure to the right to set up a histogram with random variable values from List 1 and probabilities from List 3.

5. Press $\boxed{\text{WINDOW}}$ and adjust the values. `Xmin` is the smallest value of the random variable, `Xmax` is the largest value, and `Xscl` is the difference between consecutive values of the random variable. `Ymin` is 0, and `Ymax` should be a value slightly larger than the greatest probability value in the distribution.

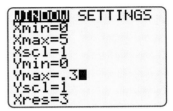

6. Press $\boxed{\text{GRAPH}}$ to display the histogram.

Expected Value

If we were to roll a pair of dice a large number of times, what would be the average value of the random variable? In Example 1, we found the probability distribution for Y to be that shown in Table 8.12.

TABLE 8.12

y	0	1	2	3	4	5
$P(Y = y)$	$\frac{6}{36}$	$\frac{10}{36}$	$\frac{8}{36}$	$\frac{6}{36}$	$\frac{4}{36}$	$\frac{2}{36}$

To determine the average value of the random variable, we multiply each value of the random variable by its associated probability and sum the results. The end result is referred to as the **expected value** $E(Y)$ of the random variable.

$$E(Y) = 0\left(\frac{6}{36}\right) + 1\left(\frac{10}{36}\right) + 2\left(\frac{8}{36}\right) + 3\left(\frac{6}{36}\right) + 4\left(\frac{4}{36}\right) + 5\left(\frac{2}{36}\right)$$

$$= 0 + \frac{10}{36} + \frac{16}{36} + \frac{18}{36} + \frac{16}{36} + \frac{10}{36}$$

$$= \frac{70}{36}$$

$$= 1.944$$

Although the value 1.944 cannot occur in a single dice roll, it is expected that the long-run average of the values will be 1.944.

> **EXPECTED VALUE OF A RANDOM VARIABLE**
>
> The **expected value** of a random variable X with probability distribution
>
> | x_i | x_1 | x_2 | $\ldots$ | x_n |
> |---|---|---|---|---|
> | $P(X = x_i)$ | p_1 | p_2 | $\ldots$ | p_n |
>
> is given by
>
> $$E(x) = x_1 p_1 + x_2 p_2 + \cdots + x_n p_n$$
>
> where p_i is the probability that x_i occurs.

EXAMPLE 2

Finding a Probability Distribution

A group of 14 applicants to Berkeley Law School in 2003 reported LSAT scores between 169 and 179. Of these applicants, 12 were accepted to the law school and 2 were rejected. (**Source:** www.lawschoolnumbers.com.) A random sample of 5 applicants is selected from the 14-applicant pool.

Let $X = $ the number of applicants in the sample who were rejected. Find the probability distribution of X and determine the expected number of rejected applicants in the sample.

SOLUTION The number of ways to select 5 applicants from a pool of 14 applicants is

$$C(14, 5) = \frac{14!}{9!5!}$$

$$= 2002$$

Since only 2 applicants were rejected, any sample will contain either 0, 1, or 2 rejected applicants. The number of samples with 0 rejected applicants is given by $C(12, 5)\, C(2, 0)$, the number with 1 rejected applicant is given by $C(12, 4)\, C(2, 1)$, and the number with 2 rejected applicants is given by $C(12, 3)\, C(2, 2)$. Therefore,

$$p_0 = \frac{C(12, 5)\, C(2, 0)}{C(14, 5)} \qquad\qquad p_1 = \frac{C(12, 4)\, C(2, 1)}{C(14, 5)}$$

$$= \frac{792 \cdot 1}{2002} \qquad\qquad\qquad = \frac{495 \cdot 2}{2002}$$

$$= 0.40 \qquad\qquad\qquad\qquad = 0.49$$

$$p_2 = \frac{C(12, 3)\, C(2, 2)}{C(14, 5)}$$

$$= \frac{220 \cdot 1}{2002}$$

$$= 0.11$$

The probability distribution of X is given by

x	0	1	2
$P(X = x)$	0.40	0.49	0.11

The expected value of X is given by

$$E(X) = 0(0.40) + 1(0.49) + 2(0.11)$$
$$= 0.71$$

If a random sample of 5 applicants was repeatedly selected from the 14-applicant pool, we expect that the average number of applicants in the sample who were rejected would be 0.71.

EXAMPLE 3

Using a Probability Distribution in a Real-World Context

The average premium for homeowner's insurance nationwide in 1999 was $487. In that year, 8.63 percent of homeowners filed payable insurance claims. The average payout was $3773. (**Source:** www.iii.org.) Assuming that there was an 8.63 percent probability that an insured homeowner received a payment for damages in 1999, determine the expected return of a homeowner who had a policy in 1999.

SOLUTION In 1999, there was an 8.63 percent probability that a homeowner received a payment. The average payment amount was $3773. However, since the insured owner had to pay a $487 premium, the net return was $3286 ($3773 − $487).

In 1999, there was a 91.37 percent probability that a homeowner did **not** receive a payment. For these homeowners, the net return was −$487.

Let X = the average net insurance return. The probability distribution of X is given by

x	−487	3,286
$P(X = x)$	0.9137	0.0863

The expected value is given by

$$E(X) = -487(0.9137) + 3286(0.0863)$$
$$= -\$161.39$$

If the insurance payment data remained constant over a number of years, the average net loss to an insured homeowner is expected to be $161.39 per year.

EXAMPLE 4

Using a Probability Distribution in a Real-World Context

The National Thoroughbred Racing Association publishes wagering information for people interested in betting on a horse race. The association reports that a $2.00 winning ticket on an entry with 10 to 1 odds will pay $22.00 to the winner. (**Source:** www.ntra.com.) A gambler repeatedly wagers on entries with 10 to 1 odds. In the long run, what is her expected net return per ticket?

SOLUTION Let X = the net return. If the gambler wins, her net gain is $20, since the ticket cost $2. If she loses, her net loss is $2. The probability distribution of X is given by

x	−2	20
$P(X = x)$	$\frac{10}{11}$	$\frac{1}{11}$

since the odds of losing are 10 to 1. The expected value is given by

$$E(X) = -2\left(\frac{10}{11}\right) + 20\left(\frac{1}{11}\right)$$

$$= 0$$

In the long run, she should expect to earn $0 per entry.

A game with an expected value of $0 is referred to as a **fair game.** The payoff amount for the wager in Example 4 is considered to be fair, since the expected value of the net return was $0. If the value had been negative, the game would have been in favor of the racetrack. If the value had been positive, the game would have been in favor of the person placing the wager.

EXAMPLE 5 Using a Probability Distribution to Make Informed Decisions

A newlywed couple debating who should do the dishes decides to settle the dispute by rolling a pair of dice. He proposes that if none of the face values are odd, he will do the dishes. Otherwise, she must do them. Should she accept his proposal?

SOLUTION We must determine who is favored by the game. We define the random variable X to be the number of dice with an odd face value. The random variable may take on the value of 0, 1, or 2. He wins if $X = 0$, and she wins if $X = 1$ or $X = 2$. Since each die has three odd faces and three even faces, there is a 50 percent probability that a die face will show an odd number and a 50 percent probability that the die face will show an even number. We have

$$P(\text{both odd}) = (0.5)(0.5) = 0.25$$
$$P(\text{first odd, second even}) = (0.5)(0.5) = 0.25$$
$$P(\text{first even, second odd}) = (0.5)(0.5) = 0.25$$
$$P(\text{both even}) = (0.5)(0.5) = 0.25$$

Therefore,

x	0	1	2
$P(X = x)$	0.25	0.50	0.25

We can see from the table that there is a 75 percent probability that she will win. The expected value of the game is

$$E(X) = 0(0.25) + 1(0.5) + 2(0.25)$$
$$= 1$$

She should readily accept his offer, since she is likely to win the game and thus not have to do the dishes.

EXAMPLE 6 Using a Probability Distribution to Make Informed Decisions

In February 2004, the collision and comprehensive loss insurance coverage offered by Progressive Max Insurance Company on a 1990 Toyota pickup truck (driven by a 35-year-old male with an accident-free history) was $105 per six-month period. (**Source:** www.progressive.com.) This quote required a $250 deductible. The estimated value of the vehicle was $2000. Assuming that there is a 0.2 percent probability that the vehicle will sustain a collision or a comprehensive loss during the six-month period, determine whether or not the driver should carry collision and comprehensive loss insurance coverage.

SOLUTION If an insured driver is in a crash, his total loss will include the six-month premium and the $250 deductible ($105 + $250 = $365). Using this result together with the additional data given in the problem, we construct Table 8.13.

TABLE 8.13

	Insurance Coverage	No Insurance Coverage
Collision or comprehensive loss ($p = 0.002$)	−$365	−$2,000
No collision or comprehensive loss ($p = 0.998$)	−$105	$0

We will first determine the expected loss for a driver with insurance coverage.

$$E(X) = 0.002(-365) + 0.998(-105)$$
$$= -105.52$$

The expected six-month loss for a driver with comprehensive and collision coverage is $105.52.

We will next determine the expected loss for a driver without insurance coverage.

$$E(X) = 0.002(-2000) + 0.998(0)$$
$$= -4$$

The expected six-month loss for the driver without collision and comprehensive coverage is $4.00. Since $4.00 < $105.52, the driver may want to decline collision and comprehensive loss coverage, since the expected loss is lower when collision and comprehensive loss coverage is not carried.

8.4 Summary

In this section you learned how to calculate the probability distribution of a random variable, draw a histogram, and determine the expected value of a random variable. You used these concepts in a variety of consumer decision-making applications.

8.4 Exercises

In Exercises 1–10, determine the probability distribution for the random variable X and draw its associated histogram.

1. Two coins are tossed.

   $X =$ the number of heads shown

2. Five cards are pulled from a deck of cards containing 13 hearts, 13 clubs, 13 diamonds, and 13 spades.

   $X =$ the number of cards that are hearts

3. Five cards are pulled from a deck of cards containing 4 kings and 48 other cards.

   $X =$ the number of cards that are kings

4. Four marbles are randomly selected from a bucket of marbles containing five red marbles, three blue marbles, and two green marbles.

   $X =$ the number of green marbles

5. A sample of 6 batteries is pulled from an order of 1000 batteries with a known defect rate of 1.3 percent.

   $X =$ the number of defective
   batteries in the sample

6. Three six-sided dice are rolled.

   $X =$ the number of dice showing
   an even number

7. A sample of 10 law school applicants is randomly selected from a pool of 500 applicants. The law school has a known acceptance rate of 65 percent.

   $X =$ the number of applicants in the
   sample who were accepted

8. Three people are playing the game Rock-Paper-Scissors. (At each round of the game, each person indicates either Rock, Paper, or Scissors.)

   $X =$ the number of Rocks

9. Four marbles are randomly selected from a bucket of marbles containing three blue marbles, three yellow marbles, two red marbles, and one green marble.

   $X =$ the number of red marbles

10. Five people are randomly selected from a group containing ten convicts and four police officers.

    $X =$ the difference between the number
    of police officers and convicts

In Exercises 11–23, apply the concepts of probability distribution and expected value, as appropriate.

11. **Raffle Tickets** A campus club sells 250 raffle tickets at a price of $2 each. The raffle offers one $50 prize, two $25 prizes, and five $10 prizes. What is the expected value of a raffle ticket?

12. **Raffle Tickets** A campus club sells 500 raffle tickets at a price of $2 each. The raffle offers one $100 prize, two $50 prizes, and five $25 prizes. What is the expected value of a raffle ticket?

13. **Income Tax Returns** Based on data from 1988 through 1998, the probability of an income tax return being examined by an Internal Revenue Service tax auditor is estimated to be 0.44 percent. Based on the data from the same time period, the average tax penalty for audited income tax returns was $2707. (**Source:** *Statistical Abstract of the United States, 2001,* Table 471, p. 313.)

    For an additional $25 fee, an accountant offers to pay any tax penalties of his clients who are audited by the IRS. Would you recommend the audit-guarantee service? Justify your answer.

14. **Income Tax Returns** An accountant determines that 0.20 percent of the income tax returns she processes are audited by the IRS. The average tax penalty for her audited returns is $1240. She guarantees her clients that she will pay any tax penalties on audited returns. How much must she charge for the guarantee service in order to break even?

15. **Test-Taking Strategy** Each question on a multiple-choice test has four possible choices. A correct answer earns 1 point, an incorrect answer loses 0.5 point, and a blank answer earns 0 points. If the student can't eliminate any of the choices for a particular question, should he guess? Explain.

16. **Test-Taking Strategy** Each question on a multiple-choice test has five possible choices. A correct answer earns 1 point, an incorrect answer loses 0.25 point, and a blank answer earns 0 points. If a student can eliminate one of the incorrect choices on a question and then guesses from among the remaining choices, what is the expected value of the points earned on that question?

17. **Test-Taking Strategy** Each question on a multiple-choice test has four possible choices. A correct answer earns 1 point, an incorrect answer loses 0.5 point, and a blank answer earns 0 points. A student determines that she can eliminate one of the incorrect choices on a question. If she guesses from among the remaining three choices, what is the expected value of the points earned on that question?

18. **Projected Earnings** A student works two part-time jobs. She typically works 30 hours per month as a waitress, earning an average of $8.00 per hour, and 50 hours per month as a tutor, earning $6.50 per hour. What are her expected average earnings per hour during a given month?

19. **Auto Insurance Policy Choices** In 2004, United Services Automobile Association charged the author $89.85 to insure his minivan against collision and comprehensive loss for a period of six months. (**Source:** USAA insurance policy.) The author estimated the value of the minivan to be $5300 and the probability of suffering a collision or comprehensive loss in the next six months to be 0.1 percent. Should the author continue to buy collision and comprehensive coverage on the minivan? Justify your answer.

20. **Auto Insurance Policy Choices** An automobile insurance policy's comprehensive and collision coverage costs $620 for a six-month period. The value of the insured vehicle is estimated to be $23,400. The insured driver estimates that there is a 1 percent probability that he will destroy his car in a collision in the next six months. Should he continue to buy collision and comprehensive loss coverage? Justify your answer.

21. **Homeowners' Insurance** In 2000, the average consumer expenditure on homeowners' insurance was $508 per year. In that year, 7.72 percent of insured homeowners filed payable claims. The average payout was $4168 per claim. (**Source:** www.iii.org.) Assuming that the consumer insurance expenditure, filing frequency, and payout amount remain constant for future years, determine the expected value of a homeowners' insurance policy and interpret the meaning of the result.

22. **Lottery Winnings** The Florida Mega Money lottery game sells tickets for $1 each. The estimated probability of winning various cash prizes is shown in the following table.

Prize Amount	Probability of Winning
$500,000	$\frac{1}{2,986,523}$
$1,200	$\frac{1}{142,216}$
$350	$\frac{1}{18,667}$
$50	$\frac{1}{890}$
$25	$\frac{1}{639}$
$3	$\frac{1}{77}$
$2	$\frac{1}{31}$

Source: www.flalottery.com.

What is the expected value of a Mega Money lottery ticket?

23. **Lottery Winnings** On February 28, 2004, the value of the Florida Lotto jackpot was estimated to be $3 million. Lottery tickets cost $1 each. The estimated probability of winning various cash prizes is shown in the following table.

Prize Amount	Probability of Winning
Jackpot	$\frac{1}{22,957,481}$
$5,000	$\frac{1}{81,411}$
$70	$\frac{1}{1,417}$
$5	$\frac{1}{72}$

Source: www.flalottery.com.

What is the expected value of a Florida Lotto ticket when the jackpot is valued at $3 million? $5 million? $10 million?

In Exercises 24–30, calculate the expected rate of change in the share price of the given company. Assume that the probability of an increase or decrease in the share price is equal to the percentage of time that the share price actually increased or decreased, respectively.

24. **Investment Performance** The share price of American Eagle Outfitters, Inc., varied between February 2 and March 3, 2004. The price increased on 71.4 percent of the trading days and decreased on 28.6 percent of the trading days. On days when the price increased, the

average increase was $0.45 per day. On days when the price decreased, the average change in price was $-\$0.28$ per day. Based on these data, what was the expected change in the share price on any day in the given time period?

25. **Investment Performance** The share price of Big Dog Holdings, Inc., varied between February 2 and March 3, 2004. The price increased on 47.6 percent of the trading days, remained the same 19.1 percent of the trading days, and decreased on 33.3 percent of the trading days. On days when the price increased, the average increase was $0.16 per day. On days when the price decreased, the average change in price was $-\$0.14$ per day. Based on these data, what was the expected change in the share price on any day in the given time period?

26. **Investment Performance** The share price of Pepsico, Inc., varied between 1977 and 2002. The annual closing price increased during 68 percent of the years and decreased during 32 percent of the years. In years when the share price increased, the average annual increase was $3.28 per year. In years when the share price decreased, the average annual change in price was $-\$1.95$ per year. Based on these data, what was the expected change in the share price in any year in the given time period?

27. **Investment Performance** The share price of Scholastic Corp. varied between 1992 and 2002. The annual closing price increased during 72.7 percent of the years and decreased during 27.3 percent of the years. In years when the share price increased, the average annual increase was $8.81 per year. In years when the share price decreased, the average annual change in price was $-\$11.50$ per year. Based on these data, what was the expected change in the share price in any year in the given time period?

28. **Investment Performance** The share price of Gap, Inc., varied between February 2 and March 3, 2004. The price increased on 42.9 percent of the trading days and decreased on 57.1 percent of the trading days. On days when the price increased, the average increase was $0.44 per day. On days when the price decreased, the average change in price was $-\$0.15$ per day. Based on these data, what was the expected change in the share price on any day in the given time period?

29. **Investment Performance** The share price of the Coca Cola Company varied between February 2 and March 3, 2004. The price increased on 33.3 percent of the trading days, remained constant on 4.8 percent of the trading days, and decreased on 61.9 percent of the trading days. On days when the price increased, the average increase was $0.47 per day. On days when the price decreased, the average change in price was $-\$0.29$ per day. Based on these data, what was the expected rate of change in the share price on any day in the given time period?

30. **Investment Performance** The share price of the Coca Cola Company varied between 1965 and 2002. The annual closing price increased during 75 percent of the years and decreased during 25 percent of the years. In years when the share price increased, the average annual increase was $2.53 per year. In years when the share price decreased, the average annual change in price was $-\$2.50$ per year. Based on these data, what was the expected change in the share price in any year in the given time period?

Exercises 31–35 are intended to challenge your understanding of probability distributions and expected values of random variables.

31. **Life Insurance Premiums** Term life insurance is used to insure a person's life for a fixed period of time. In February 2004, First Colony Life advertised a $500,000 30-year term life insurance policy for a 36-year-old woman in excellent health. The $440 annual premium guaranteed that if the woman died before February 2034, her designated beneficiaries would receive $500,000. (**Source:** vitalterm.lifelinkcorp.com.)

   The woman anticipates that there is a 1.5 percent probability she will die before her policy expires. Assuming that the total cost of the policy is $C = 440t$, where t is the number of years she pays on the policy, what is the expected value of the life insurance policy? (*Hint:* The expected value will be in terms of t.)

32. Refer to the expected value in Exercise 31. Determine for which whole-number values of t the expected value of the life insurance policy will be positive and interpret the meaning of the result.

33. **Dice Roll** A fair six-sided die has six equally likely outcomes: 1, 2, 3, 4, 5, 6. A fair four-sided die has four equally likely outcomes: 1, 2, 3, 4. The two dice are rolled. A random variable X is defined to be the larger of the face values shown on the two dice. Determine the expected value of X.

34. **Insurance Rates** An insurance company estimates that the probability that a client will file a legitimate claim in a given year is 4 percent. The company predicts that the average payout for each legitimate claim will be $2600. In order to be profitable, the company must earn an average of $200 per client per year, after deducting the amount of money spent on claims. How much should the company charge policyholders for the insurance?

35. **Insurance Rates** An insurance company estimates that the probability that a client will file a legitimate claim in a given year is 3 percent. In order to be profitable, the company must earn an average of $200 per client per year, after deducting the amount of money spent on claims. The company is currently charging each client $350 per year. In order to reach its profitability goal, what level does it need to keep its average payout below?

8.5 Measures of Central Tendency and Dispersion

- Calculate the mean, median, mode, and standard deviation of a data set

GETTING STARTED Every year, the U.S. government makes direct payments to the agriculture industry. The amount of money paid to the industry varies widely from year to year. By using a variety of statistical measures, we can anticipate future agricultural funding requirements. In this section, we will investigate an area of mathematics referred to as *statistics*. We will introduce the statistical measures of *mean*, *median*, *mode*, and *standard deviation* and apply them to real-life scenarios such as the agricultural payments.

Measures of Central Tendency

Table 8.14 shows the amount of money the government paid directly to the agriculture industry between 1990 and 2000.

TABLE 8.14 Direct Government Payments to Agriculture Industry

Year (t)	Payments (P) (billions of dollars)	Year (t)	Payments (P) (billions of dollars)
1990	9.3	1996	7.3
1991	8.2	1997	7.5
1992	9.2	1998	12.4
1993	13.4	1999	21.5
1994	7.9	2000	22.9
1995	7.3		

Source: *Statistical Abstract of the United States, 2001*, Table 807, p. 527.

What is the typical amount of money that the government pays to the agriculture industry in a given year? We can answer this question using each of the three common **measures of central tendency** that are used in statistical analysis: mean, median, and mode. The mean finds the "typical" payment by dividing the sum of all payments by the number of payments. Many people refer to this as the *average* payment.

MEAN

The *mean*, $\bar{x}$, (read "x bar") of a group of n numbers $x_1, x_2, \ldots, x_n$ is given by

$$\bar{x} = \frac{x_1 + x_2 + \cdots + x_n}{n}$$

To calculate the mean of the payments, we add up the payment amounts and divide by the number of years.

$$\bar{x} = \frac{9.3 + 8.2 + 9.2 + 13.4 + 7.9 + 7.3 + 7.3 + 7.5 + 12.4 + 21.5 + 22.9}{11}$$

$$= \frac{126.9}{11}$$

$$\approx 11.5 \text{ billion dollars per year}$$

Over the 11-year period, the mean (or average) annual payment amount was $11.5 billion.

The median finds the "typical" payment by giving the middle value of all payments listed in order from smallest to largest.

MEDIAN

The *median* of a group of n numbers $x_1, x_2, \ldots, x_n$ is found by listing the numbers in numerical order and locating the value(s) in the center of the list. If n is odd, there will be a single value at the center of the list. If n is even, there will be two values at the center of the list, and the median is the mean of these two values.

To calculate the median of the payments, we must first list the payment amounts in numerical order.

$$7.3, \quad 7.3, \quad 7.5, \quad 7.9, \quad 8.2, \quad \underset{\underset{\text{Median}}{\uparrow}}{9.2}, \quad 9.3, \quad 12.4, \quad 13.4, \quad 21.5, \quad 22.9$$

Since there are an odd number of values in the list, the centermost value will be the median. The median amount of money spent by the government annually on direct agricultural payments is $9.2 billion.

The mode finds the "typical" payment by choosing the payment amount that occurs most often.

MODE

The *mode* of a group of n numbers $x_1, x_2, \ldots, x_n$ is the value that is repeated most often. If all values are repeated the same number of times, there is no mode.

To calculate the mode of the payments, we identify the payment amount that is repeated most often. All of the payment amounts occur exactly once with the exception of 7.3, which occurs twice. The mode of the payments is $7.3 billion.

Which of the measures of central tendency is the most useful? It depends. For the agricultural payment data, we have

$$\text{Mean} = 11.5$$

$$\text{Median} = 9.2$$

$$\text{Mode} = 7.3$$

Although the payment amount has ranged between $7.3 and $22.9 billion, the average amount of the annual payment is $11.5 billion. If we expect that future payments will oscillate within this range, the mean may be a good predictor for future payment amounts.

Observe that the payment amounts in 1999 and 2000 are substantially larger than the payments in all of the other years. If we believe that this dramatic increase in payment size is an anomaly and not indicative of a future trend, then the median may be the best measure to use. Unlike the mean, the median is not dramatically affected by a data value that is substantially different from the other values in the data set.

The mode for the agricultural payments was 7.3. This value occurred twice, while all other values occurred only once. If this value had appeared substantially more often (say 50 percent of all data values were 7.3), we might have concluded that the mode would be the best predictor of future payment amounts. However, since 7.3 occurred only one more time than all of the other data values, we do not expect it to be a good predictor of future payment amounts.

EXAMPLE 1

Calculating and Interpreting Measures of Central Tendency

The amount of money spent on aerobic shoes in the United States between 1991 and 2000 is shown in Table 8.15.

TABLE 8.15 Spending on Aerobic Shoes

Year (t)	Sales (S) (millions of dollars)
1991	600
1992	590
1993	500
1994	356
1995	372
1996	401
1997	380
1998	334
1999	275
2000	272

Source: *Statistical Abstract of the United States, 2001*, Table 1251, p. 766.

Calculate the mean, median, and mode for the annual amount of money spent on aerobic shoe sales and interpret the meaning of each result.

SOLUTION

$$\bar{x} = \frac{600 + 590 + 500 + 356 + 372 + 401 + 380 + 334 + 275 + 272}{10}$$

$$= \frac{4080}{10}$$

$$= 408$$

From 1991 through 2000, the mean (average) annual amount of money spent on aerobic shoes was $408 million.

To find the median, we sequence the sales values in numerical order and seek to find the centermost value.

$$272, 275, 334, 356, 372, 380, 401, 500, 590, 600$$

Since there is an even number of terms in the sequence, there are two values occupying the centermost positions. To determine the median, we find the mean of these two values.

$$\text{Median} = \frac{372 + 380}{2}$$

$$= 376$$

In five of the ten years, the amount of money spent on aerobic shoe sales was less than $376 million. In the remaining five years, the amount of money spent was more than $376 million.

Since all of the data values occur with the same frequency, there is no mode.

EXAMPLE 2

Calculating Measures of Central Tendency

In 2003–2004, the Scottsdale Community College Mathematics Department had 15 full-time faculty members. For each member of the department, we recorded the number of years that the member had been employed by the college. We summarized our results in the frequency distribution in Table 8.16.

TABLE 8.16

Years Employed (x)	2	4	7	8	9	14	15	20	21	29
Frequency (number of math faculty)	3	2	2	1	1	1	2	1	1	1

Source: Scottsdale Community College General Catalog and Student Handbook.

From the frequency distribution, we see that three members had been employed for two years, two members had been employed for four years, two members had been employed for seven years, and so on.

Calculate the mean, median, and mode for the number of years a department member had been employed by the college and interpret the meaning of each result. Then draw a histogram of the frequency distribution, indicating the location of the median and mean.

SOLUTION Calculating the mean from a frequency distribution is somewhat different from calculating the mean from a data set. The frequency value tells us the number of times we must write the corresponding value of x (e.g., since the frequency of 2 is 3, we write the 2 three times in the numerator sum).

$$\bar{x} = \frac{(2 + 2 + 2) + (4 + 4) + (7 + 7) + 8 + 9 + 14 + (15 + 15) + 20 + 21 + 29}{15}$$

$$= \frac{159}{15}$$

$$= 10.6 \text{ years}$$

The mean amount of time a member of the department had been employed by the college was 10.6 years. Does this signify that half of the members had worked fewer than 10.6 years and the other half of the members had worked more? No. We can see from the frequency distribution that more than half (9 of the 15 members) had worked fewer than 10.6 years.

The outlying value of 29 is largely responsible for increasing the size of the mean. The value 29 is 8 years away from the next closest data value (21), while all of the other values in the data set are within 3 units of the next closest data value. If the value 29 had been removed from the data set, the mean would have been 9.3. (This is 1.3 units lower than the mean of the entire data set.) In contrast, if one of the values of 7 had been removed from the data set, the mean would have been 10.9. (This is 0.3 unit higher than the mean of the entire data set.)

If we want to split the math faculty into two similarly sized groups based upon seniority (the number of years at the college), we need to use the median. Listing the data values in numerical order yields

$$2, 2, 2, 4, 4, 7, 7, 8, 9, 14, 15, 15, 20, 21, 29$$

The eighth term is the middle value and is equal to 8. That is, the median is 8. Seven of the faculty members had worked at the college for fewer than eight years, and seven of the faculty members had worked at the college for more than eight years. To create two similarly sized groups, we could add the one math faculty member who had worked at the college for eight years to either of the groups.

The mode of the number of years math faculty had worked at the college is 2, since this is the value with the largest frequency in our frequency distribution. If the math faculty were broken into groups based upon the year they were hired, the largest group would be those who had worked at the college for two years.

A histogram of the frequency distribution is shown in Figure 8.20.

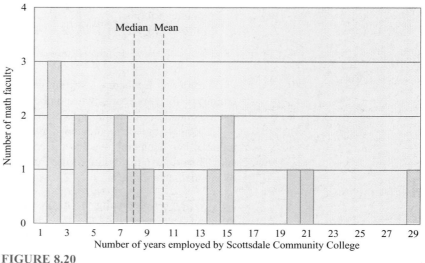

FIGURE 8.20

Of the three measures of central tendency, the mean is used the most often. It is easy to compute, and it tends to be a very reliable measure. That is, repeated samples taken from the same population are likely to have similar means. However, the mean is easily influenced by extreme values, as was shown in Example 2.

The median, on the other hand, is not significantly influenced by extreme values. However, the median is often difficult to calculate for large data sets, since calculating the median requires sequencing the values in numerical order. The sorting function of a spreadsheet is often used to sequence the values of a large data set.

The mode is easy to find but may not yield a unique value for each data set. For example, the data set {2, 2, 3, 3, 4, 4, 5, 6} has mode 2, 3, and 4! The values 2, 3, and 4 each occur twice, while the other values in the list occur only once.

Measures of Dispersion

While measures of central tendency identify the "middle" or "average" of a data set, **measures of dispersion** identify how "spread out" the data items are. For example, consider the sets of data in Table 8.17.

TABLE 8.17

List A	List B
4	1
5	1
6	1
6	1
7	7
7	7
8	12
8	12
9	14
10	14
Mean = 7	Mean = 7
Median = 7	Median = 7

Both data sets have the same mean and median. However, the data sets have very different frequency distributions, as shown in the histograms in Figures 8.21 and 8.22).

FIGURE 8.21

FIGURE 8.22

The data values from List A are clustered close to the mean. On the other hand, most of the data values from List B are far away from the mean. We can quantify the *dispersion* of the data by looking at the *deviations* from the mean. A **deviation from the mean** is the difference found by subtracting the mean from each number in the sample (Table 8.18).

TABLE 8.18

List A	Deviation from the Mean	List B	Deviation from the Mean
4	-3	1	-6
5	-2	1	-6
6	-1	1	-6
6	-1	1	-6
7	0	7	0
7	0	7	0
8	1	12	5
8	1	12	5
9	2	14	7
10	3	14	7
Mean = 7		Mean = 7	

To find a measure of dispersion, we might try finding the mean of the deviations. Unfortunately, the sum of the deviations from the mean will always equal zero because of the canceling of positive and negative values. To offset this effect, we sum the *squares* of the deviations. The data set that is more widely dispersed will result in the larger value.

List A:

$$\text{Sum of squares} = (-3)^2 + (-2)^2 + (-1)^2 + (-1)^2 + (0)^2 + (0)^2 + (1)^2 + (1)^2 + (2)^2 + (3)^2$$
$$= 9 + 4 + 1 + 1 + 0 + 0 + 1 + 1 + 4 + 9$$
$$= 30$$

List B:

$$\text{Sum of squares} = (-6)^2 + (-6)^2 + (-6)^2 + (-6)^2 + (0)^2 + (0)^2 + (5)^2 + (5)^2 + (7)^2 + (7)^2$$
$$= 36 + 36 + 36 + 36 + 0 + 0 + 25 + 25 + 49 + 49$$
$$= 292$$

To determine the mean of the sum of the squares of the deviations, we would divide each sum by the number of terms in the data set, n. However, statisticians typically divide the sum of the squares by $n - 1$ instead of by n. For large values of n, the results are essentially the same. (Using $n - 1$ instead of n makes some other statistical processes work out nicely.) Since $n = 10$ for both List A and List B, we will divide the sums of the squares of the deviations by 9. The result is referred to as the **sample variance.** The symbol s^2 is used to represent the sample variance.

$$\text{List A sample variance} = \frac{30}{9}$$

$$s^2 = 3.333$$

$$\text{List B sample variance} = \frac{292}{9}$$

$$s^2 = 32.44$$

It is often convenient to write the sample variance using *sigma notation*. The Greek letter sigma (Σ) is used to represent the concept of summing a number of values. The expression

$$(x_1 - \bar{x})^2 + (x_2 - \bar{x})^2 + \cdots + (x_n - \bar{x})^2$$

may be written as $\displaystyle\sum_{i=1}^{n}(x_i - \bar{x})^2$. The variable i is an index value that moves through whole-number values beginning at 1 and continuing until it reaches n. Each time the index value changes, another term is added to the summation. For example,

$$\sum_{i=1}^{n}(x_i - \bar{x})^2 = (x_1 - \bar{x})^2 + \sum_{i=2}^{n}(x_i - \bar{x})^2$$

$$= (x_1 - \bar{x})^2 + (x_2 - \bar{x})^2 + \sum_{i=3}^{n}(x_i - \bar{x})^2$$

$$= (x_1 - \bar{x})^2 + (x_2 - \bar{x})^2 + (x_3 - \bar{x})^2 + \sum_{i=4}^{n}(x_i - \bar{x})^2$$

$$\vdots$$

$$= (x_1 - \bar{x})^2 + (x_2 - \bar{x})^2 + \cdots + (x_{n-1} - \bar{x})^2 + \sum_{i=n}^{n}(x_i - \bar{x})^2$$

$$= (x_1 - \bar{x})^2 + (x_2 - \bar{x})^2 + \cdots + (x_{n-1} - \bar{x})^2 + (x_n - \bar{x})^2$$

Summation notation may feel cumbersome at first; however, as you practice, you will become more comfortable with it. The sample variance is typically written using summation notation.

SAMPLE VARIANCE

The *sample variance*, s^2, of a group of n numbers $x_1, x_2, \ldots, x_n$ with mean $\bar{x}$ is given by

$$s^2 = \frac{\displaystyle\sum_{i=1}^{n}(x_i - \bar{x})^2}{n - 1}$$

The sample variance may be calculated more easily by using the equivalent form

$$s^2 = \frac{\left[\displaystyle\sum_{i=1}^{n}(x_i)^2\right] - n(\bar{x})^2}{n - 1}$$

Converting from $s^2 = \dfrac{\displaystyle\sum_{i=1}^{n}(x_i - \bar{x})^2}{n - 1}$ to $s^2 = \dfrac{\left[\displaystyle\sum_{i=1}^{n}(x_i)^2\right] - n(\bar{x})^2}{n - 1}$ is not

immediately obvious. We will demonstrate the equivalency of the two expressions using $n = 2$. A similar approach may be used to show that the two expressions are equivalent for any value of n. For $n = 2$, the sample variance equation may be written as follows:

$$s^2 = \frac{\displaystyle\sum_{i=1}^{2}(x_i - \bar{x})^2}{2 - 1}$$

$$= \frac{(x_1 - \bar{x})^2 + (x_2 - \bar{x})^2}{1} \qquad \text{Convert from sigma notation}$$

$$= (x_1 - \bar{x})^2 + (x_2 - \bar{x})^2$$

$$= (x_1 - \bar{x})(x_1 - \bar{x}) + (x_2 - \bar{x})(x_2 - \bar{x})$$

$$= [(x_1)^2 - 2x_1\bar{x} + (\bar{x})^2] + [(x_2)^2 - 2x_2\bar{x} + (\bar{x})^2] \qquad \text{Calculate products}$$

$$= (x_1)^2 + (x_2)^2 - 2x_1\bar{x} - 2x_2\bar{x} + 2(\bar{x})^2 \qquad \text{Reorder the terms}$$

$$= (x_1)^2 + (x_2)^2 - 2\bar{x}(x_1 + x_2) + 2(\bar{x})^2 \qquad \text{Factor out } 2\bar{x}$$

$$= (x_1)^2 + (x_2)^2 - 2\bar{x}(2\bar{x}) + 2(\bar{x})^2 \qquad \text{Since } \bar{x} = \frac{x_1 + x_2}{2} \Rightarrow x_1 + x_2 = 2\bar{x}$$

$$= \left[\sum_{i=1}^{2}(x_i)^2\right] - 4(\bar{x})^2 + 2(\bar{x})^2 \qquad \text{Rewrite with sigma notation}$$

$$= \left[\sum_{i=1}^{2}(x_i)^2\right] - 2(\bar{x})^2 \qquad \text{Group like terms}$$

$$= \frac{\left[\displaystyle\sum_{i=1}^{2}(x_i)^2\right] - 2(\bar{x})^2}{2 - 1}$$

This alternative form of writing the sample variance is a useful shortcut for calculating the sample variance.

Since the sample variance was derived from the square of the deviations from the mean, we must calculate the square root of the sample variance to return to the same units as the data. The result is called the **sample standard deviation.**

SAMPLE STANDARD DEVIATION

The *sample standard deviation*, s, of a group of n numbers $x_1, x_2, \ldots, x_n$ with mean $\bar{x}$ is given by

$$s = \sqrt{\frac{\left[\displaystyle\sum_{i=1}^{n}(x_i)^2\right] - n(\bar{x})^2}{n - 1}}$$

The sample standard deviation is the most common measure of dispersion. If the data points tend to be bunched up close to the mean, the standard deviation will be relatively small. If the data points are spread out from the mean, the standard deviation will be relatively large.

Earlier in our discussion, we derived the following results for List A and List B:

$$\text{List A sample variance} = \frac{30}{9}$$

$$s^2 = 3.333$$

$$\text{List B sample variance} = \frac{292}{9}$$

$$s^2 = 32.44$$

It follows that

$$\text{List A sample standard deviation} = \sqrt{\frac{30}{9}}$$

$$s = 1.826$$

$$\text{List B sample standard deviation} = \sqrt{\frac{292}{9}}$$

$$s = 5.696$$

Note that the more widely dispersed data set has a much larger standard deviation.

The Russian mathematician Pafnuty L. Chebyshev (1821–1894) discovered a fascinating relationship between the sample standard deviation of a data set and the elements of the data set. This relationship is referred to as Chebyshev's rule.

CHEBYSHEV'S RULE

For any set of data, at least $\frac{k^2 - 1}{k^2}$ data points fall within k standard deviations of the mean. That is, at least $\frac{k^2 - 1}{k^2}$ data points lie in the interval $[\bar{x} - ks, \bar{x} + ks]$.

According to Chebyshev's Rule, at least $\frac{3}{4}$ of the data points of any data set fall within two sample standard deviations of the mean. For List A, we have

$$\bar{x} - 2s = 7 - 2(1.826) \qquad \bar{x} + 2s = 7 + 2(1.826)$$
$$= 3.348 \qquad \text{and} \qquad = 10.652$$

So at least 75 percent $\left(\frac{3}{4}\right)$ of the data points of List A lie within the interval $[3.348, 10.652]$. In fact, all 10 of the data points of List A $\{4, 5, 6, 6, 7, 7, 8, 8, 9, 10\}$ lie within this interval.

For List B, we have

$$\bar{x} - 2s = 7 - 2(5.696) \qquad \bar{x} + 2s = 7 + 2(5.696)$$
$$= -4.392 \qquad \text{and} \qquad = 18.392$$

So at least 75 percent $\left(\frac{3}{4}\right)$ of the data points of List B lie within the interval $[-4.392, 18.392]$. In fact, all 10 of the data points of List B $\{1, 1, 1, 1, 7, 7, 12, 12, 14, 14\}$ lie within this interval.

EXAMPLE **3**

Finding the Mean and Standard Deviation

From 1965 through 1999, there were 3753 successful space launches worldwide (Table 8.19). (A successful space launch entails attainment of Earth orbit or Earth escape.)

TABLE 8.19   Worldwide Space Launches

Time Period (t)	Successful Launches (N)
1965–1969	586
1970–1974	555
1975–1979	607
1980–1984	605
1985–1989	550
1990–1994	466
1995–1999	384

Source: *Statistical Abstract of the United States, 2001*, Table 793, p. 519.

Determine the mean and sample standard deviation for the number of successful launches in a five-year period. How many of the data points lie within one sample standard deviation of the mean?

SOLUTION

$$\bar{x} = \frac{586 + 555 + 607 + 605 + 550 + 466 + 384}{7}$$

$$= \frac{3753}{7}$$

$$\approx 536$$

An average of 536 successful space launches occurred during each five-year period.

To calculate the sample standard deviation, we will first calculate the component parts of the equation and then substitute in the simplified results. We first determine $\sum_{i=1}^{n}(x_i)^2$.

$$\sum_{i=1}^{7}(x_i)^2 = (586)^2 + (555)^2 + (607)^2 + (605)^2 + (550)^2 + (466)^2 + (384)^2$$

$$= 2{,}053{,}007$$

We next find $n(\bar{x})^2$.

$$7(\bar{x})^2 = 7\left(\frac{3753}{7}\right)^2$$

$$\approx 2{,}012{,}144$$

Substituting these results into the sample standard deviation equation yields

$$s = \sqrt{\frac{\left[\sum_{i=1}^{7}(x_i)^2\right] - 7(\bar{x})^2}{7-1}}$$

$$= \sqrt{\frac{2,053,007 - 2,012,144}{6}}$$

$$= \sqrt{\frac{40,863}{6}}$$

$$\approx 83$$

The sample standard deviation (rounded to the nearest whole number) is equal to 83 successful launches.

To determine the interval of values within one standard deviation of the mean, we calculate the following:

$$\bar{x} - s = 536 - 83 \qquad \text{and} \qquad \bar{x} + s = 536 + 83$$
$$= 453 \qquad\qquad\qquad\qquad = 619$$

All data points within one standard deviation of the mean lie in the interval [453, 619]. Six of the seven points lie within this interval. Only the value 384 is not within one standard deviation of the mean.

The TI-83 Plus quickly and easily calculates the mean, median, and standard deviation of a data set, as shown in the following Technology Tip.

TECHNOLOGY TIP

Calculating the Mean, Median, and Standard Deviation

1. Press [STAT] [ENTER] to open the Statistics Editor. Enter the values from the list in L1.

2. Press [STAT] and use the blue arrows to move to the CALC menu.

(Continued)

3. Press [ENTER] to place the command
   1-Var Stats on the home screen and
   press [ENTER] again to calculate statistics
   for the data set. The mean of the data set
   is $\bar{x} \approx 536$. The sample standard deviation
   (given by $Sx = 82.52 \ldots$) is $s \approx 83$.

```
1-Var Stats
x̄=536.1428571
Σx=3753
Σx²=2053007
Sx=82.5256093
σx=76.40386784
↓n=7
```

4. To find the median, press the blue down
   arrow to display the rest of the statistical
   results. The median is shown as
   Med=555. The median is 555.

```
1-Var Stats
↑n=7
minX=384
Q₁=466
Med=555
Q₃=605
maxX=607
```

The technology tip also shows the **population standard deviation** σ. The Greek letter σ is read "sigma." It is the lowercase form of Σ. The population standard deviation is the square root of the **population variance.** The population variance is given by

$$\sigma^2 = \frac{\left[\displaystyle\sum_{i=1}^{n}(x_i)^2\right] - n(\bar{x})^2}{n}$$

and the population standard deviation is given by

$$\sigma = \sqrt{\frac{\left[\displaystyle\sum_{i=1}^{n}(x_i)^2\right] - n(\bar{x})^2}{n}}$$

These equations are nearly identical to those used to calculate the sample variance and sample standard deviation. The only difference is that the denominator of the rational expression is n instead of $n - 1$. We will use the population standard deviation extensively as we discuss normal distributions in the next section.

EXAMPLE 4

Finding the Mean and Standard Deviation

On September 25, 2004, travelocity.com advertised a variety of fares for a one-way flight on December 25, 2004, from Phoenix, Arizona, to Orlando, Florida, as shown by the frequency distribution in Table 8.20.

TABLE 8.20

Fare	Frequency
$152	5
$159	1
$184	1
$197	1
$210	2
$242	3
$254	1
$284	1
$301	2
$307	1
$314	1
$322	1

Source: www.travelocity.com.

Calculate the mean and the sample standard deviation for the data set. Then determine the number of data values within two standard deviations of the mean.

SOLUTION   We can determine the total number of flights by summing the frequencies.

$$\text{Total number of flights} = 5 + 1 + 1 + 1 + 2 + 3 + 1 + 1 + 2 + 1 + 1 + 1$$
$$= 20$$

Because of the relatively large size of the data set, we will use technology to calculate the mean and standard deviation, as demonstrated in the previous Technology Tip (see Figure 8.23).

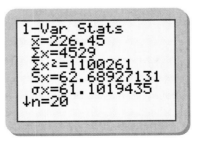

FIGURE 8.23

We see that $\bar{x} = 226.45$. The mean (average) ticket price is $226.45. The sample standard deviation is $62.69. The equation $\Sigma x = 4529$ means that the sum of the 20 fares is $4529. The equation $n = 20$ indicates that 20 data points have been entered into the calculator. (It is good practice to count the number of data points manually as well to ensure that you didn't inadvertently omit entering one of the data points.)

We are asked to find the number of data points that are within two standard deviations of the mean.

$$\bar{x} - 2s = 226.45 - 2(62.69) \qquad \bar{x} + 2s = 226.45 + 2(62.69)$$
$$= 226.45 - 125.38 \qquad\qquad\quad\; = 226.45 + 125.38$$
$$= 101.07 \qquad\qquad\qquad\qquad = 351.83$$

Since all 20 fares are between \$101.07 and \$351.83, all of the fares lie within two standard deviations of the mean.

8.5 Summary

In this section, you learned how to use the statistical measures of mean, median, mode, and standard deviation in analyzing a variety of real-life data sets. You also discovered that Chebyshev's rule is useful for estimating the dispersion of the points in any data set.

8.5 Exercises

In Exercises 1–15, calculate the mean, median, and mode for each data set and interpret the meaning of each result. Create a frequency distribution for each data set. Then draw a histogram of the frequency distribution, indicating the location of the median and mean.

1. **University of Oklahoma Football** As of the end of the 2003 season, 16 football players had scored 30 or more touchdowns during their careers at the University of Oklahoma. The number of touchdowns scored was as follows: 57, 53, 51, 43, 41, 37, 37, 36, 35, 35, 34, 32, 32, 31, 30, and 30. (**Source:** www.soonerstats.com.)

2. **Space Shuttle Columbia** The space shuttle Columbia flew 30 missions between 1984 and 2001; however, the number of missions flown each year varied as shown in the following table.

Year	Number of Missions
1984	2
1985	4
1986	0
1987	0
1988	1
1989	2
1990	2
1991	2
1992	2
1993	2
1994	2
1995	2
1996	0
1997	2
1998	2
1999	2
2000	1
2001	2

Source: www.pao.ksc.nasa.gov.

The space shuttle Columbia and her crew were lost during entry on February 1, 2003, 16 minutes before the scheduled landing time.

3. 🌐 **Yale University Football Ranking**   In the early 1900s, Yale University was a football powerhouse. The following table shows how one ranking system rated the Yale team from 1900 to 1909.

Year	Ranking
1900	1
1901	4
1902	2
1903	3
1904	7
1905	2
1906	4
1907	6
1908	9
1909	4

Source: www.kiko13.com.

4. 🌐 **Death Rates**   Death rates in the United States between 1950 and 1995 have fluctuated as shown in the following table.

Year	Deaths (per 1000 people)
1950	9.6
1955	9.3
1960	9.5
1965	9.4
1970	9.5
1975	8.8
1980	8.8
1985	8.8
1990	8.6
1995	8.8

Source: *Statistical Abstract of the United States, 2001,* Table 68, p. 59.

5. 🌐 **Births to Teenagers**   Between 1990 and 1999, the percentage of all births that were to teenage mothers ranged from 12.3 percent to 13.1 percent, as shown in the following table.

Year	Births to Teenagers (percent of all births)
1990	12.8
1991	12.9
1992	12.7
1993	12.8
1994	13.1
1995	13.1
1996	12.9
1997	12.7
1998	12.5
1999	12.3

Source: *Statistical Abstract of the United States, 2001,* Table 76, p. 63.

6. 🌐 **Births to Unmarried Women**   Between 1990 and 1999, the percentage of all births that were to unmarried women ranged from 26.6 percent to 33.0 percent, as shown in the following table.

Year	Births to Unmarried Mothers (percent of all births)
1990	26.6
1991	28.0
1992	30.1
1993	31.0
1994	32.6
1995	32.2
1996	32.4
1997	32.4
1998	32.8
1999	33.0

Source: *Statistical Abstract of the United States, 2001,* Table 76, p. 63.

7. 🌐 **Tropical Storms and Hurricanes**
   Between 1980 and 2000, the number of
   tropical storms and hurricanes in the North
   Atlantic ranged between 4 and 19 per year, as
   shown in the following table.

Year	Tropical Storms and Hurricanes
1980	11
1981	11
1982	5
1983	4
1984	12
1985	11
1986	6
1987	7
1988	12
1989	11
1990	14
1991	8
1992	7
1993	8
1994	7
1995	19
1996	13
1997	7
1998	14
1999	12
2000	15

Source: *Statistical Abstract of the United States,
2001*, Table 368, p. 221.

8. 🌐 **U.S. Marines**   Between 1980 and 2000,
   the number of U.S. Marines ranged
   between 173 and 200 thousand, as shown in the
   following table.

Year	U.S. Marines (thousands)
1980	188
1981	191
1982	192
1983	194
1984	196
1985	198
1986	199
1987	200
1988	197
1989	197
1990	197
1991	194
1992	185
1993	178
1994	174
1995	175
1996	175
1997	174
1998	173
1999	173
2000	173

Source: *Statistical Abstract of the
United States, 2001*, Table 500,
p. 329.

9. **Homes of More than 2400 Square Feet** Between 1985 and 1999, the percentage of new homes built that had 2400 or more square feet ranged from 17 percent to 34 percent, as shown in the following table.

Year	Percentage of New Homes with 2400+ Square Feet (percent)
1985	17
1986	18
1987	21
1988	25
1989	26
1990	29
1991	28
1992	29
1993	29
1994	28
1995	28
1996	30
1997	31
1998	32
1999	34

Source: *Statistical Abstract of the United States, 2001*, Table 938, p. 597.

10. **Homes of Less than 1200 Square Feet** Between 1985 and 1999, the percentage of new homes built that had 1200 or less square feet ranged from 7 percent to 20 percent, as shown in the following table.

Year	Percentage of New Homes with Less than 1200 Square Feet (percent)
1985	20
1986	17
1987	14
1988	12
1989	13
1990	11
1991	12
1992	10
1993	9
1994	9
1995	10
1996	9
1997	8
1998	7
1999	7

Source: *Statistical Abstract of the United States, 2001*, Table 938, p. 597.

11. **Fuel Production** Between 1990 and 1999, the U.S. percentage of the world's fuel production ranged from 19 percent to 23 percent, as shown in the following table.

Year	U.S. Percentage of World Fuel Production (percent)
1990	19
1991	20
1992	20
1993	19
1994	21
1995	20
1996	21
1997	21
1998	22
1999	23

Source: *Statistical Abstract of the United States, 2001*, Table 872, p. 559.

12. **Milled Rice Production** Between 1990 and 2000, the U.S. percentage of the world's milled rice production ranged from 1.4 percent to 1.8 percent, as shown in the following table.

Year	U.S. Percentage of World Milled Rice Production (percent)
1990	1.5
1991	1.4
1992	1.6
1993	1.5
1994	1.8
1995	1.5
1996	1.4
1997	1.5
1998	1.5
1999	1.6
2000	1.5

Source: *Statistical Abstract of the United States, 2001*, Table 821, p. 534.

13. **Citrus Fruit Consumption** Between 1980 and 2000, the per capita consumption of citrus fruit in the United States ranged from 19 to 28 pounds, as shown in the following table.

Year	Per Capita Citrus Fruit Consumption (pounds)
1980	26
1981	24
1982	23
1983	28
1984	23
1985	22
1986	24
1987	24
1988	25
1989	24
1990	21
1991	19
1992	24
1993	26
1994	25
1995	24
1996	25
1997	27
1998	27
1999	21
2000	26

Source: *Statistical Abstract of the United States, 2001*, Table 832, p. 539.

14. **Potato Consumption** Between 1980 and 1999, the per capita consumption of potatoes in the United States ranged from 46 to 51 pounds, as shown in the following table.

Year	Per Capita Potato Consumption (pounds)
1980	51
1981	46
1982	47
1983	50
1984	48
1985	46
1986	49
1987	48
1988	50
1989	50
1990	47
1991	50
1992	49
1993	51
1994	50
1995	50
1996	51
1997	49
1998	48
1999	48

Source: *Statistical Abstract of the United States, 2001,* Table 832, p. 539.

15. **Top Ten Movies** In 2003, the movie with the highest gross theater income was *The Lord of the Rings: Return of the King*, distributed by New Line studio. The following table shows the number of New Line movies that ranked in the 10 top-grossing movies for each of the given years.

Year	Number of New Line Movies in the 10 Top Grossing Movies
1993	0
1994	2
1995	1
1996	0
1997	0
1998	1
1999	1
2000	0
2001	2
2002	2
2003	2

Source: www.boxofficemojo.com.

In Exercises 16–30, calculate the mean and the sample standard deviation for each data set. Then determine the percentage of data points within two standard deviations of the mean. (You may want to apply Chebyshev's rule, as appropriate, to substantiate your results.)

16. **Home Prices** On March 29, 2004, Realtor.com advertised five homes in Ivins, Utah, in the $200,000–$225,000 price range. The amount of square footage offered by each home was 3400, 3400, 2738, 4332, and 4332 square feet, respectively. (Source: www.realtor.com.)

17. **Home Prices** On March 29, 2004, Realtor.com advertised four homes in the Bronx and in the city of Yonkers, New York, in the $200,000–225,000 price range. The amount of square footage offered by each home was 1600, 1500, 1116, and 1497 square feet, respectively. (Source: www.realtor.com.)

18. **Air Fares** On April 2, 2004, travelocity.com advertised a variety of fares for a one-way flight on April 8, 2004, from Seattle to Phoenix, as shown by the following frequency distribution.

Fare	Frequency
$93	4
$94	6
$101	3
$104	4
$154	1
$162	1
$164	3

Source: www.travelocity.com.

19. **Arnold Schwarzenegger Movies** Between 1991 and 2003, Arnold Schwarzenegger was featured in at least 10 movies. The gross income from theaters for each movie ranged from $34.6 million to $146.3 million, as shown in the following table.

Movie	Gross Income (millions)
Terminator 3: Rise of the Machines	$143.7
Collateral Damage	$40.1
The 6th Day	$34.6
End of Days	$66.9
Batman & Robin	$107.3
Jingle All the Way	$60.6
Eraser	$101.3
Junior	$36.8
True Lies	$146.3
Last Action Hero	$50.0

Source: www.boxofficemojo.com.

20. **Drew Barrymore Movies** Between 1998 and 2003, Drew Barrymore was featured in at least 12 movies. The gross income from theaters for each movie ranged from $0.5 million to $125.3 million, as shown in the following table.

Movie	Gross Income (millions)
Duplex	$9.7
Charlie's Angels 2: Full Throttle	$100.8
Confessions of a Dangerous Mind	$16.0
Riding in Cars with Boys	$30.2
Freddy Got Fingered	$14.3
Donnie Darko	$0.5
Charlie's Angels	$125.3
Titan A.E. (voice)	$22.8
Never Been Kissed	$55.5
Home Fries	$10.5
Ever After: A Cinderella Story	$65.7
The Wedding Singer	$80.2

Source: www.boxofficemojo.com.

21. **Birth Rates** Birth rates in the United States between 1950 and 1995 have varied widely, as shown in the following table.

Year	Births (per 1,000 people)
1950	24.1
1955	25.0
1960	23.7
1965	19.4
1970	18.4
1975	14.6
1980	15.9
1985	15.8
1990	16.7
1995	14.8

Source: *Statistical Abstract of the United States, 2001*, Table 68, p. 59.

22. **Infant Death Rates**   Infant death rates in the United States declined dramatically between 1950 and 1995, as shown in the following table.

Year	Infant Deaths (per 1,000 live births)
1950	29.2
1955	26.4
1960	26.0
1965	24.7
1970	20.0
1975	16.1
1980	12.6
1985	10.6
1990	9.2
1995	7.6

Source: *Statistical Abstract of the United States, 2001*, Table 68, p. 59.

23. **Divorce Rates**   Divorce rates in the United States between 1950 and 1995 have fluctuated as shown in the following table.

Year	Divorces (per 1,000 people)
1950	2.6
1955	2.3
1960	2.2
1965	2.5
1970	3.5
1975	4.8
1980	5.2
1985	5.0
1990	4.7
1995	4.1

Source: *Statistical Abstract of the United States, 2001*, Table 68, p. 59.

24. **Marriage Rates**   Marriage rates in the United States between 1950 and 1995 have fluctuated as shown in the following table.

Year	Marriages (per 1,000 people)
1950	11.1
1955	9.3
1960	8.5
1965	9.3
1970	10.6
1975	10.0
1980	10.6
1985	10.1
1990	9.8
1995	7.6

Source: *Statistical Abstract of the United States, 2001*, Table 68, p. 59.

25. **Women's Basketball Champions**   On April 6, 2004, the University of Connecticut women's basketball team defeated Tennessee and won the national title for the third year in a row. In the final six games leading up to the championship, Connecticut beat its opponents by notable point margins. The following table shows the team played and the difference between the winning and losing scores for each of these games.

Team	Difference Between Winning and Losing Scores
Penn	36
Auburn	26
U.C. Santa Barbara	8
Penn State	17
Minnesota	9
Tennessee	9

Source: www.uconnhuskies.com.

26. **Air Pollution** Between 1990 and 1999, the national ambient air carbon monoxide concentration decreased, as shown in the following table.

Year	Carbon Monoxide Pollution (in parts per million)
1990	5.8
1991	5.7
1992	5.3
1993	5.0
1994	5.1
1995	4.6
1996	4.3
1997	3.9
1998	3.8
1999	3.7

Source: *Statistical Abstract of the United States, 2001*, Table 355, p. 216.

27. **Tornadoes** Between 1980 and 1997, the number of tornadoes ranged between 656 and 1298 per year, as shown in the following table.

Year	Tornadoes	Year	Tornadoes
1980	866	1989	856
1981	783	1990	1,133
1982	1,046	1991	1,132
1983	931	1992	1,298
1984	907	1993	1,176
1985	684	1994	1,082
1986	764	1995	1,235
1987	656	1996	1,170
1988	702	1997	1,148

Source: *Statistical Abstract of the United States, 2001*, Table 368, p. 221.

28. **Tropical Storms and Hurricanes** Between 1980 and 2000, the number of tropical storms and hurricanes in the North Atlantic ranged between 4 and 19 per year, as shown in the following table.

Year	Tropical Storms and Hurricanes
1980	11
1981	11
1982	5
1983	4
1984	12
1985	11
1986	6
1987	7
1988	12
1989	11
1990	14
1991	8
1992	7
1993	8
1994	7
1995	19
1996	13
1997	7
1998	14
1999	12
2000	15

Source: *Statistical Abstract of the United States, 2001*, Table 368, p. 221.

29. **Real Networks Net Revenue** The net revenue of Real Networks, Inc., between 1999 and 2003 is shown in the following table.

Year	Net Revenue (thousands of dollars)
1999	131,242
2000	241,538
2001	188,905
2002	182,679
2003	202,377

Source: Real Networks Inc 2003 Report, p. 14.

30. **Education and Earnings** The average annual earnings of full-time, male workers in 1999 is shown in the following table based on their level of educational attainment.

Educational Attainment	Average Earnings
Less than 9th grade	$18,743
9th–12th grade (no diploma)	$18,908
High school graduate (or equivalent)	$30,414
Some college (no degree)	$33,614
Associate degree	$40,047
Bachelor's degree	$57,706
Master's degree	$68,367
Professional	$120,352
Doctorate	$97,357

Source: *Statistical Abstract of the United States, 2001*, Table 677, p. 441.

Exercises 31–35 are intended to challenge your understanding of measures of central tendency and dispersion.

31. **Starting Salaries for Law Graduates** A professional football player with a multimillion-dollar contract pursues a law degree in the off-season. In the year after the player obtains his degree, which measure of first-year salary of new law school graduates will be most accurate: mean, median, or mode? Explain.

32. Describe the relationship between the mean of a data set and the expected value of a random variable.

33. **Dice Game Activity** Roll a pair of dice 10 times, each time noting the sum of the face values of the dice. Calculate the mean, median, mode, and standard deviation for the data set. Do you think the results would be close to the same if the dice were rolled 100 times? 1000 times? Explain.

34. Using a spreadsheet to simulate the rolling of dice, the author "rolled" a pair of dice 1000 times and summed their face values. The frequency of each event is recorded in the following table.

Sum of Face Values	Frequency
2	31
3	49
4	90
5	111
6	127
7	160
8	136
9	106
10	96
11	57
12	37

Calculate the mean, median, mode, and standard deviation for the data set. Do these results support the conclusions you made in Exercise 33? Explain.

35. Two data sets each have the following statistics:

$$\text{Mean} = 4$$
$$\text{Median} = 4$$
$$\text{Mode} = none$$
$$\text{Standard deviation} = 1$$

Are the two data sets identical? Justify your answer.

8.6 Normal Distributions

- Convert a normal distribution to a standard normal distribution
- Determine the area of a shaded region beneath a normal curve
- Calculate the probability of attaining a value within a designated number of standard deviations of the mean of a normal distribution

GETTING STARTED   Many colleges use placement tests such as the SAT as a factor in determining scholarship recipients. They assume that students who outperform their peers on the SAT are likely to excel in their academic studies at the college. Many large data sets representing human characteristics or performance, such as SAT scores, may be modeled by *normal distributions.*

In this section we will introduce normal distributions. Normal distributions are one of the most common types of probability distributions occurring in the real world. Normal distributions are *continuous* probability distributions, since the data values of the distribution may assume any one of an infinite number of values within a given interval. Recall that not all probability distributions are continuous. Consider the histogram of the discrete probability distribution for the sum of two dice given in Section 8.4 (Figure 8.24). The sum of two dice may take on only whole-number values between 2 and 12. It is impossible for the sum of two dice to equal any other value (e.g., 3.463).

FIGURE 8.24

This distribution is a *discrete* probability distribution, since there are a finite number of possible outcomes (in this case, 11 outcomes).

In contrast, the birth weight of a boy is a *continuous distribution* because the boy's weight may take on any number of values within a given interval. Is it possible for a newborn boy to weigh 6.21347 pounds? Absolutely! Although a weight scale may be unable to detect the difference between a newborn weighing 6.21000 pounds and a newborn weighing 6.21347 pounds, the fact remains that the newborns have different weights. If we were to draw the graph of the continuous distribution for the weights of newborn boys, it would look something like the continuous curve shown in Figure 8.25.

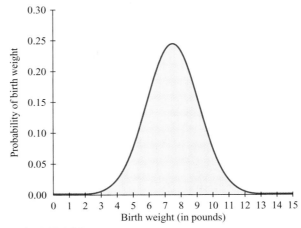

FIGURE 8.25

As shown in the graph, most newborn boys weigh between 5.5 and 9.5 pounds. The median birth weight of a newborn boy is about 7.5 pounds. (**Source:** National Center for Health Statistics growth chart at www.cdc.gov.)

Although discrete probability distributions are not continuous distributions, continuous curves are often used to model discrete probability distributions. The normal curve is one of the most common types of continuous curves used to model discrete data sets.

NORMAL DISTRIBUTION

A **normal distribution** is a continuous probability distribution with the following properties:

1. The graph of the distribution is a bell-shaped curve with a horizontal asymptote at the horizontal axis. (The graph is called a *normal curve.*)

2. The mean, median, and mode of the distribution are equal to one another.

3. The graph of the distribution is symmetric about the vertical line that passes through the population mean μ. (μ is the Greek letter mu.)

4. The distribution may be completely determined by knowing the population standard deviation σ and population mean μ of the distribution.

5. The area below the graph of the distribution and above the horizontal axis is equal to 1.

Figures 8.26 to 8.28 show the graphs of three different normal distributions. Each of the distributions has a mean of $\mu = 0$; however, they have different standard deviations. Notice that as the standard deviation σ gets larger, the normal curve becomes flatter.

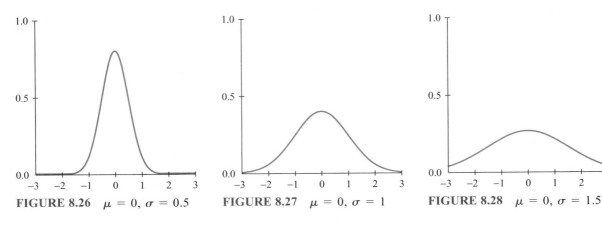

FIGURE 8.26 $\mu = 0, \sigma = 0.5$ **FIGURE 8.27** $\mu = 0, \sigma = 1$ **FIGURE 8.28** $\mu = 0, \sigma = 1.5$

The graph shown in Figure 8.27 is the normal curve for the *standard normal distribution.*

STANDARD NORMAL DISTRIBUTION

The **standard normal distribution** is the normal distribution with $\mu = 0$ and $\sigma = 1$.

Any normal distribution may be converted to a standard normal distribution by calculating the *z-scores* for the distribution. The **z-score** for a particular value x is given by $z = \frac{x - \mu}{\sigma}$. The *z*-score shows how many standard deviations x is away from the mean. For example, suppose a normal distribution has $\mu = 1.2$ and $\sigma = 0.4$. The *z*-score for $x = 2.2$ is

$$z = \frac{2.2 - 1.2}{0.4}$$

$$= \frac{1.0}{0.4}$$

$$= 2.5$$

$x = 2.2$ is 2.5 standard deviations away from the mean. We can easily confirm the result by adding 2.5 standard deviations to the mean.

$$\mu + 2.5\sigma = 1.2 + 2.5(0.4)$$

$$= 1.2 + 1.0$$

$$= 2.2$$

EXAMPLE 1

Finding the z-Score for a Normal Distribution

Determine the *z*-score for $x = 15$ given a normal distribution with $\mu = 20$ and $\sigma = 3$.

SOLUTION

$$z = \frac{x - \mu}{\sigma}$$

$$= \frac{15 - 20}{3}$$

$$= -\frac{5}{3}$$

$$\approx -1.67$$

The value $x = 15$ is approximately 1.67 standard deviations below the mean.

Area Under a Normal Curve

All normal distributions have the amazing property that the area of the shaded region between the normal curve and the horizontal axis on the interval $[a, b]$ is equal to the probability that a randomly selected data value will fall within the interval $[a, b]$. Consequently, the probability that a z-score lies within n standard deviations of the mean is determined by the area of the shaded region between the standard normal curve and the horizontal axis on the interval $[-n, n]$. Since any normal distribution may be converted to a standard normal distribution, the area under the standard normal curve on an interval $[0, z]$ is typically listed in a table displaying various values of z (Table 8.21). By using the table, we can quickly determine the probability that a z-score lies within n standard deviations of the mean.

TABLE 8.21

The area A under the standard normal curve on the interval $[0, z]$ is given in the table.

z	A	z	A	z	A	z	A	z	A	z	A	z	A	z	A
0.00	0.0000	0.21	0.0832	0.42	0.1628	0.63	0.2357	0.84	0.2995	1.05	0.3531	1.26	0.3962	1.47	0.4292
0.01	0.0040	0.22	0.0871	0.43	0.1664	0.64	0.2389	0.85	0.3023	1.06	0.3554	1.27	0.3980	1.48	0.4306
0.02	0.0080	0.23	0.0910	0.44	0.1700	0.65	0.2422	0.86	0.3051	1.07	0.3577	1.28	0.3997	1.49	0.4319
0.03	0.0120	0.24	0.0948	0.45	0.1736	0.66	0.2454	0.87	0.3078	1.08	0.3599	1.29	0.4015	1.50	0.4332
0.04	0.0160	0.25	0.0987	0.46	0.1772	0.67	0.2486	0.88	0.3106	1.09	0.3621	1.30	0.4032	1.51	0.4345
0.05	0.0199	0.26	0.1026	0.47	0.1808	0.68	0.2517	0.89	0.3133	1.10	0.3643	1.31	0.4049	1.52	0.4357
0.06	0.0239	0.27	0.1064	0.48	0.1844	0.69	0.2549	0.90	0.3159	1.11	0.3665	1.32	0.4066	1.53	0.4370
0.07	0.0279	0.28	0.1103	0.49	0.1879	0.70	0.2580	0.91	0.3186	1.12	0.3686	1.33	0.4082	1.54	0.4382
0.08	0.0319	0.29	0.1141	0.50	0.1915	0.71	0.2611	0.92	0.3212	1.13	0.3708	1.34	0.4099	1.55	0.4394
0.09	0.0359	0.30	0.1179	0.51	0.1950	0.72	0.2642	0.93	0.3238	1.14	0.3729	1.35	0.4115	1.56	0.4406
0.10	0.0398	0.31	0.1217	0.52	0.1985	0.73	0.2673	0.94	0.3264	1.15	0.3749	1.36	0.4131	1.57	0.4418
0.11	0.0438	0.32	0.1255	0.53	0.2019	0.74	0.2704	0.95	0.3289	1.16	0.3770	1.37	0.4147	1.58	0.4429
0.12	0.0478	0.33	0.1293	0.54	0.2054	0.75	0.2734	0.96	0.3315	1.17	0.3790	1.38	0.4162	1.59	0.4441
0.13	0.0517	0.34	0.1331	0.55	0.2088	0.76	0.2764	0.97	0.3340	1.18	0.3810	1.39	0.4177	1.60	0.4452
0.14	0.0557	0.35	0.1368	0.56	0.2123	0.77	0.2794	0.98	0.3365	1.19	0.3830	1.40	0.4192	1.61	0.4463
0.15	0.0596	0.36	0.1406	0.57	0.2157	0.78	0.2823	0.99	0.3389	1.20	0.3849	1.41	0.4207	1.62	0.4474
0.16	0.0636	0.37	0.1443	0.58	0.2190	0.79	0.2852	1.00	0.3413	1.21	0.3869	1.42	0.4222	1.63	0.4484
0.17	0.0675	0.38	0.1480	0.59	0.2224	0.80	0.2881	1.01	0.3438	1.22	0.3888	1.43	0.4236	1.64	0.4495
0.18	0.0714	0.39	0.1517	0.60	0.2257	0.81	0.2910	1.02	0.3461	1.23	0.3907	1.44	0.4251	1.65	0.4505
0.19	0.0753	0.40	0.1554	0.61	0.2291	0.82	0.2939	1.03	0.3485	1.24	0.3925	1.45	0.4265	1.66	0.4515
0.20	0.0793	0.41	0.1591	0.62	0.2324	0.83	0.2967	1.04	0.3508	1.25	0.3944	1.46	0.4279	1.67	0.4525

(Continued)

(Continued)

z	A	z	A	z	A	z	A	z	A	z	A	z	A	z	A
1.68	0.4535	1.97	0.4756	2.26	0.4881	2.55	0.4946	2.84	0.4977	3.13	0.4991	3.42	0.4997	3.71	0.4999
1.69	0.4545	1.98	0.4761	2.27	0.4884	2.56	0.4948	2.85	0.4978	3.14	0.4992	3.43	0.4997	3.72	0.4999
1.70	0.4554	1.99	0.4767	2.28	0.4887	2.57	0.4949	2.86	0.4979	3.15	0.4992	3.44	0.4997	3.73	0.4999
1.71	0.4564	2.00	0.4772	2.29	0.4890	2.58	0.4951	2.87	0.4979	3.16	0.4992	3.45	0.4997	3.74	0.4999
1.72	0.4573	2.01	0.4778	2.30	0.4893	2.59	0.4952	2.88	0.4980	3.17	0.4992	3.46	0.4997	3.75	0.4999
1.73	0.4582	2.02	0.4783	2.31	0.4896	2.60	0.4953	2.89	0.4981	3.18	0.4993	3.47	0.4997	3.76	0.4999
1.74	0.4591	2.03	0.4788	2.32	0.4898	2.61	0.4955	2.90	0.4981	3.19	0.4993	3.48	0.4997	3.77	0.4999
1.75	0.4599	2.04	0.4793	2.33	0.4901	2.62	0.4956	2.91	0.4982	3.20	0.4993	3.49	0.4998	3.78	0.4999
1.76	0.4608	2.05	0.4798	2.34	0.4904	2.63	0.4957	2.92	0.4982	3.21	0.4993	3.50	0.4998	3.79	0.4999
1.77	0.4616	2.06	0.4803	2.35	0.4906	2.64	0.4959	2.93	0.4983	3.22	0.4994	3.51	0.4998	3.80	0.4999
1.78	0.4625	2.07	0.4808	2.36	0.4909	2.65	0.4960	2.94	0.4984	3.23	0.4994	3.52	0.4998	3.81	0.4999
1.79	0.4633	2.08	0.4812	2.37	0.4911	2.66	0.4961	2.95	0.4984	3.24	0.4994	3.53	0.4998	3.82	0.4999
1.80	0.4641	2.09	0.4817	2.38	0.4913	2.67	0.4962	2.96	0.4985	3.25	0.4994	3.54	0.4998	3.83	0.4999
1.81	0.4649	2.10	0.4821	2.39	0.4916	2.68	0.4963	2.97	0.4985	3.26	0.4994	3.55	0.4998	3.84	0.4999
1.82	0.4656	2.11	0.4826	2.40	0.4918	2.69	0.4964	2.98	0.4986	3.27	0.4995	3.56	0.4998	3.85	0.4999
1.83	0.4664	2.12	0.4830	2.41	0.4920	2.70	0.4965	2.99	0.4986	3.28	0.4995	3.57	0.4998	3.86	0.4999
1.84	0.4671	2.13	0.4834	2.42	0.4922	2.71	0.4966	3.00	0.4987	3.29	0.4995	3.58	0.4998	3.87	0.4999
1.85	0.4678	2.14	0.4838	2.43	0.4925	2.72	0.4967	3.01	0.4987	3.30	0.4995	3.59	0.4998	3.88	0.4999
1.86	0.4686	2.15	0.4842	2.44	0.4927	2.73	0.4968	3.02	0.4987	3.31	0.4995	3.60	0.4998	3.89	0.4999
1.87	0.4693	2.16	0.4846	2.45	0.4929	2.74	0.4969	3.03	0.4988	3.32	0.4995	3.61	0.4998	3.90	0.5000
1.88	0.4699	2.17	0.4850	2.46	0.4931	2.75	0.4970	3.04	0.4988	3.33	0.4996	3.62	0.4999	3.91	0.5000
1.89	0.4706	2.18	0.4854	2.47	0.4932	2.76	0.4971	3.05	0.4989	3.34	0.4996	3.63	0.4999	3.92	0.5000
1.90	0.4713	2.19	0.4857	2.48	0.4934	2.77	0.4972	3.06	0.4989	3.35	0.4996	3.64	0.4999	3.93	0.5000
1.91	0.4719	2.20	0.4861	2.49	0.4936	2.78	0.4973	3.07	0.4989	3.36	0.4996	3.65	0.4999	3.94	0.5000
1.92	0.4726	2.21	0.4864	2.50	0.4938	2.79	0.4974	3.08	0.4990	3.37	0.4996	3.66	0.4999	3.95	0.5000
1.93	0.4732	2.22	0.4868	2.51	0.4940	2.80	0.4974	3.09	0.4990	3.38	0.4996	3.67	0.4999	3.96	0.5000
1.94	0.4738	2.23	0.4871	2.52	0.4941	2.81	0.4975	3.10	0.4990	3.39	0.4997	3.68	0.4999	3.97	0.5000
1.95	0.4744	2.24	0.4875	2.53	0.4943	2.82	0.4976	3.11	0.4991	3.40	0.4997	3.69	0.4999	3.98	0.5000
1.96	0.4750	2.25	0.4878	2.54	0.4945	2.83	0.4977	3.12	0.4991	3.41	0.4997	3.70	0.4999	3.99	0.5000

EXAMPLE 2

Finding and Interpreting the Meaning of the Area Under a Standard Normal Curve

Use Table 8.21 to determine the area beneath the standard normal curve on the interval $[0, 1.49]$. Then determine the probability that a randomly selected data value will fall within 1.49 standard deviations of the mean.

SOLUTION Referring to Table 8.21, we see that when $z = 1.49$, $A = 0.4319$. Since the standard normal distribution is symmetric about $x = 0$, the area between the curve and the horizontal axis on the interval $[-1.49, 1.49]$ is equal to $2A$. Therefore, the probability that a randomly selected z-score is within 1.49 standard deviations of the mean is

$$P(-1.49 \leq z \leq 1.49) = 2A$$
$$= 2(0.4319)$$
$$= 0.8638$$

There is an 86.38 percent likelihood that a randomly selected z-score lies within 1.49 standard deviations of the mean. We represent this graphically by shading the corresponding region of the normal curve (Figure 8.29).

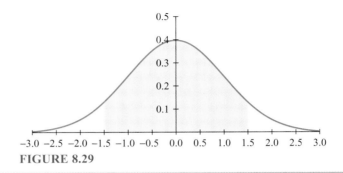

FIGURE 8.29

The built-in features of the TI-83 Plus calculator eliminate the need to refer to a z-score table. The calculator can draw and calculate the area of any region beneath a normal curve, as detailed in the following two Technology Tips.

TECHNOLOGY **TIP**

Calculating the Area Beneath a Normal Curve

1. Press [2nd] [VARS] to open the DISTR menu.

2. Move the cursor to 2:normalcdf(and press [ENTER]. The normalcdf function requires the following four parameters: a, b, μ, and σ. (a and b are the starting and ending values of the interval, respectively.)

3. After entering the parameters, press [ENTER] to display the area beneath the normal curve on the interval $[a, b]$. The area beneath the standard normal curve on $[-1.49, 1.49]$ is 0.8638.

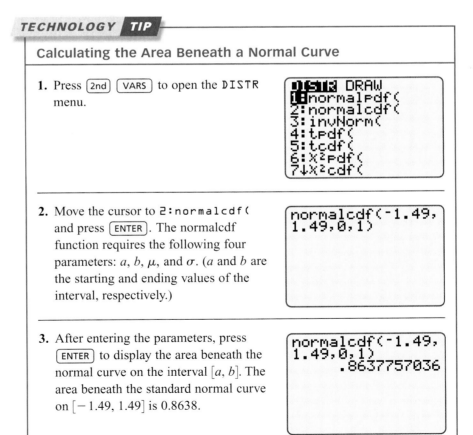

TECHNOLOGY **TIP**

Drawing the Area Beneath a Normal Curve

1. Press WINDOW and set Xmin to $\mu - 3\sigma$ and Xmax to $\mu + 3\sigma$. For example, if $\mu = 4$ and $\sigma = 3$, then Xmin=-5 and Xmax=13.

```
WINDOW
Xmin=-5
Xmax=13
Xscl=1
Ymin=-.1
Ymax=.2
Yscl=1
Xres=1
```

2. Press 2nd VARS to open the DISTR menu and move the cursor to the DRAW menu. Select the 1:ShadeNorm(function and press ENTER.

```
DISTR DRAW
1:ShadeNorm(
2:Shade_t(
3:ShadeX²(
4:ShadeF(
```

3. The ShadeNorm function requires the following four parameters: a, b, μ, and σ. To shade the region within one standard deviation of the mean, we let $a = \mu - \sigma$ and $b = \mu + \sigma$. In this case, $a = 1$ and $b = 7$.

```
ShadeNorm(1,7,4,
3)
```

4. Press ENTER to display the shaded region.

```
Area=.682689
low=1      up=7
```

EXAMPLE **3**

Using Technology to Analyze a Normal Distribution

Using a graphing calculator, determine the probability that a randomly selected value from a normal distribution with $\mu = 2.32$ and $\sigma = 0.48$ lies within 1.5 standard deviations of the mean. Then draw the shaded region of the normal distribution.

SOLUTION We must first determine $a = \mu - 1.5\sigma$ and $b = \mu + 1.5\sigma$.

$$a = \mu - 1.5\sigma \qquad\qquad b = \mu + 1.5\sigma$$
$$= 2.32 - 1.5(0.48) \qquad\qquad = 2.32 + 1.5(0.48)$$
$$= 1.60 \qquad\qquad\qquad\quad = 3.04$$

Using the normalcdf function on the TI-83 Plus yields the result shown in Figure 8.30.

FIGURE 8.30

The probability that a randomly selected value lies within 1.5 standard deviations of the mean is 86.64 percent.

In order to set the window, we find the values three standard deviations away from the mean. Doing so will ensure that the majority of the normal curve will appear in the window, even though only a portion of the region beneath the curve will be shaded.

$$\begin{aligned} \text{Xmin} &= \mu - 3\sigma & & & \text{Xmax} &= \mu + 3\sigma \\ &= 2.32 - 3(0.48) & & \text{and} & &= 2.32 + 3(0.48) \\ &= 0.88 & & & &= 3.76 \end{aligned}$$

Using the ShadeNorm function, we draw the shaded region (see Figure 8.31).

FIGURE 8.31

In Example 3, we drew the graph of the normal distribution over the interval $[\mu - 3\sigma, \mu + 3\sigma]$. From the image we see that nearly the entire graph of the normal distribution was contained within this domain. In fact, for all normal distributions, the majority of values will fall within three standard deviations of the mean.

EMPIRICAL RULE

All normal distributions have the following properties.

- Approximately 68.3 percent of values lie within one standard deviation of the mean.
- Approximately 95.4 percent of values lie within two standard deviations of the mean.
- Approximately 99.7 percent of values lie within three standard deviations of the mean.

EXAMPLE 4 **Finding the Area Under a Standard Normal Curve**

What percentage of the total area under the standard normal curve lies between the z-scores of $z = 1.94$ and $z = 3.08$?

SOLUTION From the z-score table, we see that the area below the curve on the interval $[0, 1.94]$ is 0.4738. The area below the curve on the interval $[0, 3.08]$ is 0.4990. Since we are interested in the area below the curve on $[1.94, 3.08]$, we calculate the difference between the two values.

$$0.4990 - 0.4738 = 0.0252$$

Therefore, 2.52 percent of the total area below the standard normal curve is between the z-scores of $z = 1.94$ and $z = 3.08$.

EXAMPLE 5 **Using a Normal Distribution to Forecast Unknown Results**

An IQ test has a mean of 100 and a standard deviation of 15. Assuming that test scores are normally distributed, what percentage of the population is expected to score above 110?

SOLUTION We first compute the z-score.

$$z = \frac{x - \mu}{\sigma}$$

$$= \frac{110 - 100}{15}$$

$$= \frac{10}{15}$$

$$\approx 0.67$$

The score 110 is 0.67 standard deviation above the mean. Referring to the z-score table, we see that the area between 0 and 0.67 is 0.2486. Since 50 percent of all scores are above the mean, the percentage of the population expected to score above 110 is given by

$$0.5000 - 0.2486 = 0.2514$$

Therefore, 25.14 percent of the population is expected to score above 110 on the IQ test.

EXAMPLE   6

Using a Normal Distribution to Forecast Unknown Results

In 2003, the mean SAT verbal score for college-bound seniors was 507 with a standard deviation of 111. (**Source:** www.collegeboard.com.) Assuming that the scores are normally distributed, what score would a student have to earn to be in the top 10 percent of all test takers?

SOLUTION   We are looking for a test score x and a z-score z such that

$$A = 0.40 \quad \text{and} \quad z = \frac{x - 507}{111}$$

(We pick $A = 0.40$ because we already know that 50 percent of the scores lie below the mean. We are looking for a score above the mean that has 40 percent of the scores above the mean below it. The resulting score will have a total of 90 percent of the scores below it, since 50 percent + 40 percent = 90 percent.)

Looking in the z-score table, we see that when $z = 1.28, A = 0.3997$, and when $z = 1.29, A = 0.4015$. Since the first value of A is closer to 0.40, we pick $z = 1.28$.

$$1.28 = \frac{x - 507}{111}$$

$$142.08 = x - 507$$

$$x = 649.08$$

$$\approx 650 \qquad \text{Since an SAT score must be a multiple of 10}$$

Assuming that the scores are normally distributed, a student earning 650 or better would be in the top 10 percent of college-bound seniors who took the test.

EXAMPLE   7

Using a Normal Distribution to Forecast Unknown Results

During a well-baby exam, a nurse informed a mother that her 16-month-old son was at the 95th percentile for height (that is, 95 percent of toddlers his age were shorter than he was). The average height of a 16-month-old toddler is about 31.5 inches. (**Source:** www.cdc.gov.) The woman's son was 33.5 inches tall. Approximate the value of σ for the height of a 16-month-old boy.

SOLUTION   We know that 50 percent of the heights lie below the 31.5-inch mean. We need to first determine what z-score yields a value of $A = 0.45$, since 50 percent + 45 percent = 95 percent. From the z-score table, we see that when $z = 1.64, A = 0.4495$, and when $z = 1.65, A = 0.4505$. We estimate that when $z = 1.645, A = 0.4500$. We know that $z = \frac{x - \mu}{\sigma}$. Therefore,

$$z = \frac{x - \mu}{\sigma}$$

$$1.645 = \frac{33.5 - 31.5}{\sigma}$$

$$1.645 = \frac{2}{\sigma}$$

$$\sigma = \frac{2}{1.645}$$

$$\sigma = 1.216 \text{ inches}$$

The population standard deviation σ for the height of a 16-month-old boy is about 1.2 inches.

8.6 Summary

In this section, you learned how to use normal distributions. You saw that any normal distribution may be converted into a standard normal distribution by calculating the corresponding z-scores. You also learned how to use a z-score table.

8.6 Exercises

In Exercises 1–10, calculate the z-score for the given values of x, μ, and σ.

1. $x = 4, \mu = 5, \sigma = 1.1$

2. $x = 34, \mu = 25, \sigma = 4$

3. $x = -2, \mu = 0, \sigma = 0.9$

4. $x = 124, \mu = 111, \sigma = 10.5$

5. $x = 0.21, \mu = 0.42, \sigma = 0.84$

6. $x = 22, \mu = 49, \sigma = 10$

7. $x = 0.94, \mu = 0.49, \sigma = 0.17$

8. $x = 1.57, \mu = 1.34, \sigma = 0.32$

9. $x = -0.29, \mu = -1.19, \sigma = 0.76$

10. $x = 789, \mu = 256, \sigma = 121$

In Exercises 11–20, use the z-score table to determine the area beneath the standard normal curve on the interval [0, z]. Then determine the probability that a randomly selected data value will fall within z standard deviations of the mean.

11. $z = 2.13$ **12.** $z = 1.50$

13. $z = 2.39$ **14.** $z = 3.29$

15. $z = 0.21$ **16.** $z = 1.78$

17. $z = 0.67$ **18.** $z = 1.35$

19. $z = 0.42$ **20.** $z = 3.61$

In Exercises 21–30, use a graphing calculator to shade the region within one standard deviation of the mean for a normal distribution with the indicated values of μ and σ.

21. $\mu = 5.2, \sigma = 2.6$ **22.** $\mu = 0.2, \sigma = 0.6$

23. $\mu = 9.8, \sigma = 3.5$ **24.** $\mu = 5.7, \sigma = 0.1$

25. $\mu = 7.9, \sigma = 2.7$ **26.** $\mu = 5.8, \sigma = 9.7$

27. $\mu = -1.2, \sigma = 1.7$ **28.** $\mu = -0.3, \sigma = 1.5$

29. $\mu = -1.7, \sigma = 7.4$ **30.** $\mu = 9.1, \sigma = 0.2$

In Exercises 31–35, determine the percentage of the total area between the normal curve and the horizontal axis that is between the given z-scores.

31. $z = 0.21$ and $z = 1.25$

32. $z = 1.98$ and $z = 3.96$

33. $z = -1.44$ and $z = 1.45$

34. $z = -0.25$ and $z = 2.97$

35. $z = 2.99$ and $z = 3.81$

In Exercises 36–45, apply normal distribution concepts to answer the questions.

36. **SAT Math Scores**   In 2003, the mean SAT math score for college-bound seniors was 519 with a standard deviation of 115. (**Source:** www.collegeboard.com.) Assuming that the scores are normally distributed, what score would a student have to earn to be in the top 10 percent of all test takers on the math portion of the exam? (Round your answer to the nearest multiple of 10.)

37. **SAT Math Scores—Utah** In 2003, the mean SAT math score for college-bound Utah women was 543 points with a standard deviation of 105 points. (**Source:** www.collegeboard.com.) Assuming that the scores are normally distributed, what score would a woman have to earn to be in the top 10 percent of all female test takers in Utah on the math portion of the exam? (Round your answer to the nearest multiple of 10.)

38. **Quality Control: Canned Tuna** Four cans of Bumble Bee® Chunk Light Tuna were weighed on a kitchen scale. The weight of the cans varied from 185 to 195 grams. Assuming that the weight of the cans is normally distributed with $\mu = 190$ grams and $\sigma = 3.54$, what is the probability than a can of tuna weighs less than 180 grams?

39. **Quality Control: Canned Beans** Five cans of S&W® Cut Green Beans were weighed on a kitchen scale. The weight of the cans varied from 440 to 455 grams. Assuming that the weight of the cans is normally distributed with $\mu = 448$ grams and $\sigma = 5.10$, what is the probability that a can of green beans weighs less than 440 grams?

40. **Quality Control: Boxed Pudding** Six boxes of Hain Pure Foods® Super Pudding™ were weighed on a kitchen scale. The weight of the boxes varied from 80 to 85 grams. Assuming that the weight of the boxes is normally distributed with $\mu = 81.3$ grams and $\sigma = 1.97$, what is the probability that a box of pudding weighs more than 89 grams?

41. **Quality Control: Canned Albacore** Six cans of Kirkland Signature™ Solid White Albacore were weighed on a kitchen scale. The weight of the cans varied from 185 to 187 grams. Assuming that the weight of the cans is normally distributed with $\mu = 185.7$ grams and $\sigma = 0.94$, what is the probability than a can of albacore weighs more than 190 grams?

42. Exercises 38–41 dealt with the weight variations of packaged foodstuffs. Explain how a manufacturer could use analytical results such as those presented in the indicated exercises to improve its business.

43. **Weight of a Boy** According to a growth chart published by the Centers for Disease Control, the average weight of a

24-month-old boy is approximately 27.5 pounds. The chart further indicates that 5 percent of all 24-month-old boys weigh less than 23.5 pounds. (**Source:** www.cdc.gov.) Assuming that the weights of 24-month-old boys are normally distributed, what weight range is within two standard deviations of the mean?

44. **Weight of a Girl** According to a growth chart published by the Centers for Disease Control, the average weight of a 36-month-old girl is approximately 30.5 pounds. The chart further indicates that 90 percent of all 36-month-old girls weigh less than 36 pounds. (**Source:** www.cdc.gov.) Assuming that the weights of 36-month-old girls are normally distributed, what is the standard deviation for the normal distribution?

45. **Well-Baby Exam** During a well-baby exam, pediatricians typically compare a patient's measurements with growth charts such as those published by the CDC. Why do you think that pediatricians would be interested in comparing a child to charts showing weight-for-age and height-for-age percentiles?

Exercises 46–50 are intended to challenge your understanding of normal distributions.

46. Two different normal distributions have the same mean μ; however, the standard deviation of one of the distributions is σ_1, while the standard deviation of the other is σ_2. What will be the relationship between the z-scores of each of the distributions for a given value of x?

47. Two different normal distributions have the same standard deviation σ; however, the mean of the first distribution is μ_1, while the mean of the second distribution is μ_2. What is the relationship between the z-scores of the distributions for a given value of x?

48. The equation of a graph of a normal curve is given by

$$f(x) = \frac{1}{\sigma\sqrt{2\pi}} e^{-\frac{1}{2}\left(\frac{x-\mu}{\sigma}\right)^2}$$

The graph reaches a maximum value at $x = \mu$. If $f(\mu) = k$, then determine the value of σ in terms of k.

49. Will the sum of two normal distributions be a normal distribution? Show the work leading to your conclusion.

50. Give the equation of a normal distribution with $\sigma = 1$ but $\mu \neq 0$. What is the relationship between the graph of this distribution and the standard normal curve?

Chapter 8 Review Exercises

Section 8.1 *In Exercises 1–3, determine the indicated probability.*

1. $P(A \cap B) = 0.25$ and $P(B) = 0.35$. Find $P(A|B)$.

2. Find $P(H'|D')$.

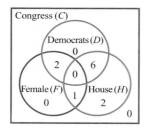

3. Twenty-two students, including twelve seniors, entered a school art contest. Five students, three of whom were seniors, won awards. What is the probability that a student won an award, given that the student was a senior?

In Exercises 4–5, use a probability tree to determine the solution.

4. In 2003, California had 55 members of the U.S. Congress, including 2 senators and 53 representatives. Of the 35 Democrats, 2 were senators. (**Source:** clerk.house.gov.) What is the likelihood that a randomly selected member of Congress is a Democratic representative?

5. In 2003, California had 55 members of the U.S. Congress, including 2 senators and 53 representatives. Of the 35 Democrats, 2 were senators. (**Source:** clerk.house.gov.) What is the likelihood that a randomly selected member of Congress is *not* a Democratic representative?

In Exercises 6–7, determine if the events E and F are independent or dependent.

6. Two coins are pulled from a bag containing one nickel, one dime, and one penny.

$E = $ the first coin is a dime
$F = $ the second coin is a nickel

7. Two coins are pulled from a bag containing five nickels and one dime.

$E = $ the first coin is a nickel
$F = $ the second coin is a dime

Section 8.2 *In Exercises 8–9, complete the table by applying Bayes' theorem.*

8.

	Percent
Prevalence	5
Sensitivity	92
Specificity	76
Predictive-value positive	
Predictive-value negative	

9.

	Percent
Prevalence	0.1
Sensitivity	25
Specificity	99
Predictive-value positive	
Predictive-value negative	

In Exercises 10–11, apply Bayes' theorem to the real-life application.

10. **Hepatitis B Test** ACON® HbsAg is a one-step hepatitis B surface antigen test strip produced by ACON Laboratories. According to the test package insert, the test has a sensitivity of at least 99 percent and a specificity of 97 percent. (**Source:** www.aconlab.com.)

    The incidence of chronic hepatitis B infection is estimated to be 1 out of 100 people in the United States. What is the probability that someone who tests negative for hepatitis B actually has the illness?

11. **Hepatitis C Test** ACON HCV Rapid Test Device (Serum/Plasma) is a one-step hepatitis C virus test device produced by ACON Laboratories. According to the package insert, the test was compared with a leading commercial HCV test that used a different method. The comparison yielded the following results.

		Other HCV Method	
		Positive HCV	Negative HCV
ACON HCV	Positive	145	2
	Negative	0	144

Source: www.aconlab.com.

    Calculate and interpret the meaning of the sensitivity, specificity, predictive-value positive, and predictive value negative for the ACON HCV test.

Section 8.3 *In Exercises 12–13, draw the state transition diagram that corresponds with the given transition matrix. Identify which states, if any, are absorbing states.*

12. $P = \begin{bmatrix} 0 & 1 & 0 \\ 1 & 0 & 0 \\ 0.5 & 0.25 & 0.25 \end{bmatrix}$

13. $P = \begin{bmatrix} 0 & 0.1 & 0.9 \\ 0.3 & 0.7 & 0 \\ 0.8 & 0 & 0.2 \end{bmatrix}$

In Exercises 14–15, write the transition matrix P that corresponds with the state transition diagram.

14.

15.

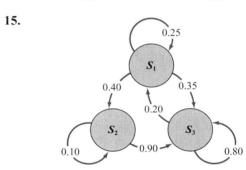

In Exercises 16–17, do the following: (i) identify the states of the Markov chain, (ii) draw the transition diagram, (iii) write the transition matrix P, (iv) identify the initial state distribution vector D, and (v) calculate DP and interpret the meaning of the result.

16. **Market Share** Assume that an advertising company has 15 percent of the market share for advertising. Suppose that the probability that a client will change to a competitor is 55 percent and that the probability that a competitor's client will change to the advertising company is 45 percent.

17. **Market Share** Suppose that two toothpaste brands dominate the market. The first brand, A, has 38 percent of the market share. The second brand, B, has 32 percent of the market share. The probability that a Brand A customer will become a brand B customer by the end of the year is 20 percent. The probability that a Brand A customer will become a customer of another non-B brand by the end of the year is 12 percent. The probability that a customer of Brand B will become a Brand A customer by the end of the year is 18 percent. The probability that a Brand B customer will become a customer of another non-A brand by the end of the year is 22 percent. There is a 50 percent probability that a customer using a brand other than Brand A or Brand B will remain with her or his current brand. There is a 20 percent probability that the customers will become a Brand A customer by the end of the year.

Section 8.4   *In Exercises 18–19, determine the probability distribution for the random variable X and draw its associated histogram.*

18. A coin is tossed three times in a row

X = the number of heads in the longest consecutive sequence of heads

19. Three six-sided dice are tossed.

X = the number of different values shown

In Exercises 20–22, apply the concepts of probability distribution and expected value, as appropriate.

20. Raffle Tickets   A campus club sells 420 raffle tickets at a price of $2 each. The raffle offers one $100 prize, two $50 prizes, and ten $10 prizes. What is the expected value of a raffle ticket?

21. True/False Tests   A true/false test is scored as follows: A correct answer adds 5 points to the overall score. An incorrect answer deducts 5 points from the overall score. A blank answer adds 0 points to the overall score.

A person taking the test doesn't know the answer to one of the questions. Should she guess? Justify your answer.

22. Income Tax Returns   An accountant determines that 0.10 percent of the income tax returns she processes are audited by the IRS. The average tax penalty for her audited returns is $1650. She guarantees her clients that she will pay any tax penalties on audited returns. How much must she charge for the guarantee service in order to break even?

Section 8.5   *In Exercises 23–25, calculate the mean, median, mode, and standard deviation and interpret the meaning of the results.*

23.   **Women's Basketball Champions**   On April 6, 2004, the University of Connecticut women's basketball team defeated

Tennessee and won the national title for the third year in a row. In the final six games leading up to the championship, Connecticut outscored each of its opponents. The table shows the number of points scored by Connecticut in each of its final six games.

Losing Team	Points Scored by Connecticut
Penn	91
Auburn	79
U.C. Santa Barbara	63
Penn State	66
Minnesota	67
Tennessee	70

Source: www.uconnhuskies.com.

24.   **Golf Champion**   Up until 2004, Phil Mickelson had been classified by some as the best player never to win a major tournament. That changed at the 2004 Augusta National golf tournament. As a result of a remarkable performance on the last nine holes of the fourth round, Phil Mickelson surpassed his opponents to clinch his first major victory. The table details Mickelson's performance in the fourth round.

Difference Between Score and Par For the Hole	Frequency
−1	6
0	9
1	3

Source: www.golf.com.

25.   **Weekend Theater Sales**   The movie *My Big Fat Greek Wedding* was somewhat unknown when it was first introduced in April 2002; however, it steadily grew in popularity among the public. The table shows the gross weekend theater sales of the movie from some of its top-grossing weekends.

Weekend in 2002	Gross Weekend Theater Sales
Aug. 23–25	$7,261,842
Aug. 30–Sept. 2	$14,809,546
Sept. 6–8	$10,372,316
Sept. 13–15	$10,772,146
Sept. 20–22	$9,748,969
Sept. 27–29	$9,434,602
Oct. 4–6	$8,223,801
Oct. 11–13	$8,453,159
Oct. 18–20	$7,145,309
Oct. 25–27	$6,209,500

Source: www.boxofficemojo.com.

Section 8.6 *In Exercises 26–27, calculate the z-score for the given values of x, μ, and σ.*

26. $x = 12.1$, $\mu = 10.9$, $\sigma = 6.8$

27. $x = 9.4$, $\mu = 9.5$, $\sigma = 0.6$

In Exercises 28–29, use the z-score table to determine the percentage of values of a normal distribution lying within the indicated number of standard deviations of the mean.

28. $z = 2.13$

29. $z = 0.25$

In Exercises 30–31, use a graphing calculator to shade the region within two standard deviations of the mean for a normal distribution with the indicated values of μ and σ.

30. $\mu = 4.5$, $\sigma = 1.5$

31. $\mu = 5.22$, $\sigma = 0.11$

In Exercises 32–33, determine the percentage of the total area between the normal curve and the horizontal axis that is between the given z-scores.

32. $z = -0.91$ and $z = 1.42$

33. $z = 1.00$ and $z = 2.00$

In Exercise 34, apply normal distribution concepts to answer the question.

34. **Fresh Whole Carrot Weight** A particular bag of fresh whole carrots contains carrots ranging in weight from 55 grams to 110 grams. Assuming that the individual weights of the carrots used by the producer to fill each bag of carrots are normally distributed with $\mu = 76$ and $\sigma = 18.4$, what percentage of carrots in the bag are likely to weigh 110 grams or more?

Make It Real

What to do

1. Collect a set of 10 or more data points from an area of personal interest.
2. Calculate the mean, median, mode, and standard deviation for the data set.
3. Explain the real-world meaning of each of the measures of central tendency.
4. Describe how your results could be used to improve business or society.

Example project

Age of Grandchild	Frequency
1	1
2	1
3	1
4	1
5	1
7	2
8	1
10	2

Mean: 5.7
The average age of a grandchild is 5.7 years.

Median: 6
Five grandchildren are less than six years old and five grandchildren are more than six years old.

Mode: 7 and 10
The most common age of a grandchild is seven years old and ten years old.
Standard deviation: 3.199

If the grandparents plan to give each grandchild $1000 in the year the child turns 18, they could use these results for financial planning. From the mode, they can determine the years in which the most money will be required. From the mean, they can determine that the average age of a grandchild will be 18 in about 12 years. The median tells them that half of the grandchildren will turn 18 in the next 12 years.

Chapter
9

The Derivative

It is impossible to determine how quickly a person is running from a single photograph, since speed is calculated as a change in distance over a change in time. Nevertheless, we may estimate a person's speed at a particular instant in time by determining the distance traveled over a small interval of time (e.g., one second). A runner's speed may be classified as a rate of change in distance. A key component of calculus is the study of rates of change.

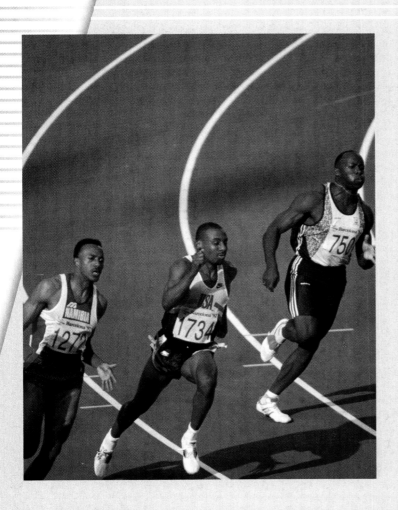

9.1 Average Rate of Change

- Calculate the average rate of change of a function over an interval

GETTING **STARTED**   Colleges and universities periodically raise their tuition rates in order to cover rising staffing and facilities costs. As a result, it is often difficult for students to know how much money they should save to cover future tuition costs. By calculating the *average rate of change* in the tuition price over a period of years, we can estimate projected increases in tuition costs. In this section, we will demonstrate how to calculate the average rate of change in the value of a function over a specified interval [a, b]. (The interval notation [a, b] is equivalent to $a \le x \le b$.)

THE DIFFERENCE QUOTIENT: AN AVERAGE RATE OF CHANGE

The average rate of change of a function $y = f(x)$ over an interval $[a, b]$ is

$$\frac{f(b) - f(a)}{b - a}$$

This expression is referred to as the **difference quotient.** For a linear function, the difference quotient gives the slope of the line.

In calculating the difference quotient, we answer the question, "Over the interval $[a, b]$, on average, how much does a one-unit increase in the x value change the y value of the function?"

EXAMPLE   **1**

Calculating an Average Rate of Change

The quarterly cost of tuition for full-time resident students at Green River Community College is shown in Table 9.1.

TABLE 9.1

Years (since 1994–1995) (t)	Quarterly Tuition Cost (dollars) $[f(t)]$	Change in Tuition Cost from Prior Year (dollars)
0	432	
1	450	18
2	467	17
3	486	19
4	505	19
5	528	23
6	547	19
7	581	34

Source: Green River Community College.

What is the average rate of change in the quarterly tuition cost from the 1994–1995 academic year to the 2001–2002 academic year? Rounded to the nearest dollar, what do you estimate the 2002–2003 quarterly tuition cost will be?

SOLUTION One way to calculate the average rate of change is to average the annual rates of change over the entire period. To calculate the average, we sum each of the seven rate of change data values and divide by the number of data values.

UNITS

$$\text{Average rate of change} = \frac{(18 + 17 + 19 + 19 + 23 + 19 + 34) \text{ dollars}}{7 \text{ years}}$$

$$= \frac{149}{7} \text{ dollars per year}$$

$$= 21.29 \text{ per year}$$

It is significant to note that the same result may be obtained more rapidly by applying the difference quotient formula. For the period 1994–1995 to 2001–2002, the interval $[a, b] = [0, 7]$. The average rate of change of the tuition is

UNITS

$$\frac{f(b) - f(a)}{b - a} = \frac{f(7) - f(0)}{7 - 0}$$

$$= \frac{581 - 432 \text{ dollars}}{7 \text{ years}}$$

$$= \frac{149 \text{ dollars}}{7 \text{ years}}$$

$$= 21.29 \text{ dollars per year}$$

From 1994–1995 to 2001–2002, the tuition increased by an average of approximately $21 per year. We estimate that the 2002–2003 quarterly tuition will be $21 more than the 2001–2002 tuition, or $602.

When determining the meaning of an average rate of change in a real-life problem, it is essential to find the units of measurement of the result. Fortunately, the units are easily determined. **The units of the rate of change are the units of the output divided by the units of the input.** In Example 1, the units of the output were *dollars* and the units of the input were *years*. Consequently, the units of the average rate of change were *dollars* divided by *years*, or *dollars per year*.

EXAMPLE 2

Calculating an Average Rate of Change

A company's stock price often fluctuates dramatically from day to day. If an investor is preoccupied with daily price changes, she may sell too early or buy too late to get a good return on her investment. One way an investor can gauge the change in a company's stock price is to calculate the average rate of change in the price of the stock over a period of time.

Table 9.2 shows the closing stock price of Apple Computer Corporation over a nine-month period.

TABLE 9.2

Months (since August 2001) (t)	Stock Price at the End of the First Trading Day of the Month (Dollars) $[f(t)]$
August 2001 ($t = 0$)	19.06
September ($t = 1$)	18.25
October ($t = 2$)	15.54
November ($t = 3$)	18.59
December ($t = 4$)	21.05
January 2002 ($t = 5$)	23.30
February ($t = 6$)	24.41
March ($t = 7$)	24.29
April ($t = 8$)	24.46

Source: www.quicken.com.

What was the average rate of change in the stock price between

(a) August 2001 and October 2001?
(b) October 2001 and April 2002?
(c) August 2001 and April 2002?

SOLUTION

(a) In the table, August is month 0 and October is month 2. The average rate of change of the stock over the interval $[0, 2]$ is

$$\frac{f(b) - f(a)}{b - a} = \frac{f(2) - f(0)}{2 - 0}$$

$$= \frac{15.54 - 19.06 \text{ dollars}}{2 \text{ months}}$$

$$= -1.76 \text{ dollars per month}$$

Between August and October 2001, the share price of Apple Computer stock decreased by an average of $1.76 per month.

(b) In the table, October is month 2 and April is month 8. The average rate of change of the stock over the interval $[2, 8]$ is

$$\frac{f(b) - f(a)}{b - a} = \frac{f(8) - f(2)}{8 - 2}$$

$$= \frac{24.46 - 15.54 \text{ dollars}}{6 \text{ months}}$$

$$= 1.49 \text{ dollars per month}$$

Between October 2001 and April 2002, the share price of Apple Computer stock increased by an average of $1.49 per month.

c. In the table, August is month 0 and April is month 8. The average rate of change of the stock over the interval $[0, 8]$ is

$$\frac{f(8) - f(0)}{8 - 0} = \frac{24.46 - 19.06 \text{ dollars}}{8 \text{ months}}$$

$$= 0.68 \text{ dollar per month}$$

Between October 2001 and April 2002, the share price of Apple Computer stock increased by an average of $0.68 per month.

Some investors choose to buy or sell a stock once it hits a certain limit price. In fact, many investment companies allow investors to activate a limit order for a period of 30 days. If the stock price hits the limit price within the 30-day period, the order is executed. If the stock price fails to hit the limit price during the 30-day period, the order expires with no trading action. Observing the overall trend in a stock's price helps the investor to pick a limit price and predict whether the order will be executed.

EXAMPLE 3

Calculating an Average Rate of Change

The population of Washington state may be modeled by the function $P(t) = 0.8981(1.019)^t$, where P is the population in millions of people and t is the number of years since 1900. (**Source:** Modeled from data at www.ofm.wa.gov.) What is the average rate of change in the population from 1990 to 2000?

SOLUTION We observe that in the model, t is the number of years since 1900. So for the model, $t = 90$ represents 1990 and $t = 100$ represents 2000. The average rate of change in the population over the interval $[90, 100]$ is

$$\frac{P(100) - P(90)}{100 - 90} = \frac{5.90 - 4.89 \text{ million people}}{10 \text{ years}}$$

$$= 0.101 \text{ million people per year}$$

Between 1990 and 2000, the population of Washington state increased by an average of 101,000 people per year.

Graphical Interpretation of the Difference Quotient

A line connecting any two points on a graph is referred to as a **secant line.** Graphically speaking, the difference quotient for a function $y = f(x)$ is the slope of the secant line connecting $(a, f(a))$ and $(b, f(b))$ (Figure 9.1).

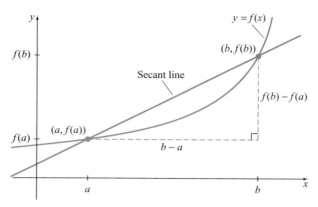

FIGURE 9.1 Secant line slope $= \dfrac{f(b) - f(a)}{b - a}$

EXAMPLE 4

Finding the Slope of a Secant Line

The graph of the function $f(x) = 3x^2 - 6x + 5$ is shown in Figure 9.2. Calculate the slope of the secant line of f that passes through $(1, 2)$ and $(2, 5)$.

FIGURE 9.2

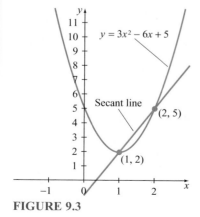

FIGURE 9.3

SOLUTION We first plot the points $(1, 2)$ and $(2, 5)$ and draw the line connecting them (Figure 9.3).

This line is the secant line of the graph between $x = 1$ and $x = 2$. The slope of the secant line is given by

$$m = \frac{f(b) - f(a)}{b - a}$$

$$= \frac{f(2) - f(1)}{2 - 1}$$

$$= \frac{5 - 2}{1}$$

$$= 3$$

The slope of the secant line is 3. That is, on the interval $[1, 2]$, a one-unit increase in x results in a three-unit increase in y, on average.

EXAMPLE 5   **Calculating the Slope of a Secant Line**

Calculate the slope of the secant line of the function f shown in Figure 9.4.

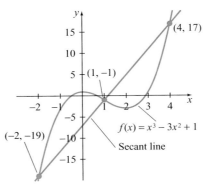

FIGURE 9.4

SOLUTION From the graph, we see that the line intersects the graph of f at $(-2, -19)$, $(1, -1)$, and $(4, 17)$. The slope of the given secant line may be determined by using any pair of these points.

$$
\begin{aligned}
m &= \frac{f(4) - f(-2)}{4 - (-2)} \\
&= \frac{17 - (-19)}{6} \\
&= \frac{36}{6} \\
&= 6
\end{aligned}
$$

For this function f, the average rate of change in f over the interval $[-2, 1]$ is the same as the average rate of change in f over the intervals $[1, 4]$ and $[-2, 4]$.

9.1 Summary

In this section, you learned how to calculate the average rate of change of a function over a specified interval. You also learned that the units of the average rate of change are the units of the *output* divided by the units of the *input*. You saw that, graphically speaking, the difference quotient is the slope of a secant line. In the next section, we will demonstrate how to use the difference quotient to estimate an *instantaneous* rate of change.

9.1 Exercises

In Exercises 1–10, calculate the average rate of change of the function over the given interval.

1. $f(x) = 2x - 5$ over the interval $[3, 5]$

2. $g(x) = x^2 - 5x$ over the interval $[-2, 4]$

3. $v(x) = -x^3$ over the interval $[-1, 1]$

4. $g(t) = 3^t - \dfrac{4}{t}$ over the interval $[2, 4]$

5. $v(m) = m^2 - m$ over the interval $[-3, 4]$

6. $h(s) = \ln(s)$ over the interval $[4, 7]$

7. $z = \dfrac{\ln(x)}{x}$ over the interval $[1, 5]$

8. $f(t) = \dfrac{2t - 5}{t + 5}$ over the interval $[-2, 2]$

9. $q(x) = \sqrt{x + 2}$ over the interval $[0, 6]$

10. $w(t) = 2t^{-2} + 5$ over the interval $[1, 4]$

In Exercises 11–20, calculate the average rate of change in the designated quantity over the given interval(s).

11. **Air Temperature** Temperature between 11:00 a.m. and 3:00 p.m.

Time of Day	Temperature (°F)
11:00 a.m.	68
1:00 p.m.	73
3:00 p.m.	75

12. **Internet Access** People with Internet access between 1997 and 2000.

Year	Number of People with Internet Access at Home or Work (in thousands)
1997	46,305
1998	62,273
1999	83,677
2000	112,949

Source: www.census.gov.

13. **Dow Jones Industrial Average** Dow Jones Industrial Average between 1996 and 1998 and between 1997 and 1999.

Year	Dow Jones Industrial Average Closing Value at the End of the Year (points)
1996	6,448.30
1997	7,908.30
1998	9,181.40
1999	11,497.10

Source: www.census.gov.

14. **Nasdaq Composite Index** Nasdaq Composite Index closing value between 1996 and 1998 and between 1997 and 1999.

Year	Nasdaq Composite Index Closing Value at the End of the Year (points)
1996	1,291.0
1997	1,570.4
1998	2,192.7
1999	4,069.3

Source: www.census.gov.

15. **Reading Scores** A third-grade student's reading score between the first and third quarter.

Quarter	Reading Score (words per minute)
1	69
2	107
3	129

Source: Author's data.

16. **Owner-Occupied Homes** The percentage of U.S. housing units that are owner-occupied when the number of housing units increased from 94.2 million (in 1990) to 107.0 million (in 2001).

Number of Housing Units (millions)	Percent of Housing Units That Are Owner-Occupied
94.2	63.9
100.0	64.7
104.9	66.8
105.7	67.4
107.0	67.8

Source: www.census.gov.

17. **Newspaper Subscriptions** Daily newspaper subscriptions as the number of cable TV subscribers increased from 50.5 million (in 1990) to 67.7 million (in 2000).

Cable TV Subscribers (millions)	Daily Newspaper Circulation (millions)
50.5	62.3
60.9	58.2
66.7	56.0
67.7	55.8

Source: www.census.gov.

18. **Personal Income** Disposable personal income between 1990 and 2000.

Year	Disposable Personal Income (dollars per person)
1990	17,176
1995	20,358
1999	23,708
2000	24,889

Source: www.census.gov.

19. **Poverty Level** Percent of people below poverty level when the unemployment rate decreased from 5.6 percent (in 1990) to 4.0 percent (in 2000).

Unemployment Rate (percent)	People Below Poverty Level (percent)
5.6	13.5
4.2	11.8
4.0	11.3

Source: www.census.gov.

20. **Net Farm Income** Net farm income when land in farms decreased from 987 million acres (in 1990) to 943 million acres (in 2000).

Land in Farms (millions of acres)	Farm Income (billions of dollars)
987	44.6
943	46.4

Source: www.census.gov.

In Exercises 21–25, graph each function. Then use the difference quotient, $\dfrac{f(b) - f(a)}{b - a}$, to calculate the slope of the secant line through the points $(1, f(1))$ and $(3, f(3))$.

21. $f(x) = 2^x$

22. $f(x) = 5 - x^2$

23. $f(x) = 5$

24. $f(x) = 3x + 1$

25. $f(x) = (x - 2)^2$

In Exercises 26–30, use the difference quotient, $\dfrac{f(b) - f(a)}{b - a}$, *to calculate the slope of the secant line through the points* $(1, f(1))$ *and* $(3, f(3))$ *for the given graph of f.*

26.

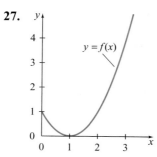

27.

28.

29.

30.

In Exercises 31–40, calculate the average rate of change in the state populations (in people per month) between April 1, 2000, and July 1, 2001.

31. **Population**

Montana	
Date	Population (in thousands)
April 2000	902.2
July 2001	904.4

Source: www.census.gov.

32. **Population**

North Dakota	
Date	Population (in thousands)
April 2000	642.2
July 2001	634.4

Source: www.census.gov.

33. **Population**

Massachusetts	
Date	Population (in thousands)
April 2000	6,349.1
July 2001	6,379.3

Source: www.census.gov.

34. **Population**

Utah	
Date	Population (in thousands)
April 2000	2,233.2
July 2001	2,269.8

Source: www.census.gov.

35. **Population**

Missouri	
Date	Population (in thousands)
April 2000	5,595.2
July 2001	5,629.7

Source: www.census.gov.

36. **Population**

Louisiana	
Date	Population (in thousands)
April 2000	4,469.0
July 2001	4,465.4

Source: www.census.gov.

37. **Population**

West Virginia	
Date	Population (in thousands)
April 2000	1,808.3
July 2001	1,801.9

Source: www.census.gov.

38. **Population**

Wyoming	
Date	Population (in thousands)
April 2000	493.8
July 2001	494.4

Source: www.census.gov.

39. **Population**

Pennsylvania	
Date	Population (in thousands)
April 2000	12,281.1
July 2001	12,287.2

Source: www.census.gov.

40. **Population**

Arizona	
Date	Population (in thousands)
April 2000	5,130.6
July 2001	5,307.3

Source: www.census.gov.

Exercises 41–45 are intended to challenge your understanding of average rates of change.

41. Let $f(x) = x^2 - 4x$. Compare and contrast the average rate of change of $f(x)$ and of $2(f(x))$ on the interval $[1, 4]$.

42. Given $f(x) = x^2 - 4$, determine the average rate of change of f over the interval $[a, a + h]$.

43. The slope of the secant line passing through $(2, f(2))$ and $(2 + h, f(2 + h))$ is 4. Determine the vertical distance between $(2, f(2))$ and $(2.01, f(2.01))$.

44. The slope of a secant line of the function $f(x) = x^2$ passes through the points $(0, 0)$ and (h, h^2). Determine the slope of the secant line when $h = 1$, $h = 0.1$, $h = 0.01$, and $h = 0.001$. As h gets smaller, what value does the slope of the secant line appear to be approaching?

45. A secant line of the function $f(x) = x^{1/3}$ passes through the points $(0, 0)$ and $(h, h^{1/3})$. Determine the slope of the secant line when $h = 1$, $h = 0.1$, $h = 0.01$, and $h = 0.001$. As h gets smaller, what value does the slope of the secant line appear to be approaching?

9.2 Limits and Instantaneous Rates of Change

- Estimate the instantaneous rate of change of a function at a point
- Use derivative notation and terminology to describe instantaneous rates of change

GETTING STARTED In the July 1988 Olympic trials, Florence Griffith Joyner shattered the world record for the 100-meter dash with a time of 10.49 seconds. How fast was she running (in meters per second) when the picture at the left was taken?

From a single photo, it is impossible for us to determine her speed. However, if we knew how long it took her to reach various checkpoints during the race, we could approximate her speed at the finish line. In this section, we will demonstrate how to use the difference quotient to estimate and the derivative to calculate an *instantaneous rate of change*.

While an average rate of change is calculated over an interval $[a, b]$, an instantaneous rate of change is calculated at a single value a. For example, the average highway speed of a person over a 200-mile trip may have been 59 miles per hour. However, when he passed a state patrol car exactly 124 miles into the trip, he was speeding at 84 miles per hour. His *average* speed on the interval $[0, 200]$ was 59 miles per hour; however, his *instantaneous* speed at $d = 124$ was 84 miles per hour.

EXAMPLE 1

Estimating an Instantaneous Rate of Change

Suppose that a runner in the 100-meter dash recorded the times shown in Table 9.3.

TABLE 9.3

Time (seconds) (t)	Total Distance Traveled (meters) $[D(t)]$
0	0
5.24	50
9.46	90
9.97	95
10.39	99
10.49	100

Estimate her speed when she crossed the finish line.

SOLUTION Her average speed over various distances may be calculated using the difference quotient, $\dfrac{D(b) - D(a)}{b - a}$.

Her average speed over the 100-meter distance was

UNITS

$$\text{Average speed} = \frac{100 - 0 \text{ meters}}{10.49 - 0 \text{ seconds}}$$

$$= 9.53 \text{ meters per second}$$

Her average speed over the last 50 meters was

UNITS

$$\text{Average speed} = \frac{100 - 50 \text{ meters}}{10.49 - 5.24 \text{ seconds}}$$

$$= \frac{50 \text{ meters}}{5.25 \text{ seconds}}$$

$$= 9.52 \text{ meters per second}$$

Her average speed over the last 10 meters was

UNITS

$$\text{Average speed} = \frac{100 - 90 \text{ meters}}{10.49 - 9.46 \text{ seconds}}$$

$$= \frac{10 \text{ meters}}{1.03 \text{ seconds}}$$

$$= 9.71 \text{ meters per second}$$

Her average speed over the last 5 meters was

UNITS

$$\text{Average speed} = \frac{100 - 95 \text{ meters}}{10.49 - 9.97 \text{ seconds}}$$

$$= \frac{5 \text{ meters}}{0.52 \text{ second}}$$

$$= 9.62 \text{ meters per second}$$

And her average speed over the last meter was

UNITS

$$\text{Average speed} = \frac{100 - 99 \text{ meters}}{10.49 - 10.39 \text{ seconds}}$$

$$= \frac{1 \text{ meter}}{0.10 \text{ second}}$$

$$= 10.0 \text{ meters per second}$$

Although each calculation yielded a different result, all of these difference quotients estimate the runner's finish-line speed. Which of the estimates do you think is most accurate?

The last calculation best estimates her finish-line speed because it measures the change in distance over the smallest interval of time: 0.1 second. (Reducing the time interval to an even smaller amount of time, say 0.01 second, would further improve the estimate.) We estimate that the runner's speed when she crossed the finish line was 10.0 meters per second.

To estimate the *instantaneous rate of change* of a function $y = f(x)$ at a point $(a, f(a))$, we calculate the average rate of change of the function over a very small interval $[a, b]$. If we let the variable h represent the distance between

$x = a$ and $x = b$, then $b = a + h$. Consequently, the difference quotient may be rewritten as

$$\frac{f(b) - f(a)}{b - a} = \frac{f(a + h) - f(a)}{(a + h) - a}$$

$$= \frac{f(a + h) - f(a)}{h}$$

THE DIFFERENCE QUOTIENT AS AN ESTIMATE OF AN INSTANTANEOUS RATE OF CHANGE

The *instantaneous rate of change* of a function $y = f(x)$ at a point $(a, f(a))$ may be *estimated* by calculating the **difference quotient of f at a**,

$$\frac{f(a + h) - f(a)}{h}$$

using an h arbitrarily close to 0. (If $h = 0$, the difference quotient is undefined.)

EXAMPLE 2

Estimating an Instantaneous Rate of Change

The average income of an elementary or secondary teacher may be modeled by

$$S(t) = 0.00713t^3 - 0.111t^2 + 1.42t + 31.6, \quad 0 \le t \le 10$$

where $S(t)$ is the annual salary in thousands of dollars and t is the number of years since 1990. (**Source:** Modeled from *Statistical Abstract of the United States, 2001,* Table 237, p. 151.) According to the model, how quickly was the average income of a teacher increasing in 1999?

SOLUTION In 1999, $t = 9$. Using the difference quotient

$$\frac{S(9 + h) - S(9)}{h}$$

and selecting increasingly small values of h, we generate Table 9.4.

TABLE 9.4

h	$\dfrac{S(9 + h) - S(9)}{h}$
1.000	1.2432
0.100	1.1628
0.010	1.1554
0.001	1.1547

We conclude that in 1999 the average annual salary of a teacher was increasing by about $1.155 thousand ($1155) per year.

The following Technology Tip details how to generate a table of values for the difference quotient using the TI-83 Plus.

TECHNOLOGY TIP

Calculating the Difference Quotient for Different Values of h

1. Enter the function $S(t)$ as Y1 by pressing the [Y=] button and typing the equation.

2. Press the [MATH] button, scroll to Ø:Solver..., and press [ENTER]. (*Hint:* It is quicker to scroll up than to scroll down.)

3. If the Solver already contains an equation, you may see something similar to the graphic shown. Press the blue up arrow and then press [CLEAR] to delete the equation.

```
X^2+3*A+G=0
 X=2
 A=46
 G=7
 bound={-1E99,1...
```

4. Notice that the equation is set equal to zero. We may rewrite the difference quotient $D = \dfrac{Y_1(x+h) - Y_1(x)}{h}$ as

$$0 = \frac{Y_1(x+h) - Y_1(x)}{h} - D.$$ Type in the second equation using the [VARS]; Y-VARS; 1:Function menu sequence to enter the function Y1. To enter the variable H, press the [ALPHA] [^] key sequence. To enter the variable D, press the [ALPHA] [X⁻¹] key sequence. Press [ENTER].

```
EQUATION SOLVER
eqn:0=((Y₁(X+H)-
Y₁(X))/H)-D
```

5. Enter the value of x where you want to evaluate the difference quotient and enter a small positive value of H. Move your cursor to the D variable and press the [ALPHA] [ENTER] key sequence to solve for the difference quotient D. Repeat for various values of H.

```
((Y₁(X+H)-Y₁(...=0
 X=9
 H=.001
▪D=1.154671517
 bound={-1E99,1...
▪left-rt=0
```

EXAMPLE 3

Using Technology to Estimate an Instantaneous Rate of Change

Based on data from 1990–2000, the cost of 1/2 gallon of prepackaged ice cream may be modeled by

$$C(t) = 0.0142t^2 - 0.0272t + 2.53 \text{ dollars}$$

where t is the number of years since 1990. (**Source:** Modeled from *Statistical Abstract of the United States, 2001*, Table 706, p. 468.) Use the Solver on your calculator to estimate the instantaneous rate of change of the cost in 1999.

SOLUTION   We begin by entering $C(t)$ into the [Y=] editor.

```
Plot1 Plot2 Plot3
\Y1■0.0142X^2-0.
0272X+2.53
\Y2=
\Y3=
\Y4=
\Y5=
\Y6=
```

In 1999, $t = 9$. Entering $x = 9$ and $H = 0.001$ and solving for D yields

```
(Y1(X+H)-Y1(X...=0
X=9
H=.001
■D=.2284142
bound={-1ε99,1...
■left-rt=0
```

Repeating this process for various values of H, we create Table 9.5.

TABLE 9.5

h	$\dfrac{C(9 + h) - C(9)}{h}$
0.001	0.22841420
0.0001	0.22840142
0.00001	0.22840014
0.000001	0.22840000

In 1999, the cost of 1/2 gallon of prepackaged ice cream was increasing by about \$0.23 per year. That is, from 1999 to 2000, the price was expected to increase by about 23 cents.

Limits

In Examples 1 through 3, we estimated the instantaneous rate of change at a point by calculating the average rate of change over a "short" interval by picking "small" values of h. The terms *short* and *small* are vague. Numerically, what does "small" mean? Mathematicians struggled with this dilemma for years before developing the concept of the **limit.** We will explore the limit concept graphically before giving a formal definition.

Consider the graph of the function $f(x) = -x^2 + 4$ over the interval $[-3, 3]$ (Figure 9.5).

We ask the question, "As x gets close to 2, to what value does $f(x)$ get close?"

Observe from the graph of f that as the value of x moves from 0 to 2, the value of $f(x)$ moves from 4 to 0. We represent this behavior symbolically with the notation

$$\lim_{x \to 2^-} f(x) = 0$$

which is read, "the limit of $f(x)$ as x approaches 2 *from the left* is 0." This is commonly referred to as a **left-hand limit** because we approach $x = 2$ through values to the *left* of 2.

Observe from the graph of f that as the value of x moves from 3 to 2, the value of $f(x)$ moves from -5 to 0. We write

$$\lim_{x \to 2^+} f(x) = 0$$

which is read, "the limit of $f(x)$ as x approaches 2 *from the right* is 0." This is commonly referred to as a **right-hand limit** because we approach $x = 2$ through values to the *right* of 2.

On a number line, $-\infty$ lies to the left of $+\infty$. For the left-hand limit, $\lim_{x \to a^-} f(x)$, the minus sign is used to indicate that we are approaching $x = a$ from the direction of $-\infty$. For the right-hand limit, $\lim_{x \to a^+} f(x)$, the plus sign is used to indicate that we are approaching $x = a$ from the direction of $+\infty$.

The left- and right-hand limit behavior can also be seen from a table of values for $f(x)$ (Table 9.6).

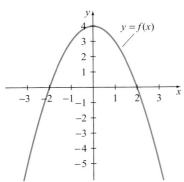

FIGURE 9.5

TABLE 9.6

	x	$f(x)$	
	0.00	4.000	
	1.00	3.000	$f(x)$ gets close to 0
Left of $x = 2$	1.90	0.390	as x nears 2
	1.99	0.040	
	2.00	**0.000**	
	2.01	-0.040	
Right of $x = 2$	2.10	-0.410	$f(x)$ gets close to 0 as x nears 2
	3.00	-5.000	

When the left- and right-hand limits of $f(x)$ approach the same finite value, we say that "the limit exists." In this case, the left- and right-hand limits of $f(x)$ neared the same value ($y = 0$) as x approached 2. We say that "the limit of $f(x)$ as x approaches 2 is 0" and write

$$\lim_{x \to 2} f(x) = 0$$

Sometimes the left- and right-hand limits of a function at a point are not equal. Consider the graph of the piecewise function

$$f(x) = \begin{cases} -x^2 + 2 & x \le 1 \\ x + 1 & x > 1 \end{cases}$$

on the interval $[-3, 3]$ (see Figure 9.6).

We ask the question, "As x gets close to 1, to what value does $f(x)$ get close?"

Observe from the graph that as the value of x moves from 0 to 1, the value of $f(x)$ moves from 2 to 1. We write

$$\lim_{x \to 1^-} f(x) = 1$$

That is, the left-hand limit of $f(x)$ as x approaches 1 is 1. Similarly, as the value of x moves from 2 to 1, the value of $f(x)$ moves from 3 to 2. We write

$$\lim_{x \to 1^+} f(x) = 2$$

That is, the right-hand limit of $f(x)$ as x approaches 1 is 2. Since for this function, the left- and right-hand limits are not equal, we say that "the limit of $f(x)$ as x approaches 1 does not exist" or, simply, "the limit does not exist." This can also be seen from a table of values for $f(x)$ (Table 9.7).

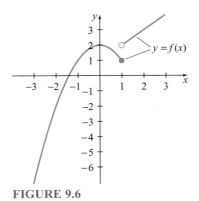

FIGURE 9.6

TABLE 9.7

	x	$f(x)$	
↓	0.00	2.000	
Left of $x = 1$	0.90	1.190	$f(x)$ gets close to 1 as x nears 1
↓	0.99	1.020	
	1.00	1.000	
↑	1.01	2.010	
Right of $x = 1$	1.10	2.100	$f(x)$ gets close to 2 as x nears 1
↑	2.00	3.000	
	3.00	4.000	

EXAMPLE 4   **Finding Left-hand and Right-hand Limits**

Given Table 9.8, determine $\lim_{x \to 0^-} f(x)$ and $\lim_{x \to 0^+} f(x)$.

TABLE 9.8

x	$f(x)$
-1.00	0.84147
-0.10	0.99833
-0.01	0.99998
0.00	Undefined
0.01	0.99998
0.10	0.99833
1.00	0.84147

SOLUTION We observe that as x moves from -1 to -0.01, the value of $f(x)$ nears 1. That is,

$$\lim_{x \to 0^-} f(x) = 1$$

Similarly, as x moves from 1 to 0.01, the value of $f(x)$ nears 1. That is,

$$\lim_{x \to 0^+} f(x) = 1$$

The limit of $f(x)$ as x approaches 0 is 1, since the left- and right-hand limits of $f(x)$ are both equal to 1. That is,

$$\lim_{x \to 0} f(x) = 1$$

It is important to note that even though $f(x)$ is undefined at $x = 0$, as x gets close to 0, $f(x)$ gets close to 1.

As illustrated in Example 4, one of the most powerful features of limits is that the limit of a function may exist at a point where the function itself is undefined. This feature will be used often when we introduce the limit definition of the derivative later in this section.

THE LIMIT OF A FUNCTION

If $f(x)$ is defined for all values of x near c, then

$$\lim_{x \to c} f(x) = L$$

means that as x approaches c, $f(x)$ approaches L.
    We say that the limit exists if

1. L is a finite number and

2. Approaching c from the left or right yields the same value of L.

The theory surrounding limits is rich and worthy of study; however, in this text, we are most interested in the limit of the difference quotient as h approaches zero. That is,

$$\lim_{h \to 0} \frac{f(a + h) - f(a)}{h}$$

Recall that the difference quotient represents the average rate of change of $f(x)$ over the interval $[a, a + h]$. When we place the limit on the difference quotient, we are symbolically asking, "As the distance between the two x values (a and $a + h$) gets smaller, what happens to the average rate of change of $f(x)$ on the interval $[a, a + h]$?" The limit of the difference quotient as h approaches zero (if the limit exists) is the **instantaneous rate of change in $f(x)$ at $x = a$**. Note that even though the difference quotient is undefined when $h = 0$, the limit may still exist.

EXAMPLE 5 Calculating an Instantaneous Rate of Change

Let $f(x) = x^2$. What is the instantaneous rate of change of $f(x)$ when $x = 3$?

SOLUTION We can calculate the instantaneous rate of change by taking the limit of the difference quotient as h approaches zero. The instantaneous rate of change of $f(x)$ at $x = 3$ is given by

$$\lim_{h \to 0} \frac{f(3 + h) - f(3)}{h} = \lim_{h \to 0} \frac{(3 + h)^2 - (3)^2}{h} \qquad \text{Since } f(x) = x^2$$

$$= \lim_{h \to 0} \frac{(9 + 6h + h^2) - 9}{h}$$

$$= \lim_{h \to 0} \frac{6h + h^2}{h}$$

$$= \lim_{h \to 0} \frac{h(6 + h)}{h}$$

$$= \lim_{h \to 0} (6 + h) \text{ for } h \neq 0 \qquad \text{Since } \frac{h}{h} = 1 \text{ for } h \neq 0$$

As h nears 0, what happens to the value of $6 + h$? Let's pick values of h near 0 (Table 9.9).

TABLE 9.9

h	$6 + h, h \neq 0$
-0.100	5.900
-0.010	5.990
-0.001	5.999
0.000	Undefined
0.001	6.001
0.010	6.010
0.100	6.100

As seen from the table, even though the difference quotient is undefined when $h = 0$, the value of the simplified difference quotient, $6 + h$, gets close to 6 as h approaches 0. In fact, by picking sufficiently small values of h, we can get as close to 6 as we would like. So the instantaneous rate of change of $f(x)$ when $x = 3$ is 6.

Observe that we can attain the same result by plugging in $h = 0$ after canceling out the h in the denominator of the difference quotient. That is,

$$\lim_{h \to 0}(6 + h) = 6 + 0$$
$$= 6$$

Throughout the rest of this chapter, we will substitute in $h = 0$ after eliminating the h in the denominator of the difference quotient. This process will simplify our computations while still giving the correct result.

The limit of the difference quotient as h approaches zero is used widely throughout calculus and is called **the derivative.**

THE DERIVATIVE OF A FUNCTION AT A POINT

The **derivative** of a function $y = f(x)$ at a point $(a, f(a))$ is

$$f'(a) = \lim_{h \to 0} \frac{f(a + h) - f(a)}{h}$$

$f'(a)$ is read "f prime of a" and is the *instantaneous rate of change* of the function f at the point $(a, f(a))$.

EXAMPLE 6

Calculating the Derivative of a Function at a Point

Given $f(x) = 3x + 1$, find $f'(2)$.

SOLUTION

$$f'(2) = \lim_{h \to 0} \frac{f(2 + h) - f(2)}{h}$$

$$= \lim_{h \to 0} \frac{[3(2 + h) + 1] - [3(2) + 1]}{h} \qquad \text{Since } f(2 + h) = [3(2 + h) + 1] \text{ and } f(2) = 3(2) + 1$$

$$= \lim_{h \to 0} \frac{(6 + 3h + 1) - (7)}{h}$$

$$= \lim_{h \to 0} \frac{(3h + 7) - (7)}{h}$$

$$= \lim_{h \to 0} \frac{3h}{h}$$

$$= \lim_{h \to 0} 3 \qquad \text{Since } \frac{h}{h} = 1$$

$$= 3$$

In this case, the difference quotient turned out to be a constant value of 3, so taking the limit of the difference quotient as h approached 0 did not alter the value of the difference quotient.

For linear functions, the slope of the line is the instantaneous rate of change of the function at any value of x. Consequently, the derivative of a linear function will always be a constant value that is equal to the slope of the line.

Recall that in Example 1, we estimated the runner's finish-line speed by finding the average rate of change in distance over a small interval of time. In other words, the distance traveled divided by the amount of time used to travel the distance yielded the average speed. One of the drawbacks of using the term *speed* is that it does not indicate direction. For example, consider two race cars traveling at 75 mph on the same road (Figure 9.7).

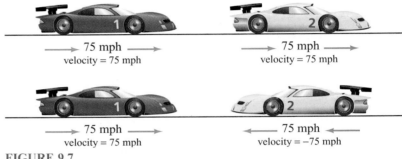

$\longrightarrow$ 75 mph $\longrightarrow$
velocity = 75 mph

$\longrightarrow$ 75 mph $\longrightarrow$
velocity = 75 mph

$\longrightarrow$ 75 mph $\longrightarrow$
velocity = 75 mph

$\longleftarrow$ 75 mph $\longleftarrow$
velocity = −75 mph

FIGURE 9.7

If the cars are traveling in the same direction, we don't foresee any problems. However, if the cars are traveling in opposite directions, a head-on collision may be imminent. The direction of the cars is a key piece of information that is not captured by the term *speed*. Fortunately, the term *velocity* combines the concepts of speed and direction. A negative sign placed on the speed is used to distinguish a difference in direction. In Figure 9.7, a positive velocity indicates that the car is traveling from left to right. A negative velocity indicates that the car is traveling from right to left.

When talking about the velocity of objects that are rising or falling, it is customary to use a positive velocity to indicate that the object is rising and a negative velocity to indicate that the object is falling.

It is common to model the vertical position of a free-falling object on earth with the equation $s(t) = -16t^2 + v_0 t + s_0$ feet, where v_0 is the velocity of the object (in feet per second) and s_0 is the vertical position of the object (in feet) at time $t = 0$ seconds. (Although this model neglects air resistance, it does give us an idea of what is happening physically.) To determine the velocity of the object at time t seconds, we simply need to find $s'(t)$.

EXAMPLE 7

Calculating the Velocity of a Falling Object

A water balloon is dropped from the top of a 20-foot roof. How fast is it falling 1 second into its flight? How fast is it falling when it hits the ground?

SOLUTION The *initial velocity* of the balloon, v_0, is 0 feet per second because the balloon was dropped, not thrown. The *initial position* of the balloon, s_0, is 20 feet, since the balloon is being dropped from 20 feet above the ground. Therefore, the balloon's position may be modeled by

$$s(t) = -16t^2 + v_0 t + s_0$$
$$= -16t^2 + 0t + 20$$
$$= -16t^2 + 20$$

To calculate the velocity of the balloon at time $t = 1$ second, we must find $s'(1)$.

$$s'(1) = \lim_{h \to 0} \frac{s(1 + h) - s(1)}{h}$$

$$= \lim_{h \to 0} \frac{[-16(1 + h)^2 + 20] - [-16(1)^2 + 20]}{h} \qquad \text{Since } s(1 + h) = -16(1 + h)^2 + 20$$

$$= \lim_{h \to 0} \frac{[-16(1 + 2h + h^2) + 20] - [-16 + 20]}{h}$$

$$= \lim_{h \to 0} \frac{(-16 - 32h - 16h^2 + 20) - (4)}{h}$$

$$= \lim_{h \to 0} \frac{(-32h - 16h^2 + 4) - (4)}{h}$$

$$= \lim_{h \to 0} \frac{-32h - 16h^2}{h}$$

$$= \lim_{h \to 0} \frac{h(-32 - 16h)}{h} \qquad \text{Factor } h \text{ out of the numerator}$$

$$= \lim_{h \to 0} (-32 - 16h) \qquad \text{Since } \frac{h}{h} = 1 \text{ for } h \neq 0$$

$$= -32 - 16(0) \qquad \text{Plug in } h = 0$$

$$= -32 \text{ feet per second}$$

So 1 second after the balloon is dropped, it is traveling toward the ground at a speed of 32 feet per second. (As previously noted, a positive velocity typically indicates that an object is moving upward, while a negative velocity signifies that an object is moving downward.)

To determine how fast the balloon is falling when it hits the ground, we must first determine when it will hit the ground. Since $s(t)$ is the height of the balloon above the ground, we must determine when $s(t) = 0$.

$$s(t) = -16t^2 + 20$$
$$0 = -16t^2 + 20$$
$$16t^2 = 20$$
$$t^2 = \frac{20}{16}$$
$$t^2 = 1.25$$
$$t \approx 1.1 \text{ seconds}$$

We estimate that the balloon will hit the ground in 1.1 seconds. To determine the velocity of the balloon at that time, we calculate $s'(1.1)$.

$$s'(1.1) = \lim_{h \to 0} \frac{s(1.1 + h) - s(1.1)}{h}$$

$$= \lim_{h \to 0} \frac{[-16(1.1 + h)^2 + 20] - [-16(1.1)^2 + 20]}{h}$$

$$= \lim_{h \to 0} \frac{[-16(1.21 + 2.2h + h^2) + 20] - [-16(1.21) + 20]}{h}$$

$$= \lim_{h \to 0} \frac{(-19.36 - 35.2h - 16h^2 + 20) - (-19.36 + 20)}{h}$$

$$= \lim_{h \to 0} \frac{(-35.2h - 16h^2 + 0.64) - (0.64)}{h}$$

$$= \lim_{h \to 0} \frac{-35.2h - 16h^2}{h}$$

$$= \lim_{h \to 0} \frac{h(-35.2 - 16h)}{h} \qquad \text{Factor } h \text{ out of the numerator}$$

$$= \lim_{h \to 0} (-35.2 - 16h) \qquad \text{Since } \frac{h}{h} = 1 \text{ for } h \neq 0$$

$$= -35.2 - 16(0) \qquad \text{Plug } in \ h = 0$$

$$= -35.2 \text{ feet per second}$$

UNITS

We estimate that the balloon will be falling at a rate of a little over 35 feet per second when it strikes the ground.

EXAMPLE 8

Finding and Interpreting the Meaning of the Derivative of a Function at a Point

The population of Washington state may be modeled by the function

$$P(t) = 0.8981(1.019)^t \text{ million people}$$

where t is the number of years since 1900. (**Source:** Modeled from www.ofm.wa.gov data.) Find and interpret the meaning of $P'(104)$.

SOLUTION Since t is the number of years since 1900, $t = 104$ is the year in 2004. $P'(104)$ is the instantaneous rate of change in the population in 2004, given in millions of people per year.

$$P'(104) = \lim_{h \to 0} \frac{P(104 + h) - P(104)}{h}$$

$$= \lim_{h \to 0} \frac{[0.8981(1.019)^{104 + h}] - [0.8981(1.019)^{104}]}{h}$$

$$= \lim_{h \to 0} \frac{0.8981[(1.019)^{104 + h} - (1.019)^{104}]}{h} \qquad \text{Factor out } 0.8981$$

$$= \lim_{h \to 0} \frac{0.8981[(1.019)^{104}(1.019)^h - (1.019)^{104}]}{h} \qquad \text{Since } (1.019)^{104 + h} = (1.019)^{104}(1.019)^h$$

$$= \lim_{h \to 0} \frac{(0.8981)(1.019)^{104}\left[(1.019)^h - 1\right]}{h} \qquad \text{Factor out } (1.019)^{104}$$

$$= \lim_{h \to 0} \frac{6.360\left[(1.019)^h - 1\right]}{h}$$

Unlike in Examples 5, 6, and 7, we are unable to eliminate the h in the denominator algebraically and calculate the exact value of $P'(104)$. Nevertheless, by picking a small value for h (say $h = 0.001$), we can estimate the instantaneous rate of change.

 UNITS

$$\frac{6.360[(1.019)^{0.001} - 1]}{0.001} = 0.1197 \text{ million people per year}$$

$$= 119.7 \text{ thousand people per year}$$

$$\approx 120 \text{ thousand people per year}$$

According to the model, the population of Washington was increasing by approximately 120 thousand people per year in 2004.

Although we were unable to obtain the exact value of the derivative of the exponential function in Example 8, we will develop the theory in later sections that will allow us to calculate the exact value of the derivative of an exponential function.

9.2 Summary

In this section, you learned how to use the difference quotient to estimate an instantaneous rate of change. You also discovered that the derivative may be used to calculate the exact value of an instantaneous rate of change of a function at a point.

9.2 Exercises

In Exercises 1–10, use the difference quotient
$$\frac{f(a + h) - f(a)}{h} \text{ (with } h = 0.1, h = 0.01, \text{ and}$$
$h = 0.001) \text{ to estimate the instantaneous rate of change of the function at the given input value.}$

1. $f(x) = x^2; x = 2$

2. $h(x) = -x^2 + 1; x = 0$

3. $s(t) = -16t^2 + 64; t = 2$

4. $s(t) = -16t^2 + 64t + 32; t = 0$

5. $w(t) = 4t + 2; t = 5$

6. $P(t) = t^2 - 2t + 1; t = 1$

7. $P(t) = 5; t = 25$

8. $g(x) = 5x - x^2; x = -1$

9. $P(r) = 500(1 + r)^2; r = 0.07$

10. $P(r) = 100(1 + r)^2; r = 0.12$

In Exercises 11–20, use the derivative to calculate the instantaneous rate of change of the function at the given input value. (In each exercise, you can eliminate the h algebraically.) Compare your answers to the solutions of Exercises 1–10.

11. $f(x) = x^2; x = 2$ **12.** $h(x) = -x^2 + 1; x = 0$

13. $s(t) = -16t^2 + 64$; $t = 2$

14. $s(t) = -16t^2 + 64t + 32$; $t = 0$

15. $w(t) = 4t + 2$; $t = 5$

16. $P(t) = t^2 - 2t + 1$; $t = 1$

17. $P(t) = 5$; $t = 25$

18. $g(x) = 5x - x^2$; $x = -1$

19. $P(r) = 500(1 + r)^2$; $r = 0.07$

20. $P(r) = 100(1 + r)^2$; $r = 0.12$

In Exercises 21–30, use the difference quotient (with $h = 0.1$, $h = 0.01$, and $h = 0.001$) to estimate the instantaneous rate of change of the function at the given input value. You may find it helpful to apply the Technology Tip demonstrated in the section.

21. $f(x) = 2x^{-3}$; $x = 3$

22. $P(t) = 50(1.02)^t$; $t = 10$

23. $P(t) = 230(0.9)^t$; $t = 25$

24. $g(x) = -2(3)^x$; $x = -1$

25. $P(r) = 500(1 + r)^{10}$; $r = 0.07$

26. $P(r) = 100(1 + r)^5$; $r = 0.12$

27. $y = \ln(x)$; $x = 2$

28. $y = x \ln(x)$; $x = 1$

29. $g(x) = e^{3x}$; $x = 1$

30. $g(x) = 8 - 2x$; $x = 5$

In Exercises 31–40, determine the instantaneous rate of change of the function at the indicated input value. (You may find it helpful to apply the Technology Tip demonstrated in the section.) Then explain the real-life meaning of the result.

31. 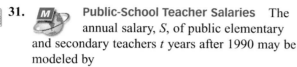 **Public-School Teacher Salaries** The annual salary, S, of public elementary and secondary teachers t years after 1990 may be modeled by

$$S(t) = 0.00713t^3 - 0.111t^2 + 1.42t + 31.6$$

thousand dollars. (**Source:** Modeled from *Statistical Abstract of the United States, 2001*, Table 237, p. 151.) Find and interpret the meaning of $S'(10)$.

32. **Apparel and Textile Industry Pay** The annual payroll, P, of the apparel and

textile manufacturing industry between 1992 and 1997 may be modeled by

$$P = -1560x + 24{,}850$$

billion dollars, where x is the number of paid employees (in thousands). (**Source:** Modeled from *Statistical Abstract of the United States, 2001*, Table 722, p. 482.) Find and interpret the meaning of $P'(15)$.

33. **Prescription Drug Spending** The amount of money spent in the United States on prescription drugs may be modeled by

$$D(t) = 0.003667t^3 - 0.0986t^2 + 0.8976t + 2.829$$

billion dollars, where t is the number of years after 1960. (**Source:** Modeled from *Statistical Abstract of the United States, 2001*, Table 119, p. 91.) Find and interpret the meaning of $D'(45)$.

34. **Video-Game Sales** The annual sales, S, of the U.S. video-game industry from 1997 to 2001 may be modeled by

$$S(t) = 0.1958t^4 - 1.275t^3 + 2.254t^2 - 0.0750t + 5.100$$

billion dollars, where t is the number of years since 1997. (**Source:** Modeled from data at www.npdfunworld.com.) Find and interpret the meaning of $S'(5)$.

35. **Video-Game Sales** In the first quarter of 2001, U.S. video-game industry sales totaled $1.6 billion. Sales increased by 20 percent in the first quarter of 2002. The number one video game in the first quarter of 2002 was Grand Theft Auto 3. (**Source:** www.npdfunworld.com.) If this pattern continues, video game industry sales in the first quarter, S, may be modeled by

$$S = 1.6(1.2)^t$$

billion dollars, where t is the number of years since 2001. Find and interpret the meaning of $S'(6)$.

36. **Nursing Home Care** The amount of money spent annually in the United States on nursing home care from 1960 to 2000 may be modeled by

$$P(t) = 0.07214t^2 - 0.4957t + 1.029$$

billion dollars, where t is the number of years after 1960. (**Source:** Modeled from *Statistical Abstract of the United States, 2001,* Table 119, p. 91). Find and interpret the meaning of $D'(45)$.

37. **M** **College Tuition** The cost of full-time resident tuition at Green River Community College from 1994 to 2001 may be modeled by

$$E(t) = 0.1869t^3 - 1.081t^2 + 19.57t + 431.6$$

dollars, where t is the number of years since 1994. (**Source:** Modeled from Green River Community College data.) Find and interpret the meaning of $E'(7)$.

38. **M** **Demand for VHS Tapes** From 2000 to 2002, the worldwide demand for blank VHS tapes fell. The demand for the tapes may be modeled by

$$V(t) = -20t^2 - 67t + 1409$$

million tapes, where t is the number of years since 2000. (**Source:** Modeled from International Recording Media Association data.) Find and interpret the meaning of $V'(2)$.

39. **M** **Demand for CD-R Disks** From 2000 to 2002, the worldwide demand for CD-R disks increased substantially. The demand for the disks may be modeled by

$$C(t) = -252.5t^2 + 1252.5t + 2730$$

million disks, where t is the number of years since 2000. (**Source:** Modeled from International Recording Media Association data.) Find and interpret the meaning of $C'(2)$.

40. **M** **VHS Tapes and CD-R Disks** From 2000 to 2002, the worldwide demand for blank VHS tapes fell, while the demand for CD-R disks increased. The demand for CD-R disks may be modeled by

$$C(v) = -0.03550v^2 + 85.45v - 47{,}200$$

million disks, where v is the number of blank VHS tapes sold (in millions). (**Source:** Modeled from International Recording Media Association data.) Find and interpret the meaning of $C'(500)$. Do you believe that the sales of blank VHS tapes and CD-R disks are related? Justify your answer.

Exercises 41–45 deal with the velocity of a free-falling object on earth. As explained in the section, the vertical position of a free-falling object may be modeled by $s(t) = -16t^2 + v_0 t + s_0$ feet, where v_0 is the velocity of the object and s_0 is the vertical position of the object at time $t = 0$ seconds.

41. **M** **Velocity of a Dropped Object** A can of soda is dropped from a diving board 40 feet above the bottom of an empty pool. How fast is the can traveling when it reaches the bottom of the pool?

42. **M** **Velocity of a Dropped Object** The Space Needle in Seattle, Washington, is a popular tourist attraction. The observation deck is approximately 520 feet above the ground. Dropping anything from that height is extremely dangerous to people on the ground. How fast would a penny dropped from the observation deck be traveling when it hit the ground?

43. **M** **Velocity of a Ball** A small rubber ball is thrown into the air by a child at a velocity of 20 feet per second. The child releases the ball 4 feet above the ground. What is the velocity of the ball after 1 second?

44. **M** **Velocity of a Diving Ring** A diving ring is thrown from a diving board at a velocity of -10 feet per second toward the surface of a diving pool. If the surface of the pool is 40 feet below the point of release, what will be the velocity of the ring when it hits the surface of the water?

45. **M** **Velocity of a Juggling Pin** A juggler tosses a bowling pin into the air at a velocity of 15 feet per second. If she catches the pin at the same point she released it, what will be the velocity of the pin when she catches it?

Exercises 46–50 are intended to challenge your understanding of instantaneous rates of change and derivatives.

46. The derivative of a continuous function f is undefined at the point $[a, f(a)]$. Draw a possible graph of f, including the point $[a, f(a)]$.

47. A function f has the property that $f'(a) = f'(b)$. Does $a = b$? Explain.

48. The instantaneous rate of change of a function f is equal to 3 for all values of x. How many x-intercepts does the graph of f have?

49. The instantaneous rate of change of a function f is equal to 0 for all values of x. How many x-intercepts does the graph of f have?

50. The Mean Value Theorem states that if f is continuous and smooth (no breaks, no sharp points) on an interval $[a, b]$, then there exists a value c with $a < c < b$ such that

$$f'(c) = \frac{f(b) - f(a)}{b - a}$$

Given $f(x) = x^2$ on the interval $[2, 4]$, find a value c with $2 < c < 4$ such that

$$f'(c) = \frac{f(4) - f(2)}{4 - 2}$$

9.3 The Derivative as a Slope: Graphical Methods

- Find the equation of the tangent line of a curve at a given point
- Numerically approximate derivatives from a table of data
- Graphically interpret average and instantaneous rates of change
- Use derivative notation and terminology to describe instantaneous rates of change

GETTING STARTED   Based on data from 1990 to 2000, the average price for a half gallon of prepackaged ice cream may be modeled by

$$C(t) = 0.0142t^2 - 0.0272t + 2.53 \text{ dollars}$$

where t is the number of years since 1990. (**Source:** Modeled from *Statistical Abstract of the United States, 2001*, Table 706, p. 468.)

To help project future revenue, a creamery wants to know what the average annual increase in the price of ice cream was from 1990 to 2000 and at what rate the price will be increasing at the end of 2000. We will answer these questions by calculating the average and instantaneous rates of change in the price of ice cream.

In Sections 9.1 and 9.2, you learned how to calculate the average rate of change of a function over an interval and the instantaneous rate of change of a function at a point. In this section, we will revisit these concepts from a graphical standpoint. We will also demonstrate how to use tangent-line approximations to estimate the value of a function.

As shown in Section 9.1, the difference quotient formula, $\frac{f(a + h) - f(a)}{h}$, gives the average rate of change in the value of the function between the points $(a, f(a))$ and $(a + h, f(a + h))$. If we let $a = c - h$, then the difference quotient formula becomes $\frac{f(c) - f(c - h)}{h}$. We will use this modified form of the difference quotient in our exploration of the price of ice cream, since we will be approaching $t = c$ through values to the left of $t = c$. Recall from Section 9.1 that, graphically speaking, the difference quotient is the **slope of the secant line** connecting the two points on the graph of a function. According to the ice cream price model, $C(0) = 2.530$ and $C(10) = 3.678$. We're interested in the slope of the secant line between $(0, 2.530)$ and $(10, 3.678)$. The slope of this line represents the average rate of change in the price of a half gallon of ice cream between 1990 and 2000 (Figure 9.8).

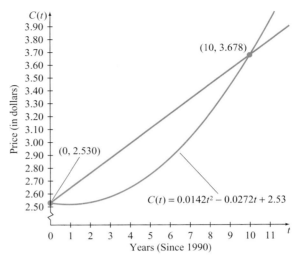

FIGURE 9.8

The ice cream graph secant line between $(0, 2.530)$ and $(10, 3.678)$ has the slope

$$m = \frac{C(10) - C(0)}{10} \frac{\text{dollars}}{\text{years}}$$

$$= \frac{3.678 - 2.530}{10} \frac{\text{dollars}}{\text{years}}$$

$$= 0.115 \text{ dollar per year}$$

Between 1990 and 2000, the average price of a half gallon of ice cream increased by $0.115 (about 12 cents) per year. Does this mean that from 2000 to 2001, the price will increase by about 12 cents? No. Looking at the graph of the model, we notice that the price of ice cream is rising at an increasing rate as time progresses (the steeper the graph, the greater the magnitude of the rate of change). We can approximate the instantaneous rate of change at the end of 2000 ($t = 10$) by calculating the slope of a secant line through $(10, 3.678)$ and a "nearby" point, as shown in Figure 9.9.

The ice cream graph secant line between $(5, 2.749)$ and $(10, 3.678)$ has the slope

$$m = \frac{C(10) - C(5)}{5} \frac{\text{dollars}}{\text{years}}$$

$$= \frac{3.678 - 2.749}{5} \frac{\text{dollars}}{\text{years}}$$

$$= 0.186 \text{ dollar per year}$$

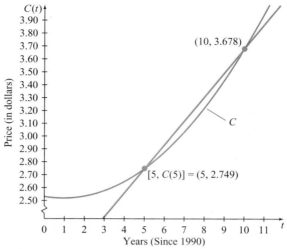

FIGURE 9.9

As shown in Figure 9.10, the ice cream graph secant line between $(9, 3.435)$ and $(10, 3.678)$ has the slope

$$m = \frac{C(10) - C(9)}{10 - 9} \frac{\text{dollars}}{\text{years}}$$

$$= \frac{3.678 - 3.435}{1} \frac{\text{dollars}}{\text{years}}$$

$$= 0.243 \text{ dollar per year}$$

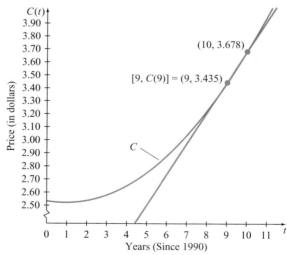

FIGURE 9.10

By zooming in (Figure 9.11), we can see that between $t = 9$ and $t = 10$, the secant line and the graph of the function are extremely close together.

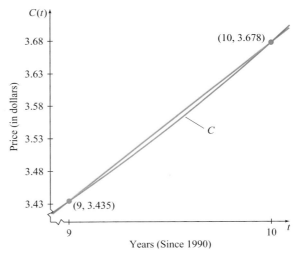

FIGURE 9.11

The ice cream graph secant line between (9.9, 3.653) and (10, 3.678) has the slope

$$m = \frac{C(10) - C(9.9)}{10 - 9.9} \frac{\text{dollars}}{\text{years}}$$

$$= \frac{3.678 - 3.652}{0.1} \frac{\text{dollars}}{\text{years}}$$

$$= 0.26 \text{ dollar per year}$$

Through (9.9, 3.652) and (10, 3.678), the slope of the secant line is $0.26 per year.

Again zooming in, we see that the secant line and the graph of the function are nearly identical between $t = 9.9$ and $t = 10$ (Figure 9.12). In fact, we are unable to visually distinguish between the two.

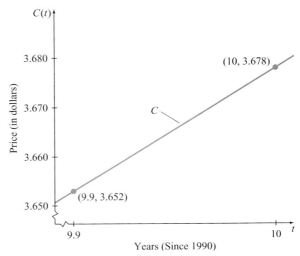

FIGURE 9.12

As t gets closer and closer to 10, the slope of the secant line will approach $C'(10)$. This occurs because $\dfrac{C(10) - C(10 - h)}{h} \approx C'(10)$ for small values of h. Since h is the horizontal distance between the points, h becomes increasingly small as t nears 10. Using the methods covered in Section 9.2, we determine algebraically that $C'(10) = 0.2568$. At the end of 2000, ice cream prices were increasing at a rate of roughly $0.26 per year. In other words, we anticipate that ice cream prices increased by *approximately* $0.26 between 2000 and 2001.

As the preceding example demonstrates, to find the instantaneous rate of change of a function $y = f(x)$ at a point $P = (a, f(a))$, we can find the limit of the slope of the secant line through P and a nearby point Q as the point Q gets closer and closer to P. Graphically speaking, we select values $Q_1, Q_2, Q_3, \ldots,$ with each consecutive value of Q_i being a point on the curve that is closer to the point P than the one before it. The limit of the slope of the secant lines (imagine the points P and Q finally coinciding) is the line *tangent* to the curve at the point $P = (a, f(a))$ (Figure 9.13).

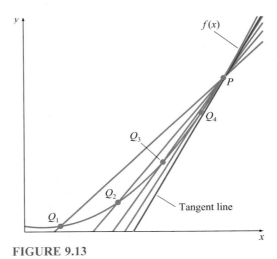

FIGURE 9.13

The **tangent line** to a graph f at a point $(a, f(a))$ is the line that passes through $(a, f(a))$ and has slope $f'(a)$.

THE GRAPHICAL MEANING OF THE DERIVATIVE

The derivative of a function f at a point $(a, f(a))$ is the **slope of the tangent line** to the graph of f at that point. The slope of the tangent line at a point is also referred to as the **slope of the curve** at that point.

EXAMPLE 1

Finding the Equation of a Tangent Line

Find the equation of the tangent line to the graph of $f(x) = x^2$ that passes through $(2, 4)$. Then graph the tangent line and the graph of f.

SOLUTION The slope of the tangent line is $f'(2)$.

$$f'(2) = \lim_{h \to 0} \frac{f(2+h) - f(2)}{h}$$

$$= \lim_{h \to 0} \frac{(2+h)^2 - 2^2}{h} \qquad \text{Since } f(x) = x^2$$

$$= \lim_{h \to 0} \frac{4 + 4h + h^2 - 4}{h}$$

$$= \lim_{h \to 0} \frac{4h + h^2}{h}$$

$$= \lim_{h \to 0} \frac{h(4+h)}{h}$$

$$= \lim_{h \to 0} (4 + h)$$

$$= 4 + 0$$

$$= 4$$

The slope of the tangent line is 4. Using the slope-intercept form of a line, we have $y = 4x + b$. Substituting in the point $(2, 4)$, we get

$$4 = 4(2) + b$$

$$4 = 8 + b$$

$$b = -4$$

The equation of the tangent line is $y = 4x - 4$.

We generate a table of values for $y = 4x - 4$ and $f(x) = x^2$. Then we graph the results (Figure 9.14).

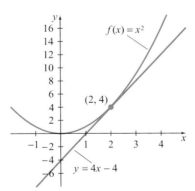

x	y	$f(x)$
-1	-8	1
0	-4	0
1	0	1
2	4	4
3	8	9
4	12	16

FIGURE 9.14

EXAMPLE 2

Finding the Equation of a Tangent Line

Find the equation of the tangent line of the graph of $f(x) = -x^2 + 2x + 2$ that passes through $(1, 3)$. Then plot the tangent line and the graph of f.

SOLUTION The slope of the tangent line is $f'(1)$. Because of the complex nature of f, we will calculate $f(1 + h)$ and $f(1)$ individually, then substitute the simplified values into the derivative formula.

$$f(1 + h) = -(1 + h)^2 + 2(1 + h) + 2 \qquad\qquad f(1) = -(1)^2 + 2(1) + 2$$
$$= -(1 + 2h + h^2) + 2 + 2h + 2 \qquad\qquad\quad = -1 + 2 + 2$$
$$= -1 - 2h - h^2 + 2h + 4 \qquad\qquad\qquad\quad = 3$$
$$= -h^2 + 3$$

$$f'(1) = \lim_{h \to 0} \frac{f(1 + h) - f(1)}{h}$$
$$= \lim_{h \to 0} \frac{(-h^2 + 3) - 3}{h}$$
$$= \lim_{h \to 0} \frac{-h^2}{h}$$
$$= \lim_{h \to 0} (-h)$$
$$= 0$$

The slope of the tangent line is 0. Using the slope-intercept form of a line, we have $y = 0x + b$. Substituting in the point $(1, 3)$, we get

$$3 = 0(1) + b$$
$$b = 3$$

The equation of the tangent line is $y = 3$. We generate a table of values for $y = 3$ and $f(x) = -x^2 + 2x + 2$. Then we graph both functions (Figure 9.15).

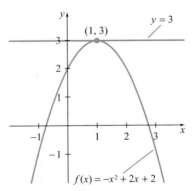

x	y	$f(x)$
-1	3	-1
0	3	2
1	3	3
2	3	2
3	3	-1

FIGURE 9.15

As shown in Example 2, a function $y = f(x)$ will have a horizontal tangent line at $(a, f(a))$ if $f'(a) = 0$.

Tangent-Line Approximations

In Examples 1 and 2, you may have noticed that the tangent line lies very near to the graph of f for values of x near a. In fact, if we zoom in to the region immediately surrounding $(a, f(a))$, the graph of f and the tangent line to the graph of f at $(a, f(a))$ appear nearly identical. Zooming in on the graph in Example 2, we can see that for values of x near 1, the tangent line y value is a good approximation of the actual function value (see Figure 9.16).

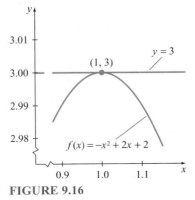

FIGURE 9.16

Since it is frequently easier to calculate the values of the tangent line than the values of the function, sometimes the tangent line is used to estimate the value of the function. For example, suppose we wanted to estimate $f(0.9)$ given $f(x) = -x^2 + 2x + 2$. Since $x = 0.9$ is near $x = 1$, we may use the equation of the tangent line, $y = 3$, to estimate $f(0.9)$. That is, $f(0.9) \approx 3$. The actual value is

$$f(0.9) = -(0.9)^2 + 2(0.9) + 2$$
$$= -0.81 + 1.8 + 2$$
$$= 2.99$$

Our tangent-line estimate ($y = 3$) was remarkably close to the actual value of the function ($f(0.9) = 2.99$) and required no additional computations.

EXAMPLE 3

Using a Tangent Line Approximation to Forecast DVD Player Demand

Based on data from 1998 to 2001, the number of DVD players shipped in North America may be modeled by

$$P(t) = 995t^2 + 271t - 245 \text{ thousand players}$$

where t is the number of years since the end of 1997. (Model based on data from DVD Entertainment Group.) According to the model, 16,759 thousand DVD players were shipped in 2001.

Determine how quickly DVD player shipments were increasing at the end of 2001. Then use a tangent line to estimate how many DVD players were shipped in 2002 and compare the estimate to the actual value predicted by the model.

SOLUTION Since $t = 4$ corresponds with the year 2001, the instantaneous rate of change in DVD player shipments at the end of 2001 is given by

UNITS

$$P'(4) = \lim_{h \to 0} \frac{P(4 + h) - P(4)}{h} \quad \frac{\text{thousand players}}{\text{years}}$$

We already know that $P(4) = 16,759$. We'll calculate $P(4 + h)$ and then substitute the simplified value into the derivative formula.

$$P(4 + h) = 995(4 + h)^2 + 271(4 + h) - 245$$
$$= 995(16 + 8h + h^2) + 1084 + 271h - 245$$
$$= 15,920 + 7960h + 995h^2 + 271h + 839$$
$$= 995h^2 + 8231h + 16,759$$

$$P'(4) = \lim_{h \to 0} \frac{P(4 + h) - P(4)}{h}$$
$$= \lim_{h \to 0} \frac{(995h^2 + 8231h + 16,759) - 16,759}{h}$$
$$= \lim_{h \to 0} \frac{995h^2 + 8231h}{h}$$
$$= \lim_{h \to 0} \frac{h(995h + 8231)}{h}$$

$$= \lim_{h \to 0} (995h + 8231)$$

$$= 995(0) + 8231$$

$$= 8231 \, \frac{\text{thousand players}}{\text{year}}$$

At the end of 2001, DVD player shipments were increasing at a rate of 8231 thousand (8.231 million) players per year. That is, between 2001 and 2002, the number of DVD players shipped annually was expected to increase by *approximately* 8.231 million.

The slope-intercept form of the tangent line is $y = 8231t + b$. Using the point (4, 16,759), we determine

$$16{,}759 = 8231(4) + b$$

$$16{,}759 = 32{,}924 + b$$

$$b = -16{,}165$$

The tangent-line equation is $y = 8231t - 16{,}165$. At $t = 5$, we have

$$y = 8231(5) - 16{,}165$$

$$= 41{,}155 - 16{,}165$$

$$= 24{,}990$$

Using the tangent-line equation, we estimate that 24.990 million DVD players were shipped in 2002. According to the model, the actual number of DVD players shipped was slightly greater.

$$P(5) = 995(5)^2 + 271(5) - 245$$

$$= 25{,}985$$

The model indicates that 25.985 million DVD players were shipped in 2002.

Why was there a discrepancy between the two estimates in Example 3? Look at the graph of the model and the tangent line (Figure 9.17).

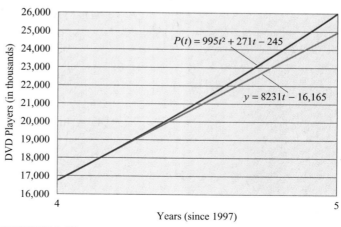

FIGURE 9.17

Although both functions were equal at $t = 4$, by the time t reached 5, the value of the model exceeded the tangent-line estimate by about 1,000 thousand DVD players. Although tangent-line estimates of a function's value are not exact, they are often good enough for their intended purpose.

Numerical Derivatives

TABLE 9.10

x	$f(x)$
0	0
1	1
2	4
3	9
4	16

Often we encounter real-life data in tables or charts. Is it possible to calculate a derivative from a table of data? We'll investigate this question by looking at a table of data for $f(x) = x^2$ (see Table 9.10). For this function, $f'(2) = 4$.

We can estimate $f'(2)$ by calculating the slope of the secant line through points whose x values are equidistant from $x = 2$. That is,

$$f'(2) \approx \frac{f(3) - f(1)}{3 - 1}$$

$$= \frac{9 - 1}{2}$$

$$= \frac{8}{2}$$

$$= 4$$

In this case, our estimate was equal to the tangent-line slope. Let's look at the situation graphically (see Figure 9.18).

When estimating a derivative numerically, we typically select the two closest data points that are horizontally equidistant from our point of interest. Doing so often yields a line that is parallel to the tangent line. Picking two points that are equidistant from the point of interest will tend to give the best estimate of the derivative and can be used as long as the point of interest is not an endpoint. (If the point of interest is an endpoint, we find the slope of the secant line between the endpoint and the next closest point.) If we assume that each output in a table of data represents $f(a)$ for a corresponding input a, then we can symbolically represent the process of numerically estimating a derivative as follows.

FIGURE 9.18

NUMERICAL ESTIMATE OF THE DERIVATIVE

The derivative of a function f at a point $(a, f(a))$ may be approximated from a table by

$$f'(a) \approx \frac{f(a + h) - f(a - h)}{(a + h) - (a - h)}$$

$$\approx \frac{f(a + h) - f(a - h)}{2h}$$

where h is the horizontal distance between a and $a + h$.

If a is the largest domain value in the data set, then

$$f'(a) \approx \frac{f(a) - f(a - h)}{(a) - (a - h)}$$

$$\approx \frac{f(a) - f(a - h)}{h}$$

If a is the smallest domain value in the data set, then

$$f'(a) \approx \frac{f(a + h) - f(a)}{(a + h) - (a)}$$

$$\approx \frac{f(a + h) - f(a)}{h}$$

In both cases, h is the distance between a and the next closest domain value.

EXAMPLE **4**

Estimating the Derivative from a Table of Data

Use Table 9.11 to estimate how quickly home sales prices in the southern United States were increasing at the end of 1990. [That is, estimate the slope of the tangent line at $(10, 99.0)$.]

TABLE 9.11   Median Sales Price of a New One-Family House in the Southern United States

Years Since 1980 (t)	Price (thousands of dollars) (P)
0	59.6
5	75.0
10	99.0
15	124.5
20	148.0

Source: *Statistical Abstract of the United States, 2001,* Table 940, p. 598.

SOLUTION   We'll estimate the slope of the tangent line at $(10, 99.0)$ by calculating the slope of the secant line between $(5, 75.0)$ and $(15, 124.5)$.

$$f'(10) \approx \frac{f(15) - f(5)}{15 - 5} \frac{\text{thousand dollars}}{\text{years}}$$

$$\approx \frac{124.5 - 75.0}{10}$$

$$\approx 4.95 \text{ thousand dollars per year}$$

We estimate that home prices in the southern United States were increasing at a rate of 5.0 thousand dollars per year at the end of 1990. (Since the original data were accurate to one decimal place, we rounded our result to one decimal place.)

EXAMPLE 5

Estimating the Derivative from a Table of Data

Given the data in Table 9.12, how quickly was the percentage of households leasing vehicles increasing in 1995 and 1998?

TABLE 9.12   Percentage of Households Leasing Vehicles

Years Since 1989 (t)	Percent (P)
0	2.5
3	2.9
6	4.5
9	6.4

Source: *Statistical Abstract of the United States, 2001,* Table 1086, p. 680.

SOLUTION We want to find $f'(6)$ and $f'(9)$.

$$f'(6) \approx \frac{f(9) - f(3)}{9 - 3} \frac{\text{percentage points}}{\text{years}}$$

$$\approx \frac{6.4 - 2.9}{6}$$

$$\approx 0.583 \text{ percentage point per year}$$

In 1995, the percentage of households with leased vehicles was increasing by about 0.6 percentage point per year.

Since an endpoint occurs at $t = 9$, we will estimate $f'(9)$ by calculating the slope of the secant line between the endpoint and the next closest point.

$$f'(9) \approx \frac{f(9) - f(6)}{9 - 6} \frac{\text{percentage points}}{\text{years}}$$

$$\approx \frac{6.4 - 4.5}{3}$$

$$\approx 0.633 \text{ percentage point per year}$$

In 1998, the percentage of households with leased vehicles was also increasing by about 0.6 percentage point per year.

9.3 Summary

In this section, you learned that the derivative at a point is the slope of the tangent line to the graph at that point. You discovered that the tangent line may be used to approximate the value of a function. You also learned how to estimate a derivative from a table of data.

9.3 Exercises

In Exercises 1–10, determine the equation of the tangent line of the function at the given point. Then graph the tangent line and the function together.

1. $f(x) = x^2 - 4x$; $(1, -3)$

2. $f(x) = -x^2 + 6$; $(2, 2)$

3. $g(x) = x^2 + 2x + 1$; $(0, 1)$

4. $g(x) = x^2 - 4$; $(3, 5)$

5. $g(x) = x^2 - 4x - 5$; $(4, -5)$

6. $h(x) = x^3$; $(2, 8)$

7. $h(x) = x^3$; $(0, 0)$

8. $h(x) = x^3$; $(1, 1)$

9. $f(x) = (x - 3)^2$; $(3, 0)$

10. $f(x) = (x + 2)^2$; $(-1, 1)$

In Exercises 11–20, answer the questions by calculating the slope of the tangent line and the tangent-line equation, as appropriate.

11. **M** **Motor Fuel Consumption** Based on data from 1980 to 1999, the amount of motor fuel consumed annually by pickups, SUVs, and vans may be modeled by

$$F(t) = 0.0407t^2 + 0.8091t + 23.33$$

billion gallons, where t is the number of years since 1980. (**Source:** Modeled from *Statistical Abstract of the United States, 2001*, Table 1105, p. 691.)

How quickly was fuel consumption increasing in 1999? What was the estimated fuel consumption in 2000? (Use a tangent-line approximation.)

12. **M** **High School Principal Salaries** Based on data from 1980 to 2000, the average salary of a high school principal may be modeled by

$$S(t) = 2499t + 30,039$$

dollars, where t is the number of years since 1980. (**Source:** Modeled from *Statistical Abstract of the United States, 2001*, Table 238, p. 152.)

How quickly were high school principals' salaries increasing in 2000? What was the average high school principal's salary in 2001? (Calculate the tangent-line approximation for the 2001 salary

and compare it to the model calculation for the 2001 salary.)

13. **M** **New One-Family Home Size** Based on data from 1970 to 1999, the average number of square feet in new one-family homes may be modeled by

$$H(t) = -0.1669t^2 + 29.95t + 1490$$

square feet, where t is the number of years after 1970. (**Source:** Modeled from *Statistical Abstract of the United States, 2001*, Table 938, p. 597.)

At what rate was home size increasing in 1999? What was the estimated size of a new home in 2000? (Use a tangent-line approximation.)

14. **M** **New One-Family Home Size** How could a real estate developer or home builder use the results of Exercise 13?

15. **M** **Student-to-Teacher Ratio** Based on data from 1960 to 1998, the student-to-teacher ratio at private elementary and secondary schools may be modeled by

$$R(t) = 0.01484t^2 - 1.000t + 31.60$$

students per teacher, where t is the number of years since 1960. (**Source:** Modeled from *Statistical Abstract of the United States, 2001*, Table 235, p. 150.)

According to the model, how quickly was the student-to-teacher ratio changing in 1998? What was the estimated student-to-teacher ratio for 1999? (Use a tangent-line approximation.)

16. **M** **Student-to-Teacher Ratio** Private schools succeed financially by convincing families that children taught at their school will obtain a better educational experience than that which is available at other public and private schools in the surrounding area. How could a private elementary school use the model and results of Exercise 15 in its marketing plan?

17. **M** **Number of Banks** Based on data from 1984 to 2000, the number of different banks (not branches of banks) in the United States may be modeled by

$$B(t) = 2.970t^3 - 68.39t^2 - 170.9t + 18,172$$

banks, where t is the number of years since 1984. (**Source:** Modeled from *Statistical Abstract of the United States, 2001*, Table 1173, p. 728.)

How quickly was the number of banks decreasing in 2000? What was the estimated number of banks in 2001? (Use a tangent-line approximation.)

18. **Decrease in Number of Banks** Using the model from Exercise 17, calculate how quickly the number of banks was decreasing in 1999. Comparing the rate of decrease in the year 1999 and in the year 2000, what can you conclude about the rate of decrease in the number of U.S. banks?

19. **Debit Cards** Based on data from 1990 to 1999, the number of debit cards in the United States may be modeled by

$$D(t) = 0.1563t^3 - 2.268t^2 + 14.84t + 164.0$$

million cards, where t is the number of years since 1990. (**Source:** Modeled from *Statistical Abstract of the United States, 2001*, Table 1189, p. 734.)

At the end of 1999, how quickly was the number of debit cards increasing? What was the estimated number of debit cards in 2000? (Use a tangent-line approximation.)

20. **Radio Broadcasting Wages** Based on data from 1992 to 1998, the average wage of a worker in the radio broadcasting industry may be modeled by

$$W(t) = 164.3t^2 + 778.6t + 23{,}500$$

dollars, where t is the number of years since 1992. (**Source:** Modeled from *Statistical Abstract of the United States, 2001*, Table 1123, p. 703.)

How quickly were wages of radio broadcasting employees increasing at the end of 1998? What was the average salary in 1999? (Use a tangent-line approximation.)

In Exercises 21–30, estimate the specified derivative by using the data in the table. Then interpret the result.

21. **Television Broadcasting Wages**

Average Annual Wage per Worker: Television Broadcasting

Year Since 1992 (t)	Annual Wage (dollars) [$W(t)$]
0	41,400
1	42,200
2	43,700
3	47,200
4	51,100
5	51,000
6	54,600

Source: *Statistical Abstract of the United States, 2001*, Table 1123, p.703.

Estimate $W'(5)$.

22. Using the table in Exercise 21, estimate $W'(6)$.

23. **Daily Newspapers**

Different Daily Newspapers

Years Since 1970 (t)	Newspapers [$N(t)$]
0	1,748
5	1,756
10	1,745
15	1,676
20	1,611
25	1,533
30	1,480

Source: *Statistical Abstract of the United States, 2001*, Table 1130, p. 706.

Estimate $N'(25)$.

24. Use the table in Exercise 23 to estimate $N'(30)$.

25. **Tuition at Green River Community College**

Full-Time Resident Tuition

Years Since 1994 (t)	Quarterly Tuition (dollars) [E(t)]
0	432
1	450
2	467
3	486
4	505
5	528
6	547
7	581

Source: Green River Community College.

Estimate $E'(6)$ and $E'(7)$. What can you conclude about the rate of increase in tuition at Green River Community College?

26. **Bread Prices**

Whole Wheat Bread Cost

Years Since 1990 (t)	Cost (dollars per loaf) [C(t)]
3	1.12
4	1.12
5	1.15
6	1.30
7	1.30
8	1.32
9	1.36
10	1.36

Source: *Statistical Abstract of the United States, 2001*, Table 706, p. 468.

Estimate $C'(4)$ and $C'(5)$.

27. **Baseball Game Attendance**

Major League Baseball Attendance

Years Since 1994 (t)	American League Games (millions of people) [A(t)]
0	24.2
1	25.4
2	29.7
3	31.3
4	31.9
5	31.8

Source: *Statistical Abstract of the United States, 2001*, Table 1241, p. 759.

Estimate $A'(3)$.

28. **Baseball Game Attendance**

Major League Baseball Attendance

Years Since 1994 (t)	National League Games (millions of people) [N(t)]
0	25.8
1	25.1
2	30.4
3	31.9
4	38.4
5	38.3

Source: *Statistical Abstract of the United States, 2001*, Table 1241, p. 759.

Estimate $N'(3)$.

29. **Major League Baseball** Using the tables from Exercises 27 and 28, estimate the overall rate of change in major league baseball game attendance in 1997. (*Hint:* First make a table with the combined attendance of both leagues.)

30. **Movie Prices**

Movie Ticket Prices

Years Since 1975 (t)	Price (dollars) [$P(t)$]
0	2.05
5	2.69
10	3.55
15	4.23
20	4.35
24	5.08

Source: *Statistical Abstract of the United States, 2001*, Table 1244, p. 761.

Estimate $P'(24)$.

Exercises 31–35 are intended to challenge your understanding of derivatives.

31. Given $f(x) = x^2$, find two pairs of points $(a, f(a))$ and $(b, f(b))$ such that $f'(2) = \frac{f(b) - f(a)}{b - a}$. What do both soltuions have in common?

32. Given $f(x) = x^2 - 2x$, find two pairs of points $(a, f(a))$ and $(b, f(b))$ such that $f'(2) = \frac{f(b) - f(a)}{b - a}$. What do both solutions have in common?

33. Given $f(x) = x^3$, calculate $f'(-1)$ and $f'(1)$. What can you conclude about the graph of f at $(-1, -1)$ and $(1, 1)$?

34. Given $f'(3) = 4$ and $g(x) = x^2 - f(x)$, find $g'(3)$.

35. Given $p(x) = f(x) + g(x)$, what is the relationship between $p'(0), f'(0)$, and $g'(0)$?

9.4 The Derivative as a Function: Algebraic Method

- Use the limit definition of the derivative to find the derivative of a function
- Use derivative notation and terminology to describe instantaneous rates of change

GETTING STARTED   Based on data from 1990 to 2000, the cumulative number of homicides resulting from a romantic triangle may be modeled by $R(t) = -20.45t^2 + 471.8t\ t - 152.5$ homicides between the start of 1991 and the end of year t, where t is the number of years since 1990. (**Source:** Modeled from *Crime in the United States 2000*, Uniform Crime Report, FBI.)

How quickly was the cumulative number of homicides increasing at the end of 1997, 1998, and 1999? Although we could calculate the derivative at $t = 7, t = 8$, and $t = 9$, we can save time by finding the *derivative function* and then substituting in the various values of t.

In this section, we will introduce the derivative function and show how to find it algebraically. Your skill in finding the derivative at a point will prove to be especially useful in this section. We'll begin with a couple of simple examples before returning to the romantic triangle problem.

EXAMPLE **1**   **Calculating the Derivative of a Function at Multiple Points**

Determine the instantaneous rate of change of $f(x) = 3x^2$ at $(1, 3), (3, 27)$, and $(10, 300)$.

SOLUTION Although we could calculate $f'(1), f'(3),$ and $f'(10)$ individually, it will be more efficient to find the derivative function itself and then substitute in the different values of x.

We begin with the derivative formula; however, instead of substituting a specific value for a, we replace a with the variable x.

$$f'(x) = \lim_{h \to 0} \frac{f(x + h) - f(x)}{h}$$

$$= \lim_{h \to 0} \frac{[3(x + h)^2] - (3x^2)}{h} \qquad \text{Since } f(x) = 3x^2$$

$$= \lim_{h \to 0} \frac{[3(x^2 + 2hx + h^2)] - (3x^2)}{h}$$

$$= \lim_{h \to 0} \frac{(3x^2 + 6hx + 3h^2) - (3x^2)}{h}$$

$$= \lim_{h \to 0} \frac{6hx + 3h^2}{h}$$

$$= \lim_{h \to 0} \frac{h(6x + 3h)}{h}$$

$$= \lim_{h \to 0} (6x + 3h)$$

$$= 6x + 3(0)$$

$$= 6x$$

The result $f'(x) = 6x$ is the derivative function for $f(x) = 3x^2$. It can be used to calculate the instantaneous rate of change of f at any point $(a, f(a))$.

$$f'(1) = 6(1)$$
$$= 6$$

The instantaneous rate of change of f at $(1, 3)$ is 6.

$$f'(3) = 6(3)$$
$$= 18$$

The instantaneous rate of change of f at $(3, 27)$ is 18.

$$f'(10) = 6(10)$$
$$= 60$$

The instantaneous rate of change of f at $(10, 300)$ is 60.

As demonstrated in Example 1, the techniques used to find the derivative function are virtually identical to the procedures used to find the derivative at a point. However, knowing the derivative function allows us to calculate the derivative at a number of different points more quickly than calculating the derivative at each point separately. The derivative function for a function f is typically called the **derivative of f.**

THE DERIVATIVE FUNCTION

The **derivative** of a function f is given by

$$f'(x) = \lim_{h \to 0} \frac{f(x + h) - f(x)}{h}$$

if the limit exists.

EXAMPLE 2 Finding the Derivative of a Function

Find the derivative of $f(x) = 4x^3 - 2x + 1$. Then calculate the slope of the tangent line at $x = 0$, $x = 1$, and $x = 2$.

SOLUTION

$$f'(x) = \lim_{h \to 0} \frac{f(x + h) - f(x)}{h}$$

$$= \lim_{h \to 0} \frac{[4(x + h)^3 - 2(x + h) + 1] - (4x^3 - 2x + 1)}{h} \qquad \text{Since } f(x) = 4x^3 - 2x + 1$$

$$= \lim_{h \to 0} \frac{[4(x^3 + 3hx^2 + 3h^2x + h^3) - 2x - 2h + 1] - (4x^3 - 2x + 1)}{h}$$

$$= \lim_{h \to 0} \frac{(4x^3 + 12hx^2 + 12h^2x + 4h^3 - 2x - 2h + 1) - (4x^3 - 2x + 1)}{h}$$

$$= \lim_{h \to 0} \frac{(12hx^2 + 12h^2x + 4h^3 - 2h)}{h}$$

$$= \lim_{h \to 0} \frac{h(12x^2 + 12hx + 4h^2 - 2)}{h}$$

$$= \lim_{h \to 0} (12x^2 + 12hx + 4h^2 - 2)$$

$$= 12x^2 + 12(0)x + 4(0)^2 - 2 \qquad \text{Replace } h \text{ with } 0$$

$$= 12x^2 - 2$$

The derivative of $f(x) = 4x^3 - 2x + 1$ is $f'(x) = 12x^2 - 2$.

Graphically speaking, the derivative is the slope of the tangent line. Therefore,

$$f'(0) = 12(0)^2 - 2$$
$$= -2$$

means that at $x = 0$ the slope of the tangent line is -2. Similarly,

$$f'(1) = 12(1)^2 - 2$$
$$= 10$$

means that at $x = 1$ the slope of the tangent line is 10. Likewise,

$$f'(2) = 12(2)^2 - 2$$
$$= 46$$

means that at $x = 2$ the slope of the tangent line is 46.

EXAMPLE 3   **Calculating the Instantaneous Rate of Change of a Function at Multiple Points**

Based on data from 1990 to 2000, the cumulative number of homicides resulting from a romantic triangle may be modeled by

$$R(t) = -20.45t^2 + 471.8t - 152.5 \text{ homicides}$$

between the start of 1991 and the end of year t, where t is the number of years since 1990. (**Source:** Modeled from *Crime in the United States 2000*, Uniform Crime Report, FBI.)

How quickly was the cumulative number of homicides increasing at the end of 1997, 1998, and 1999?

SOLUTION   We will begin by finding the derivative $R'(t)$.

$$R'(t) = \lim_{h \to 0} \frac{R(t + h) - R(t)}{h}$$

Because of the complex nature of $R(t)$, we will first calculate $R(t + h)$ and then substitute the result into the derivative formula.

$$R(t + h) = -20.45(t + h)^2 + 471.8(t + h) - 152.5$$

$$= -20.45(t^2 + 2ht + h^2) + 471.8t + 471.8h - 152.5$$

$$= -20.45t^2 - 40.90ht - 20.45h^2 + 471.8t + 471.8h - 152.5$$

We already know that $R(t) = -20.45t^2 + 471.8t - 152.5$. Substituting both of these quantities into the derivative formula yields

$$R'(t) = \lim_{h \to 0} \frac{R(t + h) - R(t)}{h}$$

$$= \lim_{h \to 0} \frac{(-20.45t^2 - 40.90ht - 20.45h^2 + 471.8t + 471.8h - 152.5) - (-20.45t^2 + 471.8t - 152.5)}{h}$$

$$= \lim_{h \to 0} \frac{-40.90ht - 20.45h^2 + 471.8h}{h}$$

$$= \lim_{h \to 0} \frac{h(-40.90t - 20.45h + 471.8)}{h} \qquad \text{Every term in the numerator without an } h \text{ cancelled out}$$

$$= \lim_{h \to 0} (-40.90t - 20.45h + 471.8) \qquad \text{Since } \frac{h}{h} = 1 \text{ for } h \neq 0$$

$$= -40.90t - 20.45(0) + 471.8$$

$$= -40.90t + 471.8$$

So $R'(t) = -40.90t + 471.8$ homicides per year. We can now compute the instantaneous rate of change in the cumulative number of homicides in 1997, 1998, and 1999.

$$R'(7) = -40.90(7) + 471.8$$

$$= 185.5$$

$$\approx 186 \text{ homicides per year}$$

$$R'(8) = -40.90(8) + 471.8$$
$$= 144.6$$
$$\approx 145 \text{ homicides per year}$$
$$R'(9) = -40.90(9) + 471.8$$
$$= 103.7$$
$$\approx 104 \text{ homicides per year}$$

The cumulative number of homicides resulting from a romantic triangle was increasing at a rate of 186 homicides per year in 1997, 145 homicides per year in 1998, and 104 homicides per year in 1999. According to the model, although the cumulative number of homicides continued to increase, the rate at which these homicides were increasing slowed between the end of 1997 and the end of 1999.

EXAMPLE 4

Finding the Derivative of a Function

Find the derivative of $g(t) = 2t^3 - 4t + 3$.

SOLUTION

We must find $g'(t) = \lim\limits_{h \to 0} \dfrac{g(t + h) - g(t)}{h}$. We'll first find $g(t + h)$ and then substitute the result into the derivative formula.

$$g(t + h) = 2(t + h)^3 - 4(t + h) + 3$$
$$= 2(t^3 + 3t^2h + 3th^2 + h^3) - 4t - 4h + 3$$
$$= 2t^3 + 6t^2h + 6th^2 + 2h^3 - 4t - 4h + 3$$

$$g'(t) = \lim_{h \to 0} \frac{g(t + h) - g(t)}{h}$$
$$= \lim_{h \to 0} \frac{(2t^3 + 6t^2h + 6th^2 + 2h^3 - 4t - 4h + 3) - (2t^3 - 4t + 3)}{h}$$
$$= \lim_{h \to 0} \frac{(6t^2h + 6th^2 + 2h^3 - 4h)}{h} \qquad \text{Every term in the numerator}$$
$$\qquad\qquad\qquad\qquad\qquad\qquad \text{without an } h \text{ cancelled out}$$
$$= \lim_{h \to 0} \frac{h(6t^2 + 6th + 2h^2 - 4)}{h}$$
$$= \lim_{h \to 0} (6t^2 + 6th + 2h^2 - 4)$$
$$= 6t^2 + 6t(0) + 2(0)^2 - 4$$
$$= 6t^2 - 4$$

The derivative of $g(t) = 2t^3 - 4t + 3$ is $g'(t) = 6t^2 - 4$.

Estimating Derivatives

For polynomial functions, all terms in the numerator of the derivative formula without an h will cancel out. This allows us always to eliminate the h in the denominator. However, with some other types of functions, the h in the denominator cannot be eliminated algebraically. In this case, we can estimate the derivative function by substituting in a small positive value (i.e. 0.001) for h. The closer the value of h is to zero, the more accurate the estimate of the derivative will be.

EXAMPLE 5

Calculating the Instantaneous Rate of Change of a Function at Multiple Points

The per capita consumption of bottled water in the United States may be modeled by

$$W = 2.593(1.106)^t \text{ gallons}$$

where t is the number of years since the end of 1980. (**Source:** Modeled from Statistical Abstract of the United States, 2001, Table 204, p. 130.)

Determine how quickly bottled water consumption was increasing at the end of 2002, 2004, and 2006.

SOLUTION We must evaluate $W'(t)$ at $t = 22$, 24, and 26.

$$W'(t) = \lim_{h \to 0} \frac{W(t+h) - W(t)}{h}$$

$$= \lim_{h \to 0} \frac{[2.593(1.106)^{t+h}] - [2.593(1.106)^t]}{h}$$

$$= \lim_{h \to 0} \frac{[2.593(1.106)^t(1.106)^h] - [2.593(1.106)^t]}{h} \qquad \text{Since } (1.106)^{t+h} = (1.106)^t(1.106)^h$$

$$= \lim_{h \to 0} \frac{2.593(1.106)^t[(1.106)^h - 1]}{h} \qquad \text{Factor out } 2.593(1.106)^t$$

$$= 2.593(1.106)^t \cdot \lim_{h \to 0} \frac{[(1.106)^h - 1]}{h}$$

We can move the expression $2.593(1.106)^t$ to the other side of the limit because it does not contain an h. Since

$$\lim_{h \to 0} \frac{[(1.106)^h - 1]}{h} \approx \frac{[(1.106)^{0.001} - 1]}{0.001}$$

$$\approx 0.1008$$

we have

$$W'(t) = 2.593(1.106)^t \cdot \lim_{h \to 0} \frac{[(1.106)^h - 1]}{h}$$

$$\approx 2.593(1.106)^t \cdot (0.1008)$$

$$\approx 0.2614(1.106)^t$$

We will now evaluate the derivative function at $t = 22$, 24, and 26.

$$W'(22) \approx 0.2614(1.106)^{22}$$

$$\approx 2.398$$

According to the model, bottled water consumption was increasing by 2.398 gallons per year at the end of 2002.

$$W'(24) \approx 0.2614(1.106)^{24}$$

$$\approx 2.934$$

Bottled water consumption was increasing by 2.934 gallons per year at the end of 2004.

$$W'(26) \approx 0.2614(1.106)^{26}$$
$$\approx 3.589$$

Bottled water consumption is expected to be increasing at a rate of 3.589 gallons per year at the end of 2006.

9.4 Summary

In this section, you learned how to find the derivative function algebraically. You discovered that it is often easier to expand and simplify $f(x + h)$ before substituting it into the derivative formula. Additionally, you found that when the h in the denominator cannot be eliminated algebraically, you can estimate the derivative numerically.

9.4 Exercises

In Exercises 1–10, find the derivative of the function.

1. $f(x) = x^2 - 4x$

2. $f(x) = -x^2 + 6$

3. $g(x) = x^2 + 2x + 1$

4. $g(x) = x^2 - 4$

5. $g(x) = x^2 - 4x - 5$

6. $w(x) = x^3$

7. $h(x) = x^3 + 2$

8. $h(x) = x^3 + x$

9. $f(t) = (t - 3)^2$

10. $f(t) = (t + 2)^2$

In Exercises 11–20, find the slope of the tangent line of the function at $x = 1$, $x = 3$, and $x = 5$.

11. $g(x) = 2x^2 + x - 1$

12. $g(x) = 3x^2 - 4x - 5$

13. $f(x) = x^2 - 2x$

14. $f(x) = 4x - 5$

15. $h(x) = -5$

16. $h(x) = -2x + 4$

17. $W(x) = -4x + 9$

18. $W(x) = x^3 - 1$

19. $S(x) = 3x^2 - 2x + 1$

20. $S(x) = -4x^2 + 9$

In Exercises 21–25, estimate the derivative of the function. When you are unable to eliminate the h in the denominator of the derivative formula algebraically, use $h = 0.001$.

21. $P(x) = 3^x$

22. $R(x) = 2 \cdot 3^x$

23. $C(x) = -3 \cdot 4^x$

24. $P(x) = 2.03^x$

25. $R(x) = 5.042 \cdot (0.98)^x$

In Exercises 26–35, use the derivative function to answer the questions.

26. **New Home Size** Based on data from 1970 to 1999, the average size of a new one-family home may be modeled by

$$H(t) = -0.1669t^2 + 29.95t + 1490$$

square feet, where t is the number of years since 1970. (**Source:** Model based on *Statistical Abstract of the United States, 2001*, Table 938, p. 597.)

Was the average size of a new home growing faster at the end of 1989 or the end of 1999?

27. **New Home Prices** Based on data from 1980 to 2000, the median sales price of a new one-family home in the southern United States may be modeled by

$$P(t) = 0.0506t^2 + 3.515t + 58.49$$

thousand dollars, where t is the number of years since 1980. (**Source:** Modeled from *Statistical Abstract of the United States, 2001*, Table 940, p. 598.)

According to the model, was the median sales price increasing more quickly at the end of 1990 or the end of 2000?

28. **M** **New Home Prices**   Based on data from 1980 to 2000, the median sales price of a new one-family home in the northeastern United States may be modeled by

$$P(t) = -0.01242t^2 + 8.279t + 68.27$$

thousand dollars, where t is the number of years since 1980. (**Source:** Modeled from *Statistical Abstract of the United States, 2001,* Table 940, p. 598.)

According to the model, was the median sales price increasing more quickly at the end of 1990 or the end of 2000?

29. New Home Prices   Using the models from Exercises 27 and 28, determine whether median new home prices were increasing faster in the South or the Northeast in 1999.

30. **M** **Restaurant Sales**   Based on data from 1980 to 2000, full-service restaurant sales may be modeled by

$$S(t) = 12.77t^2 - 276.9t + 5222t + 38,203$$

million dollars, where t is the number of years since 1980. (**Source:** Modeled from *Statistical Abstract of the United States, 2001,* Table 1268, p. 775.)

Were restaurant sales growing faster at the end of 1984 or the end of 1994? Explain.

31. **M** **Gaming Software Sales**   Based on data from 1990 to 1999, electronic gaming software factory sales may be modeled by

$$S(t) = 39.79t^2 - 48.41t + 2378$$

million dollars, where t is the number of years since the end of 1990. (**Source:** Modeled from *Statistical Abstract of the United States, 2001,* Table 1005, p. 634.)

Were software sales growing faster at the end of 1997 or the end of 1999?

32. **M** **Certified Organic Cropland**   Based on data from 1992 to 1996, the amount of certified organic cropland may be modeled by

$$A(t) = 20.57t^2 - 24.51t + 410.3$$

thousand acres, where t is the number of years since the end of 1992. (**Source:** Modeled from *Statistical Abstract of the United States, 2001,* Table 805, p. 526.)

In addition to showing that the amount of organic farmland increased between 1992 and 1996, how could you use the model to convince organic farming investors that there is still room for growth in the organic farming market?

33. **M** **Breakfast Cereal Consumption**   Based on data from 1980 to 1999, the per capita consumption of breakfast cereal may be modeled by

$$C(t) = -0.004718t^3 + 0.1165t^2 - 0.3585t + 12.17$$

pounds of cereal, where t is the number of years since the end of 1980. (**Source:** Modeled from *Statistical Abstract of the United States, 2001,* Table 202, p. 129.)

Was the per capita consumption of breakfast cereal changing more rapidly in 1989 or in 1999?

34. **M** **Federal and State Prison Rate**   Based on data from 1980 to 1998, the federal and state prison rate may be modeled by

$$R(t) = 0.3750t^2 + 11.70t + 138.7$$

prisoners per 100,000 people, where t is the number of years since the end of 1980. (**Source:** Modeled from *Statistical Abstract of the United States, 2001,* Table 332, p. 200.)

State lawmakers are debating whether or not the state should invest more money in a crime prevention program. An opponent of the measure argues that between 1980 and 1998, the prison rate increased by an average of 18 prisoners per hundred thousand people per year. She contends that additional money should not be spent on crime prevention unless the crime rate is increasing by at least 25 prisoners per 100,000 people per year. How could a proponent of the crime prevention measure use the model to convince her that crime prevention legislation should be passed now? (Assume that the debate takes place in 2000.)

35. **M** **Private College Enrollment**   Based on data from 1980 to 1998, the number of students enrolled at private colleges may be modeled by

$$S(t) = 0.9195t^2 + 23.25t + 2652$$

thousand students, where t is the number of years since the end of 1980. (**Source:** Modeled from *Statistical Abstract of the United States, 2001,* Table 205, p. 133.)

How much more rapidly were enrollments increasing in 1998 than they were increasing in 1995?

Exercises 36–40 are intended to challenge your understanding.

36. Given $f(x) = |x|$, find $f'(x)$.

37. Given $f'(x) = 3x$ and $g(x) = x^2 + 3 + f(x)$, find $g'(x)$.

38. Given $f(x) = x^2 - 2x$, determine where $f'(x) = 0$.

39. Given $f(x) = x^3 - 3x$, determine where $f'(x) = 0$.

40. Graph $f(x) = x^3 - 3x$. What does the result of Exercise 39 tell you about the graph of f?

9.5 Interpreting the Derivative

- Interpret the meaning of the derivative in the context of a word problem

GETTING STARTED   Many of us feel inundated by the advertisements we are sent through the mail. Don't expect this to let up anytime soon: Spending on direct-mail advertising has risen every year since 1990. The amount of money spent on direct-mail advertising may be modeled by

$$A(t) = 70.54t^2 + 1488t + 22{,}828 \text{ million dollars}$$

where t is the number of years since 1990. (**Source:** Modeled from *Statistical Abstract of the United States, 2001,* Table 1272, p. 777.)

According to the model, $A(9) = 41{,}934$ and $A'(9) = 2758$. But what does this mean? In this section, we will discuss how to interpret the meaning of a derivative in the context of a real-life problem.

Recall that the units of the derivative are the units of the output divided by the units of the input. In this case, the units of A' are $\dfrac{\text{millions of dollars}}{\text{year}}$ or millions of dollars per year. Note that $t = 9$ corresponds to the year 1999. We conclude that in 1999, 41,934 million dollars were spent on direct mail advertising, and spending was increasing by 2758 million dollars per year. In other words, according to the model, 41,934 million dollars were spent on direct mail advertising in 1999 and spending increased by *about* 2758 million dollars between 1999 and 2000. The term *about, approximately,* or *roughly* must be used when using the second interpretation of the derivative, since we are using a tangent-line approximation to estimate the increase over the next year. (For a graphical discussion of tangent-line approximations, refer to Section 9.3.)

INTERPRETING THE DERIVATIVE

Let $f(x)$ be a function. The meaning of $f'(a) = c$ may be written in either of the following two ways:

- When $x = a$, the value of the function f is increasing (decreasing) by c units of output per unit of input.
- The value of the function f will increase (decrease) by *about* c units of output between a units of input and $a + 1$ units of input.

EXAMPLE 1

Interpreting the Meaning of the Derivative

Based on data from 1969 to 2002, the number of students taking the AP Calculus AB exam may be modeled by

$$E(t) = 157.8t^2 - 770.6t + 10{,}268 \text{ students}$$

where t is the number of years since 1969. (**Source:** Modeled from College Board data.)

Interpret the meaning of $E(33) = 156{,}682$ and $E'(33) = 9644$. Then estimate $E(34)$.

SOLUTION Since t is the number of years since 1969, $t = 33$ corresponds to 2002. $E(33) = 156{,}682$ means that 156,682 students took the exam in 2002.

The units of the derivative are $\frac{\text{students}}{\text{year}}$. $E'(33) = 9644$ means that in 2002, the number of students taking the exam was increasing by 9644 students per year. In other words, the number of students taking the exam will increase by *about* 9644 students between 2002 and 2003.

We use a tangent-line approximation to estimate $E(34)$. Since

$$E'(33) \approx \frac{E(34) - E(33)}{34 - 33}$$

$$E'(33) \approx \frac{E(34) - E(33)}{1}$$

$$E'(33) \approx E(34) - E(33)$$

$$E(34) \approx E(33) + E'(33)$$

$$\approx 156{,}682 + 9644$$

$$\approx 166{,}326$$

We estimate that 166,326 students took the exam in 2003.

Each year the College Board hires high school and college educators from around the globe to score the free-response portion of the Calculus AB exam. If an average grader can score 270 exams during the week-long grading, $\frac{9644}{270} = 35.7 \approx 36$ additional graders were needed to score the exams in 2003.

EXAMPLE 2

Interpreting the Meaning of the Derivative

Based on data from 1940 to 2000, the average monthly Social Security benefit for men may be modeled by

$$P(t) = 0.3486t^2 - 5.505t + 35.13 \text{ dollars}$$

where t is the number of years since 1940. (**Source:** Modeled from Social Security Administration data.)

Interpret the meaning of $P(60) = 959.79$ and $P'(60) = 36.33$. Then use a tangent-line approximation to estimate $P(61)$.

SOLUTION Since t is the number of years since 1940, $t = 60$ corresponds to 2000. $P(60) = 959.79$ means that (according to the model) the average Social Security benefit for men in 2000 was $959.79.

The units of the derivative are $\frac{\text{dollars}}{\text{year}}$. $P'(60) = 36.33$ means that in 2000, the average Social Security benefit for men was increasing by \$36.33 per year. In other words, the average Social Security benefit for men was expected to increase by *about* \$36.33 between 2000 and 2001.

We estimate the benefit in 2001 by approximating $P(61)$.

$$P(61) \approx P(60) + P'(60)$$
$$\approx 959.79 + 36.33$$
$$\approx 996.12$$

In 2001, we estimate that the average Social Security benefit for men was \$996.12.

EXAMPLE 3

Using a Tangent Line Approximation to Estimate a Function Value

The average weight of a boy between the ages of 2 and 13 years old may be modeled by

$$W(a) = 0.215a^2 + 2.993a + 23.78 \text{ pounds}$$

where a is the age of the boy in years. (**Source:** Modeled from www.babybag.com data.)

Interpret the meaning of $W(10) = 75.21$ and $W'(10) = 7.29$. Then use a tangent-line approximation to estimate $W(11)$.

SOLUTION Since a is the age of the boy, $a = 10$ corresponds to a 10-year-old boy. $W(10) = 75.21$ means that the average weight of a 10-year-old boy is 75.21 pounds.

The units of the derivative are $\frac{\text{pounds}}{\text{year of his age}}$. $W'(10) = 7.29$ means that the average weight of a 10-year-old boy is increasing by 7.29 pounds per year of his age. In other words, the average weight of a boy will increase by *about* 7.29 pounds between his tenth and eleventh years.

We use a tangent-line approximation to estimate $W(11)$.

$$W(11) \approx W(10) + W'(10)$$
$$\approx 75.21 + 7.29$$
$$\approx 82.5$$

We estimate that the average weight of an 11-year-old boy is 82.5 pounds.

EXAMPLE 4

Interpreting the Meaning of the Derivative

Based on data from 1970 to 1999, the percentage of new one-family homes that are 1200 square feet or less may be modeled by

$$H(t) = 0.02676t^2 - 1.753t + 35.93 \text{ percentage points}$$

where t is the number of years since 1970. (**Source:** Modeled from *Statistical Abstract of the United States, 2001*, Table 938, p. 597.)

Interpret the meaning of $H(14) = 16.6$ and $H'(14) = -1.00$.

> **SOLUTION** Since t is the number of years since 1970, $t = 14$ corresponds to 1984. $H(14) = 16.6$ means that in 1984, 16.6 percent of new homes were 1200 square feet or less in size.
>
> The units of the derivative are $\frac{\text{percentage points}}{\text{year}}$. $H'(14) = -1.00$ means that in 1984, the percentage of new homes that are 1200 square feet or less in size was decreasing by 1 percentage point per year. In other words, the percentage of new homes that are 1200 square feet or less in size was expected to decrease by *about* 1 percentage point (from 16.6 percent to 15.6 percent) from 1984 to 1985.

9.5 Summary

In this section, you learned how to interpret the meaning of the derivative verbally in its real-world context. You also discovered two different ways of representing the concept.

9.5 Exercises

In Exercises 1–30, interpret the real-life meaning of the indicated values. Answer additional questions as appropriate.

1. **M** **Body Weight** The weight of a 2-year-old to 13-year-old girl may be modeled by

$$W(a) = 0.289a^2 + 2.464a + 23.10$$

pounds, where a is the age of the girl. (**Source:** Modeled from www.babybag.com data.)

Interpret the meaning of $W(10) = 76.64$ and $W'(10) = 8.24$. Then estimate $W(11)$.

2. **M** **Body Weight** Compare the results of Example 3 and Exercise 1. Were boys or girls expected to gain more weight between their tenth and eleventh years? Explain.

3. **M** **Per Capita Income** Based on data from 1959 to 1989, the annual per capita income of residents of Washington state may be modeled by

$$P(t) = -2.98t^2 + 327.4t + 7881$$

dollars, where t is the number of years since 1959. (**Source:** Modeled from www.census.gov.)

Interpret the meaning of $P(25) = 14,203$ and $P'(25) = 178.4$. Then estimate $P(26)$.

4. **M** **Per Capita Income** Using the model from Exercise 3, $P'(20) = 208.2$, $P'(25) = 178.4$, $P'(30) = 148.6$, and $P'(35) = 118.8$. Based on these results, what conclusions can you draw about per capita income in the state of Washington?

5. **M** **Nasdaq Market Volume** Based on data from 1980 to 2000, the average daily volume of the Nasdaq market may be modeled by

$$V(t) = 0.06754t^4 - 1.936t^3 + 17.97t^2 - 41.27t + 39.03$$

million shares, where t is the number of years since 1980. (**Source:** Model based on *Statistical Abstract of the United States, 2001*, Table 1205, p. 741.)

Interpret the meaning of $V(18) = 918$ and $V'(18) = 299$. Then estimate $V(19)$.

6. **M** **Nasdaq Market** Using the model from Exercise 5, $V'(18) = 299$, $V'(19) = 398$, and $V'(20) = 516$. Based on these results, what conclusions can you draw about the average daily volume of the Nasdaq market?

7. New York Stock Exchange Volume
Based on data from 1980 to 2000, the annual volume of the New York Stock Exchange may be modeled by

$$V(t) = 4.807t^4 - 98.56t^3 + 541.3t^2 + 2936t + 10,612$$

million shares, where t is the number of years since 1980. (**Source:** Modeled from *Statistical Abstract of the United States, 2001*, Table 1207, p. 742.)
Interpret the meaning of $V(18) = 168,659$ and $V'(18) = 38,760$. Then estimate $V(19)$.

8. New York Stock Exchange Using the model from Exercise 7, $V'(18) = 38,760$, $V'(19) = 48,650$, and $V'(20) = 60,140$. Based on these results, what conclusions can you draw about the annual volume of the New York Stock Exchange?

9. Radio Broadcasting Wages Based on data from 1992 to 1998, the average annual wage of an employee in the radio broadcasting industry may be modeled by

$$W(t) = 164.3t^2 + 778.6t + 23,500$$

dollars, where t is the number of years since 1992. (**Source:** Modeled from *Statistical Abstract of the United States, 2001*, Table 1123, p. 703.)
Interpret the meaning of $W(6) = 34,086$ and $W'(6) = 2750$. Estimate $W(7)$.

10. Restaurant Sales Based on data from 1989 to 1998, restaurant sales may be modeled by

$$S(r) = 0.9291r - 160.0$$

billion dollars, where r is the number of restaurants (and other eating places) in thousands. (**Source:** Modeled from *Statistical Abstract of the United States, 2001*, Table 1268, p. 775.)
Interpret the meaning of $S(400) = 212$ and $S'(400) = 0.9291$. (*Hint:* Convert the units of the derivative into thousands of dollars per restaurant.)

11. Billboard Advertising Based on data from 1990 to 2000, the amount of money spent on billboard advertising may be modeled by

$$A(t) = 9.396t^2 - 11.58t + 1063$$

million dollars, where t is the number of years since 1990. (**Source:** Modeled from *Statistical Abstract of the United States, 2001*, Table 1272, p. 777.)
Interpret the meaning of $A(10) = 1887$ and $A'(10) = 176$.

12. Cassette Tapes Based on data from 1992 to 2000, the net number of cassette tapes shipped by recording industry manufacturers may be modeled by

$$C(t) = 0.6527t^3 - 8.348t^2 - 10.30t + 367.5$$

million cassettes, where t is the number of years since 1992. (**Source:** Modeled from Recording Industry Association of America data.)
Interpret the meaning of $C(8) = 85.0$ and $C'(8) = -18.5$.

13. Compact Discs Based on data from 1992 to 2000, the value of CDs shipped by recording industry manufacturers may be modeled by

$$V(x) = 0.008086x^2 + 3.308x + 2735$$

million dollars, where x is the net number of CDs shipped in millions. (*Net* means "after returns.")
(**Source:** Modeled from Recording Industry Association of America data.)
Interpret the meaning of $V(900) = 12,261.9$ and $V'(900) = 17.9$.

14. DVDs Based on data from 1998 to 2001, the value of DVD videos shipped by recording industry manufacturers may be modeled by

$$V(x) = -0.2173x^2 + 25.84x - 0.02345$$

million dollars, where x is the net number of DVDs shipped in millions. (*Net* means "after returns.") (**Source:** Modeled from Recording Industry Association of America data.)
Interpret the meaning of $V(7.5) = 181.6$ and $V'(7.5) = 22.6$.

15. Summer Olympics Based on data from 1904 to 1992, the number of women in the Summer Olympics may be modeled by

$$W(t) = 0.0002344t^4 - 0.03440t^3 + 1.701t^2 - 19.69t + 70.80$$

women, where t is the number of years since 1900.
(**Source:** Modeled from www.olympicwomen.co.uk data.)
Interpret the meaning of $W(92) = 2662$ and $W'(92) = 150$.

16. Summer Olympics Refer to the results of Exercise 15. Do you think $W(96)$ will be greater than 2662 and $W'(96)$ will exceed 150? Defend your conclusions.

17. [M] **U.S. Population** Based on data from 1790 to 2000, the population of the United States may be modeled by

$$P(t) = 0.006702t^2 - 24.11t + 21{,}696$$

million people, where t is the calendar year.
(**Source:** Modeled from U.S. Bureau of the Census data.)
Interpret the meaning of $P(2000) = 284$ and $P'(2000) = 2.70$.

18. [M] **Homicide Rate** Based on data from 1990 to 2000, the homicide rate (deaths per 100,000 people) in the United States may be modeled by

$$H(t) = 0.01129t^3 - 0.2002t^2 + 0.4826t + 9.390$$

people, where t is the number of years since 1990.
(**Source:** Modeled from *Crime in the United States 2000*, Uniform Crime Report, FBI).
Interpret the meaning of $H(10) = 5.5$ and $H'(10) = -0.1$.

19. [M] **Deadly Alcohol-Related Brawls** Based on data from 1991 to 2000, the number of homicides resulting from an alcohol-related brawl may be modeled by

$$H(t) = -0.3551t^3 + 9.867t^2 - 104.7t + 600.4$$

deaths, where t is the number of years since 1990.
(**Source:** Modeled from *Crime in the United States 2000*, Uniform Crime Report, FBI.)
Interpret the meaning of $H(10) = 185$ and $H'(10) = -14$.

20. [M] **Body Height** The average height of a girl between the ages of 2 and 13 years may be modeled by

$$H(a) = -0.0392a^2 + 2.987a + 29.69$$

inches, where a is the age of the girl. (**Source:** Modeled from www.babybag.com data.)
Interpret the meaning of $H(5) = 43.6$ and $H'(5) = 2.6$.

21. [M] **Company Revenue** Based on data from 1999 to 2001, the gross revenue from sales of Johnson & Johnson and its subsidiaries may be modeled by

$$R(t) = 665t^2 + 1150t + 27{,}357$$

million dollars, where t is the number of years since the end of 1999. (**Source:** Modeled from Johnson & Johnson 2001 Annual Report.)
Interpret the meaning of $R(2) = 32{,}317$ and $R'(2) = 3810$.

22. [M] **Company Costs** Based on data from 1999 to 2001, the cost of goods sold for Johnson & Johnson and its subsidiaries may be modeled by

$$C(t) = 103t^2 + 315t + 8539$$

million dollars, where t is the number of years since the end of 1999. (**Source:** Modeled from Johnson & Johnson 2001 Annual Report.)
Interpret the meaning of $C(2) = 9581$ and $C'(2) = 727$.

23. [M] **Company Profit** Based on data from 1999 to 2001, the operating profit of Frito-Lay North America may be modeled by

$$P(t) = -47.5t^2 + 283.5t + 1679$$

million dollars, where t is the number of years since the end of 1999. (**Source:** Modeled from PepsiCo 2001 Annual Report.)
Interpret the meaning of $P(2) = 2056$ and $P'(2) = 93.5$.

24. [M] **Employee Earnings** Based on data from 1991 to 2001, the average hourly earnings for a Ford Motor Company employee may be modeled by

$$W(t) = -0.003931t^4 + 0.1005t^3 - 0.8295t^2 + 3.188t + 16.48$$

dollars, where t is the number of years since the end of 1990. (**Source:** Ford Motor Company 2001 Annual Report).
Interpret the meaning of $W(11) = 27.39$ and $W'(11) = 0.49$.

25. [M] **River Flow** Based on data from November 22 to November 24, 2002, the flow of the Snoqualmie River near Snoqualmie Falls, Washington, may be modeled by

$$f(t) = 25t^2 - 145t + 490$$

cubic feet per second, where t is the number of 24-hour periods since noon on November 22, 2002. (**Source:** Modeled from www.dreamflows.com data.)
Interpret the meaning of $f(2) = 300$ and $f'(2) = -45$. How could a river rafting company use this type of information?

26. [M] **Personal Income: California** Based on data from 1993 to 2000, the per capita personal income of California may be modeled by

$$P(t) = 123.73t^2 - 445.58t + 22{,}847$$

dollars, where t is the number of years since the end of 1993. (**Source:** Modeled from Bureau of Economic Analysis data.)

Interpret the meaning of $P(5) = 23,712$ and $P'(5) = 791.7$.

27. **Personal Income: Colorado** Based on data from 1993 to 2000, the per capita personal income of Colorado may be modeled by

$$P(t) = 80.994t^2 + 890.16t + 22,158$$

dollars, where t is the number of years since the end of 1993. (**Source:** Modeled from Bureau of Economic Analysis data.)

Interpret the meaning of $P(5) = 28,634$ and $P'(5) = 1700.1$.

28. **Personal Income: Connecticut** Based on data from 1993 to 2000, the per capita personal income of Connecticut may be modeled by

$$P(t) = -19.652t^3 + 296.96t^2 + 502.89t + 29,285$$

dollars, where t is the number of years since the end of 1993. (**Source:** Modeled from Bureau of Economic Analysis data.)

Interpret the meaning of $P(5) = 36,767$ and $P'(5) = 1998.6$.

29. **Income Analysis** If a high per capita personal income is considered to be one indicator of a strong economy, compare and contrast the income and projected income growth in California, Colorado, and Connecticut between 1998 and 1999, using the results of Exercises 26–28.

30. **Price of a VCR** Based on data from 1997 to 2003, the average price of a VCR may be modeled by

$$P(t) = \frac{123.0}{1 + 0.2649e^{0.9733t}} + 60$$

dollars, where t is the number of years since the end of 1997. (**Source:** Modeled from Consumer Electronics Association data.)

Interpret the meaning of $P'(1) = -29.00$, $P'(3) = -16.82$, and $P'(5) = -3.285$.

Exercises 31–35 are intended to challenge your understanding of the meaning of derivatives.

31. **Market Analysis** The DVD player was introduced in 1997. Based on the results

of Exercise 30, what effect did the introduction of the DVD player have on the price of a VCR? As t gets larger, what value do you think $P'(t)$ will approach?

32. **Company Sales Income** Based on data from 1993 to 2002, the annual sales income of Starbucks Corporation may be modeled by

$$S(t) = 29.23t^2 + 79.33t + 177.4$$

million dollars, where t is the number of years since the end of 1993. According to the model, in what year (between 1993 and 2002) was annual sales income increasing most rapidly? Justify your answer.

33. **AIDS Deaths in the United States** Based on data from 1981 to 2001, the number of adult and adolescent AIDS deaths in the United States may be modeled by a piecewise function f, whose graph is shown in the following figure along with a scatter plot of the number of deaths as a function of years since the end of 1981.

The highest number of AIDS deaths (50,876) occurred in 1995. (**Source:** Centers for Disease Control and Prevention.)

Estimate from the graph the year in which f' was the greatest and the year in which f' was the most negative.

34. Let f be any continuous, smooth function (no breaks or sharp points) that is defined on the interval $(-\infty, \infty)$. If $f(a) \leq f(x)$ for all x, what is the value of $f'(a)$?

35. Let f be any continuous, smooth function (no breaks or sharp points) that is defined on the interval $(-\infty, \infty)$. If $f(a) \geq f(x)$ for all x, what is the value of $f'(a)$?

Chapter 9 Review Exercises

Section 9.1 *In Exercises 1–2, calculate the average rate of change of the function over the given interval.*

1. $f(x) = 6x - 12; [1, 8]$

2. $g(x) = x^2 - 2x + 1; [0, 2]$

In Exercises 3–4, calculate the average rate of change in the designated quantity over the given interval.

3. Change in per capita income between 1969 and 1989.

United States Per Capita Income, 1959–1989

Year (t)	Income (dollars) (P)
1959	7,259
1969	9,816
1979	12,229
1989	14,420

Source: www.census.gov.

4. Change in per capita income between 1959 and 1989.

Washington Per Capita Income, 1959–1989

Year Since 1959 (t)	Income (dollars) (P)
0	7,978
10	10,565
20	13,528
30	14,923

Source: www.census.gov.

In Exercises 5–6, use the difference quotient to calculate the slope of the secant line through the points $(1, f(1))$ and $(3, f(3))$.

5. $f(x) = 3^x$

6. The graph of f is shown.

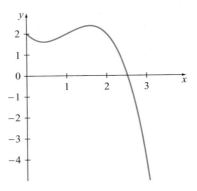

Section 9.2 *In Exercises 7–8, use the difference quotient (with $h = 0.1$, $h = 0.01$, and $h = 0.001$) to estimate the instantaneous rate of change of the function at the given input value.*

7. $f(x) = 2x^2 + 3; x = 1$

8. $h(x) = -x^2 + 5x - 2; x = 4$

In Exercises 9–10, use the derivative to calculate the instantaneous rate of change of the function at the given input value. (In each exercise, you can eliminate h algebraically.) Compare your answers to the solutions of Exercises 7–8.

9. $f(x) = 2x^2 + 3; x = 1$

10. $h(x) = -x^2 + 5x - 2; x = 4$

In Exercises 11–12, use the difference quotient (with $h = 0.1$, $h = 0.01$, and $h = 0.001$) to estimate the instantaneous rate of change of the function at the given input value. You may find it helpful to apply the Technology Tip demonstrated in Section 9.2.

11. $f(x) = 2x^{-3} + x; x = 4$

12. $P(t) = 16(1.04)^t; t = 20$

In Exercises 13–14, determine the instantaneous rate of change of the function at the indicated input value. (You may find it helpful to apply the Technology Tip demonstrated in Section 9.2.) Explain the real-life meaning of the result.

13. **Olympic Runners**   Based on the run times of the top seven runners of the 100-meter men's race in the 2000 Olympics, a runner's time may be modeled by

$$T(p) = 0.003056p^3 - 0.04274p^2 + 0.2178p + 9.694$$

seconds, where p is the place (rank) of the runner. (**Source:** Modeled from www.sporting-heroes.net data.)

Find and interpret the meaning of $T'(7)$.

14. **Presidential Election Funding**   Based on data from 1980 to 1996 the amount of federal funds used for the presidential election may be modeled by

$$F(t) = 0.03867t^3 - 0.6804t^2 + 6.604t + 62.62$$

million dollars, where t is the number of years since 1980. (**Source:** Modeled from *Statistical Abstract of the United States, 2001;* Table 409, p. 255.)

Find and interpret the meaning of $F'(16)$.

Section 9.3   In Exercises 15–16, determine the equation of the line tangent to the function at the given point. Then graph the tangent line and the function together.

15. $f(x) = 2x^2 - 2x; (1, 0)$

16. $f(x) = -x^2 + 6x; (2, 8)$

In Exercises 17–18, answer the questions by calculating the slope of the tangent line and the tangent-line equation, as appropriate.

17. **Social Security Benefit**   Based on data from 1940 to 2000, the average monthly Social Security benefit for a retired man may be modeled by

$$P(t) = 0.3486t^2 - 5.505t + 35.13$$

dollars, where t is the number of years since 1940.

How quickly was the average monthly Social Security benefit for a retired man increasing in 2000? What was the average monthly Social Security benefit for a retired man in 2001? (Use a tangent-line approximation.)

18. **Weight of an Average Girl**   The weight of a 2-year-old to 13-year-old girl may be modeled by

$$W(t) = 0.289a^2 + 2.464a + 23.10$$

pounds, where a is the age of the girl. (**Source:** Modeled from www.babybag.com data.)

How quickly is the average girl gaining weight when she is 12 years old? What is the estimated weight of a 13-year-old girl? (Use a tangent-line approximation.)

In Exercises 19–20, estimate the specified derivative by using the data in the table. Then interpret the result.

19. **Highway Fatalities**

Highway Fatalities

Years Since 1970 (t)	Fatalities (F)
0	52,627
10	51,091
20	44,599
25	41,817
30	41,821

Source: U.S. Department of Transportation, Bureau of Transportation Statistics.

Estimate $F'(15)$.

20. **Aviation Fatalities**

General Aviation Fatalities

Years Since 1970 (t)	Fatalities (F)
0	1,310
10	1,239
20	767
25	734
30	592

Source: U.S. Department of Transportation, Bureau of Transportation Statistics.

Estimate $F'(25)$.

Section 9.4 *In Exercises 21–22, find the derivative of the function.*

21. $f(x) = 2x^2 - 4$ **22.** $f(x) = -x^2 + 3x$

In Exercises 23–24, find the slope of the line tangent to the function at $x = 1$, $x = 3$, and $x = 5$.

23. $g(x) = 3x^2 + 2x - 5$

24. $g(x) = x^2 - 6x + 9$

In Exercises 25–26, estimate the derivative of the function, using $h = 0.001$ if you are unable to eliminate the h in the denominator of the derivative formula algebraically.

25. $P(x) = 2^x$ **26.** $R(x) = 3 \cdot 4^x$

In Exercises 27–28, use the derivative function to answer the questions.

27. **Major League Baseball Salaries** Based on data from 1980 to 2000, the average annual salary of a major league baseball player may be modeled by
$$B(t) = 2.699t^2 + 20.76t + 158.4$$
thousand dollars, where t is the number of years since 1980. (**Source:** Modeled from *Statistical Abstract of the United States, 2001*, Table 1241, p. 759.)

Was the average player salary growing faster at the end of 1989 or the end of 1999?

28. **Women's NCAA Basketball** Based on data from 1985 to 1999, the number of people attending women's NCAA basketball games may be modeled by
$$W(t) = 17.1t^2 + 359t + 135$$
thousand people, where t is the number of years since 1985. (**Source:** Modeled from *Statistical Abstract of the United States, 2001*, Table 1241, p. 759).

Was attendance increasing faster at the end of 1995 or the end of 1999?

Section 9.5 *In Exercises 29–30, interpret the real-life meaning of the function and of the derivative evaluated at the given point.*

29. **McDonald's Sales** Based on data from 1990 to 2001, franchised sales at McDonald's fast-food restaurants may be modeled by
$$S(t) = 1233.5t + 12{,}213$$
million dollars, where t is the number of years since 1990. (**Source:** Modeled from McDonald's July 26, 2002, Financial Report.)

Interpret the meaning of $S(12) = 27{,}015$ and $S'(12) = 1233.5$.

30. **Church Membership** Based on data from 1978 to 2000, the number of congregations (wards and branches) in the United States of the Church of Jesus Christ of Latter-Day Saints may be modeled by
$$C(t) = 8679(1.023)^t$$
congregations, where t is the number of years since 1978. (**Source:** Modeled from data compiled by Mark Davies, Brigham Young University.)

Interpret the meaning of $C(14) = 11{,}932$ and $C'(14) = 271$.

Make It Real

What to do

1. Collect a minimum of six data points related to a public or privately owned business (for example, the profit or revenue of a company over a six-year period).
2. Model the data with a polynomial function.
3. Find the derivative of the model function.
4. Evaluate the derivative at three different points.
5. Interpret the real-life meaning of the results of Step 4.
6. Discuss the benefits and the drawbacks of using the model.
7. Explain how company management could use the results of your analysis.

Where to look for data

Company Web Sites

Many company web site addresses are of the form *www.companyname.com*, where *"companyname"* is the name of the company. Look for the area of the web site with investor information. Some web sites to try are

www.mcdonalds.com
www.levistrauss.com
www.albertsons.com
www.gm.com
www.kelloggs.com

Company Financial Reports

Financial reports are typically mailed annually to investors who own shares in a company.

Financial Newspapers and Magazines

The *Wall Street Journal, BusinessWeek, the New York Times, Forbes* magazine, and others contain a wealth of business information.

Differentiation Techniques

A mathematical model may be used to forecast a company's revenue at a given point in time. However, business executives and investors aren't interested only in the dollar amount of a company's revenue; they are also interested in the direction in which revenue is heading. As a rate of change, the derivative of the revenue function shows whether revenues are increasing or decreasing.

10.1 Basic Derivative Rules

- Use various forms of derivative notation
- Use the Constant Rule, Power Rule, Constant Multiple Rule, and Sum and Difference Rule to find the derivative of a function

GETTING **STARTED** Home video game systems have become increasingly popular over the last two decades. Based on data from 1990 to 1999, electronic game system factory sales may be modeled by

$$S(t) = 8.451t^3 - 111.9t^2 + 470.0t + 965.0 \text{ million dollars}$$

where t is the number of years since 1990. How quickly were game system sales increasing in 1999? How quickly were sales expected to increase in 2000?

In the previous chapter, you learned how to find the derivative function by using the limit definition of the derivative. Unfortunately, calculating the difference quotient for a complex function (such as the electronic gaming function) is cumbersome and prone to error. In this section, we will introduce some alternative forms of derivative notation and then demonstrate several shortcuts that will greatly enhance our efficiency in calculating derivatives.

Derivative Notation

Up to this point, we have used $f'(x)$ to represent the derivative of a function $f(x)$. Due in part to the fact that calculus was first developed by two people working independently (Newton and Leibniz), there are multiple ways to represent the same concept.

DERIVATIVE NOTATION

The derivative of a function $y = f(x)$ may be represented by any of the following:

- $f'(x)$, read "f prime of x"
- y', read "y prime"
- $\dfrac{dy}{dx}$, read "dee why dee ex" or "the derivative of y with respect to x"

Note that $\dfrac{d}{dx}$ means "find the derivative with respect to x," whereas $\dfrac{dy}{dx}$ is "the derivative of y with respect to x." The form $\dfrac{dy}{dx}$ is referred to as **Leibniz notation** for the derivative. Throughout this section, we will use various forms of notation for the derivative. Additionally, we will use the term **differentiate** to mean "find the derivative of" and the term **differentiation** to refer to the process of finding the derivative. A function whose derivative exists for all values of x in its domain is said to be **differentiable.**

The Constant Rule

Recall that the derivative represents the slope of the tangent line of the function. Consider the horizontal line, $f(x) = c$. The slope of the tangent line of f at (a, c) is

$$f'(a) = \lim_{h \to 0} \frac{f(a + h) - f(a)}{h}$$

$$= \lim_{h \to 0} \frac{c - c}{h}$$

$$= \lim_{h \to 0} \frac{0}{h}$$

$$= 0 \qquad \qquad \text{Since dividing 0 by any nonzero number yields 0}$$

CONSTANT RULE

The derivative of a constant function $y = c$ is

$$y' = 0$$

EXAMPLE 1 **Finding the Derivative of a Constant Function**

Find the derivative of $y = 5$.

SOLUTION

$$y' = 0$$

The Power Rule

Recall that a power function is a function of the form $f(x) = x^n$, where x is a variable and n is a constant. In Exercises 9.4, you found that the power function $y = x^3$ had derivative $y' = 3x^2$. Similarly, the derivative of $y = x^4$ is $y' = 4x^3$, and the derivative of $y = x^5$ is $y' = 5x^4$. Although the derivative methods previously introduced allow us to find the derivative of any power function, the Power Rule provides a remarkably quick and easy way to calculate the derivative.

POWER RULE

The derivative of a function $y = x^n$, where x is a variable and n is a nonzero constant, is

$$y' = nx^{n-1}$$

EXAMPLE 2

Finding the Derivative of a Power Function

Find $\dfrac{dy}{dx}$ given $y = x^3$.

SOLUTION

$$\frac{dy}{dx} = 3x^{3-1}$$

$$= 3x^2$$

EXAMPLE 3

Finding the Derivative of a Power Function

Differentiate $g(x) = x^{-5}$.

SOLUTION

$$g'(x) = -5x^{-5-1}$$

$$= -5x^{-6}$$

EXAMPLE 4

Finding the Derivative of a Power Function

Find y' given $y = \dfrac{1}{x}$.

SOLUTION Initially, this function doesn't look like a power function. However, recall that $y = \dfrac{1}{x}$ is equivalent to $y = x^{-1}$, which is of the form $y = x^n$.

$$y' = -1x^{-1-1}$$

$$= -x^{-2}$$

$$= -\frac{1}{x^2}$$

EXAMPLE 5

Finding the Derivative of a Power Function with a Rational Exponent

Find $f'(x)$ given $f(x) = \sqrt[3]{x}$.

SOLUTION Again, $f(x) = \sqrt[3]{x}$ doesn't appear to be of the form $y = x^n$. However, any function of the form $f(x) = \sqrt[n]{x}$ may be rewritten as $f(x) = x^{1/n}$. Therefore, $f(x) = \sqrt[3]{x}$ is equivalent to $f(x) = x^{1/3}$, which is of the form $y = x^n$.

$$f'(x) = \frac{1}{3}x^{1/3-1}$$

$$= \frac{1}{3}x^{-2/3}$$

This result may be written in a variety of alternative forms, such as the following:

$$f'(x) = \frac{1}{3x^{2/3}}$$

$$f'(x) = \frac{1}{3\sqrt[3]{x^2}}$$

When working with functions with rational exponents, it is helpful to recall that

$$y = \sqrt[n]{x^m}$$
$$= x^{m/n}$$
$$= (x^m)^{1/n}$$
$$= (x^{1/n})^m$$

Each of these forms is a correct way to represent the function $y = \sqrt[n]{x^m}$.

Constant Multiple Rule

The Constant Multiple Rule is used when a constant is multiplied by a function. This rule allows us to factor out the constant, take the derivative of the function, and then multiply the result by the factored-out constant.

CONSTANT MULTIPLE RULE

The derivative of a differentiable function $g(x) = kf(x)$ with constant k is given by

$$g'(x) = kf'(x)$$

EXAMPLE 6 **Finding a Derivative of a Power Function**

Differentiate $g(x) = 3x^2$.

SOLUTION

$$g'(x) = \frac{d}{dx}(3x^2)$$

$$= 3 \cdot \frac{d}{dx}(x^2) \qquad \text{Constant Multiple Rule}$$

$$= 3(2x) \qquad \text{Power Rule}$$

$$= 6x$$

EXAMPLE 7

Finding the Derivative of a Power Function

Find the derivative of $y = 4x^3$.

SOLUTION

$$y' = \frac{d}{dx}(4x^3)$$

$$= 4 \cdot \frac{d}{dx}(x^3) \qquad \text{Constant Multiple Rule}$$

$$= 4(3x^2) \qquad \text{Power Rule}$$

$$= 12x^2$$

Sum and Difference Rule

The Sum and Difference Rule allows us to calculate the derivative of a function with multiple terms by adding together the derivatives of each of the terms.

SUM AND DIFFERENCE RULE

The derivative of a differentiable function $h(x) = f(x) \pm g(x)$ is given by

$$h'(x) = f'(x) \pm g'(x)$$

EXAMPLE 8

Finding the Derivative of a Sum of Functions

Differentiate $h(x) = 3x^{-2} + 5x$.

SOLUTION

$$h'(x) = \frac{d}{dx}(3x^{-2} + 5x)$$

$$= \frac{d}{dx}(3x^{-2}) + \frac{d}{dx}(5x) \qquad \text{Sum and Difference Rule}$$

$$= 3 \cdot \frac{d}{dx}(x^{-2}) + 5 \cdot \frac{d}{dx}(x) \qquad \text{Constant Multiple Rule}$$

$$= 3(-2x^{-3}) + 5(1) \qquad \text{Power Rule}$$

$$= -6x^{-3} + 5$$

Although we showed multiple steps in this example, as you become skilled in using the rules, you will frequently be able to calculate the derivative in one or two steps.

EXAMPLE 9

Finding the Derivative of a Sum of Functions

Find $\dfrac{dy}{dx}$ given $y = 5x^3 - 6x^2 + 1$.

SOLUTION

$$\frac{dy}{dx} = \frac{d}{dx}(5x^3 - 6x^2 + 1)$$

$$= \frac{d}{dx}(5x^3) - \frac{d}{dx}(6x^2) + \frac{d}{dx}(1) \qquad \text{By the Sum and Difference Rule}$$

$$= 15x^2 - 12x + 0 \qquad\qquad \text{By the Constant Multiple, Constant,}$$
$$\qquad\qquad\qquad\qquad\qquad\qquad \text{and Power Rules}$$

$$= 15x^2 - 12x$$

We will now return to the electronic games sales function introduced at the beginning of this section.

EXAMPLE 10

Using Derivative Rules to Quickly Calculate an Instantaneous Rate of Change

Based on data from 1990 to 1999, electronic game system factory sales may be modeled by

$$S(t) = 8.451t^3 - 111.9t^2 + 470.0t + 965.0 \text{ million dollars}$$

where t is the number of years since 1990. (**Source:** Modeled from *Statistical Abstract of the United States, 2001*, Table 1005, p. 634.)

How quickly were game system sales increasing in 1999? How quickly were sales expected to be increasing in 2000?

SOLUTION We are asked to evaluate the derivative of the function at $t = 9$ and $t = 10$.

$$S'(t) = \frac{d}{dt}(8.451t^3 - 111.9t^2 + 470.0t + 965.0)$$

$$= \frac{d}{dt}(8.451t^3) - \frac{d}{dt}(111.9t^2) + \frac{d}{dt}(470.0t) + \frac{d}{dt}(965.0) \qquad \text{Sum and Difference Rule}$$

$$= 3(8.451)t^2 - 2(111.9)t + 470.0 + 0 \qquad\qquad \text{Power, Constant Multiple, and}$$
$$\qquad\qquad\qquad\qquad\qquad\qquad\qquad\qquad\qquad \text{Constant Rules}$$

$$= 25.353t^2 - 223.8t + 470.0$$

$$S'(9) = 25.353(9)^2 - 223.8(9) + 470.0$$

$$= 2053.593 - 2014.2 + 470.0$$

$$\approx 509.4$$

$$S'(10) = 25.353(10)^2 - 223.8(10) + 470.0$$

$$= 2535.3 - 2238.0 + 470.0$$

$$= 767.3$$

In 1999, game system sales were increasing by \$509.4 million per year. In 2000, game system sales were expected to be increasing at a rate of \$767.3 million per year.

EXAMPLE 11

Using Derivative Rules to Quickly Calculate an Instantaneous Rate of Change

In Example 10, we saw that game system sales were increasing rapidly between 1990 and 1999. Game software sales were also increasing rapidly over the same time period. Based on data from 1990 to 1999, electronic game software factory sales may be modeled by

$$S(t) = 39.79t^2 - 48.41t + 2378 \text{ million dollars}$$

where t is the number of years since 1990. (**Source:** Modeled from *Statistical Abstract of the United States, 2001*, Table 1005, p. 634.)

Were electronic game software sales increasing faster in 1995 or 1999?

SOLUTION We need to evaluate $S'(t)$ at $t = 5$ and $t = 9$.

$$S'(t) = \frac{d}{dt}(39.79t^2 - 48.41t + 2378)$$

$$= 2(39.79)t - 48.41 + 0 \qquad \text{Constant Multiple, Power, Constant,}$$
$$\text{and Sum and Difference Rules}$$

$$= 79.58t - 48.41$$

$$S'(5) = 79.58(5) - 48.41$$
$$= 397.9 - 48.41$$
$$= 349.49$$

$$S'(9) = 79.58(9) - 48.41$$
$$= 716.22 - 48.41$$
$$= 667.81$$

In 1995, game software sales were increasing by $349.5 million per year. In 1999, sales were increasing by $667.8 million per year. Sales were increasing much faster in 1999 than in 1995.

10.1 Summary

In this section you learned how to use the Constant Rule, the Power Rule, the Constant Multiple Rule, and the Sum and Difference Rule. You also learned alternative ways of representing the derivative.

10.1 Exercises

In Exercises 1–10, use the Constant Rule, the Power Rule, the Constant Multiple Rule, and the Sum and Difference Rule (as appropriate) to find the derivative of the function.

1. $f(x) = 5$

2. $g(x) = 2x + 4$

3. $v(t) = 5t^3 - 10t^{-2}$

4. $s(t) = 7t^4 - 12t^2 + 2t + 43$

5. $g(x) = -2x + 4x^3$

6. $s(n) = 44n^2 - n$

7. $f(t) = 4t^{-1}$

8. $h(t) = 15t^5 - 75t^4 + 300t^3$

9. $g(t) = t^3 + 3t^2 + 6t + 6$

10. $f(x) = 0$

In Exercises 11–20, find the equation of the tangent line at the indicated domain value.

11. $y = \sqrt{x}; \; x = 1$

12. $y = 6.9x^4 - 3.2x^2 + 9.1; \; x = 0$

13. $y = -2.6x^5 + 3x^{-2}; \; x = -1$

14. $y = -7.62x^2 - 5.21x + 6.11; \; x = 1$

15. $y = 1235.3t^3 + 551.23t - 1203.9; \; t = 0$

16. $y = -t^{-1} + 2t^{-2} - 3t^{-3}; \; t = 4$

17. $f(n) = 9n^{1.3} + 5n^{2.1} + 92n^{-0.2}; \; n = 1$

18. $w(t) = 3.2t - t^3; \; t = 5$

19. $g(x) = x^{-3} + 2.22x^3 - 12.3; \; x = 1$

20. $r(x) = \frac{1}{2}x^{1.5}; \; x = 0$

In Exercises 21–30, answer the question by evaluating the derivative using the derivative rules introduced in this section.

21. **Organic Cropland** Based on data from 1992 to 1996, the amount of certified organic cropland in the United States may be modeled by

$$A(t) = 20.57t^2 - 24.51t + 410.3$$

thousand acres, where t is the number of years since 1992. (**Source:** Modeled from *Statistical Abstract of the United States, 2001*, Table 805, p. 526.)

    How quickly was the amount of organic cropland increasing in 1996?

22. **Traffic Fatalities** Based on data from 1970 to 2000, the number of highway traffic fatalities may be modeled by

$$F(t) = 1.554t^3 - 72.90t^2 + 427.0t + 52{,}615$$

fatalities, where t is the number of years since 1970. (**Source:** Modeled from U.S. Department of Transportation, Bureau of Transportation Statistics data.)

    At what rate was the number of highway fatalities changing in 1995?

23. **Average Height of a Girl** The average height of a 2- to 13-year-old girl may be modeled by

$$H(a) = -0.0392a^2 + 2.987a + 29.69$$

inches, where a is the age of the girl. (**Source:** Modeled from www.babybag.com data.)

    Does the average height of a 4-year-old girl increase faster than the average height of an 11-year-old girl?

24. **Average Height of a Boy** The average height of a 2- to 13-year-old boy may be modeled by

$$H(a) = -0.0507a^2 + 2.997a + 30.44$$

inches, where a is the age of the boy. (**Source:** Modeled from www.babybag.com data.)

    Does the average height of a 4-year-old boy increase faster than the average height of an 11-year-old boy?

25. Using the models in Exercises 23 and 24, does a 9-year-old boy or a 9-year-old girl grow at a quicker rate?

26. **Average Weight of a Girl** The average weight of a 2- to 13-year-old girl may be modeled by

$$W(a) = 0.289a^2 + 2.464a + 23.10$$

pounds, where a is the age of the girl. (**Source:** Modeled from www.babybag.com data.)

    Does the average weight of a 4-year-old girl increase faster than that of an 11-year-old girl?

27. **Average Weight of a Boy** The average weight of a 2- to 13-year-old boy may be modeled by

$$W(a) = 0.215a^2 + 2.993a + 23.78$$

pounds, where a is the age of the boy. (**Source:** Modeled from www.babybag.com data.)

    Does the average weight of a 4-year-old boy increase faster than the average weight of an 11-year-old boy?

28. Using the models in Exercises 26 and 27, does a 9-year-old boy or a 9-year-old girl gain weight at a quicker rate?

29. **Billboard Advertising** Based on data from 1990 to 2000, the amount of money spent on billboard advertising may be modeled by

$$A(t) = 34.99t^2 + 94.91t + 8963$$

million dollars, where t is the number of years since 1990. (**Source:** Modeled from *Statistical Abstract of the United States, 2001,* Table 1272, p. 777.)

Was billboard advertising expected to increase more rapidly in 1999 or in 2000?

30. 🅼 **Television Advertising** Based on data from 1990 to 2000, the amount of money spent on broadcast TV advertising may be modeled by

$$A(t) = 10.22t^4 - 221.5t^3 + 1616t^2 - 2488t + 26,671$$

million dollars, where t is the number of years since 1990. (**Source:** Modeled from *Statistical Abstract of the United States, 2001,* Table 1272, p. 777.)

At what rate was television advertising expected to increase in 2000?

Exercises 31–35 are intended to challenge your understanding of the basic rules of differentiation.

31. Let f, g, and h be differentiable functions with $h(x) = f(x) \cdot g(x)$. May the Constant Multiple Rule be used to find $h'(x)$? Explain.

32. Let $f(x) = 2^x$. Is $f'(x) = x \cdot 2^{x-1}$? Defend your conclusions.

33. Use the limit definition of the derivative to prove that if s is differentiable and $s(x) = f(x) + g(x)$, then $s'(x) = f'(x) + g'(x)$.

34. Use the limit definition of the derivative to prove that if f is differentiable, k is a constant, and $f(x) = k \cdot g(x)$, then $f'(x) = k \cdot g'(x)$.

35. Given $f(x) = x^{1/3}$, for what values of x is $f'(x)$ undefined?

10.2 The Product and Quotient Rules

- Use the Product Rule and Quotient Rule to find the derivative of a function

GETTING STARTED Based on data from 1985 to 1999, the per capita consumption of chicken may be modeled by

$$C(t) = 0.002623t^4 - 0.07545t^3 + 0.6606t^2 - 0.4688t + 36.92$$

pounds of *boneless, trimmed* chicken, where t is the number of years since 1985. (**Source:** Modeled from *Statistical Abstract of the United States, 2001,* Table 202, p. 129.)

Based on data from 1985 to 2000, the average retail price of *fresh whole* chicken may be modeled by

$$P(t) = 0.0007346t^2 - 0.009701t + 0.7803$$

dollars per pound, where t is the number of years since 1985. (**Source:** Modeled from *Statistical Abstract of the United States, 2001,* Table 706, p. 468.) A whole chicken loses about 30 percent of its weight when it is boned and trimmed.

Was per capita spending on chicken increasing or decreasing in 2000? By how much? We will use the Product Rule to answer this question in Example 4 of this section.

In this section, we will demonstrate how to find the derivative of a product by using the Product Rule and the derivative of a quotient by using the Quotient Rule.

> **PRODUCT RULE**
>
> The derivative of a function $h(x) = f(x) \cdot g(x)$ is given by
>
> $$h'(x) = f'(x) \cdot g(x) + g'(x) \cdot f(x)$$

A common error among beginning calculus students is to assume that $\dfrac{d}{dx}[f(x) \cdot g(x)] = f'(x) \cdot g'(x)$. We can easily convince ourselves that this is not true by calculating the derivative of $f(x) = x^2$ using the Power Rule and then by using the erroneous rule.

Using the Power Rule,

$$\frac{d}{dx}(x^2) = 2x$$

Using the erroneous rule,

$$\frac{d}{dx}(x^2) = \frac{d}{dx}(x \cdot x)$$

$$= \frac{d}{dx}(x) \cdot \frac{d}{dx}(x)$$

$$= 1 \cdot 1$$

$$= 1$$

Since $2x \neq 1$, we know that the erroneous rule is not valid. Let's calculate the derivative using the Product Rule.

Using the Product Rule,

$$\frac{d}{dx}(x^2) = \frac{d}{dx}(x \cdot x)$$

$$= \frac{d}{dx}(x) \cdot x + \frac{d}{dx}(x) \cdot x$$

$$= 1 \cdot x + 1 \cdot x$$

$$= 2x$$

The Product Rule yielded the same result as the Power Rule.

EXAMPLE 1   **Finding the Derivative of a Product of Functions**

Find the derivative of $h(x) = (2x + 4)(5x^3 - 3x)$.

SOLUTION

$$h'(x) = \frac{d}{dx}[(2x + 4)(5x^3 - 3x)]$$

$$= \left[\frac{d}{dx}(2x + 4)\right] \cdot (5x^3 - 3x) + \left[\frac{d}{dx}(5x^3 - 3x)\right] \cdot (2x + 4) \qquad \text{Product Rule}$$

$$= (2) \cdot (5x^3 - 3x) + (15x^2 - 3) \cdot (2x + 4) \qquad \text{Power, Constant Multiple, and Constant Rules}$$
$$= (10x^3 - 6x) + (30x^3 + 60x^2 - 6x - 12)$$
$$= 40x^3 + 60x^2 - 12x - 12$$

We can check our work by multiplying out the factors of $h(x)$ and differentiating the result using the Power Rule.

$$h(x) = (2x + 4)(5x^3 - 3x)$$
$$= 10x^4 + 20x^3 - 6x^2 - 12x$$
$$h'(x) = 40x^3 + 60x^2 - 12x - 12$$

The result is the same as that obtained by using the Product Rule. We are confident that we have done the problem correctly.

In the previous two examples, we saw that there are multiple ways to calculate the derivative of a function. In general, we will use the simplest techniques possible to find the derivative. However, in order to demonstrate the Product Rule, in this section we may use the Product Rule in places where other methods are typically preferred.

EXAMPLE 2 **Finding the Derivative of a Product of Functions**

Differentiate $y = (2x^3 - 6x + 5)(x^4 - 2x^2 + 1)$.

SOLUTION

$$\frac{dy}{dx} = \frac{d}{dx}(y)$$

$$= \frac{d}{dx}\left[(2x^3 - 6x + 5)(x^4 - 2x^2 + 1)\right]$$

$$= \left[\frac{d}{dx}(2x^3 - 6x + 5)\right](x^4 - 2x^2 + 1) + \left[\frac{d}{dx}(x^4 - 2x^2 + 1)\right](2x^3 - 6x + 5)$$

$$= (6x^2 - 6)(x^4 - 2x^2 + 1) + (4x^3 - 4x)(2x^3 - 6x + 5)$$

$$= (6x^6 - 12x^4 + 6x^2 - 6x^4 + 12x^2 - 6) + (8x^6 - 24x^4 + 20x^3 - 8x^4 + 24x^2 - 20x)$$

$$= 14x^6 - 50x^4 + 20x^3 + 42x^2 - 20x - 6$$

EXAMPLE 3 **Finding the Derivative of a Product of Three Functions**

Differentiate $R(t) = (5t - 1)(4t + 3)(2t + 6)$.

SOLUTION This function is different because it is the product of three factors instead of two. Nevertheless, the Product Rule may be generalized to work here. If we have a function of the form $w(x) = f(x) \cdot g(x) \cdot h(x)$, then $w'(x) = f'(x)g(x)h(x) + g'(x)f(x)h(x) + h'(x)f(x)g(x)$. To find each term of the derivative, we calculate the derivative of each factor and then multiply it by the remaining factors. The derivative of $w(x)$ is the sum of the individual terms.

$$
\begin{aligned}
R'(t) &= 5(4t + 3)(2t + 6) + 4(5t - 1)(2t + 6) + 2(5t - 1)(4t + 3) \\
&= 5(8t^2 + 24t + 6t + 18) + 4(10t^2 + 30t - 2t - 6) + 2(20t^2 + 15t - 4t - 3) \\
&= 5(8t^2 + 30t + 18) + 4(10t^2 + 28t - 6) + 2(20t^2 + 11t - 3) \\
&= (40t^2 + 150t + 90) + (40t^2 + 112t - 24) + (40t^2 + 22t - 6) \\
&= 120t^2 + 284t + 60
\end{aligned}
$$

EXAMPLE 4   **Applying the Product Rule in a Real-World Context**

Based on data from 1985 to 1999, the per capita consumption of chicken may be modeled by

$$C(t) = 0.0026t^4 - 0.075t^3 + 0.66t^2 - 0.47t + 37$$

pounds of *boneless, trimmed* chicken, where t is the number of years since 1985. (**Source:** Modeled from *Statistical Abstract of the United States, 2001*, Table 202, p. 129.)

Based on data from 1985 to 2000, the average retail price of *fresh whole* chicken may be modeled by

$$P(t) = 0.00073t^2 - 0.0097t + 0.78$$

dollars per pound, where t is the number of years since 1985. (**Source:** Modeled from *Statistical Abstract of the United States, 2001*, Table 706, p. 468.) A whole chicken loses about 30 percent of its weight when it is boned and trimmed.

Was per capita spending on chicken increasing or decreasing in 2000? At what rate?

SOLUTION We first need to find a function to model per capita spending on chicken. The units of $C(t)$ are pounds of *boneless, trimmed* chicken and the units of $P(t)$ are dollars per pound of *fresh whole chicken*. Since 30% of a whole chicken's weight is lost when it is boned and trimmed, 1 pound of *fresh whole* chicken yields 0.7 pound of *boneless, trimmed* chicken. We define a new function $S(t) = C(t)P(t)$. What are the units of $S(t)$? We have

$$(C \text{ pounds of boneless chicken})\left(\frac{P \text{ dollars}}{\text{pound of whole chicken}}\right)$$

Dividing the pounds of boneless chicken by 0.7 will yield the number of pounds of whole chicken that were used to get C pounds of boneless chicken and allow us to cancel units. Thus

UNITS

$$\left(\frac{C}{0.7} \text{ pounds of whole chicken}\right)\left(\frac{P \text{ dollars}}{\text{pound of whole chicken}}\right) = \frac{10}{7} CP \text{ dollars}$$

Therefore, the units of $S(t)$ are dollars.

$$
\begin{aligned}
S(t) &= \frac{10}{7} C(t) \cdot P(t) \\
&= \frac{10}{7}(0.0026t^4 - 0.075t^3 + 0.66t^2 - 0.47t + 37)(0.00073t^2 - 0.0097t + 0.78)
\end{aligned}
$$

We are asked to calculate the instantaneous rate of change of $S(t)$ in 2000. That is, we are to find $S'(15)$. We know from the Product Rule that

UNITS

$$S'(t) = \frac{10}{7}[C'(t) \cdot P(t) + P'(t) \cdot C(t)] \frac{\text{dollars}}{\text{year}}$$

Although we could calculate $S'(15)$ algebraically, the complexity of $S(t)$ would make the process extremely tedious. In cases such as this, it is appropriate to use the power of a graphing calculator, as demonstrated in the following Technology Tip. Using the Technology Tip, we determine that

$$S'(15) \approx 5.33 \frac{\text{dollars}}{\text{year}}$$

In 2000, per capita spending on chicken was increasing by $5.33 per year.

TECHNOLOGY TIP

Calculating a Derivative with *nDeriv(*

1. Enter the function equation using the Y= editor. (In this case, we entered C in Y_1, P in Y_2, and S in Y_3. We input the variables Y_1 and Y_2 into equation Y_3 by using the VARS Y-VARS; 1:Function key sequence and selecting the respective functions.)

2. Press 2ND MODE to quit and return to the home screen. Then press MATH and select 8:nDeriv(. Then press ENTER.

3. The nDeriv(function requires three input values: the function name, the variable of differentiation, and the value of the variable. Enter the appropriate values.

4. Press ENTER to display the result.

nDeriv(Y₃,X,15
              5.334334044

Although any quotient of two functions may be rewritten as a product, using the **Quotient Rule** to find the derivative of a quotient is often easier than using the Product Rule.

QUOTIENT RULE

The derivative of a function $h(x) = \dfrac{f(x)}{g(x)}$ is given by

$$h'(x) = \frac{f'(x) \cdot g(x) - g'(x) \cdot f(x)}{[g(x)]^2}$$

EXAMPLE 5   **Finding the Derivative of a Quotient of Functions**

Find $\dfrac{dy}{dx}$ for $y = \dfrac{x^2 + 1}{x}$.

SOLUTION

$$\frac{dy}{dx} = \frac{(2x)(x) - (1)(x^2 + 1)}{(x)^2}$$

$$= \frac{2x^2 - x^2 - 1}{x^2}$$

$$= \frac{x^2 - 1}{x^2}$$

EXAMPLE 6   **Finding the Derivative of a Quotient of Functions**

Differentiate $y = \dfrac{6x + 1}{3x - 1}$.

SOLUTION

$$y' = \frac{(6)(3x - 1) - (3)(6x + 1)}{(3x - 1)^2}$$

$$= \frac{18x - 6 - 18x - 3}{(3x - 1)^2}$$

$$= \frac{-9}{(3x - 1)^2}$$

10.2 Summary

In this section, you learned how to use the Product and Quotient Rules to find the derivative of a function that is written as a product or quotient of factors. You also learned how technology may be used to differentiate functions with complex factors.

10.2 Exercises

In Exercises 1–10, use the Product or the Quotient Rule to find the derivative of the function. Don't simplify the result.

1. $f(x) = 2x(3x + 4)$ **2.** $g(x) = 5x(9x - 2)$

3. $f(t) = (2t - 6)(10t + 5)$

4. $s(t) = 2t^3(3t^2 + 7t)$

5. $w(n) = (3n^2 + 8)(n^3 - n)$

6. $D(p) = (p^2 - 7p)(1 - p^3)$

7. $Q(t) = (t^2 + 2t + 1)(t^2 + 4t + 4)$

8. $P(t) = \dfrac{t^2 + 5t + 6}{t^2 + 5t + 4}$ **9.** $f(x) = \dfrac{x^3 - 1}{x^4 - x^3}$

10. $g(x) = \dfrac{x^3 + 5x^2}{x^2 - 4x + 2}$

In Exercises 11–20, determine the slope of the graph at the indicated domain value.

11. $h(t) = \dfrac{t^2 + 5t + 6}{t}$; $t = -2$

12. $g(t) = \dfrac{t^3 - 12t^2}{t}$; $t = 1$

13. $s(t) = \dfrac{t^4 - 6t^2 + 8}{2t}$; $t = -1$

14. $p(t) = \dfrac{7t^2 - 2t + 9}{t}$; $t = 2$

15. $f(x) = 2x(3x + 4)(5x + 7)$; $x = 0$

16. $R(p) = 5p(3p + 1)(4p - 1)$; $p = 1$

17. $r(x) = (2x + 11)(x - 4)(7x + 3)$; $x = 4$

18. $T(n) = (n^2 + 4)(2n - 8)(n + 9)$; $n = 3$

19. $f(x) = (x + 1)(x - 1)(x^2 + 1)$; $x = -1$

20. $g(x) = (x - 2)(x + 2)(x^2 + 4)$; $x = 2$

In Exercises 21–25, use the Technology Tip to answer the questions from the real-life scenarios.

21.   **Employer Labor Costs** Based on data from 1995 to 1999, the average annual earnings in the lumber and wood products manufacturing industry may be modeled by
$$S(t) = -63.57t^2 + 1253t + 25{,}066$$

dollars per employee, and the average number of employees in the lumber and wood products manufacturing industry may be modeled by
$$E(t) = 3.143t^2 + 5.029t + 772.5$$
thousand employees, where t is the number of years since 1995. (**Source:** Modeled from *Statistical Abstract of the United States, 2001*, Table 979, p. 622.)

In 1999, at what rate was employer spending on lumber and wood products manufacturing industry employee earnings increasing?

22.   **Employer Labor Costs** Based on data from 1995 to 1999, the average annual earnings in the rubber and plastics manufacturing industry may be modeled by
$$E(t) = 1196t + 29{,}825$$

dollars per employee, and the average number of employees in the rubber and plastics manufacturing industry may be modeled by
$$N(t) = -2.917t^3 + 15.29t^2 - 10.73t + 962.9$$

thousand employees, where t is the number of years since 1995. (**Source:** Modeled from *Statistical Abstract of the United States, 2001*, Table 979, p. 622.)

In 1999, at what rate was employer spending on rubber and plastics manufacturing industry employee earnings increasing?

23.   **Employer Labor Costs** Based on data from 1995 to 1999, the average annual earnings in the paper and allied products manufacturing industry may be modeled by
$$E(t) = 1335t + 39{,}408$$

dollars per employee, and the average number of employees in the same industry may be modeled by
$$N(t) = -t^3 + 5.571t^2 - 12.29t + 684.9$$

thousand employees, where t is the number of years since 1995. (**Source:** Modeled from *Statistical Abstract of the United States, 2001*, Table 979, p. 622.)

In 1999, at what rate was employer spending on paper and allied products manufacturing industry employee earnings increasing?

24. **Employer Labor Costs**   Based on data from 1995 to 1999, the average annual earnings in the apparel and other textile products manufacturing industry may be modeled by

$$E(t) = 1118t + 18{,}729$$

dollars per employee, and the average number of employees in the same industry may be modeled by

$$N(t) = -3.5t^3 + 20.21t^2 - 85.86t + 918.2$$

thousand employees, where t is the number of years since 1995. (**Source:** Modeled from *Statistical Abstract of the United States, 2001*, Table 979, p. 622.)

In 1999, at what rate was employer spending on apparel and other textile products manufacturing industry employee earnings changing?

25. **Employer Labor Costs**   Based on data from 1995 to 1999, the average annual earnings in the printing and publishing industry may be modeled by

$$E(t) = 88.36t^2 + 1291t + 34{,}528$$

dollars per employee, and the average number of employees in the printing and publishing industry may be modeled by

$$N(t) = -4.417t^3 + 24.93t^2 \\ - 25.30t + 1450$$

thousand employees, where t is the number of years since 1995. (**Source:** Modeled from *Statistical Abstract of the United States, 2001*, Table 979, p. 622.)

In 1999, at what rate was employer spending on printing and publishing industry employee earnings increasing?

Exercises 26–30 are intended to challenge your understanding of the Product Rule. Solve the problems without using the Technology Tip.

26. **Apple Farming**   Historically, many apple farmers spaced trees 40 feet by 40 feet apart (27 trees per acre). Trees typically took 25 years to reach their maximum production of 500 bushels per acre. (A bushel is about 44 pounds.) In recent years, agriculturists have created dwarf and semidwarf varieties that allow trees to be spaced 10 feet by 10 feet apart. (**Source:** USDA.)

Suppose that a farmer has an apple orchard with 40 trees per acre. The orchard yields 10 bushels per tree. The farmer estimates that for each additional tree planted per acre, the average yield per tree is reduced by 0.1 bushel.

If $y = f(x)$ is the total number of bushels of apples produced per acre when an additional x trees per acre are planted, calculate and interpret the meaning of $f'(25)$, $f'(30)$, and $f'(35)$.

27. Apple Farming   Suppose that a farmer has an apple orchard with 30 trees per acre. The orchard yields 12 bushels per tree. The farmer estimates that for each additional tree planted per acre, the average yield per tree is reduced by 0.1 bushel.

If $y = f(x)$ is the total number of bushels of apples produced per acre when an additional x trees per acre are planted, calculate and interpret the meaning of $f'(40)$, $f'(45)$, and $f'(50)$.

28. Based on the results of Exercises 26 and 27, do you think that planting a particular number of trees will maximize the yield per acre? Justify your answer.

29. Apple Supplier Prices   A fruit farmer sells apples to a grocery store chain. The amount of apples the store buys depends linearly upon the price per pound that the farmer charges. The farmer estimates that for every $0.02 per pound increase in the price, the store will reduce its order by 44 pounds. The store currently orders 440 pounds per week and pays $0.18 per pound. What price should the farmer charge in order to maximize her revenue from apple sales?

30. Apple Retailer Prices   A grocery store has priced apples at $0.65 per pound and sells 1000 pounds per week. The amount of apples the store sells depends linearly upon the price per pound that the store charges. The store manager estimates that for every $0.05 per pound increase in the price, the store will reduce its sales by 100 pounds. What price should the store charge for its apples in order to maximize revenue from apple sales?

10.3 The Chain Rule

- Use the Chain Rule to find the derivative of a function

Based on data from 1995 to 1999, the average annual earnings of an employee in the lumber and wood products industry may be modeled by

$$S(n) = -0.7685n^2 + 1295n - 516{,}596 \text{ dollars}$$

where n is the number of employees in thousands. (**Source:** Modeled from *Statistical Abstract of the United States, 2001*, Table 979, p. 622.)

Based on data from 1995 to 1999, the number of employees in the lumber and wood products industry may be modeled by

$$n(t) = 3.143t^2 + 5.029t + 772.5 \text{ thousand employees}$$

where t is the number of years since 1995. (**Source:** Modeled from *Statistical Abstract of the United States, 2001*, Table 979, p. 622.)

How quickly was the average annual earnings of an employee increasing in 1998? We will answer this question using the Chain Rule.

In this section, we will demonstrate how to find the composition of two functions. We will then demonstrate how to use the Chain Rule to find the derivative of a composition of functions.

Composition of Functions

A composition of two functions occurs when the outputs of one function are the inputs of a second function. Let's consider a common example. Let f represent the freezer function and b represent the blender function.

$$w \longrightarrow f \longrightarrow f(w) \longrightarrow b \longrightarrow b(f(w))$$

FIGURE 10.1

$b(f(w))$ means that first w is plugged into the function f, and then the result is substituted into b. In our example, water was placed in the freezer, and the result, ice, was put in the blender.

EXAMPLE 1

Finding a Composition of Functions

Let $f(t) = 3t + 2$ and $g(t) = t^2 + 1$. Find $f(g(t))$.

SOLUTION

$$
\begin{aligned}
f(g(t)) &= f(t^2 + 1) && \text{The output of the function } g \text{ is the input for } f \\
&= 3(t^2 + 1) + 2 && \text{The variable in } f \text{ is replaced with the output of } g \\
&= 3t^2 + 3 + 2 \\
&= 3t^2 + 5
\end{aligned}
$$

Is $f(g(t))$ the same as $g(f(t))$? Let's return to the freezer and blender example and see.

$$w \longrightarrow b \longrightarrow b\,(w) \longrightarrow f \longrightarrow f(b(w))$$

FIGURE 10.2

When we reversed the order of the functions, we got a different result. *Ice* is not the same as *chopped ice*. There are certain functions for which the two different compositions are equal, but, in general, $f(g(t)) \neq g(f(t))$.

EXAMPLE 2

Finding a Composition of Functions

Let $f(t) = 3t + 2$ and $g(t) = t^2 + 1$. Find $g(f(t))$.

SOLUTION

$$
\begin{aligned}
g(f(t)) &= g(3t + 2) \\
&= (3t + 2)^2 + 1 \\
&= 9t^2 + 12t + 4 + 1 \\
&= 9t^2 + 12t + 5
\end{aligned}
$$

Note that although Examples 1 and 2 used the same functions f and g, $f(g(t)) \neq g(f(t))$.

The Chain Rule

The Chain Rule is used whenever we have to find the derivative of a composition of functions.

CHAIN RULE

The derivative of a function $h(x) = f(g(x))$ is given by

$$h'(x) = f'(g(x)) \cdot g'(x)$$

The Chain Rule tells us to find $f'(x)$, substitute $g(x)$ in for the variable x, and then multiply the result by $g'(x)$.

EXAMPLE 3

Finding a Derivative by Using the Chain Rule

Given $h(x) = (2x + 1)^2 - 4$, find $h'(x)$.

SOLUTION Because there is a nonconstant function $(2x + 1)$ inside the parentheses, we know that we will need the Chain Rule. The function $h(x) = (2x + 1)^2 - 4$ is a function of the form $h(x) = f(g(x))$. We must determine the functions f and g. Often, selecting the function inside the parentheses as g gives the desired result. In this case, we pick $g(x) = 2x + 1$. Then

$$f(g(x)) = (2x + 1)^2 - 4$$
$$f(g(x)) = [g(x)]^2 - 4$$

To find the function f, we simply replace the $g(x)$ in the equation with the variable x. Thus we have

$$f(x) = x^2 - 4$$

We'll first find $f'(x)$ and then use it to calculate $f'(g(x))$.

$$f'(x) = \frac{d}{dx}(x^2 - 4)$$
$$= 2x$$
$$f'(g(x)) = 2 \cdot g(x) \qquad \text{By substitution}$$
$$= 2(2x + 1)$$

Next we'll find $g'(x)$.

$$g'(x) = \frac{d}{dx}(2x + 1)$$
$$= 2$$

The derivative of $f(g(x))$ is

$$f'(g(x)) \cdot g'(x) = [2(2x + 1)] \cdot (2)$$
$$= 4(2x + 1)$$
$$= 8x + 4$$

We'll rework the problem using a slightly different but somewhat easier approach.

$$h(x) = (2x + 1)^2 - 4$$

$$\frac{dh}{dx} = \frac{d}{dx}[(2x + 1)^2] - \frac{d}{dx}(4) \qquad \text{Sum and Difference Rule}$$

To use the Chain Rule to differentiate $(2x + 1)^2$, we treat the expression $2x + 1$ inside the parentheses as if it were a single variable and use the Power Rule. Then we multiply the result by the derivative of $2x + 1$, the expression inside the parentheses.

$$\frac{dh}{dx} = 2(2x + 1)^{2-1} \cdot \frac{d}{dx}(2x + 1) - 0 \qquad \text{Chain, Power, and Constant Rules}$$

$$= 2(2x + 1) \cdot (2)$$

$$= 4(2x + 1)$$

$$= 8x + 4$$

The second technique shown in Example 3 is a demonstration of the Generalized Power Rule.

GENERALIZED POWER RULE

Let $u = g(x)$. The derivative of a function $f(x) = u^n$ is given by

$$f'(x) = nu^{n-1}u'$$

The Generalized Power Rule is a special application of the Chain Rule. We'll use this rule in the next two examples.

EXAMPLE 4 **Finding a Derivative by Using the Generalized Power Rule**

Differentiate $f(x) = 4(2x^2 + x)^3$.

SOLUTION We pick $u = 2x^2 + x$. Then $f(x) = 4(2x^2 + x)^3$ may be rewritten as $f(x) = 4u^3$. From the Generalized Power Rule, we know that

$$\frac{d}{dx}(f(x)) = \frac{d}{dx}(4u^3)$$

$$= 4\frac{d}{dx}(u^3) \qquad \text{Constant Multiple Rule}$$

$$f'(x) = 12u^2u' \qquad \text{Generalized Power Rule}$$

We need to write the result in terms of x, not u. Since $u = 2x^2 + x$ and $u' = 4x + 1$, the derivative may be rewritten as

$$f'(x) = 12u^2u'$$
$$= 12(2x^2 + x)^2(4x + 1)$$
$$= 12[x(2x + 1)]^2(4x + 1)$$
$$= 12(x^2)(2x + 1)^2(4x + 1)$$
$$= 12x^2(2x + 1)^2(4x + 1)$$

We will often write derivatives in factored form. In real-life optimization problems, we are often interested in the values of x that make the derivative equal to zero. These values are easiest to find if the function is written in factored form.

EXAMPLE 5 — **Finding a Derivative by Using the Generalized Power Rule**

Find $\dfrac{dy}{dx}$ given $y = \dfrac{1}{2x^3 + 4x + 1}$.

SOLUTION We first rewrite $y = \dfrac{1}{2x^3 + 4x + 1}$ as $y = (2x^3 + 4x + 1)^{-1}$. If we pick $u = 2x^3 + 4x + 1$, then $y = (2x^3 + 4x + 1)^{-1}$ may be rewritten as $y = u^{-1}$. From the Generalized Power Rule, we know that

$$\frac{d}{dx}(y) = \frac{d}{dx}(u^{-1})$$

$$\frac{dy}{dx} = -u^{-2}u' \qquad \text{Generalized Power Rule}$$

$$= -\frac{u'}{u^2}$$

Since $u = 2x^3 + 4x + 1$, $u' = 6x^2 + 4$. Thus

$$\frac{dy}{dx} = -\frac{u'}{u^2}$$

$$= -\frac{6x^2 + 4}{(2x^3 + 4x + 1)^2}$$

$$= -\frac{2(3x^2 + 2)}{(2x^3 + 4x + 1)^2}$$

As demonstrated in the next example, it is necessary to apply the Chain Rule repeatedly for some functions.

EXAMPLE 6 — **Finding a Derivative by Using the Chain Rule Repeatedly**

Find the derivative of $S(t) = [3(2t + 1)^2 + 5]^3$.

SOLUTION We let $u = 3(2t + 1)^2 + 5$. Notice that this function also contains an expression inside parentheses. If we let $v = 2t + 1$, then $u = 3v^2 + 5$.

$$S(t) = u^3$$

$$\frac{dS}{dt} = 3u^2 u' \qquad\qquad\qquad \text{Generalized Power Rule}$$

$$= 3u^2 \frac{d}{dt}(u) \qquad\qquad \text{Since } u' = \frac{d}{dt}(u)$$

$$= 3(3v^2 + 5)^2 \cdot \frac{d}{dt}(3v^2 + 5) \qquad \text{Since } u = 3v^2 + 5$$

$$= 3(3v^2 + 5)^2 \cdot 6vv' \qquad\qquad \text{Generalized Power Rule}$$

$$= 3(3v^2 + 5)^2 \cdot 6v \cdot \frac{d}{dt}(v) \qquad \text{Since } v' = \frac{d}{dt}(v)$$

$$= 3[3(2t + 1)^2 + 5]^2 \cdot 6(2t + 1) \cdot \frac{d}{dt}(2t + 1) \qquad \text{Since } v = 2t + 1$$

$$= 3[3(2t + 1)^2 + 5]^2 \cdot 6(2t + 1) \cdot (2)$$

At this point, we have calculated the derivative. If we want to rewrite the derivative in its simplest factored form, we continue as follows:

$$\frac{dS}{dt} = 3\,[3(2t + 1)^2 + 5]^2 \cdot 6(2t + 1) \cdot (2)$$

$$= 36\,[3(2t + 1)^2 + 5]^2\,(2t + 1)$$

$$= 36\,[3(4t^2 + 4t + 1) + 5]^2\,(2t + 1)$$

$$= 36\,(12t^2 + 12t + 3 + 5)^2\,(2t + 1)$$

$$= 36\,(12t^2 + 12t + 8)^2\,(2t + 1)$$

$$= 36\,[4(3t^2 + 3t + 2)]^2\,(2t + 1)$$

$$= 36\,(4)^2\,(3t^2 + 3t + 2)^2\,(2t + 1)$$

$$= 576\,(3t^2 + 3t + 2)^2\,(2t + 1)$$

Writing the derivative in factored form required a significant number of additional steps. Although there are some instances that require us to write derivatives in factored form, we will consider the additional steps optional at this point.

Chain Rule: Alternative Form

The Chain Rule may also be expressed using Leibniz notation.

CHAIN RULE: ALTERNATIVE FORM

If $y = f(x)$ and $x = g(t)$, then $y = f(g(t))$ and

$$\frac{dy}{dt} = \frac{dy}{dx} \cdot \frac{dx}{dt}$$

The units of $\frac{dy}{dt}$ are obtained by multiplying the units of $\frac{dy}{dx}$ by the units of $\frac{dx}{dt}$.

EXAMPLE 7

Finding a Derivative by Using the Alternative Form of the Chain Rule

Based on data from 1989 to 1998, restaurant sales in the United States may be modeled by

$$S(r) = 0.9291r - 160.0 \text{ billion dollars}$$

where r is the number of restaurants (in thousands).

The number of restaurants may be modeled by

$$r(t) = 0.3405t^2 + 6.910t + 330.9 \text{ thousand restaurants}$$

where t is the number of years since 1989. (**Source:** Modeled from *Statistical Abstract of the United States, 2001*, Table 1268, p. 775.)

How quickly were restaurant sales increasing in 1998?

SOLUTION We are asked to find $\dfrac{dS}{dt}$. Recall that the alternative form of the Chain Rule states that $\dfrac{dS}{dt} = \dfrac{dS}{dr} \cdot \dfrac{dr}{dt}$. We will first find $\dfrac{dS}{dr}$ and $\dfrac{dr}{dt}$.

$$S(r) = 0.9291r - 160.0$$

UNITS
$$\frac{dS}{dr} = 0.9291 \; \frac{\text{billion dollars}}{\text{thousand restaurants}}$$

$$r(t) = 0.3405t^2 + 6.910t + 330.9$$

UNITS
$$\frac{dr}{dt} = 0.681t + 6.910 \; \frac{\text{thousand restaurants}}{\text{year}}$$

UNITS
$$\frac{dS}{dt} = \frac{dS}{dr} \cdot \frac{dr}{dt}$$

$$= \left(0.9291 \; \frac{\text{billion dollars}}{\text{thousand restaurants}} \right) \cdot \left(0.681t + 6.910 \; \frac{\text{thousand restaurants}}{\text{year}} \right)$$

$$= (0.9291) \cdot (0.681t + 6.910) \frac{\text{billion dollars}}{\text{thousand restaurants}} \; \frac{\text{thousand restaurants}}{\text{year}}$$

$$= (0.9291) \cdot (0.681t + 6.910) \text{ billion dollars per year}$$

$$\approx 0.6327t + 6.420 \text{ billion dollars per year}$$

We need to determine the rate of change in sales for 1998 ($t = 9$).

$$\frac{dS}{dt} \bigg|_{t=9} = 0.6327(9) + 6.420$$

$$\approx 12.11 \text{ billion dollars per year}$$

(Note: The notation $\dfrac{dS}{dt} \bigg|_{t=9}$ means the value of $\dfrac{dS}{dt}$ when $t = 9$.)

In 1998, restaurant sales in the United States were increasing at an estimated rate of 12.11 billion dollars per year.

We'll now return to the problem introduced at the beginning of this section.

EXAMPLE 8

Finding a Derivative by Using the Alternative Form of the Chain Rule

Based on data from 1995 to 1999, the average annual earnings of an employee in the lumber and wood products industry may be modeled by

$$S(n) = -0.7685n^2 + 1295n - 516{,}596 \text{ dollars}$$

where n is the number of employees in thousands. (**Source:** Modeled from *Statistical Abstract of the United States, 2001,* Table 979, p. 622.)

Based on data from 1995 to 1999, the number of employees in the lumber and wood products industry may be modeled by

$$n(t) = 3.143t^2 + 5.029t + 772.5 \text{ thousand employees}$$

where t is the number of years since 1995. (**Source:** Modeled from *Statistical Abstract of the United States, 2001,* Table 979, p. 622.)

How quickly was the average annual earnings of an employee increasing in 1998?

SOLUTION We are asked to find $\dfrac{dS}{dt}$. By the Chain Rule, $\dfrac{dS}{dt} = \dfrac{dS}{dn} \cdot \dfrac{dn}{dt}$. We will first find $\dfrac{dS}{dn}$ and $\dfrac{dn}{dt}$.

$$S(n) = -0.7685n^2 + 1295n - 516{,}596$$

UNITS ▶

$$\frac{dS}{dn} = -1.537n + 1295 \; \frac{\text{dollars}}{\text{thousand employees}}$$

$$n(t) = 3.143t^2 + 5.029t + 772.5$$

UNITS ▶

$$\frac{dn}{dt} = 6.286t + 5.029 \; \frac{\text{thousand employees}}{\text{year}}$$

UNITS ▶

$$\frac{dS}{dt} = \frac{dS}{dn} \cdot \frac{dn}{dt}$$

$$= \left(-1.537n + 1295 \; \frac{\text{dollars}}{\text{thousand employees}}\right) \cdot \left(6.286t + 5.029 \; \frac{\text{thousand employees}}{\text{year}}\right)$$

$$= (-1.537n + 1295) \cdot (6.286t + 5.029)\frac{\text{dollars}}{\text{thousand employees}} \; \frac{\text{thousand employees}}{\text{year}}$$

$$= [-1.537(3.143t^2 + 5.029t + 772.5) + 1295] \cdot (6.286t + 5.029) \text{ dollars per year}$$

$$\approx (-4.8308t^2 - 7.7296t - 1187.3325 + 1295) \cdot (6.286t + 5.029)$$

$$= (-4.8308t^2 - 7.7296t + 107.6675) \cdot (6.286t + 5.029) \text{ dollars per year}$$

We must find $\dfrac{dS}{dt}\bigg|_{t=3}$, since $t = 3$ in the year 1998.

$$\frac{dS}{dt}\bigg|_{t=3} = [-4.8308(3)^2 - 7.7296(3) + 107.6675] \cdot [6.286(3) + 5.029]$$

$$= (-43.4772 - 23.1888 + 107.6675) \cdot (23.887)$$

$$= (41.0015) \cdot (23.887)$$

$$\approx 979.40 \text{ dollars per year}$$

According to the model, the average annual earnings of a lumber and wood products industry employee was increasing at a rate of roughly $979 per year in 1998.

10.3 Summary

In this section you learned how to compose two functions. You also discovered how to use the Chain Rule to find the derivative of a composition of functions.

10.3 Exercises

In Exercises 1–5, find the function $f(g(x))$. Then simplify the result.

1. $f(x) = 2x + 3, g(x) = x^2 + 4$

2. $f(x) = -2x^2 + 7x, g(x) = x - 10$

3. $f(x) = 7x - 4, g(x) = x^2 - 2x + 1$

4. $f(x) = x^2 - 1, g(x) = x^2 + 1$

5. $f(x) = 3x^2 - 11x, g(x) = x^3$

In Exercises 6–10, the function is of the form $h(x) = f(g(x))$. Determine $f(x)$ and $g(x)$.

6. $h(x) = 5(x^2 - 4)^3 + 1$

7. $h(x) = (x + 1)^2 - 6(x + 1) + 9$

8. $h(x) = e^{2x-4}$ **9.** $h(x) = 2\ln(x^2 + 4)$

10. $h(x) = 4(x^3 - 1)^2 - x^3 + 1$

In Exercises 11–20, use the Chain Rule in finding the derivative of the function.

11. $f(x) = 5(3x + 1)^2$

12. $g(t) = 4(6t - 7)^3 + 2$

13. $h(t) = 3(t^3 - t^2 + 1)^4$

14. $S(n) = (n^3 - 3n^2)^2$

15. $f(n) = 5(n^2 + 2n + 1)^2 + 6$

16. $g(x) = [4(3x + 5)^2 + x]^2$

17. $f(x) = [-5(4x - 1)^2 + 2x]^2$

18. $s(t) = (2t + 1)^2(6t)$

19. $f(t) = (5t^2 + t)(3t + 2)^2$

20. $h(x) = (4x + 1)^2(-x + 2)^2$

In Exercises 21–25, determine the slope of the graph at the indicated domain value.

21. $g(x) = \dfrac{x^2 + 3}{x^2 - x}; x = 2$

22. $r(x) = \dfrac{2x^2 - 8}{2x^2 + 8}; x = 0$

23. $R(x) = \dfrac{3x + 1}{2x + 5}; x = 2$

24. $P(t) = (3t - 1)^{-3}(5t - 2); t = 0.4$

25. $f(x) = [5(4x + 1)^2 + 2]^2; x = -1$

In Exercises 26–30, find the equation of the tangent line at the indicated domain value.

26. $R(n) = [5(-n + 5)^3 - 2n]^2; n = 5$

27. $s(t) = (3t + 1)^2[2(-4t - 1) + 6]^{-2}; t = 0$

28. $g(x) = [5x(x^2 + 1)^3 + (x + 1)^3]^2; x = -1$

29. $p(x) = \{[(x + 1)^2 + 2]^3 + 3\}^4; x = -1$

30. $r(t) = \{[(2t + 1)^2 - 1]^3 + 2\}^4; t = 0$

In Exercises 31–32, use the Chain Rule to answer the questions.

31. **Average Height of a Girl** The average height of a 2- to 13-year-old girl may be modeled by

$$F(m) = 1.071m - 3.554 \text{ inches}$$

where m is the average height of a boy of the same age.

The average height of a 2- to 13-year-old boy may be modeled by

$$m(a) = -0.0507a^2 + 2.997a + 30.44 \text{ inches}$$

where a is the age of the boy in years. (**Source:** Modeled from www.babybag.com data.)

According to the models, how quickly is the height of a girl increasing when she is 10 years old?

32. **Average Weight of a Boy** The average weight of a 2- to 13-year-old boy may be modeled by

$$M(f) = 0.9099f + 5.065 \text{ pounds}$$

where f is the average weight of a girl of the same age.

The average weight of a 2- to 13-year-old girl may be modeled by

$$f(a) = 0.289a^2 + 2.464a + 23.10 \text{ pounds}$$

where a is the age of the girl in years. (**Source:** Modeled from www.babybag.com data.)

According to the models, how quickly is the average weight of a boy increasing when he is 10 years old?

Exercises 33–35 are intended to challenge your understanding of the Chain Rule.

33. Find the derivative of
$$f(x) = [(x^2 + 1)^2 + 1]^2 + 1$$

34. Differentiate $g(t) = (t^2 + 1)^3(t^2 + 3)^4$

35. Given that $\dfrac{dy}{dt} = 5$ people per day,

$\dfrac{dt}{dx} = 2$ days per year, and

$\dfrac{dp}{dy} = 4$ sales per person, calculate

$\dfrac{dp}{dx}$, including units.

10.4 Exponential and Logarithmic Rules

- Use the Exponential Rule and Logarithmic Rule to find the derivative of a function

GETTING STARTED Despite the infamous "Got Milk?" advertising campaign, the per capita milk beverage consumption decreased from nearly 28 gallons per year in 1980 to less than 24 gallons per year in 1999. The annual per capita milk beverage consumption may be modeled by

$$M(t) = 27.76(0.9914)^t \text{ gallons}$$

where t is the number of years since 1980. Does the four-gallon decrease over the 19-year period indicate that consumer consumption of milk will continue to drop? According to the model, at what rate was consumption decreasing in 2000? Questions such as these may be answered using the Exponential Rule for derivatives.

In this section, we will introduce the Exponential Rule and the Logarithmic Rule for derivatives. These rules will allow us to calculate the exact derivative of these types of functions quickly and accurately instead of relying upon a numerical estimate as we have done previously.

Exponential Rule

Recall that an exponential function may be written in the form $y = ab^x$, where a and b are constants and b is positive and not equal to 1. Since the variable of an exponential function is in the exponent, the Power Rule cannot be used to find the derivative of this type of function.

EXPONENTIAL RULE

The derivative of a function $f(x) = b^x$ is given by

$$f'(x) = (\ln b)b^x$$

This rule is not entirely obvious. To better understand its origin, let's consider the limit definition of the derivative of an exponential function.

$$f'(x) = \lim_{h \to 0} \frac{f(x + h) - f(x)}{h}$$

$$= \lim_{h \to 0} \frac{b^{x+h} - b^x}{h} \qquad \text{Since } f(x) = b^x$$

$$= \lim_{h \to 0} \frac{b^x b^h - b^x}{h} \qquad \text{Algebraic rules of exponents}$$

$$= \lim_{h \to 0} \frac{b^x(b^h - 1)}{h}$$

$$= b^x \lim_{h \to 0} \frac{(b^h - 1)}{h} \qquad \text{Since } b^x \text{ is independent of the value of } h$$

Although we cannot simplify $\lim_{h \to 0} \frac{(b^h - 1)}{h}$ algebraically, it turns out that $\lim_{h \to 0} \frac{(b^h - 1)}{h} = \ln(b)$. Consequently,

$$f'(x) = b^x \lim_{h \to 0} \frac{(b^h - 1)}{h}$$

$$= (\ln b)b^x$$

Exponential functions frequently use the number e as the base. Recall that $e \approx 2.7182818$.

EXAMPLE 1

Finding the Derivative of an Exponential Function

Differentiate $y = e^x$.

SOLUTION

$$\frac{dy}{dx} = (\ln e)e^x$$

$$= 1 \cdot e^x \qquad \text{Since } \ln e = 1$$

$$= e^x$$

The derivative of e^x is e^x! This quirky result makes the function $y = e^x$ one of the most popular functions in calculus. For the function $f(x) = e^x$, the instantaneous rate of change in the function at a point $(a, f(a))$ is the same as $f(a)$. In other words, the slope of the tangent line to the graph of $f(x) = e^x$ at $x = a$ is $f(a)$.

EXPONENTIAL RULE: SPECIAL CASE

The derivative of a function $f(x) = e^x$ is given by

$$f'(x) = e^x$$

EXAMPLE 2

Finding the Derivative of an Exponential Function

Find the derivative of $g(x) = 2^x$.

SOLUTION

$$g'(x) = (\ln 2)2^x$$

EXAMPLE 3

Finding the Derivative of an Exponential Function

Calculate $\dfrac{d}{dx}(3 \cdot 4^x)$.

SOLUTION

$$\frac{d}{dx}(3 \cdot 4^x) = 3 \cdot \frac{d}{dx}(4^x) \qquad \text{Constant Multiple Rule}$$

$$= 3(\ln 4)4^x \qquad \text{Exponential Rule}$$

EXAMPLE 4

Finding a Derivative by Using the Exponential and Chain Rules

Find y' given $y = 3^{x^2+x}$.

SOLUTION This function is different because there is a function in the exponent rather than a single variable. Consequently, we'll have to use the Chain Rule in conjunction with the Exponential Rule to find the derivative.

$$y' = (\ln 3)3^{x^2+x} \cdot \frac{d}{dx}(x^2 + x) \qquad \text{Exponential and Chain Rules}$$

$$= (\ln 3)3^{x^2+x} \cdot (2x + 1) \qquad \text{Power Rule}$$

$$= (2x + 1)(\ln 3)3^{x^2+x}$$

EXAMPLE 5

Finding a Derivative by Using the Exponential and Chain Rules

Differentiate $y = e^{5x-6}$.

SOLUTION

$$\frac{dy}{dx} = (\ln e)e^{5x-6} \cdot \frac{d}{dx}(5x - 6) \qquad \text{Exponential and Chain Rules}$$

$$= (1)e^{5x-6} \cdot (5) \qquad \text{Constant, Constant Multiple, and Power Rules}$$

$$= 5e^{5x-6}$$

In general, a function of the form $y = e^u$, where u is some function of x, will have the derivative $y' = e^u \cdot u'$. This is referred to as the Generalized Exponential Rule; it is a special application of the Chain Rule together with the Exponential Rule.

GENERAL EXPONENTIAL RULE

Let $u = g(x)$. The derivative of a function $f(x) = b^u$ is given by

$$f'(x) = (\ln b)b^u u'$$

If $b = e$, the rule simplifies to

$$f'(x) = e^u u'$$

EXAMPLE 6

Finding a Derivative by Using Multiple Derivative Rules

Find the derivative of $P(t) = 2te^{3t}$.

SOLUTION Although this function looks relatively simple, there is a lot going on. To find the derivative, we must use the Constant Multiple Rule, the Power Rule, the Product Rule, and the Generalized Exponential Rule.

$$\frac{dP}{dt} = \frac{d}{dt}(2t) \cdot e^{3t} + \frac{d}{dt}(e^{3t}) \cdot (2t) \qquad \text{Product Rule}$$

$$= 2e^{3t} + (e^{3t} \cdot 3) \cdot (2t) \qquad \text{Constant Multiple, Generalized Exponential, and Power Rules}$$

$$= 2e^{3t} + 6te^{3t}$$

$$= e^{3t}(2 + 6t)$$

$$= 2e^{3t}(3t + 1)$$

EXAMPLE 7

Finding a Derivative by Using the General Exponential Rule

The annual per capita milk beverage consumption may be modeled by

$$M(t) = 27.76(0.9914)^t \text{ gallons}$$

where t is the number of years since 1980. According to the model, at what rate was consumption decreasing in 2000?

SOLUTION

$$\frac{dM}{dt} = 27.76[\ln{(0.9914)}](0.9914)^t$$

$$\approx -0.2398(0.9914)^t \text{ gallons per year} \qquad \text{Evaluate ln (0.9914) and multtiply by 27.76}$$

In 2000, $t = 20$, so

$$\frac{dM}{dt}\bigg|_{t=20} = -0.2398(0.9914)^{20} \text{ gallons per year}$$

$$\approx -0.2398(0.8414)$$

$$\approx -0.2018$$

In 2000, the per capita milk beverage consumption was decreasing by about two-tenths (0.2) of a gallon per year.

It is important to note that although the per capita consumption of milk was decreasing, overall milk beverage sales still could have increased as a result of an increase in the U.S. population.

Logarithmic Rule

What is the derivative of a function of the form $y = \log_b x$? The answer is not obvious. We will state the rule here and show how the formula was determined when we discuss implicit differentiation in a subsequent section.

LOGARITHMIC RULE

The derivative of a function $f(x) = \log_b x$ is given by

$$f'(x) = \frac{1}{\ln b} \cdot \frac{1}{x}$$

EXAMPLE 8 Finding the Derivative of a Logarithmic Function

Differentiate $y = \ln x$.

SOLUTION Recall that $\ln x = \log_e x$.

$$y' = \frac{1}{\ln e} \cdot \frac{1}{x}$$

$$= \frac{1}{1} \cdot \frac{1}{x}$$

$$= \frac{1}{x}$$

The result of Example 8 will be used extensively, so we will classify it as a special case of the Logarithmic Rule.

LOGARITHMIC RULE: SPECIAL CASE

The derivative of a function $f(x) = \ln x$ is given by

$$f'(x) = \frac{1}{x}$$

EXAMPLE **9**

Finding the Derivative of a Logarithmic Function

Find the derivative of $y = 3 \log t$.

SOLUTION Recall that $\log t$ means $\log_{10} t$.

$$\frac{dy}{dt} = \frac{d}{dt}(3 \log t)$$

$$= 3 \cdot \frac{d}{dt}(\log_{10} t) \qquad \text{Constant Multiple Rule}$$

$$= 3 \cdot \frac{1}{\ln 10} \cdot \frac{1}{t} \qquad \text{Logarithmic Rule}$$

$$= \frac{3}{\ln 10} \cdot \frac{1}{t}$$

EXAMPLE **10**

Finding the Derivative by Using the Logarithmic and Chain Rules

Find $\dfrac{dy}{dx}$ for $y = \log(x^2)$.

SOLUTION Since y is the composition of $f(x) = \log x$ and $g(x) = x^2$, we must use the Chain Rule in conjunction with the Logarithmic Rule.

$$\frac{dy}{dx} = \left(\frac{1}{\ln 10} \cdot \frac{1}{x^2} \right) \cdot \frac{d}{dx}(x^2) \qquad \text{Logarithmic and Chain Rules}$$

$$= \left(\frac{1}{\ln 10} \cdot \frac{1}{x^2} \right) \cdot (2x) \qquad \text{Power Rule}$$

$$= \frac{1}{\ln 10} \cdot \frac{2x}{x^2}$$

$$= \frac{2}{(\ln 10)x}$$

EXAMPLE 11

Finding a Derivative by Using Multiple Rules

Differentiate $P(t) = e^{2t} \ln(3t)$.

SOLUTION We must use the Product Rule, the Chain Rule, the Exponential Rule, the Constant Multiple Rule, and the Power Rule in finding the derivative.

$$P'(t) = \frac{d}{dt}(e^{2t})\ln(3t) + \frac{d}{dt}[\ln (3t)]e^{2t} \qquad \text{Product Rule}$$

$$= (e^{2t} \cdot 2)\ln(3t) + \left(\frac{1}{3t} \cdot 3\right)e^{2t} \qquad \begin{array}{l}\text{Generalized Exponential, Logarithmic,} \\ \text{Constant Multiple, and Power Rules}\end{array}$$

$$= 2e^{2t}\ln(3t) + \left(\frac{3}{3t}\right)e^{2t}$$

$$= e^{2t}\left[2\ln(3t) + \frac{1}{t}\right]$$

EXAMPLE 12

Applying the Logarithmic Rule in a Real-World Context

The year in which the per capita consumption of bottled water will reach w gallons may be modeled by

$$T(w) = -9.249 + 9.839(\ln w) \text{ years since 1980}$$

(**Source:** *Statistical Abstract of the United States, 2001,* Table 204, p 130.) Calculate and interpret the meaning of $T'(20)$.

SOLUTION

UNITS

$$T'(w) = 9.839 \cdot \frac{1}{w} \qquad \begin{array}{l}\text{Constant and} \\ \text{Logarithmic Rules}\end{array}$$

$$T'(20) = \frac{9.839}{20} \frac{\text{years since 1980}}{\text{gallon}}$$

$$= 0.4920 \text{ year per gallon}$$

When the annual per capita consumption of bottled water is 20 gallons, it will take about 0.4920 year (roughly 6 months) for the annual per capita consumption to increase to 21 gallons.

EXAMPLE 13

Applying the Logarithmic Rule in a Real-World Context

Based on data from 1989 to 1999, the percentage of auto fatalities that are not alcohol-related may be modeled by

$$F(t) = 49.96 + 5.112 \ln t \text{ percent}$$

where t is the number of years since 1989. (**Source:** *Statistical Abstract of the United States, 2001,* Table 1099, p 688.) According to the model, how quickly was the percentage of non-alcohol-related fatalities increasing in 2000?

SOLUTION

UNITS

$$\frac{dF}{dt} = \frac{d}{dt}[49.96 + 5.112(\ln t)]$$

$$= 0 + 5.112 \cdot \frac{1}{t} \qquad \text{Constant, Constant Multiple, and Logarithmic Rules}$$

$$= \frac{5.112}{t} \frac{\text{percentage points}}{\text{year}}$$

In 2000, $t = 11$, so

$$\left.\frac{dF}{dt}\right|_{t=11} = \frac{5.112}{11} \frac{\text{percentage points}}{\text{year}}$$

$$\approx 0.4647 \text{ percentage point per year}$$

In 2000, the percentage of *non-alcohol-related* auto fatalities was increasing by 0.46 percentage point per year. This means that *alcohol-related* auto fatalities were declining by about 0.46 percentage point per year in 2000, a good sign.

10.4 Summary

In this section, you learned how to use the Exponential Rule and the Logarithmic Rule for derivatives. You discovered that these rules allow you to calculate the exact derivative of exponential and logarithmic functions instead of having to rely on a numerical estimate.

10.4 Exercises

In Exercises 1–10, use the Exponential Rule or the Logarithmic Rule, as appropriate, to find the derivative of the function.

1. $y = 4^x$

2. $g(t) = 3(5)^t$

3. $s(t) = 4 \ln t$

4. $f(t) = 3 \log t$

5. $y = 5e^x + \ln x$

6. $P(n) = 4n^2(2^n)$

7. $f(x) = 5e^x \ln x$

8. $g(x) = e^{2x-1}$

9. $y = 2^{x^2+5x}$

10. $h(x) = \ln(x^3 + 3x)$

In Exercises 11–20, determine the slope of the function at the indicated domain value.

11. $S(n) = \log(n^2 - 4); n = 3$

12. $P(x) = xe^x; x = 0$

13. $w(t) = \log(2^t); t = -1$

14. $f(x) = 4 \log(3^x); x = 0$

15. $H(t) = 3^{\ln t}; t = e$

16. $y = 2^{\log 5t}; t = 0.2$

17. $f(t) = 3 \log(5^t); t = 0$

18. $y = \log(5x) + 4^{2x+5}; x = 1$

19. $C(n) = 3ne^{n^2+2n-1}; n = 1$

20. $g(t) = 5t(e^{t-2}); t = 3$

In Exercises 21–25, determine the equation of the tangent line at the indicated domain value.

21. $f(t) = 2^{3 \ln t}; t = 1$

22. $y = 5x^2(2^{3x}); x = 0$

23. $g(x) = [3 \log(4x)](3^x); x = 1$

24. $h(x) = \ln(3e^{5x}); x = -1$

25. $f(x) = \frac{\ln x}{x}; x = e$

In Exercises 26–30, answer the questions by using the Exponential Rule or the Logarithmic Rule, as appropriate.

26. **M** **Teacher Salaries** Based on data from 1990 to 2000, the average salary of a public elementary or secondary school teacher may be modeled by

$$S(t) = 32.05(1.027)^t \text{ thousand dollars}$$

where t is the number of years since 1990.
(**Source:** Modeled from *Statistical Abstract of the United States, 2001*, Table 237, p. 151.)

At what rate (in thousands of dollars per year) were salaries increasing in 2000?

27. **M** **College Tuition** Based on data from 1994 to 2001, the quarterly tuition of a resident student at Green River Community College may be modeled by

$$E(t) = 430.6(1.042)^t \text{ dollars}$$

where t is the number of years since 1994.
(**Source:** Modeled from Green River Community College data.)

How quickly was tuition increasing in 2001?

28. **M** **NBA Player Salaries** Based on data from 1980 to 1998, the average salary of a National Basketball Association (NBA) player may be modeled by

$$P(t) = 161(1.17)^t \text{ thousand dollars}$$

where t is the number of years since 1980.
(**Source:** Modeled from *Statistical Abstract of the United States, 2001*, Table 1324, p. 829.)

Calculate and interpret the meaning of $P'(18)$.

29. **M** **VCR Usage** Based on data from 1985 to 1999, the percentage of homes with a VCR may be modeled by

$$V(t) = 22.24 + 24.47 \ln t \text{ percent}$$

where t is the number of years since 1984.
(**Source:** Modeled from *Statistical Abstract of the United States, 2001*, Table 1126, p. 705.)

Calculate and interpret the meaning of $V'(15)$.

30. **M** **Army Personnel** Based on data from 1980 to 2000, the number of active-duty Army personnel may be modeled by

$$A = \frac{302.2}{1 + 0.0001047e^{0.7731t}} + 480 \text{ thousand people}$$

where t is the number of years since 1980.
(**Source:** Modeled from *Statistical Abstract of the United States, 2001*, Table 500, p. 329.)

At what rate was the number of active-duty army personnel changing in 2000?

Exercises 31–35 are intended to challenge your understanding of the Exponential and Logarithmic Rules.

31. Differentiate $f(x) = 2^{\ln(x^2)}$.

32. Find $\dfrac{dy}{dx}$ given $y = 2^x \ln(2x)$.

33. Find $g'(t)$ given $g(t) = 2^{2t} \ln(2^{2t})$.

34. Differentiate $h(t) = 3^{2^{4t}}$.

35. Find $\dfrac{dx}{dt}$ given $x(t) = \ln[\ln(3^{2^t})]$

10.5 Implicit Differentiation

- Use implicit differentiation to differentiate functions and nonfunctions

GETTING STARTED The intent of this section is to give you the skills necessary to work the related-rate problems in the next chapter. We will show you how to use *implicit differentiation* to find the derivative of functions and nonfunctions. Knowing how to do implicit differentiation is a key skill that is necessary for our discussion of related rates.

The equation of a circle may be written as $x^2 + y^2 = r^2$, where r is the radius of the circle. A circle is not a function because it fails the Vertical Line Test. However, we can draw tangent lines to the graph of a circle. Since the derivative is the slope of the tangent line, we should be able to find the derivative at any point on the circle. However, since most x values have two different y values, the derivative will be a function of both x and y.

Consider $x^2 + y^2 = 1$ and its associated graph Figure 10.3. What is the slope of each of the tangent lines shown? In other words, what is the derivative at the indicated points?

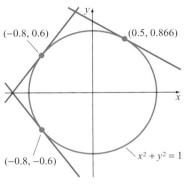

FIGURE 10.3

We will find the derivative by using **implicit differentiation.** We begin by taking the derivative of both sides of the equation with respect to x. Since y is a function of x, $\dfrac{d}{dx}(y) = \dfrac{dy}{dx}$. This fact will become important as we move through the problem.

$$x^2 + y^2 = 1$$

$$\frac{d}{dx}(x^2 + y^2) = \frac{d}{dx}(1)$$

$$\frac{d}{dx}(x^2) + \frac{d}{dx}(y^2) = 0 \qquad \text{Sum and Difference and Constant Rules}$$

$$2x + 2y \cdot \frac{d}{dx}(y) = 0 \qquad \text{Power and Chain Rules}$$

$$2x + 2y \cdot \frac{dy}{dx} = 0$$

$$2y\frac{dy}{dx} = -2x$$

$$\frac{dy}{dx} = \frac{-2x}{2y}$$

$$\frac{dy}{dx} = -\frac{x}{y}$$

Note that the derivative is a function of both x and y. We can now calculate the slope of the tangent line at each of the points shown.

$$\text{At } (-0.8, 0.6), \frac{dy}{dx} = -\frac{-0.8}{0.6} \approx 1.333.$$

$$\text{At } (-0.8, -0.6), \frac{dy}{dx} = -\frac{-0.8}{-0.6} \approx -1.333.$$

$$\text{At } (0.5, 0.866), \frac{dy}{dx} = -\frac{0.5}{0.866} \approx -0.5774.$$

HOW TO **Steps of Implicit Differentiation**

1. Differentiate both sides of the equation with respect to x.
$$\left(\text{Recall } \frac{d}{dx}(y) = \frac{dy}{dx}. \right)$$

2. Algebraically isolate the $\frac{dy}{dx}$ term.

EXAMPLE 1   **Finding $\dfrac{dy}{dx}$ by Using Implicit Differentiation**

Differentiate $xy = 1$ using implicit differentiation.

SOLUTION

$$\frac{d}{dx}(xy) = \frac{d}{dx}(1)$$

$$\frac{d}{dx}(x) \cdot y + \frac{d}{dx}(y) \cdot x = 0 \qquad \text{Product Rule}$$

$$1 \cdot y + \frac{dy}{dx} \cdot x = 0 \qquad \text{Power Rule}$$

$$y + x\frac{dy}{dx} = 0$$

$$x\frac{dy}{dx} = -y$$

$$\frac{dy}{dx} = -\frac{y}{x}$$

If we first solve the equation for y and then differentiate, we get a very different-looking result.

$$xy = 1$$

$$y = \frac{1}{x}$$

$$\frac{d}{dx}(y) = \frac{d}{dx}\left(\frac{1}{x}\right)$$

$$\frac{dy}{dx} = \frac{d}{dx}(x^{-1})$$

$$= -x^{-2} \qquad \text{Power Rule}$$

$$= -\frac{1}{x^2}$$

Is this answer equivalent to our implicit differentiation result? Let's see.

$$\frac{dy}{dx} = -\frac{y}{x}$$

$$= -\frac{1}{x}\,y$$

$$= -\frac{1}{x} \cdot \frac{1}{x} \qquad \text{Since } y = \tfrac{1}{x}$$

$$= -\frac{1}{x^2}$$

The two equations are equivalent. When using implicit differentiation, it is common for correct answers to look very different from each other.

EXAMPLE 2   **Finding $\dfrac{dy}{dx}$ by Using Implicit Differentiation**

Differentiate $(x + y)^2 = 9$.

SOLUTION

$$\frac{d}{dx}(x + y)^2 = \frac{d}{dx}(9)$$

$$2(x + y) \cdot \frac{d}{dx}(x + y) = 0 \qquad \text{Constant and Chain Rules}$$

$$2(x + y) \cdot \left(1 + \frac{dy}{dx}\right) = 0 \qquad \text{Power Rule}$$

$$\left(1 + \frac{dy}{dx}\right) = \frac{0}{2(x + y)}$$

$$1 + \frac{dy}{dx} = 0$$

$$\frac{dy}{dx} = -1$$

For all values of x and y on the graph of $(x + y)^2 = 9$, the slope of the tangent line is -1.

Let's solve the original equation for y.

$$(x + y)^2 = 9$$

$$\sqrt{(x + y)^2} = \sqrt{9}$$

$$(x + y) = \pm 3$$

$$y = -x \pm 3$$

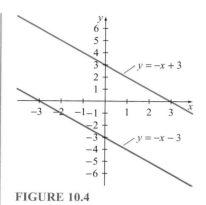

FIGURE 10.4

When we graph $y = -x - 3$ and $y = -x + 3$ simultaneously, we see a pair of parallel lines with slope -1 (Figure 10.4). This is consistent with the implicit differentiation result.

EXAMPLE 3   Finding $\dfrac{dy}{dx}$ by Using Implicit Differentiation

Differentiate $x = y^2$.

SOLUTION

$$\frac{d}{dx}(x) = \frac{d}{dx}(y^2)$$

$$1 = 2y \cdot \frac{d}{dx}(y) \qquad \text{Power and Chain Rules}$$

$$1 = 2y \cdot \frac{dy}{dx}$$

$$\frac{dy}{dx} = \frac{1}{2y}$$

EXAMPLE 4   Finding $\dfrac{dy}{dx}$ by Using Implicit Differentiation

Find the derivative of $y^2 + 2y + 1 = x$ with respect to x.

SOLUTION

$$\frac{d}{dx}(y^2 + 2y + 1) = \frac{d}{dx}(x)$$

$$2y \cdot \frac{d}{dx}(y) + 2 \cdot \frac{d}{dx}(y) + 0 = 1 \qquad \text{Power, Chain, and Constant Rules}$$

$$2y \cdot \frac{dy}{dx} + 2 \cdot \frac{dy}{dx} = 1$$

$$\frac{dy}{dx}(2y + 2) = 1$$

$$\frac{dy}{dx} = \frac{1}{(2y + 2)}$$

EXAMPLE 5

Proving $\dfrac{d}{dx}(\log_b(x)) = \dfrac{1}{\ln(b) \cdot x}$

Prove that the derivative of $y = \log_b(x)$ is $y' = \dfrac{1}{\ln(b) \cdot x}$.

SOLUTION We will do some clever manipulations with the function and then use the Exponential Rule to find the derivative.

Recall that according to the definition of logarithms, the following two equations are equivalent:

$$y = \log_b(x) \qquad \text{and}$$
$$b^y = x$$

Differentiating the second equation with respect to x yields

$$\frac{d}{dx}(b^y) = \frac{d}{dx}(x)$$

$$(\ln b)b^y \cdot \frac{d}{dx}(y) = 1 \qquad\qquad \text{Chain, Exponential, and Power Rules}$$

$$(\ln b)b^y \cdot \frac{dy}{dx} = 1 \qquad\qquad \text{Chain Rule}$$

$$\frac{dy}{dx} = \frac{1}{(\ln b)b^y}$$

$$= \frac{1}{(\ln b)x} \qquad\qquad \text{Since } b^y = x$$

EXAMPLE 6

Finding $\dfrac{dy}{dx}$ by Using Implicit Differentiation

Find $\dfrac{dy}{dx}$ given $3xy^2 = 4xy$.

SOLUTION

$$\frac{d}{dx}(3xy^2) = \frac{d}{dx}(4xy)$$

$$\frac{d}{dx}(3x) \cdot y^2 + \frac{d}{dx}(y^2) \cdot 3x = \frac{d}{dx}(4x) \cdot y + \frac{d}{dx}(y) \cdot 4x \qquad \text{Product Rule}$$

$$(3)y^2 + \left(2y\frac{dy}{dx}\right) \cdot 3x = (4)y + \frac{dy}{dx} \cdot 4x \qquad \begin{array}{l}\text{Constant Multiple, Power, and Chain}\\\text{Rules}\end{array}$$

$$3y^2 + 6xy\frac{dy}{dx} = 4y + 4x\frac{dy}{dx}$$

$$6xy\frac{dy}{dx} - 4x\frac{dy}{dx} = 4y - 3y^2$$

$$\frac{dy}{dx}(6xy - 4x) = 4y - 3y^2$$

$$\frac{dy}{dx} = \frac{4y - 3y^2}{6xy - 4x}$$

10.5 Summary

In this section, you learned how to use implicit differentiation to find the derivative of functions and nonfunctions. Knowing how to do implicit differentiation will help you understand related rates in the next chapter.

10.5 Exercises

In Exercises 1–10, use implicit differentiation to find $\dfrac{dy}{dx}$. *Then evaluate the derivative function at the designated point.*

1. $2x^2 + y^2 = 32;\ (4, 0)$ **2.** $x^2 - y^2 = 9;\ (5, 4)$

3. $x^2y - 3 = y^2;\ (2, 1)$ **4.** $4y^2 = x^2;\ (4, 2)$

5. $xy = 6;\ (2, 3)$

6. $x^2y - xy^2 = 20;\ (5, 4)$

7. $9x - y^3 = 0;\ (3, 3)$

8. $xy^2 + xy + y = -1;\ (3, -1)$

9. $20y - x^2y^2 = 75;\ (1, 5)$

10. $xy^3 - x^3y = 30;\ (-2, -3)$

In Exercises 11–20, use implicit differentiation to find $\dfrac{dy}{dx}$.

11. $y^2 = x$ **12.** $xe^y = y$

13. $ye^y = 2$ **14.** $2^y = \ln x$

15. $\ln y = xy$ **16.** $x^2y - y^2 + y = x$

17. $2xy - y^2 = 9$ **18.** $x(x^2 - y^2) = y$

19. $\ln(xy^2) = y$ **20.** $ye^x = xy^2$

In Exercises 21–25, determine at what points, if any, $\dfrac{dy}{dx} = 0.$

21. $x^2 + y^2 = 1$ **22.** $3x^2 = y \ln y$

23. $x^2 + 2y^2 = y$ **24.** $x^3 - y^2 = xy$

25. $x^2 + xy = y^2 - 5$

Exercises 26–30 are intended to challenge your understanding of implicit differentiation.

26. Given $x^2 = y^2$, calculate $\dfrac{dy}{dx}$ and $\dfrac{dx}{dy}$. What is the relationship between $\dfrac{dy}{dx}$ and $\dfrac{dx}{dy}$?

27. Given $V = \pi r^2h$, write $\dfrac{dr}{dt}$ in terms of $\dfrac{dV}{dt}, \dfrac{dh}{dt}, h,$ and r. (*Hint:* Differentiate both sides with respect to t.)

28. Find $\dfrac{dy}{dx}$ given $ye^x = xe^y$.

29. Given the equation $ye^x = xe^y$, find the solution to the equation $\dfrac{dy}{dx} = 0$, if it exists.

30. Find $\dfrac{dy}{dx}$, if possible, given $e^{xy} = \ln(xy)$.

Chapter 10 Review Exercises

Section 10.1   *In Exercises 1–4, use the Constant Rule, the Power Rule, the Constant Multiple Rule, and the Sum and Difference Rule (as appropriate) to find the derivative of the function.*

1. $f(x) = 9$

2. $g(x) = -12x + 14$

3. $v(t) = 7t^3 - 4t^{-3}$

4. $s(t) = -2t^4 + t^2 - t + 17$

In Exercises 5–6, answer the question by evaluating the derivative.

5.   **Military Personnel**   Based on data from 1990 to 1999, the number of active-duty military personnel may be modeled by

$$A = 2.5626t^2 - 58.90t + 1321$$

thousand people, where t is the number of years since 1990. (**Source:** Modeled from *Statistical Abstract of the United States, 2001*, Table 499, p. 328.)

According to the model, at what rate was the number of active-duty military personnel changing in 1999 and in 2000?

6.   **Prison Rate**   Based on data from 1980 to 1998, the federal and state prison rate may be modeled by

$$R = 0.3750t^2 + 11.70t + 138.7$$

prisoners per 100,000 people, where t is the number of years since 1980. (**Source:** Modeled from *Statistical Abstract of the United States, 2001*, Table 332, p. 200.)

Was the prison rate increasing more rapidly in 1988 or in 1998?

Section 10.2   *In Exercises 7–10, use the Product Rule to find the derivative of the function. Simplify the result.*

7. $f(x) = 3x^2(2x - 4)$

8. $g(x) = -15x(5x^2 - 2)$

9. $f(t) = (2t^2 - 4)(-t + 4)$

10. $s(t) = t^3(3t^2 + 7t)(t - 3)$

In Exercises 11–12, use the Product Rule in answering the questions from the real-life scenarios. You may find it helpful to use your calculator.

11.   **Prison Rate**   Based on data from 1980 to 1998, the federal and state prison rate may be modeled by

$$R = 0.3750t^2 + 11.70t + 138.7$$

prisoners per 100,000 people, where t is the number of years since 1980. (**Source:** Modeled from *Statistical Abstract of the United States, 2001*, Table 332, p. 200.)

The number of people in the United States may be modeled by

$$P = 0.525t^2 + 1.695t + 226.5$$

hundred thousand people, where t is the number of years since 1980. (**Source:** Modeled from www.census.gov.)

In 2000, at what rate was the number of prisoners increasing (in prisoners per year)?

12.   **Milk Consumption**   Based on data from 1980 to 1999, the per capita milk beverage consumption in the United States may be modeled by

$$M = -0.219t + 27.7$$

gallons per person, where t is the number of years since 1980. (**Source:** Modeled from *Statistical Abstract of the United States, 2001*, Table 202, p. 129.)

The number of people in the United States may be modeled by

$$P = 0.525t^2 + 1.695t + 226.5$$

hundred thousand people, where t is the number of years since 1980. (**Source:** Modeled from www.census.gov.)

In 1999, at what rate (in thousands of gallons per year) was milk beverage consumption changing?

Section 10.3   *In Exercises 13–14, find the function $f(g(x))$. Then simplify the result.*

13. $f(x) = -3x + 4, g(x) = 2x^2 - x$

14. $f(x) = -4x^2 + x$, $g(x) = 2x + 3$

In Exercises 15–16, use the Chain Rule to find the derivative of the function. Then write the result in factored form.

15. $f(x) = 2(9x + 5)^3$ **16.** $g(x) = 3(x^2 - 2x)^2$

In Exercises 17–18, use a composition of functions and the Chain Rule to answer the questions.

17. **M** **Advertising Expenditures**   Based on data from 1990 to 2000, the amount of money spent on advertising on cable television may be modeled by

$$C(m) = 0.00009027m^2 - 0.0261m - 1130$$

million dollars, where m is the amount of money (in millions of dollars) spent on advertising in magazines.

The amount of money spent on advertising in magazines may be modeled by

$$m(t) = 42.29t^2 + 158.3t + 6575$$

million dollars, where t is the number of years since 1990. (**Source:** Modeled from *Statistical Abstract of the United States, 2001,* Table 1272, p. 777.)

In 2000, at what rate was cable television advertising spending changing?

18. **M** **Advertising Expenditures**   Based on data from 1990 to 2000, the amount of money spent on advertising on radio may be modeled by

$$R(b) = 12.92b - 4984$$

million dollars, where b is the amount of money (in millions of dollars) spent on advertising on billboards.

The amount of money spent on advertising on billboards may be modeled by

$$b(t) = 9.396t^2 - 11.58t + 1063$$

million dollars, where t is the number of years since 1990. (**Source:** Modeled from *Statistical Abstract of the United States, 2001,* Table 1272, p. 777.)

In 2000, at what rate was radio advertising spending changing?

Section 10.4   *In Exercises 19–22, use the Exponential Rule or the Logarithmic Rule, as appropriate, to find the derivative of the function.*

19. $y = 9^x$ **20.** $g(t) = 20(0.5)^t$

21. $f(x) = \ln(x^2 + 3)$ **22.** $s(t) = \ln(2^{3t-4})$

In Exercises 23–24, answer the questions by using the Exponential Rule or the Logarithmic Rule, as appropriate.

23. **M** **Hawaii Population**   The projected population of Hawaii may be modeled by

$$P = 1175(1.014)^t$$

thousand people, where t is the number of years since 1995. (**Source:** Modeled from www.census.gov.)

In 2010, at what rate is the population of Hawaii expected to be changing?

24. **Church Growth**   Based on data from 1990 and 2001, the number of Lutherans in the United States may be modeled by

$$L = 9110(1.005)^t$$

thousand people, where t is the number of years since 1990. (**Source:** Modeled from American Religious Identification Survey data.)

In 2001, at what rate was the number of Lutherans in the United States increasing?

Section 10.5   *In Exercises 25–29, use implicit differentiation to find $\dfrac{dy}{dx}$. Then evaluate the derivative function at the designated point.*

25. $xe^{3y} = 7$; $(7, 0)$

26. $\log(xy) = 2$; $(4, 25)$

27. $x^2y^2 - 1 = 0$; $(0.5, 2)$

28. $2xy^2 - y = 10$; $(1, -2)$

29. $4xy - 2^{xy} = 4$; $(-3, -1)$

Make It Real

What to do

1. Collect a minimum of four data points for a rate of change (for example, a wage in dollars per hour over time).

2. Collect a minimum of four data points with the same domain (input values) as the data set in (1). The product of the range values (outputs) of each data set should make sense in a real-world context. For example, $\frac{\text{dollars}}{\text{hours}} \cdot \text{hour} = \text{dollars}$.

3. Model each of the data sets in (1) and (2) with a polynomial, exponential, or logarithmic function.

4. Multiply the two models together to come up with a third model.

5. Find the derivative of the third model.

6. Evaluate and interpret the derivative at two different points.

Example

Fresh Citrus Consumption

Years (Since 1999)	Citrus Consumption (pounds per person)
0	20.3
1	23.4
2	23.8
3	22.4

Source: Statistical Abstract of the United States, 2004–05, Table 200, page 132.

$C(t) = -1.125t^2 + 4.045t + 20.35$ pounds per person t years since the end of 1999. (Model coefficients are rounded to 4 significant digits.)

U.S. Population

Years (Since 1999)	People (in thousands)
0	279,295
1	282,388
2	285,321
3	288,205

Source: Statistical Abstract of the United States, 2004–05, Table 2, page 7.

$P(t) = 2967t + 279,400$ thousand people t years since the end of 1999. (Model coefficients are rounded to 4 significant digits.)

(Continued)

The total consumption of fresh citrus fruits in the United States may be modeled by

$$T(t) = C(t)P(t)$$

$$= \left(-1.125t^2 + 4.045t + 20.35 \frac{\text{pounds}}{\text{person}}\right)(2967t + 279{,}400 \text{ thousand people})$$

$$= (-1.125t^2 + 4.045t + 20.35)(2967t + 279{,}400) \text{ thousand pounds}$$

The rate at which the total consumption of citrus fruits is changing is given by

$$T'(t) = (-2.250t + 4.045)(2967t + 279{,}400) + (2967)(-1.125t^2 + 4.045t + 20.35) \frac{\text{thousand pounds}}{\text{year}}$$

We will calculate the rate of change in 1999 ($t = 0$) and 2002 ($t = 3$).

$$T'(0) = [-2.250(0) + 4.045][2967(0) + 279{,}400] + (2967)[-1.125(0)^2 + 4.045(0) + 20.35]$$
$$\approx 11{,}910{,}000 \text{ thousand pounds per year (accurate to 4 significant digits)}$$

According to the model, the total consumption of citrus fruits was increasing at a rate of 11.910 billion pounds per year in 1999.

$$T'(3) = [-2.250(3) + 4.045][2967(3) + 279{,}400 + (2967)[-1.125(3)^2 + 4.045(3) + 20.35]$$
$$\approx -713{,}500 \text{ thousand pounds per year (accurate to 4 significant digits)}$$

According to the model, the total consumption of citrus fruits was decreasing at a rate of 0.713 billion pounds per year in 2000.

Chapter
11

Derivative Applications

Businesses survive by being profitable. A savvy business owner effectively analyzes the factors that contribute to the financial success or failure of her business. Using mathematical models, she may forecast what prices, production levels, shipment schedules, and other such elements will result in maximum profits. Although no mathematical model is perfect at forecasting future results, a model can help a business owner make informed decisions.

11.1 Maxima and Minima

- Find relative and absolute extrema
- Use the First Derivative Test to find relative extrema

GETTING STARTED Based on data from 1984 to 2001, the rate at which tuition and fees at community colleges in Washington state are increasing may be modeled by

$$R(t) = -0.7473t^2 + 14.89t + 9.024 \text{ dollars per year}$$

where t is the number of years since the 1984–1985 school year. In what year were tuition and fees at Washington state community colleges increasing the fastest? (**Source:** Modeled from Washington State Higher Education Coordinating Board, Higher Education Statistics, September 2001.) Questions such as these may be answered using the concepts of relative and absolute extrema.

In this section, we will informally discuss continuity. In addition, we will discuss how to find *relative extrema* and *absolute extrema*.

Continuity

The notion of continuity is best understood graphically. Loosely speaking, a function is said to be **continuous** if its graph can be drawn by a pencil without lifting the pencil from the page. If there is a break in the graph, the graph is said to be **discontinuous.** This loose definition of continuity will be sufficient for our purposes. (The formal definition of continuity, which relies heavily on the concept of limits, may be obtained from a traditional calculus text.)

Many functions are continuous. For example, linear, polynomial, exponential, and logarithmic functions are all continuous. Frequently, functions that are discontinuous have domain restrictions or are defined piecewise. Consider the piecewise function $f(x) = \begin{cases} 2x + 6 & \text{if } x < 1 \\ 3x + 2 & \text{if } x \geq 1 \end{cases}$ shown in Figure 11.1. Since there is a break in the graph at $x = 1$, the graph is said to be discontinuous.

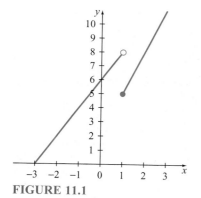

FIGURE 11.1

Relative and Absolute Extrema

An **extremum** is a maximum or minimum value of a function. The plural of extremum is **extrema.**

RELATIVE EXTREMA

- A **relative maximum** of a function f occurs at a point $(c, f(c))$ if $f(x) \leq f(c)$ for all x in some interval (a, b) containing c. (That is, $a < c < b$.)

- A **relative minimum** of a function f occurs at a point $(c, f(c))$ if $f(x) \geq f(c)$ for all x in some interval (a, b) containing c. (That is, $a < c < b$.)

According to the definition, a relative extremum may not occur at an endpoint. We will explore the concept of relative extrema by looking at the graph of

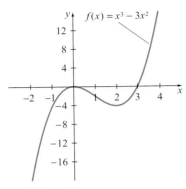

FIGURE 11.2

$f(x) = x^3 - 3x^2$ on the domain $[-2, 4]$ (Figure 11.2).

Graphically, we are looking for the peaks and valleys of the graph. In this case, a peak occurs at $(0, 0)$ and a valley occurs at $(2, -4)$. A relative maximum occurs at the point $(0, 0)$, since $y = 0$ is larger than all y values of the function *nearby* $x = 0$. (When we use the term *nearby*, we mean an arbitrarily small open interval (a, b) surrounding $x = 0$.) A relative minimum occurs at the point $(2, -4)$, since $y = -4$ is smaller than all y values of the function nearby $x = 2$.

ABSOLUTE EXTREMA

- An **absolute maximum** of a function f occurs at a point $(c, f(c))$ if $f(x) \leq f(c)$ for all x in the domain of f.
- An **absolute minimum** of a function f occurs at a point $(c, f(c))$ if $f(x) \geq f(c)$ for all x in the domain of f.

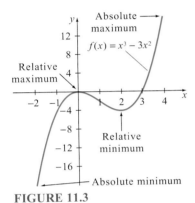

FIGURE 11.3

It is important to note that unlike relative extrema, *absolute extrema may occur at endpoints.*

To find the absolute maximum of the function $f(x) = x^3 - 3x^2$ on the domain $[-2, 4]$, we look for the largest y value. Since f is continuous (there are no breaks in the graph of f), we need only check the relative maxima and the y values of the endpoints. The graph of f has a relative maximum at $(0, 0)$ and endpoints at $(-2, -20)$ and $(4, 16)$. Since $(4, 16)$ has the largest y value, $y = 16$ is the absolute maximum.

To find the absolute minimum, we are looking for the smallest y value. Since f is continuous, we need only check the relative minima and the endpoints. The graph of f has a relative minimum at $(2, -4)$ and endpoints at $(-2, -20)$ and $(4, 16)$. Since $(-2, -20)$ has the smallest (most negative) y value, $y = -20$ is the absolute minimum (Figure 11.3).

Precisely determining relative extrema is difficult to do graphically, especially when extrema occur at irrational points such as $\left(\pi, \sqrt{2}\right)$. Fortunately, by using the derivative, we can quickly find relative extrema.

The derivative of $f(x) = x^3 - 3x^2$ is $f'(x) = 3x^2 - 6x$. Recall that the derivative at a point represents the slope of the tangent line (or slope of the graph) at that point. If the slope is positive, the graph is increasing. If the slope is negative, the graph is decreasing. If the slope is zero, the graph is neither increasing nor decreasing but remains flat. In other words, the graph has a horizontal tangent line if and only if $f'(x) = 0$. For $f(x) = x^3 - 3x^2$, we have the results shown in Table 11.1.

TABLE 11.1

x	$f(x)$	$f'(x)$	Slope of Tangent Line	Graph of f
-2	-20	24	Positive	Increasing
-1	-4	9	Positive	Increasing
0	0	0	Zero	Flat
1	-2	-3	Negative	Decreasing
2	-4	0	Zero	Flat
3	0	9	Positive	Increasing
4	16	24	Positive	Increasing

Observe that $f'(x) = 0$ where the relative extrema occurred $[(0,0)$ and $(2, -4)]$.

THE VALUE OF THE DERIVATIVE AT A RELATIVE EXTREMUM

Let f be a function with a continuous graph. If a relative extremum of f occurs at $(c, f(c))$, then $f'(c) = 0$ or $f'(c)$ is undefined.

The converse is not true. That is, if $f'(c) = 0$ or $f'(c)$ is undefined, then a relative extremum is not guaranteed to occur at $(c, f(c))$. We will demonstrate this in Example 2.

EXAMPLE   **1**

Determining from a Graph Where $f'(x) = 0$

The graph of $f(x) = x^3 - 12x$ on the interval $[-3, 3]$ is shown (Figure 11.4). Determine the points on the graph where $f'(x) = 0$.

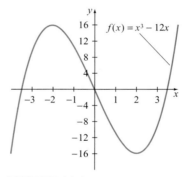

FIGURE 11.4

SOLUTION   Since the graph is a smooth curve (no sharp points), the derivative will equal zero at the relative extrema. It appears that a relative maximum occurs when $x = -2$ and a relative minimum occurs when $x = 2$. We will now use the derivative to confirm our observation.

$$f(x) = x^3 - 12x$$
$$f'(x) = 3x^2 - 12$$
$$f'(2) = 3(2)^2 - 12$$
$$= 0$$
$$f'(-2) = 3(-2)^2 - 12$$
$$= 0$$

The results confirm our graphical estimate. At the point $(-2, 16)$ and the point $(2, -16)$, $f'(x) = 0$.

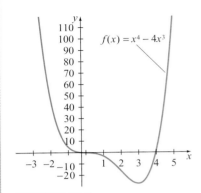

$f(x) = x^4 - 4x^3$

FIGURE 11.5

EXAMPLE 2

Determining from a Graph Where $f'(x) = 0$

The graph of $f(x) = x^4 - 4x^3$ on the interval $[-3, 5]$ is shown in Figure 11.5. Determine the points on the graph where $f'(x) = 0$.

SOLUTION The graph is a continuous, smooth curve. We are looking for places on the graph where the function has a horizontal tangent line $[f'(x) = 0]$. It looks like the graph has a horizontal tangent line at $x = 0$ and $x = 3$. We will confirm this result by evaluating $f'(x)$ at $x = 0$ and $x = 3$.

$$f(x) = x^4 - 4x^3$$
$$f'(x) = 4x^3 - 12x^2$$
$$= 4x^2(x - 3)$$
$$f'(0) = 4(0)^2(0 - 3)$$
$$= 0$$
$$f'(3) = 4(3)^2(3 - 3)$$
$$= 0$$

So $f'(x) = 0$ at $(0, 0)$ and $(3, -27)$.

As discussed previously, the fact that $f'(c) = 0$ does not guarantee that a relative extremum occurs at $(c, f(c))$. In this example, a relative minimum occurred at $(3, -27)$, but a relative extremum did **not** occur at $(0, 0)$.

In Example 3, we will use the absolute value function. Recall that the absolute value function is formally defined as a piecewise function.

$$|x| = \begin{cases} x & \text{if } x \geq 0 \\ -x & \text{if } x < 0 \end{cases}$$

The graph of the absolute value function looks like a "V" (Figure 11.6).

Since the graph is decreasing for $x < 0$, the slope of the tangent line of f is negative for $x < 0$. Since the graph is increasing for $x > 0$, the slope of the tangent line of f is positive for $x > 0$. But what is the slope of the tangent line at $x = 0$? Recall that the derivative of a function f at $x = a$ is formally defined as

$$f'(a) = \lim_{h \to 0} \frac{f(a + h) - f(a)}{h}$$

For $f(x) = |x|$, the derivative of the function at $x = 0$ is

$$f'(0) = \lim_{h \to 0} \frac{|0 + h| - |0|}{h}$$
$$= \lim_{h \to 0} \frac{|h|}{h}$$

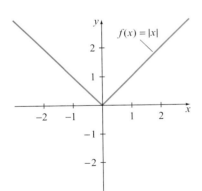

$f(x) = |x|$

FIGURE 11.6

As h gets close to 0, what value does $\frac{|h|}{h}$ approach? We first observe that $h \neq 0$ because division by 0 is not defined. From the piecewise definition of the absolute value function, we know

$$\frac{|h|}{h} = \begin{cases} \dfrac{h}{h} & \text{if } h > 0 \\ \text{undefined} & \text{if } h = 0 \\ \dfrac{-h}{h} & \text{if } h < 0 \end{cases}$$

Simplifying the equation yields

$$\frac{|h|}{h} = \begin{cases} 1 & \text{if } h > 0 \\ \text{undefined} & \text{if } h = 0 \\ -1 & \text{if } h < 0 \end{cases}$$

For positive values of h, $\frac{|h|}{h} = 1$. For negative values of h, $\frac{|h|}{h} = -1$. Since $\frac{|h|}{h}$ doesn't approach a constant value as h nears 0, $\lim_{h \to 0} \frac{|h|}{h}$ does not exist. That is, for $f(x) = |x|$, $f'(0)$ is undefined. In other words, the derivative of the absolute value function does not exist at $x = 0$. In general, the derivative of any function does not exist wherever the graph of a function has a "sharp point."

EXAMPLE 3 Finding the Relative Extrema of a Function

The graph of $f(x) = -|x| + 2$ is shown in Figure 11.7. Find the relative extrema of f on the interval $[-3, 3]$.

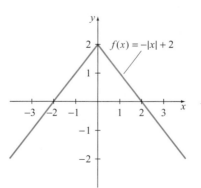

FIGURE 11.7

SOLUTION From the graph, we can see that the function has a relative maximum at $(0, 2)$. Therefore, $f'(x) = 0$ or $f'(x)$ is undefined at that point. At any sharp point on a graph, the derivative is undefined. Therefore, $f'(0)$ is undefined. Does $f'(x) = 0$ anywhere? From the graph, we see that f is increasing for all $x < 0$. Therefore, $f'(x)$ is positive for $x < 0$. Similarly, f is decreasing for all $x > 0$. Therefore, $f'(x)$ is negative for $x > 0$. Hence, $f'(x) \neq 0$ anywhere. Thus the only relative extremum occurs at $(0, 2)$.

Critical Values

In Example 2, we saw that relative extrema may occur where $f'(x) = 0$. In Example 3, we saw that relative extrema may also occur where $f'(x)$ is undefined. These two situations represent two different types of *critical values* of a function.

CRITICAL VALUES

Let f be a continuous function on the interval (a, b) with $a < c < b$. If

- $f'(c) = 0$ or
- $f'(c)$ is undefined

then c is called a **critical value** of f.

The two types of critical values may be further classified as **stationary values** $[f'(c) = 0]$ and **singular values** $[f'(c)$ is undefined$]$. The corresponding point $(c, f(c))$ is called a **critical point, stationary point,** or **singular point** of f, respectively.

EXISTENCE OF RELATIVE EXTREMA

Let f be a continuous function on the interval (a, b). All relative extrema of f occur at critical values of f.

EXAMPLE 4

Finding Critical Values of a Function

Find the critical values of $f(x) = x^3 - 3x$ on the interval $[-2, 4]$.

SOLUTION To find the stationary values, we set the derivative equal to zero and solve for x.

$$f'(x) = 3x^2 - 3$$
$$0 = 3x^2 - 3 \qquad \text{Set } f'(x) = 0$$
$$0 = 3(x^2 - 1)$$
$$0 = 3(x - 1)(x + 1)$$
$$x - 1 = 0 \qquad x + 1 = 0 \qquad \text{Set each factor equal to 0 and solve for } x$$
$$x = 1 \qquad\qquad x = -1$$

The stationary values are $x = 1$ and $x = -1$. Since the derivative is defined for all values of x, there are no singular values.

EXAMPLE 5

Finding Critical Points of a Function

Find the critical points of $f(x) = x^{1/3}$ on the interval $[-2, 2]$.

SOLUTION

$$f'(x) = \frac{1}{3}x^{-2/3}$$

$$= \frac{1}{3x^{2/3}}$$

$f'(x) \neq 0$ for all values of x, since a fraction can equal zero only if its numerator is zero. Therefore, f does not have any stationary points.

A rational expression $\dfrac{g(x)}{h(x)}$ is undefined when $h(x)$ is equal to zero. What value of x makes the denominator of $f'(x)$ equal zero?

$0 = 3x^{2/3}$	Set the denominator equal to 0
$0 = x^{2/3}$	Divide by 3
$(0)^3 = (x^{2/3})^3$	Cube both sides of the equation
$0 = x^2$	$(x^{2/3})^3 = x^{2/3 \cdot 3} = x^2$
$x = 0$	Take the square root

Therefore, $f'(0)$ is undefined. A singular point may occur at $x = 0$. We need only confirm that the function f is defined at $x = 0$. Since $f(0) = 0$, $x = 0$ is in the domain of f and is indeed a singular value of f. The point $(0, 0)$ is a singular point of f.

The First Derivative Test

Is there a nongraphical way to tell whether a relative maximum or a relative minimum occurs at a critical value? Fortunately, yes. Recall that when a function is increasing, its derivative is positive. When a function is decreasing, its derivative is negative. If the derivative changes sign at a critical value, a relative extremum occurs at that value. The First Derivative Test will allow us to determine where relative extrema occur for *nonconstant* functions.

If f is a constant function $[f(x) = k$ for some constant $k]$, then $f'(x) = 0$ for all x and all points of the function are critical points. Furthermore, since $k \leq f(x) \leq k$ for all x, both relative maxima and relative minima occur at all points of f. Since most real-life data models are not constant functions, we will focus our discussion on *nonconstant* functions.

FIRST DERIVATIVE TEST

Let $(c, f(c))$ be a critical point of a non-constant function f.

- If f' changes from positive to negative at $x = c$, then a relative maximum occurs at $(c, f(c))$.
- If f' changes from negative to positive at $x = c$, then a relative minimum occurs at $(c, f(c))$.
- If f' doesn't change sign at $x = c$, then a relative extremum does not occur at $(c, f(c))$.

It is often helpful to develop a *sign chart* when determining where relative extrema occur. We will demonstrate this with an example and then detail the specific steps of the process.

EXAMPLE 6

Finding Relative and Absolute Extrema

Find the relative and absolute extrema of $f(x) = x^4 - 4x^3$ on the interval $[-1, 5]$.

SOLUTION We will first find the relative extrema by using the First Derivative Test.

$$f'(x) = 4x^3 - 12x^2$$
$$= 4x^2(x - 3)$$

$4x^2 = 0$	$x - 3 = 0$	Set each factor equal to zero
$x = 0$	$x = 3$	Solve for x

Setting the derivative equal to zero, we see that $x = 0$ and $x = 3$ are critical values of f. We begin by drawing a number line with the critical values clearly marked.

We will evaluate $f'(x)$ at an x value in each of the following three intervals: $[-1, 0), (0, 3),$ and $(3, 5]$. (These intervals may be equivalently expressed as $-1 \le x < 0, 0 < x < 3,$ and $3 < x \le 5$, respectively.) We are not concerned with the actual value of the derivative. We simply want to know if the derivative is positive or negative.

From the interval $[-1, 0)$, we pick $x = -1$.

$$f'(-1) = 4(-1)^2(-1-3) \qquad \text{The product is equivalent to } (4)(1)(-4)$$
$$f'(-1) < 0 \qquad \text{Since there are an odd number of negative factors}$$

On the interval $[-1, 0)$, the derivative is negative. We update the chart to reflect our finding.

From the interval $(0, 3)$, we pick $x = 1$.

$$f'(1) = 4(1)^2(1-3) \qquad \text{The product is equivalent to } (4)(1)(-2)$$
$$f'(-1) < 0 \qquad \text{Since there are an odd number of negative factors}$$

On the interval $(0, 3)$, the derivative is negative. We update the chart to reflect our finding.

From the interval $(3, 5]$, we pick $x = 4$.

$$f'(4) = 4(4)^2(4-3) \qquad \text{The product is equivalent to } (4)(16)(1)$$
$$f'(4) > 0 \qquad \text{Since there are no negative factors}$$

On the interval $(3, 5]$, the derivative is positive. We update the chart to reflect our finding.

$$f' \xleftarrow{\quad - \quad \underset{0}{+} \quad - \quad \underset{3}{+} \quad + \quad}$$

At the critical value $x = 0$, the derivative did not change sign, so a relative extremum does not occur there. Graphically speaking, the graph of f is decreasing through the whole interval $[-1, 3)$. At the critical value $x = 3$, the derivative changed sign from negative to positive, so a relative minimum occurs there.

$$f' \xleftarrow{\quad - \quad \underset{0}{+} \quad - \quad \underset{3}{+} \quad \text{min } + \quad}$$

Graphically speaking, the graph of f changed from decreasing to increasing at $x = 3$.

To find absolute extrema, we consider the relative extrema and evaluate the function f at the endpoints (see Table 11.2).

TABLE 11.2

x	$f(x)$	
-1	5	
3	-27	Relative minimum, absolute minimum
5	125	Absolute maximum

The largest value of $f(x)$ in the table is the absolute maximum, and the smallest value is the absolute minimum. In this case, the relative and absolute minimum occurs at $(3, -27)$ and is equal to -27. The absolute maximum occurs at $(5, 125)$ and is equal to 125. There is no relative maximum. (Recall from our definition of a relative extremum that a relative extremum may not occur at an endpoint. This definition allows us to use critical values to find all relative extrema.)

HOW TO Constructing a Derivative Sign Chart

1. Label a number line with the critical values of the function f.
2. Write f' next to the number line.
3. Evaluate f' at a value in each of the number line intervals.
4. Record a "+" if the derivative is positive and a "−" if the derivative is negative.
5. Use the First Derivative Test to determine where relative maxima and minima occur and record the results on the chart.

EXAMPLE 7

Finding Relative and Absolute Extrema

Find the absolute and relative extrema of $g(x) = x^3 - 2x^2 - 5x + 6$ on the interval $[-3, 4]$.

SOLUTION

$$g'(x) = 3x^2 - 4x - 5$$

We set $g'(x) = 0$.

$$3x^2 - 4x - 5 = 0$$

This is a quadratic equation in the form $ax^2 + bx + c = 0$. Since the derivative doesn't factor, we must use the Quadratic Formula to find the stationary points (or use a calculator or computer).

$$x = \frac{-b \pm \sqrt{b^2 - 4ac}}{2a}$$

$$= \frac{-(-4) \pm \sqrt{(-4)^2 - 4(3)(-5)}}{2(3)}$$

$$= \frac{4 \pm \sqrt{76}}{6}$$

$$= \frac{4 \pm 2\sqrt{19}}{6}$$

$$= \frac{2}{3} \pm \frac{\sqrt{19}}{3}$$

Since $\frac{2}{3} + \frac{\sqrt{19}}{3} \approx 2.120$ and $\frac{2}{3} - \frac{\sqrt{19}}{3} \approx -0.7863$, the stationary values are $x \approx -0.7863$ and $x \approx 2.120$. The derivative is defined for all values of x, so there are no singular values. We now construct a sign chart.

To determine the sign of the derivative $g'(x) = 3x^2 - 4x - 5$ in each interval, we evaluate the derivative at $x = -1$, $x = 0$, and $x = 3$.

$$g'(-1) = 3(-1)^2 - 4(-1) - 5$$
$$= 3 + 4 - 5$$
$$g'(-1) > 0$$

$$g'(0) = 3(0)^2 - 4(0) - 5$$
$$= 0 - 0 - 5$$
$$g'(0) < 0$$

$$g'(3) = 3(3)^2 - 4(3) - 5$$
$$= 27 - 12 - 5$$
$$g'(3) > 0$$

We record the sign of the derivative in each interval on the sign chart.

A relative maximum occurs at $x = \frac{2}{3} - \frac{\sqrt{19}}{3} \approx -0.7863$ and a relative minimum occurs at $x = \frac{2}{3} + \frac{\sqrt{19}}{3} \approx 2.120$.

To find the absolute extrema, we consider the relative extrema and evaluate the function g at the endpoints, as shown in Table 11.3.

TABLE 11.3

x	$g(x)$	
-3	-24	Absolute minimum
-0.7863	8.209	Relative maximum
2.120	-4.061	Relative minimum
4	18	Absolute maximum

The absolute minimum occurs at $(-3, -24)$, although there is a relative minimum at $(2.120, -4.061)$. The absolute maximum occurs at $(4, 18)$, although there is a relative maximum at $(-0.7863, 8.209)$.

It is often helpful to use a graphing calculator to calculate values of the derivative, especially when evaluating the derivative at other than whole-number values. The following Technology Tip will detail how to use your calculator to evaluate the derivative at a point quickly.

TECHNOLOGY **TIP**

Evaluating the Derivative at a Point

1. Enter the function $f(x)$ as Y1 by pressing the [Y=] button and typing the equation.

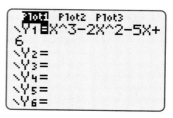

2. Graph the function over the specified domain. (You may need to press [WINDOW] and adjust the Xmin and Xmax settings. Press [Zoom] and 0:ZoomFit then [ENTER] to automatically adjust the y values so that the entire graph will appear on screen.)

3. Press [2nd] then [TRACE] to bring up the [CALCULATE] menu. Select item 6:dy/dx and press [ENTER].

(Continued)

4. Type in the *x* value of the point where you want to evaluate the derivative or use the blue arrows to select the point graphically. Then press (ENTER).

5. The value of the derivative is displayed at the bottom of the screen. (You can assume that a value like "1E-6" is equal to zero. 1E-6 = 1×10^{-6} = 0.000001. The error occurs because the calculator uses numerical methods to differentiate.)

EXAMPLE 8

Interpreting Extrema in a Real-World Context

Based on data from 1980 to 1999, the per capita consumption of breakfast cereal may be modeled by

$$C(t) = -0.004718t^3 + 0.1165t^2 - 0.3585t + 12.17 \text{ pounds per person}$$

where *t* is the number of years since 1980. (**Source:** *Statistical Abstract of the United States, 2001,* Table 202, p. 129.) Find the relative and absolute extrema of $C(t)$ on the interval $[0, 19]$ and interpret the meaning of the results.

SOLUTION We must first find the critical values of $C(t)$.

$$C'(t) = -0.014154t^2 + 0.2330t - 0.3585$$
$$0 = -0.014154t^2 + 0.2330t - 0.3585 \qquad \text{Set } C'(t) \text{ equal to } 0$$

Recall that the quadratic formula states that the solutions to a quadratic equation of the form $at^2 + bt + c = 0$ are given by

$$t = \frac{-b \pm \sqrt{b^2 - 4ac}}{2a}$$

$$= \frac{-(0.2330) \pm \sqrt{(0.2330)^2 - 4(-0.014154)(-0.3585)}}{2(-0.014154)}$$

$$= \frac{-0.2330 \pm 0.1844}{-0.028308}$$

$$t = \frac{-0.2330 + 0.1844}{-0.028308} \qquad t = \frac{-0.2330 - 0.1844}{-0.028308}$$

$$\approx 1.7 \qquad\qquad\qquad \approx 14.7$$

The critical values are $t \approx 1.7$ and $t \approx 14.7$. Using these values together with the interval test values of $t = 0$, $t = 10$, and $t = 18$, we construct the first derivative sign chart.

C has a relative minimum at $t \approx 1.7$ and a relative maximum at $t \approx 14.7$. Evaluating $C(t) = -0.004718t^3 + 0.1165t^2 - 0.3585t + 12.17$ at the critical values and the endpoints, we construct a table of values, as shown in Table 11.4.

$C'(t)$ ← $\dfrac{-\quad \text{min of } C \quad + \quad \text{max of } C \quad -}{1.7 \qquad\qquad 14.7}$

TABLE 11.4

t	C(t)	
0	12.2	
1.7	11.9	Relative and absolute minimum
14.7	17.1	Relative and absolute maximum
19	15.1	

According to the model, the per capita consumption of breakfast cereal was decreasing between the end of 1980 ($t = 0$) and mid-September 1982 ($t = 1.7$). It reached a relative low of 11.9 pounds per person in mid-September 1982. From mid-September 1982 to mid-September 1995, the per capita consumption of cereal was increasing. It reached a relative high of 17.1 pounds per person in mid-September 1995. Between mid-September 1995 and the end of 1999, the per capita consumption of cereal decreased. The highest per capita consumption of cereal between 1980 and 1999 was 17.1 pounds per person in the one-year period ending in mid-September 1995, and the lowest per capita consumption of cereal between 1980 and 1999 was 11.9 pounds per person in the one-year period ending in mid-September 1982.

EXAMPLE 9

Interpreting Extrema in a Real-World Context

Based on data from 1980 to 1999, the annual per capita bottled water consumption may be modeled by

$$W(t) = 2.593(1.106)^t \text{ gallons}$$

where t is the number of years since the end of 1980. (**Source:** *Statistical Abstract of the United States, 2001*, Table 204, p. 130.) In what year between 1980 and 1999 was per capita bottled water consumption the greatest?

SOLUTION We are asked to find the absolute maximum of $W(t) = 2.593(1.106)^t$ on the interval $[0, 19]$. We first observe that W is an exponential function. Recall that the derivative of an exponential function $y = ab^x$ is $y' = a \ln(b)b^x$. Since $a \neq 0$ and $\ln(b) \neq 0$, $y' = 0$ only when $b^x = 0$. But no value of x can make $b^x = 0$. Similarly, no value of x can make b^x undefined. Therefore, an exponential function $y = ab^x$ doesn't have any critical values. Thus exponential functions do not have any relative extrema. So to determine when the per capita bottled water consumption was the greatest, we only need to evaluate $W(t) = 2.593(1.106)^t$ at the endpoints, as shown in Table 11.5.

TABLE 11.5

t	W(t)	
0	2.6	Absolute minimum
19	17.6	Absolute maximum

According to the model, per capita bottled water consumption reached a maximum of 17.6 gallons at the end of 1999.

Forecasting Maximum Revenue

Based on data from 1985 to 1999, the number of movie theater admissions purchased may be modeled by

$$A(p) = -1036.5p^3 + 12,915p^2 - 53,497p + 74,871 \text{ million admissions}$$

where p is the price of a movie ticket in constant 1990 dollars. (That is, prices have been adjusted for inflation.) (**Source:** *Modeled from Statistical Abstract of the United States, 2001*, Table 1244, p. 761 and Table 694, p. 455.)

Assuming that prices (in constant 1990 dollars) will continue to remain between $3.45 and $4.76, at what price will the revenue from movie ticket sales be the greatest?

SOLUTION   Revenue is the product of the price and the number of admissions sold.

$$R(p) = pA$$
$$= p(-1,036.5p^3 + 12,915p^2 - 53,497p + 74,871)$$
$$= -1036.5p^4 + 12,915p^3 - 53,497p^2 + 74,871p \text{ million dollars of revenue}$$

The change in revenue is the derivative of the revenue function.

UNITS

$$R'(p) = -4146p^3 + 38,745p^2 - 106,994p + 74,871 \; \frac{\text{revenue in millions of dollars}}{\text{price in dollars}}$$

The graph of $R'(p)$ on the interval $[3.45, 4.76]$ is shown in Figure 11.8.

Using the Technology Tip following this example, we determined that the horizontal intercepts of $R'(p)$ occur at $p = 3.77$ and $p = 4.52$. In other words, $R'(3.77) = 0$ and $R'(4.52) = 0$. That is, the horizontal intercepts of $R'(p)$ are the critical points of $R(p)$.

From the graph of $R'(p)$, we can readily construct the sign chart of $R'(p)$. When the graph of $R'(p)$ is below the horizontal axis, $R'(p) < 0$. When the graph of $R'(p)$ is above the horizontal axis, $R'(p) > 0$.

A relative minimum occurs at $p = 3.77$, and a relative maximum occurs at $p = 4.52$. We evaluate the function R at the endpoints and the relative extrema, as shown in Table 11.6.

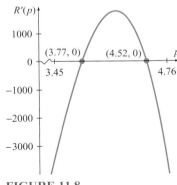

FIGURE 11.8

```
        − min    +   max −
R'(p) ←————+————————+————→
        3.77        4.52
```

TABLE 11.6

p	$R(p)$	
3.45	5,053	
3.77	4,556	Relative and absolute minimum
4.52	5,456	Relative and absolute maximum
4.76	5,053	

According to the model, the maximum revenue of $5,456 million is achieved when the ticket price is $4.52. Recall that the price and revenue are in constant 1990 dollars. A movie theater would need to adjust these figures for inflation if it were to use the model as a guide in setting ticket prices today.

TECHNOLOGY **TIP**

Finding the *x*-intercepts of a Function

1. Enter the function $f(x)$ as Y1 by pressing the (Y=) button and typing the equation.

2. Graph the function over the specified domain. (Press (WINDOW) and adjust the Xmin and Xmax settings. Press (ZOOM) then 0:ZoomFit then (ENTER) to automatically adjust the *y* values so that the entire graph will fit on the screen.)

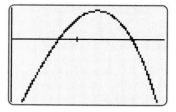

3. Press (2nd) then (TRACE) to bring up the CALCULATE menu. Select item 2:zero and press (ENTER).

4. The calculator asks Left Bound?. Enter an *x* value that is smaller than the *x*-intercept. Or, if you prefer, use the blue arrows to select a point to the left of the *x*-intercept visually.

5. The calculator asks Right Bound?. Enter an *x* value that is larger than the *x*-intercept. Or, if you prefer, use the blue arrows to select a point to the right of the *x*-intercept visually.

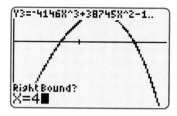

6. The calculator asks Guess?. You may enter a guess or simply press (ENTER). The calculator returns the coordinates of the *x*-intercept. Repeat Steps 3 through 6 to find additional *x*-intercepts.

11.1 Summary

In the section, you learned how to find the relative and absolute extrema of a function. You discovered that the First Derivative Test is an excellent tool for finding the location of relative extrema algebraically.

11.1 Exercises

In Exercises 1–10, determine the points on the graph where $f'(x) = 0$ or $f'(x)$ is undefined.

1. $f(x) = x^2 - 2x$

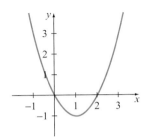

2. $f(x) = |x - 1| + 2x$

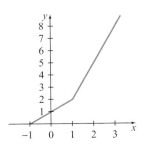

3. $f(x) = x^3 + 1$

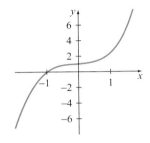

4. $f(x) = -x^3 + 3x^2 - 3x + 5$

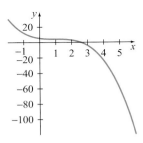

5. $f(x) = -x^2 + 4x - 4$

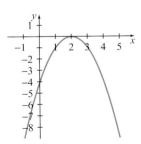

6. $f(x) = x^3 - 9x^2 + 24x + 3$

7. $f(x) = -|x - 2| + 4$

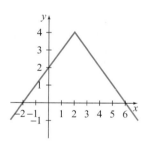

8. $f(x) = |x + 1| - |2x|$

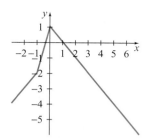

9. $f(x) = x^5 - 5x^4 + 200$

10. $f(x) = 5$

In Exercises 11–20, find the critical values of the function. Then classify the critical values as stationary values or singular values.

11. $f(x) = x^2 - 4x$ **12.** $g(t) = t^3 - 1$

13. $h(t) = t^3 - 6t^2$ **14.** $f(x) = x^{1/5} + 4$

15. $g(x) = 3x^{1/3} - x$

16. $R(p) = p^4 - 2p^2 + 1$

17. $h(x) = \dfrac{x}{x^2 + 1}$ **18.** $g(t) = \dfrac{1}{t^2}$

19. $C(p) = 4p^3 - 12p + 7$

20. $f(x) = -x + \sqrt{x}$

In Exercises 21–35, construct a sign chart for the derivative. Then determine the relative and absolute extrema of the function on the specified domain. Use the First Derivative Test as appropriate. (Note: Exercises 21–30 use the same functions as Exercises 11–20.)

21. $f(x) = x^2 - 4x;\ [-1, 5]$

22. $g(t) = t^3 - 1;\ [-2, 2]$

23. $h(t) = t^3 - 6t^2;\ [-1, 7]$

24. $f(x) = x^{1/5} + 4;\ [-1, 1]$

25. $g(x) = 3x^{1/3} - x;\ [-3, 3]$

26. $R(p) = p^4 - 2p^2 + 1;\ [0, 4]$

27. $h(x) = \dfrac{x}{x^2 + 1};\ [-2, 2]$

28. $g(t) = \dfrac{1}{t^2};\ [-2, 2]$

29. $C(p) = 4p^3 - 12p + 7;\ [-3, 3]$

30. $f(x) = -x + \sqrt{x};\ [0, 2]$

31. $f(x) = x^3 - 4x;\ [-3, 3]$

32. $g(t) = e^t - t;\ [-2, 2]$

33. $h(t) = t \ln(t);\ [0.1, 3]$

34. $p(n) = n(2^n);\ [-2, 1]$

35. $g(x) = e^x - \ln x;\ [0.1, 3]$

In Exercises 36–40, the number of tickets sold, $N(p)$, is given as a function of the ticket price p. Determine what ticket price will maximize the revenue from ticket sales.

36. $N(p) = -200p + 1000$

37. $N(p) = -60p + 3000$

38. $N(p) = 2500e^{-0.1p}$

39. $N(p) = 625e^{-0.025p}$

40. $N(p) = 8500e^{-0.04p}$

In Exercises 41–45, use the methods discussed in this section to answer the questions.

41. **Margarine Consumption** Based on data from 1980 to 1999, the per capita consumption of margarine may be modeled by

$$M(t) = 0.0002868t^4 - 0.01149t^3 + 0.1329t^2 - 0.5098t + 11.30$$

pounds per year, where t is the number of years since 1980. (**Source:** Modeled from *Statistical Abstract of the United States, 2001*, Table 202, p. 129.)

Between 1980 and 1999, when was the per capita consumption of margarine increasing?

42. **Gaming Software Sales** Based on data from 1990 to 1999, electronic gaming *software* sales may be modeled by

$$S(t) = 39.79t^2 - 48.41t + 2378$$

million dollars, where t is the number of years since 1990. (**Source:** Modeled from *Statistical Abstract of the United States, 2001*, Table 1005, p. 634.)

In what year (between 1990 and 1999) did gaming *software* sales reach their highest point and in what year did they reach their lowest point?

43. **Gaming Hardware Sales** Based on data from 1990 to 1999, electronic gaming *hardware* sales may be modeled by

$$S(t) = 8.451t^3 - 111.9t^2 + 470.0t + 965.0$$

million dollars, where t is the number of years since 1990. (**Source:** Modeled from *Statistical Abstract of the United States, 2001*, Table 1005, p. 634.)

In what year (between 1990 and 1999) did gaming *hardware* sales reach their highest point, and in what year did they reach their lowest point?

44. **Lumber Industry Earnings** Based on data from 1995 to 1999, the average earnings of a lumber and wood products industry employee may be modeled by

$$S(n) = -0.7685n^2 + 1295n - 516{,}596$$

dollars, where n is the number of employees in thousands. (**Source:** Modeled from *Statistical Abstract of the United States, 2001*, Table 979, p. 622.)

Assuming that the employee union wants to maximize the average wage of its members, what is the optimal number of employees to have working in the lumber and wood products industry (according to the model)?

45. **DVD Market Value** Based on data from 1998 to 2001, the total dollar value of DVDs shipped by manufacturers may be modeled by

$$V(x) = -0.2173x^2 + 25.84x - 0.02345$$

million dollars, where x is the number of DVDs shipped in millions. (**Source:** Modeled from Recording Industry Association of America data.)

According to the model, how many DVDs should be shipped in order to maximize the value of the manufacturers' DVD shipments?

Exercises 46–50 are intended to challenge your understanding of extrema.

46. A continuous function f has the property that $f'(2) > 0$, $f'(3) = 0$, and $f'(4) < 0$. Does f have a relative maximum at $x = 3$? Justify your answer.

47. Give an example of a nonlinear function that does not have any critical points.

48. Give an example of a continuous function that has a singular point but no relative extrema.

49. The graph of a polynomial function of degree n changes from increasing to decreasing (or vice versa) at most $n - 1$ times. What is the maximum number of stationary points the function can have? Defend your conclusions.

50. A continuous function f' has two x-intercepts: $x = 2$ and $x = 5$. The function f has exactly two relative extrema. (*Note:* $\dfrac{d}{dx}(f) = f'$.)

If $f'(3) = 1$, determine the x values of the relative maximum and relative minimum of f.

11.2 Applications of Maxima and Minima

■ Analyze and interpret revenue, cost, and profit functions

GETTING STARTED The equation that expresses the price of an item as a function of the number of items sold is referred to as a **demand equation.** Based on data from 1992 to 2000, the wholesale demand equation for audio compact discs may be modeled by

$$p = -0.000002549x^2 + 0.001036x + 13.12$$

where x is the number of CDs shipped by manufacturers (in millions) and p is the price per CD (in constant 1992 dollars). The financial data have been adjusted for inflation. (**Source:** Modeled from Recording Industry Association of America data.)

How many CDs should be sold in order to maximize revenue? Questions such as these can be answered using the notion of the derivative.

In this section, we will further illustrate applications of maxima and minima. Specifically, we will look at the business concepts of revenue, cost, and profit as well as marginal revenue, marginal cost, and marginal profit. We will also explore area and volume problems.

Revenue, Cost, and Profit

A company's **revenue** is the total amount of money it brings in. The company's **cost** is the total amount of money it spends; this includes both fixed costs and variable costs. **Fixed costs** are those costs that the company incurs regardless of production. **Variable costs** are typically expenses that vary based on the level of production. **Profit** is the difference between the company's revenue and its cost. The **break-even point** is the production level that results in revenue equaling cost.

Marginal revenue is an approximation of the additional revenue generated if one more unit is produced and sold. **Marginal cost** is the approximate cost incurred by producing one more unit, and **marginal profit** is the approximate profit resulting from the production and sale of one more unit. In order to maximize profits, product manufacturers should continue to increase production as long as the marginal profit is positive. If they continue production when the marginal profit turns negative, they will actually lose money by producing more items.

Marginal profit is determined by taking the derivative of the profit function. This relationship between the derivative and marginal profit is seen by looking at the difference quotient of the profit function $P(x)$. Suppose we wanted to estimate the additional profit earned by selling one more item after 100 items had been sold. We know that $P(101) - P(100)$ is the additional profit earned by selling the 101st item. Note that

$$P(101) - P(100) = \frac{P(101) - P(100)}{101 - 100}$$

$$\approx \lim_{h \to 0} \frac{P(100 + h) - P(100)}{(100 + h) - 100}$$

$$\approx \lim_{h \to 0} \frac{P(100 + h) - P(100)}{h}$$

$$\approx P'(100)$$

It may seem illogical to evaluate the derivative at $x = 100$ when we could calculate the exact amount of the additional profit earned by producing and selling the 101st item by evaluating $P(101) - P(100)$. However, if P is a complicated function, it is often easier to evaluate the derivative at a single value instead of calculating P at two different values.

In a similar manner, we can see that marginal cost is the derivative of the cost function and marginal revenue is the derivative of the revenue function. We'll begin by working a straightforward example to demonstrate the concepts of marginal revenue and marginal profit.

EXAMPLE 1

Forecasting Maximum Profit

The demand equation for a certain brand of squirt gun is given by

$$p = -x + 15$$

where x is the number of squirt guns sold (in thousands) and p is the price per squirt gun (in dollars).

The cost to produce x thousand squirt guns is

$$C(x) = 5x \text{ thousand dollars}$$

(a) If the company is currently producing 7000 squirt guns per year, should it increase or decrease squirt gun production? Explain using the concepts of marginal revenue and marginal profit.

(b) Determine the production level that will maximize profit.

SOLUTION

(a) The revenue function $R(x)$ is the product of the price p and the number of squirt guns sold.

UNITS

$$R(x) = \left(p \, \frac{\text{dollars}}{\text{squirt gun}}\right) \cdot (x \text{ thousand squirt guns})$$

$$= (-x + 15)x \, \frac{(\text{dollars})(\text{thousand squirt guns})}{\text{squirt gun}}$$

$$= -x^2 + 15x \text{ thousand dollars}$$

The marginal revenue is the derivative of the revenue function.

UNITS

$$R'(x) = -2x + 15 \, \frac{\text{thousand dollars}}{\text{thousand squirt guns}}$$

Evaluating the marginal revenue at a production level of 7 thousand squirt guns yields

$$R'(7) = -2(7) + 15$$

$$= 1 \, \frac{\text{thousand dollars}}{\text{thousand squirt guns}}$$

Increasing production from 7 thousand squirt guns to 8 thousand squirt guns will increase revenue by about $1000.

Profit is the difference between revenue and cost. The profit function is

$$P(x) = R(x) - C(x)$$
$$= (-x^2 + 15x) - (5x)$$
$$= -x^2 + 10x \text{ thousand dollars}$$

The marginal profit function is

$$P'(x) = -2x + 10 \frac{\text{thousand dollars}}{\text{thousand squirt guns}}$$

Evaluating the marginal profit at a production level of 7 thousand squirt guns yields

$$P'(7) = -2(7) + 10$$
$$= -4 \text{ thousand dollars per thousand squirt guns}$$

Increasing production from 7 thousand squirt guns to 8 thousand squirt guns will decrease profit by about $4000.

Although marginal revenue is positive at a production level of 7 thousand squirt guns, marginal profit is negative at that production level. This means that any additional revenue brought in won't be enough to cover the cost of producing the extra squirt guns. The company should reduce production.

(b) The company wants to maximize profit. When profit is maximized, the marginal profit will be equal to 0. Recall that the profit function is

$$P(x) = -x^2 + 10x \text{ thousand dollars}$$

and the marginal profit function is the derivative

$$P'(x) = -2x + 10 \text{ thousand dollars per thousand squirt guns}$$

Setting the marginal profit to zero yields

$$0 = -2x + 10$$
$$2x = 10$$
$$x = 5 \text{ thousand squirt guns}$$

Does a maximum or a minimum of the profit function occur at the critical value $x = 5$? Observe that the graph of $P(x) = -x^2 + 10x$ is a concave down parabola. A concave down parabola has one relative maximum and no relative minima. The relative maximum is also an absolute maximum. Consequently, an absolute maximum of the profit function $P(x) = -x^2 + 10x$ occurs at the critical value $x = 5$. At a production level of five thousand squirt guns, the profit will be maximized. The maximum profit is

$$P(5) = -(5)^2 + 10(5)$$
$$= 25 \text{ thousand dollars}$$

We can verify the accuracy of our result by evaluating the profit function at other production levels on either side of $x = 5$.

$$P(4) = -(4)^2 + 10(4)$$
$$= 24 \text{ thousand dollars}$$

$$P(7) = -(7)^2 + 10(7)$$

$$= 21 \text{ thousand dollars}$$

Increasing or decreasing the production level from 5000 squirt guns will decrease profit.

EXAMPLE 2

Forecasting Maximum Revenue

Based on data from 1992 to 2000, the wholesale demand equation for audio compact disks may be modeled by

$$p = -0.000002549x^2 + 0.001036x + 13.12$$

where x is the number of CDs shipped by manufacturers (in millions) and p is the price per CD (in constant 1992 dollars). The financial data have been adjusted for inflation. (**Source:** Modeled from Recording Industry Association of America data.)

(a) What is the revenue function for the wholesale shipment of CDs?
(b) At a production level of 1 billion CDs, what is the marginal revenue?
(c) At a production level of 2 billion CDs, what is the marginal revenue?
(d) How many CDs should be shipped in order to maximize revenue?

SOLUTION

(a) Revenue is the product of an item's price and the number of items sold.

$$R(x) = (-0.000002549x^2 + 0.001036x + 13.12)x$$

$$= -0.000002549x^3 + 0.001036x^2 + 13.12x \text{ million dollars}$$

(b) Marginal revenue is the derivative of the revenue function.

UNITS

$$R'(x) = \frac{d}{dx}(-0.000002549x^3 + 0.001036x^2 + 13.12x)$$

$$= -0.000007647x^2 + 0.002072x + 13.12 \frac{\text{millions of dollars}}{\text{millions of CDs}}$$

$$R'(1000) = -0.000007647(1000)^2 + 0.002072(1000) + 13.12$$

$$= 7.55 \text{ million dollars per million CDs}$$

Increasing production from 1000 million to 1001 million CDs will increase revenue by about 7.55 million dollars.

(c) Evaluating marginal revenue at a production level of 2000 million CDs yields

$$R'(2000) = -0.000007647(2000)^2 + 0.002072(2000) + 13.12$$

$$= -13.32 \text{ million dollars per million CDs}$$

Increasing production from 2000 million to 2001 million CDs will decrease revenue by about 13.32 million dollars.

(d) We will use a sign chart and the First Derivative Test to determine the production level that will maximize revenue.

$$R'(x) = -0.000007647x^2 + 0.002072x + 13.12$$

The derivative is defined for all values of x, so there are no singular values. We'll find the stationary values by setting the derivative equal to zero and solving the equation with the Quadratic Formula.

$$0 = -0.000007647x^2 + 0.002072x + 13.12$$

$$0 = -7.647x^2 + 2072x + 13{,}120{,}000 \qquad \text{Multiply equation by 1,000,000}$$

$$x = \frac{-b \pm \sqrt{b^2 - 4ac}}{2a}$$

$$= \frac{-2072 \pm \sqrt{(2072)^2 - 4(-7.647)(13{,}120{,}000)}}{2(-7.647)}$$

$$= \frac{-2072 \pm \sqrt{405{,}607{,}744}}{2(-7.647)}$$

$$= \frac{-2072 \pm 20{,}140}{-15.294}$$

$$\approx -1181;\ 1452$$

The revenue function has two critical values: $x \approx -1181$ and $x \approx 1452$. In the context of the problem, it doesn't make sense to talk about a production level of -1181 million CDs. Consequently, we'll throw out that critical value. We're only interested in what happens when 1452 million CDs are produced. Using the data from parts (b) and (c) of this problem, we can construct the sign chart.

A relative maximum exists at $x = 1452$. Is this an absolute maximum? The domain of the revenue function is not explicitly stated; however, it is implied. Clearly, $x \geq 0$, since it doesn't make sense to talk about a negative quantity of CDs being shipped. To determine the maximum domain value, we analyze the graph of the cubic function $R(x) = -0.000002549x^3 + 0.001036x^2 + 13.12x$ (Figure 11.9).

Using a graphing calculator, we determined that the revenue function has an x-intercept at $x = 2481.03$. As seen on the graph, the revenue function R is negative for values of x greater than 2481.03. Since it doesn't make sense to have negative revenue, $x \leq 2481.03$. Therefore, the implied domain of the function is $[0, 2481.03]$. Redrawing $R(x)$ on the implied domain yields Figure 11.10.

From the graph and equation of $R(x)$, we confirm the results of our earlier analysis, as shown in Table 11.7.

FIGURE 11.9

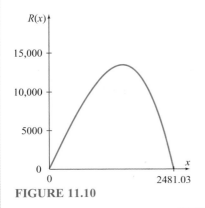

FIGURE 11.10

TABLE 11.7

x	$R(x)$	
0	0	Absolute minimum
1452	13,431	Relative and absolute maximum
2481.03	0	Absolute minimum

According to the model, 1452 million CDs should be shipped in order to maximize revenue. The maximum revenue is forecasted to be 13,431 million dollars (in constant 1992 dollars).

EXAMPLE 3

Forecasting Maximum Profit

Based on data from 1992 to 2000, the wholesale demand for cassette tapes may be modeled by

$$p = 8.954 - \frac{156.2}{x} \text{ dollars per cassette (in constant 1992 dollars)}$$

where x is the number of cassettes shipped by manufacturers (in millions). The financial data have been adjusted for inflation. In order for the price to be positive, x must equal at least 17.5 million cassettes. (**Source:** Modeled from Recording Industry Association of America data.)

The cassette cost function for cassette tape retailers is the same as the revenue function for cassette tape manufacturers, since retailers purchase their cassettes from the manufacturers.

(a) Find the product cost function for cassette tape retailers.

(b) Assuming that the average retail price of a cassette is $p = 11 - \frac{x}{400}$, find the revenue function for cassette tape retailers.

(c) Find the profit function for cassette tape retailers.

(d) Determine what quantity of cassettes will maximize the profit of cassette tape retailers.

SOLUTION

(a) The cost function for cassette tape retailers is

$$C(x) = \left(8.954 - \frac{156.2}{x}\right)x$$

$$= 8.954x - 156.2 \text{ million dollars}$$

with the domain $[17.5, \infty)$.

(b) The revenue function for cassette tape retailers is

$$R(x) = \left(11 - \frac{x}{400}\right)x$$

$$= 11x - \frac{x^2}{400} \text{ million dollars}$$

(c) The profit function for cassette tape retailers is

$$P(x) = R(x) - C(x)$$

$$= \left(11x - \frac{x^2}{400}\right) - (8.954x - 156.2)$$

$$= 2.046x - \frac{x^2}{400} + 156.2 \text{ with domain } [17.5, \infty)$$

(d) The marginal profit function is

$$P'(x) = 2.046 - \frac{2x}{400}$$

$$= 2.046 - \frac{x}{200} \text{ with domain } [17.5, \infty)$$

We want to know when the marginal profit is zero.

$$0 = 2.046 - \frac{x}{200}$$

$$\frac{x}{200} = 2.046$$

$$x = 409.2$$

Evaluating the derivative on either side of the critical value $x = 409.2$, we get

$$P'(0) = 2.046$$

$$P'(500) = 2.046 - \frac{500}{200}$$

$$= -0.454$$

The profit function has a relative maximum when 409.2 million cassettes are sold, as shown in the sign chart and Table 11.8.

TABLE 11.8

x	$P(x)$
17.5	191.2
409.2	574.8
1,000	−297.8

Relative and absolute maximum

We do not have a right-hand endpoint, since the domain of the function is $[17.5, \infty)$. However, we can calculate the profit at a large value of x greater than 409.2 and compare it to $P(409.2)$. Additionally, since the graph of the profit function is a concave down parabola, we know that there will be an absolute maximum at the vertex. The vertex is $(409.2, 574.8)$. According to the model, the maximum profit of $574.8 million occurs when 409.2 million cassettes are sold.

EXAMPLE 4

Minimizing Inventory Costs

Some authors turn to self-publishing when they are unable to convince a publisher to produce their work. Authors such as Deepak Chopra, Virginia Woolf, and James Joyce have self-published. (**Source:** The Self Publishing Manual by Dan Poynter.)

    An author anticipates that she will sell 10,000 copies of her book annually. The setup cost for each print run is $300. The storage cost for each printed book is $1.50 per year. Assuming that customer demand for the book is constant for the 250 business days of a year, determine the number and size of each print run that will minimize her inventory cost. (Inventory cost is the sum of the setup and storage costs.)

SOLUTION The author expects to sell 10,000 copies in the 250 business days of the year. The number of books sold per day is $\frac{10,000}{250} = 40$. She could print

40 books a day; however, her annual setup cost would be $75,000! She could print all 10,000 books at once; however, her annual storage cost would be $7500. We seek to determine the size of the print run that will minimize her annual inventory cost.

Let x be the number of books in a print run. Since the demand is 40 books per day, the number of books in storage will decrease linearly from x books to 0 books at a rate of 40 books per day. Because the number of books in storage decreases linearly, the average number of books in storage is equal to the average of the starting and ending quantities. That is, the average number of books in storage is given by $\frac{x + 0}{2} = \frac{x}{2}$. On the same day that the number of books from the first print run reaches zero, the x books from the second print run are delivered.

For example, if 400 books were produced in a print run, then the number of books in inventory would be as shown in Figure 11.11.

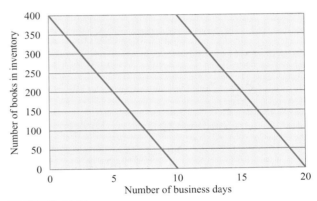

FIGURE 11.11

It takes 10 days to sell all of the books from a print run. The average number of books in inventory over each 10-day period is $\frac{400 + 0}{2} = 200$ books. Since this value is the same for every 10-day period, the average inventory over the course of the year is 200 books when 400 books are produced per print run.

Since the average number of books in storage over the course of the year is $\frac{x}{2}$ and the storage cost is $1.50 per book per year, the annual storage cost is

$$S(x) = 1.5\left(\frac{x}{2}\right)$$

$$= 0.75x \text{ dollars}$$

Since she needs to produce 10,000 books annually, the total number of print runs is equal to $\frac{10,000}{x}$, where x is the number of books in a print run. Since the setup cost of each print run is $300, her annual setup cost is

$$C(x) = 300\left(\frac{10,000}{x}\right)$$

$$= 3,000,000x^{-1} \text{ dollars}$$

The author's annual inventory cost is the sum of the storage and the setup costs.

$$I(x) = S(x) + C(x)$$
$$= 0.75x + 3,000,000x^{-1}$$

Differentiating I and setting it equal to 0 yields

$$0 = 0.75 - 3,000,000x^{-2}$$
$$\frac{3,000,000}{x^2} = 0.75$$
$$3,000,000 = 0.75x^2$$
$$4,000,000 = x^2$$
$$x = 2000$$

The critical value of $I(x)$ is $x = 2000$. We construct a sign chart for $I'(x)$.

The inventory cost function is minimized when 2000 books are ordered. The author will need to make five 2000-book orders during the year. Her minimum annual inventory cost is given by

$$I(2000) = 0.75(2000) + 3,000,000(2000)^{-1}$$
$$= \$3000$$

Area and Volume

Shipping companies such as United Parcel Service (UPS) and the U.S. Postal Service (USPS) classify packages by size and weight. USPS advises customers that Parcel Post™ is the best value when shipping a single package or a small number of packages. A Parcel Post package "can weigh up to 70 pounds and measure up to 130 inches in length and girth combined." (**Source:** U.S. Postal Service.) The length of a package is the length of the longest side, and the girth of a package is the distance around its thickest part. Businesses are often interested in fitting the maximum amount of their product into a box whose dimensions meet the postal service guidelines. They can save a substantial amount of money in shipping costs by conforming to the postal service standard.

EXAMPLE   5   ### Maximizing Volume

A compact disc in a plastic jewel case weighs 3.75 ounces and has dimensions $5.6'' \times 4.9'' \times 0.41''$. A box 22 inches long can fit four CDs placed end to end with an additional $2.4''$ of space to ensure ease of packing. The box will be shipped Parcel Post.

(a) What are the dimensions of the 22-inch rectangular box that conforms to the Parcel Post criteria that has maximum volume?

(b) How many CDs could be packed into the box without exceeding the 70-pound weight restriction?

SOLUTION   To help us visualize what is going on in the problem, we draw a picture of the box (Figure 11.12).

FIGURE 11.12

(a) The volume of a box is $V = lwh$. We want to maximize $V = 22wh$; however, we can't take the derivative of V because it is a function of two variables, w and h. We must find a *secondary equation* that will allow us to write one of the variables in terms of the other.

The girth of the package is

$$g = w + h + w + h$$
$$= 2w + 2h$$

A Parcel Post package can measure up to 130 inches in length and girth combined ($l + g = 130$).

$$l + g = 130$$
$$22 + (2w + 2h) = 130$$
$$2w + 2h = 108$$
$$w + h = 54$$
$$w = 54 - h$$

Now that we have w in terms of h, we may write V in terms of a single variable.

$$V = 22wh$$
$$= 22(54 - h)h$$
$$= 1188h - 22h^2$$

Do we have any domain restrictions on h? Yes. Clearly, $h \geq 0$ and $w \geq 0$, since neither the height nor the width of the package can be negative. Recall that $w = 54 - h$. Thus we may write $w \geq 0$ in terms of h.

$$w \geq 0$$
$$54 - h \geq 0$$
$$54 \geq h$$
$$h \leq 54$$

The domain of the volume function $V = 1188h - 22h^2$ is $[0, 54]$. We must maximize V on this interval.

$$V'(h) = 1188 - 44h$$
$$0 = 1188 - 44h \qquad \text{Set } V'(h) = 0$$
$$44h = 1188$$
$$h = 27$$

The critical value of the volume function is $h = 27$. Since V is a quadratic equation with a negative coefficient on h^2, the graph of V is a concave down parabola, and a relative and absolute maximum occurs at $h = 27$ (see Table 11.9).

TABLE 11.9

h	$V(h)$	
0	0	
27	16,038	Relative and absolute maximum
54	0	

We need to find the value of w in order to determine the dimensions of the box.

$$w = 54 - h$$
$$= 54 - 27$$
$$= 27$$

A rectangular box of size $22'' \times 27'' \times 27''$ will have a volume of 16,038 cubic inches. This is the maximum volume of a box 22 inches long that meets the Parcel Post criteria.

(b) We'll first determine how many CDs can fit across the width of the box.

$$\frac{\text{Width of box}}{\text{Width of CD}} = \frac{27 \text{ inches}}{5.6 \text{ inches per CD}}$$
$$= 4.8 \text{ CDs}$$

Since it doesn't make sense to talk about a fraction of a CD, we'll round this figure to 4 CDs. The width of a row 4 CDs wide is $5.6(4) = 22.4$ inches. There is an additional 4.6 inches of space available for packing material.

We'll then determine how many CDs can fit across the length of the box.

$$\frac{\text{Length of box}}{\text{Length of CD}} = \frac{22 \text{ inches}}{4.9 \text{ inches per CD}}$$
$$= 4.5 \text{ CDs}$$

Since it doesn't make sense to talk about a fraction of a CD, we'll round this figure to 4 CDs. The length of a row 4 CDs long is $4.9(4) = 19.6$ inches. There is an additional 2.4 inch of space available for packing material. (This confirms the information given at the start of the problem.)

We'll next determine how high we can stack the CDs.

$$\frac{\text{Height of box}}{\text{Height of CD}} = \frac{27 \text{ inches}}{0.41 \text{ inches per CD}}$$
$$\approx 65 \text{ CDs}$$

We can stack the CDs 65 units high. Visually, a stack 4 CDs wide, 4 CDs long, and 24 CDs high in the box would look like Figure 11.13.

FIGURE 11.13

A total of $4 \cdot 4 \cdot 65 = 1040$ CDs can fit in the box.

The total weight of the CDs will be $1040 \cdot 3.75 = 3900$ ounces. There are 16 ounces in a pound. The weight of the CDs is

$$3900 \text{ ounces} \cdot \frac{1 \text{ pound}}{16 \text{ ounces}} \approx 244 \text{ pounds}$$

Since the weight of the package exceeds the 70-pound limit, we will need to put fewer CDs in the box. Since 70 pounds is equivalent to 1120 ounces, the combined weight of the CDs and packing material can be at most 1120 ounces.

$$\frac{1120 \text{ ounces}}{3.75 \frac{\text{ounces}}{\text{CD}}} = 298\frac{2}{3} \text{ CDs}$$

We can pack at most 298 CDs into the box. Depending upon the weight of the packing materials, we may have to remove additional CDs to remain within the weight limit.

EXAMPLE 6

Minimizing Use of Resources

Moving companies are hired by businesses and private individuals to pack, load, move, and unload household goods. As part of their service, movers bring cardboard boxes that can be easily constructed and easily broken down. Boxes are purchased by the moving company from a box manufacturer.

In an effort to cut costs, a box manufacturer wants to design a closable box with a square base that has a volume of 8 cubic feet and uses the least amount of cardboard. The box must be at least 24 inches in height and will be constructed according to the design shown in Figure 11.14. What are the dimensions of the box that uses the least material?

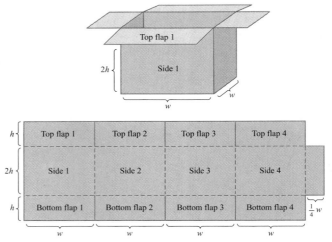

FIGURE 11.14

SOLUTION The surface area of the cardboard material is

$$A = (h + 2h + h)(w + w + w + w) + \frac{1}{4}w(2h)$$

$$= (4h)(4w) + \frac{1}{2}hw$$

$$= 16hw + \frac{1}{2}hw$$

$$= 16.5hw \text{ square feet}$$

Since we want to minimize the amount of cardboard used, we must minimize the surface area. We are unable to find A' because it is a function of two variables. We must use a secondary equation to write one variable in terms of the other.

The volume of the box is

$$V = w \cdot w \cdot 2h$$

$$= 2hw^2$$

Since the box has a volume of 8 cubic feet, we have

$$8 = 2hw^2$$

$$h = \frac{8}{2w^2}$$

$$= 4w^{-2}$$

Substituting this result into the surface area equation, we get

$$A = 16.5hw$$

$$= 16.5(4w^{-2})w$$

$$= 66w^{-1}$$

Do we have any restrictions on w? Clearly, $w > 0$, since the box must have a positive width. (If $w = 0$, the surface area function is undefined.) Additionally, the box is required to have a height of at least 24 inches (2 feet). That is,

$$2h \geq 2$$

$$h \geq 1$$

Since $h = 4w^{-2}$, we can write $h \geq 1$ in terms of w.

$$h \geq 1$$
$$4w^{-2} \geq 1$$
$$4 \geq w^2$$
$$4 - w^2 \geq 0$$

Graphically speaking, we want to know when the graph of $f(w) = 4 - w^2$ is on or above the horizontal axis. [This is when $f(w) \geq 0$.] Rewriting $f(w)$ in the standard form of a quadratic equation $[f(w) = aw^2 + bw + c]$ yields $f(w) = -w^2 + 4$. Since $a < 0$, the parabola is concave down. Since $c = 4$, the graph has a vertical intercept of $(0, 4)$. Since $f(-2) = 0$ and $f(2) = 0$, $(-2, 0)$ and $(2, 0)$ are the horizontal intercepts of $f(w) = -w^2 + 4$ (Figure 11.15).

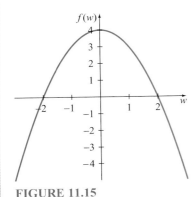

FIGURE 11.15

From the graph, we see that when $-2 \leq w \leq 2$, $f(w) = -w^2 + 4 \geq 0$. Recall that since w is the width, we also required $w > 0$. Thus the domain of the surface area function is $(0, 2]$.

To minimize the surface area, we will differentiate A.

$$A(w) = 66w^{-1}$$
$$A'(w) = -66w^{-2}$$
$$= -\frac{66}{w^2}$$

$A'(w)$ is negative for all values of w in the domain of A, so there are no critical values. Since the derivative is negative, the surface area function is decreasing over its entire domain. Consequently, the smallest area will occur at the rightmost endpoint, $w = 2$. This may be easily confirmed by looking at a table of values for A (Table 11.10).

TABLE 11.10

w	$A(w)$	
0	Undefined	
0.5	132	
1.0	66	
1.5	44	
2.0	33	Absolute minimum

When $w = 2$, $h = 1$, since

$$h = \frac{4}{w^2}$$
$$= \frac{4}{(2)^2}$$
$$= \frac{4}{4}$$
$$= 1$$

The box has dimensions $w \times w \times 2h$, so the optimal dimensions are $2' \times 2' \times 2'$.

FIGURE 11.16

EXAMPLE 7 **Minimizing Use of Resources**

Many bulk food commodities are packaged in #10 cans. These cylindrical metal cans are 6.875 inches tall and have a 7.25-inch diameter. There is a 0.125-inch lip on the top and on the bottom of the can, which reduces the height of the interior of the can (from lid to lid) to 6.625 inches (Figure 11.16). The volume of a cylindrical can is $V = \pi r^2 h$, where r is the radius of the can and h is the interior height of the can.

A cylindrical can may be constructed from a flat sheet of metal of length $2\pi r$ and two circular lids each with surface area $A = \pi r^2$. The height of the sheet of metal is $h + 0.5$ (Figure 11.17).

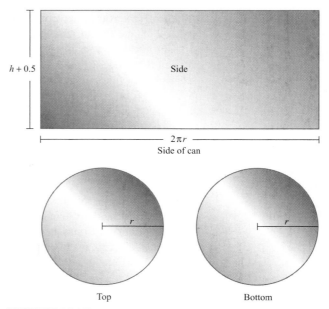

FIGURE 11.17

The additional 0.5 inch of metal is required to create the lip at the top and the bottom of the can. (The metal initially extends upward about 0.25 inch but is folded in half when the can is sealed, creating the 0.125-inch top lip. We assume that the 0.125-inch bottom lip was created in a similar manner.)

(a) What is the volume of a #10 can?
(b) How much metal is required to construct a #10 can?
(c) Is it possible to construct a can with the same volume as a #10 can that requires less metal? If so, what is the minimum amount of metal required?

SOLUTION

(a) The diameter is 7.25 inches. Since the radius r is equal to half of the diameter,

$$r = 0.5(7.25)$$
$$= 3.625$$

The volume V of a cylinder with radius r and height h is given by $V = \pi r^2 h$.

$$V = \pi r^2 h$$
$$= \pi (3.625)^2 (6.625)$$
$$= 273.5 \text{ cubic inches}$$

The volume of the can is 273.5 cubic inches.

(b) The amount of metal required to construct the sides of the can is

$$S = 2\pi r(h + 0.5)$$
$$= 2\pi(3.625)(6.625 + 0.5)$$
$$= 162.3 \text{ square inches}$$

The amount of metal required to construct each lid is

$$L = \pi r^2$$
$$= \pi(3.625)^2$$
$$= 41.28 \text{ square inches}$$

The total amount of metal required is

$$A = 2\pi r(h + 0.5) + \pi r^2 + \pi r^2$$
$$= 162.3 + 41.28 + 41.28$$
$$= 244.9 \text{ square inches}$$

(c) The surface area of a cylindrical can with a 0.125-inch lip at the top and bottom of the can is

$$A = 2\pi r(h + 0.5) + \pi r^2 + \pi r^2$$
$$= 2\pi rh + \pi r + 2\pi r^2$$
$$= \pi r(2h + 1 + 2r)$$

In order to minimize A, we must find A'. However, A is presently a function of two variables. We must use a secondary equation to solve for one variable in terms of the other.

The volume of the can is $V = \pi r^2 h$. We want the can to have the same volume as a #10 can. So

$$V = \pi r^2 h$$
$$273.5 = \pi r^2 h \qquad \text{The volume of a #10 can is 273.5 cubic inches}$$
$$h = \frac{273.5}{\pi r^2}$$

We may now rewrite the surface area equation in terms of a single variable, r.

$$A = \pi r(2h + 1 + 2r)$$
$$= \pi r\left[2\left(\frac{273.5}{\pi r^2}\right) + 1 + 2r\right]$$
$$= \pi r\left(\frac{547}{\pi r^2} + 1 + 2r\right)$$
$$= \frac{547}{r} + \pi r + 2\pi r^2$$
$$= 547r^{-1} + \pi r + 2\pi r^2$$
$$A'(r) = -547r^{-2} + \pi + 4\pi r$$
$$0 = -\frac{547}{r^2} + \pi + 4\pi r$$

We will use the graphing calculator to find the x-intercept(s) of the graph of $A'(r)$ (Figure 11.18).

FIGURE 11.18

The only critical value is $r = 3.436$. We can visually see that $A'(r) < 0$ for values of $r < 3.436$. Likewise, $A'(r) > 0$ for values of $r > 3.436$. Since the sign of $A'(r)$ changes from negative to positive at $r = 3.436$, a relative minimum of A occurs at $r = 3.436$. The corresponding value of h is

$$h = \frac{273.5}{\pi(3.436)^2}$$

$$= 7.374 \text{ inches}$$

Using these values of h and r, we calculate the surface area.

$$A(r) = 547r^{-1} + \pi r + 2\pi r^2$$
$$A(3.436) = 547(3.436)^{-1} + \pi(3.436) + 2\pi(3.436)^2$$
$$= 244.2 \text{ square inches}$$

Remarkably, the #10 can very nearly minimizes the amount of metal required (244.9 square inches versus the minimal 244.2 square inches).

EXAMPLE 8

Minimizing Landscaping Costs

A landscape designer offers an Economy garden package to her clients. The rectangular garden has flowers along the front of the garden and shrubs along the back and sides (Figure 11.19).

FIGURE 11.19

The designer charges $25 per linear foot for the shrubs and $6 per linear foot for the flowers. A client wants to install a 400-square-foot garden. What will be the dimensions of the least expensive garden?

SOLUTION We begin by labeling the figure. We let l represent the length of the garden and w represent the width of the garden (Figure 11.20). The area of the garden is given by

$$A = lw$$
$$400 = lw$$

FIGURE 11.20

Assuming that the shrubs extend across the back of the garden and fully down both sides, the cost of the shrubs (in dollars) is

$$S = 25(w + l + w)$$
$$= 25l + 50w$$

Assuming that the flowers extend across the front of the garden, the cost of the flowers (in dollars) is

$$F = 6l$$

The combined cost for the garden is

$$C = (25l + 50w) + 6l$$
$$= 31l + 50w$$

We want to minimize this cost. We can't differentiate C because it contains two input variables, l and w. However, recall that

$$lw = 400$$
$$l = \frac{400}{w}$$

Therefore, the cost function may be rewritten as

$$C = 31l + 50w$$

$$= 31\left(\frac{400}{w}\right) + 50w$$

$$= 12{,}400w^{-1} + 50w$$

Differentiating the cost function and setting it equal to zero yields

$$C'(w) = -12{,}400w^{-2} + 50$$

$$0 = -12{,}400w^{-2} + 50 \qquad \text{Set } C'(w) = 0$$

$$\frac{12{,}400}{w^2} = 50$$

$$12{,}400 = 50w^2$$

$$w^2 = 248$$

$$w \approx 15.75 \qquad\qquad \text{Since } w > 0, \text{ we omit the solution} \\ w \approx -15.75$$

Applying the First Derivative Test, we determine that a relative minimum occurs at $w = 15.75$. (*Note:* $C'(15) < 0$ and $C'(16) > 0$.) To find the corresponding value of l, we substitute the value of w into the area equation.

$$l = \frac{400}{w}$$

$$= \frac{400}{15.75}$$

$$= 25.40$$

The dimensions that will minimize the cost are 25.40 feet by 15.75 feet. The minimum cost is

$$C = 31l + 50w$$

$$= 31(25.40) + 50(15.75)$$

$$= \$1574.90$$

The shape of the garden (drawn to scale) is shown in Figure 11.21. The client may want to change the dimensions; however, any change in dimensions will result in a higher cost.

FIGURE 11.21 25.4 feet $\times$ 15.75 feet

11.2 Summary

In this section, you learned how to find marginal revenue, marginal cost, and marginal profit. You also learned how to use the derivative to optimize specified areas and volumes.

11.2 Exercises

In Exercises 1–30, use the derivative techniques demonstrated in this section to determine the answer to each question.

1. **M** **Company Profit** In 2001, the Kellogg Company introduced a new breakfast cereal: Special K® Red Berries cereal. It quickly achieved an impressive 1 percent market share during its first six months on the market and helped to boost corporate sales. Based on data from 1999 to 2001, the net sales of the Kellogg Company may be modeled by

$$R(t) = 964.1t^2 - 993.6t + 6984$$

million dollars, where t is the number of years since 1999. The cost of goods sold may be modeled by

$$C(t) = 399.8t^2 - 379.9t + 3325.1$$

million dollars, where t is the number of years since 1999. (**Source:** Modeled from Kellogg Company 2001 Annual Report, pp. 7, 27.)

(a) Find the equation for the gross profit. (This is net sales minus the cost of goods sold.)

(b) Between 1999 and 2001, in which year was gross profit minimized?

2. **Carton Design** A carton of Peer Foods Sliced Canadian Style Bacon holds four 3-pound packages of sliced Canadian bacon. The dimensions of a carton are 11″ × 19.75″ × 4″. (**Source:** www.peerfoods.com.)

(a) What is the volume of the carton?

(b) Is it possible to construct a carton with the same volume but less surface area? If so, find the dimensions of the carton with minimal surface area.

(c) What other issues should Peer Foods consider before moving to a different sized carton?

3. **Box Design** Peer Foods sells its bacon in two different case sizes. The Hotel Pack case has dimensions

10.1875″ × 8.5″ × 7.5625″. The Single Slice Pack case has dimensions 19.75″ × 11″ × 4″. (**Source:** www.peerfoods.com.)

(a) Calculate the surface area and volume of each of the cases.

(b) Find the dimensions of the box with a square base that has the same surface area as the Hotel Pack case but has the greatest possible volume.

(c) Find the dimensions of the box with a square base that has the same surface area as the Single Slice Pack case but has the greatest possible volume.

(d) Based on the results of (b) and (c), what conclusion can you draw regarding the dimensions of a box that has the maximum possible volume for its surface area? (*Hint:* Divide the height of each box by its width and look for a pattern.)

4. **M** **Company Revenue** Uline Shipping Supply Specialists sell 40″ × 48″ hardwood pallets. The price they charge per pallet is reduced as the order quantity increases. The price of hardwood pallets is given by the piecewise function

$$p = \begin{cases} -5.5x + 124.5 & 1 \le x \le 2 \\ 105 & 3 \le x \le 10 \end{cases}$$

dollars, where x is the number of groups of 10 pallets. (**Source:** www.uline.com.)

(a) Write the equation for revenue as a function of the number of 10-pallet groups sold.

(b) Determine the marginal revenue when 10, 20, 30, and 40 pallets are sold.

(c) Explain the real-world meaning of the results of (b).

5. **M** **Company Revenue** In its 2001 Annual Report, the Coca-Cola Company reported that "Our worldwide unit case volume

increased 4 percent in 2001, on top of a 4 percent increase in 2000. The increase in unit case volume reflects consistent performance across certain key operations despite difficult global economic conditions. Our business system sold 17.8 billion unit cases in 2001." (**Source:** Coca-Cola Company 2001 Annual Report, p. 46.) A unit case is equivalent to 24 eight-ounce servings of finished beverage.

(a) Find a model for the number of unit cases sold, s, as a function of t, where t is the number of years since 1999.

(b) Based on data from 1999 to 2001, the net operating revenue of the Coca-Cola Company may be modeled by

$$R(s) = -0.4029s^2 + 14.44s - 109.2$$

billion dollars, where s is the number of unit cases sold (in billions). Find the marginal revenue function.

(c) According to the model, at what unit case production level will revenue be maximized?

(d) Rewrite the revenue function as a function of t (years since 1999) instead of as a function of s (billions of unit cases sold). (*Hint:* Use a composition of functions.)

(e) According to the model, in what year will revenue be maximized?

6. **M** **Company Profit** Based on data from 1999 to 2001, the net operating revenues of the Coca-Cola Company may be modeled by

$$R(t) = -201t^2 + 806t + 19,284$$

million dollars and the cost of goods sold may be modeled by

$$C(t) = -177.5t^2 + 372.5t + 6009$$

million dollars, where t is the number of years since 1999.

(a) Find the gross profit function.

(b) According to the model, in what year is the gross profit projected to reach a maximum?

(c) Does the result of part (b) seem reasonable? Explain.

7. **M** **Company Costs** Frito-Lay North America, a subsidiary of PepsiCo, produces Doritos, Cheetos, Fritos corn chips, and a variety of other salty, sweet, or grain-based snacks. Based on data from 1999 to 2001, the net sales (revenue) of Frito-Lay North America may be modeled by

$$R(t) = -168t^2 + 907t + 8232$$

million dollars and the operating profit (earnings before interest and taxes) may be modeled by

$$P(t) = -47.5t^2 + 283.5t + 1679$$

million dollars, where t is the number of years since 1999. (**Source:** Modeled from 2001 PepsiCo Annual Report, pp. 23, 44.)

(a) In what year are net sales projected to reach a maximum?

(b) Find the cost function for Frito-Lay North America.

(c) According to the model, in what year are costs expected to reach a maximum?

(d) Compare the results of (a) and (c). Do the results seem reasonable?

8. **M** **Company Revenue Versus Profit** Based on data from 1999 to 2001, the net sales (revenue) of Pepsi-Cola North America may be modeled by

$$R(t) = -65.5t^2 - 749.5t + 2605$$

million dollars and the operating profit may be modeled by

$$P(t) = 6t^2 + 76t + 751$$

million dollars, where t is the number of years since 1999. (**Source:** Modeled from PepsiCo 2001 Annual Report, p. 44.)

(a) Find the marginal revenue and marginal profit functions.

(b) At what rate was revenue increasing in 2001?

(c) At what rate was profit increasing in 2001?

(d) Project the rate of change in revenue and in profit in 2014. Does this projection seem reasonable? Explain.

9. **M** **Employee Wages** Based on data from 1991 to 2001, the average wage of a Ford Motor Company employee may be modeled by

$$W(t) = -0.003931t^4 + 0.1005t^3 - 0.8295t^2 + 3.188t + 16.48$$

dollars per hour, where t is the number of years since 1990. (**Source:** Modeled from Ford Motor Company 2001 Annual Report, p. 71.)

Was the average wage changing more rapidly in 1995 or 2000?

10. **Pool Size Optimization** A pool builder makes two types of economy-priced pools: square and circular. He estimates the price of the job by multiplying the perimeter or circumference of the pool (in linear feet) by a fixed price per linear foot. A homeowner wants her pool to have the

maximum amount of water surface area for the lowest possible price.

(a) Should the homeowner have a square or a circular pool built? Explain. (The area of a square with a perimeter p is $S = \frac{1}{16}p^2$. The area of a circle with circumference c is $A = \frac{1}{4\pi}c^2$.)

(b) Shortly before construction, the homeowner decides to increase the 100-foot distance around the pool by 1 foot. For both shapes of pool, use the derivative to approximate how much the surface area of the pool will increase by increasing the perimeter (circumference) by 1 foot.

(c) Does your result in (b) confirm your conclusion in (a)? Explain.

11. **Apple Farming** Historically, many apple farmers spaced trees 40 feet by 40 feet apart (27 per acre). Trees typically took 25 years to reach their maximum production of 500 bushels per acre. (A bushel is about 44 pounds.) In recent years, agriculturists have created dwarf and semi-dwarf varieties that allow trees to be spaced 10 feet by 10 feet apart. (**Source:** USDA.)

Suppose that a farmer has an apple orchard with 40 trees per acre. The orchard yields 10 bushels per tree. The farmer estimates that for each additional tree planted (per acre), the average yield per tree is reduced by 0.1 bushel.

If $y = f(x)$ is the total number of bushels of apples produced per acre when an additional x trees per acre are planted, determine how many additional trees should be planted in order to maximize the number of bushels of apples produced.

12. **Apple Farming** Suppose that a farmer has an apple orchard with 30 trees per acre. The orchard yields 12 bushels per tree. The farmer estimates that for each additional tree planted (per acre), the average yield per tree is reduced by 0.1 bushel.

If $y = f(x)$ is the total number of bushels of apples produced per acre when an additional x trees per acre are planted, determine how many additional trees should be planted in order to maximize the number of bushels of apples produced.

13. **Apple Supplier Prices** A fruit farmer sells apples to a grocery store chain. The amount of apples the store buys depends linearly upon the price per pound the farmer charges. The farmer estimates that for every $0.02 per pound increase in the price, the store will reduce its order by 50 pounds. The store presently orders 500 pounds per week and pays $0.18 per pound. What price should the farmer charge in order to maximize her revenue from apple sales?

14. **Apple Retailer Prices** A grocery store has priced apples at $0.70 per pound and sells 1000 pounds per week. The amount of apples the store sells depends linearly upon the price per pound the store charges. The store manager estimates that for every $0.04 per pound increase in the price, the store will reduce its sales by 100 pounds. What price should the store charge for its apples in order to maximize revenue from apple sales?

15. **Company Sales** Based on data from 1993 to 2002, the annual sales of Starbucks Corporation may be modeled by

$$S(t) = 29.23t^2 + 79.33t + 177.4$$

million dollars, where t is the number of years since 1993.

According to the model, how much more rapidly were sales increasing in 2001 than in 1999?

16. **Company Sales** Based on data from 1991 to 2001, the franchised sales of McDonalds Corporation may be modeled by

$$S(t) = -41.293t^2 + 1729.0t + 11{,}139$$

million dollars, where t is the number of years since 1990.

Calculate the instantaneous rate of change of sales in 1999 and 2001. Explain the financial significance of the result.

17. **Landscape Design** A landscaping company offers its clients the Rectangular Garden package at a reduced rate.

The Rectangular Garden has shrubs along the back and sides of the garden and flowers along the front of the garden. (See figure.)

The designer charges $25 per linear foot for the shrubs and $8 per linear foot for the flowers. A client wants to install a 500-square-foot garden. What will be the dimensions of the least expensive garden?

18. **Landscape Design** A landscaping company offers its clients the Rectangular Garden package at a reduced rate.

   The Rectangular Garden has shrubs along the back and sides of the garden and flowers along the front of the garden. (See figure.)

   The designer charges $20 per linear foot for the shrubs and $5 per linear foot for the flowers. A client wants to install a 250-square-foot garden. What will be the dimensions of the least expensive garden?

19. **Inventory Cost** An author anticipates that she will sell 6,000 copies of her book annually. The setup cost for each print run is $300. The storage cost for each printed book is $2.50 per year. Assuming that customer demand for the book is constant for the 250 business days of a year, determine the size of each print run that will minimize her inventory cost.

20. **Inventory Cost** An author anticipates that she will sell 5,000 copies of her book annually. The setup cost for each print run is $300. The storage cost for each printed book is $1.92 per year. Assuming that customer demand for the book is constant for the 250 business days of a year, determine the size of each print run that will minimize her inventory cost.

21. **Inventory Cost** An author anticipates that he will sell 500 copies of his book annually. The setup cost for each print run is $300. The storage cost for each printed book is $1.20 per year. Assuming that customer demand for the book is constant for the 250 business days of a year, determine the size of each print run that will minimize his inventory cost.

22. **Inventory Cost** An author anticipates that he will sell 12,000 copies of his book annually. The setup cost for each print run is $300. The storage

cost for each printed book is $1.25 per year. Assuming that customer demand for the book is constant for the 250 business days of a year, determine the size of each print run that will minimize his inventory cost.

23. **Fleet Vehicle Sales** A business that owns or leases 10 or more vehicles may qualify for auto manufacturer fleet purchase incentives. Incentives vary from $300 to $7000 on new vehicles. In August 2003, the manufacturer's fleet incentive for a 2004 Chevrolet Cavalier was $1400. (**Source:** www.fleet-central.com.)

   An auto dealer offers an additional discount to fleet buyers who purchase one or more new Cavaliers. To encourage sales, the dealer reduces the after-incentive price of each car by $x\%$, where x is the total number of cars purchased. Assuming that the pre-incentive price of a 2004 Chevrolet Cavalier is $14,400, how many vehicles would the dealer need to sell in order to maximize revenue?

24. **Fleet Vehicle Sales** In August 2003, the manufacturer's fleet incentive for a 2004 Cadillac Seville was $2000. (**Source:** www.fleet-central.com.)

   An auto dealer offers an additional discount to fleet buyers who purchase one or more new Cadillac Sevilles. To encourage sales, the dealer reduces the after-incentive price of each car by $2x\%$, where x is the total number of cars purchased. Assuming that the pre-incentive price of a 2004 Cadillac Seville is $41,000, how many vehicles would the dealer need to sell in order to maximize revenue?

25. **Fleet Vehicle Sales** In August 2003, the manufacturer's fleet incentive for a 2004 Buick Century was $1800. (**Source:** www.fleet-central.com.)

   An auto dealer offers an additional discount to fleet buyers who purchase one or more new Buick Century vehicles. To encourage sales, the dealer reduces the after-incentive price of each car by $200x$ dollars, where x is the total number of cars purchased. Assuming that the pre-incentive price of a 2004 Buick Century is $19,000, how many vehicles would the dealer need to sell in order to maximize revenue?

26. **Fleet Vehicle Sales** In August 2003, the manufacturer's fleet incentive for a 2004 Pontiac Bonneville was $900. (**Source:** www.fleet-central.com.)

An auto dealer offers an additional discount to fleet buyers who purchases one or more new Pontiac Bonneville vehicles. To encourage sales, the dealer reduces the after-incentive price of each car by $250x$ dollars, where x is the total number of cars purchased. Assuming that the pre-incentive price of a 2004 Pontiac Bonneville is $24,000, how many vehicles would the dealer need to sell in order to maximize revenue?

27. Pricing Analysis Based on the results of Exercises 23 to 26, what additional restrictions (if any) should the dealer place on his advertised discount in order to ensure that selling additional vehicles won't reduce his revenue?

28. Profit Lost Due to Waste A fruit vendor purchases x pounds of fruit from her supplier. She estimates that $0.2x\%$ of each pound of produce she buys spoils before it is purchased by a consumer. For each pound of fruit she sells, she makes a profit of $0.40. How many pounds of fruit should she buy in order to maximize profit?

29. Profit Lost Due to Waste A fruit vendor purchases x pounds of fruit from his supplier. He estimates that $0.2x\%$ of each pound of produce he buys spoils before it is purchased by a consumer. For each pound of fruit he sells, he makes a profit of $0.50. How many pounds of fruit should he buy in order to maximize profit?

30. Profit Analysis For many commodities, the marginal profit fluctuates. Based on the results of Exercises 28 and 29, what effect does the profit

per pound have on the quantity of fruit that maximizes profit?

Exercises 31–36 are intended to challenge your skill in finding and interpreting the meaning of extrema. Each of the questions deals with a parametric function model of the position of a roller coaster train over a six-second interval. The x value of the model represents the horizontal position of a roller coaster train (in meters) at time t seconds. The y value of the model represents the vertical position of the roller coaster train (in meters) at time t seconds. The model is

$$x = t^3 - 9t^2 + 23t$$
$$y = 6t - t^2$$

31. Calculate and interpret the meaning of $\dfrac{dx}{dt}$ and $\dfrac{dy}{dt}$.

32. Determine when $\dfrac{dx}{dt} = 0$ and $\dfrac{dy}{dt} = 0$.

33. Given that the domain of the function is $0 \leq t \leq 6$, find the relative and absolute extrema of x and y.

34. Over the six-second interval, what was the slowest horizontal speed of the train?

35. Over the six-second interval, what was the slowest vertical speed of the train?

36. Do you think that the roller coaster would be safe to ride? Defend your conclusion.

11.3 Concavity and the Second Derivative

- Use the Second Derivative Test to find relative extrema
- Determine the concavity of the graph of a function

GETTING STARTED Based on data from 1992 to 1998, the average annual earnings of a prepackaged software retailer may be modeled by

$$W(t) = 49.62t^4 - 637.12t^3 + 2895t^2 - 3768t + 32,150 \text{ dollars}$$

where t is the number of years since the end of 1992. (**Source:** Modeled from *Statistical Abstract of the United States, 2001*, Table 1123, p. 703.)

In what year were annual earnings increasing the fastest, and in what year were they increasing the slowest? Did the rate of increase in earnings continually grow, or did the rate of increase in earnings shrink at some point? Questions such as these may be answered using the *second derivative*.

In this section, we will introduce the second derivative and discuss the graphical concepts of concavity and inflection points. We will use these concepts to investigate the relationship between position, velocity, and acceleration. We will also introduce the Second Derivative Test as an alternative means of finding relative extrema. We will conclude with curve-sketching techniques.

Concavity

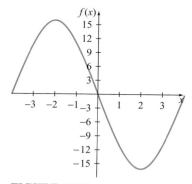

f(x)

FIGURE 11.22

The term **concavity** refers to the curvature of a graph. A graph is said to be **concave up** if it curves upward and **concave down** if it curves downward. A simple rhyme is helpful in remembering the meaning of the terms.

Concave up is like a cup.

Concave down is like a frown.

We will explore the concept of concavity by looking at the graph of the function $f(x) = x^3 - 12x$ on the interval $[-3, 3]$ (Figure 11.22).

From $x = -3$ to $x = 0$, the graph is curved downward (looks like a frown). At $x = 0$, the concavity appears to change. From $x = 0$ to $x = 3$, the graph is curved upward (looks like a cup). Is there a way to determine the concavity of a graph algebraically? Let's see. We'll begin by finding the derivative of the function and then the derivative of the derivative. The derivative of the derivative is called the **second derivative** and is commonly denoted by y'' (read "y double prime") or $f''(x)$ (read "f double prime of x").

$$f(x) = x^3 - 12x$$
$$f'(x) = 3x^2 - 12$$
$$f''(x) = 6x$$

We will generate a table of values for each of the functions (see Table 11.11).

TABLE 11.11

x	$f(x)$	$f'(x)$	$f''(x)$
-3	9	15	-18
-2	16	0	-6
-1	11	-9	-3
0	0	-12	0
1	-11	-9	3
2	-16	0	6
3	-9	15	18

We can learn much about the graph of the function from this table.

The graph of the function has x-intercepts when $f(x) = 0$. The x-intercept of the function on the domain $[-3, 3]$ occurs at $x = 0$, since $f(0) = 0$.

The graph of the function has relative extrema when $f'(x) = 0$ and the derivative changes sign. A relative maximum occurs at $x = -2$, since $f'(-2) = 0$ and the derivative changes from positive to negative at $x = -2$. A relative minimum occurs at $x = 2$, since $f'(2) = 0$ and the derivative changes from negative to positive at $x = 2$.

We observe that on the interval $[-3, 0)$, $f''(x) < 0$, and on the interval $(0, 3]$, $f''(x) > 0$. At $x = 0$, $f''(x) = 0$. Observe that the graph of f is concave down when $f''(x) < 0$ and is concave up when $f''(x) > 0$. The graph changes concavity from down to up at $x = 0$. The point on the graph where a function changes concavity is called an **inflection point.**

CONCAVITY AND INFLECTION POINTS OF THE GRAPH OF A FUNCTION

1. A continuous function f is *concave up* at a point $(c, f(c))$ if $f''(c) > 0$.
2. A continuous function f is *concave down* at a point $(c, f(c))$ if $f''(c) < 0$.
3. A continuous function f has an *inflection point* at $(c, f(c))$ if
   - $f''(c) = 0$ or $f''(c)$ is undefined **and**
   - $f''(c)$ changes sign at $x = c$

By referring back to the earlier rhyme, we can come up with a clever strategy to remember how to determine the concavity of a graph.

When we're positive, we are up (like the cup).

When we're negative, we're down and we frown.

Admittedly, the rhyme is a bit silly; however, its sheer wackiness will make the concept easier to remember.

A common error made by learners in their search for inflection points is to assume that if $f''(c) = 0$, then $(c, f(c))$ is an inflection point. Although $f''(c) = 0$ is a necessary condition for an inflection point, it is not a sufficient condition. Consider the function $f(x) = x^4$ with its associated second derivative $f''(x) = 12x^2$. Although $f''(0) = 0$, the point $(0, 0)$ is not an inflection point of f since the graph of f is concave up on both sides of the point (Figure 11.23).

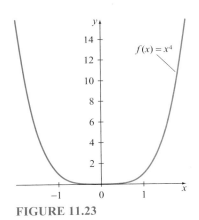

FIGURE 11.23

EXAMPLE 1

Determining the Concavity of a Graph

The graph of $f(x) = x^4 - 4x^3$ on the interval $[-3, 5]$ is shown in Figure 11.24. Determine where the graph is concave up and where the graph is concave down. Identify all inflection points.

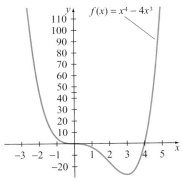

FIGURE 11.24

SOLUTION The graph appears to be concave up from $[-3, 0]$, concave down from $[0, 2]$, and concave up again from $[2, 5]$. It is difficult to determine the exact location of inflection points visually, so we will proceed algebraically to verify our visual conclusions.

$$f(x) = x^4 - 4x^3$$
$$f'(x) = 4x^3 - 12x^2$$
$$f''(x) = 12x^2 - 24x$$

In order to identify potential inflection points, we must set $f''(x)$ equal to zero.

$$0 = 12x^2 - 24x$$
$$0 = 12x(x - 2)$$
$$12x = 0 \quad \text{or} \quad x - 2 = 0$$
$$x = 0 \quad\quad\quad x = 2$$

Inflection points may occur at $x = 0$ or $x = 2$; however, we must verify that $f''(x)$ changes sign at each of these points. We will do this using a sign chart for f''. We evaluate $f''(x)$ using points on either side of the potential inflection points.

$$f''(x) = 12x\,(x - 2)$$
$$f''(-1) = 12(-1)(-1 - 2) > 0$$
$$f''(1) = 12(1)(1 - 2) < 0$$
$$f''(3) = 12(3)(3 - 2) > 0$$

We update the sign chart for f'' with the results.

Since $f''(x)$ changes sign at $x = 0$ and $x = 2$, inflection points occur at each of these points. Since $f''(x) > 0$ on $[-3, 0)$ and on $(2, 5]$, it is concave up on those intervals. Since $f''(x) < 0$ on $(0, 2)$, it is concave down on that interval. Our algebraic analysis confirms our graphical estimation.

EXAMPLE **2**   ### Determining the Concavity of a Graph

Determine the concavity of the graph $f(x) = x^3 + 3x^2 + 3x + 1$ on the interval $[-2, 2]$. Identify the location of any inflection points.

SOLUTION

$$f(x) = x^3 + 3x^2 + 3x + 1$$
$$f'(x) = 3x^2 + 6x + 3$$
$$f''(x) = 6x + 6$$

We set $f''(x) = 0$ and solve.

$$0 = 6x + 6$$
$$-6x = 6$$
$$x = -1$$

An inflection point may occur at $x = -1$. We'll evaluate the second derivative on either side of $x = -1$ and construct a sign chart for f''.

$$f''(-2) = 6(-2) + 6$$
$$= -6$$

$$f''(0) = 6(0) + 6$$
$$= 6$$

$f(x)$ is concave down on the interval $[-2, -1)$ and concave up on the interval $(-1, 2]$. An inflection point occurs at the point $(-1, 0)$ (Figure 11.25).

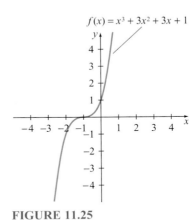

$$f(x) = x^3 + 3x^2 + 3x + 1$$

FIGURE 11.25

EXAMPLE 3

Using the Second Derivative to Determine Maximum and Minimum Rates of Change

Based on data from 1992 to 1998, the average annual earnings of a prepackaged software retailer may be modeled by

$$W(t) = 49.62t^4 - 637.12t^3 + 2895t^2 - 3768t + 32{,}150 \text{ dollars}$$

where t is the number of years since 1992. (**Source:** Modeled from *Statistical Abstract of the United States, 2001*, Table 1123, p. 703.)

In what year were annual earnings increasing the fastest, and in what year were they increasing the slowest? Did the rate of increase in earnings continually grow, or did the rate of increase in earnings shrink at some point?

SOLUTION The rate of change of the earnings function is $W'(t)$.

$$W'(t) = 198.48t^3 - 1911.36t^2 + 5790t - 3768 \ \frac{\text{dollars}}{\text{year}}$$

This function tells us how fast or how slowly earnings are increasing. We want to determine the extrema of $W'(t)$. To do this, we differentiate $W'(t)$.

UNITS

$$W''(t) = 595.44t^2 - 3822.72t + 5790 \ \frac{\text{dollars per year}}{\text{year}}$$

We set $W''(t)$ equal to zero and solve using the quadratic formula.

$$t = \frac{-b \pm \sqrt{b^2 - 4ac}}{2a}$$

$$= \frac{3822.72 \pm \sqrt{(-3822.72)^2 - 4(595.44)(5790)}}{2(595.44)}$$

$$= \frac{3822.72 \pm 907.08}{1190.88}$$

$$= 2.448; \ 3.972$$

W'' sign chart:

	max of W'		min of W'	
+		−		+

2.448   3.972

We evaluate $W''(t)$ on either side of the critical points of $W'(t)$ and construct a sign chart for $W''(t)$.

The relative extrema of $W'(t)$ are in fact the inflection points of $W(t)$. Inflection points tell us when the growth rate of $W(t)$ is accelerating or slowing down. Let's interpret the financial meaning of our results.

$$W'(2.448) = 1863 \text{ dollars per year}$$

$$W'(3.972) = 1513 \text{ dollars per year}$$

Let's look again at the sign chart of W''.

Recall that W' is the growth rate of employee earnings (in dollars per year). Between the end of 1992 ($t = 0$) and the middle of 1995 ($t = 2.448$), the growth rate of employee earnings was increasing. However, between the middle of 1995 and the end of 1996 ($t = 3.972$), the growth rate of employee earnings was decreasing. That is, although earnings were increasing, they were increasing at a slower rate. In the middle of 1995 ($t = 2.448$), the earnings growth rate was $1863 per year, but by the end of 1996 ($t = 3.972$), the earnings growth rate had fallen to $1513 per year. Between the end of 1996 and the end of 1998, the growth rate of employee earnings was again increasing. We can verify our conclusions by generating a table of values for W, W', and W''. Tables such as Table 11.12 may be quickly generated using the Technology Tip following this example.

TABLE 11.12

Years Since 1992 (t)	$W(t)$ (dollars)	$W'(t)$ (dollars per year)	$W''(t)$ (dollars per year per year)
0	32,150	−3,768	5,790
1	30,690	309	2,563
2	31,891	1,754	526
3	33,718	1,759	−319
4	35,325	1,513	26
5	37,058	2,208	1,562
6	40,452	5,035	4,290

In what year were annual earnings increasing the fastest, and in what year were they increasing the slowest? Earnings were diminishing at a rate of $3768 per year in 1992. This was the slowest rate of growth between 1992 and 1998. Earnings were increasing at a rate of $5035 per year in 1998. This is the fastest rate of growth between 1992 and 1998.

Did the rate of increase in earnings grow continually, or did the rate of increase in earnings shrink at some point? Notice that the values of $W'(t)$ increased every year except between 1995 ($t = 3$) and 1996 ($t = 4$). The fact that $W''(3)$ was negative told us that the rate of increase in earnings was decreasing in 1995.

TECHNOLOGY **TIP**

Creating a Table of Values

1. Use the $\boxed{Y=}$ editor to enter the function(s) to be evaluated. In this case we enter W, W', and W''.

2. Press $\boxed{2nd}$ $\boxed{WINDOW}$ to open Table Setup. `TblStart=` is the x value where we want the table to begin, and Δ`Tbl` is the distance between consecutive x values. We will start the table at $x = 0$ and space values one unit apart.

```
TABLE SETUP
 TblStart=0
 △Tbl=1
Indpnt: AUTO Ask
Depend: AUTO Ask
```

3. Press $\boxed{2nd}$ $\boxed{GRAPH}$ to display the table.

4. To see additional function values, use the blue arrows to navigate between rows and columns.

5. If you prefer to select the x values of the function(s) manually, press $\boxed{2nd}$ $\boxed{WINDOW}$ and select `Indpnt:Ask`.

```
TABLE SETUP
 TblStart=0
 △Tbl=1
Indpnt: Auto ASK
Depend: AUTO Ask
```

6. Press $\boxed{2nd}$ $\boxed{GRAPH}$ to bring up the table, then enter the desired x values.

```
  X   │  Y1  │  Y2
 0    │ 32150│ -3768
 2.448│ 32710│ 1863.4
 3    │ 33718│ 1758.7
 3.972│ 35283│ 1512.6
 5    │ 37058│ 2208
 ▬▬▬▬ │      │
────────────────────
X=
```

The Second Derivative Test

As we've seen, the second derivative may be used to determine the concavity of the graph and find the location of inflection points. The second derivative may also be used to find relative extrema. Consider the function $f(x) = x^3 - 12x$. Differentiating the function yields

$$f'(x) = 3x^2 - 12$$
$$0 = 3(x^2 - 4) \qquad \text{Set } f'(x) = 0 \text{ to find critical values}$$
$$0 = 3(x - 2)(x + 2)$$
$$\begin{array}{ccc} x - 2 = 0 & & x + 2 = 0 \\ & \text{or} & \\ x = 2 & & x = -2 \end{array}$$

The function f has critical values $x = -2$ and $x = 2$, since $f'(-2) = 0$ and $f'(2) = 0$. Graphically speaking, the graph of $f(x) = x^3 - 12x$ is flat (has a horizontal tangent line) at $x = -2$ and $x = 2$.

The second derivative of $f(x) = x^3 - 12x$ is $f''(x) = 6x$. We evaluate f'' at each of the critical values of f.

$$f''(x) = 6x$$
$$f''(-2) = 6(-2)$$
$$= -12$$
$$f''(2) = 6(2)$$
$$= 12$$

Since $f''(-2) < 0$, the graph of f is concave down at $x = -2$. However, since f also has a horizontal tangent line at $x = -2$, a relative maximum occurs there.

Since $f''(2) > 0$, the graph of f is concave up at $x = 2$. However, since f also has a horizontal tangent line at $x = 2$, a relative minimum occurs there (Figure 11.26).

These results are summarized in the Second Derivative Test.

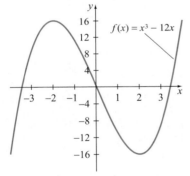

FIGURE 11.26

THE SECOND DERIVATIVE TEST

Let f be a continuous function with $f'(c) = 0$. If

- $f''(c) > 0$, then a relative minimum of f occurs at $x = c$.
- $f''(c) < 0$, then a relative maximum of f occurs at $x = c$.
- $f''(c) = 0$, then the test is inconclusive.

EXAMPLE **4**

Finding Relative Extrema with the Second Derivative Test

Use the Second Derivative Test to find the relative extrema of the function $f(x) = x^5 - 5x^4$.

SOLUTION

$$f'(x) = 5x^4 - 20x^3$$
$$0 = 5x^3(x - 4) \qquad \text{Set } f'(x) = 0 \text{ and find critical values}$$

$$5x^3 = 0 \qquad \text{or} \qquad x - 4 = 0$$
$$x = 0 \qquad\qquad\qquad x = 4$$

The critical values are $x = 0$ and $x = 4$. To apply the Second Derivative Test, we calculate $f''(x)$ and evaluate it at the critical values, $x = 0$ and $x = 4$. Writing the second derivative in factored form will make it easier to evaluate.

$$f''(x) = 20x^3 - 60x^2$$
$$= 20x^2(x - 3)$$
$$f''(0) = 20(0)^2(0 - 3)$$
$$= 0$$
$$f''(4) = 20(4)^2(4 - 3)$$
$$= 320$$

Since $f''(4) > 0$, a relative minimum occurs at $x = 4$. The relative minimum is $(4, -256)$.

At $x = 0$, the Second Derivative Test is inconclusive. What is happening at $x = 0$? Let's evaluate $f'(x)$ at points on either side of the critical value $x = 0$ to see if the slope of f changes sign at $x = 0$. (This is the First Derivative Test.)

$$f'(-1) = 5(-1)^3(-1 - 4)$$
$$= 25$$

Since $f'(-1) > 0$, the graph of f is increasing at $x = -1$.

$$f'(1) = 5(1)^3(1 - 4)$$
$$= -15$$

Since $f'(1) < 0$, the graph of f is decreasing at $x = 1$. Since f' changes from positive to negative at $x = 0$, a relative maximum occurs at $x = 0$. The relative maximum is $(0, 0)$. The graph of the function (Figure 11.27) confirms our conclusion.

Although calculations are often simpler when we use the Second Derivative Test, we may have to revert back to the First Derivative Test if the Second Derivative Test yields an inconclusive result.

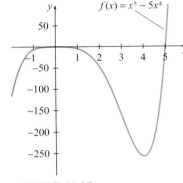

FIGURE 11.27

Sometimes we want to maximize the value of the first derivative or the second derivative instead of the value of the original function. In Example 5, we will maximize the derivative using the First Derivative Test. The notation gets a bit tricky because the critical points of f' are found by setting f'' equal to zero.

EXAMPLE 5

Using the Second Derivative in a Real-World Context

Based on data from 1991 to 2001, the average wage of a Ford Motor Company employee may be modeled by

$$W(t) = -0.003931t^4 + 0.1005t^3 - 0.8295t^2 + 3.188t + 16.48 \text{ dollars per hour}$$

where t is the number of years since 1990. (**Source:** Modeled from Ford Motor Company 2001 Annual Report, p. 71.) According to the model, did the rate of increase in employee hourly wages (in dollars) ever diminish between 1991 and 2001? What was the minimum and what was the maximum *annual increase* in the average hourly wage?

SOLUTION The rate of increase in hourly wages is given by

$$W'(t) = -0.015724t^3 + 0.3015t^2 - 1.659t + 3.188 \; \frac{\text{dollars}}{\text{year}}$$

We want to know if W' ever decreased between 1991 and 2001. That is, we want to know when $W'' < 0$.

$$W''(t) = -0.047172t^2 + 0.603t - 1.659 \; \frac{\text{dollars per year}}{\text{year}}$$

We set $W''(t) = 0$ and solve using the Quadratic Formula.

$$0 = -0.047172t^2 + 0.603t - 1.659$$

$$t = \frac{-0.603 \pm \sqrt{(0.603)^2 - 4(-0.047172)(-1.659)}}{2(-0.047172)}$$

$$= \frac{-0.603 \pm 0.2249}{-0.094344}$$

$$= 4.008, \; 8.775$$

The critical values of W' are $t = 4.008$ and $t = 8.775$. We'll evaluate $W''(t)$ on either side of the critical values of W'.

$$W''(0) = -1.659$$
$$W''(5) = 0.1767$$
$$W''(10) = -0.3462$$

From 1991 ($t = 1$) to early 1995 ($t = 4.008$), the annual rate of increase in hourly wages was decreasing (since $W'' < 0$). Although the hourly wage continued to rise during those years, each annual increase was less than that of the year before.

From early 1995 ($t = 4.008$) to late 1999 ($t = 8.775$), the annual rate of increase in hourly wages was increasing (since $W'' > 0$). The hourly wage continued to rise during those years, and each annual increase was more than that of the year before.

From late 1999 ($t = 8.775$) to the end of 2001 ($t = 11$), the annual rate of increase in hourly wages was decreasing (since $W'' < 0$). Although the hourly wage continued to rise during those years, each annual increase was less than that of the year before.

The maximum annual increase in hourly wage will occur when W' reaches its maximum value. From the sign chart for W'', we see that a relative maximum of W' occurs when $t = 8.775$. We'll need to check the endpoints to see if this is an absolute maximum.

The minimum annual increase in hourly wage will occur when W' reaches its minimum value. From the sign chart for W'', we see that a relative minimum of W' occurs when $t = 4.008$. We'll need to check the endpoints to see if this is an absolute minimum. (See Table 11.13.)

TABLE 11.13

Years Since 1990 (t)	Dollars per Year [$W'(t)$]	
1	1.81	Absolute maximum of $W'(t)$
4.008	0.37	Relative and absolute minimum of $W'(t)$
8.775	1.22	Relative maximum of $W'(t)$
11	0.49	

According to the model, the average hourly wage was increasing most slowly in early 1995 ($t = 4.008$). At that time, it was increasing at a rate of $0.37 per year.

According to the model, the average hourly wage was increasing most rapidly at the end of 1991 ($t = 1$). At that time, it was increasing at a rate of $1.81 per year.

Point of Diminishing Returns

In business, we are often interested in knowing when the rate of change in revenue or profit from the sale of an item reaches a maximum value. The point at which this occurs is called the **point of diminishing returns.** In terms of calculus, we want to know when the derivative $R'(x)$ of a revenue function $R(x)$ attains a maximum.

EXAMPLE 6

Finding the Point of Diminishing Returns

Based on data from 1990 to 2000, the revenue from sales of newspaper advertising may be modeled by

$$A(t) = -32.08t^3 + 656.2t^2 - 1641t + 31,914 \text{ million dollars}$$

where t is the number of years since the end of 1990. (**Source:** Modeled from *Statistical Abstract of the United States, 2001,* Table 1272, p. 777.) Find the point of diminishing returns for newspaper advertising revenue and determine the rate at which advertising revenue is changing at that point.

SOLUTION The rate of change in newspaper advertising is $A'(t)$.

$$A(t) = -32.08t^3 + 656.2t^2 - 1641t + 31,914$$

$$A'(t) = -96.24t^2 + 1312.4t - 1641 \text{ million dollars per year}$$

To determine when $A'(t)$ attains a maximum, we must find $A''(t)$.

$$A''(t) = -192.48t + 1312.4 \text{ million dollars per year per year}$$

To find the critical values of $A'(t)$, we set $A''(t) = 0$.

$$0 = -192.48t + 1312.4$$

$$192.48t = 1312.4$$

$$t = \frac{1312.4}{192.48}$$

$$t \approx 6.8$$

At $t = 6.8$, $A'(t)$ attains a relative maximum. Thus the point of diminishing returns occurs at $t = 6.8$. The point of diminishing returns is $(6.8, A(6.8)) = (6.8, 41,011)$. According to the model, revenue from newspaper advertising was $41,011 million in late 1997 ($t = 6.8$).

To determine the rate of change in revenue at the point of diminishing returns, we evaluate $A'(t)$ at the critical value. (We also evaluate $A'(t)$ at the implied endpoints as one strategy for verifying the accuracy of our work; see Table 11.14.)

TABLE 11.14

t	$A'(t) = -96.24t^2 + 1{,}312.4t - 1641$	
0	$-1{,}641$	Absolute minimum of $A'(t)$
6.8	$2{,}833$	Relative and absolute maximum of $A'(t)$
10	$1{,}859$	

Newspaper advertising revenue was increasing most rapidly in late 1997 ($t = 6.8$). At that time, revenue was increasing at a rate of $2833 million per year.

Position, Velocity, and Acceleration

The velocity of an object is the rate of change in its position over time. The acceleration of an object is the rate of change in its velocity over time. These relationships between position, velocity, and acceleration are nicely captured using the derivative concept.

> **RELATIONSHIP BETWEEN POSITION, VELOCITY, AND ACCELERATION**
>
> Let $s(t)$ be the function that describes the position of an object at time t. Then
> - $v(t) = s'(t)$, where $v(t)$ is the velocity of the object at time t.
> - $a(t) = v'(t) = s''(t)$, where $a(t)$ is the acceleration of the object at time t.

EXAMPLE 7 **Using a Position Function to Determine Velocity and Acceleration**

The author tracked his mileage as he drove through a residential area and into a cemetery. Every 15 seconds he recorded the mileage (accurate to 0.05 mile). Based on 1.5 minutes of data, his distance from a stoplight at the bottom of a hill is given by the position function

$$s(t) = -0.5333t^4 + 1.333t^3 - 0.7667t^2 + 0.3167t \text{ miles}$$

where t is in minutes.

Determine the velocity and acceleration functions. Then calculate the author's velocity and acceleration at 1 minute into his timed trip and at 1.25 minutes into his trip.

SOLUTION

$$v(t) = s'(t)$$

$$v(t) = \frac{d}{dt}(-0.5333t^4 + 1.333t^3 - 0.7667t^2 + 0.3167t)$$

$$v(t) = -2.1332t^3 + 3.999t^2 - 1.5334t + 0.3167 \frac{\text{miles}}{\text{minute}}$$

$$a(t) = v'(t)$$

$$a(t) = \frac{d}{dt}(-2.1332t^3 + 3.999t^2 - 1.5334t + 0.3167)$$

UNITS →

$$a(t) = -6.3996t^2 + 7.998t - 1.5334 \frac{\text{miles per minute}}{\text{minute}}$$

To calculate the velocity and acceleration one minute into the trip, we evaluate $v(1)$ and $a(1)$.

$$v(1) = -2.1332(1)^3 + 3.999(1)^2 - 1.5334(1) + 0.3167 \frac{\text{miles}}{\text{minute}}$$

$$= 0.6491 \frac{\text{miles}}{\text{minute}}$$

UNITS →

$$= 0.6491 \frac{\text{miles}}{\text{minute}} \cdot \frac{60 \text{ minutes}}{1 \text{ hour}}$$

$$= 38.946 \text{ miles per hour}$$

$$\approx 39 \text{ miles per hour}$$

$$a(1) = -6.3996(1)^2 + 7.998(1) - 1.5334 \frac{\text{miles per minute}}{\text{minute}}$$

$$= 0.065 \frac{\text{miles per minute}}{\text{minute}}$$

UNITS →

$$= 0.065 \frac{\text{miles}}{\text{minute}} \cdot \frac{1}{\text{minute}}$$

$$= 0.065 \frac{\text{miles}}{\text{minute}} \cdot \frac{1}{\text{minute}} \cdot \frac{60 \text{ minutes}}{1 \text{ hour}}$$

$$= 3.9 \frac{\text{miles}}{\text{hour}} \cdot \frac{1}{\text{minute}}$$

$$\approx 4 \text{ miles per hour per minute}$$

At 1 minute into his trip, the author was traveling at approximately 39 miles per hour and was accelerating at a rate of 4 miles per hour per minute. That is, if he maintained his current rate of acceleration for the next minute, his speed would increase by about 4 miles per hour.

To calculate the velocity and acceleration 1.25 minutes into the trip, we evaluate $v(1.25)$ and $a(1.25)$.

$$v(1.25) = -2.1332(1.25)^3 + 3.999(1.25)^2 - 1.5334(1.25) + 0.3167 \frac{\text{miles}}{\text{minute}}$$

$$= 0.4820 \frac{\text{miles}}{\text{minute}}$$

$$= 0.4820 \, \frac{\text{miles}}{\text{minute}} \cdot \frac{60 \, \text{minutes}}{1 \, \text{hour}}$$

$$\approx 29 \text{ miles per hour}$$

$$a(1.25) = -6.3996(1.25)^2 + 7.998(1.25) - 1.5334 \, \frac{\text{miles per minute}}{\text{minute}}$$

$$= -1.535 \, \frac{\text{miles per minute}}{\text{minute}}$$

UNITS

$$= -1.535 \, \frac{\text{miles}}{\text{minute}} \cdot \frac{1}{\text{minute}} \cdot \frac{60 \, \text{minutes}}{1 \, \text{hour}}$$

$$= -92.10 \, \frac{\text{miles}}{\text{hour}} \cdot \frac{1}{\text{minute}}$$

$$\approx -92 \text{ miles per hour per minute}$$

At 1 minute and 15 seconds into his trip, he was traveling at approximately 29 miles per hour and was decelerating at a rate of 92 miles per hour per minute. That is, if he maintained his current rate of deceleration for the next minute, his speed would decrease by about 92 miles per hour. The rapid rate of deceleration indicates that he is braking, probably in preparation for turning into the cemetery. (Since he was traveling at 29 mph, we know that he won't be able to maintain the high rate of deceleration for an entire minute. Otherwise, his velocity would turn negative, indicating that he was moving away from the cemetery and back toward the stop light.)

Curve Sketching

Although nowadays it is customary to graph functions using a graphing calculator, knowing how to graph functions by hand will greatly increase your understanding of calculus. If you know the x- and y-intercepts, the relative and absolute extrema, and the concavity and inflection points of a function, it is relatively easy to come up with a good sketch of the curve. We will demonstrate the curve-sketching process in Example 8.

EXAMPLE 8 **Sketching a Polynomial Function**

Sketch the graph of $f(x) = x^3 - 2x^2 - 4x + 8$ on the interval $[-3, 4]$.

SOLUTION We will set $f(x) = 0$ to find x-intercepts, $f'(x) = 0$ to find critical values, and $f''(x) = 0$ to determine where inflection points may occur.

x- and y-intercepts

$$f(x) = x^3 - 2x^2 - 4x + 8$$

$$0 = (x^3 - 2x^2) - (4x - 8) \qquad \text{Set } f(x) = 0 \text{ and group terms}$$

$$0 = x^2(x - 2) - 4(x - 2) \qquad \text{Factor each group}$$

$$0 = (x - 2)(x^2 - 4) \qquad \text{Factor } (x - 2) \text{ out of each term}$$

$$0 = (x - 2)(x - 2)(x + 2) \qquad \text{Factor } (x^2 - 4)$$

$$0 = (x - 2)^2(x + 2) \qquad \text{Group like terms}$$

$$\begin{array}{ccc} x - 2 = 0 & & x + 2 = 0 \\ & \text{or} & \\ x = 2 & & x = -2 \end{array}$$

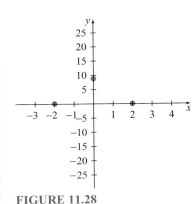

FIGURE 11.28

The graph of $f(x) = x^3 - 2x^2 - 4x + 8$ has x-intercepts at $(-2, 0)$ and $(2, 0)$. To find the y-intercept, we evaluate $f(x) = x^3 - 2x^2 - 4x + 8$ at $x = 0$.

$$f(0) = (0)^3 - 2(0)^2 - 4(0) + 8$$
$$= 8$$

The y-intercept is $(0, 8)$. We plot these points (Figure 11.28).

Relative and Absolute Extrema

To find critical values, we find $f'(x)$ and set it equal to zero.

$$f(x) = x^3 - 2x^2 - 4x + 8$$
$$f'(x) = 3x^2 - 4x - 4 \qquad \text{Differentiate } f(x)$$
$$0 = 3x^2 - 4x - 4 \qquad \text{Set } f'(x) = 0$$
$$0 = (3x + 2)(x - 2) \qquad \text{Factor}$$

$$3x + 2 = 0 \qquad\qquad x - 2 = 0$$
$$3x = -2 \qquad \text{or} \qquad x = 2$$
$$x = -\frac{2}{3}$$

$f(x)$ has critical values at $x = -\frac{2}{3}$ and $x = 2$. We construct a sign chart for $f'(x)$ to determine where the relative extrema occur.

We evaluate $f(x)$ at the critical values and the endpoints (see Table 11.15).

TABLE 11.15

x	$f(x) = x^3 - 2x^2 - 4x + 8$	
-3	-25	Absolute minimum
$-\frac{2}{3}$	$9\frac{13}{27} \approx 9.5$	Relative maximum
2	0	Relative minimum
4	24	Absolute maximum

We plot the corresponding points on the graph of f (Figure 11.29).

From the sign chart of $f'(x)$, we can also determine the intervals on which f is increasing or decreasing. Since $f'(x) > 0$ on $\left[-3, -\frac{2}{3}\right)$ and $(2, 4]$, f is increasing on those intervals. Similarly, since $f'(x) < 0$ on the interval $\left(-\frac{2}{3}, 2\right)$, f is decreasing on that interval. We won't update the graph of f with this information yet, since we still don't know the concavity of the graph on these intervals.

Inflection Points

To find the inflection points of f, we set $f''(x) = 0$.

$$f'(x) = 3x^2 - 4x - 4$$
$$f''(x) = 6x - 4 \qquad \text{Differentiate } f'(x)$$
$$0 = 6x - 4 \qquad \text{Set } f''(x) = 0$$
$$4 = 6x$$
$$x = \frac{2}{3} \qquad \text{Simplify } \frac{4}{6} \text{ to } \frac{2}{3}$$

FIGURE 11.29

inflection
point of f

$f''(x)$

$\frac{2}{3}$

An inflection point *may* occur at $x = \frac{2}{3}$. We'll construct a sign chart for $f''(x)$ and see if the concavity of f changes at $x = \frac{2}{3}$.

Since $f''(x) < 0$ on the interval $\left[-3, \frac{2}{3}\right)$, the graph of f is concave down on that interval. Since $f''(x) > 0$ on the interval $\left(\frac{2}{3}, 4\right]$, the graph of f is concave up on that interval. An inflection point occurs at $x = \frac{2}{3}$.

TABLE 11.16

x	$f(x) = x^3 - 2x^2 - 4x + 8$
$\frac{2}{3}$	$4\frac{20}{27} \approx 4.7$

Inflection point

We update the graph with the inflection point (Figure 11.30).

Referring to our earlier observations regarding the increasing/decreasing behavior and concavity of f, we finish the graph (Figure 11.31).

FIGURE 11.30

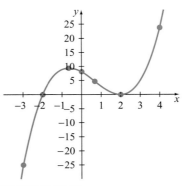

FIGURE 11.31

We summarize the curve-sketching steps in the following box.

HOW TO   Curve Sketching

To graph a function $f(x)$ on an interval $[a, b]$, complete the following steps. After each step, graph the corresponding point(s).

1. Find the x-intercepts of $f(x)$ by setting $f(x) = 0$ and solving for x.
2. Find the y-intercept of $f(x)$ by evaluating $f(0)$.
3. Find the relative extrema and increasing/decreasing behavior of $f(x)$ by constructing a sign chart for $f'(x)$.

(Continued)

4. Find the absolute extrema of $f(x)$ by evaluating $f(x)$ at each critical value and at the endpoints $x = a$ and $x = b$.

5. Find the inflection points and concavity of $f(x)$ by constructing a sign chart for $f''(x)$.

6. Connect the points, paying attention to the increasing/decreasing behavior and concavity of $f(x)$.

EXAMPLE 9

Sketching a Function

Using the curve-sketching techniques of calculus, graph the function $f(x) = \frac{\ln(x)}{2x}$ on the domain $[1, 8]$.

SOLUTION

x- and y-intercepts

$$f(x) = \frac{\ln(x)}{2x}$$

$$0 = \frac{\ln(x)}{2x} \qquad \text{Set } f(x) = 0$$

$$0 = \ln(x) \qquad \text{Multiply both sides by } 2x$$

$$e^0 = x \qquad \text{Rewrite as an exponential function}$$

$$x = 1 \qquad \text{Since } e^0 = 1$$

The function has an x-intercept at $(1, 0)$. Since $x = 0$ is not in the domain of the function, the function does not have a y-intercept.

Relative Extrema and Increasing/Decreasing Behavior

$$f(x) = \frac{\ln(x)}{2x}$$

$$f'(x) = \frac{\left(\frac{1}{x}\right)(2x) - (2)[\ln(x)]}{(2x)^2} \qquad \text{Apply the Quotient Rule}$$

$$= \frac{2 - 2\ln(x)}{4x^2}$$

$$0 = \frac{2 - 2\ln(x)}{4x^2} \qquad \text{Set } f'(x) = 0$$

$$0 = 2 - 2\ln(x) \qquad \text{Set numerator equal to 0}$$

$$2 = 2\ln(x)$$

$$1 = \ln(x)$$

$$e^1 = e^{\ln(x)} \qquad \text{Exponentiate both sides}$$

$$e = x \qquad \text{Since } e^{\ln(x)} = x$$

The critical value of the function is $x = e$. [Although the value $x = 0$ makes $f'(x)$ undefined, $x = 0$ is not a critical value because it is not in the domain of $f(x)$.]

We construct a sign chart to determine the relative extrema and the increasing/decreasing behavior of f.

The graph of f is increasing on the interval $[1, e)$ and decreasing on the interval $(e, 8]$. A relative maximum occurs at $x = e$. We evaluate f at the critical value and endpoints (see Table 11.17).

TABLE 11.17

x	$f(x) = \dfrac{\ln(x)}{2x}$	
1	0	Absolute minimum
e	$\dfrac{1}{2e} \approx 0.18$	Relative maximum, absolute maximum
8	$\dfrac{\ln(8)}{16} \approx 0.13$	

We create a graph of the points (Figure 11.32).

Inflection Points and Concavity

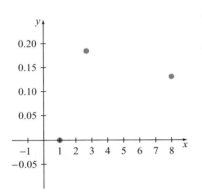

FIGURE 11.32

$$f'(x) = \frac{2 - 2\ln(x)}{4x^2}$$

$$f''(x) = \frac{\left(-\dfrac{2}{x}\right)(4x^2) - 8x[2 - 2\ln(x)]}{(4x^2)^2}$$ 　Apply the Quotient Rule

$$= \frac{-8x - 16x + 16x\ln(x)}{16x^4}$$

$$= \frac{-24x + 16x\ln(x)}{16x^4}$$

$$= \frac{-8x[3 - 2\ln(x)]}{16x^4}$$ 　Factor the numerator

$$0 = \frac{-8x[3 - 2\ln(x)]}{16x^4}$$ 　Set $f'(x) = 0$

$$0 = -8x[3 - 2\ln(x)]$$ 　Multiply both sides by $16x^4$

$$-8x = 0 \quad \text{or} \quad 3 - 2\ln(x) = 0$$
$$x = 0 \qquad\qquad 3 = 2\ln(x)$$
$$\frac{3}{2} = \ln(x)$$
$$x = e^{3/2}$$

Since $x = 0$ is not in the domain of the function, $x = 0$ is not a point of inflection. We construct a sign chart for $f''(x)$.

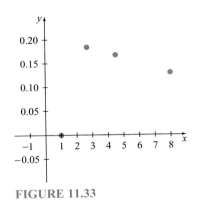

The graph of f is concave down on $[1, e^{3/2})$ and concave up on $(e^{3/2}, 8]$. An inflection point occurs at $x = e^{3/2} \approx 4.5$ (see Table 11.18).

TABLE 11.18

x	$f(x) = \dfrac{\ln(x)}{2x}$	
$e^{3/2}$	$\dfrac{3}{4e^{3/2}} \approx 0.17$	Inflection point

We update the graph with the inflection point (Figure 11.33).

Paying attention to the increasing/decreasing behavior and concavity of f, we finish the graph (Figure 11.34).

FIGURE 11.33

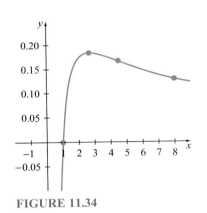

FIGURE 11.34

11.3 Summary

In this section, you learned how to use the second derivative in discussing the graphical concepts of concavity and inflection points. You saw how the Second Derivative Test may be used as an alternative means of finding relative extrema. You also discovered that velocity is the derivative of the position function and that acceleration is the second derivative of the position function. You also learned some curve sketching techniques.

11.3 Exercises

In Exercises 1–10, indicate on the graph where the function is concave up and where the function is concave down. Mark the graph where you think inflection points occur.

1. $f(x) = x^3 - 3x^2$

2. $y = \dfrac{\ln(x)}{x}$

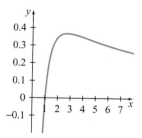

3. $f(x) = x^4 - 12x^2$

4. $g(x) = x^2$

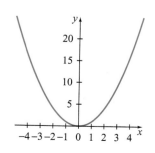

5. $y = x^4 - 8x^3 + 18x^2$

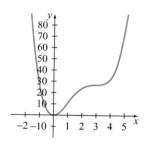

6. $f(x) = x^4 + 2x^3 - 12x$

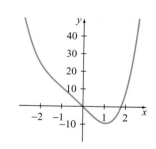

7. $y = -x^4 - 2x^3 + 12x^2$

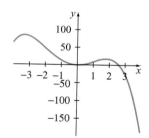

8. $f(x) = x^5 - 5x^4$

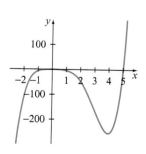

9. $f(x) = \dfrac{x^2}{x^2 + 1}$

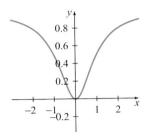

10. $g(x) = \dfrac{3}{x^2 + 1}$

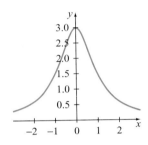

In Exercises 11–20, use the second derivative to find the inflection points of the function algebraically. You may find it helpful to refer to the graphs of the functions in Exercises 1–10.

11. $f(x) = x^3 - 3x^2$

12. $y = \dfrac{\ln(x)}{x}$

13. $f(x) = x^4 - 12x^2$

14. $g(x) = x^2$

15. $y = x^4 - 8x^3 + 18x^2$

16. $f(t) = t^4 + 2t^3 - 12t$

17. $y = -x^4 - 2x^3 + 12x^2$

18. $f(x) = x^5 - 5x^4$

19. $h(t) = \dfrac{t^2}{t^2 + 1}$

20. $g(x) = \dfrac{3}{x^2 + 1}$

In Exercises 21–30, find the stationary points of the function. Then use the Second Derivative Test to determine where relative extrema occur. If the Second Derivative Test is inconclusive, use the First Derivative Test. You may find it helpful to refer to the graphs of the functions in Exercises 1–10 and your work in Exercises 11–20.

21. $f(x) = x^3 - 3x^2$

22. $y = \dfrac{\ln(x)}{x}$

23. $f(r) = r^4 - 12r^2$

24. $g(x) = x^2$

25. $y = x^4 - 8x^3 + 18x^2$

26. $s(n) = n^4 + 2n^3 - 12n$

27. $y = -x^4 - 2x^3 + 12x^2$

28. $f(t) = t^5 - 5t^4$

29. $f(x) = \dfrac{x^2}{x^2 + 1}$

30. $g(x) = \dfrac{3}{x^2 + 1}$

In Exercises 31–40, use the curve-sketching techniques introduced in the section to graph each function by hand.

31. $f(x) = \dfrac{1}{3}x^3 - x^2$ on $[-2, 4]$

32. $g(x) = x^4 + 4x^3$ on $[-2, 5]$

33. $f(x) = -x^3 + 3x + 1$ on $[-2, 2]$

34. $f(x) = -x^4 + 2x^2$ on $[-2, 2]$

35. $g(x) = -x^3 + 12x$ on $[-4, 4]$

36. $f(x) = x^4 - 2x^2 + 1$ on $[-2, 2]$

37. $g(x) = \dfrac{1}{5}x^5 - x^4$ on $[-3, 5]$

38. $f(x) = \ln(x)/x$ on $[1, 10]$

39. $g(x) = 2^x - x\ln(2)$ on $[-2, 2]$

   Hint: This function doesn't have any x-intercepts.

40. $h(x) = 2^{-x} - 1$ on $[-2, 4]$

In Exercises 41–50, answer the questions by using the Second Derivative Test, as appropriate. In determining where absolute extrema occur, remember to check the endpoints.

41. **West Coast Sailor Wages** Based on data from 1975 to 2000, the average monthly wage (in addition to room and board) of a sailor working on the West Coast may be modeled by

$$W(t) = 0.1446t^2 - 5.765t + 883.7$$

dollars, where t is the number of years since 1975. (**Source:** Modeled from *Statistical Abstract of the United States, 2001*, Table 1072, p. 674.)

What was the maximum and what was the minimum increase in monthly wages, and in what year (between 1975 and 2000) did each occur? (Round to the nearest year.) *Hint:* A decrease in wages is a negative increase.

42. **East Coast Sailor Wages** Based on data from 1975 to 2000, the average monthly wage (in addition to room and board) of a sailor working on the East Coast may be modeled by

$$E(t) = 0.1643t^2 - 5.658t + 112.3$$

dollars, where t is the number of years since 1975. (**Source:** Modeled from *Statistical Abstract of the United States, 2001*, Table 1072, p. 674.)

What was the maximum and what was the minimum increase in monthly wages, and in what year (between 1975 and 2000) did each occur? (Round to the nearest year.)

43. **Salary Negotiations** How could an East Coast sailors' labor union use the information in Exercises 41 and 42 in negotiating future wage increases?

44. **Homicide Rate** Based on data from 1990 to 2000, the homicide rate may be modeled by

$$H(t) = 0.01129t^3 - 0.2002t^2 + 0.4826t + 9.390$$

deaths per 100,000 people, where t is the number of years since 1990. (**Source:** Modeled from *FBI Uniform Crime Reports, 2000*.)

Between 1990 and 2000, in what year did the homicide rate increase most rapidly, and in what year did it decrease most rapidly?

45. **Internet Usage** Based on usage data from 1995 to 1999 and Census Bureau projections of future usage, the number of hours that the average person spends on the Internet annually may be modeled by

$$L(t) = -0.4514t^3 + 6.125t^2 + 6.171t + 1.994$$

hours, where t is the number of years since 1995. (**Source:** Modeled from *Statistical Abstract of the United States, 2001*, Table 1125, p. 704.)

In what year (between 1995 and 2004) was Internet usage expected to increase most slowly, and in what year was it expected to increase most rapidly?

46. Based on the results of Exercise 45, do you think that the rate of change in annual Internet usage will ever exceed 35 hours per year? Justify your answer.

47. **Manufacturing Employees** Based on data from 1995 to 1999, the number of employees in the leather and leather products manufacturing industry may be modeled by

$$A(t) = -0.6667t^3 + 4.5t^2 - 14.83t + 106$$

thousand employees, where t is the number of years since 1995. (**Source:** Modeled from *Statistical Abstract of the United States, 2001*, Table 979, p. 622.)

In what year did the number of employees in the leather and leather products industry drop most rapidly? In what year did the number of employees increase most rapidly (or decrease most slowly)?

48. **Political Influence** People for the Ethical Treatment of Animals (PETA) is a nonprofit organization that believes that animals are not ours to eat or wear. (**Source:** www.peta.org.)

Based on the results of Exercise 47, do you think PETA is having an influence on the leather and leather products industry? Justify your answer.

49. **Love Triangles** Based on data from 1991 to 2000, the cumulative number of homicides since the *start* of 1991 and the *end* of year t that are due to a romantic triangle may be modeled by

$$R(t) = -20.45t^2 + 471.8t - 152.5$$

homicides, where t is the number of years since the end of 1990. (**Source:** Modeled from *Crime in the United States 2000*, Uniform Crime Report, FBI .)

In what year between 1991 and 2000 did the cumulative number of such homicides increase most rapidly? How many homicides due to a romantic triangle occurred that year?

50. **M** **Breakfast Cereal Consumption** Based on data from 1980 to 1999, the per capita consumption of breakfast cereal may be modeled by

$$C(t) = -0.004718t^3 + 0.1165t^2 - 0.3585t + 12.17$$

pounds, where t is the number of years since 1980. (**Source:** Modeled from *Statistical Abstract of the United States, 2001*, Table 202, p. 129.)

In what year (between 1980 and 1999) was the per capita breakfast cereal consumption increasing most rapidly, and in what year was it decreasing most rapidly?

Exercises 51–55 deal with position, velocity, and acceleration. The vertical position of a free-falling object on Earth may be modeled by $s(t) = -16t^2 + v_0 t + s_0$ feet above the earth after being airborne for t seconds. v_0 is the velocity of the object, and s_0 is the position of the object at time $t = 0$. Use this information and the techniques shown in the section to answer the questions.

51. A rock is *thrown* upward into the air at a rate of 32 feet per second. After how many seconds of flight will it attain its maximum height above the ground?

52. When the rock in Exercise 51 attains its maximum height, what will be its velocity?

53. Given $s(t) = -16t^2 + v_0 t + s_0$, calculate $s''(t)$ and interpret the meaning of the result.

54. A pebble is *dropped* into the ocean from the top of a 240-foot cliff. What will be the pebble's velocity when it hits the surface of the ocean?

55. When will the pebble in Exercise 54 attain its maximum speed?

In Exercises 56–57, find the point of diminishing returns for the given function.

56. **M** **Electronic Game Unit Sales** Based on data from 1996 to 2000, the annual sales of computer and video game units may be modeled by

$$S(t) = \frac{121.8}{1 + 17.23e^{-1.794t}} + 100$$

units, where t is the number of years since the end of 1996. (**Source:** Modeled from Interactive Digital Software Association State of the Industry Report, 2000–2001.)

Find the point of diminishing returns and interpret its real-life meaning.

57. **M** **Box Office Movie Sales** The movie *My Big Fat Greek Wedding* was first released on April 19, 2002. Based on the first 52 weeks of sales, the cumulative gross box office sales of the movie may be modeled by

$$S(t) = \frac{240,800,000}{1 + 125.4e^{-0.2086t}}$$

dollars, where t is in weeks with $t = 1$ corresponding to the first week the movie was released.

Find the point of diminishing returns and interpret its real-life meaning.

Exercises 58–63 are intended to challenge your skill in finding and interpreting the meaning of concavity and the second derivative. Each of the questions deals with a parametric function model of the position of a roller-coaster train over a six-second interval $[0, 6]$. The x value of the model represents the horizontal position of a roller-coaster train (in meters) at time t seconds. The y value of the model represents the vertical position of the roller-coaster train (in meters) at time t seconds. The model is

$$x = t^3 - 9t^2 + 23t$$
$$y = 6t - t^2$$

58. Find the second derivative of x and of y with respect to t. (These may be represented as x'' and y''.)

59. Determine when $x'' = 0$ and $y'' = 0$.

60. Given that the domain of the function is $0 \le t \le 6$, find the points of inflection of x and y.

61. Over the six-second interval $[0, 6]$, what was the quickest horizontal acceleration of the train?

62. Over the same six-second interval, what was the slowest horizontal acceleration of the train?

63. Do you think that the roller coaster would be safe to ride? Defend your conclusion.

11.4 Related Rates

■ Solve related-rate problems

On March 24, 1989, the oil tanker *Exxon Valdez* ran aground on Bligh Reef in Prince William Sound, Alaska. In the ensuing spill, 11 million gallons of crude oil flowed into the water. About 250,000 seabirds; 2,650 sea otters; 1,000 cormorants, and hundreds of loons, harbor seals, and bald eagles were killed. Although much larger spills have occurred internationally, the *Exxon Valdez* spill was one of the most environmentally destructive. Because of the location of the accident, approximately 1,300 miles of the 9,000 miles of shoreline in the region were contaminated. Despite cleanup efforts, there were still pockets of oil remaining a decade after the accident.

When an oil spill occurs, many questions immediately arise. How quickly is the oil spilling into the water? How rapidly is the area covered by the oil expanding? If the oil continues to spill at its current rate, how much area will be covered an hour from now? These questions may be addressed using the concept of *related rates*.

In this section, we will demonstrate how related rates can be used to measure how the rate of change in one variable affects the rate of change in another variable. We will use the implicit differentiation techniques covered in Section 10.5 extensively in this section.

The area of a circular oil spill with radius r feet is given by the formula $A = \pi r^2$. If the area of the circle is increasing at a rate of 100 square feet per minute, how quickly is the radius changing? Since the radius rate of change function is the derivative of the radius function, we are looking for $\dfrac{dr}{dt}$ with units $\dfrac{\text{feet}}{\text{minute}}$. The rate of change in the area is given by $\dfrac{dA}{dt}$. The units of $\dfrac{dA}{dt}$ are $\dfrac{\text{square feet}}{\text{minute}}$. Notice that both A and r are differentiated with respect to t. Using implicit differentiation, we have

$$A = \pi r^2$$

$$\frac{d}{dt}(A) = \frac{d}{dt}(\pi r^2)$$

$$\frac{dA}{dt} = \pi \frac{d}{dt}(r^2)$$

$$\frac{dA}{dt} = \pi \left(2r \frac{dr}{dt} \right)$$

$$\frac{dA}{dt} = 2\pi r \frac{dr}{dt}$$

$$\frac{dr}{dt} = \frac{1}{2\pi r} \frac{dA}{dt}$$

Thus the rate of change in the radius is the rate of change in the area divided by the product of 2π and the current radius. For this problem, we indicated that the area was increasing at a rate of 100 square feet/minute. That is, $\dfrac{dA}{dt} = 100$.

Therefore, the rate of change in the radius is

$$\frac{dr}{dt} = \frac{1}{2\pi r} \frac{dA}{dt}$$

$$= \frac{1}{2\pi r}(100)$$

$$= \frac{50}{\pi r}$$

When the radius is 1 foot, $\dfrac{dr}{dt} = \dfrac{50}{\pi(1)} \approx 15.9$ feet per minute. When the radius is 2 feet, $\dfrac{dr}{dt} = \dfrac{50}{\pi(2)} \approx 7.96$ feet per minute. When the radius is 1000 feet, $\dfrac{dr}{dt} = \dfrac{50}{\pi(1000)} \approx 0.01592$ feet per minute. Assuming that the shape of the spill remains circular, we can calculate the rate of change in the radius for any value of the radius.

Related rates are most often used when comparing the changes in two related quantities over time. In the next several examples, we will illustrate how related rates are used.

EXAMPLE 1

Interpreting the Meaning of $\dfrac{dV}{dt}, \dfrac{dr}{dt},$ and $\dfrac{dh}{dt}$

The volume of space occupied by a liquid in a cylindrical can is given by $V = \pi r^2 h$, where r is the radius of the can and h is the depth of the liquid in the can. Differentiate V with respect to t and interpret the meaning of $\dfrac{dV}{dt}, \dfrac{dr}{dt},$ and $\dfrac{dh}{dt}$.

Assume that the radius and height are measured in centimeters and that time is measured in seconds. Assuming that the radius of the can is fixed, simplify the result for $\dfrac{dV}{dt}$.

SOLUTION

$$\frac{d}{dt}(V) = \frac{d}{dt}(\pi r^2 h)$$

$$\frac{dV}{dt} = \pi \frac{d}{dt}(r^2 h) \qquad\qquad \text{Constant Multiple Rule}$$

$$= \pi \left(2r\frac{dr}{dt} \cdot h + \frac{dh}{dt} \cdot r^2 \right) \qquad \text{Product and Chain Rules}$$

$$= \pi r \left(2h\frac{dr}{dt} + r\frac{dh}{dt} \right)$$

$\dfrac{dr}{dt}$ is the rate of change in the radius in centimeters per second.

$\dfrac{dh}{dt}$ is the rate of change in depth of the liquid in centimeters per second.

$\dfrac{dV}{dt}$ is the rate of change in the volume of the liquid in the can in cubic centimeters per second. It is a function of the radius, the depth, and the rates of change in the radius and the depth. Notice that if the can has a fixed radius, then $\dfrac{dr}{dt} = 0$ and

$$\frac{dV}{dt} = \pi r \left[2h(0) + r\frac{dh}{dt} \right]$$

$$= \pi r^2 \frac{dh}{dt}$$

EXAMPLE 2

Using Related Rates in a Real-World Context

The volume of a cylindrical can is given by $V = \pi r^2 h$, where r is the radius and h is the height of the interior of the can from lid to lid. In Section 11.2, we learned that a #10 can has a radius of 3.625 inches (9.208 centimeters). Suppose that the can is being filled with rice and that the rice is being poured into the can at a constant rate of 20 cups per minute. At what rate is the height of the rice in the can changing? Will it take more or less than a minute to fill the can?

SOLUTION You may immediately ask, "Why do we care?" Admittedly, we may not care if we have to fill only one can. However, if we worked in a cannery and wanted to program a machine to fill thousands of cans, the problem would immediately become meaningful. Failure to program the machine with the correct fill rate would result in overfilled cans or insufficiently filled cans. Both of these would hamper the cannery's effort to create a uniform product with minimum waste.

Since we are filling #10 cans, the volume equation becomes

$$V = \pi (9.208)^2 h$$

$$= 84.79\pi h \text{ cubic centimeters}$$

and the rate of change in the volume is given by

$$\frac{dV}{dt} = 84.79\pi \frac{dh}{dt} \frac{\text{cubic centimeters}}{\text{minute}}$$

One of the challenges of the standard measurement system (inches, feet, miles, cups, gallons, etc.) is that conversion between quantities (e.g., cups to cubic inches) is difficult. For this reason, we will convert standard measurements to their metric equivalents. One cup is equivalent to 236.6 cubic centimeters. Since the volume is changing at a rate of 20 cups per minute, we have

UNITS

$$\frac{dV}{dt} = 20 \, \frac{\text{cups}}{\text{minute}} \cdot \frac{236.6 \text{ cubic centimeters}}{1 \text{ cup}}$$

$$= 4732 \, \frac{\text{cubic centimeters}}{\text{minute}}$$

We must find $\dfrac{dh}{dt}$.

$$\frac{dV}{dt} = 84.79\pi \frac{dh}{dt}$$

$$4732 = 266.4 \frac{dh}{dt}$$

$$\frac{dh}{dt} = \frac{4732}{266.4}$$

$$= 17.76 \text{ centimeters per minute}$$

The height of the rice in the can is increasing by about 17.76 centimeters per minute. Since the interior height of a #10 can is 16.83 centimeters (6.625 inches), it will take less than a minute to fill the can. (To be precise, it will take 57 seconds.)

EXAMPLE 3

Using Related Rates in a Real-World Context

A circular above-ground swimming pool has a 12-foot radius and may be filled to a maximum depth of 4 feet. By measuring the amount of time it took to fill a bucket, the author determined that water leaves his outside faucet at a rate of 9 gallons per minute. How quickly is the depth of the water in the pool changing when the pool is being filled from a hose connected to the faucet? How long will it take to fill the pool?

SOLUTION We will first convert the dimensions to metric measurements. We know that 1 foot ≈ 30.48 centimeters and 1 gallon ≈ 3785 cubic centimeters.

The radius of the pool is $12(30.48) = 365.76$ centimeters. The rate at which the water is leaving the faucet is $9(3785) = 34{,}065$ cubic centimeters per minute.

The pool is cylindrical, so its volume is $V = \pi r^2 h$, where r is the radius and h is the depth of the water in the pool. Since the radius of the pool will not change, we have

$$V = \pi(365.76)^2 h$$

$$= 420{,}283.45h \text{ cubic centimeters}$$

Differentiating with respect to t, we get

$$\frac{dV}{dt} = 420{,}283.45 \frac{dh}{dt} \, \frac{\text{cubic centimeters}}{\text{minute}}$$

But we know that $\dfrac{dV}{dt} = 34{,}065$ cubic centimeters per minute, so

$$34{,}065 = 420{,}283.45\dfrac{dh}{dt}$$

UNITS

$$\dfrac{dh}{dt} = 0.08105\,\dfrac{\text{centimeters}}{\text{minute}}$$

$$= 0.08105\,\dfrac{\text{centimeters}}{\text{minute}} \cdot \dfrac{60\ \cancel{\text{minutes}}}{1\ \text{hour}}$$

$$= 4.863\ \text{centimeters per hour}$$

The pool depth is increasing at a rate of 4.863 centimeters per hour. Since 4 feet = 121.92 centimeters, the amount of time it will take to fill the pool is

UNITS

$$\dfrac{121.92\ \text{centimeters}}{4.863\ \text{centimeters per hour}} = 25.07\ \text{hours}$$

It will take about 25 hours to fill the pool.

EXAMPLE 4   **Calculating a Change in Wind Chill Temperature**

Wind chill temperature is the temperature in calm air that has the same chilling effect on a person as that of a particular combination of temperature and wind. For example, when the temperature is 35° Fahrenheit (°F) and the wind speed is 40 miles per hour, it feels like it is 3°F. The wind chill temperature may be calculated according to the following formula:

$$F = 91.4 - \left(0.474677 - 0.020425w + 0.303107\sqrt{w}\right)(91.4 - T)\ \text{degrees Fahrenheit}$$

where w is the wind speed (in mph) and T is the temperature (in °F). (**Source:** National Climatic Data Center.)

If the temperature is currently 20° and is dropping at a rate of 3° per hour, and if the wind is currently blowing at 16 mph and is increasing at a rate of 2 mph per hour, how quickly is the wind chill temperature changing?

SOLUTION We must first differentiate the wind chill equation with respect to time.

$$\dfrac{d}{dt}(F) = \dfrac{d}{dt}\left[91.4 - \left(0.474677 - 0.020425w + 0.303107\sqrt{w}\right)(91.4 - T)\right]$$

$$\dfrac{dF}{dt} = 0 - \left[-0.020425\dfrac{dw}{dt} + 0.5(0.303107w^{-0.5})\dfrac{dw}{dt}\right](91.4 - T)$$

$$+ \left(-\dfrac{dT}{dt}\right)\left[-\left(0.474677 - 0.020425w + 0.303107\sqrt{w}\right)\right] \qquad \text{Product Rule}$$

$$\dfrac{dF}{dt} = \left[0.020425\dfrac{dw}{dt} - 0.5(0.303107w^{-0.5})\dfrac{dw}{dt}\right](91.4 - T)$$

$$+ \left(\dfrac{dT}{dt}\right)\left(0.474677 - 0.020425w + 0.303107\sqrt{w}\right)$$

The current wind speed is 16 mph, so $w = 16$.

$\dfrac{dw}{dt}$ is the rate of change in the wind and equals 2 mph per hour.

The current temperature is 20°, so $T = 20$.

$\dfrac{dT}{dt}$ is the rate of change in the temperature and equals -3 degrees per hour.

Substituting these values into the derivative equation, we get

$$\frac{dF}{dt} = \{0.020425(2) - 0.5[0.303107(16)^{-0.5}](2)\}[91.4 - (20)]$$

$$+ (-3)[0.474677 - 0.020425(16) + 0.303107\sqrt{16}]$$

$$= (-0.03493)(71.4) - 4.081$$

$$= -2.494 - 4.081$$

$$= -6.575$$

The wind chill index is dropping by about 6.6 degrees per hour. That is, although the temperature is dropping by only 3 degrees per hour, it feels like it is dropping by 6.6 degrees per hour because of the wind.

Because of the number of terms and the complexity of the coefficients, the type of calculation shown in Example 4 is time-consuming and prone to error. For this reason, we recommend using the following Technology Tip to speed up calculations and increase accuracy.

TECHNOLOGY **TIP**

Using the Equation Solver

1. Press $\boxed{\text{MATH}}$ `0:Solver...` to bring up the Equation Solver. Immediately press the blue up arrow. Press $\boxed{\text{CLEAR}}$ to delete any equation that may be visible.

```
EQUATION SOLVER
eqn:0=
```

2. Enter the equation you need to solve, using the alpha keys for variables. Since the calculator requires the equation to be set equal to zero, we must subtract $\dfrac{dF}{dt}$ from both sides of our equation. For our equation, we let $\dfrac{dF}{dt} = A$, $\dfrac{dT}{dt} = B$, and $\dfrac{dw}{dt} = C$.

```
EQUATION SOLVER
eqn:0=(.020425*C
-.5*(.303107*W^-
.5)*C)*(91.4-T)+
B*(.474677-.0204
25W+.303107*W^.5
)-A
```

(Continued)

3. Press ENTER to display the variable input menu. Edit the variables to reflect the known values from our problem.

```
(.020425*C-.5…=0
C=2
W=16
T=20
B=-3
A=■
bound={-1E99,1…
left-rt=0
```

4. Move the cursor to the line containing A= and press ALPHA ENTER to solve the equation for A.

```
(.020425*C-.5…=0
C=2
W=16
T=20
B=-3
•A=-6.57468495
bound={-1E99,1…
•left-rt=0
```

EXAMPLE 5

Calculating a Change in Wind Chill Temperature

If the temperature is currently 30° and dropping at a rate of 5° per hour, and the wind is currently blowing at 20 mph and is decreasing at a rate of 10 mph per hour, how quickly is the wind chill temperature changing? Use the Technology Tip and the equation for $\dfrac{dF}{dt}$ from Example 4 to answer the question.

SOLUTION

The current wind speed is 20 mph, so $w = 20$.

$C = \dfrac{dw}{dt}$ is the rate of change in the wind and equals -10 mph per hour.

The current temperature is 30°, so $T = 30$.

$B = \dfrac{dT}{dt}$ is the rate of change in the temperature and equals -5 degrees per hour.

We need to solve for $A = \dfrac{dF}{dt}$. Since we have already entered the equation (from Example 4) in our calculator, we'll use the Technology Tip to solve the problem (see Figure 11.35).

The wind chill temperature is increasing by $1.1579 \approx 1.2$ degrees per hour. Although the temperature is dropping 5 degrees per hour, it feels like it is warming by 1.2 degrees per hour because the wind speed is decreasing.

```
(.020425*C-.5…=0
C=-10
W=20
T=30
B=-5
•A=1.1579596286…
bound={-1E99,1…
•left-rt=0
```

FIGURE 11.35

11.4 Summary

In this section, you learned how related rates can be used to measure how the rate of change in one variable affects the rate of change in another variable. You discovered that technology can simplify the arithmetic computations of related-rate problems.

11.4 Exercises

In Exercises 1–10, differentiate the function with respect to t. Explain the physical meaning of each of the rates. (For each of the problems, A is area, V is volume, r is radius, h is height, l is length, w is width, b is base, and t is time. Use inches, square inches, cubic inches, and minutes as the units of length, area, volume, and time, respectively.)

1. $A = lw$ (area of a rectangle)

2. $A = 2\pi rh + 2\pi r^2$ (surface area of a cylinder including top and bottom)

3. $V = lwh$ (volume of a box)

4. $V = \frac{4}{3}\pi r^3$ (volume of a sphere)

5. $A = 2lw + 2wh + 2lh$ (surface area of a box)

6. $V = \frac{1}{3}\pi r^2 h$ (volume of a cone)

7. $A = 4\pi r^2$ (surface area of a sphere)

8. $A = \frac{1}{2}bh$ (area of a triangle)

9. $A = \pi r^2$ (area of a circle)

10. $A = 2\pi r^2$ (surface area of a hemisphere)

In Exercises 11–15, solve the related-rate problem using the techniques demonstrated in this section.

11. **Swimming Pool Depth**   A circular above-ground swimming pool has a 9-foot radius and may be filled to a maximum depth of 4 feet. The pool is being filled with a faucet that releases water at a rate of 10 gallons per minute. How quickly is the depth of the water in the pool changing when the pool is being filled from a hose connected to the faucet? How long will it take to fill the pool?

12. **Water Pressure**   A circular above-ground swimming pool has a 9-foot radius and may be filled to a maximum depth of 4 feet. If the depth of the water in the pool is rising at a rate of 0.5 inch per minute, at what rate is the water leaving the hose that is filling the pool?

13. **Water Depth**   A cylindrical can has a 5-inch radius and an 8-inch height and is to be filled with water. At what constant rate must the depth of the water increase if the can is to be filled in exactly 30 seconds?

14. **Pizza Size**   The owner of a pizzeria is contemplating changing the size of his large pizza. He intends to decrease the size of the large pizza from a 16-inch diameter to a 14-inch diameter by gradually reducing the diameter by $\frac{1}{2}$ inch each month. (He figures that consumers won't notice the change if it is gradual.) At what rate will the surface area of the pizza be changing two months into the reduction period?

15. **Oil Spill Spread**   The surface area of a circular oil spill is increasing at a rate of 100 square feet per hour. How quickly is the radius increasing when the radius is 157 feet?

In Exercises 16–20, solve the related-rate problems using the Technology Tip demonstrated in this section. The wind chill temperature may be estimated using the following model equation:

$$F = 91.4 - (0.474677 - 0.020425w + 0.303107\sqrt{w})$$
$$\cdot\,(91.4 - T) \text{ degrees Fahrenheit}$$

where w is the wind speed (in mph) and T is the temperature (in °F). (**Source:** National Climatic Data Center.)

16. **Wind Chill**   If the temperature is currently 0°F and is dropping at a rate of 5°F per hour, and the wind is currently blowing at 18 mph and is increasing at a rate of 2 mph per hour, how quickly is the wind chill temperature changing?

17. **Wind Chill**   If the temperature is currently −10°F and is warming at a rate of 8°F per hour, and the wind is currently blowing at 25 mph and is increasing at a rate of 3 mph per hour, how quickly is the wind chill temperature changing?

18. **Wind Chill**   If the temperature is currently −2°F and is decreasing at a rate of 4°F per hour, and the wind is currently blowing at 55 mph and is decreasing at a rate of 5 mph per hour, how quickly is the wind chill temperature changing?

19. **Wind Chill**   If the temperature is currently 40°F and is increasing at a rate of 4°F per hour, and the wind is currently blowing at 5 mph and is not increasing its speed, how quickly is the wind chill temperature changing?

20. **Wind Chill**   If the temperature is currently 32°F and is not changing, and the wind is currently blowing at 25 mph and is increasing its speed by 4 mph per hour, how quickly is the wind chill temperature changing?

Chapter 11 Review Exercises

Section 11.1 *In Exercises 1–2, determine the points on the graph where $f'(x) = 0$ or $f'(x)$ is undefined.*

1. $f(x) = |x - 2| - (x - 2)^2 + 2$

2. $f(x) = x^5 - 4x^3 + 10$

In Exercises 3–6, find the critical points of the function.

3. $f(x) = x^5 - 4x^3 + 10$

4. $g(x) = 5x^{1/5}$ **5.** $y = -4x^2 + 4$

6. $y = -2x + 4$

In Exercises 7–10, construct a sign chart for the derivative. Then determine the relative and absolute extrema of the function on the specified domain. Use the First Derivative Test as appropriate. (These functions are the same as those in Exercises 3–6.)

7. $f(x) = x^5 - 4x^3 + 10; [-2, 2]$

8. $g(x) = 5x^{1/5}; [-2, 2]$

9. $y = -4x^2 + 4; [-3, 3]$

10. $y = -2x + 4; [0, 4]$

In Exercises 11–12, use the concepts of relative and absolute extrema to find the answer to the question.

11. **Exxon Valdez Oil Spill** Based on data from the *fourth* to the fifty-sixth day after the accident, the spread of the oil from the site of the *Exxon Valdez* oil spill may be modeled by

$$D(t) = 0.004108t^3 - 0.4432t^2 + 21.07t - 42.07$$

miles from the spill site t days after the accident. Note that the model is invalid for $t < 4$. (**Source:** Modeled from www.oilspill.state.ak.us/facts/spillmap.html.)

According to the model, what was the greatest distance the oil had spread between the fourth and the fifty-sixth day?

12. **Industry Work Force Size** Based on data from 1995 to 1999, the number of employees in the printing and publishing industry may be modeled by

$$E(t) = -4.417t^3 + 24.93t^2 - 25.30t + 1450$$

thousand employees, where t is the number of years since 1995. (**Source:** Modeled from *Statistical Abstract of the United States, 2001*, Table 979, p. 622.)

According to the model, when did the industry have its maximum and minimum number of employees? How many people were employed at these times?

Section 11.2 *In Exercises 13–14, use the derivative and other optimization techniques to find the optimal value requested.*

13. **Company Revenue and Cost** Gatorade/Tropicana North America, a subsidiary of PepsiCo, produces fruit juices and other flavored beverages. Based on data from 1999 to 2001, the net sales (revenue) of Gatorade/Tropicana North America may be modeled by

$$R(t) = -107t^2 + 496t + 3452$$

million dollars and the operating profit (earnings before interest and taxes) may be modeled by

$$P(t) = -18.5t^2 + 85.5t + 433$$

million dollars, where t is the number of years since 1999. (**Source:** Modeled from 2001 PepsiCo Annual Report, p. 44.)

(a) In what year are net sales projected to reach a maximum?

(b) Find the cost function for Gatorade/Tropicana North America.

(c) According to the model, in what year are costs expected to reach a maximum?

(d) Compare the results of (a) and (c). Do the results seem reasonable?

14. **Ice Cream Cone Design** The volume of an ice cream cone (minus the ice cream) is given by

$$V = \frac{1}{3}\pi r^2 h$$

cubic inches, where r is the radius of the top of the cone and h is the height of the cone (both in inches). The surface area of the exterior of the cone is given by

$$A = \pi r \sqrt{r^2 + h^2}$$

square inches.

For a cone with a surface area of $47.12(15\pi)$ square inches, what are the dimensions of the cone with maximum volume? Should ice cream cones be constructed with these dimensions? Explain.

Section 11.3 *In Exercises 15–16, indicate on the graph where the function is concave up and where the function is concave down. Mark the graph where you think inflection points occur.*

15. $f(x) = \dfrac{x^2 + 9}{x^2 + 10}$

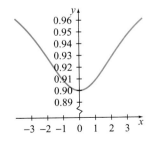

16. $y = x^3 - x^2 - 8x + 12$

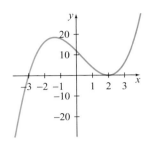

In Exercises 17–18, use the second derivative to find the inflection points of the function algebraically. You may find it helpful to refer to the graphs of the functions in Exercises 15–16.

17. $f(x) = \dfrac{x^2 + 9}{x^2 + 10}$

18. $y = x^3 - x^2 - 8x + 12$

In Exercises 19–20, find the stationary points of the function. Then use the Second Derivative Test to determine where relative extrema occur. If the Second Derivative Test is inconclusive, use the First Derivative Test. You may find it helpful to refer to the graphs of the functions in Exercises 15–16 and your work in Exercises 17–18.

19. $f(x) = \dfrac{x^2 + 9}{x^2 + 10}$

20. $y = x^3 - x^2 - 8x + 12$

In Exercises 21–22, use the second derivative to determine where the maximum and minimum rates of change occur for the specified function. Be sure to evaluate the rate of change function at the endpoints to ensure that you have found the absolute extrema.

21. **Exxon Valdez Oil Spill** Based on data from the fourth to the fifty-sixth day after the accident, the spread of the oil from the site of the *Exxon Valdez* oil spill may be modeled by

$$D(t) = 0.004108t^3 - 0.4432t^2 + 21.07t - 42.07$$

miles from the spill site t days after the accident. (**Source:** Modeled from www.oilspill.state.ak.us/facts/ spillmap.html.)

On what day was the oil spreading at the quickest rate? At what rate was it spreading?

22.   **Minivan Speed**   Based on 2.5 minutes of data, the distance of the author's minivan from a stoplight at the beginning of a highway may be modeled by

$$M(t) = -0.0591t^3 + 0.3399t^2 + 0.5019t + 0.005594$$

miles, where t is the amount of time (in minutes) that the minivan has been moving away from the stoplight.

     Rounded to the nearest mile per hour, what were the highest and lowest speeds of the minivan?

Section 11.4   *In Exercises 23–24, differentiate the function with respect to t. Explain the physical meaning of each of the rates. (A is area and s is the length of a side. Lengths are measured in centimeters, areas in square centimeters, and time in seconds.)*

23. $A = 5s^2$ (area of an equilateral cross)

24. $A = \dfrac{\sqrt{3}}{4}s^2$ (area of an equilateral triangle)

In Exercises 25–30, solve the related-rate problem.

25. Swimming Pool Depth   A circular above-ground swimming pool has a 4-foot radius and may be filled to a maximum depth of 3 feet. A woman is filling the pool with a faucet that releases water at a rate of 10 gallons per minute. How quickly is the depth of the water in the pool changing 2 minutes after she begins to fill the pool?

26. Swimming Pool Depth   How rapidly is the depth of the water in the pool in Exercise 25 increasing 5 minutes, 10 minutes, and 15 minutes after the woman begins to fill the pool?

27. Swimming Pool Depth   A rectangular above-ground swimming pool has a 12-foot length and a 6-foot width and may be filled to a maximum depth of 4 feet. A teen is filling the pool with a faucet that releases water at a rate of 9 gallons per minute. How quickly is the depth of the water in the pool changing 2 minutes after he begins to fill the pool?

28. Swimming Pool Depth   How rapidly is the depth of the water in the pool in Exercise 27 increasing 5 minutes, 10 minutes, and 15 minutes after the teen begins to fill the pool?

29. Oil Spill Spread   The surface area of a circular oil spill is increasing at a rate of 250 square feet per hour. How quickly is the radius increasing when it is 2500 feet?

30. Oil Spill Spread   The radius of a circular oil spill is increasing at a rate of 10 feet per hour. How quickly is the surface area changing when the radius is 200 feet?

Make It Real

What to do

1. Recruit a friend, neighbor, or family member to go on a drive with you.
2. Drive to the base of a freeway (or major highway) on-ramp.
3. As you drive up the on-ramp and onto the freeway, have your friend record the odometer reading every 15 seconds for at least 2 minutes. If the odometer is between tenths of a mile, round to the nearest twentieth. (That is, if it is between 2.0 and 2.1, round to 2.05.)
4. Find the equation of the function that best fits the data.
5. Use differentiation to find the velocity and acceleration functions.
6. Determine your maximum velocity and maximum acceleration.

How to get a good data set

1. As you're driving, try to avoid any rapid changes in your acceleration. (This will make a smoother scatter plot.)
2. Collect several minutes of data. You can eliminate extra points at the end of your data set if the model doesn't fit those points well.
3. Use as many data points as possible, yet aim for a model coefficient of determination (r^2) of 0.99 or better.
4. Avoid excessive speeds, erratic driving, and driving under the influence.
5. If you see flashing lights behind you, pull over to the right side of the road and abort the project!.

Chapter 12

The Integral

On March 24, 1989, the *Exxon Valdez* oil tanker ran aground in Prince William Sound. The ensuing oil spill was one of the most widely publicized environmental disasters in history. If a function modeling the variable rate at which oil is spilling from a tanker is known, the Fundamental Theorem of Calculus may be used to calculate the total amount of oil spilled over a specified time period.

12.1 Indefinite Integrals

- Use integration rules to find the antiderivative of a function

GETTING STARTED A speedometer measures a car's velocity. If we record our car's velocity at 15-second intervals, we will generate a table of data that may be used to find a model for the velocity of the car. Can we determine how far the car traveled from the velocity equation alone? Yes. By using a process called *integration*, we can determine the distance equation if we know the velocity equation.

In this section, we will introduce integration: the process of finding the equation of a function that has a given derivative. We will also define and demonstrate how to use basic rules of integration.

We will begin by looking at three functions: $f(x) = x^3 - x^2 - 8x + 17$, $g(x) = x^3 - x^2 - 8x + 12$, and $h(x) = x^3 - x^2 - 8x + 7$. The equations of these functions look remarkably similar. In fact, their equations differ only by a constant. Similarly, the basic shape of the graphs is the same; they differ only in their vertical placement on the coordinate system (see Figure 12.1).

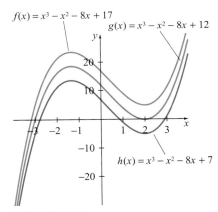

FIGURE 12.1

Let's differentiate each of the functions.

$$f'(x) = 3x^2 - 2x - 8$$

$$g'(x) = 3x^2 - 2x - 8$$

$$h'(x) = 3x^2 - 2x - 8$$

The functions all have the same derivative! In fact, any function of the form

$$f(x) = x^3 - x^2 - 8x + C \ (C \text{ is any constant})$$

will have the derivative

$$f'(x) = 3x^2 - 2x - 8$$

Thus, an infinite number of functions share the same derivative. The graphs of all of these functions will have the same basic shape, but their vertical placement on the coordinate system will differ.

Often we are given the rate of change function (derivative) and asked to find the equation of the function that has the given derivative. As previously illustrated, there is an infinite number of functions that share the same derivative. To represent all such functions, we will write the letter C in the place of the constant term.

EXAMPLE 1

Finding the Antiderivative of a Function

Find the general form of the function that has the derivative $f(x) = 2x$.

SOLUTION We are looking for a function $F(x)$ such that $\frac{d}{dx}[F(x)] = f(x)$.

Let's consider the function $F(x) = x^2 + C$.

$$\frac{d}{dx}[F(x)] = \frac{d}{dx}(x^2 + C)$$

$$= 2x + 0$$

$$= 2x$$

Therefore, any function of the form $F(x) = x^2 + C$ has the derivative $f(x) = 2x$. Specific functions include $F(x) = x^2 + 2$, $F(x) = x^2 - 12$, $F(x) = x^2 + 256$, and so on.

DEFINITION OF TERMS

Integration: The process of finding the general form of a function that has a given derivative. Also referred to as *antidifferentiation*.

Indefinite integral: The general form of a function that has a given derivative. Also referred to as the *antiderivative*.

Integral sign: The symbol $\int$, which indicates that integration is to be performed.

Integrand: The function written to the right of the integral sign. This is the function that is to be integrated.

The notation $\int f(x)\,dx$ is read "the integral of $f(x)$ with respect to x" or "the antiderivative of $f(x)$ with respect to x." Just as $\frac{d}{dx}[f(x)]$ means "find the derivative of $f(x)$ with respect to x," $\int f(x)\,dx$ means "find the general form of the function that has derivative $f(x)$ with respect to x." The integral sign $\int$ must always be written together with a *differential* (e.g., dx) to indicate the variable of integration. (We discuss differentials in more detail in the next section.) A function f is said to be **integrable** if there exists a function F such that $F' = f$.

EXAMPLE   2   **Finding the Antiderivative of a Function**

Find $\int 3x^2\, dx$.

SOLUTION   We are to find the general equation of the function whose derivative is $f(x) = 3x^2$. Recall that $\dfrac{d}{dx}(x^3) = 3x^2$. Therefore, the indefinite integral of $f(x) = 3x^2$ is $F(x) = x^3 + C$.

We can check our work by differentiating the function F with respect to x.

$$\frac{d}{dx}[F(x)] = \frac{d}{dx}(x^3 + C)$$
$$= 3x^2 + 0$$
$$= 3x^2$$

You may have noticed that we have shifted our notation slightly from our earlier treatment of derivatives. Since we are given the rate of change function as the original function, we are using the notation $f(x)$ instead of $f'(x)$ to represent the rate of change function. We use the notation $F(x)$ to represent the function whose derivative is $f(x)$. That is, $F'(x) = f(x)$. Admittedly, this shift in notation is a bit confusing. However, since the notation shift is so widely adopted by people who use calculus, we will also use this commonly accepted notation. Making the transition may be difficult at first, but by the end of the chapter, you will be using the new notation with ease.

Basic Integration Rules

As you might expect, the rules for derivatives and the rules for integrals are intimately related. In fact, many of the rules even share the same names.

POWER RULE FOR INTEGRALS

Let $f(x) = x^n$, where x is a variable and n is a constant with $n \neq -1$. Then

$$\int x^n\, dx = \frac{x^{n+1}}{n+1} + C, \quad n \neq -1$$

The rule can be easily verified by differentiating the resultant function.

$$\frac{d}{dx}\left(\frac{x^{n+1}}{n+1} + C\right) = \frac{1}{n+1}\frac{d}{dx}(x^{n+1}) + 0 \qquad \text{Constant Multiple Rule}$$

$$= \frac{n+1}{n+1}x^{n+1-1} \qquad\qquad \text{Power Rule}$$

$$= x^n$$

Notice that $n \neq -1$. If $n = -1$, the denominator of $\dfrac{x^{n+1}}{n+1} + C$ is equal to 0, which makes the expression undefined. We will deal with the case of $n = -1$ later.

> **CONSTANT MULTIPLE RULE FOR INTEGRALS**
>
> Let $f(x) = a \cdot g(x)$, where x is a variable and a is a constant. Then
>
> $$\int [a \cdot g(x)]\,dx = a \cdot \int g(x)\,dx$$
> $$= a \cdot G(x) + C$$
>
> where $g(x) = G'(x)$.

EXAMPLE 3

Integrating a Function

Integrate $f(x) = 6x^2$.

SOLUTION

$$\int 6x^2\,dx = 6\int x^2\,dx \qquad \text{Constant Multiple Rule}$$

$$= 6\left(\frac{x^3}{3} + C_1\right) \qquad \text{Power Rule}$$

$$= \frac{6x^3}{3} + 6C_1$$

$$= 2x^3 + C$$

Notice that we rewrote $6C_1$ as C. Since C_1 is a constant, multiplying it by 6 will still result in a constant. The same would be true if we were to add two different constants C_1 and C_2. The result would still be a constant, which we could write as C. Therefore, when doing integration, we will write a single constant C to represent a multiple of a constant or a sum of constants.

Remember that the derivative represents the slope of the tangent line. The derivative of a linear function $f(x) = mx + b$ is $f'(x) = m$. That is, for a linear function, the derivative of the function is the slope of the line. Similarly, the antiderivative of a constant function $f(x) = m$ is the linear function $F(x) = mx + b$. This relationship is summarized in the Constant Rule for Integrals.

> **CONSTANT RULE FOR INTEGRALS**
>
> Let k be a constant. Then
>
> $$\int k\,dx = kx + C.$$

The Sum and Difference Rule is another extremely useful rule to use when working with integrals.

SUM AND DIFFERENCE RULE FOR INTEGRALS

Let $f(x)$ and $g(x)$ be integrable functions of x. Then

$$\int [f(x) \pm g(x)]dx = \int f(x)\,dx \pm \int g(x)\,dx$$

EXAMPLE 4

Antidifferentiating a Function

Find the antiderivative of $h(x) = 2x + 4$.

SOLUTION

$$
\begin{aligned}
\int (2x + 4)\,dx &= \int 2x\,dx + \int 4\,dx && \text{Sum and Difference Rule} \\
&= 2\int x\,dx + 4\int 1\,dx && \text{Constant Multiple Rule} \\
&= 2\left(\frac{x^2}{2}\right) + 4(x) + C && \text{Power and Constant Rules} \\
&= x^2 + 4x + C
\end{aligned}
$$

EXPONENTIAL RULE FOR INTEGRALS

Let $a > 0$ and $a \neq 1$. Then

$$\int a^x\,dx = \frac{1}{\ln(a)}a^x + C$$

The Exponential Rule may be easily verified by differentiating $y = \frac{1}{\ln(a)}a^x + C$.

$$
\begin{aligned}
\frac{d}{dx}\left(\frac{1}{\ln(a)}a^x + C\right) &= \frac{d}{dx}\left(\frac{1}{\ln(a)}a^x\right) + \frac{d}{dx}(C) && \text{Sum and Difference Rule} \\
&= \frac{1}{\ln(a)}\left[\frac{d}{dx}(a^x)\right] + 0 && \text{Constant and Constant} \\
& && \text{Multiple Rules} \\
&= \frac{1}{\ln(a)}[\ln(a)a^x] && \text{Exponential Rule} \\
&= \frac{\ln(a)}{\ln(a)}(a^x) \\
&= a^x
\end{aligned}
$$

It immediately follows from the Exponential Rule for Integrals that

$$
\begin{aligned}
\int e^x\,dx &= \frac{1}{\ln(e)}e^x + C \\
&= \frac{1}{1}e^x + C \\
&= e^x + C
\end{aligned}
$$

EXAMPLE 5

Integrating an Exponential Function

Find $\int 2 \cdot 3^x dx$.

SOLUTION

$$\int 2 \cdot 3^x dx = 2 \int 3^x dx \qquad \text{Constant Multiple Rule for Integrals}$$

$$= 2 \cdot \frac{1}{\ln(3)} (3^x) + C \qquad \text{Exponential Rule for Integrals}$$

$$= \frac{2}{\ln(3)} (3^x) + C$$

Recall that $x^{-1} = \frac{1}{x}$. We saw previously that we cannot use the Power Rule for Integrals to integrate $f(x) = x^{-1}$. However, the function is integrable.

INTEGRAL RULE FOR $\frac{1}{x}$

If $x \neq 0$, then

$$\int \frac{1}{x} dx = \ln|x| + C$$

The absolute value of x is necessary to ensure that the two functions $f(x) = \frac{1}{x}$ and $F(x) = \ln|x| + C$ have the same domain ($\ln x$ is defined only for $x > 0$, but $\ln|x|$ is defined for all $x \neq 0$). We can convince ourselves that the rule holds true by differentiating $\ln|x| + C$.

$$\frac{d}{dx}(\ln|x| + C) = \frac{d}{dx}(\ln|x|) + \frac{d}{dx}(C) \qquad \text{Sum and Difference Rule}$$

$$= \frac{d}{dx}\left(\begin{cases} \ln(x) & x > 0 \\ \ln(-x) & x < 0 \end{cases}\right) + 0 \qquad \begin{array}{l}\text{Definition of absolute value} \\ \text{Constant Rule}\end{array}$$

$$= \begin{cases} \dfrac{1}{x}, x > 0 \\ \dfrac{1}{-x}(-1), x < 0 \end{cases} \qquad \begin{array}{l}\text{Logarithm Rule} \\ \\ \text{Logarithm and Chain Rules}\end{array}$$

$$= \begin{cases} \dfrac{1}{x}, x > 0 \\ \dfrac{1}{x}, x < 0 \end{cases}$$

$$= \frac{1}{x}, x \neq 0$$

We see that for all nonzero values of x, $\dfrac{d}{dx}(\ln|x| + C) = \dfrac{1}{x}$.

Integrating a Function of the Form $y = \dfrac{c}{x}$

Integrate $\dfrac{3}{x}$ with respect to x.

SOLUTION

$$\int \left(\frac{3}{x}\right) dx = 3 \int \left(\frac{1}{x}\right) dx \qquad \text{Constant Multiple Rule for Integrals}$$

$$= 3\ln|x| + C \qquad \text{Integral Rule for } \frac{1}{x}$$

Indefinite Integral Applications

Recall that the units of $\dfrac{dy}{dx}$ are the units of y over the units of x. Since $\int \left(\dfrac{dy}{dx}\right) dx = y + C$, the units of the antiderivative of $\dfrac{dy}{dx}$ are the units of y. In real-world applications, we are often given enough information to figure out the specific value of C, as demonstrated in Example 7.

Using Integration to Find a Position Function

Based on 2.5 minutes of data, the velocity of the author's minivan as he drove from a stoplight onto a highway may be modeled by $V(t) = -0.1773t^2 + 0.6798t + 0.5019$ miles per minute, where t is the number of minutes since he left the stoplight. Find the position function $D(t)$ that shows his distance from the stoplight at time t. Find the total distance traveled during the first minute and during the first two minutes.

SOLUTION The position function $D(t)$ is the antiderivative of the velocity function $V(t)$.

$$\int [V(t)] dt = \int (-0.1773t^2 + 0.6798t + 0.5019)\, dt$$

$$D(t) = -0.1773\frac{t^3}{3} + 0.6798\frac{t^2}{2} + 0.5019t + C \qquad \substack{\text{Constant Multiple, Power, and} \\ \text{Sum and Difference Rules}}$$

$$= -0.0591t^3 + 0.3399t^2 + 0.5019t + C \text{ miles}$$

You may ask, "What is the meaning of C in this context?" Let's see. $D(0)$ represents the distance traveled after 0 minutes. Intuitively, we know that after 0 minutes, the minivan hasn't traveled at all, so $D(0) = 0$.

$$D(0) = -0.0591(0)^3 + 0.3399(0)^2 + 0.5019(0) + C$$

$$= C$$

But $D(0) = 0$, so $C = 0$. $D(t) = -0.0591t^3 + 0.3399t^2 + 0.5019t$ is the distance function for the minivan. According to our distance model, $D(1) = 0.78$ and $D(2) = 1.89$. That is, the total distance traveled during the first minute is 0.78 mile. The total distance traveled during the first two minutes is 1.89 miles.

EXAMPLE 8 **Using Antidifferentiation to Find a Cost Function**

Have you ever dreamed of being a published author? Golden Pillar Publishing (www.goldenpillarpublishing.com) publishes books for aspiring authors. In 2003, the marginal cost for printing an 8.25″ × 11″ soft-cover book was

$$c(p) = 0.018 \text{ dollar per page}$$

where p is the number of pages in the book. Including the setup costs, a 1000-page book has a \$19.30 production cost. Find the book production cost function.

SOLUTION Recall that marginal cost is the derivative of the cost function. Therefore, the antiderivative of the marginal cost is the cost function.

$$\int c(p) \, dp = \int 0.018 \, dp$$
$$C(p) = 0.018p + k \qquad \text{Constant Rule for Integrals}$$

(Note: We used k as a constant, since the function was named C.)
The cost of a 1000-page book is \$19.30, so

$$C(1000) = 0.018(1000) + k$$
$$19.30 = 18 + k$$
$$k = 1.30$$

The constant $k = 1.30$ represents the fixed setup cost. The cost to produce a book with p pages is $C(p) = 0.018p + 1.30$ dollars.

12.1 Summary

In this section you learned how to find the equation of some functions that have a given derivative through the process of integration. You also discovered how to use basic integration rules to integrate a variety of functions.

12.1 Exercises

In Exercises 1–20, find the general antiderivative of the function.

1. $f(x) = 2$

2. $g(x) = -2x + 9$

3. $r(t) = t^{-1} - 3t$

4. $f(t) = \dfrac{1.21}{t}$

5. $v(t) = 0.5t + 20$

6. $a(t) = -32$

7. $s(t) = t^3 - 3t^2 + 3t - 1$

8. $s(x) = 2^x$

9. $h(x) = \dfrac{2.3}{x} + 1$

10. $f(x) = 0$

11. $p(x) = 3^x - x^3$

12. $r(x) = 200 - \dfrac{225}{x}$

13. $f(t) = 4t^{-2} + 2t^{-1} + 1$

14. $g(x) = 2^x - x^2$

15. $h(x) = 2(3^x) - 3(2^x)$

16. $q = 2500 - \dfrac{1000}{p^2}$

17. $q = \dfrac{2000}{p} - \dfrac{500}{p^2}$

18. $f(x) = 3(2^x) + 4x^{-1}$

19. $y = 5t^{-2} + 2^t$

20. $h(x) = \dfrac{4x^3 - 15}{x^2}$

In Exercises 21–35, calculate the indefinite integral.

21. $\int (3x - 5)\, dx$ **22.** $\int (-x^2 + 4x)\, dx$

23. $\int \dfrac{t - 2}{t}\, dt$ (*Hint:* Divide each term in the

numerator by the denominator before integrating.)

24. $\int \dfrac{25u - 1}{u}\, du$

25. $\int (3x^{-2} - 4x^3 + 2)\, dx$

26. $\int (4t^{-1} - 12)\, dt$

27. $\int (5t^{-2} - 16t^2 - 9)\, dt$

28. $\int [5(2^t) - 3t^2]\, dt$

29. $\int \dfrac{5u^2 - 4u + 1}{u^2}\, du$

30. $\int (4t^{-2} - 2^t)\, dt$

31. $\int \left(\dfrac{400}{x^2} + \dfrac{200}{x} + 50 \right) dx$

32. $\int (9^t - 3^t)\, dt$ **33.** $\int \left(\dfrac{4}{x} - \dfrac{5}{x^2} \right) dx$

34. $\int \left(20 - \dfrac{5u + 2}{u} \right) du$

35. $\int (4t + 4^t)\, dt$

In Exercises 36–39, apply the concept of integration in solving the real-world applications.

36. **Publishing Costs** Golden Pillar Publishing (www.goldenpillarpublishing .com) publishes books for aspiring authors. In 2003, the marginal printing cost for a 7.5″ × 9.25″ soft-cover book was

$$c(p) = 0.013$$

dollar per page, where p is the number of pages in the book. Including setup costs, a 1000-page book has a $13.90 production cost. Find the book production cost function.

37. **Publishing Costs** Golden Pillar Publishing (www.goldenpillarpublishing. com) publishes books for aspiring authors. In 2003, the marginal printing cost for a 6″ × 9″ hard-cover book was

$$c(p) = 0.013$$

dollar per page, where p is the number of pages in the book. Including setup costs, a 1000-page book has an $18.00 production cost. Find the book production cost function.

38. **Cost Analysis** The marginal printing cost of the soft-cover and hard-cover books in Exercises 36 and 37 was the same, yet their cost functions were different. Why?

39. **Minivan Position** Based on 1.5 minutes of data, the velocity of the author's minivan as he drove from a stoplight into a sparsely populated residential area may be modeled by

$$V(t) = -2.133t^3 + 3.999t^2 - 0.1533t + 0.3167$$

miles per minute, where t is the number of minutes since he left the stoplight. Find the position function $D(t)$ that shows his distance from the stoplight at time t.

Exercises 40–45 are intended to challenge your understanding of the basic integration rules.

40. Let $F_1(x)$ and $F_2(x)$ be specific antiderivatives of $f(x)$. What is the graphical interpretation of $|F_1(x) - F_2(x)|$?

41. Let $F_1(x)$ and $F_2(x)$ be specific antiderivatives of $f(x)$, with $F_1(x) \neq F_2(x)$. Do the graphs of $F_1(x)$ and $F_2(x)$ ever intersect?

42. Let $f(x)$ be a polynomial of degree n with $f(x) = F'(x)$. At most, how many x-intercepts does $F(x)$ have?

43. Let $F'(x) = 3 + \dfrac{1}{x^2}$. Is $F(x)$ an increasing or a decreasing function? Explain.

44. Let $F'(x) = 3x^2$ and $F(0) = 0$. On what interval(s) is $F(x) \geq 0$?

45. Let $F'(x) = 4 - 2x$ and $F(0) = -4$. On what interval(s) is $F(x) \geq 0$?

12.2 Integration by Substitution

- Use the method of substitution to integrate functions

The process of *integration by substitution* is a clever way to integrate a function that initially doesn't appear to be integrable. (Recall that a function f is said to be integrable if there exists a function F such that $F' = f$.) Because of the relatively complex nature of the process, we will defer illustrations of how this process may be used in real-life applications to later sections.

In this section, we will introduce the concept of differentials. We will then demonstrate how to do integration by substitution.

Differentials

Recall that $\dfrac{d}{dx}(y) = \dfrac{dy}{dx}$. The derivative $\dfrac{dy}{dx}$ is a single term, not the quotient of dy and dx. However, it would be useful for us to be able to treat $\dfrac{dy}{dx}$ as if it were a fraction. For this purpose, we introduce the concept of *differentials*. We define the **differential** dx to be an arbitrarily small real number and

$$dy = y'\,dx$$

Observe that dividing both sides of the equation $dy = y'\,dx$ by dx yields

$$\frac{dy}{dx} = y'$$

which is consistent with our earlier notation for the derivative. Thus, by using differentials, we can treat dy and dx in $\dfrac{dy}{dx}$ as separate entities and use the rules for fractions in manipulating the terms.

EXAMPLE 1

Calculating the Differential *dy*

Find dy given $y = x^2$.

SOLUTION
$$dy = 2x\,dx$$

EXAMPLE 2

Calculating the Differential *dy*

Find dy given $y = 2x^3 - 3x$.

SOLUTION
$$dy = (6x^2 - 3)\,dx$$

We may also use differentials with other variables, such as u and t, as demonstrated in the next two examples.

EXAMPLE 3

Calculating the Differential *du*

Find *du* given $u = e^{3t} + 2$.

SOLUTION

$$du = 3e^{3t}dt$$

EXAMPLE 4

Calculating the Differential *du*

Find *du* given $u = \ln(t)$.

SOLUTION

$$du = \frac{1}{t}dt$$

Integration by Substitution

The technique of **integration by substitution** is used to rewrite a function that doesn't appear to be readily integrable in such a way that it can be integrated using the basic integration rules.

EXAMPLE 5

Integrating by Substitution

Integrate $f(x) = \frac{2\ln(x)}{x}$.

SOLUTION We must find $\int \left[\frac{2\ln(x)}{x} \right] dx$. This function cannot be readily integrated using any of the rules covered in Section 12.1. This is because $f(x) = \frac{2\ln(x)}{x}$ is the product of $2\ln(x)$ and $\frac{1}{x}$, and we haven't yet discussed how to integrate a product. (Contrary to what you might think, $\int [f(x)\,g(x)]\,dx \neq \int f(x)\,dx \cdot \int g(x)\,dx$.

We will let $u = \ln(x)$. We take the derivative of both sides and use the concept of differentials to find *du*.

$$u = \ln(x)$$

$$\frac{d}{dx}(u) = \frac{d}{dx}[\ln(x)]$$

$$\frac{du}{dx} = \frac{1}{x}$$

$$du = \frac{1}{x}dx$$

We will now rewrite the integral in terms of *u*.

$$\int 2\underbrace{[\ln(x)]}_{u}\underbrace{\left(\frac{1}{x}dx\right)}_{du} = \int 2u\,du$$

But we know that

$$\int 2u\,du = u^2 + C$$

(Although the differential du represents an arbitrarily small number and the du of an integral means "with respect to u," the two uses of the notation du may be interchanged. A more in-depth explanation may be obtained from an advanced calculus text.)

Since the variable u was something that we introduced into the problem, we must rewrite the solution in terms of x. Recall that $u = \ln(x)$.

$$u^2 + C = [\ln(x)]^2 + C$$

Thus, we have

$$\int \frac{2\ln(x)}{x}\,dx = [\ln(x)]^2 + C$$

It is not always possible to find a substitution of variables that makes a difficult function integrable. However, if the integral can be rewritten in any of the following forms, it may be integrated using the basic integration rules.

- $\int u^n\,du$

- $\int e^u\,du$

- $\int \frac{1}{u}\,du$

EXAMPLE 6

Integrating by Substitution

Find the antiderivative of $f(x) = 2x(x^2 + 5)^{10}$.

SOLUTION We need to determine $\int 2x(x^2 + 5)^{10}\,dx$. The function is a product of two functions, so we hope to be able to write it in the form $u^n\,du$. We'll try $u = x^2 + 5$. Then

$$u = x^2 + 5$$

$$\frac{du}{dx} = 2x$$

$$du = 2x\,dx$$

Let's see if we can rewrite the integral in terms of u and du. Recall that $u = x^2 + 5$ and $du = 2x\,dx$.

$$\int 2x(x^2 + 5)^{10}\,dx = \int [(x^2 + 5)^{10}](2x\,dx) \qquad \text{Switch the order of the factors and group}$$

$$= \int u^{10}\,du \qquad \text{Since } u = x^2 + 5 \text{ and } du = 2x\,dx$$

$$= \frac{u^{11}}{11} + C$$

Rewriting the result in terms of x, we have

$$\int 2x(x^2 + 5)^{10}\,dx = \frac{(x^2 + 5)^{11}}{11} + C$$

We can check our result by differentiating the antiderivative and comparing it to $f(x)$.

$$\frac{d}{dx}\left[\frac{(x^2 + 5)^{11}}{11} + C\right] = \frac{1}{11}\frac{d}{dx}[(x^2 + 5)^{11}] + 0 \qquad \text{Constant and Constant Multiple Rules}$$

$$= 11 \cdot \frac{1}{11}(x^2 + 5)^{10} \cdot 2x \qquad \text{Power and Chain Rules}$$

$$= 2x(x^2 + 5)^{10}$$

The derivative of the antiderivative is equal to $f(x)$, so we are confident that we integrated the function correctly.

EXAMPLE **7**   **Integrating by Substitution**

Integrate $g(t) = e^{2t-4}$.

SOLUTION We must find $\int e^{2t-4}\,dt$. We hope to be able to rewrite the integral as $\int e^u du$. We'll let $u = 2t - 4$. Then

$$\frac{du}{dt} = 2$$

$$du = 2\,dt$$

Looking at the integrand (the function to be integrated), we see that we have dt but not $2\,dt$. Can we solve $du = 2\,dt$ for dt? Yes!

$$2\,dt = du$$

$$dt = \frac{1}{2}du$$

Substituting, we get

$$\int e^{2t-4}dt = \int e^u \left(\frac{1}{2}du\right) \qquad \text{Since } u = 2t - 4 \text{ and } dt = \tfrac{1}{2}du$$

$$= \frac{1}{2}\int e^u\,du \qquad \text{Constant Multiple Rule for Integrals}$$

$$= \frac{1}{2}e^u + C \qquad \text{Exponential Rule}$$

$$= \frac{1}{2}e^{2t-4} + C \qquad \text{Rewrite in terms of } t$$

EXAMPLE **8**   **Integrating by Substitution**

Find the general form of a function that has the derivative $f(x) = \dfrac{2x + 3}{x^2 + 3x + 4}$.

SOLUTION We must find $\int \dfrac{2x + 3}{x^2 + 3x + 4}\,dx$. We observe that the numerator is the derivative of the denominator. Consequently, we hope to be able to rewrite the integral in the form $\int \dfrac{1}{u}\,du$. We'll let $u = x^2 + 3x + 4$. Then

$$\frac{du}{dx} = 2x + 3$$

$$du = (2x + 3)dx$$

Rewriting the integral in terms of u and solving, we get

$$\int \frac{2x + 3}{x^2 + 3x + 4} \, dx = \int \frac{(2x + 3)dx}{x^2 + 3x + 4}$$

$$= \int \frac{du}{u} \qquad \text{Since } u = x^2 + 3x + 4$$
$$\text{and } du = (2x + 3)dx$$

$$= \int \frac{1}{u} \, du$$

$$= \ln|u| + C \qquad \text{Integral Rule for } \frac{1}{u}$$

$$= \ln|x^2 + 3x + 4| + C \qquad \text{Rewrite in terms of } x$$

All functions of the form $F(x) = \ln|x^2 + 3x + 4| + C$ have the derivative $f(x) = \dfrac{2x + 3}{x^2 + 3x + 4}$.

In each of the previous examples, we selected the "correct" u each time. However, it is very common to attempt integration by substitution with the "wrong" function for u. If you rewrite a function in terms of u and the resulting function isn't integrable, go back and select a different function to be u. This is part of the process of integration by substitution and should not be viewed as an error. As you gain experience with these types of problems, you will more easily recognize what is the best choice for u. In Example 9, we will demonstrate a type of problem that requires an especially clever selection of u.

EXAMPLE 9   **Integrating by Substitution a Function Containing a Radical**

Find $\int x\sqrt{x - 1} \, dx$.

SOLUTION   We begin by rewriting the integrand with a rational exponent.

$$\int x(x - 1)^{1/2} \, dx$$

Let's pick $u = x - 1$. Then

$$\frac{du}{dx} = 1$$

$$du = dx$$

We make the substitution.

$$\int xu^{1/2} \, du$$

What do we do with the extra x? The integrand must be written in terms of u. (We can't use the Constant Multiple Rule to move the x to the left-hand side of the integral sign because x is a variable, not a constant.) Recall that $u = x - 1$; therefore, $x = u + 1$. Thus we may rewrite the integral as

$$\int xu^{1/2}\,du = \int (u+1)u^{1/2}\,du$$

Since we have completely rewritten the integral in terms of u, we may continue. We will multiply out the integrand and then integrate each term.

$$\int \left(u^{3/2} + u^{1/2}\right)du = \int u^{3/2}\,du + \int u^{1/2}\,du$$

$$= \frac{u^{5/2}}{\frac{5}{2}} + \frac{u^{3/2}}{\frac{3}{2}} + C \qquad\qquad \text{Power Rule for Integrals}$$

$$= \frac{2}{5}u^{5/2} + \frac{2}{3}u^{3/2} + C$$

$$= \frac{2}{5}(x-1)^{5/2} + \frac{2}{3}(x-1)^{3/2} + C$$

Thus $\int x\sqrt{x-1}\,dx = \frac{2}{5}(x-1)^{5/2} + \frac{2}{3}(x-1)^{3/2} + C.$

EXAMPLE 10

Determining When Integration by Substitution Won't Work

Integrate $f(x) = xe^x$.

SOLUTION We must find $\int xe^x\,dx$. If we pick $u = x$, then $du = dx$ and the integral becomes

$$\int ue^u\,du$$

Observe that this is identical to the original function written, except that it is written in terms of u. Using the integral techniques we've covered, we don't know how to integrate a function in this form.

Let's try $u = e^x$. Then

$$u = e^x$$
$$\ln(u) = \ln(e^x)$$
$$\ln(u) = x$$
$$x = \ln(u)$$

Since $\dfrac{du}{dx} = e^x$, $du = e^x dx$. The resultant integral is

$$\int xe^x\,dx = \int [\ln(u)]\,du$$

We haven't yet learned how to integrate $\ln(u)$, so we are unable to find the solution using this choice of u.

As a final attempt, we'll try $u = xe^x$. Then $du = (1 \cdot e^x + e^x \cdot x)dx$, by the Product Rule. If we attempt to make a substitution, we'll have a mixture of u and x in the integrand, leaving us with a function that we can't integrate

$$\left(\int xe^x dx = \int u\frac{du}{(1 \cdot e^x + e^x \cdot x)}\right).$$

The function $f(x) = xe^x$ cannot be integrated using integration by substitution. Nevertheless, there is a method that can be used to integrate $f(x) = xe^x$. In Section 13.1, we will show how to integrate this function using integration by parts.

It is important to note that, despite the many advances in calculus, there are some functions that we still don't know how to integrate. Nevertheless, we must attempt to integrate a function using all known methods before we can conclude that it is not integrable.

12.2 Summary

In this section you learned how to work with differentials. You used integration by substitution to find the antiderivative of a function that didn't initially appear to be integrable. You also learned that there are functions that we can't integrate simply by reversing the derivative rules.

12.2 Exercises

In Exercises 1–40, integrate the functions. Some integrals will require integration by substitution, while others may be integrated using basic integration rules. All of the functions may be integrated using the techniques we've covered.

1. $f(x) = 2x(x^2 + 3)^5$ **2.** $g(x) = 6x^2(x^3 + 7)^2$

3. $h(x) = 2x(3x + 5)$

4. $f(t) = (2t + 3)(t^2 + 3t + 1)^3$

5. $s(x) = e^x(e^x + 1)$ **6.** $v(t) = 3^t + t^{-1}$

7. $f(x) = 1.2x^2 - 2.4x + 0.6$

8. $w(t) = 10te^{5t^2 - 4}$ **9.** $p(x) = (x - 2)e^{x^2 - 4}$

10. $y = 3e^{2x}$ **11.** $y = 3xe^{3x^2 - 6}$

12. $f(x) = (2x - 9)e^{x^2 - 9x + 21}$

13. $g(t) = 4e^{3t}$ **14.** $y = 3(2^x)$

15. $s(x) = \dfrac{\ln(x^2)}{x}$ **16.** $y = 4x(x^2 + 1)^3$

17. $h(t) = \dfrac{2t - 1}{t^2 - t + 2}$ **18.** $g(x) = \dfrac{x - 4}{x^2 - 8x + 9}$

19. $f(x) = \dfrac{6x^2}{x^3 - 9}$ **20.** $h(t) = \dfrac{2t - 1}{t^2 - t + 2}$

21. $p(x) = \dfrac{1}{x \ln x}$ **22.** $s(x) = \dfrac{\ln x}{x}$

23. $y = 3x\sqrt{x - 2}$ **24.** $f(x) = x\sqrt{x + 2}$

25. $g(x) = 2x\sqrt{x^2 + 1}$

26. $h(x) = (x - 1)\sqrt{x^2 - 2x}$

27. $f(t) = t\sqrt{t^2 - 1}$ **28.** $s(x) = \dfrac{x}{\sqrt{x^2 - 1}}$

29. $f(x) = \dfrac{3x^2 - 2x + 1}{x^3 - x^2 + x - 1}$

30. $g(t) = \dfrac{e^{\sqrt{t}}}{2\sqrt{t}}$ **31.** $f(x) = \dfrac{x}{2x - 1}$

32. $g(x) = \dfrac{e^x - \ln 2(2^x)}{e^x - 2^x}$

33. $h(x) = \dfrac{2x}{4x^2 - 5}$ **34.** $v(t) = \ln 2(3^t)$

35. $a(t) = (4t - 1)(t^2 - 2t)$

36. $q = \dfrac{1000}{p} - p\sqrt{p - 2}$

37. $q = \dfrac{4p}{2p^2 + 1} - 4p\sqrt{2p^2 + 1}$

38. $s(t) = 6t(t^2 + 1)^3$

39. $s(t) = (4t - 2)(t^2 - t)^{-1}$

40. $v(x) = x(2^{x^2 - 1})$

Exercises 41–45 are intended to challenge your understanding of integration by substitution.

41. Determine if $f(x) = 2x^3(x^2 - 1)^4$ may be integrated using integration by substitution.

42. Determine if $f(x) = \dfrac{\ln(x) + 1}{x \ln(x)}$ may be integrated using integration by substitution.

43. Determine if $f(x) = \dfrac{x}{\ln(x)}$ may be integrated using integration by substitution.

44. Determine if $f(x) = \dfrac{\ln(x^2)}{4x}$ may be integrated using integration by substitution.

45. Determine if $f(x) = 3x^2(x^2 - 1)^7$ may be integrated using integration by substitution.

12.3 Using Sums to Approximate Area

- Estimate the area between a curve and the x-axis by using left- and right-hand sums

GETTING STARTED   River rafting or kayaking is an exciting sport that requires both physical skill and an intellectual understanding of the hydraulics of a river. One factor that enthusiasts must consider before navigating a river is the flow rate of the water. The flow rate indicates how much water (in cubic feet) will pass a fixed point in a second. The flow rate fluctuates constantly, ever changing the intensity of the river. For the inexperienced, flow-rate values mean little. One way to add conceptual meaning to the numbers is to determine how much water has passed a given point over a period of time. As will be explained in the section, we can do this by estimating the area between a graph of the flow rate and the horizontal axis.

In this section, we will demonstrate how left- and right-hand sums may be used to approximate the area between the graph of a function and the x-axis. This will prepare us to link the notion of area with the concept of the definite integral in Section 12.4. The relationship between the two concepts is truly remarkable.

Suppose that the flow rate $f(x)$ of a river is a constant 3000 cubic feet per second. Let's consider the graph of the flow-rate function $f(x) = 3$, shown in Figure 12.2.

The graph is a horizontal line with y-intercept $(0, 3)$. We ask the question, "What is the area of the region between the graph of the function and the x-axis on the interval $[1, 3]$?

FIGURE 12.2

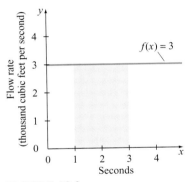

FIGURE 12.3

That is, what is the area of the shaded region in Figure 12.3? We quickly recognize that the region is a rectangle. To calculate its area, we multiply its height by its width.

$$A = 3 \cdot 2$$
$$= 6$$

The area of the region is 6 square units. What does this mean in the context of the problem? Let's recalculate A, this time keeping track of the physical meaning of the length and width of the rectangle.

$$A = \left(3 \, \frac{\text{thousand cubic feet}}{\text{second}} \right)(2 \text{ seconds})$$
$$= 6 \text{ thousand cubic feet}$$

Between the first and third second after we started timing, 6000 cubic feet of water passed our water flow measurement station. In general, if $f(x)$ is a positive rate of change function, then the area between the graph of $f(x)$ and the x-axis on the interval $[a, b]$ represents the total change in $F(x)$ [an antiderivative of $f(x)$] from $x = a$ to $x = b$.

When the shaded region is a familiar geometric shape (triangle, rectangle, and so on), it is easy to calculate the area using known formulas. However, what do we do if the shaded region is not a familiar geometric shape? Can we still calculate the area? In Example 1, we will estimate the area of an irregularly shaped region by drawing a series of rectangles. As you will see, the accuracy of our estimate will increase as we increase the number of rectangles.

EXAMPLE **1**

Using Riemann Sums to Estimate an Area

Estimate the area of the region between $f(x) = \sqrt{x} + 2$ and the x-axis on the interval $[0, 4]$.

SOLUTION We begin by drawing the graph of $f(x) = \sqrt{x} + 2$ (see Figure 12.4).

We will first estimate the area by drawing a single rectangle whose upper left corner touches the graph of f (Figure 12.5).

The area of the rectangle is $A = 2 \cdot 4 = 8$ square units. Since the entire region below the graph is not completely shaded, 8 square units is an underestimate of the actual area. To improve our accuracy, we will draw two rectangles that are each two units wide. The upper left-hand corner of each rectangle will touch the graph (see Figure 12.6).

FIGURE 12.4

FIGURE 12.5

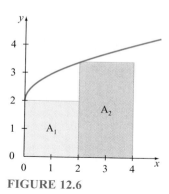

FIGURE 12.6

The area of the first rectangle is easily calculated. $A_1 = 2 \cdot 2 = 4$ square units. We can easily see that the second rectangle has a width of 2 units, but it is difficult to tell from the graph what the height of the rectangle is. However, since the upper left-hand corner of the rectangle is on the graph of f, we can use the equation of $f(x)$ to calculate the height of the rectangle. Since the rectangle touches the graph when $x = 2$, the height of the rectangle is

$$f(2) = \sqrt{2} + 2$$
$$\approx 1.414 + 2$$
$$\approx 3.414 \text{ units}$$

Consequently, the approximate area of the second rectangle is

$$A_2 = 3.414 \cdot 2$$
$$= 6.828$$

We can estimate the area of the region by adding together the areas of the two rectangles.

$$A = A_1 + A_2$$
$$= 4 + 6.828$$
$$= 10.828 \text{ square units}$$

This estimate is substantially better than our first estimate because it has increased the size of the shaded region below the graph. However, it is still an underestimate. How can we increase the accuracy of our estimate? Add more rectangles. This time we'll use four rectangles of width 1 unit, as shown in Figure 12.7.

The width of each rectangle is 1 unit. Since the upper left-hand corner of each rectangle touches the graph of f, the heights of the rectangles are $f(0)$, $f(1)$, $f(2)$, and $f(3)$, respectively. The areas of the rectangles are given by

$$A_1 = f(0) \cdot 1$$
$$= 2 \cdot 1$$
$$= 2 \text{ square units}$$

$$A_2 = f(1) \cdot 1$$
$$= 3 \cdot 1$$
$$= 3 \text{ square units}$$

$$A_3 = f(2) \cdot 1$$
$$= \left(\sqrt{2} + 2\right) \cdot 1$$
$$= 3.414 \text{ square units}$$

$$A_4 = f(3) \cdot 1$$
$$= \left(\sqrt{3} + 2\right) \cdot 1$$
$$= 3.732 \text{ square units}$$

The area of the entire shaded region is the sum of the individual areas.

$$A = A_1 + A_2 + A_3 + A_4$$
$$= 2 + 3 + 3.414 + 3.732$$
$$= 12.146 \text{ square units}$$

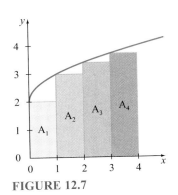

FIGURE 12.7

How can we improve our estimate? Increase the number of rectangles. This time we'll use eight rectangles of width 0.5 unit (see Figure 12.8).

The width of each rectangle is 0.5 unit. The heights of the rectangles are found by evaluating $f(x)$ at $x = 0.0$, $x = 0.5$, $x = 1.0$, $x = 1.5$, $x = 2.0$, $x = 2.5$, $x = 3.0$, and $x = 3.5$ (see Table 12.1).

FIGURE 12.8

TABLE 12.1

x	$f(x) = \sqrt{x} + 2$
0.0	2.000
0.5	2.707
1.0	3.000
1.5	3.225
2.0	3.414
2.5	3.581
3.0	3.732
3.5	3.871

The area of the shaded region is given by

$$A = A_1 + A_2 + A_3 + A_4 + A_5 + A_6 + A_7 + A_8$$
$$= 2(0.5) + 2.707(0.5) + 3(0.5) + 3.225(0.5) + 3.414(0.5) + 3.581(0.5)$$
$$+ 3.732(0.5) + 3.871(0.5)$$
$$= (2.000 + 2.707 + 3.000 + 3.225 + 3.414 + 3.581 + 3.732 + 3.871)(0.5)$$
$$= (25.530)(0.5)$$
$$= 12.765 \text{ square units}$$

We will next estimate the area using 16 rectangles. We know that our area estimate will continue to improve as we increase the number of rectangles; however, as the number of rectangles increases and their width decreases, it becomes increasingly difficult to draw the rectangles. We need a better method.

Since the width of the entire region is 4 units, the width of each of the 16 rectangles will be $\frac{4}{16} = 0.25$ unit. We'll make a table for $f(x)$ with the x-values starting at $x = 0$ and continuing to $x = 4$, with values spaced 0.25 unit apart (see Table 12.2).

TABLE 12.2

x	$f(x) = \sqrt{x} + 2$
0.00	2.000
0.25	2.500
0.50	2.707
0.75	2.866
1.00	3.000
1.25	3.118
1.50	3.225
1.75	3.323
2.00	3.414
2.25	3.500
2.50	3.581
2.75	3.658
3.00	3.732
3.25	3.803
3.50	3.871
3.75	3.936
4.00	4.000

We know that the height of each left-hand rectangle is determined by evaluating $f(x)$ at the x-value on the leftmost edge of the rectangle. The leftmost edge of the first rectangle occurs at $x = 0$. The leftmost edge of the second rectangle occurs at $x = 0.25$. Continuing on down, we see that the leftmost edge of the sixteenth rectangle occurs at $x = 3.75$. The heights of the rectangles are highlighted in the red box in Table 12.3.

TABLE 12.3

Rectangle Number	x	$f(x) = \sqrt{x} + 2$
1	0.00	2.000
2	0.25	2.500
3	0.50	2.707
4	0.75	2.866
5	1.00	3.000
6	1.25	3.118
7	1.50	3.225
8	1.75	3.323
9	2.00	3.414
10	2.25	3.500
11	2.50	3.581
12	2.75	3.658
13	3.00	3.732
14	3.25	3.803
15	3.50	3.871
16	3.75	3.936
	4.00	4.000

Since each rectangle has the same width, we may add up all of the heights of the rectangles and multiply them by the width instead of having to multiply the height of each rectangle individually by its width. (Both methods will yield the same result, but the first method requires fewer computations.)

$$A = (2 + 2.5 + 2.707 + 2.866 + 3 + 3.118 + 3.225 + 3.323 + 3.414$$
$$+ 3.5 + 3.581 + 3.658 + 3.732 + 3.803 + 3.871 + 3.936)(0.25)$$
$$= (52.234)(0.25)$$
$$\approx 13.06 \text{ square units}$$

This method is easier than drawing the rectangles; however, even it can become cumbersome when there are large numbers of rectangles. As it turns out, the exact area of the region is $13\frac{1}{3}$ square units. Our final estimate was fairly close to the exact value.

Each of the estimates in Example 1 is called a **left-hand sum,** since the upper *left* corner of each of the rectangles used in the sum touched the graph of f. The notion of using sums is credited to the famous mathematician Bernhard Riemann. Consequently, these sums are commonly called **Riemann sums.**

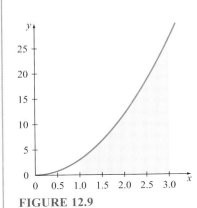

FIGURE 12.9

EXAMPLE 2 Using Riemann Sums to Estimate an Area

Estimate the area of the region between $f(x) = 3x^2$ and the x-axis on the interval $[1, 3]$ using a left-hand sum with $n = 4$.

SOLUTION We are asked to estimate the area of the shaded region in Figure 12.9 by using a left-hand sum.

Note that the shaded region is constrained to the interval $[1, 3]$.

Since $n = 4$, we are to use 4 rectangles. The width of each rectangle is

$$\frac{3 - 1}{4} = \frac{2}{4}$$

$$= 0.5 \text{ unit}$$

(In general, if we want to use n rectangles on a region between $x = a$ and $x = b$, the width of each rectangle is $\frac{b - a}{n}$.)

We will create a table of data for $f(x)$ starting at $x = 1$ and continuing to $x = 3$ with x-values spaced 0.5 unit apart (see Table 12.4).

TABLE 12.4

Rectangle Number	x	$f(x) = 3x^2$
1	1.0	3.00
2	1.5	6.75
3	2.0	12.00
4	2.5	18.75
	3.0	27.00

FIGURE 12.10

The left-hand sum estimate of the area is

$$A = (3.00 + 6.75 + 12.00 + 18.75)(0.5)$$

$$= 20.25 \text{ square units}$$

By drawing the rectangles, as shown in Figure 12.10, we can see that this is an underestimate of the actual area.

EXAMPLE 3 Using Riemann Sums to Estimate an Area

Estimate the area of the region between $f(x) = \frac{1}{x}$ and the x-axis on the interval $[1, 2]$ using a left-hand sum with $n = 5$.

SOLUTION We are asked to estimate the area of the shaded region in Figure 12.11 by using a left-hand sum with five rectangles.

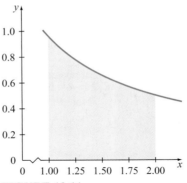

FIGURE 12.11

The width of each rectangle is $\frac{2 - 1}{5} = 0.2$ unit. We will create a table of values for $f(x)$ starting at $x = 1$ and continuing to $x = 2$ with x-values spaced 0.2 unit apart (see Table 12.5).

TABLE 12.5

Rectangle Number	x	$f(x) = \dfrac{1}{x}$	
1	1.0	1.000	
2	1.2	0.833	
3	1.4	0.714	
4	1.6	0.625	
5	1.8	0.556	
	2.0	0.500	

The left-hand sum estimate of the area is

$$A = (1.000 + 0.833 + 0.714 + 0.625 + 0.556)(0.2)$$
$$= (3.728)(0.2)$$
$$= 0.746 \text{ square unit}$$

By drawing the rectangles as shown in Figure 12.12, we can see that this is an overestimate of the actual area.

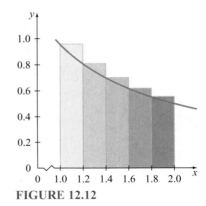

FIGURE 12.12

HOW TO Finding the Left-Hand Sum for Small Values of n ($n \leq 10$)

Let f be a function defined on the interval $[a, b]$ whose graph lies above the x-axis. The left-hand sum estimate for the area between the graph of f and the x-axis may be found using the following steps.

1. Calculate the width of each rectangle, $\Delta x = \dfrac{b - a}{n}$. "$\Delta x$" is read "delta x" and is the distance between consecutive x values.
2. Create a table of values for $f(x)$ starting at $x = a$ and ending at $x = b$, with intermediate x-values spaced Δx units apart.
3. Add up the first to the penultimate (second to last) values of $f(x)$ listed in the table. (This is the sum of the heights of the rectangles.)
4. Multiply the result of Step 3 by Δx to get the left-hand sum approximation of the area.

(*Note:* This method may be used for any value of n; however, for values of n larger than 10, the process becomes extremely tedious.)

EXAMPLE 4

Using a Left-Hand Sum to Estimate an Area

Estimate the area between the graph of $f(x) = -x^2 + 4x$ and the x-axis on the interval $[0, 4]$ using a left-hand sum with eight rectangles.

SOLUTION We are asked to estimate the area of the region shown in Figure 12.13.

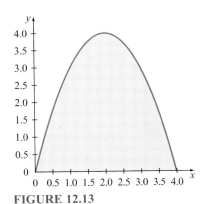

FIGURE 12.13

The width of each rectangle is

$$\Delta x = \frac{4 - 0}{8}$$

$$= \frac{1}{2}$$

$$= 0.5$$

We will create a table of values starting at $x = 0$ and continuing to $x = 4$ with the x-values spaced 0.5 unit apart (see Table 12.6).

TABLE 12.6

Rectangle Number	x	$f(x) = -x^2 + 4x$
1	0.0	0.00
2	0.5	1.75
3	1.0	3.00
4	1.5	3.75
5	2.0	4.00
6	2.5	3.75
7	3.0	3.00
8	3.5	1.75
	4.0	0.00

The sum of the heights of the rectangles is

$$0 + 1.75 + 3 + 3.75 + 4 + 3.75 + 3 + 1.75 = 21$$

We now multiply the sum of the heights by the width of a rectangle.

$$A = 21(0.5)$$
$$= 10.5 \text{ square units}$$

This left-hand sum estimate is close to the exact area $(10\frac{2}{3}$ square units$)$.

As you may have guessed, we can also calculate a right-hand sum. The right-hand sum is the sum of the areas of rectangles whose upper right-hand corner touches the graph of f. The process for finding the right-hand sum is similar to the process for finding the left-hand sum, as will be shown in Example 5.

EXAMPLE 5

Using a Right-Hand Sum to Estimate an Area

Estimate the area between the graph of $f(x) = x^3 - 3x^2 + 3x$ and the x-axis on the interval $[0, 2]$ using a right-hand sum with four rectangles.

SOLUTION We are asked to estimate the area of the shaded region in Figure 12.14.

For the right-hand sum, the upper right-hand corner of each rectangle will touch the graph of $f(x)$, as shown in Figure 12.15. The width of each rectangle will be

$$\Delta x = \frac{2 - 0}{4}$$
$$= 0.5$$

FIGURE 12.14

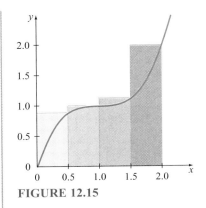

FIGURE 12.15

We will create a table of values for $f(x)$ starting at $x = 0$ and continuing to $x = 2$ with the x-values spaced 0.5 unit apart (see Table 12.7). For the right-hand sum, we select the second through the last term in the table. These are the respective heights of the rectangles.

TABLE 12.7

Rectangle Number	x	$f(x) = x^3 - 3x^2 + 3x$	
	0	0.000	
1	0.5	0.875	
2	1.0	1.000	
3	1.5	1.125	
4	2.0	2.000	

We will sum the heights together and multiply them by the width of a rectangle.

$$A = (0.875 + 1.000 + 1.125 + 2.000)\,(0.5)$$
$$= (5)\,(0.5)$$
$$= 2.5 \text{ square units}$$

The right-hand sum estimate for the area is 2.5 square units (using four rectangles).

The process for finding the right-hand sum is identical to the process for finding the left-hand sum, with the exception of Step 3. For the left-hand sum, you add up the first through the penultimate term in the table. For the right-hand sum, you add up the second through the last term.

HOW TO Finding the Right-Hand Sum for Small Values of n ($n \le 10$)

Let f be a function defined on the interval $[a, b]$ whose graph lies above the x-axis. The right-hand sum estimate for the area between the graph of f and the x-axis may be found using the following steps.

1. Calculate the width of each rectangle, $\Delta x = \dfrac{b - a}{n}$.
2. Create a table of values for $f(x)$ starting at $x = a$ and ending at $x = b$, with intermediate x-values spaced Δx units apart.
3. Add up the second through the last values of $f(x)$ listed in the table. (This is the sum of the heights of the rectangles.)
4. Multiply the result of Step 3 by Δx to get the right-hand sum approximation of the area.

(*Note:* This method may be used for any value of n; however, for values of n larger than 10, the process becomes extremely tedious.)

The table method for finding the left- and right-hand sums works well for small values of n. However, for large values of n (say $n = 100{,}000$), the table method is not practical. In Section 12.4, we will demonstrate how summation notation may be used for left- and right-hand sums with large values of n.

You may wonder, "Is the left-hand or the right-hand sum a better estimate of the area?" Actually, it varies from function to function. Often you can get one of the best area estimates by averaging the left- and right-hand sums.

EXAMPLE 6

Using Riemann Sums to Estimate an Area

Estimate the area between the graph of $f(x) = x^3 - 5x^2 + 6x + 1$ and the x-axis on the interval $[0, 4]$ with $n = 4$ using a left-hand sum, a right-hand sum, and the average of the sums.

SOLUTION We are asked to estimate the area of the shaded region in Figure 12.16.

We generate the table of values for $f(x)$ given in Table 12.8.

FIGURE 12.16

TABLE 12.8

x	$f(x) = x^3 - 5x^2 + 6x + 1$
0	1
1	3
2	1
3	1
4	9

The left-hand sum is given by

$$LHS = (1 + 3 + 1 + 1)(1)$$
$$= 6 \text{ square units}$$

The right-hand sum is given by

$$RHS = (3 + 1 + 1 + 9)(1)$$
$$= 14 \text{ square units}$$

Averaging the two sums, we get

$$\frac{LHS + RHS}{2} = \frac{6 + 14}{2}$$
$$= 10 \text{ square units}$$

Our estimate is fairly close to the exact area of $9\frac{1}{3}$ square units. (In Section 12.5, we will show you how to calculate the exact area.)

EXAMPLE 7

Using Riemann Sums to Estimate Distance Traveled

The velocity of the author's minivan as he drove from a traffic light onto a highway may be modeled by

$$V(t) = -0.1773t^2 + 0.6798t + 0.5019 \text{ miles per minute}$$

where t is time in minutes. Estimate the area between the velocity graph and the t-axis on the interval $[0.0, 2.5]$ using five rectangles. Then interpret the real-world meaning of the result.

SOLUTION We are asked to estimate the area of the shaded region in Figure 12.17 using five rectangles.

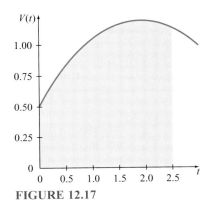

FIGURE 12.17

We construct a table of values for $V(t)$ starting at $t = 0.0$ and ending at $t = 2.5$ with t-values spaced $\Delta t = \dfrac{2.5 - 0}{5} = 0.5$ minute apart (see Table 12.9).

TABLE 12.9

t	$V(t) = -0.1773t^2 + 0.6798t + 0.5019$
0.0	0.502
0.5	0.797
1.0	1.004
1.5	1.123
2.0	1.152
2.5	1.093

Since the height of the rectangles represents velocity in miles per minute and the width of the rectangles represents minutes, the units of the area will be

$$\frac{\text{miles}}{\text{minute}} \cdot \cancel{\text{minute}} = \text{miles}$$

We first find the left-hand sum.

$$LHS = (0.502 + 0.797 + 1.004 + 1.123 + 1.152)(0.5)$$
$$= (4.578)(0.5)$$
$$\approx 2.289 \text{ miles}$$

Then we find the right-hand sum.

$$RHS = (0.797 + 1.004 + 1.123 + 1.152 + 1.093)(0.5)$$
$$= (5.169)(0.5)$$
$$\approx 2.585 \text{ miles}$$

Now we average the sums.

$$\frac{LHS + RHS}{2} = \frac{2.289 + 2.585}{2}$$
$$= 2.437 \text{ miles}$$
$$\approx 2.45 \text{ (rounded to the nearest twentieth of a mile)}$$

We estimate that in the first 2.5 minutes of the highway trip, the author traveled a distance of 2.45 miles. (We rounded to twentieths of a mile because the raw data were recorded accurate to the nearest twentieth of a mile.) According to the raw data, the actual distance traveled by the author was 2.45 miles. In this case, our rounded estimate was right on!

Although the TI-83 Plus doesn't come preloaded with a Riemann sum program, there are several excellent programs available at www.ticalc.org. One outstanding program is *Riemann.8xp*, written by Mike Miller of Corban College. We transferred the downloaded program from our PC to our calculator using the TI-CONNECT software. In the following Technology Tip, we will repeat Example 7 using the Riemann.8xp program. (Users of the TI-83 calculator should download the file Rieman83.83P.)

TECHNOLOGY TIP

Finding Riemann Sums with Riemann.8xp

1. Press the [PRGM] key to bring up the list of programs. Select the program RIEMANN. Press [ENTER] twice.

   ```
   EXEC EDIT NEW
   1:RIEMAN83
   2:RIEMANN
   ```

2. The introduction and credit screen is displayed. Press [ENTER].

   ```
   Riemann Sums
   Finds area under a function
   on a given interval using
   Riemann Sums. Increasing
   the number of partitions
   leads to better estimates.
                    Press Enter
   From the Mind of Mike, 2003
   ```

(Continued)

3. To find a Riemann sum of a function f on an interval $[a, b]$ with n rectangles, enter the following:

$f(x)$: (Enter the function f.)
Lower Bound: a
Upper Bound: b
Partitions: n

4. The calculator draws the graph of f and draws vertical lines at $x = a$ and $x = b$. Press (ENTER).

5. The calculator displays an option screen. To calculate the left-hand sum, select 2:LEFT SUM and press (ENTER).

6. The calculator graphs the function, draws the left-hand rectangles, and displays the value of the left-hand sum. Press (ENTER) to return to the option screen.

7. Select 3:RIGHT SUM. The calculator graphs the function, draws the right-hand rectangles, and displays the value of the right-hand sum. Press (ENTER) to return to the option screen.

8. Select 5:TRAPEZOID SUM to calculate the average of the left- and right-hand sums. Press (ENTER) to return to the option screen.

(Continued)

9. Select 1:Settings to change the function, interval, or number of rectangles. Otherwise, select 7:QUIT to quit the program.

EXAMPLE 8

Using Riemann Sums to Estimate Water Volume

Based on data from November 22 to 24, 2002, the flow rate of the Rogue River near Agness, Oregon, at noon may be modeled by

$$f(t) = 25t^2 - 75t + 1800 \text{ cubic feet per second}$$

where t is the number of 24-hour periods since noon on November 22, 2002. (**Source:** waterdata.usgs.gov.)

How much water (in gallons) passed by this point near Agness, Oregon, between noon on November 22 and noon on November 24, 2002?

SOLUTION   We must initially find the area of the shaded region in Figure 12.18.

FIGURE 12.18

We decide to use four rectangles and generate the appropriate table of values (Table 12.10).

TABLE 12.10

t	$f(t) = 25t^2 - 75t + 1800$
0.0	1,800
0.5	1,769
1.0	1,750
1.5	1,744
2.0	1,750

This problem is a bit tricky, since the flow rate is given in cubic feet per second and the time intervals are 24-hour periods. Additionally, we're asked to give our answer in gallons, not cubic feet. We'll need to do some unit conversion before reaching our final answer. Since the flow rate is given in cubic feet per second and t is in terms of 24-hour periods, the units of the area of each rectangle will be

$$\frac{\text{cubic feet}}{\text{second}} \cdot \text{24-hour periods}$$

We'll convert these to the appropriate units after finding the left- and right-hand sums.

$$\begin{aligned}
LHS &= (1800 + 1769 + 1750 + 1744)(0.5) \\
&= (7063)(0.5) \\
&= 3531 \text{ (accurate to four significant figures)} \\
RHS &= (1769 + 1750 + 1744 + 1750)(0.5) \\
&= (7013)(0.5) \\
&= 3506 \text{ (accurate to four significant figures)}
\end{aligned}$$

$$\frac{LHS + RHS}{2} = \frac{3531 + 3506}{2}$$

UNITS

$$= 3519 \, \frac{\text{cubic feet}}{\text{second}} \cdot \text{24-hour periods}$$

$$= 3519 \text{ cubic feet} \cdot \frac{\text{24-hour periods}}{\text{second}}$$

How many seconds are there in a 24-hour period?

UNITS

$$\begin{aligned}
\text{24-hour periods} &= 24 \cdot 1 \text{ hour} \\
&= 24 \cdot 60 \text{ minutes} \\
&= 1440 \cdot 1 \text{ minute} \\
&= 1440 \cdot 60 \text{ seconds} \\
&= 86{,}400 \text{ seconds}
\end{aligned}$$

Thus we have

UNITS

$$= 3519 \text{ cubic feet} \cdot \frac{86{,}400 \text{ ~~seconds~~}}{\text{~~second~~}}$$

$$= 304{,}041{,}600 \text{ cubic feet}$$

Since 1 gallon equals 0.13368 cubic foot, we have

UNITS

$$304{,}041{,}600 \text{ ~~cubic feet~~} \cdot \frac{1 \text{ gallon}}{0.13368 \text{ ~~cubic feet~~}} \approx 2{,}274{,}398{,}564 \text{ gallons}$$

Between noon on November 22 and noon on November 24, we estimate that more than 2,274 million gallons of water passed by the measurement station on the Rogue River near Agness, Oregon. This is enough water to fill more than 267 Olympic-size swimming pools!

As noted earlier in the section, if $f(x)$ is a rate of change function, then the area between the graph of $f(x)$ and the x-axis on the interval $[a, b]$ represents the total change in $F(x)$ [an antiderivative of $f(x)$] from $x = a$ to $x = b$. We will discuss this in greater detail in Section 12.5.

12.3 Summary

In this section, you learned how to use left- and right-hand sums to approximate the area between the graph of a function and the horizontal axis. A solid understanding of these concepts will help you grasp the material presented in the next section.

12.3 Exercises

In Exercises 1–5, draw the rectangles used to calculate the left-hand sum estimate of the area between the graph of the function and the horizontal axis on the specified interval. (In each case, use four rectangles.) Then calculate the left-hand sum.

1. $f(x) = 2; [1, 5]$

2. $g(x) = -2x + 9; [0, 4]$

3. $h(x) = 2^x + x; [0, 2]$

4. $f(t) = \dfrac{3}{t}; [1, 3]$

5. $v(t) = 0.5t + 20; [3, 7]$

In Exercises 6–15, use the left-hand sum to estimate the area between the graph of the function and the horizontal axis on the specified interval. For each exercise, calculate the sum with $n = 2$, $n = 4$, and $n = 10$ rectangles.

6. $s(t) = -3t^2 + 3t; [0, 1]$

7. $s(t) = t^3 - 3t^2 + 3t - 1; [1, 2]$

8. $g(x) = -x^2 + 4; [0, 2]$

9. $h(x) = \dfrac{2}{x} + 1; [1, 5]$

10. $f(x) = -2x + 20; [0, 10]$

11. $s(x) = \ln(x); [e, e^2]$

12. $y = x^2 - 2^x; [2, 4]$

13. $y = 4t^2 - 1; [2, 4]$

14. $q = 20 - \dfrac{1000}{p}; [100, 200]$

15. $f(t) = \dfrac{t + 2}{t - 2}; [3, 11]$

In Exercises 16–35, use left- and right-hand sums (with $n = 4$) to estimate the area between the graph of the function and the horizontal axis on the specified interval. In each exercise, calculate the left-hand sum, the right-hand sum, and the average of the two sums. The exact area, A, is given so that you can compare your estimates to the actual area.

16. $f(x) = 6x + 1$ on $[2, 4]$; $A = 38$

17. $g(x) = x^2 - 2x + 2$ on $[3, 5]$; $A = 20\frac{2}{3}$

18. $h(x) = 3x^2 + 3$ on $[0, 2]$; $A = 14$

19. $f(x) = -(x - 2)^3 + 2$ on $[1, 2]$; $A = 2.25$

20. $g(x) = x^3 - 10x^2 + 33x - 36$ on $[5, 9]$; $A = 250\frac{2}{3}$

21. $h(x) = x^3 - 3x^2 + 2x + 1$ on $[0, 2]$; $A = 2$

22. $f(x) = x^4 - 5x^2 + 8$ on $[0, 3]$; $A = 27.6$

23. $f(x) = x^3 - 9x^2 + 23x$ on $[2, 6]$; $A = 64$

24. $g(x) = x^2 - 7x + 16$ on $[3, 5]$; $A = 8\frac{2}{3}$

25. $h(x) = x^3 - 9x^2 + 26x$ on $[1, 5]$; $A = 96$

26. $y = \ln 4(4^x) - \ln 2(2^x)$ on $[0, 1]$; $A = 2$

27. $f(x) = 4x - x^4$ on $[0, 1]$; $A = 1.8$

28. $g(x) = x^4 - 4x^3$ on $[5, 6]$; $A = 259.2$

29. $h(x) = e^{2x} - e^x$ on $[0, \ln 2]$; $A = 0.5$

30. $f(t) = e^{3t} - e^t$ on $[0, \ln 4]$; $A = 18$

31. $s(t) = \ln(t)$ on $[1, e]$; $A = 1$

32. $p(x) = \ln 3(3^x)$ on $[0, 1]$; $A = 2$

33. $y = e^x$ on $[\ln 2, \ln 5]$; $A = 3$

34. $y = 3x^2 - \ln 2(2^x)$ on $[3, 8]$; $A = 237$

35. $y = 5x^4 - \ln 5(5^x)$ on $[2, 4]$; $A = 392$

In Exercises 36–40, estimate the area between the graph of the function and the horizontal axis. Then interpret the physical meaning of the result.

36. **M** **Minivan Position** The velocity of the author's minivan as he drove from a traffic light onto a highway may be modeled by

$$V(t) = -0.1773t^2 + 0.6798t + 0.5019$$

miles per minute, where t is time in minutes. Use left- and right-hand sums with $n = 8$ to estimate how far he traveled in the first 2.0 minutes of his trip.

37. **M** **River Flow** Between 5:45 a.m. and 7:45 a.m. on November 22, 2002, the stream flow rate of the Columbia River below the Priest Rapids Dam dropped dramatically. The flow rate during that time period may be modeled by

$$F(t) = 14.50t^4 - 80.15t^3 + 168.3t^2 - 180.8t + 158.9$$

cubic feet per second, where t is the number of hours since 5:45 a.m. (**Source:** waterdata.usgs.gov.)

(a) Convert the units of the flow-rate function from cubic feet per second to cubic feet per hour.

(b) Use left- and right-hand sums with $n = 8$ to approximate the total amount of water that flowed past the flow-rate measurement station between 5:45 a.m. and 7:45 a.m.

(c) Why do you think the flow rate decreased so substantially?

38. **M** **River Flow** Based on data from November 22 to 24, 2002, the flow rate of the Wenatchee River at Peshastin, Washington, at noon may be modeled by

$$f(t) = -100t + 1300$$

cubic feet per second, where t is the number of 24-hour periods since noon on November 22, 2002. (**Source:** waterdata.usgs.gov.)

How much water (in gallons) passed by Peshastin between noon on November 22 and noon on November 24, 2002?

39. **M** **River Flow** Based on data from November 22 to 24, 2002, the flow rate of the Snake River near Irwin, Idaho, at noon may be modeled by

$$f(t) = -50t^2 + 100t + 1050$$

cubic feet per second, where t is the number of 24-hour periods since noon on November 22, 2002. (**Source:** waterdata.usgs.gov.)

How much water (in gallons) passed by this point near Irwin, Idaho, between noon on November 22 and noon on November 24, 2002? (Use $n = 8$.)

40. **M** **River Flow** Based on data from November 22 to 24, 2002, the flow rate of the North Snoqualmie River near Snoqualmie Falls, Washington, at noon may be modeled by

$$f(t) = 25t^2 - 145t + 490$$

cubic feet per second, where t is the number of 24-hour periods since noon on November 22, 2002. (**Source:** waterdata.usgs.gov.)

How much water (in gallons) passed by this point near Snoqualmie Falls between noon on November 22 and noon on November 24, 2002? (Use $n = 8$.)

Exercises 41–45 are intended to challenge your understanding of Riemann sums.

41. Give an example of a positive function f on an interval $[a, b]$ that has the property that the left-hand sum and right-hand sum approximation of the area between the graph of f and the x-axis are equal for all values of Δx.

42. The left-hand sum and right-hand sum of the function $f(x) = -x^2 + 4x$ on the interval $[0, 4]$ are both equal to 10 when four rectangles are used. Does this mean that the area between the graph of $f(x) = -x^2 + 4x$ and the x-axis is equal to 10? Explain.

43. A positive function f is an increasing function. (That is, $f(a) \le f(b)$ whenever $a \le b$.) Which sum will best approximate the area between the graph of f and the x-axis on the interval $[a, b]$: the left-hand sum, the right-hand sum, or the average of the left- and right-hand sums? Explain.

44. Calculate the left-hand sum for the function $f(x) = |-2x + 2|$ on the interval $[0, 2]$ using $n = 1$, $n = 2$, $n = 3$, $n = 4$, $n = 5$, and $n = 6$. For which value(s) of n does the left-hand sum best approximate the actual area between the graph of f and the x-axis on the interval $[a, b]$?

45. For a nonnegative function f on the interval $[a, b]$, does reducing the width of the rectangles used in a Riemann sum ever *worsen* the Riemann sum estimate of the area between the graph of f and the x-axis on the interval $[a, b]$? Explain. (*Hint:* Consider the function $f(x) = |-2x + 2|$ using left-hand sums with $n = 2$ and $n = 3$.)

12.4 The Definite Integral

- Apply definite integral properties
- Calculate the exact area between a curve and the x-axis by using definite integrals

Based on data from 1999 to 2001, the marginal revenue of the Coca-Cola Company may be modeled by

$$M = -0.8058s + 14.44 \text{ dollars per unit case}$$

where s is the number of unit cases (in billions). A unit case is equivalent to 24 eight-ounce servings of finished beverage. In 1999, 16.5 billion unit cases were sold. In 2001, 17.8 billion unit cases were sold. (**Source:** Modeled from Coca-Cola Company 2001 Annual Report, pp. 46, 57.) By how much did the revenue of the Coca-Cola Company grow as the number of unit cases sold increased from 16.5 billion to 17.8 billion? If you were a financial consultant for Coca-Cola, would you encourage the company to reduce its prices so that it could increase unit case sales? We can address questions such as these using the notion of the definite integral.

In this section, we will continue our discussion of area estimates by introducing *summation notation* and the *definite integral*. We will also show you several helpful definite integral properties.

Summation Notation

We use *summation notation* in order to represent the sum of a large number of terms easily. The notation may feel a bit awkward at first; however, as you become skilled in using it, you will come to appreciate its usefulness. Let's return to the table we introduced in Example 5 of Section 12.3 (Table 12.7), reproduced here as Table 12.11.

TABLE 12.11

x	$f(x) = x^3 - 3x^2 + 3x$
0	0.000
0.5	0.875
1.0	1.000
1.5	1.125
2.0	2.000

Recall that for the left-hand sum, the sum of the rectangle heights may be written as $f(0) + f(0.5) + f(1.0) + f(1.5)$. Notice that the x values are $\Delta x = 0.5$ unit apart. An alternative way to write this sum is

$$\sum_{i=0}^{3} f(0 + 0.5i)$$

The Greek letter *sigma* (Σ) tells us that we are to sum the values of the form $f(0 + 0.5i)$. The i is the *index of summation*. It is a variable, and it is always equal to a whole number. The numbers above and below Σ are the *limits of summation*. In this case, the initial value of i is 0 and the ending value is 3.

We will first substitute the initial value of i into $f(0 + 0.5i)$.

$$f[0 + 0.5(0)] = f(0)$$

We will then increase i from 0 to 1 and substitute $i = 1$ into $f(0 + 0.5i)$.

$$f[0 + 0.5(1)] = f(0.5)$$

The summation symbol Σ tells us to sum these two values together. So far we have

$$f(0) + f(0.5)$$

However, the summation tells us that we must continue to increase the index until $i = 3$, each time summing the values.

$$f[0 + 0.5(2)] = f(1)$$
$$f[0 + 0.5(3)] = f(1.5)$$

Therefore, $\displaystyle\sum_{i=0}^{3} f(0 + 0.5i) = f(0) + f(0.5) + f(1.0) + f(1.5)$.

In the next two examples we will temporarily move away from our discussion of heights and areas in order to give you practice using summation notation.

EXAMPLE 1 Using Summation Notation

Calculate $\displaystyle\sum_{i=1}^{10} i^2$.

SOLUTION We are to sum terms of the form i^2 by substituting in the values 1 through 10 for i.

$$\sum_{i=1}^{10} i^2 = (1)^2 + (2)^2 + (3)^2 + (4)^2 + (5)^2 + (6)^2 + (7)^2 + (8)^2 + (9)^2 + (10)^2$$

$$= 1 + 4 + 9 + 16 + 25 + 36 + 49 + 64 + 81 + 100$$

$$= 385$$

EXAMPLE 2 Using Summation Notation

Calculate $\displaystyle\sum_{i=0}^{6} (i - 3)$.

SOLUTION We must sum terms of the form $i - 3$ by substituting in the values 0 through 6 for i.

$$\sum_{i=0}^{6} (i - 3) = (0 - 3) + (1 - 3) + (2 - 3) + (3 - 3) + (4 - 3) + (5 - 3) + (6 - 3)$$

$$= (-3) + (-2) + (-1) + 0 + 1 + 2 + 3$$

$$= 0$$

We can now describe the left- and right-hand sums in terms of summation notation.

SUMMATION NOTATION FOR THE LEFT-HAND SUM

Let f be a function defined on the interval $[a, b]$. If the graph of f is above the x-axis, then the area of the region between the graph of f and the x-axis may be approximated by

$$\sum_{i=0}^{n-1} f(x_i) \Delta x = f(x_0) \Delta x + f(x_1) \Delta x + \cdots + f(x_{n-1}) \Delta x$$

where $\Delta x = \dfrac{b - a}{n}$ and

$$x_0 = a$$
$$x_1 = a + \Delta x$$
$$\vdots$$
$$x_{n-1} = a + (n - 1) \Delta x$$

$\sum_{i=0}^{n-1} f(x_i) \Delta x$ is the left-hand sum estimate of the area using n rectangles.

SUMMATION NOTATION FOR THE RIGHT-HAND SUM

Let f be a function defined on the interval $[a, b]$. If the graph of f is above the x-axis, then the area of the region between the graph of f and the x-axis may be approximated by

$$\sum_{i=1}^{n} f(x_i) \Delta x = f(x_1) \Delta x + f(x_2) \Delta x + \cdots + f(x_n) \Delta x$$

where $\Delta x = \dfrac{b - a}{n}$ and

$$x_1 = a + \Delta x$$
$$x_2 = a + 2 \cdot \Delta x$$
$$\vdots$$
$$x_n = b$$

$\sum_{i=1}^{n} f(x_i) \Delta x$ is the right-hand sum estimate of the area using n rectangles.

In general, a Riemann sum does not require that each rectangle have the same width. However, by making the rectangles have the same width, we reduce the number of calculations needed to find the solution.

Let's work an area example using summation notation.

EXAMPLE **3**   **Using Summation Notation in Calculating a Riemann Sum**

Use a left-hand sum with $n = 6$ to estimate the area between the graph of $f(x) = x^2 + 2$ and the x-axis on the interval $[1, 3]$.

SOLUTION We are asked to estimate the area of the shaded region in Figure 12.19 using a left-hand sum with six rectangles.

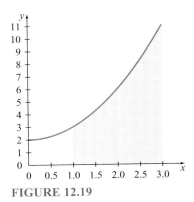

FIGURE 12.19

We will first find the width of each rectangle.

$$\Delta x = \frac{3 - 1}{6}$$

$$= \frac{1}{3} \text{ unit}$$

The left-hand sum is given by

$$\sum_{i=0}^{n-1} f(x_i) \, \Delta x = \sum_{i=0}^{6-1} f(x_i) \left(\frac{1}{3}\right)$$

$$= \sum_{i=0}^{5} [(x_i)^2 + 2]\left(\frac{1}{3}\right)$$

$$= \sum_{i=0}^{5} [(1 + i \cdot \Delta x)^2 + 2]\left(\frac{1}{3}\right) \qquad \text{Since } x_i = a + i\Delta x \text{ and } a = 1$$

$$= \sum_{i=0}^{5} \{[1 + i\left(\frac{1}{3}\right)]^2 + 2\}\left(\frac{1}{3}\right) \qquad \text{Since } \Delta x = \frac{1}{3}$$

$$= \{[1 + 0\left(\frac{1}{3}\right)]^2 + 2\}\left(\frac{1}{3}\right) + \{[1 + 1\left(\frac{1}{3}\right)]^2 + 2\}\left(\frac{1}{3}\right) + \{[1 + 2\left(\frac{1}{3}\right)]^2 + 2\}\left(\frac{1}{3}\right)$$

$$\quad + \{[1 + 3\left(\frac{1}{3}\right)]^2 + 2\}\left(\frac{1}{3}\right) + \{[1 + 4\left(\frac{1}{3}\right)]^2 + 2\}\left(\frac{1}{3}\right) + \{[1 + 5\left(\frac{1}{3}\right)]^2 + 2\}\left(\frac{1}{3}\right)$$

$$= \{[(1)^2 + 2] + [\left(\frac{4}{3}\right)^2 + 2] + [\left(\frac{5}{3}\right)^2 + 2] + [\left(\frac{6}{3}\right)^2 + 2] + [\left(\frac{7}{3}\right)^2 + 2] + [\left(\frac{8}{3}\right)^2 + 2]\}\left(\frac{1}{3}\right)$$

$$= [3 + \left(\frac{34}{9}\right) + \left(\frac{43}{9}\right) + (6) + \left(\frac{67}{9}\right) + \left(\frac{82}{9}\right)]\left(\frac{1}{3}\right)$$

$$\approx 11.37$$

Admittedly, formulating the sum using summation notation was more complicated than our table method. However, as we use an increasingly large number of rectangles, using summation notation will be the easiest way to represent the left- and right-hand sums symbolically.

The Definite Integral

What would happen if we used infinitely many rectangles to estimate the area between the graph of a function and the horizontal axis? Would the estimate be exact? As we have seen, when we increase the number of rectangles, the width of each rectangle is reduced and our area estimate improves. The definite integral captures the idea of infinitely many rectangles.

THE DEFINITE INTEGRAL

Let f be a continuous function defined on the interval $[a, b]$. The **definite integral of f from a to b** is given by

$$\int_a^b f(x)\,dx = \lim_{n \to \infty}\{[f(x_1) + f(x_2) + \cdots + f(x_n)]\Delta x\}$$

where $x_1, x_2, \ldots, x_n$ are the right-hand endpoints of subintervals of length

$$\Delta x = \frac{b-a}{n}.$$

The use of infinitely many rectangles may also be represented using summation notation in conjunction with limit notation. We have $\displaystyle\int_a^b f(x)\,dx = \lim_{n \to \infty} \sum_{i=1}^{n} f(x_i)\,\Delta x$

and $\displaystyle\int_a^b f(x)\,dx = \lim_{n \to \infty} \sum_{i=0}^{n-1} f(x_i)\,\Delta x$. That is, when we use infinitely many rectangles, the left- and right-hand sums become equal. When we write the symbol $\displaystyle\lim_{n \to \infty}$, we mean that n is made to be infinitely large.

For the definite integral $\displaystyle\int_a^b f(x)\,dx$, a is called the **lower limit of integration** and b is called the **upper limit of integration.** As with indefinite integrals, $f(x)$ is called the **integrand.** If $\displaystyle\int_a^b f(x)\,dx$ is defined, then f is said to be **integrable** on $[a, b]$.

The connection between the indefinite integral $\displaystyle\int f(x)\,dx$ and the definite integral $\displaystyle\int_a^b f(x)\,dx$ is not immediately obvious. For a nonnegative function $f(x)$, $\displaystyle\int f(x)\,dx$ is a family of functions with the same derivative, while $\displaystyle\int_a^b f(x)\,dx$ is the area between the graph of f and the x-axis. However, despite the dramatically different meanings of indefinite and definite integrals, the two are closely related, as will be discussed in Section 12.5.

In our interpretation of the definite integral, we required f to be a nonnegative function. In general, the expression $\displaystyle\int_a^b f(x)\,dx$ is the sum of the shaded regions above the x-axis minus the sum of the shaded regions below the x-axis. In other words, $\displaystyle\int_a^b f(x)\,dx$ is the **sum of the signed areas.**

Consider the graph of $f(x) = x^3 - 4x^2 + 3x$ on the interval $[0, 3]$ (Figure 12.20).

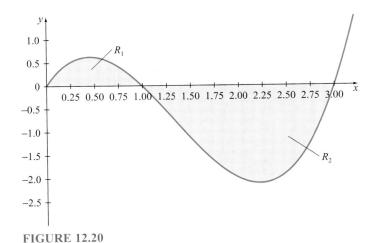

FIGURE 12.20

The first region, R_1, has area $\frac{5}{12}$ square unit. The second region, R_2, has area $2\frac{2}{3}$ square units. Therefore, $\int_0^3 (x^3 - 4x^2 + 3x)\, dx = \frac{5}{12} - 2\frac{2}{3} = -2\frac{1}{4}$. The fact that the number is negative indicates that more area lies below the x-axis than above it.

In considering the relationship between $\int_a^b f(x)\, dx$ and the left- and right-hand sums, it is helpful to think of $f(x)$ as the height of an infinitely narrow rectangle and dx as the width of that rectangle. The integral sign $\int$ is a somewhat distorted S, meaning "sum," and tells us to sum the areas of infinitely many adjacent rectangles of width dx between $x = a$ and $x = b$. (From our earlier discussion of differentials, we know that we can choose the value of dx to be as small as we like.) The height of each rectangle varies as x moves from a to b in steps of length dx.

THE MEANING OF $\int_a^b f(x)\, dx$ FOR A NONNEGATIVE FUNCTION f

Let f be a nonnegative function on $[a, b]$. The definite integral

$$\int_a^b f(x)\, dx$$

is the area of the region between the graph of f and the x-axis.

EXAMPLE 4   **Interpreting the Graphical Meaning of a Definite Integral**

Interpret the graphical meaning of $\int_0^3 (-x^2 + 3x)\, dx = 4.5$.

SOLUTION   We know that $\int_0^3 (-x^2 + 3x)\, dx$ represents the sum of the signed areas between the graph of $y = -x^2 + 3x$ and the x-axis. However, since $y \geq 0$

for all values of x in the interval $[0, 3]$, $\int_0^3 (-x^2 + 3x)\,dx$ gives the area of the shaded region in Figure 12.21.

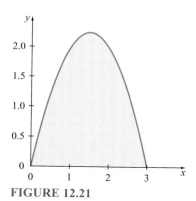

FIGURE 12.21

The area of the shaded region is 4.5 square units.

EXAMPLE 5

Interpreting the Meaning of a Definite Integral in an Applied Setting

Based on data from 1985 to 1999, the rate of change in the per capita consumption of breakfast cereal may be modeled by

$$R(t) = -0.014154t^2 + 0.233t - 0.3585 \text{ pounds per person per year}$$

where t is the number of years since the end of 1980. The graph of R is depicted in Figure 12.22.

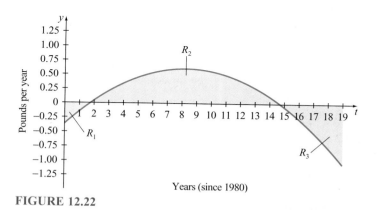

FIGURE 12.22

The areas of the three shaded regions are 0.296 square unit, 5.214 square units, and 2.034 square units, respectively. Calculate and interpret the meaning of

$$\int_0^{19} (-0.014154t^2 + 0.233t - 0.3585)\,dt.$$

SOLUTION $\int_0^{19} (-0.014154t^2 + 0.233t - 0.3585)\,dt$ is the sum of the signed areas. Since regions R_1 and R_3 are below the t-axis, their signed areas are negative.

Therefore,

$$\int_0^{19} (-0.014154t^2 + 0.233t - 0.3585)\,dt = -0.296 + 5.214 - 2.034 = 2.884$$

Recall that the units of the integral are found by multiplying the units of the independent variable by the units of the dependent variable. That is,

$$\text{years since 1980} \cdot \frac{\text{pounds}}{\text{year since 1980}} = \text{pounds}$$

So from the end of 1980 through 1999, the annual per capita breakfast cereal consumption increased by a net of 2.884 pounds. That is, taking into account the increases and decreases in cereal consumption rates, annual breakfast cereal consumption increased by 2.884 pounds per person from the end of 1980 through 1999.

Definite Integral Properties

Some of the definite integral properties are directly related to the indefinite integral rules. You will readily note the similarities.

DEFINITE INTEGRAL PROPERTIES

For integrable functions f and g, the following properties hold:

- $$\int_a^a f(x)\,dx = 0$$

- $$\int_a^b f(x)\,dx = -\int_b^a f(x)\,dx$$

- $$\int_a^b k \cdot f(x)\,dx = k \int_a^b f(x)\,dx \text{ for constant } k$$

- $$\int_a^b [f(x) \pm g(x)]\,dx = \int_a^b f(x)\,dx \pm \int_a^b g(x)\,dx$$

- $$\int_a^b f(x)\,dx = \int_a^c f(x)\,dx + \int_c^b f(x)\,dx \text{ for } a \le c \le b$$

EXAMPLE 6

Applying Integration Properties

Given $\int_{-2}^0 (2x + 3)\,dx = 2$, determine $\int_0^{-2} (2x + 3)\,dx$.

SOLUTION Switching the limits of integration changes the sign of the definite integral. Therefore, $\int_0^{-2} (2x + 3)\,dx = -2$.

EXAMPLE 7 **Applying Integration Properties**

Given $\int_2^4 2x \, dx = 12$ and $\int_2^4 (2x + 5) \, dx = 22$, find $\int_2^4 5 \, dx$.

SOLUTION We know that $\int_2^4 2x \, dx + \int_2^4 5 \, dx = \int_2^4 (2x + 5) \, dx$. Therefore,

$$12 + \int_2^4 5 \, dx = 22$$

$$\int_2^4 5 \, dx = 10$$

EXAMPLE 8 **Applying Integration Properties**

Given $\int_{-1}^1 3x^2 \, dx = 2$ and $\int_1^3 3x^2 \, dx = 26$, find $\int_{-1}^3 3x^2 \, dx$.

SOLUTION The only difference between the three integrals is the limits of integration. Since the upper limit of integration of $\int_{-1}^1 3x^2 \, dx$ is the same as the lower limit of integration of $\int_1^3 3x^2 \, dx$, the two integrals may be written together as a single integral $\int_{-1}^3 3x^2 \, dx$. The value of this integral is found by summing the values of the two integrals from which it was created. That is,

$$\int_{-1}^3 3x^2 \, dx = \int_{-1}^1 3x^2 \, dx + \int_1^3 3x^2 \, dx$$

$$= 2 + 26$$

$$= 28$$

EXAMPLE 9 **Interpreting the Meaning of a Definite Integral in an Applied Setting**

The rate of change in the annual per capita consumption of chicken from the end of 1985 through 1999 may be modeled by

$$R(t) = 0.010492t^3 - 0.22635t^2 + 1.3212t - 0.4688 \text{ pounds per person per year}$$

where t is the number of years since the end of 1985. (**Source:** Modeled from *Statistical Abstract of the United States, 2001,* Table 202, p. 129.)

The sum of the signed areas of the shaded regions of the graph in Figure 12.23 is $\int_0^{15} R(t) \, dt$ and equals 19.75.

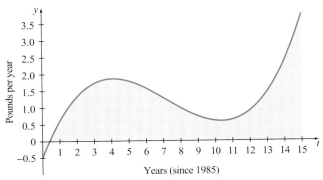

FIGURE 12.23

Given $\int_{10}^{15} R(t)\,dt = 7.60$, calculate and interpret the meaning of $\int_{0}^{10} R(t)\,dt$.

SOLUTION $\int_{0}^{10} R(t)\,dt$ is the net change in the annual per capita consumption of chicken from the end of 1985 through the end of 1995 in pounds per person. We know that $\int_{0}^{15} R(t)\,dt = 19.75$ and $\int_{10}^{15} R(t)\,dt = 7.60$, so

$$\int_{0}^{10} R(t)\,dt + \int_{10}^{15} R(t)\,dt = \int_{0}^{15} R(t)\,dt$$

$$\int_{0}^{10} R(t)\,dt = \int_{0}^{15} R(t)\,dt - \int_{10}^{15} R(t)\,dt$$

$$= 19.75 - 7.60$$

$$= 12.15$$

From the end of 1985 through 1995, the annual per capita consumption of chicken increased by 12.15 pounds. That is, at the end of 1995, on average, a person was eating 12.15 pounds more chicken per year than he or she was at the end of 1985.

For poultry producers, this information is great news. For beef producers, it may be a cause for concern. Over the same time period, annual per capita beef consumption dropped by 10.2 pounds. (**Source:** *Statistical Abstract of the United States, 2001,* Table 202, p. 129.)

EXAMPLE 10

Using a Definite Integral to Calculate an Accumulated Change

Based on data from 1999 to 2001, the marginal revenue of the Coca-Cola Company may be modeled by

$$M = -0.8058s + 14.44 \text{ dollars per unit case}$$

where s is the number of unit cases (in billions). A unit case is equivalent to 24 eight-ounce servings of finished beverage. In 1999, 16.5 billion unit cases were sold. In 2001, 17.8 billion unit cases were sold. (**Source:** Modeled from Coca-Cola Company 2001 Annual Report, pp. 46, 57.) By how much did the revenue of Coca-Cola Company grow as the number of unit cases sold increased from 16.5 billion to 17.8 billion?

SOLUTION   We begin by drawing a graph of the marginal revenue function on the interval $[16.5, 17.8]$ and shading the region between the graph and the horizontal axis (see Figure 12.24).

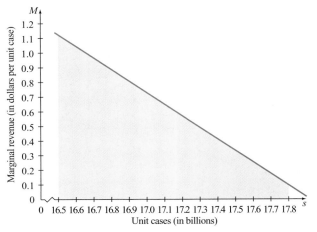

FIGURE 12.24

We are asked to find $\displaystyle\int_{16.5}^{17.8} (-0.8058s + 14.44)\,ds$. Since the marginal revenue function is positive on the interval $[16.5, 17.8]$, $\displaystyle\int_{16.5}^{17.8} (-0.8058s + 14.44)\,ds$ is the area of the shaded region. We recognize that the shaded region is a trapezoid.

The area of a trapezoid is given by $A = \dfrac{1}{2}(b_1 + b_2)h$, where h is the height of the trapezoid and b_1 and b_2 are the respective lengths of the bases. Our graph is a trapezoid turned on its side. We can calculate the length of each base by evaluating M at $s = 16.5$ and $s = 17.8$. The height of the trapezoid is the length of the interval. We have $M(16.5) = 1.144$ and $M(17.8) = 0.097$.

The length of the interval is $17.8 - 16.5 = 1.3$. Therefore, the area of the trapezoid is

$$A = \frac{1}{2}(b_1 + b_2)h$$

$$= \frac{1}{2}(1.144 + 0.097)(1.3)$$

$$= 0.8067$$

Thus $\int_{16.5}^{17.8} (-0.8058s + 14.44)\, ds \approx 0.8$ billion dollars. (The raw data were accurate to the nearest tenth of a billion dollars.) Increasing sales from 16.5 billion to 17.8 billion unit cases increased revenue by $0.8 billion.

When working definite integral problems, it is often helpful to use a graphing calculator to verify the accuracy of our results. The following Technology Tip describes how to find a definite integral graphically.

TECHNOLOGY TIP

Finding a Definite Integral Graphically

1. Enter the function $f(x)$ as Y1 by pressing the Y= button and typing the equation.

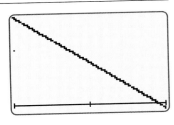

2. Graph the function over the specified domain. (You may need to press WINDOW and adjust the Xmin and Xmax settings. Press ZOOM then 0 to automatically adjust the y-values so that the entire graph will appear on the screen.)

3. Press 2nd then TRACE to bring up the CALCULATE menu. Select item

   7: $\int f(x)\, dx$ and press ENTER.

4. The calculator asks, Lower Limit? Enter the value of the lower limit of integration. Then press ENTER. (*Warning:* The lower limit must be within the domain of the viewing rectangle.)

(Continued)

5. The calculator asks, `Upper Limit?` Enter the value of the upper limit of integration. Then press ⌈ENTER⌉. (*Warning:* The upper limit must be within the domain of the viewing rectangle.)

6. The calculator shades the region between the graph of the function and the *x*-axis and then displays the value of the definite integral. In this case, we found

$$\int_{16.5}^{17.8} (-0.8058s + 14.44)\, ds = 0.8067.$$

12.4 Summary

In this section, you learned how to use summation notation. You discovered that the definite integral is the sum of the signed areas. You also learned how to manipulate definite integrals using definite integral properties.

12.4 Exercises

In Exercises 1–10, calculate the sum.

1. $\displaystyle\sum_{i=1}^{6} i$

2. $\displaystyle\sum_{i=3}^{7} (i^2 - i)$

3. $\displaystyle\sum_{i=0}^{5} (-1)^i i$

4. $\displaystyle\sum_{i=0}^{4} [(3i - 1) \cdot 4]$

5. $\displaystyle\sum_{i=2}^{3} (4i^2 - 2i - 3)$

6. $\displaystyle\sum_{i=1}^{4} [4(i^2 - 2i)]$

7. $\displaystyle\sum_{i=1}^{4} f(1 + i \cdot \Delta x)(\Delta x)$ given that

$f(x) = x^2$ and $\Delta x = 0.25$

8. $\displaystyle\sum_{i=0}^{3} f(1 + i \cdot \Delta x)(\Delta x)$ given that

$f(x) = x^2$ and $\Delta x = 0.25$

9. $\displaystyle\sum_{i=1}^{4} f(2 + i \cdot \Delta x)(\Delta x)$ given that

$f(x) = 3x - 1$ and $\Delta x = 0.25$

10. $\displaystyle\sum_{i=0}^{3} f(2 + i \cdot \Delta x)(\Delta x)$ given that

$f(x) = 3x - 1$ and $\Delta x = 0.25$

In Exercises 11–20, draw the graph of the integrand function and shade the region corresponding to the definite integral. Then calculate the definite integral using geometric formulas for area.

11. $\displaystyle\int_{5}^{7} 1 \, dx$

12. $\displaystyle\int_{2}^{5} 3 \, dx$

13. $\displaystyle\int_{0}^{4} x \, dx$

14. $\displaystyle\int_{0}^{3} (-x + 3) \, dx$

15. $\displaystyle\int_{1}^{2} (-2x + 5) \, dx$

16. $\displaystyle\int_{1}^{3} (4x - 4) \, dx$

17. $\displaystyle\int_{-1}^{1} (2x + 2) \, dx$

18. $\displaystyle\int_{-1}^{1} (3x + 4) \, dx$

19. $\displaystyle\int_{-2}^{-1} (-2x + 1) \, dx$

20. $\displaystyle\int_{-2}^{3} (-x + 5) \, dx$

In Exercises 21–30, use the definite integral properties to find the numeric value of each definite integral given that $\int_{2}^{4} f(x)\,dx = 5$,

$$\int_{2}^{6} f(x)\,dx = 9,\ \int_{2}^{4} g(x)\,dx = 2,\ \text{and}\ \int_{2}^{6} g(x)\,dx = 1.$$

21. $\int_{2}^{4} [2f(x)]\,dx$ **22.** $\int_{2}^{6} [3f(x)]\,dx$

23. $\int_{2}^{6} [f(x) + g(x)]\,dx$ **24.** $\int_{2}^{6} [f(x) - g(x)]\,dx$

25. $\int_{4}^{6} f(x)\,dx$ **26.** $\int_{4}^{6} g(x)\,dx$

27. $\int_{4}^{6} [f(x) + g(x)]\,dx$ **28.** $\int_{4}^{6} [2f(x) - g(x)]\,dx$

29. $\int_{6}^{2} [3f(x) - 4g(x)]\,dx$

30. $\int_{6}^{2} [f(x) + g(x)]\,dx$

In Exercises 31–35, write the definite integral that represents the sum of the signed areas of the shaded regions of the graph.

31. $f(x) = -x^2 + 5x - 6$

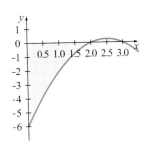

32. $f(x) = x^3 - 4x^2$

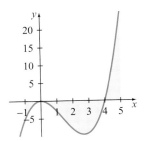

33. $f(x) = -x^3 + 4x + 20$

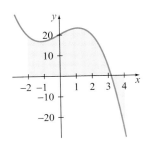

34. $f(x) = \dfrac{x + 1}{x^2 + 1}$

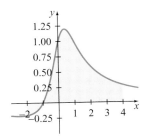

35. $f(x) = -|x - 3| + 3$

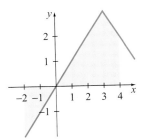

In Exercises 36–40, shade the regions of the graph that will be used in calculating the definite integral.

36. $\int_{2}^{4} (x - 3)\,dx$

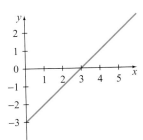

37. $\displaystyle\int_{1}^{3} (-x^2 + 4x - 3)\, dx$

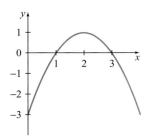

38. $\displaystyle\int_{0}^{2} \frac{x^2 - 1}{x^2 + 1}\, dx$

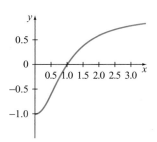

39. $\displaystyle\int_{-1}^{1} e^{-x^2 + 1}\, dx$

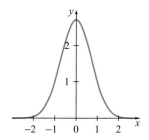

40. $\displaystyle\int_{0}^{2} \ln\left(\frac{1}{x^2 + 1}\right) dx$

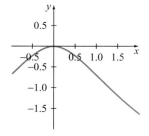

In Exercises 41–45, use the properties of definite integrals to find the solution.

41.   **Bottled Water Consumption**   Based on data from 1980 to 1999, the rate of change in the annual per capita consumption of bottled water may be modeled by

$$R(t) = 0.2612(1.106)^t$$

gallons per person per year, where t is the number of years since the end of 1980. (**Source:** Modeled from *Statistical Abstract of the United States, 2001*, Table 204, p. 130.) The graph of $R(t)$ is shown.

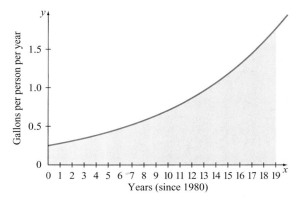

The area of the shaded region of the graph is $\displaystyle\int_{0}^{19} [0.2612(1.106)^t]\, dt$ and is equal to 15.0.

Additionally, $\displaystyle\int_{0}^{10} [0.2612(1.106)^t]\, dt = 4.5$.

Calculate and interpret the meaning of $\displaystyle\int_{10}^{19} [0.2612(1.106)^t]\, dt$.

42.   **Margarine Consumption**   Based on data from 1980 to 1999, the rate of change in the annual per capita consumption of margarine may be modeled by

$$R(t) = 0.001147t^3 - 0.03445t^2 + 0.2658t - 0.5098$$

pounds per person per year, where t is the number of years since the end of 1980. (**Source:** Modeled from *Statistical Abstract of the United States, 2001*, Table 202, p. 129.) The graph of $R(t)$ is shown.

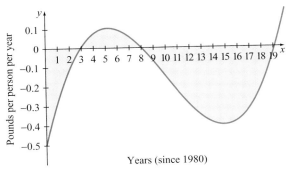

Years (since 1980)

The sum of the signed areas of the shaded regions is $\int_0^{19}\left(\begin{array}{c}0.001147t^3 - 0.03445t^2 \\ + 0.2658t - 0.5098\end{array}\right)dt$ and is equal to -3.10. Additionally,

$$\int_0^8\left(\begin{array}{c}0.001147t^3 - 0.03445t^2 \\ + 0.2658t - 0.5098\end{array}\right)dt = -0.28.$$

Calculate and interpret the meaning of

$$\int_8^{19}\left(\begin{array}{c}0.001147t^3 - 0.03445t^2 \\ + 0.2658t - 0.5098\end{array}\right)dt.$$

43. **M** **Mozzarella Cheese Consumption** Based on data from 1980 to 1999, the rate of change in the annual per capita consumption of mozzarella cheese may be modeled by

$$R(t) = 0.0009778t^3 - 0.02958t^2 + 0.2392t - 0.06311$$

pounds per person per year, where t is the number of years since the end of 1980. (**Source:** Modeled from *Statistical Abstract of the United States, 2001*, Table 202, p. 129.) The graph of $R(t)$ is shown.

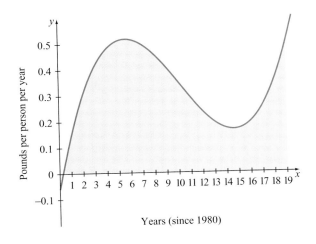

Years (since 1980)

The sum of the signed areas of the shaded regions is $\int_0^{19}\left(\begin{array}{c}0.0009778t^3 - 0.02958t^2 \\ + 0.2392t - 0.06311\end{array}\right)dt$ and is equal to 6.20. Additionally,

$$\int_{15}^{19}\left(\begin{array}{c}0.0009778t^3 - 0.02958t^2 \\ + 0.2392t - 0.06311\end{array}\right)dt = 1.14.$$

Calculate and interpret the meaning of

$$\int_0^{15}\left(\begin{array}{c}0.0009778t^3 - 0.02958t^2 \\ + 0.2392t - 0.06311\end{array}\right)dt.$$

44. **M** **Flounder Production** Based on data from 1990 to 1999, the rate of change in flounder steak and filet production in the continental United States may be modeled by

$$R(t) = 0.2842t - 4.755$$

million pounds per year, where t is the number of years since the end of 1990. (**Source:** Modeled from *Statistical Abstract of the United States, 2001*, Table 863, p. 553.)

Graph R and shade the region whose area is given by $\int_0^9 (0.2842t - 4.755)\,dt$. Then calculate $\int_0^9 (0.2842t - 4.755)\,dt$ and interpret its real-world meaning. (*Hint:* Use the area of a trapezoid formula to calculate the area between the graph of the function and the horizontal axis.)

45. **M** **AP Calculus Participants** Based on data from 1969 to 2002, the rate of change in the number of AP Calculus AB exam participants may be modeled by

$$R(t) = 315.70t - 770.64$$

people per year, where t is the number of years since the end of 1969. (**Source:** Modeled from The College Board data.)

Graph R and shade the region whose area is given by $\int_{23}^{33} (315.70t - 770.64)\,dt$. Then calculate $\int_{23}^{33} (315.70t - 770.64)\,dt$ and interpret its real-world meaning. (*Hint:* Use the area of a trapezoid formula to calculate the value of the definite integral.)

Exercises 46 to 50 are intended to challenge your understanding of Riemann sums and definite integrals.

46. A function f on $[a, b]$ has the property that

$$\sum_{i=0}^{3} f(x_i) \Delta x = \sum_{i=1}^{4} f(x_i) \Delta x.$$ (That is, the left-hand sum is equal to the right-hand sum.) Does this mean that $\int_{a}^{b} f(x) dx = \sum_{i=0}^{3} f(x_i) \Delta x$? Explain.

47. Suppose that $f(x)$ is *increasing* on the interval $[a, b]$. Is the following statement true for all values of n?

$$\sum_{i=0}^{n-1} f(x_i) \Delta x \le \int_{a}^{b} f(x) dx \le \sum_{i=1}^{n} f(x_i) \Delta x$$

Justify your answer.

48. Suppose that $f(x)$ is *decreasing* on the interval $[a, b]$. Is the following statement true for all values of n?

$$\sum_{i=0}^{n-1} f(x_i) \Delta x \le \int_{a}^{b} f(x) dx \le \sum_{i=1}^{n} f(x_i) \Delta x$$

Justify your answer.

49. Find three different functions f that have the property that $\int_{2}^{4} f(x) dx = 0$.

50. Give an example of an integrable function f that has the property that $\int_{a}^{b} f(x) dx = f(b) - f(a)$.

12.5 The Fundamental Theorem of Calculus

- Apply the Fundamental Theorem of Calculus
- Find the function for accumulated change given a rate of change function

GETTING STARTED   Based on data from 1993 to 2002, the rate of change in the sales income of the Starbucks Corporation may be modeled by

$$R(t) = 58.46t + 79.33 \text{ million dollars per year}$$

where t is the number of years since 1993.

According to the model, by how much did the sales income increase between the end of 1996 and the end of 2002?

Although we may use the methods previously covered to answer this question, in this section we will demonstrate how the Fundamental Theorem of Calculus can greatly simplify our computations. This theorem is one of the most powerful tools in calculus and gives us a remarkably easy way to calculate the sum of the signed areas between the graph of a function and the x-axis.

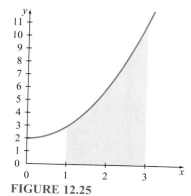

FIGURE 12.25

In Example 3 of Section 12.4, we used a left-hand sum with six rectangles to estimate the area between the graph of $f(x) = x^2 + 2$ and the x-axis on the interval $[1, 3]$ (see Figure 12.25).

The left-hand sum approximation of the area was approximately 11.37 square units, which we noted was an underestimate of the actual area.

The right-hand sum approximation of the area between $f(x) = x^2 + 2$ and the x-axis on $[1, 3]$ using six rectangles is 14.04 square units. This is an overestimate of the actual area. Averaging the two sums, we estimate that the shaded region has an approximate area of 12.70 square units.

Let's find the antiderivative of the function f.

$$\int f(x)\,dx = \int (x^2 + 2)\,dx$$

$$F(x) = \frac{x^3}{3} + 2x + C$$

We'll evaluate the function F at the endpoints of the interval.

$$F(3) = \frac{(3)^3}{3} + 2(3) + C$$

$$= 15 + C$$

$$F(1) = \frac{(1)^3}{3} + 2(1) + C$$

$$= 2\frac{1}{3} + C$$

We next find the difference between the two values.

$$F(3) - F(1) = 15 + C - \left(2\frac{1}{3} + C\right)$$

$$= 12\frac{2}{3}$$

$$\approx 12.67$$

Notice that the constant C was eliminated in the computation. This will happen anytime we calculate the difference between two different values of the antiderivative. This value, 12.67, is extremely close to our estimate of 12.70. In fact, the exact area of the region is $12\frac{2}{3}$ square units. Thus, the definite integral $\int_1^3 f(x)\,dx$ not only represents the area of the region between $f(x)$ and the x-axis on the interval $[1, 3]$, it also represents the total change in $F(x)$ from $x = 1$ to $x = 3$, where $F'(x) = f(x)$. This remarkable result is a key component of the Fundamental Theorem of Calculus.

FUNDAMENTAL THEOREM OF CALCULUS

Let f be a continuous function on $[a, b]$. Then

$$\int_a^b f(x)\,dx = F(b) - F(a)$$

where F is any antiderivative of f.

Since F may be *any* antiderivative of f, we will choose F to be the antiderivative of f that has the constant $C = 0$. This choice will make all of our computations easier. We'll work several examples to illustrate the power of the Fundamental Theorem of Calculus.

EXAMPLE **1**

Using the Fundamental Theorem of Calculus to Calculate a Definite Integral

Calculate $\displaystyle\int_3^7 (2x - 4)\, dx$.

SOLUTION We must first find an antiderivative of $f(x) = 2x - 4$ and then evaluate the antiderivative at $x = 7$ and $x = 3$.

$$\int_3^7 (2x - 4)\, dx = (x^2 - 4x)\Big|_3^7$$

An antiderivative of $f(x) = 2x - 4$ is $F(x) = x^2 - 4x$. The vertical bar tells us that we must evaluate this antiderivative at $x = 7$ and $x = 3$ and then find the difference of the two values.

$$(x^2 - 4x)\Big|_3^7 = [(7)^2 - 4(7)] - [(3)^2 - 4(3)]$$
$$= (49 - 28) - (9 - 12)$$
$$= 21 - (-3)$$
$$= 24$$

Therefore, $\displaystyle\int_3^7 (2x - 4)\, dx = 24$. Since $f(x) = 2x - 4$ is positive on the interval $[3, 7]$, the total area between the graph of f and the x-axis on $[3, 7]$ is 24 square units.

EXAMPLE **2**

Using the Fundamental Theorem of Calculus to Calculate a Definite Integral

Calculate $\displaystyle\int_{-1}^1 (x^3 - x)\, dx$.

SOLUTION

$$\int_{-1}^1 (x^3 - x)\, dx = \left(\frac{x^4}{4} - \frac{x^2}{2}\right)\Big|_{-1}^1$$
$$= \left[\frac{(1)^4}{4} - \frac{(1)^2}{2}\right] - \left[\frac{(-1)^4}{4} - \frac{(-1)^2}{2}\right]$$
$$= \left(\frac{1}{4} - \frac{1}{2}\right) - \left(\frac{1}{4} - \frac{1}{2}\right)$$
$$= 0$$

This result is due to the fact that the area of the shaded region above the x-axis and the area of the shaded region below the x-axis are equal.

EXAMPLE **3**

Using the Fundamental Theorem of Calculus to Find the Area of a Region

Find the area of the region between the graph of the function $f(x) = x^3 - x$ and the x-axis over the interval $[-1, 1]$.

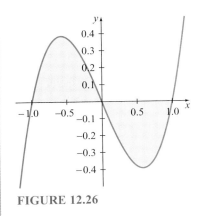

FIGURE 12.26

We are asked to find the combined area of the regions enclosed by the graph of f and the x-axis. We begin by graphing f and shading the enclosed regions (see Figure 12.26).

The graph of f bordering the first region intersects the x-axis at $x = -1$ and $x = 0$. The area of the region is given by

$$\int_{-1}^{0} (x^3 - x)\, dx = \frac{x^4}{4} - \frac{x^2}{2}\Big|_{-1}^{0}$$

$$= \left[\frac{(0)^4}{4} - \frac{(0)^2}{2}\right] - \left[\frac{(-1)^4}{4} - \frac{(-1)^2}{2}\right]$$

$$= 0 - \left(\frac{1}{4} - \frac{1}{2}\right)$$

$$= \frac{1}{4}$$

The graph of f bordering the second region intersects the x-axis at $x = 0$ and $x = 1$. The region lies below the x-axis, so $\int_{0}^{1} (x^3 - x)\, dx$ will be a negative number equal to -1 times the area.

$$\int_{0}^{1} (x^3 - x)\, dx = \frac{x^4}{4} - \frac{x^2}{2}\Big|_{0}^{1}$$

$$= \left[\frac{(1)^4}{4} - \frac{(1)^2}{2}\right] - \left[\frac{(0)^4}{4} - \frac{(0)^2}{2}\right]$$

$$= \left(\frac{1}{4} - \frac{1}{2}\right) - 0$$

$$= -\frac{1}{4}$$

$$= -1 \cdot \frac{1}{4}$$

The area of the second region is also $\frac{1}{4}$ square unit. Therefore, the area bounded by the graph of f and the x-axis is

$$A = \frac{1}{4} + \frac{1}{4}$$

$$= \frac{1}{2} \text{ square unit}$$

Notice in Example 3 that we used separate definite integrals to calculate the area above the x-axis and the area below the x-axis. Using a single definite integral would have led to an erroneous result. If we calculate $\int_{-1}^{1} (x^3 - x)\, dx$, we get the sum of the *signed* areas, not the actual area.

$$\int_{-1}^{1} (x^3 - x)\, dx = \frac{1}{4}x^4 - \frac{1}{2}x^2 \Big|_{-1}^{1}$$

$$= \left[\frac{1}{4}(1)^4 - \frac{1}{2}(1)^2\right] - \left[\frac{1}{4}(-1)^4 - \frac{1}{2}(-1)^2\right]$$

$$= \left(\frac{1}{4} - \frac{1}{2}\right) - \left(\frac{1}{4} - \frac{1}{2}\right)$$

$$= 0$$

A common error among beginning calculus students is to assume that the definite integral *always* yields the area between the graph of the function and the x-axis. As was just illustrated, this is not the case.

Although we are often given an interval over which to find the area, we are sometimes required to find the interval ourselves. When we see the phrase "Find the area of the region bounded by $f(x)$ and the x-axis," we are being asked to calculate the area of the region(s) enclosed by the graph of f and the x-axis. As long as f is a continuous function, we can determine the area by setting up definite integrals with the limits of integration representing each consecutive pair of x-intercepts, as demonstrated in Example 4.

EXAMPLE 4

Using the Fundamental Theorem of Calculus to Find the Area of a Region

Find the area of the region bounded by $f(x) = 4x^3 - 2x$ and the x-axis.

SOLUTION We must first determine where f crosses the x-axis. To do this, we set $f(x)$ equal to zero and solve.

$$0 = 4x^3 - 2x$$

$$0 = 4x\left(x^2 - \frac{1}{2}\right)$$

$$0 = 4x\left(x - \sqrt{\frac{1}{2}}\right)\left(x + \sqrt{\frac{1}{2}}\right)$$

The function has x-intercepts at $x = -\sqrt{\frac{1}{2}}$, $x = 0$, and $x = \sqrt{\frac{1}{2}}$. Let's graph the function so that we can see the location of the bounded regions (Figure 12.27).

On the interval $\left(-\infty, -\sqrt{\frac{1}{2}}\right)$, the graph of f and the x-axis do not enclose a region. The first enclosed region occurs on the interval $\left(-\sqrt{\frac{1}{2}}, 0\right)$. The second enclosed region occurs on the interval $\left(0, \sqrt{\frac{1}{2}}\right)$. On the interval $\left(\sqrt{\frac{1}{2}}, \infty\right)$, the graph of f and the x-axis do not enclose a region. The area we're looking for is

$$A = \left| \int_{-\sqrt{1/2}}^{0} (4x^3 - 2x)\, dx \right| + \left| \int_{0}^{\sqrt{1/2}} (4x^3 - 2x)\, dx \right|$$

Notice that we placed absolute values on each of the definite integrals. This will guarantee that the value returned for each region is that region's area instead of its *signed* area.

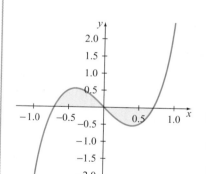

FIGURE 12.27

$$A = \left| \int_{-\sqrt{1/2}}^{0} (4x^3 - 2x)\, dx \right| + \left| \int_{0}^{\sqrt{1/2}} (4x^3 - 2x)\, dx \right|$$

$$= \left| (x^4 - x^2)\, \big|_{-\sqrt{1/2}}^{0} \right| + \left| (x^4 - x^2)\, \big|_{0}^{\sqrt{1/2}} \right|$$

$$= \left| \left\{ [(0)^4 - (0)^2] - \left[\left(-\sqrt{\frac{1}{2}}\right)^4 - \left(-\sqrt{\frac{1}{2}}\right)^2 \right] \right\} \right| + \left| \left\{ \left[\left(\sqrt{\frac{1}{2}}\right)^4 - \left(\sqrt{\frac{1}{2}}\right)^2 \right] - [(0)^4 - (0)^2] \right\} \right|$$

$$= \left| \left[(0-0) - \left(\frac{1}{4} - \frac{1}{2} \right) \right] \right| + \left| \left[\left(\frac{1}{4} - \frac{1}{2} \right) - (0-0) \right] \right|$$

$$= \left| 0 - \left(-\frac{1}{4} \right) \right| + \left| \left(-\frac{1}{4} - 0 \right) \right|$$

$$= \frac{1}{2} \text{ square unit}$$

The area of the region bounded by $f(x) = 4x^3 - 2x$ and the x-axis is $\frac{1}{2}$ square unit.

The relationship between a function and its antiderivative becomes especially meaningful if the function represents a rate of change.

ACCUMULATED CHANGE OF A FUNCTION

Let f be the rate of change function (derivative) of F on $[a, b]$. Then $\int_a^b f(x) \, dx = F(b) - F(a)$ is the accumulated change in F over the interval $[a, b]$.

EXAMPLE 5   **Using the Fundamental Theorem of Calculus to Calculate an Accumulated Change In Revenue**

Based on data from 1999 to 2001, the marginal revenue of the Coca-Cola Company may be modeled by

$$m(s) = -0.8058s + 14.44 \text{ dollars per unit case}$$

where s is the number of unit cases (in billions). A unit case is equivalent to 24 eight-ounce servings of finished beverage. In 1999, 16.5 billion unit cases were sold. In 2001, 17.8 billion unit cases were sold. (**Source:** Modeled from Coca-Cola Company 2001 Annual Report, pp. 46, 57.) By how much did the revenue of Coca-Cola Company grow as the number of unit cases sold increased from 16.5 billion to 17.8 billion?

SOLUTION Marginal revenue is the rate of change in the revenue function. The accumulated change in revenue is given by

$$\int_{16.5}^{17.8} (-0.8058s + 14.44) \, ds = (-0.4029s^2 + 14.44s) \Big|_{16.5}^{17.8}$$

$$= [-0.4029(17.8)^2 + 14.44(17.8)] - [-0.4029(16.5)^2 + 14.44(16.5)]$$

$$= (-127.65 + 257.03) - (-109.69 + 238.26)$$

$$= (129.38) - (128.57)$$

$$= 0.81$$

Increasing the number of cases sold from 16.5 billion to 17.8 billion increased revenue by 0.8 billion dollars. (The raw data were accurate to the nearest tenth of a billion dollars.)

Let's look at the functions m and M graphically. The graph of m is a line, as shown in Figure 12.28. As previously shown in Section 12.4, the area of the shaded region between the graph of m and the horizontal axis is 0.8.

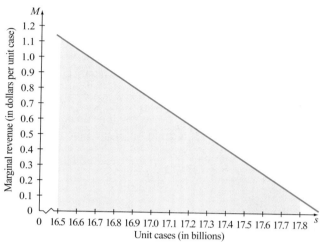

FIGURE 12.28

The graph of M is a parabola (Figure 12.29).

FIGURE 12.29

The vertical distance between the point $[17.8, M(17.8)]$ and the point $[16.5, M(16.5)]$ is given by $M(17.8) - M(16.5)$.

$$M(17.8) - M(16.5) = 129.38 - 128.57$$
$$= 0.81$$

Amazingly, the area between the graph of m and the horizontal axis on the interval $[16.5, 17.8]$ is easily calculated by finding the difference $M(17.8) - M(16.5)$, where M is an antiderivative of m.

A common error of beginning calculus students is to assume that the revenue earned by the company was 0.8 billion dollars. The value 0.8 billion dollars is the accumulated *change* in revenue, not the revenue value itself.

EXAMPLE 6

Using the Fundamental Theorem of Calculus to Calculate an Accumulated Change in Sales

Based on data from 1993 to 2002, the rate of change in the sales income of the Starbucks Corporation may be modeled by

$$r(t) = 58.46t + 79.33 \text{ million dollars per year}$$

where t is the number of years since 1993.

According to the model, by how much did the sales income increase between the end of 1996 and the end of 2002?

SOLUTION   We are asked to find $\displaystyle\int_3^9 (58.46t + 79.33)\,dt.$

$$\int_3^9 (58.46t + 79.33)\,dt = (29.23t^2 + 79.33t)\Big|_3^9$$

$$= [29.23(9)^2 + 79.33(9)] - [29.23(3)^2 + 79.33(3)]$$

$$= 3081.6 - 501.06$$

$$\approx 2580 \text{ million dollars}$$

Between the end of 1996 and the end of 2002, the sales income increased by 2580 million dollars. (Based on the raw data, the actual change was 2592 million. Our model estimate was off by about $12 million. Although that may seem like a lot, our model estimate was within 0.5 percent of the actual value.)

Changing Limits of Integration

When the integrand function in a definite integral requires integration by substitution, the limits of integration must be changed along with the variable. This technique will be demonstrated in Examples 7 and 8.

EXAMPLE 7

Changing the Limits of Integration of a Definite Integral

Calculate $\displaystyle\int_0^1 x^2(x^3 + 1)^3\,dx.$

SOLUTION   Let $u = x^3 + 1$. Then

$$\frac{d}{dx}(u) = \frac{d}{dx}(x^3 + 1)$$

$$\frac{du}{dx} = 3x^2$$

$$du = 3x^2 dx$$

$$\frac{du}{3} = x^2 dx$$

Therefore,

$$\int_0^1 x^2(x^3 + 1)^3 \, dx = \int_a^b u^3 \left(\frac{du}{3}\right)$$

$$= \frac{1}{3} \int_a^b u^3 \, du$$

What are the values of a and b? Recall that the limits of integration of the original function were $x = 0$ and $x = 1$. We will substitute these values of x into the equation relating u and x to find the new limits of integration. Since $u = x^3 + 1$, the new limits of integration are $a = 0^3 + 1$ and $b = 1^3 + 1$. Therefore,

$$\int_0^1 x^2(x^3 + 1)^3 \, dx = \frac{1}{3} \int_1^2 u^3 \, du$$

$$= \frac{1}{3} \frac{u^4}{4} \Big|_1^2$$

$$= \frac{2^4}{12} - \frac{1^4}{12}$$

$$= 1.25$$

EXAMPLE 8

Changing the Limits of Integration of a Definite Integral

Calculate $\displaystyle\int_1^3 \frac{\ln(x)}{x} \, dx$.

SOLUTION   At first, the function does not appear to be integrable; however, if we write $f(x) = \frac{\ln(x)}{x}$ as $f(x) = \ln(x) \cdot \frac{1}{x}$, we recognize that f is of the form $u \cdot u'$. Let $u = \ln x$. Then

$$\frac{d}{dx}(u) = \frac{d}{dx}(\ln x)$$

$$\frac{du}{dx} = \frac{1}{x}$$

$$du = \frac{1}{x} \, dx$$

Therefore, $\displaystyle\int_1^3 \left(\ln x \cdot \frac{1}{x}\right) dx = \int_a^b u \, du$. What are the values of a and b? Recall that the limits of integration of the original function were $x = 1$ and $x = 3$. We will substitute these values of x into the equation relating u and x to find the new limits of integration. Since $u = \ln(x)$, the new limits of integration are $a = \ln(1)$ and $b = \ln(3)$. Since $\ln(1) = 0$, $a = 0$. Therefore,

$$\int_1^3 \frac{\ln(x)}{x}\, dx = \int_0^{\ln(3)} u\, du$$

$$= \frac{u^2}{2}\Big|_0^{\ln(3)}$$

$$= \frac{[\ln(3)]^2}{2} - \frac{(0)^2}{2}$$

$$= \frac{[\ln(3)]^2}{2}$$

$$\approx 0.6035$$

In Section 12.4, we showed how to use the TI-83 Plus to determine the value of a definite integral graphically. In the following Technology Tip, we will demonstrate an alternative way to calculate the definite integral.

TECHNOLOGY **TIP**

Calculating a Definite Integral with *fnInt*

1. Press [MATH] to bring up the MATH menu. Select 9:fnInt. Press [ENTER].

```
MATH NUM CPX PRB
4↑³√(
5: ×√
6:fMin(
7:fMax(
8:nDeriv(
9:fnInt(
0:Solver…
```

2. The calculator program *fnInt* requires four values: the function, the variable of integration, the lower limit, and the upper limit. Enter each of the values and press [ENTER]. To calculate $\int_1^3 \frac{\ln(x)}{x}\, dx$, we enter *fnInt* $(\ln(x)/x, x, 1, 3)$.

```
fnInt(ln(X)/X,X,
1,3)
```

3. The calculator displays the value of the definite integral.

```
fnInt(ln(X)/X,X,
1,3)
        .6034744804
```

12.5 Summary

In this section, you learned how to use one of the most powerful tools in calculus: the Fundamental Theorem of Calculus. You used this theorem to calculate a bounded area quickly and to determine the accumulated change of a rate of change function. In future sections, you will continue to use this powerful tool.

12.5 Exercises

In Exercises 1–20, use the Fundamental Theorem of Calculus to calculate the definite integral.

1. $\int_{5}^{7} 1 \, dx$

2. $\int_{2}^{5} 3 \, dx$

3. $\int_{0}^{4} x \, dx$

4. $\int_{0}^{3} (-x + 3) \, dx$

5. $\int_{1}^{2} (-3x^2 + 5) \, dx$

6. $\int_{1}^{3} (4t^2 - 4t) \, dt$

7. $\int_{-1}^{1} (8x^3 + 2x) \, dx$

8. $\int_{-1}^{1} (5t^4 - 4t^3) \, dt$

9. $\int_{1}^{3} \left(\frac{1}{x}\right) dx$

10. $\int_{1}^{3} \left(\frac{4}{x}\right) dx$

11. $\int_{2}^{5} (2^t) \, dt$

12. $\int_{-3}^{3} (3^t) \, dt$

13. $\int_{-2}^{4} (5 \cdot 2^x) \, dx$

14. $\int_{-4}^{-2} (7 \cdot 3^x) \, dx$

15. $\int_{0}^{1} \left(\frac{2x}{x^2 + 1}\right) dx$

16. $\int_{1}^{4} \left(\frac{e^x}{e^x + 1}\right) dx$

17. $\int_{0}^{2} (x^2 - 1)(x^3 - 3x)^3 \, dx$

18. $\int_{-1}^{1} (3x + 4) \, dx$

19. $\int_{-2}^{-1} (-2x + 1) \, dx$

20. $\int_{-2}^{3} (-x + 5) \, dx$

In Exercises 21–30, calculate the area of the region bounded by the graph of the function and the x-axis. You may find it helpful to graph the function before attempting each exercise.

21. $h(x) = x^3 + x^2 - 12x$

22. $g(x) = -x^2 + 4$

23. $f(x) = x^3 - x^2$

24. $g(x) = x^4 - 4x^2$

25. $f(x) = x^3 - 1$ between $x = -1$ and $x = 1$

26. $h(x) = x^2 - 4$ between $x = -1$ and $x = 1$

27. $g(x) = 1 - 3^x$ between $x = 0$ and $x = 2$

28. $f(x) = e^x - x^2$ between $x = 0$ and $x = 2$

29. $h(x) = x - x^2$ between $x = -1$ and $x = 2$

30. $g(x) = x^3 - \frac{1}{x}$ between $x = 1$ and $x = 2$

In Exercises 31–35, use the Fundamental Theorem of Calculus to find the answer to the question.

31. **M** **AP Calculus Exam**   Based on data from 1969 to 2002, the rate of change in the number of AP Calculus AB exam participants may be modeled by

$$R(t) = 315.70t - 770.64$$

people per year, where t is the number of years since the end of 1969. (**Source:** Modeled from The College Board data.)

If the model is an accurate predictor of the future, by how much will the number of people taking the exam each year increase between the end of 2002 and the end of 2012?

32. **M** **Bottled Water Consumption**   Based on data from 1980 to 1999, the rate of change in the annual per capita consumption of bottled water may be modeled by

$$R(t) = 0.2612(1.106)^t$$

gallons per person per year, where t is the number of years since the end of 1980. (**Source:** Modeled from *Statistical Abstract of the United States, 2001*, Table 204, p. 130.)

Between the end of 1995 and the end of 1999, by how much did per capita bottled water consumption increase?

33. **Minivan Position** The velocity of the author's minivan as he drove from a traffic light onto a highway may be modeled by

$$V(t) = -0.1773t^2 + 0.6798t + 0.5019$$

miles per minute, where t is time in minutes. How far did he travel in the first 1.5 minutes of his trip?

34. **McDonald's Sales** Based on data from 1990 to 2001, the rate of change of franchised sales of McDonald's Corporation may be modeled by

$$R(t) = -82.586t + 1729.0$$

million dollars per year, where t is the number of years since the end of 1990. (**Source:** Modeled from www.mcdonalds.com data.)

Between the end of 1990 and the end of 2001, by how much did McDonald's franchised sales increase?

35. **Mozzarella Cheese Consumption** Based on data from 1980 to 1999, the rate of change in the annual per capita consumption of mozzarella cheese may be modeled by

$$R(t) = 0.0009778t^3 - 0.02958t^2 + 0.2392t - 0.06311$$

pounds per person per year, where t is the number of years since the end of 1980. (**Source:** Modeled from *Statistical Abstract of the United States, 2001*, Table 202, p. 129.)

Between the end of 1990 and the end of 1999, by how much did per capita mozzarella cheese consumption change?

Exercises 36–40 are intended to challenge your understanding of the Fundamental Theorem of Calculus.

36. A function $F(x)$ has the properties that $F'(x) = 3x^2 + 2x + 1$ and $F(1) = 5$. Find $F(2)$.

37. A function $F(x)$ has the properties that $F'(x) = \dfrac{1}{x}$ and $F(1) = 1 + \ln(4)$. Find $F(4)$.

38. The velocity of a car is modeled by the function $v(t) = 66$ miles per hour, where t is the number of hours that the car has been traveling. Calculate $\displaystyle\int_{0.5}^{1} v(t)\, dt$ and interpret the meaning of the result.

39. Calculate the area of the region bounded by the graph of $g(x) = 2^x$ and the graph of $h(x) = x^2$. (*Hint:* The graphs of g and h intersect in exactly three places.)

40. Find a function $f(x) \neq 0$ such that $\displaystyle\int_{-a}^{a} f(x)\, dx = 0$ on any interval $[-a, a]$. Then discuss the relationship between $F(-a)$ and $F(a)$ given that $F'(x) = f(x)$.

Chapter 12 Review Exercises

Section 12.1 *In Exercises 1–4, find the general antiderivative of the function.*

1. $g(x) = -2x + 5$

2. $f(x) = 3^x + x - 1$

3. $h(x) = 4x - \dfrac{3}{x}$

4. $g(x) = x^4 - 4x^3$

In Exercises 5–8, calculate the indefinite integral.

5. $\displaystyle\int (4x^3 - 5x - 1)\, dx$

6. $\displaystyle\int \frac{u + 2}{u}\, du$

7. $\displaystyle\int \frac{2t + 1}{2t}\, dt$

8. $\displaystyle\int (x^2 - 2x + 3)\, dx$

In Exercises 9–10, apply the concept of integration in solving the real-world applications.

9. **Publishing Costs** Golden Pillar Publishing (www.goldenpillarpublishing.com) publishes books for aspiring authors. In 2003, the marginal printing cost for a 8.25" × 11" soft-cover book was

$$c(p) = 0.018$$

dollars per page, where p is the number of pages in the book. Including setup costs, a 100-page book had a $3.10 production cost. Find the book production cost function.

10. **Free Fall**   The Giant Drop is one of the most popular amusement park rides manufactured by Intamin. Riders are hoisted to a height of 227 feet and then dropped into a free fall. (**Source:** www.sixflags.com.)

The velocity of an object that is dropped into a free fall may be modeled by

$$v(t) = -32t$$

feet per second, where t is the number of seconds since the object was dropped. A negative velocity indicates that the object is moving toward the ground. Given that position is the antiderivative of velocity, find the position function for the free-fall portion of the Great Drop thrill ride.

Section 12.2   *In Exercises 11–15, integrate the functions. Some integrals will require integration by substitution, while others may be integrated using basic integration rules.*

11. $f(x) = x(x^2 - 9)^5$ **12.** $f(t) = 2t\sqrt{t - 4}$

13. $h(x) = \dfrac{2\ln x}{x}$ **14.** $f(x) = \dfrac{2x}{x^2 + 1}$

15. $g(t) = \dfrac{6 + t}{2t}$

Section 12.3   *In Exercises 16–20, use the left-hand sum to estimate the area between the graph of the function and the horizontal axis on the specified interval. For each exercise, calculate the sum with $n = 2$, $n = 4$, and $n = 10$ rectangles.*

16. $f(x) = x^2 - 1; [1, 5]$

17. $g(x) = -4x + 16; [0, 4]$

18. $h(x) = \ln x; [1, 2]$ **19.** $f(t) = t - \dfrac{1}{t}; [3, 5]$

20. $g(t) = \sqrt{t + 2}; [0, 1]$

In Exercises 21–25, use left- and right-hand sums (with $n = 4$) to estimate the area between the graph of the function and the horizontal axis on the specified interval. In each exercise, calculate the left-hand sum, the right-hand sum, and the average of the two sums. The exact area, A, is given so that you can compare your estimates to the actual area.

21. $f(t) = 3t^2 - 2t + 1$ on $[0, 1]; A = 1$

22. $g(x) = x - \sqrt{x}$ on $[1, 4]; A = 2\dfrac{5}{6}$

23. $h(x) = 2x + \dfrac{1}{x}$ on $[1, 2]; A = 3 + \ln 2 \approx 3.693$

24. $g(x) = \sqrt{x + 4}$ on $[0, 5]; A = 12\dfrac{2}{3}$

25. $g(x) = -3x^2 + 3$ on $[0, 1]; A = 2$

In Exercises 26–27, estimate the area between the graph of the function and the horizontal axis. Then interpret the physical meaning of the result.

26. **Free Fall**   The Giant Drop is one of the most popular amusement park rides manufactured by Intamin. Riders are hoisted to a height of 227 feet and then dropped into a free fall. (**Source:** www.sixflags.com.)

The velocity of an object that is dropped into a free fall may be modeled by

$$v(t) = -32t$$

feet per second, where t is the number of seconds since the object was dropped. A negative velocity indicates that the object is moving toward the ground. Estimate the area of the region between the graph of v and the horizontal axis on the interval $[0, 2]$ and interpret the meaning of the result. (*Hint:* The function v is entirely negative on the interval. Consequently, the left- and right-hand sums will yield a negative value. Since the entire region is below the horizontal axis, you can simply drop the negative sign to get the area estimate.)

27. **River Flow**   Between 5:00 p.m. and 9:45 p.m. on November 22, 2002, the stream flow rate of the Columbia River below the Priest Rapids Dam increased dramatically. The flow rate during that time period may be modeled by

$$f(t) = 714.05t^3 - 5282.6t^2 + 21,738t + 61,656$$

cubic feet per second, where t is the number of hours since 5:00 p.m. (**Source:** waterdata.usgs.gov.)

(a) Convert the units of the flow-rate function from cubic feet per second to cubic feet per hour.

(b) Estimate the area of the region between the graph of f and the horizontal axis on the interval $[0, 4.75]$ and interpret its meaning.

(c) Depending upon the pool dimensions, an Olympic-size swimming pool can hold between 94,000 and 114,000 cubic feet of water. How many 100,000-cubic-foot swimming pools could be filled with the volume of water that flowed past the checkpoint below the Priest Rapids Dam on the Columbia River between 5:00 p.m. and 9:45 p.m.?

Section 12.4 *In Exercises 28–30, calculate the sum.*

28. $\displaystyle\sum_{i=1}^{6} i^2 - 2i + 1$

29. $\displaystyle\sum_{i=3}^{7} (i - 4)^2$

30. $\displaystyle\sum_{i=3}^{7} 4(2i - 9)^2$

In Exercises 31–35, draw the graph of the integrand function. Then calculate the definite integral using geometric formulas for area.

31. $\displaystyle\int_{2}^{5} 2x \, dx$

32. $\displaystyle\int_{0}^{2} (-3x + 6) \, dx$

33. $\displaystyle\int_{0}^{6} (-|x - 3| + 3) \, dx$

34. $\displaystyle\int_{1}^{2} (-2x + 4) \, dx$

35. $\displaystyle\int_{4}^{6} 2 \, dx$

In Exercises 36–40, use the definite integral properties to find the numeric value of each definite integral, given that $\displaystyle\int_{1}^{3} f(x) \, dx = 2,$ $\displaystyle\int_{1}^{4} f(x) \, dx = 5,$ $\displaystyle\int_{1}^{3} g(x) \, dx = 1,$ *and* $\displaystyle\int_{3}^{4} g(x) \, dx = 3.$

36. $\displaystyle\int_{3}^{4} [2f(x)] \, dx$

37. $\displaystyle\int_{1}^{4} [3g(x)] \, dx$

38. $\displaystyle\int_{3}^{4} [2f(x) + 4g(x)] \, dx$

39. $\displaystyle\int_{1}^{4} [f(x) - g(x)] \, dx$

40. $\displaystyle\int_{1}^{4} [4f(x) - 5g(x)] \, dx$

In Exercises 41–42, shade the regions of the graph that will be used in calculating the definite integral.

41. $\displaystyle\int_{1}^{2} \left(\frac{x}{x^2 + 1}\right) dx$

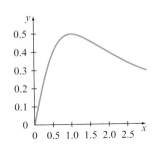

42. $\displaystyle\int_{1}^{2} \left(\frac{x^2 + 2}{x^2 + 1}\right) dx$

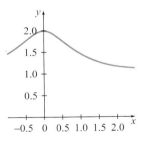

In Exercises 43–44, use the properties of definite integrals to find the solution.

43. **M** **Milk Consumption** Based on data from 1985 to 1999, the rate of change in the annual per capita consumption of milk as a beverage may be modeled by

$$R(t) = 0.001326t^2 - 0.02192t - 0.1541$$

gallons per person per year, where t is the number of years since the end of 1985. (**Source:** Modeled from *Statistical Abstract of the United States, 2001*, Table 204, p. 130.) The graph of $R(t)$ is shown.

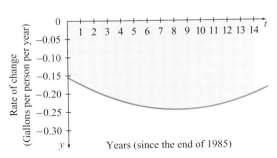

The signed area of the shaded region of the graph is

$$\int_{0}^{14} (0.001326t^2 - 0.02192t - 0.1541) \, dt$$

and is equal to -3.093. Additionally,

$$\int_{0}^{8} (0.001326t^2 - 0.02192t - 0.1541) \, dt$$
$$= -1.708$$

(a) Calculate and interpret the meaning of

$$\int_{8}^{14} (0.001326t^2 - 0.02192t - 0.1541) \, dt$$

(b) In 1993, California milk processors launched the "Got milk?" advertising campaign in an effort to turn around the 20-year trend of declining milk consumption. (**Source:** www.gotmilk.com.) Considering the results of part (a) together with the graph of R, do you think that the "Got milk?" advertising campaign has been successful?

44. **M** **Personal Income** Based on data from 1993 to 2000, the rate of change in the annual per capita personal income in New Mexico may be modeled by

$$R(t) = 8.4546t^2 + 45.944t + 723.00$$

dollars per person per year, where t is the number of years since the end of 1993. (**Source:** Modeled from Bureau of Economic Analysis data.) The graph of $R(t)$ is shown.

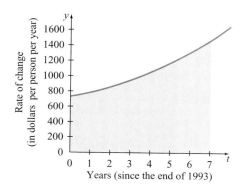

The signed area of the shaded region of the graph is

$$\int_0^7 (8.4546t^2 + 45.944t + 723.00)\, dt$$

and is equal to 7153. Additionally,

$$\int_4^7 (8.4546t^2 + 45.944t + 723.00)\, dt = 3713$$

Calculate and interpret the meaning of

$$\int_0^4 (8.4546t^2 + 45.944t + 723.00)\, dt.$$

Section 12.5 *In Exercises 45–50, use the Fundamental Theorem of Calculus to calculate the definite integral.*

45. $\displaystyle\int_2^3 (3x + 1)\, dx$

46. $\displaystyle\int_{-2}^2 (x^2)\, dx$

47. $\displaystyle\int_{-1}^1 (-x^3)\, dx$

48. $\displaystyle\int_0^1 (3e^x + 3x^2)\, dx$

49. $\displaystyle\int_2^3 \frac{2x}{x^2 + 1}\, dx$

50. $\displaystyle\int_1^2 \frac{\ln x}{x}\, dx$

In Exercises 51–55, calculate the area of the region bounded by the graph of the function and the x-axis. You may find it helpful to graph the function before attempting the exercise.

51. $h(x) = x^3 - 8$ between $x = 1$ and $x = 5$

52. $g(x) = 2.5 - \dfrac{1}{x} - x$ between $x = 0.5$ and $x = 4$

53. $f(x) = \dfrac{2x}{x^2 + 1}$ between $x = -3$ and $x = 3$

54. $f(t) = 2^t - t^2$ between $t = 0$ and $t = 4$

55. $g(x) = (x^2 - 1)(x^3 - 3x)^3$ between $x = -\sqrt{3}$ and $x = \sqrt{3}$.

In Exercises 56–57, use the Fundamental Theorem of Calculus to determine the answer to the question.

56. **M** **Personal Income** Based on data from 1993 to 2000, the rate of change in the annual per capita personal income in Louisiana may be modeled by

$$R(t) = -28.940t + 883.18$$

dollars per person per year, where t is the number of years since the end of 1993. (**Source:** Modeled from Bureau of Economic Analysis data.)

Between the end of 1995 and the end of 2000, by how much did the annual per capita personal income increase?

57. **M** **Personal Income** Based on data from 1993 to 2000, the rate of change in the annual per capita personal income in Florida may be modeled by

$$R(t) = -18.914t^2 + 209.16t + 645.4$$

dollars per person per year, where t is the number of years since the end of 1993. (**Source:** Modeled from Bureau of Economic Analysis data.)

Between 1995 and 2000, by how much did the annual per capita personal income increase?

Make It Real

What to do

1. Collect at least five years' worth of data that are reported annually (for example, annual income or annual tuition cost).

2. Find a model for the data. This model represents the annual rate of change function.

3. Find the antiderivative of the model equation.

4. Use the Fundamental Theorem of Calculus to determine the projected accumulated change in the annual rate of change function over a period of at least four years, including at most one future year.

5. Interpret the meaning of the results from Step 4 and explain why the information is useful.

Example

Data Points

Johnson and Johnson Pharmaceutical Operating Profit

Years (Since 1997) (t)	Operating Profit (millions of dollars) (P)
0	2,332
1	3,114
2	3,735
3	4,394
4	4,928

Meaning

From the end of 1998 through 2002, the cumulative operating profit earned by Johnson and Johnson is projected to increase by 17,350 million dollars. When making investment choices, long-term investors often look at cumulative profit over time instead of focusing on fluctuations in annual profits.

Fundamental Theorem

$$A(5) - A(1) = 20{,}070 - 2719.2$$
$$\approx 17{,}350$$

Model

$$P = -32.71t^2 + 778.1t + 2341$$

Antiderivative

$$A = -10.90t^3 + 389.1t^2 + 2341t$$

Advanced Integration Techniques and Applications

Fashion trends come and go. When a new product is first introduced to the market, sales are often slow. However, as the popularity of the product increases, sales increase rapidly. After a period of time, the market becomes saturated with the product, and sales taper off. The cumulative sales of a product that demonstrates the type of sales growth just described may often be modeled by a logistic function.

13.1 Integration by Parts
- Use the integration by parts method to find the antiderivative of a function

13.2 Area Between Two Curves
- Calculate the area of the region between two continuous curves

13.3 Differential Equations and Applications
- Solve introductory-level differential equations
- Apply Newton's Law of Heating and Cooling
- Analyze and interpret the real-life meaning of differential equations

13.4 Differential Equations: Limited Growth and Logistic Models
- Find the equation of a limited growth model
- Find the equation of a logistic growth model
- Analyze and interpret the real-life meaning of differential equations

13.1 Integration by Parts

- Use the integration by parts method to find the antiderivative of a function

GETTING STARTED   As we saw in Chapter 12, many derivative rules have a corresponding integral rule. Although there is no product rule for integrals, the *integration by parts method* is directly related to the product rule for derivatives.

In this section, we will demonstrate how to use the method of integration by parts to integrate a function. This technique is among the more challenging methods to use, so we will use it only after basic integration methods and the substitution method have failed.

Recall that the Product Rule for derivatives states that

$$\frac{d}{dx}[u(x) \cdot v(x)] = u'(x) \cdot v(x) + v'(x) \cdot u(x)$$

where u and v are differentiable functions. If we integrate both sides of the equation with respect to x, we get

$$\int \left\{ \frac{d}{dx}[u(x) \cdot v(x)] \right\} dx = \int [u'(x) \cdot v(x) + v'(x) \cdot u(x)]dx$$

$$u(x) \cdot v(x) = \int [u'(x) \cdot v(x)]dx + \int [v'(x) \cdot u(x)]dx$$

$$= \int \left[\frac{du}{dx} \cdot v(x) \right] dx + \int \left[\frac{dv}{dx} \cdot u(x) \right] dx$$

$$u(x) \cdot v(x) = \int v(x) \, du + \int u(x) \, dv$$

For notational ease, we will write $u(x)$ as u and $v(x)$ as v, keeping in mind that both u and v are functions of x.

$$uv = \int v \, du + \int u \, dv$$

We'll solve this equation for $\int u \, dv$.

$$\int u \, dv = uv - \int v \, du$$

This equation is the basis of the integration by parts method.

INTEGRATION BY PARTS

Let u and v be differentiable functions of x. Then

$$\int u \, dv = uv - \int v \, du$$

Consider the integral $\int (xe^x) \, dx$. We attempted to integrate this function by substitution in Section 12.2, but we were unable to do so. We will now attempt to integrate the function using integration by parts. In order to use this method, we must first identify a function u and a function v' such that $uv' = xe^x$. We typically will pick the function with the simplest derivative to be u or the most complex function that may be easily integrated to be v'. In this case, we pick

$$u(x) = x \qquad \text{and} \qquad v'(x) = e^x$$

Next we'll differentiate u with respect to x and, using the concept of differentials, solve for du.

$$\frac{d}{dx}(u) = \frac{d}{dx}(x)$$

$$\frac{du}{dx} = 1$$

$$du = 1 \, dx$$

Next we'll integrate both sides of the equation $v'(x) = e^x$ with respect to x.

$$v'(x) = e^x$$

$$\int v'(x) \, dx = \int e^x dx$$

$$v(x) = e^x \qquad \text{We will add in the constant } C \text{ at the end of the problem}$$

We'll now return to the formula $\int u \, dv = uv - \int v \, du$ and plug in the correspon-

ding pieces. (Recall that, by definition, $dv = v'(x) \, dx$. In this case, $dv = e^x dx$.)

$$\int u \, dv = uv - \int v \, du$$

$$\int (x)(e^x dx) = (x) \cdot (e^x) - \int (e^x)(dx)$$

$$= xe^x - \int e^x \, dx$$

$$= xe^x - e^x + C$$

(In this example, as in previous examples, the constant C represents the sum of all constants generated throughout the integration process.)

Let's check our work by differentiating $F(x) = xe^x - e^x + C$.

$$\frac{d}{dx}[F(x)] = \frac{d}{dx}(xe^x - e^x + C)$$

$$f(x) = [1 \cdot e^x + (e^x) \cdot (x)] - e^x + 0$$

$$= e^x + xe^x - e^x$$

$$= xe^x$$

The result checks out, so we did the problem correctly.

Now that we've demonstrated the basic theory that surrounds the integration by parts method, we will refine the process to increase our efficiency. We'll demonstrate the streamlined process in Example 1 and then detail the steps of the process.

EXAMPLE 1   **Integrating by Parts**

Integrate $f(x) = 3x(2^x)$ with respect to x.

SOLUTION We are asked to find $\int 3x(2^x) \, dx$. Since $\frac{d}{dx}(3x)$ is a constant, we will select $u = 3x$. Consequently, $dv = 2^x dx$. Observe that $\int 3x(2^x)\,dx$ may be rewritten as $\int u\,dv$. We are now ready to do integration by parts.

$u = 3x$	$dv = 2^x\,dx$	Choose u and dv
$du = 3\,dx$	$v = \dfrac{2^x}{\ln(2)}$	Differentiate u and integrate dv with respect to x

$$\int 3x(2^x)\,dx = \int u\,dv$$
$$= uv - \int v\,du$$
$$= (3x)\left[\frac{2^x}{\ln(2)}\right] - \int \frac{2^x}{\ln(2)}(3\,dx)$$
$$= \frac{3x(2^x)}{\ln(2)} - \frac{3}{\ln(2)}\int 2^x\,dx \qquad \text{Constant Multiple Rule}$$
$$= \frac{3x(2^x)}{\ln(2)} - \frac{3}{\ln(2)}\left[\frac{2^x}{\ln(2)}\right] + C \qquad \text{Exponential Rule}$$
$$= \left[\frac{2^x}{\ln(2)}\right]\left[3x - \frac{3}{\ln(2)}\right] + C \qquad \text{Factor out } \frac{2^x}{\ln(2)}$$

Therefore, $\int 3x(2^x)\,dx = \left[\frac{2^x}{\ln(2)}\right]\left[3x - \frac{3}{\ln(2)}\right] + C.$

HOW TO Detailed Steps for the Integration by Parts Method

Let u and v be differentiable functions of x. To integrate $\int u\,dv$, do the following:

1. Identify u and dv. The product of these two factors must equal the entire integrand coupled with the associated dx.
2. Differentiate u and integrate dv with respect to x.
3. Write the expression $uv - \int v\,du$ in terms of x.
4. Integrate $\int v\,du$ to eliminate the integral sign.
5. Simplify.

We'll demonstrate these steps in the next several examples.

EXAMPLE 2 Integrating by Parts

Integrate $f(x) = x\ln(x)$.

SOLUTION We must find $\int x \ln(x)\, dx$. We'll pick $u = \ln(x)$ and $dv = x\, dx$.

$u = \ln(x)$	$dv = x\, dx$	
$du = \dfrac{1}{x}\, dx$	$v = \dfrac{x^2}{2}$	Differentiate u and integrate dv with respect to x

The top of the table: Choose u and dv

$$uv - \int v\, du = \ln(x) \cdot \frac{x^2}{2} - \int \left(\frac{x^2}{2} \cdot \frac{1}{x} \right) dx$$

$$= \frac{1}{2} x^2 \ln(x) - \frac{1}{2} \int x\, dx \qquad \text{Constant Multiple Rule}$$

$$= \frac{1}{2} x^2 \ln(x) - \frac{1}{2} \left(\frac{1}{2} x^2 \right) + C \qquad \text{Power Rule}$$

$$= \frac{1}{2} x^2 \left[\ln(x) - \frac{1}{2} \right] + C \qquad \text{Factor out } \frac{1}{2} x^2$$

Therefore, $\int x \ln(x)\, dx = \frac{1}{2} x^2 \left[\ln(x) - \frac{1}{2} \right] + C$.

EXAMPLE 3

Integrating by Parts

Simplify $\int (2x + 1)(x - 3)^3 dx$.

SOLUTION We'll pick $u = 2x + 1$ and $dv = (x - 3)^3 dx$.

$u = 2x + 1$	$dv = (x - 3)^3 dx$	
$du = 2\, dx$	$v = \dfrac{(x - 3)^4}{4}$	By the substitution method

$$uv - \int v\, du = (2x + 1) \left[\frac{(x - 3)^4}{4} \right] - \int \left[\frac{(x - 3)^4}{4} \right] (2\, dx)$$

$$= \frac{(2x + 1)(x - 3)^4}{4} - \frac{2}{4} \int (x - 3)^4 dx \qquad \text{Constant Multiple Rule}$$

$$= \frac{1}{4} (2x + 1)(x - 3)^4 - \frac{1}{2} \int (x - 3)^4 dx \qquad \text{Simplify fractions}$$

$$= \frac{1}{4} (2x + 1)(x - 3)^4 - \frac{1}{2} \left[\frac{(x - 3)^5}{5} \right] + C \qquad \text{Power Rule}$$

$$= \frac{1}{4} (2x + 1)(x - 3)^4 - \frac{1}{10} (x - 3)^5 + C \qquad \text{Simplify fractions}$$

$$= \frac{5}{20} (2x + 1)(x - 3)^4 - \frac{2}{20} (x - 3)^5 + C \qquad \text{Rewrite fractions}$$

$$= \frac{1}{20} (x - 3)^4 [5(2x + 1) - 2(x - 3)] + C \qquad \text{Factor out } \frac{1}{20} (x - 3)^4$$

$$= \frac{1}{20}(x - 3)^4(10x + 5 - 2x + 6) + C \qquad \text{Simplify}$$

$$= \frac{1}{20}(x - 3)^4(8x + 11) + C \qquad \text{Group like terms}$$

Therefore, $\int (2x + 1)(x - 3)^3 dx = \frac{1}{20}(x - 3)^4(8x + 11) + C$.

EXAMPLE 4

Integrating by Parts

Simplify $\int (x^2)(x + 1)^3 dx$.

SOLUTION Let $u = x^2$ and $dv = (x + 1)^3 \, dx$.

$u = x^2$	$dv = (x + 1)^3 \, dx$
$du = 2x \, dx$	$v = \dfrac{(x + 1)^4}{4}$

By the substitution method

$$uv - \int v \, du = (x^2)\left[\frac{(x + 1)^4}{4}\right] - \int \left[\frac{(x + 1)^4}{4}\right](2x \, dx)$$

$$= \frac{1}{4}x^2(x + 1)^4 - \frac{1}{2}\int x(x + 1)^4 \, dx \qquad \text{Constant Multiple Rule}$$

Observe that $\int x(x + 1)^4 dx$ is not readily integrable. We'll integrate this piece separately using integration by substitution and then substitute the result back into the equation.

Let $w = x + 1$. Then $dw = dx$ and $x = w - 1$. Thus

$$\int x(x + 1)^4 dx = \int (w - 1)(w^4) \, dw$$

$$= \int (w^5 - w^4) \, dw$$

$$= \frac{w^6}{6} - \frac{w^5}{5} + C \qquad \text{Power Rule}$$

$$= \frac{5w^6}{30} - \frac{6w^5}{30} + C \qquad \text{Get a common denominator}$$

$$= \frac{1}{30}w^5(5w - 6) + C \qquad \text{Factor out } \frac{1}{30}w^5$$

$$= \frac{1}{30}(x + 1)^5[5(x + 1) - 6] + C \qquad \text{Replace } w \text{ with } x + 1$$

$$= \frac{1}{30}(x + 1)^5(5x + 5 - 6) + C \qquad \text{Simplify}$$

$$= \frac{1}{30}(x + 1)^5(5x - 1) + C \qquad \text{Group like terms}$$

We'll substitute this result back into the integration by parts equation.

$$\frac{1}{4}x^2(x+1)^4 - \frac{1}{2}\int x(x+1)^4\,dx = \frac{1}{4}x^2(x+1)^4 - \frac{1}{2}\left[\frac{1}{30}(x+1)^5(5x-1)\right] + C$$

$$= \frac{15}{60}x^2(x+1)^4 - \frac{1}{60}(x+1)^5(5x-1) + C$$

$$= \frac{1}{60}(x+1)^4[15x^2 - (x+1)(5x-1)] + C$$

$$= \frac{1}{60}(x+1)^4[15x^2 - (5x^2+4x-1)] + C$$

$$= \frac{1}{60}(x+1)^4(10x^2-4x+1) + C$$

Therefore, $\displaystyle\int (x^2)(x+1)^3\,dx = \frac{1}{60}(x+1)^4(10x^2-4x+1) + C$. Phew!

Although we used the substitution method to integrate the second integral in Example 4, we could have used integration by parts to integrate the second integral. In Example 5, we will apply the integration by parts method multiple times.

EXAMPLE   **5**

Integrating by Parts

Integrate $f(x) = x^2e^x$.

SOLUTION   We're asked to find $\displaystyle\int x^2e^x\,dx$. We'll pick $u = x^2$ and $dv = e^x\,dx$.

$u = x^2$	$dv = e^x\,dx$
$du = 2x\,dx$	$v = e^x$

$$uv - \int v\,du = (x^2)(e^x) - \int 2xe^x\,dx$$

$$= x^2e^x - 2\int xe^x\,dx$$

The remaining integral, $\displaystyle\int xe^x\,dx$, is not readily integrable. We will integrate this function using integration by parts. We will select new values for u and dv; however, it is important to note that although the variables are the same, the functions are not equal to the u and v identified previously.

$u = x$	$dv = e^x\,dx$
$du = dx$	$v = e^x$

$$x^2e^x - 2\int xe^x\,dx = x^2e^x - 2\left(xe^x - \int e^x\,dx\right)$$

$$= x^2e^x - 2(xe^x - e^x) + C$$

$$= x^2e^x - 2xe^x + 2e^x + C$$

$$= e^x(x^2 - 2x + 2) + C$$

Integration by parts can also be used to integrate functions that formerly did not appear integrable. In Example 6, we will integrate the natural log function.

EXAMPLE 6

Integrating the Natural Logarithm Function

Simplify $\int \ln(x)\, dx$.

SOLUTION Let $u = \ln(x)$ and $dv = dx$.

$u = \ln(x)$	$dv = dx$
$du = \dfrac{1}{x}dx$	$v = x$

$$uv - \int v\, du = [\ln(x)](x) - \int x\left(\frac{1}{x}dx\right)$$
$$= x\ln(x) - \int 1\, dx$$
$$= x\ln(x) - x + C$$
$$= x[\ln(x) - 1] + C$$

Therefore, $\int \ln(x)\, dx = x[\ln(x) - 1] + C$.

EXAMPLE 7

Using Integration to Forecast Motor Vehicle Deaths

Based on data from 1980, 1985, 1990, and 1995, the number of motor vehicle deaths in South Carolina may be modeled by

$$m(t) = 843.3 + 65.92\ln(t)$$

where t is the number of years since 1979. According to the model, how many motor vehicle deaths occurred in South Carolina between 1980 and 1995?

SOLUTION Since $m(t)$ is the annual death rate, we must calculate $\int_{1}^{16}[843.3 + 65.92\ln(t)]\, dt$ to determine the cumulative number of deaths between 1980 and 1995.

$$\text{Deaths} = \int_{1}^{16}[843.3 + 65.92\ln(t)]\, dt$$

$$= 843.3t\Big|_{1}^{16} + 65.92\int_{1}^{16}\ln(t)\, dt \qquad \text{Power, Constant Multiple Rules}$$

$$= 843.3t\Big|_{1}^{16} + 65.92\{t[\ln(t) - 1]\}\Big|_{1}^{16} \qquad \text{Integral of natural log function}$$

$$= [843.3(16) - 843.3(1)]$$
$$+ 65.92\{16[\ln(16) - 1]\} - 65.92\{1[\ln(1) - 1]\} \qquad \text{Fundamental Theorem of Calculus}$$

$$= 12{,}649.5 + 1935.5$$

$$= 14{,}585$$

According to the model, the cumulative number of motor vehicle deaths in South Carolina between 1980 and 1995 was 14,585 deaths.

13.1 Summary

In this section, you learned how to use the method of integration by parts to integrate a function. You discovered that this technique is among the more challenging integration methods and should be used only after basic integration methods and the substitution method have failed.

13.1 Exercises

In Exercises 1–10, integrate the function using integration by parts.

1. $f(x) = xe^{-x}$

2. $g(x) = 5x \ln(x)$

3. $h(t) = (3t - 5) \ln(t)$

4. $f(t) = \dfrac{6}{t} + \ln(t)$

5. $h(t) = t^3 \ln(t)$

6. $h(t) = \dfrac{\ln(t)}{t} - 3$

7. $f(x) = 4xe^{2x+1}$

8. $f(x) = \ln(x^2)$

9. $h(x) = e^{2x}(2x - 1)$

10. $f(t) = (3t + 4)e^{2t}$

In Exercises 11–25, integrate the function using the simplest possible method.

11. $g(t) = \dfrac{3^t}{3^t - 1}$

12. $h(t) = (2t^2 - 9)(e^t)$

13. $f(x) = 3x(e^x - x)$

14. $g(x) = (4x - 1)[\ln(x)]$

15. $h(x) = 2x(x^2 - 9)$

16. $f(t) = (t - 1)^5(2t - 7)$

17. $g(t) = (3t^2)(t^3 - 1)$

18. $h(x) = x^6(3x - 4)$

19. $g(x) = 3^x(x^2 - 3)$

20. $g(x) = \dfrac{[\ln(x)]^2}{x}$

21. $h(x) = 3xe^{3x^2}$

22. $g(t) = t - t\ln(t)$

23. $f(t) = e^{2t}t^2$

24. $h(t) = (2t - 3) \ln(t)$

25. $f(x) = \dfrac{2^{\ln x}}{x}$

In Exercises 26–30, use the integration by parts method to determine the answer to the question.

26. Textbook Sales   A new college textbook edition typically generates most of its sales in the year of its publication. Sales drop off in subsequent years as a result of competition from the used book market. Suppose that the annual sales of a particular textbook may be modeled by

$$S(t) = 30,000 \, te^{-1.5t}$$

textbooks, where t is the number of years since the edition was published. According to the model, how many textbooks will be sold in the first three years of the edition?

27. **M** **Real Networks Licensing Revenue**   Based on data from 1999 to 2003, the annual net revenue from *software licensing fees* for Real Networks, Inc., may be modeled by

$$R(t) = 1208(t - 0.88)e^{-1.34t} + 56.7$$

million dollars, where t is the number of years since the end of 1998. (**Source:** Modeled from Real Networks, Inc., 2003 Report, p. 14.)

According to the model, what was the accumulated net revenue from software licensing fees between the end of 2000 and the end of 2003?

28. **M** **Real Networks Advertising Revenue**   Based on data from 1999 to 2003, the annual net revenue from *advertising* for Real Networks, Inc., may be modeled by

$$A(t) = 1430(t - 0.95)e^{-1.89t} + 6$$

million dollars, where t is the number of years since the end of 1998. (**Source:** Modeled from Real Networks, Inc., 2003 Report, p. 14.)

According to the model, what was the accumulated net revenue from advertising between the end of 2000 and the end of 2003?

29. **Electronic Arts Profit** Based on data from 2000 to 2004, the annual gross profit for Electronic Arts, Inc., may be modeled by

$$P(t) = 274(t - 0.6)e^{-0.232t} + 596$$

million dollars, where t is the number of years since the end of 1998. (**Source:** Modeled from Electronic Arts 2004 Annual Report, p. 19.)

According to the model, what was the accumulated profit between the end of 2000 and the end of 2004?

30. **Real Networks Operating Expenses** Based on data from 1999 to 2003, the annual total operating expenses for Real Networks, Inc., may be modeled by

$$C(t) = 5970(t - 1.04)e^{-1.73t} + 150$$

million dollars, where t is the number of years since the end of 1998. (**Source:** Modeled from Real Networks, Inc., 2003 Report, p. 14.)

According to the model, what was the accumulated operating expense between the end of 1999 and the end of 2003?

Exercises 31–35 are intended to challenge your understanding of integration by parts.

31. Integrate $h(x) = 4x^2e^x$.

32. Integrate $f(t) = 3^t\sqrt{t}$.

33. Integrate $g(t) = 6t^{0.5}\ln(t)$.

34. Explain the relationship between the integration by parts method for integrals and the Product Rule for derivatives.

35. Explain an effective strategy for determining when to use the integration by parts method.

13.2 Area Between Two Curves

- Calculate the area of the region between two continuous curves

GETTING STARTED Electronic Arts, Inc., is arguably the most dominant force in the electronic gaming software market. Sixteen of the games the company shipped in 2002 sold more than one million copies each. Its 2002 title "Harry Potter and the Sorcerer's Stone"™ generated more than $206 million in net revenues!

The rate of change in the net revenues at Electronic Arts, Inc., may be modeled by

$$r(t) = 162.840t^3 - 823.299t^2 + 961.788t + 65.839 \text{ million dollars per year}$$

and the rate of change in cost of goods sold may be modeled by

$$c(t) = 66.288t^3 - 328.464t^2 + 355.674t + 61.435 \text{ million dollars per year}$$

where t is the number of years since the end of fiscal year 1998. (**Source:** Modeled from Electronic Arts, Inc., Annual Report; March 31, 2002.) By how much did the company's gross profits increase between the end of fiscal year 1998 and the end of fiscal year 2002? This problem may be interpreted graphically by calculating the area of the region between the graph of r and the graph of c.

In this section, we will discuss how to calculate the area of the bounded region between two graphs. We will also explain the meaning of the definite integral of the difference between two rate of change functions.

AREA BETWEEN TWO CURVES

If the graph of f lies above the graph of g on an interval $[a, b]$, then the area of the region between the two graphs from $x = a$ to $x = b$ is given by

$$\int_a^b [f(x) - g(x)]dx$$

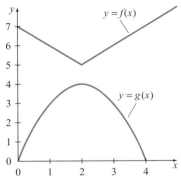

FIGURE 13.1

This result is relatively easy to verify graphically. Consider the graphs of the functions f and g shown in Figure 13.1. We see that on the interval $[0, 4]$, $f(x) > g(x)$. That is, the graph of f lies above the graph of g.

Consider $A = \int_1^3 [f(x) - g(x)]dx$. By the Sum and Difference Rule for integrals, we know that this equation is equivalent to

$$A = \int_1^3 f(x)\, dx - \int_1^3 g(x)\, dx$$
$$= (\text{area between } f \text{ and } x\text{-axis on } [1, 3])$$
$$- (\text{area between } g \text{ and } x\text{-axis on } [1, 3])$$

The shaded region in Figure 13.2 is the area between the graph of f and the x-axis on the interval $[1, 3]$.

The shaded region in Figure 13.3 is the area between the graph of g and the x-axis on the interval $[1, 3]$.

FIGURE 13.2

FIGURE 13.3

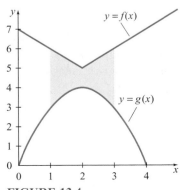

FIGURE 13.4

Subtracting the second area from the first area results in the area between the graphs of f and g on the interval $[1, 3]$ (Figure 13.4).

We'll do a few examples before returning to the Electronic Arts profit question.

Finding the Area of an Enclosed Region

Calculate the area of the region between the graphs of $f(x) = x^2 - 2x$ and $g(x) = x + 2$ on the interval $[0, 3]$.

SOLUTION We begin by graphing the functions together and shading the appropriate region (see Figure 13.5).

We see that the graph of $g(x) = x + 2$ lies above the graph of $f(x) = x^2 - 2x$ on the interval $[0, 3]$. We will calculate the area of the region bounded by the two graphs.

$$\int_0^3 [g(x) - f(x)] \, dx = \int_0^3 [(x + 2) - (x^2 - 2x)] \, dx$$

$$= \int_0^3 (-x^2 + 3x + 2) \, dx$$

$$= \left(-\frac{1}{3}x^3 + \frac{3}{2}x^2 + 2x \right) \Big|_0^3$$

$$= \left[-\frac{1}{3}(3)^3 + \frac{3}{2}(3)^2 + 2(3) \right] - (0)$$

$$= -9 + \frac{27}{2} + 6$$

$$= \frac{21}{2}$$

$$= 10.5$$

The area of the region between the two graphs is 10.5 square units.

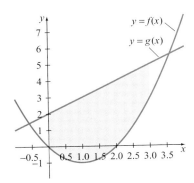

FIGURE 13.5

Finding the Area of an Enclosed Region

Calculate the area of the bounded region between the graphs of $f(x) = -x^2 + 2x + 1$ and $g(x) = x^2 - 2x + 1$.

SOLUTION We begin by graphing the functions together and shading the bounded region (Figure 13.6).

It appears that the graphs intersect at $x = 0$ and $x = 2$; however, we must confirm this algebraically.

$$f(x) = g(x)$$
$$-x^2 + 2x + 1 = x^2 - 2x + 1$$
$$0 = 2x^2 - 4x$$
$$2x(x - 2) = 0$$
$$x = 0, x = 2$$

Our algebraic solution confirms our graphical observation.

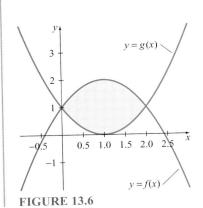

FIGURE 13.6

We see that the graph of f lies above the graph of g on the interval $[0, 2]$. We will calculate the area of the "eye-shaped" region between the two graphs.

$$
\begin{aligned}
A &= \int_0^2 [f(x) - g(x)]\,dx \\
&= \int_0^2 [(-x^2 + 2x + 1) - (x^2 - 2x + 1)]\,dx \\
&= \int_0^2 (-x^2 + 2x + 1 - x^2 + 2x - 1)\,dx \\
&= \int_0^2 (-2x^2 + 4x)\,dx \\
&= \left(-\frac{2}{3}x^3 + 2x^2 \right)\Big|_0^2 \\
&= \left[-\frac{2}{3}(2)^3 + 2(2)^2 \right] - 0 \\
&= \frac{8}{3}
\end{aligned}
$$

The area of the region between the two graphs is $2\frac{2}{3}$ square units.

EXAMPLE 3

Finding the Area of an Enclosed Region

Find the area of the bounded region between $f(x) = x^3 - 3x^2 + 3x$ and $g(x) = -x^2 + 4x - 2$.

SOLUTION We begin by graphing the functions together and shading the bounded regions (see Figure 13.7).

The graphs intersect in three different places: $x = -1$, $x = 1$, and $x = 2$. We can confirm our graphical observation by evaluating both f and g at these points and ensuring that the function values are equal at these x values.

FIGURE 13.7

TABLE 13.1

x	$f(x) = x^3 - 3x^2 + 3x$	$g(x) = -x^2 + 4x - 2$
-1	-7	-7
1	1	1
2	2	2

Our computational results, shown in Table 13.1, confirm our graphical observation.

There are two bounded regions. The first bounded region occurs on the interval $[-1, 1]$. On this interval, the graph of f lies above the graph of g. The second bounded region occurs on the interval $[1, 2]$. On this interval, the graph of g lies above the graph of f. The combined area of the regions is given by

$$A = \int_{-1}^{1} [f(x) - g(x)] \, dx + \int_{1}^{2} [g(x) - f(x)] \, dx$$

$$= \int_{-1}^{1} [f(x) - g(x)] \, dx - \int_{1}^{2} [f(x) - g(x)] \, dx \qquad \text{Factor out a } -1 \text{ to make the integrands equal}$$

Since both integrands are equal, we will simplify the expression $f(x) - g(x)$ and substitute the result into each integrand.

$$f(x) - g(x) = x^3 - 3x^2 + 3x - (-x^2 + 4x - 2)$$
$$= x^3 - 3x^2 + 3x + x^2 - 4x + 2$$
$$= x^3 - 2x^2 - x + 2$$

$$A = \int_{-1}^{1} [f(x) - g(x)] \, dx - \int_{1}^{2} [f(x) - g(x)] \, dx$$

$$= \int_{-1}^{1} (x^3 - 2x^2 - x + 2) \, dx - \int_{1}^{2} (x^3 - 2x^2 - x + 2) \, dx$$

$$A = \left(\frac{1}{4} x^4 - \frac{2}{3} x^3 - \frac{1}{2} x^2 + 2x \right) \Big|_{-1}^{1} - \left(\frac{1}{4} x^4 - \frac{2}{3} x^3 - \frac{1}{2} x^2 + 2x \right) \Big|_{1}^{2}$$

$$= \left\{ \left[\frac{1}{4}(1)^4 - \frac{2}{3}(1)^3 - \frac{1}{2}(1)^2 + 2(1) \right] - \left[\frac{1}{4}(-1)^4 - \frac{2}{3}(-1)^3 - \frac{1}{2}(-1)^2 + 2(-1) \right] \right\}$$

$$- \left\{ \left[\frac{1}{4}(2)^4 - \frac{2}{3}(2)^3 - \frac{1}{2}(2)^2 + 2(2) \right] - \left[\frac{1}{4}(1)^4 - \frac{2}{3}(1)^3 - \frac{1}{2}(1)^2 + 2(1) \right] \right\}$$

$$= \left[\left(\frac{13}{12} \right) - \left(-\frac{19}{12} \right) \right] - \left[\left(\frac{8}{12} \right) - \left(\frac{13}{12} \right) \right]$$

$$= \frac{37}{12} = 3 \frac{1}{12}$$

The combined area of the bounded regions between the two graphs is $3\frac{1}{12}$ square units.

Difference of Accumulated Changes

In Chapter 12, we saw that the Fundamental Theorem of Calculus had special meaning when working with a rate of change function f. If the units of f were miles per hour and t was in terms of hours, then $\int_{a}^{b} f(t) \, dt$ indicated that the total change in distance (in miles) between $t = a$ hours and $t = b$ hours was $F(b) - F(a)$. This notion may be applied to the difference of two rate of change functions as detailed in the following box.

DIFFERENCE OF TWO ACCUMULATED CHANGES

If f and g are continuous rate of change functions defined on the interval $[a, b]$, then the difference of the accumulated change of f from $x = a$ to $x = b$ and the accumulated change of g from $x = a$ to $x = b$ is given by

$$\int_{a}^{b} [f(x) - g(x)] \, dx$$

EXAMPLE 4

Calculating an Accumulated Change

Based on data from 1998 to 2002, the rate of change in the annual net revenues of Electronic Arts, Inc., may be modeled by

$$r(t) = 162.840t^3 - 823.299t^2 + 961.788t + 65.839 \text{ million dollars per year}$$

and the rate of change in annual cost of goods sold may be modeled by

$$c(t) = 66.288t^3 - 328.464t^2 + 355.674t + 61.435 \text{ million dollars per year}$$

where t is the number of years since the end of fiscal year 1998. (**Source:** Modeled from Electronic Arts, Inc., Annual Report, March 31, 2002.) By how much did the company's annual net revenues, cost of goods sold, and gross profits change between the end of fiscal year 1998 and the end of fiscal year 2002?

SOLUTION The total change in the annual net revenues of Electronic Arts, Inc., between the end of 1998 and the end of 2002 may be determined by calculating the sum of the signed areas of the shaded regions between the graph of r and the horizontal axis on the interval [0, 4] (see Figure 13.8).

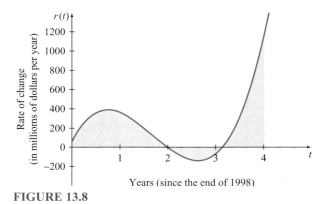

FIGURE 13.8

In other words, the change in the annual net revenues is given by

$$\int_0^4 r(t)\,dt = \int_0^4 (162.840t^3 - 823.299t^2 + 961.788t + 65.839)\,dt$$

$$= (40.710t^4 - 274.433t^3 + 480.894t^2 + 65.839t)\Big|_0^4$$

$$= [40.710(4)^4 - 274.433(4)^3 + 480.894(4)^2 + 65.839(4)] - (0)$$

$$= 10{,}421.76 - 17{,}563.712 + 7694.304 + 263.356$$

$$= 815.708 \text{ million dollars}$$

Between the end of fiscal year 1998 and the end of fiscal year 2002, annual net revenues increased by about $815,708,000.

The total change in the annual cost of goods sold between the end of fiscal year 1998 and the end of fiscal year 2002 may be determined by calculating the sum of the signed areas of the shaded regions between the graph of c and the horizontal axis on the interval [0, 4] (see Figure 13.9).

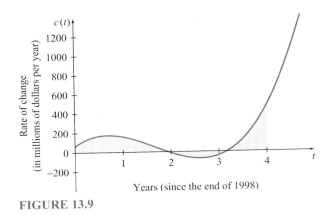

FIGURE 13.9

In other words, the total change in the annual cost of goods sold is given by

$$\int_0^4 c(t)\, dt = \int_0^4 (66.288t^3 - 328.464t^2 + 355.674t + 61.435)\, dt$$

$$= (16.572t^4 - 109.488t^3 + 177.837t^2 + 61.435t)\Big|_0^4$$

$$= [16.572(4)^4 - 109.488(4)^3 + 177.837(4)^2 + 61.435(4)] - (0)$$

$$= 4242.432 - 7007.232 + 2845.392 + 245.740$$

$$= 326.332 \text{ million dollars}$$

Between the end of fiscal year 1998 and the end of fiscal year 2002, the annual cost of goods sold increased by about $326,332,000.

The change in annual gross profit between the end of fiscal year 1998 and the end of fiscal year 2002 is given by

$$\int_0^4 [r(t) - c(t)]\, dt$$

since the rate of change in profit is the difference of the rate of change in revenue and the rate of change in cost. However,

$$\int_0^4 [r(t) - c(t)]\, dt = \int_0^4 r(t)\, dt - \int_0^4 c(t)\, dt$$

Therefore,

$$\int_0^4 [r(t) - c(t)]\, dt = 815.708 - 326.332$$

$$= 489.376 \text{ million dollars}$$

Between the end of fiscal year 1998 and the end of fiscal year 2002, the annual profit increased by about $489,376,000.

We have drawn the graphs of r and c together and shaded the bounded regions between the two graphs, as shown in Figure 13.10. These regions represent the accumulated change in annual profit.

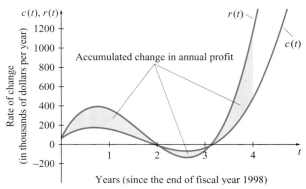

FIGURE 13.10

Note that although annual profits increased from the start of fiscal year 1998 through the end of fiscal year 2000 and from early in fiscal year 2002 through the end of fiscal year 2002, they decreased between the end of fiscal year 2000 and early in fiscal year 2002.

EXAMPLE 5

Calculating an Accumulated Change

Based on data from 1959 to 1989, the rate of change in the annual per capita income in Washington state may be modeled by

$$w(t) = -5.96t + 327.4 \text{ dollars per year}$$

and the rate of change in the per capita income in the United States may be modeled by

$$u(t) = 239.0 \text{ dollars per year}$$

where t is the number of years since the end of 1959. (**Source:** Modeled from www.census.gov.) In 1959, the per capita income in Washington state exceeded the national per capita income.

From the end of 1959 until the end of 1989, by how much did the pay gap between the Washington state and the national per capita income change?

SOLUTION We can calculate $\int_0^{30} w(t)\,dt$ and $\int_0^{30} u(t)\,dt$ independently and then calculate their difference, as was demonstrated in Example 4. However, it is often easier to calculate $\int_0^{30} [w(t) - u(t)]\,dt$.

$$w(t) - u(t) = (-5.96t + 327.4) - (239.0)$$
$$= -5.96t + 88.4$$

$$\int_0^{30} [w(t) - u(t)]\,dt = \int_0^{30} (-5.96t + 88.4)\,dt$$

$$= -2.98t^2 + 88.4t \Big|_0^{30}$$

$$= [-2.98(30)^2 + 88.4(30)] - [-2.98(0)^2 + 88.4(0)]$$
$$= -30$$

Between 1959 and 1989, the pay gap between the annual per capita income in Washington state and the national average annual per capita income changed by $30. Since the Washington state per capita income exceeded the national per capita income in 1959, the pay gap between the two was *reduced* by $30 between 1959 and 1989. (If the Washington state per capita income had been less than the national per capita income in 1959, the pay gap would have *increased* by $30 between 1959 and 1989.)

Let's look at the problem graphically. We begin by graphing both u and w (see Figure 13.11). The sum of the signed areas between the graph of u and the graph of w represents the change in the pay gap.

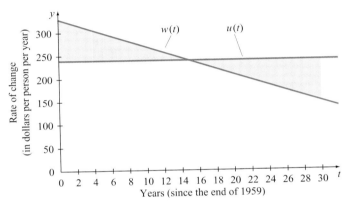

FIGURE 13.11

The sum of the signed areas of the shaded regions shown in the graph is -30.

Another way to look at the problem graphically is to graph the function $g(t) = w(t) - u(t)$, which represents the rate of change in the pay gap between the Washington state and the national per capita income (see Figure 13.12).

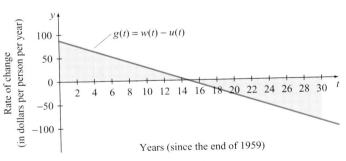

FIGURE 13.12

When $g(t) > 0$, the Washington state per capita income is growing faster than the national per capita income. When $g(t) < 0$, the Washington state per capita income is growing more slowly than the national per capita income. The sum of the signed areas of the shaded regions shown in the graph is -30. This means that, over the 30-year period from 1959 to 1989, the gap between the Washington state per capita income and the national per capita income was reduced by $30.

13.2 Summary

In this section, you learned how to calculate the area of the bounded region between two graphs. You also learned that the definite integral of the difference between two rate of change functions is equal to the difference between the accumulated changes in each of the individual functions.

13.2 Exercises

In Exercises 1–20, calculate the combined area of the region(s) bounded by the graphs of the two functions. (On some of the exercises, you may find it helpful to use your graphing calculator to locate the intersection points of the graphs.)

1. $f(x) = x^2; g(x) = x$

2. $f(x) = x^2 - 3x + 2; g(x) = 2$

3. $f(x) = x^2 - 4; g(x) = -x^2 + 4$

4. $f(x) = 2x^2 - 1; g(x) = 1$

5. $f(x) = x^2 - 4x - 5; g(x) = -3x - 5$

6. $f(x) = 2x^2 - 1; g(x) = x^3 - 1$

7. $f(x) = -x; g(x) = -x^2 + 3x$

8. $f(x) = x; g(x) = x^3 - x$

9. $f(x) = x^3 - 6x^2 + 5x; g(x) = 6x^2 + 50x$

10. $f(x) = 2x^2 + 2x + 4; g(x) = x^3 + 2x + 4$

11. $f(x) = x^2 - 2x; g(x) = x^3 - 3x^2$

12. $f(x) = x - 1; g(x) = x^2 - 2x$

13. $f(x) = x^3 - x; g(x) = -x^3 + x$

14. $f(x) = 2^x + x; g(x) = 3^x - x$

15. $f(x) = 3^x; g(x) = x^3$

16. $f(x) = 2^x; g(x) = x^3 + 1$

17. $f(x) = 2^x; g(x) = x^2$

18. $f(x) = 3\ln(x); g(x) = 2x - 4$

19. $f(x) = 2 + \ln(x); g(x) = 5x^2$

20. $f(x) = x^2 - 4; g(x) = \dfrac{\ln(x)}{x}$

In Exercises 21–25, graph the functions together on the same axes. Then determine the solution by using the integral techniques demonstrated in this section. (You may integrate each function separately and then calculate the difference, or, if you prefer, you may calculate the difference between the functions and then integrate the result.)

21. **Company Cost and Profit** Based on data from 1999 to 2001, the annual gross revenue of the Johnson & Johnson company and its subsidiaries may be modeled by
$$R(t) = 665t^2 + 1150t + 27{,}357$$
million dollars, and the annual cost of goods sold may be modeled by
$$C(t) = 103t^2 + 315t + 8539$$
million dollars, where t is the number of years since the end of 1999. (**Source:** Modeled from Johnson & Johnson 2001 Annual Report, p. 6.)

According to the models, what was the total profit earned between 1999 and 2001?

22. **Company Sales and Profit** Based on data from 1999 to 2001, the annual net sales of Gatorade/Tropicana North America may be modeled by
$$R(t) = -107t^2 + 496t + 3452$$
million dollars, and the annual operating profit may be modeled by
$$P(t) = -18.5t^2 + 85.5t + 433$$
million dollars, where t is the number of years since the end of 1999. (**Source:** Modeled from PepsiCo 2001 Annual Report, p. 44.)

According to the models, what were the total costs of the company between 1999 and 2001? (Assume that *annual net sales* is the same as *annual revenue*.)

23. 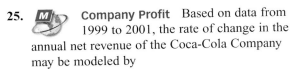 **Prison vs. College Population** Based on data from 1980 to 1998, the rate of change in the number of adults in prison, in jail, on probation, or on parole may be modeled by

$$r(t) = 238.1$$

thousand people per year, where t is the number of years since 1980. (**Source:** Modeled from *Statistical Abstract of the United States, 2001, Table 335, p. 202.*)

    Based on data from 1980 to 1998, the rate of change in the number of students enrolled in private colleges may be modeled by

$$s(t) = 1.839t + 23.25$$

thousand students per year, where t is the number of years since 1980. (**Source:** Modeled from *Statistical Abstract of the United States, 2001, Table 205, p. 133.*)

    Determine the accumulated change in the difference of the two populations from the end of 1980 through 1998. Interpret the meaning of the result.

24. 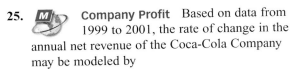 **Company Operating Costs** Based on data from 1999 to 2001, the rate of change in the annual net sales of Pepsi-Cola North America may be modeled by

$$R(t) = -131t - 749.5$$

million dollars per year, and the rate of change in the annual operating profit may be modeled by

$$P(t) = 12t + 76$$

million dollars per year, where t is the number of years since the end of 1999. (**Source:** Modeled from PepsiCo 2001 Annual Report, pp. 23 and 24.)

    Determine the accumulated change in annual operating costs from the end of 1999 through 2001.

25. 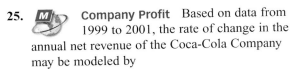 **Company Profit** Based on data from 1999 to 2001, the rate of change in the annual net revenue of the Coca-Cola Company may be modeled by

$$r(t) = -402t + 806$$

million dollars, and the rate of change in the annual cost of goods sold may be modeled by

$$c(t) = -355t + 372.5$$

million dollars, where t is the number of years since the end of 1999. (**Source:** Modeled from Coca-Cola Company 2001 Annual Report.)

    Determine the accumulated change in annual profit from the end of 1999 through 2001.

Exercises 26–30 are intended to challenge your understanding of calculating the area between curves. In each exercise, we are interested in the region that is bordered by all of the given functions.

26. What is the combined area of the regions bordered by all three functions $f(x) = x^2$, $g(x) = -x^2 + 4$, and $h(x) = 2x$?

27. What is the combined area of the regions bordered by the three functions $f(x) = 2^x$, $g(x) = 2^{-x}$, and $h(x) = -x + 4$?

28. What is the combined area of the regions bordered by the three functions $f(x) = x^2$, $g(x) = x$, and $h(x) = -x^2 + 1$?

29. What is the area of the region bordered by the three functions $f(x) = x^2$, $g(x) = 5x - 6$, and $h(x) = -x^2$?

30. What is the area of the region bordered by the four functions $f(x) = x$, $g(x) = -x + 4$, $h(x) = (x - 2)^2$, and $j(x) = 1.5$?

13.3 Differential Equations and Applications

- Solve introductory-level differential equations
- Apply Newton's Law of Heating and Cooling
- Analyze and interpret the real-life meaning of differential equations

GETTING STARTED In February 1992, a 79-year-old woman sustained third-degree burns on over 6 percent of her body when she spilled a cup of McDonald's coffee on herself. She requested that McDonald's pay the associated medical expenses, and when the company denied her request, she filed suit. During the trial, a McDonald's quality assurance manager testified that the company required that coffee in the pot be held at between 180 and 190 degrees Fahrenheit (°F). Liquids at 180 degrees will cause third-degree burns in two to seven seconds. If the temperature of the coffee had been 25 degrees cooler, the woman might have avoided serious injury. The jury initially awarded the woman $2.7 million in damages; however, a judge later reduced this amount to $480,000. Ultimately, the woman and McDonald's settled the case for an undisclosed amount. (**Sources:** www.lectlaw.com; www.consumerrights.net.)

How long does it take a 186-degree cup of coffee to cool to 155 degrees? Questions such as these may be answered using differential equations.

In this section, we will introduce differential equations and show how integral calculus may be used to solve them. Additionally, we'll discuss Newton's Law of Heating and Cooling and other applications of differential equations.

A **first-order differential equation** is an equation that relates an unknown function and its first derivative. In the first few examples, we will look at some simple first-order differential equations and their solutions.

EXAMPLE 1   Solving a Separable Differential Equation

Find a function $F(t)$ given that $\dfrac{dF}{dt} = F$.

SOLUTION   We are looking for a function that is its own derivative.

$$\frac{dF}{dt} = F$$

$$\frac{dF}{F} = dt \qquad \text{Multiply by } dt \text{ and divide by } F$$

$$\frac{1}{F} dF = dt \qquad \text{Rewrite } \frac{dF}{F} \text{ as } \frac{1}{F} dF$$

$$\int \frac{1}{F} dF = \int dt \qquad \text{Take the integral of both sides}$$

$$\ln|F| = t + C \qquad \text{Integrate}$$

$$e^{t+C} = |F| \qquad \text{Recall } \ln|y| = x \text{ is equivalent to } e^x = |y|$$

$$F = \pm e^t \cdot e^C \qquad \text{Rules of exponents and definition of absolute value}$$

$$= \pm e^C e^t$$

Letting $k = e^C$, we get the *general solution* of the differential equation:

$$F(t) = \pm k e^t$$

Any function of this form is a solution to the differential equation $\dfrac{dF}{dt} = F$. For example, $F(t) = 2e^t$, $F(t) = -3e^t$, and $F(t) = e^t$ are all *particular solutions* to the differential equation.

EXAMPLE 2

Finding a Particular Solution for a Separable Differential Equation

Find the particular solution of the differential equation $\dfrac{dF}{dt} = F^2$ given $F(0) = 4$.

SOLUTION

$$\frac{dF}{dt} = F^2$$

$$\frac{dF}{F^2} = dt \qquad \text{Divide by } F^2 \text{ and multiply by } dt$$

$$\frac{1}{F^2}dF = dt \qquad \text{Rewrite } \frac{dF}{F^2} \text{ as } \frac{1}{F^2}dF$$

$$F^{-2}dF = dt \qquad \text{Rewrite } \frac{1}{F^2}dF \text{ as } F^{-2}dF$$

$$\int F^{-2}dF = \int dt \qquad \text{Take the integral of both sides}$$

$$\frac{F^{-1}}{-1} = t + C \qquad \text{Integrate with Power Rule}$$

$$F^{-1} = -t - C \qquad \text{Multiply both sides by } -1$$

$$\frac{1}{F} = -t - C \qquad \text{Rewrite } F^{-1} \text{ as } \frac{1}{F}$$

$$F = \frac{1}{-t - C} \qquad \text{Divide by } -t - C \text{ and multiply by } F$$

This is the *general solution* to the differential equation. However, since we know that $F(0) = 4$, we will be able to find the *particular solution*.

$$F = \frac{1}{-t - C}$$

$$4 = \frac{1}{-0 - C} \qquad \text{Substitute } F = 4 \text{ and } t = 0$$

$$4 = -\frac{1}{C}$$

$$C = -\frac{1}{4}$$

Therefore, the particular solution is $F(t) = \dfrac{1}{-t + \frac{1}{4}}$. We can verify the accuracy of our work by differentiating this function.

$$\frac{d}{dt}[F(t)] = \frac{d}{dt}\left(\frac{1}{-t + \frac{1}{4}}\right)$$

$$\frac{dF}{dt} = \frac{d}{dt}\left(-t + \frac{1}{4}\right)^{-1}$$

$$= -1\left(-t + \frac{1}{4}\right)^{-2}(-1)$$

$$= \left(-t + \frac{1}{4}\right)^{-2}$$

$$= \frac{1}{\left(-t + \frac{1}{4}\right)^{2}}$$

$$= \left(\frac{1}{-t + \frac{1}{4}}\right)^{2}$$

$$= F^{2}$$

Since the result is equivalent to the differential equation we were given at the start of the problem, we believe that our calculated particular solution, $F(t) = \frac{1}{-t + \frac{1}{4}}$, is correct. To verify, we check to see that $F(0) = 4$.

$$F(0) = \frac{1}{-(0) + \frac{1}{4}}$$

$$= 4$$

We used the **separation of variables** process to solve the differential equations in Examples 1 and 2. We will now formally describe the process.

SOLVING SEPARABLE FIRST-ORDER DIFFERENTIAL EQUATIONS

A first-order differential equation is said to be **separable** if it may be written as

$$\frac{dy}{dx} = \frac{f(x)}{g(y)}$$

for some functions $f(x)$ and $g(y)$. The solution to the differential equation is obtained by moving the x and y variables to opposite sides of the equal sign and integrating. That is,

$$\int g(y)\, dy = \int f(x)\, dx$$

It is essential that all x terms be grouped with the dx and all y terms be grouped with the dy.

EXAMPLE 3   **Finding a General Solution to a Separable Differential Equation**

The rate of change of an investment account earning compound interest may be given by the first-order differential equation $\frac{dA}{dt} = kA$, where k is a positive constant. (If $k = 0$, the value of the investment remains constant. If k is negative, the investment is losing value.) Find the general solution to the differential equation.

SOLUTION We will solve the differential equation using the separation of variables process.

$$\frac{dA}{dt} = kA$$

$$\frac{dA}{A} = k\,dt \qquad\qquad\qquad \text{Separate the variables}$$

$$\int \frac{1}{A}\,dA = \int k\,dt \qquad\qquad \text{Take the integral of both sides}$$

$$\ln|A| = kt + C \qquad\qquad \text{Integrate}$$

$$\ln A = kt + C \qquad\qquad A, \text{ the account value, is assumed to be positive}$$

$$A = e^{kt+C} \qquad\qquad \text{Since, in general, } \ln A = b \text{ is equivalent to } A = e^{b}$$

$$= e^{kt} \cdot e^{C} \qquad\qquad \text{Rules of exponents}$$

$$= Pe^{kt} \text{ where } P = e^{C}$$

The general solution to the differential equation is $A = Pe^{kt}$.

We immediately recognize the result from Example 3 as the continuous compound interest formula. The standard compound interest formula $A = P\left(1 + \frac{r}{n}\right)^{nt}$ may be converted to the form $A = Pe^{kt}$ by letting

$$\left(1 + \frac{r}{n}\right)^{n} = e^{k}$$

$$k = \ln\left[\left(1 + \frac{r}{n}\right)^{n}\right]$$

A DIFFERENTIAL EQUATION FOR CONTINUOUS COMPOUND INTEREST

The value of an investment account earning continuous compound interest is

$$A = Pe^{kt} \text{ dollars}$$

where A is the value of the investment after t years, P is the initial value of the investment (in dollars), and k is the continuous interest rate.
The rate of change in the value of the investment is given by

$$\frac{dA}{dt} = kA \;\frac{\text{dollars}}{\text{year}}$$

EXAMPLE 4

Finding a Particular Solution to a Differential Equation

The rate of change of an investment account earning compound interest is given by $\frac{dA}{dt} = kA$, where k is a positive constant. The initial account value was $1000. At the end of the third year, the account value is $1120. Find the particular solution to the differential equation.

SOLUTION The general solution to the differential equation $\dfrac{dA}{dt} = kA$ is $A = Pe^{kt}$. We also know that $A(0) = 1000$ and $A(3) = 1120$.

$$A = Pe^{kt}$$
$$1000 = Pe^{k(0)}$$
$$1000 = P \cdot 1$$
$$P = 1000$$

Thus $A = 1000e^{kt}$. We will find the value of k by substituting in the second point, $(3, 1120)$.

$$A = 1000e^{kt}$$
$$1120 = 1000e^{k(3)}$$
$$1.12 = e^{3k}$$
$$\ln(1.12) = \ln(e^{3k})$$
$$0.1133 = 3k$$
$$k = 0.03778$$

Therefore, $A = 1000e^{0.03778t}$ is the particular solution to the differential equation. This equation may also be written as $A = 1000(1.0385)^t$, since $e^{0.03778} = 1.0385$. The account is earning interest at a rate of 3.778 percent compounded continuously or a rate of 3.85 percent compounded annually.

Newton's Law of Heating and Cooling

The world-renowned scientist and scholar Sir Isaac Newton determined that the rate of change in an object's temperature is proportional to the difference between the constant temperature of the environment surrounding the object and the object's temperature. This observation is summarized in **Newton's Law of Heating and Cooling.**

NEWTON'S LAW OF HEATING AND COOLING

Let T be the temperature of an object at time t and A be the temperature of the environment surrounding the object (ambient temperature). Then

$$\frac{dT}{dt} = k(T - A)$$

where k is a constant that varies depending upon the physical properties of the object.

According to Newton's Law of Heating and Cooling,

$$\frac{dT}{dt} = k(T - A)$$

We will use substitution to rewrite the equation.

We define the function $y = T - A$ and differentiate this function with respect to t.

$$\frac{d}{dt}(y) = \frac{d}{dt}(T - A)$$

$$\frac{dy}{dt} = \frac{d}{dt}(T) - \frac{d}{dt}(A)$$

$$= \frac{dT}{dt} - \frac{dA}{dt}$$

$$= \frac{dT}{dt} - 0 \qquad \qquad \frac{dA}{dt} = 0 \text{ since } A \text{ is a constant}$$

$$= \frac{dT}{dt}$$

So $\frac{dy}{dt} = \frac{dT}{dt}$. Rewriting the equation, $\frac{dT}{dt} = k(T - A)$, in terms of y, we get

$$\frac{dy}{dt} = ky$$

since $\frac{dT}{dt} = \frac{dy}{dt}$ and $T - A = y$. Separating the variables, we rewrite the equation as

$$\frac{1}{y}dy = k\,dt$$

Integrating both sides of the equation, we get

$$\int \frac{1}{y}dy = \int(k\,dt)$$

$$\ln|y| = kt + C$$

Recall that if $x = \ln|y|$, then $e^x = |y|$. Consequently,

$$\ln|y| = kt + C$$

$$|y| = e^{kt+C}$$

$$y = \pm(e^{kt} \cdot e^C)$$

Now we'll back substitute $y = T - A$.

$$T - A = (e^{kt} \cdot e^C)$$

$$T = A + (e^{kt} \cdot e^C)$$

$$T = A + Se^{kt} \text{ where } S = e^C$$

Thus the general solution to the differential equation $\frac{dT}{dt} = k(T - A)$ is

$T = A + Se^{kt}$.

EXAMPLE 5

Using a Differential Equation to Model Water Temperature

To test Newton's Law of Heating and Cooling, the author heated a cup of water to 186°F and placed it in a 76°F room. The temperature of the water was recorded at two-minute time intervals as shown in Table 13.2.

TABLE 13.2 Hot Water Cooling in 76°F Room

Minutes (t)	Temperature (degrees Fahrenheit) (T)
0	186
2	175
4	167
6	159
8	153
10	147
12	142
14	138

Use Newton's Law of Heating and Cooling to find a model for the water temperature.

SOLUTION We know that the general solution to the differential equation given in Newton's Law of Heating and Cooling is $T = A + Se^{kt}$. We'll substitute in values from the table to determine the numeric values of k and C. We know that $A = 76$, since the ambient temperature is 76 degrees Fahrenheit. Additionally, we know that at time $t = 0$, $T = 186$.

$$T = 76 + Se^{kt}$$
$$186 = 76 + Se^{k(0)}$$
$$110 = Se^0$$
$$110 = S$$

Our equation $T = A + Se^{kt}$ may now be rewritten as

$$T(t) = 76 + 110e^{kt}$$

We'll use the table values $t = 14$ and $T = 138$ to find the value of k. (We can select any pair of table values. Since we're developing a model, selecting a different pair of table values would slightly alter the model equation.)

$$138 = 76 + 110e^{k(14)}$$
$$62 = 110e^{14k}$$
$$0.5636 = e^{14k}$$
$$\ln(0.5636) = 14k \qquad \text{Log both sides}$$
$$k = -0.04095$$

Since k is negative, the temperature of the water is dropping at a continuous rate of $0.04095°F$ per minute. The temperature of the water may be modeled by

$$T(t) = 76 + 110e^{-0.04095t} \text{ degrees Fahrenheit}$$

Let's see how well this function fits the table of data.

TABLE 13.3 Hot Water Cooling in 76°F Room

Minutes (t)	Actual Temperature (degrees Fahrenheit) (T)	Model Temperature (degrees Fahrenheit) (M)
0	186	186
2	175	177
4	167	169
6	159	162
8	153	155
10	147	149
12	142	143
14	138	138

Newton's Law of Heating and Cooling gave a remarkably good estimate for the measured temperature. The discrepancies between the two may be due to measurement device error or other physical phenomena. As with any mathematical model, some error is to be expected.

EXAMPLE 6

Using a Differential Equation to Calculate Cooling Rates

According to the model developed in Example 5, at what rate was the water cooling 4, 8, and 12 minutes into the cooling period?

SOLUTION The model for the water temperature was given by

$$T(t) = 76 + 110e^{-0.04095t} \text{ degrees}$$

t seconds after the temperature of the water was measured to be 186 degrees. The rate of change in the temperature is given by the differential equation

$$\frac{dT}{dt} = k(T - A)$$

$$= -0.04095(T - 76)$$

We're asked to evaluate this function at $t = 4$, $t = 8$, and $t = 12$.

$$\frac{dT}{dt}\bigg|_{t=4} = -0.04095[T(4) - 76]$$

$$= -0.04095(169.4 - 76)$$

$$= -3.82$$

Four minutes into the cooling period, the water was cooling at a rate of 3.82 degrees per minute.

$$\frac{dT}{dt}\bigg|_{t=8} = -0.04095[T(8) - 76]$$
$$= -0.04095(155.3 - 76)$$
$$= -3.25$$

Eight minutes into the cooling period, the water was cooling at a rate of 3.25 degrees per minute.

$$\frac{dT}{dt}\bigg|_{t=12} = -0.04095[T(12) - 76]$$
$$= -0.04095(143.3 - 76)$$
$$= -2.76$$

Twelve minutes into the cooling period, the water was cooling at a rate of 2.76 degrees per minute.

Assuming that coffee cools at the same rate as water, we estimate that it would take a cup of McDonald's coffee eight minutes to cool from 186 degrees to 155 degrees Fahrenheit. However, the size and shape of the coffee cup, the addition of cream and sugar, and other environmental variables could dramatically affect the accuracy of our estimate.

13.3 Summary

In this section, you learned how to use integral calculus to solve first-order differential equations. You also learned how to apply differential equations in Newton's Law of Heating and Cooling and in the continuous compound interest formula.

13.3 Exercises

In Exercises 1–10, find the general solution of the separable differential equation.

1. $\dfrac{dy}{dx} = \dfrac{x}{y}$

2. $\dfrac{dA}{dt} = 3A^2$

3. $\dfrac{dy}{dx} = \sqrt{y}$

4. $\dfrac{dA}{dt} = \dfrac{3A}{t}$

5. $\dfrac{dy}{dx} = \dfrac{y^2}{x^2}$

6. $\dfrac{dA}{dt} = 4(20 - A)$

7. $\dfrac{dy}{dx} = 0.25(y - 4)$

8. $\dfrac{dy}{dx} = 4y$

9. $\dfrac{dy}{dx} = 4y - y^2$

(*Hint:* $\displaystyle\int \frac{dy}{y(M - y)} = \frac{1}{M} \ln \left| \frac{y}{M - y} \right|$.)

10. $\dfrac{dA}{dt} = 6 - 3A$

In Exercises 11–20, find the particular solution for the differential equations.

(Hint: $\int \dfrac{dy}{y(M-y)} = \dfrac{1}{M}\ln\left|\dfrac{y}{M-y}\right|$.)

11. $\dfrac{dy}{dx} = \dfrac{2x}{y}$; $y(1) = 4$

12. $\dfrac{dA}{dt} = 4A^2$; $A(0) = -1$

13. $\dfrac{dy}{dx} = 3y$; $y(0) = -1$ **14.** $\dfrac{dA}{dt} = \dfrac{2A}{t}$; $A(1) = 1$

15. $\dfrac{dy}{dx} = 25(y-5)$; $y(0) = 0$

16. $\dfrac{dA}{dt} = 4 - 2A$; $A(0) = 0$

17. $\dfrac{dy}{dx} = y(4-y)$; $y(0) = \dfrac{1}{2}$

18. $\dfrac{dA}{dt} = A^2 - 4A$; $A(0) = 8$

19. $\dfrac{dy}{dx} = 2y - y^2$; $y(0) = 100$

20. $\dfrac{dA}{dt} = A(10 - A)$; $A(0) = 5$

In Exercises 21–28, use Newton's Law of Heating and Cooling to find the equation of the model that best fits the data. Then answer the given questions.

21. **Water Temperature** The author placed a cup of ice water in a 76°F room and recorded the temperature of the water at 10-minute intervals as shown in the table.

Cold Water Warming in a 76°F Room

Minutes (t)	Temperature (degrees Fahrenheit) (T)
0	34.3
10	40.3
20	43.8
30	48.2
40	51.8
50	54.7

(a) Find a model for the temperature of the water at time t minutes.

(b) Determine the rate of change in the water's temperature at 10 minutes and 50 minutes into the warming period.

22. Water Temperature How long will it take to freeze a 60°F cup of water in a 10°F freezer? (At 32°F, water freezes.) (Since the result will vary depending upon the quantity of water and the shape of the container, assume that for the container in question, the water temperature will be 46°F at the end of the eighth minute.)

23. **Cake Temperature** Cakes are typically baked at 350°F. If a pan filled with cake batter is moved from a 75°F kitchen into a 350°F oven, how long will it take for the cake batter to heat up to 200°F? (Since the result will vary depending upon the quantity of the batter and the shape of the pan, assume that for the container in question, the batter temperature will be 100°F at the end of the third minute.)

24. **Cake Temperature** A cake is removed from a 350°F oven and placed on a cooling rack in a 75°F room. If the temperature of the cake was 350°F when it was removed from the oven and its temperature is 250°F two minutes later, how long will it take for the cake to cool to 100°F?

25. Time of Death When the forensic investigator arrives upon the scene of a homicide, one of the most important things to determine is the time of death of the victim. Suppose that the victim's internal body temperature was 84.5°F when the body was discovered in a 70°F room. Thirty minutes later, the person's body temperature had dropped to 83.6°F. If the person's internal body temperature was 98°F at the time of death, how long had the person been dead when the forensic investigator arrived on the scene?

26. According to the model developed in Exercise 25, at what rate was the corpse's temperature dropping 1 hour, 2 hours, and 4 hours into the cooling period?

27. Time of Death Suppose that a homicide victim's internal body temperature was 59.1°F when the body was discovered in a 30°F room. Thirty minutes later, the person's body temperature had dropped to 56.9°F. If the person's internal body temperature was 98°F at the time of death, how long had the person been dead when the forensic investigator arrived on the scene?

28. According to the model developed in Exercise 27, at what rate was the corpse cooling 4 hours, 6 hours, and 12 hours into the cooling period?

Exercises 29–32 deal with the notion of continuous growth.

29. Investment Value A $1000 investment has a continuous interest rate of 10 percent. Write an equation for the value of the investment after t years.

30. Investment Value The value of an investment is modeled by $A = 2000e^{0.08t}$ dollars, where t is the number of years since 2006. What will be the annual rate of change in the value of the investment when the account value is $3000? $4000? $5000?

31. Investment Value The rate of change of an investment account earning continuous compound interest is given by
$$\frac{dA}{dt} = kA$$
where k is a positive constant. The initial account value was $2500. At the end of the third year, the account value was $4200. Find the particular solution to the differential equation.

32. Investment Value The rate of change of an investment account earning continuous compound interest is given by
$$\frac{dA}{dt} = kA$$
where k is a positive constant. The initial account value was $420. At the end of the fifth year, the account value was $630. Find the particular solution to the differential equation.

Exercises 33–35 are intended to challenge your understanding of differential equations.

33. Show that the derivative of $A = Pe^{kt}$ is equivalent to kA.

34. Find the general solution of the differential equation.
$$2y + y'x^2 = 5$$

35. Find the general solution of the differential equation.
$$3x^2y^2 + 2yy'x^3 = 0$$

13.4 Differential Equations: Limited Growth and Logistic Models

- Find the equation of a limited growth model
- Find the equation of a logistic growth model
- Analyze and interpret the real-life meaning of differential equations

GETTING STARTED   On September 11, 2001, many of us woke up to the horrific news of the terrorist attack on the World Trade Center. News bulletins announcing the attack flooded the radio and television channels. Initially, the news of the attack spread very rapidly; however, as the day progressed, the news spread more slowly, since most people had already received word of the attack. The spread of information, such as the news of the September 11 attack, may often be modeled mathematically. In this section, we will continue our discussion of differential equations and show how they can be used to find limited growth and logistic models.

Limited Growth Model

Sociologists often assume that the rate at which news spreads via mass media is proportional to the number of people who have not yet heard the news. This type of growth is called **limited growth.** For a limited growth function y with a maximum value of M, the rate of change, $\frac{dy}{dt}$, is proportional to the difference between the present value of y and M. That is, $\frac{dy}{dt} = k(M - y)$.

EXAMPLE 1

Using Differential Equations to Forecast the Spread of Information

At 8:46 a.m. on September 11, 2001, the hijacked American Airlines Flight 11 slammed into the North Tower of the World Trade Center in New York City. Eighteen minutes later, United Airlines Flight 175 crashed into the South Tower of the World Trade Center. By the end of the day, thousands of people were dead as a result of the deadliest terrorist attack in American history.

Suppose that in a community of 10,000 people, 3000 people were watching television or listening to the radio when the news broke on September 11, 2001. For the purpose of the model, we'll assume that all 3000 people heard the news simultaneously *five* minutes after the first attack. (It takes time for news crews to reach the site of any newsworthy event.) Find the equation for the limited growth model that models the spread of the news of the attack. Then determine how long it took for 90 percent of the community to hear the news and at what rate people were hearing the news at that time.

SOLUTION The maximum number of people that can be informed is 10,000, so $M = 10,000$. Additionally, we know that when $t = 5$, $y = 3000$. We have

$$\frac{dy}{dt} = k(M - y)$$

$$\frac{dy}{dt} = k(10,000 - y)$$

We can use the separation of variables process to find the equation of the general solution and then use the point (5, 3000) to find the particular solution.

$$\frac{dy}{dt} = k(10,000 - y)$$

$$\frac{dy}{(10,000 - y)} = k\,dt$$

Let $u = 10,000 - y$; then $du = -dy$. Rewriting the left-hand side of the equation in terms of u, we get

$$\frac{-du}{u} = k\,dt$$

$$-\int \frac{1}{u}\,du = \int k\,dt$$

$$-\ln|u| = kt + C$$

$$-\ln u = kt + C \qquad \text{Since } u > 0$$

$$\ln u = -kt - C$$

$$u = e^{-kt - C}$$

$$10,000 - y = e^{-kt}e^{-C} \qquad \text{Since } u = 10,000 - y$$

$$y = 10,000 - Se^{-kt} \qquad \text{Where } S = e^{-C}$$

At $t = 0$, nobody had heard the news of the attack because the information hadn't yet been broadcasted to the community. (It wasn't until five minutes after the crash that the first 3000 people in the community heard the news.) We can use this information to find the value of S.

$$y(0) = 10{,}000 - Se^{-k(0)}$$
$$0 = 10{,}000 - Se^0$$
$$S = 10{,}000$$

A general solution to the differential equation is

$$y = 10{,}000 - 10{,}000e^{-kt}$$
$$= 10{,}000(1 - e^{-kt})$$

We'll now use the point (5, 3000) to find the particular solution.

$$3000 = 10{,}000(1 - e^{-k(5)})$$
$$0.3 = 1 - e^{-5k}$$
$$-0.7 = -e^{-5k}$$
$$0.7 = e^{-5k}$$
$$\ln(0.7) = -5k$$
$$k = \frac{\ln(0.7)}{-5}$$
$$= 0.07133$$

The limited growth model that fits the data is given by

$$y = 10{,}000(1 - e^{-0.07133t})$$

Since 90 percent of 10,000 is 9000, we need to determine when 9000 people had heard the news.

$$9000 = 10{,}000(1 - e^{-0.07133t})$$
$$0.9 = 1 - e^{-0.07133t}$$
$$-0.1 = -e^{-0.07133t}$$
$$0.1 = e^{-0.07133t}$$
$$\ln(0.1) = -0.07133t$$
$$t = \frac{\ln(0.1)}{-0.07133}$$
$$= 32.28$$

We estimate that 32 minutes after the attack, 90 percent of the people in the community had heard the news. We'll now determine at what rate the word was spreading.

$$\frac{dy}{dt} = k(10{,}000 - y)$$
$$= 0.07133(10{,}000 - y)$$
$$\left.\frac{dy}{dt}\right|_{t=32.28} = 0.07133(10{,}000 - 9000) \qquad \text{[Note: } y(32.28) = 9000\text{]}$$
$$= 71.33$$

Approximately 32 minutes after the attack, the news was spreading at a rate of roughly 71 people per minute.

In Example 1, we solved the given differential equation by using the separation of variables method. Using this same approach, we can derive the general solution to the differential equation $\dfrac{dy}{dt} = k(M - y)$.

LIMITED GROWTH

Assume that the rate of growth of a function y with maximum value M is proportional to the difference between the present value of y and M. That is,

$$\frac{dy}{dt} = k(M - y)$$

The solution to the differential equation with initial condition

$$y(0) = 0$$

is given by

$$y = M(1 - e^{-kt})$$

and has the graph shown in Figure 13.13.

FIGURE 13.13

A limited growth model may be used to model the cumulative number of people infected by a disease during a bioterrorist attack once quarantine efforts are underway. In the journal article *Modeling Potential Responses to Smallpox as a Bioterrorist Weapon*, scientists at the Centers for Disease Control and Prevention and at Don Millar and Associates demonstrated the catastrophic effect of a smallpox attack. Table 13.4 shows how rapidly the cumulative number of smallpox cases would increase should there be no intervention.

TABLE 13.4 Estimates of Cumulative Total Smallpox Cases After 365 Days with No Intervention

Number Initially Infected[a]	Number Infected per Infectious Person[b]	Cumulative Total Number of Smallpox Cases, Days Postrelease[c]			
		30 Days	90 Days	180 Days	365 Days
10	1.5	31	214	2,190	224 thousand
10	3.0	64	4,478	2.2 million	774 billion
1,000	1.5	3,094	21,372	219,006	22 million
1,000	3.0	6,397	447,794	222 million	77 trillion

[a]Number initially infected refers to those who are exposed during a release so that they subsequently become infectious to others. This scenario excludes those who are exposed but either do not become ill (i.e., are immune or are not exposed to an infectious dose) or do not become infectious (residual immunity from prior vaccination may be sufficient to prevent onward transmission).
[b]The number of persons infected per infectious person is the transmission rate.
[c]Assumes an unlimited supply of smallpox-susceptible persons.

Source: *Emerging Infectious Diseases,* vol. 7, no. 6, November–December 2001, p. 959.

Figure 13.14 shows the effectiveness of adequate intervention once smallpox is detected. If 25 percent of the people infected are quarantined daily, the cumulative number of people infected will continue to increase at an increasing rate; however, if 50 percent of the people infected are quarantined daily, the cumulative number of people infected will increase at a decreasing rate. The graphs in the bottom right-hand corner of the figure show that the cumulative number of people infected may be modeled by a limited growth function when 50 percent of the infected individuals are quarantined daily.

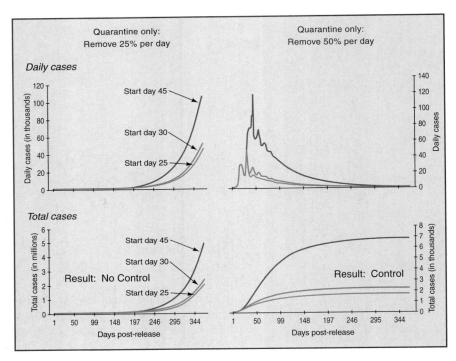

Figure 4. Daily and total cases of smallpox after quarantining infectious persons at two daily rates and three post-release start dates. The graphs demonstrate that if quarantine is the only intervention used, a daily removal rate of ≥ 50% is needed to stop transmission within 365 days post-release. At a 25% daily removal rate of infectious persons by quarantine, a cohort of all those entering the first day of overt symptoms (i.e., rash) is entirely removed within 17 days (18 to 20 days post-incubation) after the first day of overt symptoms, with 90% removed within 9 days. At a 50% daily removal of infectious persons by quarantine, a cohort of all those entering their first day of overt symptoms (i.e., rash) is entirely removed within 7 days (8 to 10 days post-incubation) after the first day of overt symptoms, with 90% removed within 4 days. The daily rate of removal (quarantine) relates only to the removal of those who are infectious (i.e., overtly symptomatic). The rate does not include any persons who may be quarantined along with overtly symptomatic patients, such as unvaccinated household contacts. Data generated by assuming 100 persons initially infected and a transmission rate of 3 persons infected per infectious person. For clarity, the graphs of daily cases do not include the assumed 100 initially infected persons. The graphs of total cases include the 100 initially infected.

FIGURE 13.14

Source: *Emerging Infectious Diseases,* vol. 7, no. 6, November–December 2001, p. 959.

EXAMPLE **2**

Using Differential Equations to Forecast the Effect of a Bioterrorist Attack

Reading from the graph in the bottom right corner of Figure 13.14, we can construct a limited growth model. It appears that a maximum of 7000 people will be infected with smallpox as a result of a bioterrorist attack. It also appears that 99 days after the attack, 4750 people are infected. Do the following:

(a) Determine the equation of the limited growth model. (This models the cumulative number of people infected if 50 percent of the infected people are quarantined daily beginning on the 45th day after the attack.)

(b) Determine the rate at which the disease is spreading on the 50th and 100th days after the attack.

SOLUTION

(a) The equation of the limited growth model is given by

$$y = M(1 - e^{-kt})$$

Since a maximum of 7000 people will be infected, $M = 7000$.

$$y = 7000(1 - e^{-kt})$$

Additionally, we know that $(99, 4750)$ lies on the graph of the function.

$$4750 = 7000(1 - e^{-k(99)})$$
$$0.6786 = 1 - e^{-k(99)}$$
$$-0.3214 = -e^{-99k}$$
$$0.3214 = e^{-99k}$$
$$\ln(0.3214) = -99k$$
$$k = \frac{\ln(0.3214)}{-99}$$
$$= 0.01147$$

Thus $y = 7000(1 - e^{-0.01147t})$ models the number of people infected t days after an attack. (*Note:* Since we are modeling the equation from the graph, our model lacks precision. In addition, the smallpox graph doesn't appear to pass through $(0, 0)$, and our model does. However, despite the limitations of our model, it does give us a rough idea of what would happen in a bioterrorist smallpox attack.)

(b) We can use the differential equation to estimate the rate at which the disease is spreading.

$$\frac{dy}{dt} = k(M - y)$$
$$= 0.01147(7000 - y)$$
$$\left.\frac{dy}{dt}\right|_{t=50} = 0.01147[7000 - y(50)]$$
$$= 0.01147(7000 - 3055)$$
$$= 45$$

On the 50th day, we estimate that the disease is spreading at a rate of 45 people per day. That is, approximately 45 additional people are infected between the 50th and 51st days.

$$\left.\frac{dy}{dt}\right|_{t=100} = 0.01147[7000 - y(100)]$$
$$= 0.01147(7000 - 4777)$$
$$= 25$$

On the 100th day, we estimate that the disease is spreading at a rate of 25 people per day. That is, approximately 25 additional people are infected between the 100th and 101st days.

Commercial fish growers seek to create a fish habitat that will maximize the size of each fish in the least amount of time. The Von Bertalanffy Limited Growth model is widely used in the fish industry. The model assumes that there is a maximal length, L_∞, that a fish will attain under optimal conditions and that the rate of change in the length of the fish, $\dfrac{dL}{dt}$, is proportional to the difference between the maximal length and the current length. That is, $\dfrac{dL}{dt} = k(L_\infty - L)$. Solving this differential equation, we can find the general form of the Von Bertalanffy Limited Growth model.

$$\frac{dL}{dt} = k(L_\infty - L)$$

$$\frac{dL}{L_\infty - L} = k\,dt \qquad\qquad \text{Separation of variables}$$

$$\int \frac{1}{L_\infty - L}\,dL = \int k\,dt$$

$$-\ln|L_\infty - L| = kt + C \qquad\qquad \text{Since } \frac{d}{dL}[-\ln(L_\infty - L)] = \frac{1}{L_\infty - L}$$

$$\ln(L_\infty - L) = -kt - C \qquad\qquad \text{Since } L_\infty - L > 0$$

$$L_\infty - L = e^{-k(t+C)}$$

$$L = L_\infty - e^{-k(t+C)}$$

The solution to the differential equation is commonly written as $L = L_\infty(1 - e^{-k(t-t_0)})$, where t_0 is a negative constant that varies from species to species. The constant t_0 has the property that $L(t_0) = 0$. Our solution may be converted from $L = L_\infty - e^{-k(t+C)}$ to $L = L_\infty(1 - e^{-k(t-t_0)})$ by letting $C = \dfrac{-\ln(L_\infty)}{k} - t_0$. (No matter what the value of C is, we can always find a t_0 that makes this equality true.)

EXAMPLE   3

Forecasting Fish Size with a Von Bertalanffy Limited Growth Model

Turkish scientists conducted a study on the Eastern Black Sea between 1991 and 1996 and determined the values of k, L_∞, and t_0 for economically important fish species in the region. (**Source:** www.tagem.gov.tr.) The whiting fish species has the Von Bertalanffy growing constants given in Table 13.5.

Whiting	k	L_∞	t_0
Female	0.11	43.3 cm	-1.91 years
Male	0.136	34.2 cm	-2.02 years

Source: www.fishbase.org.

(a) According to the model, how long is a two-year-old female whiting?
(b) According to the model, how long is a two-year-old male whiting?
(c) Does the two-year-old female whiting grow faster than the two-year-old male?

SOLUTION

(a) We know that $L(t) = L_\infty(1 - e^{-k(t-t_0)})$. We have $L_\infty = 43.3$, $k = 0.11$, and $t_0 = -1.91$. Substituting these values into the Von Bertalanffy Limited Growth model yields $L(t) = 43.3(1 - e^{-0.11[t-(-1.91)]})$. We determine the length of a two-year-old female whiting by evaluating this function at $t = 2$.

$$L(2) = 43.3(1 - e^{-0.11[2-(-1.91)]})$$
$$= 43.3(1 - e^{-0.11(3.91)})$$
$$= 43.3(0.3496)$$
$$\approx 15.1$$

We estimate the length of a two-year-old female whiting to be about 15.1 cm.

(b) We know that $L(t) = L_\infty(1 - e^{-k(t-t_0)})$. We have $L_\infty = 34.2$, $k = 0.136$, and $t_0 = -2.02$. Substituting these values into the Von Bertalanffy Limited Growth model yields $L(t) = 34.2(1 - e^{-0.136[t-(-2.02)]})$. We determine the length of a two-year-old male whiting by evaluating this function at $t = 2$.

$$L(2) = 34.2(1 - e^{-0.136[2-(-2.02)]})$$
$$= 34.2(1 - e^{-0.136(4.02)})$$
$$= 34.2(0.4212)$$
$$\approx 14.4$$

We estimate the length of a two-year-old male whiting to be about 14.4 cm.

(c) Finally, we need to evaluate each of the differential equations at $t = 2$.
Recall that the general form of the differential equation is $\dfrac{dL}{dt} = k(L_\infty - L)$.

Female:

$$\left.\frac{dL}{dt}\right|_{t=2} = 0.11[43.3 - L(2)]$$
$$= 0.11(43.3 - 15.1)$$
$$= 3.1$$

A two-year-old female whiting grows at a rate of 3.1 cm per year.

Male:

$$\left.\frac{dL}{dt}\right|_{t=2} = 0.136[34.2 - L(2)]$$
$$= 0.136(34.2 - 14.4)$$
$$= 2.7$$

A two-year-old male whiting grows at a rate of 2.7 cm per year.

From our calculations we conclude that the two-year-old female whiting grows faster than the two-year-old male whiting.

Logistic Growth

When a successful product is introduced to the market, cumulative product sales often increase very slowly initially; however, as the popularity of the product increases, sales increase rapidly. Then, as the market becomes saturated with the product, cumulative sales growth again slows. This type of growth is called **logistic growth.**

LOGISTIC GROWTH

Assume that the rate of growth of a function y with maximum value M is proportional to the product of the present value of y and the difference between the present value of y and M. That is,

$$\frac{dy}{dt} = ky(M - y)$$

The solution to this differential equation with initial condition

$$y(0) = \frac{M}{1 + S}$$

is given by

$$y = \frac{M}{1 + Se^{-kMt}}$$

The graph of the solution is an s-shaped (sigmoidal) graph (see Figure 13.15).

FIGURE 13.15

For example, consider the digital video disc (DVD) player. The DVD player was introduced into the market in the first quarter of 1997. By the end of 2002, nearly 50 percent of U.S. homes had DVD capability. (**Source:** DVD Entertainment Group.) Looking at a scatter plot of the cumulative number of DVD players sold (Figure 13.16), it initially looks as if the cumulative number of DVD players sold is increasing exponentially.

FIGURE 13.16

However, when we attempt to model the data with an exponential function (Figure 13.17), we see that the exponential function ends up growing much more rapidly than the data.

FIGURE 13.17

However, if we model the data with a logistic function (Figure 13.18), we see that DVD player sales appear to exhibit logistic growth behavior.

FIGURE 13.18

Using logistic regression on the TI-83 Plus, we determine that the logistic model for the given DVD player data is given by

$$D(t) = \frac{68{,}571}{1 + 288.8e^{-0.28708t}} \text{ thousand DVD players}$$

where t is the number of quarters since the end of the first quarter of 1997.

Plotting the graph of D together with the scatter plot of the data (Figure 13.19), we see that the solution fits the data remarkably well.

FIGURE 13.19

How rapidly were DVD player sales increasing at the end of the first quarter of 2000? We could differentiate $D(t) = \dfrac{68,571}{1 + 288.8e^{-0.28708t}}$ and evaluate the derivative at $t = 12$. However, it may be easier to calculate $D'(12)$ using the differential equation.

$$\frac{dD}{dt} = kD(M - D)$$

Since $-0.28708t = -Mkt$, $k = \dfrac{0.28708}{M}$. Since $M = 68,571$, $k = \dfrac{0.28708}{68,571}$. The differential equation may now be written as follows:

$$\frac{dD}{dt} = \left(\frac{0.28708}{68,571} \right) D(68,571 - D)$$

$$= 0.28708D - 0.000004187D^2$$

$$\left. \frac{dD}{dt} \right|_{t=12} = 0.28708D(12) - 0.000004187[D(12)]^2$$

$$D'(12) = 0.28708(6713.1) - 0.000004187(6713.1)^2$$

$$= 1927.20 - 188.69$$

$$\approx 1739$$

At the end of the first quarter of 2000, the cumulative number of DVD players sold was increasing at a rate of 1739 thousand DVD players per quarter. That is, from the end of the first quarter of 2000 to the end of the second quarter of 2000, we estimate that 1739 thousand DVD players were sold.

How rapidly will DVD player sales be increasing at the end of the first quarter of 2006? That is, what is $D'(36)$?

$$\left. \frac{dD}{dt} \right|_{t=36} = 0.28708D(36) - 0.000004187[D(36)]^2$$

$$D'(36) = 0.28708(67,934) - 0.000004187(67,934)^2$$

$$= 19,502 - 19,323$$

$$= 179$$

At the end of the first quarter of 2006, we forecast that the cumulative number of DVD players sold will be increasing at a rate of 179 thousand DVD players per quarter. That is, from the end of the first quarter of 2006 to the end of the second quarter of 2006, about 179 thousand DVD players will be sold. We are more skeptical of this estimate than of the 2000 estimate, since the first quarter of 2006 ($t = 36$) is so far outside the raw data set ($1 \leq t \leq 22$).

In practice, it is easiest to create a logistic model by using logistic regression. The following Technology Tip demonstrates how to find a logistic model for a set of data.

TECHNOLOGY *TIP*

Logistic Regression

1. Enter the data using the Statistics Menu List Editor. (Refer to Section 1.3 if you've forgotten how to do this.)

2. Bring up the Statistics CALC menu, select item B:Logistic, and press ENTER.

3. If you want to automatically paste the regression equation into the Y= editor, press the key sequence VARS Y-VARS; 1:Function; 1:Y1 and press ENTER. Otherwise press ENTER.

EXAMPLE 4

Using a Differential Equation to Forecast the Price of a VCR

Based on data from 1997 to 2002, the average price of a video cassette recorder (VCR) may be modeled by

$$P(t) = \frac{110.0}{1 + 0.1955e^{1.150t}} + 65 \text{ dollars}$$

where t is the number of years since the end of 1997. The dramatic drop in price may be attributed in part to the introduction of the DVD player to the market in the first quarter of 1997. (**Source:** Modeled from Consumer Electronics Association data.) How quickly was the average price of a VCR changing at the end of 1999 and the end of 2001?

SOLUTION The function $P(t)$ is a logistic function shifted vertically 65 units. We'll define a new function $N(t) = P(t) - 65$. We have

$$N(t) = \frac{110.0}{1 + 0.1955e^{1.150t}}$$

This function is a solution to the differential equation

$$\frac{dN}{dt} = kN(110.0 - N)$$

$$= \left(\frac{-1.150}{110.0}\right)N(110.0 - N)$$

$$= -1.15N + 0.01045N^2$$

Since $\dfrac{dN}{dt} = \dfrac{dP}{dt}$,

$$\frac{dP}{dt} = -1.15N + 0.01045N^2$$

$$\left.\frac{dP}{dt}\right|_{t=2} = -1.15N(2) + 0.01045[N(2)]^2$$

$$= -1.15(37.29) + 0.01045(37.29)^2$$

$$= -28.35$$

At the end of 1999, the price of a VCR was dropping at a rate of about $28.35 per year.

$$\left.\frac{dP}{dt}\right|_{t=4} = -1.15N(4) + 0.01045[N(4)]^2$$

$$= -1.15(5.379) + 0.01045(5.379)^2$$

$$= -5.88$$

At the end of 2001, the price of a VCR was dropping at a rate of about $5.88 per year. The introduction of the DVD player had a dramatic impact on the price of a VCR.

In the past several examples, we have demonstrated the use of both limited and logistic growth models. Both types of models approach a constant value as the domain values grow large; however, near the origin the models behave quite differently. A limited growth model grows very rapidly at first, while a logistic model initially grows very slowly. Consequently, when choosing a model to use, it is important to consider the expected growth behavior of the function at the start of the time period in addition to looking at the shape of the scatter plot.

13.4 Summary

In this section, you learned how to use differential equations to find limited growth and logistic growth models. You also practiced interpreting the results of differential equations.

13.4 Exercises

In Exercises 1–6, determine the equation of the limited growth model for each species of fish and then answer the associated questions. You may find it helpful to refer to Example 3 in this section.

1. **Red Mullet**

Red Mullet	k	L_∞	t_0
Female	0.23	25.6 cm	−1.38 years
Male	0.21	22.2 cm	−2.08 years

Source: www.fishbase.org.

(a) How long is a male, three-year-old red mullet?

(b) How long is a female, three-year-old red mullet?

(c) Is the three-year-old male or the three-year-old female red mullet growing faster?

2. **Turbot**

Turbot	k	L_∞	t_0
Female	0.115	103.4 cm	−0.93 year
Male	0.143	77.3 cm	−1.22 years

(a) At what rate does a 50.0-cm-long male turbot grow?

(b) At what rate does a 50.0-cm-long female turbot grow?

(c) Which is older, a 50.0-cm-long male turbot or a 50.0-cm-long female turbot? Explain.

3. **Flounder**

Flounder	k	L_∞	t_0
Female	0.197	44.98 cm	−1.43 years
Male	0.22	33.0 cm	−2.24 years

Source: www.fishbase.org.

(a) At what rate does a 20.0-cm-long male flounder grow?

(b) At what rate does a 20.0-cm-long female flounder grow?

(c) Which is older, a 20-cm-long male flounder or a 20-cm-long female flounder? Explain.

4. **Picarel Fish**

Pickerel	k	L_∞	t_0
Female	0.232	17.6 cm	−1.15 years
Male	0.793	20.6 cm	0.08 years

(a) How long is a male, four-year-old picarel?

(b) How long is a female, four-year-old picarel?

(c) Is the four-year-old male or the four-year-old female picarel growing faster?

5. **Copper Shark**

Copper Shark	k	L_∞	t_0
	0.038	385.0 cm	−3.477

Source: www.fishbase.org.

(a) How long is a three-year-old copper shark?

(b) At what rate is the length of a three-year-old copper shark changing?

6.   **Cowcod**

Cowcod	k	L_∞	t_0
	0.052	86.9 cm	−1.94

(a) How long is a two-year-old cowcod?

(b) At what rate is the length of a two-year-old cowcod changing?

In Exercises 7–12, use the logistic growth model and associated differential equation to answer the questions.

7. **Pediatric AIDS** Based on data from 1992 to 2001, the estimated number of new pediatric AIDS cases in the United States in year t may be modeled by

$$P(t) = \frac{919.9}{1 + 0.0533e^{0.07758t}} + 90$$

cases, where t is the number of years since the end of 1992. (**Source:** Modeled from Centers for Disease Control and Prevention data.)

The rate of change of the pediatric AIDS function P is the same as the rate of change of the function $N(t) = \dfrac{919.9}{1 + 0.0533e^{0.07758t}}$.

(a) How many new pediatric AIDS cases are estimated to have occurred in 2000?

(b) What is the differential equation whose solution is $N(t)$?

(c) At what rate was the estimated number of new pediatric AIDS cases changing at the end of 2000?

(d) Based on the results of part (c) and the graph of $P(t)$, does it look like efforts to reduce pediatric AIDS in the United States are generating positive results? Explain.

8. **Adult and Adolescent AIDS** Based on data from 1981 to 1995, the number of adult and adolescent deaths due to AIDS in a given year in the United States may be modeled by

$$P = \frac{53{,}955}{1 + 38.834e^{-0.45127t}}$$

deaths, where t is the number of years since the end of 1981. (**Source:** Modeled from Centers for Disease Control and Prevention data.)

(a) According to the model, how many adult and adolescent AIDS deaths occurred in the United States in 1995?

(b) According to the model, at what rate were adult and adolescent AIDS deaths increasing at the end of 1995?

(c) The number of adult and adolescent AIDS deaths in the United States has decreased every year since 1995, in part because of the increased availability of drug treatments and AIDS prevention efforts. In 2001, there were 8063 deaths. If the number of AIDS deaths had followed the logistic model, how many deaths would have occurred in 2002?

9. **School Internet Access** Based on data from 1994 to 2000, the rate of change in the percentage of public-school classrooms with Internet access may be modeled by

$$\frac{dy}{dt} = 0.01077y(85.88 - y)$$

percentage points per year, where t is the number of years since the end of 1994. In 2000, 77 percent of public-school classrooms had Internet access. (**Source:** Modeled from *Statistical Abstract of the United States, 2001,* Table 243, p. 155.)

(a) Find the particular solution to the differential equation and interpret its real-world meaning.

(b) At what rate was the percentage of public schools with Internet access changing at the end of 2000?

10. **National Internet Usage** Based on data from 1995 to 1999 and U.S. Census Bureau projections for 2000 to 2004, the rate of change in annual per capita Internet usage may be modeled by

$$\frac{dy}{dt} = 0.002882y(232.2 - y)$$

hours per year, where t is the number of years since the end of 1995. (**Source:** *Statistical Abstract of the United States, 2001,* Table 1125, p. 704.) In 1999, the annual per capita Internet usage was 99 hours.

(a) Find the particular solution to the differential equation and interpret its real-world meaning.

(b) At what rate was the annual per capita Internet usage changing at the end of 2000?

11. **Hotel/Motel Occupancy** Based on data from 1990 to 1999, the average hotel/motel room rate may be modeled by

$$R(t) = \frac{27.80}{(1 + 47.23e^{-0.6425t})} + 57$$

dollars per day, where t is the number of years since the end of 1990. (**Source:** *Statistical Abstract of the United States, 2001,* Table 1266, p. 774.)

The rate of change of the room rate function R is the same as the rate of change of the function

$$N(t) = \frac{27.80}{(1 + 47.23e^{-0.6425t})}$$

(a) Find the differential equation whose solution is N.

(b) Determine at what rate the hotel/motel room rate was changing at the end of 1997 and at the end of 1999.

12. **Information Technology** Based on data from 1990 to 2000, the percentage of the economy attributed to the information technology sector may be modeled by

$$P(t) = \frac{2.956}{1 + 30.15e^{-0.5219t}} + 7.9$$

percent, where t is the number of years since the end of 1990. (**Source:** Modeled from *Statistical Abstract of the United States, 2001,* Table 1122, p. 703.)

The rate of change in the information technology percentage of the economy function P is the same as the rate of change of the function

$$N(t) = \frac{2.956}{1 + 30.15e^{-0.5219t}}$$

(a) Find the differential equation whose solution is N.

(b) Determine at what rate the information technology percentage of the economy was changing at the end of 1995 and at the end of 2000.

In Exercises 13–14, use a limited growth function to model the data. Then answer the given questions.

13. **Spread of Information** Suppose that in a city of 350,000 people, 40,000 people were watching television or listening to the radio when the news of a bioterrorist attack was first broadcast. For the purpose of the model, assume that all 40,000 people heard the news simultaneously five minutes after the attack. Model the spread of the news of the attack. How long did it take for 75 percent of the city to hear the news, and at what rate was the news spreading at that time?

14. **Video Game Sales** Blizzard Entertainment released the engaging real-time strategy game "Warcraft III: Reign of Chaos" on July 3, 2002. By July 22, 2002, more than a million copies of the game had been sold. (**Source:** www.pcgameworld.com.)

Model the "Warcraft III" game sales, assuming that at the end of the twentieth day after the game was released, 1 million copies of the game had been sold. Furthermore, assume that a total of 8 million copies of the game will be sold over the life of the game. According to the model, how long did it take for 4 million copies to be sold, and at what rate were sales increasing at that time?

In Exercises 15–20, use logistic regression to find the logistic model for the data. Then answer the given questions.

15. **Deadly Fights over Money**

Homicides Due to Arguments over Money or Property

Years since 1990 (t)	Homicides (H)
1	520
2	483
3	445
4	387
5	338
6	328
7	287
8	241
9	213
10	206

Source: *Crime in the United States 1995, 2000,* Uniform Crime Report, FBI.

According to the model, at what rate were money-related homicides decreasing in 2001?

16.   **Cassette Tape Market Share**

Years Since 1993 (t)	Percent of Music Market (percentage points) (P)
0	38.0
1	32.1
2	25.1
3	19.3
4	18.2
5	14.8
6	8.0
7	4.9
8	3.4
9	2.4

Source: Recording Industry of America.

According to the model, in what year will cassette tape market share drop below 1 percent? At what rate will the cassette tape market share be decreasing at that time?

   17.   **Movie Box Office Sales**

My Big Fat Greek Wedding

Weekend	Week #	Cumulative Gross Box Office Sales (dollars)
Apr. 19–21 (2002)	1	597,362
June 28–30	11	19,340,988
Sept. 6–8	21	95,824,732
Nov. 15–17	31	199,574,370
Jan. 24–26 (2003)	41	236,448,697
Apr. 4–6	51	241,437,427

Source: www.boxofficeguru.com.

(a) According to the model, were cumulative box office sales for *My Big Fat Greek Wedding* increasing at a higher rate in Week 1 or Week 51?

(b) IFC Films was the distributor for *My Big Fat Greek Wedding*. (**Source:** www.boxofficeguru.com.) If you were a marketing consultant to IFC Films, what would you tell the company about forecasted box office sales beyond Week 51?

  18.   **Resource Value**

Value of Fabricated Metals Shipments

Years Since 1992 (t)	Shipment Value (millions of dollars) (V)
0	170,403
1	177,967
2	194,113
3	212,444
4	222,995
5	242,812
6	253,720
7	256,900
8	258,960

Source: *Statistical Abstract of the United States, 2001*, Table 982, p. 624.

(*Hint:* Before creating the model, align the data by subtracting 170,000 from each value of V. After doing logistic regression, add back the 170,000 to the resultant model equation.)

According to the model, was the value of fabricated metals shipments increasing more rapidly in 1995 or in 1997?

19.   **TV Homes with Cable**

Years Since 1970 (t)	TV Homes with Cable (percent) (C)
0	6.7
5	12.6
10	19.9
15	42.8
16	45.6
17	47.7
18	49.4
19	52.8
20	56.4
21	58.9
22	60.2
23	61.4
24	62.4
25	63.4
26	65.3
27	66.5
28	67.2
29	67.5

Source: *Statistical Abstract of the United States, 2001*, Table 1126, p. 705.

According to the model, what percentage of TV homes will have cable in 2006, and at what rate will that percentage be increasing?

20.   **U.S. Army Personnel**

Years Since 1980 (t)	Personnel (thousands) (P)
0	777
2	780
4	780
6	781
8	772
10	732
12	610
14	541
16	491
18	484
20	482

Source: *Statistical Abstract of the United States, 2001*, Table 500, p. 329.

(*Hint:* Before creating the model, align the data by subtracting 480 from each value of P. After doing logistic regression, add back the 480 to the resultant model equation.)

According to the model, how many people were in the U.S. Army in 2005, and at what rate was that number changing?

Exercises 21–23 are intended to challenge your understanding of differential equations.

21. Prove that if $\dfrac{dy}{dt} = k(M - y)$ and $y(0) = 0$, then $y = M(1 - e^{-kt})$.

22. Prove that if $y = \dfrac{M}{1 + Se^{-kMt}}$, then $\dfrac{dy}{dt} = ky(M - y)$.

23. Show that if $D(t) = \dfrac{100}{1 + 200e^{-0.1t}}$, then $D'(t) = 0.001[D(t)]\{100 - [D(t)]\}$.

Chapter 13 Review Exercises

Section 13.1 *In Exercises 1–10, integrate the function using the simplest possible method.*

1. $f(x) = 2xe^{-x}$
2. $g(x) = -4x\ln(x)$
3. $h(x) = 3x^2e^x$
4. $g(t) = 2t\ln(t)$
5. $f(x) = \dfrac{[\ln(x)]^3}{x}$
6. $h(x) = (2x - 1)e^{x^2 - x + 1}$
7. $g(x) = (5x - 3)[\ln(x)]$
8. $f(t) = (t - 1)^4(2t - 5)$ 9. $h(x) = 4xe^{4x}$
10. $g(t) = 2t - t\ln(t)$

Section 13.2 *In Exercises 11-15, calculate the combined area of the region(s) bounded by the graphs of the two functions. (On some of the exercises, you may find it helpful to use your graphing calculator to locate the intersection points of the graphs.)*

11. $f(x) = x^2 - 1; g(x) = -x^2 + 17$
12. $f(x) = x^2 - 1; g(x) = x^3 - 3x^2 - x + 3$
13. $f(x) = 2^x; g(x) = x^2$
14. $f(x) = 3^x; g(x) = 3x^3$
15. $f(x) = \ln(x); g(x) = 0.1x$

In Exercises 16–20, graph each of the functions. Then determine the solution by using the integral techniques demonstrated in Section 13.2.

16. **Adult and Adolescent AIDS** Based on data from 1981 to 1992, the rate of change in the annual number of adult and adolescent AIDS cases diagnosed in the United States may be modeled by
$$R(t) = 75.48t^3 - 1237t^2 + 6652t - 4882$$
cases per year, and the rate of change in the annual number of adult and adolescent deaths in the United States due to AIDS may be modeled by
$$D(t) = -107.3t^2 + 1641t - 975.3$$
deaths per year, where t is the number of years since the end of 1981. (**Source:** Modeled from Centers for Disease Control and Prevention data.)

Determine the accumulated change in the difference of the two populations from the end of 1981 through 1992. Interpret the meaning of the result.

17. **Company Profit** Based on data from 1999 to 2001, the rate of change in the revenue of the Johnson & Johnson company and its subsidiaries may be modeled by
$$r(t) = 1330t + 1150$$
million dollars per year, and the rate of change in the cost of goods sold may be modeled by
$$c(t) = 206t + 315$$
million dollars, where t is the number of years since the end of 1999. (**Source:** Modeled from Johnson & Johnson 2001 Annual Report.)

Determine the accumulated change in annual profit from the end of 1999 through 2001.

18. **Public- and Private-School Enrollment** Based on U.S. National Center for Education Statistics enrollment projections for 2000 to 2010, the rate of change in the annual number of students enrolled in public schools may be modeled by
$$g(t) = 0.4971t^2 - 3.022t + 51.17$$
thousand students per year, and the rate of change in the number of students enrolled in private schools may be modeled by
$$p(t) = -19.99t + 280.5$$
thousand students per year, where t is the number of years since the end of 2000. (**Source:** Modeled from *Statistical Abstract of the United States, 2001,* Table 205, p. 133.) In 2000, 58,821,000 students were enrolled in public schools and 9,303,000 students were enrolled in private schools.

Determine the accumulated change in the annual enrollment gap between private and public schools over the period 2000 to 2010. Interpret the meaning of the result.

19. **Public- and Private-School Teachers** Based on U.S. National Center for Educational Statistics projections for 2000 to

2010, the rate of change in the number of public-school teachers may be modeled by

$$g(t) = -1.4242t + 16.11$$

thousand teachers per year, and the rate of change in the number of private-school teachers may be modeled by

$$p(t) = -0.8158t + 1.453$$

thousand teachers per year, where t is the number of years since the end of 2000. (**Source:** Modeled from *Statistical Abstract of the United States, 2001*, Table 207, p. 134.) In 2000, there were 2,850,000 public-school teachers and 402,000 private-school teachers.

Determine the predicted accumulated change in the annual gap between private- and public-school teachers over the period 2000 to 2010. Interpret the meaning of the result.

20. **M** **Wages of Clothing Producers** Based on data from 1980 to 2000, the rate of change in the average hourly wage of a men's and boys' furnishings production worker may be modeled by

$$m(t) = 0.007032t + 0.1446$$

dollars per year, and the rate of change in the average hourly wage of a women's and misses' outerwear production worker may be modeled by

$$w(t) = 0.006394t + 0.1333$$

dollars per year, where t is the number of years since 1980. (**Source:** Modeled from *Statistical Abstract of the United States, 2001*, Table 609, p. 394.) In 1980, the hourly wage of a men's and boys' furnishings production worker was $4.23, and the average hourly wage of a women's and misses outerwear production worker was $4.61.

Determine the accumulated change in the hourly wage gap between men's and boys' furnishings and women's and misses outerwear production workers over the period 1980 to 2000.

Section 13.3 *In Exercises 21–25, find the general solution of the separable differential equation. Assume that k is a constant. (Hint:*

$$\int \frac{dy}{y(M - y)} = \frac{1}{M} \ln\left(\frac{y}{M - y}\right).)$$

21. $\dfrac{dy}{dx} = \dfrac{y}{x}$

22. $\dfrac{dA}{dt} = -5A^2$

23. $\dfrac{dA}{dt} = 2(16 - A)$

24. $\dfrac{dy}{dx} = 0.1(y - 10)$

25. $\dfrac{dA}{dt} = 4A - A^2$

In Exercises 26–30, find the particular solution for the differential equations.

26. $\dfrac{dy}{dx} = \dfrac{2x}{y}$; $y(0) = -1$

27. $\dfrac{dA}{dt} = -5A^2$; $A(0) = -\dfrac{1}{2}$

28. $\dfrac{dA}{dt} = 2(16 - A)$; $A(0) = 15$

29. $\dfrac{dy}{dx} = 0.1(y - 10)$; $y(0) = 11$

30. $\dfrac{dA}{dt} = 4A - A^2$; $A(0) = 1$

In Exercise 31, use Newton's Law of Heating and Cooling.

31. **Time of Death** A crime scene investigator finds that the temperature of a corpse in a 70°F room is 80°F. Thirty minutes later, the temperature of the corpse is 77°F. Assuming that the woman's temperature was 98.6°F when she died, find a function for the temperature of the body t minutes after the woman died. Then use the model to estimate how long the woman had been dead when the crime scene investigator first took her temperature.

Section 13.4 *In Exercises 32–33, find the model indicated and answer the associated questions.*

32. **Textbook Sales** Cumulative textbook sales may often be modeled by a limited growth model. A publisher's marketing analyst predicted that a new edition of a textbook would sell 12,900 copies in its first year and 20,500 copies over the life of the edition.

(a) Find the limited growth model for the cumulative number of sales of the edition.

(b) Calculate the cumulative number of books sold by the end of the second year.

(c) Determine the rate of change in the cumulative number of textbook sales at the end of the second year.

33. **Electronic Game Sales**   Based on data from 1996 to 2000, the number of copies of computer and video games sold worldwide may be modeled by

$$S(t) = \frac{121.8}{1 + 17.23e^{-1.794t}} + 100$$

million units, where t is the number of years since the end of 1996. (**Source:** Modeled from Interactive Digital Software Association State of the Industry Report, 2000-2001.)

The rate of change in the number of game units sold, S, is the same as the rate of change of the function

$$N(t) = \frac{121.8}{1 + 17.23e^{-1.794t}}$$

(a) Find the differential equation whose solution is N.

(b) Determine at what rate computer and video game sales were changing in 1997 and in 1999.

In Exercises 34–35, determine the equation of the limited growth model for each species of fish and then answer the associated questions.

34. **Rainbow Trout in Australia**

Rainbow Trout	k	L_∞	t_0
	0.340	52.0 cm	−0.08 year

(a) How long is a three-year-old rainbow trout?

(b) At what rate is the length of a three-year-old rainbow trout changing?

35. **Albacore Tuna off California**

Albacore Tuna	k	L_∞	t_0
	0.233	109.0 cm	−2.31 years

Source: www.fishbase.org.

(a) How long is a two-year-old albacore tuna?

(b) At what rate is the length of a two-year-old albacore tuna changing?

In Exercises 36–37, use logistic regression to find the logistic model for the data. Then answer the given questions.

36. **Number of Different Banks**

Years Since 1984 (t)	Banks (B)
0	17,900
1	18,033
2	17,876
3	17,325
4	16,562
5	15,829
6	15,192
7	14,517
8	13,891
9	13,261
10	12,641
11	12,002
12	11,478
13	10,923
14	10,463
15	10,221
16	9,908

Source: *Statistical Abstract of the United States, 2001,* Table 1173, p. 728.

(*Hint:* Align the data by subtracting 9900 from B. After finding the model using logistic regression, add back the 9900.)

According to the model, at what rate was the number of banks declining in 2000?

37. **U.S. Homicide Rate**

Years Since 1990 (t)	Homicides per 100,000 People (H)
0	9.4
1	9.8
2	9.3
3	9.5
4	9
5	8.2
6	7.4
7	6.8
8	6.3
9	5.7
10	5.5

Source: FBI Uniform Crime Reports, 2000.

(*Hint:* Align the data by subtracting 5.4 from H. After finding the model using logistic regression, add back the 5.4.)

According to the model, at what rate was the number of homicides per hundred thousand people declining in 2000?

Make It Real

What to do

1. Identify a public company that you would like to work for in the future.
2. Obtain a copy of the company's annual financial report.
3. Using five or more years of data, model the revenue of the company and the costs of the company as a function of time.
4. Graph the revenue and cost models on the same axes.
5. Calculate the accumulated change in revenue and the accumulated change in costs over a five-year period using the models.
6. Shade the region that represents the accumulated change in profit.
7. Using the results from Steps 5 and 6, calculate the accumulated change in profit over the five-year period

Where to look for data

Publicly held companies are required to release financial data to the public through annual reports. Many of these annual reports are available online and may be accessed through the investor information page on the company's website. Here are links to the homepages of some prominent companies.

Safeway
www.safeway.com

McDonald's
www.mcdonalds.com

Kellogg's
www.kelloggs.com

Revlon
www.revlon.com

Chapter 14

Multivariable Functions and Partial Derivatives

The amount of moisture in the air affects how hot it feels to a person. If the air temperature is 96°F and the relative humidity is 75 percent, it feels like it is 132°! The apparent temperature (what it feels like) is a multivariable function of the actual temperature and the relative humidity. Partial derivatives may be used to determine what effect a one-degree increase in the actual temperature will have on the apparent temperature if the relative humidity remains constant.

14.1 Multivariable Functions

- Analyze and interpret multivariable mathematical models
- Evaluate multivariable functions

GETTING STARTED Businesses seek to manage their labor and capital so as to maximize their production efficiency. Since financial resources are limited, executives have to determine how much money to invest in their employees and how much money to invest in equipment and buildings. Economists Charles Cobb and Paul Douglas formulated the multivariable Cobb-Douglas production function to model the output of small- and large-scale economies.

In this section, we will discuss how to evaluate and graph *multivariable functions*. Additionally, we'll look at several applications of multivariable functions, including the Cobb-Douglas production function.

Functions requiring two or more inputs to generate a single output are called **multivariable functions.** For example, the heat index is a function of the temperature and the humidity. The wind chill factor is a function of temperature and wind speed. The volume of a cylindrical can is a function of its radius and its height. The market share of a business is a function of the market shares of each of its individual competitors. The value of a savings account is a function of the amount invested, the interest rate, the compounding frequency, and the amount of time the money is invested. Functions with two input variables may be represented using a three-dimensional graph.

FUNCTION OF TWO VARIABLES

A **function f of two variables** is a rule that associates each ordered pair of inputs (x, y) with a single output $z = f(x, y)$. The graph of a function of two variables is a three-dimensional **surface.**

In this context, both x and y are **independent variables** and z is the **dependent variable.** The **domain** of the function is the set of all ordered pairs (x, y). The **range** of the function is the set of corresponding values of $f(x, y)$.

Consider the volume of a cylindrical can. The equation for the volume is given by $V = \pi r^2 h$, where r is the radius and h is the height of the can. We may alternatively write $f(h, r) = \pi r^2 h$, where $V = f(h, r)$. What is the volume of a can with a five-inch height and a two-inch radius?

$$f(h, r) = \pi r^2 h$$
$$f(5, 2) = \pi (2)^2 (5)$$
$$= 20\pi$$
$$\approx 62.83$$

The volume of a cylindrical can with a five-inch height and a two-inch radius is approximately 62.83 cubic inches.

We may generate a table of values for V by constructing a table with a column for each of the variables (see Table 14.1).

TABLE 14.1

h	r	$V = \pi r^2 h$
1	1	π
1	2	4π
1	3	9π
2	1	2π
2	2	8π
2	3	18π

Notice that each value of h may be paired with multiple values of r. A more concise way to record the information is to draw a table with the rows containing the values of h and the columns containing the values of r (see Table 14.2). The value of V is contained in the body of the table.

TABLE 14.2

$h \downarrow$ $r \rightarrow$	1	2	3
1	π	4π	9π
2	2π	8π	18π

To graph V, we must use a three-dimensional grid with an axis for each of the variables. One of the easiest ways to visualize this is to look at an interior corner of a classroom. The edge between one wall and the floor represents the h-axis. The edge between the other wall and the floor represents the r-axis. The edge joining the two walls represents the V-axis. The origin, $(0, 0, 0)$, is the point of intersection of the three axes (see Figure 14.1).

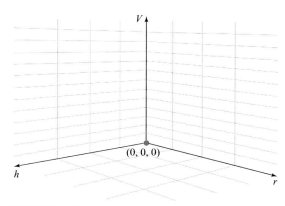

FIGURE 14.1

The hV plane is the "wall" formed by the h-axis and the V-axis. The rV plane is the "wall" formed by the r-axis and the V-axis. The hr plane is the "floor" formed by the h-axis and the r-axis. (Using the standard variables x, y, and z, the xz plane and the yz plane form the "walls" and the xy plane is the "floor.")

A point in this coordinate system is an ordered triple of the form (h, r, V). To plot the point $(2, 1, 2\pi)$, we move 2 units from the origin along the h-axis, move 1 unit from the origin in the direction of the r-axis, and finally move 2π units up in the direction of the V-axis, as shown in Figure 14.2.

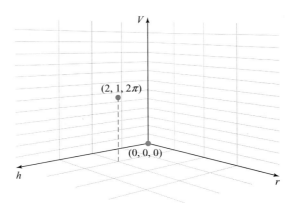

FIGURE 14.2

To plot the point $(1, 2, 4\pi)$, we move 1 unit from the origin along the h-axis, move 2 units from the origin in the direction of the r-axis, and finally move 4π units up in the direction of the V-axis, as shown in Figure 14.3.

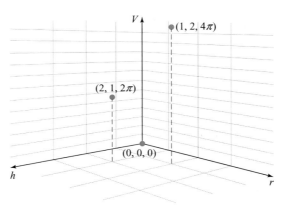

FIGURE 14.3

(If h, r, or V were negative, we would move in the opposite direction. However, since the equation represents volume as a function of lengths, all variables will be nonnegative.)

Visualizing what the entire surface will look like is often difficult, since we're trying to represent a three-dimensional object on a two-dimensional face of a sheet of paper. However, many technologies and techniques have been developed to make it easier to do so. An Internet search on the key words "3D Graphing Freeware" will reveal a variety of free graphing utilities that may be used to draw the surface. Additionally, purchased graphing programs such as Maple, Mathematica, and Autograph can create beautiful 3D surfaces that can be easily rotated.

To help you better visualize the surface, the graph of the cylinder volume function is shown from two different angles in Figure 14.4.

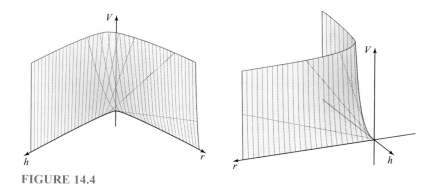

FIGURE 14.4

EXAMPLE **1**

Graphing a Multivariable Function

Generate a table of data for the multivariable function $f(x, y) = x^2 + y^2$ and use a graphing utility to draw the surface.

SOLUTION To calculate the values in Table 14.3, we substitute each ordered pair (x, y) into the function f. For example,

$$f(x, y) = x^2 + y^2$$
$$f(-3, -2) = (-3)^2 + (-2)^2$$
$$= 9 + 4$$
$$= 13$$

TABLE 14.3

x ↓	y →						
	-3	-2	-1	0	1	2	3
-3	18	13	10	9	10	13	18
-2	13	8	5	4	5	8	13
-1	10	5	2	1	2	5	10
0	9	4	1	0	1	4	9
1	10	5	2	1	2	5	10
2	13	8	5	4	5	8	13
3	18	13	10	9	10	13	18

Notice that the low point of this surface occurs at the origin, $(0, 0, 0)$. As we move away from the origin in any direction, the value of z increases. Two different views of the graph of the surface are shown in Figure 14.5.

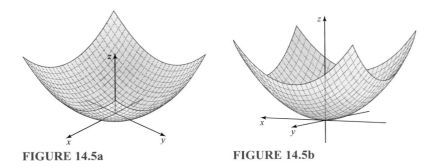

FIGURE 14.5a   **FIGURE 14.5b**

Figure 14.5a is a graph in standard position. The x-axis points to the left, the y-axis points to the right, and the z-axis points upward. In Figure 14.5b, we have rotated the axes so that we can see the surface from a different point of view.

EXAMPLE 2

Graphing a Multivariable Function

Generate a table of data for the multivariable function $f(x, y) = y^3 - yx^2$ and use a graphing utility to draw the surface.

SOLUTION

TABLE 14.4

x ↓	$y \rightarrow$						
	-3	-2	-1	0	1	2	3
-3	0	10	8	0	-8	-10	0
-2	-15	0	3	0	-3	0	15
-1	-24	-6	0	0	0	6	24
0	-27	-8	-1	0	1	8	27
1	-24	-6	0	0	0	6	24
2	-15	0	3	0	-3	0	15
3	0	10	8	0	-8	-10	0

The graph of the function on the specified domain is shown in Figure 14.6 in standard position and with the axes rotated.

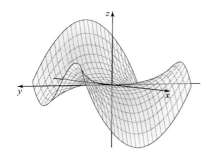

FIGURE 14.6

EXAMPLE 3

Using a Multivariable Function to Forecast the Value of an Investment

A $1000 investment is to be made into an account paying 6 percent compound interest. The value of the investment compounded n times per year after t years is given by

$$A(n, t) = 1000\left(1 + \frac{0.06}{n}\right)^{nt}$$

Generate a table of data for the multivariable function using $n = 1, 2, 4, 12, 365$ and $t = 0, 1, 2, 3$. Graph the function and interpret the meaning of $(12, 2, 1127.16)$.

SOLUTION

TABLE 14.5

n $\downarrow$	$t \rightarrow$			
	0	1	2	3
1	1000.00	1060.00	1123.60	1191.02
2	1000.00	1060.90	1125.51	1194.05
4	1000.00	1061.36	1126.49	1195.62
12	1000.00	1061.68	1127.16	1196.68
365	1000.00	1061.83	1127.49	1197.20

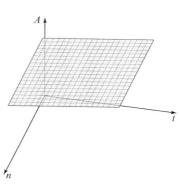

FIGURE 14.7

When we graph the function using a window with $0 \le A \le 1200$, the graph looks like a plane parallel to the nt plane (Figure 14.7).

However, when we use the window $1000 \le A \le 1200$, we get a better look at what is happening (Figure 14.8).

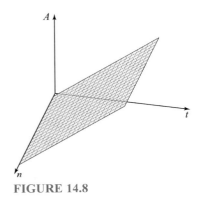

FIGURE 14.8

The compounding frequency, n, has a greater effect on the value of the investment as t increases. The meaning of $(12, 2, 1127.16)$ is that a $1000 investment earning 6 percent interest compounded monthly ($t = 12$) at the end of two years ($n = 2$) will have a value of $1127.16.

EXAMPLE 4

Using a Multivariable Function to Forecast Business Sales

Based on data from 1998 to 2001, the net sales of Kellogg Company and subsidiaries may be modeled by

$$S(a, e) = -0.36661a + 0.16556e + 4671.3 \text{ million dollars}$$

where a is the amount of money spent on advertising (in millions of dollars) and e is the number of employees working for the company. (**Source:** Modeled from Kellogg Company 2001 Annual Report, p. 26.) In 2001, the company had 26,424 employees, spent $519.2 million on advertising, and had $8853.3 million in net sales. If the company were to increase advertising spending to $700 million and decrease the number of employees to 25,000, what would be the predicted net sales?

SOLUTION We are asked to calculate $S(700, 25{,}000)$.

$$S(700, 25{,}000) = -0.36661(700) + 0.16556(25{,}000) + 4671.3$$
$$= 8553.7$$

Net sales are predicted to be $8553.7 million when advertising spending is $700 million and the number of employees is 25,000.

Let's look a little more closely at the coefficients of the model equation. The units of the output of the function are *millions of sales dollars*. Consequently, the units of each term in the sum must also be *millions of sales dollars*. The units of a are *millions of advertising dollars*. What are the units of the coefficient -0.36661? The units are *sales dollars per advertising dollar*, since

$$\frac{\text{sales dollars}}{\text{advertising dollar}} \cdot \text{millions of } \cancel{\text{advertising dollars}} = \text{millions of sales dollars}$$

According to the model, for every advertising dollar, sales decrease by 0.36661 million dollars ($366,610). What are the units of the coefficient 0.16556? The units are *millions of sales dollars per employee*, since

UNITS

$$\frac{\text{millions of sales dollars}}{\text{employee}} \cdot \text{employees} = \text{millions of sales dollars}$$

According to the model, adding an employee will increase sales by 0.16556 million dollars ($165,560).

Although the coefficients of the model indicate that increasing advertising spending will decrease net sales, we know that if Kellogg stops advertising altogether ($a = 0$), its net sales will eventually decrease, not increase. However, since the coefficient of a in the model is negative, the model does seem to suggest that the company might want to look more closely at its advertising expenditures to ensure that it is getting the desired return on its investment.

The **Cobb-Douglas production function,**

$$f(L, C) = kL^m C^n$$

is widely used in economics. The model assumes that the output of a company, industry, or country is a function of its labor L and its capital C. (Capital may be interpreted as the dollar value of the money invested in the company, including equipment, material, and building costs.) The model requires that m, n, and k are positive constants and $m + n = 1$.

EXAMPLE 5

Using a Cobb-Douglas Production Model

Based on data from 2000 and 2001, a Cobb-Douglas production model for the Coca-Cola Company is given by

$$P(L, C) = 0.02506L^{0.0481} C^{0.9519} \text{ billion unit cases of liquid beverage}$$

where L is the labor cost (including sales, payroll, and other taxes) in millions of dollars and C is the cost of capital expenditures (property, plant, and equipment) in millions of dollars. One unit case of beverage equals 24 eight-ounce servings. (**Source:** Modeled from Coca-Cola Company 2000 and 2001 Annual Reports.) According to the model, would it be better to spend $800 million on capital expenditures and $200 million on labor costs or $700 million on capital expenditures and $300 million on labor costs? Explain.

SOLUTION We are asked to calculate $P(200, 800)$ and $P(300, 700)$.

$$P(200, 800) = 0.02506(200)^{0.0481}(800)^{0.9519}$$

$$= 18.8$$

$$P(300, 700) = 0.02506(300)^{0.0481}(700)^{0.9519}$$

$$= 16.8$$

Spending $200 million on labor and $800 million on capital expenditures is expected to produce 18.8 billion unit cases of beverage, while spending $300 million on labor and $700 million on capital expenditures is expected to produce 16.8 billion unit cases of beverage. The model suggests that it may be better to spend $800 million on capital expenditures and $200 million on labor; however, there may be additional constraints to take into consideration. For example, if the Coca-Cola Company were to purchase additional equipment with its capital, it might need to hire additional people to operate the new equipment. Hiring additional people would result in a corresponding increase in the labor cost.

EXAMPLE 6

Using a Multivariable Function to Find the Surface Area of a Box

A box manufacturer designed a closeable box with a square base, as shown in Figure 14.9. Find the surface area equation for the unassembled box. How many square inches of material are required to construct a box 6 inches wide, 6 inches long, and 18 inches high?

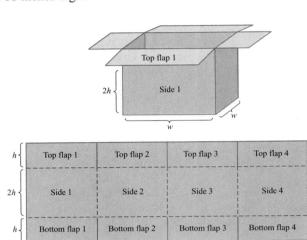

FIGURE 14.9

SOLUTION   The surface area of the unassembled box is given by

$$A(h, w) = (h + 2h + h)(w + w + w + w) + \frac{1}{4}w(2h)$$

$$= (4h)(4w) + \frac{1}{2}hw$$

$$= 16hw + \frac{1}{2}hw$$

$$= 16.5hw \text{ square inches}$$

The height of the box is $2h$. So

$$2h = 18$$

$$h = 9$$

The width and length of the box are both 6 inches, so $w = 6$.

$$A(h, w) = 16.5(9)(6)$$

$$= 891$$

The surface area of the unassembled box is 891 square inches.

14.1 Summary

In this section, you learned how to evaluate multivariable functions and saw how functions of two variables may be represented graphically. You also discovered some applications of multivariable functions.

14.1 Exercises

In Exercises 1–10, generate a table of data for the multivariable function using $x = -2, 0, 2$ and $y = -1, 0, 1$. Then use a graphing utility to graph the function. Compare the coordinates of the points in the table to the coordinates of the points on the graph of the function.

1. $z = 3x - 2y$

2. $z = xy - y^2$

3. $z = 4xy + 4$

4. $f(x, y) = x^2 + 2xy + y^2$

5. $f(x, y) = e^{-xy}$

6. $f(x, y) = x^2 - 2xy + y^2$

7. $f(x, y) = \dfrac{x^2 + 1}{y^2 + 1}$

8. $f(x, y) = \dfrac{xy}{y^2 + 1}$

9. $z = (x + y)^3$

10. $z = \ln(x^2 y^2 + 1)$

In Exercises 11–20, use the multivariable function to answer the given question.

11. **Cobb-Douglas Function** Based on data from 1999 and 2000, a Cobb-Douglas model for Ford Motor Company is given by

$$P(L, C) = 0.5319 L^{0.4558} C^{0.5442}$$

thousand vehicles, where L is the total labor cost and C is the capital expenditure cost, both in millions of dollars. In 2000, Ford Motor Company spent $25,783 million on labor and produced 7424 thousand vehicles. (**Source:** Modeled from Ford Motor Company 2001 Annual Report.)

According to the model, if Ford spends $20,000 million on labor and $9000 million on capital expenditures, how many vehicles will it produce?

12. **Housing Costs** The U.S. Department of Housing and Urban Development (HUD) published a research study in January 2002 comparing the cost of building a steel-frame home to the cost of building a wood-frame home. Two homes with identical floor plans were built side by side in Beaufort, South Carolina.

The material and labor costs of constructing each home are given in the following table.

	Material Cost	Labor Cost
Steel Frame	$97,447.52	$32,228.94
Wood Frame	$91,589.38	$25,236.83

Source: HUD.

Find an equation for the total cost of constructing w identical wood-frame homes and s steel-frame homes in Beaufort, South Carolina. How much will it cost to construct ten wood-frame and five steel-frame homes?

13. **Personal Data Assistant Software** SplashData is a leading developer of virtual wallet applications for Palm OS handhelds. As of December 2002, SplashData offered consumers six products: SplashWallet Suite ($39.95), SplashClock ($5.00), SplashID ($19.95), SplashMoney ($14.95), SplashPhoto ($19.95), and SplashShopper ($19.95). (**Source:** www.splashdata.com.)

Find the formula for the SplashData revenue function. How much revenue will SplashData generate if it sells 20 SplashWallet Suites, 50 SplashClocks, 10 SplashIDs, 100 SplashMoneys, 40 SplashPhotos, and 70 SplashShoppers?

14. **Future Value of Investment** Find the equation for the future value of an investment of P dollars into an account paying 6 percent interest compounded monthly t years after the investment is made. What is the value of a $2500 investment after 10 years?

15. **Surface Area** Referring to Example 6 in the section, find a formula for the exterior surface area of the assembled box with a square base. (Assume that the top and bottom of the box are closed.) What is the exterior surface area of the box with a 6-inch width, 6-inch length, and 18-inch height?

16. **Surface Area Analysis** In Example 6 we calculated the surface area of an unassembled box, and in Exercise 15 you calculated the exterior surface area of the assembled box. Of what practical value are each of these results?

17. **Body Mass Index** The Body Mass Index (BMI) is a helpful indicator of obesity or underweight in adults. A high BMI is predictive of death from cardiovascular disease. The Centers for Disease Control and Prevention indicate that a healthy BMI for adults is between 18.5 and 24.9. They use the following guidelines.

Underweight	BMI less than 18.5
Overweight	BMI of 25.0 to 29.9
Obese	BMI of 30.0 or more

The following formula is used to calculate body mass index.

$$BMI = \frac{703 \cdot W}{H^2}$$

where W is your weight in pounds and H is your height in inches. What is your BMI?

18. **Relative Humidity** The relative humidity of the air alters our perception of the temperature. As the humidity increases, the air feels hotter to us. The heat index (apparent temperature) is a function of air temperature and relative humidity, as shown in the table below.

When exposed to apparent temperatures exceeding 105°F for prolonged periods of time, people are at high risk of sunstroke, a potentially fatal condition. The heat index values were devised for shady, light wind conditions. Exposure to full sunshine can increase apparent temperature by up to 15°F. What is the apparent temperature when the air temperature is 90°F and the relative humidity is 80 percent?

	Relative Humidity (%)												
	40	45	50	55	60	65	70	75	80	85	90	95	100
110	136												
108	130	137											
106	124	130	137										
104	119	124	131	137									
102	114	119	124	130	137								
100	109	114	118	124	129	136							
98	105	109	113	117	123	128	134						
96	101	104	108	112	116	121	126	132					
94	97	100	103	106	110	114	119	124	129	135			
92	94	96	99	101	105	108	112	116	121	126	131		
90	91	93	95	97	100	103	106	109	113	117	122	127	132
88	88	89	91	93	95	98	100	103	106	110	113	117	121
86	85	87	88	89	91	93	95	97	100	102	105	108	112
84	83	84	85	86	88	89	90	92	94	96	98	100	103
82	81	82	83	84	84	85	86	88	89	90	91	93	95
80	80	80	81	81	82	82	83	84	84	85	86	86	87

Air Temperature (°F) — left axis label. Heat Index (Apparent Temperature) — interior label.

Source: National Weather Service.

19. **M** **Wind Chill** Wind alters our perception of the temperature. As the wind strength increases, the air feels colder to us. This phenomenon is referred to as *wind chill*. Wind chill temperature is a function of wind speed and air temperature and is modeled by

$$W(T, V) = 35.74 + 0.6215T$$
$$- 35.75V^{0.16} + 0.4275TV^{0.16}$$

degrees Fahrenheit, where T is the actual air temperature (in degrees Fahrenheit) and V is the velocity of the wind (in miles per hour). (Source: National Climatic Data Center.) When the air is 10°F and the wind is blowing at 40 mph, what is the wind chill temperature?

20. **Ice Cream Cones** The GE-500 Semi-Automatic is a machine that is designed to create a variety of sizes of cones, ranging from a diameter of 20 millimeters (mm) to a diameter of 56 mm. The number of cones that can be made in an hour depends upon the diameter of the cone. The production rate of 20-mm cones is 4400 per hour, while the production rate of 42-mm cones is 2000 per hour. (Source: www.maneklalexports.com.)

Let s be the number of hours spent producing 20-mm cones and l be the number of hours spent producing 42-mm cones. Find a formula for the cumulative number of cones produced as a function of s and l. How many cones are produced if 1.5 hours are spent making 20-mm cones and 3 hours are spent making 42-mm cones?

Exercises 21–23 are intended to challenge your understanding of multivariable functions.

21. What is the domain of the multivariable function?

$$f(x, y) = \frac{\sqrt{x^2 - 1}}{y}$$

22. What are the domain and range of the multivariable function?

$$f(x, y) = 3 + \sqrt{x - 2} + \sqrt{y^2 - 4}$$

23. The amount of a person's take-home pay is a multivariable function. Identify *four* different variables that contribute to the amount of a person's take-home pay.

14.2 Partial Derivatives

- Analyze and interpret multivariable mathematical models
- Calculate first- and second-order partial derivatives of multivariable functions
- Find the equation of a cross section of a graph of a surface
- Determine the meaning of partial rates of change

GETTING STARTED The wind chill temperature is a function of the air temperature and the wind velocity. If the wind speed is increasing at a rate of 2 mph per hour, how quickly is the wind chill temperature changing? Questions such as these may be answered by using *partial derivatives*.

In this section, we will demonstrate how to calculate partial derivatives. These partial rates of change allow us to determine the effect of a change in one of the input variables on the output.

Recall that the operator $\dfrac{d}{dx}$ means "take the derivative with respect to x." The operator $\dfrac{\partial}{\partial x}$ means "take the partial derivative with respect to x." When calculating a partial derivative, all variables except the variable of differentiation are treated as constants. Consider the function $z = x^2 - xy$.

$$\frac{\partial}{\partial x}(z) = \frac{\partial}{\partial x}(x^2 - xy)$$

$$\frac{\partial z}{\partial x} = 2x - y$$

Observe that we treated the y in xy as if it were a constant. That is, when differentiating $x^2 - xy$, we treated the expression as if it were $x^2 - cx$ (where c is a constant). Since $\frac{d}{dx}(x^2 - cx) = 2x - c$, $\frac{\partial z}{\partial x} = 2x - y$. Let's now take the partial derivative of the function with respect to y.

$$\frac{\partial}{\partial y}(z) = \frac{\partial}{\partial y}(x^2 - xy)$$

$$\frac{\partial z}{\partial y} = -x$$

Since the variable of differentiation is y, we treated x as a constant. That is, when differentiating $x^2 - xy$, we treated the expression as if it were $c^2 - cy$ (where c is a constant). Since $\frac{d}{dy}(c^2 - cy) = -c$, $\frac{\partial}{\partial y}(x^2 - xy) = -x$. Observe that $\frac{\partial z}{\partial x} \neq \frac{\partial z}{\partial y}$. The partial derivative of a function varies based upon the variable of differentiation.

PARTIAL DERIVATIVES

The **partial derivatives** of a function $f(x, y)$ are given by

$\dfrac{\partial f}{\partial x}$ (read "the partial of f with respect to x")

$\dfrac{\partial f}{\partial y}$ (read "the partial of f with respect to y")

To calculate $\dfrac{\partial f}{\partial x}$, differentiate the function treating the y variable as a constant.

To calculate $\dfrac{\partial f}{\partial y}$, differentiate the function treating the x variable as a constant.

EXAMPLE 1

Finding Partial Derivatives

Find the partial derivatives of $f(x, y) = 3xy + y^2 + 5$.

SOLUTION

$$\begin{aligned}
\frac{\partial f}{\partial x} &= \frac{\partial}{\partial x}(3xy + y^2 + 5) \\
&= \frac{d}{dx}(3cx + c^2 + 5) \qquad c \text{ is a constant representing } y \\
&= 3c \\
&= 3y \qquad\qquad\qquad \text{Replace } c \text{ with } y
\end{aligned}$$

Replacing the variable y with the constant c is an optional step used to help you apply the partial differentiation rules correctly. As you become skilled at applying the rules, the step will become unnecessary.

$$\frac{\partial f}{\partial y} = \frac{\partial}{\partial y}(3xy + y^2 + 5)$$

$$= \frac{d}{dy}(3cy + y^2 + 5) \qquad c \text{ is a constant representing } x$$

$$= 3c + 2y$$

$$= 3x + 2y \qquad \text{Replace } c \text{ with } x$$

PARTIAL DERIVATIVES (ALTERNATIVE NOTATION)

The **partial derivatives** of a function $f(x, y)$ may alternatively be written as

f_x (read "the partial of f with respect to x")

f_y (read "the partial of f with respect to y")

EXAMPLE 2

Finding Partial Derivatives

Find the partial derivatives of $f(x, y) = x^2 y^2$.

SOLUTION

$$f_x = \frac{\partial}{\partial x}(x^2 y^2)$$

$$= \frac{d}{dx}(x^2 c^2) \qquad c \text{ is a constant representing } y$$

$$= 2xc^2$$

$$= 2xy^2 \qquad \text{Replace } c \text{ with } y$$

$$f_y = \frac{\partial}{\partial y}(x^2 y^2)$$

$$= \frac{d}{dy}(c^2 y^2) \qquad c \text{ is a constant representing } x$$

$$= 2c^2 y$$

$$= 2x^2 y \qquad \text{Replace } c \text{ with } x$$

The units of $\frac{\partial f}{\partial x}$ are the units of f divided by the units of x. Similarly, the units of $\frac{\partial f}{\partial y}$ are the units of f divided by the units of y.

EXAMPLE 3

Interpreting the Meaning of the Partial Derivatives of the Compound Interest Function

Find the partial derivatives of the compound interest function $A(r, t) = 1000\left(1 + \frac{r}{4}\right)^{4t}$. Then evaluate the partial derivatives at the point $(0.12, 3, 1425.76)$ and interpret the meaning of the result.

SOLUTION We will first find the partial derivatives.

$$\frac{\partial}{\partial t}(A) = \frac{\partial}{\partial t}\left[1000\left(1 + \frac{r}{4}\right)^{4t}\right]$$

$$\frac{\partial A}{\partial t} = 1000 \cdot \ln\left(1 + \frac{r}{4}\right) \cdot \left(1 + \frac{r}{4}\right)^{4t} \cdot 4 \qquad \text{Exponential, Constant Multiple, and Chain Rules}$$

$$A_t = 4000 \ln\left(1 + \frac{r}{4}\right) \cdot \left(1 + \frac{r}{4}\right)^{4t}$$

$$\frac{\partial A}{\partial r} = 1000 \cdot 4t\left(1 + \frac{r}{4}\right)^{4t-1} \cdot \left(\frac{1}{4}\right) \qquad \text{Power, Constant Multiple, and Chain Rules}$$

$$A_r = 1000t\left(1 + \frac{r}{4}\right)^{4t-1}$$

Next, we will evaluate each of the partial derivatives at the point (0.12, 3, 1425.76).

$$A_t(0.12, 3) = 4000 \ln\left(1 + \frac{0.12}{4}\right) \cdot \left(1 + \frac{0.12}{4}\right)^{4(3)}$$

$$= 4000 \cdot \ln(1.03)(1.03)^{12}$$

$$= 168.58$$

The units of the partial derivative are the units of the output (dollars) divided by the units of t (years). An investment account earning 12 percent interest (compounded quarterly) with an initial investment of $1000 is increasing in value at a rate of $168.58 per year at the end of the third year. That is, from the end of the third year to the end of the fourth year, the account value will increase by about $168.58.

$$A_r = 1000t\left(1 + \frac{r}{4}\right)^{4t-1}$$

$$A_r(0.12, 3) = 1000(3)\left(1 + \frac{0.12}{4}\right)^{4(3)-1}$$

$$= 3000(1.03)^{11}$$

$$= 4152.70$$

The units of the partial derivative are the units of the output (dollars) divided by the units of r (100 percentage points). (Observe that $0.12 \cdot 100$ percentage points $= 12\%$.)

$$4152.70\frac{\text{dollars}}{100 \text{ percentage points}} = 41.527 \text{ dollars per percentage point}$$

An investment account earning 12 percent interest per year (compounded quarterly) with an initial investment of $1000 is increasing in value at a rate of $41.53 per percentage point increase in the interest rate at the end of the third year. That is, increasing the interest rate from 12 percent per year to 13 percent per year at the end of the third year will increase the account value at the end of the fourth year by approximately $41.53 over what it would have been without the interest rate increase.

EXAMPLE 4 Interpreting the Meaning of Partial Derivatives

The distance of a cue ball from the eight ball in the game of billiards may be expressed in terms of two perpendicular components parallel to the side or end of the pool table, respectively, as shown in Figure 14.10.

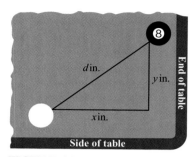

FIGURE 14.10

By the Pythagorean Theorem, $d = \sqrt{x^2 + y^2}$. Find the partial derivative of the distance function. Then evaluate each partial derivative at the point $(3, 4, 5)$ and interpret the meaning of the result.

SOLUTION We will first determine the equations of the partial derivatives.

$$d = \sqrt{x^2 + y^2}$$
$$= (x^2 + y^2)^{1/2}$$

$$d_x = \frac{1}{2}(x^2 + y^2)^{-1/2}(2x) \qquad \text{Chain Rule treating } y \text{ as a constant}$$
$$= x(x^2 + y^2)^{-1/2}$$

$$d_y = \frac{1}{2}(x^2 + y^2)^{-1/2}(2y) \qquad \text{Chain Rule treating } x \text{ as a constant}$$
$$= y(x^2 + y^2)^{-1/2}$$

Evaluating these functions at $(3, 4, 5)$, we determine

$$d_x(3, 4) = (3)(3^2 + 4^2)^{-1/2}$$
$$= \frac{3}{5}$$
$$= 0.6$$

$$d_y(3, 4) = (4)(3^2 + 4^2)^{-1/2}$$
$$= \frac{4}{5}$$
$$= 0.8$$

When the cue ball is five inches away from the eight ball, rolling toward the eight ball one inch (along the path depicted in Figure 14.10) will change the cue ball's distance from the end of the pool table by 0.6 inch and its distance from the side of the pool table by 0.8 inch.

EXAMPLE 5 **Using Partial Derivatives to Forecast Changes in the Wind Chill Temperature**

Wind chill temperature is a function of wind speed and air temperature and is modeled by

$$W(T, V) = 35.74 + 0.6215T - 35.75V^{0.16} + 0.4275TV^{0.16}$$

degrees Fahrenheit, where T is the actual air temperature (in degrees Fahrenheit) and V is the velocity of the wind (in miles per hour). (**Source:** National Climatic Data Center.) If the wind speed is 20 mph, how much will a 1-mph change in wind speed alter the wind chill temperature if the current temperature remains at 30 degrees? If the current temperature is 30 degrees, how much will a 1-degree increase in temperature alter the wind chill temperature if the current wind speed remains at 20 mph?

SOLUTION We begin by calculating the partial derivatives of W.

$$W(T, V) = 35.74 + 0.6215T - 35.75V^{0.16} + 0.4275TV^{0.16}$$

UNITS ➤

$$\frac{\partial W}{\partial T} = 0.6215 + 0.4275V^{0.16} \quad \frac{\text{degree of wind chill temperature}}{\text{degree of actual temperature}}$$

$$\frac{\partial W}{\partial V} = 0.16(-35.75)V^{-0.84} + 0.16(0.4275TV^{-0.84})$$

$$= -5.72V^{-0.84} + 0.0684TV^{-0.84}$$

UNITS ➤

$$= 0.0684V^{-0.84}(-83.63 + T) \quad \frac{\text{degree of wind chill temperature}}{\text{miles per hour}}$$

We will now evaluate each of the partial derivatives at $(30, 20)$.

$$\frac{\partial W}{\partial T}(30, 20) = 0.6215 + 0.4275(20)^{0.16}$$

UNITS ➤

$$= 1.312 \quad \frac{\text{degree of wind chill temperature}}{\text{degree of actual temperature}}$$

Increasing the temperature from 30 to 31 degrees will increase the wind chill temperature by about 1.3 degrees if the wind remains at 20 mph. (Since the partial derivative does not contain the variable T, the wind chill temperature will increase by about 1.3 degrees for every 1 degree increase in temperature regardless of the initial temperature.)

$$\frac{\partial W}{\partial V}(30, 20) = 0.0684V^{-0.84}(-83.63 + T)$$

$$= 0.0684(20)^{-0.84}(-83.63 + 30)$$

UNITS ➤

$$= -0.2962 \quad \frac{\text{degree of wind chill temperature}}{\text{miles per hour}}$$

Increasing the wind speed from 20 to 21 miles per hour will decrease the wind chill temperature by about 0.3 degree when the temperature is 30 degrees. The colder the temperature is, the more substantial the impact a change in wind speed will have on the wind chill temperature.

Cross Sections of a Surface

When an apple is sliced in half, the exposed surface is called a **cross section** of the apple. If the cross section is dipped in paint and pressed on a sheet of paper, the resultant image is a two-dimensional figure (see Figure 14.11).

FIGURE 14.11

The shape of the cross section varies depending upon the point at which the apple is sliced. Figure 14.12 shows three different cross sections of the same apple.

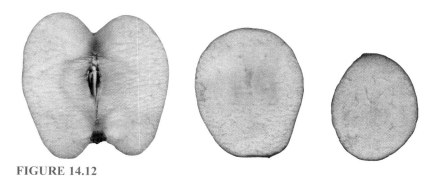

FIGURE 14.12

The equation for the border of the cross section can be derived from the original multivariable equation. (The shape of an apple may be modeled by a multivariable piecewise equation. However, because of the complexity of the equation, we will focus on simpler models as we analyze cross sections.)

Consider the graph of the function $f(x, y) = -x^3 + 4y^2$, shown in Figure 14.13.

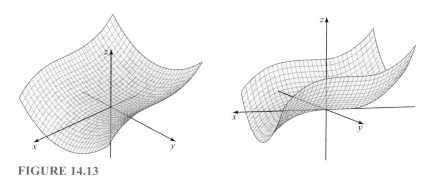

FIGURE 14.13

If we want to know what the cross section of the surface looks like at $y = 0$, we calculate $f(x, 0)$. The equation of the cross-section graph is given by

$$f(x, 0) = -x^3 + 4(0)^2$$
$$= -x^3$$

Slicing the surface at $y = 0$ yields the graph shown in Figure 14.14.

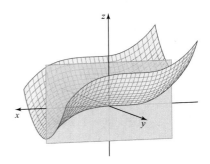

FIGURE 14.14

Note: Looking at the graph from this angle, the x-values go from positive to negative as we read the graph from left to right.

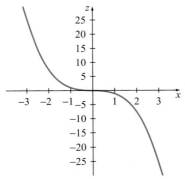

FIGURE 14.15

Graphing the cross section in two dimensions yields the curve $z = -x^3$ (Figure 14.15).

The slope of the cross-section graph is $\dfrac{dz}{dx} = -3x^2$. What is the relationship between the slope of the cross-section graph and the partial derivative? Let's find f_x and evaluate it at $(x, 0)$.

$$f_x = -3x^2$$

Observe that $f_x(x, 0) = -3x^2$. The slope equation for the cross-section graph is the same as the equation of the partial derivative of f with respect to x evaluated at $(x, 0)$. Let's look at another cross section and see if the relationship holds true. This time we'll slice the surface at $y = 2$ (see Figure 14.16).

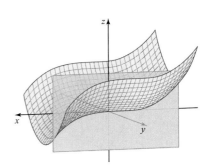

FIGURE 14.16

Note: Looking at the graph from this angle, the x-values go from positive to negative as we read the graph from left to right.

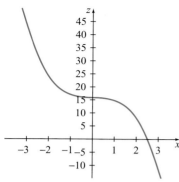

FIGURE 14.17

The equation of the cross-section graph when $y = 2$ is given by $f(x, 2)$.

$$f(x, 2) = -x^3 + 4(2)^2$$
$$= -x^3 + 16$$

Graphing the cross-section graph in two dimensions yields the graph shown in Figure 14.17.

The slope of the cross section graph is $\dfrac{dz}{dx} = -3x^2$. We find f_x and evaluate it at $(x, 2)$.

$$f_x = -3x^2$$

Observe that $f_x(x, 2) = -3x^2$. The slope equation for the cross-section graph is the same as the equation of the partial derivative of f with respect to x evaluated at $(x, 2)$.

GEOMETRIC INTERPRETATION OF PARTIAL DERIVATIVES

The partial derivatives of a function $f(x, y)$ represent rates of change of the graph of the surface.

- $f_x(x, y)$ is the rate of change of the graph of the function $f(x, y)$ when y is held constant. It is referred to as the *slope in the x direction*.
- $f_y(x, y)$ is the rate of change of the graph of the function $f(x, y)$ when x is held constant. It is referred to as the *slope in the y direction*.

Consider the graph of $f(x, y) = -x^2 + 4x - y^2 + y + xy + 100$ (see Figure 14.18). The point $(3, 1, 106)$ lies on the graph.

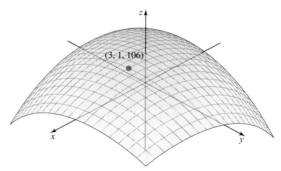

(3, 1, 106)

FIGURE 14.18

Let's calculate the partial derivatives.

$$f_x(x, y) = -2x + 4 + y$$
$$f_y(x, y) = -2y + 1 + x$$

Recall that $f_x(x, y)$ is the slope in the x-direction. Thus, $f_x(3, 1)$ is the slope of the graph in the x-direction at the point $(3, 1, 106)$.

$$f_x(x, y) = -2x + 4 + y$$
$$f_x(3, 1) = -2(3) + 4 + 1$$
$$= -1$$

There is a line tangent to the graph at $(3, 1, 106)$ that has slope -1. The line is of the form $z = -1x + b$, since the line is in the x-direction. Substituting in $x = 3$ and $z = 106$, we can find the exact equation of the line.

$$z = -1x + b$$
$$106 = -3 + b$$
$$b = 109$$
$$\text{So } z = -x + 109$$

The equation of the tangent line to the graph of f in the x-direction at the point $(3, 1, 106)$ is $z = -x + 109$. We draw the line on the graph of f (see Figure 14.19).

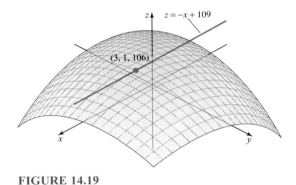

FIGURE 14.19

Recall that $f_y(x, y)$ is the slope in the y-direction. Thus, $f_y(3, 1)$ is the slope of the graph in the y-direction at the point $(3, 1, 106)$.

$$f_y(x, y) = -2y + 1 + x$$
$$f_y(3, 1) = -2(1) + 1 + 3$$
$$= 2$$

There is a line tangent to the graph at $(3, 1, 106)$ that has slope 2. The line is of the form $z = 2y + b$, since the line is in the y-direction. Substituting in $y = 1$ and $z = 106$, we can find the exact equation of the line.

$$z = 2y + b$$
$$106 = 2(1) + b$$
$$106 = 2 + b$$
$$b = 104$$
$$\text{So } z = 2y + 104$$

So the equation of the tangent line to the graph of f in the y-direction at the point $(3, 1, 106)$ is $z = 2y + 104$. We draw the line on the graph of f (see Figure 14.20).

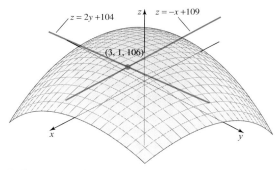

FIGURE 14.20

It is important to note that in three dimensions, the graph of the equation $z = 2y + 104$ is really the *plane* containing the green line that we have drawn, not the line itself. However, by requiring $x = 3$, we constrain the plane to the green line. Similarly, the graph of $z = -x + 109$ is the graph of the *plane* containing the blue line we have drawn, not the line itself. However, by requiring $y = 1$, we constrain the plane to the blue line. Without these additional constraints on x and y, respectively, the graph would look like Figure 14.21.

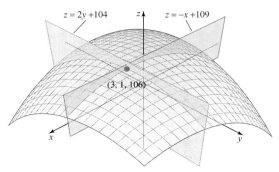

FIGURE 14.21

EXAMPLE 6

Finding Tangent Lines of a Multivariable Function

Given the function $f(x, y) = x^2 - xy^2$, determine the equations of the tangent lines in the x-direction and in the y-direction at the point $(3, 2, -3)$.

SOLUTION The rate of change of f in the x-direction is given by

$$f_x(x, y) = 2x - y^2$$

Thus the slope of the tangent line in the x-direction at $(3, 2, -3)$ is

$$f_x(3, 2) = 2(3) - (2)^2$$
$$= 2$$

Since y is being held constant, the equation will have the form $z = 2x + b$. We find the value of b by substituting in the point $(3, 2, -3)$.

$$z = 2x + b$$
$$-3 = 2(3) + b$$
$$b = -9$$
$$\text{So } z = 2x - 9$$

Thus the equation of the tangent line in the x-direction (when y is constrained to $y = 2$) is $z = 2x - 9$.

The rate of change of f in the y direction is given by

$$f_y(x, y) = -2xy$$

Thus the slope of the tangent line in the y-direction at $(3, 2, -3)$ is

$$f_y(3, 2) = -2(3)(2)$$
$$= -12$$

Since x is being held constant, the equation will have the form $z = -12y + b$. We find the value of b by substituting in the point $(3, 2, -3)$.

$$z = -12y + b$$
$$-3 = -12(2) + b$$
$$b = 21$$
$$\text{So } z = -12y + 21$$

Thus the equation of the tangent line in the y-direction (when x is constrained to $x = 3$) is $z = -12y + 21$.

We can confirm our results graphically by using a graphing utility (see Figure 14.22).

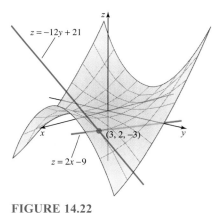

FIGURE 14.22

Second-Order Partial Derivatives

Recall that the derivative of a derivative of a function is called the second derivative of the function. Likewise, a partial derivative of a partial derivative of a function is called a **second-order partial derivative.**

SECOND-ORDER PARTIAL DERIVATIVES

The **second-order partial derivatives** of a function $f(x, y)$ may be written as

$f_{xx} = \dfrac{\partial}{\partial x}\left(\dfrac{\partial f}{\partial x}\right)$ (read "the second partial of f with respect to x")

$f_{yy} = \dfrac{\partial}{\partial y}\left(\dfrac{\partial f}{\partial y}\right)$ (read "the second partial of f with respect to y")

$f_{xy} = \dfrac{\partial}{\partial y}\left(\dfrac{\partial f}{\partial x}\right)$ (read "the mixed second partial derivative of f with respect to x then y")

$f_{yx} = \dfrac{\partial}{\partial x}\left(\dfrac{\partial f}{\partial y}\right)$ (read "the mixed second partial derivative of f with respect to y then x")

For all functions with continuous first and second partial derivatives, $f_{xy} = f_{yx}$. For all multivariable functions in this text, $f_{xy} = f_{yx}$.

EXAMPLE 7

Finding First and Second Partial Derivatives

The graph of the function $f(x, y) = x^2 + y^2 + 1$ is shown in Figure 14.23. Calculate the value of the first and second partial derivatives of the function at the point $(0, 0, 1)$.

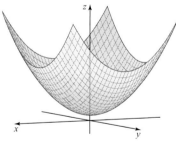

FIGURE 14.23

SOLUTION We have

$$f_x = 2x$$
$$f_y = 2y$$
$$f_{xx} = 2$$
$$f_{yy} = 2$$
$$f_{xy} = 0$$
$$f_{yx} = 0$$

Evaluating each of the partial derivatives at $(0, 0, 1)$, we get

$$f_x(0, 0) = 2(0)$$
$$= 0$$
$$f_y(0, 0) = 2(0)$$
$$= 0$$

The second partial derivatives remain constant for all values of x and y. Thus

$$f_{xx}(0, 0) = 2$$
$$f_{yy}(0, 0) = 2$$
$$f_{xy}(0, 0) = 0$$
$$f_{yx}(0, 0) = 0$$

EXAMPLE 8

Finding First and Second Partial Derivatives

The graph of the function $f(x, y) = x^3 - xy^2$ is shown in Figure 14.24. Calculate the first and second partial derivatives of the function.

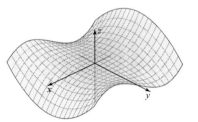

FIGURE 14.24

SOLUTION

$$f_x = 3x^2 - y^2$$
$$f_y = -2xy$$
$$f_{xx} = 6x$$
$$f_{yy} = -2x$$
$$f_{xy} = -2y$$
$$f_{yx} = -2y$$

In the next section, we will demonstrate the key role that second-order partial derivatives play in identifying the location of the maxima and minima of a surface.

14.2 Summary

In this section, you learned how to calculate partial derivatives. You saw that these partial rates of change allow you to determine the effect on the output of a change in one of the input variables.

14.2 Exercises

In Exercises 1–10, calculate the first and second partial derivatives of each of the functions.

1. $f(x, y) = 3xy$ **2.** $f(x, y) = x^2y - xy^2$

3. $f(x, y) = 4xy - y^2$ **4.** $f(x, y) = x^3 - y^3$

5. $W = t\sqrt{v}$ **6.** $W = v^2t^{1.2} + v$

7. $V = \pi r^2 h$ **8.** $V = lw^2$

9. $z = x \ln(xy)$ **10.** $z = xy \ln(y)$

In Exercises 11–17, calculate and interpret the practical meaning of each of the partial derivatives.

11. Volume of a Cylinder The volume of a cylinder is given by $V(h, r) = \pi r^2 h$, with h and r measured in inches. What is the practical meaning of $V_r(3, 2)$ and $V_h(3, 2)$?

12. Volume of a Box The volume of a rectangular box with a square base is given by $V(l, w) = lw^2$, with l and w measured in inches. What is the practical meaning of $V_l(4, 5)$ and $V_w(4, 5)$?

13. Wind Chill Temperature Wind chill temperature is a function of wind speed and air temperature and is modeled by

$$W(T, V) = 35.74 + 0.6215T$$
$$- 35.75V^{0.16} + 0.4275TV^{0.16}$$

degrees Fahrenheit, where T is the actual air temperature (in degrees Fahrenheit) and V is the velocity of the wind (in miles per hour). (Source: National Climatic Data Center.) What is the practical meaning of $W_T(27, 5)$ and $W_V(27, 5)$?

14. Wind Chill Temperature Wind chill temperature is a function of wind speed and air temperature and is modeled by

$$W(T, V) = 35.74 + 0.6215T$$
$$- 35.75V^{0.16} + 0.4275TV^{0.16}$$

degrees Fahrenheit, where T is the actual air temperature (in degrees Fahrenheit) and V is the velocity of the wind (in miles per hour). (Source: National Climatic Data Center.) What is the practical meaning of $W_T(-30, 15)$ and $W_V(-30, 15)$?

15. Body Mass Index The following formula is used to calculate the body mass index:

$$B(H, W) = \frac{703 \cdot W}{H^2}$$

where W is weight in pounds and H is height in inches. What is the practical meaning of $B_H(67, 155)$ and $B_W(67, 155)$?

16. Body Mass Index The following formula is used to calculate body mass index:

$$B(H, W) = \frac{703 \cdot W}{H^2}$$

where W is weight in pounds and H is height in inches. What is the practical meaning of $B_H(67, 100)$ and $B_W(67, 100)$?

17. Price-to-Earnings Ratio The price-to-earnings ratio (P/E) is the most common measure of how expensive a stock is, and is given by

$$R(E, P) = \frac{P}{E}$$

where P is the price of a share of the stock and E is the earnings per share. What is the practical meaning of $R_E(4, 32)$ and $R_P(4, 32)$?

In Exercises 18–20, use partial derivatives to determine the answer to each question.

18. Price-to-Earnings Ratio The price-to-earnings ratio (P/E) is the most common measure of how expensive a stock is, and is given by

$$R(E, P) = \frac{P}{E}$$

where P is the price of a share of the stock and E is the earnings per share.

On December 28, 2002, Electronic Arts stock was selling for $51.50 per share and had a published earnings per share of $0.71. (**Source:** www.quicken.com.) If the stock price increased by $1 or the earnings per share increased by $1, what effect would this have on the P/E ratio?

19. **M** **Company Sales** Based on data from 1998 to 2001, the sales of the Kellogg Company may be modeled by

$$S(a, e) = -0.36661a + 0.16556e + 4671.3$$

million dollars, where a is the amount of money spent on advertising (millions of dollars) and e is the number of employees. In 2001, the company spent $519.2 million on advertising and had 26,424 employees. (**Source:** Modeled from Kellogg Company 2001 Annual Report, p. 26.)

According to the model, what effect would adding one employee in 2002 have on sales? What effect would increasing advertising spending by $1 million have on sales?

20. **M** **Cobb-Douglas Model** Based on data from 1999 and 2000, a Cobb-Douglas model for Ford Motor Company is given by

$$P(L, C) = 0.5319L^{0.4558}C^{0.5442}$$

thousand vehicles, where L is the total labor cost and C is the capital expenditure cost, both in millions of dollars. In 2000, Ford Motor Company spent $25,783 million on labor and produced 7424 thousand vehicles. (**Source:** Modeled from Ford Motor Company 2001 Annual Report.)

According to the model, if Ford increased labor spending by $1 million or increased capital expenditures by $1 million in 2001, what effect would each change have on the production of vehicles?

In Exercises 21–25, a function graph with a cross section drawn at $x = 0$ and $y = -1$ is shown. Determine the equations of the tangent lines of the function in the x-direction and the y-direction at the point $(0, -1, f(0, -1))$.

21. $f(x, y) = x^4 - 4y^2$

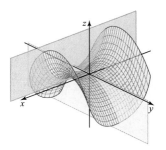

22. $f(x, y) = x^2 - xy + y^2$

23. $f(x, y) = e^{-xy}$

24. $f(x, y) = \dfrac{x - y}{x^2 + y^2 + 1}$

25. $f(x, y) = e^{x^2 + y^2}$

Exercises 26–30 are intended to challenge your understanding of partial derivatives.

26. Determine the rate of change in the *x*-direction and the *y*-direction for the function

$$f(x, y) = \frac{e^x - e^y}{e^{xy}} \text{ at the point } (2, 2, 0).$$

27. Give an example of a function $f(x, y)$ with the property that $f_x(x, y) > 0$, $f_y(x, y) > 0$, and $f_{xy}(x, y) = 0$ for every (x, y) in the domain of f.

28. Give an example of a function $f(x, y)$ with the property that $f(x, y) > 0$, $f_{xx}(x, y) > 0$, $f_{yy}(x, y) > 0$, and $f_{xy}(x, y) = 0$ for every (x, y) in the domain of f.

29. A function $f(x, y) = ax^2 + bxy + cy^2$ has the property that $f_{xx}(x, y) = 4$, $f_{yy}(x, y) = -8$, and $f_{xy}(x, y) = 3$. Determine the values of a, b, and c. Then determine the values of x and y for which $f_x(x, y)$ and $f_y(x, y)$ are simultaneously equal to 1.

30. A function $f(x, y) = ax^2 + bxy + cy^2$ has the property that $f_x(1, 2) = 4$, $f_y(1, 2) = -12$, and $f_{xy}(1, 2) = 0$. Determine the values of a, b, and c.

14.3 Multivariable Maxima and Minima

- Analyze and interpret multivariable mathematical models
- Find critical points of multivariable functions
- Locate relative extrema and saddle points of two-variable function graphs

GETTING STARTED   An investor has $3000 to invest. She wants to get a $10,000 return on her investment without an excessive amount of risk. To achieve her goal, she has decided to invest a portion of her money in a certificate of deposit paying 6 percent interest compounded annually and the rest of her money in an aggressive growth fund that is expected to earn a maximum of 16 percent annually. How much money should she invest in each account if she wants to assume a moderate amount of risk? Questions such as these can be investigated using the techniques associated with finding *multivariable extrema* and *saddle points*.

In this section, we will demonstrate the techniques used to find relative extrema and saddle points of two-variable functions. The techniques used will be similar to those covered in our discussion of single-variable optimization in Chapter 10.

We will first define relative extrema of multivariable functions.

RELATIVE EXTREMA OF MULTIVARIABLE FUNCTIONS

Consider the graph of a surface with equation $z = f(x, y)$. The value $f(a, b)$ is

- a relative minimum if $f(a, b) \leq f(x, y)$ for all (x, y) in some region surrounding (a, b).
- a relative maximum if $f(a, b) \geq f(x, y)$ for all (x, y) in some region surrounding (a, b).

Recall that relative extrema of single-variable functions could not occur at endpoints. Similarly, relative extrema of two-variable functions may not occur on the border of the domain region.

To help you visualize what relative extrema in three-dimensions look like, we will label the relative extrema on the graphs of several multivariable functions (see Figure 14.25).

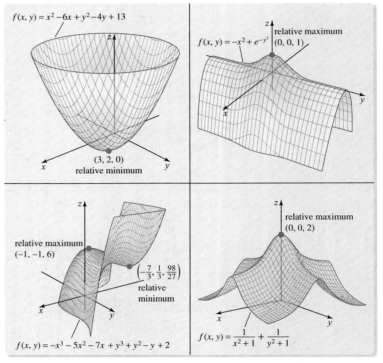

FIGURE 14.25

As was the case in two dimensions, relative extrema have horizontal tangent lines provided that the extrema don't occur at a sharp point on the graph.

In addition to relative extrema, a three-dimensional surface may have a *saddle point.* A **saddle point** is a point (a, b, c) that is a relative maximum of one cross-section graph and a relative minimum of another cross-section graph. The portion of the graph surrounding the saddle point looks like a "saddle." Consider the graph of $f(x, y) = -x^2 + y^2 + 2$, shown in Figure 14.26.

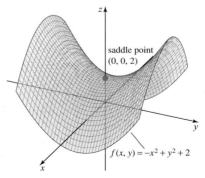

FIGURE 14.26

If we slice the graph with the plane $x = 0$, we see a cross section that looks like a concave up parabola with a vertex at $(0, 0, 2)$ (see Figure 14.27).

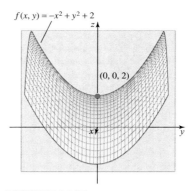

FIGURE 14.27

The smallest possible value of z (in the cross section) is reached at the vertex. Similarly, if we slice the graph with the plane $y = 0$, we see a cross section that looks like a concave down parabola with a vertex at $(0, 0, 2)$ (see Figure 14.28).

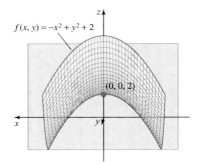

FIGURE 14.28

The largest possible value of z (in the cross section) is reached at the vertex. Since the same point is a maximum value of one cross section and a minimum value of a different cross section, the point $(0, 0, 2)$ is a saddle point. Later in the section, we will discuss how to find saddle points using partial derivatives.

Recall from our discussion of single-variable functions that a value c in the domain of f was called a critical value of f if $f'(c) = 0$ or $f'(c)$ was undefined. All relative extrema occurred at critical values.

A similar definition is used for two-variable functions.

CRITICAL POINT OF A TWO-VARIABLE FUNCTION

An ordered pair (c, d) in the domain of f is a **critical point** of f if

$$f_x(c, d) = 0 \text{ and } f_y(c, d) = 0$$

(Both statements must be true in order for (c, d) to be a critical point.)

When we refer to the critical point (c, d), we mean that (c, d) is the pair of input values whose output is $f(c, d)$.

Let's find the critical point of $f(x, y) = x^2 + y^2 + 1$.

$$f_x = 2x$$
$$f_y = 2y$$

Setting each partial derivative equal to zero and solving, we find that

$$f_x = 2x \qquad f_y = 2y$$
$$0 = 2x \qquad 0 = 2y$$
$$x = 0 \qquad y = 0$$

Therefore, the ordered pair $(0, 0)$ is a critical point of f. The domain region of the function is defined by $-\infty < x < \infty$ and $-\infty < y < \infty$. Since $(0, 0)$ is in the interior of the region, it is a relative extremum candidate.

To determine whether a maximum or minimum occurs at a critical point, we will apply the Second Derivative Test for Two-Variable Functions.

SECOND DERIVATIVE TEST FOR TWO-VARIABLE FUNCTIONS (D-TEST)

To determine if a relative extremum of a function f occurs at a critical point (c, d), first calculate

$$D(c, d) = f_{xx}(c, d) \cdot f_{yy}(c, d) - [f_{xy}(c, d)]^2$$

Then

- f has a **relative maximum** at (c, d) if $D(c, d) > 0$ and $f_{xx}(c, d) < 0$.
- f has a **relative minimum** at (c, d) if $D(c, d) > 0$ and $f_{xx}(c, d) > 0$.
- f has a **saddle point** at (c, d) if $D(c, d) < 0$.
- The test is inconclusive if $D(c, d) = 0$.

To apply the D-Test, we must first find the second partial derivatives. For the function $f(x, y) = x^2 + y^2 + 1$, we have the first partial derivatives $f_x = 2x$ and $f_y = 2y$. Therefore,

$$f_{xx} = 2$$
$$f_{yy} = 2$$
$$f_{xy} = 0$$

We evaluate each of the second partial derivatives at the critical point $(0, 0)$. In this case, the second partials are constant for all values of x and y. Thus

$$f_{xx}(0, 0) = 2$$
$$f_{yy}(0, 0) = 2$$
$$f_{xy}(0, 0) = 0$$

We now calculate D.

$$D(0, 0) = 2 \cdot 2 - 0^2$$
$$= 4$$

Since $D > 0$, either a maximum or a minimum occurs at $(0, 0)$. Since $f_{xx}(0, 0) > 0$, a relative minimum occurs at $(0, 0)$. Evaluating f at $(0, 0)$, we determine that

$$f(0, 0) = 0^2 + 0^2 + 1$$
$$= 1$$

The relative minimum occurs at $(0, 0, 1)$.

EXAMPLE **1**

Locating Relative Extrema and Saddle Points

Find the location of the relative extrema and saddle points of $f(x, y) = x^2 + y^2 + xy - 2$. The graph of the surface is shown in Figure 14.29.

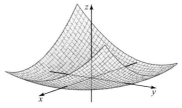

FIGURE 14.29

SOLUTION The partial derivatives of f are

$$f_x = 2x + y$$
$$f_y = 2y + x$$

We set each partial derivative equal to zero and solve.

$$0 = 2x + y \qquad 0 = 2y + x$$
$$y = -2x \qquad y = -0.5x$$

Since both equalities must be true at a critical point, we set the functions equal to each other and solve.

$$-2x = -0.5x$$
$$1.5x = 0$$
$$x = 0$$

Back substituting into the equation $y = -2x$, we determine that

$$y = -2(0)$$
$$= 0$$

The critical point of the function is $(0, 0)$. We now find the second partial derivatives.

$$f_{xx} = 2$$
$$f_{yy} = 2$$
$$f_{xy} = 1$$

Since the second partials are constant, evaluating each second partial at the critical point will result in the same values.

$$D(0, 0) = 2 \cdot 2 - 1 \cdot 1$$
$$= 3$$

Since $D(0, 0) > 0$ and $f_{xx} > 0$, a relative minimum occurs at $(0, 0)$. The value of the function at $(0, 0)$ is $f(0, 0)$.

$$f(0, 0) = (0)^2 + (0)^2 + (0)(0) - 2$$
$$= -2$$

The relative minimum occurs at $(0, 0, -2)$. There are no relative maxima or saddle points.

EXAMPLE 2

Locating Relative Extrema and Saddle Points

Find the location of the relative extrema and saddle points of $f(x, y) = e^{-xy}$. The graph of the surface is shown from two different viewpoints in Figure 14.30.

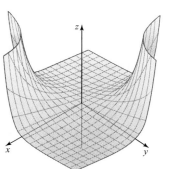

FIGURE 14.30

SOLUTION We'll first find the partial derivatives of f.

$$f_x = -ye^{-xy}$$
$$f_y = -xe^{-xy}$$

To find the critical points, we'll set each partial equal to zero and solve.

$$0 = -ye^{-xy}$$

We solve the equation by setting each factor to zero.

$$-y = 0 \qquad e^{-xy} = 0$$
$$y = 0$$

Since $e^{-xy} \neq 0$ for all values of x and y, the only value that makes the equation zero is $y = 0$.

We'll now set f_y equal to zero.

$$0 = -xe^{-xy}$$

We solve the equation by setting each factor to zero.

$$x = 0 \qquad e^{-xy} = 0$$

Since $e^{-xy} \neq 0$ for all values of x and y, the only value that makes the equation zero is $x = 0$. Thus, the critical point of the function is $(0, 0)$.

We'll now find the second partials and calculate the value of D at $(0, 0)$.

$$f_{xx} = y^2 e^{-xy}$$
$$f_{yy} = x^2 e^{-xy}$$
$$f_{xy} = -1(e^{-xy}) + (-xe^{-xy})(-y)$$
$$= -e^{-xy}(1 - xy)$$
$$f_{xx}(0, 0) = 0$$
$$f_{yy}(0, 0) = 0$$
$$f_{xy}(0, 0) = -e^{(0)(0)}[1 - (0)(0)]$$
$$= -1$$
$$D(0, 0) = 0 \cdot 0 - (-1)^2$$
$$= -1$$

Since $D < 0$, a saddle point occurs at the critical point $(0, 0)$. The coordinates of the saddle point are $(0, 0, 1)$. There are no relative extrema.

EXAMPLE 3

Finding Relative Extrema and Saddle Points

Find the location of the relative extrema and saddle points of $f(x, y) = 0.1x^2 y^2$. Two different views of the graph of the surface are shown in Figure 14.31.

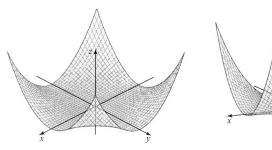

FIGURE 14.31

SOLUTION The partial derivatives of f are given by

$$f_x = 0.2xy^2$$
$$f_y = 0.2x^2 y$$

Setting the partial derivatives equal to zero and solving yields

$$0 = 0.2xy^2 \qquad\qquad 0 = 0.2x^2 y$$
$$0 = xy^2 \qquad\qquad 0 = x^2 y$$
$$x = 0 \quad \text{or} \quad y = 0 \qquad x = 0 \quad \text{or} \quad y = 0$$

Any point of the form $(a, 0)$ or $(0, b)$ is a critical point of the function (a and b are real numbers). The function f has an infinite number of critical points.

The second partial derivatives are given by

$$f_{xx} = 0.2y^2$$
$$f_{yy} = 0.2x^2$$
$$f_{xy} = 0.4xy$$
$$f_{yx} = 0.4xy$$

$$D = (0.2y^2)(0.2x^2) - (0.4xy)^2$$
$$= 0.04x^2y^2 - 0.16x^2y^2$$
$$= -0.12x^2y^2$$

Observe that $D(a, 0) = 0$ and $D(0, b) = 0$ for all values of a and b.

The D-Test is inconclusive for all critical points. However, by evaluating the function at points near the critical points, we can determine what is happening there. We'll generate a data table (Table 14.6) by increasing and decreasing each domain variable on either side of the critical points.

TABLE 14.6

x $\downarrow$	$y \rightarrow$				
	−2	−1	0	1	2
−2	1.6	0.4	0	0.4	1.6
−1	0.4	0.1	0	0.1	0.4
0	0	0	0	0	0
1	0.4	0.1	0	0.1	0.4
2	1.6	0.4	0	0.4	1.6

The values highlighted in Table 14.6 are the function values of the critical points of f. Observe that as we move away from the highlighted values vertically or horizontally, the value of f increases or remains constant. Since $f(a, 0) \leq f(x, y)$ and $f(0, b) \leq f(x, y)$ for all values of a, b, x, and y, relative minima occur at all of the critical points. There are no relative maxima or saddle points.

EXAMPLE 4

Using Multivariable Functions to Forecast Investment Growth

An investor has $3000 to invest in two separate accounts. The first account is a certificate of deposit guaranteed to earn 6 percent annually. The second account is an aggressive growth fund expected to earn between −4 percent and 16 percent annually. The combined dollar value of the return on the investment accounts after 20 years is given by

$$V = x(1.06)^{20} + (3000 - x)(1 + y)^{20} - 3000$$

where x is the amount of money invested in the certificate of deposit and y is the growth rate (as a decimal) of the aggressive growth fund. (If the growth rate is 8 percent, $y = 0.08$.) A graph of the function is shown in Figure 14.32. The green plane is a graph of the function $z = 0$ and is drawn to provide a point of reference. When the graph of V is above the green plane, the investor makes money. When the graph of V is below the green plane, the investor loses money.

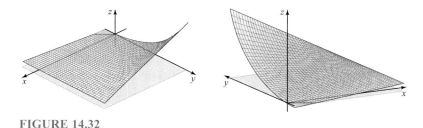

FIGURE 14.32

Find all critical points of V and determine the location of relative extrema and saddle points.

SOLUTION

$$V_x = (1.06)^{20} - (1 + y)^{20}$$
$$V_y = 20(3000 - x)(1 + y)^{19}$$

We set each partial derivative equal to zero.

$$0 = (1.06)^{20} - (1 + y)^{20}$$
$$(1 + y)^{20} = (1.06)^{20}$$
$$1 + y = 1.06$$
$$y = 0.06$$
$$0 = 20(3000 - x)(1 + y)^{19}$$
$$3000 - x = 0 \quad \text{or} \quad (1 + y)^{19} = 0$$
$$x = 3000 \qquad\qquad y = -1$$

Recall that in order to be a critical point, the ordered pair must satisfy both equations. Since $-1 \neq 0.06$, we throw out the value $y = -1$. Thus the only critical point is $(3000, 0.06)$.

We'll now find the second partials.

$$V_{xx} = 0$$
$$V_{yy} = 380(3000 - x)(1 + y)^{18}$$
$$V_{xy} = -20(1 + y)^{19}$$
$$D = (0)[380(3000 - x)(1 + y)^{18}] - [-20(1 + y)^{19}]^2$$
$$= -400(1 + y)^{38}$$
$$D(3000, 0.06) = -400(1 + 0.06)^{38}$$
$$D(3000, 0.06) < 0 \qquad\qquad \text{We only need to know the sign of } D$$

Since $D < 0$, a saddle point occurs at $(3000, 0.06, 6621.41)$. However, from the context of the problem, we know that $0 \leq x \leq 3000$. Therefore, this point sits on the border of the domain region. As with relative extrema, a saddle point may not sit on the border of the domain region. We conclude that the function does not have any relative extrema or saddle points on the domain region defined by $0 \leq x \leq 3000$ and $-0.04 \leq y \leq 0.16$.

Let's look at the table of data for the function in the region surrounding the critical point given in Example 4 (Table 14.7).

TABLE 14.7

x ↓	$y \rightarrow$						
	−0.04	0.00	0.04	0.06	0.08	0.12	0.16
2,600	5,515.35	5,738.55	6,215.00	6,621.41	7,202.94	9,197.07	13,122.86
2,700	5,791.87	5,959.27	6,316.60	6,621.41	7,057.55	8,553.15	11,497.49
2,800	6,068.38	6,179.98	6,418.20	6,621.41	6,912.17	7,909.24	9,872.13
2,900	6,344.89	6,400.69	6,519.81	6,621.41	6,766.79	7,265.32	8,246.77
3,000	6,621.41	6,621.41	6,621.41	6,621.41	6,621.41	6,621.41	6,621.41

From the table, we see that investing all of the money in the certificate of deposit or earning exactly 6 percent on the amount of money invested in the aggressive growth account will yield a combined return of \$6621.41 on the \$3000 investment. As the amount of money invested in the certificate of deposit (x) decreases, the amount of the return becomes more volatile. If \$2600 is invested in the certificate of deposit and \$400 is invested in the aggressive growth account, the combined return could be as low as \$5515.35 or as high as \$13,122.86. Many investors are willing to risk losing \$1106.06 (\$6621.41 − \$5515.35) for the potential to earn an additional \$6501.45 (\$13,122.86 − \$6621.41). Investors who are willing to risk all \$3000 on the aggressive growth account could lose as much as \$1673.99 of their initial investment or earn as much as \$55,382.28 on top of their initial investment. Most financial advisers encourage investors to diversify their investments among a variety of investment accounts with varying levels of risk in order to achieve a return that meets the investor's goals.

EXAMPLE 5   **Using Multivariable Functions to Forecast Product Sales**

A software company produces two versions of its lead product: one for the Windows operating system and the other for the Macintosh operating system. Through market research, the company determines that at a price of p dollars, approximately q copies of the Windows version will be sold, and at a price of s dollars, approximately v copies of the Macintosh version will be sold. The relationship between the price of each item and sales may be modeled by

$$q = \frac{100{,}000}{p} + 100 - 0.2p \quad \text{and} \quad v = \frac{20{,}000}{s} + 50 - 0.1s$$

Determine the price the company should charge for each version in order to maximize the combined revenue from sales.

SOLUTION   The revenue from the sale of q copies of the Windows version is given by

$$pq = p\left(\frac{100{,}000}{p} + 100 - 0.2p\right)$$
$$= 100{,}000 + 100p - 0.2p^2$$

Revenue from the sale of v copies of the Macintosh version is given by

$$sv = s\left(\frac{20{,}000}{s} + 50 - 0.1s\right)$$
$$= 20{,}000 + 50s - 0.1s^2$$

The combined revenue from the sales of the two products is given by

$$R(p, s) = 100{,}000 + 100p - 0.2p^2 + 20{,}000 + 50s - 0.1s^2$$
$$= 100p - 0.2p^2 + 50s - 0.1s^2 + 120{,}000$$

From the graph of R on the region defined by $0 \le p \le 500$ and $0 \le s \le 500$ (see Figure 14.33), we can see that the revenue function has a relative maximum.

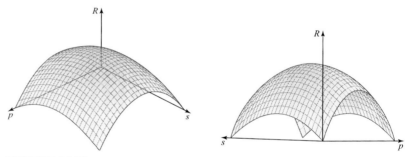

FIGURE 14.33

To find the location of the relative maximum, we first find the partial derivatives.

$$R_p = 100 - 0.4p$$
$$R_s = 50 - 0.2s$$

We then set each partial derivative equal to zero and solve.

$$0 = 100 - 0.4p \qquad 0 = 50 - 0.2s$$
$$0.4p = 100 \qquad 0.2s = 50$$
$$p = 250 \qquad s = 250$$

The critical point is $(250, 250)$. The second partial derivatives are

$$R_{pp} = -0.4$$
$$R_{ss} = -0.2$$
$$R_{ps} = 0$$

$$D(250, 250) = (-0.4)(-0.2) - 0$$
$$= 0.08$$

Since $D > 0$ and $R_{pp} < 0$, a relative maximum occurs at $(250, 250)$. To maximize revenue, the company should charge \$250 for each version of the software. The maximum combined revenue is \$138,750, with \$112,500 coming from the Windows version and \$26,250 coming from the Macintosh version.

EXAMPLE 6

Finding Relative Extrema and Saddle Points

Find the relative extrema and saddle points of $f(x, y) = x^4 - 4x^3 + 4x^2 - y^2$. Two different views of the graph are shown in Figure 14.34.

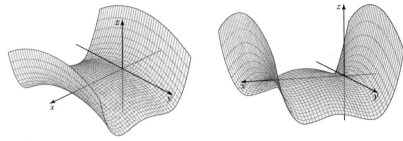

FIGURE 14.34

SOLUTION   We will first find the critical points of f. We begin by finding the partial derivatives of f.

$$f_x = 4x^3 - 12x^2 + 8x$$
$$f_y = -2y$$

We then set each partial derivative equal to 0 and solve.

$$4x^3 - 12x^2 + 8x = 0 \qquad\qquad -2y = 0$$
$$4x(x^2 - 3x + 2) = 0 \qquad\qquad y = 0$$
$$4x(x - 1)(x - 2) = 0$$
$$x = 0, 1, 2$$

We must pair up each x-value with each y-value to form the critical points. The critical points are $(0, 0)$, $(1, 0)$, and $(2, 0)$.

Next we will find the second partial derivatives and perform the D-Test on each critical point.

$$f_{xx} = 12x^2 - 24x + 8$$
$$= 4(3x^2 - 6x + 2)$$
$$f_{yy} = -2$$
$$f_{xy} = 0$$
$$D = (12x^2 - 24x + 8)(-2) - (0)(0)$$
$$= -8(3x^2 - 6x + 2)$$

When there are many critical points, it is often helpful to construct a table of data (see Table 14.8).

TABLE 14.8

Critical Point	$f_{xx} = 4(3x^2 - 6x + 2)$	$D = -8(3x^2 - 6x + 2)$	Graphical Interpretation
$(0, 0)$	8	-16	Saddle point
$(1, 0)$	-4	8	Relative maximum
$(2, 0)$	8	-16	Saddle point

FIGURE 14.35

In order to understand better what a saddle point looks like, we'll zoom in on the region surrounding the critical point at $(0, 0)$ (see Figure 14.35).

Looking at the graph from the xz plane (Figure 14.36), it looks as if a relative minimum occurs at the critical point.

FIGURE 14.36

However, looking at the graph from the yz plane (Figure 14.37), it looks as if a relative maximum occurs at the critical point.

When a graph has a saddle point, the point will look like a minimum from one plane and a maximum from the other plane. (The planes will not necessarily be the xz plane and yz plane.)

To determine the actual coordinates of each of the extrema and saddle points, we evaluate the function at the critical points (see Table 14.9).

FIGURE 14.37

TABLE 14.9

x	y
$\downarrow$	0
0	0
1	1
2	0

From the table, we see the relative maximum occurs at $(1, 0, 1)$. Saddle points occur at $(0, 0, 0)$ and $(2, 0, 0)$. These results are consistent with our earlier conclusions.

14.3 Summary

In this section, you learned the techniques used to find relative extrema and saddle points of two-variable functions. In the next section, we will look at finding absolute extrema subject to function constraints.

14.3 Exercises

In Exercises 1–25, find the relative extrema and saddle points of the function.

1. $f(x, y) = x^2 + 4y^2 - 10$

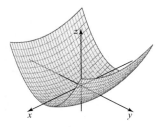

2. $f(x, y) = x^2 - 4xy$

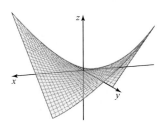

3. $f(x, y) = -x^2 - y^3 + 3y - x$

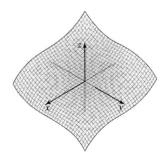

4. $f(x, y) = x^4 - 4x^2 + y^4 - 9y^2 + 1$

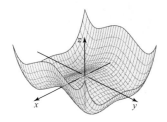

5. $f(x, y) = x^2y - y^2 + x$

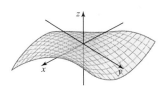

6. $f(x, y) = x^4 - 4x + y^4 - 9y^2 + 1$

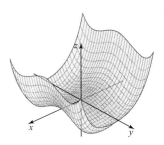

7. $f(x, y) = x^4 - 4x + y^3 - y^2$

8. $f(x, y) = x^3 - 2x + y^3 - 2y$

9. $f(x, y) = x^3 - 6x + xy - y^2$

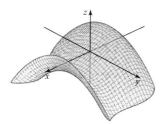

10. $f(x, y) = -\dfrac{x^2 - 4}{y^2 + 4}$

11. $f(x, y) = \dfrac{x^2 + 1}{y^2 + 1}$

12. $f(x, y) = e^{-x^2 - y^2}$

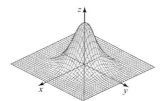

13. $f(x, y) = \dfrac{-1}{x^2 + 1} + y^4 - 4y^2$

14. $f(x, y) = e^{xy - 4}$

15. $f(x, y) = x^3 - 3x + y^3 - 12y$

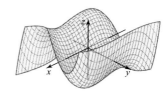

16. $f(x, y) = x^2 - 4x + y^2 - y - xy$

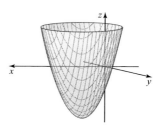

17. $f(x, y) = x^2 - x - 3y^2 + 3y$

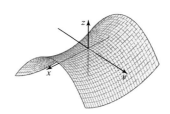

18. $f(x, y) = \ln(x^2 + 1) + \ln(y^2 + 1)$

19. $f(x, y) = \dfrac{x^2 + y^2 + 1}{e^{x^2 + y^2}}$

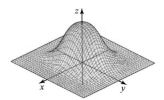

20. $f(x, y) = \dfrac{(x - 1)^2 + (y - 2)^2}{e^{(x-1)^2 + (y-2)^2}}$

21. $f(x, y) = \dfrac{x^2 - y^2}{e^{x^2 + y^2}}$

22. $f(x, y) = \dfrac{x^2 - 4}{y^2 + 1} - 1$

23. $f(x, y) = e^{-x^2 + 2x - y^2 + y + 1}$

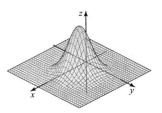

24. $f(x, y) = \sqrt{x^2 - 2x + y^2 - 4y + 5}$

25. $f(x, y) = x^3y + xy^3 - 4xy$

14.4 Constrained Maxima and Minima and Applications

- Analyze and interpret multivariable mathematical models
- Use the Lagrange Multiplier Method to find constrained maxima and minima

GETTING STARTED Based on data from 1999 and 2000, a Cobb-Douglas model for Ford Motor Company is given by

$$P(L, C) = 0.5319L^{0.4558}C^{0.5442}$$

thousand vehicles, where L is the total labor cost and C is the capital expenditure cost, both in millions of dollars. In 2000, Ford Motor Company spent $25,783 million on labor and produced 7424 thousand vehicles. (**Source:** Modeled from Ford Motor Company 2001 Annual Report.) Assuming that Ford had $34,131 million to spend on labor and capital in 2000, how much should it have spent on labor and how much should it have spent on capital expenditures in order to maximize vehicle production? Questions such as this may be answered by solving a constrained optimization problem. (See Example 3.)

In this section, we will demonstrate how to use the *Lagrange Multiplier Method* to solve constrained multivariable optimization problems. This technique may be used in two-variable and multivariable optimization problems. We will first demonstrate the technique with an example and then formally define the process.

EXAMPLE 1

Minimizing a Multivariable Function with the Lagrange Multiplier Method

Minimize $f(x, y) = x^2 - 2y$ subject to the constraint $-2x + y = 4$ (see Figure 14.38).

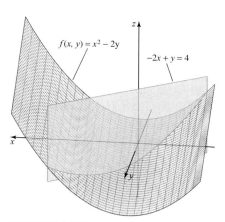

$f(x, y) = x^2 - 2y$

$-2x + y = 4$

FIGURE 14.38

SOLUTION We are to minimize $f(x, y) = x^2 - 2y$ subject to the constraint $-2x + y = 4$.

As shown in Figure 14.38, the graph of the constraint is a plane that cuts through the graph of $f(x, y) = x^2 - 2y$. We are to minimize f along the intersection of the graph of f and the graph of the constraint.

We have

$$-2x + y = 4$$
$$-2x + y - 4 = 0$$

We will call the constraint function $g(x, y)$. That is,

$$g(x, y) = -2x + y - 4$$

Observe that $g(x, y) = 0$. We will now introduce a new function $F(x, y, \lambda)$. The third input variable of the function is called the **Lagrange multiplier** and is represented by the Greek letter λ (read "lambda"). The function $F(x, y, \lambda)$ is called the **Lagrange function** and is equal to the function to be maximized plus λ times the constraint function. That is,

$$\begin{aligned} F(x, y, \lambda) &= f(x, y) + \lambda g(x, y) \\ &= x^2 - 2y + \lambda(-2x + y - 4) \\ &= x^2 - 2y - 2\lambda x + \lambda y - 4\lambda \end{aligned}$$

We will next take the partial derivative of F with respect to each of the input variables.

$$F_x = 2x - 2\lambda$$
$$F_y = -2 + \lambda$$
$$F_\lambda = -2x + y - 4$$

By design, $F_\lambda = g(x, y) = 0$. We will now set the other two partial derivatives equal to zero and solve for λ.

$$\begin{array}{ll} F_x = 2x - 2\lambda & \qquad F_y = -2 + \lambda \\ 0 = 2x - 2\lambda & \qquad 0 = -2 + \lambda \\ 2\lambda = 2x & \qquad \lambda = 2 \\ \lambda = x & \end{array}$$

Now we set the lambdas equal to each other and solve.

$$x = 2$$

So $x = 2$. We will substitute this result back into the constraint equation.

$$\begin{aligned} g(x, y) &= 0 \\ -2x + y - 4 &= 0 \\ -2x + y &= 4 \\ -2(2) + y &= 4 \\ -4 + y &= 4 \\ y &= 8 \end{aligned}$$

The critical point of the *objective function* subject to the constraint is $(2, 8)$. (Recall that the objective function is the function that is to be maximized or minimized.) It is important to note that the D-Test works only for unconstrained optimization and cannot be used to solve a constrained problem. Although the

Lagrange Multiplier Method finds the critical points of the constrained objective function, it does not tell us whether maxima or minima (or neither) occur at these points. Let's construct a table of values for f at points near the critical point that also satisfy the constraint equation $-2x + y = 4$ (see Table 14.10). Points satisfying the constraint equation nearby $(2, 8)$ include $(0, 4)$, $(1, 6)$, $(3, 10)$, and $(4, 12)$.

TABLE 14.10

x	y	$f(x, y)$
0	4	-8
1	6	-11
2	8	-12
3	10	-11
4	12	-8

The constrained minimum of f is $(2, 8, -12)$.

The approach used in Example 1 is referred to as the Lagrange Multiplier Method. The steps of the method are detailed in the following box.

HOW TO **Lagrange Multiplier Method for Solving Constrained Optimization Problems**

To optimize an objective function $f(x, y)$ subject to the constraint $g(x, y) = 0$, do the following:

1. Define $F(x, y, \lambda) = f(x, y) + \lambda g(x, y)$.
2. Set the partial derivatives of F equal to zero

$$F_x = 0 \qquad F_y = 0 \qquad F_\lambda = 0$$

   and solve to find the critical points.

3. The solution to the constrained optimization problem (if it exists) will occur at one of these critical points.

EXAMPLE 2 **Maximizing a Multivariable Function with the Lagrange Multiplier Method**

The United States Postal Service classifies a package as Parcel Post™ if its length plus its girth is less than or equal to 130 inches. (**Source:** U.S. Postal Service.) (The girth is the distance around the package at its thickest point.) What are the dimensions of the rectangular packing box with a square base that has the greatest volume and meets the requirements for Parcel Post?

SOLUTION We are looking for the dimensions of the rectangular box with maximum volume (see Figure 14.39).

FIGURE 14.39

The volume of the box is a multivariable function given by

$$V(l, w) = lw^2$$

We have the constraint that the length plus the girth must not exceed 130 inches. A package with maximum volume will use all of the 130 inches allowed. Since the length of the package is l and the girth of the package is $4w$ we have

$$130 = l + 4w$$

$$0 = l + 4w - 130$$

We will call the constraint function $g(l, w)$. That is,

$$g(l, w) = l + 4w - 130$$

Observe that $g(l, w) = 0$.

We will now find the Lagrange function $F(l, w, \lambda)$. Recall that it is equal to the function to be maximized plus λ times the constraint function. That is,

$$F(l, w, \lambda) = V(l, w) + \lambda g(l, w)$$
$$= lw^2 + \lambda(l + 4w - 130)$$
$$= lw^2 + \lambda l + 4\lambda w - 130\lambda$$

We will next take the partial derivative of F with respect to each of the input variables.

$$F_l = w^2 + \lambda$$
$$F_w = 2lw + 4\lambda$$
$$F_\lambda = l + 4w - 130$$

By design, $F_\lambda = g(l, w) = 0$. We will now set the other two partial derivatives equal to zero and solve for λ.

$$0 = w^2 + \lambda \qquad 0 = 2lw + 4\lambda$$
$$\lambda = -w^2 \qquad 4\lambda = -2lw$$
$$\lambda = -0.5lw$$

Now we set the lambdas equal to each other and solve.

$$-w^2 = -0.5lw$$
$$w^2 - 0.5lw = 0$$
$$w(w - 0.5l) = 0$$

So $w = 0$ or $w = 0.5l$. We will substitute each of these results back into the constraint equation.

$$g(l, w) = 0 \qquad\qquad g(l, w) = 0$$
$$l + 4w - 130 = 0 \qquad\qquad l + 4w - 130 = 0$$
$$l + 4(0) - 130 = 0 \quad \text{or} \quad l + 4(0.5l) - 130 = 0$$
$$l = 130 \qquad\qquad 3l = 130$$
$$l \approx 43.33$$

We now back substitute these results to solve for w.

$$g(l, w) = 0 \qquad\qquad g(l, w) = 0$$
$$l + 4w - 130 = 0 \qquad\qquad l + 4w - 130 = 0$$
$$130 + 4w - 130 = 0 \quad \text{or} \quad 43.33 + 4w - 130 = 0$$
$$4w = 0 \qquad\qquad 4w = 86.67$$
$$w = 0 \qquad\qquad w \approx 21.67$$

Since it doesn't make sense to have a box with width 0 inches, we throw out the extraneous solution $(130, 0)$. The box with maximum volume is 43.33 inches long, 21.67 inches high, and 21.67 inches wide.

As stated before, the Lagrange Multiplier Method finds the critical points of the constrained objective function, but it does not tell us whether maxima or minima (or neither) occur at these points. However, in most applied problems, we can figure out the status of the points from the context of the problem. In Example 2, we threw out the critical point $(130, 0)$ because it didn't make sense to talk about the volume of a box with width 0. Since conceptually we knew that there had to be a box with maximum volume, we concluded that the other critical point would yield the optimal solution.

Locating critical points is sometimes a challenge. However, approaching the task in a methodical way makes it easier. Consider the following technique for finding critical points.

HOW TO **Hints for Finding the Critical Points**

1. Solve $F_x = 0$ and $F_y = 0$ for λ.
2. Eliminate λ by setting the resultant equations equal to each other.
3. Substitute the result from Step 2 into the equation $g(x, y) = 0$ and solve for x and y.

EXAMPLE 3 **Using the Lagrange Multiplier Method in a Real-World Setting**

Based on data from 1999 and 2000, a Cobb-Douglas model for Ford Motor Company is given by

$$P(L, C) = 0.5319L^{0.4558}C^{0.5442}$$

thousand vehicles, where L is the total labor cost and C is the capital expenditure cost, both in millions of dollars. In 2000, Ford Motor Company spent \$25,783 million on labor and produced 7424 thousand vehicles. (**Source:** Modeled from Ford Motor Company 2001 Annual Report.) According to the model, how much should Ford have spent on labor and how much should it have spent on capital if its goal was to maximize production? According to its 2001 annual report, Ford had \$34,131 million to spend on labor and capital in 2000.

SOLUTION We want to maximize $P(L, C) = 0.5319L^{0.4558}C^{0.5442}$ subject to the constraint $L + C = 34{,}131$.

We define $g(L, C) = L + C - 34{,}131$. Then

$$
\begin{aligned}
F(L, C) &= P(L, C) + \lambda g(L, C) \\
&= 0.5319L^{0.4558}C^{0.5442} + \lambda(L + C - 34{,}131) \\
&= 0.5319L^{0.4558}C^{0.5442} + \lambda L + \lambda C - 34{,}131\lambda
\end{aligned}
$$

Next we find the partial derivatives of F.

$$
\begin{aligned}
F_L &= 0.24244L^{-0.5442}C^{0.5442} + \lambda \\
F_C &= 0.28946L^{0.4558}C^{-0.4558} + \lambda \\
F_\lambda &= L + C - 34{,}131
\end{aligned}
$$

We set the first two partials equal to zero and solve for λ.

$$
\begin{aligned}
0 &= 0.24244L^{-0.5442}C^{0.5442} + \lambda \\
\lambda &= -0.24244L^{-0.5442}C^{0.5442}
\end{aligned}
$$

$$
\begin{aligned}
0 &= 0.28946L^{0.4558}C^{-0.4558} + \lambda \\
\lambda &= -0.28946L^{0.4558}C^{-0.4558}
\end{aligned}
$$

We then set the equations equal to each other and solve.

$$
-0.24244L^{-0.5442}C^{0.5442} = -0.28946L^{0.4558}C^{-0.4558}
$$

$$
\frac{L^{0.4558}}{L^{-0.5442}} = \frac{-0.24244\,C^{0.5442}}{-0.28946\,C^{-0.4558}}
$$

$$
L = 0.83756C
$$

We substitute this result into the third equation, $F_\lambda = L + C - 34{,}131$. Since $F_\lambda = g(L, C)$, $F_\lambda = 0$. Thus $L + C - 34{,}131 = 0$ and

$$
\begin{aligned}
L + C - 34{,}131 &= 0 \\
0.83756C + C - 34{,}131 &= 0 \qquad \text{Since } L = 0.83756C \\
1.83756C &= 34{,}131 \\
C &= 18{,}574
\end{aligned}
$$

We know that $L + C = 34{,}131$; thus

$$
\begin{aligned}
L + 18{,}574 &= 34{,}131 \\
L &= 15{,}557
\end{aligned}
$$

According to our model, when \$15,557 million is invested in labor and \$18,574 is invested in capital expenditures, vehicle production will be maximized. Since $P(15{,}557, 18{,}574) = 9113$, we conclude that a maximum of 9113 thousand vehicles could have been produced in 2000.

EXAMPLE 4

Maximizing Volume with the Lagrange Multiplier Method

A box designer wants to construct a closeable box with a square base, as shown in Figure 14.40. The box will be used to store products in a warehouse.

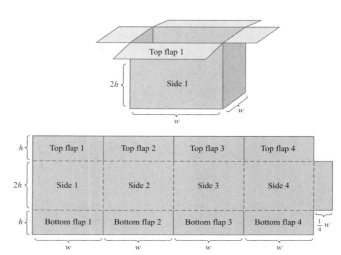

FIGURE 14.40

The designer requires that the sum of the height and width of the piece of cardboard used to make the box be at most 120 inches. What are the dimensions of the box with the maximum volume?

SOLUTION   We are asked to maximize the volume of the box subject to the constraint. The volume of the box is

$$V(h, w) = (2h)(w)(w)$$
$$= 2hw^2$$

The constraint is given by

$$(h + 2h + h) + (w + w + w + w + 0.25w) = 120$$
$$4h + 4.25w = 120$$
$$4h + 4.25w - 120 = 0$$

We'll call this function $g(h, w)$.

$$g(h, w) = 4h + 4.25w - 120$$

The Lagrange function $F(h, w, \lambda)$ is given by

$$F(h, w, \lambda) = V(h, w) + \lambda g(h, w)$$
$$= 2hw^2 + \lambda(4h + 4.25w - 120)$$
$$= 2hw^2 + 4h\lambda + 4.25w\lambda - 120\lambda$$

We'll now find the partial derivatives of F.

$$F_h = 2w^2 + 4\lambda$$
$$F_w = 4hw + 4.25\lambda$$
$$F_\lambda = 4h + 4.25w - 120$$

Observe that $F_\lambda = g(h, w) = 0$. We will now set the other two partial derivatives equal to zero and solve for λ.

$$0 = 2w^2 + 4\lambda \qquad\qquad 0 = 4hw + 4.25\lambda$$

$$4\lambda = -2w^2 \qquad\qquad 4.25\lambda = -4hw$$

$$\lambda = -0.5w^2 \qquad\qquad \lambda = -\frac{4}{4.25}hw$$

$$= -\frac{1}{2}w^2 \qquad\qquad = -\frac{16}{17}hw$$

We now set the equations equal to each other and solve.

$$-\frac{1}{2}w^2 = -\frac{16}{17}hw$$

$$w^2 = \frac{32}{17}hw$$

$$w^2 - \frac{32}{17}hw = 0$$

$$w\left(w - \frac{32}{17}h\right) = 0$$

So $w = 0$ or $w = \frac{32}{17}h$. In the context of the problem, we know that $w = 0$ will not maximize the volume. So $w = \frac{32}{17}h$. We will substitute this result into the equation of $g(h, w)$.

$$4h + 4.25w - 120 = 0$$

$$4h + 4.25\left(\frac{32}{17}h\right) - 120 = 0$$

$$4h + 8h = 120$$

$$12h = 120$$

$$h = 10$$

$$w = \frac{32}{17}h$$

$$= \frac{32}{17}(10)$$

$$\approx 18.82$$

Note that the box has height $2h$ and width w. The dimensions of the box with maximum volume are 18.82 inches $\times$ 18.82 inches $\times$ 20.00 inches. The volume of this box is 7084 cubic inches.

Although the problem in Example 4 could have been solved using optimization methods from Chapter 10, there is value in the Lagrange Multiplier Method, especially when working with functions of more than two variables. In Example 5, we will repeat the problem given in Example 4; however, this time we will allow the box to have a rectangular base instead of a square base.

EXAMPLE 5 **Maximizing Volume with the Lagrange Multiplier Method**

A box designer wants to construct a closeable box as shown in Figure 14.41. The designer requires that the sum of the height and width of the piece of cardboard used to make the box be 120 inches. What are the dimensions of the box with the maximum volume?

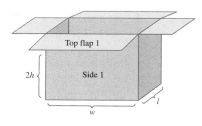

FIGURE 14.41

SOLUTION We are to maximize $V(h, l, w) = 2hlw$ subject to the constraint $4h + 2l + 2.25w = 120$.

We have $g(h, l, w) = 4h + 2l + 2.25w - 120$. Note that $g(h, l, w) = 0$. The Lagrange function is given by

$$
\begin{aligned}
F(h, l, w, \lambda) &= V(h, l, w) + \lambda g(h, l, w) \\
&= 2hlw + \lambda(4h + 2l + 2.25w - 120) \\
&= 2hlw + 4h\lambda + 2l\lambda + 2.25w\lambda - 120\lambda
\end{aligned}
$$

The partial derivatives are

$$
\begin{aligned}
F_h &= 2lw + 4\lambda \\
F_l &= 2hw + 2\lambda \\
F_w &= 2hl + 2.25\lambda \\
F_\lambda &= 4h + 2l + 2.25w - 120
\end{aligned}
$$

We will set the partial derivatives F_h, F_l, and F_w equal to 0 and solve each one for λ.

$$
\begin{array}{ll}
0 = 2lw + 4\lambda & \qquad 0 = 2hw + 2\lambda \\
4\lambda = -2lw & \qquad 2\lambda = -2hw \\
\lambda = -\dfrac{1}{2}lw & \qquad \lambda = -hw
\end{array}
$$

$$
\begin{aligned}
0 &= 2hl + 2.25\lambda \\
2.25\lambda &= -2hl \\
\lambda &= -\frac{8}{9}hl
\end{aligned}
$$

We set the first two equations equal to each other to get h in terms of l.

$$-\frac{1}{2}lw = -hw$$

$$-\frac{1}{2}lw + hw = 0$$

$$w\left(-\frac{1}{2}l + h\right) = 0$$

So $w = 0$ or $h = \frac{1}{2}l$. Since we are looking for the maximum volume, we throw out $w = 0$.

Next we set the second two equations equal to each other to get w in terms of l.

$$-hw = -\frac{8}{9}hl$$

$$hw - \frac{8}{9}hl = 0$$

$$h\left(w - \frac{8}{9}l\right) = 0$$

So $h = 0$ or $w = \frac{8}{9}l$. Since we are looking for the maximum volume, we throw out $h = 0$.

We substitute the values $w = \frac{8}{9}l$ and $h = \frac{1}{2}l$ into $g(l, w, h) = 0$ to solve for l.

$$4h + 2l + 2.25w - 120 = 0$$

$$4\left(\frac{1}{2}l\right) + 2l + 2.25\left(\frac{8}{9}l\right) - 120 = 0$$

$$2l + 2l + \frac{18}{9}l - 120 = 0$$

$$6l = 120$$

$$l = 20$$

Then we back substitute to solve for w and h.

$$w = \frac{8}{9}l$$

$$= \frac{8}{9}(20)$$

$$= \frac{160}{9}$$

$$\approx 17.78$$

$$h = \frac{1}{2}(20)$$

$$= 10$$

Recall that the height of the box is $2h$. The dimensions of the box with maximum volume are 20 inches × 17.78 inches × 10 inches. The maximum volume is 7112 cubic inches.

Graphical Interpretation of Constrained Optimization Problems

Graphically speaking, a constraint "slices" the graph of the surface, creating a cross section. A constrained maximum is the largest value of the objective function when the domain is restricted to the cross section. Similarly, a constrained minimum is the smallest value of the objective function when the domain is restricted to the cross section. Using a graphing utility to graph the objective function and its constraint helps us to identify the location of constrained extrema visually. However, we will check our work using the Lagrange Multiplier Method. It is easy to overlook extrema when relying on graphical methods alone, as will be illustrated in Example 6.

EXAMPLE 6

Maximizing a Multivariable Function with the Lagrange Multiplier Method

Maximize $f(x, y) = -x^3 + 6x^2 - y^2 + 10$ subject to the constraint $10x - y = 50$.

SOLUTION The graph of the function f is shown in Figure 14.42, together with the graph of the constraint equation.

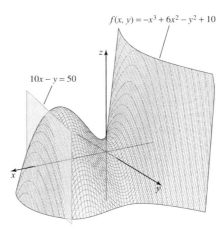

$f(x, y) = -x^3 + 6x^2 - y^2 + 10$

$10x - y = 50$

FIGURE 14.42

The constrained maximum is the highest point along the curve of intersection between the two graphs.

We let $g(x, y) = 10x - y - 50$. The Lagrange function is

$$F(x, y, \lambda) = f(x, y) + \lambda g(x, y)$$
$$= -x^3 + 6x^2 - y^2 + 10 + \lambda(10x - y - 50)$$
$$= -x^3 + 6x^2 - y^2 + 10 + 10x\lambda - y\lambda - 50\lambda$$

The partial derivatives are

$$F_x = -3x^2 + 12x + 10\lambda$$
$$F_y = -2y - \lambda$$
$$F_\lambda = 10x - y - 50$$

We set F_x and F_y equal to zero and solve for λ.

$$0 = -3x^2 + 12x + 10\lambda \qquad 0 = -2y - \lambda$$
$$-10\lambda = -3x^2 + 12x \qquad\qquad \lambda = -2y$$
$$\lambda = 0.3x^2 - 1.2x$$

We set the equations equal to each other and solve for y in terms of x.

$$0.3x^2 - 1.2x = -2y$$
$$y = -0.15x^2 + 0.6x$$

Since $g(x, y) = 0$,

$$10x - y - 50 = 0$$
$$10x - (-0.15x^2 + 0.6x) - 50 = 0$$
$$0.15x^2 + 9.4x - 50 = 0$$
$$15x^2 + 940x - 5000 = 0$$
$$3x^2 + 188x - 1000 = 0$$

We solve for x by using the Quadratic Formula.

$$x = \frac{-b \pm \sqrt{b^2 - 4ac}}{2a}$$
$$= \frac{-(188) \pm \sqrt{(188)^2 - 4(3)(-1000)}}{2(3)}$$
$$= \frac{-188 \pm \sqrt{47{,}344}}{6}$$
$$= \frac{-188 \pm 217.6}{6}$$
$$= 4.931 \qquad \text{or} \qquad -67.60$$

$$y = -0.15x^2 + 0.6x \qquad\qquad y = -0.15x^2 + 0.6x$$
$$= -0.15(4.931)^2 + 0.6(4.931) \qquad = -0.15(-67.60)^2 + 0.6(-67.60)$$
$$= -0.6886 \qquad\qquad\qquad = -726.0$$

The critical points are $(4.931, -0.6886)$ and $(-67.60, -726.0)$. Evaluating f at each of the critical points, we get

$$f(4.931, -0.6886) = 35.52 \qquad \text{and} \qquad f(-67.60, -726.0) = -190{,}732$$

The constrained maximum occurs at $(4.931, -0.6886, 35.52)$. We also discovered a constrained minimum. (The minimum is not shown in the original graph of the function.)

14.4 Summary

In this section, you learned how to use the Lagrange Multiplier Method to solve constrained multivariable optimization problems. You discovered that graphing utilities are helpful but not foolproof in identifying the location of constrained extrema.

14.4 Exercises

In Exercises 1–10, solve the constrained optimization problem using the Lagrange Multiplier Method.

1. Minimize $f(x, y) = x^2 - 2y$ subject to the constraint $-2x + y = 5$.

2. Maximize $f(x, y) = -x^2 - y^2 + 16$ subject to the constraint $x - 2y = 4$.

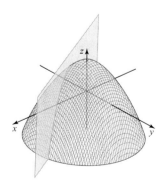

3. Maximize $f(x, y) = x^2 - y^2$ subject to the constraint $y = 3x - 1$.

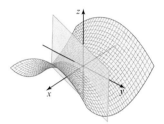

4. Minimize $f(x, y) = xy$ subject to the constraint $2x - y = 4$.

5. Maximize *and* minimize $f(x, y) = -x^3 + 4x^2 + 2y^2$ subject to the constraint $y = x$.

6. Minimize $f(x, y) = x^2 + 3y^2$ subject to the constraint $2x + y = -4$.

7. Minimize $f(x, y) = 4xy$ subject to the constraint $5x - y = 2$.

8. Maximize $f(x, y) = -4xy$ subject to the constraint $-2x + y = -4$.

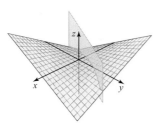

9. Maximize *and* minimize $f(x, y) = x^3 - 20x + 2y^3 + y^2$ subject to the constraint $y = 2x + 3$.

10. Maximize $f(x, y) = x^2y + 2y^2$ subject to the constraint $x - 4y = 24$.

In Exercises 11–15, use the Lagrange Multiplier Method to find the answer to the question.

11. Box Design   A box designer wants to construct a closeable box with a square base as shown in the figure.

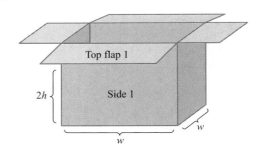

The designer requires that the sum of the height and width of the piece of cardboard used to make the box be at most 180 inches. What are the dimensions of the box with the maximum volume?

12. Cylindrical Can Volume   The volume of a cylindrical can is given by $V(h, r) = \pi r^2 h$, where r is the radius and h is the height of the can, both measured in inches. Because of manufacturing equipment limitations, the sum of the diameter and the height of the can cannot exceed 12 inches. Find the dimensions of the can with maximum volume.

13. **Cobb-Douglas Model** Based on data from 1999 and 2000, a Cobb-Douglas model for Ford Motor Company is given by

$$P(L, C) = 0.5319L^{0.4558}C^{0.5442}$$

thousand vehicles, where L is the total labor cost and C is the capital expenditure cost, both in millions of dollars. In 2000, Ford Motor Company spent $25,783 million on labor and produced 7424 thousand vehicles. (**Source:** Modeled from Ford Motor Company 2001 Annual Report.)

Assuming that the model is valid for future years, how much should Ford spend on labor and how much should it spend on capital in order to maximize vehicle production if its annual budget for labor and capital expenditures is $40,000 million?

14. Landscape Design A landscape architect is designing a garden for a client as shown in the following figure.

She intends to line the flat edges of the garden with shrubs and the curved edge of the garden with flowers. The shrubs cost $25 a linear foot, and the flowers cost $20 a linear foot, including labor. The client is willing to pay up to $1000 and wants as large a garden as possible. What are the dimensions of the garden? (*Hint:* The area of the garden is given by

$$A(x, y) = 2xy + \frac{1}{2}\pi y^2.)$$

15. **Cobb-Douglas Model** Based on data from 2000 to 2001, a Cobb-Douglas model for the Coca-Cola Company is given by

$$P(L, C) = 0.02506L^{0.0481}C^{0.9519}$$

billion unit cases of beverage, where L is the amount of money spent on labor (including sales, payroll, and other taxes) and C is the amount of money spent on capital expenditures, both in millions of dollars. (One unit case of beverage equals 24 eight-ounce servings.) In 2001, Coca-Cola spent a combined total of $917 million on labor and capital expenditures and produced 17.8 billion unit cases of beverage. (**Source:** Coca-Cola 2001 Annual Report.) According to the model, what is the maximum number of unit cases that can be produced with a labor and capital budget of $917 million?

Chapter 14 Review Exercises

Section 14.1 *In Exercises 1–3, generate a table of data for the multivariable function using $x = -2, 0, 2$ and $y = -1, 0, 1$. Then use a graphing utility to graph the function. Compare the coordinates of the points in the table to the coordinates of the points on the graph of the function.*

 1. $z = 4x - 5y$

 2. $z = 2xy - 3y^2$

 3. $z = xy + 4y$

In Exercises 4–5, use the multivariable function to answer the given question.

4. **Price-to-Earnings Ratio** The price-to-earnings ratio (P/E) is the most common measure of how expensive a stock is and is given by

$$R = \frac{P}{E}$$

where P is the price of a share of the stock and E is the earnings per share. Growth stocks typically have high P/Es compared to the overall market. Investors are willing to pay more for these stocks because they expect the company's earnings and stock price to rise. Stocks with low P/Es are often considered to be overlooked value stocks.

As of December 28, 2002, the share prices of four major companies were

Apple Computer:	$14.06
General Electric:	$24.70
Microsoft:	$52.97
Electronic Arts:	$51.50

Their earnings per share were

Apple Computer:	$0.18
General Electric:	$1.41
Microsoft:	$1.41
Electronic Arts:	$0.71

(**Source:** www.quicken.com.)

Calculate the P/E ratio for each of the companies. According to your calculations, which of the companies would you classify as "growth" stocks?

5. [M] **Debt/Asset Ratio**   The debt/asset ratio shows the proportion of a company's assets that are financed through debt and is calculated by

$$R = \frac{L}{A}$$

where L is the total value of the company's liabilities and A is the total value of its assets. If most of the company's assets are financed through equity (stock ownership interest in a company), the ratio is less than 1. If most of the company's assets are financed through debt, the ratio is greater than 1. Companies, and individuals, with high debt/asset ratios may be in danger if creditors start to demand repayment of debt.

As of December 28, 2002, the total debt of each of three major companies was

Apple Computer:	$316.0 million
General Electric:	$254,522.0 million
Microsoft:	$0.0 million

Their reported debt/equity ratios were

Apple Computer:	0.08
General Electric:	4.25
Microsoft:	0.00

(**Source:** www.quicken.com.)

Determine the value of the assets of Apple Computer and General Electric. Then interpret the meaning of the debt/equity ratio for Microsoft.

Section 14.2   *In Exercises 6–10, calculate the first and second partial derivatives of each of the functions.*

6. $f(x, y) = 6xy$

7. $f(x, y) = 2x^2y - 4xy^2$ **8.** $f(x, y) = 5xy - 5y^2$

9. $z = xy \ln(y)$ **10.** $z = e^x \ln(y)$

In Exercises 11–12, interpret the practical meaning of each of the partial derivatives.

11. Cylindrical Can Volume   The volume of a cylindrical can is given by $V(h, r) = \pi r^2 h$, where r is the radius and h is the height of the can, both measured in inches. Find V_h and V_r. Then evaluate the partial derivatives at $(4, 5)$ and interpret the meaning of the results.

12. Landscape Design   The area of the garden shown in the figure is given by

$$A(x, y) = 2xy + \frac{1}{2}\pi y^2.$$

Find A_x and A_y. Then evaluate the partial derivatives at $(6, 4)$ and interpret the meaning of the results.

In Exercises 13–14, a function graph with a cross section drawn at $x = 0$ and $y = -1$ is shown. Determine the equations of the tangent lines of the function in the x-direction and y-direction at the point $(0, -1, f(0, -1))$.

13. $f(x, y) = x - y + 2$

14. $f(x, y) = x^2 - 4y + 2$

Section 14.3   *In Exercises 15–16, find the relative extrema and saddle points of the function.*

15. $f(x, y) = x^2 - xy + 4y^2 + 10$

16. $f(x, y) = -x^3 + 3x + 4y - y^2$

Section 14.4   *In Exercises 17–18, solve the constrained optimization problem using the Lagrange Multiplier Method.*

17. Maximize $f(x, y) = -x^3 + 3x + 4y - y^2$ subject to the constraint $-2x + y = 0$.

18. Maximize $f(x, y) = -x^3 + 3x + 4y^3$ subject to $-2x + y = -6$.

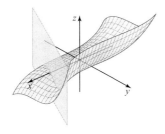

In Exercises 19–20, use the Lagrange Multiplier Method to find the answer to the question.

19. Landscape Design   A landscape architect is designing a garden for a client as shown in the following figure.

She intends to line the flat edges of the garden with shrubs and the curved edge of the garden with flowers. The shrubs cost $30 a linear foot, and the flowers cost $40 a linear foot, including labor. The client is willing to pay up to $1200 and wants as large a garden as possible. What are the dimensions of the garden?

20. **Cobb-Douglas Model**   Based on data from 2000 to 2001, a Cobb-Douglas model for the Coca-Cola Company is given by

$$P(L, C) = 0.02506L^{0.0481}C^{0.9519}$$

billion unit cases of beverage, where L is the amount of money spent on labor (including sales, payroll, and other taxes) and C is the amount of money spent on capital expenditures, both in millions of dollars. (One unit case of beverage equals 24 eight-ounce servings.) In 2001, Coca-Cola spent a combined total of $917 million on labor and capital expenditures and produced 17.8 billion unit cases of beverage. (**Source:** Coca-Cola 2001 Annual Report.)

Assuming that the model works for future years, what is the maximum number of unit cases that can be produced with a labor and capital budget of $998 million?

Make It Real

What to do

1. Go to www.weather.com and enter your zip code to access your local forecast.

2. Use the wind chill formula to calculate the current wind chill temperature. The wind chill formula is given by

$$W(T, V) = 35.74 + 0.6215T - 35.75V^{0.16} + 0.4275TV^{0.16}$$

where T is the temperature (in degrees Fahrenheit) and V is the wind speed (in miles per hour).

3. Compare the calculated wind chill temperature to the "Feels Like" temperature displayed on the forecast.

4. Estimate how much the "Feels Like" temperature will change if the wind speed increases by 1 mph.

Local weather forecast

50°F

Fair   **Feels Like**
48°F

UV Index:	1 Minimal
Dew Point:	35°F
Humidity:	57%
Visibility:	Unlimited
Pressure:	30.25 inches and falling
Wind:	From the West Southwest at 6 mph

As reported a Seattle-Tacoma, WA
Last Updated Tuesday, January 7, 2003,
at 11:56 AM Pacific Standard Time
(Tuesday, 2:56 PM EST).

Source: www.weather.com.
Reprinted by permission.

Answers to Odd-Numbered Exercises

CHAPTER 1

Section 1.1 (page 12)

1. Your weight is a function of your age since at any instant in time you have only one weight.

3. Temperature is a function of the time of day.

5. The number of salmon in a catch is not a function of the number of fish caught.

7. $C(4) = 159.80$
   The total cost of four pairs of shoes is $159.80.

9. $H(2) = 56$
   Two seconds after he jumped, the cliff diver is 56 feet above the water.

11. $E(4) = 0.06$
    In the fourth quarter since December 1999, shares in the tortilla company earned $0.06 per share.

13. $P(4) \approx 18.72$
    On November 8, 2001, the closing stock price of the computer company was approximately $18.72.

15. It appears that near $x = 1$, the graph goes vertical. If it does so, a vertical line drawn at that point would touch the graph in multiple locations. However, if the graph doesn't actually go vertical near $x = 1$, then it is a function. One drawback of reading a graph is that it is sometimes difficult to tell if the graph goes vertical or not.

17. The graph is a function.

19. The graph is a function.

21. 23.

25. $y \approx 4$; $y = 4$

27. $y \approx 2$; y is undefined.

29. The domain is all real numbers.

31. The domain is all real numbers.

33. The domain is all real numbers except $t = 1$. That is, $\{t \mid t \neq 1\}$.

35. The domain is all real numbers greater than or equal to -1. That is $\{a \mid a \geq -1\}$.

37. The domain is the set of real numbers greater than or equal to -3. That is, $\{x \mid x \geq -3\}$.

39. The domain is the set of whole numbers. That is, $\{n \mid n \text{ is a whole number}\}$.

41. Using our current calendar system, the domain of the function is the set of whole numbers between 1 and the current year.

43. The domain is all real numbers except $x = -1$ and $x = 1$.

45. Yes. Even though the domain value $x = 1$ is listed twice in the table, it is linked with the same range value both times, $y = 6$. Similarly, $x = 4$ is listed twice in the table, but each domain value is linked to the range value $y = 2$.

Section 1.2 (page 27)

1. $m = -1$ 3. $m = -0.2$ 5. $m = 0$

7. y-intercept: $(0, 10)$
   x-intercept: $(-2, 0)$

9. y-intercept: $(0, 11)$
   x-intercept: $(-5.5, 0)$

11. y-intercept: $(0, -4)$
    x-intercept: $\left(\frac{4}{3}, 0\right)$

13. The slope-intercept form of the line is $y = -x + 7$. The standard form of the line is $x + y = 7$. A point-slope form of the line is $y - 5 = -1(x - 2)$.

15. The slope-intercept form of the line is $y = -0.2x + 3.64$. The standard form of the line is $5x + 25y = 91$. A point-slope form of the line is $y - 3.4 = -0.2(x - 1.2)$.

17. The slope-intercept form of the line is
$y = 0x + 2$ and is commonly written as $y = 2$.
The standard form of the line is $0x + y = 2$ and
is also often written as $y = 2$. A point-slope form
of the line is $y - 2 = 0(x + 2)$.

19. $y = 4x - 2$

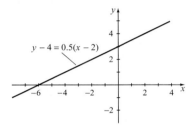

21. $y - 4 = 0.5(x - 2)$

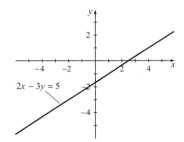

23. $2x - 3y = 5$

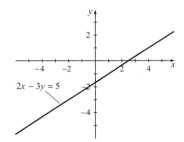

25. $y = -\frac{2}{3}x + \frac{4}{3}$

27. The slope-intercept form of the line is
$y = \frac{2}{3}x + 3$.

29. The equation of the line is $x = 3$.

31. The data represents a linear function.
$m = 993.2$ year 2000 dollars per year
Between 1989 and 1999, the U.S. average
personal income (in year 2000 dollars) increased
by an average of $993.20 per year.

33. The data represents a linear function.
$m = 1183.01$ dollars per month
Between September 2001 and October 2001, the
employee's take-home pay increased at a rate of
$1183.01 per month.

35. The data represents a linear function.
$m = \$0.0375$ per pound
It costs an average of $0.0375 per pound to
dispose of clean wood.

37. In order to consume 8 grams of fiber, you would
need to eat 3 servings $\left(2\frac{1}{4} \text{ cups}\right)$ of Wheaties
along with the large banana.

39. The equation of the vertical line is $x = 4$.

41. The slope-intercept form is $y = mx + b$, and
the point-slope form is $y - y_1 = m(x - x_1)$.
For a vertical line, the slope m is undefined and
thus may not be substituted into either of the
two forms.

43. Vertical lines are not functions, since they fail the
Vertical Line Test. However, all nonvertical lines
are functions.

45. $x = 3$

47. The x-intercept is $\left(\frac{c}{a}, 0\right)$.

The y-intercept is $\left(0, \frac{c}{b}\right)$.

The slope of the line is $m = -\frac{a}{b}$.

49. b must equal 23.   **51.** (a, b)

Section 1.3 (page 45)

1. (a) The scatter plot shows that the data are near-
linear.

(b) The equation of the line of best fit is
$y = -6.290x + 110.9$.

(c) $m = -6.290$ means that the Harbor Capital Appreciation Fund share price is dropping at a rate of $6.29 per month.

The y-intercept means that in month 0 of 2000, the fund price was $110.94. Since the months of 2000 begin with 1, not 0, the y-intercept does not represent the price at the end of January. It could, however, be interpreted as being the price at the end of December 1999.

(d) This model is a useful tool to show the trend in the stock price between October and December 2000. Since stock prices tend to be volatile, we are somewhat skeptical of the accuracy of data values outside of that domain. (Answers may vary.)

3. (a) The scatter plot shows that the data are near-linear.

(b) The equation of the line of best fit is
$y = 1165.9x + 80{,}887$.

(c) $m = 1165.9$ means that Washington State public university enrollment is increasing by about 1166 students per year.

The y-intercept means that, according to the model, Washington State public university enrollment was 80,887 in 1990.

(d) The model fits the data extremely well, as shown by the graph of the line of best fit.

This model could be used by Washington State legislators and university administrators in budgeting and strategic planning. (Answers may vary).

5. (a) The scatter plot shows that although the data are not linear, they are nonincreasing.

(b) The equation of the line of best fit is
$y = -3.4x + 39.6$.

(c) $m = -3.4$ means that the per capita income ranking of North Carolina is changing at a rate of 3.4 places per year. That is, the state is moving up in the rankings by about 3 places per year.

The y-intercept means that in 1995 (year 0), North Carolina was ranked 40th out of the 50 states. (Only positive whole number rankings make sense.)

(d) This model is not a highly accurate representation of the data, as shown by the graph of the line of best fit, so it should be used with caution.

However, an incumbent government official could use the model in a 2000 reelection campaign as evidence that the state's economy had improved during his or her tenure in office. The official might also use the model to claim that the trend of improvement will continue if he or she is reelected. (Answers may vary.)

7. (a)

$$C(x) = \begin{cases} 15.25 & 0 \le x \le 320 \\ 0.04274x + 1.036 & 340 \le x \le 760 \\ 0.04409x & 780 \le x \end{cases}$$

For values of $x \ge 780$, $C(x)$ is directly proportional to x.

(b) $15.25

(c) $23.26

(d) The cost per pound is lowest when 780 or more pounds of trash are disposed of. As a construction company, we would try to keep the weight of our trash deliveries at or above 780 pounds. (Answers may vary.)

9. (a) $C(t) = \begin{cases} 29.99 & 0 \le t \le 200 \\ 0.4t - 50.01 & t > 200 \end{cases}$

   (b) For the Qwest plan, we have

$$C(300) = \$82.49$$

For the Sprint plan, we have

$$C(300) = \$69.99$$

The Sprint plan is the best deal for a customer who uses 300 anytime minutes.

11. (a) Let t be the production year of a Toyota Land Cruiser 4-Wheel Drive and V be the value of the vehicle in 2001.

$$V = 4505t - 8,966,350$$

   (b) \$7610

   (c) The model substantially underestimated the value of a 1992 Land Cruiser.

13. (a) Let F be the number of grams of fat in x Chef Salads. We have $F = 8x$.

The number of fat grams is directly proportional to the number of Chef Salads.

   (b) less than 10 salads

   (c) Up to 7.75 salads

   (d) Up to 5.875 salads

15. Let s be the number of segments and C be the maximum total cost of the ticket (in dollars) from Seattle to Phoenix. We have

$$C = 10s + 222$$

Note that $s \ge 1$.

17. a is the price of a cup of coffee, b is the price of a bagel, and c is the price of a muffin.

19. The correlation coefficient $r = 0$ indicates that the model doesn't fit the data well.

In addition, we can see from the scatter plot that the model doesn't fit well.

21. The y-intercept form of a line passing through the origin is

$$y = mx$$

This implies that y is directly proportional to x.

If the linear model $y = mx$ fits the original data well, it is likely that the dependent variable of the original data is directly proportional to the independent variable. However, if the linear model $y = mx$ does not fit the data well, the dependent

and independent variables of the original data are not directly proportional.

23. The line of best fit is $y = 1$. This model fits the table of data perfectly.

```
LinReg
 y=ax+b
 a=0
 b=1
 r²=
 r=
```

However, from the calculator display, we see that r is undefined.

The correlation coefficient is formally defined as

$$r = \frac{n(\Sigma xy) - (\Sigma x)(\Sigma y)}{\sqrt{n(\Sigma x^2) - (\Sigma x)^2}\,\sqrt{n(\Sigma y^2) - (\Sigma y)^2}}$$

$$r = \frac{0}{0}$$

Because we can't divide by zero and there is a zero in the denominator of the expression, the correlation coefficient r is undefined.

25. $y = \begin{cases} -2x + 10 & -2 \le x \le 4 \\ 0.5x & 4 < x \le 10 \end{cases}$

CHAPTER 1 REVIEW EXERCISES

Section 1.1 *(page 49)*

1. $C(2) = 99.90$
The cost to buy two pairs of shoes is \$99.90.

3. $H(2) = 36$
The cliff diver is 36 feet above the water 2 seconds after he jumps from a 100-foot cliff.

5. The domain is all real numbers.

7. The domain is all real numbers except $r = 1$ and $r = -1$.

9. At $t = 2, P \approx 21.03$. The stock price of the computer company at the end of the day two days after December 16, 2001, was about \$21.03.

11. The graph represents a function.

Section 1.2 *(page 50)*

13. $m = \dfrac{2}{3}$ **15.** $m = -11$

17. The y-intercept is $(0, 18)$.
The x-intercept is $(6, 0)$.

19. $y = -1x + 7$
In standard form, $x + y = 7$.

Section 1.3 *(page 50)*

21. (a) Let x be the number of large orders of French fries consumed. The amount of fat consumed is given by $F(x) = 29x$ grams.
(b) Let y be the number of Big N' Tasty sandwiches consumed. The amount of fat consumed is given by $G(y) = 32y$ grams.
(c) Only one combination meal may be eaten.

23. (a) Let t be the production year of the Mercedes-Benz Roadster two-door SL500 and $v(t)$ be the value of the car in 2001.

$$v(t) = 7312.5t - 14,558,975$$

(b) $58,712.50
(c) The linear model was extremely effective at accurately predicting the value of the 1999 Mercedes-Benz Roadster two-door SL500. The $212.50 difference between the predicted value of the model and the NADA guide average value was negligible.

CHAPTER 2

Section 2.1 *(page 73)*

1. $(2, 1)$

3. $(-0.2118, 5)$

5. The lines are parallel and thus do not intersect. So, a solution does not exist.

7. $(0, -9)$

9. The lines do not intersect at a common point. So, a solution does not exist.

11. $x = 2$, $y = 1$ **13.** $x \approx -0.2118$, $y = 5$

15. There is no solution.   **17.** $x = 0$, $y = -9$

19. The three lines do not intersect at a common point, so there is no solution.

21. By early 1996 ($t = 1.354$), the diabetes infection rates had become equal.

23. Graphing the three models simultaneously yields the following:

Since the functions do not intersect at a common point, the cost of the plans will never be equal.

25. At the end of 2022.

27. Near the end of 1996, the sports team industry and independent artist industry payrolls were equal. Since the original models were based upon just two years of data (1998 and 1999), we are uncertain as to whether the payroll of each industry can, in general, be modeled well by a linear function.

If the linear model fits additional data points well near $t = 0$, $t = -2.170$ is likely to be a good estimate for the time when the payrolls of both industries were equal.

29. According to the model, the per capita consumption of fish exceeded the per capita consumption of chicken before the latter part of 1956, 14.20 years before the end of 1970. Since the value $t = -14.20$ is so far outside of the data set ($0 \le t \le 29$), we are somewhat skeptical of this result.

31. $x = 2$, $y = 1$, $z = 5$ **33.** $x = 1$, $y = 0$, $z = 0$

35. Any point of the form $\left(-\frac{9}{8}t - \frac{45}{4}, \frac{1}{8}t + \frac{45}{4}, t\right)$ is a solution.

37. $x = 9$, $y = 6$, $z = -4$

39. $x = 1$, $y = 0$, $z = -1$

41. $x = 1$, $y = 0$, $z = 1$, $w = 0$

43. The company has $100 left over after paying the dye supplier for 250 bottles of sky blue dye and paying the lease on the building. It is mathematically possible for it to generate sufficient revenue to pay the dye supplier and the building lease.

45. No. Whenever the number of variables exceeds the number of equations, there are 0 or infinitely many solutions.

Section 2.2 *(page 89)*

1. 2×3　　**3.** 1×3　　**5.** 1×1

7. $\begin{aligned} x + 3y &= 12 \\ 4x + y &= 7 \\ -x + 6y &= 9 \end{aligned}$　　**9.** $\begin{aligned} 3x - y + 9z &= 13 \\ -3x + y &= -31 \end{aligned}$

11. The matrix is not in reduced row echelon form because the leading 1 in R_2 is to the *right* of the leading 1 in R_3.

13. The matrix is not in reduced row echelon form because the leading entry in R_2 is not a 1.

15. The matrix is in reduced row echelon form.

17. The matrix is not in reduced row echelon form because the leading 1 in R_1 is to the *right* of the leading 1 in R_2.

19. The matrix is not in reduced row echelon form because the leading 1 in R_2 is not the only nonzero entry in its corresponding column.

21. $\begin{bmatrix} 2 & 5 & 2 \\ 3 & -5 & 3 \end{bmatrix}$ reduces to $\begin{bmatrix} 1 & 0 & 1 \\ 0 & 1 & 0 \end{bmatrix}$

The solution is $x = 1$ and $y = 0$.

23. $\begin{bmatrix} -2 & 6 & 4 & 10 \\ 4 & -12 & 2 & -20 \\ 3 & 4 & -1 & 11 \end{bmatrix}$ reduces to

$\begin{bmatrix} 1 & 0 & 0 & 1 \\ 0 & 1 & 0 & 2 \\ 0 & 0 & 1 & 0 \end{bmatrix}$

So $x = 1$, $y = 2$, and $z = 0$.

25. $\begin{bmatrix} 3 & -2 & 1 & 6 \\ 11 & -20 & -1 & 0 \\ 0 & 1 & 1 & 3 \end{bmatrix}$ reduces to $\begin{bmatrix} 1 & 0 & 0 & 2 \\ 0 & 1 & 0 & 1 \\ 0 & 0 & 1 & 2 \end{bmatrix}$

So $x = 2$, $y = 1$, and $z = 2$.

27. $\begin{bmatrix} 1 & -1 & -5 \\ 9 & 1 & 25 \\ 29 & 1 & 65 \end{bmatrix}$ reduces to $\begin{bmatrix} 1 & 0 & 2 \\ 0 & 1 & 7 \\ 0 & 0 & 0 \end{bmatrix}$

The solution is $x = 2$ and $y = 7$.

29. $\begin{bmatrix} 2 & -4 & 16 \\ 9 & 1 & -4 \\ -3 & 6 & -24 \end{bmatrix}$ reduces to $\begin{bmatrix} 1 & 0 & 0 \\ 0 & 1 & -4 \\ 0 & 0 & 0 \end{bmatrix}$

The solution is $x = 0$ and $y = -4$.

31. $\begin{bmatrix} 2 & -2 & 0 \\ 3 & 1 & 4 \\ 5 & -1 & 5 \end{bmatrix}$ reduces to $\begin{bmatrix} 1 & 0 & 0 \\ 0 & 1 & 0 \\ 0 & 0 & 1 \end{bmatrix}$

The system is inconsistent because $0x + 0y \neq 1$.

33. $\begin{bmatrix} 1 & 1 & 1 & 6 \\ 2 & -1 & 1 & 3 \end{bmatrix}$ reduces to $\begin{bmatrix} 1 & 0 & \frac{2}{3} & 3 \\ 0 & 1 & \frac{1}{3} & 3 \end{bmatrix}$

We have $x + \frac{2}{3}z = 3$ and $y + \frac{1}{3}z = 3$. This is a dependent system. All solutions are of the form

$$x = -\frac{2}{3}t + 3$$

$$y = -\frac{1}{3}t + 3$$

$$z = t$$

for any real number t.

35. $\begin{bmatrix} 3 & 5 & 1 & 6 \\ 6 & 10 & 2 & 10 \end{bmatrix}$ reduces to $\begin{bmatrix} 1 & \frac{5}{3} & \frac{1}{3} & 0 \\ 0 & 0 & 0 & 1 \end{bmatrix}$

This system is inconsistent because $0x + 0y + 0z \neq 2$.

37. $\begin{bmatrix} 1 & 1 & 1 & 6 \\ 2 & -1 & 1 & 3 \\ 4 & -2 & 3 & 9 \end{bmatrix}$ reduces to $\begin{bmatrix} 1 & 0 & 0 & 1 \\ 0 & 1 & 0 & 2 \\ 0 & 0 & 1 & 3 \end{bmatrix}$

The solution is $x = 1$, $y = 2$, and $z = 3$.

39. $\begin{bmatrix} 3 & 2 & 3 & 6 \\ 2 & -5 & 1 & -11 \\ 4 & 2 & 3 & 3 \end{bmatrix}$ reduces to

$\begin{bmatrix} 1 & 0 & 0 & -3 \\ 0 & 1 & 0 & \frac{30}{17} \\ 0 & 0 & 1 & \frac{65}{17} \end{bmatrix}$

The solution is $x = -3$, $y = \frac{30}{17}$, and $z = \frac{65}{17}$.

41.

$x = 2.0289$, $y = 1.5656$, $z = 1.1895$

43.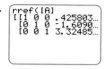

$x = 0.42580$, $y = -1.6090$, $z = 3.3249$

45.

$x = -0.81028$, $y = 1.1639$, $z = 1.2098$

47.

The system is dependent. We have $x - z = -2$ and $y + 2z = 3$. All solutions are of the form

$$x = t - 2$$
$$y = -2t + 3$$
$$z = t$$

for any real number t.

49.

$x = 1$, $y = 1$, $z = 1$

Section 2.3 *(page 99)*

1. (a) We have $S = 7.62t + 284$, where S is the projected number of students enrolled in second grade t years after 1998.

We have $T = 9.99t + 284$, where T is the projected number of students enrolled in third grade t years after 1998.

(b) $S = T$ when $t = 0$

(c) According to the models, the same number of students (284) was projected to be enrolled in second grade and third grade in 1998. The model projections slightly underestimated the actual second grade enrollment and overestimated the actual third grade enrollment.

(d) The model results could be used for strategic planning in the allocation of classrooms and teachers. The projections would also be helpful in determining when to build a new elementary school. According to the models, the number of students enrolled in second grade is expected to increase at a rate of about 8 students per year, while third grade enrollment is expected to increase at an approximate rate of 10 students per year. (Answers will vary.)

3. (a) We have $C = 0.9818t + 23.74$, where C is the annual per capita consumption of chicken (in pounds) t years after 1970.

We have $F = 0.1372t + 11.72$, where F is the annual per capita consumption of fish and shellfish (in pounds) t years after 1970.

(b) $C = F$ when $t \approx -14.2$

(c) According to the models, the per capita consumption of chicken and of fish and shellfish was equal in late 1956 (14.2 years before the end of 1970). At that time, the annual per capita consumption was approximately 9.8 pounds.

We are somewhat skeptical of this result and question the accuracy of the interpretations, since the value $t = -14.2$ is so far outside of the original data set. For the original data set, $0 \le t \le 29$.

(d) Despite the apparent inaccuracy of the interpretation of the intersection point, the models do have practical value. The annual per capita chicken consumption is increasing at a rate of 0.98 pound per year. The annual per capita fish and shellfish consumption is increasing at a rate of 0.14 pound per year. Poultry farmers and seafood producers could use these figures as a baseline when attempting to determine the effects of nationwide advertising campaigns. For example, suppose that poultry farmers launch a nationwide chicken advertising campaign at the start of 2007 and, at the end of 2007, the annual per capita chicken consumption has increased by 2 pounds. Since the model only projects a 0.98-pound increase, the producers conclude that the advertising campaign may have had an effect on chicken consumption. However, other variables would have to be analyzed to ensure that the increase was in fact due to the advertising and not some other market condition. (Answers will vary.)

5. (a) We have $N = 2.969t + 24.23$, where N is the number of people (in millions) who attended a National League baseball game t years after 1994.

We have $A = 1.689t + 24.83$, where A is the number of people (in millions) who attended an American League baseball game t years after 1994.

(b) $A = N$ when $t \approx 0.4688$

(c) According to the models, in mid-1995 (0.47 year after the end of 1994), the baseball game attendance of the two different leagues was equal. We are skeptical of this result for a couple of reasons. First, the linear models did not fit the data well. Second, the baseball season doesn't last the entire year. Therefore, we must be careful about drawing conclusions based on decimal estimates from the model. Round to the nearest year, we could estimate that annual baseball game attendance at games in the two leagues was equal in 1994 or 1995.

(d) Because of the relative inaccuracy of the models, we should be wary of any conclusions drawn as a result. The slopes of the two functions, $m_N = 2.969$ and $m_A = 1.689$, show that National League game attendance increased by roughly 3.0 million people per year and American League game attendance increased by roughly 1.7 million people per year. Representatives of Major League Baseball could use these results as an impetus for research to determine why National League game attendance is growing faster than American League game attendance. (Answers will vary.)

7. (a) The percentage of births to unmarried women in the United States may be modeled by

$$U = \frac{5}{6}t + 18$$

where t is the number of years since the end of 1980.

The percentage of births to unmarried women in the United Kingdom may be modeled by

$$K = \frac{13}{9}t + 12$$

where t is the number of years since the end of 1980.

(b) $U = K$ when $t \approx 9.82$

(c) According to the models, the percentage of all births in the United States and the United Kingdom that were to unmarried women were equal near the end of 1990 ($t = 9.82$). At that time, approximately 26 percent of all births were to unmarried women.

(d) Although the percentage of births to unmarried women in the United States exceeded the percentage of births to unmarried women in the United Kingdom by 6 percentage points at the end of 1980, by the end of 1998 the percentage of births to unmarried women in the United States trailed the percentage of births to unmarried women in the United Kingdom by 5 percentage points. Assuming that both countries have the goal of limiting the number of births to unmarried women, the countries could compare their policies and pregnancy planning services in an effort to determine why the United States had greater success than the United Kingdom. (Answers will vary.)

9. (a) The cost (in cents) to send a 1-ounce letter in the United States t years after 1985 may be modeled by $L = 0.7409t + 23.12$.

The cost (in cents) to send a postcard in the United States t years after 1985 may be modeled by $P = 0.3912t + 14.81$.

(b) $L = P$ when $t \approx -23.76$

(c) The models indicate that the cost to send a 1-ounce letter and the cost to send a postcard were equal in early 1962 ($t = -23.76$). We are skeptical of these results because $t = -23.76$ lies so far outside of the original data set ($0 \le t \le 16$).

Additionally, in our lifetime the cost of sending a letter has always exceeded the cost of sending a postcard.

(d) Postal service companies (e.g., the U.S. Postal Service, UPS, FedEx) could use the model to help determine future postage rates. The postage cost of a 1-ounce letter has increased by approximately 3 cents every four years, while the postage cost of a postcard has increased by about 2 cents every five years. (Answers may vary.)

11. (a) The average annual earnings (in dollars) of a paper manufacturing employee t years after the end of 1995 may be modeled by $E = 1335.1t + 39{,}408$.

The average annual earnings (in dollars) of a printing and publishing employee t years after the end of 1995 may be modeled by
$$P = 1644.7t + 34{,}351$$

(b) $E = P$ when $t \approx 16.333$

(c) According to the models, in early 2012 ($t = 16.333$), the annual earnings of paper manufacturing employees and of printing and publishing employees are expected to be equal. Although the results make sense in their real-world context and the models fit the data very well, the solution $t = 16.333$ is so far outside of the original data set ($0 \le t \le 4$) that we must be cautious in using the result.

(d) If current trends continue, printing and publishing employees will eventually make more than paper manufacturing employees. Printers and publishing companies could use this information to show the strength of their industry as they recruit prospective employees. (Answers will vary.)

13. (a) The number of clothing and accessories stores t years after the end of 1998 may be modeled by $C(t) = -929t + 152{,}603$

The number of food and beverage stores t years after the end of 1998 may be modeled by $F(t) = 3854t + 147{,}652$

(b) $C = F$ when $t \approx 1.035$

(c) According to the models, shortly after the end of 1999, the number of food and beverage stores surpassed the number of clothing and accessories stores.

(d) Entrepreneurs and investors could use this information to forecast the growth in the two industries. The fact that the number of food and beverage stores is increasing while the number of clothing and accessories stores is decreasing would be important to consider when making an investment decision. (Answers may vary.)

15. (a) The number of men's clothing stores t years after the end of 1998 may be modeled by $M(t) = -416t + 11{,}861$

The number of children's clothing stores t years after the end of 1998 may be modeled by $C(t) = 168t + 5165$

(b) $M = C$ when $t \approx 11.47$

(c) According to the models, roughly 6 months into 2010, the number of men's clothing stores and the number of children's clothing stores will be equal. The fact that the intersection point is so far outside of the original data set calls into question the accuracy of the projection.

(d) Clothing producers could use these models to determine in what specialty area to focus their production efforts. (Answers may vary.)

17. (a) The retail sales (in millions of dollars) of building materials and supplies stores t years after the end of 1992 may be modeled by $B(t) = 14{,}675.0t + 157{,}819$

The retail sales (in millions of dollars) of department stores t years after the end of 1992 may be modeled by $D(t) = 6889.85t + 187{,}399$

(b) $B = D$ when $t \approx 3.80$

(c) According to the models, roughly 10 months into 1996, the annual retail sales of building materials and supplies stores and the annual retail sales of department stores was equal.

(d) Investors could use these models in deciding in which industries to invest. Although the department store sales initially exceeded the building supplies sales, department store sales were increasing at a much slower rate than building supplies store sales. As a result, building supplies store sales ultimately surpassed department store sales. (Answers may vary.)

19. He should invest $818.18 in the Bond Market account, $272.73 in the Growth account, and $1909.09 in the Stock account.

21. It should buy 30 Corollas, 10 Camrys, and 15 Prius cars.

23. The quadratic model for the Johnson & Johnson cost of goods sold is
$$C(t) = 103t^2 + 315t + 8539$$
where t is the number of years after 1999 and C is in millions of dollars.

25. The quadratic model for the Gatorade/Tropicana North America net sales is
$$R(t) = -107t^2 + 496t + 3452$$
where t is the number of years after 1999 and R is in millions of dollars.

27. In order to use up all of the buns, patties, and cheese slices, the family should make 60 hamburgers, 16 cheeseburgers, and 28 double cheeseburgers.

29. The breeder should purchase 9 bags of venison meal and rice dog food, 0 bags of natural dog food, and 4 bags of lamb dog food.

31. In order to use up all of the supplies, the company should assemble 20 compact, 40 standard, and 30 deluxe first-aid kits.

33. If 1000 floor tickets, 5000 lower tickets, and 6000 upper tickets are sold, the desired revenue is obtained.

35. In order to use all of the available labor, the company should produce 40 rockers, 80 armchairs, and 40 benches.

37. There are 14 pennies, 22 nickels, 22 dimes, and 6 quarters.

39. We pick $x_4 = 150$ and $x_4 = 200$ and generate a table of values showing each solution.

x_4	x_1	x_2	x_3
150	100	50	0
200	150	100	50

(Answers may vary.)

CHAPTER 2 REVIEW EXERCISES

Section 2.1 *(page 106)*

1. $(1, 7)$

3. $(5, -6)$

5. The lines are parallel and do not intersect. So, there is no solution.

7. $(1, 0, 2)$

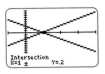

9. $x = 1, y = 7$ **11.** $x = 5, y = -6$

13. There is no solution.   **15.** $x = 1, y = 0.2$

Section 2.2 *(page 107)*

17. The leading entry in row 2 is not the only nonzero entry in its corresponding column. Also, the leading entry in row 3 is not to the right of the leading entry in the row above it.

19. All rows of zeros are not at the bottom of the matrix. Also the leading entry in row 3 is not to the right of the leading entry in the row above it.

21. Reduced row echelon form

23. Reduced row echelon form

25. $\begin{bmatrix} 2.1 & -1 & \vdots & -8 \\ 3.4 & 1 & \vdots & 8 \end{bmatrix}$ reduces to $\begin{bmatrix} 1 & 0 & \vdots & 0 \\ 0 & 1 & \vdots & 8 \end{bmatrix}$

$x = 0, y = 8$

27. $\begin{bmatrix} 4 & -8 & \vdots & -16 \\ 1 & 1 & \vdots & 5 \end{bmatrix}$ reduces to $\begin{bmatrix} 1 & 0 & \vdots & 2 \\ 0 & 1 & \vdots & 3 \end{bmatrix}$

$x = 2, y = 3$

29. $\begin{bmatrix} 1 & 1 & 1 & \vdots & 6 \\ 2 & -1 & 1 & \vdots & 3 \\ 3 & 0 & 2 & \vdots & 9 \end{bmatrix}$ reduces to $\begin{bmatrix} 1 & 0 & \frac{2}{3} & \vdots & 3 \\ 0 & 1 & \frac{1}{3} & \vdots & 3 \\ 0 & 0 & 0 & \vdots & 0 \end{bmatrix}$

$x = -\frac{2}{3}z + 3$ and $y = -\frac{1}{3}z + 3$

Let $z = t$. Then $x = -\frac{2}{3}t + 3$ and $y = -\frac{1}{3}t + 3$.

31. $\begin{bmatrix} 1 & -1 & 3 & \vdots & 5 \\ 2 & 4 & 1 & \vdots & -1 \end{bmatrix}$ reduces to

$\begin{bmatrix} 1 & 0 & \frac{13}{6} & \vdots & \frac{19}{6} \\ 0 & 1 & -\frac{5}{6} & \vdots & -\frac{11}{6} \end{bmatrix}$

Let $z = t$. Then $x = -\frac{13}{6}t + \frac{19}{6}$ and $y = \frac{5}{6}t - \frac{11}{6}$.

33. $\begin{bmatrix} 1 & 0 & 0 & \vdots & 0.3289 \\ 0 & 1 & 0 & \vdots & -0.7230 \\ 0 & 0 & 1 & \vdots & 0.5717 \end{bmatrix}$

35. $\begin{bmatrix} 1 & 0 & 0.5066 & \vdots & 0.8054 \\ 0 & 1 & -0.6303 & \vdots & -0.3195 \\ 0 & 0 & 0 & \vdots & 0 \end{bmatrix}$

Section 2.3 *(page 107)*

37. (a) $K(t) = \dfrac{13}{9}t + 12$

$I(t) = \dfrac{4}{3}t + 40$

(b) $I = K$ when $t = 252$

(c) According to the model, the number of births to unmarried mothers in Iceland and the United Kingdom will be the same 252 years after the end of 1980. According to the model, 376 percent of births will be to unmarried mothers. This doesn't make sense, since the maximum percentage of births that can be to unmarried mothers is 100 percent.

(d) The models could be used by sociologists and government policymakers to forecast the unmarried birth rate in years subsequent to 1998. However, as shown in part (c), the results should be used with caution.

39. (a) $M(t) = 153.72t + 27,110$
$W(t) = 461.28t + 1283.0$

(b) $M = W$ when $t = 83.97$

(c) According to the models, NCAA basketball attendance at men's and women's games will be equal approximately 84 years after the end of 1985. In that year, attendance is projected to be 40,019 thousand for both men's and women's basketball. The result makes sense, but the point of intersection is so far outside of the original data set that we are skeptical of the result.

(d) The models could be used by the NCAA as a baseline in monitoring fan support of men's and women's basketball programs.

41. The quadratic model is
$S(t) = 29.9t^2 + 76.6t + 178.4$

According to the model, in 2004 Starbucks will generate \$4639 million in sales.

43. The company should assemble 15 compact, 35 standard, and 25 deluxe first-aid kits.

CHAPTER 3

Section 3.1 *(page 121)*

1. $A + B = \begin{bmatrix} -4 & 6 \\ 10 & 12 \\ -4 & 3 \end{bmatrix}$

3. Since B and C do not have the same dimensions, they may not be added together.

5. $D + C = \begin{bmatrix} 3 & 0 & 4 \\ -1 & 9 & 8 \\ 14 & 6 & 2 \end{bmatrix}$

7. $3B = \begin{bmatrix} -15 & 0 \\ 21 & 24 \\ -27 & 3 \end{bmatrix}$

9. $-2C + 2D = \begin{bmatrix} -2 & -12 & 8 \\ 2 & 2 & 12 \\ 8 & 20 & 4 \end{bmatrix}$

11.
```
1.2[A]
[[1.44   7.56 ;...
 [-10.92 5.04 2...
 [1.08   -2.4 ....
```

$1.2A = \begin{bmatrix} 1.44 & 7.56 & 0.48 \\ -10.92 & 5.04 & 2.04 \\ 1.08 & -2.4 & 0.36 \end{bmatrix}$

(*Hint:* Use the blue arrow keys on the calculator to display the third column of the matrix.)

13.
```
1.2[A]-2.3[B]
[[-1.78   8.25  ...
 [-4.48  -7.61  ...
 [-20.08 -22.18...
```

$1.2A - 2.3B = \begin{bmatrix} -1.78 & 8.25 & -0.44 \\ -4.48 & -7.61 & -14.29 \\ -20.08 & -22.18 & -4.24 \end{bmatrix}$

15.
```
-1.1[A]+2.9[B]
[[2.74  -7.8  ;...
 [1.89  11.33 1...
 [25.69 27.14 5...
```

$-1.1A + 2.9B = \begin{bmatrix} 2.74 & -7.8 & 0.72 \\ 1.89 & 11.33 & 18.72 \\ 25.69 & 27.14 & 5.47 \end{bmatrix}$

17.

$$-8.7A + 8.7B = \begin{bmatrix} 1.74 & -57.42 & 0 \\ 54.81 & 11.31 & 46.98 \\ 72.21 & 92.22 & 14.79 \end{bmatrix}$$

19.

$$7.8A + 9.9B = \begin{bmatrix} 23.22 & 46.17 & 7.08 \\ -98.7 & 87.21 & 83.55 \\ 98.1 & 69.54 & 22.14 \end{bmatrix}$$

21.

Combined	LSL	RB	MJ
Consumer loan	2,700	0	2,950
Car loan	0	27,700	0
Credit card	0	8,400	0
Mortgage	0	82,500	0

23.

2005–2006 Pay Scale	240 Credits ($)	300 Credits ($)	360 Credits ($)
Level 1	40,208	44,341	48,475
Level 2	42,365	46,498	50,631
Level 3	44,521	48,654	52,787
Level 4	46,678	50,811	54,943

25. The 2006 prices are projected to be as follows:

	Albertsons	A&W	Henry Weinhards
Root beer, 6-pack	$1.79	$1.88	$5.06
Cream soda, 12-pack	$3.58	$3.76	$10.13
Club soda, 2 liters	$1.11		

27. Average Markup

	Accord	Civic
2000 model	$2,300	$1,875
2001 model	$2,400	$1,975

29. Surplus Energy

Years Since 1960	Natural Gas (quadrillion BTUs)	Coal (quadrillion BTUs)
0	0.27	0.98
10	−0.13	2.34
20	−0.48	3.18
30	−0.94	3.21
40	−3.59	0.25

31.

Surplus Energy

Years Since 1960	Nuclear Electric Power (quadrillion BTUs)	Coal (quadrillion BTUs)
0	0	0.98
10	0	2.34
20	0	3.18
30	0	3.21
40	0	0.25

33. **Total Renewable Energy Consumption 1997–1999**

Renewable Energy Type	Energy Consumed (quadrillion BTUs)
Conventional hydroelectric power	10.91
Geothermal energy	1.00
Biomass	9.48
Solar energy	0.22
Wind energy	0.10

35. **Difference in Energy Consumption**

Renewable Energy Type	Energy Consumed (quadrillion BTUs)
Conventional hydroelectric power	−0.22
Geothermal energy	0.00
Biomass	0.35
Solar energy	0.01
Wind energy	0.01

The amount of conventional hydroelectric power consumed in 1999 was less than the three-year average. The amount of geothermal energy consumed in 1999 was equal to the three-year average. The amount of biomass, solar, and wind energy consumed in 1999 exceeded the three-year average.

37. $B = \begin{bmatrix} -2 & 74 & -4 \\ 27 & -20 & -50 \\ -38 & 1 & \frac{11}{3} \end{bmatrix}$

39. $B = \begin{bmatrix} \frac{3}{2} & -11 & 1 \\ -\frac{15}{4} & 3 & 8 \\ \frac{11}{2} & 0 & -\frac{7}{12} \end{bmatrix}$

Section 3.2 *(page 138)*

1. $[0]$ **3.** $A + B = \begin{bmatrix} 5 & 0 \\ 0 & 0 \end{bmatrix}$

5. $4A = \begin{bmatrix} 12 & 16 \\ -4 & -8 \end{bmatrix}$

7. $4A - 2B = \begin{bmatrix} 8 & 24 \\ -6 & -12 \end{bmatrix}$

9. $AB = \begin{bmatrix} 10 & -4 \\ -4 & 0 \end{bmatrix}$

11. We cannot calculate CD, since C is a 3×2 matrix and D is a 3×3 matrix.

13. $CA = \begin{bmatrix} 20 & 24 \\ 17 & 22 \\ 18 & 22 \end{bmatrix}$

15. $CB = \begin{bmatrix} 20 & -24 \\ 13 & -22 \\ 17 & -22 \end{bmatrix}$

17. $B^{-1} = \begin{bmatrix} \frac{1}{4} & \frac{1}{2} \\ -\frac{1}{8} & \frac{1}{4} \end{bmatrix}$

19. $AA^{-1} = \begin{bmatrix} 1 & 0 \\ 0 & 1 \end{bmatrix}$

21. The matrix is invertible because $\det(A) = 1$.
$$A^{-1} = \begin{bmatrix} 2 & -3 \\ -1 & 2 \end{bmatrix}$$

23. The matrix is singular because $\det(C) = 0$.

25. The matrix is invertible because $\det(E) = 1.1$.
$$E^{-1} = \begin{bmatrix} \frac{34}{11} & \frac{7}{11} \\ -\frac{40}{11} & -\frac{5}{11} \end{bmatrix}$$

27. The matrix is singular because $\det(B) = 0$.

29. The matrix is invertible because $\det(D) = 1.99$.
$$D^{-1} = \begin{bmatrix} \frac{100}{199} & \frac{110}{199} \\ -\frac{90}{199} & \frac{100}{199} \end{bmatrix}$$

31. From 1995 through 1999, the total amount of money spent on employee wages and salaries was 109,129,532 thousand dollars, or roughly $109 billion.

33. The first smoothie has 68 mg vitamin C, 53 mg calcium, and 7.5 g fiber. The second smoothie has 115.5 mg vitamin C, 86 mg calcium, and 7.2 g fiber.

35. The tropical fruit smoothie has 74.1 mg vitamin C, 41 mg calcium, and 13.8 g fiber. The pina colada smoothie has 34.1 mg vitamin C, 28 mg calcium, and 11.4 g fiber.

37. $RP = \begin{bmatrix} -52 & -7 \end{bmatrix}$

From 1996 to 1998, the number of military personnel dropped by 52 thousand and the number of army personnel dropped by 7 thousand.

$$PC = \begin{bmatrix} 565 \\ 520 \end{bmatrix}$$

In 1996, the number of military personnel not in the army was 565 thousand. In 1998, the number of military personnel not in the army was 520 thousand.

39. The accumulated sales for 1999–2001 was $22,792.2 million.

The accumulated cost of goods sold for 1999–2001 was $10,780.6 million.

The accumulated profit for 1999–2001 was $12,011.6 million.

41. $A = \begin{bmatrix} 0 & 0 \\ 0 & 0 \end{bmatrix}$, $B = \begin{bmatrix} 1 & 1 \\ 2 & 2 \end{bmatrix}$, and $C = \begin{bmatrix} 1 & 0 \\ 1 & 0 \end{bmatrix}$

(Answers may vary.)

43. We have

$$AA^{-1} = \begin{bmatrix} a & 0 & 0 \\ 0 & b & 0 \\ 0 & 0 & c \end{bmatrix} \begin{bmatrix} \frac{1}{a} & 0 & 0 \\ 0 & \frac{1}{b} & 0 \\ 0 & 0 & \frac{1}{c} \end{bmatrix}$$

$$= \begin{bmatrix} 1 & 0 & 0 \\ 0 & 1 & 0 \\ 0 & 0 & 1 \end{bmatrix}$$

$$= I$$

and

$$A^{-1}A = \begin{bmatrix} \frac{1}{a} & 0 & 0 \\ 0 & \frac{1}{b} & 0 \\ 0 & 0 & \frac{1}{c} \end{bmatrix} \begin{bmatrix} a & 0 & 0 \\ 0 & b & 0 \\ 0 & 0 & c \end{bmatrix}$$

$$= \begin{bmatrix} 1 & 0 & 0 \\ 0 & 1 & 0 \\ 0 & 0 & 1 \end{bmatrix}$$

$$= I$$

Therefore, the matrix A^{-1} is the inverse of A. A is singular if a, b, or c is equal to zero.

45. Let $A = \begin{bmatrix} a & b & c \\ b & d & e \\ c & e & f \end{bmatrix}$. Then

$$A^2 = \begin{bmatrix} a & b & c \\ b & d & e \\ c & e & f \end{bmatrix} \begin{bmatrix} a & b & c \\ b & d & e \\ c & e & f \end{bmatrix}$$

$$= \begin{bmatrix} a^2 + b^2 + c^2 & ab + bd + ce & ac + be + cf \\ ab + bd + ce & b^2 + d^2 + e^2 & bc + de + ef \\ ac + be + cf & bc + de + ef & c^2 + e^2 + f^2 \end{bmatrix}$$

The matrix A^2 is symmetric.

Section 3.3 *(page 157)*

1. $A^{-1} = \begin{bmatrix} 1.2 & -1.4 & -0.2 \\ -0.4 & 0.8 & -0.1 \\ 0.2 & -0.4 & 0.3 \end{bmatrix}$

3. $A^{-1} = \begin{bmatrix} 0 & 2 & -1 \\ -1 & 5 & -2 \\ 0.5 & -4 & 2 \end{bmatrix}$

5. $A^{-1} = \begin{bmatrix} 0 & 0.2 & -0.8 \\ 1 & -1.2 & 4.8 \\ -0.5 & 0.6 & -1.9 \end{bmatrix}$

7. A is singular.

9. A is singular.

11. $\begin{bmatrix} 1 & 1 & 1 \\ 2 & -1 & 1 \\ 4 & -2 & 3 \end{bmatrix} \begin{bmatrix} x \\ y \\ z \end{bmatrix} = \begin{bmatrix} 6 \\ 3 \\ 9 \end{bmatrix}$; $x = 1$, $y = 2$, $z = 3$

13. $\begin{bmatrix} 3 & 2 & 3 \\ 2 & -5 & 1 \\ 4 & 2 & 3 \end{bmatrix} \begin{bmatrix} x \\ y \\ z \end{bmatrix} = \begin{bmatrix} 6 \\ -11 \\ 3 \end{bmatrix}$; $x = -3$, $y = \dfrac{30}{17}$,

$z = \dfrac{65}{17}$

15. $\begin{bmatrix} 1 & 1 & 1 \\ 2 & -1 & 1 \\ 4 & -2 & 3 \end{bmatrix} \begin{bmatrix} x \\ y \\ z \end{bmatrix} = \begin{bmatrix} 3 \\ 13 \\ 28 \end{bmatrix}$; $x = 4$, $y = -3$,

$z = 2$

17. $\begin{bmatrix} 3 & -1 & 2 \\ 4 & 3 & 5 \\ 6 & 8 & 4 \end{bmatrix} \begin{bmatrix} x \\ y \\ z \end{bmatrix} = \begin{bmatrix} 12 \\ 45 \\ 64 \end{bmatrix}$; $x = 2$, $y = 4$,

$z = 5$

19. $\begin{bmatrix} 2 & 5 & 3 \\ 1 & 4 & 2 \\ 0 & 2 & 4 \end{bmatrix} \begin{bmatrix} x \\ y \\ z \end{bmatrix} = \begin{bmatrix} 0 \\ -1 \\ 2 \end{bmatrix}$; $x = 1$, $y = -1$,

$z = 1$

21. The matrix A is singular.

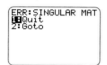

ERR:SINGULAR MAT
1:Quit
2:Goto

23. [C]⁻¹
[[0 0 ...
[0 .2380952381...
[2 -.571428571...

$C^{-1} \approx \begin{bmatrix} 0 & 0 & 3.333 \\ 0 & 0.2381 & -1.429 \\ 2 & -0.5714 & -19.24 \end{bmatrix}$

25. [E]⁻¹
[[-1.458333333 ...
[-.25 ...
[-.125 ...
[.8333333333 ...

$E^{-1} \approx \begin{bmatrix} -1.458 & 0.9722 & 0.09722 & 0.05556 \\ -0.25 & 0.5 & -0.25 & 0 \\ -0.125 & 0.08333 & 0.2083 & -0.1667 \\ 0.8333 & -0.5556 & -0.05556 & 0.1111 \end{bmatrix}$

27. [B]⁻¹
[[-.0667481663 ...
[.104400978 ...
[-.0596577017 ...

$B^{-1} \approx \begin{bmatrix} -0.06675 & 0.1858 & -0.1296 \\ 0.1044 & -0.03423 & 0.1002 \\ -0.05966 & 0.01956 & 0.2285 \end{bmatrix}$

29. The matrix D is singular.

ERR:SINGULAR MAT
1:Quit
2:Goto

31.

	Protein Blend	Low-Fat Mix	Low-Carb Mix
Protein	23.03	21.47	16.02
Carbohydrates	38.99	40.37	25.12
Fat	70.47	67.95	77.60

33. To earn a 6% return, $1539.17 should be invested in the Money Market fund and $460.83 should be invested in the Large Cap Value fund.

To earn an 8% return, $924.73 should be invested in the Money Market fund and $1075.27 should be invested in the Large Cap Value fund.

To earn a 10% return, $310.29 should be invested in the Money Market fund and $1689.71 should be invested in the Large Cap Value fund.

35. If a student has earned 40.91 credits or less, it is mathematically possible for the student to earn a cumulative GPA of 3.5 by the time he completes 90 credits. If the student has earned more than 40.91 credits, it is mathematically impossible for the student to earn a cumulative GPA of 3.5 credits by the time he completes 90 credits. Only the student with 30 credits could conceivably increase his GPA to a 3.5 by the time he earns 90 credits.

37. $A^{-1} = \begin{bmatrix} 1 & \frac{1}{2}a & -\frac{1}{2}a & -\frac{1}{2}a \\ 0 & \frac{1}{2} & \frac{1}{2} & -\frac{1}{2} \\ 0 & \frac{1}{2} & -\frac{1}{2} & \frac{1}{2} \\ 0 & -\frac{1}{2} & \frac{1}{2} & \frac{1}{2} \end{bmatrix}$

39. The statement is true.

41. The statement is false.

43. We have $AA^{-1} = I$ and $BB^{-1} = I$. If both A and B are $n \times n$ matrices, AB is defined. We must show that if AB is defined, $(AB)(AB)^{-1} = I$.

$$
\begin{aligned}
(AB)(AB)^{-1} &= (AB)(B^{-1}A^{-1}) \\
&= A(BB^{-1})A^{-1} \\
&= AIA^{-1} \\
&= AA^{-1} \\
&= I
\end{aligned}
$$

Thus, if A and B are invertible, then AB is invertible.

45. $A = \begin{bmatrix} 1 & 0 & 0 \\ 0 & 1 & 0 \\ 0 & 0 & 1 \end{bmatrix}$ and $A = \begin{bmatrix} 1 & 0 & 0 \\ 0 & 0 & 1 \\ 0 & 1 & 0 \end{bmatrix}$

Section 3.4 *(page 165)*

1. t_{11} means that 0.2 unit of Sector A is required to produce 1 unit in Sector A. t_{12} means that 0.5 unit of Sector A is required to produce 1 unit in Sector B. t_{21} means that 0.4 unit of Sector B is required to produce 1 unit in Sector A. t_{22} means that 0.8 unit of Sector B is required to produce 1 unit in Sector B.

3. t_{12} means that 0.01 unit of oil products is required to produce 1 unit of mining products. t_{21} means that 0.00 units of mining products are required to produce 1 unit of oil products. t_{23} means that 0.00 units of mining products are required to produce 1 unit of support products. t_{31} means that 0.02 unit of support products is required to produce 1 unit of oil products.

5. From the technology matrix, we see that $0.10 of agriculture products and $0.30 of manufacturing products are required to produce $1 of manufacturing products.

7. $I - T = \begin{bmatrix} 0.8 & -0.1 \\ -0.6 & 0.7 \end{bmatrix}$

$(I - T)^{-1} = \begin{bmatrix} 1.4 & 0.2 \\ 1.2 & 1.6 \end{bmatrix}$

9. $408 million of agriculture products and $864 million of manufacturing products are required to meet the internal and external demands of the economy.

11. Roughly $69.9 billion of textile products and $54.1 billion of apparel products are required to meet the internal and external demands of the economy.

13. Roughly $431.4 billion of vehicle products and $163.4 billion of transportation products are required to meet the internal and external demands of the economy.

15. Roughly $131.7 billion of metals, $232.5 billion of products, and $232.6 billion of machinery are required to meet the internal and external demands of the economy.

17. Roughly $143.3 billion of computer products, $284.6 billion of management, and $431.2 billion of support are required to meet the internal and external demands of the economy.

19. Roughly $507.0 billion of bank products, $224.5 billion of securities, $455.5 billion of insurance products, and $71.6 billion of funds products are required to meet the internal and external demands of the economy.

21. The sum of each column is greater than 1. This implies that it costs more than $1 of input to produce $1 of output in each of the given industries. An economy with input costs exceeding output value cannot survive.

23. Since output in Sectors B and C doesn't require any input from Sector A, the strike should not affect production in Sectors B and C.

25. The sum of each column must be positive and less than 1.

CHAPTER 3 REVIEW EXERCISES

Section 3.1 *(page 167)*

1. $A + B = \begin{bmatrix} 7 & 9 \\ 0 & -8 \\ 9 & 9 \end{bmatrix}$

3. $2A = \begin{bmatrix} 4 & 6 \\ -6 & -8 \\ 0 & 14 \end{bmatrix}$

5. $2A + 3B = \begin{bmatrix} 19 & 24 \\ 3 & -20 \\ 27 & 20 \end{bmatrix}$

7. $5.3B = \begin{bmatrix} 18.02 & -1.59 & -2.12 \\ -20.14 & 29.68 & 38.16 \\ 11.66 & 13.78 & 10.6 \end{bmatrix}$

9. $4.1A + 0.1B = \begin{bmatrix} 5.26 & 24.98 & -1.68 \\ -37.69 & 17.78 & 7.69 \\ 3.91 & 12.56 & 13.73 \end{bmatrix}$

11. Average Markup

	Celica	MR2 Spyder
2000 model	$2275	$2550
2001 model	$2400	$2625

Section 3.2 *(page 168)*

13. $BA = \begin{bmatrix} 25 & 42 \\ 12 & 20 \end{bmatrix}$

15. $DC = \begin{bmatrix} 30 & 27 \\ -10 & -8 \\ 16 & 17 \end{bmatrix}$

17. AC is undefined.

19. $B^{-1} = \begin{bmatrix} -1 & 2.5 \\ 1 & -2 \end{bmatrix}$

21. $\det(A) = 36$

Since $\det(A) \neq 0$, the matrix is invertible.

23. $B^{-1} = \begin{bmatrix} \frac{1}{2} & -\frac{31}{32} & -\frac{901}{80} \\ 0 & \frac{5}{16} & \frac{27}{8} \\ 0 & 0 & -2 \end{bmatrix}$

Section 3.3 *(page 168)*

25. $A^{-1} = \begin{bmatrix} 1 & -\frac{5}{3} & \frac{1}{6} \\ 2 & -4 & \frac{1}{2} \\ -1 & 2 & 0 \end{bmatrix}$

27. $A^{-1} = \begin{bmatrix} \frac{5}{3} & -\frac{8}{3} & -1 \\ -\frac{14}{3} & \frac{23}{3} & 2 \\ \frac{10}{3} & -\frac{16}{3} & -1 \end{bmatrix}$

29. $x = 8,\ y = -6,\ z = 12$

31. To earn a 7% return, $1661.31 should be invested in the Growth Fund and $338.69 in the Capital Appreciation Fund.

To earn a 9% return, $1077.37 should be invested

in the Growth Fund and $922.63 in the Capital Appreciation Fund.

To earn an 11% return, $493.43 should be invested in the Growth Fund and $1506.57 in the Capital Appreciation Fund.

33. Her flower cost is $18.41 for Type 1, $24.94 for Type 2, and $25.81 for Type 3.

The retail price is $27.62 for Type 1, $37.41 for Type 2, and $38.72 for Type 3.

Section 3.4 *(page 169)*

35. Roughly $231.5 billion of machine products and $431.3 billion of auto products are required to meet the internal and external demands of the economy.

CHAPTER 4

Section 4.1 *(page 183)*

1. $2x + y \leq 6;\ P = (2, 4)$

The point $(2, 4)$ is not in the solution region.

3. $x + 5y \leq 10\ ;\ P = (0, 0)$

The point $(0, 0)$ is in the solution region.

5. $-2x + 4y \geq -2\ ;\ P = (1, 2)$

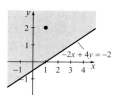

The point $(1, 2)$ is in the solution region.

7. $5x - 4y \leq 0; P = (1, 0)$

The point $(1, 0)$ is not in the solution region.

9. $2x - y \geq 8; P = (-3, 2)$

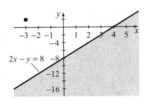

The point $(-3, 2)$ is not in the solution region.

11. $-4x + y \geq 2$
$-2x + y \geq 1$
$x \leq 0$

13. $-2x + 6y \leq 8$
$4x - 12y \leq -6$

15. $3x - 2y \leq 4$
$11x - 20y \geq 2$

17. $x - y \leq -5$
$9x + y \leq 25$

19. $6x + 2y \leq 10$
$-x - 2y \geq -5$
$x \geq 0$
$y \geq 0$

21. $2x - 4y \geq 16$
$9x + y \leq -4$
$-3x + 6y \leq -24$

Note: The first and third linear equations describe the same line.

23. $8x - y \geq 3$
$x + 2y \leq 11$
$9x + y \leq 14$

25. $-5x + y \geq 0$
$2x + y \leq 4$
$y \leq 1$
$x \leq 1$

27. Let x be the number of granola bars and y be the number of cereal bars. We have

Sodium: $130x + 65y \leq 2400$

Fat: $3.5x + 1.5y \leq 80$

and the corresponding solution region for nonnegative x and nonnegative y

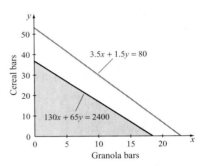

Any serving combination contained within the solution region would meet the dietary guidelines. For example, 8 granola bars and 20 cereal bars is a solution.

29. Let x be the number of hours spent managing the copy center and y be the number of hours spent designing brochures. We have

Hours worked: $x + y \leq 50$

Minimum copy center hours: $x \geq 35$

Maximum copy center hours: $x \leq 45$

Earnings: $900 + 25y \geq 1100$

Nonnegative constraints: $x \geq 0, y \geq 0$

and the corresponding solution region

If the employee is able to consistently work 42 hours or less each week managing the copy center, he will be able to meet his goals. However, any week that the man works more than 42 hours at the copy center, he will be unable to meet his weekly income goal unless he exceeds his 50-hour-per-week workload limit.

31.
$$2x + 3y \leq 6$$
$$-2x + 4y \geq 4$$
$$-5x + y \leq 15$$
$$x \leq 5$$
$$y \geq 2$$

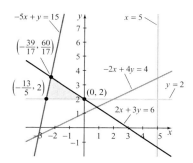

33. $-x + y \leq 0$
$$-x - y \geq -4$$
$$y \geq 2$$

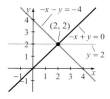

There is a single solution to the system of inequalities: $(2, 2)$. This may be loosely referred to as the "corner point" of the region.

35. $x \geq 1$
$$y \leq 3$$
$$y \geq \frac{2}{3}x - \frac{1}{3}$$
$$y \geq 1$$

37. $y \leq -\frac{1}{4}x + 5$
$$y \leq -4x + 20$$
$$x \geq 0$$
$$x \leq 5$$

39. $x \geq 0$
$$x \leq 0$$
$$y \geq 0$$
$$y \leq 0$$

The solution region is bounded and consists of a single solution: $(0, 0)$. This may be loosely referred to as a corner point.

Section 4.2 *(page 199)*

1. The optimal solution is $(0.6, 1.6)$, and the optimal value is 13.

3. The optimal solution is $(0, 16)$, and the optimal value is 16.

5. The optimal solution is $(2, 4)$, and the optimal value is 42.

7. The optimal solution is $(3, 0)$, and the optimal value is 6.

9. The feasible region is empty, so the linear programming problem has no solution.

11. The optimal solution is $(0, 1)$, and the optimal value is 7.

13. The optimal solution is $(2, 4)$, and the optimal value is 58.

15. The optimal solution is $(4, 5)$, and the optimal value is 54.

17. The optimal solution is $(5, 0)$, and the optimal value is 100.

19. Since we may make y as big as we like and still remain in the feasible region, z will never attain a maximum value. The linear programming problem has no solution.

21. Let x be the number of cartons of medium peaches and y be the number of cartons of small peaches. The wholesale cost is the objective function and is given by $z = 9x + 10y$. We are to minimize the function subject to the following constraints:

Minimum demand: $60x + 70y \geq 420$

Maximum demand: $60x + 70y \leq 630$

Nonnegative constraints: $x \geq 0, \ y \geq 0$

The optimal solution is $(0, 6)$ with optimal value $z = 60$. The minimum wholesale cost occurs when six cartons of small peaches are purchased. The cost is $60.

23. Let x be the number of shares in the Stock account and y be the number of shares in the Growth account.

$$\text{Minimize} \quad P = 6x + 7y$$

$$\text{Subject to} \quad \begin{cases} 174x + 55y \geq 2000 \\ x \geq 0, y \geq 0 \end{cases}$$

The optimal solution is $(11.49, 0)$ with optimal value $P = 68.94$. In order to minimize risk, he should invest all of his money in the Stock account.

25. Let x be the number of bags of venison dog food and y be the number of bags of lamb dog food.

$$\text{Minimize} \quad P = 21.99x + 22.99y$$

$$\text{Subject to} \quad \begin{cases} 2x + 3y \geq 30 \\ 4x - 9y \geq 0 \\ x \geq 0, y \geq 0 \end{cases}$$

The optimal solution is $(9, 4)$ with optimal value $P = 289.87$. Buying 9 bags of venison dog food and 4 bags of lamb dog food yields the minimum cost.

27. Let x be the number of rockers and y be the number of benches produced.

$$\text{Maximize} \quad P = 314x + 57y$$

$$\text{Subject to} \quad \begin{cases} 4x + y \leq 360 \\ 7x + 3y \leq 730 \\ x + y \leq 150 \\ x \geq 0, y \geq 0 \end{cases}$$

The optimal solution is $(90, 0)$ with optimal value $P = 28,260$. Producing 90 rockers and 0 benches will yield the maximum profit of $28,260.

29. Let x be the number of 29-passenger vehicles and y be the number of 49-passenger vehicles.

$$\text{Minimize} \quad P = 1000x + 1800y$$

$$\text{Subject to} \quad \begin{cases} 27x + 45y \geq 135 \\ 2x + 4y \leq 16 \\ x \geq 0, y \geq 0 \end{cases}$$

The optimal solution is $(5, 0)$ with optimal value $P = 5000$. In order to minimize transportation costs, the PTA should charter five 29-passenger vehicles. (Only 10 chaperones will be needed for the trip.)

31. Let x be the number of *CSI* spots and y be the number of *Law and Order* spots.

$$\text{Maximize} \quad P = 14,834x + 10,557y$$

$$\text{Subject to} \quad \begin{cases} x \geq 10 \\ y \geq 10 \\ 31x + 25y \leq 870 \end{cases}$$

The optimal solution is $(20, 10)$ with optimal value $P = 402,250$. In order to maximize the number of viewers seeing the commercial spot, the company should purchase 20 spots on *CSI* and 10 spots on *Law and Order*. The number of viewers is projected to be 402,250,000. (*Note:* This calculation counts a viewer each time a person sees the spot. That is, if someone views both shows and sees all 30 spots, that person is counted as 30 viewers.)

33. Let x be the number of shares in Harley-Davidson, Inc., and y be the number of shares in Polaris Industries, Inc.

$$\text{Maximize} \quad P = 2.56x + 2.50y$$

$$\text{Subject to} \quad \begin{cases} 62.7x + 50.5y \leq 4000 \\ 0.40x + 0.92y \geq 60 \\ x \geq 0, y \geq 0 \end{cases}$$

The optimal solution is $(0, 79.2)$ with optimal value $P = 198$. In order to maximize earnings, the investor should purchase 79.2 shares of Polaris Industries. (Because of round-off error, this will cost only $3999.60. If we round the required number of shares to 79.208, the cost is $4000.00.)

35. Let x be the number of shares of Polaris Industries and y be the number of shares of Winnebago.

$$\text{Maximize} \quad P = 0.92x + 0.20y$$

$$\text{Subject to} \quad \begin{cases} 50.50x + 33.42y \leq 17,000 \\ 2.5x + 1.8y \geq 900 \\ x \geq 0, y \geq 0 \end{cases}$$

The optimal solution is $(71.0, 401.4)$ with optimal value $P = 145.60$. In order to maximize dividends, the investor should buy 71.0 shares of Polaris Industries and 401.4 shares of Winnebago Industries. (Because of round-off error, this actually results in an investment of $17,000.29. If we went out more decimal places on the share amounts, we could get a more accurate answer. The investor may choose to simply work with the rounded amounts instead of worrying about the few extra cents.)

37. Yes. For example, the feasible region described by the following constraints contains exactly one point: $(0, 0)$.

$$x + y \leq 0$$

$$x \geq 0, y \geq 0$$

The maximum and minimum value of any objective function will be zero.

39. Yes. For example,

$$\text{Minimize} \quad z = x + y$$
$$\text{Subject to} \quad \{y \geq x$$

This feasible region has no corner points. In fact, any linear programming problem with exactly one inequality as a constraint will not have any corner points.

Section 4.3 *(page 230)*

1. The problem is a standard maximization problem.

3. The problem is not a standard maximization problem. The objective function is not a linear function.

5. The problem is not a standard maximization problem. It is missing the nonnegative constraints: $x \geq 0, y \geq 0, z \geq 0$.

7. The problem is not a standard maximization problem. The constant term of the first linear constraint is negative (-24).

9. The problem is not a standard maximization problem. It is missing the nonnegative constraints: $x \geq 0, y \geq 0, z \geq 0$.

11. The corner point $(3, 4)$ yields the maximum objective function value of 59. The feasible region is shown in the following graph.

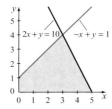

Evaluating the objective function $P = 9x + 8y$ at each of the corner points yields

x	y	P
0	0	0
0	1	8
3	4	59
5	0	45

13. The corner point $(4, 0)$ yields the maximum objective function value of 20. The feasible region is shown in the following graph.

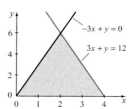

Evaluating the objective function $P = 5x - 2y$ at each of the corner points yields

x	y	P
0	0	0
2	6	-2
4	0	20

15. The corner point $(2, 4)$ yields the maximum objective function value of 24. The feasible region is shown in the following graph.

Evaluating the objective function $P = 4x + 4y$ at each of the corner points yields

x	y	P
0	0	0
0	3	12
2	4	24
5	0	20

17. The corner point $(2, 3)$ yields the maximum objective function value of 27. The feasible region is shown in the following graph.

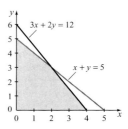

Evaluating the objective function $P = 6x + 5y$ at each of the corner points yields

x	y	P
0	0	0
0	5	25
2	3	27
4	0	24

19. The corner point $(5, 0)$ yields the maximum objective function value of 35. The feasible region is shown in the following graph.

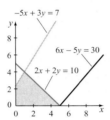

Evaluating the objective function $P = 7x - 5y$ at each of the corner points yields

x	y	P
0	0	0
0	$\dfrac{7}{3}$	$\dfrac{-35}{3}$
1	4	-13
5	0	35

21. The solution $(0, 0, 60)$ yields a maximum objective function value of 360.

23. The solution $(0, 0, 40)$ yields a maximum objective function value of 480.

25. The solution $(45, 0, 0)$ yields a maximum objective function value of 225.

27. The solution $(0, 0, 4.6)$ yields a maximum objective function value of 46.

29. The solution $(0, 4, 0)$ yields a maximum objective function value of 4.

31. Let x be the number of batches of Cranberry Cooler, y be the number of batches of Nancy's Party Punch, and z be the number of batches of Jimmy Wallbanger. We are to

$$\text{Maximize} \quad P = 1x + 9y + 1z$$

$$\text{Subject to} \quad \begin{cases} \text{Cranberry: } 4x + 32y \le 128 \\ \text{Lemon-lime: } 2x + 6z \le 64 \\ \text{Nonnegative: } x \ge 0, y \ge 0, z \ge 0 \end{cases}$$

We can make the maximum $46\frac{2}{3}$ servings by making 4 batches of Nancy's Party Punch and $10\frac{2}{3}$ batches of Jimmy Wallbanger.

33. Let x be the number of hamburgers, y be the number of cheeseburgers, and z be the number of double cheeseburgers. We are to

$$\text{Maximize} \quad P = x + y + 2z$$

$$\text{Subject to} \quad \begin{cases} \text{Beef patties: } x + y + 2z \le 132 \\ \text{Buns: } x + y + z \le 104 \\ \text{Cheese slices: } y + 2z \le 72 \\ \text{Nonnegative: } x \ge 0, y \ge 0, z \ge 0 \end{cases}$$

The optimal solution is $(60, 0, 36)$. The maximum number of beef patties that can be used is 132 patties. This occurs when 60 hamburgers, no cheeseburgers, and 36 double cheeseburgers are made.

35. Let x be the number of 50-packs, y be the number of 100-packs, and z be the number of 600-packs. We are to

$$\text{Maximize the revenue} \quad R = 50x + 100y + 600z$$

$$\text{Subject to} \quad \begin{cases} \text{Quantity: } 50x + 100y + 600z \le 1000 \\ \text{Budget: } 10x + 18y + 96z \le 175 \\ \text{Nonnegative: } x \ge 0, y \ge 0, z \ge 0 \end{cases}$$

We also have the additional constraints $x \le 17$, $y \le 9$, and $z \le 1$. The optimal solution is $(0, 4, 1)$ with optimal value $R = 1000$. She should order no 50-packs, four 100-packs, and one 600-pack. Her maximum revenue will be $1000.

37. (a) The corner points are $(0, 0)$, $(0, 4)$, $(2, 6)$, $(4, 5)$, and $(6, 0)$.

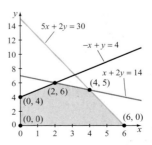

(b) The optimal solution is $(4, 5)$ with optimal value $z = 36$.

(c) The basic feasible solution $(0, 0)$ is the first corner point of the region.

The feasible solution associated with the second tableau is $(0, 4)$.

The feasible solution associated with the third tableau is $(2, 6)$.

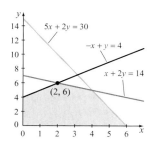

The feasible solution associated with the final tableau is $(4, 5)$.

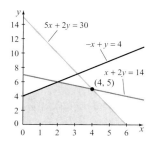

An alternative correct solution moves from $(0, 0)$ to $(6, 0)$ to $(4, 5)$. This occurs if you pick the x-column as the pivot column in the initial simplex tableau.

(d) The simplex method starts at $(0, 0)$ and moves from corner point to corner point along the border of the feasible region while simultaneously increasing the value of the objective function.

Section 4.4 *(page 249)*

1. The problem is a standard minimization problem.

3. The objective function $P = 6xy$ is not of the form $P = ax + by$. Therefore, the problem is not a standard minimization problem.

5. The required nonnegative constraints $x \geq 0$, $y \geq 0$, and $z \geq 0$ are not given. Therefore, the problem is not a standard minimization problem.

7. The constraint $-6x - y + 4z \geq -24$ is not of the form $Ax + By + Cz + \cdots \geq M$ with M nonnegative. Therefore, the problem is not a standard minimization problem.

9. The required nonnegative constraints $x \geq 0$ and $y \geq 0$ are not given. Therefore, the problem is not a standard minimization problem.

11. $A^T = \begin{bmatrix} 2 & 5 & 8 \\ 2 & 9 & 1 \\ 1 & 7 & 0 \\ 4 & 3 & 1 \end{bmatrix}$ 13. $A^T = \begin{bmatrix} 1 & 6 \\ 2 & 8 \\ 4 & 1 \end{bmatrix}$

15. $A^T = \begin{bmatrix} 9 & 7 \\ 3 & 2 \\ -1 & -2 \\ 4 & 1 \end{bmatrix}$

17. $A^T = \begin{bmatrix} 9 & -3 & 4 & 5 \\ 2 & -2 & 0 & 6 \\ 18 & 6 & 0 & 1 \end{bmatrix}$

19. $A^T = \begin{bmatrix} 0 & -5 & 8 & -1 & 4 \\ 2 & 4 & 1 & 2 & 2 \\ 1 & 7 & 0 & 1 & 1 \end{bmatrix}$

21. **(i)** The dual problem is

Maximize $P = 6x + 11y$

Subject to $\begin{cases} 3x + 2y \leq 1 \\ x + 3y \leq 2 \\ x \geq 0, y \geq 0 \end{cases}$

(ii) The final simplex tableau for the dual problem is

$$\begin{array}{ccccc} x & y & s & t & P \\ \begin{bmatrix} 1.5 & 1 & 0.5 & 0 & 0 & | & 0.5 \\ -3.5 & 0 & -1.5 & 1 & 0 & | & 0.5 \\ \hline 10.5 & 0 & 5.5 & 0 & 1 & | & 5.5 \end{bmatrix} \end{array}$$

(iii) Reading from the s and t columns in the bottom row, we see that the objective function of the *minimization* problem has the optimal solution $x = 5.5$ and $y = 0$ with optimal value $P = 5.5$.

(iv)

x	y	$P = x + 2y$
0	6	6
1	3	7
5.5	0	5.5

23. (i) The dual problem is

$$\text{Maximize} \quad P = 30x + 15y$$

$$\text{Subject to} \quad \begin{cases} 5x + 10y \le 9 \\ 2x - 5y \le 6 \\ x \ge 0, y \ge 0 \end{cases}$$

(ii) The final simplex tableau for the dual problem is

$$\begin{array}{ccccc|c} x & y & s & t & P & \\ 1 & 2 & 0.2 & 0 & 0 & 1.8 \\ 0 & -9 & -0.4 & 1 & 0 & 2.4 \\ 0 & 45 & 6 & 0 & 1 & 54 \end{array}$$

(iii) Reading from the s and t columns in the bottom row, we see that the objective function of the *minimization* problem has the optimal solution $x = 6$ and $y = 0$ with optimal value $P = 54$.

(iv)

x	y	$P = 9x + 6y$
4	5	66
6	0	54

25. (i) The dual problem is

$$\text{Maximize} \quad P = 10x + 25y$$

$$\text{Subject to} \quad \begin{cases} -5x + 5y \le 5 \\ 2x + y \le 20 \\ x \ge 0, y \ge 0 \end{cases}$$

(ii) The final simplex tableau for the dual problem is

$$\begin{array}{ccccc|c} x & y & s & t & P & \\ 0 & 1 & \frac{2}{15} & \frac{1}{3} & 0 & \frac{22}{3} \\ 1 & 0 & -\frac{1}{15} & \frac{1}{3} & 0 & \frac{19}{3} \\ 0 & 0 & \frac{8}{3} & \frac{35}{3} & 1 & \frac{740}{3} \end{array}$$

(iii) Reading from the s and t columns in the bottom row, we see that the objective function of the *minimization* problem has the optimal solution $x = \frac{8}{3}$ and $y = \frac{35}{3}$ with optimal value $P = \frac{740}{3} = 246\frac{2}{3}$.

(iv)

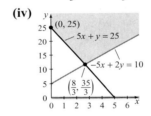

x	y	$P = 5x + 20y$
0	25	500
$\frac{8}{3} = 2\frac{2}{3}$	$\frac{35}{3} = 11\frac{2}{3}$	$\frac{740}{3} = 246\frac{2}{3}$

27. (i) The dual problem is

$$\text{Maximize} \quad P = 5x + 11y$$

$$\text{Subject to} \quad \begin{cases} x + 3y + 4z \le 3 \\ x + 2y - z \le 2 \\ x \ge 0, y \ge 0, z \ge 0 \end{cases}$$

(ii) One final simplex tableau for the dual problem is

$$\begin{array}{cccccc|c} x & y & z & s & t & P & \\ -\frac{1}{11} & 0 & 1 & \frac{2}{11} & -\frac{3}{11} & 0 & 0 \\ \frac{5}{11} & 1 & 0 & \frac{1}{11} & \frac{4}{11} & 0 & 1 \\ 0 & 0 & 0 & 1 & 4 & 1 & 11 \end{array}$$

(iii) Reading from the s and t columns in the bottom row, we see that the objective function of the *minimization* problem has the optimal solution $x = 1$ and $y = 4$ with optimal value $P = 11$.

(iv)

x	y	$P = 3x + 2y$
1	4	11
5	0	15

29. (i) The dual problem is

Maximize $P = 4x + 2y + 6z$

Subject to $\begin{cases} x + y + 2z \le 6 \\ x - y - z \le 5 \\ x \ge 0, y \ge 0, z \ge 0 \end{cases}$

(ii) The final simplex tableau for the dual problem is

$$\begin{bmatrix} x & y & z & s & t & P & \\ 0 & \frac{2}{3} & 1 & \frac{1}{3} & -\frac{1}{3} & 0 & \frac{1}{3} \\ 1 & -\frac{1}{3} & 0 & \frac{1}{3} & \frac{2}{3} & 0 & \frac{16}{3} \\ \hline 0 & \frac{2}{3} & 0 & \frac{10}{3} & \frac{2}{3} & 1 & \frac{70}{3} \end{bmatrix}$$

(iii) Reading from the s and t columns in the bottom row, we see that the objective function of the *minimization* problem has the optimal solution $x = \frac{10}{3}$ and $y = \frac{2}{3}$ with optimal value $P = \frac{70}{3}$.

(iv)

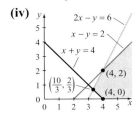

x	y	$P = 6x + 5y$
$\frac{10}{3}$	$\frac{2}{3}$	$\frac{70}{3} = 23\frac{1}{3}$
4	0	24
4	2	34

31. Let x be the number of bags of chicken dog food, y be the number of bags of lamb dog food, and z be the number of bags of turkey dog food. We must

Minimize $P = 8x + 8y + 9z$

Subject to $\begin{cases} x + y + z \ge 15 \\ -2x + y - z \ge 0 \\ x \ge 0, y \ge 0, z \ge 0 \end{cases}$

The optimal solution is $x = 0$, $y = 15$, $z = 0$ with optimal value $P = 120$. The breeder should buy 15 bags of the lamb variety of dog food to obtain the minimum cost of \$120. (An alternative optimal solution is $x = 5$, $y = 10$, $z = 0$, and $P = 120$.)

33. Let x be the number of bags of Nature's Recipe brand dog food, y be the number of bags of Nutro Max brand dog food, and z be the number of bags of PETsMART brand dog food. We must

Minimize $P = 22x + 13y + 23z$

Subject to $\begin{cases} 8x + 7y + 12z \ge 70 \\ -20x + 7y \ge 0 \\ -16x + 7y \ge 0 \\ x \ge 0, y \ge 0, z \ge 0 \end{cases}$

The optimal solution is $x = 0$, $y = 10$, $z = 0$ with optimal value $P = 130$. The optimal solution indicates the breeder should buy 10 bags of the Nutro Max brand of dog food to obtain a minimum cost of \$130.

35. Let x be the number of cases shipped from Monroe to Birmingham, y be the number of cases shipped from Monroe to Scottsboro, z be the number of cases shipped from Shelbyville to Birmingham, and w be the number of cases shipped from Shelbyville to Scottsboro.

We need to minimize the shipment cost function, $C = 0.68x + 0.65y + 0.57z + 0.32w$, subject to

$$\begin{aligned} x + y &\ge 600 \\ w + z &\ge 400 \\ x + z &\ge 700 \\ y + w &\ge 300 \\ x \ge 0, y \ge 0, z &\ge 0, w \ge 0 \end{aligned}$$

The optimal solution is $x = 600$, $y = 0$, $z = 100$, $w = 300$ and $P = 561$. In order to minimize shipping costs, 600 cases should be shipped from Monroe to Birmingham, 0 cases should be shipped from Monroe to Scottsboro, 100 cases should be shipped from Shelbyville to Birmingham, and 300 cases should be shipped from Shelbyville to Scottsboro. The minimal shipping cost is \$561.

37. Let x be the number of cases shipped from Monroe to Calhoun, y be the number of cases shipped from Monroe to Scottsboro, z be the number of cases shipped from Shelbyville to Calhoun, and w be the number of cases shipped from Shelbyville to Scottsboro.

We need to minimize the shipment cost function, $C = 0.39x + 0.65y + 0.50z + 0.32w$, subject to

$$x + y \geq 400$$
$$w + z \geq 200$$
$$x + z \geq 300$$
$$y + w \geq 300$$
$$x \geq 0, y \geq 0, z \geq 0, w \geq 0$$

The optimal solution is $x = 300$, $y = 100$, $z = 0$, $w = 200$ with optimal value $P = 246$. In order to minimize shipping costs, 300 cases should be shipped from Monroe to Calhoun, 100 cases should be shipped from Monroe to Scottsboro, 0 cases should be shipped from Shelbyville to Calhoun, and 200 cases should be shipped from Shelbyville to Scottsboro. The minimal shipping cost is \$246.

39. Let x be the number of times they went shopping and y be the number of times they went to a baseball game. We must

Minimize $P = 200x + 250y$

Subject to $\begin{cases} 3x + 4y \geq 10 \\ 11x + 14y \geq 36 \\ x \qquad\quad \geq 2 \\ x \geq 0, y \geq 0 \end{cases}$

The optimal solution is $x = 2$ and $y = 1$ with optimal value $P = 650$.

Section 4.5 *(page 272)*

1.
$$2x + y + z + s = 10$$
$$x + z + t = 3$$
$$x - 2z - u = 1$$
$$-4x - 2y - 3z + P = 0$$

3.
$$-12x + 8y + 2z + s = 15$$
$$8x + 5y - 2z + t = 10$$
$$3x - 2y + 6z - u = 5$$
$$-3x + y - 9z + P = 0$$

5.
$$4x - 3y - 2z + s = 12$$
$$-2x + 4y + 6z + t = 12$$
$$3x - 2y + 4z - u = 12$$
$$-10x - 9y - z + P = 0$$

7.

	x	y	s	t	u	P	
*	9	4	-1	0	0	0	10
	-2	5	0	1	0	0	8
*	1	1	0	0	-1	0	12
	5	4	0	0	0	1	0

9.

	x	y	s	t	u	P	
*	0	6	-2	0	0	0	16
	5	1	0	1	0	0	1
*	0	3	0	4	-4	0	6
	0	10	0	5	0	3	9

11. The optimal solution is $(5, 0)$ with optimal value $P = 10$. The points corresponding to each of the tableaus are shown in the figure below.

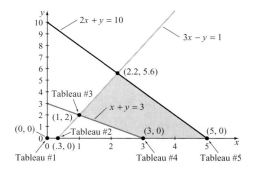

13. The optimal solution is $(2, 4)$ with optimal value $P = -22$. The points corresponding to each of the tableaus are shown in the figure below.

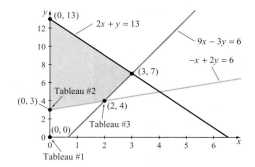

15. The optimal solution is $(0, 6)$ with optimal value $P = 24$. The points corresponding to each of the tableaus are shown in the figure below.

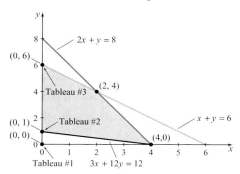

17. The optimal solution is $(2, 4)$ with optimal value $C = -30$. Since $C = -P$, P has the optimal value $P = 30$. The points corresponding to each of the tableaus are shown in the figure below.

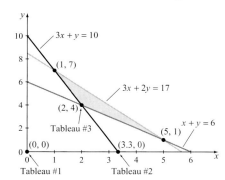

19. The feasible region is unbounded and the objective function C cannot be maximized. Consequently, the objective function $P = -C$ cannot be minimized. There is no solution to the linear programming problem.

 The points corresponding to each of the tableaus are shown in the figure below.

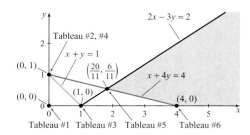

21. The optimal solution is $(5, 5, 0)$ with optimal value $P = 25$.

23. The optimal solution is $(0, 0, 8)$ with optimal value $P = 8$.

25. The optimal solution is $(0, 0, 6)$ with optimal value $P = 12$.

27. The optimal solution is $(7, 0, 0)$ with optimal value $P = 42$.

29. The optimal solution for $C = x + y - 4z$ is $(0, 0, 4)$ with optimal value $C = -16$. The optimal solution for $P = -x - y + 4z$ is also $(0, 0, 4)$ with optimal value $P = 16$.

31. The feasible region is unbounded and there is no optimal solution for $C = -9x + 3y + 3z$. Thus, $P = 9x - 3y - 3z$ does not have an optimal solution.

33. The optimal solution for $C = -6x + 2y - 4z$ is $(0, 0, 1)$ with optimal value $C = -4$. Therefore, the optimal solution for $P = 6x - 2y + 4z$ is $(0, 0, 1)$ with optimal value $P = 4$.

35. The optimal solution is $(0, \frac{22}{3}, 1)$ with optimal value $P = 24$.

37. Let x be the number of 50-packs, y be the number of 100-packs, and z be the number of 600-packs. We are to

Maximize $\quad P = 1.50(50x + 100y + 600z)$
$$= 75x + 150y + 900z$$

Subject to $\begin{cases} 50x + 100y + 600z \geq 900 \\ \text{At least 900 batteries must be purchased} \\ 10x + 18y + 96z \leq 150 \\ \text{At most \$150 may be spent} \\ z \leq 1 \\ \text{At most one 600-pack may be purchased} \\ x \geq 0, y \geq 0, z \geq 0 \end{cases}$

The optimal solution is $(0, 3, 1)$ with optimal value $P = 1350$. In order to maximize revenue, she should buy three 100-packs and one 600-pack. Her maximum revenue will be \$1350.

39. Let x be the number of batches of rice pudding, y be the number of batches of tapioca pudding, and z be the number of batches of vanilla pudding. We want to

Maximize $\quad P = 24x + 18y + 12z$

Subject to $\begin{cases} 12x + 12y + 6z \leq 108 \\ \text{At most 108 cups of milk may be used} \\ 1.5x + 1.5y + 1.5z \leq 150 \\ \text{At most 150 cups of sugar may be used} \\ 9x + 9y + 6z \leq 84 \\ \text{At most 84 eggs may be used} \\ x + y \geq 2 \\ \text{The sum of the batches of rice pudding} \\ \text{and tapioca pudding must be at least two} \\ x \geq 0, y \geq 0, z \geq 0 \end{cases}$

The optimal solution is $(9, 0, 0)$ with optimal value $P = 216$. In order to maximize the number of servings, the deli should make 9 batches of rice pudding. No other pudding variety needs to be made.

41. Let x be the number of $\frac{1}{2}$-cup servings of pinto beans, y be the number of $\frac{1}{2}$-cup servings of kidney beans, and z be the number of $\frac{1}{2}$-cup servings of black beans. We want to *minimize*

$$P = y$$

This is equivalent to *maximizing*

$$C = -y$$

We have the following constraints:

$x + y + z \le 6$	At most 6 1/2-cup servings
$7x + 11y + 2z \ge 39$	At least 39 grams of fiber is needed
$6x + 8y + 6z \ge 42$	At least 42 grams of protein is needed

$x \ge 0, y \ge 0, z \ge 0$

The optimal solution for $C = -y$ is $(0, 3, 3)$ with optimal value $C = -3$. The optimal solution for $P = y$ is $(0, 3, 3)$ with optimal value $P = 3$.

We should use three $\frac{1}{2}$-cup servings of kidney beans and three $\frac{1}{2}$-cup servings of black beans. In other words, we should use 1.5 cups of kidney beans and 1.5 cups of black beans.

43. Let x be the number of students transported from the North region to School 1, y be the number of students transported from the North region to School 2, z be the number of students transported from the South region to School 1, and w be the number of students transported from the South region to School 2.

The total transportation cost is given by

$$P = 120x + 180y + 100z + 150w.$$

We are to *minimize* this amount. In other words, we are to

Maximize $C = -P$.

Subject to

$x + z \le 500$	School 1 holds at most 500 students
$y + w \le 420$	School 2 holds at most 420 students
$x + y \ge 400$	The North region has at least 400 students
$z + w \ge 430$	The South region has at least 430 students

$x \ge 0, y \ge 0, z \ge 0, w \ge 0$

The optimal solution for $C = -120x - 180y - 100z - 150w$ is $(400, 0, 100, 330)$ with optimal value $C = -107500$. The optimal solution for $P = 120x + 180y + 100z + 150w$ is $(400, 0, 100, 330)$ with optimal value $P = 107{,}500$.

The minimum transportation cost is $107,500, which occurs when 400 children are bused from the North region to School 1, 100 children are bused from the South region to School 1, and 330 children are bused from the South region to School 2.

45. Let x be the number of desks shipped from Phoenix to Kingman, y be the number of desks shipped from Phoenix to Flagstaff, z be the number of desks shipped from Las Vegas to Kingman, and w be the number of desks shipped from Las Vegas to Flagstaff.

The total transportation cost is given by

$$P = 1.85x + 1.45y + 1.05z + 2.51w$$

We are to *minimize* this amount. In other words, we are to

Maximize $C = -P$.

Subject to

$x + y \le 500$	Phoenix can ship at most 500 desks
$z + w \le 200$	Las Vegas can ship at most 200 desks
$x + z \ge 250$	Kingman needs at least 250 desks
$y + w \ge 400$	Flagstaff needs at least 400 desks

$x \ge 0, y \ge 0, z \ge 0, w \ge 0$

The optimal solution for $C = -1.85x - 1.45y - 1.05z - 2.51w$ is $(50, 400, 200, 0)$ with optimal value $C = -882.5$. The optimal solution for $P = 1.85x + 1.45y + 1.05z + 2.51w$ is $(50, 400, 200, 0)$ with optimal value $P = 882.5$.

The minimum transportation cost is $882.50, which occurs when 50 desks are shipped from Phoenix to Kingman, 400 desks are shipped from

Phoenix to Flagstaff, and 200 desks are shipped from Las Vegas to Kingman.

47. The optimal solution for $C = -x - y - z - w$ is $(4, 0, 2, 3)$ with optimal value $C = -9$. The optimal solution for $P = x + y + z + w$ is $(4, 0, 2, 3)$ with optimal value $P = 9$.

49.

$$\text{Maximize} \quad P = x$$

$$\text{Subject to} \quad \begin{cases} x + y + z \le 2 \\ x + y + z \ge 5 \\ x + 2y \le 0 \\ x \ge 0, y \ge 0, z \ge 0 \end{cases}$$

There are no values of x, y, and z that can simultaneously satisfy the first two constraints. If the sum of the variables is less than 2, the first constraint is satisfied but the second constraint is not. If the sum of the variables is greater than 5, the second constraint is satisfied but the first constraint is not. Thus the feasible region is empty. (Answers will vary).

CHAPTER 4 REVIEW EXERCISES

Section 4.1 *(page 276)*

1. $P = (2, 9)$ is a solution.

3. $P = (8, 9)$ is a solution.

5.

$-3x + 2y = 1$
$-4x + 3y = 2$

7.

$6x - 2y = -6$
$2x + 4y = 8$

9. Let x be the number of hours she spends working in the copy center and y be the number of hours

she spends designing brochures. We have the following constraints.

$x + y \le 50$	She works at most 50 hours per week
$x \ge 40$	She must work at least 40 hours in the copy center
$x \le 50$	She may work at most 50 hours in the copy center
$30y \ge 200$	She must earn at least \$200 making brochures
$x \ge 0, y \ge 0$	

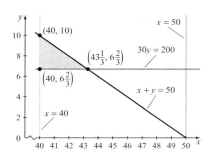

In order to meet her income and workload goals, she can work at most 43 hours and 20 minutes a week at the copy center. Since she has no control over the amount of time she works at the copy center, some weeks she will meet her goals and other weeks she will not. In the weeks she doesn't meet her goals, she will have to either work extra hours or live on less money. The struggle between standard of living and quality of life is a dilemma for many workers.

Section 4.2 *(page 277)*

11. The optimal solution is $(2, 4)$ with optimal value $z = 34$.

13. The optimal solution is $\left(2\frac{2}{3}, 0\right)$ with optimal value $z = 13\frac{1}{3}$.

Section 4.3 *(page 277)*

15. The problem is not a standard maximization problem because it is missing the nonnegative constraints $x \ge 0$ and $y \ge 0$. Additionally, the second constraint $-x + -y \le -20$ is not of the form $Ax + By \le M$ with M nonnegative.

17. The optimal solution is $(3, 4)$ with optimal value $P = 34$. We verify our solution graphically.

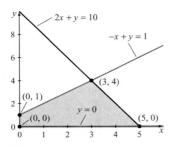

We evaluate the objective function $P = -2x + 10y$ at each of the corner points.

x	y	$P = -2x + 10y$
0	0	0
0	1	10
3	4	34
5	0	−10

19. The optimal solution is $(60, 0, 0)$ with optimal value $P = 360$.

21. Let x be the number of 100-packs, y be the number of 600-packs, and z be the number of 1200-packs. She earns \$100 for each 100-pack sold, \$600 for each 600-pack sold, and \$1200 for each 1200-pack sold. She must

Maximize $P = 100x + 600y + 1200z$

Subject to
$$\begin{cases} 100x + 600y + 1200z \leq 1800 \\ \quad \text{She wants at most 1800 batteries} \\ 18x + 96y + 180z \leq 300 \\ \quad \text{She wants to spend at most \$300} \\ x \geq 0, y \geq 0, z \geq 0 \end{cases}$$

Furthermore, to increase the likelihood that we will get a whole-number solution, we determine the maximum number of each size pack we could buy with the \$300 and add these as additional constraints.

$x \leq 16$ She can buy at most sixteen \$18 100-packs with \$300

$y \leq 3$ She can buy at most three \$96 600-packs with \$300

$z \leq 1$ She can buy at most one \$180 1200-pack with \$300

The optimal solution is $(0, 1, 1)$ with optimal value $P = 1800$. She should order one 1200-pack and one 600-pack in order to get the maximum revenue of \$1800.

Section 4.4 *(page 278)*

23. This is a standard minimization problem.

25. The second constraint is not of the form $Ax + By \geq M$ with M nonnegative and cannot be made into the correct form by multiplying by -1. Thus the problem is not a standard minimization problem.

27. $A^T = \begin{bmatrix} 12 & 14 & 16 \\ 10 & 12 & 14 \\ 8 & 10 & 12 \end{bmatrix}$

29. (i) The dual problem is

Maximize $P = 20x + 11y$

Subject to $\begin{cases} 4x + 2y \leq 4 \\ 5x + 3y \leq 2 \\ x \geq 0, y \geq 0 \end{cases}$

(ii) The final tableau is

$$\begin{array}{ccccc} x & y & s & t & P \\ \begin{bmatrix} 0 & -2 & 5 & -4 & 0 \\ 5 & 3 & 0 & 1 & 0 \\ 0 & 1 & 0 & 4 & 1 \end{bmatrix} & & & & \begin{matrix} 12 \\ 2 \\ 8 \end{matrix} \end{array}$$

(iii) Reading from the s and t columns in the bottom row, we see that the objective function of the *minimization* problem has the optimal solution $x = 0$ and $y = 4$ with optimal value $P = 8$.

(iv)

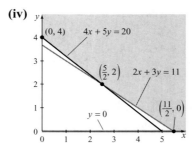

x	y	$P = 4x + 2y$
0	4	8
$\frac{5}{2}$	2	14
$\frac{11}{2}$	0	22

31. (i) The dual problem is

$$\text{Maximize} \quad P = 26x + 25y$$

$$\text{Subject to} \quad \begin{cases} 5x + 10y \le 2 \\ 2x - 5y \le 3 \\ x \ge 0, y \ge 0 \end{cases}$$

(ii) The final tableau for the dual problem is

$$\begin{array}{ccccc} x & y & s & t & P \\ \begin{bmatrix} 5 & 10 & 1 & 0 & 0 & | & 2 \\ 0 & -45 & -2 & 5 & 0 & | & 11 \\ 0 & 27 & \frac{26}{5} & 0 & 1 & | & \frac{52}{5} \end{bmatrix} \end{array}$$

(iii) Reading from the s and t columns in the bottom row, we see that the objective function of the *minimization* problem has the optimal solution $x = \frac{26}{5} = 5.2$ and $y = 0$ with optimal value $P = \frac{52}{5} = 10.4$.

(iv)

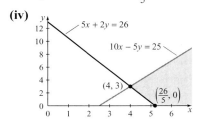

x	y	$P = 2x + 3y$
4	3	17
$\frac{26}{5}$	0	$\frac{52}{5} = 10.4$

33. Let x be the number of bags of Authority brand dog food, y be the number of bags of Bil-Jac brand dog food, and z be the number of bags of Iams brand dog food.

We must

$$\text{Minimize} \quad P = 20x + 19y + 9z$$

$$\text{Subject to} \begin{cases} 33x + 18y + 8z \ge 330 \\ \quad \text{At least 330 pounds of} \\ \quad \text{dog food is needed} \\ 0.66x - 0.54y - 0.32z \ge 0 \\ \quad \text{At least 30\% protein} \\ 0.18y + 0.08z \ge 0 \\ \quad \text{At least 3\% fiber} \\ x \ge 0, y \ge 0, z \ge 0 \end{cases}$$

The optimal solution is $(10, 0, 0)$ with optimal value $P = 200$. In order to minimize costs, the kennel should buy 10 bags of the Authority brand

dog food. Using the rounded prices, we calculate that the cost will be \$200.

Section 4.5 *(page 279)*

35.
$$2x + 4y + z + s = 10$$
$$x + 2z + t = 3$$
$$3x - 2z - u = 1$$
$$-2x - 3y - z + P = 0$$

37. The optimal solution is $(1, 0, 2)$ with optimal value $P = 24$.

CHAPTER 5

Section 5.1 *(page 299)*

1. Concave up, y-intercept $(0, 1)$, vertex $(1, 0)$

3. Concave up, y-intercept $(0, 0)$, vertex $(-0.5, -0.75)$

5. Concave up, y-intercept $(0, 2.1)$, vertex $(0.25, 1.925)$

7. $y = -x^2 + 4x - 1$ **9.** $y = x^2 - 10x + 5$

11. (a) $W(t) = 36.92t^2 - 43.60t + 2072$

(b)

(c) As shown in the scatter plot, the model appears to fit the data well. It seems reasonable that women's basketball game attendance will continue to increase; consequently, we believe this model is good.

13. (a) $D(t) = 896.67t^2 + 723.33t - 540$

(b)

(c) As shown in the scatter plot, the model appears to fit the data well. Since the scatter plot fits the data well and we expect that DVD shipments will continue to increase, we believe this model is a good model.

15. (a) $C(v) = -0.03550v^2 + 85.451v - 47,197$

(b)

(c) As shown in the scatter plot, the model fits the data perfectly. Although the scatter plot fits the data perfectly, we believe that the model has some definite limitations. According to the model, as the number of blank VHS tapes drops below 1195, the number of CD-R disks will also begin to decrease. This doesn't seem reasonable. We anticipate that CD-R disk sales will continue to increase even as VHS tape sales decrease.

17. (a) (See part d.)
(b) Since the data set consists of three nonlinear points, a quadratic model will fit the data perfectly.
(c) The quadratic model is
$I(t) = -10.31t^2 + 972.4t + 22{,}530$
(d)

(e) Since the model fits the data and since we expect personal incomes to continue to increase, we anticipate that the model will be relatively accurate in forecasting per capita personal incomes between 1980 and 2010.

19. (a)

(b) The scatter plot does not seem to resemble a parabola. Consequently, we do not think that a quadratic function will fit the data well.

21. (a) (See part d.)
(b) The scatter plot roughly resembles a parabola, so we will model the data with a quadratic function.
(c) The quadratic model is
$M(t) = 0.02628t^2 + 0.3340t + 47.46$
(d)

(e) Since the model fits the data and since it seems reasonable to assume that the upward trend in conventional mortgages will continue, we anticipate that the model will be relatively accurate in forecasting the percentage of new privately owned one-

family houses financed with a conventional mortgage between 1977 and 2005. (Note: 2005 is arbitrarily selected as a year *near* the last data point.)
(f) When $t \approx 5.4$, $M(t) = 50$. Since t is the number of years since the end of 1970, we estimate that in mid-1976 ($t = 5.4$), 50 percent of mortgages were conventional mortgages.

23. (a)

(b) The scatter plot does not seem to resemble a parabola. Consequently, we do not think that a quadratic function will fit the data well.

25. (a) (See part d.)
(b) The scatter plot seems to resemble a portion of a parabola. We think that a quadratic model will fit the data.
(c) The quadratic model is
$H(t) = -0.01293t^2 + 1.398t + 57.60$
(d)

(e) We believe the model is a good model because it seems reasonable to assume that the percentage of homes with a garage will continue to increase.
(f) Since $H(33.6) \approx 90$, we estimate that in mid-2004 ($t = 33.6$), 90 percent of new homes had garages.

27. (a) (See part d.)
(b) Although the scatter plot appears to be near-linear, we can still construct a quadratic model. When quadratic models are used to model near-linear functions, the value of a is relatively close to zero. (If we had our choice of which type of model to construct, we would construct a linear model because it is simpler.)
(c) The quadratic model is
$P(t) = -0.02143t^2 + 8.279t + 68.27$
(d)

(e) Since the model fits the scatter plot fairly well, and since home sales prices typically increase over time, we believe that our model will relatively accurately predict future home prices in the northeastern United States.

29. (a) (See part d.)
(b) The scatter plot looks concave up. Consequently, we expect that a parabola will fit the data.
(c) The quadratic model is
$P(t) = 0.0417t^2 + 0.06833t + 2.455$
(d)

(e) Since the model fits the scatter plot fairly well, and since auto leasing appears to be increasing in popularity, we believe that our model will relatively accurately predict the percentage of households leasing vehicles.
(f) Since $P(7) \approx 5$, we estimate that at the end of 1996, 5 percent of households were leasing vehicles.

31. (a) (See part d.)
(b) The scatter plot looks like a concave up parabola, so we believe that a quadratic model will fit the data well.
(c) The quadratic model is
$S(t) = 29.23t^2 + 79.33t + 177.4$
(d)

(e) Since the model fits the data extremely well, and since we expect Starbucks sales to continue to increase, we believe we have a good model.
(f) Since $S(11.6) \approx 5000$, we expect that midway through fiscal year 2005 ($t = 11.6$), Starbucks sales will reach $5000 million.

33. (a) (See part d.)
(b) The scatter plot looks more or less concave up and increasing. A quadratic model will probably be a good choice for the data set.
(c) The quadratic model is
$P(t) = 0.0007346t^2 + 0.009701t$
$+ 0.7803$

(d)

(e) Since the model fits the data relatively well, and since we expect the price of chicken to continue to increase (because of inflation), we believe we have a good model.

35. (a) (See part d.)
(b) The scatter plot is generally concave up and increasing. We believe that a quadratic model will fit the data well.
(c) The quadratic model is
$S(t) = 0.9195t^2 + 23.25t + 2652$
(d)

(e) Since the model fits the data relatively well, and since it seems reasonable to anticipate that private college enrollment will continue to increase, we believe that the model is a good model.

37. (a) (See part d.)
(b) The scatter plot appears concave up and increasing. A quadratic model may fit the data well.
(c) The quadratic model is
$P(t) = 123.73t^2 + 445.58t + 22,847$
(d)

(e) Since the model fits the data relatively well, and since it seems reasonable to anticipate that personal incomes will continue to increase, we believe the model is a good model.

39. (a) (See part d.)
(b) The scatter plot appears near-linear. Nevertheless, we can come up with a decent quadratic model for the data.
(c) The quadratic model is
$P(t) = -14.470t^2 + 883.18t + 17,620$
(d)

(e) Since the model fits the data relatively well, and since it seems reasonable to anticipate that personal incomes will continue to increase, we believe that the model is a good model.

41. (a) (See part d.)
  (b) The scatter plot appears slightly concave down and increasing. A quadratic model should fit the data well.
  (c) The quadratic model is
  $P(t) = -32.71t^2 + 778.1t + 2341$
  (d)

  (e) Since the model fits the data well, and since it seems reasonable that the company's profits will continue to increase, we believe that the model is a good model.
  (f) Since $P(6.5) \approx 6000$, we estimate that midway through 2004 ($t = 6.5$), Johnson & Johnson profits reached $6000 million.

43. (a) (See part d.)
  (b) The scatter plot is concave up and increasing. A quadratic function may fit the data well.
  (c) The quadratic model is
  $F(t) = 12.66t^2 + 48.52t + 2102$
  (d)

  (e) Since the model fits the data well, and since it seems reasonable that the payroll will continue to grow, we believe that the model is a good model.

45. (a) (See part d.)
  (b) The scatter plot appears more or less concave down and increasing. A quadratic model may fit the data well.
  (c) The quadratic model is
  $F(t) = -0.04079t^2 + 1.453t + 401.6$
  (d)

  (e) Since the model fits the data well, and since it seems reasonable that the number of

private-school teachers will continue to increase, we believe that the model is a good model.

47. (a) (See part d.)
  (b) The scatter plot appears more or less concave up and increasing. A quadratic model may fit the data well.
  (c) The quadratic model is
  $W(t) = 0.003197t^2 + 0.1333t + 4.606$
  (d)

  (e) Since the model fits the data well, and since it seems reasonable that wages will continue to increase, we believe that the model is a good model.

49. (a) (See part d.)
  (b) The scatter plot appears concave up. A quadratic model may fit the data well.
  (c) The quadratic model is
  $A(t) = 9.396t^2 - 11.58t + 1063$
  (d)

  (e) Since the model fits the data well, and since it seems reasonable that advertising expenditures will continue to increase, we believe that the model is a good model.
  (f) Since $A(10.6) \approx 2000$, we estimate that in mid-2001 ($t = 10.6$), billboard advertising expenditures reached $2000 million ($2 billion).

51. (Answers may vary.)
  $y = ax(x - 4)$
  $y = x^2 - 4x$ when $a = 1$
  $y = -x^2 + 4x$ when $a = -1$

53. No. Although any set of three *nonlinear* points defines a unique quadratic function, the three points given all lie on the same line, $y = 2x + 1$.

55. Since the function is nonnegative for all values of x, the graph of the parabola must be entirely above the x axis. Consequently, the graph must be concave up. (If it were concave down, it would cross the x axis.) Since the graph is concave up, $a > 0$.

Section 5.2 *(page 323)*

1. The quartic function that best fits the data is given by $S(t) = -0.53891t^4 + 15.050t^3 - 103.86t^2 + 48.261t + 12,393$. $S(16) = 12,904$ means that there were 12,904 thousand students enrolled in high school in 2000.

3. The quadratic function that best fits the data is $R(t) = 1.874t^2 - 1.425t + 2165$. $R(21.5) \approx 3000$ means that the number of public elementary and secondary teachers reached 3 million (3000 thousand) in mid-2002 ($t = 21.5$).

5. The cubic function that best fits the data is $C(t) = -0.5278t^3 + 5.905t^2 - 4.234t + 4.048$. $C(7) \approx 83$ means that 83 percent of public school classrooms had Internet access in 2001. $C(8) \approx 78$ means that 78 percent of public school classrooms had Internet access in 2002. The 2001 estimate seems reasonable, since we expect Internet access in schools to become increasingly common. The 2002 estimate does not seem reasonable, since it forecasts a decline in Internet access.

7. The cubic model that best fits the data is $C(t) = 0.1563t^3 - 2.268t^2 + 14.84t + 164.0$. $C(15) \approx 404$ means that there will be 404 million debt cards by the end of 2005. Our model estimate is substantially higher than the consulting firm's 270 million card projection.

9. The quartic function that best fits the data is given by $S(t) = 4.807t^4 - 98.56t^3 + 541.3t^2 + 2936t + 10,610$. $S(25) \approx 760,000$ means that 760 billion shares were traded in 2005. Because of the rapid growth in the market in recent years, this estimate seems reasonable. However, the recession in the early 2000s may have affected the number of shares sold.

11. The quartic function that best fits the data is $W(t) = 100t^4 - 1518.5t^3 + 8244.4t^2 - 17,294t + 42,667$. $W(6.5) \approx 40,000$, meaning that in mid-1999 ($t = 6.5$), the average annual wage per worker for computer and equipment retailers first reached \$40,000.

13. The quadratic function of best fit is $W(t) = 1421.4t^2 - 2321.4t + 56,429$. $W(7.6) \approx 120,000$, meaning that in mid-2000 ($t = 7.6$), the average wage per worker in the prepackaged software development industry reached \$120,000.

15. The cubic model that best fits the software wholesaler's wages is $W(t) = 113.89t^3 - 251.19t^2 + 277.78t + 52,571$. We estimate that at the end of 2002, the average wage of both software wholesalers and software retailers was \$145,000.

17. The quartic model that best fits the data is $W(t) = 51.515t^4 - 701.52t^3 + 3119.7t^2 - 2450.4t + 41,563$. $W(7) \approx 60,000$, meaning that the television broadcasting average wage will first exceed \$60,000 at the end of 1999. This estimate may be a bit high, but it is not unreasonable. Since $t = 7$ is close to the end of the data set ($t = 6$) and \$60,000 is somewhat close to \$54,600, we are comfortable with the estimate.

19. The cubic model that best fits the data is $W(t) = -0.005860t^3 + 0.2430t^2 - 0.06623t + 6.541$. $C(32) \approx 61.2$, meaning that 61.2 percent of TV homes had cable at the end of 2002. We would generally expect the percentage of homes with cable to continue to increase, as it has over the 29 years of the data set. For this reason, the estimate seems unreasonable. However, if people are replacing cable with satellite dishes or some other technology, it is possible that the percentage of homes with cable could decline.

21. A polynomial of degree n may have at most $n - 1$ bends.

23. $f(x) = f(-x)$

25. $y = x^4 - 2x^3 - x^2 + 2x$

Section 5.3 *(page 343)*

1. The graph is decreasing and concave up. The y-intercept is $(0, 4)$.

3. The graph is increasing and concave up. The y-intercept is $(0, 0.5)$.

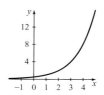

5. The graph is increasing and concave up. The y-intercept is (0, 0.4).

7. The graph is decreasing and concave down. The y-intercept is (0, −1.2).

9. The graph is decreasing and concave up. The y-intercept is (0, 3).

11. $y = 2(3)^x$

13. $y = 5(2)^x$

15. $y = 64\left(\dfrac{1}{2}\right)^x$

17. $y = 0.25(4)^x$

19. $y = 256\left(\dfrac{1}{2}\right)^x$

21. $I(t) = 82.67(1.061)^t$ (rounded model)
$I(25) = 364.5$ (using unrounded model)
At the end of 2005, dental prices are expected to be 364.5 percent of the 1984 price.

23. $I(t) = 106.1(0.9657)^t$ (rounded model)
$I(25) = 44.4$ (using unrounded model)
At the end of 2005, the price of television set is expected to be 44.4 percent of the 1984 price.

25. $I(t) = 86.75(1.051)^t$ (rounded model)
$I(25) = 302.7$ (using unrounded model)
At the end of 2005, the price of admission to an entertainment venue is expected to be 302.7 percent of its 1984 price.

27. $B(t) = 235(1.0232)^t$
where B is the balance (in dollars) and t is the number of years from now. In just over 2 years

and 8 months ($t \approx 2.7$), the balance is projected to reach $250.

29. $p(t) = 59.90(1.28)^t$
where p is the Dave Matthews Band concert tickets price and t is the number of years since September 2004. A concert ticket that costs $59.90 in September 2004 is expected to cost $98.14 in September 2006.

31. $x = 8.827$ 33. $x = 14.21$ 35. $x = 2$

37. (a) The predicted value of the investment is given by $V(t) = 2000(1.0867)^t + 1000(1.0693)^t$ where V is the investment value (in dollars) and t is the number of years from now.
(b) $V(20) = \$14{,}368.66$
Before changing the amount of money invested in each account, the investor should consider her risk tolerance. (Answers may vary.)

39. (a) Growth account: $699.40
Inflation-linked Bond account: $1163.20
(b) The combined value of the accounts is projected to double by June 2017 ($t = 14.43$).

Section 5.4 *(page 357)*

1. $y = 2$ 3. $y = 6$ 5. $y = -2$

7. $y = \log_4(x)$ is concave down and increasing.

9. $y = \log_{0.7} x$ is concave up and decreasing.

11. $x = 6$ 13. $x = -1$ 15. $x = 3$

17. $x = \dfrac{1}{16}$ 19. $x = 243$ 21. $4\log(2)$

23. $\log(2x^4)$ 25. $\log(jam)$

27. $\log\left(\dfrac{1}{324}x^{-16}\right)$ **29.** $-\log(x)$

31. $2\ln(x)$ **33.** $2\ln(x)$ **35.** $\ln(27)$

37. $\ln\left(\dfrac{1}{64}\right)$ **39.** $\ln(729x^5)$

41. $T(i) = -73.69 + 16.71\ln(i)$
$T(125) = 7.0$
According to the model, the Consumer Price Index reached 125 seven years after 1980. That is, at the end of 1987, the price of dental services were 125 percent of their 1984 price.

43. $T(i) = 129.8 - 27.76\ln(i)$
$T(125) = -4.2$
According to the model, the Consumer Price Index was at 125 4.2 years before the end of 1980. That is, in late 1976, the price of a television set was 125 percent of its 1984 price.

45. $T(i) = -88.48 + 19.84\ln(i)$
$T(125) = 7.3$
According to the model, the Consumer Price Index was at 125 7.3 years after the end of 1980. That is, in early 1988, the price of admission to an entertainment venue was 125 percent of its 1984 price.

47. We know that $y = \log_b(x)$ is equivalent to $x = b^y$. From the definition of $y = \log_b(x)$, we also know that b is positive.

We know that raising a positive number to any power yields a positive number. So b^y will always be positive. But $x = b^y$, so x must always be positive. Hence, the domain of $y = \log_b(x)$ is all positive real numbers.

49. No.

Section 5.5 *(page 370)*

1. $S(t) = 8.451t^3 - 111.9t^2$
$\qquad\quad + 470.0t + 965.0$
Since $S(12) \approx 5099$, we forecast that gaming hardware sales will be \$5099 million in 2002.

3. $C(t) = 0.3214t^2 + 0.2500t + 37.36$
Since $C(8) \approx 60$, we forecast that the 2005–06 tuition cost per credit will be \$60.

5. $S(t) = 1165.9t + 80{,}887$
Since $S(20) \approx 104{,}200$, we forecast that the 2010 public university enrollment will be 104,200.

7. $C(t) = \dfrac{1106}{1 + 0.1259e^{0.5491t}}$
Since $S(13) \approx 7$, we forecast that there will be 7 pediatric AIDS cases in the United States in 2005.

9. $A(t) = \dfrac{38.56}{1 + 14.52e^{-0.8032t}} + 16$
Since $A(10) \approx 54$, we forecast that there were 54 air carrier accidents in 2002.

11. $C(t) = 0.60(1.03)^t$ dollars, where t is the number of years from now.

13. $P(t) = 3t + 87$ dollars, where t is the number of years from now.

15. The logistic model for monthly product sales is
$$S(t) = \dfrac{200.67}{1 + 25.85e^{-0.4973t}}$$

17. The logistic model for the data is
$$P(t) = \dfrac{6930}{1 + 166.2e^{-0.4019t}} + 2600$$
The maximum projected population of the city is $6930 + 2600 = 9530$. According to the logistic model, the 2003 population was 6258, since $P(13) \approx 6258$.

The exponential model that best fits the aligned data is
$$P(t) = 67.59(1.377)^t + 2600$$
According to the exponential model, the 2003 population was 6910, since $P(13) \approx 6910$. Although this estimate is still below the actual 2003 population (7480), it is substantially better than the logistic model estimate. (Answers may vary.)

19.
$$T(x) = \begin{cases} 0.1x & \text{if } 0 \le x \le 7000 \\ 0.15x - 350 & \text{if } 7000 < x \le 28{,}400 \\ 0.25x - 3190 & \text{if } 28{,}400 < x \le 68{,}800 \end{cases}$$

21. In addition to looking at how well the model fits the data, we must consider the expected future behavior of the thing being modeled. Since the model is being used to forecast future events, we must make sure that the model's prediction seems reasonable based on what we know about the thing being modeled. (Answers may vary.)

23. Businesses can benefit financially by accurately predicting the future. Although mathematical models are imperfect at forecasting the future, they offer an educated guess as to what might happen. In short, they increase the likelihood of an accurate prediction. (Answers may vary.)

25. Because of the dramatic variability in the data set, we don't recognize any trends or patterns in the scatter plot. None of the standard mathematical models will fit this data set well. Our best guess might be to identify the range of y values and state that we would expect future data values to fall within that range. (Answers may vary.)

CHAPTER 5 REVIEW EXERCISES

Section 5.1 *(page 375)*

1. The quadratic model that best fits the data is
$$A(t) = 42.293t^2 + 158.32t + 6574.5$$
Since $A(12) \approx 14{,}564$, we estimate that \$14,564 million was spent on magazine advertising in 2002.

3. Early in the second quarter of 2001 $(t \approx 10.3)$.

5. $A(t) = 34.992t^2 + 94.909t + 8963.2$
Since $A(11.8) \approx 15{,}000$, we estimate that \$15 billion was spent on Yellow Pages advertising in the one-year period that ended in late 2002 $(t = 11.8)$.

Section 5.2 *(page 376)*

7. $A(t) = -32.085t^3 + 656.16t^2$
$\qquad - 1640.7t + 31{,}914$
Since $A(11) \approx 50{,}557$, we estimate that \$50,557 million was spent on newspaper advertising in 2001 $(t = 11)$.

9. $A(t) = 10.222t^4 - 221.51t^3 + 1616.4t^2$
$\qquad - 2487.8t + 26{,}671$
Since $A(11.0) \approx 50{,}000$, we estimate that \$50 billion was spent on broadcast TV advertising in 2001 $(t = 11)$. Broadcast TV annual advertising expenditures beyond 2001 are expected to exceed \$50 billion.

11. $E(t) = 0.001228t^3 - 0.03338t^2$
$\qquad + 0.6439t + 9.844$
Since $E(23) \approx 21.94$, we estimate that average hourly earnings of a manufacturing industry employee in Michigan in 2003 $(t = 23)$ was \$21.94.

13. The manufacturing wage in Florida is roughly 65 percent of the Michigan manufacturing wage. Thus labor costs in Florida will be substantially less than labor costs in Michigan. For this reason, a new manufacturing business may prefer to start up in Florida as opposed to Michigan. However, additional factors such as business tax rate, infrastructure, availability and cost of raw materials, etc., should also be considered. (Answers may vary.)

Section 5.3 *(page 379)*

15. The function $y = -0.3(2.8)^x$ is concave down and decreasing. The y-intercept is $(0, -0.3)$.

17. $y = 9(2)^x$

19. $N(t) = 22{,}925(1.0727)^t$
$N(10) \approx 46{,}248$ means that there will be 46,248 McDonalds restaurants in 2007.

21. $E(t) = 4500(1.023)^t$
$E(5) \approx 5042$ means that five years from now, my monthly household expenses are expected to be \$5042.

23. $x \approx 1.723$

Section 5.4 *(page 379)*

25. $y = -1$

27. The graph of $y = \log_{0.4}(x)$ is decreasing and concave up.

29. $x = 4$ 31. $x = 0.04$

33. $\log(2x^2)$ 35. $-\log(3)$

37. $F(t) = 49.96 + 5.112 \ln t$. Since $F(10) \approx 61.7$, we estimate that in 1999, 61.7 percent of auto accident fatalities were not alcohol-related.

Section 5.5 *(page 380)*

39. $P(t) = \dfrac{57.81}{1 + 0.5443e^{0.3900t}}$

CHAPTER 6

Section 6.1 *(page 389)*

1. $x = 4$

3. $x = 2$

5. $x \approx 1.585$

7. $t \approx -0.614$

9. $t \approx 0.5195$

11. $t \approx 24$; At the end of 2012 ($t = 24$), the number of congregations is expected to reach 15,000.

13. $t \approx 36.47$; By mid-2015 ($t = 36.47$), the amount of land in farms is projected to drop to 910 million acres.

15. $t \approx 20.28$; By early 2001 ($t = 20.28$), the annual per capita bottled water consumption was projected to reach 20 gallons.

17. $t \approx 2.59$; In mid-1993, the number of people who considered themselves nondenominational Christians was expected to equal the number of people who consider themselves Lutherans. At that time, both groups were expected to have 9228 thousand American adult members.

19. $t \approx 76.9$; The populations of Ohio and Illinois are projected to be equal by late 2072 ($t = 76.9$). At that time, the population of each state is expected to be 14,052 thousand people.

21. $x \approx 0.7510$.

23. The salaries of public elementary and secondary school teachers and public university assistant professors are projected to be equal in early 1995 ($t = 4.07$) and in early 2035 ($t = 44.1$).

25. At the end of 2003 ($t = 4$), the number of unit cases sold is expected to equal the number of dollars of net operating revenue.

27. $t \approx 9.298$; It will take more than 9 years for the investment value to reach $1000.

29. (a) The account values are expected to be equal after about 4 years and 4 months ($t = 4.323$).
(b) The investments will each be valued at $697.36, with a combined value of $1394.72.
(c) The combined value of the investments after 30 years will be $5113.30.
(d) $5080.07
(e) (Answers may vary.)

31. Yes. Although the combined value of the accounts initially decreases, the combined value increases after the end of the second year.

33. $x \approx 0.7304$

35. The equation will have a solution for all $a \leq 0$.

37. The equation has at least one solution: $x = 0$.

39. When $a = b$, the graphs intersect in infinitely many places.

Section 6.2 *(page 402)*

1. $612.50

3. $5137.50

5. $632.16

7. Even though the stated rate of ING DIRECT (4.8%) was higher than the stated rate of Ascencia Bank (4.7%), the difference in compounding methods made Ascencia Bank the better value. ING DIRECT yielded $620 at maturity while Ascencia Bank yielded $632.16.

9. At maturity, the accounts will have the same value.

11. $APY = 3.55$ percent

13. $APY = 3.49$ percent

15. $APY = 2.38$ percent

17. The account values will never be equal to each other in the future.

19. The account values will be equal in 22.3 years.

21. Annual rate: 7.50 percent

23. Continuous rate: 3.24 percent

25. Annual rate: 11.65 percent

27. Continuous rate: 15.87 percent

29. 735,777.99 percent

31. 8478.33 percent

33. (a) 4.13 percent per year
(b) A loaf of white bread will cost $2.06 in 2008.

35. (a) 4.33 percent per year
(b) In 2003 ($t = 19.37$)

37. 9.05 percent

39. 4.16 percent

41. 4.55 percent

43. 5.89 percent

45. $r = 0$ percent or $r = 200$ percent

47. 791.6 percent

Section 6.3 *(page 414)*

1. About 3 years ($t \approx 3.0$)

3. $76.55 monthly

5. $80.19 monthly

7. $307,934.23

9. $243,470.19

11. $163,879.49

13. $32,460.06

15. About 13.5 years ($t \approx 13.5$)

17. $100.00 per month   **19.** $988.56 per month

21. 5.66 percent   **23.** 10.91 percent

25. It will take them 17 months to save up the $9750 down payment, assuming that interest is prorated between quarters. If interest is not prorated between quarters, they will have to wait until the end of the quarter (18 months). At that time, they will have $10,469.34.

27. 72 weeks   **29.** 2.5 years

31. 7.529 years (assuming that interest is prorated)

33. When $t = 30.26$, the account values are expected to be equal. The twins will be 48 years old and about 3 months into their 49th year when the accounts reach the same value.

35. (Answers may vary.)   **37.** $1261.62

39. The graphs intersect at (0, 0) and (56.104041, 554,549.98). Each account started out with the same initial value: $0. Shortly after the 56th year of the investments, both accounts simultaneously attained the same value: $554,549.98.

41. Account A

Section 6.4 *(page 433)*

1. $819.91   **3.** $954.29   **5.** $680.83

7. $309.32   **9.** $13,148.39

11. Adding in the $930 down payment, the total cost for the timeshare is $14,607.02.

13. $1935.27   **15.** $743.97   **17.** $1611.33

19. About 26 years ($t \approx 25.9$)

21. With $1000 monthly payments, the loan balance will never reach $100,000. This is because the payments don't cover the interest on the loan.

23. 2 years and 3 months (27 payments)

25. 4 years and 9 months (57 payments)

27. 4 years (four annual payments)

29. 7 years (84 monthly payments)

31. $2188.51   **33.** $597.13   **35.** $1031.83

37. $416.86 per month   **39.** $4676.32 per month

CHAPTER 6 REVIEW EXERCISES

Section 6.1 *(page 436)*

1. $x = 2$ **3.** $x = 2$

5. In 2123 and any year thereafter

Section 6.2 *(page 436)*

7. $5028.38   **9.** Stonebridge Bank

11. $APY = 2.60$ percent

13. 38 years and 7 months ($t \approx 38.6$)

15. Continuous rate: 9.76 percent

Section 6.3 *(page 437)*

17. $55,883.41   **19.** $246,464.02

21. $1572.91 monthly

Section 6.4 *(page 438)*

23. 152 months (12 years and 8 months)

25. Yes.   **27.** $3219.72

CHAPTER 7

Section 7.1 *(page 447)*

1. $A' = \{1, 5, 8, 9\}$

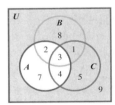

3. $B \cup C = \{1, 2, 3, 4, 5, 8\}$

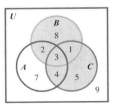

5. $(A \cap B)' = \{1, 4, 5, 7, 8, 9\}$

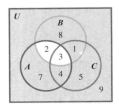

7. $A \cap A' = \{\}$ **9.** $(A \cup B) \cap C' = \{2, 7, 8\}$

        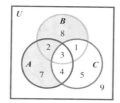

11. D' **13.** M' **15.** $S \cap M$ **17.** $A \subset B$

19. $\{x | 0 < x < 10\}$

21. $\{x | x = k^2 \text{ for some integer } k\}$

23. $\{x | x \text{ is a factor of } 24\}$

The following Venn diagram was used for Exercises 25–36.

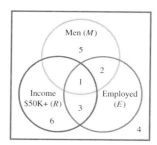

25. $M \cap R = \{1\}$; This is the set of men with a household income above $50,000.

27. $E' \cap R = \{6\}$; This is the set of unemployed people with a household income above $50,000.

29. $(E \cap M)' \cap R = \{3, 6\}$; This is the set of people with a household income above $50,000 who aren't employed men.

31. $(E' \cup R') \cap (M' \cup R) = \{4, 6\}$; This is the set of people who are unemployed or have a household income not greater than $50,000 who are also women or have a household income above $50,000. In simpler terms, it is the set of people who are unemployed women.

33. $(E \cap R') \cup M = \{1, 2, 5\}$; This is the set of people who are employed with a household income not greater than $50,000 together with all of the men.

35. $(M \cap R')' = \{1, 3, 4, 6\}$; This is the set of people who are not men with a household income of at most $50,000. In other terms, it is the set of men who have a household income greater than $50,000 together with the set of all women.

37. $F \cup E$ is the group of full-time instructors together with the group of instructors with 20 or more years of teaching experience. There are 94.9 thousand full-time instructors. There are 6.4 thousand part-time instructors with 20 or more years of teaching experience. There are a total of 101.3 thousand people in $F \cup E$.

39. $F' \cap E'$ is the set of part-time instructors with less than 20 years of teaching experience. There are 122.2 thousand part-time instructors with less than 10 years' experience and 24.5 thousand part-time instructors with 10–19 years' experience. There are a total of 146.7 thousand people in $F' \cap E'$.

41. $(F \cap E')'$ is the group of people who aren't full-time instructors with less than 20 years of teaching experience. In other terms, it is the set of all part-time instructors together with the full-time instructors with 20 or more years of teaching experience. There are 153.1 thousand part-time instructors, and there are 23.2 thousand full-time instructors with 20 or more years of teaching experience. There are 176.3 thousand people in $(F \cap E')'$. (*Note:* Because of the rounding in the table, you may also correctly end up with 176.2 thousand people.)

43. $(A \cap B)'$ $A' \cup B'$

    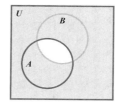

Since both statements yield the same Venn diagram, the statements are equivalent. That is,

$$(A \cap B)' = A' \cup B'$$

$(A \cup B)'$ $A' \cap B'$

    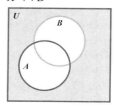

Since both statements yield the same Venn diagram, the statements are equivalent. That is,

$$(A \cup B)' = A' \cap B'$$

Since A and B represented arbitrary sets, DeMorgan's Law is true for any sets A and B.

45. $(A \cap B') \cap (A' \cap B) = \varnothing$

Section 7.2 *(page 456)*

1. $n(B) = 7$ **3.** $n(B') = 8$

5. $n(A' \cap B') = 2$ **7.** $n((A \cap B)') = 11$

9. $n(A' \cap B) = 3$

11. $n(A \times B) = 6$

$$A \times B = \begin{Bmatrix} \text{red apple, green apple,} \\ \text{yellow apple, red tomato,} \\ \text{green tomato, yellow tomato} \end{Bmatrix}$$

13. $n(A \times B) = 8$

$$A \times B = \begin{Bmatrix} \text{chocolate ice cream, chocolate pie,} \\ \text{chocolate shake, chocolate cookie,} \\ \text{vanilla ice cream, vanilla pie,} \\ \text{vanilla shake, vanilla cookie} \end{Bmatrix}$$

15. $n(A \times B) = 8$

$$A \times B = \begin{Bmatrix} (H, 1), (H, 2), (H, 3), (H, 6), \\ (T, 1), (T, 2), (T, 3), (T, 6) \end{Bmatrix}$$

17. 32 schedules   **19.** 45 frozen dairy desserts

21. 136 different dining/lodging options

23. 10 different meal options

25. 11 dish choices

27. The outcomes with a single element from each of the sets R, C, and M.

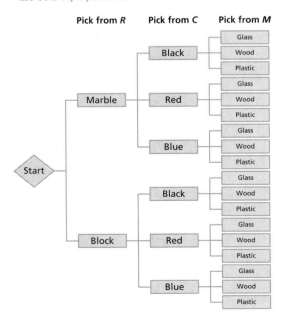

29. The outcomes with a single element from each of the sets R, C, M, and S.

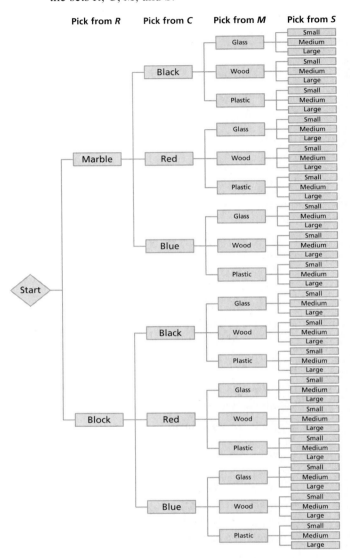

31. $n(D \cap F) = 155$ **33.** $n(B' \cap F) = 6770$

35. $n((N' \cap M) \cup F) = 68{,}969$

37. $A \neq B$ **39.** 12 sandwiches

Section 7.3 *(page 467)*

1. $P(5, 2) = 20$ **3.** $P(5, 5) = 120$

5. $P(9, 0) = 1$ **7.** $C(7, 5) = 21$

9. $C(4, 1) = 4$

11. $P(15, 7) = 32{,}432{,}400$

13. $P(19, 7) = 253{,}955{,}520$

15. $P(369, 3) = 49,835,664$

17. $C(20, 3) = 1140$

19. $C(39, 6) = 3,262,623$

21. 80,089,128

23. 11,232,000 license plates

25. 604,800 phone numbers

27. 720 zip codes   **29.** 420 schedules

31. 240 pizzas   **33.** 4050 pizzas

35. 35,048 call signs   **37.** 120 numbers

39. r^3 ways

Section 7.4 *(page 477)*

1. **Experiment:** Pick a person from Ellensburg and note the gender
Trial: 1
Outcomes: *M, F*
Sample space: $\{M, F\}$

3. **Experiment:** Roll a pair of dice and calculate the difference between the larger number and the smaller number
Trial: 1
Outcomes: 0, 1, 2, 3, 4, 5
Sample space: $\{0, 1, 2, 3, 4, 5\}$

5. **Experiment:** Pick three marbles from a bag and note their color
Trial: 1
Outcomes: RRR, RRW, RRB, RWW, RBB, RBW, WWW, WWB, WBB
Sample space:
$\left\{ \begin{array}{l} \text{RRR, RRW, RRB, RWW, RBB,} \\ \text{RBW, WWW, WWB, WBB} \end{array} \right\}$

7. **Experiment:** Pick a student in a two-credit class and note the number of credits on the student's class schedule
Trial: 1
Outcomes: 2, 3, 4, . . . , 20, 21
Sample space: $\{2, 3, 4, \ldots, 20, 21\}$

9. **Experiment:** Pick a box of cereal and note its brand
Trial: 10
Outcomes: Kellogg's, General Mills, Other Brand
Sample space: {Kellogg's, General Mills, Other Brand}

11. The sample space is {(Miguel, Andrea), (Miguel, Miki), (Miguel, Sara), (Miguel, Sam), (Sam, Andrea), (Sam, Miki), (Sam, Sara), (Andrea, Miki), (Andrea, Sara), (Miki, Sara)}

(a) *Event:* At least one of the finalists is male.
{(Miguel, Andrea), (Miguel, Miki), (Miguel, Sara), (Miguel, Sam), (Sam, Andrea), (Sam, Miki), (Sam, Sara)}

(b) *Event:* Both finalists are female.
{(Andrea, Miki), (Andrea, Sara), (Miki, Sara)}

(c) *Event:* No more than one finalist is male.
{(Miguel, Andrea), (Miguel, Miki), (Miguel, Sara), (Sam, Andrea), (Sam, Miki), (Sam, Sara), (Andrea, Miki), (Andrea, Sara), (Miki, Sara)}

13.

$$S = \left\{ \begin{array}{cccccc} (1,1) & (1,2) & (1,3) & (1,4) & (1,5) & (1,6) \\ (2,1) & (2,2) & (2,3) & (2,4) & (2,5) & (2,6) \\ (3,1) & (3,2) & (3,3) & (3,4) & (3,5) & (3,6) \\ (4,1) & (4,2) & (4,3) & (4,4) & (4,5) & (4,6) \\ (5,1) & (5,2) & (5,3) & (5,4) & (5,5) & (5,6) \\ (6,1) & (6,2) & (6,3) & (6,4) & (6,5) & (6,6) \end{array} \right\}$$

(a) $E = \left\{ \begin{array}{ccc} (1,1) & (1,3) & (1,5) \\ (2,2) & (2,4) & (2,6) \\ (3,1) & (3,3) & (3,5) \\ (4,2) & (4,4) & (4,6) \\ (5,1) & (5,3) & (5,5) \\ (6,2) & (6,4) & (6,6) \end{array} \right\}$

(b) $E = \left\{ \begin{array}{cccc} (1,4) & (2,3) & (3,2) & (4,1) \\ (4,6) & (5,5) & (6,4) & \end{array} \right\}$

(c) $E = \left\{ \begin{array}{cccc} (1,1) & (1,3) & (1,4) & (1,6) \\ (2,2) & (2,3) & (2,5) & (2,6) \\ (3,1) & (3,2) & (3,4) & (3,5) \\ (4,1) & (4,3) & (4,4) & (4,6) \\ (5,2) & (5,3) & (5,5) & (5,6) \\ (6,1) & (6,2) & (6,4) & (6,5) \end{array} \right\}$

15. $S = \{$TTTT, TTTF, TTFT, TFTT, FTTT, TTFF, TFTF, FTTF, TFFT, TFTT, FFTT, TFFF, FTFF, FFTF, FFFT, FFFF$\}$

(a) $E = \{$TTTT, TTTF, TTFT, TFTT, FTTT$\}$

(b) $E = \{$TTTT, TTTF, TTFT, TFTT, FTTT, TTFF, TFTF, FTTF, TFFT, FTFT, FFTT$\}$

(c) $E = \{$FFFF$\}$

17. $S = \{DD, DR, RD, DI, ID, IR, RI, II, RR\}$

   (a) $E = \{RR\}$

   (b) $E = \{DD, DI, ID\}$

   (c) $E = \{IR, RI, II, RR\}$

19. $S = \{$FFFF, FFFM, FFMF, FMFF, MFFF, FFMM, FMFM, MFFM, MFMF, MMFF, FMMF, FMMM, MFMM, MMFM, MMMF, MMMM$\}$

   (a) $E = \{$FFFF, FFFM, FFMF, FMFF, MFFF$\}$

   (b) $E = \{$FFFF, FFMF, FMFF, MFFF, MFMF, MMFF, FMMF, MMMF$\}$

   (c) $E = \{$MFFF, MFMF, MMFF, MMMF$\}$

21. $n(S) = 10$

   (a) $P(E) = 0.7$

      Odds for E: 7:3

      Odds against E: 3:7

   (b) $P(E) = 0.3$

      Odds for E: 3:7

      Odds against E: 7:3

   (c) $P(E) = 0.9$

      Odds for E: 9:1

      Odds against E: 1:9

23. $N(S) = 36$

   (a) $P(E) = 0.5$

      Odds for E: 1:1

      Odds against E: 1:1

   (b) $P(E) = \frac{7}{36} \approx 0.1944$

      Odds for E: 7:29

      Odds against E: 29:7

   (c) $P(E) = \frac{2}{3} \approx 0.6667$

      Odds for E: 2:1

      Odds against E: 1:2

25. $n(S) = 16$

   (a) $P(E) = 0.3125$

      Odds for E: 5:11

      Odds against E: 11:5

   (b) $P(E) = 0.6875$

      Odds for E: 11:5

      Odds against E: 5:11

   (c) $P(E) = 0.0625$

      Odds for E: 1:15

      Odds against E: 15:1

27. $n(S) = 9$

   (a) $P(E) = \frac{1}{9} \approx 0.1111$

      Odds for E: 1:8

      Odds against E: 8:1

   (b) $P(E) = \frac{3}{9} \approx 0.3333$

      Odds for E: 3:6

      Odds against E: 6:3

   (c) $P(E) = \frac{4}{9} \approx 0.4444$

      Odds for E: 4:5

      Odds against E: 5:4

29. $n(S) = 16$

   (a) $P(E) = \frac{5}{16} = 0.3125$

      Odds for E: 5:11

      Odds against E: 11:5

   (b) $P(E) = \frac{8}{16} = 0.5$

      Odds for E: 1:1

      Odds against E: 1:1

   (c) $P(E) = \frac{4}{16} = 0.25$

      Odds for E: 1:3

      Odds against E: 3:1

31. **(a)** $P(E) = \frac{6}{20} = 0.3$

   (b) $P(E) = \frac{2}{20} = 0.1$

   (c) $P(E) = \frac{2}{20} = 0.1$

*In our solution, we treated 35 and 53 as different groups even though the same two people belonged to each group. If you had treated 35 and 53 as the same group (combination versus permutation), the sample space would have had 10 elements.

   The probability for each of the events would remain the same.

33. $P(E) = \frac{108}{216} = 0.50$ **35.** $P(E) = \frac{12}{36} \approx 0.33$

Section 7.5 *(page 489)*

1. $E \cap F = \{$50-cent piece, silver dollar$\}$

   $E \cap F$ is the set of coins whose value is both a multiple of 5 and a multiple of 50.

   $E \cup F = \left\{\begin{array}{l}\text{nickel, dime, quarter,} \\ \text{50-cent piece, silver dollar}\end{array}\right\}$

   $E \cup F$ is the set of coins whose value is a multiple of 5 or a multiple of 50.

   $E' = \{$penny$\}$

   E' is the set of coins whose value is *not* a multiple of 5.

3. $E \cap F = \left\{ \begin{array}{l} x \,|\, x \text{ is a student on} \\ \text{scholarship with a} \\ \text{cumulative GPA of} \\ \text{at least 3.5} \end{array} \right\}$

$E \cap F$ is the set of students on scholarship who have a cumulative GPA of 3.5 or higher.

$E \cup F = \left\{ \begin{array}{l} x \,|\, x \text{ is a student who is} \\ \text{on scholarship or has a} \\ \text{cumulative GPA of at} \\ \text{least 3.5} \end{array} \right\}$

$E \cup F$ is the set of students on scholarship together with the set of students who have a cumulative GPA of 3.5 or higher.

$E' = \left\{ \begin{array}{l} x \,|\, x \text{ is a student who is} \\ \text{not on scholarship} \end{array} \right\}$

E' is the set of students who are not on scholarship.

5. $E \cap F = \left\{ \begin{array}{l} x \,|\, x \text{ is a European nation} \\ \text{that is a permanent member} \\ \text{of the UN Security Council} \end{array} \right\}$

$E \cap F$ is the set of European nations that are permanent members of the security council.

$E \cup F = \left\{ \begin{array}{l} x \,|\, x \text{ is a European nation or} \\ \text{is a permanent member} \\ \text{of the UN Security Council} \end{array} \right\}$

$E \cup F$ is the set of European nations on the Security Council together with the permanent members of the Security Council.

$E' = \left\{ \begin{array}{l} x \,|\, x \text{ is a non-European nation} \\ \text{on the UN Security Council} \end{array} \right\}$

E' is the set of all non-European nations on the UN Security Council.

7. $P(E \cup F) = \frac{19}{36} \approx 0.528$ and

$P(E \cap F) = \frac{2}{36} \approx 0.056$

9. $P(E \cup F) = \frac{22}{36} \approx 0.611$ and

$P(E \cap F) = \frac{5}{36} \approx 0.139$

11. $P(E) = 0.8$ **13.** $P(E) = 0.2$

15. $P(E) = 0.6$ **17.** $P(E) = \frac{1}{7} \approx 0.143$

19. $P(E) = \frac{34}{35} \approx 0.971$

21. An employer who was interested in giving all of her employees an equal opportunity to work could track the work assignments of each employee and periodically compare the work assignment to the various probabilities. For example, if the probability of having an all-women work team was 20 percent and several months of employee data showed that 50 percent of the work teams contained only women, the employer might be concerned that she is being biased against men. (Answers may vary.)

23. $E = \{x \,|\, x \text{ is a Republican}\}$
$F = \{x \,|\, x \text{ is a Democrat}\}$
$G = \{x \,|\, x \text{ is an Independent}\}$

$$P(E) = 0.49$$
$$P(F) = 0.50$$
$$P(G) = 0.01$$

25. $E = \{x \,|\, x \text{ is an America citizen}\}$
$F = \{x \,|\, x \text{ has a student visa}\}$
$G = \{x \,|\, x \text{ is an illegal alien}\}$

$$P(E) = \frac{20}{33} \approx 0.606$$
$$P(F) = \frac{11}{33} \approx 0.333$$
$$P(G) = \frac{2}{33} \approx 0.061$$

27. $E = \{x \,|\, x \text{ has a master's degree}\}$
$F = \{x \,|\, x \text{ has a bachelor's degree}\}$
$G = \{x \,|\, x \text{ has a high school diploma}\}$

$$P(E) = \frac{13}{40} = 0.325$$
$$P(F) = \frac{17}{40} = 0.425$$
$$P(G) = \frac{10}{40} = 0.25$$

*Note: The company is tracking the *highest* earned degree of each employee. Although someone with a master's degree also has a bachelor's degree and high school diploma, that person is counted only in the category of his or her highest degree.

29. The elements of the sample space that yield each product are indicated here, together with the probability of obtaining the product:

$1: (1, 1) \Rightarrow P(E) = \dfrac{1}{24}$

$2: (2, 1), (1, 2) \Rightarrow P(E) = \dfrac{1}{12}$

$3: (3, 1), (1, 3) \Rightarrow P(E) = \dfrac{1}{12}$

$4: (4, 1), (1, 4), (2, 2) \Rightarrow P(E) = \dfrac{1}{8}$

$5: (5, 1) \Rightarrow P(E) = \dfrac{1}{24}$

$6: (6, 1), (2, 3), (3, 2) \Rightarrow P(E) = \dfrac{1}{8}$

$8: (2, 4), (4, 2) \Rightarrow P(E) = \dfrac{1}{12}$

$9: (3, 3) \Rightarrow P(E) = \dfrac{1}{24}$

$10: (5, 2) \Rightarrow P(E) = \dfrac{1}{24}$

$12: (6, 2), (3, 4), (4, 3) \Rightarrow P(E) = \dfrac{1}{8}$

$15: (5, 3) \Rightarrow P(E) = \dfrac{1}{24}$

$16: (4, 4) \Rightarrow P(E) = \dfrac{1}{24}$

$18: (6, 3) \Rightarrow P(E) = \dfrac{1}{24}$

$20: (5, 4) \Rightarrow P(E) = \dfrac{1}{24}$

$24: (6, 4) \Rightarrow P(E) = \dfrac{1}{24}$

31. 50%

33. The elements of the sample space that yield each value are indicated here, together with the probability of obtaining the value:

$0: (1, 1), (1, 2), (1, 3), (1, 4), (1, 5),$
$(1, 6), (2, 1), (3, 1), (4, 1), (5, 1),$
$(6, 1) \Rightarrow P(E) = \dfrac{11}{36} \approx 0.306$

$2: (2, 2) \Rightarrow P(E) = \dfrac{1}{36} \approx 0.028$

$3: (2, 3), (3, 2) \Rightarrow P(E) = \dfrac{2}{36} \approx 0.056$

$4: (2, 4), (4, 2) \Rightarrow P(E) = \dfrac{2}{36} \approx 0.056$

$5: (2, 5), (5, 2) \Rightarrow P(E) = \dfrac{2}{36} \approx 0.056$

$6: (2, 6), (6, 2), (3, 3) \Rightarrow P(E) = \dfrac{3}{36} \approx 0.083$

$8: (3, 4), (4, 3) \Rightarrow P(E) = \dfrac{2}{36} \approx 0.056$

$10: (5, 3), (3, 5) \Rightarrow P(E) = \dfrac{2}{36} \approx 0.056$

$12: (6, 3), (3, 6), (4, 4) \Rightarrow P(E) = \dfrac{3}{36} \approx 0.083$

$15: (5, 4), (4, 5) \Rightarrow P(E) = \dfrac{2}{36} \approx 0.056$

$18: (6, 4), (4, 6) \Rightarrow P(E) = \dfrac{2}{36} \approx 0.056$

$20: (5, 5) \Rightarrow P(E) = \dfrac{1}{36} \approx 0.028$

$24: (6, 5), (5, 6) \Rightarrow P(E) = \dfrac{2}{36} \approx 0.056$

$30: (6, 6) \Rightarrow P(E) = \dfrac{1}{36} \approx 0.028$

35. Obtaining an odd value is more likely than obtaining an even value.

CHAPTER 7 REVIEW EXERCISES

Section 7.1 *(page 491)*

1. $(A \cap B)' = \{1, 3, 5, 6, 7, 8\}$

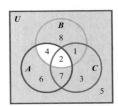

3. $(A \cap B) \cap C' = \{4\}$

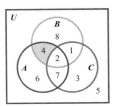

5. $M \cap E$ is the set of employed men.
$M \cap E = \{1, 2\}$

7. $E' \cap D$ is the set of unemployed people with graduate degrees. $E' \cap D = \{5\}$

Section 7.2 *(page 491)*

9. $n(B') = 10$ **11.** $n(A' \cap B') = 3$

13. 34 dishes

15. 175 three-course meals with strawberry shortcake

17. The outcomes with a single element from each of the sets *R*, *C*, and *S*.

Section 7.3 *(page 492)*

19. $P(9, 2) = 72$ **21.** $C(6, 2) = 15$

23. $P(25, 7) = 2,422,728,000$

25. 48 one-topping pizzas

27. 880 three-topping pizzas

Section 7.4 *(page 492)*

29. Experiment: Pick an entry and note the last number of the entry
Trial: 1
Outcomes: 0, 1, 2, 3, 4, 5, 6, 7, 8, 9
Sample space: {0, 1, 2, 3, 4, 5, 6, 7, 8, 9}

31. $S = \left\{ \begin{array}{l} (\text{Adi, Berto}), (\text{Adi, Marsha}), \\ (\text{Adi, Mariah}), (\text{Adi, Liz}), \\ (\text{Berto, Marsha}), (\text{Berto, Mariah}), \\ (\text{Berto, Liz}), (\text{Marsha, Mariah}), \\ (\text{Marsha, Liz}), (\text{Mariah, Liz}) \end{array} \right\}$

(a) *Event:* At least one of the finalists is male.

$E = \left\{ \begin{array}{l} (\text{Adi, Berto}), (\text{Adi, Marsha}), \\ (\text{Adi, Mariah}), (\text{Adi, Liz}), \\ (\text{Berto, Marsha}), (\text{Berto, Mariah}), \\ (\text{Berto, Liz}) \end{array} \right\}$

(b) *Event:* Both finalists are female.

$E = \left\{ \begin{array}{l} (\text{Marsha, Mariah}), \\ (\text{Marsha, Liz}), (\text{Mariah, liz}) \end{array} \right\}$

(c) *Event:* No more than one finalist is male.

$E = \left\{ \begin{array}{l} (\text{Adi, Marsha}), (\text{Adi, Mariah}), \\ (\text{Adi, Liz}), (\text{Berto, Marsha}), \\ (\text{Berto, Mariah}), (\text{Berto, Liz}), \\ (\text{Marsha, Mariah}), (\text{Marsha, Liz}), \\ (\text{Mariah, Liz}) \end{array} \right\}$

Section 7.5 *(page 492)*

33. $P(E) = \dfrac{7}{10} = 0.7$ **35.** $P(E) = \dfrac{5}{10} = 0.5$

CHAPTER 8

Section 8.1 *(page 504)*

1. 0.9   **3.** 0.1   **5.** 0.0546   **7.** 0.126

9. 0.25   **11.** $0.3\overline{3}$ **13.** $0.6\overline{6}$ **15.** 1

17. 0.414   **19.** 0.6   **21.** 0.0857

23. $0.41\overline{6}$ **25.** $0.53\overline{3}$ **27.** 0.358

29. 0.25; decreases to 0.22   **31.** Independent

33. Dependent   **35.** Independent   **37.** No

Section 8.2 *(page 521)*

1. Prevalence   14.7%
Sensitivity   86.2%
Specificity   97.9%
PVP   87.4%
PVN   97.6%
Of the entire population, 14.7% have the disease. Of those who have the disease, 86.2% test positive. Of those without the disease, 97.9% test negative. Of those who test positive, 87.4% have the disease. Of those who test negative, 97.6% don't have the disease.

3. Prevalence   19.2%
Sensitivity   86.3%
Specificity   93.8%
PVP   76.6%
PVN   96.6%
Of all e-mails, 19.2% are spam. Of those that are spam, 86.3% test positive. Of those that are not spam, 93.8% test negative. Of those that test positive, 76.6% are spam. Of those that test negative, 96.6% aren't spam.

5. Prevalence   72.7%
Sensitivity   83.3%
Specificity   66.7%
PVP   87.0%
PVN   60.0%
Of the entire population, 72.7% pass the class. Of those who pass the class, 83.3% study. Of those who fail the class, 66.7% don't study. Of those who study, 87.0% pass the class. Of those who don't study, 60.0% don't pass the class.

7. Prevalence   31.2%
Sensitivity   47.5%
Specificity   94.6%
PVP   80.0%
PVN   79.9%
Of all the homes, 31.2% don't need repairs. Of those that don't need repairs, 47.5% are newer homes. Of those that need repairs, 94.6% are older homes. Of those that are newer homes, 80.0% don't need repairs. Of those that are older homes, 79.9% need repairs.

9. Prevalence   20.4%
Sensitivity   40.0%
Specificity   74.7%
PVP   28.9%
PVN   82.9%
Of the entire population, 20.4% got a speeding ticket. Of those who got a speeding ticket, 40.0% drive a red car. Of those who didn't get a speeding ticket, 74.7% don't drive a red car. Of those who drive a red car, 28.9% got a speeding ticket. Of those who don't drive a red car, 82.9% didn't get a speeding ticket.

11. Prevalence   0.2%
Sensitivity   98%
Specificity   76%
PVP   **0.81%**
PVN   **99.99%**

13. Prevalence   20%
Sensitivity   50%
Specificity   75%
PVP   **33.3%**
PVN   **85.7%**

15. Prevalence   5%
Sensitivity   85%
Specificity   **70%**
PVP   13%
PVN   **98.9%**

17. Prevalence   2%
Sensitivity   92%
Specificity   **92.0%**
PVP   19%
PVN   **99.8%**

19. Prevalence   10%
Sensitivity   **91.1%**
Specificity   98%
PVP   **83.5%**
PVN   99%

21. 93.3% sensitivity: If a person has skin cancer, there is a 93.3% chance that the test will be positive. 97.8% specificity: If a person doesn't have skin cancer, there is a 97.8% chance that the test will be negative. 54% *PVP*: If a test is positive, there is a 54% chance that the person has skin cancer.
    99.98% *PVN*: If a test is negative, there is a 99.8% chance that the person has no skin cancer.

23. 0.0147% *PVP*: If the test is positive, there is a 0.0147% chance that the person has invasive aspergillosis. 99.9996% *PVN*: If the test is

negative, there is a 99.9996% chance that the person does not have invasive aspergillosis.

25. 44.8%

27. 100%; 0.037%

29. 50.1%; 16.7%

31. 6.9%

33. 3.3%

35. 0.00%; 99.92%

37. $1 -$ sensitivity

39. Accuracy of the tools; having a positive way of identifying people who have the disease for which you are testing; size of the sample (answers may vary)

Section 8.3 *(page 538)*

1. $p_{13} = 0$ means that the probability of moving from state 1 to state 3 is 0%.
$p_{21} = 0$ means that the probability of moving from state 2 to state 1 is 0%.
$p_{33} = 0.7$ means that the probability of moving from state 3 to state 3 is 70%.

3. $p_{13} = 0.1$ means that the probability of moving from state 1 to state 3 is 10%.
$p_{21} = 0.2$ means that the probability of moving from state 2 to state 1 is 20%.
$p_{33} = 0.0$ means that the probability of moving from state 3 to state 3 is 0%.

5. $p_{13} = 0$ means that the probability of moving from state 1 to state 3 is 0%.
$p_{21} = 0$ means that the probability of moving from state 2 to state 1 is 0%.
$p_{33} = 0.5$ means that the probability of moving from state 3 to state 3 is 50%.

7. S_1 and S_2 are absorbing states.

$$P = \begin{bmatrix} 1 & 0 \\ 0 & 1 \end{bmatrix}$$

9. S_1 is an absorbing state.

$$P = \begin{bmatrix} 1.00 & 0.00 \\ 0.05 & 0.95 \end{bmatrix}$$

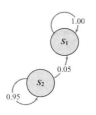

11. $P = \begin{bmatrix} 0.7 & 0.0 & 0.3 \\ 0.1 & 0.4 & 0.5 \\ 0.2 & 0.8 & 0.0 \end{bmatrix}$

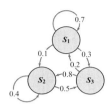

13. $P = \begin{bmatrix} 0.1 & 0.2 & 0.7 \\ 0.2 & 0.3 & 0.5 \\ 0.3 & 0.4 & 0.3 \end{bmatrix}$

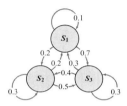

15. S_3 and S_4 are absorbing states.

$$P = \begin{bmatrix} 0 & 0.5 & 0.5 & 0 \\ 0 & 0 & 0.5 & 0.5 \\ 0 & 0 & 1 & 0 \\ 0 & 0 & 0 & 1 \end{bmatrix}$$

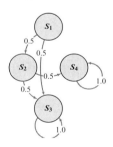

17. $P = \begin{bmatrix} 0 & 1 \\ 1 & 0 \end{bmatrix}$

19. $P = \begin{bmatrix} 0.95 & 0.05 & 0.00 \\ 0.00 & 0.99 & 0.01 \\ 0.00 & 0.00 & 1.00 \end{bmatrix}$

21. $D = \begin{bmatrix} \frac{3}{11} & \frac{8}{11} \end{bmatrix}$ **23.** $D = \begin{bmatrix} \frac{4}{13} & \frac{9}{13} \end{bmatrix}$

25. $D = \begin{bmatrix} \frac{25}{61} & \frac{36}{61} \end{bmatrix}$ **27.** $D = \begin{bmatrix} \frac{26}{115} & \frac{48}{115} & \frac{41}{115} \end{bmatrix}$

29. $D = \begin{bmatrix} \frac{1}{3} & \frac{1}{3} & \frac{1}{3} \end{bmatrix}$

31. $P^2 = \begin{bmatrix} 0.0144 & 0.7744 & 0.2112 \\ 0.2112 & 0.0144 & 0.7744 \\ 0.7744 & 0.2112 & 0.0144 \end{bmatrix}$

$P^5 = \begin{bmatrix} 0.0982 & 0.5411 & 0.3607 \\ 0.3607 & 0.0982 & 0.5411 \\ 0.5411 & 0.3607 & 0.0982 \end{bmatrix}$

$P^7 = \begin{bmatrix} 0.3950 & 0.1600 & 0.4450 \\ 0.4450 & 0.3950 & 0.1600 \\ 0.1600 & 0.4450 & 0.3950 \end{bmatrix}$

For the matrix P^k, the entry p_{ij} is the probability of moving from state i to state j in exactly k steps.

33. $P^2 = \begin{bmatrix} 0.3750 & 0.3125 & 0.3125 \\ 0.3125 & 0.3750 & 0.3125 \\ 0.3125 & 0.3125 & 0.3750 \end{bmatrix}$

$P^5 = \begin{bmatrix} 0.3340 & 0.3330 & 0.3330 \\ 0.3330 & 0.3340 & 0.3330 \\ 0.3330 & 0.3330 & 0.3340 \end{bmatrix}$

$P^7 = \begin{bmatrix} 0.3334 & 0.3333 & 0.3333 \\ 0.3333 & 0.3334 & 0.3333 \\ 0.3333 & 0.3333 & 0.3334 \end{bmatrix}$

For the matrix P^k, the entry p_{ij} is the probability of moving from state i to state j in exactly k steps.

35. $P^2 = \begin{bmatrix} 0.1759 & 0.8241 & 0 \\ 0.1230 & 0.877 & 0 \\ 0 & 0 & 1 \end{bmatrix}$

$P^5 = \begin{bmatrix} 0.1304 & 0.8696 & 0 \\ 0.1298 & 0.8702 & 0 \\ 0 & 0 & 1 \end{bmatrix}$

$P^7 = \begin{bmatrix} 0.1299 & 0.8701 & 0 \\ 0.1299 & 0.8701 & 0 \\ 0 & 0 & 1 \end{bmatrix}$

For the matrix P^k, the entry p_{ij} is the probability of moving from state i to state j in exactly k steps.

37. It is not possible for a player to lose 25 rounds in a row. The possible outcomes for any sequence of three coin tosses are

$$\begin{Bmatrix} \text{HHH, HHT, HTH, HTT,} \\ \text{THH, THT, TTH, TTT} \end{Bmatrix}$$

The outcomes correspond with the indicated win-loss sequences: win-win, win-lose, lose-win, lose-win, win-win, win-lose, win-win, win-win. Every outcome of three consecutive coin tosses will result in at least one win.

The probability that a player who lost the first round won the 25th round is 75%.

39. 82%

41. (i) Let S_1 = Mercury Interactive customer and S_2 = competitor's customer.

(ii)

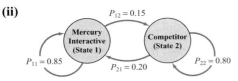

(iii) $P = \begin{bmatrix} 0.85 & 0.15 \\ 0.20 & 0.80 \end{bmatrix}$

(iv) $D = \begin{bmatrix} 0.552 & 0.448 \end{bmatrix}$

(v) $DP = \begin{bmatrix} 0.5588 & 0.4412 \end{bmatrix}$

We estimate that Mercury Interactive had a 55.9% market share of the automated software quality tools market in 2003.

43. (i) Let S_1 = price higher than previous day and S_2 = price lower than previous day.

(ii)

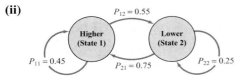

(iii) $P = \begin{bmatrix} 0.45 & 0.55 \\ 0.75 & 0.25 \end{bmatrix}$

(iv) $D = \begin{bmatrix} \dfrac{11}{20} & \dfrac{9}{20} \end{bmatrix}$

(v) $DP = \begin{bmatrix} 0.585 & 0.415 \end{bmatrix}$

According to the Markov chain model, we estimate that at the end of the next 20 trading days, 58.5% of the share prices will have been higher than the previous day's price and 41.5% of the share prices will have been lower than the previous day's price.

45. (i) Let S_1 = weight training, S_2 = cardiovascular, and S_3 = not work out.

(ii)

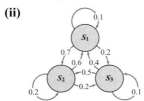

(iii) $P = \begin{bmatrix} 0.1 & 0.7 & 0.2 \\ 0.6 & 0.2 & 0.2 \\ 0.4 & 0.5 & 0.1 \end{bmatrix}$

(iv) $D = \begin{bmatrix} 0.30 & 0.25 & 0.45 \end{bmatrix}$

(v) $DP = \begin{bmatrix} 0.36 & 0.485 & 0.155 \end{bmatrix}$

According to the Markov chain model, we estimate that tomorrow 36% of the members will do weight lifting, 48.5% will do cardiovascular exercises, and 15.5% won't work out.

47. (a) $D_1 = \begin{bmatrix} 0.054 & 0.946 \end{bmatrix}$

5.4% of civilians 16 years old or older were unemployed and 94.6% were employed in 1996.

$D_2 = \begin{bmatrix} 0.040 & 0.960 \end{bmatrix}$

4.0% of civilians 16 years old or older were unemployed and 96.0% were employed in 2000.

(b) The model yielded the same results as the raw data. The model fits the data well.

(c) $D_3 = \begin{bmatrix} 0.031 & 0.969 \end{bmatrix}$

According to the model, 3.1% of civilians 16 years old and older were unemployed in 2004.

(d) The steady state distribution vector is

$$D = \begin{bmatrix} 0.012 & 0.988 \end{bmatrix}$$

The model forecasts that the long-term unemployment rate will be 1.2%.

49. (a) $D_1 = \begin{bmatrix} 0.103 & 0.897 \end{bmatrix}$

10.3% of private-sector workers were union members in 1995 and 89.7% were not.

$D_2 = \begin{bmatrix} 0.090 & 0.910 \end{bmatrix}$

9.0% of private-sector workers were union members in 2000 and 91.0% were not.

(b) The model yielded the same results as the raw data. The model fits the data well.

(c) $D_3 = \begin{bmatrix} 0.079 & 0.921 \end{bmatrix}$

According to the model, 7.9% of private-sector workers were union members in 2005 and 92.1% were not.

(d) The steady state distribution vector is

$$D = \begin{bmatrix} 0.032 & 0.968 \end{bmatrix}$$

The model forecasts that the long-term percentage of private-sector workers who are union members will be 3.2%.

(e) According to the model, union enrollment is forecasted to decline dramatically. In response to these results, union leaders could increase recruitment efforts and implement additional worker-friendly policies to increase membership. (Answers may vary.)

51. (a) $D_3 = [0.539 \quad 0.461]$

The percentage of abused children who were neglected in 1999 is estimated to be 53.9%.

(b) 53.8%

(c) Reducing child abuse may seem like a daunting task. Nevertheless, there is much that can be done by each of us. On a broad scale, any initiatives that strengthen positive family relationships will help to reduce child abuse.

Prevent Child Abuse America has indicated that supporting children and parents in your family and community reduces the likelihood of child abuse and neglect. Simple things such as offering to babysit; donating clothing, furniture, and toys to other families; being kind to new parents; and being a good neighbor go a long way in helping to prevent child abuse. (**Source:** www.preventchildabuse.org.) (Answers may vary.)

53. (a) Yes. The model fits the data perfectly.

$D_1 = [0.26 \quad 0.74]$ and $D_2 = [0.32 \quad 0.68]$

(b) The transition matrix indicates that a supermarket with a pharmacy has a 12.5% probability of *not* having a pharmacy four years later. It seems reasonable that a supermarket would replace an unprofitable service, such as a deli, bakery, or bank. The remaining probabilities in the matrix all seem reasonable.

(c) 50%.

(d) This model seems reasonable. There will always be supermarkets for which an in-store pharmacy will not make good business sense. We don't have any information that leads us to believe that more than 50% of supermarkets will have pharmacies in the future. (Answers may vary.)

55. (a) Yes. The model fits the data perfectly.

$$D_1 = [0.460 \quad 0.540] \quad \text{and}$$
$$D_2 = [0.450 \quad 0.550]$$

(b) The transition matrix indicates that a supermarket with a service fish department has a 65.8% probability of *not* having a service fish department four years later. This seems a bit higher than we would expect. Likewise, the transition matrix indicates that there is a 54.2% probability that a store without a service fish department will have such a department four years later. In light of the high probability of service fish

departments going out of business within four years, it seems unlikely that there would be a high percentage of stores without service fish departments that were clamoring to add a service fish department.

(c) 45.2%

57. $d_2 = 0$ and $d_3 = 1 - d_1$

59. Observe that

$$P^2 = \begin{bmatrix} 0 & 1 & 0 \\ 0 & 0 & 1 \\ 1 & 0 & 0 \end{bmatrix}$$

$$P^3 = \begin{bmatrix} 1 & 0 & 0 \\ 0 & 1 & 0 \\ 0 & 0 & 1 \end{bmatrix}$$

Since P^3 is the identity matrix

$$\begin{aligned} P^4 &= P^3 \cdot P \\ &= IP \\ &= P \end{aligned}$$

Consider $k = 3n + 1$.

$$\begin{aligned} P^k &= P^{3n+1} \\ &= P^{3n} \cdot P \\ &= (P^3)^n P \\ &= (I)^n P \\ &= IP \\ &= P \end{aligned}$$

In general, if $k = 3n + 1$ (for nonnegative integer values of n) then $P^k = P$.

61.

$$d_1 = d_0 P$$
$$= [x \quad 1 - x] \begin{bmatrix} a & 1 - a \\ b & 1 - b \end{bmatrix}$$
$$= [ax + (1 - x)b \quad (1 - a)x + (1 - b)(1 - x)]$$

Since $d_1 = [y \quad 1 - y]$, we know

$$y = ax + b(1 - x)$$

$$d_2 = d_1 P$$
$$= [y \quad 1 - y] \begin{bmatrix} a & 1 - a \\ b & 1 - b \end{bmatrix}$$
$$= [ay + (1 - y)b \quad (1 - a)y + (1 - b)(1 - y)]$$

Since $d_2 = [z \quad 1 - z]$, we know

$$z = ay + b(1 - y)$$

We have a system of equations in a and b.

$$ax + b(1 - x) = y$$
$$ay + b(1 - y) = z$$

Writing the system as an augmented matrix yields

$$\begin{bmatrix} x & 1-x & y \\ y & 1-y & z \end{bmatrix}$$

$$\begin{bmatrix} x & 1-x & y \\ 0 & x-y & xz-y^2 \end{bmatrix} \quad xR_2 - yR_1$$

$$\begin{bmatrix} x^2 - xy & 0 & x(xz+y-y^2-z) \\ 0 & x-y & xz-y^2 \end{bmatrix}$$
$$(x-y)R_1 - (1-x)R_2$$

$$\begin{bmatrix} 1 & 0 & \dfrac{xz+y-y^2-z}{x-y} \\ 0 & 1 & \dfrac{xz-y^2}{x-y} \end{bmatrix} \quad \begin{array}{l} \frac{1}{x^2-xy}R_1 \\ \\ \frac{1}{x-y}R_2 \end{array}$$

Thus $a = \dfrac{xz+y-y^2-z}{x-y}$ and $b = \dfrac{xz-y^2}{x-y}$.

Section 8.4   *(page 554)*

1.

X	P(X)
0	$\frac{1}{4}$
1	$\frac{1}{2}$
2	$\frac{1}{4}$

3.

X	P(X)
0	0.659
1	0.299
2	0.040
3	0.001
4	0.00002

5.

X	P(X)
0	0.9243
1	0.07342
2	0.00224
3	3.3×10^{-5}
4	2.5×10^{-7}
5	9×10^{-10}
6	1×10^{-12}

7.

X	P(X)
0	2.3×10^{-5}
1	4.6×10^{-4}
2	0.004
3	0.020
4	0.068
5	0.154
6	0.240
7	0.254
8	0.175
9	0.071
10	0.013

9.

X	P(X)
0	$\frac{35}{126} = 0.278$
1	$\frac{70}{126} = 0.556$
2	$\frac{21}{126} = 0.167$

11. $X =$ net amount won

X	P(X)
-2	$\frac{242}{250} = 0.968$
48	$\frac{1}{250} = 0.004$
23	$\frac{2}{250} = 0.008$
8	$\frac{5}{250} = 0.020$

$$E(X) = (-2)(0.968) + (48)(0.004)$$
$$+ (23)(0.008) + (8)(0.020)$$
$$= -1.40$$

The expected value is $-\$1.40$.

13. *With audit protection*
$X = $ cost

X	P(X)
25	0.9956
0	0.0044

$E(X) = 24.89$

Without audit protection
$X = $ cost

X	P(X)
0	0.9956
2,707	0.0044

$E(X) = 11.91$
Since the expected cost without audit protection is less than the expected cost with audit protection, we would not recommend the audit-guarantee service. However, since the audit protection is relatively cheap, you may choose to purchase it anyway to eliminate the possibility of being hit with a large tax penalty.

15. $X = $ points awarded if you guess

X	P(X)
1	$\frac{1}{4} = 0.25$
−0.5	$\frac{3}{4} = 0.75$

$E(X) = -0.125$
No. If he guesses, we expect he will lose 0.125 point. If he doesn't guess, no points are lost.

17. $X = $ points awarded if she guesses

X	P(X)
1	$\frac{1}{3} \approx 0.333$
−0.5	$\frac{2}{3} \approx 0.667$

$E(X) = 0$
The expected value is 0 points.

19. *With insurance*
$X = $ cost to author

X	P(X)
−89.95	0.001
−89.95	0.999

$E(X) = -89.95$

Without insurance
$X = $ cost to author

X	P(X)
−5,300	0.001
0	0.999

$E(x) = -5.30$
No. The expected value of not having insurance is less than the expected value of having insurance. However, if the author wants to eliminate the risk of a $5300 loss, he may want to buy the insurance anyway.

21. $X = $ net income from insurance

X	P(X)
−508	0.9228
3,660	0.0772

$E(X) = -186.23$
The expected value is −$186.23. Since the homeowner considers any payout as a gain, the long-term average annual cost for "peace of mind" is expected to be $186.23.

23. *$3 million jackpot*
$E(X) \approx -0.69$

$5 million jackpot
$E(X) \approx -0.60$

$10 million jackpot
$E(X) \approx -0.38$

25. $0.03 per day   **27.** $3.27 per year

29. −$0.02 per day

31. $E(X) = -5502 - 6.6t$

33. $3.91\overline{6}$ **35.** $x \leq 5000$

Section 8.5 (page 572)

1.   Mean: 38.38
Median: 35.5
Mode: 30, 32, 35, 37

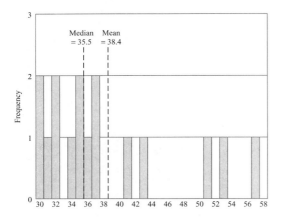

Of those players scoring more than 30 touchdowns, the mean number of touchdowns scored was 38.38. Half of the players had scored more than 35.5 touchdowns and half of the players had scored fewer than 35.5 touchdowns. The number of touchdowns scored most frequently were 30, 32, 35, and 37.

3.   Mean: 4.2
Median: 4
Mode: 4

Between 1900 and 1909, the mean rank of the Yale football team was 4.2. Placing its rankings in numerical order, the middlemost ranking was 4. The most frequent ranking was also 4.

5.   Mean: 12.78
Median: 12.8
Mode: 12.7, 12.8, 12.9, 13.1

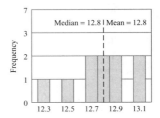

Between 1990 and 1999, the mean percentage of births to teenage mothers was 12.78%. Placing the percentages in numerical order, the middlemost percentage was 12.8%. The percentages occurring most frequently were 12.7%, 12.8%, 12.9%, and 13.1%.

7.   Mean: 10.19
Median: 11
Mode: 7, 11

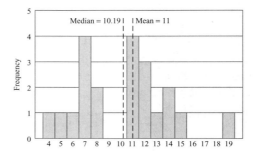

Between 1980 and 2000, the mean number of tropical storms and hurricanes each year was 10.19. Placing the number of tropical storms and hurricanes in numerical order, the middlemost number was 11. The number of tropical storms and hurricanes occurring most often in a year was 7 and 11.

9.   Mean: 27
Median: 28
Mode: 28, 29

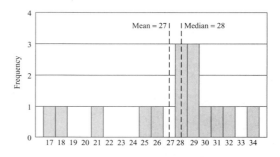

Between 1990 and 1999, the mean percentage of new homes with 2400 or more square feet was 27%. Placing the percentages in numerical order, the middlemost percentage was 28%. The percentages occurring most often were 28% and 29%.

11. Mean: 20.6
Median: 20.5
Mode: 20, 21

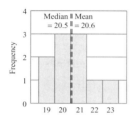

Between 1990 and 1999, the mean U.S. percentage of worldwide fuel production was 20.6%. Placing the percentages in numerical order, the middlemost percentage was 20.5%. The percentages occurring most often were 20% and 21%.

13. Mean: 24.19
Median: 24
Mode: 24

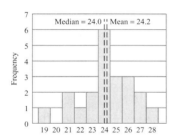

Between 1980 and 2000, the mean per capita citrus fruit consumption was 24.19%. Placing the percentages in numerical order, the middlemost percentage was 24%. The percentage occurring most often was 24%.

15. Mean: 1
Median: 1
Mode: 0, 2

Between 1993 and 2003, the mean number of top ten grossing movies of New Line studio was 1. Placing the number of top ten grossing movies in numerical order, the middlemost number was 1. The number of top ten grossing movies that occurred most often was 0 and 2.

17. $\bar{x} = 1428.25$ square feet
$s = 213.6$ square feet
100% within 2 standard deviations

19. $\bar{x} = 78.76$ million dollars
$s = 42.95$ million dollars
100% within 2 standard deviations

21. $\bar{x} = 18.84$ births per 1000 people
$s = 4.03$ births per 1000 people
100% within 2 standard deviations

23. $\bar{x} = 3.69$ divorces per 1000 people
$s = 1.21$ divorces per 1000 people
100% within 2 standard deviations

25. $\bar{x} = 17.5$ points
$s = 11.4$ points
100% within 2 standard deviations

27. $\bar{x} = 976.1$ tornadoes
$s = 206.0$ tornadoes
100% within 2 standard deviations

29. $\bar{x} \approx 189,348$ thousand dollars
$s \approx 39,732$ thousand dollars
100% within 2 standard deviations

31. Median. The mean will be skewed because of the huge salary the football player makes.

33. Answers will vary.

35. No. The sets $\{3, 4, 5\}$ and $\{3, 3, 4, 4, 5, 5\}$ both satisfy the requirements but are different sets.

Section 8.6 *(page 592)*

1. -0.91 **3.** -2.22 **5.** -0.25

7. 2.65   **9.** 1.18

11. 0.4834, 0.9668   **13.** 0.4916, 0.9832

15. 0.0832, 0.1664 (from z-score table)

17. 0.2486, 0.4972 (from z-score table)

19. 0.1628, 0.3256 (from z-score table)

21. $\mu = 5.2, \sigma = 2.6$ **23.** $\mu = 9.8, \sigma = 3.5$

25. $\mu = 7.9, \sigma = 2.7$ **27.** $\mu = -1.2, \sigma = 1.7$

29. $\mu = -1.7, \sigma = 7.4$

31. 31.12%   **33.** 85.15%   **35.** 0.13%

37. $x \geq 680$ **39.** ≈ 0.0582 **41.** ≈ 0.0000

43. 22.7 pounds to 32.3 pounds

45. The doctor would want to know that the baby is growing at a rate that is not so unusual as to be a problem.

47. $z_1 = z_2 + \dfrac{\mu_2 - \mu_1}{\sigma}$

49. The graphs of the normal distributions

$f(x) = \dfrac{1}{\sqrt{2\pi}} e^{-1/2(x-4)^2}$ and $g(x) = \dfrac{1}{\sqrt{2\pi}} e^{-1/2(x-8)^2}$

are shown in the figure below.

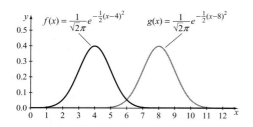

The graph of $h(x) = f(x) + g(x)$ (shown below) is clearly not a normal curve.

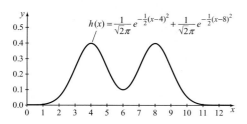

Therefore, the sum of two normal distributions is not guaranteed to be a normal distribution.

CHAPTER 8 REVIEW EXERCISES

Section 8.1 *(page 594)*

1. $P(A|B) = 0.714$

3. Let S be the set of seniors and W be the set of winners. $P(W|S) = 0.25$

5. 40.0%

7. The events are dependent.

Section 8.2 *(page 594)*

9.

	%
Prevalence	0.1
Sensitivity	25
Specificity	99
Predictive-value positive	**2.44**
Predictive-value negative	**99.9**

11. Sensitivity = 100%
There is a 100% probability that a person who tested positive with the other HCV method also tested positive with ACON HCV.

Specificity = 98.6%
There is a 98.6% probability that a person who tested negative with the other HCV method also tested negative with ACON HCV.

Predictive-value positive = 98.6%
There is a 98.6% probability that a person who tested positive with ACON HCV also tested positive with the other HCV method.

Predictive-value negative = 100%
There is a 100% probability that a person who tested negative with ACON HCV also tested negative with the other HCV method.

Section 8.3 *(page 595)*

13. $P = \begin{bmatrix} 0 & 0.1 & 0.9 \\ 0.3 & 0.7 & 0 \\ 0.8 & 0 & 0.2 \end{bmatrix}$

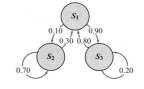

15. $P = \begin{bmatrix} 0.25 & 0.40 & 0.35 \\ 0 & 0.10 & 0.90 \\ 0.20 & 0 & 0.80 \end{bmatrix}$

17. (i) Let S_1 = Brand A; S_2 = Brand B, and S_3 = other brand.

(ii)

(iii) $P = \begin{bmatrix} 0.68 & 0.20 & 0.12 \\ 0.18 & 0.60 & 0.22 \\ 0.20 & 0.30 & 0.50 \end{bmatrix}$

(iv) $D = \begin{bmatrix} 0.38 & 0.32 & 0.30 \end{bmatrix}$

(v) $DP = \begin{bmatrix} 0.376 & 0.358 & 0.266 \end{bmatrix}$

At the end of the year, Brand A is expected to have 37.6% of the market share; Brand B, 35.8% of the market share; and other brands, 26.6% of the market share.

Section 8.4 *(page 596)*

19.

X	$P(X)$
1	$\dfrac{6}{216} \approx 0.028$
2	$\dfrac{90}{216} \approx 0.417$
3	$\dfrac{120}{216} \approx 0.556$

21. X = points awarded if she guesses

X	$P(X)$
-5	0.50
5	0.50

$$E(X) = 0$$

The expected value if she guesses is 0 points. If she doesn't guess, she will earn 0 points. According to the expected value, if she were to guess on every question, we would expect her overall score to be near 0. From a probability standpoint, either option is a good choice, since guessing and not guessing yield the same expected value. We would encourage her to think through the problem thoroughly, look back at other problems for clues, and then take the risk and guess.

Section 8.5 *(page 596)*

23. Mean = 72.67

The average number of points per game was 72.67 points.

Median = 68.50

When the six scores are lined up in numerical order, the average (mean) of the middle two values is 68.50 points.

Mode = none

No score was repeated.

Standard deviation = 10.5

Based on Chebyshev's rule, we anticipate that at least 75% of the scores lie within 21 points (2 standard deviations) of the mean.

25. Mean = 9,243,119

The average gross sales per weekend was $9,243,119.

Median = 8,943,880.5

When the 10 sales values are lined up in numerical order, the average (mean) of the middle two values is $8,943,880.50.

Mode = none

No weekend sales amount was repeated.

Standard deviation = 2,451,367.52

Based on Chebyshev's rule, we anticipate that at least 75% of the weekend sales lie within $4,902,735.04 (2 standard deviations) of the mean.

Section 8.6 *(page 597)*

27. $z = -0.1667$ **29.** 19.74%

31.   **33.** 13.59%

CHAPTER 9 EXERCISES

Section 9.1 *(page 606)*

1. 2   **3.** -1 **5.** 0

7. $\dfrac{\ln(5)}{20} \approx 0.0805$ **9.** $\dfrac{\sqrt{2}}{6} \approx 0.2357$

11. $\dfrac{7}{4}$ degrees per hour

13. Between 1996 and 1998, 1366.55 dollars per year; between 1997 and 1999, 1794.40 dollars per year

15. 30 words per minute per quarter

17. -0.378 newspaper subscription per cable TV subscriber

19. 1.375 people below poverty level per unemployed person

21. $f(x) = 2^x$

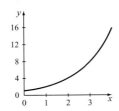

$$\dfrac{f(3) - f(1)}{3 - 1} = 3$$

23. $f(x) = 5$

$$\dfrac{f(3) - f(1)}{3 - 1} = 0$$

25. $f(x) = (x - 2)^2$

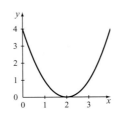

$$\dfrac{f(3) - f(1)}{3 - 1} = 0$$

27. $\dfrac{f(3) - f(1)}{3 - 1} = 2$ **29.** $\dfrac{f(3) - f(1)}{3 - 1} = -1$

31. ≈ 147 people per month

33. ≈ 2013 people per month

35. 2300 people per month

37. ≈ -427 people per month

39. ≈ 407 people per month

41. The average rate of change for $2f(x)$ is twice as big as the rate for change for $f(x)$.

43. The vertical distance between $(2, f(2))$ and $(2.01, f(2.01))$ is 0.04.

45.

h	Secant slope $= h^{-2/3}$
1.0	1.0
0.1	≈ 4.642
0.01	≈ 21.544
0.001	≈ 100

As h gets very small, the secant slope will get very large. As h approaches 0, the slope of the secant line appears to approach ∞.

Section 9.2 *(page 623)*

1.

h	Difference Quotient
0.1	4.1
0.01	4.01
0.001	4.001
Instantaneous rate ≈ 4.0	

3.

h	Difference Quotient
0.1	-65.6
0.01	-64.16
0.001	-64.016
Instantaneous rate ≈ -64.0	

5.

h	Difference Quotient
0.1	4
0.01	4
0.001	4
Instantaneous rate ≈ 4.0	

7.

h	Difference Quotient
0.1	0
0.01	0
0.001	0
Instantaneous rate ≈ 0	

9.

h	Difference Quotient
0.1	1,120
0.01	1,075
0.001	1,070.5
Instantaneous rate $\approx 1,070$	

11. 4 **13.** -64 **15.** 4 **17.** 0 **19.** 1070

21.

h	Difference Quotient
0.1	-0.06940
0.01	-0.07358
0.001	-0.07402
Instantaneous rate ≈ -0.74	

23.

h	Difference Quotient
0.1	-1.7305
0.01	-1.7388
0.001	-1.7396
Instantaneous rate ≈ -1.740	

25.

h	Difference Quotient
0.1	14,198.385
0.01	9,588.682
0.001	9,231.052
Instantaneous rate $\approx 9,200$	

27.

h	Difference Quotient
0.1	0.4879
0.01	0.4988
0.001	0.4999
Instantaneous rate ≈ 0.50	

29.

h	Difference Quotient
0.1	70.271
0.01	61.170
0.001	60.347
Instantaneous rate ≈ 60.0	

31. $1340 per year. In 2000, teacher salaries are increasing at the rate of $1340 per year.

33. $14.3 billion per year. In 2005, the amount of money spent on prescription drugs will be increasing at the rate of $14.3 billion per year.

35. $0.871 billion per year. Video game sales in the first quarter of 2007 will be increasing at a rate of $0.871 billion per year.

37. $31.91 per year. In 2001, the cost of full-time resident tuition at Green River Community College was increasing at a rate of $31.91 per year.

39. 242.5 million discs per year. In 2002, the demand for CD-R discs was increasing at a rate of 242.5 million discs per year.

41. ≈ -51 feet per second

43. ≈ -12 feet per second

45. ≈ -15 feet per second

47. No. Consider $f(x) = x^3$. For this function, $f'(-1) = f'(1) = 3$. However, $-1 \neq 1$. Therefore, different values of x may have the same derivative.

49. Since $f'(x) = 0$, $f(x) = c$ for some constant c. If $c = 0$, there are infinitely many x-intercepts. Otherwise, there are no x-intercepts.

Section 9.3 *(page 638)*

1.

3.

5.

$g(x) = x^2 - 4x - 5$

$(4, -5)$

Tangent line: $y = 4x - 21$

7.

$h(x) = x^3$

$(0, 0)$

Tangent line: $y = 0$

9.

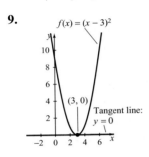

$f(x) = (x - 3)^2$

$(3, 0)$

Tangent line: $y = 0$

11. 2.356 billion gallons per year; 55.75 billion gallons

13. 20.27 square feet per year; 2238.5 square feet

15. Increasing at the rate of 0.128 student per teacher per year; 15.157 students per teacher

17. 78.42 banks per year; 10,016 banks

19. 11.997 million debit cards per year; 239.79 million debit cards

21. $1750 per year. In 1997, the annual wage was increasing at a rate of $1750 per year.

23. -13.1 newspapers per year. In 1995, the number of different daily newspapers in the United States was decreasing at a rate of about 13 newspapers per year.

25. $26.50 per year; $34 per year. The rate of increase is itself increasing.

27. 1.1 million people per year. In 1997, the attendance at American League baseball games was increasing at a rate of 1.1 million people per year.

29. Increasing at a rate of 5.1 million people per year

31. Possible pairs: (1, 1) and (3, 9) or (0, 0) and (4, 16) (answers may vary). Each pair is horizontally equidistant from the point (2, 4).

33. $f'(-1) = 3$ and $f'(1) = 3$. The graph is increasing at the same rate at both points.

35. $p'(0) = f'(0) + g'(0)$

Section 9.4 *(page 647)*

1. $f'(x) = 2x - 4$ **3.** $g'(x) = 2x + 2$

5. $g'(x) = 2x - 4$ **7.** $h'(x) = 3x^2$

9. $f'(t) = 2t - 6$

11.

x	$g'(x)$
1	5
3	13
5	21

13.

x	$f'(x)$
1	0
3	4
5	8

15.

x	$h'(x)$
1	0
3	0
5	0

17.

x	$W'(x)$
1	-4
3	-4
5	-4

19.

x	$S'(x)$
1	4
3	16
5	28

21. $P'(x) \approx 1.099(3^x)$

23. $C'(x) \approx -4.162(4^x)$

25. $R'(x) \approx -0.102(0.98^x)$

27. 2000   **29.** Northeast

31. 1999   **33.** 1999

35. 5.517 thousand students per year

37. $g'(x) = 5x$ **39.** $x = \pm 1$

Section 9.5 *(page 652)*

1. According to the model, the weight of a 10-year-old girl is 76.64 pounds. A 10-year-old girl will gain weight at a rate of 8.24 pounds per year. An 11-year-old girl will weigh about 84.88 pounds.

3. According to the model, the annual per capita income in Washington state in 1984 was $14,204. The income in 1984 was increasing at a rate of $178.40 per year. The income in 1985 was about $14,382.40.

5. According to the model, the average daily volume of the Nasdaq market in 1998 was 918 million shares. In 1998, this volume was growing at the rate of 299 million shares per year. The volume in 1999 was about 1217 million shares.

7. According to the model, the annual volume of the New York Stock Exchange in 1998 was 168,659 million shares. This volume was growing at the rate of 38,760 million shares per year. This volume in 1999 was about 207,419 million shares.

9. According to the model, the average annual wage of an employee in the radio broadcasting industry in 1998 was $34,086. The wage in 1998 was increasing at a rate of $2750 per year. The wage in 1999 was about $36,836.

11. According to the model, the amount of money spent on billboard advertising in 2000 was $1887 million. This amount was increasing at a rate of $176 million per year in 2000.

13. According to the model, when the net number of CDs shipped is 900 million, the value of the CDs is $12,261.9 million. When the net number of CDs shipped is 900 million, the value of the CDs is increasing at a rate of $17.9 million per million CDs shipped.

15. According to the model, the number of women in the Summer Olympics in 1992 was 2662 and was growing at a rate of 150 women per year.

17. According to the model, the population of the United States in the year 2000 was 284 million people and was growing at a rate of 2.70 million people per year.

19. According to the model, the number of homicides resulting from an alcohol-related brawl in the year 2000 was 185. In 2000, this number was decreasing at a rate of 14 homicides per year.

21. According to the model, the gross revenue from sales of Johnson & Johnson and its subsidiaries at the end of 2001 was $32,317 million and was increasing at a rate of $3810 million per year.

23. According to the model, the operating profit of Frito-Lay North America at the end of 2001 was $2056 million and was increasing at the rate of $93.5 million per year.

25. According to the model, the flow of the Snoqualmie River at noon on November 24, 2002, was 300 cubic feet per second and was decreasing at 45 cubic feet per second per 24-hour period. The river rafting company could predict when the flow was fast, slow, safe, unsafe, and so on.

27. According to the model, the per capita personal income of Colorado at the end of 1998 was $28,634 and was increasing at a rate of $1700.10 per year.

29.

	Per Capita Income in 1998 (dollars)	Projected Growth in Income (dollars)	Projected per Capita Income in 1999 (dollars)
California	23,712	791.7	24,503.7
Colorado	28,634	1,700.1	30,334.1
Connecticut	36,767	1,998.6	38,765.6

31. The VCR price decreased; $P'(t)$ will approach 0 dollars per year.

33. 1989; 1996   35. 0

CHAPTER 9 REVIEW EXERCISES

Section 9.1 *(page 656)*

1. 6   3. $230.2 per year   5. 12

Section 9.2 *(page 656)*

7.

h	Difference Quotient
0.1	4.2
0.01	4.02
0.001	4.002
Instantaneous rate ≈ 4.0	

9. 4

11.

h	Difference Quotient
0.1	0.9777
0.01	0.9767
0.001	0.9766
Instantaneous rate ≈ 0.977	

13. $T'(7) \approx 0.0687$ second per rank. In the 100-meter men's race in the 2000 Olympics, the seventh-place runner's time is increasing at the rate of 0.0687 seconds per place (rank).

Section 9.3 *(page 657)*

15. $y = 2x - 2$

$f(x) = 2x^2 - 2x$

$(1, 0)$

Tangent line: $y = 2x - 2$

17. $\approx \$36.33$ per year; $\approx \$996.12$

19. -649.2 fatalities per year. In 1985, the number of highway fatalities was decreasing at a rate of about 649 fatalities per year.

Section 9.4 *(page 658)*

21. $f'(x) = 4x$

23. 8 at $x = 1$; 20 at $x = 3$; 32 at $x = 5$

25. $P'(x) \approx 0.693(2^x)$ **27.** 1999

Section 9.5 *(page 658)*

29. According to the model, McDonald's sales were $27,015 million in 2002 and were increasing at a rate of $1233.5 million per year.

CHAPTER 10

Section 10.1 *(page 667)*

1. 0   **3.** $15t^2 + 20t^{-3}$ **5.** $-2 + 12x^2$

7. $-4t^{-2}$ **9.** $3t^2 + 6t + 6$

11. $y = 0.5x + 0.5$ **13.** $y = -7x - 1.4$

15. $y = 551.23t - 1203.9$

17. $y = 3.8n + 102.2$ **19.** $y = 3.66x - 12.74$

21. 140.05 thousand acres per year

23. Yes   **25.** 9-year-old girl

27. No   **29.** 2000

31. No. Consider $f(x) = g(x) = x$. $h(x) = x^2$ and $h'(x) = 2x$. If we use the Constant Multiple Rule, we get

$$h'(x) = f(x)g'(x)$$
$$= x(1)$$
$$= x$$

This is a contradiction.

33.
$$s'(x) = \lim_{h \to 0} \frac{s(x + h) - s(x)}{h}$$
$$= \lim_{h \to 0} \frac{[f(x + h) + g(x + h)] - [f(x) + g(x)]}{h}$$
$$= \lim_{h \to 0} \frac{f(x + h) - f(x) + g(x + h) - g(x)}{h}$$
$$= \lim_{h \to 0} \frac{f(x + h) - f(x)}{h} + \lim_{h \to 0} \frac{g(x + h) - g(x)}{h}$$
$$= f'(x) + g'(x)$$

35. 0

Section 10.2 *(page 675)*

1. $2(3x + 4) + 3(2x)$

3. $2(10t + 5) + 10(2t - 6)$

5. $6n(n^3 - n) + (3n^2 - 1)(3n^2 + 8)$

7. $(2t + 2)(t^2 + 4t + 4) + (2t + 4)(t^2 + 2t + 1)$

9. $\dfrac{3x^2(x^4 - x^3) - (4x^3 - 3x^2)(x^3 - 1)}{(x^4 - x^3)^2}$

11. $-\dfrac{1}{2}$ **13.** -5.5 **15.** 56

17. 589   **19.** -4

21. ≈ 1504.35 million dollars per year

23. ≈ 178.74 million dollars per year

25. ≈ 1369.96 million dollars per year

27. $f'(40) = 1$ means that when 70 trees per acre are planted, planting an additional tree will *increase* the overall yield by one bushel of apples per acre.

$f'(45) = 0$ means that when 75 trees per acre are planted, planting an additional tree will not change the overall yield of apples per acre.

$f'(50) = -1$ means that when 80 trees per acre are planted, planting an additional tree will *decrease* the overall yield by one bushel of apples per acre.

29. $0.19 per pound

Section 10.3 *(page 685)*

1. $2x^2 + 11$ **3.** $7x^2 - 14x + 3$

5. $3x^6 - 11x^3$

7. $f(x) = x^2 - 6x + 9$; $g(x) = x + 1$

9. $f(x) = 2\ln(x)$; $g(x) = x^2 + 4$

11. $30(3x + 1)$

13. $12(t^3 - t^2 + 1)^3(3t^2 - 2t)$

15. $20(n^2 + 2n + 1)(n + 1)$ or $20(n + 1)^3$

17. $4(-80x^2 + 42x - 5)(-80x + 21)$

19. $(3t + 2)(5t + 2)(12t + 1)$

21. $-\dfrac{13}{4}$ **23.** $\dfrac{13}{81}$ **25.** $-11{,}280$

27. $y = \dfrac{5}{8}t + \dfrac{1}{16}$ **29.** $y = 14{,}641$

31. 2.124 inches per year

33. $8x[(x^2 + 1)^2 + 1](x^2 + 1)$

35. 40 sales per year

Section 10.4 *(page 693)*

1. $\ln(4)4^x$ **3.** $\dfrac{4}{t}$

5. $5e^x + \dfrac{1}{x}$ **7.** $5e^x\left[\ln(x) + \dfrac{1}{x}\right]$

9. $\ln(2)(2^{x^2+5x})(2x + 5)$

11. $\dfrac{6}{5\ln(10)}$ **13.** $\log(2)$

15. $\dfrac{3\ln(3)}{e}$ **17.** $3\log(5)$ **19.** $15e^2$

21. $y = 3\ln(2)t - 3\ln(2) + 1$

23. $y \approx 9.862x - 4.443$ **25.** $y = \dfrac{1}{e}$

27. 23.63 dollars per year

29. ≈ 1.63 percentage points per year. In 1999, the percentage of homes with a VCR was increasing at a rate of 1.63 percentage points per year.

31. $\dfrac{\ln(4)4^{\ln(x)}}{x}$

33. $2^{2t+1}\ln(2)[\ln(2)(2t) + 1]$

35. $\dfrac{1}{t}$

Section 10.5 *(page 700)*

1. Undefined   **3.** -2

5. $-\dfrac{3}{2}$ **7.** $\dfrac{1}{3}$

9. 5   **11.** $\dfrac{1}{2y}$

13. $\dfrac{dy}{dx} = 0$ **15.** $\dfrac{y^2}{1 - xy}$

17. $-\dfrac{y}{x - y}$ **19.** $\dfrac{y}{xy - 2x}$

21. $(0, 1)$ and $(0, -1)$ **23.** $(0, 0)$ and $\left(0, \dfrac{1}{2}\right)$

25. $(1, -2)$ and $(-1, 2)$

27. $\dfrac{dr}{dt} = \dfrac{\dfrac{dV}{dt} - \pi r^2 \dfrac{dh}{dt}}{2\pi rh}$

29. We have $\dfrac{dy}{dx} = \dfrac{e^x y - e^y}{xe^y - e^x}$.

Although the coordinate $(1, 1)$ is the only coordinate that makes the numerator of $\dfrac{dy}{dx}$ equal to zero, it also makes the denominator equal zero. So $\dfrac{dy}{dx}$ is undefined at $(1, 1)$. Thus $\dfrac{dy}{dx} \neq 0$ everywhere.

CHAPTER 10 REVIEW EXERCISES

Section 10.1 *(page 701)*

1. 0   **3.** $21t^2 + 12t^{-4}$

5. 1999: Decreasing at a rate of 12.773 thousand people per year

2000: Decreasing at a rate of 7.648 thousand people per year

Section 10.2 *(page 701)*

7. $18x^2 - 24x$ **9.** $-6t^2 + 16t + 4$

11. $\approx 24{,}422$ prisoners per year

Section 10.3 *(page 701)*

13. $-6x^2 + 3x + 4$ **15.** $54(9x + 5)^2$

17. 2,219.31 million dollars per year

Section 10.4 *(page 702)*

19. $\ln(9)9^x$ **21.** $\dfrac{2x}{x^2 + 3}$

23. 20.1 thousand people per year

Section 10.5 *(page 702)*

25. $-\dfrac{1}{21}$ **27.** -4 **29.** $\dfrac{-2 \ln(2) + 1}{6 \ln(2) - 3}$

CHAPTER 11

Section 11.1 *(page 721)*

1. $f'(x) = 0$ at $(1, -1)$

3. $f'(x) = 0$ at $(0, 1)$ **5.** $f'(x) = 0$ at $(2, 0)$

7. $f'(x)$ is undefined at $(2, 4)$

9. $f'(x) = 0$ at $(0, 200)$ and $(4, -56)$

11. $x = 2$; stationary   **13.** $t = 0, 4$; stationary

15. $x = 0$; singular
$x = \pm 1$; stationary

17. $x = \pm 1$; stationary   **19.** $p = \pm 1$; stationary

21.

$f'(x)$ — $\dfrac{-\quad\text{min}\quad+}{2}$ →

x	$f(x)$ on $[-1, 5]$	
-1	5	Absolute maximum
2	-4	Relative and absolute minimum
5	5	Absolute maximum

23.

$h'(t)$ — $\dfrac{+\quad\text{max}\quad-\quad\text{min}\quad+}{0 \qquad 4}$ →

t	$h(t)$ on $[-1, 7]$	
-1	-7	
0	0	Relative maximum
4	-32	Relative and absolute minimum
7	49	Absolute maximum

25.

$g'(x)$ — $\dfrac{-\quad\text{min}\quad+\quad+\quad\text{max}\quad-}{-1 \quad 0 \quad 1}$ →

x	$g(x)$ on $[-3, 3]$	
-3	-1.327	
-1	-2	Relative and absolute minimum
0	0	
1	2	Relative and absolute maximum
3	1.327	

27.

$h'(x)$ — $\dfrac{-\quad\text{min}\quad+\quad\text{max}\quad-}{-1 \qquad\quad 1}$ →

x	$h(x)$ on $[-2, 2]$	
-2	-0.4	
-1	-0.5	Relative and absolute minimum
1	0.5	Relative and absolute maximum
2	0.4	

29.

$C'(p)$ — $\dfrac{+\quad\text{max}\quad-\quad\text{min}\quad+}{-1 \qquad\quad 1}$ →

p	$C(p)$ on $[-3, 3]$	
-3	-65	Absolute minimum
-1	15	Relative maximum
1	-1	Relative minimum
3	79	Absolute maximum

31.

$f'(x)$ — $\dfrac{+\quad\text{max}\quad-\quad\text{min}\quad+}{-\sqrt{\tfrac{4}{3}} \qquad \sqrt{\tfrac{4}{3}}}$ →

x	$f(x)$ on $[-3, 3]$	
-3	-15	Absolute minimum
$-\sqrt{\dfrac{4}{3}}$	≈ 3.08	Relative maximum
$\sqrt{\dfrac{4}{3}}$	≈ -3.08	Relative minimum
3	15	Absolute maximum

33.

$h'(t)$ — $\dfrac{-\quad\text{min}\quad+}{\tfrac{1}{e}}$ →

t	$h(t)$ on $[0.1, 3]$	
0.1	≈ -0.2303	
$\dfrac{1}{e}$	≈ -0.3679	Relative and absolute minimum
3	≈ 3.2958	Absolute maximum

35.
$g'(x)$ $\xleftarrow{\quad - \quad | \text{min} \quad + \quad}$
0.567

x	$g(x)$ on $[0.1, 3]$	
0.1	≈ -3.408	
0.567	≈ -2.330	Relative and absolute minimum
3	≈ 18.987	Absolute maximum

37. $25   **39.** $40

41. From about 1983 through 1988

43. Highest: 1999
Lowest: 1990

45. 59.457 million   **47.** $y = e^x$

49. Each time $p(x)$ changes from increasing to decreasing (or vice versa) $p'(x)$ will change sign. Since $p'(x)$ is continuous, $p'(x)$ will equal zero between sign changes. Since there are at most $n - 1$ sign changes, there are at most $n - 1$ stationary points.

Section 11.2 *(page 743)*

1. (a) $P(t) = 564.3t^2 - 613.7t + 3658.9$
(b) 2000

3. (a)

	Hotel Pack	Single Slice
Surface area	455.8 in^2	680.5 in^2
Volume	654.9 in^3	869.0 in^3

(b) ≈ 8.716 in $\times$ 8.716 in $\times$ 8.716 in
(c) ≈ 10.65 in $\times$ 10.65 in $\times$ 10.65 in
(d) The box is a cube, since the length, width, and height are all equal.

5. (a) $s(t) = 17.8(1.04)^{t-2}$ billion unit cases [equivalent to $s(t) \approx 16.5(1.04)^t$ billion unit cases]
(b) $R'(s) = -0.8058s + 14.44$ dollars per case sold
(c) 17.92 billion cases
(d) $R(s(t)) = -0.4029[17.8(1.04)^{t-2}]^2 + 14.44[17.8(1.04)^{t-2}] - 109.2$
(e) Early 2002

7. (a) 2002
(b) $C(t) = -120.5t^2 + 623.5t + 6553$ million dollars; t is the number of years since the end of 1999.

(c) 2002
(d) They are reasonable, since the maximum number of sales could incur additional costs.

9. 2000   **11.** 30 trees   **13.** $0.19 per pound

15. $116.92 million per year

17. ≈ 27.5 feet $\times$ 18.2 feet   **19.** 1200 books

21. 500 books   **23.** 50 vehicles

25. 43 vehicles

27. Limit the number of cars sold to the amount that maximized revenue.

29. 250 pounds

31. $\dfrac{dx}{dt} = 3t^2 - 18t + 23$ meters per second is the rate of horizontal change in position; $\dfrac{dy}{dt} = 6 - 2t$ meters per second is the rate of vertical change in position.

33.

t	$x(t)$	
0	0	Absolute minimum
1.85	18.08	Relative maximum
4.15	11.92	Relative minimum
6	30	Absolute maximum

t	$y(t)$	
0	0	Absolute minimum
3	9	Relative maximum
6	0	Absolute minimum

35. 0 meters per second (*Note:* speed is the absolute value of velocity.)

Section 11.3 *(page 766)*

1. $f(x) = x^3 - 3x^2$

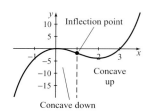

3. $f(x) = x^4 - 12x^2$

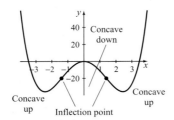

5. $y = x^4 - 8x^3 + 18x^2$

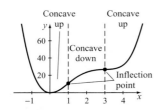

7. $y = x^4 - 2x^3 + 12x^2$

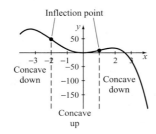

9. $f(x) = \dfrac{x^2}{x^2 + 1}$

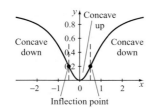

11. $(1, -2)$

13. $\left(-\sqrt{2}, -20\right); \left(\sqrt{2}, -20\right)$

15. $(1, 11); (3, 27)$ **17.** $(-2, 48); (1, 9)$

19. $\left(-\sqrt{\tfrac{1}{3}}, \tfrac{1}{4}\right); \left(\sqrt{\tfrac{1}{3}}, \tfrac{1}{4}\right)$

21. $(0, 0)$ relative maximum; $(2, -4)$ relative minimum

23. $\left(-\sqrt{6}, -36\right)$ relative minimum; $(0, 0)$ relative maximum; $\left(\sqrt{6}, -36\right)$ relative minimum

25. $(0, 0)$ relative minimum; $(3, 27)$ inflection point

27. $(0, 0)$ relative minimum; $(-3.312, 83.967)$ relative maximum; $(1.812, 16.721)$ relative maximum

29. $(0, 0)$ relative minimum

31. $A\left(-2, -6\tfrac{2}{3}\right)$ absolute minimum on $[-2, 4]$
$B(0, 0)$ relative maximum; x- and y-intercept
$C\left(1, -\tfrac{2}{3}\right)$ inflection point
$D\left(2, -\tfrac{4}{3}\right)$ relative minimum
$E(3, 0)$ x-intercept
$F\left(4, 5\tfrac{1}{3}\right)$ absolute maximum on $[-2, 4]$

33. $A(-2, 3)$ absolute maximum on $[-2, 2]$
$B(-1.532, 0)$ x-intercept
$C(-1, -1)$ relative and absolute minimum on $[-2, 2]$
$D(-.347, 0)$ x-intercept
$E(0, 1)$ inflection point and y-intercept
$F(1, 3)$ relative and absolute maximum on $[-2, 2]$
$G(2, -1)$ absolute minimum on $[-2, 2]$

35. $A(-4, 16)$ absolute maximum on $[-4, 4]$
$B(-3.464, 0)$ x-intercept
$C(-2, -16)$ relative and absolute minimum on $[-4, 4]$
$D(0, 0)$ x- and y-intercept; inflection point
$E(2, 16)$ relative and absolute maximum on $[-4, 4]$
$F(3.464, 0)$ x-intercept
$G(4, -16)$ absolute minimum on $[-4, 4]$

37.

$A(-3, -129.6)$ absolute minimum on $[-3, 5]$

$B(0, 0)$ x- and y-intercept; relative and absolute maximum on $[-3, 5]$

$C(3, -32.4)$ inflection point

$D(4, -51.2)$ relative minimum

$E(5, 0)$ x-intercept; absolute maximum on $[-3, 5]$

39.

$A(-2, 1.636)$

$B(0, 1)$ x-intercept; relative and absolute minimum on $[-2, 2]$

$C(2, 2.614)$ absolute maximum on $[-2, 2]$

41. Minimum increase: 1975, $-\$5.765$ per year
Maximum increase: 2000, $\$1.465$ per year

43. Wages should continue to increase to compensate for diminishing wages during the years directly after 1975.

45. 1995; 2000   **47.** 1995; 1998

49. 1991; 299   **51.** 1

53. $s''(t) = -32$ feet per second squared, acceleration due to the force of gravity

55. 3.9 seconds

57. $(23.16, 12.56)$; in the 23rd week, the box office sales were increasing most rapidly at about 12.56 million dollars per week.

59. $x'' = 0$ at $t = 3$; $y'' \neq 0$

61. 18 square meters per second   **63.** Yes

Section 11.4 *(page 777)*

1. $\dfrac{dA}{dt} = \dfrac{dl}{dt}w + \dfrac{dw}{dt}l$

$\dfrac{dA}{dt}$ = change in area with respect to time in square inches per minute

$\dfrac{dl}{dt}$ = change in length with respect to time in inches per minute

$\dfrac{dw}{dt}$ = change in width with respect to time in inches per minute

3. $\dfrac{dV}{dt} = \dfrac{dl}{dt}wh + \dfrac{dw}{dt}lh + \dfrac{dh}{dt}lw$

$\dfrac{dV}{dt}$ = change in volume with respect to time in cubic inches per minute

$\dfrac{dl}{dt}, \dfrac{dw}{dt}$, same as in Exercise 1.

$\dfrac{dh}{dt}$ = change in height with respect to time in inches per minute.

5. $\dfrac{dA}{dt} = 2\left(\dfrac{dl}{dt}w + \dfrac{dw}{dt}l + \dfrac{dw}{dt}h + \dfrac{dh}{dt}w \right.$
$\left. + \dfrac{dl}{dt}h + \dfrac{dh}{dt}l \right)$

Physical meanings same as in Exercises 1 and 3

7. $\dfrac{dA}{dt} = 8\pi r \dfrac{dr}{dt}$

$\dfrac{dA}{dt}$, same as in Exercise 1

$\dfrac{dr}{dt}$ = change in radius with respect to time in inches per minute

9. $\dfrac{dA}{dt} = 2\pi r \dfrac{dr}{dt}$

Physical meanings same as in Exercise 7

11. 3.782 inches per hour; 12.69 hours

13. $0.2\overline{6}$ inches per second or $0.677\overline{3}$ centimeters per second

15. 0.1014 foot per hour

17. 8.83 degrees per hour

19. 4.20 degrees per hour

CHAPTER 11 REVIEW EXERCISES

Section 11.1 *(page 778)*

1. $f'(x) = 0$ at $(1.5, 2.25)$ and $(2.5, 2.25)$
$f'(x)$ undefined at $(2, 2)$

3. $\left(-\sqrt{\dfrac{12}{5}}, \approx 15.95 \right)$; $(0, 10)$; $\left(\sqrt{\dfrac{12}{5}}, \approx 4.051 \right)$

5. $(0, 4)$

7.

f'(x) + 0 − 0 − 0 +

$-\sqrt{\frac{12}{5}}$ 0 $\sqrt{\frac{12}{5}}$

x	$f(x)$ on $[-2, 2]$	
-2	10	
$-\sqrt{12/5}$	≈ 15.95	Relative and absolute maximum
0	10	
$\sqrt{12/5}$	≈ 4.051	Relative and absolute minimum
2	10	

9.

y' + 0 −
    0

x	y on $[-3, 3]$	
-3	-32	Absolute minimum
0	4	Relative and absolute maximum
3	-32	Absolute minimum

11. 469.4 miles

Section 11.2 *(page 778)*

13. **(a)** 2002
(b) $C(t) = -88.5t^2 + 410.5t + 3019$
(c) 2002
(d) The same year is reasonable because more sales could generate more costs.

Section 11.3 *(page 779)*

15.
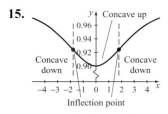
Inflection point

17. $\left(-\sqrt{\dfrac{10}{3}}, 0.925\right), \left(\sqrt{\dfrac{10}{3}}, 0.925\right)$

19. $(0, .9)$ stationary point and relative and absolute minimum

21. 4; 17.72 miles per day

Section 11.4 *(page 780)*

23. $\dfrac{dA}{dt} = 10s\dfrac{ds}{dt}$

$\dfrac{dA}{dt}$ = change in area with respect to time in square centimeters per second

$\dfrac{ds}{dt}$ = change in side length with respect to time in centimeters per second

25. 0.32 inch per minute

27. 0.20 inch per minute

29. 0.0159 foot per hour

CHAPTER 12

Section 12.1 *(page 790)*

1. $2x + C$

3. $\ln|t| - \dfrac{3t^2}{2} + C$

5. $\dfrac{t^2}{4} + 20t + C$

7. $\dfrac{t^4}{4} - t^3 + \dfrac{3t^2}{2} - t + C$

9. $2.3\ln|x| + x + C$

11. $\dfrac{3^x}{\ln(3)} - \dfrac{x^4}{4} + C$

13. $\dfrac{-4}{t} + 2\ln|t| + t + C$

15. $\dfrac{2(3^x)}{\ln(3)} - \dfrac{3(2^x)}{\ln(2)} + C$ **17.** $2000\ln|p| + \dfrac{500}{p} + C$

19. $-5t^{-1} + \dfrac{2^t}{\ln(2)} + C$ **21.** $\dfrac{3x^2}{2} - 5x + C$

23. $t - 2\ln|t| + C$

25. $-3x^{-1} - x^4 + 2x + C$

27. $-5t^{-1} - 16\dfrac{t^3}{3} - 9t + C$

29. $5u - 4\ln|u| - u^{-1} + C$

31. $-400x^{-1} + 200\ln|x| + 50x + C$

33. $4\ln|x| + 5x^{-1} + C$ **35.** $2t^2 + \dfrac{4^t}{\ln(4)} + C$

37. $0.013p + 5$ dollars, p = number of pages

39. $-0.53325t^4 + 1.333t^3 - 0.07665t^2 + 0.3167t$

41. No

43. $F'(x) > 0$ for all x so, $F(x)$ is increasing

45. $x = 2$ only

Section 12.2 *(page 798)*

1. $\dfrac{(x^2 + 3)^6}{6} + C$ **3.** $2x^3 + 5x^2 + C$

5. $\dfrac{(e^x + 1)^2}{2} + C$

7. $0.4x^3 - 1.2x^2 + 0.6x + C$

9. $\dfrac{1}{2}e^{x^2 - 4x} + C$ **11.** $\dfrac{1}{2}e^{3x^2 - 6} + C$

13. $\dfrac{4}{3}e^{3t} + C$ **15.** $[\ln(x)]^2 + C$

17. $\ln|t^2 - t + 2| + C$ **19.** $2\ln|x^3 - 9| + C$

21. $\ln|\ln(x)| + C$

23. $\dfrac{6}{5}(x - 2)^{5/2} + 4(x - 2)^{3/2} + C$

25. $\dfrac{2}{3}(x^2 + 1)^{3/2} + C$ **27.** $\dfrac{1}{3}(t^2 - 1)^{3/2} + C$

29. $\ln|x^3 - x^2 + x + 1| + C$

31. $\dfrac{1}{4}(2x - 1 + \ln|2x - 1|) + C$

33. $\dfrac{1}{4}\ln|4x^2 - 5| + C$ **35.** $t^4 - 3t^3 + t^2 + C$

37. $\ln(2p^2 + 1) - \dfrac{2}{3}(2p^2 + 1)^{3/2} + C$

39. $2\ln|t^2 - t| + C$ **41.** Yes; $u = x^2 - 1$

43. No   **45.** No

Section 12.3 *(page 816)*

1. $LHS = 8$ **3.** $LHS = 5.121$

5. $LHS = 89.0$

7.

n	LHS
2	0.0625
4	0.1406
10	0.2025

9.

n	LHS
2	9.333
4	8.167
10	7.564

11.

n	LHS
2	6.119
4	6.779
10	7.151

13.

n	LHS
2	50.00
4	61.00
10	67.92

15.

n	LHS
2	27.20
4	21.41
10	18.41

17.

LHS	17.75
RHS	23.75
Avg.	20.75

19.

LHS	2.39
RHS	2.14
Avg.	2.27

21.

LHS	2.00
RHS	2.00
Avg.	2.00

23.

LHS	60.00
RHS	72.00
Avg.	66.00

25.

LHS	90.00
RHS	102.00
Avg.	96.00

27.

LHS	1.40
RHS	2.15
Avg.	1.78

29.

LHS	0.339
RHS	0.686
Avg.	0.512

31.

LHS	0.776
RHS	1.205
Avg.	0.990

33.

LHS	2.67
RHS	3.36
Avg.	3.01

35.

LHS	324.69
RHS	441.86
Avg.	383.28

37. (a) $3600F(t)$ cubic feet per hour

(b)

LHS	685,115 cubic feet	$\approx$5,125,037 gallons
RHS	597,275 cubic feet	$\approx$4,467,946 gallons
Avg.	641,195 cubic feet	$\approx$4,796,492 gallons

(c) More hydroelectric power is needed at 5:45 a.m. than at 7:45 a.m. (Answers may vary.)

39.

LHS	2,165.6 cubic feet	16,200 gallons
RHS	2,165.6 cubic feet	16,200 gallons
Avg.	2,165.6 cubic feet	16,200 gallons

41. $f(x) = 3$ on $[0, 2]$ (Answers may vary.)

43. The left-hand sum will be an underestimate, and the right-hand sum will be an overestimate of the area. An average of the left- and right-hand sums will be the best estimate.

45. Yes. In the case of $f(x) = |-2x + 2|$ on the interval $[0, 2]$, the exact area is 2.0. Using $n = 2$, the left-hand sum estimate yields the exact area. However, using $n = 3$, the left-hand sum estimate yields 2.2, which is greater than the exact area. Increasing the number of rectangles worsened the area estimate.

Section 12.4 *(page 830)*

1. 21   **3.** -3 **5.** 36

7. 2.71875   **9.** 6.875

11. Area $= 2$ **13.** Area $= 8$

15. Area $= 2$ **17.** Area $= 4$

19. Area $= 4$

21. 10   **23.** 10   **25.** 4   **27.** 3   **29.** -23

31. $\displaystyle\int_{0}^{3} (-x^2 + 5x - 6)\, dx$

33. $\displaystyle\int_{-2}^{4} (-x^3 + 4x + 20)\, dx$

35. $\displaystyle\int_{-2}^{4} (-|x - 3| + 3)\, dx$

37.   **39.**

41. 10.5 gallons per person. From the end of 1990 to the end of 1999, the consumption of bottled water increased by 10.5 gallons per person.

43. 5.06 pounds per person. From the end of 1980 to the end of 1995, the consumption of mozzarella cheese increased by 5.06 pounds per person.

45. According to the model, the number of students participating in the AB Calculus exam increased by roughly 80,690 students from the end of 1992 to the end of 2002. That is, the number of students taking the exam in 2002 is calculated to be 80,690 students higher than the number of students taking the exam in 1992.

47. This inequality is equivalent to the left-hand sum $\leq \displaystyle\int_{a}^{b} f(x)\, dx \leq$ right-hand sum, which is true for an increasing function.

49. $f(x) = 0$, $f(x) = (x - 3)$, $f(x) = (x - 3)^3$ (Answers may vary.)

Section 12.5 *(page 844)*

1. $\displaystyle\int_5^7 1\,dx = 2$ **3.** $\displaystyle\int_0^4 x\,dx = 8$

5. $\displaystyle\int_1^2 (-3x^2 + 5)\,dx = -2$

7. $\displaystyle\int_{-1}^1 (8x^3 + 2x)\,dx = 0$

9. $\displaystyle\int_1^3 \left(\frac{1}{x}\right)dx = \ln(3) \approx 1.099$

11. $\displaystyle\int_2^5 (2^t)\,dt \approx 40.40$

13. $\displaystyle\int_{-2}^4 (5 \cdot 2^x)\,dx \approx 113.61$

15. $\displaystyle\int_0^1 \left(\frac{2x}{x^2+1}\right)dx = \ln(2) \approx 0.693$

17. $\displaystyle\int_0^2 (x^2-1)(x^3-3x)^3\,dx = \frac{4}{3} \approx 1.33$

19. $\displaystyle\int_{-2}^{-1} (-2x+1)\,dx = 4$

21. Area $= 78\dfrac{1}{12} \approx 78.08$

23. Area $= \dfrac{1}{12} \approx 0.08$ **25.** Area $= 2$

27. Area ≈ 5.28 **29.** Area $= 1\dfrac{5}{6} \approx 1.83$

31. $\displaystyle\int_{33}^{43} (315.70t - 770.64)\,dt \approx 112{,}260$
According to the model, 112,260 more people will take the exam in 2012 than in 2002.

33. $\displaystyle\int_0^{1.5} V(t)\,dt \approx 1.318$ miles

35. $\displaystyle\int_{10}^{19} (0.0009778t^3 - 0.02958t^2 + 0.2392t$
$- 0.06311)\,dt = 2.3$ pounds per person
Between the end of 1990 and the end of 1999, the per capita mozzarella cheese consumption increased by 2.3 pounds per person.
37. $F(4) = 2\ln(4) + 1 \approx 3.77$
39. Area ≈ 3.45

CHAPTER 12 REVIEW EXERCISES

Section 12.1 *(page 845)*

1. $G(x) = -x^2 + 5x + C$

3. $H(x) = 2x^2 - 3\ln|x| + C$

5. $x^4 - \dfrac{5}{2}x^2 - x + C$ **7.** $t + \dfrac{1}{2}\ln|t| + C$

9. $0.018p + 1.30$ dollars, where p is the number of pages

Section 12.2 *(page 846)*

11. $\dfrac{1}{12}(x^2 - 9)^6 + C$ **13.** $[\ln(x)]^2 + C$

15. $3\ln|t| + \dfrac{1}{2}t + C$

Section 12.3 *(page 846)*

17.

n	LHS
2	48.0
4	40.0
10	35.2

19.

n	LHS
2	6.417
4	6.954
10	7.276

21.

LHS	0.906
RHS	1.156
Avg.	1.031

23.

LHS	3.5095
RHS	3.8845
Avg.	3.697

25.

LHS	2.34375
RHS	1.59375
Avg.	1.96875

27. (a) $3600f(t)$
(b) The total amount of water that flowed past the Priest Rapids Dam between 5:00 p.m. and 9:45 p.m. on November 22, 2002, is estimated to be 1,584,925,729 cubic feet.
(c) 15,849 swimming pools

Section 12.4 *(page 847)*

29. 15   **31.** Area $= 21$

33. Area = 9

35. Area = 4

37. 12   **39.** 1

41.

43. **(a)** -1.385 gallons per person. From the end of 1993 to the end of 1999, the per capita consumption of milk decreased by 1.385 gallons.

 (b) Answers may vary.

Section 12.5 *(page 848)*

45. 8.5   **47.** 0   **49.** 0.693   **51.** 132.5

53. 4.605   **55.** $\dfrac{16}{3}$ **57.** ≈ 5821.04 dollars per person

CHAPTER 13

Section 13.1 *(page 858)*

1. $-e^{-x}(x + 1) + C$

3. $\dfrac{3}{2}t^2 \ln(t) - \dfrac{3}{4}t^2 - 5t \ln(t) + 5t + C$

5. $\dfrac{1}{4}t^4 \ln(t) - \dfrac{1}{16}t^4 + C$

7. $e^{2x+1}(2x - 1) + C$ **9.** $e^{2x}(x - 1) + C$

11. $\dfrac{\ln|3t - 1|}{\ln(3)} + C$ **13.** $3xe^x - 3e^x - x^3 + C$

15. $\dfrac{1}{2}x^4 - 9x^2 + C$ **17.** $\dfrac{1}{2}t^6 - t^3 + C$

19. $\dfrac{3^x}{\ln(3)}\left(x^2 - \dfrac{2x}{\ln(3)} + \dfrac{2}{[\ln(3)]^2} - 3\right) + C$

21. $\dfrac{1}{2}e^{3x^2} + C$ **23.** $\dfrac{1}{3}e^{2t^3} + C$

25. $\dfrac{2^{\ln(x)}}{\ln(2)} + C$

27. ≈ 280.05 million dollars

29. ≈ 3773.7 million dollars

31. $e^x(4x^2 - 8x + 8) + C$

33. $4t^{3/2} \ln(t) - \dfrac{8}{3}t^{3/2} + C$

35. When the integrand is the product or quotient of two different types of functions (i.e. polynomial, exponential, etc.), the integration by parts method is often needed.

Section 13.2 *(page 868)*

1. $\dfrac{1}{6}$ **3.** $\dfrac{64}{3}$ **5.** $\dfrac{1}{6}$ **7.** $10\dfrac{2}{3}$

9. 5980.50   **11.** ≈ 7.647 **13.** 1

15. 0.0975   **17.** ≈ 3.460 **19.** ≈ 0.0858

21. $\approx 40{,}804.7$ million dollars

23. $\approx 3{,}569{,}382$ people. According to the model, the number of adults in prison, in jail, on probation, or on parole increased by 3,569,382 more people than the number of students enrolled in private colleges from the end of 1980 through 1998.

25. 773 million dollars. According to the model, the annual profit increased by 773 million dollars from the end of 1990 through 2001.

27. 13.168   **29.** 1.5

Section 13.3 *(page 878)*

1. $y = \pm\sqrt{x^2 + C}$ **3.** $y = \left(\dfrac{x}{2} + C\right)^2$

5. $y = \dfrac{x}{1 + Cx}$ **7.** $y = \pm Ce^{0.25x} + 4$

9. $y = \dfrac{4Ce^{4x}}{1 + Ce^{4x}}$ **11.** $y = \sqrt{2x^2 + 14}$

13. $y = -e^{3x}$ **15.** $y = -5e^{25x} + 5$

17. $y = \dfrac{4e^{4x}}{7 + e^{4x}}$ **19.** $y = \dfrac{100e^{2x}}{50e^{2x} - 49}$

21. **(a)** $T(t) = 76 - 41.7e^{-0.01344t}$
  (b) $0.4899°F/min; 0.2862°F/min$

23. ≈ 19.08 min

25. About 5 hours 8 minutes

27. About 5 hours 24 minutes

29. $y = 1000e^{0.1t}$ **31.** $A(t) = 2500e^{0.17293t}$

33. $\dfrac{dA}{dt} = Pe^{kt}(k) = kA$ **35.** $y = \dfrac{\pm C}{\sqrt{|x|^3}}$

Section 13.4 *(page 893)*

1. **(a)** 14.56 cm
  (b) 16.25 cm
  (c) Female

3. **(a)** 2.86 cm/year
  (b) 4.921 cm/year
  (c) Male; male is 20 cm at $t \approx 1.994$ years and female is 20 cm at $t \approx 1.555$ years

5. **(a)** 84.00 cm
  (b) 11.438 cm/year

7. **(a)** 927
  (b) $\dfrac{dN}{dt} = -0.000084335N(919.9 - N)$
  (c) ≈ -5.86 cases per year. According to the model, the number of new pediatric AIDS cases was dropping at a rate of approximately 6 cases per year at the end of 2000.
  (d) Yes. It appears that the number of new cases will continue to drop.

9. **(a)** $y = \dfrac{85.88}{1 + 29.65e^{-0.9249t}}$. This equation represents the percentage of public schools with access to the Internet, where t is the number of years since the end of 1994.

(b) Increasing at the rate of 7.4 percentage points per year

11. **(a)** $\dfrac{dN}{dt} = 0.02311N(27.80 - N)$
  (b) Increasing at \$4.03 per year and \$1.98 per year, respectively

13. 57.1 minutes; 2124 people per minute

15. 22 homicides per year

17. **(a)** Week 1
  (b) The model predicts that the cumulative gross box office sales will level out at about 242 million dollars after Week 51. Therefore, new sales are diminishing.

19. About 69.75%; about 0.15 percentage point per year

21. Proof:

$$\frac{dy}{dt} = k(M - y)$$

$$\frac{dy}{(M - y)} = k\,dt$$

$$\int \frac{dy}{M - y} = \int k\,dt$$

$$-\ln(M - y) = kt + C \qquad \text{Assuming } M > y$$

$$\ln(M - y) = -kt - C$$

$$e^{-kt - C} = M - y$$

$$y = M - e^{-kt}e^{-C}$$

Since $y(0) = 0$, we have

$$0 = M - e^{-k \cdot 0}e^{-C}$$

$$M = e^{-C}$$

So

$$y = M - e^{-kt}(M)$$
$$= M(1 - e^{-kt})$$

23. Proof:

$$D(t) = \frac{100}{1 + 200e^{-0.1t}}$$
$$= 100(1 + 200e^{-0.1t})^{-1}$$
$$D'(t) = -100[(1 + 200e^{-0.1t})^{-2}]$$
$$[200e^{-0.1t}(-0.1)] \qquad \text{Chain Rule}$$
$$= \frac{2000e^{-0.1t}}{(1 + 200e^{-0.1t})^2}$$

We will now check to see if $D'(t) = 0.001[D(t)]\{100 - [D(t)]\}$ yields the same result.

$$D'(t) = 0.001[D(t)]\{100 - [D(t)]\}$$
$$= 0.001\left(\frac{100}{1 + 200e^{-0.1t}}\right)\left(100 - \frac{100}{1 + 200e^{-0.1t}}\right)$$

$$= 10\left(\frac{1}{1 + 200e^{-0.1t}}\right)\left(1 - \frac{1}{1 + 200e^{-0.1t}}\right)$$

$$= 10\left(\frac{1}{1 + 200e^{-0.1t}}\right)\left(\frac{(1 + 200e^{-0.1t}) - 1}{1 + 200e^{-0.1t}}\right)$$

$$= 10\left(\frac{1 + 200e^{-0.1t} - 1}{(1 + 200e^{-0.1t})^2}\right)$$

$$= \frac{2000e^{-0.1t}}{(1 + 200e^{-0.1t})^2}$$

Since both approaches yielded the same result, $D'(t) = 0.001[D(t)]\{100 - [D(t)]\}$.

CHAPTER 13 REVIEW EXERCISES

Section 13.1 *(page 898)*

1. $-2e^{-x}(x + 1) + C$

3. $e^x(3x^2 - 6x + 6) + C$ **5.** $\frac{1}{4}[\ln(x)]^4 + C$

7. $\frac{5}{2}x^2 \ln(x) - \frac{5}{4}x^2 - 3x \ln(x) + 3x + C$

9. $e^{4x}\left(x - \frac{1}{4}\right) + C$

Section 13.2 *(page 898)*

11. 72   **13.** ≈ 3.46 **15.** ≈ 29.264

17. According to the model, the accumulated change in profit from the end of 1999 through 2001 is 3918 million dollars.

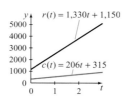

19. The gap between public- and private-school teachers is predicted to increase by 116.15 thousand teachers between the year 2000 and the year 2010.

Section 13.3 *(page 899)*

21. $y = \pm k|x|$ **23.** $A(t) = 16 \pm ke^{-2t}$

25. $A(t) = \frac{4ke^{4t}}{1 + ke^{4t}}$ **27.** $A(t) = \frac{1}{5t - 2}$

29. $y = 10 + e^{0.1x}$

31. $T(t) = 70 + 28.6e^{-0.01189t}$; about 88 minutes

Section 13.4 *(page 899)*

33. (a) $\dfrac{dN}{dt} = 0.01473N(121.8 - N)$

(b) 41.91; 14.87 million units per year

35. (a) 69.07 cm

(b) 9.3 cm per year

37. 0.134 homicide per 100,000 people per year (about 1.34 homicides per million people per year)

CHAPTER 14

Section 14.1 *(page 913)*

1. $z = 3x - 2y$

x ↓	$y \rightarrow$		
	-1	0	1
-2	-4	-6	-8
0	2	0	-2
2	8	6	4

3. $z = 4xy + 4$

x ↓	$y \rightarrow$		
	-1	0	1
-2	12	4	-4
0	4	4	4
2	-4	4	12

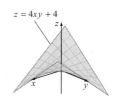

5. $f(x, y) = e^{-xy}$

x ↓	$y \rightarrow$		
	-1	0	1
-2	0.135	1.00	7.39
0	1.00	1.00	1.00
2	7.39	1.00	0.135

$z = e^{-xy}$

Note: In the graph, each square of the grid has dimensions $\frac{1}{4} \times \frac{1}{4}$.

7. $f(x, y) = \dfrac{x^2 + 1}{y^2 + 1}$

x ↓	$y \to$ -1	0	1
-2	2.5	5.0	2.5
0	0.5	1.0	0.5
2	2.5	5.0	2.5

$z = \dfrac{x^2 + 1}{y^2 + 1}$

Note: In the graph, each square of the grid has dimensions $\frac{1}{4} \times \frac{1}{4}$.

9. $z = (x + y)^3$

x ↓	$y \to$ -1	0	1
-2	-27	-8	-1
0	-1	0	1
2	1	8	27

$z = (x + y)^3$

Note: In the graph, each square of the grid has dimensions $\frac{1}{4} \times \frac{1}{4}$.

11. 6889 thousand vehicles   **13.** $4938

15. 504 square inches

17. A 187-pound man who is 71 inches tall has body mass index ≈ 26.1. (Answers will vary.)

19. $-15°F$

21. $\{(x, y) \,|\, |x| \ge 1 \text{ and } y \ne 0\}$

23. A number of variables affect a person's take-home pay, including any of the following: gross pay amount, federal income tax withholding, OASI deduction, Medicare deduction, medical insurance premium deduction, retirement plan deduction, etc. (Answers may vary.)

Section 14.2 *(page 929)*

1. $f_x = 3y$
$f_y = 3x$
$f_{xx} = 0$
$f_{xy} = 3$
$f_{yy} = 0$

3. $f_x = 4y$
$f_y = 4x - 2y$
$f_{xx} = 0$
$f_{xy} = 4$
$f_{yy} = -2$

5. $W_t = v^{0.5}$
$W_v = 0.5tv^{-0.5}$
$W_{tt} = 0$
$W_{tv} = 0.5v^{-0.5}$
$W_{vv} = -0.25tv^{-1.5}$

7. $V_r = 2\pi rh$
$V_h = \pi r^2$
$V_{rr} = 2\pi h$
$V_{rh} = 2\pi r$
$V_{hh} = 0$

9. $z_x = \ln(xy) + 1$
$z_y = xy^{-1}$
$z_{xx} = \dfrac{1}{x}$
$z_{xy} = \dfrac{1}{y}$
$z_{yy} = -xy^{-2}$

11. $V_r = 2\pi rh$
$V_h = \pi r^2$

$V_r(3, 2) = 12\pi$ means that when the height of the cylinder is 3 inches and the radius is 2 inches, increasing the radius by 1 inch while leaving the height unchanged will increase the volume by about 12π cubic inches.
$V_h(3, 2) = 4\pi$ means that when the height of the cylinder is 3 inches and the radius is 2 inches, increasing the height by 1 inch while leaving the radius unchanged will increase the volume by about 4π cubic inches.

13. $W_T = 0.6215 + 0.4275V^{0.16}$
$W_V = 0.16V^{-0.84}(-35.75 + 0.4275T)$

$W_T(27, 5) \approx 1.2$ means that when the temperature is $27°F$ and the wind speed is 5 mph, increasing the temperature by $1°F$ while leaving the wind speed unchanged will *increase* the wind chill temperature by about $1.2°F$.
$W_V(27, 5) \approx -1.0$ means that when the temperature is $27°F$ and the wind speed is 5 mph, increasing the wind speed by 1 mph while leaving the temperature unchanged will *decrease* the wind chill temperature by about $1°F$.

15. $B_H = -1406WH^{-3}$
$B_W = 703H^{-2}$

$B_H(67, 155) \approx -0.7$ means that when a person is 67 inches tall and weighs 155 pounds, increasing the person's height by 1 inch while leaving the weight constant will decrease the body mass index by about -0.7 index point.

$B_W(67, 155) \approx 0.2$ means that when a person is 67 inches tall and weighs 155 poounds, increasing the person's weight by 1 pound while leaving the height constant will increase the body mass index by about 0.2 index point.

17. $R_E = -PE^{-2}$
$R_P = E^{-1}$

$R_E(4, 32) = -2$ means that when the earnings are \$4 per share and the price is \$32 per share, increasing the earnings per share by \$1 while leaving the share price constant will decrease the price-to-earnings ratio by about 2 units.

$R_P(4, 32) = 0.25$ means that when the earnings are \$4 per share and the price is \$32 per share, increasing the share price by \$1 while leaving the earnings per share constant will increase the price-to-earnings ratio by about 0.25 unit.

19. According to the model, increasing the number of employees from 26,424 to 26,425 while leaving the amount of money spent on advertising constant would *increase* sales by about \$0.17 million.

According to the model, increasing advertising spending from \$519.2 million to \$520.2 million while leaving the number of employees constant would *decrease* sales by about \$0.37 million.

21. The tangent line in the x-direction is $z = 0x - 4$.
The tangent line in the y-direction is $z = 8y + 4$.

23. The tangent line in the x-direction is $z = x + 1$.
The tangent line in the y-direction is $z = 0y + 1$.

25. The tangent line in the x-direction is $z = 0x + e$.
The tangent line in the y-direction is
$z = -2ey - e$.

27. $f(x, y) = 2x + 3y$ (Answers will vary.)

29. $a = 2, b = 3, c = -4$. When $x = \dfrac{11}{41}$ and

$y = -\dfrac{1}{41}$, $f_x(x, y) = 1$ and $f_y(x, y) = 1$.

Section 14.3 *(page 944)*

1. $f(x, y) = x^2 + 4y^2 - 10$ has a relative minimum at $(0, 0, -10)$.

3. $f(x, y) = -x^2 - y^3 + 3y - x$ has a relative maximum at $\left(-\dfrac{1}{2}, 1, 2\dfrac{1}{4}\right)$ and a saddle point at $\left(-\dfrac{1}{2}, -1, -1\dfrac{3}{4}\right)$.

5. $f(x, y) = x^2y - y^2 + x$ has a saddle point at $\left(-1, \dfrac{1}{2}, -\dfrac{3}{4}\right)$.

7. $f(x, y) = x^4 - 4x + y^3 - y^2$ has a saddle point at $(1, 0, -3)$ and a relative minimum at $\left(1, \dfrac{2}{3}, -\dfrac{85}{27}\right) \approx (1, 0.667, -3.15)$.

9. $f(x, y) = x^3 - 6x + xy - y^2$ has a saddle point at $\left(\dfrac{4}{3}, \dfrac{2}{3}, -\dfrac{140}{27}\right)$ and a relative maximum at $\left(-\dfrac{3}{2}, -\dfrac{3}{4}, \dfrac{99}{16}\right)$.

11. $f(x, y) = \dfrac{x^2 + 1}{y^2 + 1}$ has a saddle point at $(0, 0, 1)$.

13. $f(x, y) = \dfrac{-1}{x^2 + 1} + y^4 - 4y^2$ has a relative minimum at $(0, 0, -1)$ and saddle points at $\left(0, -\sqrt{2}, -5\right)$ and $\left(0, \sqrt{2}, -5\right)$.

15. $f(x, y) = x^3 - 3x + y^3 - 12y$ has a relative maximum at $(-1, -2, 18)$, a relative minimum at $(1, 2, -18)$, and saddle points at $(-1, 2, -14)$ and $(1, -2, 14)$.

17. $f(x, y) = x^2 - x - 3y^2 + 3y$ has a saddle point at $\left(\dfrac{1}{2}, \dfrac{1}{2}, \dfrac{1}{2}\right)$.

19. $f(x, y) = \dfrac{x^2 + y^2 + 1}{e^{x^2+y^2}}$ has a relative maximum at $(0, 0, 1)$.

21. $f(x, y) = \dfrac{x^2 - y^2}{e^{x^2+y^2}}$ has relative maxima at $\left(\pm 1, 0, \dfrac{1}{e}\right)$, relative minima at $\left(0, \pm 1, -\dfrac{1}{e}\right)$, and a saddle point at $(0, 0, 0)$.

23. $f(x, y) = e^{-x^2 + 2x - y^2 + y + 1}$ has a relative maximum at $\left(1, \dfrac{1}{2}, e^{9/4}\right)$.

25. $f(x, y) = x^3y + xy^3 - 4xy$ has relative maxima at $(-1, 1, 2)$ and $(1, -1, 2)$, relative minima at $(-1, -1, -2)$ and $(1, 1, -2)$, and saddle points at $(-2, 0, 0)$, $(0, -2, 0)$, $(0, 0, 0)$, $(0, 2, 0)$, and $(2, 0, 0)$.

Section 14.4 *(page 959)*

1. A relative minimum occurs at $(2, 9, -14)$.

3. A relative maximum occurs at $\left(\frac{3}{8}, \frac{1}{8}, \frac{1}{8}\right)$.

5. A relative maximum occurs at $(4, 4, 32)$, and a relative minimum occurs at $(0, 0, 0)$.

7. A relative minimum occurs at $\left(\frac{1}{5}, -1, -\frac{4}{5}\right)$.

9. A relative minimum occurs at $\left(-\frac{50}{51}, \frac{53}{51}, \approx 21.99\right)$, and a relative maximum occurs at $(-2, -1, 31)$.

11. The box whose dimensions are 30 inches $\times$ 28.24 inches $\times$ 28.24 inches has maximum volume. The volume of this box is 23,917 cubic inches.

13. According to the model, Ford should spend $18,232 million on labor and $21,768 million on capital in order to produce the maximum number of vehicles (10,680 thousand) while remaining within the $40,000 million budget.

15. According to the model, Coca-Cola should spend $44 million on labor and $873 million on capital in order to produce the maximum number of unit cases (18.9 billion) while remaining within the $917 million budget.

CHAPTER 14 REVIEW EXERCISES

Section 14.1 *(page 961)*

1. $z = 4x - 5y$

x $\downarrow$	$y \rightarrow$ -1	0	1
-2	-3	-8	-13
0	5	0	-5
2	13	8	3

3. $z = xy + 4y$

x $\downarrow$	$y \rightarrow$ -1	0	1
-2	-2	0	2
0	-4	0	4
2	-6	0	6

5. For Apple Computer, we have $A = \$3950$ million. For General Electric, we have $A = \$59{,}887.5$ million. Microsoft's debt/equity ratio of 0 means that all of the company's assets are financed through equity. The company doesn't have any debt.

Section 14.2 *(page 962)*

7. $f_x = 4xy - 4y^2$
$f_y = 2x^2 - 8xy$
$f_{xx} = 4y$
$f_{yy} = -8x$
$f_{xy} = 4x - 8y$

9. $f_x = y \ln(y)$
$f_y = x \ln(y) + x$
$f_{xx} = 0$
$f_{yy} = \dfrac{x}{y}$
$f_{xy} = \ln(y) + 1$

11. $V_r(h, r) = 2\pi rh$
$V_r(4, 5) = 40\pi$

When the height is 4 inches and the radius is 5 inches, increasiing the radius by 1 inch will increase the volume by *about* $40\pi \approx 126$ cubic inches.

$V_h(h, r) = \pi r^2$
$V_h(4, 5) = 25\pi$

When the height is 4 inches and the radius is 5 inches, increasing the height by 1 inch will increase the volume by *about* $25\pi \approx 79$ cubic inches.

13. The tangent line in the x-direction is $z = x + 3$. The tangent line in the y-direction is $z = -y + 2$.

Section 14.3 *(page 963)*

15. $f(x, y) = x^2 - xy + 4y^2 + 10$ has a relative minimum at $(0, 0, 10)$.

Section 14.4 *(page 963)*

17. A relative maximum occurs at $(1, 2, 6)$.

19. A relative maximum occurs at $(6.60, 4.33, 86.6)$. The dimensions of the garden that maximize the area subject to the constraint are shown in the figure.

8.66ft

6.60ft   arc length = 13.6ft

Index